Student Solutions Manual
for Jon Rogawski, Colin Adams, and Robert Franzosa's

Multivariable
CALCULUS

FOURTH EDITION

ROGER LIPSETT

Brandeis University

w.h.freeman
Macmillan Learning
New York

Copyright © 2019 by W. H. Freeman and Company

ISBN-13: 978-1-319-25441-4
ISBN-10: 1-319-25441-1

Printed in the United States of America
1 2 3 4 5 6   23 22 21 20 19 18

W. H. Freeman and Company
One New York Plaza
Suite 4500
New York, NY 10004-1562

www.macmillanlearning.com

CONTENTS

Chapter 10 INFINITE SERIES (LT CHAPTER 11) **685**

10.1 Sequences (LT Section 11.1) 685
10.2 Summing an Infinite Series (LT Section 11.2) 698
10.3 Convergence of Series with Positive Terms (LT Section 11.3) 712
10.4 Absolute and Conditional Convergence (LT Section 11.4) 728
10.5 The Ratio and Root Tests and Strategies for Choosing Tests (LT Section 11.5) 735
10.6 Power Series (LT Section 11.6) 744
10.7 Taylor Polynomials (LT Section 11.7) 758
10.8 Taylor Series (LT Section 11.8) 775
Chapter Review Exercises 794

Chapter 11 PARAMETRIC EQUATIONS, POLAR COORDINATES, AND CONIC SECTIONS (LT CHAPTER 12) **813**

11.1 Parametric Equations (LT Section 12.1) 813
11.2 Arc Length and Speed (LT Section 12.2) 833
11.3 Polar Coordinates (LT Section 12.3) 842
11.4 Area and Arc Length in Polar Coordinates (LT Section 12.4) 857
11.5 Conic Sections (LT Section 12.5) 868
Chapter Review Exercises 880

Chapter 12 VECTOR GEOMETRY (LT CHAPTER 13) **893**

12.1 Vectors in the Plane (LT Section 13.1) 893
12.2 Three-Dimensional Space: Surfaces, Vectors, and Curves (LT Section 13.2) 905
12.3 Dot Product and the Angle Between Two Vectors (LT Section 13.3) 915
12.4 The Cross Product (LT Section 13.4) 931
12.5 Planes in 3-Space (LT Section 13.5) 949
12.6 A Survey of Quadric Surfaces (LT Section 13.6) 963
12.7 Cylindrical and Spherical Coordinates (LT Section 13.7) 972
Chapter Review Exercises 985

Chapter 13 CALCULUS OF VECTOR-VALUED FUNCTIONS (LT CHAPTER 14) **997**

13.1 Vector-Valued Functions (LT Section 14.1) 997
13.2 Calculus of Vector-Valued Functions (LT Section 14.2) 1010
13.3 Arc Length and Speed (LT Section 14.3) 1024
13.4 Curvature (LT Section 14.4) 1035
13.5 Motion in 3-Space (LT Section 14.5) 1061
13.6 Planetary Motion According to Kepler and Newton (LT Section 14.6) 1076
Chapter Review Exercises 1085

Chapter 14 DIFFERENTIATION IN SEVERAL VARIABLES (LT CHAPTER 15) **1095**

14.1 Functions of Two or More Variables (LT Section 15.1) 1095
14.2 Limits and Continuity in Several Variables (LT Section 15.2) 1107
14.3 Partial Derivatives (LT Section 15.3) 1115
14.4 Differentiability, Tangent Planes, and Linear Approximation (LT Section 15.4) 1126
14.5 The Gradient and Directional Derivatives (LT Section 15.5) 1138
14.6 Multivariable Calculus Chain Rules (LT Section 15.6) 1151
14.7 Optimization in Several Variables (LT Section 15.7) 1166
14.8 Lagrange Multipliers: Optimizing with a Constraint (LT Section 15.8) 1191
Chapter Review Exercises 1217

Chapter 15 MULTIPLE INTEGRATION (LT CHAPTER 16) **1235**

15.1 Integration in Two Variables (LT Section 16.1) 1235
15.2 Double Integrals over More General Regions (LT Section 16.2) 1250
15.3 Triple Integrals (LT Section 16.3) 1271
15.4 Integration in Polar, Cylindrical, and Spherical Coordinates (LT Section 16.4) 1287
15.5 Applications of Multiple Integrals (LT Section 16.5) 1311
15.6 Change of Variables (LT Section 16.6) 1328
Chapter Review Exercises 1347

Chapter 16 LINE AND SURFACE INTEGRALS (LT CHAPTER 17) **1369**

16.1 Vector Fields (LT Section 17.1) 1369
16.2 Line Integrals (LT Section 17.2) 1378

- **16.3 Conservative Vector Fields** (LT Section 17.3)   **1396**
- **16.4 Parametrized Surfaces and Surface Integrals** (LT Section 17.4)   **1404**
- **16.5 Surface Integrals of Vector Fields** (LT Section 17.5)   **1420**
- **Chapter Review Exercises**   **1438**

Chapter 17   FUNDAMENTAL THEOREMS OF VECTOR ANALYSIS (LT CHAPTER 18)   1455

- **17.1 Green's Theorem** (LT Section 18.1)   **1455**
- **17.2 Stokes' Theorem** (LT Section 18.2)   **1473**
- **17.3 Divergence Theorem** (LT Section 18.3)   **1487**
- **Chapter Review Exercises**   **1505**

10 INFINITE SERIES

10.1 Sequences (LT Section 11.1)

Preliminary Questions

1. What is a_4 for the sequence $a_n = n^2 - n$?

SOLUTION Substituting $n = 4$ in the expression for a_n gives

$$a_4 = 4^2 - 4 = 12$$

2. Which of the following sequences converge to zero?

(a) $\dfrac{n^2}{n^2 + 1}$
(b) 2^n
(c) $\left(\dfrac{-1}{2}\right)^n$

SOLUTION

(a) This sequence does not converge to zero:

$$\lim_{n \to \infty} \frac{n^2}{n^2 + 1} = \lim_{x \to \infty} \frac{x^2}{x^2 + 1} = \lim_{x \to \infty} \frac{1}{1 + \frac{1}{x^2}} = \frac{1}{1 + 0} = 1$$

(b) This sequence does not converge to zero: this is a geometric sequence with $r = 2 > 1$; hence, the sequence diverges to ∞.

(c) Recall that if $|a_n|$ converges to 0, then a_n must also converge to zero. Here,

$$\left|\left(-\frac{1}{2}\right)^n\right| = \left(\frac{1}{2}\right)^n$$

which is a geometric sequence with $0 < r < 1$; hence, $(\frac{1}{2})^n$ converges to zero. It therefore follows that $(-\frac{1}{2})^n$ converges to zero.

3. Let a_n be the nth decimal approximation to $\sqrt{2}$. That is, $a_1 = 1$, $a_2 = 1.4$, $a_3 = 1.41$, and so on. What is $\lim\limits_{n \to \infty} a_n$?

SOLUTION $\lim\limits_{n \to \infty} a_n = \sqrt{2}$

4. Which of the following sequences is defined recursively?

(a) $a_n = \sqrt{4 + n}$
(b) $b_n = \sqrt{4 + b_{n-1}}$

SOLUTION

(a) a_n can be computed directly, since it depends on n only and not on preceding terms. Therefore a_n is defined explicitly and not recursively.

(b) b_n is computed in terms of the preceding term b_{n-1}, hence the sequence $\{b_n\}$ is defined recursively.

5. Theorem 5 says that every convergent sequence is bounded. Determine if the following statements are true or false, and if false, give a counterexample:

(a) If $\{a_n\}$ is bounded, then it converges.
(b) If $\{a_n\}$ is not bounded, then it diverges.
(c) If $\{a_n\}$ diverges, then it is not bounded.

SOLUTION

(a) This statement is false. The sequence $a_n = \cos \pi n$ is bounded since $-1 \leq \cos \pi n \leq 1$ for all n, but it does not converge: since $a_n = \cos n\pi = (-1)^n$, the terms assume the two values 1 and -1 alternately, hence they do not approach one value.

(b) By Theorem 5, a converging sequence must be bounded. Therefore, if a sequence is not bounded, it certainly does not converge.

(c) The statement is false. The sequence $a_n = (-1)^n$ is bounded, but it does not approach one limit.

Exercises

1. Match each sequence with its general term:

$a_1, a_2, a_3, a_4, \ldots$	General term
(a) $\frac{1}{2}, \frac{2}{3}, \frac{3}{4}, \frac{4}{5}, \ldots$	(i) $\cos \pi n$
(b) $-1, 1, -1, 1, \ldots$	(ii) $\frac{n!}{2^n}$
(c) $1, -1, 1, -1, \ldots$	(iii) $(-1)^{n+1}$
(d) $\frac{1}{2}, \frac{2}{4}, \frac{6}{8}, \frac{24}{16} \ldots$	(iv) $\frac{n}{n+1}$

SOLUTION

(a) The numerator of each term is the same as the index of the term, and the denominator is one more than the numerator; hence $a_n = \frac{n}{n+1}, n = 1, 2, 3, \ldots$.

(b) The terms of this sequence are alternating between -1 and 1 so that the positive terms are in the even places. Since $\cos \pi n = 1$ for even n and $\cos \pi n = -1$ for odd n, we have $a_n = \cos \pi n, n = 1, 2, \ldots$.

(c) The terms a_n are 1 for odd n and -1 for even n. Hence, $a_n = (-1)^{n+1}, n = 1, 2, \ldots$

(d) The numerator of each term is $n!$, and the denominator is 2^n; hence, $a_n = \frac{n!}{2^n}, n = 1, 2, 3, \ldots$.

In Exercises 3–12, calculate the first four terms of the sequence, starting with $n = 1$.

3. $c_n = \dfrac{3^n}{n!}$

SOLUTION Setting $n = 1, 2, 3, 4$ in the formula for c_n gives

$$c_1 = \frac{3^1}{1!} = \frac{3}{1} = 3, \qquad c_2 = \frac{3^2}{2!} = \frac{9}{2},$$

$$c_3 = \frac{3^3}{3!} = \frac{27}{6} = \frac{9}{2}, \qquad c_4 = \frac{3^4}{4!} = \frac{81}{24} = \frac{27}{8}$$

5. $a_1 = 2, \quad a_{n+1} = 2a_n^2 - 3$

SOLUTION For $n = 1, 2, 3$ we have:

$$a_2 = a_{1+1} = 2a_1^2 - 3 = 2 \cdot 2^2 - 3 = 5$$
$$a_3 = a_{2+1} = 2a_2^2 - 3 = 2 \cdot 5^2 - 3 = 47$$
$$a_4 = a_{3+1} = 2a_3^2 - 3 = 2 \cdot 47^2 - 3 = 4415$$

The first four terms of $\{a_n\}$ are 2, 5, 47, 4415.

7. $b_n = 5 + \cos \pi n$

SOLUTION For $n = 1, 2, 3, 4$ we have

$$b_1 = 5 + \cos \pi = 4$$
$$b_2 = 5 + \cos 2\pi = 6$$
$$b_3 = 5 + \cos 3\pi = 4$$
$$b_4 = 5 + \cos 4\pi = 6$$

The first four terms of $\{b_n\}$ are 4, 6, 4, 6.

9. $c_n = 1 + \dfrac{1}{2} + \dfrac{1}{3} + \cdots + \dfrac{1}{n}$

SOLUTION

$$c_1 = 1$$
$$c_2 = 1 + \frac{1}{2} = \frac{3}{2}$$

$$c_3 = 1 + \frac{1}{2} + \frac{1}{3} = \frac{3}{2} + \frac{1}{3} = \frac{11}{6}$$

$$c_4 = 1 + \frac{1}{2} + \frac{1}{3} + \frac{1}{4} = \frac{11}{6} + \frac{1}{4} = \frac{25}{12}$$

11. $b_1 = 2$, $b_2 = 3$, $b_n = 2b_{n-1} + b_{n-2}$

SOLUTION We need to find b_3 and b_4. Setting $n = 3$ and $n = 4$ and using the given values for b_1 and b_2 we obtain:

$$b_3 = 2b_{3-1} + b_{3-2} = 2b_2 + b_1 = 2 \cdot 3 + 2 = 8$$

$$b_4 = 2b_{4-1} + b_{4-2} = 2b_3 + b_2 = 2 \cdot 8 + 3 = 19$$

The first four terms of the sequence $\{b_n\}$ are 2, 3, 8, 19.

13. Find a formula for the nth term of each sequence.

(a) $\frac{1}{1}, \frac{-1}{8}, \frac{1}{27}, \ldots$

(b) $\frac{2}{6}, \frac{3}{7}, \frac{4}{8}, \ldots$

SOLUTION

(a) The denominators are the third powers of the positive integers starting with $n = 1$. Also, the sign of the terms is alternating with the sign of the first term being positive. Thus,

$$a_1 = \frac{1}{1^3} = \frac{(-1)^{1+1}}{1^3}; \quad a_2 = -\frac{1}{2^3} = \frac{(-1)^{2+1}}{2^3}; \quad a_3 = \frac{1}{3^3} = \frac{(-1)^{3+1}}{3^3}$$

This rule leads to the following formula for the nth term:

$$a_n = \frac{(-1)^{n+1}}{n^3}$$

(b) Assuming a starting index of $n = 1$, we see that each numerator is one more than the index and the denominator is four more than the numerator. Thus, the general term a_n is

$$a_n = \frac{n+1}{n+5}$$

In Exercises 15–28, use Theorem 1 to determine the limit of the sequence or state that the sequence diverges.

15. $a_n = 5 - 2n$

SOLUTION We have $a_n = f(n)$ where $f(x) = 5 - 2x$; thus,

$$\lim_{n \to \infty} a_n = \lim_{x \to \infty} f(x) = \lim_{x \to \infty} (5 - 2x) = -\infty$$

The sequence diverges.

17. $b_n = \frac{5n - 1}{12n + 9}$

SOLUTION We have $b_n = f(n)$ where $f(x) = \frac{5x - 1}{12x + 9}$; thus,

$$\lim_{n \to \infty} \frac{5n - 1}{12n + 9} = \lim_{x \to \infty} \frac{5x - 1}{12x + 9} = \frac{5}{12}$$

19. $a_n = \left(\frac{1}{2}\right)^{-n}$

SOLUTION We have $a_n = f(n)$ where $f(x) = \left(\frac{1}{2}\right)^{-n} = 2^n$; thus,

$$\lim_{n \to \infty} a_n = \lim_{x \to \infty} f(x) = \lim_{x \to \infty} 2^x = \infty$$

The sequence diverges.

21. $c_n = 9^n$

SOLUTION We have $c_n = f(n)$ where $f(x) = 9^x$; thus,

$$\lim_{n \to \infty} 9^n = \lim_{x \to \infty} 9^x = \infty$$

Thus, the sequence 9^n diverges.

23. $a_n = \dfrac{n}{\sqrt{n^2+1}}$

SOLUTION We have $a_n = f(n)$ where $f(x) = \dfrac{x}{\sqrt{x^2+1}}$; thus,

$$\lim_{n \to \infty} \frac{n}{\sqrt{n^2+1}} = \lim_{x \to \infty} \frac{x}{\sqrt{x^2+1}} = \lim_{x \to \infty} \frac{\frac{x}{x}}{\frac{\sqrt{x^2+1}}{x}} = \lim_{x \to \infty} \frac{1}{\sqrt{\frac{x^2+1}{x^2}}} = \lim_{x \to \infty} \frac{1}{\sqrt{1+\frac{1}{x^2}}} = \frac{1}{\sqrt{1+0}} = 1$$

25. $a_n = \ln\left(\dfrac{12n+2}{-9+4n}\right)$

SOLUTION We have $a_n = f(n)$ where $f(x) = \ln\left(\dfrac{12x+2}{-9+4x}\right)$; thus,

$$\lim_{n \to \infty} \ln\left(\frac{12n+2}{-9+4n}\right) = \lim_{x \to \infty} \ln\left(\frac{12x+2}{-9+4x}\right) = \ln \lim_{x \to \infty} \left(\frac{12x+2}{-9+4x}\right) = \ln 3$$

27. $z_n = \dfrac{n+1}{e^n}$

SOLUTION We have $z_n = f(n)$ where $f(n) = \dfrac{n+1}{e^n}$; thus, using L'Hôpital's rule together with Theorems 1 and 2,

$$\lim_{n \to \infty} z_n = \lim_{x \to \infty} f(x) = \lim_{x \to \infty} \frac{x+1}{e^x} = \lim_{x \to \infty} \frac{1}{e^x} = 0$$

In Exercises 29–32, use Theorem 4 to determine the limit of the sequence.

29. $a_n = \sqrt{4 + \dfrac{1}{n}}$

SOLUTION We have

$$\lim_{n \to \infty} 4 + \frac{1}{n} = \lim_{x \to \infty} 4 + \frac{1}{x} = 4$$

Since $\sqrt{x}$ is a continuous function for $x > 0$, Theorem 4 tells us that

$$\lim_{n \to \infty} \sqrt{4 + \frac{1}{n}} = \sqrt{\lim_{n \to \infty} 4 + \frac{1}{n}} = \sqrt{4} = 2$$

31. $a_n = \cos^{-1}\left(\dfrac{n^3}{2n^3+1}\right)$

SOLUTION We have

$$\lim_{n \to \infty} \frac{n^3}{2n^3+1} = \frac{1}{2}$$

Since $\cos^{-1}(x)$ is continuous for all x, Theorem 4 tells us that

$$\lim_{n \to \infty} \cos^{-1}\left(\frac{n^3}{2n^3+1}\right) = \cos^{-1}\left(\lim_{n \to \infty} \frac{n^3}{2n^3+1}\right) = \cos^{-1}(1/2) = \frac{\pi}{3}$$

In Exercises 33–34 let $a_n = \dfrac{F_{n+1}}{F_n}$, where $\{F_n\}$ is the Fibonacci sequence. The sequence $\{a_n\}$ has a limit. We do not prove this fact, but investigate the value of the limit in these exercises.

33. *CAS* Estimate $\lim_{n \to \infty} a_n$ to five decimal places by computing a_n for sufficiently large n.

SOLUTION For example,

$$a_9 \approx 1.61765, \quad a_{99} \approx a_{999} \approx 1.61803$$

35. Let $a_n = \dfrac{n}{n+1}$. Find a number M such that:

(a) $|a_n - 1| \le 0.001$ for $n \ge M$.

(b) $|a_n - 1| \le 0.00001$ for $n \ge M$.

Then use the limit definition to prove that $\lim_{n \to \infty} a_n = 1$.

SOLUTION

(a) We have

$$|a_n - 1| = \left|\frac{n}{n+1} - 1\right| = \left|\frac{n - (n+1)}{n+1}\right| = \left|\frac{-1}{n+1}\right| = \frac{1}{n+1}$$

Therefore $|a_n - 1| \le 0.001$ provided $\frac{1}{n+1} \le 0.001$, that is, $n \ge 999$. It follows that we can take $M = 999$.

(b) By part (a), $|a_n - 1| \le 0.00001$ provided $\frac{1}{n+1} \le 0.00001$, that is, $n \ge 99999$. It follows that we can take $M = 99999$.

We now prove formally that $\lim_{n \to \infty} a_n = 1$. Using part (a), we know that

$$|a_n - 1| = \frac{1}{n+1} < \epsilon$$

provided $n > \frac{1}{\epsilon} - 1$. Thus, Let $\epsilon > 0$ and take $M = \frac{1}{\epsilon} - 1$. Then, for $n > M$, we have

$$|a_n - 1| = \frac{1}{n+1} < \frac{1}{M+1} = \epsilon$$

37. Use the limit definition to prove that $\lim_{n \to \infty} n^{-2} = 0$.

SOLUTION We see that

$$|n^{-2} - 0| = \left|\frac{1}{n^2}\right| = \frac{1}{n^2} < \epsilon$$

provided

$$n > \frac{1}{\sqrt{\epsilon}}$$

Thus, let $\epsilon > 0$ and take $M = \frac{1}{\sqrt{\epsilon}}$. Then, for $n > M$, we have

$$|n^{-2} - 0| = \left|\frac{1}{n^2}\right| = \frac{1}{n^2} < \frac{1}{M^2} = \epsilon$$

In Exercises 39–66, use the appropriate limit laws and theorems to determine the limit of the sequence or show that it diverges.

39. $a_n = 10 + \left(-\dfrac{1}{9}\right)^n$

SOLUTION By the Limit Laws for Sequences we have:

$$\lim_{n \to \infty}\left(10 + \left(-\frac{1}{9}\right)^n\right) = \lim_{n \to \infty} 10 + \lim_{n \to \infty}\left(-\frac{1}{9}\right)^n = 10 + \lim_{n \to \infty}\left(-\frac{1}{9}\right)^n$$

Now for all n,

$$-\left(\frac{1}{9}\right)^n \le \left(-\frac{1}{9}\right)^n \le \left(\frac{1}{9}\right)^n$$

We will use the Squeeze Theorem. The limit of the right-hand side of the above inequality is zero since it is a geometric series with ratio less than 1. The limit of the left-hand side is also zero, since

$$\lim_{n \to \infty} \left(-\left(\frac{1}{9}\right)^n\right) = -\lim_{n \to \infty} \left(\frac{1}{9}\right)^n = 0$$

Therefore

$$10 = 10 + \lim_{n \to \infty} \left(-\left(\frac{1}{9}\right)^n\right) \leq 10 + \lim_{n \to \infty} \left(-\frac{1}{9}\right)^n \leq 10 + \lim_{n \to \infty} \left(\frac{1}{9}\right)^n = 10$$

so by the Squeeze Theorem,

$$\lim_{n \to \infty} \left(10 + \left(-\frac{1}{9}\right)^n\right) = 10$$

41. $c_n = 1.01^n$

SOLUTION Since $c_n = f(n)$ where $f(x) = 1.01^x$, we have

$$\lim_{n \to \infty} 1.01^n = \lim_{x \to \infty} 1.01^x = \infty$$

so that the sequence diverges.

43. $a_n = 2^{1/n}$

SOLUTION Because 2^x is a continuous function,

$$\lim_{n \to \infty} 2^{1/n} = \lim_{x \to \infty} 2^{1/x} = 2^{\lim_{x \to \infty}(1/x)} = 2^0 = 1$$

45. $c_n = \dfrac{9^n}{n!}$

SOLUTION For $n \geq 9$, write

$$c_n = \frac{9^n}{n!} = \underbrace{\frac{9}{1} \cdot \frac{9}{2} \cdots \frac{9}{9}}_{\text{call this } C} \cdot \underbrace{\frac{9}{10} \cdot \frac{9}{11} \cdots \frac{9}{n-1}}_{\text{Each factor is less than 1}} \cdot \frac{9}{n}$$

Then clearly

$$0 \leq \frac{9^n}{n!} \leq C \frac{9}{n}$$

since each factor after the first nine is < 1. The squeeze theorem tells us that

$$\lim_{n \to \infty} 0 \leq \lim_{n \to \infty} \frac{9^n}{n!} \leq \lim_{n \to \infty} C \frac{9}{n} = C \lim_{n \to \infty} \frac{9}{n} = C \cdot 0 = 0$$

so that $\lim_{n \to \infty} c_n = 0$ as well.

47. $a_n = \dfrac{3n^2 + n + 2}{2n^2 - 3}$

SOLUTION

$$\lim_{n \to \infty} \frac{3n^2 + n + 2}{2n^2 - 3} = \lim_{x \to \infty} \frac{3x^2 + x + 2}{2x^2 - 3} = \frac{3}{2}$$

49. $a_n = \dfrac{\cos n}{n}$

SOLUTION Since $-1 \leq \cos n \leq 1$ the following holds:

$$-\frac{1}{n} \leq \frac{\cos n}{n} \leq \frac{1}{n}$$

We now apply the Squeeze Theorem for Sequences and the limits

$$\lim_{n \to \infty} -\frac{1}{n} = \lim_{n \to \infty} \frac{1}{n} = 0$$

to conclude that $\lim_{n \to \infty} \frac{\cos n}{n} = 0$.

51. $d_n = \ln 5^n - \ln n!$

SOLUTION Note that

$$d_n = \ln \frac{5^n}{n!}$$

so that

$$e^{d_n} = \frac{5^n}{n!} \quad \text{so} \quad \lim_{n \to \infty} e^{d_n} = \lim_{n \to \infty} \frac{5^n}{n!} = 0$$

by the method of Exercise 45. If d_n converged, we could, since $f(x) = e^x$ is continuous, then write

$$\lim_{n \to \infty} e^{d_n} = e^{\lim_{n \to \infty} d_n} = 0$$

which is impossible. Thus $\{d_n\}$ diverges.

53. $a_n = \left(2 + \frac{4}{n^2}\right)^{1/3}$

SOLUTION Let $a_n = \left(2 + \frac{4}{n^2}\right)^{1/3}$. Taking the natural logarithm of both sides of this expression yields

$$\ln a_n = \ln \left(2 + \frac{4}{n^2}\right)^{1/3} = \frac{1}{3} \ln \left(2 + \frac{4}{n^2}\right)$$

Thus,

$$\lim_{n \to \infty} \ln a_n = \lim_{n \to \infty} \frac{1}{3} \ln \left(2 + \frac{4}{n^2}\right)^{1/3} = \frac{1}{3} \lim_{x \to \infty} \ln \left(2 + \frac{4}{x^2}\right) = \frac{1}{3} \ln \left(\lim_{x \to \infty} \left(2 + \frac{4}{x^2}\right)\right)$$

$$= \frac{1}{3} \ln (2 + 0) = \frac{1}{3} \ln 2 = \ln 2^{1/3}$$

Because $f(x) = e^x$ is a continuous function, it follows that

$$\lim_{n \to \infty} a_n = \lim_{n \to \infty} e^{\ln a_n} = e^{\lim_{n \to \infty} (\ln a_n)} = e^{\ln 2^{1/3}} = 2^{1/3}$$

55. $c_n = \ln \left(\frac{2n+1}{3n+4}\right)$

SOLUTION Because $f(x) = \ln x$ is a continuous function, it follows that

$$\lim_{n \to \infty} c_n = \lim_{x \to \infty} \ln \left(\frac{2x+1}{3x+4}\right) = \ln \left(\lim_{x \to \infty} \frac{2x+1}{3x+4}\right) = \ln \frac{2}{3}$$

57. $y_n = \dfrac{e^n}{2^n}$

SOLUTION $\frac{e^n}{2^n} = \left(\frac{e}{2}\right)^n$ and $\frac{e}{2} > 1$. By the Limit of Geometric Sequences, we conclude that $\lim_{n \to \infty} \left(\frac{e}{2}\right)^n = \infty$. Thus, the given sequence diverges.

59. $y_n = \dfrac{e^n + (-3)^n}{5^n}$

SOLUTION

$$\lim_{n \to \infty} \frac{e^n + (-3)^n}{5^n} = \lim_{n \to \infty} \left(\frac{e}{5}\right)^n + \lim_{n \to \infty} \left(\frac{-3}{5}\right)^n$$

assuming both limits on the right-hand side exist. But by the Limit of Geometric Sequences, since

$$-1 < \frac{-3}{5} < 0 < \frac{e}{5} < 1$$

both limits on the right-hand side are 0, so that y_n converges to 0.

61. $a_n = n \sin \dfrac{\pi}{n}$

SOLUTION By the Theorem on Sequences Defined by a Function, we have

$$\lim_{n \to \infty} n \sin \frac{\pi}{n} = \lim_{x \to \infty} x \sin \frac{\pi}{x}$$

Now,

$$\lim_{x \to \infty} x \sin \frac{\pi}{x} = \lim_{x \to \infty} \frac{\sin \frac{\pi}{x}}{\frac{1}{x}} = \lim_{x \to \infty} \frac{\left(\cos \frac{\pi}{x}\right)\left(-\frac{\pi}{x^2}\right)}{-\frac{1}{x^2}} = \lim_{x \to \infty} \left(\pi \cos \frac{\pi}{x}\right)$$

$$= \pi \lim_{x \to \infty} \cos \frac{\pi}{x} = \pi \cos 0 = \pi \cdot 1 = \pi$$

Thus,

$$\lim_{n \to \infty} n \sin \frac{\pi}{n} = \pi$$

63. $b_n = \dfrac{3 - 4^n}{2 + 7 \cdot 4^n}$

SOLUTION Divide the numerator and denominator by 4^n to obtain

$$a_n = \frac{3 - 4^n}{2 + 7 \cdot 4^n} = \frac{\frac{3}{4^n} - \frac{4^n}{4^n}}{\frac{2}{4^n} + \frac{7 \cdot 4^n}{4^n}} = \frac{\frac{3}{4^n} - 1}{\frac{2}{4^n} + 7}$$

Thus,

$$\lim_{n \to \infty} a_n = \lim_{x \to \infty} \frac{\frac{3}{4^x} - 1}{\frac{2}{4^x} + 7} = \frac{\lim_{x \to \infty} \left(\frac{3}{4^x} - 1\right)}{\lim_{x \to \infty} \left(\frac{2}{4^x} + 7\right)} = \frac{3 \lim_{x \to \infty} \frac{1}{4^x} - \lim_{x \to \infty} 1}{2 \lim_{x \to \infty} \frac{1}{4^x} - \lim_{x \to \infty} 7} = \frac{3 \cdot 0 - 1}{2 \cdot 0 + 7} = -\frac{1}{7}$$

65. $a_n = \left(1 + \dfrac{1}{n}\right)^n$

SOLUTION Taking the natural logarithm of both sides of this expression yields

$$\ln a_n = \ln \left(1 + \frac{1}{n}\right)^n = n \ln \left(1 + \frac{1}{n}\right) = \frac{\ln \left(1 + \frac{1}{n}\right)}{\frac{1}{n}}$$

Thus,

$$\lim_{n \to \infty} (\ln a_n) = \lim_{x \to \infty} \frac{\ln \left(1 + \frac{1}{x}\right)}{\frac{1}{x}} = \lim_{x \to \infty} \frac{\frac{d}{dx}\left(\ln \left(1 + \frac{1}{x}\right)\right)}{\frac{d}{dx}\left(\frac{1}{x}\right)} = \lim_{x \to \infty} \frac{\frac{1}{1+\frac{1}{x}} \cdot \left(-\frac{1}{x^2}\right)}{-\frac{1}{x^2}} = \lim_{x \to \infty} \frac{1}{1 + \frac{1}{x}} = \frac{1}{1 + 0} = 1$$

Because $f(x) = e^x$ is a continuous function, it follows that

$$\lim_{n \to \infty} a_n = \lim_{n \to \infty} e^{\ln a_n} = e^{\lim_{n \to \infty}(\ln a_n)} = e^1 = e$$

In Exercises 67–70, find the limit of the sequence using L'Hôpital's Rule.

67. $a_n = \dfrac{(\ln n)^2}{n}$

SOLUTION

$$\lim_{n \to \infty} \frac{(\ln n)^2}{n} = \lim_{x \to \infty} \frac{(\ln x)^2}{x} = \lim_{x \to \infty} \frac{\frac{d}{dx}(\ln x)^2}{\frac{d}{dx}x} = \lim_{x \to \infty} \frac{\frac{2 \ln x}{x}}{1} = \lim_{x \to \infty} \frac{2 \ln x}{x}$$

$$= \lim_{x \to \infty} \frac{\frac{d}{dx} 2 \ln x}{\frac{d}{dx} x} = \lim_{x \to \infty} \frac{\frac{2}{x}}{1} = \lim_{x \to \infty} \frac{2}{x} = 0$$

69. $c_n = n(\sqrt{n^2+1} - n)$

SOLUTION

$$\lim_{n \to \infty} n\left(\sqrt{n^2+1} - n\right) = \lim_{x \to \infty} x\left(\sqrt{x^2+1} - x\right) = \lim_{x \to \infty} \frac{x\left(\sqrt{x^2+1} - x\right)\left(\sqrt{x^2+1} + x\right)}{\sqrt{x^2+1} + x}$$

$$= \lim_{x \to \infty} \frac{x}{\sqrt{x^2+1} + x} = \lim_{x \to \infty} \frac{\frac{d}{dx}x}{\frac{d}{dx}\sqrt{x^2+1} + x} = \lim_{x \to \infty} \frac{1}{1 + \frac{x}{\sqrt{x^2+1}}}$$

$$= \lim_{x \to \infty} \frac{1}{1 + \sqrt{\frac{x^2}{x^2+1}}} = \lim_{x \to \infty} \frac{1}{1 + \sqrt{\frac{1}{1+(1/x^2)}}} = \frac{1}{2}$$

In Exercises 71–74, use the Squeeze Theorem to evaluate $\lim_{n \to \infty} a_n$ by verifying the given inequality.

71. $a_n = \frac{1}{\sqrt{n^4 + n^8}}, \quad \frac{1}{\sqrt{2}n^4} \leq a_n \leq \frac{1}{\sqrt{2}n^2}$

SOLUTION For all $n > 1$ we have $n^4 < n^8$, so the quotient $\frac{1}{\sqrt{n^4+n^8}}$ is smaller than $\frac{1}{\sqrt{n^4+n^4}}$ and larger than $\frac{1}{\sqrt{n^8+n^8}}$. That is,

$$a_n < \frac{1}{\sqrt{n^4 + n^4}} = \frac{1}{\sqrt{n^4 \cdot 2}} = \frac{1}{\sqrt{2}n^2}; \text{ and}$$

$$a_n > \frac{1}{\sqrt{n^8 + n^8}} = \frac{1}{\sqrt{2n^8}} = \frac{1}{\sqrt{2}n^4}$$

Now, since $\lim_{n \to \infty} \frac{1}{\sqrt{2}n^4} = \lim_{n \to \infty} \frac{1}{\sqrt{2}n^2} = 0$, the Squeeze Theorem for Sequences implies that $\lim_{n \to \infty} a_n = 0$.

73. $a_n = (2^n + 3^n)^{1/n}, \quad 3 \leq a_n \leq (2 \cdot 3^n)^{1/n} = 2^{1/n} \cdot 3$

SOLUTION Clearly $2^n + 3^n \geq 3^n$ for all $n \geq 1$. Therefore:

$$(2^n + 3^n)^{1/n} \geq (3^n)^{1/n} = 3$$

Also $2^n + 3^n \leq 3^n + 3^n = 2 \cdot 3^n$, so

$$(2^n + 3^n)^{1/n} \leq (2 \cdot 3^n)^{1/n}$$

Thus,

$$3 \leq (2^n + 3^n)^{1/n} \leq (2 \cdot 3^n)^{1/n}$$

Because

$$\lim_{n \to \infty} (2 \cdot 3^n)^{1/n} = \lim_{n \to \infty} 2^{1/n} \cdot 3 = 3 \lim_{n \to \infty} 2^{1/n} = 3 \cdot 1 = 3$$

and $\lim_{n \to \infty} 3 = 3$, the Squeeze Theorem for Sequences guarantees

$$\lim_{n \to \infty} (2^n + 3^n)^{1/n} = 3$$

75. Which of the following statements is equivalent to the assertion $\lim_{n \to \infty} a_n = L$? Explain.

(a) For every $\epsilon > 0$, the interval $(L - \epsilon, L + \epsilon)$ contains at least one element of the sequence $\{a_n\}$.

(b) For every $\epsilon > 0$, the interval $(L - \epsilon, L + \epsilon)$ contains all but at most finitely many elements of the sequence $\{a_n\}$.

SOLUTION Statement (b) is equivalent to Definition 1 of the limit, since the assertion "$|a_n - L| < \epsilon$ for all $n > M$" means that $L - \epsilon < a_n < L + \epsilon$ for all $n > M$; that is, the interval $(L - \epsilon, L + \epsilon)$ contains all the elements a_n except (maybe) the finite number of elements $a_1, a_2, \ldots, a_M$.

Statement (a) is not equivalent to the assertion $\lim_{n \to \infty} a_n = L$. We show this, by considering the following sequence:

$$a_n = \begin{cases} \dfrac{1}{n} & \text{for odd } n \\ 1 + \dfrac{1}{n} & \text{for even } n \end{cases}$$

Clearly for every $\epsilon > 0$, the interval $(-\epsilon, \epsilon) = (L - \epsilon, L + \epsilon)$ for $L = 0$ contains at least one element of $\{a_n\}$, but the sequence diverges (rather than converges to $L = 0$). The terms in the odd places converge to 0 and the terms in the even places converge to 1. Hence, a_n does not approach one limit.

77. Show that $a_n = \dfrac{3n^2}{n^2 + 2}$ is increasing. Find an upper bound.

SOLUTION Let $f(x) = \dfrac{3x^2}{x^2+2}$. Then

$$f'(x) = \frac{6x(x^2+2) - 3x^2 \cdot 2x}{(x^2+2)^2} = \frac{12x}{(x^2+2)^2}$$

$f'(x) > 0$ for $x > 0$, hence f is increasing on this interval. It follows that $a_n = f(n)$ is also increasing. We now show that $M = 3$ is an upper bound for a_n, by writing:

$$a_n = \frac{3n^2}{n^2+2} \leq \frac{3n^2+6}{n^2+2} = \frac{3(n^2+2)}{n^2+2} = 3$$

That is, $a_n \leq 3$ for all n.

79. Give an example of a divergent sequence $\{a_n\}$ such that $\lim\limits_{n \to \infty} |a_n|$ converges.

SOLUTION Let $a_n = (-1)^n$. The sequence $\{a_n\}$ diverges because the terms alternate between $+1$ and -1; however, the sequence $\{|a_n|\}$ converges because it is a constant sequence, all of whose terms are equal to 1.

81. Using the limit definition, prove that if $\{a_n\}$ converges and $\{b_n\}$ diverges, then $\{a_n + b_n\}$ diverges.

SOLUTION We will prove this result by contradiction. Suppose $\lim_{n\to\infty} a_n = L_1$ and that $\{a_n + b_n\}$ converges to a limit L_2. Now, let $\epsilon > 0$. Because $\{a_n\}$ converges to L_1 and $\{a_n + b_n\}$ converges to L_2, it follows that there exist numbers M_1 and M_2 such that:

$$|a_n - L_1| < \frac{\epsilon}{2} \qquad \text{for all } n > M_1,$$

$$|(a_n + b_n) - L_2| < \frac{\epsilon}{2} \qquad \text{for all } n > M_2$$

Thus, for $n > M = \max\{M_1, M_2\}$,

$$|a_n - L_1| < \frac{\epsilon}{2} \quad \text{and} \quad |(a_n + b_n) - L_2| < \frac{\epsilon}{2}$$

By the triangle inequality,

$$|b_n - (L_2 - L_1)| = |a_n + b_n - a_n - (L_2 - L_1)| = |(-a_n + L_1) + (a_n + b_n - L_2)|$$
$$\leq |L_1 - a_n| + |a_n + b_n - L_2|$$

Thus, for $n > M$,

$$|b_n - (L_2 - L_1)| < \frac{\epsilon}{2} + \frac{\epsilon}{2} = \epsilon$$

that is, $\{b_n\}$ converges to $L_2 - L_1$, in contradiction to the given data. Thus, $\{a_n + b_n\}$ must diverge.

83. Theorem 1 states that if $\lim\limits_{x \to \infty} f(x) = L$, then the sequence $a_n = f(n)$ converges and $\lim\limits_{n \to \infty} a_n = L$. Show that the *converse* is false. In other words, find a function f such that $a_n = f(n)$ converges but $\lim\limits_{x \to \infty} f(x)$ does not exist.

SOLUTION Let $f(x) = \sin \pi x$ and $a_n = \sin \pi n$. Then $a_n = f(n)$. Since $\sin \pi x$ is oscillating between -1 and 1 the limit $\lim\limits_{x \to \infty} f(x)$ does not exist. However, the sequence $\{a_n\}$ is the constant sequence in which $a_n = \sin \pi n = 0$ for all n, hence it converges to zero.

85. Let $b_n = a_{n+1}$. Use the limit definition to prove that if $\{a_n\}$ converges, then $\{b_n\}$ also converges and $\lim\limits_{n \to \infty} a_n = \lim\limits_{n \to \infty} b_n$.

SOLUTION Suppose $\{a_n\}$ converges to L. Let $b_n = a_{n+1}$, and let $\epsilon > 0$. Because $\{a_n\}$ converges to L, there exists an M' such that $|a_n - L| < \epsilon$ for $n > M'$. Now, let $M = M' - 1$. Then, whenever $n > M$, $n + 1 > M + 1 = M'$. Thus, for $n > M$,

$$|b_n - L| = |a_{n+1} - L| < \epsilon$$

Hence, $\{b_n\}$ converges to L.

87. Proceed as in Example 13 to show that the sequence $\sqrt{3}, \sqrt{3\sqrt{3}}, \sqrt{3\sqrt{3\sqrt{3}}}, \ldots$ is increasing and bounded above by $M = 3$. Then prove that the limit exists and find its value.

SOLUTION This sequence is defined recursively by the formula:

$$a_{n+1} = \sqrt{3a_n}, \qquad a_1 = \sqrt{3}$$

Consider the following inequalities:

$$a_2 = \sqrt{3a_1} = \sqrt{3\sqrt{3}} > \sqrt{3} = a_1 \quad \Rightarrow \quad a_2 > a_1;$$

$$a_3 = \sqrt{3a_2} > \sqrt{3a_1} = a_2 \quad \Rightarrow \quad a_3 > a_2;$$

$$a_4 = \sqrt{3a_3} > \sqrt{3a_2} = a_3 \quad \Rightarrow \quad a_4 > a_3$$

In general, if we assume that $a_k > a_{k-1}$, then

$$a_{k+1} = \sqrt{3a_k} > \sqrt{3a_{k-1}} = a_k$$

Hence, by mathematical induction, $a_{n+1} > a_n$ for all n; that is, the sequence $\{a_n\}$ is increasing.

Because $a_{n+1} = \sqrt{3a_n}$, it follows that $a_n \geq 0$ for all n. Now, $a_1 = \sqrt{3} < 3$. If $a_k \leq 3$, then

$$a_{k+1} = \sqrt{3a_k} \leq \sqrt{3 \cdot 3} = 3$$

Thus, by mathematical induction, $a_n \leq 3$ for all n.

Since $\{a_n\}$ is increasing and bounded, it follows by the Theorem on Bounded Monotonic Sequences that this sequence is converging. Denote the limit by $L = \lim_{n \to \infty} a_n$. Using Exercise 85, it follows that

$$L = \lim_{n \to \infty} a_{n+1} = \lim_{n \to \infty} \sqrt{3a_n} = \sqrt{3 \lim_{n \to \infty} a_n} = \sqrt{3L}$$

Thus, $L^2 = 3L$, so $L = 0$ or $L = 3$. Because the sequence is increasing, we have $a_n \geq a_1 = \sqrt{3}$ for all n. Hence, the limit also satisfies $L \geq \sqrt{3}$. We conclude that the appropriate solution is $L = 3$; that is, $\lim_{n \to \infty} a_n = 3$.

Further Insights and Challenges

89. Show that $\lim_{n \to \infty} \sqrt[n]{n!} = \infty$. *Hint:* Verify that $n! \geq (n/2)^{n/2}$ by observing that half of the factors of $n!$ are greater than or equal to $n/2$.

SOLUTION We show that $n! \geq \left(\frac{n}{2}\right)^{n/2}$. For $n \geq 4$ even, we have:

$$n! = \underbrace{1 \cdots \cdots \frac{n}{2}}_{\frac{n}{2} \text{ factors}} \cdot \underbrace{\left(\frac{n}{2}+1\right) \cdots \cdots n}_{\frac{n}{2} \text{ factors}} \geq \underbrace{\left(\frac{n}{2}+1\right) \cdots \cdots n}_{\frac{n}{2} \text{ factors}}.$$

Since each one of the $\frac{n}{2}$ factors is greater than $\frac{n}{2}$, we have:

$$n! \geq \underbrace{\left(\frac{n}{2}+1\right) \cdots \cdots n}_{\frac{n}{2} \text{ factors}} \geq \underbrace{\frac{n}{2} \cdots \cdots \frac{n}{2}}_{\frac{n}{2} \text{ factors}} = \left(\frac{n}{2}\right)^{n/2}$$

For $n \geq 3$ odd, we have:

$$n! = \underbrace{1 \cdots \cdots \frac{n-1}{2}}_{\frac{n-1}{2} \text{ factors}} \cdot \underbrace{\frac{n+1}{2} \cdots \cdots n}_{\frac{n+1}{2} \text{ factors}} \geq \underbrace{\frac{n+1}{2} \cdots \cdots n}_{\frac{n+1}{2} \text{ factors}}$$

Since each one of the $\frac{n+1}{2}$ factors is greater than $\frac{n}{2}$, we have:

$$n! \geq \underbrace{\frac{n+1}{2} \cdots \cdots n}_{\frac{n+1}{2} \text{ factors}} \geq \underbrace{\frac{n}{2} \cdots \cdots \frac{n}{2}}_{\frac{n+1}{2} \text{ factors}} = \left(\frac{n}{2}\right)^{(n+1)/2} = \left(\frac{n}{2}\right)^{n/2} \sqrt{\frac{n}{2}} \geq \left(\frac{n}{2}\right)^{n/2}$$

In either case we have $n! \geq \left(\frac{n}{2}\right)^{n/2}$. Thus,

$$\sqrt[n]{n!} \geq \sqrt{\frac{n}{2}}$$

Since $\lim_{n\to\infty} \sqrt{\frac{n}{2}} = \infty$, it follows that $\lim_{n\to\infty} \sqrt[n]{n!} = \infty$. Thus, the sequence $a_n = \sqrt[n]{n!}$ diverges.

91. Given positive numbers $a_1 < b_1$, define two sequences recursively by

$$a_{n+1} = \sqrt{a_n b_n}, \qquad b_{n+1} = \frac{a_n + b_n}{2}$$

(a) Show that $a_n \leq b_n$ for all n (Figure 14).

(b) Show that $\{a_n\}$ is increasing and $\{b_n\}$ is decreasing.

(c) Show that $b_{n+1} - a_{n+1} \leq \dfrac{b_n - a_n}{2}$.

(d) Prove that both $\{a_n\}$ and $\{b_n\}$ converge and have the same limit. This limit, denoted $\text{AGM}(a_1, b_1)$, is called the **arithmetic-geometric mean** of a_1 and b_1.

(e) Estimate $\text{AGM}(1, \sqrt{2})$ to three decimal places.

FIGURE 14

SOLUTION

(a) Use induction on n. The statement is true for $n = 1$. Choose $k > 1$; then

$$b_k - a_k = \frac{a_{k-1} + b_{k-1}}{2} - \sqrt{a_{k-1} b_{k-1}} = \frac{a_{k-1} + b_{k-1} - 2\sqrt{a_{k-1} b_{k-1}}}{2} = \frac{\left(\sqrt{a_{k-1}}\right)^2 - 2\sqrt{a_{k-1}}\sqrt{b_{k-1}} + \left(\sqrt{b_{k-1}}\right)^2}{2}$$

$$= \frac{\left(\sqrt{a_{k-1}} - \sqrt{b_{k-1}}\right)^2}{2} \geq 0$$

We conclude that $b_k \geq a_k$, so by induction $b_n \geq a_n$ for all n.

(b) By part (a), $b_n \geq a_n$ for all n, so

$$a_{n+1} = \sqrt{a_n b_n} \geq \sqrt{a_n \cdot a_n} = \sqrt{a_n^2} = a_n$$

for all n. Hence, the sequence $\{a_n\}$ is increasing. Moreover, since $a_n \leq b_n$ for all n,

$$b_{n+1} = \frac{a_n + b_n}{2} \leq \frac{b_n + b_n}{2} = \frac{2b_n}{2} = b_n$$

for all n; that is, the sequence $\{b_n\}$ is decreasing.

(c) Since $\{a_n\}$ is increasing, $a_{n+1} \geq a_n$. Thus,

$$b_{n+1} - a_{n+1} \leq b_{n+1} - a_n = \frac{a_n + b_n}{2} - a_n = \frac{a_n + b_n - 2a_n}{2} = \frac{b_n - a_n}{2}$$

(d) By part (a), $a_n \leq b_n$ for all n. By part (b), $\{b_n\}$ is decreasing. Hence $b_n \leq b_1$ for all n. Combining the two inequalities we conclude that $a_n \leq b_1$ for all n. That is, the sequence $\{a_n\}$ is increasing and bounded ($0 \leq a_n \leq b_1$). By the Theorem on Bounded Monotonic Sequences we conclude that $\{a_n\}$ converges. Similarly, since $\{a_n\}$ is increasing, $a_n \geq a_1$ for all n. We combine this inequality with $b_n \geq a_n$ to conclude that $b_n \geq a_1$ for all n. Thus, $\{b_n\}$ is decreasing and bounded ($a_1 \leq b_n \leq b_1$); hence this sequence converges.

To show that $\{a_n\}$ and $\{b_n\}$ converge to the same limit, note that

$$b_n - a_n \leq \frac{b_{n-1} - a_{n-1}}{2} \leq \frac{b_{n-2} - a_{n-2}}{2^2} \leq \cdots \leq \frac{b_1 - a_1}{2^{n-1}}$$

Thus,

$$\lim_{n\to\infty}(b_n - a_n) = (b_1 - a_1)\lim_{n\to\infty}\frac{1}{2^{n-1}} = 0$$

(e) We have

$$a_{n+1} = \sqrt{a_n b_n}, \quad a_1 = 1; \quad b_{n+1} = \frac{a_n + b_n}{2}, \quad b_1 = \sqrt{2}$$

Computing the values of a_n and b_n until the first three decimal digits are equal in successive terms, we obtain:

$$a_2 = \sqrt{a_1 b_1} = \sqrt{1 \cdot \sqrt{2}} = 1.1892$$

$$b_2 = \frac{a_1 + b_1}{2} = \frac{1 + \sqrt{2}}{2} = 1.2071$$

$$a_3 = \sqrt{a_2 b_2} = \sqrt{1.1892 \cdot 1.2071} = 1.1981$$

$$b_3 = \frac{a_2 + b_2}{2} = \frac{1.1892 + 1.2071}{2} = 1.1981$$

$$a_4 = \sqrt{a_3 b_3} = 1.1981$$

$$b_4 = \frac{a_3 + b_3}{2} = 1.1981$$

Thus,

$$AGM\left(1, \sqrt{2}\right) \approx 1.198$$

93. Let $a_n = H_n - \ln n$, where H_n is the nth harmonic number:

$$H_n = 1 + \frac{1}{2} + \frac{1}{3} + \cdots + \frac{1}{n}$$

(a) Show that $a_n \geq 0$ for $n \geq 1$. *Hint:* Show that $H_n \geq \int_1^{n+1} \frac{dx}{x}$.

(b) Show that $\{a_n\}$ is decreasing by interpreting $a_n - a_{n+1}$ as an area.

(c) Prove that $\lim_{n \to \infty} a_n$ exists.

This limit, denoted γ, is known as *Euler's Constant*. It appears in many areas of mathematics, including analysis and number theory, and has been calculated to more than 100 million decimal places, but it is still not known whether γ is an irrational number. The first 10 digits are $\gamma \approx 0.5772156649$.

SOLUTION

(a) Since the function $y = \frac{1}{x}$ is decreasing, the left endpoint approximation to the integral $\int_1^{n+1} \frac{dx}{x}$ is greater than this integral; that is,

$$1 \cdot 1 + \frac{1}{2} \cdot 1 + \frac{1}{3} \cdot 1 + \cdots + \frac{1}{n} \cdot 1 \geq \int_1^{n+1} \frac{dx}{x}$$

or

$$H_n \geq \int_1^{n+1} \frac{dx}{x}$$

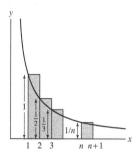

Moreover, since the function $y = \frac{1}{x}$ is positive for $x > 0$, we have:

$$\int_1^{n+1} \frac{dx}{x} \geq \int_1^n \frac{dx}{x}$$

Thus,
$$H_n \geq \int_1^n \frac{dx}{x} = \ln x \Big|_1^n = \ln n - \ln 1 = \ln n$$

and

$$a_n = H_n - \ln n \geq 0 \quad \text{for all } n \geq 1$$

(b) To show that $\{a_n\}$ is decreasing, we consider the difference $a_n - a_{n+1}$:

$$a_n - a_{n+1} = H_n - \ln n - (H_{n+1} - \ln(n+1)) = H_n - H_{n+1} + \ln(n+1) - \ln n$$

$$= 1 + \frac{1}{2} + \cdots + \frac{1}{n} - \left(1 + \frac{1}{2} + \cdots + \frac{1}{n} + \frac{1}{n+1}\right) + \ln(n+1) - \ln n$$

$$= -\frac{1}{n+1} + \ln(n+1) - \ln n$$

Now, $\ln(n+1) - \ln n = \int_n^{n+1} \frac{dx}{x}$, whereas $\frac{1}{n+1}$ is the right endpoint approximation to the integral $\int_n^{n+1} \frac{dx}{x}$. Recalling $y = \frac{1}{x}$ is decreasing, it follows that

$$\int_n^{n+1} \frac{dx}{x} \geq \frac{1}{n+1}$$

so

$$a_n - a_{n+1} \geq 0$$

(c) By parts (a) and (b), $\{a_n\}$ is decreasing and 0 is a lower bound for this sequence. Hence $0 \leq a_n \leq a_1$ for all n. A monotonic and bounded sequence is convergent, so $\lim_{n\to\infty} a_n$ exists.

10.2 Summing an Infinite Series   (LT Section 11.2)

Preliminary Questions

1. What role do partial sums play in defining the sum of an infinite series?

SOLUTION The sum of an infinite series is defined as the limit of the sequence of partial sums. If the limit of this sequence does not exist, the series is said to diverge.

2. What is the sum of the following infinite series?

$$\frac{1}{4} + \frac{1}{8} + \frac{1}{16} + \frac{1}{32} + \frac{1}{64} + \cdots$$

SOLUTION This is a geometric series with $c = \frac{1}{4}$ and $r = \frac{1}{2}$. The sum of the series is therefore

$$\frac{\frac{1}{4}}{1 - \frac{1}{2}} = \frac{\frac{1}{4}}{\frac{1}{2}} = \frac{1}{2}$$

3. What happens if you apply the formula for the sum of a geometric series to the following series? Is the formula valid?

$$1 + 3 + 3^2 + 3^3 + 3^4 + \cdots$$

SOLUTION This is a geometric series with $c = 1$ and $r = 3$. Applying the formula for the sum of a geometric series then gives

$$\sum_{n=0}^{\infty} 3^n = \frac{1}{1-3} = -\frac{1}{2}$$

Clearly, this is not valid: a series with all positive terms cannot have a negative sum. The formula is not valid in this case because a geometric series with $r = 3$ diverges.

4. Indicate whether or not the reasoning in the following statement is correct: $\sum_{n=1}^{\infty} \frac{1}{n^2} = 0$ because $\frac{1}{n^2}$ tends to zero.

SOLUTION This reasoning is not correct. Though the terms in the series do tend to zero, the general term in the sequence of partial sums,

$$S_n = 1 + \frac{1}{2^2} + \frac{1}{3^2} + \cdots + \frac{1}{n^2}$$

is clearly larger than 1. The sum of the series therefore cannot be zero.

5. Indicate whether or not the reasoning in the following statement is correct: $\sum_{n=1}^{\infty} \frac{1}{\sqrt{n}}$ converges because

$$\lim_{n \to \infty} \frac{1}{\sqrt{n}} = 0$$

SOLUTION The reasoning is not correct. Although the general term of a convergent series must tend to zero, a series whose general term tends to zero need not converge. In the case of $\sum_{n=1}^{\infty} \frac{1}{\sqrt{n}}$, the series diverges even though its general term tends to zero.

6. Find an N such that $S_N > 25$ for the series $\sum_{n=1}^{\infty} 2$.

SOLUTION The Nth partial sum of the series is:

$$S_N = \sum_{n=1}^{N} 2 = \underbrace{2 + \cdots + 2}_{N} = 2N$$

7. Does there exist an N such that $S_N > 25$ for the series $\sum_{n=1}^{\infty} 2^{-n}$? Explain.

SOLUTION The series $\sum_{n=1}^{\infty} 2^{-n}$ is a convergent geometric series with the common ratio $r = \frac{1}{2}$. The sum of the series is:

$$S = \frac{\frac{1}{2}}{1 - \frac{1}{2}} = 1$$

Notice that the sequence of partial sums $\{S_N\}$ is increasing and converges to 1; therefore $S_N \leq 1$ for all N. Thus, there does not exist an N such that $S_N > 25$.

8. Give an example of a divergent infinite series whose general term tends to zero.

SOLUTION Consider the series $\sum_{n=1}^{\infty} \frac{1}{n^{\frac{9}{10}}}$. The general term tends to zero, since $\lim_{n \to \infty} \frac{1}{n^{\frac{9}{10}}} = 0$. However, the Nth partial sum satisfies the following inequality:

$$S_N = \frac{1}{1^{\frac{9}{10}}} + \frac{1}{2^{\frac{9}{10}}} + \cdots + \frac{1}{N^{\frac{9}{10}}} \geq \frac{N}{N^{\frac{9}{10}}} = N^{1-\frac{9}{10}} = N^{\frac{1}{10}}$$

That is, $S_N \geq N^{\frac{1}{10}}$ for all N. Since $\lim_{N \to \infty} N^{\frac{1}{10}} = \infty$, the sequence of partial sums S_n diverges; hence, the series $\sum_{n=1}^{\infty} \frac{1}{n^{\frac{9}{10}}}$ diverges.

Exercises

1. Find a formula for the general term a_n (not the partial sum) of the infinite series.

(a) $\dfrac{1}{3} + \dfrac{1}{9} + \dfrac{1}{27} + \dfrac{1}{81} + \cdots$

(b) $\dfrac{1}{1} + \dfrac{5}{2} + \dfrac{25}{4} + \dfrac{125}{8} + \cdots$

(c) $\dfrac{1}{1} - \dfrac{2^2}{2\cdot 1} + \dfrac{3^3}{3\cdot 2\cdot 1} - \dfrac{4^4}{4\cdot 3\cdot 2\cdot 1} + \cdots$

(d) $\dfrac{2}{1^2+1} + \dfrac{1}{2^2+1} + \dfrac{2}{3^2+1} + \dfrac{1}{4^2+1} + \cdots$

SOLUTION

(a) The denominators of the terms are powers of 3, starting with the first power. Hence, the general term is:

$$a_n = \frac{1}{3^n}$$

(b) The numerators are powers of 5, and the denominators are the same powers of 2. The first term is $a_1 = 1$ so,

$$a_n = \left(\frac{5}{2}\right)^{n-1}$$

(c) The general term of this series is

$$a_n = (-1)^{n+1}\frac{n^n}{n!}$$

(d) Notice that the numerators of a_n equal 2 for odd values of n and 1 for even values of n. Thus,

$$a_n = \begin{cases} \dfrac{2}{n^2+1} & \text{odd } n \\ \dfrac{1}{n^2+1} & \text{even } n \end{cases}$$

The formula can also be rewritten as follows:

$$a_n = \frac{1 + \frac{(-1)^{n+1}+1}{2}}{n^2+1}$$

In Exercises 3–6, compute the partial sums S_2, S_4, and S_6.

3. $1 + \dfrac{1}{2^2} + \dfrac{1}{3^2} + \dfrac{1}{4^2} + \cdots$

SOLUTION

$$S_2 = 1 + \frac{1}{2^2} = \frac{5}{4};$$

$$S_4 = 1 + \frac{1}{2^2} + \frac{1}{3^2} + \frac{1}{4^2} = \frac{205}{144};$$

$$S_6 = 1 + \frac{1}{2^2} + \frac{1}{3^2} + \frac{1}{4^2} + \frac{1}{5^2} + \frac{1}{6^2} = \frac{5369}{3600}$$

5. $\dfrac{1}{1\cdot 2} + \dfrac{1}{2\cdot 3} + \dfrac{1}{3\cdot 4} + \cdots$

SOLUTION

$$S_2 = \frac{1}{1\cdot 2} + \frac{1}{2\cdot 3} = \frac{1}{2} + \frac{1}{6} = \frac{4}{6} = \frac{2}{3};$$

$$S_4 = S_2 + a_3 + a_4 = \frac{2}{3} + \frac{1}{3\cdot 4} + \frac{1}{4\cdot 5} = \frac{2}{3} + \frac{1}{12} + \frac{1}{20} = \frac{4}{5};$$

$$S_6 = S_4 + a_5 + a_6 = \frac{4}{5} + \frac{1}{5\cdot 6} + \frac{1}{6\cdot 7} = \frac{4}{5} + \frac{1}{30} + \frac{1}{42} = \frac{6}{7}$$

7. The series $1 + \left(\frac{1}{5}\right) + \left(\frac{1}{5}\right)^2 + \left(\frac{1}{5}\right)^3 + \cdots$ converges to $\frac{5}{4}$. Calculate S_N for $N = 1, 2, \ldots$ until you find an S_N that approximates $\frac{5}{4}$ with an error less than 0.0001.

SOLUTION

$$S_1 = 1$$

$$S_2 = 1 + \frac{1}{5} = \frac{6}{5} = 1.2$$

$$S_3 = 1 + \frac{1}{5} + \frac{1}{25} = \frac{31}{25} = 1.24$$

$$S_3 = 1 + \frac{1}{5} + \frac{1}{25} + \frac{1}{125} = \frac{156}{125} = 1.248$$

$$S_4 = 1 + \frac{1}{5} + \frac{1}{25} + \frac{1}{125} + \frac{1}{625} = \frac{781}{625} = 1.2496$$

$$S_5 = 1 + \frac{1}{5} + \frac{1}{25} + \frac{1}{125} + \frac{1}{625} + \frac{1}{3125} = \frac{3906}{3125} = 1.24992$$

Note that

$$1.25 - S_5 = 1.25 - 1.24992 = 0.00008 < 0.0001$$

In Exercises 9 and 10, use a computer algebra system to compute S_{10}, S_{100}, S_{500}, and S_{1000} for the series. Do these values suggest convergence to the given value?

9. CAS

$$\frac{\pi - 3}{4} = \frac{1}{2 \cdot 3 \cdot 4} - \frac{1}{4 \cdot 5 \cdot 6} + \frac{1}{6 \cdot 7 \cdot 8} - \frac{1}{8 \cdot 9 \cdot 10} + \cdots$$

SOLUTION Write

$$a_n = \frac{(-1)^{n+1}}{2n \cdot (2n+1) \cdot (2n+2)}$$

Then

$$S_N = \sum_{i=1}^{N} a_n$$

Computing, we find

$$\frac{\pi - 3}{4} \approx 0.0353981635$$

$$S_{10} \approx 0.03535167962$$

$$S_{100} \approx 0.03539810274$$

$$S_{500} \approx 0.03539816290$$

$$S_{1000} \approx 0.03539816334$$

It appears that $S_N \to \frac{\pi - 3}{4}$.

11. Calculate S_3, S_4, and S_5 and then find the sum of the telescoping series

$$\sum_{n=1}^{\infty} \left(\frac{1}{n+1} - \frac{1}{n+2}\right)$$

SOLUTION

$$S_3 = \left(\frac{1}{2} - \frac{1}{3}\right) + \left(\frac{1}{3} - \frac{1}{4}\right) + \left(\frac{1}{4} - \frac{1}{5}\right) = \frac{1}{2} - \frac{1}{5} = \frac{3}{10};$$

$$S_4 = S_3 + \left(\frac{1}{5} - \frac{1}{6}\right) = \frac{1}{2} - \frac{1}{6} = \frac{1}{3};$$

$$S_5 = S_4 + \left(\frac{1}{6} - \frac{1}{7}\right) = \frac{1}{2} - \frac{1}{7} = \frac{5}{14}$$

The general term in the sequence of partial sums is

$$S_N = \left(\frac{1}{2} - \frac{1}{3}\right) + \left(\frac{1}{3} - \frac{1}{4}\right) + \left(\frac{1}{4} - \frac{1}{5}\right) + \cdots + \left(\frac{1}{N+1} - \frac{1}{N+2}\right) = \frac{1}{2} - \frac{1}{N+2}$$

thus,

$$S = \lim_{N \to \infty} S_N = \lim_{N \to \infty} \left(\frac{1}{2} - \frac{1}{N+2}\right) = \frac{1}{2}$$

The sum of the telescoping series is therefore $\frac{1}{2}$.

13. Calculate S_3, S_4, and S_5 and then find the sum $\sum_{n=1}^{\infty} \frac{1}{4n^2 - 1}$ using the identity

$$\frac{1}{4n^2 - 1} = \frac{1}{2}\left(\frac{1}{2n-1} - \frac{1}{2n+1}\right)$$

SOLUTION

$$S_3 = \frac{1}{2}\left(\frac{1}{1} - \frac{1}{3}\right) + \frac{1}{2}\left(\frac{1}{3} - \frac{1}{5}\right) + \frac{1}{2}\left(\frac{1}{5} - \frac{1}{7}\right) = \frac{1}{2}\left(1 - \frac{1}{7}\right) = \frac{3}{7};$$

$$S_4 = S_3 + \frac{1}{2}\left(\frac{1}{7} - \frac{1}{9}\right) = \frac{1}{2}\left(1 - \frac{1}{9}\right) = \frac{4}{9};$$

$$S_5 = S_4 + \frac{1}{2}\left(\frac{1}{9} - \frac{1}{11}\right) = \frac{1}{2}\left(1 - \frac{1}{11}\right) = \frac{5}{11}$$

The general term in the sequence of partial sums is

$$S_N = \frac{1}{2}\left(\frac{1}{1} - \frac{1}{3}\right) + \frac{1}{2}\left(\frac{1}{3} - \frac{1}{5}\right) + \frac{1}{2}\left(\frac{1}{5} - \frac{1}{7}\right) + \cdots + \frac{1}{2}\left(\frac{1}{2N-1} - \frac{1}{2N+1}\right) = \frac{1}{2}\left(1 - \frac{1}{2N+1}\right)$$

thus,

$$S = \lim_{N \to \infty} S_N = \lim_{N \to \infty} \frac{1}{2}\left(1 - \frac{1}{2N+1}\right) = \frac{1}{2}$$

15. Find the sum of $\frac{1}{1 \cdot 3} + \frac{1}{3 \cdot 5} + \frac{1}{5 \cdot 7} + \cdots$.

SOLUTION We may write this sum as

$$\sum_{n=1}^{\infty} \frac{1}{(2n-1)(2n+1)} = \sum_{n=1}^{\infty} \frac{1}{2}\left(\frac{1}{2n-1} - \frac{1}{2n+1}\right)$$

The general term in the sequence of partial sums is

$$S_N = \frac{1}{2}\left(\frac{1}{1} - \frac{1}{3}\right) + \frac{1}{2}\left(\frac{1}{3} - \frac{1}{5}\right) + \frac{1}{2}\left(\frac{1}{5} - \frac{1}{7}\right) + \cdots + \frac{1}{2}\left(\frac{1}{2N-1} - \frac{1}{2N+1}\right) = \frac{1}{2}\left(1 - \frac{1}{2N+1}\right)$$

thus,

$$\lim_{N \to \infty} S_N = \lim_{N \to \infty} \frac{1}{2}\left(1 - \frac{1}{2N+1}\right) = \frac{1}{2}$$

and

$$\sum_{n=1}^{\infty} \frac{1}{(2n-1)(2n+1)} = \frac{1}{2}$$

In Exercises 17–22, use the nth Term Divergence Test (Theorem 4) to prove that the following series diverge.

17. $\sum_{n=1}^{\infty} \dfrac{n}{10n+12}$

SOLUTION The general term, $\dfrac{n}{10n+12}$, has limit

$$\lim_{n\to\infty} \frac{n}{10n+12} = \lim_{n\to\infty} \frac{1}{10+(12/n)} = \frac{1}{10}$$

Since the general term does not tend to zero, the series diverges.

19. $\dfrac{0}{1} - \dfrac{1}{2} + \dfrac{2}{3} - \dfrac{3}{4} + \cdots$

SOLUTION The general term $a_n = (-1)^{n-1}\dfrac{n-1}{n}$ does not tend to zero. In fact, because $\lim_{n\to\infty} \dfrac{n-1}{n} = 1$, $\lim_{n\to\infty} a_n$ does not exist. By Theorem 4, we conclude that the given series diverges.

21. $\cos\dfrac{1}{2} + \cos\dfrac{1}{3} + \cos\dfrac{1}{4} + \cdots$

SOLUTION The general term $a_n = \cos\dfrac{1}{n+1}$ tends to 1, not zero. By Theorem 4, we conclude that the given series diverges.

In Exercises 23–38, either use the formula for the sum of a geometric series to find the sum, or state that the series diverges.

23. $1 + \dfrac{1}{6} + \dfrac{1}{36} + \dfrac{1}{216} + \cdots$

SOLUTION This is a geometric series with

$$c = 1 \quad \text{and} \quad r = \frac{1}{6}$$

so its sum is

$$\frac{c}{1-r} = \frac{1}{1-\frac{1}{6}} = \frac{1}{\frac{5}{6}} = \frac{6}{5}$$

25. $\dfrac{7}{3} + \dfrac{7}{3^2} + \dfrac{7}{3^3} + \dfrac{7}{3^4} + \cdots$

SOLUTION This is a geometric series with

$$c = \frac{7}{3} \quad \text{and} \quad r = \frac{1}{3}$$

so its sum is

$$\frac{c}{1-r} = \frac{7/3}{1-\frac{1}{3}} = \frac{7}{3-1} = \frac{7}{2}$$

27. $\sum_{n=3}^{\infty} \left(\dfrac{3}{11}\right)^{-n}$

SOLUTION Rewrite this series as

$$\sum_{n=3}^{\infty} \left(\frac{11}{3}\right)^n$$

This is a geometric series with $r = \dfrac{11}{3} > 1$, so it is divergent.

29. $\sum_{n=-4}^{\infty} \left(-\frac{4}{9}\right)^n$

SOLUTION This is a geometric series with $c = 1$ and $r = -\frac{4}{9}$, starting at $n = -4$. Its sum is thus

$$\frac{cr^{-4}}{1-r} = \frac{c}{r^4 - r^5} = \frac{1}{\frac{4^4}{9^4} + \frac{4^5}{9^5}} = \frac{9^5}{9 \cdot 4^4 + 4^5} = \frac{59{,}049}{3328}$$

31. $\sum_{n=1}^{\infty} e^{-n}$

SOLUTION Rewrite the series as

$$\sum_{n=1}^{\infty} \left(\frac{1}{e}\right)^n$$

to recognize it as a geometric series with $c = \frac{1}{e}$ and $r = \frac{1}{e}$. Thus,

$$\sum_{n=1}^{\infty} e^{-n} = \frac{\frac{1}{e}}{1 - \frac{1}{e}} = \frac{1}{e - 1}$$

33. $\sum_{n=0}^{\infty} \frac{8 + 2^n}{5^n}$

SOLUTION Rewrite the series as

$$\sum_{n=0}^{\infty} \frac{8}{5^n} + \sum_{n=0}^{\infty} \frac{2^n}{5^n} = \sum_{n=0}^{\infty} 8 \cdot \left(\frac{1}{5}\right)^n + \sum_{n=0}^{\infty} \left(\frac{2}{5}\right)^n$$

which is a sum of two geometric series. The first series has $c = 8\left(\frac{1}{5}\right)^0 = 8$ and $r = \frac{1}{5}$; the second has $c = \left(\frac{2}{5}\right)^0 = 1$ and $r = \frac{2}{5}$. Thus,

$$\sum_{n=0}^{\infty} 8 \cdot \left(\frac{1}{5}\right)^n = \frac{8}{1 - \frac{1}{5}} = \frac{8}{\frac{4}{5}} = 10$$

$$\sum_{n=0}^{\infty} \left(\frac{2}{5}\right)^n = \frac{1}{1 - \frac{2}{5}} = \frac{1}{\frac{3}{5}} = \frac{5}{3}$$

and

$$\sum_{n=0}^{\infty} \frac{8 + 2^n}{5^n} = 10 + \frac{5}{3} = \frac{35}{3}$$

35. $5 - \frac{5}{4} + \frac{5}{4^2} - \frac{5}{4^3} + \cdots$

SOLUTION This is a geometric series with $c = 5$ and $r = -\frac{1}{4}$. Thus,

$$\sum_{n=0}^{\infty} 5 \cdot \left(-\frac{1}{4}\right)^n = \frac{5}{1 - \left(-\frac{1}{4}\right)} = \frac{5}{1 + \frac{1}{4}} = \frac{5}{\frac{5}{4}} = 4$$

37. $\frac{7}{8} - \frac{49}{64} + \frac{343}{512} - \frac{2401}{4096} + \cdots$

SOLUTION This is a geometric series with $c = \frac{7}{8}$ and $r = -\frac{7}{8}$. Thus,

$$\sum_{n=0}^{\infty} \frac{7}{8} \cdot \left(-\frac{7}{8}\right)^n = \frac{\frac{7}{8}}{1 - \left(-\frac{7}{8}\right)} = \frac{\frac{7}{8}}{\frac{15}{8}} = \frac{7}{15}$$

In Exercises 39–44, determine a reduced fraction that has this decimal expansion.

39. $0.222\ldots$

SOLUTION   The decimal may be regarded as a geometric series:

$$0.222\ldots = \frac{2}{10} + \frac{2}{100} + \frac{2}{1000} \cdots = \sum_{n=1}^{\infty} \frac{2}{10^n}$$

The series has first term $\frac{2}{10} = \frac{1}{5}$ and ratio $\frac{1}{10}$, so its sum is

$$0.222\ldots = \frac{1/5}{1 - 1/10} = \frac{1}{5} \cdot \frac{10}{9} = \frac{2}{9}$$

41. $0.313131\ldots$

SOLUTION   The decimal may be regarded as a geometric series:

$$0.313131\ldots = \frac{31}{100} + \frac{31}{10000} + \frac{31}{1000000} \cdots = \sum_{n=1}^{\infty} \frac{31}{10^{2n}}$$

The series has first term $\frac{31}{100}$ and ratio $\frac{1}{100}$, so its sum is

$$0.313131\ldots = \frac{31/100}{1 - 1/100} = \frac{31}{100} \cdot \frac{100}{99} = \frac{31}{99}$$

43. $0.123333333\ldots$

SOLUTION   The decimal may be regarded as a constant plus a geometric series:

$$0.123333333\ldots = 0.12 + 0.003 + 0.0003 + 0.00003 + \cdots$$

$$= \frac{12}{100} + \frac{3}{10^3} + \frac{3}{10^4} + \frac{3}{10^5} + \cdots$$

$$= \frac{12}{100} + \sum_{n=1}^{\infty} \frac{3}{10^{n+2}}$$

The series has first term $\frac{3}{1000}$ and ratio $\frac{1}{10}$, so the sum is

$$\frac{12}{100} + \frac{3/1000}{1 - \frac{1}{10}} = \frac{12}{100} + \frac{3}{900} = \frac{111}{900}$$

45. Verify that $0.999999\cdots = 1$ by expressing the left side as a geometric series and determining the sum of the series.

SOLUTION   The decimal may be regarded as a geometric series:

$$0.999999\cdots = \frac{9}{10} + \frac{9}{100} + \frac{9}{1000} + \cdots = \sum_{n=1}^{\infty} \frac{9}{10^n}$$

This series has first term $\frac{9}{10}$ and ratio $\frac{1}{10}$, so its sum is

$$0.999999\cdots = \frac{9/10}{1 - \frac{1}{10}} = \frac{9/10}{9/10} = 1$$

47. Which of the following are *not* geometric series?

(a) $\sum_{n=0}^{\infty} \frac{7^n}{29^n}$

(b) $\sum_{n=3}^{\infty} \frac{1}{n^4}$

(c) $\sum_{n=0}^{\infty} \frac{n^2}{2^n}$

(d) $\sum_{n=5}^{\infty} \pi^{-n}$

SOLUTION

(a) $\sum_{n=0}^{\infty} \frac{7^n}{29^n} = \sum_{n=0}^{\infty} \left(\frac{7}{29}\right)^n$: this is a geometric series with common ratio $r = \frac{7}{29}$.

(b) The ratio between two successive terms is

$$\frac{a_{n+1}}{a_n} = \frac{\frac{1}{(n+1)^4}}{\frac{1}{n^4}} = \frac{n^4}{(n+1)^4} = \left(\frac{n}{n+1}\right)^4$$

This ratio is not constant since it depends on n. Hence, the series $\sum_{n=3}^{\infty} \frac{1}{n^4}$ is not a geometric series.

(c) The ratio between two successive terms is

$$\frac{a_{n+1}}{a_n} = \frac{\frac{(n+1)^2}{2^{n+1}}}{\frac{n^2}{2^n}} = \frac{(n+1)^2}{n^2} \cdot \frac{2^n}{2^{n+1}} = \left(1 + \frac{1}{n}\right)^2 \cdot \frac{1}{2}$$

This ratio is not constant since it depends on n. Hence, the series $\sum_{n=0}^{\infty} \frac{n^2}{2^n}$ is not a geometric series.

(d) $\sum_{n=5}^{\infty} \pi^{-n} = \sum_{n=5}^{\infty} \left(\frac{1}{\pi}\right)^n$: this is a geometric series with common ratio $r = \frac{1}{\pi}$.

49. Prove that if $\sum_{n=1}^{\infty} a_n$ converges and $\sum_{n=1}^{\infty} b_n$ diverges, then $\sum_{n=1}^{\infty} (a_n + b_n)$ diverges. *Hint:* If not, derive a contradiction by writing

$$\sum_{n=1}^{\infty} b_n = \sum_{n=1}^{\infty} (a_n + b_n) - \sum_{n=1}^{\infty} a_n$$

SOLUTION Suppose to the contrary that $\sum_{n=1}^{\infty} a_n$ converges, $\sum_{n=1}^{\infty} b_n$ diverges, but $\sum_{n=1}^{\infty} (a_n + b_n)$ converges. Then by the Linearity of Infinite Series, we have

$$\sum_{n=1}^{\infty} b_n = \sum_{n=1}^{\infty} (a_n + b_n) - \sum_{n=1}^{\infty} a_n$$

so that $\sum_{n=1}^{\infty} b_n$ converges, a contradiction.

51. Give a counterexample to show that each of the following statements is false.

(a) If the general term a_n tends to zero, then $\sum_{n=1}^{\infty} a_n = 0$.

(b) The Nth partial sum of the infinite series defined by $\{a_n\}$ is a_N.

(c) If a_n tends to zero, then $\sum_{n=1}^{\infty} a_n$ converges.

(d) If a_n tends to L, then $\sum_{n=1}^{\infty} a_n = L$.

SOLUTION

(a) If the general term a_n tends to zero, the series may or may not converge. Even if the series converges, it may not converge to zero. For example, with the harmonic series, where $a_n = \frac{1}{n}$, we have $\lim_{n \to \infty} a_n = 0$, but $\sum_{n=1}^{\infty} a_n$ diverges.

(b) The Nth partial sum of the series $\sum_{n=1}^{\infty} a_n$ is $S_N = a_1 + a_2 + \cdots + a_N$ rather than a_N. For example, take the infinite series defined by $a_n = 1$ for all n. Then $a_N = 1$, but the Nth partial sum is N.

(c) If the general term a_n tends to zero, the series may or may not converge. See part (a) for an example.

(d) If $L \neq 0$, then the series diverges by the nth Term Divergence Test. If $L = 0$, the series may or may not converge, and even if it does converge, it may not converge to $L = 0$. For an example of the first case, take the series defined by $a_n = 1$ for all n; for the second, take the harmonic series.

53. Consider the archery competition in Example 6.

(a) Assume that Nina goes first. Let p_n represent the probability that Brook wins on his nth turn. Give an expression for p_n.

(b) Use the result from (a) and a geometric series to determine the probability that Brook wins when Nina goes first.

(c) Now assume that Brook goes first. Use a geometric series to compute the probability that Brook wins the competition.

SOLUTION

(a) The probability p_n that Brook wins on his nth turn when Nina goes first is the probability that Nina misses on her first shot, then both players miss on $n - 1$ consecutive shots, and then Brook hits. This probability is

$$p_n = 0.55 \cdot (0.48 \cdot 0.55)^{n-1} \cdot 0.52 = 0.286 \cdot 0.264^{n-1}$$

(b) p_n is a geometric series with first term ($n = 1$) of 0.286 and a ratio of 0.264, so its sum is

$$p = \sum_{n=1}^{\infty} p_n = \frac{0.286}{1 - 0.264} = \frac{0.286}{0.736} \approx 0.389$$

(c) Let q_n be the probability that Brook wins on his nth turn when he goes first. As before, this is the probability that both players miss on their first $n - 1$ turns times the probability that Brook hits on his nth turn. The probability that both players miss on any given turn is $(0.48)(0.55)$, and the probability that Brook hits on his nth turn is 0.52. Thus

$$q_n = (0.48 \cdot 0.55)^{n-1} \cdot 0.52 = 0.52 \cdot 0.264^{n-1}$$

q_n is a geometric series with first term ($n = 1$) of 0.52 and a ratio of 0.264, so its sum is

$$q = \sum_{n=1}^{\infty} q_n = \frac{0.52}{1 - 0.264} = \frac{0.52}{0.736} \approx 0.707$$

55. Compute the total area of the (infinitely many) triangles in Figure 4.

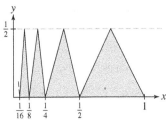

FIGURE 4

SOLUTION The area of a triangle with base B and height H is $A = \frac{1}{2}BH$. Because all of the triangles in Figure 4 have height $\frac{1}{2}$, the area of each triangle equals one-quarter of the base. Now, for $n \geq 0$, the nth triangle has a base which extends from $x = \frac{1}{2^{n+1}}$ to $x = \frac{1}{2^n}$. Thus,

$$B = \frac{1}{2^n} - \frac{1}{2^{n+1}} = \frac{1}{2^{n+1}} \quad \text{and} \quad A = \frac{1}{4}B = \frac{1}{2^{n+3}}$$

The total area of the triangles is then given by the geometric series

$$\sum_{n=0}^{\infty} \frac{1}{2^{n+3}} = \sum_{n=0}^{\infty} \frac{1}{8}\left(\frac{1}{2}\right)^n = \frac{\frac{1}{8}}{1 - \frac{1}{2}} = \frac{1}{4}$$

57. If a patient takes a dose of D units of a particular drug, the amount of the dosage that remains in the patient's bloodstream after t days is De^{-kt}, where k is a positive constant depending on the particular drug.

(a) Show that if the patient takes a dose D every day for an extended period, the amount of drug in the bloodstream approaches $R = \dfrac{De^{-k}}{1 - e^{-k}}$.

(b) Show that if the patient takes a dose D once every t days for an extended period, the amount of drug in the bloodstream approaches $R = \dfrac{De^{-kt}}{1 - e^{-kt}}$.

(c) Suppose that it is considered dangerous to have more than S units of the drug in the bloodstream. What is the minimal time between doses that is safe? *Hint:* $D + R \leq S$.

SOLUTION

(a) The total amount remaining after t days is $\sum_{n=1}^{t} De^{-kn} = D \sum_{n=1}^{t} e^{-kn}$, so that as $t \to \infty$, the amount approaches $D \sum_{n=1}^{\infty} e^{-kn}$. Since $e^{-kn} = D\left(e^{-k}\right)^n = D\left(e^{-k}\right)^{n-1} \cdot e^{-k}$, this is a geometric series with initial term De^{-k} and ratio e^{-k}, so the sum of the series is

$$R = \dfrac{De^{-k}}{1 - e^{-k}}$$

(b) In the equation for part (a), $R = D \sum_{n=1}^{\infty} e^{-kn}$, replace n by nt; this gives the series where dosages are given only on days that are a multiple of t; that is, only every t days. Then

$$R = D \sum_{n=1}^{\infty} e^{-knt} = D \sum_{n=1}^{\infty} \left(e^{-kt}\right)^n = \dfrac{De^{-kt}}{1 - e^{-kt}}$$

summing the geometric series as in part (a).

(c) Since we want the total amount in the bloodstream to be at most S, and since the amount in the bloodstream is at a maximum immediately after a dose is given, we want $D + R \leq S$. From part (b), we get

$$D + R = D + \dfrac{De^{-kt}}{1 - e^{-kt}} \leq S$$

Multiply both sides of this inequality by the positive number $1 - e^{-kt}$, giving

$$D(1 - e^{-kt}) + De^{-kt} \leq S(1 - e^{-kt}), \text{ or } D \leq S(1 - e^{-kt})$$

Simplifying gives $e^{-kt} \leq \dfrac{S-D}{S}$. Take logs and divide both sides by $-k$ (remembering to reverse the inequality on dividing by a negative number), giving

$$t \geq -\dfrac{1}{k} \log \dfrac{S - D}{S} = \dfrac{1}{k} \log \dfrac{S}{S - D}$$

So the minimum amount of time between doses is the expression on the right-hand side of the final inequality.

59. Find the total length of the infinite zigzag path in Figure 5 (each zag occurs at an angle of $\frac{\pi}{4}$).

FIGURE 5

SOLUTION Because the angle at the lower left in Figure 5 has measure $\frac{\pi}{4}$ and each zag in the path occurs at an angle of $\frac{\pi}{4}$, every triangle in the figure is an isosceles right triangle. Accordingly, the length of each new segment in the path is $\frac{1}{\sqrt{2}}$ times the length of the previous segment. Since the first segment has length 1, the total length of the path is

$$\sum_{n=0}^{\infty} \left(\dfrac{1}{\sqrt{2}}\right)^n = \dfrac{1}{1 - \frac{1}{\sqrt{2}}} = \dfrac{\sqrt{2}}{\sqrt{2} - 1} = 2 + \sqrt{2}$$

61. Show that if a is a positive integer, then

$$\sum_{n=1}^{\infty} \frac{1}{n(n+a)} = \frac{1}{a}\left(1 + \frac{1}{2} + \cdots + \frac{1}{a}\right)$$

SOLUTION By partial fraction decomposition

$$\frac{1}{n(n+a)} = \frac{A}{n} + \frac{B}{n+a}$$

clearing the denominators gives

$$1 = A(n+a) + Bn$$

Setting $n = 0$ then yields $A = \frac{1}{a}$, while setting $n = -a$ yields $B = -\frac{1}{a}$. Thus,

$$\frac{1}{n(n+a)} = \frac{\frac{1}{a}}{n} - \frac{\frac{1}{a}}{n+a} = \frac{1}{a}\left(\frac{1}{n} - \frac{1}{n+a}\right)$$

and

$$\sum_{n=1}^{\infty} \frac{1}{n(n+a)} = \sum_{n=1}^{\infty} \frac{1}{a}\left(\frac{1}{n} - \frac{1}{n+a}\right)$$

For $N > a$, the Nth partial sum is

$$S_N = \frac{1}{a}\left(1 + \frac{1}{2} + \frac{1}{3} + \cdots + \frac{1}{a}\right) - \frac{1}{a}\left(\frac{1}{N+1} + \frac{1}{N+2} + \frac{1}{N+3} + \cdots + \frac{1}{N+a}\right)$$

Thus,

$$\sum_{n=1}^{\infty} \frac{1}{n(n+a)} = \lim_{N\to\infty} S_N = \frac{1}{a}\left(1 + \frac{1}{2} + \frac{1}{3} + \cdots + \frac{1}{a}\right)$$

63. In this exercise, we resolve the paradox of Gabriel's Horn (Example 3 in Section 7.7 and Example 7 in Section 8.2). Recall that the horn is the surface formed by rotating $y = \frac{1}{x}$ for $x \geq 1$ around the x-axis. The surface encloses a finite volume and has an infinite surface area. Thus, apparently we can fill the surface with a finite volume of paint, but an infinite volume of paint is required to paint the surface.

(a) Explain that if we can fill the horn with paint, then the paint must be Magic Paint that can be spread arbitrarily thin, thinner than the thickness of the molecules in normal paint.

(b) Explain that if we use Magic Paint, then we can paint the surface of the horn with a finite volume of paint, in fact with just a milliliter of it. *Hint:* A geometric series helps here. Use half of a milliliter to paint that part of the surface between $x = 1$ and $x = 2$.

SOLUTION

(a) If we can fill the horn with paint, then clearly the surface will be painted as well. Since the surface area is infinite but the amount of paint used is finite, the thickness of the portion of the paint on the surface must be arbitrarily small. Thus it must be Magic Paint.

(b) Since we can spread the paint arbitrarily thinly, start by using half a milliliter of paint to paint the part of the surface between $x = 1$ and $x = 2$. For the part between $x = 2$ and $x = 3$, spread the paint so that we need only one-quarter of a milliliter. Between $x = 3$ and $x = 4$, use only one-eighth of a milliliter, and so forth. Then the total amount of Magic Paint used to paint the surface is

$$\frac{1}{2} + \frac{1}{2^2} + \frac{1}{2^3} + \cdots = \frac{1/2}{1-\frac{1}{2}} = 1$$

So using Magic Paint, we can in fact paint the surface with only 1 milliliter of paint.

65. Let $\{b_n\}$ be a sequence and let $a_n = b_n - b_{n-1}$. Show that $\sum_{n=1}^{\infty} a_n$ converges if and only if $\lim_{n \to \infty} b_n$ exists.

SOLUTION Let $a_n = b_n - b_{n-1}$. The general term in the sequence of partial sums for the series $\sum_{n=1}^{\infty} a_n$ is then

$$S_N = (b_1 - b_0) + (b_2 - b_1) + (b_3 - b_2) + \cdots + (b_N - b_{N-1}) = b_N - b_0$$

Now, if $\lim_{N \to \infty} b_N$ exists, then so does $\lim_{N \to \infty} S_N$ and $\sum_{n=1}^{\infty} a_n$ converges. On the other hand, if $\sum_{n=1}^{\infty} a_n$ converges, then $\lim_{N \to \infty} S_N$ exists, which implies that $\lim_{N \to \infty} b_N$ also exists. Thus, $\sum_{n=1}^{\infty} a_n$ converges if and only if $\lim_{n \to \infty} b_n$ exists.

Further Insights and Challenges

In Exercises 67–69, use the formula

$$1 + r + r^2 + \cdots + r^{N-1} = \frac{1 - r^N}{1 - r} \qquad \boxed{6}$$

67. Professor George Andrews of Pennsylvania State University observed that we can use Eq. (6) to calculate the derivative of $f(x) = x^N$ (for $N \geq 0$). Assume that $a \neq 0$ and let $x = ra$. Show that

$$f'(a) = \lim_{x \to a} \frac{x^N - a^N}{x - a} = a^{N-1} \lim_{r \to 1} \frac{r^N - 1}{r - 1}$$

and evaluate the limit.

SOLUTION According to the definition of derivative of $f(x)$ at $x = a$

$$f'(a) = \lim_{x \to a} \frac{x^N - a^N}{x - a}$$

Now, let $x = ra$. Then $x \to a$ if and only if $r \to 1$, and

$$f'(a) = \lim_{x \to a} \frac{x^N - a^N}{x - a} = \lim_{r \to 1} \frac{(ra)^N - a^N}{ra - a} = \lim_{r \to 1} \frac{a^N (r^N - 1)}{a(r - 1)} = a^{N-1} \lim_{r \to 1} \frac{r^N - 1}{r - 1}$$

By formula (6) for a geometric sum,

$$\frac{1 - r^N}{1 - r} = \frac{r^N - 1}{r - 1} = 1 + r + r^2 + \cdots + r^{N-1}$$

so

$$\lim_{r \to 1} \frac{r^N - 1}{r - 1} = \lim_{r \to 1} \left(1 + r + r^2 + \cdots + r^{N-1}\right) = 1 + 1 + 1^2 + \cdots + 1^{N-1} = N$$

Therefore, $f'(a) = a^{N-1} \cdot N = Na^{N-1}$.

69. Verify the Gregory–Leibniz formula in part (d) as follows.

(a) Set $r = -x^2$ in Eq. (6) and rearrange to show that

$$\frac{1}{1 + x^2} = 1 - x^2 + x^4 - \cdots + (-1)^{N-1} x^{2N-2} + \frac{(-1)^N x^{2N}}{1 + x^2}$$

(b) Show, by integrating over $[0, 1]$, that

$$\frac{\pi}{4} = 1 - \frac{1}{3} + \frac{1}{5} - \frac{1}{7} + \cdots + \frac{(-1)^{N-1}}{2N - 1} + (-1)^N \int_0^1 \frac{x^{2N} \, dx}{1 + x^2}$$

(c) Use the Comparison Theorem for integrals to prove that

$$0 \leq \int_0^1 \frac{x^{2N} \, dx}{1 + x^2} \leq \frac{1}{2N + 1}$$

Hint: Observe that the integrand is $\leq x^{2N}$.

(d) Prove that
$$\frac{\pi}{4} = 1 - \frac{1}{3} + \frac{1}{5} - \frac{1}{7} + \frac{1}{9} - \cdots$$

Hint: Use (b) and (c) to show that the partial sums S_N satisfy $\left|S_N - \frac{\pi}{4}\right| \leq \frac{1}{2N+1}$, and thereby conclude that $\lim_{N \to \infty} S_N = \frac{\pi}{4}$.

SOLUTION

(a) Start with Eq. (6), and substitute $-x^2$ for r:

$$1 + r + r^2 + \cdots + r^{N-1} = \frac{1 - r^N}{1 - r}$$

$$1 - x^2 + x^4 + \cdots + (-1)^{N-1} x^{2N-2} = \frac{1 - (-1)^N x^{2N}}{1 - (-x^2)}$$

$$1 - x^2 + x^4 + \cdots + (-1)^{N-1} x^{2N-2} = \frac{1}{1 + x^2} - \frac{(-1)^N x^{2N}}{1 + x^2}$$

$$\frac{1}{1 + x^2} = 1 - x^2 + x^4 + \cdots + (-1)^{N-1} x^{2N-2} + \frac{(-1)^N x^{2N}}{1 + x^2}$$

(b) The integrals of both sides must be equal. Now,

$$\int_0^1 \frac{1}{1 + x^2}\, dx = \tan^{-1} x \Big|_0^1 = \tan^{-1} 1 - \tan^{-1} 0 = \frac{\pi}{4}$$

while

$$\int_0^1 \left(1 - x^2 + x^4 + \cdots + (-1)^{N-1} x^{2N-2} + \frac{(-1)^N x^{2N}}{1 + x^2}\right) dx$$

$$= \left(x - \frac{1}{3}x^3 + \frac{1}{5}x^5 + \cdots + (-1)^{N-1} \frac{1}{2N-1} x^{2N-1}\right) + (-1)^N \int_0^1 \frac{x^{2N}\, dx}{1 + x^2}$$

$$= 1 - \frac{1}{3} + \frac{1}{5} + \cdots + (-1)^{N-1} \frac{1}{2N-1} + (-1)^N \int_0^1 \frac{x^{2N}\, dx}{1 + x^2}$$

(c) Note that for $x \in [0, 1]$, we have $1 + x^2 \geq 1$, so that

$$0 \leq \frac{x^{2N}}{1 + x^2} \leq x^{2N}$$

By the Comparison Theorem for integrals, we then see that

$$0 \leq \int_0^1 \frac{x^{2N}\, dx}{1 + x^2} \leq \int_0^1 x^{2N}\, dx = \frac{1}{2N+1} x^{2N+1} \Big|_0^1 = \frac{1}{2N+1}$$

(d) Write

$$a_n = (-1)^n \frac{1}{2n - 1}, \quad n \geq 1$$

and let S_N be the partial sums. Then

$$\left|S_N - \frac{\pi}{4}\right| = \left|(-1)^N \int_0^1 \frac{x^{2N}\, dx}{1 + x^2}\right| = \int_0^1 \frac{x^{2N}\, dx}{1 + x^2} \leq \frac{1}{2N + 1}$$

Thus $\lim_{N \to \infty} S_N = \frac{\pi}{4}$ so that

$$\frac{\pi}{4} = 1 - \frac{1}{3} + \frac{1}{5} - \frac{1}{7} + \frac{1}{9} - \cdots$$

71. The **Koch snowflake** (described in 1904 by Swedish mathematician Helge von Koch) is an infinitely jagged "fractal" curve obtained as a limit of polygonal curves (it is continuous but has no tangent line at any point). Begin with an equilateral triangle (Stage 0) and produce Stage 1 by replacing each edge with four edges of one-third the length, arranged as in Figure 9. Continue the process: At the nth stage, replace each edge with four edges of one-third the length of the edge from the $(n - 1)$st stage.

(a) Show that the perimeter P_n of the polygon at the nth stage satisfies $P_n = \frac{4}{3}P_{n-1}$. Prove that $\lim\limits_{n \to \infty} P_n = \infty$. The snowflake has infinite length.

(b) Let A_0 be the area of the original equilateral triangle. Show that $(3)4^{n-1}$ new triangles are added at the nth stage, each with area $A_0/9^n$ (for $n \geq 1$). Show that the total area of the Koch snowflake is $\frac{8}{5}A_0$.

Stage 1   Stage 2   Stage 3

FIGURE 9

SOLUTION

(a) Each edge of the polygon at the $(n-1)$st stage is replaced by four edges of one-third the length; hence the perimeter of the polygon at the nth stage is $\frac{4}{3}$ times the perimeter of the polygon at the $(n-1)$th stage. That is, $P_n = \frac{4}{3}P_{n-1}$. Thus,

$$P_1 = \frac{4}{3}P_0; \quad P_2 = \frac{4}{3}P_1 = \left(\frac{4}{3}\right)^2 P_0, \quad P_3 = \frac{4}{3}P_2 = \left(\frac{4}{3}\right)^3 P_0$$

and, in general, $P_n = \left(\frac{4}{3}\right)^n P_0$. As $n \to \infty$, it follows that

$$\lim_{n \to \infty} P_n = P_0 \lim_{n \to \infty} \left(\frac{4}{3}\right)^n = \infty$$

(b) When each edge is replaced by four edges of one-third the length, one new triangle is created. At the $(n-1)$st stage, there are $3 \cdot 4^{n-1}$ edges in the snowflake, so $3 \cdot 4^{n-1}$ new triangles are generated at the nth stage. Because the area of an equilateral triangle is proportional to the square of its side length and the side length for each new triangle is one-third the side length of triangles from the previous stage, it follows that the area of the triangles added at each stage is reduced by a factor of $\frac{1}{9}$ from the area of the triangles added at the previous stage. Thus, each triangle added at the nth stage has an area of $A_0/9^n$. This means that the nth stage contributes

$$3 \cdot 4^{n-1} \cdot \frac{A_0}{9^n} = \frac{3}{4}A_0\left(\frac{4}{9}\right)^n$$

to the area of the snowflake. The total area is therefore

$$A = A_0 + \frac{3}{4}A_0 \sum_{n=1}^{\infty} \left(\frac{4}{9}\right)^n = A_0 + \frac{3}{4}A_0 \frac{\frac{4}{9}}{1 - \frac{4}{9}} = A_0 + \frac{3}{4}A_0 \cdot \frac{4}{5} = \frac{8}{5}A_0$$

10.3 Convergence of Series with Positive Terms (LT Section 11.3)

Preliminary Questions

1. For the series $\sum\limits_{n=1}^{\infty} a_n$, if the partial sums S_N are increasing, then (choose the correct conclusion)

(a) $\{a_n\}$ is an increasing sequence.

(b) $\{a_n\}$ is a positive sequence.

SOLUTION The correct response is **(b)**. Recall that $S_N = a_1 + a_2 + a_3 + \cdots + a_N$; thus, $S_N - S_{N-1} = a_N$. If S_N is increasing, then $S_N - S_{N-1} \geq 0$. It then follows that $a_N \geq 0$; that is, $\{a_n\}$ is a positive sequence.

2. What are the hypotheses of the Integral Test?

SOLUTION The hypotheses for the Integral Test are: A function $f(x)$ such that $a_n = f(n)$ must be positive, decreasing, and continuous for $x \geq 1$.

3. Which test would you use to determine whether $\sum_{n=1}^{\infty} n^{-3.2}$ converges?

SOLUTION Because $n^{-3.2} = \frac{1}{n^{3.2}}$, we see that the indicated series is a p-series with $p = 3.2 > 1$. Therefore, the series converges.

4. Which test would you use to determine whether $\sum_{n=1}^{\infty} \frac{1}{2^n + \sqrt{n}}$ converges?

SOLUTION Because

$$\frac{1}{2^n + \sqrt{n}} < \frac{1}{2^n} = \left(\frac{1}{2}\right)^n$$

and

$$\sum_{n=1}^{\infty} \left(\frac{1}{2}\right)^n$$

is a convergent geometric series, the Direct Comparison Test would be an appropriate choice to establish that the given series converges.

5. Ralph hopes to investigate the convergence of $\sum_{n=1}^{\infty} \frac{e^{-n}}{n}$ by comparing it with $\sum_{n=1}^{\infty} \frac{1}{n}$. Is Ralph on the right track?

SOLUTION No, Ralph is not on the right track. For $n \geq 1$,

$$\frac{e^{-n}}{n} < \frac{1}{n}$$

however, $\sum_{n=1}^{\infty} \frac{1}{n}$ is a divergent series. The Direct Comparison Test therefore does not allow us to draw a conclusion about the convergence or divergence of the series $\sum_{n=1}^{\infty} \frac{e^{-n}}{n}$.

Exercises

In Exercises 1–12, use the Integral Test to determine whether the infinite series is convergent.

1. $\sum_{n=1}^{\infty} \frac{1}{(n+1)^4}$

SOLUTION Let $f(x) = \frac{1}{(x+1)^4}$. This function is continuous, positive, and decreasing on the interval $x \geq 1$, so the Integral Test applies. Moreover,

$$\int_1^{\infty} \frac{dx}{(x+1)^4} = \lim_{R \to \infty} \int_1^R \frac{dx}{(x+1)^4} = \lim_{R \to \infty} \left(-\frac{1}{3(x+1)^3}\Big|_1^R\right) = \lim_{R \to \infty} \left(\frac{1}{24} - \frac{1}{3(R+1)^3}\right) = \frac{1}{24}$$

The integral converges; hence, the series $\sum_{n=1}^{\infty} \frac{1}{(n+1)^4}$ also converges.

3. $\sum_{n=1}^{\infty} n^{-1/3}$

SOLUTION Let $f(x) = x^{-\frac{1}{3}} = \frac{1}{\sqrt[3]{x}}$. This function is continuous, positive, and decreasing on the interval $x \geq 1$, so the Integral Test applies. Moreover,

$$\int_1^{\infty} x^{-1/3}\, dx = \lim_{R \to \infty} \int_1^R x^{-1/3}\, dx = \frac{3}{2} \lim_{R \to \infty} \left(R^{2/3} - 1\right) = \infty$$

The integral diverges; hence, the series $\sum_{n=1}^{\infty} n^{-1/3}$ also diverges.

5. $\sum_{n=25}^{\infty} \dfrac{n^2}{(n^3+9)^{5/2}}$

SOLUTION Let $f(x) = \dfrac{x^2}{(x^3+9)^{5/2}}$. This function is positive and continuous for $x \geq 25$. Moreover, because

$$f'(x) = \dfrac{2x(x^3+9)^{5/2} - x^2 \cdot \tfrac{5}{2}(x^3+9)^{3/2} \cdot 3x^2}{(x^3+9)^5} = \dfrac{x(36-11x^3)}{2(x^3+9)^{7/2}}$$

we see that $f'(x) < 0$ for $x \geq 25$, so f is decreasing on the interval $x \geq 25$. The Integral Test therefore applies. To evaluate the improper integral, we use the substitution $u = x^3 + 9$, $du = 3x^2 dx$. We then find

$$\int_{25}^{\infty} \dfrac{x^2}{(x^3+9)^{5/2}}\, dx = \lim_{R\to\infty} \int_{25}^{R} \dfrac{x^2}{(x^3+9)^{5/2}}\, dx = \dfrac{1}{3}\lim_{R\to\infty} \int_{15634}^{R^3+9} \dfrac{du}{u^{5/2}}$$

$$= -\dfrac{2}{9}\lim_{R\to\infty} \left(\dfrac{1}{(R^3+9)^{3/2}} - \dfrac{1}{15634^{3/2}}\right) = \dfrac{2}{9 \cdot 15634^{3/2}}$$

The integral converges; hence, the series $\sum_{n=25}^{\infty} \dfrac{n^2}{(n^3+9)^{5/2}}$ also converges.

7. $\sum_{n=1}^{\infty} \dfrac{1}{n^2+1}$

SOLUTION Let $f(x) = \dfrac{1}{x^2+1}$. This function is positive, decreasing and continuous on the interval $x \geq 1$, hence the Integral Test applies. Moreover,

$$\int_1^{\infty} \dfrac{dx}{x^2+1} = \lim_{R\to\infty}\int_1^R \dfrac{dx}{x^2+1} = \lim_{R\to\infty}\left(\tan^{-1} R - \dfrac{\pi}{4}\right) = \dfrac{\pi}{2} - \dfrac{\pi}{4} = \dfrac{\pi}{4}$$

The integral converges; hence, the series $\sum_{n=1}^{\infty}\dfrac{1}{n^2+1}$ also converges.

9. $\sum_{n=1}^{\infty} \dfrac{1}{n(n+5)}$

SOLUTION Let $f(x) = \dfrac{1}{x(x+5)}$. This function is positive, continuous, and decreasing on the interval $x \geq 1$, so the Integral Test applies. We compute the improper integral using partial fractions:

$$\int_1^{\infty}\dfrac{dx}{x(x+5)} = \dfrac{1}{5}\lim_{R\to\infty}\int_1^R \left(\dfrac{1}{x} - \dfrac{1}{x+5}\right) dx = \dfrac{1}{5}\lim_{R\to\infty} \ln\dfrac{x}{x+5}\bigg|_1^R = \dfrac{1}{5}\lim_{R\to\infty}\left(\ln\dfrac{R}{R+5} - \ln\dfrac{1}{6}\right)$$

$$= \dfrac{1}{5}\left(\ln 1 - \ln\dfrac{1}{6}\right) = \dfrac{1}{5}\ln 6$$

The integral converges; hence, the series $\sum_{n=1}^{\infty}\dfrac{1}{n(n+5)}$ converges.

11. $\sum_{n=2}^{\infty} \dfrac{1}{n(\ln n)^{3/2}}$

SOLUTION Let $f(x) = \dfrac{1}{x(\ln x)^{3/2}}$. This function is positive and continuous for $x \geq 2$. Moreover,

$$f'(x) = -\dfrac{1}{x^2(\ln x)^3}\left(1 \cdot (\ln x)^{3/2} + x \cdot \dfrac{3}{2}(\ln x)^{1/2} \cdot \dfrac{1}{x}\right) = -\dfrac{3 + 2\ln x}{2x^2(\ln x)^{5/2}} \tag{1}$$

Since $\ln x > 0$ for $x > 1$, $f'(x)$ is negative for $x > 1$; hence, f is decreasing for $x \geq 2$. To compute the improper integral, we make the substitution $u = \ln x$, $du = \dfrac{1}{x}dx$. We obtain:

$$\int_2^\infty \frac{1}{x(\ln x)^{3/2}} dx = \lim_{R\to\infty} \int_2^R \frac{1}{x(\ln x)^{3/2}} dx = \lim_{R\to\infty} \int_{\ln 2}^{\ln R} \frac{du}{u^{3/2}}$$

$$= -\lim_{R\to\infty} 2u^{-1/2}\Big|_{u=\ln 2}^{\ln R}$$

$$= -2\lim_{R\to\infty} \left((\ln R)^{-1/2} - (\ln 2)^{-1/2}\right) = 2(\ln 2)^{-1/2}$$

The integral converges; hence, the series $\sum_{n=2}^\infty \frac{1}{n(\ln n)^{3/2}}$ also converges.

13. Show that $\sum_{n=1}^\infty \frac{1}{n^3 + 8n}$ converges by using the Direct Comparison Test with $\sum_{n=1}^\infty n^{-3}$.

SOLUTION We compare the series with the p-series $\sum_{n=1}^\infty n^{-3}$. For $n \geq 1$,

$$\frac{1}{n^3 + 8n} \leq \frac{1}{n^3}$$

Since $\sum_{n=1}^\infty \frac{1}{n^3}$ converges (it is a p-series with $p = 3 > 1$), the series $\sum_{n=1}^\infty \frac{1}{n^3 + 8n}$ also converges by the Direct Comparison Test.

15. For $\sum_{n=1}^\infty \frac{1}{n + \sqrt{n}}$, verify that for $n \geq 1$,

$$\frac{1}{n + \sqrt{n}} \leq \frac{1}{n}, \qquad \frac{1}{n + \sqrt{n}} \leq \frac{1}{\sqrt{n}}$$

Can either inequality be used to show that the series diverges? Show that $\frac{1}{n + \sqrt{n}} \geq \frac{1}{2n}$ for $n \geq 1$ and conclude that the series diverges.

SOLUTION For $n \geq 1$, $n + \sqrt{n} \geq n$ and $n + \sqrt{n} \geq \sqrt{n}$. Taking the reciprocal of each of these inequalities yields

$$\frac{1}{n + \sqrt{n}} \leq \frac{1}{n} \quad \text{and} \quad \frac{1}{n + \sqrt{n}} \leq \frac{1}{\sqrt{n}}$$

These inequalities indicate that the series $\sum_{n=1}^\infty \frac{1}{n + \sqrt{n}}$ is smaller than both $\sum_{n=1}^\infty \frac{1}{n}$ and $\sum_{n=1}^\infty \frac{1}{\sqrt{n}}$; however, $\sum_{n=1}^\infty \frac{1}{n}$ and $\sum_{n=1}^\infty \frac{1}{\sqrt{n}}$ both diverge so neither inequality allows us to show that the original series diverges.

On the other hand, for $n \geq 1$, $n \geq \sqrt{n}$, so $2n \geq n + \sqrt{n}$ and

$$\frac{1}{n + \sqrt{n}} \geq \frac{1}{2n}$$

The series $\sum_{n=1}^\infty \frac{1}{2n} = 2 \sum_{n=1}^\infty \frac{1}{n}$ diverges, since the harmonic series diverges. The Direct Comparison Test then lets us conclude that the larger series $\sum_{n=1}^\infty \frac{1}{n + \sqrt{n}}$ also diverges.

In Exercises 17–28, use the Direct Comparison Test to determine whether the infinite series is convergent.

17. $\sum_{n=1}^\infty \frac{1}{n 2^n}$

SOLUTION We compare with the geometric series $\sum_{n=1}^{\infty} \left(\frac{1}{2}\right)^n$. For $n \geq 1$,

$$\frac{1}{n2^n} \leq \frac{1}{2^n} = \left(\frac{1}{2}\right)^n$$

Since $\sum_{n=1}^{\infty} \left(\frac{1}{2}\right)^n$ converges (it is a geometric series with $r = \frac{1}{2}$), we conclude by the Direct Comparison Test that $\sum_{n=1}^{\infty} \frac{1}{n2^n}$ also converges.

19. $\sum_{n=1}^{\infty} \frac{1}{n^{1/3} + 2^n}$

SOLUTION For $n \geq 1$,

$$\frac{1}{n^{1/3} + 2^n} \leq \frac{1}{2^n}$$

The series $\sum_{n=1}^{\infty} \frac{1}{2^n}$ is a geometric series with $r = \frac{1}{2}$, so it converges. By the Comparison test, so does $\sum_{n=1}^{\infty} \frac{1}{n^{1/3} + 2^n}$.

21. $\sum_{m=1}^{\infty} \frac{4}{m! + 4^m}$

SOLUTION For $m \geq 1$,

$$\frac{4}{m! + 4^m} \leq \frac{4}{4^m} = \left(\frac{1}{4}\right)^{m-1}$$

The series $\sum_{m=1}^{\infty} \left(\frac{1}{4}\right)^{m-1}$ is a geometric series with $r = \frac{1}{4}$, so it converges. By the Direct Comparison Test we can therefore conclude that the series $\sum_{m=1}^{\infty} \frac{4}{m! + 4^m}$ also converges.

23. $\sum_{k=1}^{\infty} \frac{\sin^2 k}{k^2}$

SOLUTION For $k \geq 1$, $0 \leq \sin^2 k \leq 1$, so

$$0 \leq \frac{\sin^2 k}{k^2} \leq \frac{1}{k^2}$$

The series $\sum_{k=1}^{\infty} \frac{1}{k^2}$ is a p-series with $p = 2 > 1$, so it converges. By the Direct Comparison Test we can therefore conclude that the series $\sum_{k=1}^{\infty} \frac{\sin^2 k}{k^2}$ also converges.

25. $\sum_{n=1}^{\infty} \frac{2}{3^n + 3^{-n}}$

SOLUTION Since $3^{-n} > 0$ for all n,

$$\frac{2}{3^n + 3^{-n}} \leq \frac{2}{3^n} = 2\left(\frac{1}{3}\right)^n$$

The series $\sum_{n=1}^{\infty} 2\left(\frac{1}{3}\right)^n$ is a geometric series with $r = \frac{1}{3}$, so it converges. By the Direct Comparison Test we can therefore conclude that the series $\sum_{n=1}^{\infty} \frac{2}{3^n + 3^{-n}}$ also converges.

27. $\sum_{n=1}^{\infty} \dfrac{1}{(n+1)!}$

SOLUTION Note that for $n \geq 2$,

$$(n+1)! = 1 \cdot \underbrace{2 \cdot 3 \cdots n \cdot (n+1)}_{n \text{ factors}} \leq 2^n$$

so that

$$\sum_{n=1}^{\infty} \dfrac{1}{(n+1)!} = 1 + \sum_{n=2}^{\infty} \dfrac{1}{(n+1)!} \leq 1 + \sum_{n=2}^{\infty} \dfrac{1}{2^n}$$

But $\sum_{n=2}^{\infty} \dfrac{1}{2^n}$ is a geometric series with ratio $r = \dfrac{1}{2}$, so it converges. By the comparison test, $\sum_{n=1}^{\infty} \dfrac{1}{(n+1)!}$ converges as well.

Exercise 29–34: For all $a > 0$ and $b > 1$, the inequalities $\ln n \leq n^a$, $n^a < b^n$ are true for n sufficiently large (this can be proved using L'Hôpital's Rule). Use this, together with the Direct Comparison Test, to determine whether the series converges or diverges.

29. $\sum_{n=1}^{\infty} \dfrac{\ln n}{n^3}$

SOLUTION For n sufficiently large (say $n = k$, although in this case $n = 1$ suffices), we have $\ln n \leq n$, so that

$$\sum_{n=k}^{\infty} \dfrac{\ln n}{n^3} \leq \sum_{n=k}^{\infty} \dfrac{n}{n^3} = \sum_{n=k}^{\infty} \dfrac{1}{n^2}$$

This is a p-series with $p = 2 > 1$, so it converges. Thus $\sum_{n=k}^{\infty} \dfrac{\ln n}{n^3}$ also converges; adding back in the finite number of terms for $1 \leq n \leq k$ does not affect this result.

31. $\sum_{n=1}^{\infty} \dfrac{(\ln n)^{100}}{n^{1.1}}$

SOLUTION Choose N so that $\ln n \leq n^{0.0005}$ for $n \geq N$. Then also for $n > N$, $(\ln n)^{100} \leq (n^{0.0005})^{100} = n^{0.05}$. Then

$$\sum_{n=N}^{\infty} \dfrac{(\ln n)^{100}}{n^{1.1}} \leq \sum_{n=N}^{\infty} \dfrac{n^{0.05}}{n^{1.1}} = \sum_{n=N}^{\infty} \dfrac{1}{n^{1.05}}$$

But $\sum_{n=N}^{\infty} \dfrac{1}{n^{1.05}}$ is a p-series with $p = 1.05 > 1$, so is convergent. It follows that $\sum_{n=N}^{\infty} \dfrac{(\ln n)^{100}}{n^{1.1}}$ is also convergent; adding back in the finite number of terms for $n = 1, 2, \ldots, N - 1$ shows that $\sum_{n=1}^{\infty} \dfrac{(\ln n)^{100}}{n^{1.1}}$ converges as well.

33. $\sum_{n=1}^{\infty} \dfrac{n}{3^n}$

SOLUTION Choose N such that $n \leq 2^n$ for $n \geq N$. Then

$$\sum_{n=N}^{\infty} \dfrac{n}{3^n} \leq \sum_{n=N}^{\infty} \left(\dfrac{2}{3}\right)^n$$

The latter sum is a geometric series with $r = \dfrac{2}{3} < 1$, so it converges. Thus the series on the left converges as well. Adding back in the finite number of terms for $n < N$ shows that $\sum_{n=1}^{\infty} \dfrac{n}{3^n}$ converges.

35. Show that $\sum_{n=1}^{\infty} \sin \frac{1}{n^2}$ converges. *Hint:* Use $\sin x \leq x$ for $x \geq 0$.

SOLUTION For $n \geq 1$,

$$0 \leq \frac{1}{n^2} \leq 1 < \pi$$

therefore, $\sin \frac{1}{n^2} > 0$ for $n \geq 1$. Moreover, for $n \geq 1$,

$$\sin \frac{1}{n^2} \leq \frac{1}{n^2}$$

The series $\sum_{n=1}^{\infty} \frac{1}{n^2}$ is a p-series with $p = 2 > 1$, so it converges. By the Direct Comparison Test we can therefore conclude that the series $\sum_{n=1}^{\infty} \sin \frac{1}{n^2}$ also converges.

In Exercises 37–46, use the Limit Comparison Test to prove convergence or divergence of the infinite series.

37. $\sum_{n=2}^{\infty} \frac{n^2}{n^4 - 1}$

SOLUTION Let $a_n = \frac{n^2}{n^4 - 1}$. For large n, $\frac{n^2}{n^4 - 1} \approx \frac{n^2}{n^4} = \frac{1}{n^2}$, so we apply the Limit Comparison Test with $b_n = \frac{1}{n^2}$. We find

$$L = \lim_{n \to \infty} \frac{a_n}{b_n} = \lim_{n \to \infty} \frac{\frac{n^2}{n^4 - 1}}{\frac{1}{n^2}} = \lim_{n \to \infty} \frac{n^4}{n^4 - 1} = 1$$

The series $\sum_{n=1}^{\infty} \frac{1}{n^2}$ is a p-series with $p = 2 > 1$, so it converges; hence, $\sum_{n=2}^{\infty} \frac{1}{n^2}$ also converges. Because L exists, by the Limit Comparison Test we can conclude that the series $\sum_{n=2}^{\infty} \frac{n^2}{n^4 - 1}$ converges.

39. $\sum_{n=2}^{\infty} \frac{n}{\sqrt{n^3 + 1}}$

SOLUTION Let $a_n = \frac{n}{\sqrt{n^3 - 1}}$. For large n, $\frac{n}{\sqrt{n^3 - 1}} \approx \frac{n}{\sqrt{n^3}} = \frac{1}{\sqrt{n}}$, so we apply the Limit Comparison test with $b_n = \frac{1}{\sqrt{n}}$. We find

$$L = \lim_{n \to \infty} \frac{a_n}{b_n} = \lim_{n \to \infty} \frac{\frac{n}{\sqrt{n^3 - 1}}}{\frac{1}{\sqrt{n}}} = \lim_{n \to \infty} \frac{\sqrt{n^3}}{\sqrt{n^3 - 1}} = 1$$

The series $\sum_{n=1}^{\infty} \frac{1}{\sqrt{n}}$ is a p-series with $p = \frac{1}{2} < 1$, so it diverges; hence, $\sum_{n=2}^{\infty} \frac{1}{\sqrt{n}}$ also diverges. Because $L > 0$, by the Limit Comparison Test we can conclude that the series $\sum_{n=2}^{\infty} \frac{n}{\sqrt{n^3 - 1}}$ diverges.

41. $\sum_{n=3}^{\infty} \dfrac{3n+5}{n(n-1)(n-2)}$

SOLUTION Let $a_n = \dfrac{3n+5}{n(n-1)(n-2)}$. For large n, $\dfrac{3n+5}{n(n-1)(n-2)} \approx \dfrac{3n}{n^3} = \dfrac{3}{n^2}$, so we apply the Limit Comparison Test with $b_n = \dfrac{1}{n^2}$. We find

$$L = \lim_{n\to\infty} \dfrac{a_n}{b_n} = \lim_{n\to\infty} \dfrac{\frac{3n+5}{n(n+1)(n+2)}}{\frac{1}{n^2}} = \lim_{n\to\infty} \dfrac{3n^3 + 5n^2}{n(n+1)(n+2)} = 3$$

The series $\sum_{n=1}^{\infty} \dfrac{1}{n^2}$ is a p-series with $p = 2 > 1$, so it converges; hence, the series $\sum_{n=3}^{\infty} \dfrac{1}{n^2}$ also converges. Because L exists, by the Limit Comparison Test we can conclude that the series $\sum_{n=3}^{\infty} \dfrac{3n+5}{n(n-1)(n-2)}$ converges.

43. $\sum_{n=1}^{\infty} \dfrac{1}{\sqrt{n} + \ln n}$

SOLUTION Let

$$a_n = \dfrac{1}{\sqrt{n} + \ln n}$$

For large n, $\sqrt{n} + \ln n \approx \sqrt{n}$, so apply the Comparison Test with $b_n = \dfrac{1}{\sqrt{n}}$. We find

$$L = \lim_{n\to\infty} \dfrac{a_n}{b_n} = \lim_{n\to\infty} \dfrac{1}{\sqrt{n}+\ln n} \cdot \dfrac{\sqrt{n}}{1} = \lim_{n\to\infty} \dfrac{1}{1+\frac{\ln n}{\sqrt{n}}} = 1$$

The series $\sum_{n=1}^{\infty} \dfrac{1}{\sqrt{n}}$ is a p-series with $p = \dfrac{1}{2} < 1$, so it diverges. Because L exists, the Limit Comparison Test tells us the the original series also diverges.

45. $\sum_{n=1}^{\infty} \left(1 - \cos \dfrac{1}{n}\right)$ *Hint:* Compare with $\sum_{n=1}^{\infty} n^{-2}$.

SOLUTION Let $a_n = 1 - \cos \dfrac{1}{n}$, and apply the Limit Comparison Test with $b_n = \dfrac{1}{n^2}$. We find

$$L = \lim_{n\to\infty} \dfrac{a_n}{b_n} = \lim_{n\to\infty} \dfrac{1-\cos\frac{1}{n}}{\frac{1}{n^2}} = \lim_{x\to\infty} \dfrac{1-\cos\frac{1}{x}}{\frac{1}{x^2}} = \lim_{x\to\infty} \dfrac{-\frac{1}{x^2}\sin\frac{1}{x}}{-\frac{2}{x^3}} = \dfrac{1}{2} \lim_{x\to\infty} \dfrac{\sin\frac{1}{x}}{\frac{1}{x}}$$

As $x \to \infty$, $u = \dfrac{1}{x} \to 0$, so

$$L = \dfrac{1}{2} \lim_{x\to\infty} \dfrac{\sin\frac{1}{x}}{\frac{1}{x}} = \dfrac{1}{2} \lim_{u\to 0} \dfrac{\sin u}{u} = \dfrac{1}{2}$$

The series $\sum_{n=1}^{\infty} \dfrac{1}{n^2}$ is a p-series with $p = 2 > 1$, so it converges. Because L exists, by the Limit Comparison Test we can conclude that the series $\sum_{n=1}^{\infty} \left(1 - \cos \dfrac{1}{n}\right)$ also converges.

720 CHAPTER 10 | INFINITE SERIES (LT CHAPTER 11)

In Exercises 47–76, determine convergence or divergence using any method covered so far.

47. $\sum_{n=4}^{\infty} \dfrac{1}{n^2 - 9}$

SOLUTION Apply the Limit Comparison Test with $a_n = \dfrac{1}{n^2 - 9}$ and $b_n = \dfrac{1}{n^2}$:

$$L = \lim_{n \to \infty} \frac{a_n}{b_n} = \lim_{n \to \infty} \frac{\frac{1}{n^2-9}}{\frac{1}{n^2}} = \lim_{n \to \infty} \frac{n^2}{n^2 - 9} = 1$$

Since the p-series $\sum_{n=1}^{\infty} \dfrac{1}{n^2}$ converges, the series $\sum_{n=4}^{\infty} \dfrac{1}{n^2}$ also converges. Because L exists, by the Limit Comparison Test we can conclude that the series $\sum_{n=4}^{\infty} \dfrac{1}{n^2 - 9}$ converges.

49. $\sum_{n=1}^{\infty} \dfrac{\sqrt{n}}{4n + 9}$

SOLUTION Apply the Limit Comparison Test with $a_n = \dfrac{\sqrt{n}}{4n + 9}$ and $b_n = \dfrac{1}{\sqrt{n}}$:

$$L = \lim_{n \to \infty} \frac{a_n}{b_n} = \lim_{n \to \infty} \frac{\frac{\sqrt{n}}{4n+9}}{\frac{1}{\sqrt{n}}} = \lim_{n \to \infty} \frac{n}{4n + 9} = \frac{1}{4}$$

The series $\sum_{n=1}^{\infty} \dfrac{1}{\sqrt{n}}$ is a divergent p-series. Because $L > 0$, by the Limit Comparison Test we can conclude that the series $\sum_{n=1}^{\infty} \dfrac{\sqrt{n}}{4n + 9}$ also diverges.

51. $\sum_{n=1}^{\infty} \dfrac{n^2 - 1}{n^5 + 1}$

SOLUTION Observe that

$$\frac{n^2 - 1}{n^5 + 1} < \frac{n^2}{n^5} = \frac{1}{n^3}$$

for $n \geq 1$. The series $\sum_{n=1}^{\infty} \dfrac{1}{n^3}$ is a convergent p-series, so by the Direct Comparison Test we can conclude that the series $\sum_{n=1}^{\infty} \dfrac{n^2 - 1}{n^5 + 1}$ also converges.

53. $\sum_{n=5}^{\infty} (4/5)^{-n}$

SOLUTION

$$\sum_{n=5}^{\infty} \left(\frac{4}{5}\right)^{-n} = \sum_{n=5}^{\infty} \left(\frac{5}{4}\right)^{n}$$

which is a geometric series starting at $n = 5$ with ratio $r = \dfrac{5}{4} > 1$. Thus the series diverges.

55. $\sum_{n=2}^{\infty} \dfrac{1}{n^{3/2} \ln n}$

SOLUTION For $n \geq 3$, $\ln n > 1$, so $n^{3/2} \ln n > n^{3/2}$ and

$$\frac{1}{n^{3/2} \ln n} < \frac{1}{n^{3/2}}$$

The series $\sum_{n=1}^{\infty} \dfrac{1}{n^{3/2}}$ is a convergent p-series, so the series $\sum_{n=3}^{\infty} \dfrac{1}{n^{3/2}}$ also converges. By the Direct Comparison Test we can therefore conclude that the series $\sum_{n=3}^{\infty} \dfrac{1}{n^{3/2} \ln n}$ converges. Hence, the series $\sum_{n=2}^{\infty} \dfrac{1}{n^{3/2} \ln n}$ also converges.

57. $\sum_{k=1}^{\infty} 4^{1/k}$

SOLUTION

$$\lim_{k \to \infty} a_k = \lim_{k \to \infty} 4^{1/k} = 2^0 = 1 \neq 0$$

therefore, the series $\sum_{k=1}^{\infty} 4^{1/k}$ diverges by the Divergence Test.

59. $\sum_{n=2}^{\infty} \dfrac{1}{(\ln n)^4}$

SOLUTION By the comment preceding Exercise 29, we can choose N so that for $n \geq N$, we have $\ln n < n^{1/8}$, so that $(\ln n)^4 < n^{1/2}$. Then

$$\sum_{n=N}^{\infty} \frac{1}{(\ln n)^4} > \sum_{n=N}^{\infty} \frac{1}{n^{1/2}}$$

which is a divergent p-series. Thus the series on the left diverges as well, and adding back in the finite number of terms for $n < N$ does not affect the result. Thus $\sum_{n=2}^{\infty} \dfrac{1}{(\ln n)^4}$ diverges.

61. $\sum_{n=3}^{\infty} \dfrac{1}{n \ln n - n}$

SOLUTION For $n \geq 2$, $n \ln n - n \leq n \ln n$; therefore,

$$\frac{1}{n \ln n - n} \geq \frac{1}{n \ln n}$$

Now, let $f(x) = \dfrac{1}{x \ln x}$. For $x \geq 2$, this function is continuous, positive, and decreasing, so the Integral Test applies. Using the substitution $u = \ln x$, $du = \dfrac{1}{x} dx$, we find

$$\int_2^{\infty} \frac{dx}{x \ln x} = \lim_{R \to \infty} \int_2^R \frac{dx}{x \ln x} = \lim_{R \to \infty} \int_{\ln 2}^{\ln R} \frac{du}{u} = \lim_{R \to \infty} (\ln(\ln R) - \ln(\ln 2)) = \infty$$

The integral diverges; hence, the series $\sum_{n=2}^{\infty} \dfrac{1}{n \ln n}$ also diverges. By the Direct Comparison Test we can therefore conclude that the series $\sum_{n=2}^{\infty} \dfrac{1}{n \ln n - n}$ diverges.

63. $\displaystyle\sum_{n=1}^{\infty} \frac{1}{n^n}$

SOLUTION For $n \geq 2$, $n^n \geq 2^n$; therefore,

$$\frac{1}{n^n} \leq \frac{1}{2^n} = \left(\frac{1}{2}\right)^n$$

The series $\displaystyle\sum_{n=1}^{\infty}\left(\frac{1}{2}\right)^n$ is a convergent geometric series, so $\displaystyle\sum_{n=2}^{\infty}\left(\frac{1}{2}\right)^n$ also converges. By the Direct Comparison Test we can therefore conclude that the series $\displaystyle\sum_{n=2}^{\infty}\frac{1}{n^n}$ converges. Hence, the series $\displaystyle\sum_{n=1}^{\infty}\frac{1}{n^n}$ converges.

65. $\displaystyle\sum_{n=1}^{\infty} \frac{1 + (-1)^n}{n}$

SOLUTION Let

$$a_n = \frac{1 + (-1)^n}{n}$$

Then

$$a_n = \begin{cases} 0 & n \text{ odd} \\ \frac{2}{2k} = \frac{1}{k} & n = 2k \text{ even} \end{cases}$$

Therefore, $\{a_n\}$ consists of 0s in the odd places and the harmonic series in the even places, so $\sum_{i=1}^{\infty} a_n$ is just the sum of the harmonic series, which diverges. Thus $\sum_{i=1}^{\infty} a_n$ diverges as well.

67. $\displaystyle\sum_{n=1}^{\infty} \sin\frac{1}{n}$

SOLUTION Apply the Limit Comparison Test with $a_n = \sin\frac{1}{n}$ and $b_n = \frac{1}{n}$:

$$L = \lim_{n\to\infty} \frac{\sin\frac{1}{n}}{\frac{1}{n}} = \lim_{u\to 0} \frac{\sin u}{u} = 1$$

where $u = \frac{1}{n}$. The harmonic series diverges. Because $L > 0$, by the Limit Comparison Test we can conclude that the series $\displaystyle\sum_{n=1}^{\infty} \sin\frac{1}{n}$ also diverges.

69. $\displaystyle\sum_{n=1}^{\infty} \frac{2n+1}{4^n}$

SOLUTION For $n \geq 3$, $2n + 1 < 2^n$, so

$$\frac{2n+1}{4^n} < \frac{2^n}{4^n} = \left(\frac{1}{2}\right)^n$$

The series $\displaystyle\sum_{n=1}^{\infty}\left(\frac{1}{2}\right)^n$ is a convergent geometric series, so $\displaystyle\sum_{n=3}^{\infty}\left(\frac{1}{2}\right)^n$ also converges. By the Direct Comparison Test we can therefore conclude that the series $\displaystyle\sum_{n=3}^{\infty}\frac{2n+1}{4^n}$ converges. Finally, the series $\displaystyle\sum_{n=1}^{\infty}\frac{2n+1}{4^n}$ converges.

71. $\sum_{n=4}^{\infty} \dfrac{\ln n}{n^2 - 3n}$

SOLUTION By the comment preceding Exercise 29, we can choose $N \geq 4$ so that for $n \geq N$, $\ln n < n^{1/2}$. Then

$$\sum_{n=N}^{\infty} \dfrac{\ln n}{n^2 - 3n} \leq \sum_{n=N}^{\infty} \dfrac{n^{1/2}}{n^2 - 3n} = \sum_{n=N}^{\infty} \dfrac{1}{n^{3/2} - 3n^{1/2}}$$

To evaluate convergence of the latter series, let $a_n = \dfrac{1}{n^{3/2} - 3n^{1/2}}$ and $b_n = \dfrac{1}{n^{3/2}}$, and apply the Limit Comparison Test:

$$L = \lim_{n \to \infty} \dfrac{a_n}{b_n} = \lim_{n \to \infty} \dfrac{1}{n^{3/2} - 3n^{1/2}} \cdot n^{3/2} = \lim_{n \to \infty} \dfrac{1}{1 - 3n^{-1}} = 0$$

Thus $\sum a_n$ converges if $\sum b_n$ does. But $\sum b_n$ is a convergent p-series. Thus $\sum a_n$ converges and, by the comparison test, so does the original series. Adding back in the finite number of terms for $n < N$ does not affect convergence.

73. $\sum_{n=2}^{\infty} \dfrac{1}{n^{1/2} \ln n}$

SOLUTION By the comment preceding Exercise 29, we can choose $N \geq 2$ so that for $n \geq N$, $\ln n < n^{1/4}$. Then

$$\sum_{n=N}^{\infty} \dfrac{1}{n^{1/2} \ln n} > \sum_{n=N}^{\infty} \dfrac{1}{n^{3/4}}$$

which is a divergent p-series. Thus the original series diverges as well—as usual, adding back in the finite number of terms for $n < N$ does not affect convergence.

75. $\sum_{n=2}^{\infty} \dfrac{4n^2 + 15n}{3n^4 - 5n^2 - 17}$

SOLUTION Apply the Limit Comparison Test with

$$a_n = \dfrac{4n^2 + 15n}{3n^4 - 5n^2 - 17}, \qquad b_n = \dfrac{4n^2}{3n^4} = \dfrac{4}{3n^2}$$

We have

$$L = \lim_{n \to \infty} \dfrac{a_n}{b_n} = \lim_{n \to \infty} \dfrac{4n^2 + 15n}{3n^4 - 5n^2 - 17} \cdot \dfrac{3n^2}{4} = \lim_{n \to \infty} \dfrac{12n^4 + 45n^3}{12n^4 - 20n^2 - 68} = \lim_{n \to \infty} \dfrac{12 + 45/n}{12 - 20/n^2 - 68/n^4} = 1$$

Now, $\sum_{n=1}^{\infty} b_n$ is a p-series with $p = 2 > 1$, so converges. Since $L = 1$, we see that $\sum_{n=1}^{\infty} \dfrac{4n^2 + 15n}{3n^4 - 5n^2 - 17}$ converges as well.

77. For which a does $\sum_{n=2}^{\infty} \dfrac{1}{n(\ln n)^a}$ converge?

SOLUTION First consider the case $a > 0$ but $a \neq 1$. Let $f(x) = \dfrac{1}{x(\ln x)^a}$. This function is continuous, positive, and decreasing for $x \geq 2$, so the Integral Test applies. Now,

$$\int_2^{\infty} \dfrac{dx}{x(\ln x)^a} = \lim_{R \to \infty} \int_2^R \dfrac{dx}{x(\ln x)^a} = \lim_{R \to \infty} \int_{\ln 2}^{\ln R} \dfrac{du}{u^a} = \dfrac{1}{1-a} \lim_{R \to \infty} \left(\dfrac{1}{(\ln R)^{a-1}} - \dfrac{1}{(\ln 2)^{a-1}} \right)$$

Because

$$\lim_{R \to \infty} \dfrac{1}{(\ln R)^{a-1}} = \begin{cases} \infty, & 0 < a < 1 \\ 0, & a > 1 \end{cases}$$

we conclude the integral diverges when $0 < a < 1$ and converges when $a > 1$. Therefore

$$\sum_{n=2}^{\infty} \frac{1}{n(\ln n)^a} \text{ converges for } a > 1 \text{ and diverges for } 0 < a < 1$$

Next, consider the case $a = 1$. The series becomes $\sum_{n=2}^{\infty} \frac{1}{n \ln n}$. Let $f(x) = \frac{1}{x \ln x}$. For $x \geq 2$, this function is continuous, positive, and decreasing, so the Integral Test applies. Using the substitution $u = \ln x$, $du = \frac{1}{x} dx$, we find

$$\int_2^{\infty} \frac{dx}{x \ln x} = \lim_{R \to \infty} \int_2^R \frac{dx}{x \ln x} = \lim_{R \to \infty} \int_{\ln 2}^{\ln R} \frac{du}{u} = \lim_{R \to \infty} (\ln(\ln R) - \ln(\ln 2)) = \infty$$

The integral diverges; hence, the series also diverges.

Finally, consider the case $a < 0$. Let $b = -a > 0$ so the series becomes $\sum_{n=2}^{\infty} \frac{(\ln n)^b}{n}$. Since $\ln n > 1$ for all $n \geq 3$, it follows that

$$(\ln n)^b > 1 \quad \text{so} \quad \frac{(\ln n)^b}{n} > \frac{1}{n}$$

The series $\sum_{n=3}^{\infty} \frac{1}{n}$ diverges, so by the Direct Comparison Test we can conclude that $\sum_{n=3}^{\infty} \frac{(\ln n)^b}{n}$ also diverges. Consequently, $\sum_{n=2}^{\infty} \frac{(\ln n)^b}{n}$ diverges. Thus,

$$\sum_{n=2}^{\infty} \frac{1}{n(\ln n)^a} \text{ diverges for } a < 0$$

To summarize:

$$\sum_{n=2}^{\infty} \frac{1}{n(\ln n)^a} \text{ converges if } a > 1 \text{ and diverges if } a \leq 1$$

79. For which values of p does $\sum_{n=1}^{\infty} \frac{n^2}{(n^3 + 1)^p}$ converge?

SOLUTION Using the Integral Test, we want to evaluate

$$\int_1^{\infty} \frac{x^2}{(x^3 + 1)^p} dx$$

Use the substitution $u = x^3 + 1$, so that $\frac{1}{3} du = x^2 dx$. Then

$$\int_1^{\infty} \frac{x^2}{(x^3 + 1)^p} dx = \frac{1}{3} \int_2^{\infty} \frac{1}{u^p} du$$

But by the proof of Theorem 3 in the text, this integral converges for $p > 1$ and diverges for $p \leq 1$. Thus the series $\sum_{n=1}^{\infty} \frac{n^2}{(n^3+1)^p}$ converges for $p > 1$ and diverges for $p \leq 1$.

Approximating Infinite Sums *In Exercises 81–83, let $a_n = f(n)$, where f is a continuous, decreasing function such that $f(x) \geq 0$ and $\int_1^{\infty} f(x) dx$ converges.*

81. Show that

$$\int_1^{\infty} f(x) dx \leq \sum_{n=1}^{\infty} a_n \leq a_1 + \int_1^{\infty} f(x) dx \qquad \boxed{3}$$

SOLUTION From the proof of the Integral Test, we know that

$$a_2 + a_3 + a_4 + \cdots + a_N \leq \int_1^N f(x)\,dx \leq \int_1^\infty f(x)\,dx$$

that is,

$$S_N - a_1 \leq \int_1^\infty f(x)\,dx \quad \text{or} \quad S_N \leq a_1 + \int_1^\infty f(x)\,dx$$

Also from the proof of the Integral test, we know that

$$\int_1^N f(x)\,dx \leq a_1 + a_2 + a_3 + \cdots + a_{N-1} = S_N - a_N \leq S_N$$

Thus,

$$\int_1^N f(x)\,dx \leq S_N \leq a_1 + \int_1^\infty f(x)\,dx$$

Taking the limit as $N \to \infty$ yields Eq. (3), as desired.

83. Assume $\sum_{n=1}^\infty a_n$ converges to S. Arguing as in Exercise 81, show that

$$\sum_{n=1}^M a_n + \int_{M+1}^\infty f(x)\,dx \leq S \leq \sum_{n=1}^{M+1} a_n + \int_{M+1}^\infty f(x)\,dx \qquad \boxed{4}$$

Conclude that

$$0 \leq S - \left(\sum_{n=1}^M a_n + \int_{M+1}^\infty f(x)\,dx\right) \leq a_{M+1} \qquad \boxed{5}$$

This provides a method for approximating S with an error of at most a_{M+1}.

SOLUTION Following the proof of the Integral Test and the argument in Exercise 81, but starting with $n = M + 1$ rather than $n = 1$, we obtain

$$\int_{M+1}^\infty f(x)\,dx \leq \sum_{n=M+1}^\infty a_n \leq a_{M+1} + \int_{M+1}^\infty f(x)\,dx$$

Adding $\sum_{n=1}^M a_n$ to each part of this inequality yields

$$\sum_{n=1}^M a_n + \int_{M+1}^\infty f(x)\,dx \leq \sum_{n=1}^\infty a_n = S \leq \sum_{n=1}^{M+1} a_n + \int_{M+1}^\infty f(x)\,dx$$

Subtracting $\sum_{n=1}^M a_n + \int_{M+1}^\infty f(x)\,dx$ from each part of this last inequality then gives us

$$0 \leq S - \left(\sum_{n=1}^M a_n + \int_{M+1}^\infty f(x)\,dx\right) \leq a_{M+1}$$

85. *CAS* Use the inequalities in (4) from Exercise 83 with $M = 40{,}000$ to show that

$$1.644934066 \leq \sum_{n=1}^\infty \frac{1}{n^2} \leq 1.644934068$$

Is this consistent with Euler's result, according to which this infinite series has sum $\pi^2/6$?

SOLUTION From the given inequalities with $f(x) = \frac{1}{x^2}$, $a_n = \frac{1}{n^2}$ and $M = 40000$, we find

$$S_{40000} + \int_{40001}^{\infty} \frac{dx}{x^2} \leq \sum_{n=1}^{\infty} \frac{1}{n^2} \leq S_{40001} + \int_{40001}^{\infty} \frac{dx}{x^2}$$

Now,

$$S_{40000} = 1.6449090672;$$

$$S_{40001} = S_{40000} + \frac{1}{40001} = 1.6449090678$$

and

$$\int_{40001}^{\infty} \frac{dx}{x^2} = \lim_{R \to \infty} \int_{40001}^{R} \frac{dx}{x^2} = -\lim_{R \to \infty} \left(\frac{1}{R} - \frac{1}{40001} \right) = \frac{1}{40001} = 0.0000249994$$

Thus,

$$1.6449090672 + 0.0000249994 \leq \sum_{n=1}^{\infty} \frac{1}{n^2} \leq 1.6449090678 + 0.0000249994$$

or

$$1.6449340665 \leq \sum_{n=1}^{\infty} \frac{1}{n^2} \leq 1.6449340672$$

Since $\frac{\pi^2}{6} \approx 1.6449340668$, our approximation is consistent with Euler's result.

87. *CAS* Use a CAS and the inequalities in (5) from Exercise 83 to determine the value of $\sum_{n=1}^{\infty} n^{-5}$ to within an error less than 10^{-4}.

SOLUTION Using the inequalities in (5) with $f(x) = x^{-5}$ and $a_n = n^{-5}$, we have

$$0 \leq \sum_{n=1}^{\infty} n^{-5} - \left(\sum_{n=1}^{M+1} n^{-5} + \int_{M+1}^{\infty} x^{-5} \, dx \right) \leq (M+1)^{-5}$$

To guarantee an error less than 10^{-4}, we need $(M+1)^{-5} \leq 10^{-4}$. This yields $M \geq 10^{4/5} - 1 \approx 5.3$, so we choose $M = 6$. Now,

$$\sum_{n=1}^{7} n^{-5} = 1.0368498887$$

and

$$\int_{7}^{\infty} x^{-5} \, dx = \lim_{R \to \infty} \int_{7}^{R} x^{-5} \, dx = -\frac{1}{4} \lim_{R \to \infty} \left(R^{-4} - 7^{-4} \right) = \frac{1}{4 \cdot 7^4} = 0.0001041233$$

Thus,

$$\sum_{n=1}^{\infty} n^{-5} \approx \sum_{n=1}^{7} n^{-5} + \int_{7}^{\infty} x^{-5} \, dx = 1.0368498887 + 0.0001041233 = 1.0369540120$$

89. The following argument proves the divergence of the harmonic series $\sum_{n=1}^{\infty} 1/n$ without using the Integral Test. To begin, assume that the harmonic series converges to a value S.

(a) Prove that the following two series must also converge:

$$1 + \frac{1}{3} + \frac{1}{5} + \cdots \quad \text{and} \quad \frac{1}{2} + \frac{1}{4} + \frac{1}{6} + \cdots$$

(b) Prove that if S_1 and S_2 are the sums of the series on the left and right, respectively, then $S = S_1 + S_2$.

(c) Prove that $S_1 \geq S_2 + \frac{1}{2}$, and $S_2 = \frac{1}{2}S$. Explain how this leads to a contradiction and the conclusion that the harmonic series diverges.

SOLUTION

(a) For the first series, let $a_{2k+1} = \frac{1}{2k+1}$ for $k \geq 0$ and $a_{2k} = 0$ for $k \geq 1$. Then $a_k \leq \frac{1}{k}$, so by the Direct Comparison Test, if the harmonic series converges, so does $\sum a_k$; but clearly $\sum a_k$ is the same as the sum of the series on the left. For the series on the right, let $b_{2k} = \frac{1}{2k}$ for $k \geq 1$ and $b_{2k+1} = 0$ for $k \geq 0$. Then $b_k \leq \frac{1}{k}$, so again by the Direct Comparison Test convergence of the harmonic series implies convergence of $\sum b_k$, and therefore of the series on the right.

(b) Since $\sum a_k = S_1$ as well, and $\sum b_k = S_2$, Theorem 2 in Section 10.1 shows that $\sum (a_k + b_k) = \sum \frac{1}{k} = S_1 + S_2$. Thus $S = S_1 + S_2$.

(c) Note that

$$S_2 + \frac{1}{2} = \frac{1}{2} + \frac{1}{2} + \frac{1}{4} + \frac{1}{6} + \cdots = 1 + \frac{1}{4} + \frac{1}{6} + \cdots$$

Then each term of the underlying sequence is less than each term of the underlying sequence for S_1, so that $S_1 \geq S_2 + \frac{1}{2}$. However,

$$S_2 = \frac{1}{2} + \frac{1}{4} + \frac{1}{6} + \cdots = \frac{1}{2}\left(1 + \frac{1}{2} + \frac{1}{3} + \cdots\right) = \frac{1}{2}S$$

so that $S_2 = \frac{1}{2}S$. Since $S = S_1 + S_2$, this implies that $S_1 = \frac{1}{2}S = S_2$. This contradicts the fact that $S_1 \geq S_2 + \frac{1}{2}$, so the original assumption that the harmonic series converges must have been false.

Further Insights and Challenges

91. Kummer's Acceleration Method Suppose we wish to approximate $S = \sum_{n=1}^{\infty} 1/n^2$. There is a similar telescoping series whose value can be computed exactly (Example 2 in Section 10.2):

$$\sum_{n=1}^{\infty} \frac{1}{n(n+1)} = 1$$

(a) Verify that

$$S = \sum_{n=1}^{\infty} \frac{1}{n(n+1)} + \sum_{n=1}^{\infty} \left(\frac{1}{n^2} - \frac{1}{n(n+1)}\right)$$

Thus for M large,

$$S \approx 1 + \sum_{n=1}^{M} \frac{1}{n^2(n+1)} \qquad \boxed{6}$$

(b) Explain what has been gained. Why is (6) a better approximation to S than $\sum_{n=1}^{M} 1/n^2$?

(c) *CAS* Compute

$$\sum_{n=1}^{1000} \frac{1}{n^2}, \qquad 1 + \sum_{n=1}^{100} \frac{1}{n^2(n+1)}$$

Which is a better approximation to S, whose exact value is $\pi^2/6$?

SOLUTION

(a) Because the series $\sum_{n=1}^{\infty} \frac{1}{n^2}$ and $\sum_{n=1}^{\infty} \frac{1}{n(n+1)}$ both converge,

$$\sum_{n=1}^{\infty} \frac{1}{n(n+1)} + \sum_{n=1}^{\infty} \left(\frac{1}{n^2} - \frac{1}{n(n+1)}\right) = \sum_{n=1}^{\infty} \frac{1}{n(n+1)} + \sum_{n=1}^{\infty} \frac{1}{n^2} - \sum_{n=1}^{\infty} \frac{1}{n(n+1)} = \sum_{n=1}^{\infty} \frac{1}{n^2} = S$$

Now,

$$\frac{1}{n^2} - \frac{1}{n(n+1)} = \frac{n+1}{n^2(n+1)} - \frac{n}{n^2(n+1)} = \frac{1}{n^2(n+1)}$$

so, for M large,

$$S \approx 1 + \sum_{n=1}^{M} \frac{1}{n^2(n+1)}$$

(b) The series $\sum_{n=1}^{\infty} \frac{1}{n^2(n+1)}$ converges more rapidly than $\sum_{n=1}^{\infty} \frac{1}{n^2}$ since the degree of n in the denominator is larger.

(c) Using a computer algebra system, we find

$$\sum_{n=1}^{1000} \frac{1}{n^2} = 1.6439345667 \quad \text{and} \quad 1 + \sum_{n=1}^{100} \frac{1}{n^2(n+1)} = 1.6448848903$$

The second sum is more accurate because it is closer to the exact solution $\frac{\pi^2}{6} \approx 1.6449340668$.

10.4 Absolute and Conditional Convergence (LT Section 11.4)

Preliminary Questions

1. Give an example of a series such that $\sum a_n$ converges but $\sum |a_n|$ diverges.

SOLUTION The series $\sum \frac{(-1)^n}{\sqrt[3]{n}}$ converges by the Alternating Series Test, but the positive series $\sum \frac{1}{\sqrt[3]{n}}$ is a divergent p-series.

2. Which of the following statements is equivalent to Theorem 1?

(a) If $\sum_{n=0}^{\infty} |a_n|$ diverges, then $\sum_{n=0}^{\infty} a_n$ also diverges.

(b) If $\sum_{n=0}^{\infty} a_n$ diverges, then $\sum_{n=0}^{\infty} |a_n|$ also diverges.

(c) If $\sum_{n=0}^{\infty} a_n$ converges, then $\sum_{n=0}^{\infty} |a_n|$ also converges.

SOLUTION The correct answer is **(b)**: If $\sum_{n=0}^{\infty} a_n$ diverges, then $\sum_{n=0}^{\infty} |a_n|$ also diverges. Take $a_n = (-1)^n \frac{1}{n}$ to see that statements **(a)** and **(c)** are not true in general.

3. Indicate whether or not the reasoning in the following statement is correct: Since $\sum_{n=1}^{\infty} (-1)^n \sqrt{n}$ is an alternating series, it must converge.

SOLUTION No. Although $\sum_{n=1}^{\infty} (-1)^n \sqrt{n}$ is an alternating series, the terms $a_n = \sqrt{n}$ do not form a decreasing sequence that tends to zero. In fact, $a_n = \sqrt{n}$ is an increasing sequence that tends to ∞, so $\sum_{n=1}^{\infty} (-1)^n \sqrt{n}$ diverges by the Divergence Test.

4. Suppose that b_n is positive, decreasing, and tends to 0, and let $S = \sum_{n=1}^{\infty} (-1)^{n-1} b_n$. What can we say about $|S - S_{100}|$ if $a_{101} = 10^{-3}$? Is S larger or smaller than S_{100}?

SOLUTION From the text, we know that $|S - S_{100}| < a_{101} = 10^{-3}$. Also, the Leibniz test tells us that $S_{2N} < S < S_{2N+1}$ for any $N \geq 1$, so that $S_{100} < S$.

Exercises

1. Show that
$$\sum_{n=0}^{\infty} \frac{(-1)^n}{2^n}$$
converges absolutely.

SOLUTION The positive series $\sum_{n=0}^{\infty} \frac{1}{2^n}$ is a geometric series with $r = \frac{1}{2}$. Thus, the positive series converges, and the given series converges absolutely.

In Exercises 3–10, determine whether the series converges absolutely, conditionally, or not at all.

3. $\sum_{n=1}^{\infty} \frac{(-1)^{n-1}}{n^{1/3}}$

SOLUTION The sequence $a_n = \frac{1}{n^{1/3}}$ is positive, decreasing, and tends to zero; hence, the series $\sum_{n=1}^{\infty} \frac{(-1)^{n-1}}{n^{1/3}}$ converges by the Alternating Series Test. However, the positive series $\sum_{n=1}^{\infty} \frac{1}{n^{1/3}}$ is a divergent p-series, so the original series converges conditionally.

5. $\sum_{n=0}^{\infty} \frac{(-1)^n}{(1.001)^n}$

SOLUTION Because
$$\left| \frac{(-1)^n}{(1.001)^n} \right| = \frac{1}{1.001^n}$$
the positive series forms a geometric series with ratio $\frac{1}{1.001} < 1$; thus, the original series converges absolutely.

7. $\sum_{n=1}^{\infty} \frac{(-1)^n e^{-n}}{n^2}$

SOLUTION We have
$$\left| \frac{(-1)^n e^{-n}}{n^2} \right| = \frac{e^{-n}}{n^2} < \frac{1}{n^2}$$
since $e^{-n} < 1$ for $n \geq 1$. Therefore the positive series is bounded above by the convergent p-series $\sum_{n=1}^{\infty} \frac{1}{n^2}$, so it is convergent and the original series converges absolutely.

9. $\sum_{n=2}^{\infty} \frac{(-1)^n}{n \ln n}$

SOLUTION Let $a_n = \frac{1}{n \ln n}$. Then a_n forms a decreasing sequence (note that n and $\ln n$ are both increasing functions of n) that tends to zero; hence, the series $\sum_{n=2}^{\infty} \frac{(-1)^n}{n \ln n}$ converges by the Alternating Series Test. However, the positive series $\sum_{n=2}^{\infty} \frac{1}{n \ln n}$ diverges, so the original series converges conditionally.

11. Let $S = \sum_{n=1}^{\infty} (-1)^{n+1} \frac{1}{n^3}$.

(a) Calculate S_n for $1 \leq n \leq 10$.

(b) Use the inequality in (2) to show that $0.9 \leq S \leq 0.902$.

SOLUTION

(a)

$$S_1 = 1 \qquad\qquad S_6 = S_5 - \frac{1}{6^3} = 0.899782407$$

$$S_2 = 1 - \frac{1}{2^3} = \frac{7}{8} = 0.875 \qquad\qquad S_7 = S_6 + \frac{1}{7^3} = 0.902697859$$

$$S_3 = S_2 + \frac{1}{3^3} = 0.912037037 \qquad\qquad S_8 = S_7 - \frac{1}{8^3} = 0.900744734$$

$$S_4 = S_3 - \frac{1}{4^3} = 0.896412037 \qquad\qquad S_9 = S_8 + \frac{1}{9^3} = 0.902116476$$

$$S_5 = S_4 + \frac{1}{5^3} = 0.904412037 \qquad\qquad S_{10} = S_9 - \frac{1}{10^3} = 0.901116476$$

(b) By (2),

$$|S_{10} - S| \le a_{11} = \frac{1}{11^3}$$

so

$$S_{10} - \frac{1}{11^3} \le S \le S_{10} + \frac{1}{11^3}$$

or

$$0.900365161 \le S \le 0.901867791$$

13. Approximate $\sum_{n=1}^{\infty} \frac{(-1)^{n+1}}{n^4}$ to three decimal places.

SOLUTION Let $S = \sum_{n=1}^{\infty} \frac{(-1)^{n+1}}{n^4}$, so that $a_n = \frac{1}{n^4}$. By Eq. (2),

$$|S_N - S| \le a_{N+1} = \frac{1}{(N+1)^4}$$

To guarantee accuracy to three decimal places, we must choose N so that

$$\frac{1}{(N+1)^4} < 5 \times 10^{-4} \quad \text{or} \quad N > \sqrt[4]{2000} - 1 \approx 5.7$$

The smallest value that satisfies the required inequality is then $N = 6$. Thus,

$$S \approx S_6 = 1 - \frac{1}{2^4} + \frac{1}{3^4} - \frac{1}{4^4} + \frac{1}{5^4} - \frac{1}{6^4} = 0.946767824$$

In Exercises 15–16, find a value of N such that S_N approximates the series with an error of at most 10^{-5}. Using technology, compute this value of S_N.

15. $\sum_{n=1}^{\infty} \frac{(-1)^{n+1}}{n(n+2)(n+3)}$

SOLUTION Let $S = \sum_{n=1}^{\infty} \frac{(-1)^{n+1}}{n(n+2)(n+3)}$, so that $a_n = \frac{1}{n(n+2)(n+3)}$. By Eq. (2),

$$|S_N - S| \le a_{N+1} = \frac{1}{(N+1)(N+3)(N+4)}$$

We must choose N so that

$$\frac{1}{(N+1)(N+3)(N+4)} \le 10^{-5} \quad \text{or} \quad (N+1)(N+3)(N+4) \ge 10^5$$

SECTION 10.4 | Absolute and Conditional Convergence (LT SECTION 11.4) 731

For $N = 43$, the product on the left-hand side is 95,128, while for $N = 44$ the product is 101,520; hence, the smallest value of N which satisfies the required inequality is $N = 44$. Thus,

$$S \approx S_{44} = \sum_{n=1}^{44} \frac{(-1)^{n+1}}{n(n+2)(n+3)} = 0.0656746$$

In Exercises 17–32, determine convergence or divergence by any method.

17. $\sum_{n=0}^{\infty} 7^{-n}$

SOLUTION This is a (positive) geometric series with $r = \frac{1}{7} < 1$, so it converges.

19. $\sum_{n=1}^{\infty} \frac{1}{5^n - 3^n}$

SOLUTION Use the Limit Comparison Test with $\frac{1}{5^n}$:

$$L = \lim_{n \to \infty} \frac{1/(5^n - 3^n)}{1/5^n} = \lim_{n \to \infty} \frac{5^n}{5^n - 3^n} = \lim_{n \to \infty} \frac{1}{1 - (3/5)^n} = 1$$

But $\sum_{n=1}^{\infty} \frac{1}{5^n}$ is a convergent geometric series. Since $L = 1$, the Limit Comparison Test tells us that the original series converges as well.

21. $\sum_{n=1}^{\infty} \frac{1}{3n^4 + 12n}$

SOLUTION Use the Limit Comparison Test with $\frac{1}{3n^4}$:

$$L = \lim_{n \to \infty} \frac{(1/(3n^4 + 12n))}{1/3n^4} = \lim_{n \to \infty} \frac{3n^4}{3n^4 + 12n} = \lim_{n \to \infty} \frac{1}{1 + 4n^{-3}} = 1$$

But $\sum_{n=1}^{\infty} \frac{1}{3n^4} = \frac{1}{3} \sum_{n=1}^{\infty} \frac{1}{n^4}$ is a convergent p-series. Since $L = 1$, the Limit Comparison Test tells us that the original series converges as well.

23. $\sum_{n=1}^{\infty} \frac{1}{\sqrt{n^2 + 1}}$

SOLUTION Apply the Limit Comparison Test and compare the series with the divergent harmonic series:

$$L = \lim_{n \to \infty} \frac{\frac{1}{\sqrt{n^2+1}}}{\frac{1}{n}} = \lim_{n \to \infty} \frac{n}{\sqrt{n^2 + 1}} = 1$$

Because $L > 0$, we conclude that the series $\sum_{n=1}^{\infty} \frac{1}{\sqrt{n^2 + 1}}$ diverges.

25. $\sum_{n=1}^{\infty} \frac{3^n + (-2)^n}{5^n}$

SOLUTION The two series $\sum_{n=1}^{\infty} \frac{3^n}{5^n} = \sum_{n=1}^{\infty} \left(\frac{3}{5}\right)^n$ and $\sum_{n=1}^{\infty} \frac{(-2)^n}{5^n} = \sum_{n=1}^{\infty} \left(-\frac{2}{5}\right)^n$ are both convergent geometric series. Hence,

$$\sum_{n=1}^{\infty} \frac{3^n + (-2)^n}{5^n} = \sum_{n=1}^{\infty} \left(\frac{3}{5}\right)^n + \sum_{n=1}^{\infty} \left(-\frac{2}{5}\right)^n$$

also converges.

27. $\sum_{n=1}^{\infty} (-1)^n n^2 e^{-n^3/3}$

SOLUTION Consider the associated positive series $\sum_{n=1}^{\infty} n^2 e^{-n^3/3}$. This series can be seen to converge by the Integral Test:

$$\int_1^{\infty} x^2 e^{-x^3/3} \, dx = \lim_{R \to \infty} \int_1^R x^2 e^{-x^3/3} \, dx = -\lim_{R \to \infty} e^{-x^3/3} \Big|_1^R = e^{-1/3} + \lim_{R \to \infty} e^{-R^3/3} = e^{-1/3}$$

The integral converges, so the original series converges absolutely.

29. $\sum_{n=2}^{\infty} \frac{(-1)^n}{n^{1/2}(\ln n)^2}$

SOLUTION This is an alternating series with $a_n = \frac{1}{n^{1/2}(\ln n)^2}$. Because a_n is a decreasing sequence which converges to zero, the series $\sum_{n=2}^{\infty} \frac{(-1)^n}{n^{1/2}(\ln n)^2}$ converges by the Leibniz Test. (Note that the series converges only conditionally, not absolutely; the associated positive series is eventually greater than $\frac{1}{n^{3/4}}$, which is a divergent p-series.)

31. $\sum_{n=1}^{\infty} \frac{\ln n}{n^{1.05}}$

SOLUTION Choose N so that for $n \geq N$ we have $\ln n \leq n^{0.01}$. Then

$$\sum_{n=N}^{\infty} \frac{\ln n}{n^{1.05}} \leq \sum_{n=N}^{\infty} \frac{n^{0.01}}{n^{1.05}} = \sum_{n=N}^{\infty} \frac{1}{n^{1.04}}$$

This is a convergent p-series, so by the Comparison Test, the original series converges as well.

33. Show that

$$\frac{1}{2} - \frac{1}{2} + \frac{1}{3} - \frac{1}{3} + \frac{1}{4} - \frac{1}{4} + \cdots$$

converges by computing the partial sums. Does it converge absolutely?

SOLUTION The sequence of partial sums is

$$S_1 = \frac{1}{2}$$
$$S_2 = S_1 - \frac{1}{2} = 0$$
$$S_3 = S_2 + \frac{1}{3} = \frac{1}{3}$$
$$S_4 = S_3 - \frac{1}{3} = 0$$

and, in general,

$$S_N = \begin{cases} \frac{1}{N}, & \text{for odd } N \\ 0, & \text{for even } N \end{cases}$$

Thus, $\lim_{N \to \infty} S_N = 0$, and the series converges to 0. The positive series is

$$\frac{1}{2} + \frac{1}{2} + \frac{1}{3} + \frac{1}{3} + \frac{1}{4} + \frac{1}{4} + \cdots = 2 \sum_{n=2}^{\infty} \frac{1}{n}$$

which diverges. Therefore, the original series converges conditionally, not absolutely.

SECTION 10.4 | Absolute and Conditional Convergence (LT SECTION 11.4)

35. Assumptions Matter Show that the following series diverges:

$$\frac{1}{2} - \frac{1}{4} + \frac{1}{3} - \frac{1}{8} + \frac{1}{4} - \frac{1}{16} + \cdots + \left(\frac{1}{n} - \frac{1}{2^n}\right) + \cdots$$

(*Note:* This demonstrates that in the Alternating Series Test, we *need* the assumption that the sequence a_n is decreasing. It is not enough to assume only that a_n tends to zero.)

SOLUTION Let

$$R = \frac{1}{2} - \frac{1}{4} + \frac{1}{3} - \frac{1}{8} + \frac{1}{4} - \frac{1}{16} + \cdots + \left(\frac{1}{n+1} - \frac{1}{2^{n+1}}\right) + \cdots$$

This is an alternating series with

$$a_n = \begin{cases} \frac{1}{k+1}, & n = 2k-1 \\ \frac{1}{2^{k+1}}, & n = 2k \end{cases}$$

Note that $a_n \to 0$ as $n \to \infty$, but the sequence $\{a_n\}$ is not decreasing. We will now establish that R diverges. For sake of contradiction, suppose that R converges. The geometric series

$$\sum_{n=1}^{\infty} \frac{1}{2^{n+1}}$$

converges, so the sum of R and this geometric series must also converge; however,

$$R + \sum_{n=1}^{\infty} \frac{1}{2^{n+1}} = \sum_{n=2}^{\infty} \frac{1}{n}$$

which diverges because the harmonic series diverges. Thus, the series R must diverge.

37. Prove that if $\sum a_n$ converges absolutely, then $\sum a_n^2$ also converges. Give an example where $\sum a_n$ is only conditionally convergent and $\sum a_n^2$ diverges.

SOLUTION Suppose the series $\sum a_n$ converges absolutely. Because $\sum |a_n|$ converges, we know that

$$\lim_{n \to \infty} |a_n| = 0$$

Therefore, there exists a positive integer N such that $|a_n| < 1$ for all $n \geq N$. It then follows that for $n \geq N$,

$$0 \leq a_n^2 = |a_n|^2 = |a_n| \cdot |a_n| < |a_n| \cdot 1 = |a_n|$$

By the Direct Comparison Test we can then conclude that $\sum a_n^2$ also converges.

Consider the series $\sum_{n=1}^{\infty} \frac{(-1)^n}{\sqrt{n}}$. This series converges by the Alternating Series Test, but the corresponding positive series is a divergent p-series; that is, $\sum_{n=1}^{\infty} \frac{(-1)^n}{\sqrt{n}}$ is conditionally convergent. Now, $\sum_{n=1}^{\infty} a_n^2$ is the divergent harmonic series $\sum_{n=1}^{\infty} \frac{1}{n}$. Thus, $\sum a_n^2$ need not converge if $\sum a_n$ is only conditionally convergent.

Further Insights and Challenges

39. Use Exercise 38 to show that the following series converges:

$$\frac{1}{\ln 2} + \frac{1}{\ln 3} - \frac{2}{\ln 4} + \frac{1}{\ln 5} + \frac{1}{\ln 6} - \frac{2}{\ln 7} + \cdots$$

SOLUTION The given series has the structure of the generic series from Exercise 38 with $a_n = \frac{1}{\ln(n+1)}$. Because a_n is a positive, decreasing sequence with $\lim_{n \to \infty} a_n = 0$, we can conclude from Exercise 38 that the given series converges.

41. Show that the following series diverges:

$$1 + \frac{1}{2} + \frac{1}{3} - \frac{2}{4} + \frac{1}{5} + \frac{1}{6} + \frac{1}{7} - \frac{2}{8} + \cdots$$

Hint: Use the result of Exercise 40 to write the series as the sum of a convergent series and a divergent series.

SOLUTION Let

$$R = 1 + \frac{1}{2} + \frac{1}{3} - \frac{3}{4} + \frac{1}{5} + \frac{1}{6} + \frac{1}{7} - \frac{3}{8} + \cdots$$

and

$$S = 1 + \frac{1}{2} + \frac{1}{3} - \frac{2}{4} + \frac{1}{5} + \frac{1}{6} + \frac{1}{7} - \frac{2}{8} + \cdots$$

For sake of contradiction, suppose the series S converges. From Exercise 40, we know that the series R converges. Thus, the series $S - R$ must converge; however,

$$S - R = \frac{1}{4} + \frac{1}{8} + \frac{1}{12} + \cdots = \frac{1}{4} \sum_{k=1}^{\infty} \frac{1}{k}$$

which diverges because the harmonic series diverges. Thus, the series S must diverge.

43. We say that $\{b_n\}$ is a rearrangement of $\{a_n\}$ if $\{b_n\}$ has the same terms as $\{a_n\}$ but occurring in a different order. Show that if $\{b_n\}$ is a rearrangement of $\{a_n\}$ and $\sum_{n=1}^{\infty} a_n$ converges absolutely, then $\sum_{n=1}^{\infty} b_n$ also converges absolutely. (This result does not hold if $\sum_{n=1}^{\infty} a_n$ is only conditionally convergent.) *Hint:* Prove that the partial sums $\sum_{n=1}^{N} |b_n|$ are bounded. It can be shown further that the two series converge to the same value.

SOLUTION Suppose the series $S = \sum_{n=1}^{\infty} a_n$ converges absolutely and denote the corresponding positive series by

$$S^+ = \sum_{n=1}^{\infty} |a_n|$$

Further, let $T_N = \sum_{n=1}^{N} |b_n|$ denote the Nth partial sum of the series $\sum_{n=1}^{\infty} |b_n|$. Because $\{b_n\}$ is a rearrangement of $\{a_n\}$, we know that

$$0 \leq T_N \leq \sum_{n=1}^{\infty} |a_n| = S^+$$

that is, the sequence $\{T_N\}$ is bounded. Moreover,

$$T_{N+1} = \sum_{n=1}^{N+1} |b_n| = T_N + |b_{N+1}| \geq T_N$$

that is, $\{T_N\}$ is increasing. It follows that $\{T_N\}$ converges, so the series $\sum_{n=1}^{\infty} |b_n|$ converges, which means the series $\sum_{n=1}^{\infty} b_n$ converges absolutely.

10.5 The Ratio and Root Tests and Strategies for Choosing Tests (LT Section 11.5)

Preliminary Questions

1. Consider the geometric series $\sum_{n=0}^{\infty} cr^n$.

(a) In the Ratio Test, what do the terms $\left|\frac{a_{n+1}}{a_n}\right|$ equal?

(b) In the Root Test, what do the terms $\sqrt[n]{|a_n|}$ equal?

SOLUTION

(a) $\left|\frac{a_{n+1}}{a_n}\right| = \left|\frac{cr^{n+1}}{cr^n}\right| = |r|$

(b) $\sqrt[n]{|a_n|} = \sqrt[n]{|cr^n|} = \sqrt[n]{|c|} \cdot \sqrt[n]{|r^n|} = |r|\sqrt[n]{|c|}$

2. Consider the p-series $\sum_{n=1}^{\infty} n^{-p}$.

(a) In the Ratio Test, what do the terms $\left|\frac{a_{n+1}}{a_n}\right|$ equal?

(b) What can be concluded from the Ratio Test?

SOLUTION

(a) $\left|\frac{a_{n+1}}{a_n}\right| = \left|\frac{(n+1)^{-p}}{n^{-p}}\right| = \left(\frac{n}{n+1}\right)^p$

(b) Since

$$\lim_{n \to \infty} \left(\frac{n}{n+1}\right)^p = \left(\lim_{n \to \infty} \frac{n}{n+1}\right)^p = 1$$

the Ratio Test is inconclusive.

3. Is the Ratio Test conclusive for $\sum_{n=1}^{\infty} \frac{1}{n!}$? Is it conclusive for $\sum_{n=1}^{\infty} \frac{1}{n+1}$?

SOLUTION The general term of $\sum_{n=1}^{\infty} \frac{1}{n!}$ is $a_n = \frac{1}{n!}$; thus,

$$\left|\frac{a_{n+1}}{a_n}\right| = \frac{1}{(n+1)!} \cdot \frac{n!}{1} = \frac{1}{n+1}$$

and

$$\rho = \lim_{n \to \infty} \left|\frac{a_{n+1}}{a_n}\right| = \lim_{n \to \infty} \frac{1}{n+1} = 0 < 1$$

Consequently, the Ratio Test guarantees that the series $\sum_{n=1}^{\infty} \frac{1}{n!}$ converges.

The general term of $\sum_{n=1}^{\infty} \frac{1}{n+1}$ is $a_n = \frac{1}{n+1}$; thus,

$$\left|\frac{a_{n+1}}{a_n}\right| = \frac{1}{n+2} \cdot \frac{n+1}{1} = \frac{n+1}{n+2}$$

and

$$\rho = \lim_{n \to \infty} \left|\frac{a_{n+1}}{a_n}\right| = \lim_{n \to \infty} \frac{n+1}{n+2} = 1$$

The Ratio Test is therefore inconclusive for the series $\sum_{n=1}^{\infty} \frac{1}{n+1}$.

4. Is the Root Test conclusive for $\sum_{n=1}^{\infty} \frac{1}{2^n}$? Is it conclusive for $\sum_{n=1}^{\infty} \left(1 + \frac{1}{n}\right)^{-n}$?

SOLUTION The general term of $\sum_{n=1}^{\infty} \frac{1}{2^n}$ is $a_n = \frac{1}{2^n}$; thus,

$$\left|\frac{a_{n+1}}{a_n}\right| = \frac{1}{2^{n+1}} \cdot \frac{2^n}{1} = \frac{1}{2}$$

and

$$\rho = \lim_{n \to \infty} \left|\frac{a_{n+1}}{a_n}\right| = \frac{1}{2} < 1$$

Consequently, the Ratio Test guarantees that the series $\sum_{n=1}^{\infty} \frac{1}{2^n}$ converges.

The general term of $\sum_{n=1}^{\infty} \left(1 + \frac{1}{n}\right)^{-n}$ is

$$\left(1 + \frac{1}{n}\right)^{-n} = \left(\frac{n+1}{n}\right)^{-n} = \left(\frac{n}{n+1}\right)^n$$

thus,

$$\left|\frac{a_{n+1}}{a_n}\right| = \left(\frac{n+1}{n+2}\right)^{n+1} \cdot \left(\frac{n+1}{n}\right)^n = \frac{(n+1)^{2n+1}}{(n+2)^{n+1} n^n}$$

and

$$\rho = \lim_{n \to \infty} \left|\frac{a_{n+1}}{a_n}\right| = \lim_{n \to \infty} \frac{(n+1)^{2n+1}}{(n+2)^{n+1} n^n} = \lim_{n \to \infty} \frac{(1 + 1/n)^{2n+1}}{(1 + 2/n)^{n+1} 1^n} = 1$$

The Ratio Test is therefore inconclusive for the series $\sum_{n=1}^{\infty} \left(1 + \frac{1}{n}\right)^{-n}$.

Exercises

In Exercises 1–20, apply the Ratio Test to determine convergence or divergence, or state that the Ratio Test is inconclusive.

1. $\sum_{n=1}^{\infty} \frac{1}{5^n}$

SOLUTION With $a_n = \frac{1}{5^n}$,

$$\left|\frac{a_{n+1}}{a_n}\right| = \frac{1}{5^{n+1}} \cdot \frac{5^n}{1} = \frac{1}{5} \quad \text{and} \quad \rho = \lim_{n \to \infty} \left|\frac{a_{n+1}}{a_n}\right| = \frac{1}{5} < 1$$

Therefore, the series $\sum_{n=1}^{\infty} \frac{1}{5^n}$ converges by the Ratio Test.

3. $\sum_{n=1}^{\infty} \frac{1}{n^n}$

SOLUTION With $a_n = \frac{(-1)^{n-1}}{n^n}$,

$$\left|\frac{a_{n+1}}{a_n}\right| = \frac{1}{(n+1)^{n+1}} \cdot \frac{n^n}{1} = \frac{1}{n+1} \left(\frac{n}{n+1}\right)^n = \frac{1}{n+1} \left(1 + \frac{1}{n}\right)^{-n}$$

and

$$\rho = \lim_{n \to \infty} \left|\frac{a_{n+1}}{a_n}\right| = 0 \cdot \frac{1}{e} = 0 < 1$$

Therefore, the series $\sum_{n=1}^{\infty} \frac{(-1)^{n-1}}{n^n}$ converges by the Ratio Test.

5. $\sum_{n=1}^{\infty} \dfrac{n}{n^2+1}$

SOLUTION With $a_n = \dfrac{n}{n^2+1}$,

$$\left|\dfrac{a_{n+1}}{a_n}\right| = \dfrac{n+1}{(n+1)^2+1} \cdot \dfrac{n^2+1}{n} = \dfrac{n+1}{n} \cdot \dfrac{n^2+1}{n^2+2n+2}$$

and

$$\rho = \lim_{n\to\infty} \left|\dfrac{a_{n+1}}{a_n}\right| = 1 \cdot 1 = 1$$

Therefore, for the series $\sum_{n=1}^{\infty} \dfrac{n}{n^2+1}$, the Ratio Test is inconclusive.

We can show that this series diverges by using the Limit Comparison Test and comparing with the divergent harmonic series.

7. $\sum_{n=1}^{\infty} \dfrac{2^n}{n^{100}}$

SOLUTION With $a_n = \dfrac{2^n}{n^{100}}$,

$$\left|\dfrac{a_{n+1}}{a_n}\right| = \dfrac{2^{n+1}}{(n+1)^{100}} \cdot \dfrac{n^{100}}{2^n} = 2\left(\dfrac{n}{n+1}\right)^{100} \quad \text{and} \quad \rho = \lim_{n\to\infty}\left|\dfrac{a_{n+1}}{a_n}\right| = 2 \cdot 1^{100} = 2 > 1$$

Therefore, the series $\sum_{n=1}^{\infty} \dfrac{2^n}{n^{100}}$ diverges by the Ratio Test.

9. $\sum_{n=1}^{\infty} \dfrac{10^n}{2^{n^2}}$

SOLUTION With $a_n = \dfrac{10^n}{2^{n^2}}$,

$$\left|\dfrac{a_{n+1}}{a_n}\right| = \dfrac{10^{n+1}}{2^{(n+1)^2}} \cdot \dfrac{2^{n^2}}{10^n} = 10 \cdot \dfrac{1}{2^{2n+1}} \quad \text{and} \quad \rho = \lim_{n\to\infty}\left|\dfrac{a_{n+1}}{a_n}\right| = 10 \cdot 0 = 0 < 1$$

Therefore, the series $\sum_{n=1}^{\infty} \dfrac{10^n}{2^{n^2}}$ converges by the Ratio Test.

11. $\sum_{n=1}^{\infty} \dfrac{e^n}{n^n}$

SOLUTION With $a_n = \dfrac{e^n}{n^n}$,

$$\left|\dfrac{a_{n+1}}{a_n}\right| = \dfrac{e^{n+1}}{(n+1)^{n+1}} \cdot \dfrac{n^n}{e^n} = \dfrac{e}{n+1}\left(\dfrac{n}{n+1}\right)^n = \dfrac{e}{n+1}\left(1+\dfrac{1}{n}\right)^{-n}$$

and

$$\rho = \lim_{n\to\infty}\left|\dfrac{a_{n+1}}{a_n}\right| = 0 \cdot \dfrac{1}{e} = 0 < 1$$

Therefore, the series $\sum_{n=1}^{\infty} \dfrac{e^n}{n^n}$ converges by the Ratio Test.

13. $\sum_{n=0}^{\infty} \dfrac{n!}{6^n}$

SOLUTION With $a_n = (-1)^n \dfrac{n!}{4^n}$,

$$\left|\dfrac{a_{n+1}}{a_n}\right| = \dfrac{(n+1)!}{4^{n+1}} \cdot \dfrac{4^n}{n!} = \dfrac{n+1}{4} \quad \text{and} \quad \rho = \lim_{n\to\infty}\left|\dfrac{a_{n+1}}{a_n}\right| = \infty > 1$$

Therefore, the series $\sum_{n=0}^{\infty} (-1)^n \dfrac{n!}{4^n}$ diverges by the Ratio Test.

15. $\sum_{n=2}^{\infty} \dfrac{1}{n^{3/2} \ln n}$

SOLUTION With $a_n = \dfrac{1}{n^{3/2} \ln n}$,

$$\left| \dfrac{a_{n+1}}{a_n} \right| = \dfrac{1}{(n+1)^{3/2} \ln(n+1)} \cdot \dfrac{n^{3/2} \ln n}{1} = \left(\dfrac{n}{n+1} \right)^{3/2} \dfrac{\ln n}{\ln(n+1)}$$

and

$$\rho = \lim_{n \to \infty} \left| \dfrac{a_{n+1}}{a_n} \right| = 1 \cdot \lim_{n \to \infty} \dfrac{\ln n}{\ln(n+1)}$$

Now,

$$\lim_{n \to \infty} \dfrac{\ln n}{\ln(n+1)} = \lim_{x \to \infty} \dfrac{\ln x}{\ln(x+1)} = \lim_{x \to \infty} \dfrac{1/x}{1/(x+1)} = \lim_{x \to \infty} \dfrac{x+1}{x} = 1$$

Thus $\rho = 1$ and the Ratio Test is inconclusive for the series $\sum_{n=2}^{\infty} \dfrac{1}{n^{3/2} \ln n}$.

Using the Integral Test, or the Comparison Test against $\dfrac{1}{n^{3/2}}$, we can show that the series $\sum_{n=2}^{\infty} \dfrac{1}{n^{3/2} \ln n}$ converges.

17. $\sum_{n=1}^{\infty} \dfrac{n^2}{(2n+1)!}$

SOLUTION With $a_n = \dfrac{n^2}{(2n+1)!}$,

$$\left| \dfrac{a_{n+1}}{a_n} \right| = \dfrac{(n+1)^2}{(2n+3)!} \cdot \dfrac{(2n+1)!}{n^2} = \left(\dfrac{n+1}{n} \right)^2 \dfrac{1}{(2n+3)(2n+2)}$$

and

$$\rho = \lim_{n \to \infty} \left| \dfrac{a_{n+1}}{a_n} \right| = 1^2 \cdot 0 = 0 < 1$$

Therefore, the series $\sum_{n=1}^{\infty} \dfrac{n^2}{(2n+1)!}$ converges by the Ratio Test.

19. $\sum_{n=2}^{\infty} \dfrac{1}{2^n + 1}$

SOLUTION With $a_n = \dfrac{1}{2^n + 1}$,

$$\left| \dfrac{a_{n+1}}{a_n} \right| = \dfrac{1}{2^{n+1} + 1} \cdot \dfrac{2^n + 1}{1} = \dfrac{1 + 2^{-n}}{2 + 2^{-n}}$$

and

$$\rho = \lim_{n \to \infty} \left| \dfrac{a_{n+1}}{a_n} \right| = \dfrac{1}{2} < 1$$

Therefore, the series $\sum_{n=2}^{\infty} \dfrac{1}{2^n + 1}$ converges by the Ratio Test.

21. Show that $\sum_{n=1}^{\infty} n^k 3^{-n}$ converges for all exponents k.

SOLUTION With $a_n = n^k 3^{-n}$,

$$\left| \dfrac{a_{n+1}}{a_n} \right| = \dfrac{(n+1)^k 3^{-(n+1)}}{n^k 3^{-n}} = \dfrac{1}{3} \left(1 + \dfrac{1}{n} \right)^k$$

and, for all k,

$$\rho = \lim_{n \to \infty} \left| \frac{a_{n+1}}{a_n} \right| = \frac{1}{3} \cdot 1 = \frac{1}{3} < 1$$

Therefore, the series $\sum_{n=1}^{\infty} n^k \, 3^{-n}$ converges for all exponents k by the Ratio Test.

23. Show that $\sum_{n=1}^{\infty} 2^n x^n$ converges if $|x| < \frac{1}{2}$.

SOLUTION With $a_n = 2^n x^n$,

$$\left| \frac{a_{n+1}}{a_n} \right| = \frac{2^{n+1} |x|^{n+1}}{2^n |x|^n} = 2|x| \quad \text{and} \quad \rho = \lim_{n \to \infty} \left| \frac{a_{n+1}}{a_n} \right| = 2|x|$$

Therefore, $\rho < 1$ and the series $\sum_{n=1}^{\infty} 2^n x^n$ converges by the Ratio Test provided $|x| < \frac{1}{2}$.

25. Show that $\sum_{n=1}^{\infty} \frac{r^n}{n}$ converges if $|r| < 1$.

SOLUTION With $a_n = \frac{r^n}{n}$,

$$\left| \frac{a_{n+1}}{a_n} \right| = \frac{|r|^{n+1}}{n+1} \cdot \frac{n}{|r|^n} = |r| \frac{n}{n+1} \quad \text{and} \quad \rho = \lim_{n \to \infty} \left| \frac{a_{n+1}}{a_n} \right| = 1 \cdot |r| = |r|$$

Therefore, by the Ratio Test, the series $\sum_{n=1}^{\infty} \frac{r^n}{n}$ converges provided $|r| < 1$.

In Exercises 27–28, the following limit could be helpful: $\lim_{n \to \infty} \left(1 + \frac{1}{n}\right)^n = e$.

27. Does $\sum_{n=1}^{\infty} \frac{n!}{n^n}$ converge or diverge?

SOLUTION With $a_n = \frac{n!}{n^n}$,

$$\left| \frac{a_{n+1}}{a_n} \right| = \frac{(n+1)!}{(n+1)^{n+1}} \cdot \frac{n^n}{n!} = \left(\frac{n}{n+1} \right)^n = \left(1 + \frac{1}{n}\right)^{-n}$$

and

$$\rho = \lim_{n \to \infty} \left| \frac{a_{n+1}}{a_n} \right| = \frac{1}{e} < 1$$

Therefore, the series $\sum_{n=1}^{\infty} \frac{n!}{n^n}$ converges by the Ratio Test.

In Exercises 29–33, assume that $|a_{n+1}/a_n|$ converges to $\rho = \frac{1}{3}$. What can you say about the convergence of the given series?

29. $\sum_{n=1}^{\infty} n^3 a_n$

SOLUTION Let $b_n = n^3 a_n$. Then

$$\rho = \lim_{n \to \infty} \left| \frac{b_{n+1}}{b_n} \right| = \lim_{n \to \infty} \left(\frac{n+1}{n} \right)^3 \left| \frac{a_{n+1}}{a_n} \right| = 1^3 \cdot \frac{1}{3} = \frac{1}{3} < 1$$

Therefore, the series $\sum_{n=1}^{\infty} n^3 a_n$ converges by the Ratio Test.

31. $\sum_{n=1}^{\infty} 3^n a_n$

SOLUTION Let $b_n = 3^n a_n$. Then

$$\rho = \lim_{n \to \infty} \left| \frac{b_{n+1}}{b_n} \right| = \lim_{n \to \infty} \frac{3^{n+1}}{3^n} \left| \frac{a_{n+1}}{a_n} \right| = 3 \cdot \frac{1}{3} = 1$$

Therefore, the Ratio Test is inconclusive for the series $\sum_{n=1}^{\infty} 3^n a_n$.

33. $\sum_{n=1}^{\infty} a_n^2$

SOLUTION Let $b_n = a_n^2$. Then

$$\rho = \lim_{n \to \infty} \left| \frac{b_{n+1}}{b_n} \right| = \lim_{n \to \infty} \left| \frac{a_{n+1}}{a_n} \right|^2 = \left(\frac{1}{3} \right)^2 = \frac{1}{9} < 1$$

Therefore, the series $\sum_{n=1}^{\infty} a_n^2$ converges by the Ratio Test.

35. Show that the root test is inconclusive for the p-series $\sum_{n=1}^{\infty} \frac{1}{n^p}$.

SOLUTION With $a_n = \frac{1}{n^p} = n^{-p}$,

$$\lim_{n \to \infty} \sqrt[n]{n^{-p}} = \lim_{n \to \infty} n^{-p/n}$$

To compute this limit, we compute the limit of its logarithm and use L'Hôpital's rule:

$$\lim_{n \to \infty} \ln\left(n^{-p/n} \right) = \lim_{n \to \infty} \frac{-p \ln n}{n} = \lim_{n \to \infty} \frac{-p/n}{1} = \lim_{n \to \infty} \frac{-p}{n} = 0$$

Since the limit of the logarithm is zero, the limit of the original expression is 1, so the root test is inconclusive.

In Exercises 36–41, use the Root Test to determine convergence or divergence (or state that the test is inconclusive).

37. $\sum_{n=1}^{\infty} \frac{1}{n^n}$

SOLUTION With $a_n = \frac{1}{n^n}$,

$$\sqrt[n]{a_n} = \sqrt[n]{\frac{1}{n^n}} = \frac{1}{n} \quad \text{and} \quad \lim_{n \to \infty} \sqrt[n]{a_n} = 0 < 1$$

Therefore, the series $\sum_{n=0}^{\infty} \frac{1}{n^n}$ converges by the Root Test.

39. $\sum_{k=0}^{\infty} \left(\frac{k}{3k+1} \right)^k$

SOLUTION With $a_k = \left(\frac{k}{3k+1} \right)^k$,

$$\sqrt[k]{a_k} = \sqrt[k]{\left(\frac{k}{3k+1} \right)^k} = \frac{k}{3k+1} \quad \text{and} \quad \lim_{k \to \infty} \sqrt[k]{a_k} = \frac{1}{3} < 1$$

Therefore, the series $\sum_{k=0}^{\infty} \left(\frac{k}{3k+1} \right)^k$ converges by the Root Test.

41. $\sum_{n=4}^{\infty} \left(1 + \frac{1}{n}\right)^{-n^2}$

SOLUTION With $a_k = \left(1 + \frac{1}{n}\right)^{-n^2}$,

$$\sqrt[n]{a_n} = \sqrt[n]{\left(1 + \frac{1}{n}\right)^{-n^2}} = \left(1 + \frac{1}{n}\right)^{-n} \quad \text{and} \quad \lim_{n \to \infty} \sqrt[n]{a_n} = e^{-1} < 1$$

Therefore, the series $\sum_{k=0}^{\infty} \left(1 + \frac{1}{n}\right)^{-n^2}$ converges by the Root Test.

In Exercises 43–62, determine convergence or divergence using any method covered in the text so far.

43. $\sum_{n=1}^{\infty} \frac{2^n + 4^n}{7^n}$

SOLUTION Because the series

$$\sum_{n=1}^{\infty} \frac{2^n}{7^n} = \sum_{n=1}^{\infty} \left(\frac{2}{7}\right)^n \quad \text{and} \quad \sum_{n=1}^{\infty} \frac{4^n}{7^n} = \sum_{n=1}^{\infty} \left(\frac{4}{7}\right)^n$$

are both convergent geometric series, it follows that

$$\sum_{n=1}^{\infty} \frac{2^n + 4^n}{7^n} = \sum_{n=1}^{\infty} \left(\frac{2}{7}\right)^n + \sum_{n=1}^{\infty} \left(\frac{4}{7}\right)^n$$

also converges.

45. $\sum_{n=1}^{\infty} \frac{n}{2n+1}$

SOLUTION Since

$$\lim_{n \to \infty} \frac{n}{2n+1} = \lim_{n \to \infty} \frac{1}{2 + 1/n} = \frac{1}{2}$$

this series fails the nth Term Divergence Test, so it diverges.

47. $\sum_{n=1}^{\infty} \frac{\sin n}{n^2}$

SOLUTION Since $|\sin x| \leq 1$ for any x, we have $\left|\frac{\sin n}{n^2}\right| \leq \frac{1}{n^2}$. Since $\sum_{n=1}^{\infty} \frac{1}{n^2}$ is a p-series with $p > 1$, it converges. Then by the Direct Comparison Test,

$$\sum_{n=1}^{\infty} \left|\frac{\sin n}{n^2}\right|$$

converges as well, so that the original series converges absolutely.

49. $\sum_{n=1}^{\infty} \frac{1}{n + \sqrt{n}}$

SOLUTION Use the Direct Comparison Test. Since $n + \sqrt{n} \geq 2\sqrt{n}$ for $n \geq 1$, we have $\frac{1}{n + \sqrt{n}} \leq \frac{1}{2\sqrt{n}}$. But

$$\sum_{n=1}^{\infty} \frac{1}{2\sqrt{n}} = \frac{1}{2} \sum_{n=1}^{\infty} \frac{1}{n^{1/2}}$$

diverges, since it is a p-series with $p \leq 1$, so that $\sum_{n=1}^{\infty} \frac{1}{n + \sqrt{n}}$ diverges as well.

51. $\sum_{n=1}^{\infty} \frac{n^3}{5^n}$

SOLUTION The presence of the exponential term suggests applying the Ratio Test. With $a_n = \frac{n^3}{5^n}$,

$$\left|\frac{a_{n+1}}{a_n}\right| = \frac{(n+1)^3}{5^{n+1}} \cdot \frac{5^n}{n^3} = \frac{1}{5}\left(1 + \frac{1}{n}\right)^3 \quad \text{and} \quad \rho = \lim_{n \to \infty} \left|\frac{a_{n+1}}{a_n}\right| = \frac{1}{5} \cdot 1^3 = \frac{1}{5} < 1$$

Therefore, the series $\sum_{n=1}^{\infty} \dfrac{n^3}{5^n}$ converges by the Ratio Test.

53. $\sum_{n=2}^{\infty} \dfrac{1}{\sqrt{n^3 - n^2}}$

SOLUTION This series is similar to a *p*-series; because

$$\dfrac{1}{\sqrt{n^3 - n^2}} \approx \dfrac{1}{\sqrt{n^3}} = \dfrac{1}{n^{3/2}}$$

for large *n*, we will apply the Limit Comparison Test comparing with the *p*-series with $p = \tfrac{3}{2}$. Now,

$$L = \lim_{n \to \infty} \dfrac{\frac{1}{\sqrt{n^3 - n^2}}}{\frac{1}{n^{3/2}}} = \lim_{n \to \infty} \sqrt{\dfrac{n^3}{n^3 - n^2}} = 1$$

The *p*-series with $p = \tfrac{3}{2}$ converges and *L* exists; therefore, the series $\sum_{n=2}^{\infty} \dfrac{1}{\sqrt{n^3 - n^2}}$ also converges.

55. $\sum_{n=1}^{\infty} n^{-0.8}$

SOLUTION

$$\sum_{n=1}^{\infty} n^{-0.8} = \sum_{n=1}^{\infty} \dfrac{1}{n^{0.8}}$$

so that this is a divergent *p*-series.

57. $\sum_{n=1}^{\infty} 4^{-2n+1}$

SOLUTION Observe

$$\sum_{n=1}^{\infty} 4^{-2n+1} = \sum_{n=1}^{\infty} 4 \cdot (4^{-2})^n = \sum_{n=1}^{\infty} 4 \left(\dfrac{1}{16} \right)^n$$

is a geometric series with $r = \tfrac{1}{16}$; therefore, this series converges.

59. $\sum_{n=1}^{\infty} \sin \dfrac{1}{n^2}$

SOLUTION Here, we will apply the Limit Comparison Test, comparing with the *p*-series with $p = 2$. Now,

$$L = \lim_{n \to \infty} \dfrac{\sin \frac{1}{n^2}}{\frac{1}{n^2}} = \lim_{u \to 0} \dfrac{\sin u}{u} = 1$$

where $u = \tfrac{1}{n^2}$. The *p*-series with $p = 2$ converges and *L* exists; therefore, the series $\sum_{n=1}^{\infty} \sin \dfrac{1}{n^2}$ also converges.

61. $\sum_{n=1}^{\infty} \dfrac{(-2)^n}{\sqrt{n}}$

SOLUTION Because

$$\lim_{n \to \infty} \dfrac{2^n}{\sqrt{n}} = \lim_{x \to \infty} \dfrac{2^x}{\sqrt{x}} = \lim_{x \to \infty} \dfrac{2^x \ln 2}{\frac{1}{2\sqrt{x}}} = \lim_{x \to \infty} 2^{x+1} \sqrt{x} \ln 2 = \infty \neq 0$$

the general term in the series $\sum_{n=1}^{\infty} \dfrac{(-2)^n}{\sqrt{n}}$ does not tend toward zero; therefore, the series diverges by the Divergence Test.

Further Insights and Challenges

63. **Proof of the Root Test** Let $\sum_{n=0}^{\infty} a_n$ be a positive series, and assume that $L = \lim_{n \to \infty} \sqrt[n]{a_n}$ exists.

(a) Show that the series converges if $L < 1$. *Hint:* Choose R with $L < R < 1$ and show that $a_n \leq R^n$ for n sufficiently large. Then compare with the geometric series $\sum R^n$.

(b) Show that the series diverges if $L > 1$.

SOLUTION Suppose $\lim_{n \to \infty} \sqrt[n]{a_n} = L$ exists.

(a) If $L < 1$, let $\epsilon = \dfrac{1-L}{2}$. By the definition of a limit, there is a positive integer N such that

$$-\epsilon \leq \sqrt[n]{a_n} - L \leq \epsilon$$

for $n \geq N$. From this, we conclude that

$$0 \leq \sqrt[n]{a_n} \leq L + \epsilon$$

for $n \geq N$. Now, let $R = L + \epsilon$. Then

$$R = L + \frac{1-L}{2} = \frac{L+1}{2} < \frac{1+1}{2} = 1$$

and

$$0 \leq \sqrt[n]{a_n} \leq R \quad \text{or} \quad 0 \leq a_n \leq R^n$$

for $n \geq N$. Because $0 \leq R < 1$, the series $\sum_{n=N}^{\infty} R^n$ is a convergent geometric series, so the series $\sum_{n=N}^{\infty} a_n$ converges by the Direct Comparison Test. Therefore, the series $\sum_{n=0}^{\infty} a_n$ also converges.

(b) If $L > 1$, let $\epsilon = \dfrac{L-1}{2}$. By the definition of a limit, there is a positive integer N such that

$$-\epsilon \leq \sqrt[n]{a_n} - L \leq \epsilon$$

for $n \geq N$. From this, we conclude that

$$L - \epsilon \leq \sqrt[n]{a_n}$$

for $n \geq N$. Now, let $R = L - \epsilon$. Then

$$R = L - \frac{L-1}{2} = \frac{L+1}{2} > \frac{1+1}{2} = 1$$

and

$$R \leq \sqrt[n]{a_n} \quad \text{or} \quad R^n \leq a_n$$

for $n \geq N$. Because $R > 1$, the series $\sum_{n=N}^{\infty} R^n$ is a divergent geometric series, so the series $\sum_{n=N}^{\infty} a_n$ diverges by the Direct Comparison Test. Therefore, the series $\sum_{n=0}^{\infty} a_n$ also diverges.

65. Let $\sum_{n=1}^{\infty} \dfrac{c^n n!}{n^n}$, where c is a constant.

(a) Prove that the series converges absolutely if $|c| < e$ and diverges if $|c| > e$.

(b) It is known that $\lim_{n \to \infty} \dfrac{e^n n!}{n^{n+1/2}} = \sqrt{2\pi}$. Verify this numerically.

(c) Use the Limit Comparison Test to prove that the series diverges for $c = e$.

SOLUTION

(a) With $a_n = \frac{c^n n!}{n^n}$,

$$\left|\frac{a_{n+1}}{a_n}\right| = \frac{|c|^{n+1}(n+1)!}{(n+1)^{n+1}} \cdot \frac{n^n}{|c|^n n!} = |c|\left(\frac{n}{n+1}\right)^n = |c|\left(1 + \frac{1}{n}\right)^{-n}$$

and

$$\rho = \lim_{n\to\infty}\left|\frac{a_{n+1}}{a_n}\right| = |c|e^{-1}$$

Thus, by the Ratio Test, the series $\sum_{n=1}^{\infty} \frac{c^n n!}{n^n}$ converges when $|c|e^{-1} < 1$, or when $|c| < e$. The series diverges when $|c| > e$.

(b) The table below lists the value of $\frac{e^n n!}{n^{n+1/2}}$ for several increasing values of n. Since $\sqrt{2\pi} = 2.506628275$, the numerical evidence verifies that

$$\lim_{n\to\infty} \frac{e^n n!}{n^{n+1/2}} = \sqrt{2\pi}$$

n	100	1000	10000	100000
$\frac{e^n n!}{n^{n+1/2}}$	2.508717995	2.506837169	2.506649163	2.506630363

(c) With $c = e$, the series becomes $\sum_{n=1}^{\infty} \frac{e^n n!}{n^n}$. Using the result from part (b),

$$L = \lim_{n\to\infty} \frac{\frac{e^n n!}{n^n}}{\sqrt{n}} = \lim_{n\to\infty} \frac{e^n n!}{n^{n+1/2}} = \sqrt{2\pi}$$

Because the series $\sum_{n=1}^{\infty} \sqrt{n}$ diverges by the Divergence Test and $L > 0$, we conclude that $\sum_{n=1}^{\infty} \frac{e^n n!}{n^n}$ diverges by the Limit Comparison Test.

10.6 Power Series   (LT Section 11.6)

Preliminary Questions

1. Suppose that $\sum a_n x^n$ converges for $x = 5$. Must it also converge for $x = 4$? What about $x = -3$?

SOLUTION The power series $\sum a_n x^n$ is centered at $x = 0$. Because the series converges for $x = 5$, the radius of convergence must be at least 5 and the series converges absolutely at least for the interval $|x| < 5$. Both $x = 4$ and $x = -3$ are inside this interval, so the series converges for $x = 4$ and for $x = -3$.

2. Suppose that $\sum a_n(x - 6)^n$ converges for $x = 10$. At which of the points (a)–(d) must it also converge?

(a) $x = 8$ (b) $x = 11$ (c) $x = 3$ (d) $x = 0$

SOLUTION The given power series is centered at $x = 6$. Because the series converges for $x = 10$, the radius of convergence must be at least $|10 - 6| = 4$ and the series converges absolutely at least for the interval $|x - 6| < 4$, or $2 < x < 10$.

(a) $x = 8$ is inside the interval $2 < x < 10$, so the series converges for $x = 8$.
(b) $x = 12$ is not inside the interval $2 < x < 10$, so the series may or may not converge for $x = 12$.
(c) $x = 2$ is an endpoint of the interval $2 < x < 10$, so the series may or may not converge for $x = 2$.
(d) $x = 0$ is not inside the interval $2 < x < 10$, so the series may or may not converge for $x = 0$.

3. What is the radius of convergence of $F(3x)$ if $F(x)$ is a power series with radius of convergence $R = 12$?

SOLUTION If the power series $F(x)$ has radius of convergence $R = 12$, then the power series $F(3x)$ has radius of convergence $R = \frac{12}{3} = 4$.

4. The power series $F(x) = \sum_{n=1}^{\infty} nx^n$ has radius of convergence $R = 1$. What is the power series expansion of $F'(x)$ and what is its radius of convergence?

SOLUTION We obtain the power series expansion for $F'(x)$ by differentiating the power series expansion for $F(x)$ term-by-term. Thus,

$$F'(x) = \sum_{n=1}^{\infty} n^2 x^{n-1}$$

The radius of convergence for this series is $R = 1$, the same as the radius of convergence for the series expansion for $F(x)$.

Exercises

1. Use the Ratio Test to determine the radius of convergence R of $\sum_{n=0}^{\infty} \frac{x^n}{2^n}$. Does it converge at the endpoints $x = \pm R$?

SOLUTION With $a_n = \frac{x^n}{2^n}$,

$$\left|\frac{a_{n+1}}{a_n}\right| = \frac{|x|^{n+1}}{2^{n+1}} \cdot \frac{2^n}{|x|^n} = \frac{|x|}{2} \quad \text{and} \quad \rho = \lim_{n \to \infty} \left|\frac{a_{n+1}}{a_n}\right| = \frac{|x|}{2}$$

By the Ratio Test, the series converges when $\rho = \frac{|x|}{2} < 1$, or $|x| < 2$, and diverges when $\rho = \frac{|x|}{2} > 1$, or $|x| > 2$. The radius of convergence is therefore $R = 2$.

3. Show that the power series (a)–(c) have the same radius of convergence. Then show that (a) diverges at both endpoints, (b) converges at one endpoint but diverges at the other, and (c) converges at both endpoints.

(a) $\sum_{n=1}^{\infty} \frac{x^n}{3^n}$ **(b)** $\sum_{n=1}^{\infty} \frac{x^n}{n 3^n}$ **(c)** $\sum_{n=1}^{\infty} \frac{x^n}{n^2 3^n}$

SOLUTION

(a) With $a_n = \frac{x^n}{3^n}$,

$$\rho = \lim_{n \to \infty} \left|\frac{a_{n+1}}{a_n}\right| = \lim_{n \to \infty} \left|\frac{x^{n+1}}{3^{n+1}} \cdot \frac{3^n}{x^n}\right| = \lim_{n \to \infty} \left|\frac{x}{3}\right| = |x| \lim_{n \to \infty} \frac{1}{3} = \frac{|x|}{3}$$

Thus $\rho < 1$ when $|x| < 3$, so the radius of convergence is $R = 3$. For the endpoint $x = 3$, the series becomes

$$\sum_{n=1}^{\infty} \frac{3^n}{3^n} = \sum_{n=1}^{\infty} 1$$

which diverges by the Divergence Test. For the endpoint $x = -3$, the series becomes

$$\sum_{n=1}^{\infty} \frac{(-3)^n}{3^n} = \sum_{n=1}^{\infty} (-1)^n$$

which also diverges by the Divergence Test.

(b) With $a_n = \frac{x^n}{n 3^n}$,

$$\rho = \lim_{n \to \infty} \left|\frac{a_{n+1}}{a_n}\right| = \lim_{n \to \infty} \left|\frac{x^{n+1}}{(n+1)3^{n+1}} \cdot \frac{n 3^n}{x^n}\right| = \lim_{n \to \infty} \left|\frac{nx}{3(n+1)}\right| = |x| \lim_{n \to \infty} \frac{n}{3n+3} = \frac{|x|}{3}$$

Thus $\rho < 1$ when $|x| < 3$, so the radius of convergence is $R = 3$. For the endpoint $x = 3$, the series becomes

$$\sum_{n=1}^{\infty} \frac{3^n}{n 3^n} = \sum_{n=1}^{\infty} \frac{1}{n}$$

which is the divergent harmonic series. For the endpoint $x = -3$, the series becomes

$$\sum_{n=1}^{\infty} \frac{(-3)^n}{n 3^n} = \sum_{n=1}^{\infty} \frac{(-1)^n}{n}$$

which converges by the Alternating Series Test.

(c) With $a_n = \frac{x^n}{n^2 3^n}$,

$$\rho = \lim_{n\to\infty}\left|\frac{a_{n+1}}{a_n}\right| = \lim_{n\to\infty}\left|\frac{x^{n+1}}{(n+1)^2 3^{n+1}} \cdot \frac{n^2 3^n}{x^n}\right| = \lim_{n\to\infty}\left|\frac{n^2 x}{3(n+1)^2}\right| = |x|\lim_{n\to\infty}\frac{n^2}{3n^2+6n+3} = \frac{|x|}{3}$$

Thus $\rho < 1$ when $|x| < 3$, so the radius of convergence is $R = 3$. For the endpoint $x = 3$, the series becomes

$$\sum_{n=1}^{\infty}\frac{3^n}{n^2 3^n} = \sum_{n=1}^{\infty}\frac{1}{n^2}$$

which is a convergent p-series. For the endpoint $x = -3$, the series becomes

$$\sum_{n=1}^{\infty}\frac{(-3)^n}{n^2 3^n} = \sum_{n=1}^{\infty}\frac{(-1)^n}{n^2}$$

which converges by the Alternating Series Test.

5. Show that $\sum_{n=0}^{\infty} n^n x^n$ diverges for all $x \neq 0$.

SOLUTION With $a_n = n^n x^n$,

$$\rho = \lim_{n\to\infty}\left|\frac{a_{n+1}}{a_n}\right| = \lim_{n\to\infty}\left|\frac{(n+1)^{n+1} x^{n+1}}{n^n x^n}\right| = |x|\lim_{n\to\infty}\frac{(n+1)^{n+1}}{n^n} = |x|\lim_{n\to\infty}\left(1+\frac{1}{n}\right)^n (n+1) = \infty \text{ if } x \neq 0$$

Then $\rho < 1$ only when $x = 0$, so the radius of convergence is $R = 0$. In other words, the power series converges only for $x = 0$.

7. Use the Ratio Test to show that $\sum_{n=0}^{\infty}\frac{x^{2n}}{3^n}$ has radius of convergence $R = \sqrt{3}$.

SOLUTION With $a_n = \frac{x^{2n}}{3^n}$,

$$\rho = \lim_{n\to\infty}\left|\frac{a_{n+1}}{a_n}\right| = \lim_{n\to\infty}\left|\frac{x^{2(n+1)}}{3^{n+1}} \cdot \frac{3^n}{x^{2n}}\right| = \lim_{n\to\infty}\left|\frac{x^2}{3}\right| = \left|\frac{x^2}{3}\right|$$

Then $\rho < 1$ when $|x^2| < 3$, or $|x| < \sqrt{3}$, so the radius of convergence is $R = \sqrt{3}$.

In Exercises 9–34, find the interval of convergence.

9. $\sum_{n=0}^{\infty} nx^n$

SOLUTION With $a_n = nx^n$,

$$\rho = \lim_{n\to\infty}\left|\frac{a_{n+1}}{a_n}\right| = \lim_{n\to\infty}\left|\frac{(n+1)x^{n+1}}{nx^n}\right| = |x|\lim_{n\to\infty}\frac{n+1}{n} = |x|$$

Then $\rho < 1$ when $|x| < 1$, so that $R = 1$ and the series converges absolutely on the interval $-1 < x < 1$. For the endpoint $x = 1$, the series becomes $\sum_{n=0}^{\infty} n$, which diverges by the Divergence Test. For the endpoint $x = -1$, the series becomes $\sum_{n=1}^{\infty}(-1)^n n$, which also diverges by the Divergence Test. Thus, the series $\sum_{n=0}^{\infty} nx^n$ converges for $-1 < x < 1$ and diverges elsewhere.

11. $\sum_{n=1}^{\infty}(-1)^n \frac{x^{2n+1}}{2^n n}$

SOLUTION With $a_n = (-1)^n \frac{x^{2n+1}}{2^n n}$,

$$\rho = \lim_{n\to\infty}\left|\frac{x^{2(n+1)+1}}{2^{n+1}(n+1)} \cdot \frac{2^n n}{x^{2n+1}}\right| = \lim_{n\to\infty}\left|\frac{x^2}{2} \cdot \frac{n}{n+1}\right| = \left|\frac{x^2}{2}\right|$$

Then $\rho < 1$ when $|x| < \sqrt{2}$, so the radius of convergence is $R = \sqrt{2}$, and the series converges absolutely on the interval $-\sqrt{2} < x < \sqrt{2}$. For the endpoint $x = -\sqrt{2}$, the series becomes $\sum_{n=1}^{\infty} (-1)^n \frac{-\sqrt{2}}{n} = \sum_{n=1}^{\infty} (-1)^{n+1} \frac{\sqrt{2}}{n}$, which converges by the Leibniz test. For the endpoint $x = \sqrt{2}$, the series becomes $\sum_{n=1}^{\infty} (-1)^n \frac{\sqrt{2}}{n}$ which also converges by the Leibniz test. Thus the series $\sum_{n=1}^{\infty} (-1)^n \frac{x^{2n+1}}{2^n n}$ converges for $-\sqrt{2} \le x \le \sqrt{2}$ and diverges elsewhere.

13. $\sum_{n=4}^{\infty} \frac{x^n}{n^5}$

SOLUTION With $a_n = \frac{x^n}{n^5}$,

$$\rho = \lim_{n \to \infty} \left| \frac{a_{n+1}}{a_n} \right| = \lim_{n \to \infty} \left| \frac{x^{n+1}}{(n+1)^5} \cdot \frac{n^5}{x^n} \right| = |x| \lim_{n \to \infty} \frac{(n+1)^5}{n^5} = |x|$$

Then $\rho < 1$ when $|x| < 1$, so the radius of convergence is $R = 1$, and the series converges absolutely on $-1 < x < 1$. For the endpoint $x = 1$, the series becomes $\sum_{n=4}^{\infty} \frac{1}{n^5}$, which is a convergent p-series. For the endpoint $x = -1$, the series becomes $\sum_{n=4}^{\infty} \frac{(-1)^n}{n^5}$, which converges by the Alternating Series Test. Thus, the series $\sum_{n=4}^{\infty} \frac{x^n}{n^5}$ converges for $-1 \le x \le 1$ and diverges elsewhere.

15. $\sum_{n=0}^{\infty} \frac{x^n}{(n!)^2}$

SOLUTION With $a_n = \frac{x^n}{(n!)^2}$,

$$\rho = \lim_{n \to \infty} \left| \frac{a_{n+1}}{a_n} \right| = \lim_{n \to \infty} \left| \frac{x^{n+1}(n!)^2}{((n+1)!)^2 x^n} \right| = |x| \lim_{n \to \infty} \frac{(n!)^2}{((n+1)!)^2} = |x| \lim_{n \to \infty} \frac{1}{(n+1)^2} = 0$$

Then $\rho < 1$ for any x, so the radius of convergence is $R = \infty$ and the series converges absolutely for all x.

17. $\sum_{n=0}^{\infty} \frac{(2n)!}{(n!)^3} x^n$

SOLUTION With $a_n = \frac{(2n)! x^n}{(n!)^3}$, and assuming $x \neq 0$,

$$\rho = \lim_{n \to \infty} \left| \frac{a_{n+1}}{a_n} \right| = \lim_{n \to \infty} \left| \frac{(2(n+1))! x^{n+1}}{((n+1)!)^3} \cdot \frac{(n!)^3}{(2n)! x^n} \right| = \lim_{n \to \infty} \left| x \frac{(2n+2)(2n+1)}{(n+1)^3} \right|$$

$$= \lim_{n \to \infty} \left| x \frac{4n^2 + 6n + 2}{n^3 + 3n^2 + 3n + 1} \right| = \lim_{n \to \infty} \left| x \frac{4n^{-1} + 6n^{-1} + 2n^{-3}}{1 + 3n^{-1} + 3n^{-2} + n^{-3}} \right| = 0$$

Then $\rho < 1$ for all x, so the radius of convergence is $R = \infty$, and the series converges absolutely for all x.

19. $\sum_{n=0}^{\infty} \frac{(-1)^n x^n}{\sqrt{n^2 + 1}}$

SOLUTION With $a_n = \frac{(-1)^n x^n}{\sqrt{n^2+1}}$,

$$\rho = \lim_{n \to \infty} \left| \frac{a_{n+1}}{a_n} \right| = \lim_{n \to \infty} \left| \frac{(-1)^{n+1} x^{n+1}}{\sqrt{(n+1)^2+1}} \cdot \frac{\sqrt{n^2+1}}{(-1)^n x^n} \right| = |x| \lim_{n \to \infty} \frac{\sqrt{n^2+1}}{\sqrt{(n+1)^2+1}} = |x| \lim_{n \to \infty} \sqrt{\frac{n^2+1}{n^2+2n+2}} = |x|$$

Then $\rho < 1$ when $|x| < 1$, so the radius of convergence is $R = 1$, and the series converges absolutely on the interval $-1 < x < 1$. For the endpoint $x = 1$, the series becomes $\sum_{n=0}^{\infty} \frac{(-1)^n}{\sqrt{n^2+1}}$, which converges by the Alternating Series Test. For the endpoint $x = -1$, the series becomes $\sum_{n=0}^{\infty} \frac{1}{\sqrt{n^2+1}}$, which diverges by the Limit Comparison Test comparing with the divergent harmonic series. Thus, the series $\sum_{n=0}^{\infty} \frac{(-1)^n x^n}{\sqrt{n^2+1}}$ converges for $-1 < x \le 1$ and diverges elsewhere.

21. $\sum_{n=15}^{\infty} \frac{x^{2n+1}}{3n+1}$

SOLUTION With $a_n = \frac{x^{2n+1}}{3n+1}$,

$$\rho = \lim_{n\to\infty} \left| \frac{a_{n+1}}{a_n} \right| = \lim_{n\to\infty} \left| \frac{x^{2(n+1)+1}}{3(n+1)+1} \cdot \frac{3n+1}{x^{2n+1}} \right| = \lim_{n\to\infty} \left| \frac{x^{2n+3}(3n+1)}{(3n+4)x^{2n+1}} \right| = |x|^2 \lim_{n\to\infty} \frac{3n+1}{3n+4} = |x|^2$$

Then $\rho < 1$ when $|x| < 1$, so the radius of convergence is $R = 1$, and the series converges absolutely on the interval $-1 < x < 1$. For the endpoint $x = 1$, the series becomes $\sum_{n=15}^{\infty} \frac{1}{3n+1}$, which diverges by the Limit Comparison Test comparing with the divergent harmonic series. For the endpoint $x = -1$, the series becomes $\sum_{n=15}^{\infty} \frac{-1}{3n+1}$, which also diverges by the Limit Comparison Test comparing with the divergent harmonic series. Thus, the series $\sum_{n=15}^{\infty} \frac{x^{2n+1}}{3n+1}$ converges for $-1 < x < 1$ and diverges elsewhere.

23. $\sum_{n=2}^{\infty} \frac{x^n}{\ln n}$

SOLUTION With $a_n = \frac{x^n}{\ln n}$,

$$\rho = \lim_{n\to\infty} \left| \frac{a_{n+1}}{a_n} \right| = \lim_{n\to\infty} \left| \frac{x^{n+1}}{\ln(n+1)} \cdot \frac{\ln n}{x^n} \right| = |x| \lim_{n\to\infty} \frac{\ln n}{\ln(n+1)}$$

Using L'Hôpital's Rule, we have

$$\rho = |x| \lim_{n\to\infty} \frac{\ln n}{\ln(n+1)} = |x| \lim_{n\to\infty} \frac{1/n}{1/(n+1)} = |x| \lim_{n\to\infty} \frac{n+1}{n} = |x|$$

Then $\rho < 1$ when $|x| < 1$, so the radius of convergence is $R = 1$, and the series converges absolutely on the interval $-1 < x < 1$. For the endpoint $x = 1$, the series becomes $\sum_{n=2}^{\infty} \frac{1}{\ln n}$. Because $\frac{1}{\ln n} > \frac{1}{n}$ and $\sum_{n=2}^{\infty} \frac{1}{n}$ is the divergent harmonic series, the endpoint series diverges by the Comparison Test. For the endpoint $x = -1$, the series becomes $\sum_{n=2}^{\infty} \frac{(-1)^n}{\ln n}$, which converges by the Alternating Series Test. Thus, the series $\sum_{n=2}^{\infty} \frac{x^n}{\ln n}$ converges for $-1 \le x < 1$ and diverges elsewhere.

25. $\sum_{n=1}^{\infty} n(x-3)^n$

SOLUTION With $a_n = n(x-3)^n$,

$$\rho = \lim_{n\to\infty} \left| \frac{a_{n+1}}{a_n} \right| = \lim_{n\to\infty} \left| \frac{(n+1)(x-3)^{n+1}}{n(x-3)^n} \right| = |x-3| \lim_{n\to\infty} \left| \frac{n+1}{n} \right| = |x-3|$$

Then $\rho < 1$ when $|x-3| < 1$, so the radius of convergence is $R = 1$ centered at 3, and the series converges absolutely on the interval $2 < x < 4$. For the endpoint $x = 4$, the series becomes $\sum_{n=1}^{\infty} n$, which diverges by the Divergence Test. For the endpoint $x = 2$, the series becomes $\sum_{n=1}^{\infty}(-1)^n n$, which also diverges by the Divergence Test. Thus, the series $\sum_{n=1}^{\infty} n(x-3)^n$ converges for $2 < x < 4$ and diverges elsewhere.

27. $\sum_{n=1}^{\infty}(-1)^n n^5 (x-7)^n$

SOLUTION With $a_n = (-1)^n n^5 (x-7)^n$,

$$\rho = \lim_{n\to\infty}\left|\frac{a_{n+1}}{a_n}\right| = \lim_{n\to\infty}\left|\frac{(-1)^{n+1}(n+1)^5(x-7)^{n+1}}{(-1)^n n^5(x-7)^n}\right| = |x-7|\lim_{n\to\infty}\left|\frac{-(n+1)^5}{n^5}\right| = |x-7|\lim_{n\to\infty}\frac{(n+1)^5}{n^5} = |x-7|$$

Then $\rho < 1$ when $|x-7| < 1$, so the radius of convergence is $R = 1$ centered at 7, and the series converges absolutely on the interval $6 < x < 8$. For the endpoint $x = 6$, the series becomes $\sum_{n=1}^{\infty} n^5$, which diverges by the Divergence Test. For the endpoint $x = 8$, the series becomes $\sum_{n=1}^{\infty}(-1)^n n^5$, which also diverges by the Divergence Test. Thus, the series $\sum_{n=1}^{\infty}(-1)^n n^5(x-7)^n$ converges for $6 < x < 8$ and diverges elsewhere.

29. $\sum_{n=1}^{\infty}\frac{2^n}{3n}(x+3)^n$

SOLUTION With $a_n = \frac{2^n(x+3)^n}{3n}$,

$$\rho = \lim_{n\to\infty}\left|\frac{a_{n+1}}{a_n}\right| = \lim_{n\to\infty}\left|\frac{2^{n+1}(x+3)^{n+1}}{3(n+1)} \cdot \frac{3n}{2^n(x+3)^n}\right| = |x+3|\lim_{n\to\infty}\left|\frac{6n}{3n+3}\right| = 2|x+3|$$

Then $\rho < 1$ when $2|x+3| < 1$, or when $|x+3| < \frac{1}{2}$. Thus the radius of convergence is $\frac{1}{2}$ centered at $x = -3$, so the series converges absolutely on the interval $-\frac{7}{2} < x < -\frac{5}{2}$. For the endpoint $x = -\frac{5}{2}$, the series becomes $\sum_{n=1}^{\infty}\frac{1}{3n}$, which diverges because it is a multiple of the divergent harmonic series. For the endpoint $x = -\frac{7}{2}$, the series becomes $\sum_{n=1}^{\infty}\frac{(-1)^n}{3n}$, which converges by the Alternating Series Test. Thus, the series $\sum_{n=1}^{\infty}\frac{2^n}{3n}(x+3)^n$ converges for $-\frac{7}{2} \le x < -\frac{5}{2}$ and diverges elsewhere.

31. $\sum_{n=0}^{\infty}\frac{(-5)^n}{n!}(x+10)^n$

SOLUTION With $a_n = \frac{(-5)^n}{n!}(x+10)^n$,

$$\rho = \lim_{n\to\infty}\left|\frac{a_{n+1}}{a_n}\right| = \lim_{n\to\infty}\left|\frac{(-5)^{n+1}(x+10)^{n+1}}{(n+1)!} \cdot \frac{n!}{(-5)^n(x+10)^n}\right| = \lim_{n\to\infty}\left|5(x+10)\frac{1}{n}\right| = 0$$

Thus $\rho < 1$ for all x, so the radius of convergence is infinite, and $\sum_{n=0}^{\infty}\frac{(-5)^n}{n!}(x+10)^n$ converges for all x.

33. $\sum_{n=12}^{\infty} e^n(x-2)^n$

SOLUTION With $a_n = e^n(x-2)^n$,

$$\rho = \lim_{n\to\infty}\left|\frac{a_{n+1}}{a_n}\right| = \lim_{n\to\infty}\left|\frac{e^{n+1}(x-2)^{n+1}}{e^n(x-2)^n}\right| = \lim_{n\to\infty}|e(x-2)| = |e(x-2)|$$

750 CHAPTER 10 | INFINITE SERIES (LT CHAPTER 11)

Thus $\rho < 1$ when $|e(x-2)| < 1$, or when $|x-2| < e^{-1}$. So the radius of convergence is e^{-1} centered at $x = 2$, and the series converges absolutely on the interval $2 - e^{-1} < x < 2 + e^{-1}$. For the endpoint $x = 2 + e^{-1}$, the series becomes $\sum_{n=12}^{\infty} 1$, which diverges by the Divergence Test. For the endpoint $x = 2 - e^{-1}$, the series becomes $\sum_{n=12}^{\infty} (-1)^n$, which also diverges by the Divergence Test. Thus, the series $\sum_{n=12}^{\infty} e^n(x-2)^n$ converges for $2 - e^{-1} < x < 2 + e^{-1}$ and diverges elsewhere.

In Exercises 35–40, use Eq. (2) to expand the function in a power series with center $c = 0$ and determine the interval of convergence.

35. $f(x) = \dfrac{1}{1 - 3x}$

SOLUTION Substituting $3x$ for x in Eq. (2), we obtain

$$\frac{1}{1 - 3x} = \sum_{n=0}^{\infty} (3x)^n = \sum_{n=0}^{\infty} 3^n x^n$$

This series is valid for $|3x| < 1$, or $|x| < \frac{1}{3}$.

37. $f(x) = \dfrac{1}{3 - x}$

SOLUTION First write

$$\frac{1}{3 - x} = \frac{1}{3} \cdot \frac{1}{1 - \frac{x}{3}}$$

Substituting $\frac{x}{3}$ for x in Eq. (2), we obtain

$$\frac{1}{1 - \frac{x}{3}} = \sum_{n=0}^{\infty} \left(\frac{x}{3}\right)^n = \sum_{n=0}^{\infty} \frac{x^n}{3^n}$$

Thus,

$$\frac{1}{3 - x} = \frac{1}{3} \sum_{n=0}^{\infty} \frac{x^n}{3^n} = \sum_{n=0}^{\infty} \frac{x^n}{3^{n+1}}$$

This series is valid for $|x/3| < 1$, or $|x| < 3$.

39. $f(x) = \dfrac{1}{1 - x^3}$

SOLUTION Substituting x^3 for x in Eq. (2), we obtain

$$\frac{1}{1 - x^3} = \sum_{n=0}^{\infty} \left(x^3\right)^n = \sum_{n=0}^{\infty} x^{3n}$$

This series is valid for $|x^3| < 1$, or $|x| < 1$.

41. Differentiate the power series in Exercise 39 to obtain a power series for $g(x) = \dfrac{3x^2}{(1 - x^3)^2}$.

SOLUTION The general term in Exercise 39 is x^{3n}, whose derivative is $3nx^{3n-1}$. The term for $n = 0$ is a constant, so it vanishes when we take the derivative. The derivative of $\dfrac{1}{1 - x^3}$ is $g(x)$, so the power series about $c = 0$ for $g(x)$ is

$$\sum_{n=1}^{\infty} 3nx^{3n-1}$$

This series is valid for $|x| < 1$.

43. (a) Divide the power series in Exercise 41 by $3x^2$ to obtain a power series for $h(x) = \dfrac{1}{(1-x^3)^2}$ and use the Ratio Test to show that the radius of convergence is 1.

(b) Another way to obtain a power series for $h(x)$ is to square the power series for $f(x)$ in Exercise 39. By multiplying term by term, determine the terms up to degree 9 in the resulting power series for $(f(x))^2$ and show that they match the terms in the power series for $h(x)$ found in part (a).

SOLUTION

(a) Dividing the series in Exercise 41 by $3x^2$ gives the power series for $\dfrac{1}{(1-x^3)^2}$:

$$\sum_{n=1}^{\infty} \frac{3nx^{3n-1}}{3x^2} = \sum_{n=1}^{\infty} nx^{3n-3} = \sum_{n=0}^{\infty} (n+1)x^{3n}$$

By the Ratio Test,

$$\lim_{n \to \infty} \left| \frac{a_{n+1}}{a_n} \right| = \lim_{n \to \infty} \frac{(n+2)x^{3(n+1)}}{(n+1)x^{3n}} = x^3 \lim_{n \to \infty} \frac{n+2}{n+1} = x^3$$

so the radius of convergence is 1.

(b) Multiplying the first few terms of the power series in Exercise 41 by themselves gives

$$(1 + x^3 + x^6 + x^9 + \cdots)(1 + x^3 + x^6 + x^9 + \cdots) = 1 + 2x^3 + 3x^6 + 4x^9 + \cdots$$

which matches terms of the form $(n+1)x^{3n}$ for $n = 0, 1, 2, 3$.

45. Use the equalities

$$\frac{1}{1-x} = \frac{1}{-3-(x-4)} = \frac{-\frac{1}{3}}{1+\left(\frac{x-4}{3}\right)}$$

to show that for $|x - 4| < 3$,

$$\frac{1}{1-x} = \sum_{n=0}^{\infty} (-1)^{n+1} \frac{(x-4)^n}{3^{n+1}}$$

SOLUTION Substituting $-\frac{x-4}{3}$ for x in Eq. (2), we obtain

$$\frac{1}{1+\left(\frac{x-4}{3}\right)} = \sum_{n=0}^{\infty} \left(-\frac{x-4}{3}\right)^n = \sum_{n=0}^{\infty} (-1)^n \frac{(x-4)^n}{3^n}$$

Thus,

$$\frac{1}{1-x} = -\frac{1}{3} \sum_{n=0}^{\infty} (-1)^n \frac{(x-4)^n}{3^n} = \sum_{n=0}^{\infty} (-1)^{n+1} \frac{(x-4)^n}{3^{n+1}}$$

This series is valid for $\left| -\frac{x-4}{3} \right| < 1$, or $|x-4| < 3$.

47. Use the method of Exercise 45 to expand $1/(4 - x)$ in a power series with center $c = 5$. Determine the interval of convergence.

SOLUTION First write

$$\frac{1}{4-x} = \frac{1}{-1-(x-5)} = -\frac{1}{1+(x-5)}$$

Substituting $-(x - 5)$ for x in Eq. (2), we obtain

$$\frac{1}{1+(x-5)} = \sum_{n=0}^{\infty}(-(x-5))^n = \sum_{n=0}^{\infty}(-1)^n(x-5)^n$$

Thus,

$$\frac{1}{4-x} = -\sum_{n=0}^{\infty}(-1)^n(x-5)^n = \sum_{n=0}^{\infty}(-1)^{n+1}(x-5)^n$$

This series is valid for $|-(x-5)| < 1$, or $|x - 5| < 1$.

49. Apply integration to the expansion

$$\frac{1}{1+x} = \sum_{n=0}^{\infty}(-1)^n x^n = 1 - x + x^2 - x^3 + \cdots$$

to prove that for $-1 < x < 1$,

$$\ln(1+x) = \sum_{n=1}^{\infty}\frac{(-1)^{n-1}x^n}{n} = x - \frac{x^2}{2} + \frac{x^3}{3} - \frac{x^4}{4} + \cdots$$

SOLUTION To obtain the first expansion, substitute $-x$ for x in Eq. (2):

$$\frac{1}{1+x} = \sum_{n=0}^{\infty}(-x)^n = \sum_{n=0}^{\infty}(-1)^n x^n$$

This expansion is valid for $|-x| < 1$, or $-1 < x < 1$.

Upon integrating both sides of the above equation, we find

$$\ln(1+x) = \int\frac{dx}{1+x} = \int\left(\sum_{n=0}^{\infty}(-1)^n x^n\right)dx$$

Integrating the series term-by-term then yields

$$\ln(1+x) = C + \sum_{n=0}^{\infty}(-1)^n\frac{x^{n+1}}{n+1}$$

To determine the constant C, set $x = 0$. Then $0 = \ln(1 + 0) = C$. Finally,

$$\ln(1+x) = \sum_{n=0}^{\infty}(-1)^n\frac{x^{n+1}}{n+1} = \sum_{n=1}^{\infty}(-1)^{n-1}\frac{x^n}{n}$$

51. Let $F(x) = (x + 1)\ln(1 + x) - x$.

(a) Apply integration to the result of Exercise 49 to prove that for $-1 < x < 1$,

$$F(x) = \sum_{n=1}^{\infty}(-1)^{n+1}\frac{x^{n+1}}{n(n+1)}$$

(b) Evaluate at $x = \frac{1}{2}$ to prove

$$\frac{3}{2}\ln\frac{3}{2} - \frac{1}{2} = \frac{1}{1\cdot 2\cdot 2^2} - \frac{1}{2\cdot 3\cdot 2^3} + \frac{1}{3\cdot 4\cdot 2^4} - \frac{1}{4\cdot 5\cdot 2^5} + \cdots$$

(c) Use a calculator to verify that the partial sum S_4 approximates the left-hand side with an error no greater than the term a_5 of the series.

SOLUTION

(a) Note that

$$\int \ln(x+1)\,dx = (x+1)\ln(x+1) - x + C$$

Then integrating both sides of the result of Exercise 49 gives

$$(x+1)\ln(x+1) - x = \int \ln(x+1)\,dx = \int \sum_{n=1}^{\infty} \frac{(-1)^{n-1} x^n}{n}\,dx$$

For $-1 < x < 1$, which is the interval of convergence of the series in Exercise 49, therefore, we can integrate term by term to get

$$(x+1)\ln(x+1) - x = \sum_{n=1}^{\infty} \frac{(-1)^{n-1}}{n} \int x^n\,dx = \sum_{n=1}^{\infty} \frac{(-1)^{n-1}}{n} \cdot \frac{x^{n+1}}{n+1} + C = \sum_{n=1}^{\infty} (-1)^{n+1} \frac{x^{n+1}}{n(n+1)} + C$$

(noting that $(-1)^{n-1} = (-1)^{n+1}$). To determine C, evaluate both sides at $x = 0$ to get

$$0 = \ln 1 - 0 = 0 + C$$

so that $C = 0$ and we get finally

$$(x+1)\ln(x+1) - x = \sum_{n=1}^{\infty} (-1)^{n+1} \frac{x^{n+1}}{n(n+1)}$$

(b) Evaluating the result of part(a) at $x = \frac{1}{2}$ gives

$$\frac{3}{2} \ln \frac{3}{2} - \frac{1}{2} = \sum_{n=1}^{\infty} (-1)^{n+1} \frac{1}{n(n+1)2^{n+1}}$$

$$= \frac{1}{1 \cdot 2 \cdot 2^2} - \frac{1}{2 \cdot 3 \cdot 2^3} + \frac{1}{3 \cdot 4 \cdot 2^4} - \frac{1}{4 \cdot 5 \cdot 2^5} + \cdots$$

(c)

$$S_4 = \frac{1}{1 \cdot 2 \cdot 2^2} - \frac{1}{2 \cdot 3 \cdot 2^3} + \frac{1}{3 \cdot 4 \cdot 2^4} - \frac{1}{4 \cdot 5 \cdot 2^5} = 0.1078125$$

$$a_5 = \frac{1}{5 \cdot 6 \cdot 2^6} \approx 0.0005208$$

$$\frac{3}{2} \ln \frac{3}{2} - \frac{1}{2} \approx 0.10819766$$

and

$$\left| S_4 - \frac{3}{2} \ln \frac{3}{2} - \frac{1}{2} \right| \approx 0.0003852 < a_5$$

53. Use the result of Example 7 to show that

$$F(x) = \frac{x^2}{1 \cdot 2} - \frac{x^4}{3 \cdot 4} + \frac{x^6}{5 \cdot 6} - \frac{x^8}{7 \cdot 8} + \cdots$$

is an antiderivative of $f(x) = \tan^{-1} x$ satisfying $F(0) = 0$. What is the radius of convergence of this power series?

SOLUTION For $-1 < x < 1$, which is the interval of convergence for the power series for arctangent, we can integrate term-by-term, so integrate that power series to get

$$F(x) = \int \tan^{-1} x\,dx = \sum_{n=0}^{\infty} \int \frac{(-1)^n x^{2n+1}}{2n+1}\,dx = \sum_{n=0}^{\infty} (-1)^n \frac{x^{2n+2}}{(2n+1)(2n+2)}$$

$$= \frac{x^2}{1 \cdot 2} - \frac{x^4}{3 \cdot 4} + \frac{x^6}{5 \cdot 6} - \frac{x^8}{7 \cdot 8} + \cdots + C$$

If we assume $F(0) = 0$, then we have $C = 0$. The radius of convergence of this power series is the same as that of the original power series, which is 1.

55. Evaluate $\sum_{n=1}^{\infty} \dfrac{n}{2^n}$. *Hint:* Use differentiation to show that

$$(1-x)^{-2} = \sum_{n=1}^{\infty} nx^{n-1} \quad (\text{for } |x| < 1)$$

SOLUTION Differentiate both sides of Eq. (2) to obtain

$$\dfrac{1}{(1-x)^2} = \sum_{n=1}^{\infty} nx^{n-1}$$

Setting $x = \tfrac{1}{2}$ then yields

$$\sum_{n=1}^{\infty} \dfrac{n}{2^{n-1}} = \dfrac{1}{\left(1-\tfrac{1}{2}\right)^2} = 4$$

Divide this equation by 2 to obtain

$$\sum_{n=1}^{\infty} \dfrac{n}{2^n} = 2$$

57. Show that the following series converges absolutely for $|x| < 1$ and compute its sum:

$$F(x) = 1 - x - x^2 + x^3 - x^4 - x^5 + x^6 - x^7 - x^8 + \cdots$$

Hint: Write $F(x)$ as a sum of three geometric series with common ratio x^3.

SOLUTION To find the radius of convergence, we compute

$$\rho = \lim_{n \to \infty} \left| \dfrac{a_{n+1}}{a_n} \right| = \lim_{n \to \infty} \left| \dfrac{\pm 1 \cdot x^{n+1}}{\pm 1 \cdot x^n} \right| = \lim_{n \to \infty} |x| = |x|$$

The radius of convergence is therefore $R = 1$, and the series converges absolutely for $|x| < 1$.

By Exercise 43 of Section 4, any rearrangement of the terms of an absolutely convergent series yields another absolutely convergent series with the same sum as the original series. Following the hint, we now rearrange the terms of $F(x)$ as the sum of three geometric series:

$$F(x) = \left(1 + x^3 + x^6 + \cdots\right) - \left(x + x^4 + x^7 + \cdots\right) - \left(x^2 + x^5 + x^8 + \cdots\right)$$

$$= \sum_{n=0}^{\infty} (x^3)^n - \sum_{n=0}^{\infty} x(x^3)^n - \sum_{n=0}^{\infty} x^2(x^3)^n = \dfrac{1}{1-x^3} - \dfrac{x}{1-x^3} - \dfrac{x^2}{1-x^3} = \dfrac{1-x-x^2}{1-x^3}$$

59. Find all values of x such that $\sum_{n=1}^{\infty} \dfrac{x^{n^2}}{n!}$ converges.

SOLUTION With $a_n = \dfrac{x^{n^2}}{n!}$,

$$\left| \dfrac{a_{n+1}}{a_n} \right| = \dfrac{|x|^{(n+1)^2}}{(n+1)!} \cdot \dfrac{n!}{|x|^{n^2}} = \dfrac{|x|^{2n+1}}{n+1}$$

if $|x| \le 1$, then

$$\lim_{n \to \infty} \dfrac{|x|^{2n+1}}{n+1} = 0$$

and the series converges absolutely. On the other hand, if $|x| > 1$, then

$$\lim_{n \to \infty} \dfrac{|x|^{2n+1}}{n+1} = \infty$$

and the series diverges. Thus, $\sum_{n=1}^{\infty} \dfrac{x^{n^2}}{n!}$ converges for $-1 \le x \le 1$ and diverges elsewhere.

61. Find a power series $P(x) = \sum_{n=0}^{\infty} a_n x^n$ satisfying the differential equation $y' = -y$ with initial condition $y(0) = 1$. Then use Eq. (8) in Section 9.1 to conclude that $P(x) = e^{-x}$.

SOLUTION Let $P(x) = \sum_{n=0}^{\infty} a_n x^n$ and note that $P(0) = a_0$; thus, to satisfy the initial condition $P(0) = 1$, we must take $a_0 = 1$. Now,

$$P'(x) = \sum_{n=1}^{\infty} n a_n x^{n-1}$$

so

$$P'(x) + P(x) = \sum_{n=1}^{\infty} n a_n x^{n-1} + \sum_{n=0}^{\infty} a_n x^n = \sum_{n=0}^{\infty} [(n+1)a_{n+1} + a_n] x^n$$

In order for this series to be equal to zero, the coefficient of x^n must be equal to zero for each n; thus

$$(n+1)a_{n+1} + a_n = 0 \quad \text{or} \quad a_{n+1} = -\frac{a_n}{n+1}$$

Starting from $a_0 = 1$, we then calculate

$$a_1 = -\frac{a_0}{1} = -1;$$

$$a_2 = -\frac{a_1}{2} = \frac{1}{2};$$

$$a_3 = -\frac{a_2}{3} = -\frac{1}{6} = -\frac{1}{3!}$$

and, in general,

$$a_n = (-1)^n \frac{1}{n!}$$

Hence,

$$P(x) = \sum_{n=0}^{\infty} (-1)^n \frac{x^n}{n!}$$

The solution to the initial value problem $y' = -y$, $y(0) = 1$ is $y = e^{-x}$. Because this solution is unique, it follows that

$$P(x) = \sum_{n=0}^{\infty} (-1)^n \frac{x^n}{n!} = e^{-x}$$

63. Use the power series for $y = e^x$ to show that

$$\frac{1}{e} = \frac{1}{2!} - \frac{1}{3!} + \frac{1}{4!} - \cdots$$

Use your knowledge of alternating series to find an N such that the partial sum S_N approximates e^{-1} to within an error of at most 10^{-3}. Confirm this using a calculator to compute both S_N and e^{-1}.

SOLUTION Recall that the series for e^x is

$$\sum_{n=0}^{\infty} \frac{x^n}{n!} = 1 + x + \frac{x^2}{2!} + \frac{x^3}{3!} + \frac{x^4}{4!} + \cdots$$

Setting $x = -1$ yields

$$e^{-1} = 1 - 1 + \frac{1}{2!} - \frac{1}{3!} + \frac{1}{4!} - + \cdots = \frac{1}{2!} - \frac{1}{3!} + \frac{1}{4!} - + \cdots$$

This is an alternating series with $a_n = \frac{1}{(n+1)!}$. The error in approximating e^{-1} with the partial sum S_N is therefore bounded by

$$|S_N - e^{-1}| \le a_{N+1} = \frac{1}{(N+2)!}$$

To make the error at most 10^{-3}, we must choose N such that

$$\frac{1}{(N+2)!} \le 10^{-3} \quad \text{or} \quad (N+2)! \ge 1000$$

For $N = 4$, $(N+2)! = 6! = 720 < 1000$, but for $N = 5$, $(N+2)! = 7! = 5040$; hence, $N = 5$ is the smallest value that satisfies the Error Bound. The corresponding approximation is

$$S_5 = \frac{1}{2!} - \frac{1}{3!} + \frac{1}{4!} - \frac{1}{5!} + \frac{1}{6!} = 0.368055555$$

Now, $e^{-1} = 0.367879441$, so

$$|S_5 - e^{-1}| = 1.761 \times 10^{-4} < 10^{-3}$$

65. Find a power series $P(x)$ satisfying the differential equation

$$y'' - xy' + y = 0 \qquad \boxed{10}$$

with initial condition $y(0) = 1$, $y'(0) = 0$. What is the radius of convergence of the power series?

SOLUTION Let $P(x) = \sum_{n=0}^{\infty} a_n x^n$. Then

$$P'(x) = \sum_{n=1}^{\infty} n a_n x^{n-1} \quad \text{and} \quad P''(x) = \sum_{n=2}^{\infty} n(n-1) a_n x^{n-2}$$

Note that $P(0) = a_0$ and $P'(0) = a_1$; in order to satisfy the initial conditions $P(0) = 1$, $P'(0) = 0$, we must have $a_0 = 1$ and $a_1 = 0$. Now,

$$P''(x) - xP'(x) + P(x) = \sum_{n=2}^{\infty} n(n-1) a_n x^{n-2} - \sum_{n=1}^{\infty} n a_n x^n + \sum_{n=0}^{\infty} a_n x^n$$

$$= \sum_{n=0}^{\infty} (n+2)(n+1) a_{n+2} x^n - \sum_{n=1}^{\infty} n a_n x^n + \sum_{n=0}^{\infty} a_n x^n$$

$$= 2a_2 + a_0 + \sum_{n=1}^{\infty} [(n+2)(n+1) a_{n+2} - n a_n + a_n] x^n$$

In order for this series to be equal to zero, the coefficient of x^n must be equal to zero for each n; thus, $2a_2 + a_0 = 0$ and $(n+2)(n+1) a_{n+2} - (n-1) a_n = 0$, or

$$a_2 = -\frac{1}{2} a_0 \quad \text{and} \quad a_{n+2} = \frac{n-1}{(n+2)(n+1)} a_n$$

Since $a_1 = 0$ and a_{n+2} is a multiple of a_n for all n, we conclude that $a_n = 0$ if n is odd. As for the even coefficients, we have $a_0 = 1$, $a_2 = -\frac{1}{2}$,

$$a_4 = \frac{1}{(4)(3)} a_2 = -\frac{1}{4!}; \qquad a_6 = \frac{3}{(6)(5)} a_4 = -\frac{3}{6!}; \qquad a_8 = \frac{5}{(8)(7)} a_6 = -\frac{15}{8!}$$

and so on. Claim that in general

$$a_{2n} = -\frac{1 \cdot 3 \cdots (2n-3)}{(2n)!} \quad \text{for } n \ge 2$$

We prove this by induction on n. First, the computation above shows that it holds for $n = 2$, $n = 3$, and $n = 4$. Now assume it holds for $n = k$, and compute:

$$a_{2(k+1)} = a_{2k+2} = \frac{2k-1}{(2k+2)(2k+1)} a_{2k} = -\frac{2k-1}{(2k+2)(2k+1)} \frac{1 \cdot 3 \cdots (2k-3)}{(2k)!}$$

$$= -\frac{1 \cdot 3 \cdots (2k-1)}{(2k+2)!} = -\frac{1 \cdot 3 \cdots (2(k+1)-1)}{(2(k+1))!}$$

and the proof is complete. Thus

$$P(x) = 1 - \frac{1}{2}x^2 - \sum_{n=2}^{\infty} \frac{1 \cdot 3 \cdots (2n-3)}{(2n)!} x^{2n}$$

To compute the radius of convergence, since we are interested in the behavior as $n \to \infty$, we may assume $n \geq 4$; then treating the series as a power series in x^2 we get

$$\rho = \lim_{n \to \infty} \left| \frac{a_{2n+2}}{a_{2n}} \right| = \lim_{n \to \infty} \left| \frac{1 \cdot 3 \cdots (2n-3) \cdot (2n-1)x^{2n+2}}{(2(n+1))!} \cdot \frac{(2n)!}{1 \cdot 3 \cdots (2n-3)x^{2n}} \right|$$

$$= \lim_{n \to \infty} \left| \frac{(2n-1)x^2}{(2n+1)(2n+2)} \right|$$

$$= |x|^2 \lim_{n \to \infty} \frac{2n-1}{4n^2 + 6n + 2} = 0$$

Thus, the radius of convergence is $R = \infty$.

67. Prove that

$$J_2(x) = \sum_{k=0}^{\infty} \frac{(-1)^k}{2^{2k+2} k! (k+3)!} x^{2k+2}$$

is a solution of the Bessel differential equation of order 2:

$$x^2 y'' + xy' + (x^2 - 4)y = 0$$

SOLUTION Let $J_2(x) = \sum_{k=0}^{\infty} \frac{(-1)^k}{2^{2k+2} k! (k+2)!} x^{2k+2}$. Then

$$J_2'(x) = \sum_{k=0}^{\infty} \frac{(-1)^k (k+1)}{2^{2k+1} k! (k+2)!} x^{2k+1}$$

$$J_2''(x) = \sum_{k=0}^{\infty} \frac{(-1)^k (k+1)(2k+1)}{2^{2k+1} k! (k+2)!} x^{2k}$$

and

$$x^2 J_2''(x) + x J_2'(x) + (x^2 - 4) J_2(x) = \sum_{k=0}^{\infty} \frac{(-1)^k (k+1)(2k+1)}{2^{2k+1} k! (k+2)!} x^{2k+2} + \sum_{k=0}^{\infty} \frac{(-1)^k (k+1)}{2^{2k+1} k! (k+2)!} x^{2k+2}$$

$$- \sum_{k=0}^{\infty} \frac{(-1)^k}{2^{2k+2} k! (k+2)!} x^{2k+4} - \sum_{k=0}^{\infty} \frac{(-1)^k}{2^{2k} k! (k+2)!} x^{2k+2}$$

$$= \sum_{k=0}^{\infty} \frac{(-1)^k k(k+2)}{2^{2k} k! (k+2)!} x^{2k+2} + \sum_{k=1}^{\infty} \frac{(-1)^{k-1}}{2^{2k} (k-1)! (k+1)!} x^{2k+2}$$

$$= \sum_{k=1}^{\infty} \frac{(-1)^k}{2^{2k} (k-1)!(k+1)!} x^{2k+2} - \sum_{k=1}^{\infty} \frac{(-1)^k}{2^{2k} (k-1)!(k+1)!} x^{2k+2} = 0$$

Further Insights and Challenges

69. Suppose that the coefficients of $F(x) = \sum_{n=0}^{\infty} a_n x^n$ are *periodic*; that is, for some whole number $M > 0$, we have $a_{M+n} = a_n$. Prove that $F(x)$ converges absolutely for $|x| < 1$ and that

$$F(x) = \frac{a_0 + a_1 x + \cdots + a_{M-1} x^{M-1}}{1 - x^M}$$

Hint: Use the hint for Exercise 57.

SOLUTION Suppose the coefficients of $F(x)$ are periodic, with $a_{M+n} = a_n$ for some whole number M and all n. The $F(x)$ can be written as the sum of M geometric series:

$$F(x) = a_0 \left(1 + x^M + x^{2M} + \cdots\right) + a_1 \left(x + x^{M+1} + x^{2M+1} + \cdots\right) +$$
$$= a_2 \left(x^2 + x^{M+2} + x^{2M+2} + \cdots\right) + \cdots + a_{M-1} \left(x^{M-1} + x^{2M-1} + x^{3M-1} + \cdots\right)$$
$$= \frac{a_0}{1 - x^M} + \frac{a_1 x}{1 - x^M} + \frac{a_2 x^2}{1 - x^M} + \cdots + \frac{a_{M-1} x^{M-1}}{1 - x^M} = \frac{a_0 + a_1 x + a_2 x^2 + \cdots + a_{M-1} x^{M-1}}{1 - x^M}$$

As each geometric series converges absolutely for $|x| < 1$, it follows that $F(x)$ also converges absolutely for $|x| < 1$.

10.7 Taylor Polynomials   (LT Section 11.7)

Preliminary Questions

1. What is T_3 centered at $a = 3$ for a function f such that $f(3) = 9$, $f'(3) = 8$, $f''(3) = 4$, and $f'''(3) = 12$?

SOLUTION In general, with $a = 3$,

$$T_3(x) = f(3) + f'(3)(x - 3) + \frac{f''(3)}{2}(x - 3)^2 + \frac{f'''(3)}{6}(x - 3)^3$$

Using the information provided, we find

$$T_3(x) = 9 + 8(x - 3) + 2(x - 3)^2 + 2(x - 3)^3$$

2. The dashed graphs in Figure 7 are Taylor polynomials for a function f. Which of the two is a Maclaurin polynomial?

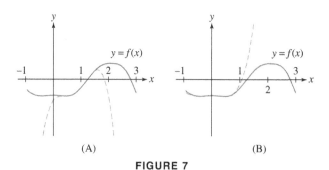

FIGURE 7

SOLUTION A Maclaurin polynomial always gives the value of $f(0)$ exactly. This is true for the Taylor polynomial sketched in (B); thus, this is the Maclaurin polynomial.

3. For which value of x does the Maclaurin polynomial T_n satisfy $T_n(x) = f(x)$, no matter what f is?

SOLUTION A Maclaurin polynomial always gives the value of $f(0)$ exactly.

4. Let T_n be the Maclaurin polynomial of a function f satisfying $|f^{(4)}(x)| \leq 1$ for all x. Which of the following statements follow from the Error Bound?

(a) $|T_4(2) - f(2)| \leq \frac{2}{3}$

(b) $|T_3(2) - f(2)| \leq \frac{2}{3}$

(c) $|T_3(2) - f(2)| \leq \frac{1}{3}$

SOLUTION For a function $f(x)$ satisfying $|f^{(4)}(x)| \leq 1$ for all x,

$$|T_3(2) - f(2)| \leq \frac{1}{24} |f^{(4)}(x)| 2^4 \leq \frac{16}{24} < \frac{2}{3}$$

Thus, **(b)** is the correct answer.

Exercises

In Exercises 1–16, calculate the Taylor polynomials T_2 and T_3 centered at $x = a$ for the given function and value of a.

1. $f(x) = \sin x, \quad a = 0$

SOLUTION First, we calculate and evaluate the needed derivatives:

$$f(x) = \sin x \qquad f(a) = 0$$
$$f'(x) = \cos x \qquad f'(a) = 1$$
$$f''(x) = -\sin x \qquad f''(a) = 0$$
$$f'''(x) = -\cos x \qquad f'''(a) = -1$$

Now,

$$T_2(x) = f(a) + f'(a)(x-a) + \frac{f''(a)}{2}(x-a)^2 = 0 + 1(x-0) + \frac{0}{2}(x-0)^2 = x; \text{ and}$$

$$T_3(x) = f(a) + f'(a)(x-a) + \frac{f''(a)}{2}(x-a)^2 + \frac{f'''(a)}{6}(x-a)^3$$

$$= 0 + 1(x-0) + \frac{0}{2}(x-0)^2 + \frac{-1}{6}(x-0)^3 = x - \frac{1}{6}x^3$$

3. $f(x) = \dfrac{1}{1+x}, \quad a = 2$

SOLUTION First, we calculate and evaluate the needed derivatives:

$$f(x) = \frac{1}{1+x} \qquad f(a) = \frac{1}{3}$$
$$f'(x) = \frac{-1}{(1+x)^2} \qquad f'(a) = -\frac{1}{9}$$
$$f''(x) = \frac{2}{(1+x)^3} \qquad f''(a) = \frac{2}{27}$$
$$f'''(x) = \frac{-6}{(1+x)^4} \qquad f'''(a) = -\frac{2}{27}$$

Now,

$$T_2(x) = f(a) + f'(a)(x-a) + \frac{f''(a)}{2!}(x-a)^2 = \frac{1}{3} - \frac{1}{9}(x-2) + \frac{2/27}{2!}(x-2)^2$$

$$= \frac{1}{3} - \frac{1}{9}(x-2) + \frac{1}{27}(x-2)^2$$

$$T_3(x) = f(a) + f'(a)(x-a) + \frac{f''(a)}{2!}(x-a)^2 + \frac{f'''(a)}{3!}(x-a)^3$$

$$= \frac{1}{3} - \frac{1}{9}(x-2) + \frac{2/27}{2!}(x-2)^2 - \frac{2/27}{3!}(x-2)^3 = \frac{1}{3} - \frac{1}{9}(x-2) + \frac{1}{27}(x-2)^2 - \frac{1}{81}(x-2)^3$$

5. $f(x) = x^4 - 2x, \quad a = 3$

SOLUTION First calculate and evaluate the needed derivatives:

$$f(x) = x^4 - 2x \qquad f(a) = 75$$
$$f'(x) = 4x^3 - 2 \qquad f'(a) = 106$$
$$f''(x) = 12x^2 \qquad f''(a) = 108$$
$$f'''(x) = 24x \qquad f'''(a) = 72$$

Now,

$$T_2(x) = f(a) + f'(a)(x-a) + \frac{f''(a)}{2}(x-a)^2 = 75 + 106(x-3) + \frac{108}{2}(x-3)^2$$
$$= 75 + 106(x-3) + 54(x-3)^2$$

$$T_3(x) = f(a) + f'(a)(x-a) + \frac{f''(a)}{2}(x-a)^2 + \frac{f'''(a)}{3!}(x-a)^3$$
$$= 75 + 106(x-3) + \frac{108}{2}(x-3)^2 + \frac{72}{3!}(x-3)^3$$
$$= 75 + 106(x-3) + 54(x-3)^2 + 12(x-3)^3$$

7. $f(x) = \sqrt{x}, \quad a = 1$

SOLUTION First calculate and evaluate the needed derivatives:

$$f(x) = \sqrt{x} \qquad f(a) = 1$$
$$f'(x) = \frac{1}{2}x^{-1/2} \qquad f'(a) = \frac{1}{2}$$
$$f''(x) = -\frac{1}{4}x^{-3/2} \qquad f''(a) = -\frac{1}{4}$$
$$f'''(x) = \frac{3}{8}x^{-5/2} \qquad f'''(a) = \frac{3}{8}$$

Now,

$$T_2(x) = f(a) + f'(a)(x-a) + \frac{f''(a)}{2!}(x-a)^2 = 1 + \frac{1}{2}(x-1) + \frac{-1/4}{2}(x-1)^2$$
$$= 1 + \frac{1}{2}(x-1) - \frac{1}{8}(x-1)^2$$

$$T_3(x) = f(a) + f'(a)(x-a) + \frac{f''(a)}{2!}(x-a)^2 + \frac{f'''(a)}{3!}(x-a)^3$$
$$= 1 + \frac{1}{2}(x-1) + \frac{-1/4}{2}(x-1)^2 + \frac{3/8}{3!}(x-1)^3$$
$$= 1 + \frac{1}{2}(x-1) - \frac{1}{8}(x-1)^2 + \frac{1}{16}(x-1)^3$$

9. $f(x) = \tan x, \quad a = 0$

SOLUTION First, we calculate and evaluate the needed derivatives:

$$f(x) = \tan x \qquad f(a) = 0$$
$$f'(x) = \sec^2 x \qquad f'(a) = 1$$
$$f''(x) = 2\sec^2 x \tan x \qquad f''(a) = 0$$
$$f'''(x) = 2\sec^4 x + 4\sec^2 x \tan^2 x \qquad f'''(a) = 2$$

Now,

$$T_2(x) = f(a) + f'(a)(x-a) + \frac{f''(a)}{2}(x-a)^2 = 0 + 1(x-0) + \frac{0}{2}(x-0)^2 = x; \text{ and}$$

$$T_3(x) = f(a) + f'(a)(x-a) + \frac{f''(a)}{2}(x-a)^2 + \frac{f'''(a)}{6}(x-a)^3$$
$$= 0 + 1(x-0) + \frac{0}{2}(x-0)^2 + \frac{2}{6}(x-0)^3 = x + \frac{1}{3}x^3$$

11. $f(x) = e^{-x} + e^{-2x}, \quad a = 0$

SOLUTION First, we calculate and evaluate the needed derivatives:

$$f(x) = e^{-x} + e^{-2x} \qquad f(a) = 2$$
$$f'(x) = -e^{-x} - 2e^{-2x} \qquad f'(a) = -3$$
$$f''(x) = e^{-x} + 4e^{-2x} \qquad f''(a) = 5$$
$$f'''(x) = -e^{-x} - 8e^{-2x} \qquad f'''(a) = -9$$

Now,

$$T_2(x) = f(a) + f'(a)(x - a) + \frac{f''(a)}{2}(x - a)^2$$

$$= 2 + (-3)(x - 0) + \frac{5}{2}(x - 0)^2 = 2 - 3x + \frac{5}{2}x^2; \text{ and}$$

$$T_3(x) = f(a) + f'(a)(x - a) + \frac{f''(a)}{2}(x - a)^2 + \frac{f'''(a)}{6}(x - a)^3$$

$$= 2 + (-3)(x - 0) + \frac{5}{2}(x - 0)^2 + \frac{-9}{6}(x - 0)^3 = 2 - 3x + \frac{5}{2}x^2 - \frac{3}{2}x^3$$

13. $f(x) = x^2 e^{-x}, \quad a = 1$

SOLUTION First, we calculate and evaluate the needed derivatives:

$$f(x) = x^2 e^{-x} \qquad f(a) = 1/e$$
$$f'(x) = (2x - x^2)e^{-x} \qquad f'(a) = 1/e$$
$$f''(x) = (x^2 - 4x + 2)e^{-x} \qquad f''(a) = -1/e$$
$$f'''(x) = (-x^2 + 6x - 6)e^{-x} \qquad f'''(a) = -1/e$$

Now,

$$T_2(x) = f(a) + f'(a)(x - a) + \frac{f''(a)}{2}(x - a)^2$$

$$= \frac{1}{e} + \frac{1}{e}(x - 1) + \frac{-1/e}{2}(x - 1)^2 = \frac{1}{e} + \frac{1}{e}(x - 1) - \frac{1}{2e}(x - 1)^2; \text{ and}$$

$$T_3(x) = f(a) + f'(a)(x - a) + \frac{f''(a)}{2}(x - a)^2 + \frac{f'''(a)}{6}(x - a)^3$$

$$= \frac{1}{e} + \frac{1}{e}(x - 1) + \frac{-1/e}{2}(x - 1)^2 + \left(\frac{-1/e}{6}\right)(x - 1)^3$$

$$= \frac{1}{e} + \frac{1}{e}(x - 1) - \frac{1}{2e}(x - 1)^2 - \frac{1}{6e}(x - 1)^3$$

15. $f(x) = \dfrac{\ln x}{x}, \quad a = 1$

SOLUTION First calculate and evaluate the needed derivatives:

$$f(x) = \frac{\ln x}{x} \qquad f(a) = 0$$

$$f'(x) = \frac{1 - \ln x}{x^2} \qquad f(a) = 1$$

$$f''(x) = \frac{-3 + 2\ln x}{x^3} \qquad f(a) = -3$$

$$f'''(x) = \frac{11 - 6\ln x}{x^4} \qquad f(a) = 11$$

so that

$$T_2(x) = f(a) + f'(a)(x-a) + \frac{f''(a)}{2!}(x-a)^2 = 0 + 1(x-1) + \frac{-3}{2!}(x-1)^2$$

$$= (x-1) - \frac{3}{2}(x-1)^2; \text{ and}$$

$$T_3(x) = f(a) + f'(a)(x-a) + \frac{f''(a)}{2!}(x-a)^2 + \frac{f'''(a)}{3!}(x-a)^3$$

$$= 0 + 1(x-1) + \frac{-3}{2!}(x-1)^2 + \frac{11}{3!}(x-1)^3$$

$$= (x-1) - \frac{3}{2}(x-1)^2 + \frac{11}{6}(x-1)^3$$

17. Show that the second Taylor polynomial for $f(x) = px^2 + qx + r$, centered at $a = 1$, is $f(x)$.

SOLUTION Calculating the derivatives at $a = 1$, we get

$$f(x) = px^2 + qx + r \qquad f(1) = p + q + r$$
$$f'(x) = 2px + q \qquad f'(1) = 2p + q$$
$$f''(x) = 2p \qquad f''(1) = 2p$$

Now,

$$T_2(x) = f(a) + f'(a)(x-a) + \frac{f''(a)}{2}(x-a)^2$$

$$= p + q + r + (2p+q)(x-1) + \frac{2p}{2}(x-1)^2$$

$$= px^2 + qx + r = f(x)$$

19. Show that the nth Maclaurin polynomial for $f(x) = e^x$ is

$$T_n(x) = 1 + \frac{x}{1!} + \frac{x^2}{2!} + \cdots + \frac{x^n}{n!}$$

SOLUTION With $f(x) = e^x$, it follows that $f^{(n)}(x) = e^x$ and $f^{(n)}(0) = 1$ for all n. Thus,

$$T_n(x) = 1 + 1(x-0) + \frac{1}{2}(x-0)^2 + \cdots + \frac{1}{n!}(x-0)^n = 1 + x + \frac{x^2}{2} + \cdots + \frac{x^n}{n!}$$

21. Show that the Maclaurin polynomials for $f(x) = \sin x$ are

$$T_{2n+1}(x) = T_{2n+2}(x) = x - \frac{x^3}{3!} + \frac{x^5}{5!} - \cdots + (-1)^n \frac{x^{2n+1}}{(2n+1)!}$$

SOLUTION Let $f(x) = \sin x$. Then

$$f(x) = \sin x \qquad f(0) = 0$$
$$f'(x) = \cos x \qquad f'(0) = 1$$
$$f''(x) = -\sin x \qquad f''(0) = 0$$
$$f'''(x) = -\cos x \qquad f'''(0) = -1$$
$$f^{(4)}(x) = \sin x \qquad f^{(4)}(0) = 0$$
$$f^{(5)}(x) = \cos x \qquad f^{(5)}(0) = 1$$
$$\vdots \qquad \vdots$$

Consequently,

$$T_{2n-1}(x) = x - \frac{x^3}{3!} + \frac{x^5}{5!} + \cdots + (-1)^{n+1} \frac{x^{2n-1}}{(2n-1)!}$$

and

$$T_{2n}(x) = x - \frac{x^3}{3!} + \frac{x^5}{5!} + \cdots + (-1)^{n+1} \frac{x^{2n-1}}{(2n-1)!} + 0 = T_{2n-1}(x)$$

In Exercises 23–30, find T_n centered at $x = a$ for all n.

23. $f(x) = \dfrac{1}{1+x}$, $a = 0$

SOLUTION We have

$$\frac{1}{1+x} = (\ln(1+x))'$$

so that from Exercise 22, letting $g(x) = \ln(1+x)$,

$$f^{(n)}(x) = g^{(n+1)}(x) = (-1)^n n!(x+1)^{-1-n} \quad \text{and} \quad f^{(n)}(0) = (-1)^n n!$$

Then

$$T_n(x) = f(0) + f'(0)x + \frac{f''(0)}{2!}x^2 + \cdots + \frac{f^{(n)}(0)}{n!}x^n$$
$$= 1 - x + \frac{2!}{2!}x^2 - \frac{3!}{3!}x^3 + \cdots + (-1)^n \frac{n!}{n!}x^n$$
$$= 1 - x + x^2 - x^3 + \cdots + (-1)^n x^n$$

25. $f(x) = e^x$, $a = 1$

SOLUTION Let $f(x) = e^x$. Then $f^{(n)}(x) = e^x$ and $f^{(n)}(1) = e$ for all n. Therefore,

$$T_n(x) = e + e(x-1) + \frac{e}{2!}(x-1)^2 + \cdots + \frac{e}{n!}(x-1)^n$$

27. $f(x) = x^{-2}$, $a = 1$

SOLUTION Taking derivatives, we get

$$f(x) = x^{-2} \qquad f(1) = 1$$
$$f'(x) = -2x^{-3} \qquad f'(1) = -2$$
$$f''(x) = 6x^{-4} \qquad f''(1) = 6$$
$$f'''(x) = -24x^{-5} \qquad f'''(1) = -24$$
$$\vdots \qquad \vdots$$
$$f^{(n)}(x) = (-1)^n n! x^{-n-2} \qquad f^{(n)}(1) = (-1)^n n!$$

Therefore,

$$T_n(x) = 1 - 2(x-1) + \frac{3!}{2!}(x-1)^2 - \frac{4!}{3!}(x-1)^3 + \cdots + (-1)^n \frac{(n+1)!}{n!}(x-1)^n$$
$$= 1 - 2(x-1) + 3(x-1)^2 - 4(x-1)^3 + \cdots + (-1)^n(n+1)(x-1)^n$$

29. $f(x) = \cos x$, $a = \dfrac{\pi}{4}$

SOLUTION Let $f(x) = \cos x$. Then

$$f(x) = \cos x \qquad f(\pi/4) = \frac{1}{\sqrt{2}}$$
$$f'(x) = -\sin x \qquad f'(\pi/4) = -\frac{1}{\sqrt{2}}$$
$$f''(x) = -\cos x \qquad f''(\pi/4) = -\frac{1}{\sqrt{2}}$$
$$f'''(x) = \sin x \qquad f'''(\pi/4) = \frac{1}{\sqrt{2}}$$

This pattern of four values repeats indefinitely. Thus,

$$f^{(n)}(\pi/4) = \begin{cases} (-1)^{(n+1)/2}\dfrac{1}{\sqrt{2}}, & n \text{ odd} \\ (-1)^{n/2}\dfrac{1}{\sqrt{2}}, & n \text{ even} \end{cases}$$

and

$$T_n(x) = \frac{1}{\sqrt{2}} - \frac{1}{\sqrt{2}}\left(x - \frac{\pi}{4}\right) - \frac{1}{2\sqrt{2}}\left(x - \frac{\pi}{4}\right)^2 + \frac{1}{6\sqrt{2}}\left(x - \frac{\pi}{4}\right)^3 \cdots$$

In general, the coefficient of $(x - \pi/4)^n$ is

$$\pm \frac{1}{(\sqrt{2})n!}$$

with the pattern of signs $+, -, -, +, +, -, -, \ldots$.

In Exercises 31–34, find T_2 and use a calculator to compute the error $|f(x) - T_2(x)|$ for the given values of a and x.

31. $y = e^x$, $a = 0$, $x = -0.5$

SOLUTION Let $f(x) = e^x$. Then $f'(x) = e^x$, $f''(x) = e^x$, $f(a) = 1$, $f'(a) = 1$ and $f''(a) = 1$. Therefore

$$T_2(x) = 1 + 1(x - 0) + \frac{1}{2}(x - 0)^2 = 1 + x + \frac{1}{2}x^2$$

and

$$T_2(-0.5) = 1 + (-0.5) + \frac{1}{2}(-0.5)^2 = 0.625$$

Using a calculator, we find

$$f(-0.5) = \frac{1}{\sqrt{e}} = 0.606531$$

so

$$|T_2(-0.5) - f(-0.5)| = 0.0185$$

33. $y = x^{-2/3}$, $a = 1$, $x = 1.2$

SOLUTION Let $f(x) = x^{-2/3}$. Then $f'(x) = -\frac{2}{3}x^{-5/3}$, $f''(x) = \frac{10}{9}x^{-8/3}$, $f(1) = 1$, $f'(1) = -\frac{2}{3}$, and $f''(1) = \frac{10}{9}$. Thus

$$T_2(x) = 1 - \frac{2}{3}(x - 1) + \frac{10}{2 \cdot 9}(x - 1)^2 = 1 - \frac{2}{3}(x - 1) + \frac{5}{9}(x - 1)^2$$

and

$$T_2(1.2) = 1 - \frac{2}{3}(0.2) + \frac{5}{9}(0.2)^2 = \frac{8}{9} \approx 0.88889$$

Using a calculator, $f(1.2) = (1.2)^{-2/3} \approx 0.88555$ so that

$$|T_2(1.2) - f(1.2)| \approx 0.00334$$

35. [GU] Compute T_3 for $f(x) = \sqrt{x}$ centered at $a = 1$. Then use a plot of the error $|f(x) - T_3(x)|$ to find a value $c > 1$ such that the error on the interval $[1, c]$ is at most 0.25.

SOLUTION We have

$$f(x) = x^{1/2} \qquad f(1) = 1$$

$$f'(x) = \frac{1}{2}x^{-1/2} \qquad f'(1) = \frac{1}{2}$$

$$f''(x) = -\frac{1}{4}x^{-3/2} \qquad f''(1) = -\frac{1}{4}$$

$$f'''(x) = \frac{3}{8}x^{-5/2} \qquad f'''(1) = \frac{3}{8}$$

Therefore

$$T_3(x) = 1 + \frac{1}{2}(x-1) - \frac{1}{4 \cdot 2!}(x-1)^2 + \frac{3}{8 \cdot 3!}(x-1)^3 = 1 + \frac{1}{2}(x-1) - \frac{1}{8}(x-1)^2 + \frac{1}{16}(x-1)^3$$

A plot of $|f(x) - T_3(x)|$ is below.

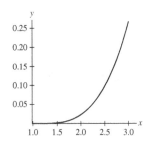

It appears that for $x \in [1, 2.9]$ that the error does not exceed 0.25. The error at $x = 3$ appears to just exceed 0.25.

37. Let T_3 be the Maclaurin polynomial of $f(x) = e^x$. Use the Error Bound to find the maximum possible value of $|f(1.1) - T_3(1.1)|$. Show that we can take $K = e^{1.1}$.

SOLUTION Since $f(x) = e^x$, we have $f^{(n)}(x) = e^x$ for all n; since e^x is increasing, the maximum value of e^x on the interval $[0, 1.1]$ is $K = e^{1.1}$. Then by the Error Bound,

$$\left|e^{1.1} - T_3(1.1)\right| \le K\frac{(1.1-0)^4}{4!} = \frac{e^{1.1} 1.1^4}{24} \approx 0.183$$

In Exercises 39–42, compute the Taylor polynomial indicated and use the Error Bound to find the maximum possible size of the error. Verify your result with a calculator.

39. $f(x) = \cos x$, $a = 0$; $|\cos 0.25 - T_5(0.25)|$

SOLUTION The Maclaurin series for $\cos x$ is

$$1 - \frac{x^2}{2!} + \frac{x^4}{4!} - \frac{x^6}{6!} + \dots$$

so that

$$T_5(x) = 1 - \frac{x^2}{2} + \frac{x^4}{24}$$

$$T_5(0.25) \approx 0.9689127604$$

In addition, $f^{(6)}(x) = -\cos x$ so that $|f^{(6)}(x)| \le 1$ and we may take $K = 1$ in the Error Bound formula. Then

$$|\cos 0.25 - T_5(0.25)| \le K\frac{0.25^6}{6!} = \frac{1}{2^{12} \cdot 6!} \approx 3.390842014 \times 10^{-7}$$

(The true value is $\cos 0.25 \approx 0.9689124217$ and the difference is in fact $\approx 3.387 \times 10^{-7}$.)

41. $f(x) = x^{-1/2}$, $a = 4$; $|f(4.3) - T_3(4.3)|$

SOLUTION We have

$$f(x) = x^{-1/2} \qquad f(4) = \frac{1}{2}$$

$$f'(x) = -\frac{1}{2}x^{-3/2} \qquad f'(4) = -\frac{1}{16}$$

$$f''(x) = \frac{3}{4}x^{-5/2} \qquad f''(4) = \frac{3}{128}$$

$$f'''(x) = -\frac{15}{8}x^{-7/2} \qquad f'''(4) = -\frac{15}{1024}$$

$$f^{(4)}(x) = \frac{105}{16}x^{-9/2}$$

so that

$$T_3(x) = \frac{1}{2} - \frac{1}{16}(x-4) + \frac{3}{256}(x-4)^2 - \frac{5}{2048}(x-4)^3$$

Using the Error Bound formula,

$$|f(4.3) - T_3(4.3)| \le K\frac{|4.3-4|^4}{4!} = \frac{27K}{80{,}000}$$

where K is a number such that $|f^{(4)}(x)| \le K$ for x between 4 and 4.3. Now, $f^{(4)}(x)$ is a decreasing function for $x > 1$, so it takes its maximum value on $[4, 4.3]$ at $x = 4$; there, its value is

$$K = \frac{105}{16}4^{-9/2} = \frac{105}{8192}$$

so that

$$|f(4.3) - T_3(4.3)| \le \frac{27\frac{105}{8192}}{80{,}000} = \frac{27 \cdot 105}{8192 \cdot 80{,}000} \approx 4.3258667 \times 10^{-6}$$

43. Calculate the Maclaurin polynomial T_3 for $f(x) = \tan^{-1} x$. Compute $T_3(\frac{1}{2})$ and use the Error Bound to find a bound for the error $\left|\tan^{-1}\frac{1}{2} - T_3(\frac{1}{2})\right|$. Refer to the graph in Figure 8 to find an acceptable value of K. Verify your result by computing $\left|\tan^{-1}\frac{1}{2} - T_3(\frac{1}{2})\right|$ using a calculator.

FIGURE 8 Graph of $f^{(4)}(x) = \dfrac{-24x(x^2-1)}{(x^2+1)^4}$, where $f(x) = \tan^{-1} x$.

SOLUTION Let $f(x) = \tan^{-1} x$. Then

$$f(x) = \tan^{-1} x \qquad\qquad f(0) = 0$$

$$f'(x) = \frac{1}{1+x^2} \qquad\qquad f'(0) = 1$$

$$f''(x) = \frac{-2x}{(1+x^2)^2} \qquad\qquad f''(0) = 0$$

$$f'''(x) = \frac{(1+x^2)^2(-2) - (-2x)(2)(1+x^2)(2x)}{(1+x^2)^4} \qquad f'''(0) = -2$$

and

$$T_3(x) = 0 + 1(x-0) + \frac{0}{2}(x-0)^2 + \frac{-2}{6}(x-0)^3 = x - \frac{x^3}{3}$$

Since $f^{(4)}(x) \le 5$ for $x \ge 0$, we may take $K = 5$ in the Error Bound; then,

$$\left|\tan^{-1}\left(\frac{1}{2}\right) - T_3\left(\frac{1}{2}\right)\right| \le \frac{5(1/2)^4}{4!} = \frac{5}{384}$$

45. [GU] Let T_2 be the Taylor polynomial at $a = 0.5$ for $f(x) = \cos(x^2)$. Use the error bound to find the maximum possible value of $|f(0.6) - T_2(0.6)|$. Plot $f^{(3)}$ to find an acceptable value of K.

SOLUTION We have

$$f(x) = \cos(x^2) \qquad f(0.5) = \cos(0.25) \approx 0.9689$$
$$f'(x) = -2x\sin(x^2) \qquad f'(0.5) = -\sin(0.25) \approx -0.2474039593$$
$$f''(x) = -4x^2\cos(x^2) - 2\sin(x^2) \qquad f''(0.5) = -\cos(0.25) - 2\sin(0.25) \approx -1.463720340$$
$$f^{(3)}(x) = 8x^3\sin(x^2) - 12x\cos(x^2)$$

so that

$$T_2(x) = 0.9689 - 0.2472039593(x - .5) - 0.73186017(x - .5)^2$$

and $T_2(0.6) \approx 0.9368534237$. A graph of $f^{(3)}(x)$ for x near 0.5 is below.

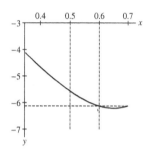

Clearly the maximum value of $|f^{(3)}(x)|$ on $[0.5, 0.6]$ is bounded by 7 (near $x = 0.5$), so we may take $K = 7$; then

$$|f(0.6) - T_2(0.6)| \leq K\frac{|0.6 - 0.5|^3}{3!} = \frac{7}{6000} \approx 0.0011666667$$

In Exercises 47–50, use the Error Bound to find a value of n for which the given inequality is satisfied. Then verify your result using a calculator.

47. $|\cos 0.1 - T_n(0.1)| \leq 10^{-7}, \quad a = 0$

SOLUTION Using the Error Bound with $K = 1$ (every derivative of $f(x) = \cos x$ is $\pm \sin x$ or $\pm \cos x$, so $|f^{(n)}(x)| \leq 1$ for all n), we have

$$|T_n(0.1) - \cos 0.1| \leq \frac{(0.1)^{n+1}}{(n+1)!}$$

With $n = 3$,

$$\frac{(0.1)^4}{4!} \approx 4.17 \times 10^{-6} > 10^{-7}$$

but with $n = 4$,

$$\frac{(0.1)^5}{5!} \approx 8.33 \times 10^{-8} < 10^{-7}$$

so we choose $n = 4$. Now,

$$T_4(x) = 1 - \frac{1}{2}x^2 + \frac{1}{24}x^4$$

so

$$T_4(0.1) = 1 - \frac{1}{2}(0.1)^2 + \frac{1}{24}(0.1)^4 = 0.995004166$$

Using a calculator, $\cos 0.1 = 0.995004165$, so

$$|T_4(0.1) - \cos 0.1| = 1.387 \times 10^{-8} < 10^{-7}$$

49. $|\sqrt{1.3} - T_n(1.3)| \leq 10^{-6}$, $a = 1$

SOLUTION Using the Error Bound, we have

$$|\sqrt{1.3} - T_n(1.3)| \leq K\frac{|1.3 - 1|^{n+1}}{(n+1)!} = K\frac{|0.3|^{n+1}}{(n+1)!}$$

where K is a number such that $|f^{(n+1)}(x)| \leq K$ for x between 1 and 1.3. For $f(x) = \sqrt{x}$, $|f^{(n)}(x)|$ is decreasing for $x > 1$, hence the maximum value of $|f^{(n+1)}(x)|$ occurs at $x = 1$. We may therefore take

$$K = |f^{(n+1)}(1)| = \frac{1 \cdot 3 \cdot 5 \cdots (2n+1)}{2^{n+1}}$$

$$= \frac{1 \cdot 3 \cdot 5 \cdots (2n+1)}{2^{n+1}} \cdot \frac{2 \cdot 4 \cdot 6 \cdots (2n+2)}{2 \cdot 4 \cdot 6 \cdots (2n+2)} = \frac{(2n+2)!}{(n+1)! 2^{2n+2}}$$

Then

$$|\sqrt{1.3} - T_n(1.3)| \leq \frac{(2n+2)!}{(n+1)! 2^{2n+2}} \cdot \frac{|.3|^{n+1}}{(n+1)!} = \frac{(2n+2)!}{[(n+1)!]^2}(0.075)^{n+1}$$

With $n = 9$

$$\frac{(20)!}{[(10)!]^2}(0.075)^{10} = 1.040 \times 10^{-6} > 10^{-6}$$

but with $n = 10$

$$\frac{(22)!}{[(11)!]^2}(0.075)^{11} = 2.979 \times 10^{-7} < 10^{-6}$$

Hence, $n = 10$ will guarantee the desired accuracy. Using technology to compute and evaluate $T_{10}(1.3)$ gives

$$T_{10}(1.3) \approx 1.140175414, \quad \sqrt{1.3} \approx 1.140175425$$

and

$$|\sqrt{1.3} - T_{10}(1.3)| \approx 1.1 \times 10^{-8} < 10^{-6}$$

51. Let $f(x) = e^{-x}$ and $T_3(x) = 1 - x + \frac{x^2}{2} - \frac{x^3}{6}$.

(a) Use the Error Bound to show that for all $x \geq 0$,

$$|f(x) - T_3(x)| \leq \frac{x^4}{24}$$

(b) [GU] Illustrate this inequality by plotting $y = f(x) - T_3(x)$ and $y = x^4/24$ together over $[0, 1]$.

SOLUTION

(a) Note that $f^{(n)}(x) = \pm e^{-x}$, so that $|f^{(n)}(x)| = f(x)$. Now, $f(x)$ is a decreasing function for $x \geq 0$, so that for any $c > 0$, $|f^{(n)}(x)|$ takes its maximum value at $x = 0$; this value is $e^0 = 1$. Thus we may take $K = 1$ in the Error Bound equation. Thus for any x,

$$|f(x) - T_3(x)| \leq K\frac{|x - 0|^4}{4!} = \frac{x^4}{24}$$

(b) A plot of $f(x) - T_3(x)$ and $\frac{x^4}{24}$ is shown below.

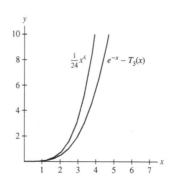

53. Let T_n be the Taylor polynomial for $f(x) = \ln x$ at $a = 1$, and let $c > 1$. Show that

$$|\ln c - T_n(c)| \le \frac{|c-1|^{n+1}}{n+1}$$

Then find a value of n such that $|\ln 1.5 - T_n(1.5)| \le 10^{-2}$.

SOLUTION With $f(x) = \ln x$, we have

$$f'(x) = x^{-1}, \ f''(x) = -x^{-2}, \ f'''(x) = 2x^{-3}, \ f^{(4)}(x) = -6x^{-4}$$

and, in general,

$$f^{(k+1)}(x) = (-1)^k k! \, x^{-k-1}$$

Notice that $|f^{(k+1)}(x)| = k!|x|^{-k-1}$ is a decreasing function for $x > 0$. Therefore, the maximum value of $|f^{(k+1)}(x)|$ on $[1, c]$ is $|f^{(k+1)}(1)|$. Using the Error Bound, we have

$$|\ln c - T_n(c)| \le K \frac{|c-1|^{n+1}}{(n+1)!}$$

where K is a number such that $|f^{(n+1)}(x)| \le K$ for x between 1 and c. We have just determined that we may take $K = |f^{(n+1)}(1)| = n!$. Then

$$|\ln c - T_n(c)| \le n! \frac{|c-1|^{n+1}}{(n+1)!} = \frac{|c-1|^{n+1}}{n+1}$$

Evaluating at $c = 1.5$ gives

$$|\ln 1.5 - T_n(1.5)| \le \frac{|1.5-1|^{n+1}}{n+1} = \frac{(0.5)^{n+1}}{n+1}$$

With $n = 3$,

$$\frac{(0.5)^4}{4} = 0.015625 > 10^{-2}$$

but with $n = 4$,

$$\frac{(0.5)^5}{5} = 0.00625 < 10^{-2}$$

Hence, $n = 4$ will guarantee the desired accuracy.

55. Verify that the third Maclaurin polynomial for $f(x) = e^x \sin x$ is equal to the product of the third Maclaurin polynomials of $f(x) = e^x$ and $f(x) = \sin x$ (after discarding terms of degree greater than 3 in the product).

SOLUTION Let $f(x) = e^x \sin x$. Then

$$f(x) = e^x \sin x \qquad\qquad f(0) = 0$$
$$f'(x) = e^x(\cos x + \sin x) \qquad f'(0) = 1$$
$$f''(x) = 2e^x \cos x \qquad\qquad f''(0) = 2$$
$$f'''(x) = 2e^x(\cos x - \sin x) \qquad f'''(0) = 2$$

and

$$T_3(x) = 0 + (1)x + \frac{2}{2!}x^2 + \frac{2}{3!}x^3 = x + x^2 + \frac{x^3}{3}$$

Now, the third Maclaurin polynomial for e^x is $1 + x + \frac{x^2}{2} + \frac{x^3}{6}$, and the third Maclaurin polynomial for $\sin x$ is $x - \frac{x^3}{6}$. Multiplying these two polynomials, and then discarding terms of degree greater than 3, yields

$$e^x \sin x \approx x + x^2 + \frac{x^3}{3}$$

which agrees with the Maclaurin polynomial obtained from the definition.

57. Find the Maclaurin polynomials T_n for $f(x) = \cos(x^2)$. You may use the fact that $T_n(x)$ is equal to the sum of the terms up to degree n obtained by substituting x^2 for x in the nth Maclaurin polynomial of $\cos x$.

SOLUTION The Maclaurin polynomials for $\cos x$ are of the form

$$T_{2n}(x) = 1 - \frac{x^2}{2} + \frac{x^4}{4!} + \cdots + (-1)^n \frac{x^{2n}}{(2n)!}$$

Accordingly, the Maclaurin polynomials for $\cos(x^2)$ are of the form

$$T_{4n}(x) = 1 - \frac{x^4}{2} + \frac{x^8}{4!} + \cdots + (-1)^n \frac{x^{4n}}{(2n)!}$$

59. Let $f(x) = 3x^3 + 2x^2 - x - 4$. Calculate T_j for $j = 1, 2, 3, 4, 5$ at both $a = 0$ and $a = 1$. Show that $T_3(x) = f(x)$ in both cases.

SOLUTION Let $f(x) = 3x^3 + 2x^2 - x - 4$. Then

$$\begin{array}{lll} f(x) = 3x^3 + 2x^2 - x - 4 & f(0) = -4 & f(1) = 0 \\ f'(x) = 9x^2 + 4x - 1 & f'(0) = -1 & f'(1) = 12 \\ f''(x) = 18x + 4 & f''(0) = 4 & f''(1) = 22 \\ f'''(x) = 18 & f'''(0) = 18 & f'''(1) = 18 \\ f^{(4)}(x) = 0 & f^{(4)}(0) = 0 & f^{(4)}(1) = 0 \\ f^{(5)}(x) = 0 & f^{(5)}(0) = 0 & f^{(5)}(1) = 0 \end{array}$$

At $a = 0$,

$$T_1(x) = -4 - x;$$
$$T_2(x) = -4 - x + 2x^2;$$
$$T_3(x) = -4 - x + 2x^2 + 3x^3 = f(x);$$
$$T_4(x) = T_3(x); \text{ and}$$
$$T_5(x) = T_3(x)$$

At $a = 1$,

$$T_1(x) = 12(x - 1);$$
$$T_2(x) = 12(x - 1) + 11(x - 1)^2;$$
$$T_3(x) = 12(x - 1) + 11(x - 1)^2 + 3(x - 1)^3 = -4 - x + 2x^2 + 3x^3 = f(x);$$
$$T_4(x) = T_3(x); \text{ and}$$
$$T_5(x) = T_3(x)$$

61. Let $s(t)$ be the distance of a truck to an intersection. At time $t = 0$, the truck is 60 m from the intersection, travels away from it with a velocity of 24 m/s, and begins to slow down with an acceleration of $a = -3$ m/s^2. Determine the second Maclaurin polynomial of s, and use it to estimate the truck's distance from the intersection after 4 s.

SOLUTION Place the origin at the intersection, so that $s(0) = 60$ (the truck is traveling away from the intersection). The second Maclaurin polynomial of $s(t)$ is

$$T_2(t) = s(0) + s'(0)t + \frac{s''(0)}{2}t^2$$

The conditions of the problem tell us that $s(0) = 60$, $s'(0) = 24$, and $s''(0) = -3$. Thus

$$T_2(t) = 60 + 24t - \frac{3}{2}t^2$$

so that after 4 seconds,

$$T_2(4) = 60 + 24 \cdot 4 - \frac{3}{2} \cdot 4^2 = 132 \text{ m}$$

The truck is 132 m past the intersection.

63. A narrow, negatively charged ring of radius R exerts a force on a positively charged particle P located at distance x above the center of the ring of magnitude

$$F(x) = -\frac{kx}{(x^2 + R^2)^{3/2}}$$

where $k > 0$ is a constant (Figure 10).

(a) Compute the third-degree Maclaurin polynomial for F.

(b) Show that $F \approx -(k/R^3)x$ to second order. This shows that when x is small, $F(x)$ behaves like a restoring force similar to the force exerted by a spring.

(c) Show that $F(x) \approx -k/x^2$ when x is large by showing that

$$\lim_{x \to \infty} \frac{F(x)}{-k/x^2} = 1$$

Thus, $F(x)$ behaves like an inverse square law, and the charged ring looks like a point charge from far away.

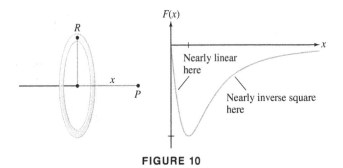

FIGURE 10

SOLUTION

(a) Start by computing and evaluating the necessary derivatives:

$$F(x) = -\frac{kx}{(x^2 + R^2)^{3/2}} \qquad F(0) = 0$$

$$F'(x) = \frac{k(2x^2 - R^2)}{(x^2 + R^2)^{5/2}} \qquad F'(0) = -\frac{k}{R^3}$$

$$F''(x) = \frac{3kx(3R^2 - 2x^2)}{(x^2 + R^2)^{7/2}} \qquad F''(0) = 0$$

$$F'''(x) = \frac{3k(8x^4 - 24x^2 R^2 + 3R^4)}{(x^2 + R^2)^{9/2}} \qquad F'''(0) = \frac{9k}{R^5}$$

so that

$$T_3(x) = F(0) + F'(0)x + \frac{F''(0)}{2!}x^2 + \frac{F'''(0)}{3!}x^3 = -\frac{k}{R^3}x + \frac{3k}{2R^5}x^3$$

(b) To degree 2, $F(x) \approx T_3(x) \approx -\frac{k}{R^3}x$ as we may ignore the x^3 term of $T_3(x)$.

(c) We have

$$\lim_{x \to \infty} \frac{F(x)}{-k/x^2} = \lim_{x \to \infty} \left(-\frac{x^2}{k} \cdot \frac{-kx}{(x^2 + R^2)^{3/2}}\right) = \lim_{x \to \infty} \frac{x^3}{(x^2 + R^2)^{3/2}}$$

$$= \lim_{x \to \infty} \frac{1}{x^{-3}(x^2 + R^2)^{3/2}} = \lim_{x \to \infty} \frac{1}{(1 + R^2/x^2)^{3/2}}$$

$$= 1$$

Thus as x grows large, $F(x)$ looks like an inverse square function.

65. Referring to Figure 12, let a be the length of the chord $\overline{AC}$ of angle θ of the unit circle. Derive the following approximation for the excess of the arc over the chord:

$$\theta - a \approx \frac{\theta^3}{24}$$

Hint: Show that $\theta - a = \theta - 2\sin(\theta/2)$ and use the third Maclaurin polynomial as an approximation.

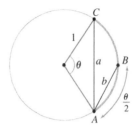

FIGURE 12 Unit circle.

SOLUTION Draw a line from the center O of the circle to B, and label the point of intersection of this line with AC as D. Then $CD = \frac{a}{2}$, and the angle COB is $\frac{\theta}{2}$. Since $CO = 1$, we have

$$\sin\frac{\theta}{2} = \frac{a}{2}$$

so that $a = 2\sin(\theta/2)$. Thus $\theta - a = \theta - 2\sin(\theta/2)$. Now, the third Maclaurin polynomial for $f(\theta) = \sin(\theta/2)$ can be computed as follows: $f(0) = 0$, $f'(x) = \frac{1}{2}\cos(\theta/2)$ so that $f'(0) = \frac{1}{2}$. $f''(x) = -\frac{1}{4}\sin(\theta/2)$ and $f''(0) = 0$. Finally, $f'''(x) = -\frac{1}{8}\cos(\theta/2)$ and $f'''(0) = -\frac{1}{8}$. Thus

$$T_3(\theta) = f(0) + f'(0)\theta + \frac{f''(0)}{2!}\theta^2 + \frac{f'''(0)}{3!}\theta^3 = \frac{1}{2}\theta - \frac{1}{48}\theta^3$$

Finally,

$$\theta - a = \theta - 2\sin\frac{\theta}{2} \approx \theta - 2T_3(\theta) = \theta - \left(\theta - \frac{1}{24}\theta^3\right) = \frac{\theta^3}{24}$$

Further Insights and Challenges

67. Show that the nth Maclaurin polynomial of $f(x) = \arcsin x$ for n odd is

$$T_n(x) = x + \frac{1}{2}\frac{x^3}{3} + \frac{1\cdot 3}{2\cdot 4}\frac{x^5}{5} + \cdots + \frac{1\cdot 3\cdot 5\cdots(n-2)}{2\cdot 4\cdot 6\cdots(n-1)}\frac{x^n}{n}$$

SOLUTION Let $f(x) = \sin^{-1} x$. Then

$$f(x) = \sin^{-1} x \qquad f(0) = 0$$

$$f'(x) = \frac{1}{\sqrt{1-x^2}} \qquad f'(0) = 1$$

$$f''(x) = -\frac{1}{2}(1-x^2)^{-3/2}(-2x) \qquad f''(0) = 0$$

$$f'''(x) = \frac{2x^2+1}{(1-x^2)^{5/2}} \qquad f'''(0) = 1$$

$$f^{(4)}(x) = \frac{-3x(2x^2+3)}{(1-x^2)^{7/2}} \qquad f^{(4)}(0) = 0$$

$$f^{(5)}(x) = \frac{24x^4 + 72x^2 + 9}{(1-x^2)^{9/2}} \qquad f^{(5)}(0) = 9$$

$$\vdots \qquad \vdots$$

$$f^{(7)}(0) = 225$$

and

$$T_7(x) = x + \frac{x^3}{3!} + \frac{9x^5}{5!} + \frac{225x^7}{7!} = x + \frac{1}{2}\frac{x^3}{3} + \frac{1\cdot 3}{2\cdot 4}\frac{x^5}{5} + \frac{1\cdot 3\cdot 5}{2\cdot 4\cdot 6}\frac{x^7}{7}$$

Thus, we can infer that

$$T_n(x) = x + \frac{1}{2}\cdot\frac{x^3}{3} + \frac{1\cdot 3}{2\cdot 4}\frac{x^5}{5} + \frac{1\cdot 3\cdot 5}{2\cdot 4\cdot 6}\frac{x^7}{7} + \cdots + \frac{1\cdot 3}{2\cdot 4}\cdots\frac{n-2}{n-1}\frac{x^n}{n}$$

69. Use Exercise 68 to show that for $x \geq 0$ and all n,

$$e^x \geq 1 + x + \frac{x^2}{2!} + \cdots + \frac{x^n}{n!}$$

Sketch the graphs of $y = e^x$, $y = T_1(x)$, and $y = T_2(x)$ on the same coordinate axes. Does this inequality remain true for $x < 0$?

SOLUTION Let $f(x) = e^x$. Then $f^{(n)}(x) = e^x$ for all n. Because $e^x > 0$ for all x, it follows from Exercise 68 that $f(x) \geq T_n(x)$ for all $x \geq 0$ and for all n. For $f(x) = e^x$,

$$T_n(x) = 1 + x + \frac{x^2}{2!} + \cdots + \frac{x^n}{n!}$$

thus,

$$e^x \geq 1 + x + \frac{x^2}{2!} + \cdots + \frac{x^n}{n!}$$

From the figure below, we see that the inequality does not remain true for $x < 0$, as $T_2(x) \geq e^x$ for $x < 0$.

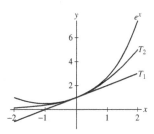

In Exercises 71–75, we estimate integrals using Taylor polynomials. Exercise 72 is used to estimate the error.

71. Find the fourth Maclaurin polynomial T_4 for $f(x) = e^{-x^2}$, and calculate $I = \int_0^{1/2} T_4(x)\,dx$ as an estimate for $\int_0^{1/2} e^{-x^2}\,dx$. A CAS yields the value $I \approx 0.461281$. How large is the error in your approximation? *Hint:* T_4 is obtained by substituting $-x^2$ in the second Maclaurin polynomial for e^x.

SOLUTION Following the hint, since the second Maclaurin polynomial for e^x is

$$1 + x + \frac{x^2}{2}$$

we substitute $-x^2$ for x to get the fourth Maclaurin polynomial for e^{x^2}:

$$T_4(x) = 1 - x^2 + \frac{x^4}{2}$$

Then

$$\int_0^{1/2} e^{-x^2}\,dx \approx \int_0^{1/2} T_4(x)\,dx = \left(x - \frac{1}{3}x^3 + \frac{1}{10}x^5\right)\bigg|_0^{1/2} = \frac{443}{960} \approx 0.4614583333$$

Using a CAS, we have $\int_0^{1/2} e^{-x^2}\,dx \approx 0.4612810064$, so the error is about 1.77×10^{-4}.

73. Let T_4 be the fourth Maclaurin polynomial for $f(x) = \cos x$.

(a) Show that

$$|\cos x - T_4(x)| \le \frac{\left(\frac{1}{2}\right)^6}{6!} \quad \text{for all} \quad x \in \left[0, \frac{1}{2}\right]$$

Hint: $T_4(x) = T_5(x)$.

(b) Evaluate $\displaystyle\int_0^{1/2} T_4(x)\,dx$ as an approximation to $\displaystyle\int_0^{1/2} \cos x\,dx$. Use Exercise 72 to find a bound for the size of the error.

SOLUTION

(a) Let $f(x) = \cos x$. Then

$$T_4(x) = 1 - \frac{x^2}{2} + \frac{x^4}{24}$$

Moreover, with $a = 0$, $T_4(x) = T_5(x)$ and

$$|\cos x - T_4(x)| \le K\frac{|x|^6}{6!}$$

where K is a number such that $|f^{(6)}(u)| \le K$ for u between 0 and x. Now $|f^{(6)}(u)| = |\cos u| \le 1$, so we may take $K = 1$. Finally, with the restriction $x \in [0, \frac{1}{2}]$,

$$|\sin x - T_4(x)| \le \frac{(1/2)^6}{6!} \approx 0.000022$$

(b)

$$\int_0^{1/2} \left(1 - \frac{x^2}{2} + \frac{x^4}{24}\right) dx = \frac{1841}{3840} \approx 0.479427$$

By (a) and Exercise 72, the error associated with this approximation is less than or equal to

$$\frac{(1/2)^6}{6!}\left(\frac{1}{2} - 0\right) = \frac{1}{92160} \approx 1.1 \times 10^{-5}$$

Note that $\displaystyle\int_0^{1/2} \cos x\,dx \approx 0.4794255$, so the actual error is roughly 1.5×10^{-6}.

75. (a) Compute the sixth Maclaurin polynomial T_6 for $f(x) = \sin(x^2)$ by substituting x^2 in $P(x) = x - x^3/6$, the third Maclaurin polynomial for $f(x) = \sin x$.

(b) Show that $|\sin(x^2) - T_6(x)| \le \dfrac{|x|^{10}}{5!}$.

Hint: Substitute x^2 for x in the Error Bound for $|\sin x - P(x)|$, noting that P is also the fourth Maclaurin polynomial for $f(x) = \sin x$.

(c) Use T_6 to approximate $\displaystyle\int_0^{1/2} \sin(x^2)\,dx$ and find a bound for the error.

SOLUTION Let $s(x) = \sin x$ and $f(x) = \sin(x^2)$. Then

(a) The third Maclaurin polynomial for $\sin x$ is

$$S_3(x) = x - \frac{x^3}{6}$$

so, substituting x^2 for x, we see that the sixth Maclaurin polynomial for $\sin(x^2)$ is

$$T_6(x) = x^2 - \frac{x^6}{6}$$

(b) Since all derivatives of $s(x)$ are either $\pm \cos x$ or $\pm \sin x$, they are bounded in magnitude by 1, so we may take $K = 1$ in the Error Bound for $\sin x$. Since the third Maclaurin polynomial $S_3(x)$ for $\sin x$ is also the fourth Maclaurin polynomial $S_4(x)$, we have

$$|\sin x - S_3(x)| = |\sin x - S_4(x)| \le K\frac{|x|^5}{5!} = \frac{|x|^5}{5!}$$

Now substitute x^2 for x in the above inequality and note from part (a) that $S_3(x^2) = T_6(x)$ to get

$$|\sin(x^2) - S_3(x^2)| = |\sin(x^2) - T_6(x)| \le \frac{|x^2|^5}{5!} = \frac{|x|^{10}}{5!}$$

(c)

$$\int_0^{1/2} \sin(x^2)\,dx \approx \int_0^{1/2} T_6(x)\,dx = \left(\frac{1}{3}x^3 - \frac{1}{42}x^7\right)\bigg|_0^{1/2} \approx 0.04148065476$$

From part (b) and Exercise 72, the error is bounded by

$$\frac{x^{10}}{5!}\left(\frac{1}{2} - 0\right) = \frac{(1/2)^{11}}{120} = \frac{1}{2048 \cdot 120} \approx 4.069010417 \times 10^{-6}$$

The true value of the integral is approximately 0.04148102427, which is consistent with the computed Error Bound.

77. Let a be any number and let

$$P(x) = a_n x^n + a_{n-1} x^{n-1} + \cdots + a_1 x + a_0$$

be a polynomial of degree n or less.
(a) Show that if $P^{(j)}(a) = 0$ for $j = 0, 1, \ldots, n$, then $P(x) = 0$, that is, $a_j = 0$ for all j. *Hint:* Use induction, noting that if the statement is true for degree $n - 1$, then $P'(x) = 0$.
(b) Prove that T_n is the only polynomial of degree n or less that agrees with f at $x = a$ to order n. *Hint:* If Q is another such polynomial, apply (a) to $P(x) = T_n(x) - Q(x)$.

SOLUTION
(a) Note first that if $n = 0$, i.e., if $P(x) = a_0$ is a constant, then the statement holds: if $P^{(0)}(a) = P(a) = 0$, then $a_0 = 0$ so that $P(x) = 0$. Next, assume the statement holds for all polynomials of degree $n - 1$ or less, and let $P(x)$ be a polynomial of degree at most n with $P^{(j)}(a) = 0$ for $j = 0, 1, \ldots, n$. If $P(x)$ has degree less than n, then we know $P(x) = 0$ by induction, so assume the degree of $P(x)$ is exactly n. Then

$$P(x) = a_n x^n + a_{n-1} x^{n-1} + \cdots + a_1 x + a_0$$

where $a_n \ne 0$; also,

$$P'(x) = n a_n x^{n-1} + (n-1) a_{n-1} x^{n-2} + \cdots + a_1$$

Note that $P^{(j+1)}(a) = (P')^{(j)}(a)$ for $j = 0, 1, \ldots, n - 1$. But then

$$0 = P^{(j+1)}(a) = (P')^{(j)}(a) \quad \text{for all } j = 0, 1, \ldots, n - 1$$

Since $P'(x)$ has degree at most $n - 1$, it follows by induction that $P'(x) = 0$. Thus $a_n = a_{n-1} = \cdots = a_1 = 0$ so that $P(x) = a_0$. But $P(a) = 0$ so that $a_0 = 0$ as well and thus $P(x) = 0$.
(b) Suppose $Q(x)$ is a polynomial of degree at most n that agrees with $f(x)$ at $x = a$ up to order n. Let $P(x) = T_n(x) - Q(x)$. Note that $P(x)$ is a polynomial of degree at most n since both $T_n(x)$ and $Q(x)$ are. Since both $T_n(x)$ and $Q(x)$ agree with $f(x)$ at $x = a$ to order n, we have

$$T_n^{(j)}(a) = f^{(j)}(a) = Q^{(j)}(a), \quad j = 0, 1, 2, \ldots, n$$

Thus

$$P^{(j)}(a) = T_n^{(j)}(a) - Q^{(j)}(a) = 0 \quad \text{for } j = 0, 1, 2, \ldots, n$$

But then by part (a), $P(x) = 0$ so that $T_n(x) = Q(x)$.

10.8 Taylor Series (LT Section 11.8)

Preliminary Questions

1. Determine $f(0)$ and $f'''(0)$ for a function f with Maclaurin series

$$T(x) = 3 + 2x + 12x^2 + 5x^3 + \cdots$$

SOLUTION The Maclaurin series for a function f has the form

$$f(0) + \frac{f'(0)}{1!}x + \frac{f''(0)}{2!}x^2 + \frac{f'''(0)}{3!}x^3 + \cdots$$

Matching this general expression with the given series, we find $f(0) = 3$ and $\frac{f'''(0)}{3!} = 5$. From this latter equation, it follows that $f'''(0) = 30$.

2. Determine $f(-2)$ and $f^{(4)}(-2)$ for a function with Taylor series

$$T(x) = 3(x+2) + (x+2)^2 - 4(x+2)^3 + 2(x+2)^4 + \cdots$$

SOLUTION The Taylor series for a function f centered at $x = -2$ has the form

$$f(-2) + \frac{f'(-2)}{1!}(x+2) + \frac{f''(-2)}{2!}(x+2)^2 + \frac{f'''(-2)}{3!}(x+2)^3 + \frac{f^{(4)}(-2)}{4!}(x+2)^4 + \cdots$$

Matching this general expression with the given series, we find $f(-2) = 0$ and $\frac{f^{(4)}(-2)}{4!} = 2$. From this latter equation, it follows that $f^{(4)}(-2) = 48$.

3. What is the easiest way to find the Maclaurin series for the function $f(x) = \sin(x^2)$?

SOLUTION The easiest way to find the Maclaurin series for $\sin\left(x^2\right)$ is to substitute x^2 for x in the Maclaurin series for $\sin x$.

4. Find the Taylor series for f centered at $c = 3$ if $f(3) = 4$ and $f'(x)$ has a Taylor expansion

$$f'(x) = \sum_{n=1}^{\infty} \frac{(x-3)^n}{n}$$

SOLUTION Integrating the series for $f'(x)$ term-by-term gives

$$f(x) = C + \sum_{n=1}^{\infty} \frac{(x-3)^{n+1}}{n(n+1)}$$

Substituting $x = 3$ then yields

$$f(3) = C = 4$$

so

$$f(x) = 4 + \sum_{n=1}^{\infty} \frac{(x-3)^{n+1}}{n(n+1)}$$

5. Let $T(x)$ be the Maclaurin series of $f(x)$. Which of the following guarantees that $f(2) = T(2)$?
(a) $T(x)$ converges for $x = 2$.
(b) The remainder $R_k(2)$ approaches a limit as $k \to \infty$.
(c) The remainder $R_k(2)$ approaches zero as $k \to \infty$.

SOLUTION The correct response is (c): $f(2) = T(2)$ if and only if the remainder $R_k(2)$ approaches zero as $k \to \infty$.

Exercises

1. Write out the first four terms of the Maclaurin series of $f(x)$ if

$$f(0) = 2, \quad f'(0) = 3, \quad f''(0) = 4, \quad f'''(0) = 12$$

SOLUTION The first four terms of the Maclaurin series of $f(x)$ are

$$f(0) + f'(0)x + \frac{f''(0)}{2!}x^2 + \frac{f'''(0)}{3!}x^3 = 2 + 3x + \frac{4}{2}x^2 + \frac{12}{6}x^3 = 2 + 3x + 2x^2 + 2x^3$$

In Exercises 3–18, find the Maclaurin series and find the interval on which the expansion is valid.

3. $f(x) = \dfrac{1}{1+10x}$

SOLUTION Substituting $-10x$ for x in the Maclaurin series for $\frac{1}{1-x}$ gives

$$\frac{1}{1+10x} = \sum_{n=0}^{\infty}(-10x)^n = \sum_{n=0}^{\infty}(-1)^n 10^n x^n$$

This series is valid for $|10x| < 1$, or $|x| < \frac{1}{10}$.

5. $f(x) = \cos 3x$

SOLUTION Substituting $3x$ for x in the Maclaurin series for $\cos x$ gives

$$\cos 3x = \sum_{n=0}^{\infty}(-1)^n \frac{(3x)^{2n}}{(2n)!} = \sum_{n=0}^{\infty}(-1)^n \frac{9^n x^{2n}}{(2n)!}$$

This series is valid for all x.

7. $f(x) = \sin(x^2)$

SOLUTION Substituting x^2 for x in the Maclaurin series for $\sin x$ gives

$$\sin x^2 = \sum_{n=0}^{\infty}(-1)^n \frac{(x^2)^{2n+1}}{(2n+1)!} = \sum_{n=0}^{\infty}(-1)^n \frac{x^{4n+2}}{(2n+1)!}$$

This series is valid for all x.

9. $f(x) = \ln(1 - x^2)$

SOLUTION Substituting $-x^2$ for x in the Maclaurin series for $\ln(1+x)$ gives

$$\ln(1 - x^2) = \sum_{n=1}^{\infty} \frac{(-1)^{n-1}(-x^2)^n}{n} = \sum_{n=1}^{\infty} \frac{(-1)^{2n-1} x^{2n}}{n} = -\sum_{n=1}^{\infty} \frac{x^{2n}}{n}$$

This series is valid for $|x| < 1$.

11. $f(x) = \tan^{-1}(x^2)$

SOLUTION Substituting x^2 for x in the Maclaurin series for $\tan^{-1} x$ gives

$$\tan^{-1}(x^2) = \sum_{n=0}^{\infty}(-1)^n \frac{(x^2)^{2n+1}}{2n+1} = \sum_{n=0}^{\infty}(-1)^n \frac{x^{4n+2}}{2n+1}$$

This series is valid for $|x| \leq 1$.

13. $f(x) = e^{x-2}$

SOLUTION $e^{x-2} = e^{-2} e^x$; thus,

$$e^{x-2} = e^{-2} \sum_{n=0}^{\infty} \frac{x^n}{n!} = \sum_{n=0}^{\infty} \frac{x^n}{e^2 n!}$$

This series is valid for all x.

15. $f(x) = \ln(1 - 5x)$

SOLUTION Substituting $-5x$ for x in the Maclaurin series for $\ln(1+x)$ gives

$$\ln(1-5x) = \sum_{n=1}^{\infty} \frac{(-1)^{n-1}(-5x)^n}{n} = \sum_{n=1}^{\infty} \frac{(-1)^{2n-1}5^n x^n}{n} = -\sum_{n=1}^{\infty} \frac{5^n x^n}{n}$$

This series is valid for $|5x| < 1$, or $|x| < \frac{1}{5}$, and for $x = -\frac{1}{5}$.

17. $f(x) = \sinh x$

SOLUTION Recall that

$$\sinh x = \frac{1}{2}(e^x - e^{-x})$$

Therefore,

$$\sinh x = \frac{1}{2}\left(\sum_{n=0}^{\infty} \frac{x^n}{n!} - \sum_{n=0}^{\infty} \frac{(-x)^n}{n!}\right) = \sum_{n=0}^{\infty} \frac{x^n}{2(n!)}(1 - (-1)^n)$$

Now,

$$1 - (-1)^n = \begin{cases} 0, & n \text{ even} \\ 2, & n \text{ odd} \end{cases}$$

so

$$\sinh x = \sum_{k=0}^{\infty} 2\frac{x^{2k+1}}{2(2k+1)!} = \sum_{k=0}^{\infty} \frac{x^{2k+1}}{(2k+1)!}$$

This series is valid for all x.

In Exercises 19–30, find the terms through degree 4 of the Maclaurin series of $f(x)$. Use multiplication and substitution as necessary.

19. $f(x) = e^x \sin x$

SOLUTION Multiply the fourth-order Taylor polynomials for e^x and $\sin x$:

$$\left(1 + x + \frac{x^2}{2} + \frac{x^3}{6} + \frac{x^4}{24}\right)\left(x - \frac{x^3}{6}\right) = x + x^2 - \frac{x^3}{6} + \frac{x^3}{2} - \frac{x^4}{6} + \frac{x^4}{6} + \text{higher-order terms}$$

$$= x + x^2 + \frac{x^3}{3} + \text{higher-order terms}$$

The terms in the Maclaurin series through degree 4 for $f(x) = e^x \sin x$ are therefore

$$x + x^2 + \frac{x^3}{3}$$

21. $f(x) = \dfrac{\sin x}{1-x}$

SOLUTION Multiply the fourth order Taylor polynomials for $\sin x$ and $\dfrac{1}{1-x}$:

$$\left(x - \frac{x^3}{6}\right)\left(1 + x + x^2 + x^3 + x^4\right) = x + x^2 - \frac{x^3}{6} + x^3 + x^4 - \frac{x^4}{6} + \text{higher-order terms}$$

$$= x + x^2 + \frac{5x^3}{6} + \frac{5x^4}{6} + \text{higher-order terms}$$

The terms of the Maclaurin series through degree 4 for $f(x) = \dfrac{\sin x}{1-x}$ are therefore

$$x + x^2 + \frac{5x^3}{6} + \frac{5x^4}{6}$$

23. $f(x) = (1+x)^{1/4}$

SOLUTION The first five generalized binomial coefficients for $a = \frac{1}{4}$ are

$$1, \quad \frac{1}{4}, \quad \frac{\frac{1}{4}\left(\frac{-3}{4}\right)}{2!} = -\frac{3}{32}, \quad \frac{\frac{1}{4}\left(\frac{-3}{4}\right)\left(\frac{-7}{4}\right)}{3!} = \frac{7}{128}, \quad \frac{\frac{1}{4}\left(\frac{-3}{4}\right)\left(\frac{-7}{4}\right)\left(\frac{-11}{4}\right)}{4!} = \frac{-77}{2048}$$

Therefore, the first four terms in the binomial series for $(1+x)^{1/4}$ are

$$1 + \frac{1}{4}x - \frac{3}{32}x^2 + \frac{7}{128}x^3 - \frac{77}{2048}x^4$$

25. $f(x) = e^x \tan^{-1} x$

SOLUTION Using the Maclaurin series for e^x and $\tan^{-1} x$, we find

$$e^x \tan^{-1} x = \left(1 + x + \frac{x^2}{2} + \frac{x^3}{6} + \cdots\right)\left(x - \frac{x^3}{3} + \cdots\right) = x + x^2 - \frac{x^3}{3} + \frac{x^3}{2} + \frac{x^4}{6} - \frac{x^4}{3} + \cdots$$

$$= x + x^2 + \frac{1}{6}x^3 - \frac{1}{6}x^4 + \cdots$$

27. $f(x) = e^{\sin x}$

SOLUTION Substituting $\sin x$ for x in the Maclaurin series for e^x and then using the Maclaurin series for $\sin x$, we find

$$e^{\sin x} = 1 + \sin x + \frac{\sin^2 x}{2} + \frac{\sin^3 x}{6} + \frac{\sin^4 x}{24} + \cdots$$

$$= 1 + \left(x - \frac{x^3}{6} + \cdots\right) + \frac{1}{2}\left(x - \frac{x^3}{6} + \cdots\right)^2 + \frac{1}{6}(x - \cdots)^3 + \frac{1}{24}(x - \cdots)^4$$

$$= 1 + x + \frac{1}{2}x^2 - \frac{1}{6}x^3 + \frac{1}{6}x^3 - \frac{1}{6}x^4 + \frac{1}{24}x^4 + \cdots$$

$$= 1 + x + \frac{1}{2}x^2 - \frac{1}{8}x^4 + \cdots$$

29. $f(x) = \cosh(x^2)$

SOLUTION From Exercise 18, the Maclaurin series for $\cosh x$ is

$$\cosh x = \sum_{k=0}^{\infty} \frac{x^{2k}}{(2k)!}$$

Substituting x^2 for x gives

$$\cosh(x^2) = \sum_{k=0}^{\infty} \frac{x^{4k}}{(2k)!}$$

Thus the terms through degree 4 are

$$1 + \frac{1}{2}x^4 + \cdots$$

In Exercises 31–40, find the Taylor series centered at c and the interval on which the expansion is valid.

31. $f(x) = \frac{1}{x}, \quad c = 1$

SOLUTION Write

$$\frac{1}{x} = \frac{1}{1 + (x-1)}$$

and then substitute $-(x-1)$ for x in the Maclaurin series for $\frac{1}{1-x}$ to obtain

$$\frac{1}{x} = \sum_{n=0}^{\infty} [-(x-1)]^n = \sum_{n=0}^{\infty} (-1)^n (x-1)^n$$

This series is valid for $|x-1| < 1$.

33. $f(x) = \dfrac{1}{1-x}$, $c = 5$

SOLUTION Write

$$\frac{1}{1-x} = \frac{1}{-4-(x-5)} = -\frac{1}{4} \cdot \frac{1}{1+\frac{x-5}{4}}$$

Substituting $-\frac{x-5}{4}$ for x in the Maclaurin series for $\frac{1}{1-x}$ yields

$$\frac{1}{1+\frac{x-5}{4}} = \sum_{n=0}^{\infty} \left(-\frac{x-5}{4}\right)^n = \sum_{n=0}^{\infty} (-1)^n \frac{(x-5)^n}{4^n}$$

Thus,

$$\frac{1}{1-x} = -\frac{1}{4} \sum_{n=0}^{\infty} (-1)^n \frac{(x-5)^n}{4^n} = \sum_{n=0}^{\infty} (-1)^{n+1} \frac{(x-5)^n}{4^{n+1}}$$

This series is valid for $\left|\frac{x-5}{4}\right| < 1$, or $|x-5| < 4$.

35. $f(x) = x^4 + 3x - 1$, $c = 2$

SOLUTION To determine the Taylor series with center $c = 2$, we compute

$$f'(x) = 4x^3 + 3, \quad f''(x) = 12x^2, \quad f'''(x) = 24x$$

and $f^{(4)}(x) = 24$. All derivatives of order five and higher are zero. Now,

$$f(2) = 21, \quad f'(2) = 35, \quad f''(2) = 48, \quad f'''(2) = 48$$

and $f^{(4)}(2) = 24$. Therefore, the Taylor series is

$$21 + 35(x-2) + \frac{48}{2}(x-2)^2 + \frac{48}{6}(x-2)^3 + \frac{24}{24}(x-2)^4$$

or

$$21 + 35(x-2) + 24(x-2)^2 + 8(x-2)^3 + (x-2)^4$$

37. $f(x) = \dfrac{1}{x^2}$, $c = 4$

SOLUTION We will first find the Taylor series for $\frac{1}{x}$ and then differentiate to obtain the series for $\frac{1}{x^2}$. Write

$$\frac{1}{x} = \frac{1}{4+(x-4)} = \frac{1}{4} \cdot \frac{1}{1+\frac{x-4}{4}}$$

Now substitute $-\frac{x-4}{4}$ for x in the Maclaurin series for $\frac{1}{1-x}$ to obtain

$$\frac{1}{x} = \frac{1}{4} \sum_{n=0}^{\infty} \left(-\frac{x-4}{4}\right)^n = \sum_{n=0}^{\infty} (-1)^n \frac{(x-4)^n}{4^{n+1}}$$

Differentiating term-by-term yields

$$-\frac{1}{x^2} = \sum_{n=1}^{\infty} (-1)^n n \frac{(x-4)^{n-1}}{4^{n+1}}$$

so that

$$\frac{1}{x^2} = \sum_{n=1}^{\infty} (-1)^{n-1} n \frac{(x-4)^{n-1}}{4^{n+1}} = \sum_{n=0}^{\infty} (-1)^n (n+1) \frac{(x-4)^n}{4^{n+2}}$$

This series is valid for $\left|\frac{x-4}{4}\right| < 1$, or $|x-4| < 4$.

39. $f(x) = \dfrac{1}{1-x^2}$, $c = 3$

SOLUTION By partial fraction decomposition

$$\frac{1}{1-x^2} = \frac{\frac{1}{2}}{1-x} + \frac{\frac{1}{2}}{1+x}$$

so

$$\frac{1}{1-x^2} = \frac{\frac{1}{2}}{-2-(x-3)} + \frac{\frac{1}{2}}{4+(x-3)} = -\frac{1}{4} \cdot \frac{1}{1+\frac{x-3}{2}} + \frac{1}{8} \cdot \frac{1}{1+\frac{x-3}{4}}$$

Substituting $-\frac{x-3}{2}$ for x in the Maclaurin series for $\frac{1}{1-x}$ gives

$$\frac{1}{1+\frac{x-3}{2}} = \sum_{n=0}^{\infty} \left(-\frac{x-3}{2}\right)^n = \sum_{n=0}^{\infty} \frac{(-1)^n}{2^n}(x-3)^n$$

while substituting $-\frac{x-3}{4}$ for x in the same series gives

$$\frac{1}{1+\frac{x-3}{4}} = \sum_{n=0}^{\infty} \left(-\frac{x-3}{4}\right)^n = \sum_{n=0}^{\infty} \frac{(-1)^n}{4^n}(x-3)^n$$

Thus,

$$\frac{1}{1-x^2} = -\frac{1}{4}\sum_{n=0}^{\infty}\frac{(-1)^n}{2^n}(x-3)^n + \frac{1}{8}\sum_{n=0}^{\infty}\frac{(-1)^n}{4^n}(x-3)^n = \sum_{n=0}^{\infty}\frac{(-1)^{n+1}}{2^{n+2}}(x-3)^n + \sum_{n=0}^{\infty}\frac{(-1)^n}{2^{2n+3}}(x-3)^n$$

$$= \sum_{n=0}^{\infty}\left(\frac{(-1)^{n+1}}{2^{n+2}} + \frac{(-1)^n}{2^{2n+3}}\right)(x-3)^n = \sum_{n=0}^{\infty}\frac{(-1)^{n+1}(2^{n+1}-1)}{2^{2n+3}}(x-3)^n$$

This series is valid for $|x-3| < 2$.

41. Use the identity $\cos^2 x = \frac{1}{2}(1+\cos 2x)$ to find the Maclaurin series for $f(x) = \cos^2 x$.

SOLUTION The Maclaurin series for $\cos 2x$ is

$$\sum_{n=0}^{\infty}(-1)^n\frac{(2x)^{2n}}{(2n)!} = \sum_{n=0}^{\infty}(-1)^n\frac{2^{2n}x^{2n}}{(2n)!}$$

so the Maclaurin series for $\cos^2 x = \frac{1}{2}(1+\cos 2x)$ is

$$\frac{1 + \left(1 + \sum_{n=1}^{\infty}(-1)^n\frac{2^{2n}x^{2n}}{(2n)!}\right)}{2} = 1 + \sum_{n=1}^{\infty}(-1)^n\frac{2^{2n-1}x^{2n}}{(2n)!}$$

43. Use the Maclaurin series for $\ln(1+x)$ and $\ln(1-x)$ to show that

$$\frac{1}{2}\ln\left(\frac{1+x}{1-x}\right) = x + \frac{x^3}{3} + \frac{x^5}{5} + \cdots$$

for $|x| < 1$. What can you conclude by comparing this result with that of Exercise 42?

SOLUTION Using the Maclaurin series for $\ln(1+x)$ and $\ln(1-x)$, we have for $|x| < 1$

$$\ln(1+x) - \ln(1-x) = \sum_{n=1}^{\infty}\frac{(-1)^{n-1}}{n}x^n - \sum_{n=1}^{\infty}\frac{(-1)^{n-1}}{n}(-x)^n$$

$$= \sum_{n=1}^{\infty}\frac{(-1)^{n-1}}{n}x^n + \sum_{n=1}^{\infty}\frac{x^n}{n} = \sum_{n=1}^{\infty}\frac{1+(-1)^{n-1}}{n}x^n$$

Since $1+(-1)^{n-1} = 0$ for even n and $1+(-1)^{n-1} = 2$ for odd n,

$$\ln(1+x) - \ln(1-x) = \sum_{k=0}^{\infty}\frac{2}{2k+1}x^{2k+1}$$

Thus,
$$\frac{1}{2}\ln\left(\frac{1+x}{1-x}\right) = \frac{1}{2}(\ln(1+x) - \ln(1-x)) = \frac{1}{2}\sum_{k=0}^{\infty}\frac{2}{2k+1}x^{2k+1} = \sum_{k=0}^{\infty}\frac{x^{2k+1}}{2k+1}$$

Observe that this is the same series we found in Exercise 42; therefore,
$$\frac{1}{2}\ln\left(\frac{1+x}{1-x}\right) = \tanh^{-1} x$$

45. Show, by integrating the Maclaurin series for $f(x) = \dfrac{1}{\sqrt{1-x^2}}$, that for $|x| < 1$,
$$\sin^{-1} x = x + \sum_{n=1}^{\infty}\frac{1\cdot 3\cdot 5\cdots(2n-1)}{2\cdot 4\cdot 6\cdots(2n)}\frac{x^{2n+1}}{2n+1}$$

SOLUTION From Example 11, we know that for $|x| < 1$
$$\frac{1}{\sqrt{1-x^2}} = \sum_{n=0}^{\infty}\frac{1\cdot 3\cdot 5\cdots(2n-1)}{2\cdot 4\cdot 6\cdots(2n)}x^{2n} = 1 + \sum_{n=1}^{\infty}\frac{1\cdot 3\cdot 5\cdots(2n-1)}{2\cdot 4\cdot 6\cdots(2n)}x^{2n}$$

so, for $|x| < 1$,
$$\sin^{-1} x = \int\frac{dx}{\sqrt{1-x^2}} = C + x + \sum_{n=1}^{\infty}\frac{1\cdot 3\cdot 5\cdots(2n-1)}{2\cdot 4\cdot 6\cdots(2n)}\frac{x^{2n+1}}{2n+1}$$

Since $\sin^{-1} 0 = 0$, we find that $C = 0$. Thus,
$$\sin^{-1} x = x + \sum_{n=1}^{\infty}\frac{1\cdot 3\cdot 5\cdots(2n-1)}{2\cdot 4\cdot 6\cdots(2n)}\frac{x^{2n+1}}{2n+1}$$

47. How many terms of the Maclaurin series of $f(x) = \ln(1+x)$ are needed to compute $\ln 1.2$ to within an error of at most 0.0001? Make the computation and compare the result with the calculator value.

SOLUTION Substitute $x = 0.2$ into the Maclaurin series for $\ln(1+x)$ to obtain:
$$\ln 1.2 = \sum_{n=1}^{\infty}(-1)^{n-1}\frac{(0.2)^n}{n} = \sum_{n=1}^{\infty}(-1)^{n-1}\frac{1}{5^n n}$$

This is an alternating series with $a_n = \dfrac{1}{n\cdot 5^n}$. Using the Error Bound for alternating series
$$|\ln 1.2 - S_N| \le a_{N+1} = \frac{1}{(N+1)5^{N+1}}$$

so we must choose N so that
$$\frac{1}{(N+1)5^{N+1}} < 0.0001 \quad\text{or}\quad (N+1)5^{N+1} > 10000$$

For $N = 3$, $(N+1)5^{N+1} = 4\cdot 5^4 = 2500 < 10{,}000$, and for $N = 4$, $(N+1)5^{N+1} = 5\cdot 5^5 = 15{,}625 > 10{,}000$; thus, the smallest acceptable value for N is $N = 4$. The corresponding approximation is:
$$S_4 = \sum_{n=1}^{4}\frac{(-1)^{n-1}}{5^n\cdot n} = \frac{1}{5} - \frac{1}{5^2\cdot 2} + \frac{1}{5^3\cdot 3} - \frac{1}{5^4\cdot 4} = 0.182266666$$

Now, $\ln 1.2 = 0.182321556$, so
$$|\ln 1.2 - S_4| = 5.489\times 10^{-5} < 0.0001$$

49. Use the Maclaurin expansion for e^{-t^2} to express the function $F(x) = \int_0^x e^{-t^2}\,dt$ as an alternating power series in x (Figure 3).

(a) How many terms of the Maclaurin series are needed to approximate the integral for $x = 1$ to within an error of at most 0.001?

(b) CAS Carry out the computation and check your answer using a computer algebra system.

SOLUTION Substituting $-t^2$ for t in the Maclaurin series for e^t yields

$$e^{-t^2} = \sum_{n=0}^{\infty} \frac{(-t^2)^n}{n!} = \sum_{n=0}^{\infty} (-1)^n \frac{t^{2n}}{n!}$$

thus,

$$\int_0^x e^{-t^2}\,dt = \sum_{n=0}^{\infty} (-1)^n \frac{x^{2n+1}}{n!(2n+1)}$$

(a) For $x = 1$,

$$\int_0^1 e^{-t^2}\,dt = \sum_{n=0}^{\infty} (-1)^n \frac{1}{n!(2n+1)}$$

This is an alternating series with $a_n = \frac{1}{n!(2n+1)}$; therefore, the error incurred by using S_N to approximate the value of the definite integral is bounded by

$$\left| \int_0^1 e^{-t^2}\,dt - S_N \right| \le a_{N+1} = \frac{1}{(N+1)!(2N+3)}$$

To guarantee the error is at most 0.001, we must choose N so that

$$\frac{1}{(N+1)!(2N+3)} < 0.001 \quad \text{or} \quad (N+1)!(2N+3) > 1000$$

For $N = 3$, $(N+1)!(2N+3) = 4! \cdot 9 = 216 < 1000$ and for $N = 4$, $(N+1)!(2N+3) = 5! \cdot 11 = 1320 > 1000$; thus, the smallest acceptable value for N is $N = 4$. The corresponding approximation is

$$S_4 = \sum_{n=0}^{4} \frac{(-1)^n}{n!(2n+1)} = 1 - \frac{1}{3} + \frac{1}{2! \cdot 5} - \frac{1}{3! \cdot 7} + \frac{1}{4! \cdot 9} = 0.747486772$$

(b) Using a computer algebra system, we find

$$\int_0^1 e^{-t^2}\,dt = 0.746824133$$

therefore

$$\left| \int_0^1 e^{-t^2}\,dt - S_4 \right| = 6.626 \times 10^{-4} < 10^{-3}$$

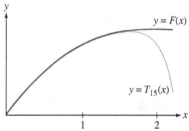

FIGURE 3 The Maclaurin polynomial $T_{15}(x)$ for $F(t) = \int_0^x e^{-t^2}\,dt$.

In Exercises 51–54, express the definite integral as an infinite series and find its value to within an error of at most 10^{-4}.

51. $\int_0^1 \cos(x^2)\,dx$

SOLUTION Substituting x^2 for x in the Maclaurin series for $\cos x$ yields

$$\cos(x^2) = \sum_{n=0}^{\infty} (-1)^n \frac{(x^2)^{2n}}{(2n)!} = \sum_{n=0}^{\infty} (-1)^n \frac{x^{4n}}{(2n)!}$$

therefore,

$$\int_0^1 \cos(x^2)\,dx = \sum_{n=0}^{\infty} (-1)^n \frac{x^{4n+1}}{(2n)!(4n+1)}\Big|_0^1 = \sum_{n=0}^{\infty} \frac{(-1)^n}{(2n)!(4n+1)}$$

This is an alternating series with $a_n = \frac{1}{(2n)!(4n+1)}$; therefore, the error incurred by using S_N to approximate the value of the definite integral is bounded by

$$\left|\int_0^1 \cos(x^2)\,dx - S_N\right| \le a_{N+1} = \frac{1}{(2N+2)!(4N+5)}$$

To guarantee the error is at most 0.0001, we must choose N so that

$$\frac{1}{(2N+2)!(4N+5)} < 0.0001 \quad \text{or} \quad (2N+2)!(4N+5) > 10000$$

For $N=2$, $(2N+2)!(4N+5) = 6! \cdot 13 = 9360 < 10000$ and for $N=3$, $(2N+2)!(4N+5) = 8! \cdot 17 = 685440 > 10000$; thus, the smallest acceptable value for N is $N=3$. The corresponding approximation is

$$S_3 = \sum_{n=0}^{3} \frac{(-1)^n}{(2n)!(4n+1)} = 1 - \frac{1}{5 \cdot 2!} + \frac{1}{9 \cdot 4!} - \frac{1}{13 \cdot 6!} = 0.904522792$$

53. $\int_0^1 e^{-x^3}\,dx$

SOLUTION Substituting $-x^3$ for x in the Maclaurin series for e^x yields

$$e^{-x^3} = \sum_{n=0}^{\infty} \frac{(-x^3)^n}{n!} = \sum_{n=0}^{\infty} (-1)^n \frac{x^{3n}}{n!}$$

therefore,

$$\int_0^1 e^{-x^3}\,dx = \sum_{n=0}^{\infty} (-1)^n \frac{x^{3n+1}}{n!(3n+1)}\Big|_0^1 = \sum_{n=0}^{\infty} \frac{(-1)^n}{n!(3n+1)}$$

This is an alternating series with $a_n = \frac{1}{n!(3n+1)}$; therefore, the error incurred by using S_N to approximate the value of the definite integral is bounded by

$$\left|\int_0^1 e^{-x^3}\,dx - S_N\right| \le a_{N+1} = \frac{1}{(N+1)!(3N+4)}$$

To guarantee the error is at most 0.0001, we must choose N so that

$$\frac{1}{(N+1)!(3N+4)} < 0.0001 \quad \text{or} \quad (N+1)!(3N+4) > 10000$$

For $N=4$, $(N+1)!(3N+4) = 5! \cdot 16 = 1920 < 10000$ and for $N=5$, $(N+1)!(3N+4) = 6! \cdot 19 = 13680 > 10000$; thus, the smallest acceptable value for N is $N=5$. The corresponding approximation is

$$S_5 = \sum_{n=0}^{5} \frac{(-1)^n}{n!(3n+1)} = 0.807446200$$

In Exercises 55–58, express the integral as an infinite series.

55. $\displaystyle\int_0^x \frac{1-\cos t}{t}\,dt,\quad \text{for all } x$

SOLUTION The Maclaurin series for $\cos t$ is

$$\cos t = \sum_{n=0}^{\infty}(-1)^n \frac{t^{2n}}{(2n)!} = 1 + \sum_{n=1}^{\infty}(-1)^n \frac{t^{2n}}{(2n)!}$$

so

$$1-\cos t = -\sum_{n=1}^{\infty}(-1)^n \frac{t^{2n}}{(2n)!} = \sum_{n=1}^{\infty}(-1)^{n+1}\frac{t^{2n}}{(2n)!}$$

and

$$\frac{1-\cos t}{t} = \frac{1}{t}\sum_{n=1}^{\infty}(-1)^{n+1}\frac{t^{2n}}{(2n)!} = \sum_{n=1}^{\infty}(-1)^{n+1}\frac{t^{2n-1}}{(2n)!}$$

Thus,

$$\int_0^x \frac{1-\cos(t)}{t}\,dt = \sum_{n=1}^{\infty}(-1)^{n+1}\frac{t^{2n}}{(2n)!\,2n}\bigg|_0^x = \sum_{n=1}^{\infty}(-1)^{n+1}\frac{x^{2n}}{(2n)!\,2n}$$

57. $\displaystyle\int_0^x \ln(1+t^2)\,dt,\quad \text{for } |x|<1$

SOLUTION Substituting t^2 for t in the Maclaurin series for $\ln(1+t)$ yields

$$\ln(1+t^2) = \sum_{n=1}^{\infty}(-1)^{n-1}\frac{(t^2)^n}{n} = \sum_{n=1}^{\infty}(-1)^n \frac{t^{2n}}{n}$$

Thus,

$$\int_0^x \ln(1+t^2)\,dt = \sum_{n=1}^{\infty}(-1)^n \frac{t^{2n+1}}{n(2n+1)}\bigg|_0^x = \sum_{n=1}^{\infty}(-1)^n \frac{x^{2n+1}}{n(2n+1)}$$

59. Which function has Maclaurin series $\displaystyle\sum_{n=0}^{\infty}(-1)^n 2^n x^n$?

SOLUTION We recognize that

$$\sum_{n=0}^{\infty}(-1)^n 2^n x^n = \sum_{n=0}^{\infty}(-2x)^n$$

is the Maclaurin series for $\frac{1}{1-x}$ with x replaced by $-2x$. Therefore,

$$\sum_{n=0}^{\infty}(-1)^n 2^n x^n = \frac{1}{1-(-2x)} = \frac{1}{1+2x}$$

61. Using Maclaurin series, determine to exactly what value the following series converges:

$$\sum_{n=0}^{\infty}(-1)^n \frac{(\pi)^{2n}}{(2n)!}$$

SOLUTION The Maclaurin series for $\cos x$ is

$$\cos x = \sum_{n=0}^{\infty}(-1)^n \frac{x^{2n}}{(2n)!}$$

substitute $x=\pi$ to get

$$\sum_{n=0}^{\infty}(-1)^n \frac{\pi^{2n}}{(2n)!} = \cos\pi = -1$$

In Exercises 61–64, use Theorem 2 to prove that the $f(x)$ is represented by its Maclaurin series for all x.

63. $f(x) = \sin(x/2) + \cos(x/3)$

SOLUTION With $f(x) = \sin(x/2) + \cos(x/3)$, we see that

$$f^{(k)}(x) = \frac{1}{2^k}g(x/2) + \frac{1}{3^k}h(x/3)$$

where $g(x)$ and $h(x)$ are each either $\pm \sin x$ or $\pm \cos x$. But then for any real number x,

$$\left|f^{(k)}(x)\right| = \left|\frac{1}{2^k}g(x/2) + \frac{1}{3^k}h(x/3)\right| \leq \left|\frac{1}{2^k}g(x/2)\right| + \left|\frac{1}{3^k}h(x/3)\right| \leq |g(x/2)| + |h(x/3)| \leq 2$$

So all derivatives of f are bounded by 2 everywhere. So if x is any real number, choose $R > |x|$. Then by Theorem 2, the Maclaurin series for f converges to f on $(-R, R)$ and therefore converges to f at x. So it converges to f for all x.

65. $f(x) = \sinh x$

SOLUTION By definition, $\sinh x = \frac{1}{2}(e^x - e^{-x})$, so if both e^x and e^{-x} are represented by their Taylor series centered at c, then so is $\sinh x$. But the previous exercise shows that e^{-x} is so represented, and the text shows that e^x is.

In Exercises 67–70, find the functions with the following Maclaurin series (refer to Table 2 on page 611).

67. $1 + x^3 + \dfrac{x^6}{2!} + \dfrac{x^9}{3!} + \dfrac{x^{12}}{4!} + \cdots$

SOLUTION We recognize

$$1 + x^3 + \frac{x^6}{2!} + \frac{x^9}{3!} + \frac{x^{12}}{4!} + \cdots = \sum_{n=0}^{\infty} \frac{x^{3n}}{n!} = \sum_{n=0}^{\infty} \frac{(x^3)^n}{n!}$$

as the Maclaurin series for e^x with x replaced by x^3. Therefore,

$$1 + x^3 + \frac{x^6}{2!} + \frac{x^9}{3!} + \frac{x^{12}}{4!} + \cdots = e^{x^3}$$

69. $1 - \dfrac{5^3 x^3}{3!} + \dfrac{5^5 x^5}{5!} - \dfrac{5^7 x^7}{7!} + \cdots$

SOLUTION Note

$$1 - \frac{5^3 x^3}{3!} + \frac{5^5 x^5}{5!} - \frac{5^7 x^7}{7!} + \cdots = 1 - 5x + \left(5x - \frac{5^3 x^3}{3!} + \frac{5^5 x^5}{5!} - \frac{5^7 x^7}{7!} + \cdots\right)$$

$$= 1 - 5x + \sum_{n=0}^{\infty} (-1)^n \frac{(5x)^{2n+1}}{(2n+1)!}$$

The series is the Maclaurin series for $\sin x$ with x replaced by $5x$, so

$$1 - \frac{5^3 x^3}{3!} + \frac{5^5 x^5}{5!} - \frac{5^7 x^7}{7!} + \cdots = 1 - 5x + \sin(5x)$$

In Exercises 71 and 72, let

$$f(x) = \frac{1}{(1-x)(1-2x)}$$

71. Find the Maclaurin series of $f(x)$ using the identity

$$f(x) = \frac{2}{1-2x} - \frac{1}{1-x}$$

SOLUTION Substituting $2x$ for x in the Maclaurin series for $\frac{1}{1-x}$ gives

$$\frac{1}{1-2x} = \sum_{n=0}^{\infty} (2x)^n = \sum_{n=0}^{\infty} 2^n x^n$$

which is valid for $|2x| < 1$, or $|x| < \frac{1}{2}$. Because the Maclaurin series for $\frac{1}{1-x}$ is valid for $|x| < 1$, the two series together are valid for $|x| < \frac{1}{2}$. Thus, for $|x| < \frac{1}{2}$,

$$\frac{1}{(1-2x)(1-x)} = \frac{2}{1-2x} - \frac{1}{1-x} = 2\sum_{n=0}^{\infty} 2^n x^n - \sum_{n=0}^{\infty} x^n$$

$$= \sum_{n=0}^{\infty} 2^{n+1} x^n - \sum_{n=0}^{\infty} x^n = \sum_{n=0}^{\infty} \left(2^{n+1} - 1\right) x^n$$

73. When a voltage V is applied to a series circuit consisting of a resistor R and an inductor L, the current at time t is

$$I(t) = \left(\frac{V}{R}\right)\left(1 - e^{-Rt/L}\right)$$

Expand $I(t)$ in a Maclaurin series. Show that $I(t) \approx \frac{Vt}{L}$ for small t.

SOLUTION Substituting $-\frac{Rt}{L}$ for t in the Maclaurin series for e^t gives

$$e^{-Rt/L} = \sum_{n=0}^{\infty} \frac{\left(-\frac{Rt}{L}\right)^n}{n!} = \sum_{n=0}^{\infty} \frac{(-1)^n}{n!}\left(\frac{R}{L}\right)^n t^n = 1 + \sum_{n=1}^{\infty} \frac{(-1)^n}{n!}\left(\frac{R}{L}\right)^n t^n$$

Thus,

$$1 - e^{-Rt/L} = 1 - \left(1 + \sum_{n=1}^{\infty} \frac{(-1)^n}{n!}\left(\frac{R}{L}\right)^n t^n\right) = \sum_{n=1}^{\infty} \frac{(-1)^{n+1}}{n!}\left(\frac{Rt}{L}\right)^n$$

and

$$I(t) = \frac{V}{R}\sum_{n=1}^{\infty} \frac{(-1)^{n+1}}{n!}\left(\frac{Rt}{L}\right)^n = \frac{Vt}{L} + \frac{V}{R}\sum_{n=2}^{\infty} \frac{(-1)^{n+1}}{n!}\left(\frac{Rt}{L}\right)^n$$

If t is small, then so is Rt/L, so the terms in the series are even smaller, and we find

$$V(t) \approx \frac{Vt}{L}$$

75. Find the Maclaurin series for $f(x) = \cos(x^3)$ and use it to determine $f^{(6)}(0)$.

SOLUTION The Maclaurin series for $\cos x$ is

$$\cos x = \sum_{n=0}^{\infty} (-1)^n \frac{x^{2n}}{(2n)!}$$

Substituting x^3 for x gives

$$\cos(x^3) = \sum_{n=0}^{\infty} (-1)^n \frac{x^{6n}}{(2n)!}$$

Now, the coefficient of x^6 in this series is

$$-\frac{1}{2!} = -\frac{1}{2} = \frac{f^{(6)}(0)}{6!}$$

so

$$f^{(6)}(0) = -\frac{6!}{2} = -360$$

77. Use substitution to find the first three terms of the Maclaurin series for $f(x) = e^{x^{20}}$. How does the result show that $f^{(k)}(0) = 0$ for $1 \le k \le 19$?

SOLUTION Substituting x^{20} for x in the Maclaurin series for e^x yields

$$e^{x^{20}} = \sum_{n=0}^{\infty} \frac{(x^{20})^n}{n!} = \sum_{n=0}^{\infty} \frac{x^{20n}}{n!}$$

the first three terms in the series are then

$$1 + x^{20} + \frac{1}{2}x^{40}$$

Recall that the coefficient of x^k in the Maclaurin series for f is $\frac{f^{(k)}(0)}{k!}$. For $1 \le k \le 19$, the coefficient of x^k in the Maclaurin series for $f(x) = e^{x^{20}}$ is zero; it therefore follows that

$$\frac{f^{(k)}(0)}{k!} = 0 \quad \text{or} \quad f^{(k)}(0) = 0$$

for $1 \le k \le 19$.

79. Does the Maclaurin series for $f(x) = (1+x)^{3/4}$ converge to $f(x)$ at $x = 2$? Give numerical evidence to support your answer.

SOLUTION The Taylor series for $f(x) = (1+x)^{3/4}$ converges to $f(x)$ for $|x| < 1$; because $x = 2$ is not contained on this interval, the series does not converge to $f(x)$ at $x = 2$. The graph below displays

$$S_N = \sum_{n=0}^{N} \binom{\frac{3}{4}}{n} 2^n$$

for $0 \le N \le 14$. The divergent nature of the sequence of partial sums is clear.

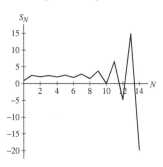

81. [GU] Let $f(x) = \sqrt{1+x}$.

(a) Use a graphing calculator to compare the graph of f with the graphs of the first five Taylor polynomials for f. What do they suggest about the interval of convergence of the Taylor series?

(b) Investigate numerically whether or not the Taylor expansion for f is valid for $x = 1$ and $x = -1$.

SOLUTION

(a) The five first terms of the Binomial series with $a = \frac{1}{2}$ are

$$\sqrt{1+x} = 1 + \frac{1}{2}x + \frac{\frac{1}{2}(\frac{1}{2}-1)}{2!}x^2 + \frac{\frac{1}{2}(\frac{1}{2}-1)(\frac{1}{2}-2)}{3!}x^3 + \frac{\frac{1}{2}(\frac{1}{2}-1)(\frac{1}{2}-2)(\frac{1}{2}-3)}{4!}x^4 + \cdots$$

$$= 1 + \frac{1}{2}x - \frac{1}{8}x^2 + \frac{9}{4}x^3 - \frac{45}{2}x^4 + \cdots$$

Therefore, the first five Taylor polynomials are

$$T_0(x) = 1;$$

$$T_1(x) = 1 + \frac{1}{2}x;$$

$$T_2(x) = 1 + \frac{1}{2}x - \frac{1}{8}x^2;$$

$$T_3(x) = 1 + \frac{1}{2}x - \frac{1}{8}x^2 + \frac{1}{8}x^3;$$

$$T_4(x) = 1 + \frac{1}{2}x - \frac{1}{8}x^2 + \frac{1}{8}x^3 - \frac{5}{128}x^4$$

The figure displays the graphs of these Taylor polynomials, along with the graph of the function $f(x) = \sqrt{1+x}$, which is shown in red.

The graphs suggest that the interval of convergence for the Taylor series is $-1 < x < 1$.

(b) Using a computer algebra system to calculate $S_N = \sum_{n=0}^{N} \binom{\frac{1}{2}}{n} x^n$ for $x = 1$ we find

$$S_{10} = 1.409931183, \quad S_{100} = 1.414073048, \quad S_{1000} = 1.414209104$$

which appears to be converging to $\sqrt{2}$ as expected. At $x = -1$ we calculate $S_N = \sum_{n=0}^{N} \binom{\frac{1}{2}}{n} \cdot (-1)^n$, and find

$$S_{10} = 0.176197052, \quad S_{100} = 0.056348479, \quad S_{1000} = 0.017839011$$

which appears to be converging to zero, though slowly.

83. Use Example 12 and the approximation $\sin x \approx x$ to show that the period T of a pendulum released at an angle θ has the following second-order approximation:

$$T \approx 2\pi \sqrt{\frac{L}{g}} \left(1 + \frac{\theta^2}{16}\right)$$

SOLUTION The period T of a pendulum of length L released from an angle θ is

$$T = 4\sqrt{\frac{L}{g}} E(k)$$

where $g \approx 9.8$ m/s^2 is the acceleration due to gravity, $E(k)$ is the elliptic integral of the first kind and $k = \sin \frac{\theta}{2}$. From Example 12, we know that

$$E(k) = \frac{\pi}{2} \sum_{n=0}^{\infty} \left(\frac{1 \cdot 3 \cdot 5 \cdots (2n-1)}{2 \cdot 4 \cdot 6 \cdots (2n)}\right)^2 k^{2n}$$

Using the approximation $\sin x \approx x$, we have

$$k = \sin\frac{\theta}{2} \approx \frac{\theta}{2}$$

moreover, using the first two terms of the series for $E(k)$, we find

$$E(k) \approx \frac{\pi}{2}\left[1 + \left(\frac{1}{2}\right)^2 \left(\frac{\theta}{2}\right)^2\right] = \frac{\pi}{2}\left(1 + \frac{\theta^2}{16}\right)$$

Therefore,

$$T = 4\sqrt{\frac{L}{g}} E(k) \approx 2\pi \sqrt{\frac{L}{g}} \left(1 + \frac{\theta^2}{16}\right)$$

In Exercises 84–87, the limits can be done using multiple L'Hôpital's Rule steps. Power series provide an alternative approach. In each case substitute in the Maclaurin series for the trig function or the inverse trig function involved, simplify, and compute the limit.

85. $\lim\limits_{x \to 0} \dfrac{\sin x - x + \frac{x^3}{6}}{x^5}$

SOLUTION Using the Maclaurin series for $\sin x$, we find

$$\sin x = \sum_{n=0}^{\infty} (-1)^n \frac{x^{2n+1}}{(2n+1)!} = x - \frac{x^3}{6} + \frac{x^5}{120} + \sum_{n=3}^{\infty} (-1)^n \frac{x^{2n+1}}{(2n+1)!}$$

Thus,

$$\sin x - x + \frac{x^3}{6} = \frac{x^5}{120} + \sum_{n=3}^{\infty} (-1)^n \frac{x^{2n+1}}{(2n+1)!}$$

and

$$\frac{\sin x - x + \frac{x^3}{6}}{x^5} = \frac{1}{120} + \sum_{n=3}^{\infty} (-1)^n \frac{x^{2n-4}}{(2n+1)!}$$

Note that the radius of convergence for this series is infinite, and recall from the previous section that a convergent power series is continuous within its radius of convergence. Thus to calculate the limit of this power series as $x \to 0$ it suffices to evaluate it at $x = 0$:

$$\lim_{x \to 0} \frac{\sin x - x + \frac{x^3}{6}}{x^5} = \lim_{x \to 0} \left(\frac{1}{120} + \sum_{n=3}^{\infty} (-1)^n \frac{x^{2n-4}}{(2n+1)!} \right) = \frac{1}{120} + 0 = \frac{1}{120}$$

87. $\lim\limits_{x \to 0} \left(\dfrac{\sin(x^2)}{x^4} - \dfrac{\cos x}{x^2} \right)$

SOLUTION We start with

$$\sin x = \sum_{n=0}^{\infty} (-1)^n \frac{x^{2n+1}}{(2n+1)!} \qquad \cos x = \sum_{n=0}^{\infty} (-1)^n \frac{x^{2n}}{(2n)!}$$

so that

$$\frac{\sin(x^2)}{x^4} = \sum_{n=0}^{\infty} (-1)^n \frac{x^{4n+2}}{(2n+1)! x^4} = \sum_{n=0}^{\infty} (-1)^n \frac{x^{4n-2}}{(2n+1)!}$$

$$\frac{\cos x}{x^2} = \sum_{n=0}^{\infty} (-1)^n \frac{x^{2n-2}}{(2n)!}$$

Expanding the first few terms gives

$$\frac{\sin(x^2)}{x^4} = \frac{1}{x^2} - \sum_{n=1}^{\infty} (-1)^n \frac{x^{4n-2}}{(2n+1)!}$$

$$\frac{\cos x}{x^2} = \frac{1}{x^2} - \frac{1}{2} + \sum_{n=2}^{\infty} (-1)^n \frac{x^{2n-2}}{(2n)!}$$

so that

$$\frac{\sin(x^2)}{x^4} - \frac{\cos x}{x^2} = \frac{1}{2} - \sum_{n=1}^{\infty} (-1)^n \frac{x^{4n-2}}{(2n+1)!} - \sum_{n=2}^{\infty} (-1)^n \frac{x^{2n-2}}{(2n)!}$$

Note that all terms under the summation signs have positive powers of x. Now, the radius of convergence of the series for both sin and cos is infinite, so the radius of convergence of this series is infinite. Recall from the previous section that a convergent power series is continuous within its radius of convergence. Thus to calculate the limit of this power series as $x \to 0$ it suffices to evaluate it at $x = 0$:

$$\lim_{x \to 0} \left(\frac{\sin(x^2)}{x^4} - \frac{\cos x}{x^2} \right) = \lim_{x \to 0} \left(\frac{1}{2} - \sum_{n=1}^{\infty} (-1)^n \frac{x^{4n-2}}{(2n+1)!} - \sum_{n=2}^{\infty} (-1)^n \frac{x^{2n-2}}{(2n)!} \right) = \frac{1}{2} + 0 = \frac{1}{2}$$

89. Use Euler's Formula to express each of the following in $a + bi$ form.
(a) $-e^{\frac{3\pi}{4}i}$
(b) $e^{2\pi i}$
(c) $3ie^{\frac{-\pi}{3}i}$

SOLUTION

(a) $-\left(\cos \frac{3\pi}{4} + i \sin \frac{3\pi}{4}\right) = -\left(-\frac{\sqrt{2}}{2} + \frac{\sqrt{2}}{2}i\right) = \frac{\sqrt{2}}{2} - \frac{\sqrt{2}}{2}i$

(b) $\cos 2\pi + i \sin 2\pi = 1$

(c) $3i\left(\cos\left(-\frac{\pi}{3}\right) + i \sin\left(-\frac{\pi}{3}\right)\right) = 3i\left(\frac{1}{2} - \frac{\sqrt{3}}{2}i\right) = \frac{3\sqrt{3}}{2} + \frac{3}{2}i$

In Exercises 90–91, use Euler's Formula to prove that the identity holds. Note the similarity between these relationships and the definitions of the hyperbolic sine and cosine functions.

91. $\sin z = \frac{e^{iz} - e^{-iz}}{2i}$

SOLUTION Using Euler's identity,

$$\frac{e^{iz} - e^{-iz}}{2i} = \frac{(\cos z + i \sin z) - (\cos(-z) + i \sin(-z))}{2i} = \frac{(\cos z + i \sin z) - (\cos z - i \sin z)}{2i} = \frac{2i \sin z}{2i} = \sin z$$

Further Insights and Challenges

93. Let $g(t) = \frac{1}{1+t^2} - \frac{t}{1+t^2}$.

(a) Show that $\int_0^1 g(t)\, dt = \frac{\pi}{4} - \frac{1}{2} \ln 2$.

(b) Show that $g(t) = 1 - t - t^2 + t^3 + t^4 - t^5 - t^6 + \cdots$.

(c) Evaluate $S = 1 - \frac{1}{2} - \frac{1}{3} + \frac{1}{4} + \frac{1}{5} - \frac{1}{6} - \frac{1}{7} + \cdots$.

SOLUTION

(a)

$$\int_0^1 g(t)\, dt = \left(\tan^{-1} t - \frac{1}{2} \ln(t^2 + 1)\right)\bigg|_0^1 = \tan^{-1} 1 - \frac{1}{2} \ln 2 = \frac{\pi}{4} - \frac{1}{2} \ln 2$$

(b) Start with the Taylor series for $\frac{1}{1+t}$:

$$\frac{1}{1+t} = \sum_{n=0}^{\infty} (-1)^n t^n$$

and substitute t^2 for t to get

$$\frac{1}{1+t^2} = \sum_{n=0}^{\infty} (-1)^n t^{2n} = 1 - t^2 + t^4 - t^6 + \ldots$$

so that

$$\frac{t}{1+t^2} = \sum_{n=0}^{\infty} (-1)^n t^{2n+1} = t - t^3 + t^5 - t^7 + \ldots$$

Finally,

$$g(t) = \frac{1}{1+t^2} - \frac{t}{1+t^2} = 1 - t - t^2 + t^3 + t^4 - t^5 - t^6 + t^7 + \ldots$$

(c) We have

$$\int g(t)\,dt = \int (1 - t - t^2 + t^3 + t^4 - t^5 - \ldots)\,dt = t - \frac{1}{2}t^2 - \frac{1}{3}t^3 + \frac{1}{4}t^4 + \frac{1}{5}t^5 - \frac{1}{6}t^6 - \cdots + C$$

The radius of convergence of the series for $g(t)$ is 1, so the radius of convergence of this series is also 1. However, this series converges at the right endpoint, $t = 1$, since

$$\left(1 - \frac{1}{2}\right) - \left(\frac{1}{3} - \frac{1}{4}\right) + \left(\frac{1}{5} - \frac{1}{6}\right) - \cdots$$

is an alternating series with general term decreasing to zero. Thus by part (a),

$$1 - \frac{1}{2} - \frac{1}{3} + \frac{1}{4} + \frac{1}{5} - \frac{1}{6} - \cdots = \frac{\pi}{4} - \frac{1}{2}\ln 2$$

In Exercises 94 and 95, we investigate the convergence of the binomial series

$$T_a(x) = \sum_{n=0}^{\infty} \binom{a}{n} x^n$$

95. By Exercise 94, $T_a(x)$ converges for $|x| < 1$, but we do not yet know whether $T_a(x) = (1 + x)^a$.

(a) Verify the identity

$$a\binom{a}{n} = n\binom{a}{n} + (n+1)\binom{a}{n+1}$$

(b) Use (a) to show that $y = T_a(x)$ satisfies the differential equation $(1 + x)y' = ay$ with initial condition $y(0) = 1$.

(c) Prove $T_a(x) = (1 + x)^a$ for $|x| < 1$ by showing that the derivative of the ratio $\dfrac{T_a(x)}{(1 + x)^a}$ is zero.

SOLUTION

(a)

$$n\binom{a}{n} + (n+1)\binom{a}{n+1} = n \cdot \frac{a(a-1)\cdots(a-n+1)}{n!} + (n+1) \cdot \frac{a(a-1)\cdots(a-n+1)(a-n)}{(n+1)!}$$

$$= \frac{a(a-1)\cdots(a-n+1)}{(n-1)!} + \frac{a(a-1)\cdots(a-n+1)(a-n)}{n!}$$

$$= \frac{a(a-1)\cdots(a-n+1)(n + (a-n))}{n!} = a \cdot \binom{a}{n}$$

(b) Differentiating $T_a(x)$ term-by-term yields

$$T_a'(x) = \sum_{n=1}^{\infty} n\binom{a}{n} x^{n-1}$$

Thus,

$$(1+x)T_a'(x) = \sum_{n=1}^{\infty} n\binom{a}{n} x^{n-1} + \sum_{n=1}^{\infty} n\binom{a}{n} x^n = \sum_{n=0}^{\infty} (n+1)\binom{a}{n+1} x^n + \sum_{n=0}^{\infty} n\binom{a}{n} x^n$$

$$= \sum_{n=0}^{\infty} \left[(n+1)\binom{a}{n+1} + n\binom{a}{n}\right] x^n = a\sum_{n=0}^{\infty} \binom{a}{n} x^n = aT_a(x)$$

Moreover,

$$T_a(0) = \binom{a}{0} = 1$$

(c)

$$\frac{d}{dx}\left(\frac{T_a(x)}{(1+x)^a}\right) = \frac{(1+x)^a T_a'(x) - a(1+x)^{a-1} T_a(x)}{(1+x)^{2a}} = \frac{(1+x)T_a'(x) - aT_a(x)}{(1+x)^{a+1}} = 0$$

Thus,
$$\frac{T_a(x)}{(1+x)^a} = C$$

for some constant C. For $x = 0$,
$$\frac{T_a(0)}{(1+0)^a} = \frac{1}{1} = 1, \text{ so } C = 1$$

Finally, $T_a(x) = (1+x)^a$.

97. Assume that $a < b$ and let L be the arc length (circumference) of the ellipse $\left(\frac{x}{a}\right)^2 + \left(\frac{y}{b}\right)^2 = 1$ shown in Figure 4. There is no explicit formula for L, but it is known that $L = 4bG(k)$, with $G(k)$ as in Exercise 96 and $k = \sqrt{1 - a^2/b^2}$. Use the first three terms of the expansion of Exercise 96 to estimate L when $a = 4$ and $b = 5$.

FIGURE 4 The ellipse $\left(\frac{x}{a}\right)^2 + \left(\frac{y}{b}\right)^2 = 1$.

SOLUTION With $a = 4$ and $b = 5$,
$$k = \sqrt{1 - \frac{4^2}{5^2}} = \frac{3}{5}$$

and the arc length of the ellipse $\left(\frac{x}{4}\right)^2 + \left(\frac{y}{5}\right)^2 = 1$ is
$$L = 20G\left(\frac{3}{5}\right) = 20\left(\frac{\pi}{2} - \frac{\pi}{2}\sum_{n=1}^{\infty}\left(\frac{1 \cdot 3 \cdots (2n-1)}{2 \cdot 4 \cdots (2n)}\right)^2 \frac{\left(\frac{3}{5}\right)^{2n}}{2n-1}\right)$$

Using the first three terms in the series for $G(k)$ gives
$$L \approx 10\pi - 10\pi\left(\left(\frac{1}{2}\right)^2 \cdot \frac{(3/5)^2}{1} + \left(\frac{1 \cdot 3}{2 \cdot 4}\right)^2 \cdot \frac{(3/5)^4}{3}\right) = 10\pi\left(1 - \frac{9}{100} - \frac{243}{40000}\right) = \frac{36157\pi}{4000} \approx 28.398$$

99. Irrationality of e Prove that e is an irrational number using the following argument by contradiction. Suppose that $e = M/N$, where M, N are nonzero integers.

(a) Show that $M!\, e^{-1}$ is a whole number.

(b) Use the power series for $f(x) = e^x$ at $x = -1$ to show that there is an integer B such that $M!\, e^{-1}$ equals
$$B + (-1)^{M+1}\left(\frac{1}{M+1} - \frac{1}{(M+1)(M+2)} + \cdots\right)$$

(c) Use your knowledge of alternating series with decreasing terms to conclude that $0 < |M!\, e^{-1} - B| < 1$ and observe that this contradicts (a). Hence, e is not equal to M/N.

SOLUTION Suppose that $e = M/N$, where M, N are nonzero integers.
(a) With $e = M/N$,
$$M!\, e^{-1} = M!\frac{N}{M} = (M-1)!N$$

which is a whole number.

(b) Substituting $x = -1$ into the Maclaurin series for e^x and multiplying the resulting series by $M!$ yields

$$M!e^{-1} = M!\left(1 - 1 + \frac{1}{2!} - \frac{1}{3!} + \cdots + \frac{(-1)^k}{k!} + \cdots\right)$$

For all $k \leq M$, $\frac{M!}{k!}$ is a whole number, so

$$M!\left(1 - 1 + \frac{1}{2!} - \frac{1}{3!} + \cdots + \frac{(-1)^k}{M!}\right)$$

is an integer. Denote this integer by B. Thus,

$$M!e^{-1} = B + M!\left(\frac{(-1)^{M+1}}{(M+1)!} + \frac{(-1)^{M+2}}{(M+2)!} + \cdots\right) = B + (-1)^{M+1}\left(\frac{1}{M+1} - \frac{1}{(M+1)(M+2)} + \cdots\right)$$

(c) The series for $M!e^{-1}$ obtained in part (b) is an alternating series with $a_n = \frac{M!}{n!}$. Using the Error Bound for an alternating series and noting that $B = S_M$, we have

$$|M!e^{-1} - B| \leq a_{M+1} = \frac{1}{M+1} < 1$$

This inequality implies that $M!e^{-1} - B$ is not a whole number; however, B is a whole number so $M!e^{-1}$ cannot be a whole number. We get a contradiction to the result in part (a), which proves that the original assumption that e is a rational number is false.

CHAPTER REVIEW EXERCISES

1. Let $a_n = \frac{n-3}{n!}$ and $b_n = a_{n+3}$. Calculate the first three terms in each sequence.

(a) a_n^2
(b) b_n
(c) $a_n b_n$
(d) $2a_{n+1} - 3a_n$

SOLUTION

(a)

$$a_1^2 = \left(\frac{1-3}{1!}\right)^2 = (-2)^2 = 4;$$

$$a_2^2 = \left(\frac{2-3}{2!}\right)^2 = \left(-\frac{1}{2}\right)^2 = \frac{1}{4};$$

$$a_3^2 = \left(\frac{3-3}{3!}\right)^2 = 0$$

(b)

$$b_1 = a_4 = \frac{4-3}{4!} = \frac{1}{24};$$

$$b_2 = a_5 = \frac{5-3}{5!} = \frac{1}{60};$$

$$b_3 = a_6 = \frac{6-3}{6!} = \frac{1}{240}$$

(c) Using the formula for a_n and the values in (b) we obtain:

$$a_1 b_1 = \frac{1-3}{1!} \cdot \frac{1}{24} = -\frac{1}{12};$$

$$a_2 b_2 = \frac{2-3}{2!} \cdot \frac{1}{60} = -\frac{1}{120};$$

$$a_3 b_3 = \frac{3-3}{3!} \cdot \frac{1}{240} = 0$$

(d)

$$2a_2 - 3a_1 = 2\left(-\frac{1}{2}\right) - 3(-2) = 5;$$

$$2a_3 - 3a_2 = 2 \cdot 0 - 3\left(-\frac{1}{2}\right) = \frac{3}{2};$$

$$2a_4 - 3a_3 = 2 \cdot \frac{1}{24} - 3 \cdot 0 = \frac{1}{12}$$

In Exercises 3–8, compute the limit (or state that it does not exist) assuming that $\lim_{n\to\infty} a_n = 2$.

3. $\lim_{n\to\infty} (5a_n - 2a_n^2)$

SOLUTION

$$\lim_{n\to\infty}\left(5a_n - 2a_n^2\right) = 5\lim_{n\to\infty} a_n - 2\lim_{n\to\infty} a_n^2 = 5\lim_{n\to\infty} a_n - 2\left(\lim_{n\to\infty} a_n\right)^2 = 5 \cdot 2 - 2 \cdot 2^2 = 2$$

5. $\lim_{n\to\infty} e^{a_n}$

SOLUTION The function $f(x) = e^x$ is continuous, hence:

$$\lim_{n\to\infty} e^{a_n} = e^{\lim_{n\to\infty} a_n} = e^2$$

7. $\lim_{n\to\infty} (-1)^n a_n$

SOLUTION Because $\lim_{n\to\infty} a_n \neq 0$, it follows that $\lim_{n\to\infty} (-1)^n a_n$ does not exist.

In Exercises 9–22, determine the limit of the sequence or show that the sequence diverges.

9. $a_n = \sqrt{n+5} - \sqrt{n+2}$

SOLUTION First rewrite a_n as follows:

$$a_n = \frac{\left(\sqrt{n+5} - \sqrt{n+2}\right)\left(\sqrt{n+5} + \sqrt{n+2}\right)}{\sqrt{n+5} + \sqrt{n+2}} = \frac{(n+5) - (n+2)}{\sqrt{n+5} + \sqrt{n+2}} = \frac{3}{\sqrt{n+5} + \sqrt{n+2}}$$

Thus,

$$\lim_{n\to\infty} a_n = \lim_{n\to\infty} \frac{3}{\sqrt{n+5} + \sqrt{n+2}} = 0$$

11. $a_n = 2^{1/n^2}$

SOLUTION The function $f(x) = 2^x$ is continuous, so

$$\lim_{n\to\infty} a_n = \lim_{n\to\infty} 2^{1/n^2} = 2^{\lim_{n\to\infty}(1/n^2)} = 2^0 = 1$$

13. $b_m = 1 + (-1)^m$

SOLUTION Because $1 + (-1)^m$ is equal to 0 for m odd and is equal to 2 for m even, the sequence $\{b_m\}$ does not approach one limit; hence this sequence diverges.

15. $b_n = \tan^{-1}\left(\frac{n+2}{n+5}\right)$

SOLUTION The function $\tan^{-1} x$ is continuous, so

$$\lim_{n\to\infty} b_n = \lim_{n\to\infty} \tan^{-1}\left(\frac{n+2}{n+5}\right) = \tan^{-1}\left(\lim_{n\to\infty} \frac{n+2}{n+5}\right) = \tan^{-1} 1 = \frac{\pi}{4}$$

17. $b_n = \sqrt{n^2 + n} - \sqrt{n^2 + 1}$

SOLUTION Rewrite b_n as

$$b_n = \frac{\left(\sqrt{n^2+n}-\sqrt{n^2+1}\right)\left(\sqrt{n^2+n}+\sqrt{n^2+1}\right)}{\sqrt{n^2+n}+\sqrt{n^2+1}} = \frac{(n^2+n)-(n^2+1)}{\sqrt{n^2+n}+\sqrt{n^2+1}} = \frac{n-1}{\sqrt{n^2+n}+\sqrt{n^2+1}}$$

Then

$$\lim_{n\to\infty} b_n = \lim_{n\to\infty} \frac{\frac{n}{n}-\frac{1}{n}}{\sqrt{\frac{n^2}{n^2}+\frac{n}{n^2}}+\sqrt{\frac{n^2}{n^2}+\frac{1}{n^2}}} = \lim_{n\to\infty}\frac{1-\frac{1}{n}}{\sqrt{1+\frac{1}{n}}+\sqrt{1+\frac{1}{n^2}}} = \frac{1-0}{\sqrt{1+0}+\sqrt{1+0}} = \frac{1}{2}$$

19. $b_m = \left(1+\dfrac{1}{m}\right)^{3m}$

SOLUTION $\displaystyle\lim_{m\to\infty} b_m = \lim_{m\to\infty}\left(1+\frac{1}{m}\right)^m = e$

21. $b_n = n(\ln(n+1) - \ln n)$

SOLUTION Write

$$b_n = n\ln\left(\frac{n+1}{n}\right) = \frac{\ln\left(1+\frac{1}{n}\right)}{\frac{1}{n}}$$

Using L'Hôpital's Rule, we find

$$\lim_{n\to\infty} b_n = \lim_{n\to\infty}\frac{\ln\left(1+\frac{1}{n}\right)}{\frac{1}{n}} = \lim_{x\to\infty}\frac{\ln\left(1+\frac{1}{x}\right)}{\frac{1}{x}} = \lim_{x\to\infty}\frac{\left(1+\frac{1}{x}\right)^{-1}\cdot\left(-\frac{1}{x^2}\right)}{-\frac{1}{x^2}} = \lim_{x\to\infty}\left(1+\frac{1}{x}\right)^{-1} = 1$$

23. Use the Squeeze Theorem to show that $\displaystyle\lim_{n\to\infty}\frac{\arctan(n^2)}{\sqrt{n}} = 0$.

SOLUTION For all x,

$$-\frac{\pi}{2} < \arctan x < \frac{\pi}{2}$$

so

$$-\frac{\pi/2}{\sqrt{n}} < \frac{\arctan(n^2)}{\sqrt{n}} < \frac{\pi/2}{\sqrt{n}}$$

for all n. Because

$$\lim_{n\to\infty}\left(-\frac{\pi/2}{\sqrt{n}}\right) = \lim_{n\to\infty}\frac{\pi/2}{\sqrt{n}} = 0$$

it follows by the Squeeze Theorem that

$$\lim_{n\to\infty}\frac{\arctan(n^2)}{\sqrt{n}} = 0$$

25. Calculate $\displaystyle\lim_{n\to\infty}\frac{a_{n+1}}{a_n}$, where $a_n = \dfrac{1}{2}3^n - \dfrac{1}{3}2^n$.

SOLUTION
(a) Because

$$\frac{1}{2}3^n - \frac{1}{3}2^n \geq \frac{1}{2}3^n - \frac{1}{3}3^n = \frac{3^n}{6}$$

and

$$\lim_{n\to\infty}\frac{3^n}{6} = \infty$$

we conclude that

$$\lim_{n\to\infty} a_n = \infty$$

(b) $\lim\limits_{n\to\infty} \dfrac{a_{n+1}}{a_n} = \lim\limits_{n\to\infty} \dfrac{\frac{1}{2}3^{n+1} - \frac{1}{3}2^{n+1}}{\frac{1}{2}3^n - \frac{1}{3}2^n} = \lim\limits_{n\to\infty} \dfrac{3^{n+2} - 2^{n+2}}{3^{n+1} - 2^{n+1}} = \lim\limits_{n\to\infty} \dfrac{3 - 2\left(\frac{2}{3}\right)^{n+1}}{1 - \left(\frac{2}{3}\right)^{n+1}} = \dfrac{3-0}{1-0} = 3$

27. Calculate the partial sums S_4 and S_7 of the series $\sum\limits_{n=1}^{\infty} \dfrac{n-2}{n^2+2n}$.

SOLUTION

$$S_4 = -\frac{1}{3} + 0 + \frac{1}{15} + \frac{2}{24} = -\frac{11}{60} = -0.183333$$

$$S_7 = -\frac{1}{3} + 0 + \frac{1}{15} + \frac{2}{24} + \frac{3}{35} + \frac{4}{48} + \frac{5}{63} = \frac{287}{4410} = 0.065079$$

29. Find the sum $\dfrac{4}{9} + \dfrac{8}{27} + \dfrac{16}{81} + \dfrac{32}{243} + \cdots$.

SOLUTION This is a geometric series with common ratio $r = \frac{2}{3}$. Therefore,

$$\frac{4}{9} + \frac{8}{27} + \frac{16}{81} + \frac{32}{243} + \cdots = \frac{\frac{4}{9}}{1 - \frac{2}{3}} = \frac{4}{3}$$

31. Use series to determine a reduced fraction that has decimal expansion $0.108108108\cdots$.

SOLUTION The decimal may be regarded as a geometric series:

$$0.108108108\ldots = \frac{108}{10^3} + \frac{108}{10^6} + \frac{108}{10^9}\cdots = \sum_{n=1}^{\infty} \frac{108}{10^{3n}}$$

The series has first term $\frac{108}{10^3}$ and ratio $\frac{1}{10^3}$, so its sum is

$$0.108108108 = \frac{108/10^3}{1 - 1/10^3} = \frac{108}{10^3} \cdot \frac{10^3}{999} = \frac{108}{999} = \frac{4}{37}$$

33. Find the sum $\sum\limits_{n=-1}^{\infty} \dfrac{2^{n+3}}{3^n}$.

SOLUTION Note

$$\sum_{n=0}^{\infty} \frac{2^{n+1}}{3^n} = 2\sum_{n=0}^{\infty} \frac{2^n}{3^n} = 2\sum_{n=0}^{\infty} \left(\frac{2}{3}\right)^n$$

therefore,

$$\sum_{n=0}^{\infty} \frac{2^{n+1}}{3^n} = 2\frac{1}{1-\frac{2}{3}} = 2 \cdot 3 = 6$$

35. Give an example of divergent series $\sum\limits_{n=1}^{\infty} a_n$ and $\sum\limits_{n=1}^{\infty} b_n$ such that $\sum\limits_{n=1}^{\infty}(a_n + b_n) = 1$.

SOLUTION Let $a_n = \left(\frac{1}{2}\right)^n + 1$, $b_n = -1$. The corresponding series diverge by the Divergence Test; however,

$$\sum_{n=1}^{\infty}(a_n + b_n) = \sum_{n=1}^{\infty}\left(\frac{1}{2}\right)^n = \frac{\frac{1}{2}}{1-\frac{1}{2}} = 1$$

37. Evaluate $S = \sum\limits_{n=3}^{\infty} \dfrac{1}{n(n+3)}$.

SOLUTION Note that

$$\frac{1}{n(n+3)} = \frac{1}{3}\left(\frac{1}{n} - \frac{1}{n+3}\right)$$

so that

$$\sum_{n=3}^{N} \frac{1}{n(n+3)} = \frac{1}{3} \sum_{n=3}^{N} \left(\frac{1}{n} - \frac{1}{n+3} \right)$$

$$= \frac{1}{3} \left(\left(\frac{1}{3} - \frac{1}{6} \right) + \left(\frac{1}{4} - \frac{1}{7} \right) + \left(\frac{1}{5} - \frac{1}{8} \right) \right.$$

$$\left. \left(\frac{1}{6} - \frac{1}{9} \right) + \cdots + \left(\frac{1}{N-1} - \frac{1}{N+2} \right) + \left(\frac{1}{N} - \frac{1}{N+3} \right) \right)$$

$$= \frac{1}{3} \left(\frac{1}{3} + \frac{1}{4} + \frac{1}{5} - \frac{1}{N+1} - \frac{1}{N+2} - \frac{1}{N+3} \right)$$

Thus

$$\sum_{n=3}^{\infty} \frac{1}{n(n+3)} = \frac{1}{3} \lim_{N \to \infty} \sum_{n=3}^{N} \left(\frac{1}{n} - \frac{1}{n+3} \right)$$

$$= \frac{1}{3} \left(\frac{1}{3} + \frac{1}{4} + \frac{1}{5} - \frac{1}{N+1} - \frac{1}{N+2} - \frac{1}{N+3} \right) = \frac{1}{3} \left(\frac{1}{3} + \frac{1}{4} + \frac{1}{5} \right) = \frac{47}{180}$$

In Exercises 39–42, use the Integral Test to determine whether the infinite series converges.

39. $\sum_{n=1}^{\infty} \frac{n^2}{n^3 + 1}$

SOLUTION Let $f(x) = \frac{x^2}{x^3+1}$. This function is continuous and positive for $x \geq 1$. Because

$$f'(x) = \frac{(x^3 + 1)(2x) - x^2(3x^2)}{(x^3 + 1)^2} = \frac{x(2 - x^3)}{(x^3 + 1)^2}$$

we see that $f'(x) < 0$ and f is decreasing on the interval $x \geq 2$. Therefore, the Integral Test applies on the interval $x \geq 2$. Now,

$$\int_2^{\infty} \frac{x^2}{x^3 + 1} \, dx = \lim_{R \to \infty} \int_2^R \frac{x^2}{x^3 + 1} \, dx = \frac{1}{3} \lim_{R \to \infty} \left(\ln(R^3 + 1) - \ln 9 \right) = \infty$$

The integral diverges; hence, the series $\sum_{n=2}^{\infty} \frac{n^2}{n^3 + 1}$ diverges, as does the series $\sum_{n=1}^{\infty} \frac{n^2}{n^3 + 1}$.

41. $\sum_{n=1}^{\infty} \frac{1}{(n+2)(\ln(n+2))^3}$

SOLUTION Let $f(x) = \frac{1}{(x+2)\ln^3(x+2)}$. Using the substitution $u = \ln(x + 2)$, so that $du = \frac{1}{x+2} dx$, we have

$$\int_0^{\infty} f(x) \, dx = \int_{\ln 2}^{\infty} \frac{1}{u^3} \, du = \lim_{R \to \infty} \int_{\ln 2}^{R} \frac{1}{u^3} \, du = \lim_{R \to \infty} \left(-\frac{1}{2u^2} \Big|_{\ln 2}^{R} \right)$$

$$= \lim_{R \to \infty} \left(\frac{1}{2(\ln 2)^2} - \frac{1}{2(\ln R)^2} \right) = \frac{1}{2(\ln 2)^2}$$

Since the integral of $f(x)$ converges, so does the series.

In Exercises 43–50, use the Direct Comparison or Limit Comparison Test to determine whether the infinite series converges.

43. $\sum_{n=1}^{\infty} \frac{1}{(n+1)^2}$

SOLUTION For all $n \geq 1$,

$$0 < \frac{1}{n+1} < \frac{1}{n} \quad \text{so} \quad \frac{1}{(n+1)^2} < \frac{1}{n^2}$$

The series $\sum_{n=1}^{\infty} \frac{1}{n^2}$ is a convergent p-series, so the series $\sum_{n=1}^{\infty} \frac{1}{(n+1)^2}$ converges by the Direct Comparison Test.

45. $\sum_{n=2}^{\infty} \frac{n^2+1}{n^{3.5}-2}$

SOLUTION Apply the Limit Comparison Test with $a_n = \frac{n^2+1}{n^{3.5}-2}$ and $b_n = \frac{1}{n^{1.5}}$. Now,

$$L = \lim_{n \to \infty} \frac{\frac{n^2+1}{n^{3.5}-2}}{\frac{1}{n^{1.5}}} = \lim_{n \to \infty} \frac{n^{3.5}+n^{1.5}}{n^{3.5}-2} = 1$$

Because L exists and $\sum_{n=1}^{\infty} \frac{1}{n^{1.5}}$ is a convergent p-series, we conclude by the Limit Comparison Test that the series $\sum_{n=2}^{\infty} \frac{n^2+1}{n^{3.5}-2}$ also converges.

47. $\sum_{n=2}^{\infty} \frac{n}{\sqrt{n^5+5}}$

SOLUTION For all $n \geq 2$,

$$\frac{n}{\sqrt{n^5+2}} < \frac{n}{n^{5/2}} = \frac{1}{n^{3/2}}$$

The series $\sum_{n=2}^{\infty} \frac{1}{n^{3/2}}$ is a convergent p-series, so the series $\sum_{n=2}^{\infty} \frac{n}{\sqrt{n^5+5}}$ converges by the Direct Comparison Test.

49. $\sum_{n=1}^{\infty} \frac{n^{10}+10^n}{n^{11}+11^n}$

SOLUTION Apply the Limit Comparison Test with $a_n = \frac{n^{10}+10^n}{n^{11}+11^n}$ and $b_n = \left(\frac{10}{11}\right)^n$. Then,

$$L = \lim n \to \infty \frac{a_n}{b_n} = \lim_{n \to \infty} \frac{\frac{n^{10}+10^n}{n^{11}+11^n}}{\left(\frac{10}{11}\right)^n} = \lim_{n \to \infty} \frac{\frac{n^{10}+10^n}{10^n}}{\frac{n^{11}+11^n}{11^n}} = \lim_{n \to \infty} \frac{\frac{n^{10}}{10^n}+1}{\frac{n^{11}}{11^n}+1} = 1$$

The series $\sum_{n=1}^{\infty} \left(\frac{10}{11}\right)^n$ is a convergent geometric series; because L exists, we may therefore conclude by the Limit Comparison Test that the series $\sum_{n=1}^{\infty} \frac{n^{10}+10^n}{n^{11}+11^n}$ also converges.

51. Determine the convergence of $\sum_{n=1}^{\infty} \frac{2^n+n}{3^n-2}$ using the Limit Comparison Test with $b_n = \left(\frac{2}{3}\right)^n$.

SOLUTION With $a_n = \frac{2^n+n}{3^n-2}$, we have

$$L = \lim_{n \to \infty} \frac{a_n}{b_n} = \lim_{n \to \infty} \frac{2^n+n}{3^n-2} \cdot \frac{3^n}{2^n} = \lim_{n \to \infty} \frac{6^n+n3^n}{6^n-2^{n+1}} = \lim_{n \to \infty} \frac{1+n\left(\frac{1}{2}\right)^n}{1-2\left(\frac{1}{3}\right)^n} = 1$$

Since $L = 1$, the two series either both converge or both diverge. Since $\sum_{n=1}^{\infty} \left(\frac{2}{3}\right)^n$ is a convergent geometric series, the Limit Comparison Test tells us that $\sum_{n=1}^{\infty} \frac{2^n+n}{3^n-2}$ also converges.

53. Let $a_n = 1 - \sqrt{1 - \frac{1}{n}}$. Show that $\lim_{n \to \infty} a_n = 0$ and that $\sum_{n=1}^{\infty} a_n$ diverges. *Hint:* Show that $a_n \geq \frac{1}{2n}$.

SOLUTION

$$1 - \sqrt{1 - \frac{1}{n}} = 1 - \sqrt{\frac{n-1}{n}} = \frac{\sqrt{n} - \sqrt{n-1}}{\sqrt{n}} = \frac{n - (n-1)}{\sqrt{n}(\sqrt{n} + \sqrt{n-1})} = \frac{1}{n + \sqrt{n^2 - n}}$$

$$\geq \frac{1}{n + \sqrt{n^2}} = \frac{1}{2n}$$

The series $\sum_{n=2}^{\infty} \frac{1}{2n}$ diverges, so the series $\sum_{n=2}^{\infty} \left(1 - \sqrt{1 - \frac{1}{n}}\right)$ also diverges by the Direct Comparison Test.

55. Consider $\sum_{n=1}^{\infty} \frac{n}{(n^2 + 1)^2}$.

(a) Show that the series converges.

(b) **CAS** Use the inequality in (4) from Exercise 83 of Section 10.3 with $M = 99$ to approximate the sum of the series. What is the maximum size of the error?

SOLUTION

(a) For $n \geq 1$,

$$\frac{n}{(n^2 + 1)^2} < \frac{n}{(n^2)^2} = \frac{1}{n^3}$$

The series $\sum_{n=1}^{\infty} \frac{1}{n^3}$ is a convergent p-series, so the series $\sum_{n=1}^{\infty} \frac{n}{(n^2 + 1)^2}$ also converges by the Direct Comparison Test.

(b) With $a_n = \frac{n}{(n^2+1)^2}$, $f(x) = \frac{x}{(x^2+1)^2}$ and $M = 99$, Eq. (4) in Exercise 83 of Section 10.3 becomes

$$\sum_{n=1}^{99} \frac{n}{(n^2 + 1)^2} + \int_{100}^{\infty} \frac{x}{(x^2 + 1)^2}\, dx \leq S \leq \sum_{n=1}^{100} \frac{n}{(n^2 + 1)^2} + \int_{100}^{\infty} \frac{x}{(x^2 + 1)^2}\, dx$$

or

$$0 \leq S - \left(\sum_{n=1}^{99} \frac{n}{(n^2 + 1)^2} + \int_{100}^{\infty} \frac{x}{(x^2 + 1)^2}\, dx\right) \leq \frac{100}{(100^2 + 1)^2}$$

Now,

$$\sum_{n=1}^{99} \frac{n}{(n^2 + 1)^2} = 0.397066274; \text{ and}$$

$$\int_{100}^{\infty} \frac{x}{(x^2 + 1)^2}\, dx = \lim_{R \to \infty} \int_{100}^{R} \frac{x}{(x^2 + 1)^2}\, dx = \frac{1}{2} \lim_{R \to \infty} \left(-\frac{1}{R^2 + 1} + \frac{1}{100^2 + 1}\right)$$

$$= \frac{1}{20002} = 0.000049995$$

thus,

$$S \approx 0.397066274 + 0.000049995 = 0.397116269$$

The bound on the error in this approximation is

$$\frac{100}{(100^2 + 1)^2} = 9.998 \times 10^{-7}$$

In Exercises 56–59, determine whether the series converges absolutely. If it does not, determine whether it converges conditionally.

57. $\sum_{n=1}^{\infty} \dfrac{(-1)^n}{n^{1.1} \ln(n+1)}$

SOLUTION Consider the corresponding positive series $\sum_{n=1}^{\infty} \dfrac{1}{n^{1.1} \ln(n+1)}$. Because

$$\dfrac{1}{n^{1.1} \ln(n+1)} < \dfrac{1}{n^{1.1}}$$

and $\sum_{n=1}^{\infty} \dfrac{1}{n^{1.1}}$ is a convergent p-series, we can conclude by the Direct Comparison Test that $\sum_{n=1}^{\infty} \dfrac{(-1)^n}{n^{1.1} \ln(n+1)}$ also converges. Thus, $\sum_{n=1}^{\infty} \dfrac{(-1)^n}{n^{1.1} \ln(n+1)}$ converges absolutely.

59. $\sum_{n=1}^{\infty} \dfrac{\cos\left(\frac{\pi}{4} + 2\pi n\right)}{\sqrt{n}}$

SOLUTION $\cos\left(\frac{\pi}{4} + 2\pi n\right) = \cos\frac{\pi}{4} = \frac{\sqrt{2}}{2}$, so

$$\sum_{n=1}^{\infty} \dfrac{\cos\left(\frac{\pi}{4} + 2\pi n\right)}{\sqrt{n}} = \dfrac{\sqrt{2}}{2} \sum_{n=1}^{\infty} \dfrac{1}{\sqrt{n}}$$

This is a divergent p-series, so the series $\sum_{n=1}^{\infty} \dfrac{\cos\left(\frac{\pi}{4} + 2\pi n\right)}{\sqrt{n}}$ diverges.

61. Catalan's constant is defined by $K = \sum_{k=0}^{\infty} \dfrac{(-1)^k}{(2k+1)^2}$.

(a) How many terms of the series are needed to calculate K with an error of less than 10^{-6}?

(b) *CAS* Carry out the calculation.

SOLUTION

(a) Using the Error Bound for an alternating series, we have

$$|S_N - K| \le \dfrac{1}{(2(N+1)+1)^2} = \dfrac{1}{(2N+3)^2}$$

For accuracy to 10^{-6}, we must choose N so that

$$\dfrac{1}{(2N+3)^2} < 10^{-6} \quad \text{or} \quad (2N+3)^2 > 10^6$$

Solving for N yields

$$N > \dfrac{997}{2} = 498.5$$

(b) Computing the sum using a CAS gives

$$K \approx \sum_{k=0}^{499} \dfrac{(-1)^k}{(2k+1)^2} = 0.915965094$$

63. Let $\sum_{n=1}^{\infty} a_n$ be an absolutely convergent series. Determine whether the following series are convergent or divergent:

(a) $\sum_{n=1}^{\infty} \left(a_n + \dfrac{1}{n^2}\right)$

(b) $\sum_{n=1}^{\infty} (-1)^n a_n$

(c) $\sum_{n=1}^{\infty} \dfrac{1}{1+a_n^2}$

(d) $\sum_{n=1}^{\infty} \dfrac{|a_n|}{n}$

SOLUTION Because $\sum_{n=1}^{\infty} a_n$ converges absolutely, we know that $\sum_{n=1}^{\infty} a_n$ converges and that $\sum_{n=1}^{\infty} |a_n|$ converges.

(a) Because we know that $\sum_{n=1}^{\infty} a_n$ converges and the series $\sum_{n=1}^{\infty} \dfrac{1}{n^2}$ is a convergent p-series, the sum of these two series, $\sum_{n=1}^{\infty} \left(a_n + \dfrac{1}{n^2} \right)$ also converges.

(b) We have,

$$\sum_{n=1}^{\infty} |(-1)^n a_n| = \sum_{n=1}^{\infty} |a_n|$$

Because $\sum_{n=1}^{\infty} |a_n|$ converges, it follows that $\sum_{n=1}^{\infty} (-1)^n a_n$ converges absolutely, which implies that $\sum_{n=1}^{\infty} (-1)^n a_n$ converges.

(c) Because $\sum_{n=1}^{\infty} a_n$ converges, $\lim_{n\to\infty} a_n = 0$. Therefore,

$$\lim_{n\to\infty} \frac{1}{1+a_n^2} = \frac{1}{1+0^2} = 1 \neq 0$$

and the series $\sum_{n=1}^{\infty} \dfrac{1}{1+a_n^2}$ diverges by the Divergence Test.

(d) $\dfrac{|a_n|}{n} \leq |a_n|$ and the series $\sum_{n=1}^{\infty} |a_n|$ converges, so the series $\sum_{n=1}^{\infty} \dfrac{|a_n|}{n}$ also converges by the Direct Comparison Test.

In Exercises 65–72, apply the Ratio Test to determine convergence or divergence, or state that the Ratio Test is inconclusive.

65. $\sum_{n=1}^{\infty} \dfrac{n^5}{5^n}$

SOLUTION With $a_n = \dfrac{n^5}{5^n}$,

$$\left| \frac{a_{n+1}}{a_n} \right| = \frac{(n+1)^5}{5^{n+1}} \cdot \frac{5^n}{n^5} = \frac{1}{5}\left(1+\frac{1}{n}\right)^5$$

and

$$\rho = \lim_{n\to\infty} \left| \frac{a_{n+1}}{a_n} \right| = \frac{1}{5} \lim_{n\to\infty} \left(1+\frac{1}{n}\right)^5 = \frac{1}{5} \cdot 1 = \frac{1}{5}$$

Because $\rho < 1$, the series converges by the Ratio Test.

67. $\sum_{n=1}^{\infty} \dfrac{1}{n2^n + n^3}$

SOLUTION With $a_n = \dfrac{1}{n2^n + n^3}$,

$$\left| \frac{a_{n+1}}{a_n} \right| = \frac{n2^n + n^3}{(n+1)2^{n+1} + (n+1)^3} = \frac{n2^n\left(1+\frac{n^2}{2^n}\right)}{(n+1)2^{n+1}\left(1+\frac{(n+1)^2}{2^{n+1}}\right)} = \frac{1}{2} \cdot \frac{n}{n+1} \cdot \frac{1+\frac{n^2}{2^n}}{1+\frac{(n+1)^2}{2^{n+1}}}$$

and

$$\rho = \lim_{n\to\infty} \left| \frac{a_{n+1}}{a_n} \right| = \frac{1}{2} \cdot 1 \cdot 1 = \frac{1}{2}$$

Because $\rho < 1$, the series converges by the Ratio Test.

69. $\sum_{n=1}^{\infty} \dfrac{2^{n^2}}{n!}$

SOLUTION With $a_n = \dfrac{2^{n^2}}{n!}$,

$$\left|\dfrac{a_{n+1}}{a_n}\right| = \dfrac{2^{(n+1)^2}}{(n+1)!} \cdot \dfrac{n!}{2^{n^2}} = \dfrac{2^{2n+1}}{n+1} \quad \text{and} \quad \rho = \lim_{n\to\infty} \left|\dfrac{a_{n+1}}{a_n}\right| = \infty$$

Because $\rho > 1$, the series diverges by the Ratio Test.

71. $\sum_{n=1}^{\infty} \left(\dfrac{n}{2}\right)^n \dfrac{1}{n!}$

SOLUTION With $a_n = \left(\dfrac{n}{2}\right)^n \dfrac{1}{n!}$,

$$\left|\dfrac{a_{n+1}}{a_n}\right| = \left(\dfrac{n+1}{2}\right)^{n+1} \dfrac{1}{(n+1)!} \cdot \left(\dfrac{2}{n}\right)^n n! = \dfrac{1}{2}\left(\dfrac{n+1}{n}\right)^n = \dfrac{1}{2}\left(1+\dfrac{1}{n}\right)^n$$

and

$$\rho = \lim_{n\to\infty} \left|\dfrac{a_{n+1}}{a_n}\right| = \dfrac{1}{2}e$$

Because $\rho = \dfrac{e}{2} > 1$, the series diverges by the Ratio Test.

In Exercises 73–76, apply the Root Test to determine convergence or divergence, or state that the Root Test is inconclusive.

73. $\sum_{n=1}^{\infty} \dfrac{1}{4^n}$

SOLUTION With $a_n = \dfrac{1}{4^n}$,

$$L = \lim_{n\to\infty} \sqrt[n]{a_n} = \lim_{n\to\infty} \sqrt[n]{\dfrac{1}{4^n}} = \dfrac{1}{4}$$

Because $L < 1$, the series converges by the Root Test.

75. $\sum_{n=1}^{\infty} \left(\dfrac{3}{4n}\right)^n$

SOLUTION With $a_n = \left(\dfrac{3}{4n}\right)^n$,

$$L = \lim_{n\to\infty} \sqrt[n]{a_n} = \lim_{n\to\infty} \sqrt[n]{\left(\dfrac{3}{4n}\right)^n} = \lim_{n\to\infty} \dfrac{3}{4n} = 0$$

Because $L < 1$, the series converges by the Root Test.

In Exercises 77–100, determine convergence or divergence using any method covered in the text.

77. $\sum_{n=1}^{\infty} \left(\dfrac{2}{3}\right)^n$

SOLUTION This is a geometric series with ratio $r = \dfrac{2}{3} < 1$; hence, the series converges.

79. $\sum_{n=1}^{\infty} e^{-0.02n}$

SOLUTION This is a geometric series with common ratio $r = \dfrac{1}{e^{0.02}} \approx 0.98 < 1$; hence, the series converges.

81. $\sum_{n=1}^{\infty} \dfrac{(-1)^{n-1}}{\sqrt{n}+\sqrt{n+1}}$

SOLUTION In this alternating series, $a_n = \dfrac{1}{\sqrt{n}+\sqrt{n+1}}$. The sequence $\{a_n\}$ is decreasing, and

$$\lim_{n\to\infty} a_n = 0$$

therefore the series converges by the Alternating Series Test.

83. $\sum_{n=2}^{\infty} \dfrac{(-1)^n}{\ln n}$

SOLUTION The sequence $a_n = \dfrac{1}{\ln n}$ is decreasing for $n \geq 10$ and

$$\lim_{n \to \infty} a_n = 0$$

therefore, the series converges by the Alternating Series Test.

85. $\sum_{n=2}^{\infty} \dfrac{n}{1 + 100n}$

SOLUTION Note that

$$\lim_{n \to \infty} \dfrac{n}{1 + 100n} = \lim_{n \to \infty} \dfrac{1}{1/n + 100} = \dfrac{1}{100}$$

Therefore the terms of the sequence do not converge to zero, so this series is divergent since it fails the Divergence Test.

87. $\sum_{n=1}^{\infty} \dfrac{\cos n}{n^{3/2}}$

SOLUTION Note that this is not a positive series, since $\cos n$ may be either positive or negative. However, $|\cos n| \leq 1$, so that

$$\left| \dfrac{\cos n}{n^{3/2}} \right| \leq \dfrac{1}{n^{3/2}}$$

But $\sum_{n=1}^{\infty} \dfrac{1}{n^{3/2}}$ is a p-series with $p = \dfrac{3}{2} > 1$, so it is convergent. Thus the original series is absolutely convergent by the Direct Comparison Test.

89. $\sum_{n=1}^{\infty} \left(\dfrac{n}{5n+2} \right)^n$

SOLUTION Since $5n + 2 > 5n$, it follows that $\dfrac{n}{5n+2} < \dfrac{n}{5n} = \dfrac{1}{5}$. Since $\sum_{n=1}^{\infty} \left(\dfrac{1}{5} \right)^n$ is a geometric series with ratio $\dfrac{1}{5}$, it converges, so the original series converges as well by the Direct Comparison Test.

91. $\sum_{n=1}^{\infty} \dfrac{1}{n\sqrt{n} + \ln n}$

SOLUTION For $n \geq 1$,

$$\dfrac{1}{n\sqrt{n} + \ln n} \leq \dfrac{1}{n\sqrt{n}} = \dfrac{1}{n^{3/2}}$$

The series $\sum_{n=1}^{\infty} \dfrac{1}{n^{3/2}}$ is a convergent p-series, so the series $\sum_{n=1}^{\infty} \dfrac{1}{n\sqrt{n} + \ln n}$ converges by the Direct Comparison Test.

93. $\sum_{n=1}^{\infty} \left(\dfrac{1}{\sqrt{n}} - \dfrac{1}{\sqrt{n+1}} \right)$

SOLUTION This series telescopes:

$$\sum_{n=1}^{\infty} \left(\dfrac{1}{\sqrt{n}} - \dfrac{1}{\sqrt{n+1}} \right) = \left(1 - \dfrac{1}{\sqrt{2}} \right) + \left(\dfrac{1}{\sqrt{2}} - \dfrac{1}{\sqrt{3}} \right) + \left(\dfrac{1}{\sqrt{3}} - \dfrac{1}{\sqrt{4}} \right) + \ldots$$

so that the nth partial sum S_n is

$$S_n = \left(1 - \dfrac{1}{\sqrt{2}} \right) + \left(\dfrac{1}{\sqrt{2}} - \dfrac{1}{\sqrt{3}} \right) + \left(\dfrac{1}{\sqrt{3}} - \dfrac{1}{\sqrt{4}} \right) + \cdots + \left(\dfrac{1}{\sqrt{n}} - \dfrac{1}{\sqrt{n+1}} \right) = 1 - \dfrac{1}{\sqrt{n+1}}$$

and then

$$\sum_{n=1}^{\infty}\left(\frac{1}{\sqrt{n}}-\frac{1}{\sqrt{n+1}}\right) = \lim_{n\to\infty} S_n = 1 - \lim_{n\to\infty}\frac{1}{\sqrt{n+1}} = 1$$

95. $\sum_{n=1}^{\infty}\frac{1}{n+\sqrt{n}}$

SOLUTION For $n \geq 101$, the sequence $\frac{1}{n-100.1}$ is positive and

$$\frac{1}{n-100.1} > \frac{1}{n}$$

Now, $\sum_{n=101}^{\infty}\frac{1}{n}$ diverges, so $\sum_{n=101}^{\infty}\frac{1}{n-100.1}$ diverges by the Direct Comparison Test. Since adding the finite sum $\sum_{n=1}^{100}\frac{1}{n-100.1}$ does not affect the convergence of the series, it follows that $\sum_{n=1}^{\infty}\frac{1}{n-100.1}$ diverges as well.

97. $\sum_{n=2}^{\infty}\frac{1}{n^{\ln n}}$

SOLUTION For $n \geq N$ large enough, $\ln n \geq 2$ so that

$$\sum_{n=N}^{\infty}\frac{1}{n^{\ln n}} \leq \sum_{n=N}^{\infty}\frac{1}{n^2}$$

which is a convergent p-series. Thus by the Comparison Test, $\sum_{n=N}^{\infty}\frac{1}{n^{\ln n}}$ also converges; adding back in the terms for $n < N$ does not affect convergence.

99. $\sum_{n=1}^{\infty}\sin^2\frac{\pi}{n}$

SOLUTION For all $x > 0$, $\sin x < x$. Therefore, $\sin^2 x < x^2$, and for $x = \frac{\pi}{n}$,

$$\sin^2\frac{\pi}{n} < \frac{\pi^2}{n^2} = \pi^2 \cdot \frac{1}{n^2}$$

The series $\sum_{n=1}^{\infty}\frac{1}{n^2}$ is a convergent p-series, so the series $\sum_{n=1}^{\infty}\sin^2\frac{\pi}{n}$ also converges by the Direct Comparison Test.

In Exercises 101–106, find the interval of convergence of the power series.

101. $\sum_{n=0}^{\infty}\frac{2^n x^n}{n!}$

SOLUTION With $a_n = \frac{2^n x^n}{n!}$,

$$\rho = \lim_{n\to\infty}\left|\frac{a_{n+1}}{a_n}\right| = \lim_{n\to\infty}\left|\frac{2^{n+1}x^{n+1}}{(n+1)!} \cdot \frac{n!}{2^n x^n}\right| = |x|\lim_{n\to\infty}\frac{2}{n+1} = 0$$

The radius of convergence is therefore $R = \infty$, and the series converges for all x.

103. $\sum_{n=0}^{\infty}\frac{n^6}{n^8+1}(x-3)^n$

SOLUTION With $a_n = \frac{n^6(x-3)^n}{n^8+1}$,

$$\rho = \lim_{n\to\infty}\left|\frac{a_{n+1}}{a_n}\right| = \lim_{n\to\infty}\left|\frac{(n+1)^6(x-3)^{n+1}}{(n+1)^8+1} \cdot \frac{n^8+1}{n^6(x-3)^n}\right|$$

$$= |x-3|\lim_{n\to\infty}\frac{(n+1)^6(n^8+1)}{((n+1)^8+1)n^6} = |x-3|\lim_{n\to\infty}\frac{n^{14}+\cdots}{n^{14}+\cdots} = |x-3|$$

Thus $\rho < 1$ when $|x-3| < 1$, so the radius of convergence is $R = 1$ centered at $x = 3$, and the series converges absolutely for $2 < x < 4$. For the endpoint $x = 4$, the series becomes $\sum_{n=0}^{\infty} \frac{n^6}{n^8 + 1}$, which converges by the Direct Comparison Test comparing with the convergent p-series $\sum_{n=1}^{\infty} \frac{1}{n^2}$. For the endpoint $x = 2$, the series becomes $\sum_{n=0}^{\infty} \frac{n^6(-1)^n}{n^8 + 1}$, which converges by the Alternating Series Test. The series $\sum_{n=0}^{\infty} \frac{n^6(x-3)^n}{n^8 + 1}$ therefore converges for $2 \le x \le 4$.

105. $\sum_{n=0}^{\infty} (nx)^n$

SOLUTION With $a_n = n^n x^n$,

$$\rho = \lim_{n \to \infty} \left| \frac{a_{n+1}}{a_n} \right| = \lim_{n \to \infty} \left| \frac{(n+1)^{n+1} x^{n+1}}{n^n x^n} \right| = |x| \lim_{n \to \infty} \frac{(n+1)^{n+1}}{n^n} = |x| \lim_{n \to \infty} \left((n+1) \left(1 + \frac{1}{n} \right)^n \right) = \infty \text{ if } x \ne 0$$

Thus $\rho = \infty$ unless $x = 0$, so the series converges only at $x = 0$.

107. Expand $f(x) = \frac{2}{4 - 3x}$ as a power series centered at $c = 0$. Determine the values of x for which the series converges.

SOLUTION Write

$$\frac{2}{4 - 3x} = \frac{1}{2} \frac{1}{1 - \frac{3}{4}x}$$

Substituting $\frac{3}{4}x$ for x in the Maclaurin series for $\frac{1}{1-x}$, we obtain

$$\frac{1}{1 - \frac{3}{4}x} = \sum_{n=0}^{\infty} \left(\frac{3}{4} \right)^n x^n$$

This series converges for $\left| \frac{3}{4}x \right| < 1$, or $|x| < \frac{4}{3}$. Hence, for $|x| < \frac{4}{3}$,

$$\frac{2}{4 - 3x} = \frac{1}{2} \sum_{n=0}^{\infty} \left(\frac{3}{4} \right)^n x^n$$

109. Let $F(x) = \sum_{k=0}^{\infty} \frac{x^{2k}}{2^k \cdot k!}$.

(a) Show that $F(x)$ has infinite radius of convergence.

(b) Show that $y = F(x)$ is a solution of

$$y'' = xy' + y, \qquad y(0) = 1, \qquad y'(0) = 0$$

(c) *CAS* Plot the partial sums S_N for $N = 1, 3, 5, 7$ on the same set of axes.

SOLUTION

(a) With $a_k = \frac{x^{2k}}{2^k \cdot k!}$,

$$\left| \frac{a_{k+1}}{a_k} \right| = \frac{|x|^{2k+2}}{2^{k+1} \cdot (k+1)!} \cdot \frac{2^k \cdot k!}{|x|^{2k}} = \frac{x^2}{2(k+1)}$$

and

$$\rho = \lim_{n \to \infty} \left| \frac{a_{k+1}}{a_k} \right| = x^2 \cdot 0 = 0$$

Because $\rho < 1$ for all x, we conclude that the series converges for all x; that is, $R = \infty$.

(b) Let

$$y = F(x) = \sum_{k=0}^{\infty} \frac{x^{2k}}{2^k \cdot k!}$$

Then

$$y' = \sum_{k=1}^{\infty} \frac{2kx^{2k-1}}{2^k k!} = \sum_{k=1}^{\infty} \frac{x^{2k-1}}{2^{k-1}(k-1)!}$$

$$y'' = \sum_{k=1}^{\infty} \frac{(2k-1)x^{2k-2}}{2^{k-1}(k-1)!}$$

and

$$xy' + y = x\sum_{k=1}^{\infty} \frac{x^{2k-1}}{2^{k-1}(k-1)!} + \sum_{k=0}^{\infty} \frac{x^{2k}}{2^k k!} = \sum_{k=1}^{\infty} \frac{x^{2k}}{2^{k-1}(k-1)!} + 1 + \sum_{k=1}^{\infty} \frac{x^{2k}}{2^k k!}$$

$$= 1 + \sum_{k=1}^{\infty} \frac{(2k+1)x^{2k}}{2^k k!} = \sum_{k=0}^{\infty} \frac{(2k+1)x^{2k}}{2^k k!} = \sum_{k=1}^{\infty} \frac{(2k-1)x^{2k-2}}{2^{k-1}(k-1)!} = y''$$

Moreover,

$$y(0) = 1 + \sum_{k=1}^{\infty} \frac{0^{2k}}{2^k k!} = 1 \quad \text{and} \quad y'(0) = \sum_{k=1}^{\infty} \frac{0^{2k-1}}{2^{k-1}(k-1)!} = 0$$

Thus, $\sum_{k=0}^{\infty} \frac{x^{2k}}{2^k k!}$ is the solution to the equation $y'' = xy' + y$ satisfying $y(0) = 1$, $y'(0) = 0$.

(c) The partial sums S_1, S_3, S_5, and S_7 are plotted in the figure below.

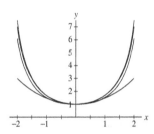

111. Use power series to evaluate $\lim\limits_{x \to 0} \dfrac{x^2 e^x}{\cos x - 1}$.

SOLUTION The relevant Maclaurin series are

$$e^x = 1 + x + \frac{x^2}{2!} + \frac{x^3}{3!} + \frac{x^4}{4!} + \cdots$$

$$\cos x = 1 - \frac{x^2}{2!} + \frac{x^4}{4!} - \frac{x^6}{6!} + \cdots$$

Hence the limit becomes

$$\lim_{x \to 0} \frac{x^2 e^x}{\cos x - 1} = \lim_{x \to 0} \frac{x^2\left(1 + x + \frac{x^2}{2!} + \frac{x^3}{3!} + \frac{x^4}{4!} + \cdots\right)}{\left(1 - \frac{x^2}{2!} + \frac{x^4}{4!} - \frac{x^6}{6!} + \cdots\right) - 1}$$

$$= \lim_{x \to 0} \frac{x^2\left(1 + x + \frac{x^2}{2!} + \frac{x^3}{3!} + \frac{x^4}{4!} + \cdots\right)}{-\frac{x^2}{2!} + \frac{x^4}{4!} - \frac{x^6}{6!} + \cdots}$$

$$= \lim_{x \to 0} \frac{1 + x + \frac{x^2}{2!} + \frac{x^3}{3!} + \frac{x^4}{4!} + \cdots}{-\frac{1}{2!} + \frac{x^2}{4!} - \frac{x^4}{6!} + \cdots}$$

$$= \frac{1}{-1/2} = -2$$

In Exercises 113–118, find the Taylor polynomial at $x = a$ for the given function.

113. $f(x) = x^3$, T_3, $a = 1$

SOLUTION We start by computing the first three derivatives of $f(x) = x^3$:

$$f'(x) = 3x^2$$
$$f''(x) = 6x$$
$$f'''(x) = 6$$

Evaluating the function and its derivatives at $x = 1$, we find

$$f(1) = 1, \ f'(1) = 3, \ f''(1) = 6, \ f'''(1) = 6$$

Therefore,

$$T_3(x) = f(1) + f'(1)(x-1) + \frac{f''(1)}{2!}(x-2)^2 + \frac{f'''(1)}{3!}(x-1)^3$$
$$= 1 + 3(x-1) + \frac{6}{2!}(x-2)^2 + \frac{6}{3!}(x-1)^3$$
$$= 1 + 3(x-1) + 3(x-2)^2 + (x-1)^3$$

115. $f(x) = x \ln(x)$, T_4, $a = 1$

SOLUTION We start by computing the first four derivatives of $f(x) = x \ln x$:

$$f'(x) = \ln x + x \cdot \frac{1}{x} = \ln x + 1$$
$$f''(x) = \frac{1}{x}$$
$$f'''(x) = -\frac{1}{x^2}$$
$$f^{(4)}(x) = \frac{2}{x^3}$$

Evaluating the function and its derivatives at $x = 1$, we find

$$f(1) = 0, \ f'(1) = 1, \ f''(1) = 1, \ f'''(1) = -1, f^{(4)}(1) = 2$$

Therefore,

$$T_4(x) = f(1) + f'(1)(x-1) + \frac{f''(1)}{2!}(x-1)^2 + \frac{f'''(1)}{3!}(x-1)^3 + \frac{f^{(4)}(1)}{4!}(x-1)^4$$
$$= 0 + 1(x-1) + \frac{1}{2!}(x-1)^2 - \frac{1}{3!}(x-1)^3 + \frac{2}{4!}(x-1)^4$$
$$= (x-1) + \frac{1}{2}(x-1)^2 - \frac{1}{6}(x-1)^3 + \frac{1}{12}(x-1)^4$$

117. $f(x) = xe^{-x^2}$, T_4, $a = 0$

SOLUTION We start by computing the first four derivatives of $f(x) = xe^{-x^2}$:

$$f'(x) = e^{-x^2} + x \cdot (-2x)e^{-x^2} = (1 - 2x^2)e^{-x^2}$$
$$f''(x) = -4xe^{-x^2} + (1 - 2x^2) \cdot (-2x)e^{-x^2} = (4x^3 - 6x)e^{-x^2}$$
$$f'''(x) = (12x^2 - 6)e^{-x^2} + (4x^3 - 6x) \cdot (-2x)e^{-x^2} = (-8x^4 + 24x^2 - 6)e^{-x^2}$$
$$f^{(4)}(x) = (-32x^3 + 48x)e^{-x^2} + (-8x^4 + 24x^2 - 6) \cdot (-2x)e^{-x^2} = (16x^5 - 80x^3 + 60x)e^{-x^2}$$

Evaluating the function and its derivatives at $x = 0$, we find

$$f(0) = 0, \ f'(0) = 1, \ f''(0) = 0, \ f'''(0) = -6, \ f^{(4)}(0) = 0$$

Therefore,

$$T_4(x) = f(0) + f'(0)x + \frac{f''(0)}{2!}x^2 + \frac{f'''(0)}{3!}x^3 + \frac{f^{(4)}(0)}{4!}x^4$$

$$= 0 + x + 0 \cdot x^2 - \frac{6}{3!}x^3 + 0 \cdot x^4 = x - x^3$$

119. Find the nth Maclaurin polynomial for $f(x) = e^{3x}$.

SOLUTION We differentiate the function $f(x) = e^{3x}$ repeatedly, looking for a pattern:

$$f'(x) = 3e^{3x} = 3^1 e^{3x}$$
$$f''(x) = 3 \cdot 3e^{3x} = 3^2 e^{3x}$$
$$f'''(x) = 3 \cdot 3^2 e^{3x} = 3^3 e^{3x}$$

Thus, for general n, $f^{(n)}(x) = 3^n e^{3x}$ and $f^{(n)}(0) = 3^n$. Substituting into the formula for the nth Taylor polynomial, we obtain:

$$T_n(x) = 1 + \frac{3x}{1!} + \frac{3^2 x^2}{2!} + \frac{3^3 x^3}{3!} + \frac{3^4 x^4}{4!} + \cdots + \frac{3^n x^n}{n!}$$

$$= 1 + 3x + \frac{1}{2!}(3x)^2 + \frac{1}{3!}(3x)^3 + \cdots + \frac{1}{n!}(3x)^n$$

121. Use the third Taylor polynomial of $f(x) = \tan^{-1} x$ at $a = 1$ to approximate $f(1.1)$. Use a calculator to determine the error.

SOLUTION We start by computing the first three derivatives of $f(x) = \tan^{-1} x$:

$$f'(x) = \frac{1}{1 + x^2}$$

$$f''(x) = -\frac{2x}{(1 + x^2)^2}$$

$$f'''(x) = \frac{-2(1 + x^2)^2 + 2x \cdot 2(1 + x^2) \cdot 2x}{(1 + x^2)^4} = \frac{2(3x^2 - 1)}{(1 + x^2)^3}$$

Evaluating the function and its derivatives at $x = 1$, we find

$$f(1) = \frac{\pi}{4}, \ f'(1) = \frac{1}{2}, \ f''(1) = -\frac{1}{2}, \ f'''(1) = \frac{1}{2}$$

Therefore,

$$T_3(x) = f(1) + f'(1)(x - 1) + \frac{f''(1)}{2!}(x - 1)^2 + \frac{f'''(1)}{3!}(x - 1)^3$$

$$= \frac{\pi}{4} + \frac{1}{2}(x - 1) - \frac{1}{4}(x - 1)^2 + \frac{1}{12}(x - 1)^3$$

Setting $x = 1.1$ yields:

$$T_3(1.1) = \frac{\pi}{4} + \frac{1}{2}(0.1) - \frac{1}{4}(0.1)^2 + \frac{1}{12}(0.1)^3 = 0.832981496$$

Using a calculator, we find $\tan^{-1} 1.1 = 0.832981266$. The error in the Taylor polynomial approximation is

$$|T_3(1.1) - \tan^{-1} 1.1| = |0.832981496 - 0.832981266| = 2.301 \times 10^{-7}$$

123. Find n such that $|e - T_n(1)| < 10^{-8}$, where T_n is the nth Maclaurin polynomial for $f(x) = e^x$.

SOLUTION Using the Error Bound, we have

$$|e - T_n(1)| \leq K \frac{|1 - 0|^{n+1}}{(n + 1)!} = \frac{K}{(n + 1)!}$$

where K is a number such that $\left|f^{(n+1)}(x)\right| = e^x \le K$ for all $0 \le x \le 1$. Since e^x is increasing, the maximum value on the interval $0 \le x \le 1$ is attained at the endpoint $x = 1$. Thus, for $0 \le u \le 1$, $e^u \le e^1 < 2.8$. Hence we may take $K = 2.8$ to obtain:

$$|e - T_n(1)| \le \frac{2.8}{(n+1)!}$$

We now choose n such that

$$\frac{2.8}{(n+1)!} < 10^{-8}$$

$$\frac{(n+1)!}{2.8} > 10^8$$

$$(n+1)! > 2.8 \times 10^8$$

For $n = 10$, $(n+1)! = 3.99 \times 10^7 < 2.8 \times 10^8$ and for $n = 11$, $(n+1)! = 4.79 \times 10^8 > 2.8 \times 10^8$. Hence, to make the error less than 10^{-8}, $n = 11$ is sufficient; that is,

$$|e - T_{11}(1)| < 10^{-8}$$

125. Verify that $T_n(x) = 1 + x + x^2 + \cdots + x^n$ is the nth Maclaurin polynomial of $f(x) = 1/(1-x)$. Show using substitution that the nth Maclaurin polynomial for $f(x) = 1/(1 - x/4)$ is

$$T_n(x) = 1 + \frac{1}{4}x + \frac{1}{4^2}x^2 + \cdots + \frac{1}{4^n}x^n$$

What is the nth Maclaurin polynomial for $g(x) = \dfrac{1}{1+x}$?

SOLUTION Let $f(x) = (1-x)^{-1}$. Then, $f'(x) = (1-x)^{-2}$, $f''(x) = 2(1-x)^{-3}$, $f'''(x) = 3!(1-x)^{-4}$, and, in general, $f^{(n)}(x) = n!(1-x)^{-(n+1)}$. Therefore, $f^{(n)}(0) = n!$ and

$$T_n(x) = 1 + \frac{1!}{1!}x + \frac{2!}{2!}x^2 + \cdots + \frac{n!}{n!}x^n = 1 + x + x^2 + \cdots + x^n$$

Upon substituting $x/4$ for x, we find that the nth Maclaurin polynomial for $f(x) = \dfrac{1}{1 - x/4}$ is

$$T_n(x) = 1 + \frac{1}{4}x + \frac{1}{4^2}x^2 + \cdots + \frac{1}{4^n}x^n$$

Substituting $-x$ for x, the nth Maclaurin polynomial for $g(x) = \dfrac{1}{1+x}$ is

$$T_n(x) = 1 - x + x^2 - x^3 + - \cdots + (-x)^n$$

In Exercises 127–136, find the Taylor series centered at c.

127. $f(x) = e^{4x}$, $c = 0$

SOLUTION Substituting $4x$ for x in the Maclaurin series for e^x yields

$$e^{4x} = \sum_{n=0}^{\infty} \frac{(4x)^n}{n!} = \sum_{n=0}^{\infty} \frac{4^n}{n!}x^n$$

129. $f(x) = x^4$, $c = 2$

SOLUTION We have

$$f'(x) = 4x^3 \quad f''(x) = 12x^2 \quad f'''(x) = 24x \quad f^{(4)}(x) = 24$$

and all higher derivatives are zero, so that

$$f(2) = 2^4 = 16 \quad f'(2) = 4 \cdot 2^3 = 32 \quad f''(2) = 12 \cdot 2^2 = 48 \quad f'''(2) = 24 \cdot 2 = 48 \quad f^{(4)}(2) = 24$$

Thus the Taylor series centered at $c = 2$ is

$$\sum_{n=0}^{4} \frac{f^{(n)}(2)}{n!}(x-2)^n = 16 + \frac{32}{1!}(x-2) + \frac{48}{2!}(x-2)^2 + \frac{48}{3!}(x-2)^3 + \frac{24}{4!}(x-2)^4$$

$$= 16 + 32(x-2) + 24(x-2)^2 + 8(x-2)^3 + (x-2)^4$$

131. $f(x) = \sin x$, $c = \pi$

SOLUTION Because $\sin x = -\sin(x - \pi)$,

$$x \sin x = (x - \pi)\sin x + \pi \sin x = -(x - \pi)\sin(x - \pi) - \pi \sin(x - \pi)$$

Substituting $x - \pi$ for x in the Maclaurin for $\sin x$ yields

$$\sin(x - \pi) = \sum_{n=0}^{\infty}(-1)^n \frac{(x - \pi)^{2n+1}}{(2n+1)!}$$

Thus,

$$x \sin x = -(x - \pi)\sum_{n=0}^{\infty}(-1)^n \frac{(x-\pi)^{2n+1}}{(2n+1)!} - \pi \sum_{n=0}^{\infty}(-1)^n \frac{(x-\pi)^{2n+1}}{(2n+1)!}$$

$$= \sum_{n=0}^{\infty}(-1)^{n+1}\frac{(x-\pi)^{2n+2}}{(2n+1)!} + \pi \sum_{n=0}^{\infty}(-1)^{n+1}\frac{(x-\pi)^{2n+1}}{(2n+1)!}$$

133. $f(x) = \dfrac{1}{1 - 2x}$, $c = -2$

SOLUTION Write

$$\frac{1}{1-2x} = \frac{1}{5 - 2(x+2)} = \frac{1}{5} \frac{1}{1 - \frac{2}{5}(x+2)}$$

Substituting $\frac{2}{5}(x+2)$ for x in the Maclaurin series for $\frac{1}{1-x}$ yields

$$\frac{1}{1 - \frac{2}{5}(x+2)} = \sum_{n=0}^{\infty} \frac{2^n}{5^n}(x+2)^n$$

hence,

$$\frac{1}{1 - 2x} = \frac{1}{5}\sum_{n=0}^{\infty} \frac{2^n}{5^n}(x+2)^n = \sum_{n=0}^{\infty} \frac{2^n}{5^{n+1}}(x+2)^n$$

135. $f(x) = \ln \dfrac{x}{2}$, $c = 2$

SOLUTION Write

$$\ln \frac{x}{2} = \ln\left(\frac{(x-2)+2}{2}\right) = \ln\left(1 + \frac{x-2}{2}\right)$$

Substituting $\frac{x-2}{2}$ for x in the Maclaurin series for $\ln(1 + x)$ yields

$$\ln \frac{x}{2} = \sum_{n=1}^{\infty} \frac{(-1)^{n+1}\left(\frac{x-2}{2}\right)^n}{n} = \sum_{n=1}^{\infty} \frac{(-1)^{n+1}(x-2)^n}{n \cdot 2^n}$$

This series is valid for $|x - 2| < 2$.

In Exercises 137–140, find the first three terms of the Maclaurin series of $f(x)$ and use it to calculate $f^{(3)}(0)$.

137. $f(x) = (x^2 - x)e^{x^2}$

SOLUTION Substitute x^2 for x in the Maclaurin series for e^x to get

$$e^{x^2} = 1 + x^2 + \frac{1}{2}x^4 + \frac{1}{6}x^6 + \dots$$

so that the Maclaurin series for $f(x)$ is

$$(x^2 - x)e^{x^2} = x^2 + x^4 + \frac{1}{2}x^6 + \dots - x - x^3 - \frac{1}{2}x^5 - \dots = -x + x^2 - x^3 + x^4 + \dots$$

The coefficient of x^3 is

$$\frac{f'''(0)}{3!} = -1$$

so that $f'''(0) = -6$.

139. $f(x) = \dfrac{1}{1 + \tan x}$

SOLUTION Substitute $-\tan x$ in the Maclaurin series for $\frac{1}{1-x}$ to get

$$\frac{1}{1 + \tan x} = 1 - \tan x + (\tan x)^2 - (\tan x)^3 + \dots$$

We have not yet encountered the Maclaurin series for $\tan x$. We need only the terms up through x^3, so compute

$$\tan'(x) = \sec^2 x \quad \tan''(x) = 2(\tan x)\sec^2 x \quad \tan'''(x) = 2(1 + \tan^2 x)\sec^2 x + 4(\tan^2 x)\sec^2 x$$

so that

$$\tan'(0) = 1 \quad \tan''(0) = 0 \quad \tan'''(0) = 2$$

Then the Maclaurin series for $\tan x$ is

$$\tan x = \tan 0 + \frac{\tan'(0)}{1!}x + \frac{\tan''(0)}{2!}x^2 + \frac{\tan'''(0)}{3!}x^3 + \dots = x + \frac{1}{3}x^3 + \dots$$

Substitute these into the series above to get

$$\frac{1}{1 + \tan x} = 1 - \left(x + \frac{1}{3}x^3\right) + \left(x + \frac{1}{3}x^3\right)^2 - \left(x + \frac{1}{3}x^3\right)^3 + \dots$$

$$= 1 - x - \frac{1}{3}x^3 + x^2 - x^3 + \text{higher degree terms}$$

$$= 1 - x + x^2 - \frac{4}{3}x^3 + \text{higher degree terms}$$

The coefficient of x^3 is

$$\frac{f'''(0)}{3!} = -\frac{4}{3}$$

so that

$$f'''(0) = -6 \cdot \frac{4}{3} = -8$$

141. Calculate $\dfrac{\pi}{2} - \dfrac{\pi^3}{2^3 3!} + \dfrac{\pi^5}{2^5 5!} - \dfrac{\pi^7}{2^7 7!} + \dots$.

SOLUTION We recognize that

$$\frac{\pi}{2} - \frac{\pi^3}{2^3 3!} + \frac{\pi^5}{2^5 5!} - \frac{\pi^7}{2^7 7!} + \dots = \sum_{n=0}^{\infty} (-1)^n \frac{(\pi/2)^{2n+1}}{(2n+1)!}$$

is the Maclaurin series for $\sin x$ with x replaced by $\pi/2$. Therefore,

$$\frac{\pi}{2} - \frac{\pi^3}{2^3 3!} + \frac{\pi^5}{2^5 5!} - \frac{\pi^7}{2^7 7!} + \dots = \sin \frac{\pi}{2} = 1$$

11 PARAMETRIC EQUATIONS, POLAR COORDINATES, AND CONIC SECTIONS

11.1 Parametric Equations (LT Section 12.1)

Preliminary Questions

1. Describe the shape of the curve $x = 3\cos t, y = 3\sin t$.

SOLUTION For all t,

$$x^2 + y^2 = (3\cos t)^2 + (3\sin t)^2 = 9(\cos^2 t + \sin^2 t) = 9 \cdot 1 = 9$$

therefore the curve is on the circle $x^2 + y^2 = 9$. Also, each point on the circle $x^2 + y^2 = 9$ can be represented in the form $(3\cos t, 3\sin t)$ for some value of t. We conclude that the curve $x = 3\cos t, y = 3\sin t$ is the circle of radius 3 centered at the origin.

2. How does $x = 4 + 3\cos t, y = 5 + 3\sin t$ differ from the curve in the previous question?

SOLUTION In this case we have

$$(x - 4)^2 + (y - 5)^2 = (3\cos t)^2 + (3\sin t)^2 = 9(\cos^2 t + \sin^2 t) = 9 \cdot 1 = 9$$

Therefore, the given equations parametrize the circle of radius 3 centered at the point $(4, 5)$.

3. What is the maximum height of a particle whose path has parametric equations $x = t^9, y = 4 - t^2$?

SOLUTION The particle's height is $y = 4 - t^2$. To find the maximum height we set the derivative equal to zero and solve:

$$\frac{dy}{dt} = \frac{d}{dt}(4 - t^2) = -2t = 0 \quad \text{or} \quad t = 0$$

The maximum height is $y(0) = 4 - 0^2 = 4$.

4. Can the parametric curve $(t, \sin t)$ be represented as a graph $y = f(x)$? What about $(\sin t, t)$?

SOLUTION In the parametric curve $(t, \sin t)$ we have $x = t$ and $y = \sin t$; therefore, $y = \sin x$. That is, the curve can be represented as a graph of a function. In the parametric curve $(\sin t, t)$ we have $x = \sin t, y = t$; therefore, $x = \sin y$. This equation does not define y as a function of x; therefore, the parametric curve $(\sin t, t)$ cannot be represented as a graph of a function $y = f(x)$.

5. (a) Describe the path of an ant that is crawling along the plane according to $c_1(t) = (f(t), f(t))$, where $f(t)$ is an increasing function.
(b) Compare that path to the path of a second ant crawling according to $c_2(t) = f(2t), f(2t)$.

SOLUTION

(a) At any time t, the ant's x-coordinate is equal to its y-coordinate, since they are both equal to $f(t)$. Thus the ant is always on the line $y = x$; since $f(t)$ is increasing, the ant is always moving up and to the right.

(b) At any time t, the ant's x-coordinate is equal to its y-coordinate, since they are both equal to $f(t)$. Thus the ant is always on the line $y = x$; since $f(t)$ is increasing, the ant is always moving up and to the right. This is the same path as in part (a), but the ant is moving at a different speed. For example, from $t = 0$ to $t = 1$, the first ant moves from $(f(0), f(0))$ to $(f(1), f(1))$, while the second moves from $(f(0), f(0))$ to $(f(2), f(2))$, passing through $(f(1), f(1))$ when $t = \frac{1}{2}$.

6. Find three different parametrizations of the graph of $y = x^3$.

SOLUTION For example, if we let $x = t$, we get the parametrization (t, t^3). Letting $x = 2t$ gives $(2t, 8t^3)$. Finally, let $y = t$, then the parametrization is $(t^{1/3}, t)$.

7. Match the derivatives with a verbal description:

(a) $\dfrac{dx}{dt}$ (b) $\dfrac{dy}{dt}$ (c) $\dfrac{dy}{dx}$

(i) Slope of the tangent line to the curve
(ii) Vertical rate of change with respect to time
(iii) Horizontal rate of change with respect to time

SOLUTION

(a) The derivative $\dfrac{dx}{dt}$ is the horizontal rate of change with respect to time.

(b) The derivative $\dfrac{dy}{dt}$ is the vertical rate of change with respect to time.

(c) The derivative $\dfrac{dy}{dx}$ is the slope of the tangent line to the curve.

Hence, (a) ↔ (iii), (b) ↔ (ii), (c) ↔ (i).

Exercises

1. Find the coordinates at times $t = 0, 2, 4$ of a particle following the path $x = 1 + t^3$, $y = 9 - 3t^2$.

SOLUTION Substituting $t = 0$, $t = 2$, and $t = 4$ into $x = 1 + t^3$, $y = 9 - 3t^2$ gives the coordinates of the particle at these times respectively. That is,

$$(t = 0) \quad x = 1 + 0^3 = 1, \ y = 9 - 3 \cdot 0^2 = 9 \quad \Rightarrow (1, 9)$$
$$(t = 2) \quad x = 1 + 2^3 = 9, \ y = 9 - 3 \cdot 2^2 = -3 \quad \Rightarrow (9, -3)$$
$$(t = 4) \quad x = 1 + 4^3 = 65, \ y = 9 - 3 \cdot 4^2 = -39 \quad \Rightarrow (65, -39)$$

3. Show that the path traced by the model rocket in Example 3 is a parabola by eliminating the parameter.

SOLUTION The path traced by the model rocket is given by the following parametric equations:

$$x = 200t, \ y = 400t - 16t^2$$

We eliminate the parameter. Since $x = 200t$, we have $t = \dfrac{x}{200}$. Substituting into the equation for y we obtain:

$$y = 400t - 16t^2 = 400 \cdot \dfrac{x}{200} - 16\left(\dfrac{x}{200}\right)^2 = 2x - \dfrac{x^2}{2500}$$

The equation $y = -\dfrac{x^2}{2500} + 2x$ is the equation of a parabola.

5. Graph the parametric curves. Include arrows indicating the direction of motion.

(a) (t, t), $-\infty < t < \infty$
(b) $(\sin t, \sin t)$, $0 \le t \le 2\pi$
(c) (e^t, e^t), $-\infty < t < \infty$
(d) (t^3, t^3), $-1 \le t \le 1$

SOLUTION

(a) For the trajectory $c(t) = (t, t)$, $-\infty < t < \infty$ we have $y = x$. Also the two coordinates tend to ∞ and $-\infty$ as $t \to \infty$ and $t \to -\infty$ respectively. The graph is shown next:

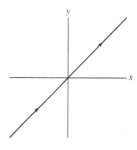

(b) For the curve $c(t) = (\sin t, \sin t)$, $0 \le t \le 2\pi$, we have $y = x$. $\sin t$ is increasing for $0 < t \le \frac{\pi}{2}$, decreasing for $\frac{\pi}{2} \le t \le \frac{3\pi}{2}$ and increasing again for $\frac{3\pi}{2} \le t < 2\pi$. Hence the particle moves from $c(0) = (0, 0)$ to $c(\frac{\pi}{2}) = (1, 1)$, then moves back to $c(\frac{3\pi}{2}) = (-1, -1)$ and then returns to $c(2\pi) = (0, 0)$. We obtain the following trajectory:

SECTION 11.1 | Parametric Equations (LT SECTION 12.1) 815

These three parts of the trajectory are shown together in the next figure:

(c) For the trajectory $c(t) = (e^t, e^t)$, $-\infty < t < \infty$, we have $y = x$. However since $\lim\limits_{t \to -\infty} e^t = 0$ and $\lim\limits_{t \to \infty} e^t = \infty$, the trajectory is the part of the line $y = x$, $0 < x$.

(d) For the trajectory $c(t) = (t^3, t^3)$, $-1 \leq t \leq 1$, we have again $y = x$. Since the function t^3 is increasing the particle moves in one direction starting at $((-1)^3, (-1)^3) = (-1, -1)$ and ending at $(1^3, 1^3) = (1, 1)$. The trajectory is shown next:

In Exercises 7–14, express in the form $y = f(x)$ by eliminating the parameter.

7. $x = t + 3$, $\quad y = 4t$

SOLUTION We eliminate the parameter. Since $x = t + 3$, we have $t = x - 3$. Substituting into $y = 4t$ we obtain
$$y = 4t = 4(x - 3) \Rightarrow y = 4x - 12$$

9. $x = t^3 - 1$, $\quad y = t^2 + 1$

SOLUTION From $x = t^3 - 1$, we get $t = \sqrt[3]{x + 1}$. Substituting into $y = t^2 + 1$ we obtain
$$y = t^2 + 1 = \left(\sqrt[3]{x + 1}\right)^2 + 1 = (x + 1)^{2/3} + 1$$

11. $x = e^{-2t}, \quad y = 6e^{4t}$

SOLUTION We eliminate the parameter. Since $x = e^{-2t}$, we have $-2t = \ln x$ or $t = -\frac{1}{2} \ln x$. Substituting in $y = 6e^{4t}$ we get

$$y = 6e^{4t} = 6e^{4 \cdot (-\frac{1}{2} \ln x)} = 6e^{-2 \ln x} = 6e^{\ln x^{-2}} = 6x^{-2} \Rightarrow y = \frac{6}{x^2}, \quad x > 0$$

13. $x = \ln t, \quad y = 2 - t$

SOLUTION Since $x = \ln t$ we have $t = e^x$. Substituting in $y = 2 - t$ we obtain $y = 2 - e^x$.

In Exercises 15–18, graph the curve and draw an arrow specifying the direction corresponding to motion.

15. $x = \frac{1}{2}t, \quad y = 2t^2$

SOLUTION Let $c(t) = (x(t), y(t)) = (\frac{1}{2}t, 2t^2)$. Then $c(-t) = (-x(t), y(t))$ so the curve is symmetric with respect to the y-axis. Also, the function $\frac{1}{2}t$ is increasing. Hence there is only one direction of motion on the curve. The corresponding function is the parabola $y = 2 \cdot (2x)^2 = 8x^2$. We obtain the following trajectory:

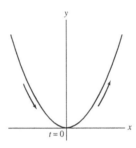

17. $x = \pi t, \quad y = \sin t$

SOLUTION We find the function by eliminating t. Since $x = \pi t$, we have $t = \frac{x}{\pi}$. Substituting $t = \frac{x}{\pi}$ into $y = \sin t$ we get $y = \sin \frac{x}{\pi}$. We obtain the following curve:

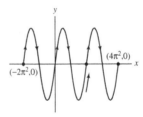

19. Match the parametrizations (a)–(d) with their plots in Figure 15, and draw an arrow indicating the direction of motion.

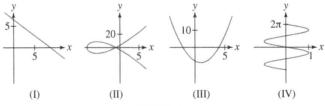

FIGURE 15

(a) $c(t) = (\sin t, -t)$
(b) $c(t) = (t^2 - 9, 8t - t^3)$
(c) $c(t) = (1 - t, t^2 - 9)$
(d) $c(t) = (4t + 2, 5 - 3t)$

SOLUTION

(a) In the curve $c(t) = (\sin t, -t)$ the x-coordinate is varying between -1 and 1 so this curve corresponds to plot IV. As t increases, the y-coordinate $y = -t$ is decreasing so the direction of motion is downward.

(IV) $c(t) = (\sin t, -t)$

(b) The curve $c(t) = (t^2 - 9, 8t - t^3)$ intersects the x-axis where $y = 8t - t^3$, or $t = 0$ and $t = \pm 2\sqrt{2}$. These correspond to the points $(-9, 0)$ and $(-1, 0)$. The y-intercepts are obtained where $x = t^2 - 9 = 0$, or $t = \pm 3$. These correspond to the points $(0, \pm 3)$. As t increases from $-\infty$ to 0, y starts out near ∞; as t gets large, towards ∞, y approaches $-\infty$. We obtain the following trajectory:

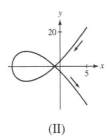

(II)

(c) The curve $c(t) = (1 - t, t^2 - 9)$ intersects the y-axis where $x = 1 - t = 0$, or $t = 1$. The y-intercept is $(0, -8)$. The x-intercepts are obtained where $t^2 - 9 = 0$ or $t = \pm 3$. These are the points $(-2, 0)$ and $(4, 0)$. Setting $t = 1 - x$ we get

$$y = t^2 - 9 = (1 - x)^2 - 9 = x^2 - 2x - 8$$

As t increases the x coordinate decreases and we obtain the following trajectory:

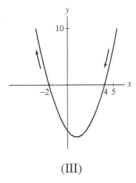

(III)

(d) The curve $c(t) = (4t + 2, 5 - 3t)$ is a straight line, since eliminating t in $x = 4t + 2$ and substituting in $y = 5 - 3t$ gives $y = 5 - 3 \cdot \frac{x-2}{4} = -\frac{3}{4}x + \frac{13}{2}$ which is the equation of a line. As t increases, the x coordinate $x = 4t + 2$ increases and the y-coordinate $y = 5 - 3t$ decreases. We obtain the following trajectory:

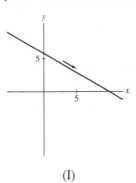

(I)

21. A particle follows the trajectory

$$x(t) = \frac{1}{4}t^3 + 2t, \qquad y(t) = 20t - t^2$$

with t in seconds and distance in centimeters.

(a) What is the particle's maximum height?

(b) When does the particle hit the ground and how far from the origin does it land?

SOLUTION

(a) To find the maximum height $y(t)$, we set the derivative of $y(t)$ equal to zero and solve:
$$\frac{dy}{dt} = \frac{d}{dt}(20t - t^2) = 20 - 2t = 0 \Rightarrow t = 10$$

The maximum height is $y(10) = 20 \cdot 10 - 10^2 = 100$ cm.

(b) The object hits the ground when its height is zero. That is, when $y(t) = 0$. Solving for t we get
$$20t - t^2 = t(20 - t) = 0 \Rightarrow t = 0, t = 20$$

$t = 0$ is the initial time, so the solution is $t = 20$. At time $t = 20$, when the object hits the ground, its x coordinate is $x(20) = \frac{1}{4} \cdot 20^3 + 2 \cdot 20 = 2040$. Thus, when it hits the ground, the object is 2040 cm far from the origin.

In Exercises 23–38, find parametric equations for the given curve.

23. $y = 9 - 4x$

SOLUTION This is a line through $P = (0, 9)$ with slope $m = -4$. Using the parametric representation of a line, as given in Example 3, we obtain $c(t) = (t, 9 - 4t)$.

25. $4x - y^2 = 5$

SOLUTION We define the parameter $t = y$. Then, $x = \frac{5 + y^2}{4} = \frac{5 + t^2}{4}$, giving us the parametrization $c(t) = \left(\frac{5 + t^2}{4}, t\right)$.

27. $(x + 9)^2 + (y - 4)^2 = 49$

SOLUTION This is a circle of radius 7 centered at $(-9, 4)$. Using the parametric representation of a circle we get $c(t) = (-9 + 7\cos t, 4 + 7\sin t)$.

29. Line of slope 8 through $(-4, 9)$

SOLUTION Using the parametric representation of a line given in Example 3, we get the parametrization $c(t) = (-4 + t, 9 + 8t)$.

31. Line through $(3, 1)$ and $(-5, 4)$

SOLUTION We use the two-point parametrization of a line with $P = (a, b) = (3, 1)$ and $Q = (c, d) = (-5, 4)$. Then $c(t) = (3 - 8t, 1 + 3t)$ for $-\infty < t < \infty$.

33. Segment joining $(1, 1)$ and $(2, 3)$

SOLUTION We use the two-point parametrization of a line with $P = (a, b) = (1, 1)$ and $Q = (c, d) = (2, 3)$. Then $c(t) = (1 + t, 1 + 2t)$; since we want only the segment joining the two points, we want $0 \le t \le 1$.

35. Circle of radius 4 with center $(3, 9)$

SOLUTION Substituting $(a, b) = (3, 9)$ and $R = 4$ in the parametric equation of the circle we get $c(t) = (3 + 4\cos t, 9 + 4\sin t)$.

37. $y = x^2$, translated so that the minimum occurs at $(-4, -8)$

SOLUTION We may parametrize $y = x^2$ by (t, t^2) for $-\infty < t < \infty$. The minimum of $y = x^2$ occurs at $(0, 0)$, so the desired curve is translated by $(-4, -8)$ from $y = x^2$. Thus a parametrization of the desired curve is $c(t) = (-4 + t, -8 + t^2)$.

In Exercises 39–42, find a parametrization $c(t)$ of the curve satisfying the given condition.

39. $y = 3x - 4$, $c(0) = (2, 2)$

SOLUTION Let $x(t) = t + a$ and $y(t) = 3x - 4 = 3(t + a) - 4$. We want $x(0) = 2$, thus we must use $a = 2$. Our line is $c(t) = (x(t), y(t)) = (t + 2, 3(t + 2) - 4) = (t + 2, 3t + 2)$.

41. $y = x^2$, $c(0) = (3, 9)$

SOLUTION Let $x(t) = t + a$ and $y(t) = x^2 = (t + a)^2$. We want $x(0) = 3$, thus we must use $a = 3$. Our curve is $c(t) = (x(t), y(t)) = (t + 3, (t + 3)^2) = (t + 3, t^2 + 6t + 9)$.

43. Find a parametrization of the top half of the ellipse $4x^2 + 5y^2 = 100$, starting at $(-5, 0)$ and ending at $(5, 0)$.

SOLUTION Dividing the given equation through by 100 gives

$$\frac{x^2}{25} + \frac{y^2}{20} = 1, \text{ or } \left(\frac{x}{5}\right)^2 + \left(\frac{y}{2\sqrt{5}}\right)^2 = 1$$

From Example 5, this curve is parametrized from $(5, 0)$ to $(-5, 0)$ by

$$c(t) = (5 \cos t, 2\sqrt{5} \sin t), \quad 0 \le t \le \pi$$

Since we want the parametrization to run in the reverse direction, we get

$$c(t) = (5\cos(\pi - t), 2\sqrt{5} \sin(\pi - t)), \text{ or } c(t) = (-5\cos t, 2\sqrt{5} \sin t), \quad 0 \le t \le \pi$$

45. Describe $c(t) = (\sec t, \tan t)$ for $0 \le t < \frac{\pi}{2}$ in the form $y = f(x)$. Specify the domain of x.

SOLUTION For $0 < t \le \frac{\pi}{2}$, both $\sec t$ and $\tan t$ are nonnegative. Using the identity $\tan^2 t = \sec^2 t - 1$, we have

$$y = \tan t \quad \Rightarrow \quad y^2 = \tan^2 t = \sec^2 t - 1 = x^2 - 1$$

Thus the equation is $y^2 = x^2 - 1$, or $y = \sqrt{x^2 - 1}$ (note that we take the positive square root because $y = \tan t \ge 0$ on the domain of t). Finally, note that the domain of x is $[0, \infty)$ since $x = \sec t \ge 0$ on the domain of t. A graph of this equation is that portion of the hyperbola $x^2 - y^2 = 1$ that lies in the first quadrant:

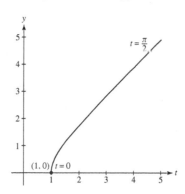

47. The graphs of $x(t)$ and $y(t)$ as functions of t are shown in Figure 16(A). Which of (I)–(III) is the plot of $c(t) = (x(t), y(t))$? Explain.

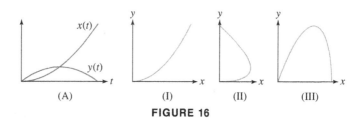

FIGURE 16

SOLUTION As seen in Figure 16(A), the x-coordinate is an increasing function of t, while $y(t)$ is first increasing and then decreasing. In Figure I, x and y are both increasing or both decreasing (depending on the direction on the curve). In Figure II, x does not maintain one tendency, rather, it is decreasing and increasing for certain values of t. The plot $c(t) = (x(t), y(t))$ is plot III.

49. Figure 18 shows a parametric curve $c(t) = (p(t), q(t))$ that models the changing population sizes of a predator (p) and its prey (q).

FIGURE 18

(a) Discuss how you expect the predator and prey populations to change as time increases at points C and D on the parametric curve.

(b) As functions of t, sketch graphs of $p(t)$ and $q(t)$ for three cycles around the parametric curve, beginning at point C.

(c) Both graphs in (b) should show oscillations between minimum and maximum values. Indicate which (predator or prey) has its peaks shortly after the other has its peaks, and explain why that makes sense in terms of an interaction between a predator and its prey.

SOLUTION

(a) At point C, the population of the prey is near its minimum, so we expect the predator population to decrease due to the absence of prey. Thus over time the population will move down and to the left, then down and to the right. At point D, both the predator and prey species will increase for a while until the predator population gets large enough that it starts to reduce the prey population.

(b) A graph is below, with the predator population shown in black and the prey population in red. Note that the prey population is near its minimum and the predator population is decreasing, as at point C.

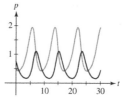

(c) From the graph in part (b), the predator population peaks shortly after the prey population peaks, since at that time food is plentiful.

In Exercises 51–58, use Eq. (6) to find dy/dx at the given point.

51. $(t^3, t^2 - 1)$, $t = -4$

SOLUTION By Eq. (6) we have

$$\frac{dy}{dx} = \frac{y'(t)}{x'(t)} = \frac{(t^2 - 1)'}{(t^3)'} = \frac{2t}{3t^2} = \frac{2}{3t}$$

Substituting $t = -4$ we get

$$\frac{dy}{dx} = \frac{2}{3t}\bigg|_{t=-4} = \frac{2}{3 \cdot (-4)} = -\frac{1}{6}$$

53. $(s^{-1} - 3s, s^3)$, $s = -1$

SOLUTION Using Eq. (6) we get

$$\frac{dy}{dx} = \frac{y'(s)}{x'(s)} = \frac{(s^3)'}{(s^{-1} - 3s)'} = \frac{3s^2}{-s^{-2} - 3} = \frac{3s^4}{-1 - 3s^2}$$

Substituting $s = -1$ we obtain

$$\frac{dy}{dx} = \frac{3s^4}{-1 - 3s^2}\bigg|_{s=-1} = \frac{3 \cdot (-1)^4}{-1 - 3 \cdot (-1)^2} = -\frac{3}{4}$$

55. $(\sin^3 \theta, \cos \theta)$, $\theta = \frac{\pi}{4}$

SOLUTION Using Eq. (6) we get

$$\frac{dy}{dx} = \frac{y'(\theta)}{x'(\theta)} = \frac{(\cos\theta)'}{(\sin^3\theta)'} = \frac{-\sin\theta}{3\sin^2\theta\cos\theta} = -\frac{1}{3\sin\theta\cos\theta}$$

Substituting $\theta = \frac{\pi}{4}$ we obtain

$$\frac{dy}{dx} = -\frac{1}{3\sin\theta\cos\theta}\bigg|_{\theta=\pi/4} = -\frac{1}{3\cdot\frac{1}{\sqrt{2}}\cdot\frac{1}{\sqrt{2}}} = -\frac{2}{3}$$

57. $(\ln t, \frac{1}{t})$, $t = 4$

SOLUTION Using Eq. (6) we get

$$\frac{dy}{dx} = \frac{y'(t)}{x'(t)} = \frac{(1/t)'}{(\ln t)'} = \frac{-1/t^2}{1/t} = -\frac{1}{t}$$

Substituting $t = 1$ we obtain

$$\frac{dy}{dx} = -\frac{1}{t}\bigg|_{t=1} = -1$$

In Exercises 59–64, find an equation $y = f(x)$ for the parametric curve and compute dy/dx in two ways: using Eq. (6) and by differentiating $f(x)$.

59. $c(t) = (2t + 1, 1 - 9t)$

SOLUTION Since $x = 2t + 1$, we have $t = \frac{x-1}{2}$. Substituting in $y = 1 - 9t$ we have

$$y = 1 - 9\left(\frac{x-1}{2}\right) = -\frac{9}{2}x + \frac{11}{2}$$

Differentiating $y = -\frac{9}{2}x + \frac{11}{2}$ gives $\frac{dy}{dx} = -\frac{9}{2}$. We now find $\frac{dy}{dx}$ using Eq. (6):

$$\frac{dy}{dx} = \frac{y'(t)}{x'(t)} = \frac{(1-9t)'}{(2t+1)'} = -\frac{9}{2}$$

61. $x = s^3$, $y = s^6 + s^{-3}$

SOLUTION We find y as a function of x:

$$y = s^6 + s^{-3} = \left(s^3\right)^2 + \left(s^3\right)^{-1} = x^2 + x^{-1}$$

We now differentiate $y = x^2 + x^{-1}$. This gives

$$\frac{dy}{dx} = 2x - x^{-2}$$

Alternatively, we can use Eq. (6) to obtain the following derivative:

$$\frac{dy}{dx} = \frac{y'(s)}{x'(s)} = \frac{\left(s^6 + s^{-3}\right)'}{(s^3)'} = \frac{6s^5 - 3s^{-4}}{3s^2} = 2s^3 - s^{-6}$$

Hence, since $x = s^3$,

$$\frac{dy}{dx} = 2x - x^{-2}$$

63. $x = 1 - e^t$, $y = t - 1$

SOLUTION Since $x = 1 - e^t$, we have $t = \ln(1 - x)$. Substituting in $y = t - 1$ we have

$$y = \ln(1 - x) - 1$$

Differentiating gives $\dfrac{dy}{dx} = -\dfrac{1}{1-x} = \dfrac{1}{x-1}$. We now find $\dfrac{dy}{dx}$ using Eq. (6):

$$\frac{dy}{dx} = \frac{y'(t)}{x'(t)} = \frac{(t-1)'}{(1-e^t)'} = \frac{1}{-e^t}$$

Since $x = 1 - e^t$, we have $-e^t = x - 1$, so this becomes

$$\frac{dy}{dx} = \frac{1}{x-1}$$

65. Find the points on the parametric curve $c(t) = (3t^2 - 2t, t^3 - 6t)$ where the tangent line has slope 3.

SOLUTION We solve

$$\frac{dy}{dx} = \frac{3t^2 - 6}{6t - 2} = 3$$

or $3t^2 - 6 = 18t - 6$, or $t^2 - 6t = 0$, so the slope is 3 at $t = 0, 6$ and the points are $(0, 0)$ and $(60, 190)$.

In Exercises 67–70, let $c(t) = (t^2 - 9, t^2 - 8t)$ (see Figure 19).

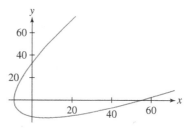

FIGURE 19 Plot of $c(t) = (t^2 - 9, t^2 - 8t)$.

67. Draw an arrow indicating the direction of motion, and determine the interval of t-values corresponding to the portion of the curve in each of the four quadrants.

SOLUTION We plot the functions $x(t) = t^2 - 9$ and $y(t) = t^2 - 8t$:

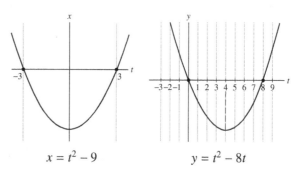

We note carefully where each of these graphs are positive or negative, increasing or decreasing. In particular, $x(t)$ is decreasing for $t < 0$, increasing for $t > 0$, positive for $|t| > 3$, and negative for $|t| < 3$. Likewise, $y(t)$ is decreasing for $t < 4$, increasing for $t > 4$, positive for $t > 8$ or $t < 0$, and negative for $0 < t < 8$. We now draw arrows on the path following the decreasing/increasing behavior of the coordinates as indicated above. We obtain:

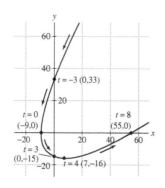

This plot also shows that:

- The graph is in the first quadrant for $t < -3$ or $t > 8$.
- The graph is in the second quadrant for $-3 < t < 0$.
- The graph is in the third quadrant for $0 < t < 3$.
- The graph is in the fourth quadrant for $3 < t < 8$.

69. Find the points where the tangent has slope $\frac{1}{2}$.

SOLUTION The slope of the tangent at t is

$$\frac{dy}{dx} = \frac{(t^2 - 8t)'}{(t^2 - 9)'} = \frac{2t - 8}{2t} = 1 - \frac{4}{t}$$

The point where the tangent has slope $\frac{1}{2}$ corresponds to the value of t that satisfies

$$\frac{dy}{dx} = 1 - \frac{4}{t} = \frac{1}{2} \Rightarrow \frac{4}{t} = \frac{1}{2} \Rightarrow t = 8$$

We substitute $t = 8$ in $x(t) = t^2 - 9$ and $y(t) = t^2 - 8t$ to obtain the following point:

$$\begin{aligned} x(8) &= 8^2 - 9 = 55 \\ y(8) &= 8^2 - 8 \cdot 8 = 0 \end{aligned} \Rightarrow (55, 0)$$

71. Let A and B be the points where the ray of angle θ intersects the two concentric circles of radii $r < R$ centered at the origin (Figure 20). Let P be the point of intersection of the horizontal line through A and the vertical line through B. Express the coordinates of P as a function of θ and describe the curve traced by P for $0 \leq \theta \leq 2\pi$.

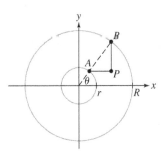

FIGURE 20

SOLUTION We use the parametric representation of a circle to determine the coordinates of the points A and B. That is,

$$A = (r\cos\theta, r\sin\theta), \quad B = (R\cos\theta, R\sin\theta)$$

The coordinates of P are therefore

$$P = (R\cos\theta, r\sin\theta)$$

In order to identify the curve traced by P, we notice that the x and y coordinates of P satisfy $\frac{x}{R} = \cos\theta$ and $\frac{y}{r} = \sin\theta$. Hence

$$\left(\frac{x}{R}\right)^2 + \left(\frac{y}{r}\right)^2 = \cos^2\theta + \sin^2\theta = 1$$

The equation

$$\left(\frac{x}{R}\right)^2 + \left(\frac{y}{r}\right)^2 = 1$$

is the equation of ellipse. Hence, the coordinates of P, $(R\cos\theta, r\sin\theta)$ describe an ellipse for $0 \leq \theta \leq 2\pi$.

In Exercises 73–76, refer to the Bézier curve defined by Eqs. (7) and (8).

73. Show that the Bézier curve with control points

$$P_0 = (1, 4), \quad P_1 = (3, 12), \quad P_2 = (6, 15), \quad P_3 = (7, 4)$$

has parametrization

$$c(t) = (1 + 6t + 3t^2 - 3t^3, 4 + 24t - 15t^2 - 9t^3)$$

Verify that the slope at $t = 0$ is equal to the slope of the segment $\overline{P_0P_1}$.

SOLUTION For the given Bézier curve we have $a_0 = 1$, $a_1 = 3$, $a_2 = 6$, $a_3 = 7$, and $b_0 = 4$, $b_1 = 12$, $b_2 = 15$, $b_3 = 4$. Substituting these values in Eq. (7)–(8) and simplifying gives

$$\begin{aligned} x(t) &= (1-t)^3 + 9t(1-t)^2 + 18t^2(1-t) + 7t^3 \\ &= 1 - 3t + 3t^2 - t^3 + 9t(1 - 2t + t^2) + 18t^2 - 18t^3 + 7t^3 \\ &= 1 - 3t + 3t^2 - t^3 + 9t - 18t^2 + 9t^3 + 18t^2 - 18t^3 + 7t^3 \\ &= -3t^3 + 3t^2 + 6t + 1 \end{aligned}$$

$$\begin{aligned} y(t) &= 4(1-t)^3 + 36t(1-t)^2 + 45t^2(1-t) + 4t^3 \\ &= 4(1 - 3t + 3t^2 - t^3) + 36t(1 - 2t + t^2) + 45t^2 - 45t^3 + 4t^3 \\ &= 4 - 12t + 12t^2 - 4t^3 + 36t - 72t^2 + 36t^3 + 45t^2 - 45t^3 + 4t^3 \\ &= 4 + 24t - 15t^2 - 9t^3 \end{aligned}$$

Then

$$c(t) = (1 + 6t + 3t^2 - 3t^3, 4 + 24t - 15t^2 - 9t^3), \quad 0 \le t \le 1$$

We find the slope at $t = 0$. Using the formula for slope of the tangent line we get

$$\frac{dy}{dx} = \frac{(4 + 24t - 15t^2 - 9t^3)'}{(1 + 6t + 3t^2 - 3t^3)'} = \frac{24 - 30t - 27t^2}{6 + 6t - 9t^2} \Rightarrow \left.\frac{dy}{dx}\right|_{t=0} = \frac{24}{6} = 4$$

The slope of the segment $\overline{P_0P_1}$ is the slope of the line determined by the points $P_0 = (1, 4)$ and $P_1 = (3, 12)$. That is, $\frac{12-4}{3-1} = \frac{8}{2} = 4$. We see that the slope of the tangent line at $t = 0$ is equal to the slope of the segment $\overline{P_0P_1}$, as expected.

75. *CAS* Find and plot the Bézier curve $c(t)$ with control points

$$P_0 = (3, 2), \quad P_1 = (0, 2), \quad P_2 = (5, 4), \quad P_3 = (2, 4)$$

SOLUTION Setting $a_0 = 3$, $a_1 = 0$, $a_2 = 5$, $a_3 = 2$, and $b_0 = 2$, $b_1 = 2$, $b_2 = 4$, $b_3 = 4$ into Eq. (7)–(8) and simplifying gives

$$\begin{aligned} x(t) &= 3(1-t)^3 + 0 + 15t^2(1-t) + 2t^3 \\ &= 3(1 - 3t + 3t^2 - t^3) + 15t^2 - 15t^3 + 2t^3 = 3 - 9t + 24t^2 - 16t^3 \end{aligned}$$

$$\begin{aligned} y(t) &= 2(1-t)^3 + 6t(1-t)^2 + 12t^2(1-t) + 4t^3 \\ &= 2(1 - 3t + 3t^2 - t^3) + 6t(1 - 2t + t^2) + 12t^2 - 12t^3 + 4t^3 \\ &= 2 - 6t + 6t^2 - 2t^3 + 6t - 12t^2 + 6t^3 + 12t^2 - 12t^3 + 4t^3 = 2 + 6t^2 - 4t^3 \end{aligned}$$

We obtain the following equation

$$c(t) = (3 - 9t + 24t^2 - 16t^3, 2 + 6t^2 - 4t^3), \quad 0 \le t \le 1$$

The graph of the Bézier curve is shown in the following figure:

77. A launched projectile follows the trajectory

$$x = at, \quad y = bt - 16t^2 \quad (a, b > 0)$$

Show that the projectile is launched at an angle $\theta = \tan^{-1}\left(\dfrac{b}{a}\right)$ and lands at a distance $\dfrac{ab}{16}$ from the origin.

SOLUTION The height of the projectile equals the value of the y-coordinate. When the projectile is launched, $y(t) = t(b - 16t) = 0$. The solutions to this equation are $t = 0$ and $t = \dfrac{b}{16}$, with $t = 0$ corresponding to the moment the projectile is launched. We find the slope m of the tangent line at $t = 0$:

$$\frac{dy}{dx} = \frac{y'(t)}{x'(t)} = \frac{b - 32t}{a} \Rightarrow m = \left.\frac{b - 32t}{a}\right|_{t=0} = \frac{b}{a}$$

It follows that $\tan\theta = \dfrac{b}{a}$ or $\theta = \tan^{-1}\left(\dfrac{b}{a}\right)$. The projectile lands at $t = \dfrac{b}{16}$. We find the distance of the projectile from the origin at this time, by substituting $t = \dfrac{b}{16}$ in $x(t) = at$. This gives

$$x\left(\frac{b}{16}\right) = \frac{ab}{16}$$

79. **CAS** Plot the astroid $x = \cos^3\theta$, $y = \sin^3\theta$ and find the equation of the tangent line at $\theta = \dfrac{\pi}{3}$.

SOLUTION The graph of the astroid $x = \cos^3\theta$, $y = \sin^3\theta$ is shown in the following figure:

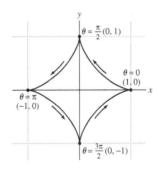

The slope of the tangent line at $\theta = \dfrac{\pi}{3}$ is

$$m = \left.\frac{dy}{dx}\right|_{\theta=\pi/3} = \left.\frac{(\sin^3\theta)'}{(\cos^3\theta)'}\right|_{\theta=\pi/3} = \left.\frac{3\sin^2\theta\cos\theta}{3\cos^2\theta(-\sin\theta)}\right|_{\theta=\pi/3} = -\left.\frac{\sin\theta}{\cos\theta}\right|_{\theta=\pi/3} = -\tan\theta\bigg|_{\pi/3} = -\sqrt{3}$$

We find the point of tangency:

$$\left(x\left(\frac{\pi}{3}\right), y\left(\frac{\pi}{3}\right)\right) = \left(\cos^3\frac{\pi}{3}, \sin^3\frac{\pi}{3}\right) = \left(\frac{1}{8}, \frac{3\sqrt{3}}{8}\right)$$

The equation of the tangent line at $\theta = \dfrac{\pi}{3}$ is, thus,

$$y - \frac{3\sqrt{3}}{8} = -\sqrt{3}\left(x - \frac{1}{8}\right) \Rightarrow y = -\sqrt{3}x + \frac{\sqrt{3}}{2}$$

81. Find the points with a horizontal tangent line on the cycloid with parametric equation (4).

SOLUTION The parametric equations of the cycloid are

$$x = t - \sin t, \quad y = 1 - \cos t$$

We find the slope of the tangent line at t:

$$\frac{dy}{dx} = \frac{(1 - \cos t)'}{(t - \sin t)'} = \frac{\sin t}{1 - \cos t}$$

The tangent line is horizontal where it has slope zero. That is,

$$\frac{dy}{dx} = \frac{\sin t}{1 - \cos t} = 0 \quad \Rightarrow \quad \begin{array}{l} \sin t = 0 \\ \cos t \neq 1 \end{array} \quad \Rightarrow \quad t = (2k - 1)\pi, \quad k = 0, \pm 1, \pm 2, \ldots$$

We find the coordinates of the points with horizontal tangent line, by substituting $t = (2k - 1)\pi$ in $x(t)$ and $y(t)$. This gives

$$x = (2k - 1)\pi - \sin(2k - 1)\pi = (2k - 1)\pi$$
$$y = 1 - \cos((2k - 1)\pi) = 1 - (-1) = 2$$

The required points are

$$((2k - 1)\pi, 2), \quad k = 0, \pm 1, \pm 2, \ldots$$

83. A *curtate cycloid* (Figure 23) is the curve traced by a point at a distance h from the center of a circle of radius R rolling along the x-axis where $h < R$. Show that this curve has parametric equations $x = Rt - h \sin t$, $y = R - h \cos t$.

FIGURE 23 Curtate cycloid.

SOLUTION Let P be a point at a distance h from the center C of the circle. Assume that at $t = 0$, the line of CP is passing through the origin. When the circle rolls a distance Rt along the x-axis, the length of the arc $\overset{\frown}{SQ}$ (see figure) is also Rt and the angle $\angle SCQ$ has radian measure t. We compute the coordinates x and y of P.

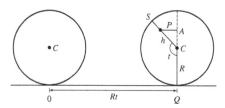

$$x = Rt - \overline{PA} = Rt - h\sin(\pi - t) = Rt - h\sin t$$
$$y = R + \overline{AC} = R + h\cos(\pi - t) = R - h\cos t$$

We obtain the following parametrization:

$$x = Rt - h\sin t, \quad y = R - h\cos t$$

85. Show that the line of slope t through $(-1, 0)$ intersects the unit circle in the point with coordinates

$$x = \frac{1 - t^2}{t^2 + 1}, \quad y = \frac{2t}{t^2 + 1} \qquad \boxed{10}$$

Conclude that these equations parametrize the unit circle with the point $(-1, 0)$ excluded (Figure 24). Show further that $t = y/(x + 1)$.

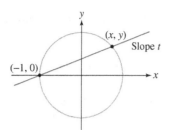

FIGURE 24 Unit circle.

SOLUTION The equation of the line of slope t through $(-1,0)$ is $y = t(x+1)$. The equation of the unit circle is $x^2 + y^2 = 1$. Hence, the line intersects the unit circle at the points (x, y) that satisfy the equations:

$$y = t(x+1) \tag{1}$$
$$x^2 + y^2 = 1 \tag{2}$$

Substituting y from equation (1) into equation (2) and solving for x we obtain

$$x^2 + t^2(x+1)^2 = 1$$
$$x^2 + t^2 x^2 + 2t^2 x + t^2 = 1$$
$$(1+t^2)x^2 + 2t^2 x + (t^2 - 1) = 0$$

This gives

$$x_{1,2} = \frac{-2t^2 \pm \sqrt{4t^4 - 4(t^2+1)(t^2-1)}}{2(1+t^2)} = \frac{-2t^2 \pm 2}{2(1+t^2)} = \frac{\pm 1 - t^2}{1+t^2}$$

So $x_1 = -1$ and $x_2 = \dfrac{1-t^2}{t^2+1}$. The solution $x = -1$ corresponds to the point $(-1,0)$. We are interested in the second point of intersection that is varying as t varies. Hence the appropriate solution is

$$x = \frac{1-t^2}{t^2+1}$$

We find the y-coordinate by substituting x in equation (1). This gives

$$y = t(x+1) = t\left(\frac{1-t^2}{t^2+1} + 1\right) = t \cdot \frac{1-t^2+t^2+1}{t^2+1} = \frac{2t}{t^2+1}$$

We conclude that the line and the unit circle intersect, besides at $(-1,0)$, at the point with the following coordinates:

$$x = \frac{1-t^2}{t^2+1}, \quad y = \frac{2t}{t^2+1} \tag{3}$$

Since these points determine all the points on the unit circle except for $(-1,0)$ and no other points, the equations in (3) parametrize the unit circle with the point $(-1,0)$ excluded.

We show that $t = \dfrac{y}{x+1}$. Using (3) we have

$$\frac{y}{x+1} = \frac{\frac{2t}{t^2+1}}{\frac{1-t^2}{t^2+1} + 1} = \frac{\frac{2t}{t^2+1}}{\frac{1-t^2+t^2+1}{t^2+1}} = \frac{\frac{2t}{t^2+1}}{\frac{2}{t^2+1}} = \frac{2t}{2} = t$$

87. Use the results of Exercise 86 to show that the asymptote of the folium is the line $x + y = -a$. *Hint:* Show that $\lim\limits_{t \to -1}(x + y) = -a$.

SOLUTION We must show that as $x \to \infty$ or $x \to -\infty$ the graph of the folium is getting arbitrarily close to the line $x + y = -a$, and the derivative $\frac{dy}{dx}$ is approaching the slope -1 of the line.

In Exercise 86 we showed that $x \to \infty$ when $t \to (-1^-)$ and $x \to -\infty$ when $t \to (-1^+)$. We first show that the graph is approaching the line $x + y = -a$ as $x \to \infty$ or $x \to -\infty$, by showing that $\lim\limits_{t \to -1-} x + y = \lim\limits_{t \to -1+} x + y = -a$.

For $x(t) = \dfrac{3at}{1 + t^3}$, $y(t) = \dfrac{3at^2}{1 + t^3}$, $a > 0$, calculated in Exercise 86, we obtain using L'Hôpital's Rule:

$$\lim_{t \to -1-}(x + y) = \lim_{t \to -1-} \frac{3at + 3at^2}{1 + t^3} = \lim_{t \to -1-} \frac{3a + 6at}{3t^2} = \frac{3a - 6a}{3} = -a$$

$$\lim_{t \to -1+}(x + y) = \lim_{t \to -1+} \frac{3at + 3at^2}{1 + t^3} = \lim_{t \to -1+} \frac{3a + 6at}{3t^2} = \frac{3a - 6a}{3} = -a$$

We now show that $\dfrac{dy}{dx}$ is approaching -1 as $t \to -1-$ and as $t \to -1+$. We use $\dfrac{dy}{dx} = \dfrac{6at - 3at^4}{3a - 6at^3}$ computed in Exercise 86 to obtain

$$\lim_{t \to -1-} \frac{dy}{dx} = \lim_{t \to -1-} \frac{6at - 3at^4}{3a - 6at^3} = \frac{-9a}{9a} = -1$$

$$\lim_{t \to -1+} \frac{dy}{dx} = \lim_{t \to -1+} \frac{6at - 3at^4}{3a - 6at^3} = \frac{-9a}{9a} = -1$$

We conclude that the line $x + y = -a$ is an asymptote of the folium as $x \to \infty$ and as $x \to -\infty$.

89. Second Derivative for a Parametrized Curve Given a parametrized curve $c(t) = (x(t), y(t))$, show that

$$\frac{d}{dt}\left(\frac{dy}{dx}\right) = \frac{x'(t)y''(t) - y'(t)x''(t)}{x'(t)^2}$$

Use this to prove the formula

$$\boxed{\frac{d^2y}{dx^2} = \frac{x'(t)y''(t) - y'(t)x''(t)}{x'(t)^3}} \quad \boxed{11}$$

SOLUTION By the formula for the slope of the tangent line we have

$$\frac{dy}{dx} = \frac{y'(t)}{x'(t)}$$

Differentiating with respect to t, using the Quotient Rule, gives

$$\frac{d}{dt}\left(\frac{dy}{dx}\right) = \frac{d}{dt}\left(\frac{y'(t)}{x'(t)}\right) = \frac{x'(t)y''(t) - y'(t)x''(t)}{x'(t)^2}$$

By the Chain Rule we have

$$\frac{d^2y}{dx^2} = \frac{d}{dx}\left(\frac{dy}{dx}\right) = \frac{d}{dt}\left(\frac{dy}{dx}\right) \cdot \frac{dt}{dx}$$

Substituting into the above equation $\left(\text{and using } \dfrac{dt}{dx} = \dfrac{1}{dx/dt} = \dfrac{1}{x'(t)}\right)$ gives

$$\frac{d^2y}{dx^2} = \frac{x'(t)y''(t) - y'(t)x''(t)}{x'(t)^2} \cdot \frac{1}{x'(t)} = \frac{x'(t)y''(t) - y'(t)x''(t)}{x'(t)^3}$$

In Exercises 91–94, use Eq. (11) to find d^2y/dx^2.

91. $x = t^3 + t^2$, $y = 7t^2 - 4$, $t = 2$

SOLUTION We find the first and second derivatives of $x(t)$ and $y(t)$:

$$x'(t) = 3t^2 + 2t \Rightarrow x'(2) = 3 \cdot 2^2 + 2 \cdot 2 = 16$$
$$x''(t) = 6t + 2 \Rightarrow x''(2) = 6 \cdot 2 + 2 = 14$$
$$y'(t) = 14t \Rightarrow y'(2) = 14 \cdot 2 = 28$$
$$y''(t) = 14 \Rightarrow y''(2) = 14$$

Using Eq. (11) we get

$$\left.\frac{d^2y}{dx^2}\right|_{t=2} = \left.\frac{x'(t)y''(t) - y'(t)x''(t)}{x'(t)^3}\right|_{t=2} = \frac{16 \cdot 14 - 28 \cdot 14}{16^3} = \frac{-21}{512}$$

93. $x = 8t + 9$, $y = 1 - 4t$, $t = -3$

SOLUTION We compute the first and second derivatives of $x(t)$ and $y(t)$:

$$x'(t) = 8 \Rightarrow x'(-3) = 8$$
$$x''(t) = 0 \Rightarrow x''(-3) = 0$$
$$y'(t) = -4 \Rightarrow y'(-3) = -4$$
$$y''(t) = 0 \Rightarrow y''(-3) = 0$$

Using Eq. (11) we get

$$\left.\frac{d^2y}{dx^2}\right|_{t=-3} = \frac{x'(-3)y''(-3) - y'(-3)x''(-3)}{x'(-3)^3} = \frac{8 \cdot 0 - (-4) \cdot 0}{8^3} = 0$$

95. Use Eq. (11) to find the t-intervals on which $c(t) = (t^2, t^3 - 4t)$ is concave up.

SOLUTION The curve is concave up where $\frac{d^2y}{dx^2} > 0$. Thus,

$$\frac{x'(t)y''(t) - y'(t)x''(t)}{x'(t)^3} > 0 \tag{1}$$

We compute the first and second derivatives:

$$x'(t) = 2t, \quad x''(t) = 2$$
$$y'(t) = 3t^2 - 4, \quad y''(t) = 6t$$

Substituting in (1) and solving for t gives

$$\frac{12t^2 - (6t^2 - 8)}{8t^3} = \frac{6t^2 + 8}{8t^3}$$

Since $6t^2 + 8 > 0$ for all t, the quotient is positive if $8t^3 > 0$. We conclude that the curve is concave up for $t > 0$.

97. Calculate the area under $y = x^2$ over $[0, 1]$ using Eq. (9) with the parametrizations (t^3, t^6) and (t^2, t^4).

SOLUTION The area A under $y = x^2$ on $[0, 1]$ is given by the integral

$$A = \int_{t_0}^{t_1} y(t)x'(t)\,dt$$

where $x(t_0) = 0$ and $x(t_1) = 1$. We first use the parametrization (t^3, t^6). We have $x(t) = t^3$, $y(t) = t^6$. Hence,

$$0 = x(t_0) = t_0^3 \Rightarrow t_0 = 0$$
$$1 = x(t_1) = t_1^3 \Rightarrow t_1 = 1$$

Also $x'(t) = 3t^2$. Substituting these values in Eq. (9) we obtain

$$A = \int_0^1 t^6 \cdot 3t^2 \, dt = \int_0^1 3t^8 \, dt = \frac{3}{9}t^9 \Big|_0^1 = \frac{3}{9} = \frac{1}{3}$$

Using the parametrization $x(t) = t^2$, $y(t) = t^4$, we have $x'(t) = 2t$. We find t_0 and t_1:

$$0 = x(t_0) = t_0^2 \Rightarrow t_0 = 0$$
$$1 = x(t_1) = t_1^2 \Rightarrow t_1 = 1 \quad \text{or} \quad t_1 = -1$$

Equation (9) is valid if $x'(t) > 0$; that is, if $t > 0$. Hence we choose the positive value, $t_1 = 1$. We now use Eq. (9) to obtain

$$A = \int_0^1 t^4 \cdot 2t \, dt = \int_0^1 2t^5 \, dt = \frac{2}{6}t^6 \Big|_0^1 = \frac{2}{6} = \frac{1}{3}$$

Both answers agree, as expected.

99. Consider the curve $c(t) = (t^2, t^3)$ for $0 \le t \le 1$.
(a) Find the area under the curve using Eq. (9).
(b) Find the area under the curve by expressing y as a function of x and finding the area using the standard method.

SOLUTION
(a) Using Eq. (9), note first that for $0 \le t \le 1$, the path of $c(t)$ is traced once, so that this equation applies. The area under the curve is then

$$\int_0^1 y(t)x'(t) \, dt = \int_0^1 t^3 \left(t^2\right)' dt = \int_0^1 2t^4 \, dt = \frac{2}{5}t^5 \Big|_0^1 = \frac{2}{5}$$

(b) To express y as a function of x, note that if $x = t^2$, then $y = t^3 = (t^2)^{3/2} = x^{3/2}$. $t = 0$ corresponds to $x = 0$, and $t = 1$ to $x = 1$, so the area under the curve is

$$\int_0^1 x^{3/2} \, dx = \frac{2}{5}x^{5/2} \Big|_0^1 = \frac{2}{5}$$

which agrees with the answer in part (a).

101. Compute the area under the parametrized curve given by $c(t) = (\sin t, \cos^2 t)$ for $0 \le t \le \pi/2$ using Eq. (9).

SOLUTION Since $\sin t$ is increasing for $0 \le t \le \frac{\pi}{2}$, Eq. (9) applies, and the area under the curve is

$$\int_0^{\pi/2} y(t)x'(t) \, dt = \int_0^{\pi/2} \cos^2 t (\sin t)' \, dt = \int_0^{\pi/2} \cos^3 t \, dt$$

$$= \int_0^{\pi/2} (1 - \sin^2 t) \cos t \, dt = \left[\sin t - \frac{1}{3}\sin^3 t\right]_0^{\pi/2} = \frac{2}{3}$$

103. Galileo tried unsuccessfully to find the area under a cycloid. Around 1630, Gilles de Roberval proved that the area under one arch of the cycloid $c(t) = (Rt - R\sin t, R - R\cos t)$ generated by a circle of radius R is equal to three times the area of the circle (Figure 26). Verify Roberval's result using Eq. (9).

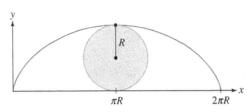

FIGURE 26 The area of one arch of the cycloid equals three times the area of the generating circle.

SOLUTION Using the given equation,

$$\int_0^{2\pi} (R - R\cos t)(Rt - R\sin t)' \, dt = \int_0^{2\pi} R^2 (1 - \cos t)^2 \, dt$$

$$= R^2 \int_0^{2\pi} (1 - 2\cos t + \cos^2 t) \, dt$$

$$= R^2 \int_0^{2\pi} \left(1 - 2\cos t + \frac{1}{2} + \frac{1}{2}\cos 2t\right) dt$$

$$= R^2 \left(\frac{3}{2} t - 2\sin t + \frac{1}{4}\sin 2t\right)\bigg|_0^{2\pi}$$

$$= 3\pi R^2$$

Further Insights and Challenges

105. Derive the formula for the slope of the tangent line to a parametric curve $c(t) = (x(t), y(t))$ using a method different from that presented in the text. Assume that $x'(t_0)$ and $y'(t_0)$ exist and $x'(t_0) \neq 0$. Show that

$$\lim_{h \to 0} \frac{y(t_0 + h) - y(t_0)}{x(t_0 + h) - x(t_0)} = \frac{y'(t_0)}{x'(t_0)}$$

Then explain why this limit is equal to the slope dy/dx. Draw a diagram showing that the ratio in the limit is the slope of a secant line.

SOLUTION Since $y'(t_0)$ and $x'(t_0)$ exist, we have the following limits:

$$\lim_{h \to 0} \frac{y(t_0 + h) - y(t_0)}{h} = y'(t_0), \quad \lim_{h \to 0} \frac{x(t_0 + h) - x(t_0)}{h} = x'(t_0) \tag{1}$$

We use Basic Limit Laws, the limits in (1) and the given data $x'(t_0) \neq 0$, to write

$$\lim_{h \to 0} \frac{y(t_0 + h) - y(t_0)}{x(t_0 + h) - x(t_0)} = \lim_{h \to 0} \frac{\frac{y(t_0+h)-y(t_0)}{h}}{\frac{x(t_0+h)-x(t_0)}{h}} = \frac{\lim_{h \to 0} \frac{y(t_0+h)-y(t_0)}{h}}{\lim_{h \to 0} \frac{x(t_0+h)-x(t_0)}{h}} = \frac{y'(t_0)}{x'(t_0)}$$

Notice that the quotient $\dfrac{y(t_0 + h) - y(t_0)}{x(t_0 + h) - x(t_0)}$ is the slope of the secant line determined by the points $P = (x(t_0), y(t_0))$ and $Q = (x(t_0 + h), y(t_0 + h))$. Hence, the limit of the quotient as $h \to 0$ is the slope of the tangent line at P, that is the derivative $\dfrac{dy}{dx}$.

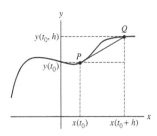

107. In Exercise 62 of Section 9.1, we described the tractrix by the differential equation

$$\frac{dy}{dx} = -\frac{y}{\sqrt{\ell^2 - y^2}}$$

Show that the parametric curve $c(t)$ identified as the tractrix in Exercise 106 satisfies this differential equation. Note that the derivative on the left is taken with respect to x, not t.

SOLUTION Note that $dx/dt = 1 - \text{sech}^2(t/\ell) = \tanh^2(t/\ell)$ and $dy/dt = -\text{sech}(t/\ell)\tanh(t/\ell)$. Thus,

$$\frac{dy}{dx} = \frac{dy/dt}{dx/dt} = \frac{-\text{sech}(t/\ell)}{\tanh(t/\ell)} = \frac{-y/\ell}{\sqrt{1 - y^2/\ell^2}}$$

Multiplying top and bottom by ℓ/ℓ gives

$$\frac{dy}{dx} = \frac{-y}{\sqrt{\ell^2 - y^2}}$$

In Exercises 108 and 109, refer to Figure 29.

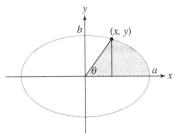

FIGURE 29 The parameter θ on the ellipse $\left(\dfrac{x}{a}\right)^2 + \left(\dfrac{y}{b}\right)^2 = 1$.

109. Show that the parametrization of the ellipse by the angle θ is

$$x = \frac{ab\cos\theta}{\sqrt{a^2\sin^2\theta + b^2\cos^2\theta}}$$

$$y = \frac{ab\sin\theta}{\sqrt{a^2\sin^2\theta + b^2\cos^2\theta}}$$

SOLUTION We consider the ellipse

$$\frac{x^2}{a^2} + \frac{y^2}{b^2} = 1$$

For the angle θ we have $\tan\theta = \frac{y}{x}$, hence,

$$y = x\tan\theta \qquad (1)$$

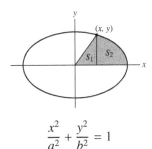

$$\frac{x^2}{a^2} + \frac{y^2}{b^2} = 1$$

Substituting in the equation of the ellipse and solving for x we obtain

$$\frac{x^2}{a^2} + \frac{x^2\tan^2\theta}{b^2} = 1$$

$$b^2x^2 + a^2x^2\tan^2\theta = a^2b^2$$

$$(a^2\tan^2\theta + b^2)x^2 = a^2b^2$$

$$x^2 = \frac{a^2b^2}{a^2\tan^2\theta + b^2} = \frac{a^2b^2\cos^2\theta}{a^2\sin^2\theta + b^2\cos^2\theta}$$

We now take the square root. Since the sign of the x-coordinate is the same as the sign of $\cos\theta$, we take the positive root, obtaining

$$x = \frac{ab\cos\theta}{\sqrt{a^2\sin^2\theta + b^2\cos^2\theta}} \qquad (2)$$

Hence by (1), the y-coordinate is

$$y = x\tan\theta = \frac{ab\cos\theta\tan\theta}{\sqrt{a^2\sin^2\theta + b^2\cos^2\theta}} = \frac{ab\sin\theta}{\sqrt{a^2\sin^2\theta + b^2\cos^2\theta}} \qquad (3)$$

Equalities (2) and (3) give the following parametrization for the ellipse:

$$c_1(\theta) = \left(\frac{ab\cos\theta}{\sqrt{a^2\sin^2\theta + b^2\cos^2\theta}}, \frac{ab\sin\theta}{\sqrt{a^2\sin^2\theta + b^2\cos^2\theta}}\right)$$

11.2 Arc Length and Speed (LT Section 12.2)

Preliminary Questions

1. What is the definition of arc length?

SOLUTION A curve can be approximated by a polygonal path obtained by connecting points

$$p_0 = c(t_0), p_1 = c(t_1), \ldots, p_N = c(t_N)$$

on the path with segments. One gets an approximation by summing the lengths of the segments. The definition of arc length is the limit of that approximation when increasing the number of points so that the lengths of the segments approach zero. In doing so, we obtain the following theorem for the arc length:

$$S = \int_a^b \sqrt{x'(t)^2 + y'(t)^2}\, dt$$

which is the length of the curve $c(t) = (x(t), y(t))$ for $a \le t \le b$.

2. Can the distance traveled by a particle ever be less than its displacement? When are they equal?

SOLUTION Suppose the displacement is d. The particle clearly must travel a distance of at least d, so that the distance must always be at least as large as the displacement. They are the same when the particle moves in a straight line, without reversing, from the start point to the end point (this is a fancy way of saying "a straight line is the shortest distance between two points.").

3. What is the interpretation of $\sqrt{x'(t)^2 + y'(t)^2}$ for a particle following the trajectory $(x(t), y(t))$?

SOLUTION The expression $\sqrt{x'(t)^2 + y'(t)^2}$ denotes the speed at time t of a particle following the trajectory $(x(t), y(t))$.

4. A particle travels along a path from $(0, 0)$ to $(3, 4)$. What is the displacement? Can the distance traveled be determined from the information given?

SOLUTION The net displacement is the distance between the initial point $(0, 0)$ and the endpoint $(3, 4)$. That is

$$\sqrt{(3-0)^2 + (4-0)^2} = \sqrt{25} = 5$$

The distance traveled can be determined only if the trajectory $c(t) = (x(t), y(t))$ of the particle is known.

5. A particle traverses the parabola $y = x^2$ with constant speed 3 cm/s. What is the distance traveled during the first minute? *Hint:* Only simple computation is necessary.

SOLUTION Since the speed is constant, the distance traveled is the following product: $L = st = 3 \cdot 60 = 180$ cm.

6. If the straight line segment given by $c(t) = (t, 3)$ for $0 \le t \le 2$ is rotated around the x-axis, what surface area results? *Hint:* Only simple computation is necessary.

SOLUTION The given line segment is a vertical segment from $(0, 3)$ to $(2, 3)$. Rotating this around the x-axis gives a circle of radius 2, which has surface area 4π.

Exercises

In Exercises 1–2, use Eq. (3) to find the length of the path over the given interval, and verify your answer using geometry.

1. $(3t - 1, 2 - 2t), \quad 0 \le t \le 5$

SOLUTION Using Eq. (3), the length is

$$s = \int_0^5 \sqrt{x'(t)^2 + y'(t)^2}\, dt = \int_0^5 \sqrt{3^2 + (-2)^2}\, dt = \int_0^5 \sqrt{13}\, dt = 5\sqrt{13}$$

To verify this using geometry, eliminate the variable t. Since $x = 3t - 1$, we have $t = \dfrac{x+1}{3}$; substituting into $y = 2 - 2t$ we get

$$y = 2 - 2t = 2 - 2 \cdot \frac{x+1}{3} = -\frac{2}{3}x + \frac{4}{3}$$

which is a straight line. $t = 0$ corresponds to $(x, y) = (-1, 2)$, while $t = 5$ corresponds to $(x, y) = (3 \cdot 5 - 1, 2 - 2 \cdot 5) = (14, -8)$. Since the path is a straight line, the length of the path is the distance between these two points, which by the Pythagorean Theorem is

$$\sqrt{(14 - (-1))^2 + (-8 - 2)^2} = \sqrt{325} = 5\sqrt{13}$$

This matches the value derived previously.

In Exercises 3–8, use Eq. (3) to find the length of the path over the given interval.

3. $(2t^2, 3t^2 - 1), \quad 0 \le t \le 4$

SOLUTION Since $x = 2t^2$ and $y = 3t^2 - 1$, we have $x' = 4t$ and $y' = 6t$. By the formula for the arc length we get

$$S = \int_0^4 \sqrt{x'(t)^2 + y'(t)^2}\, dt = \int_0^4 \sqrt{16t^2 + 36t^2}\, dt = \sqrt{52} \int_0^4 t\, dt = \sqrt{52} \cdot \frac{t^2}{2}\bigg|_0^4 = 16\sqrt{13}$$

5. $(3t^2, 4t^3), \quad 1 \le t \le 4$

SOLUTION We have $x = 3t^2$ and $y = 4t^3$. Hence $x' = 6t$ and $y' = 12t^2$. By the formula for the arc length we get

$$S = \int_1^4 \sqrt{x'(t)^2 + y'(t)^2}\, dt = \int_1^4 \sqrt{36t^2 + 144t^4}\, dt = 6\int_1^4 \sqrt{1 + 4t^2}\, t\, dt$$

Using the substitution $u = 1 + 4t^2$, $du = 8t\, dt$ we obtain

$$S = \frac{6}{8}\int_5^{65} \sqrt{u}\, du = \frac{3}{4} \cdot \frac{2}{3} u^{3/2}\bigg|_5^{65} = \frac{1}{2}(65^{3/2} - 5^{3/2}) \approx 256.43$$

7. $(\sin 3t, \cos 3t), \quad 0 \le t \le \pi$

SOLUTION We have $x = \sin 3t$, $y = \cos 3t$, hence $x' = 3\cos 3t$ and $y' = -3\sin 3t$. By the formula for the arc length we obtain:

$$S = \int_0^\pi \sqrt{x'(t)^2 + y'(t)^2}\, dt = \int_0^\pi \sqrt{9\cos^2 3t + 9\sin^2 3t}\, dt = \int_0^\pi \sqrt{9}\, dt = 3\pi$$

In Exercises 9 and 10, find the length of the path. The following identity should be helpful:

$$\frac{1 - \cos t}{2} = \sin^2 \frac{t}{2}$$

9. $(2\cos t - \cos 2t, 2\sin t - \sin 2t), \quad 0 \le t \le \frac{\pi}{2}$

SOLUTION We have $x = 2\cos t - \cos 2t$, $y = 2\sin t - \sin 2t$. Thus, $x' = -2\sin t + 2\sin 2t$ and $y' = 2\cos t - 2\cos 2t$. We get

$$x'(t)^2 + y'(t)^2 = (-2\sin t + 2\sin 2t)^2 + (2\cos t - 2\cos 2t)^2$$

$$= 4\sin^2 t - 8\sin t \sin 2t + 4\sin^2 2t + 4\cos^2 t - 8\cos t \cos 2t + 4\cos^2 2t$$

$$= 4(\sin^2 t + \cos^2 t) + 4(\sin^2 2t + \cos^2 2t) - 8(\sin t \sin 2t + \cos t \cos 2t)$$

$$= 4 + 4 - 8\cos(2t - t) = 8 - 8\cos t = 8(1 - \cos t)$$

We now use the formula for the arc length to obtain

$$S = \int_0^{\pi/2} \sqrt{x'(t)^2 + y'(t)^2} = \int_0^{\pi/2} \sqrt{8(1-\cos t)}\,dt = \int_0^{\pi/2} \sqrt{16\sin^2 \frac{t}{2}}\,dt = 4\int_0^{\pi/2} \sin \frac{t}{2}\,dt$$

$$= -8\cos \frac{t}{2}\Big|_0^{\pi/2} = -8\left(\cos \frac{\pi}{4} - \cos 0\right) = -8\left(\frac{\sqrt{2}}{2} - 1\right) \approx 2.34$$

11. Show that one arch of a cycloid generated by a circle of radius R has length $8R$.

SOLUTION Recall from earlier that the cycloid generated by a circle of radius R has parametric equations $x = Rt - R\sin t$, $y = R - R\cos t$. Hence, $x' = R - R\cos t$, $y' = R\sin t$. Using the identity $\sin^2 \frac{t}{2} = \frac{1-\cos t}{2}$, we get

$$x'(t)^2 + y'(t)^2 = R^2(1-\cos t)^2 + R^2 \sin^2 t = R^2(1 - 2\cos t + \cos^2 t + \sin^2 t)$$

$$= R^2(1 - 2\cos t + 1) = 2R^2(1 - \cos t) = 4R^2 \sin^2 \frac{t}{2}$$

One arch of the cycloid is traced as t varies from 0 to 2π. Hence, using the formula for the arc length we obtain:

$$S = \int_0^{2\pi} \sqrt{x'(t)^2 + y'(t)^2}\,dt = \int_0^{2\pi} \sqrt{4R^2 \sin^2 \frac{t}{2}}\,dt = 2R \int_0^{2\pi} \sin \frac{t}{2}\,dt = 4R \int_0^{\pi} \sin u\,du$$

$$= -4R \cos u\Big|_0^{\pi} = -4R(\cos \pi - \cos 0) = 8R$$

13. Find the length of the parabola given by $c(t) = (t, t^2)$ for $0 \le t \le 1$. See the hint for Exercise 12.

SOLUTION Since $x = t - \tanh(t)$ and $y = \text{sech}(t)$ we have $x' = 1 - \text{sech}^2(t)$ and $y' = -\text{sech}(t)\tanh(t)$. Hence,

$$x'(t)^2 + y'(t)^2 = (1 - \text{sech}^2(t))^2 + \text{sech}^2(t)\tanh^2(t)$$

$$= 1 - 2\text{sech}^2(t) + \text{sech}^4(t) + \text{sech}^2(t)\tanh^2(t)$$

$$= 1 - 2\text{sech}^2(t) + \text{sech}^2(t)(\text{sech}^2(t) + \tanh^2(t))$$

$$= 1 - 2\text{sech}^2(t) + \text{sech}^2(t) = 1 - \text{sech}^2(t) = \tanh^2(t)$$

Hence, using the formula for the arc length we get:

$$S = \int_0^A \sqrt{x'(t)^2 + y'(t)^2}\,dt = \int_0^A \sqrt{\tanh^2(t)}\,dt = \int_0^A \tanh(t)\,dt = \ln(\cosh(t))\Big|_0^A$$

$$= \ln(\cosh(A)) - \ln(\cosh(0)) = \ln(\cosh(A)) - \ln 1 = \ln(\cosh(A))$$

In Exercises 15–20, determine the speed $\frac{ds}{dt}$ at time t (assume units of meters and seconds).

15. (t^3, t^2), $t = 2$

SOLUTION We have $x(t) = t^3$, $y(t) = t^2$ hence $x'(t) = 3t^2$, $y'(t) = 2t$. The speed of the particle at time t is thus, $\frac{ds}{dt} = \sqrt{x'(t)^2 + y'(t)^2} = \sqrt{9t^4 + 4t^2} = t\sqrt{9t^2 + 4}$. At time $t = 2$ the speed is

$$\frac{ds}{dt}\Big|_{t=2} = 2\sqrt{9 \cdot 2^2 + 4} = 2\sqrt{40} = 4\sqrt{10} \approx 12.65 \text{ m/s}$$

17. $(5t + 1, 4t - 3)$, $t = 9$

SOLUTION Since $x = 5t + 1$, $y = 4t - 3$, we have $x' = 5$ and $y' = 4$. The speed of the particle at time t is

$$\frac{ds}{dt} = \sqrt{x'(t) + y'(t)} = \sqrt{5^2 + 4^2} = \sqrt{41} \approx 6.4 \text{ m/s}$$

We conclude that the particle has constant speed of 6.4 m/s.

19. (t^2, e^t), $t = 0$

SOLUTION Since $x = t^2$ and $y = e^t$, we have $x' = 2t$ and $y' = e^t$. The speed of the particle at time t is

$$\frac{ds}{dt} = \sqrt{x'(t)^2 + y'(t)^2} = \sqrt{4t^2 + e^{2t}} \text{ m/s}$$

The speed at time $t = 0$ is

$$\left.\frac{ds}{dt}\right|_{t=0} = \sqrt{4 \cdot 0^2 + e^{2 \cdot 0}} = 1 \text{ m/s}$$

21. Find the minimum speed of a particle with trajectory $c(t) = (t^3 - 4t, t^2 + 1)$ for $t \geq 0$. *Hint:* It is easier to find the minimum of the square of the speed.

SOLUTION We first find the speed of the particle. We have $x(t) = t^3 - 4t$, $y(t) = t^2 + 1$, hence $x'(t) = 3t^2 - 4$ and $y'(t) = 2t$. The speed is thus

$$\frac{ds}{dt} = \sqrt{(3t^2 - 4)^2 + (2t)^2} = \sqrt{9t^4 - 24t^2 + 16 + 4t^2} = \sqrt{9t^4 - 20t^2 + 16}$$

The square root function is an increasing function, hence the minimum speed occurs at the value of t where the function $f(t) = 9t^4 - 20t^2 + 16$ has minimum value. Since $\lim_{t \to \infty} f(t) = \infty$, f has a minimum value on the interval $0 \leq t < \infty$, and it occurs at a critical point or at the endpoint $t = 0$. We find the critical point of f on $t \geq 0$:

$$f'(t) = 36t^3 - 40t = 4t(9t^2 - 10) = 0 \Rightarrow t = 0, t = \sqrt{\frac{10}{9}}$$

We compute the values of f at these points:

$$f(0) = 9 \cdot 0^4 - 20 \cdot 0^2 + 16 = 16$$

$$f\left(\sqrt{\frac{10}{9}}\right) = 9\left(\sqrt{\frac{10}{9}}\right)^4 - 20\left(\sqrt{\frac{10}{9}}\right)^2 + 16 = \frac{44}{9} \approx 4.89$$

We conclude that the minimum value of f on $t \geq 0$ is 4.89. The minimum speed is therefore

$$\left(\frac{ds}{dt}\right)_{\min} \approx \sqrt{4.89} \approx 2.21$$

23. Find the speed of the cycloid $c(t) = (4t - 4\sin t, 4 - 4\cos t)$ at points where the tangent line is horizontal.

SOLUTION We first find the points where the tangent line is horizontal. The slope of the tangent line is the following quotient:

$$\frac{dy}{dx} = \frac{dy/dt}{dx/dt} = \frac{4\sin t}{4 - 4\cos t} = \frac{\sin t}{1 - \cos t}$$

To find the points where the tangent line is horizontal we solve the following equation for $t \geq 0$:

$$\frac{dy}{dx} = 0, \quad \frac{\sin t}{1 - \cos t} = 0 \Rightarrow \sin t = 0 \quad \text{and} \quad \cos t \neq 1$$

Now, $\sin t = 0$ and $t \geq 0$ at the points $t = \pi k$, $k = 0, 1, 2, \ldots$. Since $\cos \pi k = (-1)^k$, the points where $\cos t \neq 1$ are $t = \pi k$ for k odd. The points where the tangent line is horizontal are, therefore:

$$t = \pi(2k - 1), \quad k = 1, 2, 3, \ldots$$

The speed at time t is given by the following expression:

$$\frac{ds}{dt} = \sqrt{x'(t)^2 + y'(t)^2} = \sqrt{(4 - 4\cos t)^2 + (4\sin t)^2}$$

$$= \sqrt{16 - 32\cos t + 16\cos^2 t + 16\sin^2 t} = \sqrt{16 - 32\cos t + 16}$$

$$= \sqrt{32(1 - \cos t)} = \sqrt{32 \cdot 2\sin^2 \frac{t}{2}} = 8\left|\sin \frac{t}{2}\right|$$

That is, the speed of the cycloid at time t is

$$\frac{ds}{dt} = 8\left|\sin\frac{t}{2}\right|$$

We now substitute

$$t = \pi(2k-1), \quad k = 1, 2, 3, \ldots$$

to obtain

$$\frac{ds}{dt} = 8\left|\sin\frac{\pi(2k-1)}{2}\right| = 8|(-1)^{k+1}| = 8$$

CAS *In Exercises 25–28, plot the curve and use the Midpoint Rule with $N = 10, 20, 30$, and 50 to approximate its length.*

25. $c(t) = (\cos t, e^{\sin t})$ for $0 \leq t \leq 2\pi$

SOLUTION The curve of $c(t) = (\cos t, e^{\sin t})$ for $0 \leq t \leq 2\pi$ is shown in the figure below:

$c(t) = (\cos t, e^{\sin t}), 0 \leq t \leq 2\pi$

The length of the curve is given by the following integral:

$$S = \int_0^{2\pi} \sqrt{x'(t)^2 + y'(t)^2}\,dt = \int_0^{2\pi} \sqrt{(-\sin t)^2 + (\cos t\, e^{\sin t})^2}\,dt$$

That is, $S = \int_0^{2\pi} \sqrt{\sin^2 t + \cos^2 t\, e^{2\sin t}}\,dt$. We approximate the integral using the Mid-Point Rule with $N = 10, 20, 30, 50$. For $f(t) = \sqrt{\sin^2 t + \cos^2 t\, e^{2\sin t}}$ we obtain

$(N = 10)$: $\Delta x = \frac{2\pi}{10} = \frac{\pi}{5}, c_i = \left(i - \frac{1}{2}\right) \cdot \frac{\pi}{5}$

$$M_{10} = \frac{\pi}{5}\sum_{i=1}^{10} f(c_i) = 6.903734$$

$(N = 20)$: $\Delta x = \frac{2\pi}{20} = \frac{\pi}{10}, c_i = \left(i - \frac{1}{2}\right) \cdot \frac{\pi}{10}$

$$M_{20} = \frac{\pi}{10}\sum_{i=1}^{20} f(c_i) = 6.915035$$

$(N = 30)$: $\Delta x = \frac{2\pi}{30} = \frac{\pi}{15}, c_i = \left(i - \frac{1}{2}\right) \cdot \frac{\pi}{15}$

$$M_{30} = \frac{\pi}{15}\sum_{i=1}^{30} f(c_i) = 6.914949$$

$(N = 50)$: $\Delta x = \frac{2\pi}{50} = \frac{\pi}{25}, c_i = \left(i - \frac{1}{2}\right) \cdot \frac{\pi}{25}$

$$M_{50} = \frac{\pi}{25}\sum_{i=1}^{50} f(c_i) = 6.914951$$

27. The ellipse $\left(\frac{x}{5}\right)^2 + \left(\frac{y}{3}\right)^2 = 1$

SOLUTION We use the parametrization given in Example 4, Section 12.1, that is, $c(t) = (5\cos t, 3\sin t)$, $0 \le t \le 2\pi$. The curve is shown in the figure below:

$$c(t) = (5\cos t, 3\sin t), 0 \le t \le 2\pi$$

The length of the curve is given by the following integral:

$$S = \int_0^{2\pi} \sqrt{x'(t)^2 + y'(t)^2}\, dt = \int_0^{2\pi} \sqrt{(-5\sin t)^2 + (3\cos t)^2}\, dt$$
$$= \int_0^{2\pi} \sqrt{25\sin^2 t + 9\cos^2 t}\, dt = \int_0^{2\pi} \sqrt{9(\sin^2 t + \cos^2 t) + 16\sin^2 t}\, dt = \int_0^{2\pi} \sqrt{9 + 16\sin^2 t}\, dt$$

That is,

$$S = \int_0^{2\pi} \sqrt{9 + 16\sin^2 t}\, dt$$

We approximate the integral using the Mid-Point Rule with $N = 10, 20, 30, 50$, for $f(t) = \sqrt{9 + 16\sin^2 t}$. We obtain

$(N = 10)$: $\quad \Delta x = \dfrac{2\pi}{10} = \dfrac{\pi}{5}, c_i = \left(i - \dfrac{1}{2}\right) \cdot \dfrac{\pi}{5}$

$$M_{10} = \frac{\pi}{5} \sum_{i=1}^{10} f(c_i) = 25.528309$$

$(N = 20)$: $\quad \Delta x = \dfrac{2\pi}{20} = \dfrac{\pi}{10}, c_i = \left(i - \dfrac{1}{2}\right) \cdot \dfrac{\pi}{10}$

$$M_{20} = \frac{\pi}{10} \sum_{i=1}^{20} f(c_i) = 25.526999$$

$(N = 30)$: $\quad \Delta x = \dfrac{2\pi}{30} = \dfrac{\pi}{15}, c_i = \left(i - \dfrac{1}{2}\right) \cdot \dfrac{\pi}{15}$

$$M_{30} = \frac{\pi}{15} \sum_{i=1}^{30} f(c_i) = 25.526999$$

$(N = 50)$: $\quad \Delta x = \dfrac{2\pi}{50} = \dfrac{\pi}{25}, c_i = \left(i - \dfrac{1}{2}\right) \cdot \dfrac{\pi}{25}$

$$M_{50} = \frac{\pi}{25} \sum_{i=1}^{50} f(c_i) = 25.526999$$

29. If you unwind thread from a stationary circular spool, keeping the thread taut at all times, then the endpoint traces a curve C called the **involute** of the circle (Figure 9). Observe that $\overline{PQ}$ has length $R\theta$. Show that C is parametrized by

$$c(\theta) = (R(\cos\theta + \theta\sin\theta), R(\sin\theta - \theta\cos\theta))$$

Then find the length of the involute for $0 \le \theta \le 2\pi$.

FIGURE 9 Involute of a circle.

SOLUTION Suppose that the arc $\widehat{QT}$ corresponding to the angle θ is unwound. Then the length of the segment $\overline{QP}$ equals the length of this arc. That is, $\overline{QP} = R\theta$. With the help of the figure we can see that

$$x = \overline{OA} + \overline{AB} = \overline{OA} + \overline{EP} = R\cos\theta + \overline{QP}\sin\theta = R\cos\theta + R\theta\sin\theta = R(\cos\theta + \theta\sin\theta)$$

Furthermore,

$$y = \overline{QA} - \overline{QE} = R\sin\theta - \overline{QP}\cos\theta = R\sin\theta - R\theta\cos\theta = R(\sin\theta - \theta\cos\theta)$$

The coordinates of P with respect to the parameter θ form the following parametrization of the curve:

$$c(\theta) = (R(\cos\theta + \theta\sin\theta), R(\sin\theta - \theta\cos\theta)), \quad 0 \le \theta \le 2\pi$$

We find the length of the involute for $0 \le \theta \le 2\pi$, using the formula for the arc length:

$$S = \int_0^{2\pi} \sqrt{x'(\theta)^2 + y'(\theta)^2}\, d\theta$$

We compute the integrand:

$$x'(\theta) = \frac{d}{d\theta}(R(\cos\theta + \theta\sin\theta)) = R(-\sin\theta + \sin\theta + \theta\cos\theta) = R\theta\cos\theta$$

$$y'(\theta) = \frac{d}{d\theta}(R(\sin\theta - \theta\cos\theta)) = R(\cos\theta - (\cos\theta - \theta\sin\theta)) = R\theta\sin\theta$$

$$\sqrt{x'(\theta)^2 + y'(\theta)^2} = \sqrt{(R\theta\cos\theta)^2 + (R\theta\sin\theta)^2} = \sqrt{R^2\theta^2(\cos^2\theta + \sin^2\theta)} = \sqrt{R^2\theta^2} = R\theta$$

We now compute the arc length:

$$S = \int_0^{2\pi} R\theta\, d\theta = \left.\frac{R\theta^2}{2}\right|_0^{2\pi} = \frac{R\cdot(2\pi)^2}{2} = 2\pi^2 R$$

In Exercises 31–38, use Eq. (4) to compute the surface area of the given surface.

31. The cone generated by revolving $c(t) = (t, mt)$ about the x-axis for $0 \le t \le A$

SOLUTION Substituting $y(t) = mt$, $y'(t) = m$, $x'(t) = 1$, $a = 0$, and $b = 0$ in the formula for the surface area, we get

$$S = 2\pi\int_0^A mt\sqrt{1+m^2}\, dt = 2\pi\sqrt{1+m^2}\,m\int_0^A t\, dt = 2\pi m\sqrt{1+m^2}\cdot\left.\frac{t^2}{2}\right|_0^A = m\sqrt{1+m^2}\,\pi A^2$$

33. The surface generated by revolving the curve $c(t) = (t^2, t)$ about the x-axis for $0 \le t \le 1$

SOLUTION With $x(t) = t^2$ and $y(t) = t$, we have $x'(t) = 2t$ and $y'(t) = 1$. Using the formula for the surface area, we get

$$S = 2\pi\int_0^1 y(t)\sqrt{x'(t)^2 + y'(t)^2}\, dt = 2\pi\int_0^1 t\sqrt{1+4t^2}\, dt = 2\pi\left[\frac{1}{12}(1+4t^2)^{3/2}\right]_0^1 = \frac{\pi}{6}\left(5^{3/2} - 1\right)$$

35. The surface generated by revolving the curve $c(t) = (\sin^2 t, \cos^2 t)$ about the x-axis for $0 \le t \le \frac{\pi}{2}$

SOLUTION With $x(t) = \sin^2 t$ and $y(t) = \cos^2 t$, we have $x'(t) = 2\sin t \cos t$ and $y'(t) = -2\sin t \cos t$. Using the formula for the surface area, we get

$$S = 2\pi \int_0^{\pi/2} y(t) \sqrt{x'(t)^2 + y'(t)^2}\, dt$$

$$= 2\pi \int_0^{\pi/2} \cos^2 t \sqrt{4\sin^2 t \cos^2 t + 4\sin^2 t \cos^2 t}\, dt$$

$$= 4\pi\sqrt{2} \int_0^{\pi/2} \sin t \cos^3 t\, dt$$

$$= 4\pi\sqrt{2} \left[-\frac{1}{4} \cos^4 t \right]_0^{\pi/2}$$

$$= 4\pi\sqrt{2} \cdot \frac{1}{4} = \pi\sqrt{2}$$

37. The surface generated by revolving one arch of the cycloid $c(t) = (t - \sin t, 1 - \cos t)$ about the x-axis

SOLUTION One arch of the cycloid is traced as t varies from 0 to 2π. Since $x(t) = t - \sin t$ and $y(t) = 1 - \cos t$, we have $x'(t) = 1 - \cos t$ and $y'(t) = \sin t$. Hence, using the identity $1 - \cos t = 2\sin^2 \frac{t}{2}$, we get

$$x'(t)^2 + y'(t)^2 = (1 - \cos t)^2 + \sin^2 t = 1 - 2\cos t + \cos^2 t + \sin^2 t = 2 - 2\cos t = 4\sin^2 \frac{t}{2}$$

By the formula for the surface area we obtain:

$$S = 2\pi \int_0^{2\pi} y(t) \sqrt{x'(t)^2 + y'(t)^2}\, dt = 2\pi \int_0^{2\pi} (1 - \cos t) \cdot 2\sin \frac{t}{2}\, dt$$

$$= 2\pi \int_0^{2\pi} 2\sin^2 \frac{t}{2} \cdot 2\sin \frac{t}{2}\, dt = 8\pi \int_0^{2\pi} \sin^3 \frac{t}{2}\, dt = 16\pi \int_0^{\pi} \sin^3 u\, du$$

We use a reduction formula to compute this integral, obtaining

$$S = 16\pi \left[\frac{1}{3} \cos^3 u - \cos u \right]_0^{\pi} = 16\pi \left[\frac{4}{3} \right] = \frac{64\pi}{3}$$

39. CAS Use Simpson's Rule and $N = 30$ to approximate the surface area of the surface generated by revolving $c(t) = (t^2, e^{-t})$, $0 \le t \le 2$ about the x-axis.

SOLUTION With $x(t) = t^2$ and $y(t) = e^{-t}$, we have $x'(t) = 2t$ and $y'(t) = -e^{-t}$. Using the formula for the surface area, we get

$$S = 2\pi \int_0^2 y(t) \sqrt{x'(t)^2 + y'(t)^2}\, dt$$

$$= 2\pi \int_0^2 e^{-t} \sqrt{4t^2 + e^{-2t}}\, dt$$

Using Simpson's Rule with $n = 30$, the width of each interval is $\frac{1}{15}$, and the interval endpoints are $\frac{i}{15}$, $i = 0, 1, \ldots, 30$. The approximation is thus

$$s \approx 2\pi \cdot \frac{1/15}{3} \left(e^0 \sqrt{4 \cdot 0^2 + e^0} + 4e^{-1/15} \sqrt{4 \cdot (1/15)^2 + e^{-2/15}} + \cdots \right.$$

$$\left. + 4e^{-29/15} \sqrt{4 \cdot (29/15)^2 + e^{-58/15}} + e^{-2} \sqrt{4 \cdot 2^2 + e^{-4}} \right)$$

$$\approx 8.88635$$

Further Insights and Challenges

41. *CAS* Let $b(t)$ be the "Butterfly Curve":

$$x(t) = \sin t \left(e^{\cos t} - 2\cos 4t - \sin\left(\frac{t}{12}\right)^5 \right)$$

$$y(t) = \cos t \left(e^{\cos t} - 2\cos 4t - \sin\left(\frac{t}{12}\right)^5 \right)$$

(a) Use a computer algebra system to plot $b(t)$ and the speed $s'(t)$ for $0 \le t \le 12\pi$.

(b) Approximate the length $b(t)$ for $0 \le t \le 10\pi$.

SOLUTION

(a) Let $f(t) = e^{\cos t} - 2\cos 4t - \sin\left(\frac{t}{12}\right)^5$, then

$$x(t) = \sin t f(t)$$
$$y(t) = \cos t f(t)$$

and so

$$(x'(t))^2 + (y'(t))^2 = [\sin t f'(t) + \cos t f(t)]^2 + [\cos t f'(t) - \sin t f(t)]^2$$

Using the identity $\sin^2 t + \cos^2 t = 1$, we get

$$(x'(t))^2 + (y'(t))^2 = (f'(t))^2 + (f(t))^2$$

Thus, $s'(t)$ is the following:

$$\sqrt{\left[e^{\cos t} - 2\cos 4t - \sin\left(\frac{t}{12}\right)^5\right]^2 + \left[-\sin t e^{\cos t} + 8\sin 4t - \frac{5}{12}\left(\frac{t}{12}\right)^4 \cos\left(\frac{t}{12}\right)^5\right]^2}$$

The following figures show the curves of $b(t)$ and the speed $s'(t)$ for $0 \le t \le 10\pi$:

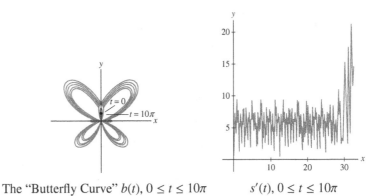

The "Butterfly Curve" $b(t)$, $0 \le t \le 10\pi$ $s'(t)$, $0 \le t \le 10\pi$

Looking at the graph, we see it would be difficult to compute the length using numeric integration; due to the high-frequency oscillations, very small steps would be needed.

(b) The length of $b(t)$ for $0 \le t \le 10\pi$ is given by the integral: $L = \int_0^{10\pi} s'(t)\,dt$ where $s'(t)$ is given in part (a). We approximate the length using the Midpoint Rule with $N = 30$. The numerical methods in Mathematica approximate the answer by 211.952. Using the Midpoint Rule with $N = 50$, we get 204.48; with $N = 500$, we get 211.6; and with $N = 5000$, we get 212.09.

43. A satellite orbiting at a distance R from the center of the earth follows the circular path $x(t) = R\cos \omega t$, $y(t) = R\sin \omega t$.

(a) Show that the period T (the time of one revolution) is $T = 2\pi/\omega$.

(b) According to Newton's Laws of Motion and Gravity,

$$x''(t) = -Gm_e \frac{x}{R^3}, \qquad y''(t) = -Gm_e \frac{y}{R^3}$$

where G is the universal gravitational constant and m_e is the mass of the earth. Prove that $R^3/T^2 = Gm_e/4\pi^2$. Thus, R^3/T^2 has the same value for all orbits (a special case of Kepler's Third Law).

SOLUTION

(a) As shown in Example 4, the circular path has constant speed of $\frac{ds}{dt} = \omega R$. Since the length of one revolution is $2\pi R$, the period T is

$$T = \frac{2\pi R}{\omega R} = \frac{2\pi}{\omega}$$

(b) Differentiating $x = R \cos \omega t$ twice with respect to t gives

$$x'(t) = -R\omega \sin \omega t$$

$$x''(t) = -R\omega^2 \cos \omega t$$

Substituting $x(t)$ and $x''(t)$ in the equation $x''(t) = -Gm_e \frac{x}{R^3}$ and simplifying, we obtain

$$-R\omega^2 \cos \omega t = -Gm_e \cdot \frac{R \cos \omega t}{R^3}$$

$$-R\omega^2 = -\frac{Gm_e}{R^2} \Rightarrow R^3 = \frac{Gm_e}{\omega^2}$$

By part (a), $T = \frac{2\pi}{\omega}$. Hence, $\omega = \frac{2\pi}{T}$. Substituting yields

$$R^3 = \frac{Gm_e}{\frac{4\pi^2}{T^2}} = \frac{T^2 Gm_e}{4\pi^2} \Rightarrow \frac{R^3}{T^2} = \frac{Gm_e}{4\pi^2}$$

11.3 Polar Coordinates (LT Section 12.3)

Preliminary Questions

1. Points P and Q with the same radial coordinate (choose the correct answer):

(a) lie on the same circle with the center at the origin.

(b) lie on the same ray based at the origin.

SOLUTION Two points with the same radial coordinate are equidistant from the origin; therefore, they lie on the same circle centered at the origin. The angular coordinate defines a ray based at the origin. Therefore, if the two points have the same angular coordinate, they lie on the same ray based at the origin.

2. Give two polar representations for the point $(x, y) = (0, 1)$, one with negative r and one with positive r.

SOLUTION The point $(0, 1)$ is on the y-axis, distant 1 unit from the origin, hence the polar representation with positive r is $(r, \theta) = \left(1, \frac{\pi}{2}\right)$. The point $(r, \theta) = \left(-1, \frac{\pi}{2}\right)$ is the reflection of $(r, \theta) = \left(1, \frac{\pi}{2}\right)$ through the origin, hence we must add π to return to the original point.

We obtain the following polar representation of $(0, 1)$ with negative r:

$$(r, \theta) = \left(-1, \frac{\pi}{2} + \pi\right) = \left(-1, \frac{3\pi}{2}\right)$$

3. Describe each of the following curves:

(a) $r = 2$ (b) $r^2 = 2$ (c) $r \cos \theta = 2$

SOLUTION

(a) Converting to rectangular coordinates we get

$$\sqrt{x^2 + y^2} = 2 \quad \text{or} \quad x^2 + y^2 = 2^2$$

This is the equation of the circle of radius 2 centered at the origin.

(b) We convert to rectangular coordinates, obtaining $x^2 + y^2 = 2$. This is the equation of the circle of radius $\sqrt{2}$, centered at the origin.

(c) We convert to rectangular coordinates. Since $x = r \cos \theta$ we obtain the following equation: $x = 2$. This is the equation of the vertical line through the point $(2, 0)$.

4. If $f(-\theta) = f(\theta)$, then the curve $r = f(\theta)$ is symmetric with respect to the (choose the correct answer):

(a) x-axis.   (b) y-axis.   (c) origin.

SOLUTION   The equality $f(-\theta) = f(\theta)$ for all θ implies that whenever a point (r, θ) is on the curve, also the point $(r, -\theta)$ is on the curve. Since the point $(r, -\theta)$ is the reflection of (r, θ) with respect to the x-axis, we conclude that the curve is symmetric with respect to the x-axis.

Exercises

1. Find polar coordinates for each of the seven points plotted in Figure 17. [Choose $r \geq 0$ and θ in $[0, 2\pi)$.]

FIGURE 17

SOLUTION   We mark the points as shown in the figure.

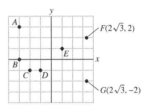

Using the data given in the figure for the x and y coordinates and the quadrants in which the point are located, we obtain:

(A): The rectangular coordinates of this point are $(-3, 3)$, so that

$$r = \sqrt{(-3)^2 + 3^2} = \sqrt{18}$$
$$\theta = \pi - \frac{\pi}{4} = \frac{3\pi}{4}$$
$$\Rightarrow (r, \theta) = \left(3\sqrt{2}, \frac{3\pi}{4}\right)$$

(B): The rectangular coordinates of this point are $(-3, 0)$, so that

$$r = \sqrt{(-3)^2 + 0^2} = 3$$
$$\theta = \pi$$
$$\Rightarrow (r, \theta) = (3, \pi)$$

(C): The rectangular coordinates of this point are $(-2, -1)$, so that

$$r = \sqrt{(-2)^2 + (-1)^2} = \sqrt{5} \approx 2.2$$
$$\theta = \tan^{-1}\left(\frac{-1}{-2}\right) = \tan^{-1}\left(\frac{1}{2}\right) \approx \pi + 0.46 \approx 3.6$$
$$\Rightarrow (r, \theta) \approx \left(\sqrt{5}, 3.6\right)$$

(D): The rectangular coordinates of this point are $(-1, -1)$, so that

$$r = \sqrt{(-1)^2 + (-1)^2} = \sqrt{2}$$
$$\theta = \pi + \frac{\pi}{4} = \frac{5\pi}{4}$$
$$\Rightarrow (r, \theta) = \left(\sqrt{2}, \frac{5\pi}{4}\right)$$

(E): The rectangular coordinates of this point are $(1, 1)$, so that

$$r = \sqrt{1^2 + 1^2} = \sqrt{2}$$
$$\theta = \tan^{-1}\left(\frac{1}{1}\right) = \frac{\pi}{4}$$
$$\Rightarrow (r, \theta) = \left(\sqrt{2}, \frac{\pi}{4}\right)$$

(F): The rectangular coordinates of this point are $(2\sqrt{3}, 2)$, so that

$$r = \sqrt{\left(2\sqrt{3}\right)^2 + 2^2} = \sqrt{16} = 4$$
$$\theta = \tan^{-1}\left(\frac{2}{2\sqrt{3}}\right) = \tan^{-1}\left(\frac{1}{\sqrt{3}}\right) = \frac{\pi}{6}$$
$$\Rightarrow (r, \theta) = \left(4, \frac{\pi}{6}\right)$$

(G): G is the reflection of F about the x axis; hence the two points have equal radial coordinates, and the angular coordinate of G is obtained from the angular coordinate of F: $\theta = 2\pi - \frac{\pi}{6} = \frac{11\pi}{6}$. Hence, the polar coordinates of G are $\left(4, \frac{11\pi}{6}\right)$.

3. Convert from rectangular to polar coordinates:

(a) $(1, 0)$ (b) $(3, \sqrt{3})$ (c) $(-2, 2)$ (d) $(-1, \sqrt{3})$

SOLUTION

(a) The point $(1, 0)$ is on the positive x axis distanced 1 unit from the origin. Hence, $r = 1$ and $\theta = 0$. Thus, $(r, \theta) = (1, 0)$.

(b) The point $\left(3, \sqrt{3}\right)$ is in the first quadrant so $\theta = \tan^{-1}\left(\frac{\sqrt{3}}{3}\right) = \frac{\pi}{6}$. Also, $r = \sqrt{3^2 + \left(\sqrt{3}\right)^2} = \sqrt{12}$. Hence, $(r, \theta) = \left(\sqrt{12}, \frac{\pi}{6}\right)$.

(c) The point $(-2, 2)$ is in the second quadrant. Hence,

$$\theta = \tan^{-1}\left(\frac{2}{-2}\right) = \tan^{-1}(-1) = \pi - \frac{\pi}{4} = \frac{3\pi}{4}$$

Also, $r = \sqrt{(-2)^2 + 2^2} = \sqrt{8}$. Hence, $(r, \theta) = \left(\sqrt{8}, \frac{3\pi}{4}\right)$.

(d) The point $\left(-1, \sqrt{3}\right)$ is in the second quadrant, hence,

$$\theta = \tan^{-1}\left(\frac{\sqrt{3}}{-1}\right) = \tan^{-1}\left(-\sqrt{3}\right) = \pi - \frac{\pi}{3} = \frac{2\pi}{3}$$

Also, $r = \sqrt{(-1)^2 + \left(\sqrt{3}\right)^2} = \sqrt{4} = 2$. Hence, $(r, \theta) = \left(2, \frac{2\pi}{3}\right)$

5. Convert from polar to rectangular coordinates:
(a) $(3, \frac{\pi}{6})$ (b) $(6, \frac{3\pi}{4})$ (c) $(0, \frac{\pi}{5})$ (d) $(5, -\frac{\pi}{2})$

SOLUTION

(a) Since $r = 3$ and $\theta = \frac{\pi}{6}$, we have:

$$x = r\cos\theta = 3\cos\frac{\pi}{6} = 3 \cdot \frac{\sqrt{3}}{2} \approx 2.6$$
$$y = r\sin\theta = 3\sin\frac{\pi}{6} = 3 \cdot \frac{1}{2} = 1.5$$
$$\Rightarrow (x, y) \approx (2.6, 1.5).$$

(b) For $\left(6, \frac{3\pi}{4}\right)$ we have $r = 6$ and $\theta = \frac{3\pi}{4}$. Hence,

$$x = r\cos\theta = 6\cos\frac{3\pi}{4} \approx -4.24$$
$$y = r\sin\theta = 6\sin\frac{3\pi}{4} \approx 4.24$$
$$\Rightarrow (x, y) \approx (-4.24, 4.24).$$

(c) Since $r = 0$, this point is the origin: $(x, y) = (0, 0)$.

(d) Since $r = 5$ and $\theta = -\frac{\pi}{2}$ we have

$$x = r\cos\theta = 5\cos\left(-\frac{\pi}{2}\right) = 5 \cdot 0 = 0$$
$$y = r\sin\theta = 5\sin\left(-\frac{\pi}{2}\right) = 5 \cdot (-1) = -5$$
$$\Rightarrow (x, y) = (0, -5)$$

7. Describe each tan shaded sector in Figure 18 by inequalities in r and θ.

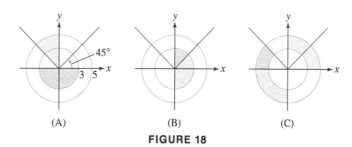

FIGURE 18

SOLUTION

(a) In the sector shown r is varying between 0 and 3 and θ is varying between π and 2π. Hence the following inequalities describe the sector:

$$0 \leq r \leq 3$$
$$\pi \leq \theta \leq 2\pi$$

(b) In the sector shown r is varying between 0 and 3 and θ is varying between $\frac{\pi}{4}$ and $\frac{\pi}{2}$. Hence, the inequalities for the sector are:

$$0 \leq r \leq 3$$
$$\frac{\pi}{4} \leq \theta \leq \frac{\pi}{2}$$

(c) In the sector shown r is varying between 3 and 5 and θ is varying between $\frac{3\pi}{4}$ and π. Hence, the inequalities are:

$$3 \leq r \leq 5$$
$$\frac{3\pi}{4} \leq \theta \leq \pi$$

9. Find an equation in polar coordinates of the line through the origin with slope $\frac{1}{\sqrt{3}}$.

SOLUTION This line makes an angle of $\theta_0 = \tan^{-1}\frac{1}{\sqrt{3}} = \frac{\pi}{6}$ with the positive x-axis, so its equation is $\theta = \frac{\pi}{6}$.

11. What is the slope of the line $\theta = \frac{3\pi}{5}$?

SOLUTION This line makes an angle $\theta_0 = \frac{3\pi}{5}$ with the positive x-axis; hence the slope of the line is $m = \tan\frac{3\pi}{5} \approx -3.1$.

In Exercises 13–18, convert to an equation in rectangular coordinates.

13. $r = 7$

SOLUTION $r = 7$ describes the points having distance 7 from the origin, that is, the circle with radius 7 centered at the origin. The equation of the circle in rectangular coordinates is

$$x^2 + y^2 = 7^2 = 49$$

15. $r = 2\sin\theta$

SOLUTION We multiply the equation by r and substitute $r^2 = x^2 + y^2$, $r\sin\theta = y$. This gives

$$r^2 = 2r\sin\theta$$
$$x^2 + y^2 = 2y$$

Moving the $2y$ and completing the square yield: $x^2 + y^2 - 2y = 0$ and $x^2 + (y-1)^2 = 1$. Thus, $r = 2\sin\theta$ is the equation of a circle of radius 1 centered at $(0, 1)$.

17. $r = \dfrac{1}{\cos\theta - \sin\theta}$

SOLUTION We multiply the equation by $\cos\theta - \sin\theta$ and substitute $y = r\sin\theta$, $x = r\cos\theta$. This gives

$$r(\cos\theta - \sin\theta) = 1$$
$$r\cos\theta - r\sin\theta = 1$$

$x - y = 1 \Rightarrow y = x - 1$. Thus,

$$r = \frac{1}{\cos\theta - \sin\theta}$$

is the equation of the line $y = x - 1$.

In Exercises 19–24, convert to an equation in polar coordinates of the form $r = f(\theta)$.

19. $x^2 + y^2 = 5$

SOLUTION We make the substitution $x^2 + y^2 = r^2$ to obtain; $r^2 = 5$ or $r = \sqrt{5}$.

21. $y = x^2$

SOLUTION Substituting $y = r \sin \theta$ and $x = r \cos \theta$ yields

$$r \sin \theta = r^2 \cos^2 \theta$$

Then, dividing by $r \cos^2 \theta$ we obtain,

$$\frac{\sin \theta}{\cos^2 \theta} = r \quad \text{so} \quad r = \tan \theta \sec \theta$$

23. $e^{\sqrt{x^2+y^2}} = 1$

SOLUTION Make the substitution $r = \sqrt{x^2 + y^2}$ to get $e^r = 1$. Taking logs of both sides gives $r = 0$. The graph of this equation consists only of the origin.

25. Match each equation with its description:

(a) $r = 2$ (i) Vertical line
(b) $\theta = 2$ (ii) Horizontal line
(c) $r = 2 \sec \theta$ (iii) Circle
(d) $r = 2 \csc \theta$ (iv) Line through origin

SOLUTION

(a) $r = 2$ describes the points 2 units from the origin. Hence, it is the equation of a circle.

(b) $\theta = 2$ describes the points P so that $\overline{OP}$ makes an angle of $\theta_0 = 2$ with the positive x-axis. Hence, it is the equation of a line through the origin.

(c) This is $r \cos \theta = 2$, which is $x = 2$, a vertical line.

(d) Converting to rectangular coordinates, we get $r = 2 \csc \theta$, so $r \sin \theta = 2$ and $y = 2$. This is the equation of a horizontal line.

27. Find the values of θ in the plot of $r = 4 \cos \theta$ corresponding to points A, B, C, D in Figure 19. Then indicate the portion of the graph traced out as θ varies in the following intervals:

(a) $0 \leq \theta \leq \frac{\pi}{2}$ (b) $\frac{\pi}{2} \leq \theta \leq \pi$ (c) $\pi \leq \theta \leq \frac{3\pi}{2}$

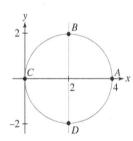

FIGURE 19 Plot of $r = 4 \cos \theta$.

SOLUTION The point A is on the x-axis hence $\theta = 0$. The point B is in the first quadrant with $x = y = 2$; hence $\theta = \tan^{-1}\left(\frac{2}{2}\right) = \tan^{-1}(1) = \frac{\pi}{4}$. The point C is at the origin. Thus,

$$r = 0 \Rightarrow 4 \cos \theta = 0 \Rightarrow \theta = \frac{\pi}{2}, \frac{3\pi}{2}$$

The point D is in the fourth quadrant with $x = 2$, $y = -2$, hence

$$\theta = \tan^{-1}\left(\frac{-2}{2}\right) = \tan^{-1}(-1) = 2\pi - \frac{\pi}{4} = \frac{7\pi}{4}$$

$0 \leq \theta \leq \frac{\pi}{2}$ represents the first quadrant; hence the points (r, θ) where $r = 4 \cos \theta$ and $0 \leq \theta \leq \frac{\pi}{2}$ are the points on the circle which are in the first quadrant, as shown below:

If we insist that $r \geq 0$, then since $\frac{\pi}{2} \leq \theta \leq \pi$ represents the second quadrant and $\pi \leq \theta \leq \frac{3\pi}{2}$ represents the third quadrant, and since the circle $r = 4\cos\theta$ has no points in the left xy-plane, then there are no points for (b) and (c). However, if we allow $r < 0$ then (b) represents the semicircle

and (c) like (a) represent

29. What are the polar equations of the lines parallel to the line $r\cos\left(\theta - \frac{\pi}{3}\right) = 1$?

SOLUTION The line $r\cos\left(\theta - \frac{\pi}{3}\right) = 1$, or $r = \sec\left(\theta - \frac{\pi}{3}\right)$, is perpendicular to the ray $\theta = \frac{\pi}{3}$ and at distance $d = 1$ from the origin. Hence, the lines parallel to this line are also perpendicular to the ray $\theta = \frac{\pi}{3}$, so the polar equations of these lines are $r = d\sec\left(\theta - \frac{\pi}{3}\right)$ or $r\cos\left(\theta - \frac{\pi}{3}\right) = d$.

31. Sketch the curve $r = \frac{1}{2}\theta$ (the spiral of Archimedes) for θ between 0 and 2π by plotting the points for $\theta = 0, \frac{\pi}{4}, \frac{\pi}{2}, \ldots, 2\pi$.

SOLUTION We first plot the following points (r, θ) on the spiral:

$$O = (0,0), \ A = \left(\frac{\pi}{8}, \frac{\pi}{4}\right), \ B = \left(\frac{\pi}{4}, \frac{\pi}{2}\right), \ C = \left(\frac{3\pi}{8}, \frac{3\pi}{4}\right), \ D = \left(\frac{\pi}{2}, \pi\right),$$

$$E = \left(\frac{5\pi}{8}, \frac{5\pi}{4}\right), \ F = \left(\frac{3\pi}{4}, \frac{3\pi}{2}\right), \ G = \left(\frac{7\pi}{8}, \frac{7\pi}{4}\right), \ H = (\pi, 2\pi)$$

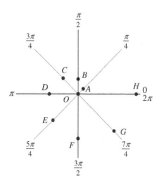

Since $r(0) = \frac{0}{2} = 0$, the graph begins at the origin and moves toward the points A, B, C, D, E, F, G, and H as θ varies from $\theta = 0$ to the other values stated above. Connecting the points in this direction we obtain the following graph for $0 \leq \theta \leq 2\pi$:

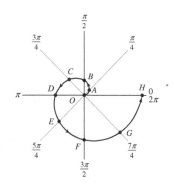

33. Sketch the cardioid curve $r = 1 + \cos\theta$.

SOLUTION Since $\cos\theta$ is periodic with period 2π, the entire curve will be traced out as θ varies from 0 to 2π. Additionally, since $\cos(2\pi - \theta) = \cos(\theta)$, we can sketch the curve for θ between 0 and π and reflect the result through the x axis to obtain the whole curve. Use the values $\theta = 0, \frac{\pi}{6}, \frac{\pi}{4}, \frac{\pi}{3}, \frac{\pi}{2}, \frac{2\pi}{3}, \frac{3\pi}{4}, \frac{5\pi}{6}$, and π:

θ	r	point
0	$1 + \cos 0 = 2$	$(2, 0)$
$\frac{\pi}{6}$	$1 + \cos\frac{\pi}{6} = \frac{2+\sqrt{3}}{2}$	$\left(\frac{2+\sqrt{3}}{2}, \frac{\pi}{6}\right)$
$\frac{\pi}{4}$	$1 + \cos\frac{\pi}{4} = \frac{2+\sqrt{2}}{2}$	$\left(\frac{2+\sqrt{2}}{2}, \frac{\pi}{4}\right)$
$\frac{\pi}{3}$	$1 + \cos\frac{\pi}{3} = \frac{3}{2}$	$\left(\frac{3}{2}, \frac{\pi}{3}\right)$
$\frac{\pi}{2}$	$1 + \cos\frac{\pi}{2} = 1$	$\left(1, \frac{\pi}{2}\right)$
$\frac{2\pi}{3}$	$1 + \cos\frac{2\pi}{3} = \frac{1}{2}$	$\left(\frac{1}{2}, \frac{2\pi}{3}\right)$
$\frac{3\pi}{4}$	$1 + \cos\frac{3\pi}{4} = \frac{2-\sqrt{2}}{2}$	$\left(\frac{2-\sqrt{2}}{2}, \frac{3\pi}{4}\right)$
$\frac{5\pi}{6}$	$1 + \cos\frac{5\pi}{6} = \frac{2-\sqrt{3}}{2}$	$\left(\frac{2-\sqrt{3}}{2}, \frac{5\pi}{6}\right)$

$\theta = 0$ corresponds to the point $(2, 0)$, and the graph moves clockwise as θ increases from 0 to π. Thus the graph is

Reflecting through the x axis gives the other half of the curve:

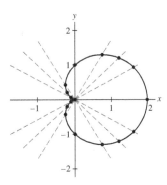

35. Figure 21 displays the graphs of $r = \sin 2\theta$ in r versus θ rectangular coordinates and in polar coordinates, where it is a "rose with four petals." Identify:

(a) The points in (B) corresponding to points A–I in (A).

(b) The parts of the curve in (B) corresponding to the angle intervals $[0, \frac{\pi}{2}], [\frac{\pi}{2}, \pi], [\pi, \frac{3\pi}{2}]$, and $[\frac{3\pi}{2}, 2\pi]$.

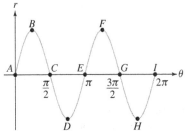

(A) Graph of r as a function of θ, where $r = \sin 2\theta$

(B) Graph of $r = \sin 2\theta$ in polar coordinates

FIGURE 21

SOLUTION

(a) The graph (I) gives the following polar coordinates of the labeled points:

$$A: \quad \theta = 0, \quad r = 0$$

$$B: \quad \theta = \frac{\pi}{4}, \quad r = \sin\frac{2\pi}{4} = 1$$

$$C: \quad \theta = \frac{\pi}{2}, \quad r = 0$$

$$D: \quad \theta = \frac{3\pi}{4}, \quad r = \sin\frac{2 \cdot 3\pi}{4} = -1$$

$$E: \quad \theta = \pi, \quad r = 0$$

$$F: \quad \theta = \frac{5\pi}{4}, \quad r = 1$$

$$G: \quad \theta = \frac{3\pi}{2}, \quad r = 0$$

$$H: \quad \theta = \frac{7\pi}{4}, \quad r = -1$$

$$I: \quad \theta = 2\pi, \quad r = 0$$

Since the maximal value of $|r|$ is 1, the points with $r = 1$ or $r = -1$ are the farthest points from the origin. The corresponding quadrant is determined by the value of θ and the sign of r. If $r_0 < 0$, the point (r_0, θ_0) is on the ray $\theta = -\theta_0$. These considerations lead to the following identification of the points in the xy plane. Notice that $A, C, G, E,$ and I are the same point.

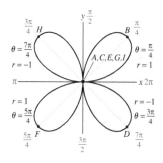

(b) We use the graph (I) to find the sign of $r = \sin 2\theta$: $0 \leq \theta \leq \frac{\pi}{2} \Rightarrow r \geq 0 \Rightarrow (r, \theta)$ is in the first quadrant. $\frac{\pi}{2} \leq \theta \leq \pi \Rightarrow r \leq 0 \Rightarrow (r, \theta)$ is in the fourth quadrant. $\pi \leq \theta \leq \frac{3\pi}{2} \Rightarrow r \geq 0 \Rightarrow (r, \theta)$ is in the third quadrant. $\frac{3\pi}{2} \leq \theta \leq 2\pi \Rightarrow r \leq 0 \Rightarrow (r, \theta)$ is in the second quadrant. That is,

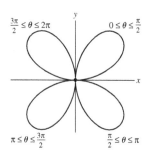

37. (a) [GU] Plot the curve $r = \frac{\theta}{2\pi - \theta}$ for $0 \leq \theta < 2\pi$.

(b) With r as in (a), compute the limits $\lim\limits_{\theta \to 2\pi^-} r\cos\theta$ and $\lim\limits_{\theta \to 2\pi^-} r\sin\theta$.

(c) Explain how the limits in (b) show that the curve approaches a horizontal asymptote as θ approaches 2π from the left. What is the asymptote?

SOLUTION

(a) The plot is below:

(b)

$$\lim_{\theta \to 2\pi^-} r\cos\theta = \lim_{\theta \to 2\pi^-} \frac{\theta \cos\theta}{2\pi - \theta}$$

$$\lim_{\theta \to 2\pi^-} r\sin\theta = \lim_{\theta \to 2\pi^-} \frac{\theta \sin\theta}{2\pi - \theta}$$

Note that in both cases the denominator approaches 0 positively. For the first limit, the numerator approaches 2π, so the limit diverges to $+\infty$. For the second limit, the numerator also approaches zero, so we use L'Hôpital's Rule:

$$\lim_{\theta \to 2\pi^-} r\sin\theta = \lim_{\theta \to 2\pi^-} \frac{\theta \sin\theta}{2\pi - \theta} = \lim_{\theta \to 2\pi^-} \frac{\sin\theta + \theta \cos\theta}{-1} = -2\pi$$

These results are confirmed by the graph, which appears to approach $y = -2\pi$ but gets arbitrarily large in the x-coordinate.

39. [GU] Plot the **cissoid** $r = 2\sin\theta \tan\theta$ and show that its equation in rectangular coordinates is

$$y^2 = \frac{x^3}{2 - x}$$

SOLUTION Using a CAS we obtain the following curve of the cissoid:

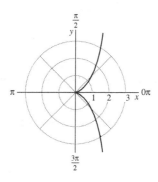

We substitute $\sin\theta = \frac{y}{r}$ and $\tan\theta = \frac{y}{x}$ in $r = 2\sin\theta \tan\theta$ to obtain

$$r = 2\frac{y}{r} \cdot \frac{y}{x}$$

Multiplying by rx, setting $r^2 = x^2 + y^2$ and simplifying, yields

$$r^2 x = 2y^2$$
$$(x^2 + y^2)x = 2y^2$$
$$x^3 + y^2 x = 2y^2$$
$$y^2(2 - x) = x^3$$

so

$$y^2 = \frac{x^3}{2 - x}$$

41. Show that

$$r = a\cos\theta + b\sin\theta$$

is the equation of a circle passing through the origin. Express the radius and center (in rectangular coordinates) in terms of a and b and write the equation in rectangular coordinates.

SOLUTION We multiply the equation by r and then make the substitution $x = r\cos\theta$, $y = r\sin\theta$, and $r^2 = x^2 + y^2$. This gives

$$r^2 = ar\cos\theta + br\sin\theta$$

$$x^2 + y^2 = ax + by$$

Transferring sides and completing the square yields

$$x^2 - ax + y^2 - by = 0$$

$$\left(x^2 - 2\cdot\frac{a}{2}x + \left(\frac{a}{2}\right)^2\right) + \left(y^2 - 2\cdot\frac{b}{2}y + \left(\frac{b}{2}\right)^2\right) = \left(\frac{a}{2}\right)^2 + \left(\frac{b}{2}\right)^2$$

$$\left(x - \frac{a}{2}\right)^2 + \left(y - \frac{b}{2}\right)^2 = \frac{a^2 + b^2}{4}$$

This is the equation of the circle with radius $\frac{\sqrt{a^2+b^2}}{2}$ centered at the point $\left(\frac{a}{2}, \frac{b}{2}\right)$. By plugging in $x = 0$ and $y = 0$ it is clear that the circle passes through the origin.

43. Use the identity $\cos 2\theta = \cos^2\theta - \sin^2\theta$ to find a polar equation of the hyperbola $x^2 - y^2 = 1$.

SOLUTION We substitute $x = r\cos\theta$, $y = r\sin\theta$ in $x^2 - y^2 = 1$ to obtain

$$r^2\cos^2\theta - r^2\sin^2\theta = 1$$

$$r^2(\cos^2\theta - \sin^2\theta) = 1$$

Using the identity $\cos 2\theta = \cos^2\theta - \sin^2\theta$ we obtain the following equation of the hyperbola:

$$r^2\cos 2\theta = 1 \quad \text{or} \quad r^2 = \sec 2\theta$$

45. Show that $\cos 3\theta = \cos^3\theta - 3\cos\theta\sin^2\theta$ and use this identity to find an equation in rectangular coordinates for the curve $r = \cos 3\theta$.

SOLUTION We use the identities $\cos(\alpha + \beta) = \cos\alpha\cos\beta - \sin\alpha\sin\beta$, $\cos 2\alpha = \cos^2\alpha - \sin^2\alpha$, and $\sin 2\alpha = 2\sin\alpha\cos\alpha$ to write

$$\cos 3\theta = \cos(2\theta + \theta) = \cos 2\theta\cos\theta - \sin 2\theta\sin\theta$$

$$= (\cos^2\theta - \sin^2\theta)\cos\theta - 2\sin\theta\cos\theta\sin\theta$$

$$= \cos^3\theta - \sin^2\theta\cos\theta - 2\sin^2\theta\cos\theta$$

$$= \cos^3\theta - 3\sin^2\theta\cos\theta$$

Using this identity we may rewrite the equation $r = \cos 3\theta$ as follows:

$$r = \cos^3\theta - 3\sin^2\theta\cos\theta \tag{1}$$

Since $x = r\cos\theta$ and $y = r\sin\theta$, we have $\cos\theta = \frac{x}{r}$ and $\sin\theta = \frac{y}{r}$. Substituting into (1) gives:

$$r = \left(\frac{x}{r}\right)^3 - 3\left(\frac{y}{r}\right)^2\left(\frac{x}{r}\right)$$

$$r = \frac{x^3}{r^3} - \frac{3y^2 x}{r^3}$$

We now multiply by r^3 and make the substitution $r^2 = x^2 + y^2$ to obtain the following equation for the curve:

$$r^4 = x^3 - 3y^2 x$$

$$(x^2 + y^2)^2 = x^3 - 3y^2 x$$

In Exercises 47–50, find an equation in polar coordinates of the line $\mathcal{L}$ with the given description.

47. The point on $\mathcal{L}$ closest to the origin has polar coordinates $(2, \frac{\pi}{9})$.

SOLUTION In Example 5, it is shown that the polar equation of the line where (r, α) is the point on the line closest to the origin is $r = d \sec(\theta - \alpha)$. Setting $(d, \alpha) = (2, \frac{\pi}{9})$ we obtain the following equation of the line:
$$r = 2 \sec\left(\theta - \frac{\pi}{9}\right)$$

49. $\mathcal{L}$ is tangent to the circle $r = 2\sqrt{10}$ at the point with rectangular coordinates $(-2, -6)$.

SOLUTION

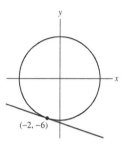

Since $\mathcal{L}$ is tangent to the circle at the point $(-2, -6)$, this is the point on $\mathcal{L}$ closest to the center of the circle which is at the origin. Therefore, we may use the polar coordinates (d, α) of this point in the equation of the line:
$$r = d \sec(\theta - \alpha) \tag{1}$$

We thus must convert the coordinates $(-2, -6)$ to polar coordinates. This point is in the third quadrant so $\pi < \alpha < \frac{3\pi}{2}$. We get
$$d = \sqrt{(-2)^2 + (-6)^2} = \sqrt{40} = 2\sqrt{10}$$
$$\alpha = \tan^{-1}\left(\frac{-6}{-2}\right) = \tan^{-1} 3 \approx \pi + 1.25 \approx 4.39$$

Substituting in (1) yields the following equation of the line:
$$r = 2\sqrt{10} \sec(\theta - 4.39)$$

51. Show that every line that does not pass through the origin has a polar equation of the form
$$r = \frac{b}{\sin\theta - a\cos\theta}$$
where $b \neq 0$.

SOLUTION Write the equation of the line in rectangular coordinates as $y = ax + b$. Since the line does not pass through the origin, we have $b \neq 0$. Substitute for y and x to convert to polar coordinates, and simplify:
$$y = ax + b$$
$$r \sin\theta = ar \cos\theta + b$$
$$r(\sin\theta - a\cos\theta) = b$$
$$r = \frac{b}{\sin\theta - a\cos\theta}$$

53. For $a > 0$, a **lemniscate curve** is the set of points P such that the product of the distances from P to $(a, 0)$ and $(-a, 0)$ is a^2. Show that the equation of the lemniscate is
$$(x^2 + y^2)^2 = 2a^2(x^2 - y^2)$$

Then find the equation in polar coordinates. To obtain the simplest form of the equation, use the identity $\cos 2\theta = \cos^2\theta - \sin^2\theta$. Use graph plotting software and plot the lemniscate for $a = 2$.

SOLUTION We compute the distances d_1 and d_2 of $P(x,y)$ from the points $(a,0)$ and $(-a,0)$, respectively. We obtain:

$$d_1 = \sqrt{(x-a)^2 + (y-0)^2} = \sqrt{(x-a)^2 + y^2}$$

$$d_2 = \sqrt{(x+a)^2 + (y-0)^2} = \sqrt{(x+a)^2 + y^2}$$

For the points $P(x,y)$ on the lemniscate we have $d_1 d_2 = a^2$. That is,

$$a^2 = \sqrt{(x-a)^2 + y^2} \sqrt{(x+a)^2 + y^2} = \sqrt{[(x-a)^2 + y^2][(x+a)^2 + y^2]}$$

$$= \sqrt{(x-a)^2(x+a)^2 + y^2(x-a)^2 + y^2(x+a)^2 + y^4}$$

$$= \sqrt{(x^2 - a^2)^2 + y^2[(x-a)^2 + (x+a)^2] + y^4}$$

$$= \sqrt{x^4 - 2a^2x^2 + a^4 + y^2(x^2 - 2xa + a^2 + x^2 + 2xa + a^2) + y^4}$$

$$= \sqrt{x^4 - 2a^2x^2 + a^4 + 2y^2x^2 + 2y^2a^2 + y^4}$$

$$= \sqrt{x^4 + 2x^2y^2 + y^4 + 2a^2(y^2 - x^2) + a^4}$$

$$= \sqrt{(x^2 + y^2)^2 + 2a^2(y^2 - x^2) + a^4}$$

Squaring both sides and simplifying yields

$$a^4 = (x^2 + y^2)^2 + 2a^2(y^2 - x^2) + a^4$$

$$0 = (x^2 + y^2)^2 + 2a^2(y^2 - x^2)$$

so

$$(x^2 + y^2)^2 = 2a^2(x^2 - y^2)$$

We now find the equation in polar coordinates. We substitute $x = r\cos\theta$, $y = r\sin\theta$, and $x^2 + y^2 = r^2$ into the equation of the lemniscate. This gives

$$(r^2)^2 = 2a^2(r^2\cos^2\theta - r^2\sin^2\theta) = 2a^2r^2(\cos^2\theta - \sin^2\theta) = 2a^2r^2\cos 2\theta$$

$$r^4 = 2a^2r^2\cos 2\theta$$

$r = 0$ is a solution, hence the origin is on the curve. For $r \neq 0$ we divide the equation by r^2 to obtain $r^2 = 2a^2\cos 2\theta$. This curve also includes the origin ($r = 0$ is obtained for $\theta = \frac{\pi}{4}$ for example); hence this is the polar equation of the lemniscate. Setting $a = 2$ we get $r^2 = 8\cos 2\theta$.

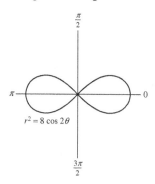

55. The Slope of the Tangent Line in Polar Coordinates Show that a polar curve $r = f(\theta)$ has parametric equations

$$x = f(\theta)\cos\theta, \quad y = f(\theta)\sin\theta$$

Then apply Theorem 1 of Section 11.1 to prove

$$\frac{dy}{dx} = \frac{f(\theta)\cos\theta + f'(\theta)\sin\theta}{-f(\theta)\sin\theta + f'(\theta)\cos\theta} \quad \boxed{2}$$

where $f'(\theta) = df/d\theta$.

SOLUTION Since $x = r\cos\theta$ and $y = r\sin\theta$ in polar coordinates, if $r = f(\theta)$, then

$$x = r\cos\theta = f(\theta)\cos\theta, \quad y = r\sin\theta = f(\theta)\sin\theta$$

for points on the curve. Therefore these parametric equations also describe the curve. Now, by the formula for the derivative we have

$$\frac{dy}{dx} = \frac{y'(\theta)}{x'(\theta)} \quad (1)$$

We differentiate the functions $x = f(\theta)\cos\theta$ and $y = f(\theta)\sin\theta$ using the Product Rule for differentiation. This gives

$$y'(\theta) = f'(\theta)\sin\theta + f(\theta)\cos\theta$$
$$x'(\theta) = f'(\theta)\cos\theta - f(\theta)\sin\theta$$

Substituting in (1) gives

$$\frac{dy}{dx} = \frac{f'(\theta)\sin\theta + f(\theta)\cos\theta}{f'(\theta)\cos\theta - f(\theta)\sin\theta} = \frac{f(\theta)\cos\theta + f'(\theta)\sin\theta}{-f(\theta)\sin\theta + f'(\theta)\cos\theta}$$

57. Use Eq. (2) to find the slope of the tangent line to $r = \theta$ at $\theta = \frac{\pi}{2}$ and $\theta = \pi$.

SOLUTION In the given curve we have $r = f(\theta) = \theta$. Using Eq. (2) we obtain the following derivative, which is the slope of the tangent line at (r, θ).

$$\frac{dy}{dx} = \frac{f(\theta)\cos\theta + f'(\theta)\sin\theta}{-f(\theta)\sin\theta + f'(\theta)\cos\theta} = \frac{\theta\cos\theta + 1\cdot\sin\theta}{-\theta\sin\theta + 1\cdot\cos\theta} \quad (1)$$

The slope, m, of the tangent line at $\theta = \frac{\pi}{2}$ and $\theta = \pi$ is obtained by substituting these values in (1). We get ($\theta = \frac{\pi}{2}$):

$$m = \frac{\frac{\pi}{2}\cos\frac{\pi}{2} + \sin\frac{\pi}{2}}{-\frac{\pi}{2}\sin\frac{\pi}{2} + \cos\frac{\pi}{2}} = \frac{\frac{\pi}{2}\cdot 0 + 1}{-\frac{\pi}{2}\cdot 1 + 0} = \frac{1}{-\frac{\pi}{2}} = -\frac{2}{\pi}$$

($\theta = \pi$):

$$m = \frac{\pi\cos\pi + \sin\pi}{-\pi\sin\pi + \cos\pi} = \frac{-\pi}{-1} = \pi$$

59. Find the polar coordinates of the points on the lemniscate $r^2 = \cos 2\theta$ in Figure 24 where the tangent line is horizontal.

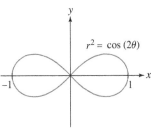

FIGURE 24

SOLUTION This curve is defined for $-\frac{\pi}{2} \le 2t \le \frac{\pi}{2}$ (where $\cos 2t \ge 0$), so for $-\frac{\pi}{4} \le t \le \frac{\pi}{4}$. For each θ in that range, there are two values of r satisfying the equation ($\pm\sqrt{\cos 2t}$). By symmetry, we need only calculate the coordinates of the points corresponding to the positive square root (i.e., to the right of the y axis). Then the equation becomes $r = \sqrt{\cos 2t}$. Now, by Eq. (2), with $f(t) = \sqrt{\cos(2t)}$ and $f'(t) = -\sin(2t)(\cos(2t))^{-1/2}$, we have

$$\frac{dy}{dx} = \frac{f(t)\cos t + f'(t)\sin t}{-f(t)\sin t + f'(t)\cos t} = \frac{\cos t\sqrt{\cos(2t)} - \sin(2t)\sin t(\cos(2t))^{-1/2}}{-\sin t\sqrt{\cos(2t)} - \sin(2t)\cos t(\cos(2t))^{-1/2}}$$

The tangent line is horizontal when this derivative is zero, which occurs when the numerator of the fraction is zero and the denominator is not. Multiply top and bottom of the fraction by $\sqrt{\cos(2t)}$, and use the identities $\cos 2t = \cos^2 t - \sin^2 t$, $\sin 2t = 2 \sin t \cos t$ to get

$$-\frac{\cos t \cos 2t - \sin t \sin 2t}{\sin t \cos 2t + \cos t \sin 2t} = -\frac{\cos t(\cos^2 t - 3 \sin^2 t)}{\sin t \cos 2t + \cos t \sin 2t}$$

The numerator is zero when $\cos t = 0$, so when $t = \frac{\pi}{2}$ or $t = \frac{3\pi}{2}$, or when $\tan t = \pm \frac{1}{\sqrt{3}}$, so when $t = \pm\frac{\pi}{6}$ or $t = \pm\frac{5\pi}{6}$. Of these possibilities, only $t = \pm\frac{\pi}{6}$ lie in the range $-\frac{\pi}{4} \le t \le \frac{\pi}{4}$. Note that the denominator is nonzero for $t = \pm\frac{\pi}{6}$, so these are the two values of t for which the tangent line is horizontal. The corresponding values of r are solutions to

$$r^2 = \cos\left(2 \cdot \frac{\pi}{6}\right) = \cos\left(\frac{\pi}{3}\right) = \frac{1}{2}$$

$$r^2 = \cos\left(2 \cdot \frac{-\pi}{6}\right) = \cos\left(-\frac{\pi}{3}\right) = \frac{1}{2}$$

Finally, the four points are $(r, t) =$

$$\left(\frac{1}{\sqrt{2}}, \frac{\pi}{6}\right), \quad \left(-\frac{1}{\sqrt{2}}, \frac{\pi}{6}\right), \quad \left(\frac{1}{\sqrt{2}}, -\frac{\pi}{6}\right), \quad \left(-\frac{1}{\sqrt{2}}, -\frac{\pi}{6}\right)$$

If desired, we can change the second and fourth points by adding π to the angle and making r positive, to get

$$\left(\frac{1}{\sqrt{2}}, \frac{\pi}{6}\right), \quad \left(\frac{1}{\sqrt{2}}, \frac{7\pi}{6}\right), \quad \left(\frac{1}{\sqrt{2}}, -\frac{\pi}{6}\right), \quad \left(\frac{1}{\sqrt{2}}, \frac{5\pi}{6}\right)$$

61. Use Eq. (2) to show that for $r = \sin\theta + \cos\theta$,

$$\frac{dy}{dx} = \frac{\cos 2\theta + \sin 2\theta}{\cos 2\theta - \sin 2\theta}$$

Then calculate the slopes of the tangent lines at points A, B, C in Figure 20.

SOLUTION In Exercise 55 we proved that for a polar curve $r = f(\theta)$ the following formula holds:

$$\frac{dy}{dx} = \frac{f(\theta)\cos\theta + f'(\theta)\sin\theta}{-f(\theta)\sin\theta + f'(\theta)\cos\theta} \tag{1}$$

For the given circle we have $r = f(\theta) = \sin\theta + \cos\theta$, hence $f'(\theta) = \cos\theta - \sin\theta$. Substituting in (1) we have

$$\frac{dy}{dx} = \frac{(\sin\theta + \cos\theta)\cos\theta + (\cos\theta - \sin\theta)\sin\theta}{-(\sin\theta + \cos\theta)\sin\theta + (\cos\theta - \sin\theta)\cos\theta} = \frac{\sin\theta\cos\theta + \cos^2\theta + \cos\theta\sin\theta - \sin^2\theta}{-\sin^2\theta - \cos\theta\sin\theta + \cos^2\theta - \sin\theta\cos\theta}$$

$$= \frac{\cos^2\theta - \sin^2\theta + 2\sin\theta\cos\theta}{\cos^2\theta - \sin^2\theta - 2\sin\theta\cos\theta}$$

We use the identities $\cos^2\theta - \sin^2\theta = \cos 2\theta$ and $2\sin\theta\cos\theta = \sin 2\theta$ to obtain

$$\frac{dy}{dx} = \frac{\cos 2\theta + \sin 2\theta}{\cos 2\theta - \sin 2\theta} \tag{2}$$

The derivative $\frac{dy}{dx}$ is the slope of the tangent line at (r, θ). The slopes of the tangent lines at the points with polar coordinates $A = \left(1, \frac{\pi}{2}\right)$ $B = \left(0, \frac{3\pi}{4}\right)$ $C = (1, 0)$ are computed by substituting the values of θ in (2). This gives

$$\left.\frac{dy}{dx}\right|_A = \frac{\cos\left(2 \cdot \frac{\pi}{2}\right) + \sin\left(2 \cdot \frac{\pi}{2}\right)}{\cos\left(2 \cdot \frac{\pi}{2}\right) - \sin\left(2 \cdot \frac{\pi}{2}\right)} = \frac{\cos\pi + \sin\pi}{\cos\pi - \sin\pi} = \frac{-1 + 0}{-1 - 0} = 1$$

$$\left.\frac{dy}{dx}\right|_B = \frac{\cos\left(2 \cdot \frac{3\pi}{4}\right) + \sin\left(2 \cdot \frac{3\pi}{4}\right)}{\cos\left(2 \cdot \frac{3\pi}{4}\right) - \sin\left(2 \cdot \frac{3\pi}{4}\right)} = \frac{\cos\frac{3\pi}{2} + \sin\frac{3\pi}{2}}{\cos\frac{3\pi}{2} - \sin\frac{3\pi}{2}} = \frac{0 - 1}{0 + 1} = -1$$

$$\left.\frac{dy}{dx}\right|_C = \frac{\cos(2 \cdot 0) + \sin(2 \cdot 0)}{\cos(2 \cdot 0) - \sin(2 \cdot 0)} = \frac{\cos 0 + \sin 0}{\cos 0 - \sin 0} = \frac{1 + 0}{1 - 0} = 1$$

Further Insights and Challenges

63. [GU] Use a graphing utility to convince yourself that the polar equations $r = f_1(\theta) = 2\cos\theta - 1$ and $r = f_2(\theta) = 2\cos\theta + 1$ have the same graph. Then explain why. *Hint:* Show that the points $(f_1(\theta + \pi), \theta + \pi)$ and $(f_2(\theta), \theta)$ coincide.

SOLUTION The graphs of $r = 2\cos\theta - 1$ and $r = 2\cos\theta + 1$ in the xy-plane coincide as shown in the graph obtained using a CAS.

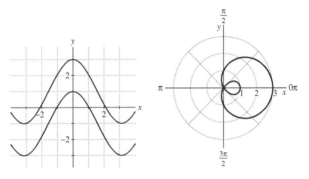

Recall that (r, θ) and $(-r, \theta + \pi)$ represent the same point. Replacing θ by $\theta + \pi$ and r by $(-r)$ in $r = 2\cos\theta - 1$ we obtain

$$-r = 2\cos(\theta + \pi) - 1$$
$$-r = -2\cos\theta - 1$$
$$r = 2\cos\theta + 1$$

Thus, the two equations define the same graph. (One could also convert both equations to rectangular coordinates and note that they come out identical.)

11.4 Area and Arc Length in Polar Coordinates (LT Section 12.4)

Preliminary Questions

1. Polar coordinates are suited to finding the area (choose one):
(a) under a curve between $x = a$ and $x = b$.
(b) bounded by a curve and two rays through the origin.

SOLUTION Polar coordinates are best suited to finding the area bounded by a curve and two rays through the origin. The formula for the area in polar coordinates gives the area of this region.

2. Is the formula for area in polar coordinates valid if $f(\theta)$ takes negative values?

SOLUTION The formula for the area

$$\frac{1}{2}\int_\alpha^\beta f(\theta)^2\, d\theta$$

always gives the actual (positive) area, even if $f(\theta)$ takes on negative values.

3. The horizontal line $y = 1$ has polar equation $r = \csc\theta$. Which area is represented by the integral $\dfrac{1}{2}\int_{\pi/6}^{\pi/2} \csc^2\theta\, d\theta$ (Figure 10)?

(a) $\square ABCD$ (b) $\triangle ABC$ (c) $\triangle ACD$

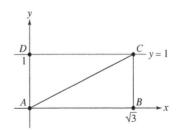

FIGURE 10

SOLUTION This integral represents an area taken from $\theta = \pi/6$ to $\theta = \pi/2$, which can only be the triangle $\triangle ACD$, as seen in part (c).

Exercises

1. Sketch the region bounded by the circle $r = 5$ and the rays $\theta = \frac{\pi}{2}$ and $\theta = \pi$, and compute its area as an integral in polar coordinates.

SOLUTION The region bounded by the circle $r = 5$ and the rays $\theta = \frac{\pi}{2}$ and $\theta = \pi$ is the shaded region in the figure. The area of the region is given by the following integral:

$$\frac{1}{2} \int_{\pi/2}^{\pi} r^2 \, d\theta = \frac{1}{2} \int_{\pi/2}^{\pi} 5^2 \, d\theta = \frac{25}{2}\left(\pi - \frac{\pi}{2}\right) = \frac{25\pi}{4}$$

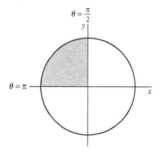

3. Calculate the area of the circle $r = 4 \sin \theta$ as an integral in polar coordinates (see Figure 4). Be careful to choose the correct limits of integration.

SOLUTION The equation $r = 4 \sin \theta$ defines a circle of radius 2 tangent to the x-axis at the origin as shown in the figure:

The circle is traced as θ varies from 0 to π. We use the area in polar coordinates and the identity

$$\sin^2 \theta = \frac{1}{2}(1 - \cos 2\theta)$$

to obtain the following area:

$$A = \frac{1}{2} \int_0^{\pi} r^2 \, d\theta = \frac{1}{2} \int_0^{\pi} (4 \sin \theta)^2 \, d\theta = 8 \int_0^{\pi} \sin^2 \theta \, d\theta = 4 \int_0^{\pi} (1 - \cos 2\theta) \, d\theta = 4\left[\theta - \frac{\sin 2\theta}{2}\right]_0^{\pi}$$

$$= 4\left(\left(\pi - \frac{\sin 2\pi}{2}\right) - 0\right) = 4\pi$$

5. Find the area of the shaded region in Figure 12. Note that θ varies from 0 to $\frac{\pi}{2}$.

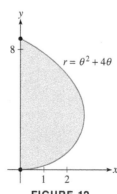

FIGURE 12

SOLUTION Since θ varies from 0 to $\frac{\pi}{2}$, the area is

$$\frac{1}{2}\int_0^{\pi/2} r^2\, d\theta = \frac{1}{2}\int_0^{\pi/2}(\theta^2+4\theta)^2\, d\theta = \frac{1}{2}\int_0^{\pi/2}\theta^4+8\theta^3+16\theta^2\, d\theta$$

$$=\frac{1}{2}\left(\frac{1}{5}\theta^5+2\theta^4+\frac{16}{3}\theta^3\right)\bigg|_0^{\pi/2} = \frac{\pi^5}{320}+\frac{\pi^4}{16}+\frac{\pi^2}{3}$$

7. Find the total area enclosed by the cardioid in Figure 14.

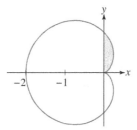

FIGURE 14 The cardioid $r = 1 - \cos\theta$.

SOLUTION We graph $r = 1 - \cos\theta$ in r and θ (cartesian, not polar, this time):

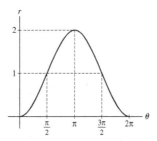

We see that as θ varies from 0 to π, the radius r increases from 0 to 2, so we get the upper half of the cardioid (the lower half is obtained as θ varies from π to 2π and consequently r decreases from 2 to 0). Since the cardioid is symmetric with respect to the x-axis we may compute the upper area and double the result. Using

$$\cos^2\theta = \frac{\cos 2\theta + 1}{2}$$

we get

$$A = 2\cdot\frac{1}{2}\int_0^{\pi} r^2\, d\theta = \int_0^{\pi}(1-\cos\theta)^2\, d\theta = \int_0^{\pi}\left(1 - 2\cos\theta + \cos^2\theta\right)\, d\theta$$

$$= \int_0^{\pi}\left(1 - 2\cos\theta + \frac{\cos 2\theta + 1}{2}\right)\, d\theta = \int_0^{\pi}\left(\frac{3}{2} - 2\cos\theta + \frac{1}{2}\cos 2\theta\right)\, d\theta$$

$$= \frac{3}{2}\theta - 2\sin\theta + \frac{1}{4}\sin 2\theta\bigg|_0^{\pi} = \frac{3\pi}{2}$$

The total area enclosed by the cardioid is $A = \frac{3\pi}{2}$.

9. Find the area of one leaf of the "four-petaled rose" $r = \sin 2\theta$ (Figure 15). Then prove that the total area of the rose is equal to one-half the area of the circumscribed circle.

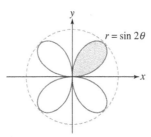

FIGURE 15 Four-petaled rose $r = \sin 2\theta$.

SOLUTION We consider the graph of $r = \sin 2\theta$ in cartesian and in polar coordinates:

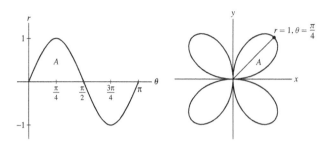

We see that as θ varies from 0 to $\frac{\pi}{4}$ the radius r is increasing from 0 to 1, and when θ varies from $\frac{\pi}{4}$ to $\frac{\pi}{2}$, r is decreasing back to zero. Hence, the leaf in the first quadrant is traced as θ varies from 0 to $\frac{\pi}{2}$. The area of the leaf (the four leaves have equal areas) is thus

$$A = \frac{1}{2}\int_0^{\pi/2} r^2\, d\theta = \frac{1}{2}\int_0^{\pi/2} \sin^2 2\theta\, d\theta$$

Using the identity

$$\sin^2 2\theta = \frac{1 - \cos 4\theta}{2}$$

we get

$$A = \frac{1}{2}\int_0^{\pi/2}\left(\frac{1}{2} - \frac{\cos 4\theta}{2}\right) d\theta = \frac{1}{2}\left(\frac{\theta}{2} - \frac{\sin 4\theta}{8}\right)\bigg|_0^{\pi/2} = \frac{1}{2}\left(\left(\frac{\pi}{4} - \frac{\sin 2\pi}{8}\right) - 0\right) = \frac{\pi}{8}$$

The area of one leaf is $A = \frac{\pi}{8} \approx 0.39$. Therefore the total area of the rose is $4 \cdot \frac{\pi}{8} = \frac{\pi}{2}$. The circumscribed circle is the unit circle; hence one half of its area is $\frac{\pi}{2}$, and the two are equal.

11. Find the area of the intersection of the circles $r = 2\sin\theta$ and $r = 2\cos\theta$.

SOLUTION A graph of the area is shown below; $r = 2\sin\theta$ is in blue, $r = 2\cos\theta$ is in red, and the area to be computed is shaded:

This region is traversed as θ varies from 0 to $\frac{\pi}{2}$, and it is clearly symmetric about the line $\theta = \frac{\pi}{4}$. The total area is thus twice the area between $\theta = 0$ and $\theta = \frac{\pi}{4}$, so it is

$$A = 2 \cdot \frac{1}{2}\int_0^{\pi/4} r^2\, d\theta = \int_0^{\pi/4} 4\sin^2\theta\, d\theta$$

Using the identity

$$\sin^2\theta = \frac{1 - \cos 2\theta}{2}$$

we get

$$A = \int_0^{\pi/4} 4\sin^2\theta\, d\theta = 2\int_0^{\pi/4}(1 - \cos 2\theta)\, d\theta = 2\left(\theta - \frac{1}{2}\sin 2\theta\right)\bigg|_{\theta=0}^{\pi/4} = \frac{\pi}{2} - 1$$

13. Find the area of region A in Figure 17.

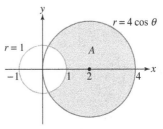

FIGURE 17

SOLUTION We first find the values of θ at the points of intersection of the two circles, by solving the following equation for $-\frac{\pi}{2} \leq x \leq \frac{\pi}{2}$:

$$4\cos\theta = 1 \Rightarrow \cos\theta = \frac{1}{4} \Rightarrow \theta_1 = \cos^{-1}\left(\frac{1}{4}\right)$$

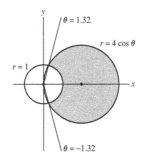

We now compute the area using the formula for the area between two curves:

$$A = \frac{1}{2}\int_{-\theta_1}^{\theta_1} \left((4\cos\theta)^2 - 1^2\right) d\theta = \frac{1}{2}\int_{-\theta_1}^{\theta_1} \left(16\cos^2\theta - 1\right) d\theta$$

Using the identity $\cos^2\theta = \frac{\cos 2\theta + 1}{2}$ we get

$$A = \frac{1}{2}\int_{-\theta_1}^{\theta_1} \left(\frac{16(\cos 2\theta + 1)}{2} - 1\right) d\theta = \frac{1}{2}\int_{-\theta_1}^{\theta_1} (8\cos 2\theta + 7) d\theta = \frac{1}{2}(4\sin 2\theta + 7\theta)\Big|_{-\theta_1}^{\theta_1}$$

$$= 4\sin 2\theta_1 + 7\theta_1 = 8\sin\theta_1\cos\theta_1 + 7\theta_1 = 8\sqrt{1-\cos^2\theta_1}\cos\theta_1 + 7\theta_1$$

Using the fact that $\cos\theta_1 = \frac{1}{4}$ we get

$$A = \frac{\sqrt{15}}{2} + 7\cos^{-1}\left(\frac{1}{4}\right) \approx 11.163$$

15. Find the area of the inner loop of the limaçon with polar equation $r = 2\cos\theta - 1$ (Figure 19).

SOLUTION We consider the graph of $r = 2\cos\theta - 1$ in cartesian and in polar, for $-\frac{\pi}{2} \leq x \leq \frac{\pi}{2}$:

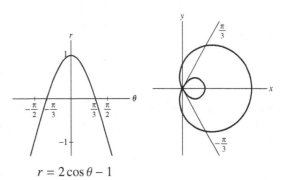

As θ varies from $-\frac{\pi}{3}$ to 0, r increases from 0 to 1. As θ varies from 0 to $\frac{\pi}{3}$, r decreases from 1 back to 0. Hence, the inner loop of the limaçon is traced as θ varies from $-\frac{\pi}{3}$ to $\frac{\pi}{3}$. The area of the shaded region is thus

$$A = \frac{1}{2}\int_{-\pi/3}^{\pi/3} r^2\, d\theta = \frac{1}{2}\int_{-\pi/3}^{\pi/3}(2\cos\theta - 1)^2\, d\theta = \frac{1}{2}\int_{-\pi/3}^{\pi/3}\left(4\cos^2\theta - 4\cos\theta + 1\right)\, d\theta$$

$$= \frac{1}{2}\int_{-\pi/3}^{\pi/3}(2(\cos 2\theta + 1) - 4\cos\theta + 1)\, d\theta = \frac{1}{2}\int_{-\pi/3}^{\pi/3}(2\cos 2\theta - 4\cos\theta + 3)\, d\theta$$

$$= \frac{1}{2}(\sin 2\theta - 4\sin\theta + 3\theta)\bigg|_{-\pi/3}^{\pi/3} = \frac{1}{2}\left(\left(\sin\frac{2\pi}{3} - 4\sin\frac{\pi}{3} + \pi\right) - \left(\sin\left(-\frac{2\pi}{3}\right) - 4\sin\left(-\frac{\pi}{3}\right) - \pi\right)\right)$$

$$= \frac{\sqrt{3}}{2} - \frac{4\sqrt{3}}{2} + \pi = \pi - \frac{3\sqrt{3}}{2} \approx 0.54$$

17. Find the area of the part of the circle $r = \sin\theta + \cos\theta$ in the fourth quadrant (see Exercise 30 in Section 11.3).

SOLUTION The value of θ corresponding to the point B is the solution of $r = \sin\theta + \cos\theta = 0$ for $-\pi \leq \theta \leq \pi$.

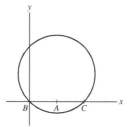

That is,
$$\sin\theta + \cos\theta = 0 \Rightarrow \sin\theta = -\cos\theta \Rightarrow \tan\theta = -1 \Rightarrow \theta = -\frac{\pi}{4}$$

At the point C, we have $\theta = 0$. The part of the circle in the fourth quadrant is traced if θ varies between $-\frac{\pi}{4}$ and 0. This leads to the following area:

$$A = \frac{1}{2}\int_{-\pi/4}^{0} r^2\, d\theta = \frac{1}{2}\int_{-\pi/4}^{0}(\sin\theta + \cos\theta)^2\, d\theta = \frac{1}{2}\int_{-\pi/4}^{0}\left(\sin^2\theta + 2\sin\theta\cos\theta + \cos^2\theta\right)\, d\theta$$

Using the identities $\sin^2\theta + \cos^2\theta = 1$ and $2\sin\theta\cos\theta = \sin 2\theta$ we get:

$$A = \frac{1}{2}\int_{-\pi/4}^{0}(1 + \sin 2\theta)\, d\theta = \frac{1}{2}\left(\theta - \frac{\cos 2\theta}{2}\right)\bigg|_{-\pi/4}^{0}$$

$$= \frac{1}{2}\left(\left(0 - \frac{1}{2}\right) - \left(-\frac{\pi}{4} - \frac{\cos\left(\frac{-\pi}{2}\right)}{2}\right)\right) = \frac{1}{2}\left(\frac{\pi}{4} - \frac{1}{2}\right) = \frac{\pi}{8} - \frac{1}{4} \approx 0.14$$

19. Find the area between the two curves in Figure 20(A).

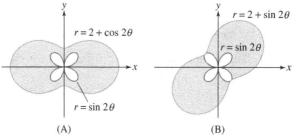

FIGURE 20

SOLUTION We compute the area A between the two curves as the difference between the area A_1 of the region enclosed in the outer curve $r = 2 + \cos 2\theta$ and the area A_2 of the region enclosed in the inner curve $r = \sin 2\theta$. That is,

$$A = A_1 - A_2$$

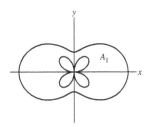

In Exercise 9 we showed that $A_2 = \frac{\pi}{2}$, hence,

$$A = A_1 - \frac{\pi}{2} \tag{1}$$

We compute the area A_1.

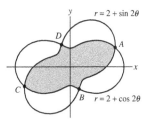

Using symmetry, the area is four times the area enclosed in the first quadrant. That is,

$$A_1 = 4 \cdot \frac{1}{2} \int_0^{\pi/2} r^2 \, d\theta = 2 \int_0^{\pi/2} (2 + \cos 2\theta)^2 \, d\theta = 2 \int_0^{\pi/2} \left(4 + 4\cos 2\theta + \cos^2 2\theta\right) d\theta$$

Using the identity $\cos^2 2\theta = \frac{1}{2}\cos 4\theta + \frac{1}{2}$ we get

$$A_1 = 2 \int_0^{\pi/2} \left(4 + 4\cos 2\theta + \frac{1}{2}\cos 4\theta + \frac{1}{2}\right) d\theta = 2 \int_0^{\pi/2} \left(\frac{9}{2} + \frac{1}{2}\cos 4\theta + 4\cos 2\theta\right) d\theta$$

$$= 2\left(\frac{9\theta}{2} + \frac{\sin 4\theta}{8} + 2\sin 2\theta\right)\bigg|_0^{\pi/2} = 2\left(\left(\frac{9\pi}{4} + \frac{\sin 2\pi}{8} + 2\sin\pi\right) - 0\right) = \frac{9\pi}{2} \tag{2}$$

Combining (1) and (2) we obtain

$$A = \frac{9\pi}{2} - \frac{\pi}{2} = 4\pi$$

21. Find the area inside both curves in Figure 21.

SOLUTION The area we need to find is the area of the shaded region in the figure.

We first find the values of θ at the points of intersection A, B, C, and D of the two curves, by solving the following equation for $-\pi \leq \theta \leq \pi$:

$$2 + \cos 2\theta = 2 + \sin 2\theta$$

$$\cos 2\theta = \sin 2\theta$$

$$\tan 2\theta = 1 \Rightarrow 2\theta = \frac{\pi}{4} + \pi k \Rightarrow \theta = \frac{\pi}{8} + \frac{\pi k}{2}$$

The solutions for $-\pi \leq \theta \leq \pi$ are

$$A: \quad \theta = \frac{\pi}{8}$$

$$B: \quad \theta = -\frac{3\pi}{8}$$

$$C: \quad \theta = -\frac{7\pi}{8}$$

$$D: \quad \theta = \frac{5\pi}{8}$$

Using symmetry, we compute the shaded area in the figure below and multiply it by 4:

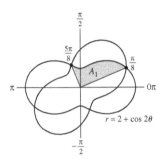

$$A = 4 \cdot A_1 = 4 \cdot \frac{1}{2} \cdot \int_{\pi/8}^{5\pi/8} (2 + \cos 2\theta)^2 \, d\theta = 2 \int_{\pi/8}^{5\pi/8} \left(4 + 4\cos 2\theta + \cos^2 2\theta\right) d\theta$$

$$= 2 \int_{\pi/8}^{5\pi/8} \left(4 + 4\cos 2\theta + \frac{1 + \cos 4\theta}{2}\right) d\theta = \int_{\pi/8}^{5\pi/8} (9 + 8\cos 2\theta + \cos 4\theta) \, d\theta$$

$$= 9\theta + 4\sin 2\theta + \frac{\sin 4\theta}{4} \bigg|_{\pi/8}^{5\pi/8} = 9\left(\frac{5\pi}{8} - \frac{\pi}{8}\right) + 4\left(\sin \frac{5\pi}{4} - \sin \frac{\pi}{4}\right) + \frac{1}{4}\left(\sin \frac{5\pi}{2} - \sin \frac{\pi}{2}\right) = \frac{9\pi}{2} - 4\sqrt{2}$$

23. (a) **GU** Plot $r(\theta) = 1 - \cos(10\theta)$ for $0 \leq \theta \leq 2\pi$.

(b) Compute the area enclosed inside the ten petals of the graph of $r(\theta)$.

(c) Explain why, for a positive integer n and $0 \leq \theta \leq 2\pi$, $r_n(\theta) = 1 - \cos(n\theta)$ traces out an n-petal flower inscribed in a circle of radius 2 centered at the origin.

(d) Show that the area enclosed inside the n petals of the graph of $r_n(\theta)$ is independent of n and equals $\frac{3}{8}A$, where A is the area of the circle of radius 2.

SOLUTION

(a) A graph of $r(\theta)$ is below:

(b) The area enclosed by the graph is (since the graph is traversed once for $0 \leq \theta \leq 2\pi$)

$$A = \frac{1}{2} \int_0^{2\pi} r^2 \, d\theta$$

$$= \frac{1}{2} \int_0^{2\pi} (1 - 2\cos 10\theta + \cos^2 10\theta) \, d\theta$$

$$= \frac{1}{2} \int_0^{2\pi} \left(1 - 2\cos 10\theta + \frac{1 + \cos 20\theta}{2}\right) d\theta$$

$$= \frac{1}{2} \int_0^{2\pi} \left(\frac{3}{2} - 2\cos 10\theta + \frac{1}{2}\cos 20\theta\right) d\theta$$

$$= \frac{1}{2}\left(\frac{3}{2}\theta - \frac{1}{5}\sin 10\theta + \frac{1}{40}\sin 20\theta\right)\bigg|_{\theta=0}^{2\pi}$$

$$= \frac{3\pi}{2}.$$

(c) $\cos n\theta$ has period $\frac{2\pi}{n}$, so over the range $0 \leq \theta \leq 2\pi$, $r_n(\theta)$ is zero $n + 1$ times. Between two such zeros, r_n varies from 0 to 2 (when $\cos n\theta = -1$), giving one petal of the rose. Since each petal has a maximum distance of 2 from the origin, the entire figure may be inscribed in a circle of radius 2 centered at the origin.

(d) Mimicking the argument from part (b), the area enclosed by the curve is

$$A = \frac{1}{2} \int_0^{2\pi} r^2 \, d\theta$$

$$= \frac{1}{2} \int_0^{2\pi} (1 - 2\cos n\theta + \cos^2 n\theta) \, d\theta$$

$$= \frac{1}{2} \int_0^{2\pi} \left(1 - 2\cos n\theta + \frac{1 + \cos 2n\theta}{2}\right) d\theta$$

$$= \frac{1}{2} \int_0^{2\pi} \left(\frac{3}{2} - 2\cos n\theta + \frac{1}{2}\cos 2n\theta\right) d\theta$$

$$= \frac{1}{2} \left(\frac{3}{2}\theta - \frac{2}{n}\sin n\theta + \frac{1}{4n}\sin 2n\theta\right)\bigg|_{\theta=0}^{2\pi}$$

$$= \frac{3\pi}{2}$$

Thus the area enclosed by the curve is independent of n. The area of a circle of radius 2 is 4π, and $\frac{3}{8} \cdot 4\pi = \frac{3\pi}{2}$.

25. Calculate the total length of the circle $r = 4\sin\theta$ as an integral in polar coordinates.

SOLUTION We use the formula for the arc length:

$$S = \int_\alpha^\beta \sqrt{f(\theta)^2 + f'(\theta)^2} \, d\theta \tag{1}$$

In this case, $f(\theta) = 4\sin\theta$ and $f'(\theta) = 4\cos\theta$, hence

$$\sqrt{f(\theta)^2 + f'(\theta)^2} = \sqrt{(4\sin\theta)^2 + (4\cos\theta)^2} = \sqrt{16} = 4$$

The circle is traced as θ is varied from 0 to π. Substituting $\alpha = 0, \beta = \pi$ in (1) yields $S = \int_0^\pi 4 \, d\theta = 4\pi$.

The circle $r = 4\sin\theta$

In Exercises 27–34, compute the length of the polar curve.

27. The length of $r = \theta^2$ for $0 \leq \theta \leq \pi$

SOLUTION We use the formula for the arc length. In this case $f(\theta) = \theta^2$, $f'(\theta) = 2\theta$, so we obtain

$$S = \int_0^\pi \sqrt{(\theta^2)^2 + (2\theta)^2} \, d\theta = \int_0^\pi \sqrt{\theta^4 + 4\theta^2} \, d\theta = \int_0^\pi \theta\sqrt{\theta^2 + 4} \, d\theta$$

We compute the integral using the substitution $u = \theta^2 + 4$, $du = 2\theta \, d\theta$. This gives

$$S = \frac{1}{2} \int_4^{\pi^2+4} \sqrt{u} \, du = \frac{1}{2} \cdot \frac{2}{3} u^{3/2} \bigg|_4^{\pi^2+4} = \frac{1}{3}\left((\pi^2 + 4)^{3/2} - 4^{3/2}\right) = \frac{1}{3}\left((\pi^2 + 4)^{3/2} - 8\right) \approx 14.55$$

29. The curve $r = \sin\theta$ for $0 \leq \theta \leq \pi$

SOLUTION We use the formula for the arc length. In this case $f(\theta) = \sin\theta$, and $f'(\theta) = \cos\theta$. Integrating gives

$$L = \int_0^\pi \sqrt{f(\theta)^2 + f'(\theta)^2}\, d\theta = \int_0^\pi \sqrt{\sin^2\theta + \cos^2\theta}\, d\theta = \int_0^\pi 1\, d\theta = \pi$$

31. $r = \sqrt{1 + \sin 2\theta}$ for $0 \le \theta \le \pi/4$

SOLUTION With $f(\theta) = \sqrt{1 + \sin 2\theta}$, we get

$$f'(\theta) = \frac{1}{2}(1 + \sin 2\theta)^{-1/2} \cdot 2\cos 2\theta = \frac{\cos 2\theta}{\sqrt{1 + \sin 2\theta}}$$

Then using the arc length formula gives

$$L = \int_0^{\pi/4} \sqrt{f(\theta)^2 + f'(\theta)^2}\, d\theta$$

$$= \int_0^{\pi/4} \sqrt{(1 + \sin 2\theta) + \frac{\cos^2 2\theta}{1 + \sin 2\theta}}\, d\theta$$

$$= \int_0^{\pi/4} \sqrt{\frac{(1 + \sin 2\theta)^2 + \cos^2 2\theta}{1 + \sin 2\theta}}\, d\theta$$

$$= \int_0^{\pi/4} \sqrt{\frac{2 + 2\sin 2\theta}{1 + \sin 2\theta}}\, d\theta$$

$$= \int_0^{\pi/4} \sqrt{2}\, d\theta = \frac{\pi}{4}\sqrt{2}$$

33. $r = \cos^2\theta$

SOLUTION Since $\cos\theta = \cos(-\theta)$ and $\cos^2(\pi - \theta) = \cos^2\theta$ the curve is symmetric with respect to the x and y-axis. Therefore, we may compute the length as four times the length of the part of the curve in the first quadrant. We use the formula for the arc length in polar coordinates. In this case, $f(\theta) = \cos^2\theta$, $f'(\theta) = 2\cos\theta(-\sin\theta)$, so we obtain

$$\sqrt{f(\theta)^2 + f'(\theta)^2} = \sqrt{\cos^4\theta + 4\cos^2\theta\sin^2\theta} = \cos\theta\sqrt{\cos^2\theta + 4\sin^2\theta}$$

$$= \cos\theta\sqrt{\cos^2\theta + \sin^2\theta + 3\sin^2\theta} = \cos\theta\sqrt{1 + 3\sin^2\theta}$$

Thus,

$$L = \int_0^{\pi/2} \sqrt{f(\theta)^2 + f'(\theta)^2}\, d\theta = \int_0^{\pi/2} \cos\theta\sqrt{1 + 3\sin^2\theta}\, d\theta$$

We compute the integral using the substitution $u = \sqrt{3}\sin\theta$ we get

$$L = \frac{1}{\sqrt{3}} \int_0^{\sqrt{3}} \sqrt{1 + u^2}\, du = \frac{1}{\sqrt{3}} \left(\frac{u}{2}\sqrt{1 + u^2} + \frac{1}{2}\ln|u + \sqrt{1 + u^2}| \right)\Big|_0^{\sqrt{3}}$$

$$= \frac{1}{\sqrt{3}}\left(\frac{\sqrt{3}}{2}\sqrt{1 + 3} + \frac{1}{2}\ln\left(\sqrt{3} + \sqrt{1+3}\right) - 0 \right) = 1 + \frac{1}{2\sqrt{3}}\ln\left(2 + \sqrt{3}\right)$$

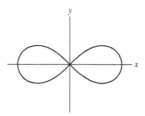

Graph of $r = \cos^2\theta$

Thus the total length equals $4L = 4 + \frac{2}{\sqrt{3}}\ln\left(2 + \sqrt{3}\right) \approx 5.52$.

In Exercises 35–38, express the length of the curve as an integral but do not evaluate it.

35. $r = e^\theta + 1, \quad 0 \leq \theta \leq \pi/2$

SOLUTION With $f(\theta) = e^\theta + 1$, we have $f'(\theta) = e^\theta$. Then using the arc length formula gives

$$L = \int_0^{\pi/2} \sqrt{f(\theta)^2 + f'(\theta)^2}\, d\theta = \int_0^{\pi/2} \sqrt{(e^\theta + 1)^2 + (e^\theta)^2}\, d\theta = \int_0^{\pi/2} \sqrt{2e^{2\theta} + 2e^\theta + 1}\, d\theta$$

37. $r = \sin^3 \theta, \quad 0 \leq \theta \leq 2\pi$

SOLUTION We have $f(t) = \sin^3 t$, $f'(t) = 3\sin^2 t \cos t$, so that

$$\sqrt{f(t)^2 + f'(t)^2} = \sqrt{\sin^6 t + 9\sin^4 t \cos^2 t} = \sin^2 t \sqrt{\sin^2 t + 9\cos^2 t}$$
$$= \sin^2 t \sqrt{\sin^2 t + \cos^2 t + 8\cos^2 t} = \sin^2 t \sqrt{1 + 8\cos^2 t}$$

Using the formula for arc length integral we get

$$L = \int_0^{2\pi} \sin^2 t \sqrt{1 + 8\cos^2 t}\, dt$$

In Exercises 39–42, use a computer algebra system to calculate the total length to two decimal places.

39. **CAS** The three-petal rose $r = \cos 3\theta$ in Figure 18

SOLUTION We have $f(\theta) = \cos 3\theta$, $f'(\theta) = -3\sin 3\theta$, so that

$$\sqrt{f(\theta)^2 + f'(\theta)^2} = \sqrt{\cos^2 3\theta + 9\sin^2 3\theta} = \sqrt{\cos^2 3\theta + \sin^2 3\theta + 8\sin^2 3\theta} = \sqrt{1 + 8\sin^2 3\theta}$$

Note that the curve is traversed completely for $0 \leq \theta \leq \pi$. Using the arc length formula and evaluating with Maple gives

$$L = \int_0^\pi \sqrt{f(\theta)^2 + f'(\theta)^2}\, d\theta = \int_0^\pi \sqrt{1 + 8\sin^2 3\theta}\, d\theta \approx 6.682446608$$

41. **CAS** The curve $r = \theta \sin \theta$ in Figure 22 for $0 \leq \theta \leq 4\pi$

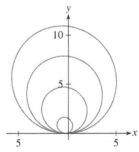

FIGURE 22 $r = \theta \sin \theta$ for $0 \leq \theta \leq 4\pi$.

SOLUTION We have $f(\theta) = \theta \sin \theta$, $f'(\theta) = \sin \theta + \theta \cos \theta$, so that

$$\sqrt{f(\theta)^2 + f'(\theta)^2} = \sqrt{\theta^2 \sin^2 \theta + (\sin \theta + \theta \cos \theta)^2} = \sqrt{\theta^2 \sin^2 \theta + \sin^2 \theta + 2\theta \sin \theta \cos \theta + \theta^2 \cos^2 \theta}$$
$$= \sqrt{\theta^2 + \sin^2 \theta + \theta \sin 2\theta}$$

using the identities $\sin^2 \theta + \cos^2 \theta = 1$ and $2\sin \theta \cos \theta = \sin 2\theta$. Thus by the arc length formula and evaluating with Maple, we have

$$L = \int_0^{4\pi} \sqrt{f(\theta)^2 + f'(\theta)^2}\, d\theta = \int_0^{4\pi} \sqrt{\theta^2 + \sin^2 \theta + \theta \sin 2\theta}\, d\theta \approx 79.56423976$$

Further Insights and Challenges

43. Suppose that the polar coordinates of a moving particle at time t are $(r(t), \theta(t))$. Prove that the particle's speed is equal to $\sqrt{(dr/dt)^2 + r^2(d\theta/dt)^2}$.

SOLUTION The speed of the particle in rectangular coordinates is:

$$\frac{ds}{dt} = \sqrt{x'(t)^2 + y'(t)^2} \tag{1}$$

We need to express the speed in polar coordinates. The x and y coordinates of the moving particles as functions of t are

$$x(t) = r(t)\cos\theta(t), \quad y(t) = r(t)\sin\theta(t)$$

We differentiate $x(t)$ and $y(t)$, using the Product Rule for differentiation. We obtain (omitting the independent variable t)

$$x' = r'\cos\theta - r(\sin\theta)\theta'$$
$$y' = r'\sin\theta - r(\cos\theta)\theta'$$

Hence,

$$x'^2 + y'^2 = (r'\cos\theta - r\theta'\sin\theta)^2 + (r'\sin\theta + r\theta'\cos\theta)^2$$
$$= r'^2\cos^2\theta - 2r'r\theta'\cos\theta\sin\theta + r^2\theta'^2\sin^2\theta + r'^2\sin^2\theta + 2r'r\theta'\sin\theta\cos\theta + r^2\theta'^2\cos^2\theta$$
$$= r'^2\left(\cos^2\theta + \sin^2\theta\right) + r^2\theta'^2\left(\sin^2\theta + \cos^2\theta\right) = r'^2 + r^2\theta'^2 \tag{2}$$

Substituting (2) into (1) we get

$$\frac{ds}{dt} = \sqrt{r'^2 + r^2\theta'^2} = \sqrt{\left(\frac{dr}{dt}\right)^2 + r^2\left(\frac{d\theta}{dt}\right)^2}$$

11.5 Conic Sections   (LT Section 12.5)

Preliminary Questions

1. Decide if the equation defines an ellipse, a hyperbola, a parabola, or no conic section at all.

(a) $4x^2 - 9y^2 = 12$
(b) $-4x + 9y^2 = 0$
(c) $4y^2 + 9x^2 = 12$
(d) $4x^3 + 9y^3 = 12$

SOLUTION

(a) This is the equation of the hyperbola $\left(\frac{x}{\sqrt{3}}\right)^2 - \left(\frac{y}{\frac{2}{\sqrt{3}}}\right)^2 = 1$, which is a conic section.

(b) The equation $-4x + 9y^2 = 0$ can be rewritten as $x = \frac{9}{4}y^2$, which defines a parabola. This is a conic section.

(c) The equation $4y^2 + 9x^2 = 12$ can be rewritten in the form $\left(\frac{y}{\sqrt{3}}\right)^2 + \left(\frac{x}{\frac{2}{\sqrt{3}}}\right)^2 = 1$, hence it is the equation of an ellipse, which is a conic section.

(d) This is not the equation of a conic section, since it is not an equation of degree two in x and y.

2. For which conic sections do the vertices lie between the foci?

SOLUTION If the vertices lie between the foci, the conic section is a hyperbola.

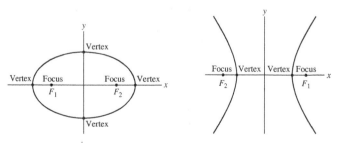

ellipse: foci between vertices   hyperbola: vertices between foci

3. What are the foci of $\left(\frac{x}{a}\right)^2 + \left(\frac{y}{b}\right)^2 = 1$ if $a < b$?

SOLUTION If $a < b$ the foci of the ellipse $\left(\frac{x}{a}\right)^2 + \left(\frac{y}{b}\right)^2 = 1$ are at the points $(0, c)$ and $(0, -c)$ on the y-axis, where $c = \sqrt{b^2 - a^2}$.

$\left(\frac{x}{a}\right)^2 + \left(\frac{y}{b}\right)^2 = 1;\ a < b$

4. What is the geometric interpretation of b/a in the equation of a hyperbola in standard position?

SOLUTION The vertices (i.e., the points where the focal axis intersects the hyperbola) are at the points $(a, 0)$ and $(-a, 0)$. The values $\pm \frac{b}{a}$ are the slopes of the two asymptotes of the hyperbola.

Hyperbola in standard position

Exercises

In Exercises 1–6, find the vertices and foci of the conic section.

1. $\left(\frac{x}{9}\right)^2 + \left(\frac{y}{4}\right)^2 = 1$

SOLUTION This is an ellipse in standard position with $a = 9$ and $b = 4$. Hence, $c = \sqrt{9^2 - 4^2} = \sqrt{65} \approx 8.06$. The foci are at $F_1 = (-8.06, 0)$ and $F_2 = (8.06, 0)$, and the vertices are $(9, 0), (-9, 0), (0, 4), (0, -4)$.

3. $\left(\frac{x}{4}\right)^2 - \left(\frac{y}{9}\right)^2 = 1$

SOLUTION This is a hyperbola in standard position with $a = 4$ and $b = 9$. Hence, $c = \sqrt{a^2 + b^2} = \sqrt{97} \approx 9.85$. The foci are at $(\pm \sqrt{97}, 0)$ and the vertices are $(\pm 2, 0)$.

5. $\left(\frac{x-4}{6}\right)^2 - \left(\frac{y+3}{5}\right)^2 = 1$

SOLUTION This is a hyperbola, but it is centered at $(4, -3)$. So first consider the same hyperbola, but centered at the origin:

$$\left(\frac{x}{6}\right)^2 - \left(\frac{y}{5}\right)^2 = 1$$

This is a hyperbola in standard position, with $a = 6$, $b = 5$, and $c = \sqrt{6^2 + 5^2} = \sqrt{61}$. The foci are then $(\pm \sqrt{61}, 0)$, and the vertices are $(\pm a, 0) = (\pm 6, 0)$. Translating back to $(4, -3)$ gives foci of $(4 \pm \sqrt{61}, -3)$ and vertices of $(4 \pm 6, -3)$, or $(10, -3)$ and $(-2, -3)$ for vertices.

In Exercises 7–10, find the equation of the ellipse obtained by translating (as indicated) the ellipse

$$\left(\frac{x-8}{6}\right)^2 + \left(\frac{y+4}{3}\right)^2 = 1$$

7. Translated with center at the origin

SOLUTION Recall that the equation

$$\frac{(x-h)^2}{a^2} + \frac{(y-k)^2}{b^2} = 1$$

describes an ellipse with center (h, k). Thus, for our ellipse to be located at the origin, it must have equation

$$\frac{x^2}{6^2} + \frac{y^2}{3^2} = 1$$

9. Translated to the right 6 units

SOLUTION Recall that the equation

$$\frac{(x-h)^2}{a^2} + \frac{(y-k)^2}{b^2} = 1$$

describes an ellipse with center (h, k). Thus, for our ellipse to be moved 6 units right, its new center is $(14, -4)$, so it must have equation

$$\frac{(x-14)^2}{6^2} + \frac{(y+4)^2}{3^2} = 1$$

In Exercises 11–14, find the equation of the given ellipse.

11. Vertices $(\pm 3, 0)$ and $(0, \pm 5)$

SOLUTION Here $b = 5 > a = 3$; since the vertices are symmetric about the origin, the ellipse is centered at the origin, so its equation is

$$\left(\frac{x}{3}\right)^2 + \left(\frac{y}{5}\right)^2 = 1$$

13. Foci $(0, \pm 10)$ and eccentricity $e = \frac{3}{5}$

SOLUTION Since the foci are on the y axis, this ellipse has a vertical major axis with center $(0, 0)$, so its equation is

$$\left(\frac{x}{b}\right)^2 + \left(\frac{y}{a}\right)^2 = 1$$

We have $a = \frac{c}{e} = \frac{10}{3/5} = \frac{50}{3}$ and

$$b = \sqrt{a^2 - c^2} = \sqrt{\frac{2500}{9} - 100} = \frac{1}{3}\sqrt{2500 - 900} = \frac{40}{3}$$

Thus the equation of the ellipse is

$$\left(\frac{x}{40/3}\right)^2 + \left(\frac{y}{50/3}\right)^2 = 1$$

In Exercises 15–20, find the equation of the given hyperbola.

15. Vertices $(\pm 3, 0)$ and foci $(\pm 5, 0)$

SOLUTION The equation is $\left(\frac{x}{a}\right)^2 - \left(\frac{y}{b}\right)^2 = 1$. The vertices are $(\pm a, 0)$ with $a = 3$ and the foci $(\pm c, 0)$ with $c = 5$. We use the relation $c = \sqrt{a^2 + b^2}$ to find b:

$$b = \sqrt{c^2 - a^2} = \sqrt{5^2 - 3^2} = \sqrt{16} = 4$$

Therefore, the equation of the hyperbola is

$$\left(\frac{x}{3}\right)^2 - \left(\frac{y}{4}\right)^2 = 1$$

17. Foci $(\pm 3, 0)$ and eccentricity $e = 3$

SOLUTION Since the foci lie on the x-axis and are symmetric about the origin, this is a hyperbola in standard position; by Theorem 4,

$$e = \frac{c}{a} \quad \Rightarrow \quad 3 = \frac{3}{a} \quad \Rightarrow \quad a = 1$$

Then $c = \sqrt{a^2 + b^2}$, so that $3 = \sqrt{1 + b^2}$ and therefore $b = \sqrt{8} = 2\sqrt{2}$. The equation of the hyperbola is

$$x^2 - \left(\frac{y}{2\sqrt{2}}\right)^2 = 1$$

19. Vertices $(-3, 0)$, $(7, 0)$ and eccentricity $e = 3$

SOLUTION The center is at $\frac{-3+7}{2} = 2$ with a horizontal focal axis, so the equation is

$$\left(\frac{x-2}{a}\right)^2 - \left(\frac{y}{b}\right)^2 = 1$$

Then $a = 7 - 2 = 5$, and $c = ae = 5 \cdot 3 = 15$. Finally,

$$b = \sqrt{c^2 - a^2} = \sqrt{15^2 - 5^2} = 10\sqrt{2}$$

so that the equation of the hyperbola is

$$\left(\frac{x-2}{5}\right)^2 - \left(\frac{y}{10\sqrt{2}}\right)^2 = 1$$

In Exercises 21–28, find the equation of the parabola with the given properties.

21. Vertex $(0, 0)$, focus $(\frac{1}{12}, 0)$

SOLUTION Since the focus is on the x-axis rather than the y-axis, and the vertex is $(0, 0)$, the equation is $x = \frac{1}{4c}y^2$. The focus is $(0, c)$ with $c = \frac{1}{12}$, so the equation is

$$x = \frac{1}{4 \cdot \frac{1}{12}}y^2 = 3y^2$$

23. Vertex $(0, 0)$, directrix $y = -5$

SOLUTION The equation is $y = \frac{1}{4c}x^2$. The directrix is $y = -c$ with $c = 5$, hence $y = \frac{1}{20}x^2$.

25. Focus $(0, 4)$, directrix $y = -4$

SOLUTION The focus is $(0, c)$ with $c = 4$ and the directrix is $y = -c$ with $c = 4$, hence the equation of the parabola is

$$y = \frac{1}{4c}x^2 = \frac{x^2}{16}$$

27. Focus $(2, 0)$, directrix $x = -2$

SOLUTION The focus is on the x-axis rather than on the y-axis and the directrix is a vertical line rather than horizontal as in the parabola in standard position. Therefore, the equation of the parabola is obtained by interchanging x and y in $y = \frac{1}{4c}x^2$. Also, by the given information $c = 2$. Hence, $x = \frac{1}{4c}y^2 = \frac{1}{4 \cdot 2}y^2$ or $x = \frac{y^2}{8}$.

In Exercises 29–38, find the vertices, foci, center (if an ellipse or a hyperbola), and asymptotes (if a hyperbola).

29. $x^2 + 4y^2 = 16$

SOLUTION We first divide the equation by 16 to convert it to the equation in standard form:

$$\frac{x^2}{16} + \frac{4y^2}{16} = 1 \Rightarrow \frac{x^2}{16} + \frac{y^2}{4} = 1 \Rightarrow \left(\frac{x}{4}\right)^2 + \left(\frac{y}{2}\right)^2 = 1$$

For this ellipse, $a = 4$ and $b = 2$ hence $c = \sqrt{4^2 - 2^2} = \sqrt{12} \approx 3.5$. Since $a > b$ we have:

- The vertices are at $(\pm 4, 0)$, $(0, \pm 2)$.
- The foci are $F_1 = (-3.5, 0)$ and $F_2 = (3.5, 0)$.
- The focal axis is the x-axis and the conjugate axis is the y-axis.
- The ellipse is centered at the origin.

31. $\left(\dfrac{x-3}{4}\right)^2 - \left(\dfrac{y+5}{7}\right)^2 = 1$

SOLUTION For this hyperbola $a = 4$ and $b = 7$ so $c = \sqrt{4^2 + 7^2} \approx 8.06$. For the standard hyperbola $\left(\dfrac{x}{4}\right)^2 - \left(\dfrac{y}{7}\right)^2 = 1$, we have:

- The vertices are $A = (4, 0)$ and $A' = (-4, 0)$.
- The foci are $F_1 = (8.06, 0)$ and $F_2 = (-8.06, 0)$.
- The focal axis is the x-axis and the conjugate axis is the y-axis.
- The center is at the midpoint of $\overline{F_1 F_2}$, that is, at the origin.
- The asymptotes $y = \pm \dfrac{b}{a} x$ are $y = \pm \dfrac{7}{4} x$.

The given hyperbola is a translation of the standard hyperbola, 3 units to the right and 5 units downward. Hence the following holds:

- The vertices are at $(7, -5)$ and $(-1, -5)$.
- The foci are at $(11.06, -5)$ and $(-5.06, -5)$.
- The focal axis is $y = -5$ and the conjugate axis is $x = 3$.
- The center is at $(3, -5)$.
- The asymptotes are $y + 5 = \pm \dfrac{7}{4}(x - 3)$.

33. $4x^2 - 3y^2 + 8x + 30y = 215$

SOLUTION Since there is no cross term, we complete the square of the terms involving x and y separately:

$$4x^2 - 3y^2 + 8x + 30y = 4\left(x^2 + 2x\right) - 3\left(y^2 - 10y\right) = 4(x+1)^2 - 4 - 3(y-5)^2 + 75 = 215$$

Hence,

$$4(x+1)^2 - 3(y-5)^2 = 144$$

$$\dfrac{4(x+1)^2}{144} - \dfrac{3(y-5)^2}{144} = 1$$

$$\left(\dfrac{x+1}{6}\right)^2 - \left(\dfrac{y-5}{\sqrt{48}}\right)^2 = 1$$

This is the equation of the hyperbola obtained by translating the hyperbola $\left(\dfrac{x}{6}\right)^2 - \left(\dfrac{y}{\sqrt{48}}\right)^2 = 1$, 1 unit to the left and 5 units upwards. Since $a = 6$, $b = \sqrt{48}$, we have $c = \sqrt{36 + 48} = \sqrt{84} \sim 9.2$. We obtain the following table:

	Standard position	Translated hyperbola
vertices	$(6, 0), (-6, 0)$	$(5, 5), (-7, 5)$
foci	$(\pm 9.2, 0)$	$(8.2, 5), (-10.2, 5)$
focal axis	The x-axis	$y = 5$
conjugate axis	The y-axis	$x = -1$
center	The origin	$(-1, 5)$
asymptotes	$y = \pm 1.15 x$	$y = -1.15x + 3.85$ $y = 1.15x + 6.15$

35. $y = 4(x - 4)^2$

SOLUTION By Exercise 34, the parabola $y = 4x^2$ has the vertex at the origin, the focus at $\left(0, \dfrac{1}{16}\right)$ and its axis is the y-axis. Our parabola is a translation of the standard parabola 4 units to the right. Hence its vertex is at $(4, 0)$, the focus is at $\left(4, \dfrac{1}{16}\right)$ and its axis is the vertical line $x = 4$.

37. $4x^2 + 25y^2 - 8x - 10y = 20$

SOLUTION Since there are no cross terms this conic section is obtained by translating a conic section in standard position. To identify the conic section we complete the square of the terms involving x and y separately:

$$4x^2 + 25y^2 - 8x - 10y = 4\left(x^2 - 2x\right) + 25\left(y^2 - \frac{2}{5}y\right)$$

$$= 4(x-1)^2 - 4 + 25\left(y - \frac{1}{5}\right)^2 - 1$$

$$= 4(x-1)^2 + 25\left(y - \frac{1}{5}\right)^2 - 5 = 20$$

Hence,
$$l4(x-1)^2 + 25\left(y - \frac{1}{5}\right)^2 = 25$$

$$\frac{4}{25}(x-1)^2 + \left(y - \frac{1}{5}\right)^2 = 1$$

$$\left(\frac{x-1}{\frac{5}{2}}\right)^2 + \left(y - \frac{1}{5}\right)^2 = 1$$

This is the equation of the ellipse obtained by translating the ellipse in standard position $\left(\frac{x}{\frac{5}{2}}\right)^2 + y^2 = 1$, 1 unit to the right and $\frac{1}{5}$ unit upward. Since $a = \frac{5}{2}$, $b = 1$ we have $c = \sqrt{\left(\frac{5}{2}\right)^2 - 1} \approx 2.3$, so we obtain the following table:

	Standard position	Translated ellipse
Vertices	$\left(\pm\frac{5}{2}, 0\right), (0, \pm 1)$	$\left(1 \pm \frac{5}{2}, \frac{1}{5}\right), \left(1, \frac{1}{5} \pm 1\right)$
Foci	$(-2.3, 0), (2.3, 0)$	$\left(-1.3, \frac{1}{5}\right), \left(3.3, \frac{1}{5}\right)$
Focal axis	The x-axis	$y = \frac{1}{5}$
Conjugate axis	The y-axis	$x = 1$
Center	The origin	$\left(1, \frac{1}{5}\right)$

In Exercises 39–42, use the Discriminant Test to determine the type of the conic section (in each case, the equation is nondegenerate). Use a graphing utility or computer algebra system to plot the curve.

39. $4x^2 + 5xy + 7y^2 = 24$

SOLUTION Here, $D = 25 - 4 \cdot 4 \cdot 7 = -87$, so the conic section is an ellipse.

41. $2x^2 - 8xy + 3y^2 - 4 = 0$

SOLUTION Here, $D = 64 - 4 \cdot 2 \cdot 3 = 40$, giving us a hyperbola.

43. Show that the "conic" $x^2 + 3y^2 - 6x + 12y + 23 = 0$ has no points.

SOLUTION Complete the square in each variable separately:

$$-23 = x^2 - 6x + 3y^2 + 12y = (x^2 - 6x + 9) + (3y^2 + 12y + 12) - 9 - 12 = (x-3)^2 + 3(y+2)^2 - 21$$

Collecting constants and reversing sides gives

$$(x-3)^2 + 3(y+2)^2 = -2$$

which has no solutions since the left-hand side is a sum of two squares and so is always nonnegative.

45. Show that $\frac{b}{a} = \sqrt{1 - e^2}$ for a standard ellipse of eccentricity e.

SOLUTION By the definition of eccentricity:

$$e = \frac{c}{a} \tag{1}$$

For the ellipse in standard position, $c = \sqrt{a^2 - b^2}$. Substituting into (1) and simplifying yields

$$e = \frac{\sqrt{a^2 - b^2}}{a} = \sqrt{\frac{a^2 - b^2}{a^2}} = \sqrt{1 - \left(\frac{b}{a}\right)^2}$$

We square the two sides and solve for $\frac{b}{a}$:

$$e^2 = 1 - \left(\frac{b}{a}\right)^2 \Rightarrow \left(\frac{b}{a}\right)^2 = 1 - e^2 \Rightarrow \frac{b}{a} = \sqrt{1 - e^2}$$

47. Explain why the dots in Figure 23 lie on a parabola. Where are the focus and directrix located?

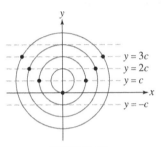

FIGURE 23

SOLUTION All the circles are centered at $(0, c)$ and the kth circle has radius kc. Hence the indicated point P_k on the kth circle has a distance kc from the point $F = (0, c)$. The point P_k also has distance kc from the line $y = -c$. That is, the indicated point on each circle is equidistant from the point $F = (0, c)$ and the line $y = -c$, hence it lies on the parabola with focus at $F = (0, c)$ and directrix $y = -c$.

49. A **latus rectum** of a conic section is a chord through a focus parallel to the directrix. Find the area bounded by the parabola $y = x^2/(4c)$ and its latus rectum (refer to Figure 8).

SOLUTION The directrix is $y = -c$, and the focus is $(0, c)$. The chord through the focus parallel to $y = -c$ is clearly $y = c$; this line intersects the parabola when $c = x^2/(4c)$ or $4c^2 = x^2$, so when $x = \pm 2c$. The desired area is then

$$\int_{-2c}^{2c} c - \frac{1}{4c}x^2\, dx = \left(cx - \frac{1}{12c}x^3\right)\bigg|_{-2c}^{2c}$$

$$= 2c^2 - \frac{8c^3}{12c} - \left(-2c^2 - \frac{(-2c)^3}{12c}\right) = 4c^2 - \frac{4}{3}c^2 = \frac{8}{3}c^2$$

In Exercises 51–54, find the polar equation of the conic with the given eccentricity and directrix, and focus at the origin.

51. $e = \frac{1}{2}$, $x = 3$

SOLUTION Substituting $e = \frac{1}{2}$ and $d = 3$ in the polar equation of a conic section we obtain

$$r = \frac{ed}{1 + e\cos\theta} = \frac{\frac{1}{2} \cdot 3}{1 + \frac{1}{2}\cos\theta} = \frac{3}{2 + \cos\theta} \Rightarrow r = \frac{3}{2 + \cos\theta}$$

53. $e = 1$, $x = 4$

SOLUTION We substitute $e = 1$ and $d = 4$ in the polar equation of a conic section to obtain

$$r = \frac{ed}{1 + e\cos\theta} = \frac{1 \cdot 4}{1 + 1 \cdot \cos\theta} = \frac{4}{1 + \cos\theta} \Rightarrow r = \frac{4}{1 + \cos\theta}$$

In Exercises 55–58, identify the type of conic, the eccentricity, and the equation of the directrix.

55. $r = \dfrac{8}{1 + 4\cos\theta}$

SOLUTION Matching with the polar equation $r = \frac{ed}{1+e\cos\theta}$ we get $ed = 8$ and $e = 4$ yielding $d = 2$. Since $e > 1$, the conic section is a hyperbola, having eccentricity $e = 4$ and directrix $x = 2$ (referring to the focus-directrix definition (11)).

57. $r = \dfrac{8}{4 + 3\cos\theta}$

SOLUTION We first rewrite the equation in the form $r = \frac{ed}{1+e\cos\theta}$, obtaining

$$r = \frac{2}{1 + \frac{3}{4}\cos\theta}$$

Hence, $ed = 2$ and $e = \frac{3}{4}$ yielding $d = \frac{8}{3}$. Since $e < 1$, the conic section is an ellipse, having eccentricity $e = \frac{3}{4}$ and directrix $x = \frac{8}{3}$.

59. Find a polar equation for the hyperbola with focus at the origin, directrix $x = -2$, and eccentricity $e = 1.2$.

SOLUTION We substitute $d = -2$ and $e = 1.2$ in the polar equation $r = \frac{ed}{1+e\cos\theta}$ and use Exercise 40 to obtain

$$r = \frac{1.2 \cdot (-2)}{1 + 1.2\cos\theta} = \frac{-2.4}{1 + 1.2\cos\theta} = \frac{-12}{5 + 6\cos\theta} = \frac{12}{5 - 6\cos\theta}$$

61. Find an equation in rectangular coordinates of the conic

$$r = \frac{16}{5 + 3\cos\theta}$$

Hint: Use the results of Exercise 60.

SOLUTION Put this equation in the form of the referenced exercise:

$$\frac{16}{5 + 3\cos\theta} = \frac{\frac{16}{5}}{1 + \frac{3}{5}\cos\theta} = \frac{\frac{16}{3} \cdot \frac{3}{5}}{1 + \frac{3}{5}\cos\theta}$$

so that $e = \frac{3}{5}$ and $d = \frac{16}{3}$. Then the center of the ellipse has x-coordinate

$$-\frac{de^2}{1 - e^2} = -\frac{\frac{16}{3} \cdot \frac{9}{25}}{1 - \frac{9}{25}} = -\frac{16}{3} \cdot \frac{9}{25} \cdot \frac{25}{16} = -3$$

and y-coordinate 0, and A' has x-coordinate

$$-\frac{de}{1 - e} = -\frac{\frac{16}{3} \cdot \frac{3}{5}}{1 - \frac{3}{5}} = -\frac{16}{3} \cdot \frac{3}{5} \cdot \frac{5}{2} = -8$$

and y-coordinate 0, so $a = -3 - (-8) = 5$, and the equation is

$$\left(\frac{x + 3}{5}\right)^2 + \left(\frac{y}{b}\right)^2 = 1$$

To find b, set $\theta = \frac{\pi}{2}$; then $r = \frac{16}{5}$. But the point corresponding to $\theta = \frac{\pi}{2}$ lies on the y-axis, and so has coordinates $\left(0, \frac{16}{5}\right)$. This point is on the ellipse, so that

$$\left(\frac{0 + 3}{5}\right)^2 + \left(\frac{\frac{16}{5}}{b}\right)^2 = 1 \quad \Rightarrow \quad \frac{256}{25 \cdot b^2} = \frac{16}{25} \quad \Rightarrow \quad \frac{256}{b^2} = 16 \quad \Rightarrow \quad b = 4$$

and the equation is

$$\left(\frac{x+3}{5}\right)^2 + \left(\frac{y}{4}\right)^2 = 1$$

63. Kepler's First Law states that planetary orbits are ellipses with the sun at one focus. The orbit of Pluto has eccentricity $e \approx 0.25$. Its **perihelion** (closest distance to the sun) is approximately 2.7 billion miles. Find the **aphelion** (farthest distance from the sun).

SOLUTION We define an xy-coordinate system so that the orbit is an ellipse in standard position, as shown in the figure.

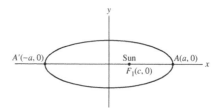

The aphelion is the length of $\overline{A'F_1}$, that is $a + c$. By the given data, we have

$$0.25 = e = \frac{c}{a} \Rightarrow c = 0.25a$$

$$a - c = 2.7 \Rightarrow c = a - 2.7$$

Equating the two expressions for c we get

$$0.25a = a - 2.7$$

$$0.75a = 2.7 \Rightarrow a = \frac{2.7}{0.75} = 3.6, \ c = 3.6 - 2.7 = 0.9$$

The aphelion is thus

$$\overline{A'F_0} = a + c = 3.6 + 0.9 = 4.5 \text{ billion miles}$$

65. Prove that if $a > b > 0$ and $c = \sqrt{a^2 - b^2}$, then a point $P = (x, y)$ on the ellipse

$$\left(\frac{x}{a}\right)^2 + \left(\frac{y}{b}\right)^2 = 1$$

satisfies $PF = ePD$ with $F = (c, 0)$, $e = \frac{c}{a}$, and vertical directrix $\mathcal{D}$ at $x = \frac{a}{e}$.

SOLUTION The derivation in the text for the first part of the hyperbola relationship can largely be worked backward, and modified for the ellipse. For an ellipse in standard position, $e = \frac{c}{a}$, so that $b^2 = a^2 - c^2 = a^2 - (ae)^2 = a^2(1 - e^2)$. Then the equation of the ellipse may be written

$$\frac{x^2}{a^2} + \frac{y^2}{a^2(1-e^2)} = 1, \text{ or } (1-e^2)x^2 + y^2 = a^2(1-e^2)$$

Let (x, y) be any point on the ellipse. Then

$$(1-e^2)x^2 + y^2 = a^2(1-e^2)$$

$$x^2 - e^2x^2 + y^2 = a^2 - a^2e^2$$

$$x^2 + a^2e^2 + y^2 = a^2 + e^2x^2$$

$$x^2 - 2aex + a^2e^2 + y^2 = a^2 - 2aex + e^2x^2$$

$$x^2 - 2cx + c^2 + y^2 = a^2 - 2aex + e^2x^2 \quad (\text{Use } c = ae.)$$

$$(x - c)^2 + y^2 = (a - ex)^2 = e^2(x - a/e)^2$$

$$\sqrt{(x-c)^2 + y^2} = \sqrt{e^2(x - a/e)^2} = e\sqrt{(x - a/e)^2}$$

The left-hand side of the last equation is the distance from (x, y) to the focus at $(c, 0)$, and the right-hand side is e times the distance from (x, y) to the directrix $\mathcal{D}$: $x = \frac{a}{e}$. Thus $PF = ePD$.

Further Insights and Challenges

67. Prove Theorem 2.

SOLUTION Let $F_1 = (c, 0)$ and $F_2 = (-c, 0)$ and let $P(x, y)$ be an arbitrary point on the hyperbola. Then for some constant a,

$$\overline{PF_1} - \overline{PF_2} = \pm 2a$$

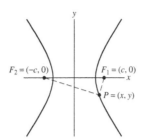

Using the distance formula we write this as

$$\sqrt{(x-c)^2 + y^2} - \sqrt{(x+c)^2 + y^2} = \pm 2a$$

Moving the second term to the right and squaring both sides gives

$$\sqrt{(x-c)^2 + y^2} = \sqrt{(x+c)^2 + y^2} \pm 2a$$

$$(x-c)^2 + y^2 = (x+c)^2 + y^2 \pm 4a\sqrt{(x+c)^2 + y^2} + 4a^2$$

$$(x-c)^2 - (x+c)^2 - 4a^2 = \pm 4a\sqrt{(x+c)^2 + y^2}$$

$$xc + a^2 = \pm a\sqrt{(x+c)^2 + y^2}$$

We square and simplify to obtain

$$x^2c^2 + 2xca^2 + a^4 = a^2\left((x+c)^2 + y^2\right)$$
$$= a^2x^2 + 2a^2xc + a^2c^2 + a^2y^2$$

$$\left(c^2 - a^2\right)x^2 - a^2y^2 = a^2\left(c^2 - a^2\right)$$

$$\frac{x^2}{a^2} - \frac{y^2}{c^2 - a^2} = 1$$

For $b = \sqrt{c^2 - a^2}$ (or $c = \sqrt{a^2 + b^2}$) we get

$$\frac{x^2}{a^2} - \frac{y^2}{b^2} = 1 \Rightarrow \left(\frac{x}{a}\right)^2 - \left(\frac{y}{b}\right)^2 = 1$$

69. Verify that if $e > 1$, then Eq. (11) defines a hyperbola of eccentricity e, with its focus at the origin and directrix at $x = d$.

SOLUTION The points $P = (r, \theta)$ on the hyperbola satisfy $\overline{PF} = e\overline{PD}$, $e > 1$. Referring to the figure we see that

$$\overline{PF} = r, \overline{PD} = d - r\cos\theta \tag{1}$$

Hence

$$r = e(d - r\cos\theta)$$

$$r = ed - er\cos\theta$$

$$r(1 + e\cos\theta) = ed \Rightarrow r = \frac{ed}{1 + e\cos\theta}$$

Remark: Equality (1) holds also for $\theta > \frac{\pi}{2}$. For example, in the following figure, we have

$$\overline{PD} = d + r\cos(\pi - \theta) = d - r\cos\theta$$

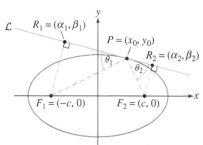

Reflective Property of the Ellipse In Exercises 70–72, we prove that the focal radii at a point on an ellipse make equal angles with the tangent line $\mathcal{L}$. Let $P = (x_0, y_0)$ be a point on the ellipse in Figure 25 with foci $F_1 = (-c, 0)$ and $F_2 = (c, 0)$, and eccentricity $e = c/a$.

FIGURE 25 The ellipse $\left(\dfrac{x}{a}\right)^2 + \left(\dfrac{y}{b}\right)^2 = 1$.

71. Points R_1 and R_2 in Figure 25 are defined so that $\overline{F_1 R_1}$ and $\overline{F_2 R_2}$ are perpendicular to the tangent line.

(a) Show, with A and B as in Exercise 70, that

$$\frac{\alpha_1 + c}{\beta_1} = \frac{\alpha_2 - c}{\beta_2} = \frac{A}{B}$$

(b) Use (a) and the distance formula to show that

$$\frac{F_1 R_1}{F_2 R_2} = \frac{\beta_1}{\beta_2}$$

(c) Use (a) and the equation of the tangent line in Exercise 70 to show that

$$\beta_1 = \frac{B(1 + Ac)}{A^2 + B^2}, \qquad \beta_2 = \frac{B(1 - Ac)}{A^2 + B^2}$$

SOLUTION

(a) From Exercise 70, the tangent line at P has equation $Ax + By = 1$. Since $R_1 = (\alpha_1, \beta_1)$ and $R_2 = (\alpha_2, \beta_2)$ lie on that tangent line, we have

$$A\alpha_1 + B\beta_1 = 1 \quad \text{and} \quad A\alpha_2 + B\beta_2 = 1$$

The slope of the line R_1F_1 is $\frac{\beta_1}{\alpha_1+c}$ and it is perpendicular to the tangent line having slope $-\frac{A}{B}$. Similarly, the slope of the line R_2F_2 is $\frac{\beta_2}{\alpha_2-c}$ and it is also perpendicular to the tangent line. Hence,

$$\frac{\alpha_1+c}{\beta_1} = \frac{A}{B} \quad \text{and} \quad \frac{\alpha_2-c}{\beta_2} = \frac{A}{B}$$

(b) Using the distance formula, we have

$$\overline{R_1F_1}^2 = (\alpha_1+c)^2 + \beta_1^2 = \beta_1^2\left(\left(\frac{\alpha_1+c}{\beta_1}\right)^2 + 1\right) \tag{1}$$

By part (a), $\frac{\alpha_1+c}{\beta_1} = \frac{A}{B}$. Substituting in (1) gives

$$\overline{R_1F_1}^2 = \beta_1^2\left(\frac{A^2}{B^2} + 1\right) \tag{2}$$

Likewise,

$$\overline{R_2F_2}^2 = (\alpha_2-c)^2 + \beta_2^2 = \beta_2^2\left(\left(\frac{\alpha_2-c}{\beta_2}\right)^2 + 1\right)$$

but since $\frac{\alpha_2-c}{\beta_2} = \frac{A}{B}$, we get that

$$\overline{R_2F_2}^2 = \beta_2^2\left(\frac{A^2}{B^2} + 1\right) \tag{3}$$

Dividing, we find that

$$\frac{\overline{R_1F_1}^2}{\overline{R_2F_2}^2} = \frac{\beta_1^2}{\beta_2^2} \quad \text{so} \quad \frac{\overline{R_1F_1}}{\overline{R_2F_2}} = \frac{\beta_1}{\beta_2}$$

as desired.

(c) From part (a),

$$A\alpha_1 + B\beta_1 = 1 \text{ and } \frac{\beta_1}{\alpha_1+c} = \frac{B}{A} \quad \Rightarrow \quad \beta_1 = \frac{B(1+Ac)}{A^2+B^2}$$

Similarly solving the equations in part (a) for β_2 yields

$$A\alpha_2 + B\beta_2 = 1 \text{ and } \frac{\beta_2}{\alpha_2-c} = \frac{B}{A} \quad \Rightarrow \quad \beta_2 = \frac{B(1-Ac)}{A^2+B^2}$$

73. Here is another proof of the Reflective Property.
(a) Figure 25 suggests that $\mathcal{L}$ is the unique line that intersects the ellipse only in the point P. Assuming this, prove that $QF_1 + QF_2 > PF_1 + PF_2$ for all points Q on the tangent line other than P.
(b) Use the Principle of Least Distance (Example 6 in Section 4.7) to prove that $\theta_1 = \theta_2$.

SOLUTION
(a) Consider a point $Q \neq P$ on the line $\mathcal{L}$ (see figure). Since $\mathcal{L}$ intersects the ellipse in only one point, the remainder of the line lies outside the ellipse, so that QR does not have zero length, and F_2QR is a triangle. Thus

$$QF_1 + QF_2 = QR + RF_1 + QF_2 = RF_1 + (QR + QF_2) > RF_1 + RF_2$$

since the sum of lengths of two sides of a triangle exceeds the length of the third side. But since point R lies on the ellipse, $RF_2 + RF_2 = PF_1 + PF_2$, and we are done.

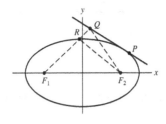

(b) Consider a beam of light traveling from F_1 to F_2 by reflection off of the line $\mathcal{L}$. By the principle of least distance, the light takes the shortest path, which by part (a) is the path through P. By Example 6 in Section 4.7, this shortest path has the property that the angle of incidence (θ_1) is equal to the angle of reflection (θ_2).

75. Show that $y = x^2/4c$ is the equation of a parabola with directrix $y = -c$, focus $(0, c)$, and the vertex at the origin, as stated in Theorem 3.

SOLUTION The points $P = (x, y)$ on the parabola are equidistant from $F = (0, c)$ and the line $y = -c$.

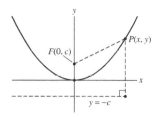

That is, by the distance formula, we have

$$\overline{PF} = \overline{PD}$$

$$\sqrt{x^2 + (y - c)^2} = |y + c|$$

Squaring and simplifying yields

$$x^2 + (y - c)^2 = (y + c)^2$$

$$x^2 + y^2 - 2yc + c^2 = y^2 + 2yc + c^2$$

$$x^2 - 2yc = 2yc$$

$$x^2 = 4yc \Rightarrow y = \frac{x^2}{4c}$$

Thus, we showed that the points that are equidistant from the focus $F = (0, c)$ and the directrix $y = -c$ satisfy the equation $y = \frac{x^2}{4c}$.

77. Derive Eqs. (13) and (14) in the text as follows. Write the coordinates of P with respect to the rotated axes in Figure 21 in polar form $\tilde{x} = r \cos \alpha$, $\tilde{y} = r \sin \alpha$. Explain why P has polar coordinates $(r, \alpha + \theta)$ with respect to the standard x- and y-axes, and derive Eqs. (13) and (14) using the addition formulas for cosine and sine.

SOLUTION If the polar coordinates of P with respect to the rotated axes are (r, α), then the line from the origin to P has length r and makes an angle of α with the rotated x-axis (the x'-axis). Since the x'-axis makes an angle of θ with the x-axis, it follows that the line from the origin to P makes an angle of $\alpha + \theta$ with the x-axis, so that the polar coordinates of P with respect to the standard axes are $(r, \alpha + \theta)$. Write (x', y') for the rectangular coordinates of P with respect to the rotated axes and (x, y) for the rectangular coordinates of P with respect to the standard axes. Then

$$x = r \cos(\alpha + \theta) = (r \cos \alpha) \cos \theta - (r \sin \alpha) \sin \theta = x' \cos \theta - y' \sin \theta$$

$$y = r \sin(\alpha + \theta) = r \sin \alpha \cos \theta + r \cos \alpha \sin \theta = (r \cos \alpha) \sin \theta + (r \sin \alpha) \cos \theta = x' \sin \theta + y' \cos \theta$$

CHAPTER REVIEW EXERCISES

1. Which of the following curves pass through the point $(1, 4)$?
(a) $c(t) = (t^2, t + 3)$
(b) $c(t) = (t^2, t - 3)$
(c) $c(t) = (t^2, 3 - t)$
(d) $c(t) = (t - 3, t^2)$

SOLUTION To check whether it passes through the point $(1, 4)$, we solve the equations $c(t) = (1, 4)$ for the given curves.

(a) Comparing the second coordinate of the curve and the point yields:

$$t + 3 = 4$$
$$t = 1$$

We substitute $t = 1$ in the first coordinate, to obtain

$$t^2 = 1^2 = 1$$

Hence the curve passes through $(1, 4)$.

(b) Comparing the second coordinate of the curve and the point yields:

$$t - 3 = 4$$
$$t = 7$$

We substitute $t = 7$ in the first coordinate to obtain

$$t^2 = 7^2 = 49 \neq 1$$

Hence the curve does not pass through $(1, 4)$.

(c) Comparing the second coordinate of the curve and the point yields

$$3 - t = 4$$
$$t = -1$$

We substitute $t = -1$ in the first coordinate, to obtain

$$t^2 = (-1)^2 = 1$$

Hence the curve passes through $(1, 4)$.

(d) Comparing the first coordinate of the curve and the point yields

$$t - 3 = 1$$
$$t = 4$$

We substitute $t = 4$ in the second coordinate, to obtain:

$$t^2 = 4^2 = 16 \neq 4$$

Hence the curve does not pass through $(1, 4)$.

3. Find parametric equations for the circle of radius 2 with center $(1, 1)$. Use the equations to find the points of intersection of the circle with the x- and y-axes.

SOLUTION Using the standard technique for parametric equations of curves, we obtain

$$c(t) = (1 + 2\cos t, 1 + 2\sin t)$$

We compare the x coordinate of $c(t)$ to 0:

$$1 + 2\cos t = 0$$
$$\cos t = -\frac{1}{2}$$
$$t = \pm\frac{2\pi}{3}$$

Substituting in the y coordinate yields

$$1 + 2\sin\left(\pm\frac{2\pi}{3}\right) = 1 \pm 2\frac{\sqrt{3}}{2} = 1 \pm \sqrt{3}$$

Hence, the intersection points with the y-axis are $(0, 1 \pm \sqrt{3})$. We compare the y coordinate of $c(t)$ to 0:

$$1 + 2\sin t = 0$$

$$\sin t = -\frac{1}{2}$$

$$t = -\frac{\pi}{6} \quad \text{or} \quad \frac{7}{6}\pi$$

Substituting in the x coordinates yields

$$1 + 2\cos\left(-\frac{\pi}{6}\right) = 1 + 2\frac{\sqrt{3}}{2} = 1 + \sqrt{3}$$

$$1 + 2\cos\left(\frac{7}{6}\pi\right) = 1 - 2\cos\left(\frac{\pi}{6}\right) = 1 - 2\frac{\sqrt{3}}{2} = 1 - \sqrt{3}$$

Hence, the intersection points with the x-axis are $(1 \pm \sqrt{3}, 0)$.

5. Find a parametrization $c(\theta)$ of the unit circle such that $c(0) = (-1, 0)$.

SOLUTION The unit circle has the parametrization

$$c(t) = (\cos t, \sin t)$$

This parametrization does not satisfy $c(0) = (-1, 0)$. We replace the parameter t by a parameter θ so that $t = \theta + \alpha$, to obtain another parametrization for the circle:

$$c^*(\theta) = (\cos(\theta + \alpha), \sin(\theta + \alpha)) \qquad (1)$$

We need that $c^*(0) = (1, 0)$, that is,

$$c^*(0) = (\cos \alpha, \sin \alpha) = (-1, 0)$$

Hence

$$\begin{aligned}\cos \alpha &= -1 \\ \sin \alpha &= 0\end{aligned} \quad \Rightarrow \quad \alpha = \pi$$

Substituting in (1) we obtain the following parametrization:

$$c^*(\theta) = (\cos(\theta + \pi), \sin(\theta + \pi))$$

7. Find a path $c(t)$ that traces the line $y = 2x + 1$ from $(1, 3)$ to $(3, 7)$ for $0 \le t \le 1$.

SOLUTION Solution 1: By one of the examples in section 12.1, the line through $P = (1, 3)$ with slope 2 has the parametrization

$$c(t) = (1 + t, 3 + 2t)$$

But this parametrization does not satisfy $c(1) = (3, 7)$. We replace the parameter t by a parameter s so that $t = \alpha s + \beta$. We get

$$c^*(s) = (1 + \alpha s + \beta, 3 + 2(\alpha s + \beta)) = (\alpha s + \beta + 1, 2\alpha s + 2\beta + 3)$$

We need that $c^*(0) = (1, 3)$ and $c^*(1) = (3, 7)$. Hence,

$$c^*(0) = (1 + \beta, 3 + 2\beta) = (1, 3)$$

$$c^*(1) = (\alpha + \beta + 1, 2\alpha + 2\beta + 3) = (3, 7)$$

We obtain the equations

$$\begin{aligned} 1 + \beta &= 1 \\ 3 + 2\beta &= 3 \\ \alpha + \beta + 1 &= 3 \\ 2\alpha + 2\beta + 3 &= 7 \end{aligned} \quad \Rightarrow \quad \beta = 0, \alpha = 2$$

Substituting in (1) gives

$$c^*(s) = (2s + 1, 4s + 3)$$

Solution 2: The segment from $(1, 3)$ to $(3, 7)$ has the following vector parametrization:

$$(1 - t)\langle 1, 3\rangle + t\langle 3, 7\rangle = \langle 1 - t + 3t, 3(1 - t) + 7t\rangle = \langle 1 + 2t, 3 + 4t\rangle$$

The parametrization is thus

$$c(t) = (1 + 2t, 3 + 4t)$$

In Exercises 9–12, express the parametric curve in the form $y = f(x)$.

9. $c(t) = (4t - 3, 10 - t)$

SOLUTION We use the given equation to express t in terms of x.

$$x = 4t - 3$$
$$4t = x + 3$$
$$t = \frac{x + 3}{4}$$

Substituting in the equation of y yields

$$y = 10 - t = 10 - \frac{x + 3}{4} = -\frac{x}{4} + \frac{37}{4}$$

That is,

$$y = -\frac{x}{4} + \frac{37}{4}$$

11. $c(t) = \left(3 - \frac{2}{t}, t^3 + \frac{1}{t}\right)$

SOLUTION We use the given equation to express t in terms of x:

$$x = 3 - \frac{2}{t}$$
$$\frac{2}{t} = 3 - x$$
$$t = \frac{2}{3 - x}$$

Substituting in the equation of y yields

$$y = \left(\frac{2}{3 - x}\right)^3 + \frac{1}{2/(3 - x)} = \frac{8}{(3 - x)^3} + \frac{3 - x}{2}$$

In Exercises 13–16, calculate dy/dx at the point indicated.

13. $c(t) = (t^3 + t, t^2 - 1), \quad t = 3$

SOLUTION The parametric equations are $x = t^3 + t$ and $y = t^2 - 1$. We use the theorem on the slope of the tangent line to find $\frac{dy}{dx}$:

$$\frac{dy}{dx} = \frac{\frac{dy}{dt}}{\frac{dx}{dt}} = \frac{2t}{3t^2 + 1}$$

We now substitute $t = 3$ to obtain

$$\left.\frac{dy}{dx}\right|_{t=3} = \frac{2 \cdot 3}{3 \cdot 3^2 + 1} = \frac{3}{14}$$

15. $c(t) = (e^t - 1, \sin t), \quad t = 20$

SOLUTION We use the theorem for the slope of the tangent line to find $\frac{dy}{dx}$:

$$\frac{dy}{dx} = \frac{\frac{dy}{dt}}{\frac{dx}{dt}} = \frac{(\sin t)'}{(e^t - 1)'} = \frac{\cos t}{e^t}$$

We now substitute $t = 20$:

$$\left.\frac{dy}{dx}\right|_{t=0} = \frac{\cos 20}{e^{20}}$$

17. **CAS** Find the point on the cycloid $c(t) = (t - \sin t, 1 - \cos t)$ where the tangent line has slope $\frac{1}{2}$.

SOLUTION Since $x = t - \sin t$ and $y = 1 - \cos t$, the theorem on the slope of the tangent line gives

$$\frac{dy}{dx} = \frac{\frac{dy}{dt}}{\frac{dx}{dt}} = \frac{\sin t}{1 - \cos t}$$

The points where the tangent line has slope $\frac{1}{2}$ are those where $\frac{dy}{dx} = \frac{1}{2}$. We solve for t:

$$\frac{dy}{dx} = \frac{1}{2}$$

$$\frac{\sin t}{1 - \cos t} = \frac{1}{2} \tag{1}$$

$$2 \sin t = 1 - \cos t$$

We let $u = \sin t$. Then $\cos t = \pm\sqrt{1 - \sin^2 t} = \pm\sqrt{1 - u^2}$. Hence

$$2u = 1 \pm \sqrt{1 - u^2}$$

We transfer sides and square to obtain

$$\pm\sqrt{1 - u^2} = 2u - 1$$

$$1 - u^2 = 4u^2 - 4u + 1$$

$$5u^2 - 4u = u(5u - 4) = 0$$

$$u = 0, \; u = \frac{4}{5}$$

We find t by the relation $u = \sin t$:

$$u = 0: \quad \sin t = 0 \Rightarrow t = 0, t = \pi$$

$$u = \frac{4}{5}: \quad \sin t = \frac{4}{5} \Rightarrow t \approx 0.93, t \approx 2.21$$

These correspond to the points $(0, 1)$, $(\pi, 2)$, $(0.13, 0.40)$, and $(1.41, 1.60)$, respectively, for $0 < t < 2\pi$.

19. Find the equation of the Bézier curve with control points

$$P_0 = (-1, -1), \quad P_1 = (-1, 1), \quad P_2 = (1, 1), \quad P_3(1, -1)$$

SOLUTION We substitute the given points in the appropriate formulas in the text to find the parametric equations of the Bézier curve. We obtain

$$x(t) = -(1 - t)^3 - 3t(1 - t)^2 + t^2(1 - t) + t^3$$
$$= -(1 - 3t + 3t^2 - t^3) - (3t - 6t^2 + 3t^3) + (t^2 - t^3) + t^3$$
$$= (-2t^3 + 4t^2 - 1)$$

$$y(t) = -(1 - t)^3 + 3t(1 - t)^2 + t^2(1 - t) - t^3$$
$$= -(1 - 3t + 3t^2 - t^3) + (3t - 6t^2 + 3t^3) + (t^2 - t^3) - t^3$$
$$= (2t^3 - 8t^2 + 6t - 1)$$

21. Find the speed (as a function of t) of a particle whose position at time t seconds is $c(t) = (\sin t + t, \cos t + t)$. What is the particle's maximal speed?

SOLUTION We use the parametric definition to find the speed. We obtain

$$\frac{ds}{dt} = \sqrt{((\sin t + t)')^2 + ((\cos t + t)')^2} = \sqrt{(\cos t + 1)^2 + (1 - \sin t)^2}$$
$$= \sqrt{\cos^2 t + 2\cos t + 1 + 1 - 2\sin t + \sin^2 t} = \sqrt{3 + 2(\cos t - \sin t)}$$

We now differentiate the speed function to find its maximum:

$$\frac{d^2 s}{dt^2} = \left(\sqrt{3 + 2(\cos t - \sin t)}\right)' = \frac{-\sin t - \cos t}{\sqrt{3 + 2(\cos t - \sin t)}}$$

We equate the derivative to zero, to obtain the maximum point:

$$\frac{d^2 s}{dt^2} = 0$$

$$\frac{-\sin t - \cos t}{\sqrt{3 + 2(\cos t - \sin t)}} = 0$$

$$-\sin t - \cos t = 0$$

$$-\sin t = \cos t$$

$$\sin(-t) = \cos(-t)$$

$$-t = \frac{\pi}{4} + \pi k$$

$$t = -\frac{\pi}{4} + \pi k$$

Substituting t in the function of speed we obtain the value of the maximal speed:

$$\sqrt{3 + 2\left(\cos -\frac{\pi}{4} - \sin -\frac{\pi}{4}\right)} = \sqrt{3 + 2\left(\frac{\sqrt{2}}{2} - \left(-\frac{\sqrt{2}}{2}\right)\right)} = \sqrt{3 + 2\sqrt{2}}$$

In Exercises 23 and 24, let $c(t) = (e^{-t}\cos t, e^{-t}\sin t)$.

23. Show that $c(t)$ for $0 \le t < \infty$ has finite length and calculate its value.

SOLUTION We use the formula for arc length, to obtain:

$$s = \int_0^\infty \sqrt{((e^{-t}\cos t)')^2 + ((e^{-t}\sin t)')^2}\,dt$$
$$= \int_0^\infty \sqrt{(-e^{-t}\cos t - e^{-t}\sin t)^2 + (-e^{-t}\sin t + e^{-t}\cos t)^2}\,dt$$
$$= \int_0^\infty \sqrt{e^{-2t}(\cos t + \sin t)^2 + e^{-2t}(\cos t - \sin t)^2}\,dt$$
$$= \int_0^\infty e^{-t}\sqrt{\cos^2 t + 2\sin t\cos t + \sin^2 t + \cos^2 t - 2\sin t\cos t + \sin^2 t}\,dt$$
$$= \int_0^\infty e^{-t}\sqrt{2}\,dt = \sqrt{2}(-e^{-t})\Big|_0^\infty = -\sqrt{2}\left(\lim_{t\to\infty} e^{-t} - e^0\right)$$
$$= -\sqrt{2}(0 - 1) = \sqrt{2}$$

25. **CAS** Plot $c(t) = (\sin 2t, 2\cos t)$ for $0 \le t \le \pi$. Express the length of the curve as a definite integral, and approximate it using a computer algebra system.

SOLUTION We use a CAS to plot the curve. The resulting graph is shown here.

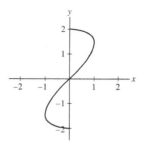

Plot of the curve $(\sin 2t, 2\cos t)$

To calculate the arc length we use the formula for the arc length to obtain

$$s = \int_0^\pi \sqrt{(2\cos 2t)^2 + (-2\sin t)^2}\, dt = 2\int_0^\pi \sqrt{\cos^2 2t + \sin^2 t}\, dt$$

We use a CAS to obtain $s = 6.0972$.

27. Convert the points $(r, \theta) = (1, \frac{\pi}{6})$, $(3, \frac{5\pi}{4})$ from polar to rectangular coordinates.

SOLUTION We convert the points from polar coordinates to cartesian coordinates. For the first point we have

$$x = r\cos\theta = 1 \cdot \cos\frac{\pi}{6} = \frac{\sqrt{3}}{2}$$

$$y = r\sin\theta = 1 \cdot \sin\frac{\pi}{6} = \frac{1}{2}$$

For the second point we have

$$x = r\cos\theta = 3\cos\frac{5\pi}{4} = -\frac{3\sqrt{2}}{2}$$

$$y = r\sin\theta = 3\sin\frac{5\pi}{4} = -\frac{3\sqrt{2}}{2}$$

29. Write $r = \dfrac{2\cos\theta}{\cos\theta - \sin\theta}$ as an equation in rectangular coordinates.

SOLUTION We use the formula for converting from polar coordinates to cartesian coordinates to substitute x and y for r and θ:

$$r = \frac{2\cos\theta}{\cos\theta - \sin\theta}$$

$$\sqrt{x^2 + y^2} = \frac{2r\cos\theta}{r\cos\theta - r\sin\theta}$$

$$\sqrt{x^2 + y^2} = \frac{2x}{x - y}$$

31. [GU] Convert the equation

$$9(x^2 + y^2) = (x^2 + y^2 - 2y)^2$$

to polar coordinates, and plot it with a graphing utility.

SOLUTION We use the formula for converting from cartesian coordinates to polar coordinates to substitute r and θ for x and y:

$$9(x^2 + y^2) = (x^2 + y^2 - 2y)^2$$

$$9r^2 = (r^2 - 2r\sin\theta)^2$$

$$3r = r^2 - 2r\sin\theta$$

$$3 = r - 2\sin\theta$$

$$r = 3 + 2\sin\theta$$

The plot of $r = 3 + 2\sin\theta$ is shown here:

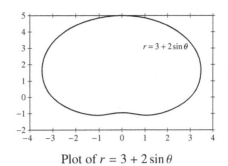

Plot of $r = 3 + 2\sin\theta$

33. Calculate the area of one petal of $r = \sin 4\theta$ (see Figure 1).

SOLUTION We use a CAS to generate the plot, as shown here.

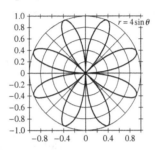

Plot of $r = \sin 4\theta$

We can see that one leaf lies between the rays $\theta = 0$ and $\theta = \dfrac{\theta}{4}$. We now use the formula for area in polar coordinates to obtain

$$A = \frac{1}{2}\int_0^{\pi/4} \sin^2 4\theta\, d\theta = \frac{1}{4}\int_0^{\pi/4}(1 - \cos 8\theta)\, d\theta = \frac{1}{4}\left(\theta - \frac{\sin 8\theta}{8}\Big|_0^{\pi/4}\right)$$

$$= \frac{\pi}{16} - \frac{1}{32}(\sin 2\pi - \sin 0) = \frac{\pi}{16}$$

35. Calculate the total area enclosed by the curve $r^2 = \cos\theta\, e^{\sin\theta}$ (Figure 2).

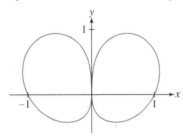

FIGURE 2 Graph of $r^2 = \cos\theta\, e^{\sin\theta}$.

SOLUTION Note that this is defined only for θ between $-\pi/2$ and $\pi/2$. We use the formula for area in polar coordinates to obtain:

$$A = \frac{1}{2}\int_{-\pi/2}^{\pi/2} r^2\, d\theta = \frac{1}{2}\int_{-\pi/2}^{\pi/2} \cos\theta\, e^{\sin\theta}\, d\theta$$

We evaluate the integral by making the substitution $x = \sin\theta\, dx = \cos\theta\, d\theta$:

$$A = \frac{1}{2}\int_{-\pi/2}^{\pi/2} \cos\theta\, e^{\sin\theta}\, d\theta = \frac{1}{2}e^x\Big|_{-1}^{1} = \frac{1}{2}\left(e - e^{-1}\right)$$

37. Find the area enclosed by the cardioid $r = a(1 + \cos\theta)$, where $a > 0$.

SOLUTION The graph of $r = a(1 + \cos\theta)$ in the $r\theta$-plane for $0 \le \theta \le 2\pi$ and the cardioid in the xy-plane are shown in the following figures:

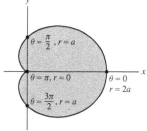

$r = a(1 + \cos\theta)$ The cardioid $r = a(1 + \cos\theta)$, $a > 0$

As θ varies from 0 to π the radius r decreases from $2a$ to 0, and this gives the upper part of the cardioid.

The lower part is traced as θ varies from π to 2π and consequently r increases from 0 back to $2a$. We compute the area enclosed by the upper part of the cardioid and the x-axis, using the following integral (we use the identity $\cos^2\theta = \frac{1}{2} + \frac{1}{2}\cos 2\theta$):

$$\frac{1}{2}\int_0^\pi r^2\, d\theta = \frac{1}{2}\int_0^\pi a^2(1 + \cos\theta)^2\, d\theta = \frac{a^2}{2}\int_0^\pi \left(1 + 2\cos\theta + \cos^2\theta\right) d\theta$$

$$= \frac{a^2}{2}\int_0^\pi \left(1 + 2\cos\theta + \frac{1}{2} + \frac{1}{2}\cos 2\theta\right) d\theta = \frac{a^2}{2}\int_0^\pi \left(\frac{3}{2} + 2\cos\theta + \frac{1}{2}\cos 2\theta\right) d\theta$$

$$= \frac{a^2}{2}\left[\frac{3\theta}{2} + 2\sin\theta + \frac{1}{4}\sin 2\theta\right]_0^\pi = \frac{a^2}{2}\left[\frac{3\pi}{2} + 2\sin\pi + \frac{1}{4}\sin 2\pi - 0\right] = \frac{3\pi a^2}{4}$$

Using symmetry, the total area A enclosed by the cardioid is

$$A = 2 \cdot \frac{3\pi a^2}{4} = \frac{3\pi a^2}{2}$$

39. **CAS** Figure 5 shows the graph of $r = e^{0.5\theta}\sin\theta$ for $0 \leq \theta \leq 2\pi$. Use a computer algebra system to approximate the difference in length between the outer and inner loops.

FIGURE 5

SOLUTION We note that the inner loop is the curve for $\theta \in [0, \pi]$, and the outer loop is the curve for $\theta \in [\pi, 2\pi]$. We express the length of these loops using the formula for the arc length. The length of the inner loop is

$$s_1 = \int_0^\pi \sqrt{(e^{0.5\theta}\sin\theta)^2 + ((e^{0.5\theta}\sin\theta)')^2}\, d\theta = \int_0^\pi \sqrt{e^\theta \sin^2\theta + \left(\frac{e^{0.5\theta}\sin\theta}{2} + e^{0.5\theta}\cos\theta\right)^2}\, d\theta$$

and the length of the outer loop is

$$s_2 = \int_\pi^{2\pi} \sqrt{e^\theta \sin^2\theta + \left(\frac{e^{0.5\theta}\sin\theta}{2} + e^{0.5\theta}\cos\theta\right)^2}\, d\theta$$

We now use the CAS to calculate the arc length of each of the loops. We obtain that the length of the inner loop is 7.5087 and the length of the outer loop is 36.121; hence the outer one is 4.81 times longer than the inner one.

In Exercises 41–44, identify the conic section. Find the vertices and foci.

41. $\left(\dfrac{x}{3}\right)^2 + \left(\dfrac{y}{2}\right)^2 = 1$

SOLUTION This is an ellipse in standard position. Its foci are $(\pm\sqrt{3^2-2^2},0)=(\pm\sqrt{5},0)$ and its vertices are $(\pm 3,0),(0,\pm 2)$.

43. $(2x+\frac{1}{2}y)^2 = 4-(x-y)^2$

SOLUTION We simplify the equation:

$$\left(2x+\frac{1}{2}y\right)^2 = 4-(x-y)^2$$

$$4x^2+2xy+\frac{1}{4}y^2 = 4-x^2+2xy-y^2$$

$$5x^2+\frac{5}{4}y^2 = 4$$

$$\frac{5x^2}{4}+\frac{5y^2}{16} = 1$$

$$\left(\frac{x}{\frac{2}{\sqrt{5}}}\right)^2+\left(\frac{y}{\frac{4}{\sqrt{5}}}\right)^2 = 1$$

This is an ellipse in standard position, with foci $\left(0,\pm\sqrt{\left(\frac{4}{\sqrt{5}}\right)^2-\left(\frac{2}{\sqrt{5}}\right)^2}\right) = \left(0,\pm\sqrt{\frac{12}{5}}\right)$ and vertices $\left(\pm\frac{2}{\sqrt{5}},0\right)$, $\left(0,\pm\frac{4}{\sqrt{5}}\right)$.

In Exercises 45–50, find the equation of the conic section indicated.

45. Ellipse with vertices $(\pm 8,0)$, foci $(\pm\sqrt{3},0)$

SOLUTION Since the foci of the desired ellipse are on the x-axis, we conclude that $a>b$. We are given that the points $(\pm 8,0)$ are vertices of the ellipse, and since they are on the x-axis, $a=8$. We are given that the foci are $(\pm\sqrt{3},0)$ and we have shown that $a>b$; hence we have that $\sqrt{a^2-b^2}=\sqrt{3}$. Solving for b yields

$$\sqrt{a^2-b^2}=\sqrt{3}$$
$$a^2-b^2=3$$
$$8^2-b^2=3$$
$$b^2=61$$
$$b=\sqrt{61}$$

Next we use a and b to construct the equation of the ellipse:

$$\left(\frac{x}{8}\right)^2+\left(\frac{y}{\sqrt{61}}\right)^2 = 1$$

47. Hyperbola with vertices $(\pm 8,0)$, asymptotes $y=\pm\frac{3}{4}x$

SOLUTION Since the asymptotes of the hyperbola are $y=\pm\frac{3}{4}x$, and the equation of the asymptotes for a general hyperbola in standard position is $y=\pm\frac{b}{a}x$, we conclude that $\frac{b}{a}=\frac{3}{4}$. We are given that the vertices are $(\pm 8,0)$, thus $a=8$. We substitute and solve for b:

$$\frac{b}{a}=\frac{3}{4}$$
$$\frac{b}{8}=\frac{3}{4}$$
$$b=6$$

Next we use a and b to construct the equation of the hyperbola:

$$\left(\frac{x}{8}\right)^2-\left(\frac{y}{6}\right)^2 = 1$$

49. Parabola with focus $(8, 0)$, directrix $x = -8$

SOLUTION This is similar to the usual equation of a parabola, but we must use y as x, and x as y, to obtain

$$x = \frac{1}{32} y^2$$

51. Find the asymptotes of the hyperbola $3x^2 + 6x - y^2 - 10y = 1$.

SOLUTION We complete the squares and simplify:

$$3x^2 + 6x - y^2 - 10y = 1$$
$$3(x^2 + 2x) - (y^2 + 10y) = 1$$
$$3(x^2 + 2x + 1 - 1) - (y^2 + 10y + 25 - 25) = 1$$
$$3(x+1)^2 - 3 - (y+5)^2 + 25 = 1$$
$$3(x+1)^2 - (y+5)^2 = -21$$
$$\left(\frac{y+5}{\sqrt{21}}\right)^2 - \left(\frac{x+1}{\sqrt{7}}\right)^2 = 1$$

We obtained a hyperbola with focal axis that is parallel to the y-axis, and is shifted -5 units on the y-axis, and -1 units in the x-axis. Therefore, the asymptotes are

$$x + 1 = \pm \frac{\sqrt{7}}{\sqrt{21}} (y + 5) \quad \text{or} \quad y + 5 = \pm \sqrt{3}(x + 1)$$

53. Show that the relation $\frac{dy}{dx} = (e^2 - 1)\frac{x}{y}$ holds on a standard ellipse or hyperbola of eccentricity e.

SOLUTION We differentiate the equations of the standard ellipse and the hyperbola with respect to x:

Ellipse:

$$\frac{x^2}{a^2} + \frac{y^2}{b^2} = 1$$
$$\frac{2x}{a^2} + \frac{2y}{b^2}\frac{dy}{dx} = 0$$
$$\frac{dy}{dx} = -\frac{b^2}{a^2}\frac{x}{y}$$

Hyperbola:

$$\frac{x^2}{a^2} - \frac{y^2}{b^2} = 1$$
$$\frac{2x}{a^2} - \frac{2y}{b^2}\frac{dy}{dx} = 0$$
$$\frac{dy}{dx} = \frac{b^2}{a^2}\frac{x}{y}$$

The eccentricity of the ellipse is $e = \frac{\sqrt{a^2 - b^2}}{a}$, hence $e^2 a^2 = a^2 - b^2$ or $e^2 = 1 - \frac{b^2}{a^2}$ yielding $\frac{b^2}{a^2} = 1 - e^2$.

The eccentricity of the hyperbola is $e = \frac{\sqrt{a^2 + b^2}}{a}$, hence $e^2 a^2 = a^2 + b^2$ or $e^2 = 1 + \frac{b^2}{a^2}$, giving $\frac{b^2}{a^2} = e^2 - 1$. Combining with the expressions for $\frac{dy}{dx}$ we get:

Ellipse:

$$\frac{dy}{dx} = -(1 - e^2)\frac{x}{y} = (e^2 - 1)\frac{x}{y}$$

Hyperbola:

$$\frac{dy}{dx} = (e^2 - 1)\frac{x}{y}$$

We thus proved that the relation $\frac{dy}{dx} = (e^2 - 1)\frac{x}{y}$ holds on a standard ellipse or hyperbola of eccentricity e.

55. Refer to Figure 25 in Section 11.5. Prove that the product of the perpendicular distances $F_1 R_1$ and $F_2 R_2$ from the foci to a tangent line of an ellipse is equal to the square b^2 of the semiminor axes.

SOLUTION We first consider the ellipse in standard position:

$$\frac{x^2}{a^2} + \frac{y^2}{b^2} = 1$$

The equation of the tangent line at $P = (x_0, y_0)$ is

$$\frac{x_0 x}{a^2} + \frac{y_0 y}{b^2} = 1$$

or
$$b^2 x_0 x + a^2 y_0 y - a^2 b^2 = 0$$

The distances of the foci $F_1 = (c, 0)$ and $F_2 = (-c, 0)$ from the tangent line are

$$\overline{F_1 R_1} = \frac{|b^2 x_0 c - a^2 b^2|}{\sqrt{b^4 x_0^2 + a^4 y_0^2}}; \quad \overline{F_2 R_2} = \frac{|b^2 x_0 c + a^2 b^2|}{\sqrt{b^4 x_0^2 + a^4 y_0^2}}$$

We compute the product of the distances:

$$\overline{F_1 R_1} \cdot \overline{F_2 R_2} = \left| \frac{\left(b^2 x_0 c - a^2 b^2\right)\left(b^2 x_0 c + a^2 b^2\right)}{b^4 x_0^2 + a^4 y_0^2} \right| = \left| \frac{b^4 x_0^2 c^2 - a^4 b^4}{b^4 x_0^2 + a^4 y_0^2} \right| \quad (1)$$

The point $P = (x_0, y_0)$ lies on the ellipse, hence:

$$\frac{x_0^2}{a^2} + \frac{y_0^2}{b^2} = 1 \Rightarrow a^4 y_0^2 = a^4 b^2 - a^2 b^2 x_0^2$$

We substitute in (1) to obtain (notice that $b^2 - a^2 = -c^2$)

$$\overline{F_1 R_1} \cdot \overline{F_2 R_2} = \frac{|b^4 x_0^2 c^2 - a^4 b^4|}{|b^4 x_0^2 + a^4 b^2 - a^2 b^2 x_0^2|} = \frac{|b^4 x_0^2 c^2 - a^4 b^4|}{|b^2(b^2 - a^2)x_0^2 + a^4 b^2|}$$

$$= \frac{|b^4 x_0^2 c^2 - a^4 b^4|}{|-b^2 x_0^2 c^2 + a^4 b^2|} = \frac{|b^2(x_0^2 c^2 - a^4)|}{|-(x_0^2 c^2 - a^4)|} = |-b^2| = b^2$$

The product $\overline{F_1 R_1} \cdot \overline{F_2 R_2}$ remains unchanged if we translate the standard ellipse.

12 VECTOR GEOMETRY

12.1 Vectors in the Plane (LT Section 13.1)

Preliminary Questions

1. Answer true or false. Every nonzero vector is:
(a) equivalent to a vector based at the origin.
(b) equivalent to a unit vector based at the origin.
(c) parallel to a vector based at the origin.
(d) parallel to a unit vector based at the origin.

SOLUTION

(a) This statement is true. Translating the vector so that it is based on the origin, we get an equivalent vector based at the origin.
(b) Equivalent vectors have equal lengths, hence vectors that are not unit vectors, are not equivalent to a unit vector.
(c) This statement is true. A vector based at the origin such that the line through this vector is parallel to the line through the given vector, is parallel to the given vector.
(d) Since parallel vectors do not necessarily have equal lengths, the statement is true by the same reasoning as in (c).

2. What is the length of $-3\mathbf{a}$ if $\|\mathbf{a}\| = 5$?

SOLUTION Using properties of the length we get

$$\|-3\mathbf{a}\| = |-3|\|\mathbf{a}\| = 3\|\mathbf{a}\| = 3 \cdot 5 = 15$$

3. Suppose that $\mathbf{v}$ has components $\langle 3, 1 \rangle$. How, if at all, do the components change if you translate $\mathbf{v}$ horizontally 2 units to the left?

SOLUTION Translating $\mathbf{v} = \langle 3, 1 \rangle$ yields an equivalent vector, hence the components are not changed.

4. What are the components of the zero vector based at $P = (3, 5)$?

SOLUTION The components of the zero vector are always $\langle 0, 0 \rangle$, no matter where it is based.

5. True or false?
(a) The vectors $\mathbf{v}$ and $-2\mathbf{v}$ are parallel.
(b) The vectors $\mathbf{v}$ and $-2\mathbf{v}$ point in the same direction.

SOLUTION

(a) The lines through $\mathbf{v}$ and $-2\mathbf{v}$ are parallel, therefore these vectors are parallel.
(b) The vector $-2\mathbf{v}$ is a scalar multiple of $\mathbf{v}$, where the scalar is negative. Therefore $-2\mathbf{v}$ points in the opposite direction as $\mathbf{v}$.

6. Explain the commutativity of vector addition in terms of the Parallelogram Law.

SOLUTION To determine the vector $\mathbf{v} + \mathbf{w}$, we translate $\mathbf{w}$ to the equivalent vector $\mathbf{w}'$ whose tail coincides with the head of $\mathbf{v}$. The vector $\mathbf{v} + \mathbf{w}$ is the vector pointing from the tail of $\mathbf{v}$ to the head of $\mathbf{w}'$.

To determine the vector $\mathbf{w} + \mathbf{v}$, we translate $\mathbf{v}$ to the equivalent vector $\mathbf{v}'$ whose tail coincides with the head of $\mathbf{w}$. Then $\mathbf{w} + \mathbf{v}$ is the vector pointing from the tail of $\mathbf{w}$ to the head of $\mathbf{v}'$. In either case, the resulting vector is the vector with the tail at the basepoint of $\mathbf{v}$ and $\mathbf{w}$, and head at the opposite vertex of the parallelogram. Therefore $\mathbf{v} + \mathbf{w} = \mathbf{w} + \mathbf{v}$.

Exercises

1. Sketch the vectors $\mathbf{v}_1, \mathbf{v}_2, \mathbf{v}_3, \mathbf{v}_4$ with tail P and head Q, and compute their lengths. Are any two of these vectors equivalent?

	$\mathbf{v}_1$	$\mathbf{v}_2$	$\mathbf{v}_3$	$\mathbf{v}_4$
P	$(2,4)$	$(-1,3)$	$(-1,3)$	$(4,1)$
Q	$(4,4)$	$(1,3)$	$(2,4)$	$(6,3)$

SOLUTION Using the definitions we obtain the following answers:

$$\mathbf{v}_1 = \overrightarrow{PQ} = \langle 4-2, 4-4 \rangle = \langle 2, 0 \rangle \qquad \mathbf{v}_2 = \langle 1-(-1), 3-3 \rangle = \langle 2, 0 \rangle$$
$$\|\mathbf{v}_1\| = \sqrt{2^2 + 0^2} = 2 \qquad \|\mathbf{v}_2\| = \sqrt{2^2 + 0^2} = 2$$

$$\mathbf{v}_3 = \langle 2-(-1), 4-3 \rangle = \langle 3, 1 \rangle \qquad \mathbf{v}_4 = \langle 6-4, 3-1 \rangle = \langle 2, 2 \rangle$$
$$\|\mathbf{v}_3\| = \sqrt{3^2 + 1^2} = \sqrt{10} \qquad \|\mathbf{v}_4\| = \sqrt{2^2 + 2^2} = \sqrt{8} = 2\sqrt{2}$$

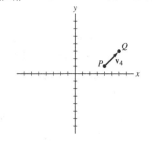

$\mathbf{v}_1$ and $\mathbf{v}_2$ are parallel and have the same length, hence they are equivalent.

3. What is the terminal point of the vector $\mathbf{a} = \langle 1, 3 \rangle$ based at $P = (2, 2)$? Sketch $\mathbf{a}$ and the vector $\mathbf{a}_0$ based at the origin and equivalent to $\mathbf{a}$.

SOLUTION The terminal point Q of the vector $\mathbf{a}$ is located 1 unit to the right and 3 units up from $P = (2, 2)$. Therefore, $Q = (2+1, 2+3) = (3, 5)$. The vector $\mathbf{a}_0$ equivalent to $\mathbf{a}$ based at the origin is shown in the figure, along with the vector $\mathbf{a}$.

In Exercises 5–8, refer to the unit vectors in Figure 21.

FIGURE 21

5. Find the components of **u**.

SOLUTION Since **u** makes an angle of $45°$ with the positive x-axis, its components are

$$\|\mathbf{u}\| \langle \cos 45°, \sin 45° \rangle = \|\mathbf{u}\| \left\langle \frac{\sqrt{2}}{2}, \frac{\sqrt{2}}{2} \right\rangle$$

7. Find the components of **w**.

SOLUTION Since **w** makes an angle of $-20°$ with the positive x-axis, its components are

$$\|\mathbf{w}\|\langle \cos(-20°), \sin(-20°) \rangle = \|\mathbf{w}\| \langle \cos 20°, -\sin 20° \rangle$$

In Exercises 9–12, find the components of $\overrightarrow{PQ}$.

9. $P = (3, 2), \quad Q = (2, 7)$

SOLUTION Using the definition of the components of a vector we have $\overrightarrow{PQ} = \langle 2 - 3, 7 - 2 \rangle = \langle -1, 5 \rangle$.

11. $P = (1, -7), \quad Q = (0, 17)$

SOLUTION Using the definition of the components of a vector we have $\overrightarrow{PQ} = \langle 0 - 1, 17 - (-7) \rangle = \langle -1, 24 \rangle$.

In Exercises 13–20, calculate.

13. $\langle 2, 1 \rangle + \langle 3, 4 \rangle$

SOLUTION Using vector algebra we have $\langle 2, 1 \rangle + \langle 3, 4 \rangle = \langle 2 + 3, 1 + 4 \rangle = \langle 5, 5 \rangle$.

15. $5 \langle 6, 2 \rangle$

SOLUTION $5 \langle 6, 2 \rangle = \langle 5 \cdot 6, 5 \cdot 2 \rangle = \langle 30, 10 \rangle$

17. $\left\langle -\frac{1}{2}, \frac{5}{3} \right\rangle + \left\langle 3, \frac{10}{3} \right\rangle$

SOLUTION The vector sum is $\left\langle -\frac{1}{2}, \frac{5}{3} \right\rangle + \left\langle 3, \frac{10}{3} \right\rangle = \left\langle -\frac{1}{2} + 3, \frac{5}{3} + \frac{10}{3} \right\rangle = \left\langle \frac{5}{2}, 5 \right\rangle$.

19. $\langle 2e, 1 - 2\pi \rangle - \langle 2e - \pi, 8 - 2\pi \rangle$

SOLUTION The vector difference is

$$\langle 2e, 1 - 2\pi \rangle - \langle 2e - \pi, 8 - 2\pi \rangle = \langle 2e - (2e - \pi), 1 - 2\pi - (8 - 2\pi) \rangle = \langle \pi, -7 \rangle$$

21. Which of the vectors (A)–(C) in Figure 22 is equivalent to $\mathbf{v} - \mathbf{w}$?

FIGURE 22

SOLUTION The vector $-\mathbf{w}$ has the same length as **w** but points in the opposite direction. The sum $\mathbf{v} + (-\mathbf{w})$, which is the difference $\mathbf{v} - \mathbf{w}$, is obtained by the parallelogram law. This vector is the vector shown in (b).

23. Sketch $2\mathbf{v}$, $-\mathbf{w}$, $\mathbf{v} + \mathbf{w}$, and $2\mathbf{v} - \mathbf{w}$ for the vectors in Figure 24.

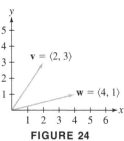

FIGURE 24

SOLUTION The scalar multiple $2\mathbf{v}$ points in the same direction as $\mathbf{v}$ and its length is twice the length of $\mathbf{v}$. It is the vector $2\mathbf{v} = \langle 4, 6 \rangle$.

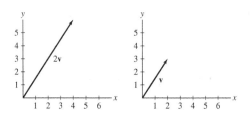

$-\mathbf{w}$ has the same length as $\mathbf{w}$ but points to the opposite direction. It is the vector $-\mathbf{w} = \langle -4, -1 \rangle$.

The vector sum $\mathbf{v} + \mathbf{w}$ is the vector:

$$\mathbf{v} + \mathbf{w} = \langle 2, 3 \rangle + \langle 4, 1 \rangle = \langle 6, 4 \rangle$$

This vector is shown in the following figure:

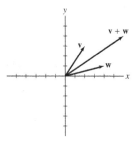

The vector $2\mathbf{v} - \mathbf{w}$ is

$$2\mathbf{v} - \mathbf{w} = 2\langle 2, 3 \rangle - \langle 4, 1 \rangle = \langle 4, 6 \rangle - \langle 4, 1 \rangle = \langle 0, 5 \rangle$$

It is shown next:

25. Sketch $\mathbf{v} = \langle 0, 2 \rangle$, $\mathbf{w} = \langle -2, 4 \rangle$, $3\mathbf{v} + \mathbf{w}$, $2\mathbf{v} - 2\mathbf{w}$.

SOLUTION We compute the vectors and then sketch them:

$$3\mathbf{v} + \mathbf{w} = 3\langle 0, 2 \rangle + \langle -2, 4 \rangle = \langle 0, 6 \rangle + \langle -2, 4 \rangle = \langle -2, 10 \rangle$$

$$2\mathbf{v} - 2\mathbf{w} = 2\langle 0, 2 \rangle - 2\langle -2, 4 \rangle = \langle 0, 4 \rangle - \langle -4, 8 \rangle = \langle 4, -4 \rangle$$

27. Sketch the vector $\mathbf{v}$ such that $\mathbf{v} + \mathbf{v}_1 + \mathbf{v}_2 = \mathbf{0}$ for $\mathbf{v}_1$ and $\mathbf{v}_2$ in Figure 25(A).

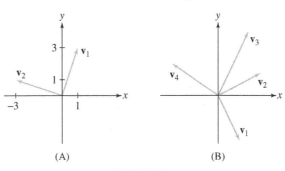

FIGURE 25

SOLUTION Since $\mathbf{v} + \mathbf{v}_1 + \mathbf{v}_2 = 0$, we have that $\mathbf{v} = -\mathbf{v}_1 - \mathbf{v}_2$, and since $\mathbf{v}_1 = \langle 1, 3 \rangle$ and $\mathbf{v}_2 = \langle -3, 1 \rangle$, then $\mathbf{v} = -\mathbf{v}_1 - \mathbf{v}_2 = \langle 2, -4 \rangle$, as seen in this picture.

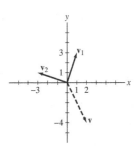

29. Let $\mathbf{v} = \overrightarrow{PQ}$, where $P = (-2, 5)$, $Q = (1, -2)$. Which of the following vectors with the given tails and heads are equivalent to $\mathbf{v}$?

(a) $(-3, 3)$, $(0, 4)$ \qquad (b) $(0, 0)$, $(3, -7)$
(c) $(-1, 2)$, $(2, -5)$ \qquad (d) $(4, -5)$, $(1, 4)$

SOLUTION Two vectors are equivalent if they have the same components. We thus compute the vectors and check whether this condition is satisfied.

$$\mathbf{v} = \overrightarrow{PQ} = \langle 1 - (-2), -2 - 5 \rangle = \langle 3, -7 \rangle$$

(a) $\langle 0 - (-3), 4 - 3 \rangle = \langle 3, 1 \rangle$ (b) $\langle 3 - 0, -7 - 0 \rangle = \langle 3, -7 \rangle$
(c) $\langle 2 - (-1), -5 - 2 \rangle = \langle 3, -7 \rangle$ (d) $\langle 1 - 4, 4 - (-5) \rangle = \langle -3, 9 \rangle$

We see that the vectors in (b) and (c) are equivalent to $\mathbf{v}$.

In Exercises 31–34, sketch the vectors $\overrightarrow{AB}$ and $\overrightarrow{PQ}$, and determine whether they are equivalent.

31. $A = (1, 1)$, $B = (3, 7)$, $P = (4, -1)$, $Q = (6, 5)$

SOLUTION We compute the vectors and check whether they have the same components:

$$\overrightarrow{AB} = \langle 3 - 1, 7 - 1 \rangle = \langle 2, 6 \rangle$$
$$\overrightarrow{PQ} = \langle 6 - 4, 5 - (-1) \rangle = \langle 2, 6 \rangle$$

$\Rightarrow$ The vectors are equivalent.

33. $A = (-3, 2)$, $B = (0, 0)$, $P = (0, 0)$, $Q = (3, -2)$

SOLUTION We compute the vectors $\overrightarrow{AB}$ and $\overrightarrow{PQ}$:

$$\overrightarrow{AB} = \langle 0 - (-3), 0 - 2 \rangle = \langle 3, -2 \rangle$$
$$\overrightarrow{PQ} = \langle 3 - 0, -2 - 0 \rangle = \langle 3, -2 \rangle$$

$\Rightarrow$ The vectors are equivalent.

In Exercises 35–38, are $\overrightarrow{AB}$ and $\overrightarrow{PQ}$ parallel? And if so, do they point in the same direction?

35. $A = (1, 1)$, $B = (3, 4)$, $P = (1, 1)$, $Q = (7, 10)$

SOLUTION We compute the vectors $\overrightarrow{AB}$ and $\overrightarrow{PQ}$:

$$\overrightarrow{AB} = \langle 3 - 1, 4 - 1 \rangle = \langle 2, 3 \rangle$$
$$\overrightarrow{PQ} = \langle 7 - 1, 10 - 1 \rangle = \langle 6, 9 \rangle$$

Since $\overrightarrow{AB} = \frac{1}{3}\langle 6, 9 \rangle$, the vectors are parallel and point in the same direction.

37. $A = (2, 2)$, $B = (-6, 3)$, $P = (9, 5)$, $Q = (17, 4)$

SOLUTION We compute the vectors $\overrightarrow{AB}$ and $\overrightarrow{PQ}$:

$$\overrightarrow{AB} = \langle -6 - 2, 3 - 2 \rangle = \langle -8, 1 \rangle$$
$$\overrightarrow{PQ} = \langle 17 - 9, 4 - 5 \rangle = \langle 8, -1 \rangle$$

Since $\overrightarrow{AB} = -\overrightarrow{PQ}$, the vectors are parallel and point in opposite directions.

In Exercises 39–42, let $R = (-2, 7)$. Calculate the following:

39. The length of $\overrightarrow{OR}$

SOLUTION Since $\overrightarrow{OR} = \langle -2, 7 \rangle$, the length of the vector is $\|\overrightarrow{OR}\| = \sqrt{(-2)^2 + 7^2} = \sqrt{53}$.

41. The point P such that $\overrightarrow{PR}$ has components $\langle -2, 7 \rangle$

SOLUTION Denoting $P = (x_0, y_0)$ we have:

$$\overrightarrow{PR} = \langle -2 - x_0, 7 - y_0 \rangle = \langle -2, 7 \rangle$$

Equating corresponding components yields:

$$\begin{aligned} -2 - x_0 &= -2 \\ 7 - y_0 &= 7 \end{aligned} \quad \Rightarrow \quad x_0 = 0, \; y_0 = 0 \quad \Rightarrow \quad P = (0, 0)$$

In Exercises 43–52, find the given vector.

43. Unit vector $\mathbf{e_v}$, where $\mathbf{v} = \langle 3, 4 \rangle$

SOLUTION The unit vector $\mathbf{e_v}$ is the following vector:

$$\mathbf{e_v} = \frac{1}{\|\mathbf{v}\|}\mathbf{v}$$

We find the length of $\mathbf{v} = \langle 3, 4 \rangle$:

$$\|\mathbf{v}\| = \sqrt{3^2 + 4^2} = \sqrt{25} = 5$$

Thus

$$\mathbf{e_v} = \frac{1}{5}\langle 3, 4 \rangle = \left\langle \frac{3}{5}, \frac{4}{5} \right\rangle$$

45. Vector of length 4 in the direction of $\mathbf{u} = \langle -1, -1 \rangle$

SOLUTION The unit vector in the direction of $\mathbf{u} = \langle -1, -1 \rangle$ is the vector:

$$\mathbf{e_u} = \frac{1}{\|\mathbf{u}\|}\mathbf{u} = \frac{1}{\sqrt{(-1)^2 + (-1)^2}}\langle -1, -1 \rangle = \frac{1}{\sqrt{2}}\langle -1, -1 \rangle = \left\langle -\frac{1}{\sqrt{2}}, -\frac{1}{\sqrt{2}} \right\rangle$$

Multiply $\mathbf{e_u}$ by 4 to obtain the desired vector:

$$4\mathbf{e_u} = 4\left\langle -\frac{1}{\sqrt{2}}, -\frac{1}{\sqrt{2}} \right\rangle = \left\langle -\frac{4}{\sqrt{2}}, -\frac{4}{\sqrt{2}} \right\rangle = \langle -2\sqrt{2}, -2\sqrt{2} \rangle$$

47. Vector of length 2 in the direction opposite to $\mathbf{v} = \mathbf{i} - \mathbf{j}$

SOLUTION We first find the unit vector in the direction of $\mathbf{v}$:

$$\mathbf{e_v} = \frac{1}{\|\mathbf{v}\|}\mathbf{v} = \frac{1}{\sqrt{1^2 + (-1)^2}}\langle 1, -1 \rangle = \frac{1}{\sqrt{2}}\langle 1, -1 \rangle = \left\langle \frac{\sqrt{2}}{2}, -\frac{\sqrt{2}}{2} \right\rangle$$

Now multiply by -2 to obtain the desired vector:

$$-2\mathbf{e_v} = -2\left\langle \frac{\sqrt{2}}{2}, -\frac{\sqrt{2}}{2} \right\rangle = \langle -\sqrt{2}, \sqrt{2} \rangle$$

49. Unit vector $\mathbf{e}$ making an angle of $\frac{4\pi}{7}$ with the x-axis

SOLUTION The unit vector $\mathbf{e}$ is the following vector:

$$\mathbf{e} = \left\langle \cos\frac{4\pi}{7}, \sin\frac{4\pi}{7} \right\rangle = \langle -0.22, 0.97 \rangle$$

51. A unit vector pointing in the direction from $(1, 1)$ to $(0, 3)$

SOLUTION We first compute the vector $\mathbf{v}$ pointing from $(1, 1)$ to $(0, 3)$. This is

$$\mathbf{v} = \langle 0, 3 \rangle - \langle 1, 1 \rangle = \langle -1, 2 \rangle$$

The desired unit vector is thus

$$\mathbf{e_v} = \frac{1}{\|\mathbf{v}\|}\mathbf{v} = \frac{1}{\sqrt{(-1)^2 + 2^2}}\langle -1, 2 \rangle = \left\langle -\frac{1}{\sqrt{5}}, \frac{2}{\sqrt{5}} \right\rangle$$

53. Find all scalars λ such that $\lambda\langle 2, 3 \rangle$ has length 1.

SOLUTION We have:

$$\|\lambda\langle 2, 3 \rangle\| = |\lambda|\|\langle 2, 3 \rangle\| = |\lambda|\sqrt{2^2 + 3^2} = |\lambda|\sqrt{13}$$

The scalar λ must satisfy

$$|\lambda|\sqrt{13} = 1$$
$$|\lambda| = \frac{1}{\sqrt{13}} \quad \Rightarrow \quad \lambda_1 = \frac{1}{\sqrt{13}}, \; \lambda_2 = -\frac{1}{\sqrt{13}}$$

55. What are the coordinates of the point P in the parallelogram in Figure 26(A)?

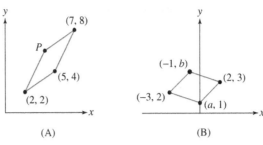

FIGURE 26

SOLUTION We denote by A, B, C the points in the figure.

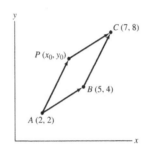

Let $P = (x_0, y_0)$. We compute the following vectors:

$$\overrightarrow{PC} = \langle 7 - x_0, 8 - y_0 \rangle$$
$$\overrightarrow{AB} = \langle 5 - 2, 4 - 2 \rangle = \langle 3, 2 \rangle$$

The vectors $\overrightarrow{PC}$ and $\overrightarrow{AB}$ are equivalent, hence they have the same components. That is:

$$\begin{aligned} 7 - x_0 &= 3 \\ 8 - y_0 &= 2 \end{aligned} \Rightarrow x_0 = 4, \ y_0 = 6 \Rightarrow P = (4, 6)$$

57. Let $\mathbf{v} = \overrightarrow{AB}$ and $\mathbf{w} = \overrightarrow{AC}$, where A, B, C are three distinct points in the plane. Match (a)–(d) with (i)–(iv). *Hint:* Draw a picture.

(a) $-\mathbf{w}$ (b) $-\mathbf{v}$ (c) $\mathbf{w} - \mathbf{v}$ (d) $\mathbf{v} - \mathbf{w}$

(i) $\overrightarrow{CB}$ (ii) $\overrightarrow{CA}$ (iii) $\overrightarrow{BC}$ (iv) $\overrightarrow{BA}$

SOLUTION

(a) $-\mathbf{w}$ has the same length as $\mathbf{w}$ and points in the opposite direction. Hence: $-\mathbf{w} = \overrightarrow{CA}$.

(b) $-\mathbf{v}$ has the same length as $\mathbf{v}$ and points in the opposite direction. Hence: $-\mathbf{v} = \overrightarrow{BA}$.

(c) By the parallelogram law we have:

$$\overrightarrow{BC} = \overrightarrow{BA} + \overrightarrow{AC} = -\mathbf{v} + \mathbf{w} = \mathbf{w} - \mathbf{v}$$

That is,

$$\mathbf{w} - \mathbf{v} = \overrightarrow{BC}$$

(d) By the parallelogram law we have:

$$\overrightarrow{CB} = \overrightarrow{CA} + \overrightarrow{AB} = -\mathbf{w} + \mathbf{v} = \mathbf{v} - \mathbf{w}$$

That is,

$$\mathbf{v} - \mathbf{w} = \overrightarrow{CB}$$

In Exercises 59–62, calculate the linear combination.

59. $3\mathbf{j} + (9\mathbf{i} + 4\mathbf{j})$

SOLUTION We have:

$$3\mathbf{j} + (9\mathbf{i} + 4\mathbf{j}) = 3\langle 0, 1\rangle + 9\langle 1, 0\rangle + 4\langle 0, 1\rangle = \langle 9, 7\rangle$$

61. $(3\mathbf{i} + \mathbf{j}) - 6\mathbf{j} + 2(\mathbf{j} - 4\mathbf{i})$

SOLUTION We have:

$$(3\mathbf{i} + \mathbf{j}) - 6\mathbf{j} + 2(\mathbf{j} - 4\mathbf{i}) = (\langle 3, 0\rangle + \langle 0, 1\rangle) - \langle 0, 6\rangle + 2(\langle 0, 1\rangle - \langle 4, 0\rangle) = \langle -5, -3\rangle$$

63. For each of the position vectors **u** with endpoints A, B, and C in Figure 27, indicate with a diagram the multiples $r\mathbf{v}$ and $s\mathbf{w}$ such that $\mathbf{u} = r\mathbf{v} + s\mathbf{w}$. A sample is shown for $\mathbf{u} = \overrightarrow{OQ}$.

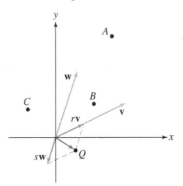

FIGURE 27

SOLUTION See the following three figures:

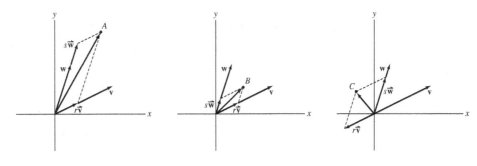

In Exercises 65 and 66, express **u** *as a linear combination* **u** $= r\mathbf{v} + s\mathbf{w}$. *Then sketch* **u, v, w,** *and the parallelogram formed by* $r\mathbf{v}$ *and* $s\mathbf{w}$.

65. $\mathbf{u} = \langle 3, -1 \rangle$; $\quad \mathbf{v} = \langle 2, 1 \rangle, \mathbf{w} = \langle 1, 3 \rangle$

SOLUTION We have

$$\mathbf{u} = \langle 3, -1 \rangle = r\mathbf{v} + s\mathbf{w} = r\langle 2, 1 \rangle + s\langle 1, 3 \rangle$$

which becomes the two equations

$$3 = 2r + s$$
$$-1 = r + 3s$$

Solving the second equation for r gives $r = -1 - 3s$, and substituting that into the first equation gives $3 = 2(-1 - 3s) + s = -2 - 6s + s$, so $5 = -5s$, so $s = -1$, and thus $r = 2$. In other words,

$$\mathbf{u} = \langle 3, -1 \rangle = 2\langle 2, 1 \rangle - 1\langle 1, 3 \rangle$$

as seen in this sketch:

67. Calculate the magnitude of the force on cables 1 and 2 in Figure 28.

FIGURE 28

SOLUTION The three forces acting on the point P are:

- The force **F** of magnitude 50 lb that acts vertically downward.
- The forces $\mathbf{F}_1$ and $\mathbf{F}_2$ that act through cables 1 and 2, respectively.

Since the point P is not in motion we have

$$\mathbf{F}_1 + \mathbf{F}_2 + \mathbf{F} = 0 \tag{1}$$

We compute the forces. Letting $\|\mathbf{F}_1\| = f_1$ and $\|\mathbf{F}_2\| = f_2$ we have:

$$\mathbf{F}_1 = f_1 \langle \cos 115°, \sin 115° \rangle = f_1 \langle -0.423, 0.906 \rangle$$
$$\mathbf{F}_2 = f_2 \langle \cos 25°, \sin 25° \rangle = f_2 \langle 0.906, 0.423 \rangle$$
$$\mathbf{F} = \langle 0, -50 \rangle$$

Substituting the forces in (1) gives

$$f_1 \langle -0.423, 0.906 \rangle + f_2 \langle 0.906, 0.423 \rangle + \langle 0, -50 \rangle = \langle 0, 0 \rangle$$

$$\langle -0.423 f_1 + 0.906 f_2, 0.906 f_1 + 0.423 f_2 - 50 \rangle = \langle 0, 0 \rangle$$

We equate corresponding components and get

$$-0.423 f_1 + 0.906 f_2 = 0$$

$$0.906 f_1 + 0.423 f_2 - 50 = 0$$

By the first equation, $f_2 = 0.467 f_1$. Substituting in the second equation and solving for f_1 yields

$$0.906 f_1 + 0.423 \cdot 0.467 f_1 - 50 = 0$$

$$1.104 f_1 = 50 \quad \Rightarrow \quad f_1 = 45.29, \ f_2 = 0.467 f_1 = 21.15$$

We conclude that the magnitude of the force on cable 1 is $f_1 = 45.29$ lb and the magnitude of the force on cable 2 is $f_2 = 21.15$ lb.

69. A plane flying due east at 200 km/h encounters a 40-km/h wind blowing in the northeast direction. The resultant velocity of the plane is the vector sum $\mathbf{v} = \mathbf{v}_1 + \mathbf{v}_2$, where $\mathbf{v}_1$ is the velocity vector of the plane and $\mathbf{v}_2$ is the velocity vector of the wind (Figure 30). The angle between $\mathbf{v}_1$ and $\mathbf{v}_2$ is $\frac{\pi}{4}$. Determine the resultant *speed* of the plane (the length of the vector $\mathbf{v}$).

FIGURE 30

SOLUTION The resultant speed of the plane is the length of the sum vector $\mathbf{v} = \mathbf{v}_1 + \mathbf{v}_2$. We place the xy-coordinate system as shown in the figure, and compute the components of the vectors $\mathbf{v}_1$ and $\mathbf{v}_2$. This gives

$$\mathbf{v}_1 = \langle v_1, 0 \rangle$$

$$\mathbf{v}_2 = \left\langle v_2 \cos \frac{\pi}{4}, v_2 \sin \frac{\pi}{4} \right\rangle = \left\langle v_2 \cdot \frac{\sqrt{2}}{2}, v_2 \cdot \frac{\sqrt{2}}{2} \right\rangle$$

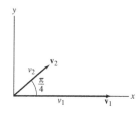

We now compute the sum $\mathbf{v} = \mathbf{v}_1 + \mathbf{v}_2$:

$$\mathbf{v} = \langle v_1, 0 \rangle + \left\langle \frac{\sqrt{2} v_2}{2}, \frac{\sqrt{2} v_2}{2} \right\rangle = \left\langle \frac{\sqrt{2}}{2} v_2 + v_1, \frac{\sqrt{2}}{2} v_2 \right\rangle$$

The resultant speed is the length of $\mathbf{v}$, that is,

$$v = \|\mathbf{v}\| = \sqrt{\left(\frac{\sqrt{2} v_2}{2}\right)^2 + \left(v_1 + \frac{\sqrt{2} v_2}{2}\right)^2} = \sqrt{\frac{v_2^2}{2} + v_1^2 + 2 \cdot \frac{\sqrt{2}}{2} v_2 v_1 + \frac{v_2^2}{2}} = \sqrt{v_1^2 + v_2^2 + \sqrt{2} v_1 v_2}$$

Finally, we substitute the given information $v_1 = 200$ and $v_2 = 40$ in the equation above, to obtain

$$v = \sqrt{200^2 + 40^2 + \sqrt{2} \cdot 200 \cdot 40} \approx 230 \text{ km/hr}$$

Further Insights and Challenges

In Exercises 70–72, refer to Figure 31, which shows a robotic arm consisting of two segments of lengths L_1 and L_2.

FIGURE 31

71. Let $L_1 = 5$ and $L_2 = 3$. Find **r** for $\theta_1 = \frac{\pi}{3}$, $\theta_2 = \frac{\pi}{4}$.

SOLUTION In Exercise 70 we showed that

$$\mathbf{r} = \langle L_1 \sin\theta_1 + L_2 \sin\theta_2, L_1 \cos\theta_1 - L_2 \cos\theta_2 \rangle$$

Substituting the given information we obtain

$$\mathbf{r} = \left\langle 5\sin\frac{\pi}{3} + 3\sin\frac{\pi}{4}, 5\cos\frac{\pi}{3} - 3\cos\frac{\pi}{4} \right\rangle = \left\langle \frac{5\sqrt{3}}{2} + \frac{3\sqrt{2}}{2}, \frac{5}{2} - \frac{3\sqrt{2}}{2} \right\rangle \approx \langle 6.45, 0.38 \rangle$$

73. Use vectors to prove that the diagonals $\overline{AC}$ and $\overline{BD}$ of a parallelogram bisect each other (Figure 32). *Hint:* Observe that the midpoint of $\overline{BD}$ is the terminal point of $\mathbf{w} + \frac{1}{2}(\mathbf{v} - \mathbf{w})$.

FIGURE 32

SOLUTION We denote by O the midpoint of $\overline{BD}$. Hence,

$$\overrightarrow{DO} = \frac{1}{2}\overrightarrow{DB}$$

Using the Parallelogram Law we have

$$\overrightarrow{AO} = \overrightarrow{AD} + \overrightarrow{DO} = \overrightarrow{AD} + \frac{1}{2}\overrightarrow{DB}$$

Since $\overrightarrow{AD} = \mathbf{w}$ and $\overrightarrow{DB} = \mathbf{v} - \mathbf{w}$ we get

$$\overrightarrow{AO} = \mathbf{w} + \frac{1}{2}(\mathbf{v} - \mathbf{w}) = \frac{\mathbf{w} + \mathbf{v}}{2} \tag{1}$$

On the other hand, $\overrightarrow{AC} = \overrightarrow{AD} + \overrightarrow{DC} = \mathbf{w} + \mathbf{v}$, hence the midpoint O' of the diagonal $\overline{AC}$ is the terminal point of $\frac{\mathbf{w}+\mathbf{v}}{2}$. That is,

$$\overrightarrow{AO'} = \frac{\mathbf{w} + \mathbf{v}}{2} \tag{2}$$

We combine (1) and (2) to conclude that O and O' are the same point. That is, the diagonal $\overline{AC}$ and $\overline{BD}$ bisect each other.

75. Prove that two nonzero vectors $\mathbf{v} = \langle a, b \rangle$ and $\mathbf{w} = \langle c, d \rangle$ are perpendicular if and only if

$$ac + bd = 0$$

SOLUTION Suppose that the vectors $\mathbf{v}$ and $\mathbf{w}$ make angles θ_1 and θ_2, which are not $\frac{\pi}{2}$ or $\frac{3\pi}{2}$, respectively, with the positive x-axis. Then their components satisfy

$$\begin{aligned} a &= \|\mathbf{v}\| \cos \theta_1 \\ b &= \|\mathbf{v}\| \sin \theta_1 \end{aligned} \quad \Rightarrow \quad \frac{b}{a} = \frac{\sin \theta_1}{\cos \theta_1} = \tan \theta_1$$

$$\begin{aligned} c &= \|\mathbf{w}\| \cos \theta_2 \\ d &= \|\mathbf{w}\| \sin \theta_2 \end{aligned} \quad \Rightarrow \quad \frac{d}{c} = \frac{\sin \theta_2}{\cos \theta_2} = \tan \theta_2$$

That is, the vectors $\mathbf{v}$ and $\mathbf{w}$ are on the lines with slopes $\frac{b}{a}$ and $\frac{d}{c}$, respectively. The lines are perpendicular if and only if their slopes satisfy

$$\frac{b}{a} \cdot \frac{d}{c} = -1 \quad \Rightarrow \quad bd = -ac \quad \Rightarrow \quad ac + bd = 0$$

We now consider the case where one of the vectors, say $\mathbf{v}$, is perpendicular to the x-axis. In this case $a = 0$, and the vectors are perpendicular if and only if $\mathbf{w}$ is parallel to the x-axis, that is, $d = 0$. So $ac + bd = 0 \cdot c + b \cdot 0 = 0$.

12.2 Three-Dimensional Space: Surfaces, Vectors, and Curves (LT Section 13.2)

Preliminary Questions

1. What is the terminal point of the vector $\mathbf{v} = \langle 3, 2, 1 \rangle$ based at the point $P = (1, 1, 1)$?

SOLUTION We denote the terminal point by $Q = (a, b, c)$. Then by the definition of components of a vector, we have

$$\langle 3, 2, 1 \rangle = \langle a - 1, b - 1, c - 1 \rangle$$

Equivalent vectors have equal components respectively, thus,

$$\begin{aligned} 3 &= a - 1 & a &= 4 \\ 2 &= b - 1 & \Rightarrow \quad b &= 3 \\ 1 &= c - 1 & c &= 2 \end{aligned}$$

The terminal point of $\mathbf{v}$ is thus $Q = (4, 3, 2)$.

2. What are the components of the vector $\mathbf{v} = \langle 3, 2, 1 \rangle$ based at the point $P = (1, 1, 1)$?

SOLUTION The components of $\mathbf{v} = \langle 3, 2, 1 \rangle$ are $\langle 3, 2, 1 \rangle$ regardless of the base point. The component of $\mathbf{v}$ and the base point $P = (1, 1, 1)$ determine the head $Q = (a, b, c)$ of the vector, as found in the previous exercise.

3. If $\mathbf{v} = -3\mathbf{w}$, then (choose the correct answer):

(a) $\mathbf{v}$ and $\mathbf{w}$ are parallel.

(b) $\mathbf{v}$ and $\mathbf{w}$ point in the same direction.

SOLUTION The vectors $\mathbf{v}$ and $\mathbf{w}$ lie on parallel lines, hence these vectors are parallel. Since $\mathbf{v}$ is a scalar multiple of $\mathbf{w}$ by a negative scalar, $\mathbf{v}$ and $\mathbf{w}$ point in opposite directions. Thus, (a) is correct and (b) is not.

4. Which of the following is a direction vector for the line through $P = (3, 2, 1)$ and $Q = (1, 1, 1)$?

(a) $\langle 3, 2, 1 \rangle$ **(b)** $\langle 1, 1, 1 \rangle$ **(c)** $\langle 2, 1, 0 \rangle$

SOLUTION Any vector that is parallel to the vector $\overrightarrow{PQ}$ is a direction vector for the line through P and Q. We compute the vector $\overrightarrow{PQ}$:

$$\overrightarrow{PQ} = \langle 1 - 3, 1 - 2, 1 - 1 \rangle = \langle -2, -1, 0 \rangle$$

The vectors $\langle 3, 2, 1 \rangle$ and $\langle 1, 1, 1 \rangle$ are not constant multiples of $\overrightarrow{PQ}$, hence they are not parallel to $\overrightarrow{PQ}$. However $\langle 2, 1, 0 \rangle = -1 \langle -2, -1, 0 \rangle = -\overrightarrow{PQ}$, hence the vector $\langle 2, 1, 0 \rangle$ is parallel to $\overrightarrow{PQ}$. Therefore, the vector $\langle 2, 1, 0 \rangle$ is a direction vector for the line through P and Q.

5. How many different direction vectors does a line have?

SOLUTION All the vectors that are parallel to a line are also direction vectors for that line. Therefore, there are infinitely many direction vectors for a line.

6. True or false? If $\mathbf{v}$ is a direction vector for a line $\mathcal{L}$, then $-\mathbf{v}$ is also a direction vector for $\mathcal{L}$.

SOLUTION True. Every vector that is parallel to $\mathbf{v}$ is a direction vector for the line L. Since $-\mathbf{v}$ is parallel to $\mathbf{v}$, it is also a direction vector for L.

7. What is the radius of the sphere $x^2 + y^2 + z^2 = 5$?

SOLUTION A sphere centered at the origin has equation $x^2 + y^2 + z^2 = r^2$, where r is the radius, so here the radius is $\sqrt{5}$.

8. Which of the following points are on the cylinder $(x - 1)^2 + y^2 = 1$?

(a) $(1, 0, 0)$ **(b)** $(0, 0, 0)$ **(c)** $(0, 0, -1)$
(d) $(0, -1, 1)$ **(e)** $(1, -1, 1)$ **(f)** $(1, 1, 0)$

SOLUTION (b), (c), (e), (f)

Exercises

1. Sketch the vector $\mathbf{v} = \langle 1, 3, 2 \rangle$ and compute its length.

SOLUTION The vector $\mathbf{v} = \langle 1, 3, 2 \rangle$ is shown in the following figure:

The length of $\mathbf{v}$ is

$$\|\mathbf{v}\| = \sqrt{1^2 + 3^2 + 2^2} = \sqrt{14}$$

3. Sketch the vector $\mathbf{v} = \langle 1, 1, 0 \rangle$ based at $P = (0, 1, 1)$. Describe this vector in the form $\overrightarrow{PQ}$ for some point Q, and sketch the vector $\mathbf{v}_0$ based at the origin equivalent to $\mathbf{v}$.

SOLUTION The vector $\mathbf{v} = \langle 1, 1, 0 \rangle$ based at $P = (0, 1, 1)$ is shown in the figure:

The head Q of the vector $\mathbf{v} = \overrightarrow{PQ}$ is at the point $Q = (0+1, 1+1, 1+0) = (1, 2, 1)$.

The vector $\mathbf{v}_0$ based at the origin and equivalent to $\mathbf{v}$ is

$$v_0 = \langle 1, 1, 0 \rangle = \overrightarrow{OS}, \text{ where } S = (1, 1, 0)$$

In Exercises 5–8, find the components of the vector $\overrightarrow{PQ}$.

5. $P = (1, 0, 1), \quad Q = (2, 1, 0)$

SOLUTION By the definition of the vector components we have

$$\overrightarrow{PQ} = \langle 2-1, 1-0, 0-1 \rangle = \langle 1, 1, -1 \rangle$$

7. $P = (4, 6, 0), \quad Q = (-\frac{1}{2}, \frac{9}{2}, 1)$

SOLUTION Using the definition of vector components we have

$$\overrightarrow{PQ} = \left\langle -\frac{1}{2} - 4, \frac{9}{2} - 6, 1 - 0 \right\rangle = \left\langle -\frac{9}{2}, -\frac{3}{2}, 1 \right\rangle$$

In Exercises 9–12, let $R = (1, 4, 3)$.

9. Calculate the length of $\overrightarrow{OR}$.

SOLUTION The length of $\overrightarrow{OR}$ is the distance from $R = (1, 4, 3)$ to the origin. That is,

$$\|\overrightarrow{OR}\| = \sqrt{(1-0)^2 + (4-0)^2 + (3-0)^2} = \sqrt{26} \approx 5.1$$

11. Find the point P such that $\mathbf{w} = \overrightarrow{PR}$ has components $\langle 3, -2, 3 \rangle$, and sketch $\mathbf{w}$.

SOLUTION Denoting $P = (x_0, y_0, z_0)$ we get

$$\overrightarrow{PR} = \langle 1 - x_0, 4 - y_0, 3 - z_0 \rangle = \langle 3, -2, 3 \rangle$$

Equating corresponding components gives

$$\begin{array}{c} 1 - x_0 = 3 \\ 4 - y_0 = -2 \\ 3 - z_0 = 3 \end{array} \quad \Rightarrow \quad x_0 = -2, \ y_0 = 6, \ z_0 = 0$$

The point P is, thus, $P = (-2, 6, 0)$.

13. Let $\mathbf{v} = \langle 4, 8, 12 \rangle$. Which of the following vectors is parallel to $\mathbf{v}$? Which point in the same direction?
(a) $\langle 2, 4, 6 \rangle$ **(b)** $\langle -1, -2, 3 \rangle$
(c) $\langle -7, -14, -21 \rangle$ **(d)** $\langle 6, 10, 14 \rangle$

908 CHAPTER 12 | VECTOR GEOMETRY (LT CHAPTER 13)

SOLUTION A vector is parallel to **v** if it is a scalar multiple of **v**. It points in the same direction if the multiplying scalar is positive. Using these properties we obtain the following answer:
(a) $\langle 2, 4, 6 \rangle = \frac{1}{2}\mathbf{v} \Rightarrow$ The vectors are parallel and point in the same direction.
(b) $\langle -1, -2, 3 \rangle$ is not a scalar multiple of **v**, hence these vectors are not parallel.
(c) $\langle -7, -14, -21 \rangle = -\frac{7}{4}\mathbf{v} \Rightarrow$ The vectors are parallel but point in opposite directions.
(d) $\langle 6, 10, 14 \rangle$ is not a constant multiple of **v**, hence these vectors are not parallel.

In Exercises 14–17, determine whether $\overrightarrow{AB}$ is equivalent to $\overrightarrow{PQ}$.

15. $A = (1, 4, 1) \quad B = (-2, 2, 0)$
$P = (2, 5, 7) \quad Q = (-3, 2, 1)$

SOLUTION We compute the two vectors:
$$\overrightarrow{AB} = \langle -2 - 1, 2 - 4, 0 - 1 \rangle = \langle -3, -2, -1 \rangle$$
$$\overrightarrow{PQ} = \langle -3 - 2, 2 - 5, 1 - 7 \rangle = \langle -5, -3, -6 \rangle$$

The components of $\overrightarrow{AB}$ and $\overrightarrow{PQ}$ are not equal, hence they are not a translate of each other, that is, the vectors are not equivalent.

17. $A = (1, 1, 0) \quad B = (3, 3, 5)$
$P = (2, -9, 7) \quad Q = (4, -7, 13)$

SOLUTION The vectors $\overrightarrow{AB}$ and $\overrightarrow{PQ}$ are the following vectors:
$$\overrightarrow{AB} = \langle 3 - 1, 3 - 1, 5 - 0 \rangle = \langle 2, 2, 5 \rangle$$
$$\overrightarrow{PQ} = \langle 4 - 2, -7 - (-9), 13 - 7 \rangle = \langle 2, 2, 6 \rangle$$

The *z*-coordinates of the vectors are not equal, hence the vectors are not equivalent.

In Exercises 18–23, calculate the linear combinations.

19. $-2 \langle 8, 11, 3 \rangle + 4 \langle 2, 1, 1 \rangle$

SOLUTION Using the operations of vector addition and scalar multiplication we have
$$-2 \langle 8, 11, 3 \rangle + 4 \langle 2, 1, 1 \rangle = \langle -16, -22, -6 \rangle + \langle 8, 4, 4 \rangle = \langle -8, -18, -2 \rangle$$

21. $\frac{1}{2} \langle 4, -2, 8 \rangle - \frac{1}{3} \langle 12, 3, 3 \rangle$

SOLUTION Using the operations on vectors we have
$$\frac{1}{2} \langle 4, -2, 8 \rangle - \frac{1}{3} \langle 12, 3, 3 \rangle = \langle 2, -1, 4 \rangle - \langle 4, 1, 1 \rangle = \langle -2, -2, 3 \rangle$$

23. $4 \langle 6, -1, 1 \rangle - 2 \langle 1, 0, -1 \rangle + 3 \langle -2, 1, 1 \rangle$

SOLUTION Using the operations of vector addition and scalar multiplication we have
$$4 \langle 6, -1, 1 \rangle - 2 \langle 1, 0, -1 \rangle + 3 \langle -2, 1, 1 \rangle = \langle 24, -4, 4 \rangle + \langle -2, 0, 2 \rangle + \langle -6, 3, 3 \rangle$$
$$= \langle 16, -1, 9 \rangle$$

In Exercises 24–27, determine whether or not the two vectors are parallel.

25. $\mathbf{u} = \langle 4, 2, -6 \rangle, \mathbf{v} = \langle 2, -1, 3 \rangle$

SOLUTION Since the first component of **u** is twice the first component of **v**, if the two vectors are to be parallel, the second component of **u** must be twice the second component of **v**. But it is not; it is -2 times the second component of **v**. Thus the two vectors are not parallel.

27. $\mathbf{u} = \langle -3, 1, 4 \rangle, \mathbf{v} = \langle 6, -2, 8 \rangle$

SOLUTION Since the first component of $\mathbf{v}$ is -2 times the first component of $\mathbf{u}$, if the two vectors are to be parallel, the third component of $\mathbf{v}$ must be -2 times the third component of $\mathbf{u}$. But it is not; it is 2 times the third component of $\mathbf{u}$. Thus the two vectors are not parallel.

In Exercises 28–31, find the given vector.

29. $\mathbf{e_w}$, where $\mathbf{w} = \langle 4, -2, -1 \rangle$

SOLUTION We first find the length of $\mathbf{w}$:

$$\|\mathbf{w}\| = \sqrt{4^2 + (-2)^2 + 1^2} = \sqrt{21}$$

Hence,

$$\mathbf{e_w} = \frac{1}{\|\mathbf{w}\|}\mathbf{w} = \left\langle \frac{4}{\sqrt{21}}, \frac{-2}{\sqrt{21}}, \frac{-1}{\sqrt{21}} \right\rangle$$

31. Unit vector in the direction opposite to $\mathbf{v} = \langle -4, 4, 2 \rangle$

SOLUTION A unit vector in the direction opposite to $\mathbf{v} = \langle -4, 4, 2 \rangle$ is the following vector:

$$-\mathbf{e_v} = -\frac{1}{\|\mathbf{v}\|}\mathbf{v}$$

We compute the length of $\mathbf{v}$:

$$\|\mathbf{v}\| = \sqrt{(-4)^2 + 4^2 + 2^2} = 6$$

The desired vector is, thus,

$$-\mathbf{e_v} = -\frac{1}{6}\langle -4, 4, 2 \rangle = \left\langle \frac{-4}{-6}, \frac{4}{-6}, \frac{2}{-6} \right\rangle = \left\langle \frac{2}{3}, -\frac{2}{3}, -\frac{1}{3} \right\rangle$$

In Exercises 33–36, describe the surface.

33. $x^2 + y^2 + (z - 2)^2 = 4$, with $z \geq 2$

SOLUTION The equation describes a sphere of radius 2 centered at $(0, 0, 2)$; the restriction that $z \geq 2$ means that we take only the upper hemisphere.

35. $x^2 + y^2 = 7$, with $|z| \leq 7$

SOLUTION The equation describes a cylinder of radius $\sqrt{7}$ with its axis lying on the z-axis. The given restriction means that we take only the portion of the cylinder lying between $z = -7$ and $z = 7$.

In Exercises 37–42, give an equation for the indicated surface.

37. The sphere of radius 3 centered at $(0, 0, -3)$

SOLUTION $x^2 + y^2 + (z - (-3))^2 = x^2 + y^2 + (z + 3)^2 = 3^2 = 9$.

39. The sphere centered at $(6, -3, 11)$ passing through $(0, 1, -4)$

SOLUTION Since the sphere is centered at $(6, -3, 11)$, its equation is of the form

$$(x - 6)^2 + (y + 3)^2 + (z - 11)^2 = r^2$$

since it passes through $(0, 1, -4)$, that point satisfies the equation, so that

$$(0 - 6)^2 + (1 + 3)^2 + (-4 - 11)^2 = 277 = r^2$$

Therefore the equation of the sphere is

$$(x - 6)^2 + (y + 3)^2 + (z - 11)^2 = 277$$

41. The cylinder passing through the origin with the vertical line through $(1, -1, 0)$ as its central axis

SOLUTION Since the axis is vertical, the equation of the cylinder has the form

$$(x - a)^2 + (y - b)^2 = r^2$$

where $(a, b, 0)$ is a point on the vertical axis and r is the radius. Therefore we have $a = 1$, $b = -1$, and the equation is

$$(x - 1)^2 + (y + 1)^2 = r^2$$

Substituting $(x, y, z) = (0, 0, 0)$ gives $(-1)^2 + 1^2 = 2 = r^2$, so finally, the equation of the cylinder is

$$(x - 1)^2 + (y + 1)^2 = 2$$

In Exercises 43–50, find a vector parametrization for the line with the given description.

43. Passes through $P = (1, 2, -8)$, direction vector $\mathbf{v} = \langle 2, 1, 3 \rangle$

SOLUTION The vector parametrization for the line is

$$\mathbf{r}(t) = \overrightarrow{OP} + t\mathbf{v}$$

Inserting the given data we get

$$\mathbf{r}(t) = \langle 1, 2, -8 \rangle + t\langle 2, 1, 3 \rangle = \langle 1 + 2t, 2 + t, -8 + 3t \rangle$$

45. Passes through $P = (4, 0, 8)$, direction vector $\mathbf{v} = 7\mathbf{i} + 4\mathbf{k}$

SOLUTION Since $\mathbf{v} = 7\mathbf{i} + 4\mathbf{k} = \langle 7, 0, 4 \rangle$ we obtain the following parametrization:

$$\mathbf{r}(t) = \overrightarrow{OP} + t\mathbf{v} = \langle 4, 0, 8 \rangle + t\langle 7, 0, 4 \rangle = \langle 4 + 7t, 0, 8 + 4t \rangle$$

47. Passes through $(1, 1, 1)$ and $(3, -5, 2)$

SOLUTION We use the equation of the line through two points P and Q:

$$\mathbf{r}(t) = (1 - t)\overrightarrow{OP} + t\overrightarrow{OQ}$$

Since $\overrightarrow{OP} = \langle 1, 1, 1 \rangle$ and $\overrightarrow{OQ} = \langle 3, -5, 2 \rangle$ we obtain

$$\mathbf{r}(t) = (1 - t)\langle 1, 1, 1 \rangle + t\langle 3, -5, 2 \rangle = \langle 1 - t, 1 - t, 1 - t \rangle + \langle 3t, -5t, 2t \rangle = \langle 1 + 2t, 1 - 6t, 1 + t \rangle$$

49. Passes through O and $(4, 1, 1)$

SOLUTION By the equation of the line through two points we get

$$\mathbf{r}(t) = (1 - t)\langle 0, 0, 0 \rangle + t\langle 4, 1, 1 \rangle = \langle 0, 0, 0 \rangle + \langle 4t, t, t \rangle = \langle 4t, t, t \rangle$$

In Exercises 51–54, find parametric equations for the lines with the given description.

51. Perpendicular to the xy-plane, passes through the origin

SOLUTION A direction vector for the line is a vector parallel to the z-axis; for instance, we may choose $\mathbf{v} = \langle 0, 0, 1 \rangle$. The line passes through the origin $(0, 0, 0)$, hence we obtain the following parametrization:

$$\mathbf{r}(t) = \langle 0, 0, 0 \rangle + t\langle 0, 0, 1 \rangle = \langle 0, 0, t \rangle$$

or $x = 0$, $y = 0$, $z = t$.

53. Parallel to the line through $(1, 1, 0)$ and $(0, -1, -2)$, passes through $(0, 0, 4)$

SOLUTION The direction vector is $\mathbf{v} = \langle 0 - 1, -1 - 1, -2 - 0 \rangle = \langle -1, -2, -2 \rangle$. Hence, using the equation of a line we obtain

$$\mathbf{r}(t) = \langle 0, 0, 4 \rangle + t\langle -1, -2, -2 \rangle = \langle -t, -2t, 4 - 2t \rangle$$

55. Which of the following is a parametrization of the line through $P = (4, 9, 8)$ perpendicular to the xz-plane (Figure 19)?

(a) $\mathbf{r}(t) = \langle 4, 9, 8 \rangle + t \langle 1, 0, 1 \rangle$
(b) $\mathbf{r}(t) = \langle 4, 9, 8 \rangle + t \langle 0, 0, 1 \rangle$
(c) $\mathbf{r}(t) = \langle 4, 9, 8 \rangle + t \langle 0, 1, 0 \rangle$
(d) $\mathbf{r}(t) = \langle 4, 9, 8 \rangle + t \langle 1, 1, 0 \rangle$

FIGURE 19

SOLUTION Since the line is perpendicular to the xz-plane, all of its points have x-coordinate equal to 4 and z-coordinate equal to 8 (see diagram). Thus only the y-coordinate varies, and the correct answer is **(c)**, $\langle 4, 9, 8 \rangle + t \langle 0, 1, 0 \rangle$.

57. Show that $\mathbf{r}_1(t)$ and $\mathbf{r}_2(t)$ define the same line, where

$$\mathbf{r}_1(t) = \langle 3, -1, 4 \rangle + t \langle 8, 12, -6 \rangle$$
$$\mathbf{r}_2(t) = \langle 11, 11, -2 \rangle + t \langle 4, 6, -3 \rangle$$

Hint: Show that $\mathbf{r}_2(t)$ passes through $(3, -1, 4)$ and that the direction vectors for $\mathbf{r}_1(t)$ and $\mathbf{r}_2(t)$ are parallel.

SOLUTION We observe first that the direction vectors of $\mathbf{r}_1(t)$ and $\mathbf{r}_2(t)$ are multiples of each other:

$$\langle 8, 12, -6 \rangle = 2 \langle 4, 6, -3 \rangle$$

Therefore $\mathbf{r}_1(t)$ and $\mathbf{r}_2(t)$ are parallel. To show they coincide, it suffices to prove that they share a point in common, so we verify that $\mathbf{r}_1(0) = \langle 3, -1, 4 \rangle$ lies on $\mathbf{r}_2(t)$ by solving for t:

$$\langle 3, -1, 4 \rangle = \langle 11, 11, -2 \rangle + t \langle 4, 6, -3 \rangle$$
$$\langle 3, -1, 4 \rangle - \langle 11, 11, -2 \rangle = t \langle 4, 6, -3 \rangle$$
$$\langle -8, -12, 6 \rangle = t \langle 4, 6, -3 \rangle$$

This equation is satisfied for $t = -2$, so $\mathbf{r}_1$ and $\mathbf{r}_2$ coincide.

59. Find two different vector parametrizations of the line through $P = (5, 5, 2)$ with direction vector $\mathbf{v} = \langle 0, -2, 1 \rangle$.

SOLUTION Two different parameterizations are

$$\mathbf{r}_1(t) = \langle 5, 5, 2 \rangle + t \langle 0, -2, 1 \rangle$$
$$\mathbf{r}_2(t) = \langle 5, 5, 2 \rangle + t \langle 0, -20, 10 \rangle$$

61. Show that the lines $\mathbf{r}_1(t) = \langle -1, 2, 2 \rangle + t \langle 4, -2, 1 \rangle$ and $\mathbf{r}_2(t) = \langle 0, 1, 1 \rangle + t \langle 2, 0, 1 \rangle$ do not intersect.

SOLUTION The two lines intersect if there exist parameter values t_1 and t_2 such that

$$\langle -1, 2, 2 \rangle + t_1 \langle 4, -2, 1 \rangle = \langle 0, 1, 1 \rangle + t_2 \langle 2, 0, 1 \rangle$$
$$\langle -1 + 4t_1, 2 - 2t_1, 2 + t_1 \rangle = \langle 2t_2, 1, 1 + t_2 \rangle$$

Equating corresponding components yields

$$-1 + 4t_1 = 2t_2$$
$$2 - 2t_1 = 1$$
$$2 + t_1 = 1 + t_2$$

The second equation implies $t_1 = \frac{1}{2}$. Substituting into the first and third equations we get

$$-1 + 4 \cdot \frac{1}{2} = 2t_2 \quad \Rightarrow \quad t_2 = \frac{1}{2}$$

$$2 + \frac{1}{2} = 1 + t_2 \quad \Rightarrow \quad t_2 = \frac{3}{2}$$

We conclude that the equations do not have solutions, which means that the two lines do not intersect.

63. Determine whether the lines $\mathbf{r}_1(t) = \langle 0, 1, 1 \rangle + t \langle 1, 1, 2 \rangle$ and $\mathbf{r}_2(s) = \langle 2, 0, 3 \rangle + s \langle 1, 4, 4 \rangle$ intersect, and if so, find the point of intersection.

SOLUTION The lines intersect if there exist parameter values t and s such that

$$\langle 0, 1, 1 \rangle + t \langle 1, 1, 2 \rangle = \langle 2, 0, 3 \rangle + s \langle 1, 4, 4 \rangle$$

$$\langle t, 1 + t, 1 + 2t \rangle = \langle 2 + s, 4s, 3 + 4s \rangle \tag{1}$$

Equating corresponding components we get

$$t = 2 + s$$

$$1 + t = 4s$$

$$1 + 2t = 3 + 4s$$

Substituting t from the first equation into the second equation we get

$$\begin{matrix} 1 + 2 + s = 4s \\ 3s = 3 \end{matrix} \quad \Rightarrow \quad s = 1, \; t = 2 + s = 3$$

We now check whether $s = 1$, $t = 3$ satisfy the third equation:

$$1 + 2 \cdot 3 = 3 + 4 \cdot 1$$

$$7 = 7$$

We conclude that $s = 1$, $t = 3$ is the solution of (1), hence the two lines intersect. To find the point of intersection we substitute $s = 1$ in the right-hand side of (1) to obtain

$$\langle 2 + 1, 4 \cdot 1, 3 + 4 \cdot 1 \rangle = \langle 3, 4, 7 \rangle$$

The point of intersection is the terminal point of this vector, that is, $(3, 4, 7)$.

65. A meteor follows a trajectory $\mathbf{r}(t) = \langle 2, 1, 4 \rangle + t \langle 3, 2, -1 \rangle$ km with t in minutes, near the surface of the earth, which is represented by the xy-plane. Determine at what time the meteor hits the ground.

SOLUTION Since the surface of the earth is the xy-plane, the z direction is the height above the ground. The meteor hits the ground when the z component of $\mathbf{r}(t)$ is zero; this happens when $4 - t = 0$, or $t = 4$. The meteor hits the ground at $t = 4$ minutes.

67. The line with vector parametrization $\mathbf{r}(t) = \langle 3, 1, -4 \rangle + t \langle -2, -2, 3 \rangle$ intersects the sphere $(x - 1)^2 + (y + 3)^2 + z^2 = 8$ in two points. Find them. *Hint*: Determine t such that the point $(x(t), y(t), z(t))$ satisfies the equation of the sphere, and then find the corresponding points on the line.

SOLUTION The line consists of points $(x(t), y(t), z(t)) = (3 - 2t, 1 - 2t, -4 + 3t)$. If such a point lies on the sphere, then it satisfies the equation of the sphere, so that

$$(x(t) - 1)^2 + (y(t) + 3)^2 + z(t)^2 = (2 - 2t)^2 + (4 - 2t)^2 + (-4 + 3t)^2 = 8$$

Expanding this equation gives

$$17t^2 - 48t + 28 = 0$$

This quadratic has the two roots $t = 2$ and $t = \frac{14}{17}$, so that the two points that lie both on the line and on the sphere (and thus are the intersection points) are

$$(3 - 2 \cdot 2, 1 - 2 \cdot 2, -4 + 3 \cdot 2) = (-1, -3, 2)$$

$$\left(3 - 2 \cdot \frac{14}{17}, 1 - 2 \cdot \frac{14}{17}, -4 + 3 \cdot \frac{14}{17}\right) = \left(\frac{23}{17}, -\frac{11}{17}, -\frac{26}{17}\right)$$

69. Find the components of the vector **v** whose tail and head are the midpoints of segments $\overline{AC}$ and $\overline{BC}$ in Figure 20. [Note that the midpoint of (a_1, a_2, a_3) and (b_1, b_2, b_3) is $\left(\dfrac{a_1+b_1}{2}, \dfrac{a_2+b_2}{2}, \dfrac{a_3+b_3}{2}\right)$.]

SOLUTION We denote by P and Q the midpoints of the segments $\overline{AC}$ and $\overline{BC}$, respectively. Thus,

$$\mathbf{v} = \overrightarrow{PQ} \tag{1}$$

We use the formula for the midpoint of a segment to find the coordinates of the points P and Q. This gives

$$P = \left(\frac{1+0}{2}, \frac{0+1}{2}, \frac{1+1}{2}\right) = \left(\frac{1}{2}, \frac{1}{2}, 1\right)$$

$$Q = \left(\frac{1+0}{2}, \frac{1+1}{2}, \frac{0+1}{2}\right) = \left(\frac{1}{2}, 1, \frac{1}{2}\right)$$

Substituting in (1) yields the following vector:

$$\mathbf{v} = \overrightarrow{PQ} = \left\langle \frac{1}{2} - \frac{1}{2}, 1 - \frac{1}{2}, \frac{1}{2} - 1 \right\rangle = \left\langle 0, \frac{1}{2}, -\frac{1}{2} \right\rangle$$

71. A box that weighs 1000 kg is hanging from a crane at the dock. The crane has a square 20 m by 20 m framework as in Figure 21, with four cables, each of the same length, supporting the box. The box hangs 10 m below the level of the framework. Find the magnitude of the force acting on each cable.

FIGURE 21

SOLUTION By symmetry, the magnitude of the force acting on all four cables is the same; denote this magnitude by f. The directions of the four forces exerted by the cables are

$$\langle 10, -10, 0 \rangle - \langle 0, 0, -10 \rangle = \langle 10, -10, 10 \rangle$$

$$\langle -10, -10, 0 \rangle - \langle 0, 0, -10 \rangle = \langle -10, -10, 10 \rangle$$

$$\langle -10, 10, 0 \rangle - \langle 0, 0, -10 \rangle = \langle -10, 10, 10 \rangle$$

$$\langle 10, 10, 0 \rangle - \langle 0, 0, -10 \rangle = \langle 10, 10, 10 \rangle$$

The sum of the forces in these four directions must balance the downward force, which is $\langle 0, 0, -1000 \rangle$. Thus we get

$$f \langle 10, -10, 10 \rangle + f \langle -10, -10, 10 \rangle + f \langle -10, 10, 10 \rangle + f \langle 10, 10, 10 \rangle = \langle 0, 0, -1000 \rangle$$

so that

$$f \langle 0, 0, 40 \rangle = \langle 0, 0, -1000 \rangle$$

which gives $f = -25$. The force has a magnitude of 25 newtons.

Further Insights and Challenges

*In Exercises 72–78, we consider the equations of a line in **symmetric form**, when $a \neq 0, b \neq 0, c \neq 0$.*

$$\frac{x - x_0}{a} = \frac{y - y_0}{b} = \frac{z - z_0}{c} \qquad \boxed{10}$$

73. Find the symmetric equations of the line through $P_0 = (-2, 3, 3)$ with direction vector $\mathbf{v} = \langle 2, 4, 3 \rangle$.

SOLUTION Using $(x_0, y_0, z_0) = (-2, 3, 3)$ and $\langle a, b, c \rangle = \langle 2, 4, 3 \rangle$ in Eq. (10) gives

$$\frac{x + 2}{2} = \frac{y - 3}{4} = \frac{z - 3}{3}$$

75. Find the symmetric equations of the line

$$x = 3 + 2t, \quad y = 4 - 9t, \quad z = 12t$$

SOLUTION If we solve each equation for t, we get:

$$t = \frac{x - 3}{2}, \quad t = \frac{4 - y}{9}, \quad t = \frac{z}{12}$$

When we set these equations equal to each other, we get:

$$\frac{x - 3}{2} = \frac{4 - y}{9} = \frac{z}{12}$$

77. Find a vector parametrization for the line $\dfrac{x}{2} = \dfrac{y}{7} = \dfrac{z}{8}$.

SOLUTION If we let t equal these three terms, as follows:

$$t = \frac{x}{2} = \frac{y}{7} = \frac{z}{8}$$

then we can break it up into three equations:

$$t = \frac{x}{2}, \quad t = \frac{y}{7}, \quad t = \frac{z}{8}$$

and solving for $x, y,$ and z gives us:

$$x = 2t, \quad y = 7t, \quad z = 8t$$

and writing this in vector form gives us

$$\mathbf{r}(t) = t \langle 2, 7, 8 \rangle$$

79. A median of a triangle is a segment joining a vertex to the midpoint of the opposite side. Referring to Figure 22(A), prove that three medians of triangle ABC intersect at the terminal point P of the vector $\frac{1}{3}(\mathbf{u} + \mathbf{v} + \mathbf{w})$. The point P is the *centroid* of the triangle. *Hint:* Show, by parametrizing the segment $\overline{AA'}$, that P lies two-thirds of the way from A to A'. It will follow similarly that P lies on the other two medians.

FIGURE 22

SOLUTION From Figure 22(A),

$$\overrightarrow{OC'} = \overrightarrow{OA} + \overrightarrow{AC'} = \mathbf{v} + \frac{1}{2}(\mathbf{w} - \mathbf{v}) = \frac{1}{2}(\mathbf{v} + \mathbf{w})$$

$$\overrightarrow{OD} = \mathbf{u}$$

The line through the points C and C' has the parametrization

$$t\mathbf{u} + (1-t)\frac{\mathbf{v} + \mathbf{w}}{2} \qquad (1)$$

Similarly, the line through B and B' has the parametrization

$$t\mathbf{w} + (1-t)\frac{\mathbf{v} + \mathbf{u}}{2} \qquad (2)$$

And the line through A and A' has the parametrization

$$t\mathbf{v} + (1-t)\frac{\mathbf{u} + \mathbf{w}}{2} \qquad (3)$$

Now, setting $t = \frac{1}{3}$ in (1), (2), and (3) yields $\frac{1}{3}(\mathbf{u} + \mathbf{v} + \mathbf{w})$. We conclude that the terminal point of this vector lies on each one of the lines, hence it is their point of intersection.

12.3 Dot Product and the Angle Between Two Vectors (LT Section 13.3)

Preliminary Questions

1. Is the dot product of two vectors a scalar or a vector?

SOLUTION The dot product of two vectors is the sum of products of scalars, hence it is a scalar.

2. What can you say about the angle between $\mathbf{a}$ and $\mathbf{b}$ if $\mathbf{a} \cdot \mathbf{b} < 0$?

SOLUTION Since the cosine of the angle between $\mathbf{a}$ and $\mathbf{b}$ satisfies $\cos\theta = \frac{\mathbf{a}\cdot\mathbf{b}}{\|\mathbf{a}\|\|\mathbf{b}\|}$, also $\cos\theta < 0$. By definition $0 \le \theta \le \pi$, but since $\cos\theta < 0$ then θ is in $[\pi/2, \pi]$. In other words, the angle between $\mathbf{a}$ and $\mathbf{b}$ is obtuse.

3. Which property of dot products allows us to conclude that if $\mathbf{v}$ is orthogonal to both $\mathbf{u}$ and $\mathbf{w}$, then $\mathbf{v}$ is orthogonal to $\mathbf{u} + \mathbf{w}$?

SOLUTION One property is that two vectors are orthogonal if and only if the dot product of the two vectors is zero. The second property is the Distributive Law. Since $\mathbf{v}$ is orthogonal to $\mathbf{u}$ and $\mathbf{w}$, we have $\mathbf{v}\cdot\mathbf{u} = 0$ and $\mathbf{v}\cdot\mathbf{w} = 0$. Therefore,

$$\mathbf{v}\cdot(\mathbf{u}+\mathbf{w}) = \mathbf{v}\cdot\mathbf{u} + \mathbf{v}\cdot\mathbf{w} = 0 + 0 = 0$$

We conclude that $\mathbf{v}$ is orthogonal to $\mathbf{u} + \mathbf{w}$.

4. Which is the projection of $\mathbf{v}$ along $\mathbf{v}$: (a) $\mathbf{v}$ or (b) $\mathbf{e}_\mathbf{v}$?

SOLUTION The projection of $\mathbf{v}$ along itself is $\mathbf{v}$, since

$$\mathbf{v}_{\|\mathbf{v}} = \left(\frac{\mathbf{v}\cdot\mathbf{v}}{\mathbf{v}\cdot\mathbf{v}}\right)\mathbf{v} = \mathbf{v}$$

5. Let $\mathbf{u}_{\|\mathbf{v}}$ be the projection of $\mathbf{u}$ along $\mathbf{v}$. Which of the following is the projection $\mathbf{u}$ along the vector $2\mathbf{v}$ and which is the projection of $2\mathbf{u}$ along $\mathbf{v}$?

(a) $\frac{1}{2}\mathbf{u}_{\|\mathbf{v}}$ \qquad (b) $\mathbf{u}_{\|\mathbf{v}}$ \qquad (c) $2\mathbf{u}_{\|\mathbf{v}}$

SOLUTION Since $u_{\|\mathbf{v}}$ is the projection of $\mathbf{u}$ along $\mathbf{v}$, we have,

$$u_{\|\mathbf{v}} = \left(\frac{\mathbf{u}\cdot\mathbf{v}}{\mathbf{v}\cdot\mathbf{v}}\right)\mathbf{v}$$

The projection of $\mathbf{u}$ along the vector $2\mathbf{v}$ is

$$\mathbf{u}_{\|2\mathbf{v}} = \left(\frac{\mathbf{u}\cdot(2\mathbf{v})}{(2\mathbf{v})\cdot(2\mathbf{v})}\right)2\mathbf{v} = \left(\frac{2(\mathbf{u}\cdot\mathbf{v})}{4(\mathbf{v}\cdot\mathbf{v})}\right)2\mathbf{v} = \left(\frac{\mathbf{u}\cdot\mathbf{v}}{\mathbf{v}\cdot\mathbf{v}}\right)\mathbf{v} = \mathbf{u}_{\|\mathbf{v}}$$

That is, $u_{\|\mathbf{v}}$ is the projection of $\mathbf{u}$ along $2\mathbf{v}$. Notice that the projection of $\mathbf{u}$ along $\mathbf{v}$ is the projection of $\mathbf{u}$ along the unit vector $\mathbf{e_v}$, hence it depends on the direction of $\mathbf{v}$ rather than on the length of $\mathbf{v}$. Therefore, the projection of $\mathbf{u}$ along $\mathbf{v}$ and along $2\mathbf{v}$ is the same vector.

For the second question,

$$(2\mathbf{u})_{\|\mathbf{v}} = \left(\frac{(2\mathbf{u}) \cdot \mathbf{v}}{\mathbf{v} \cdot \mathbf{v}}\right) \mathbf{v} = 2\left(\frac{\mathbf{u} \cdot \mathbf{v}}{\mathbf{v} \cdot \mathbf{v}}\right) \mathbf{v} = 2\mathbf{u}_{\|\mathbf{v}}$$

That is, the projection of $2\mathbf{u}$ along $\mathbf{v}$ is twice the projection of $\mathbf{u}$ along $\mathbf{v}$.

6. Which of the following is equal to $\cos \theta$, where θ is the angle between $\mathbf{u}$ and $\mathbf{v}$?

(a) $\mathbf{u} \cdot \mathbf{v}$ (b) $\mathbf{u} \cdot \mathbf{e_v}$ (c) $\mathbf{e_u} \cdot \mathbf{e_v}$

SOLUTION By the Theorems on the Dot Product and the Angle Between Vectors, we have

$$\cos \theta = \frac{\mathbf{u} \cdot \mathbf{v}}{\|\mathbf{u}\|\|\mathbf{v}\|} = \frac{\mathbf{u}}{\|\mathbf{u}\|} \cdot \frac{\mathbf{v}}{\|\mathbf{v}\|} = \mathbf{e_u} \cdot \mathbf{e_v}$$

The correct answer is (c).

Exercises

In Exercises 1–12, compute the dot product.

1. $\langle 1, 2, 1 \rangle \cdot \langle 4, 3, 5 \rangle$

SOLUTION Using the definition of the dot product we obtain

$$\langle 1, 2, 1 \rangle \cdot \langle 4, 3, 5 \rangle = 1 \cdot 4 + 2 \cdot 3 + 1 \cdot 5 = 15$$

3. $\langle 0, 1, 1 \rangle \cdot \langle -7, 41, -39 \rangle$

SOLUTION We multiply corresponding components and add to obtain

$$\langle 0, 1, 1 \rangle \cdot \langle -7, 41, -39 \rangle = 0 \cdot (-7) + 1 \cdot 41 + 1 \cdot (-39) = 2$$

5. $\langle 3, 1 \rangle \cdot \langle 4, -7 \rangle$

SOLUTION The dot product of the two vectors is the following scalar:

$$\langle 3, 1 \rangle \cdot \langle 4, -7 \rangle = 3 \cdot 4 + 1 \cdot (-7) = 5$$

7. $\mathbf{k} \cdot \mathbf{j}$

SOLUTION By the orthogonality of $\mathbf{j}$ and $\mathbf{k}$, we have $\mathbf{k} \cdot \mathbf{j} = 0$

9. $(\mathbf{i} + \mathbf{j}) \cdot (\mathbf{j} + \mathbf{k})$

SOLUTION By the distributive law and the orthogonality of $\mathbf{i}$, $\mathbf{j}$ and $\mathbf{k}$ we have

$$(\mathbf{i} + \mathbf{j}) \cdot (\mathbf{j} + \mathbf{k}) = \mathbf{i} \cdot \mathbf{j} + \mathbf{i} \cdot \mathbf{k} + \mathbf{j} \cdot \mathbf{j} + \mathbf{j} \cdot \mathbf{k} = 0 + 0 + 1 + 0 = 1$$

11. $(\mathbf{i} + \mathbf{j} + \mathbf{k}) \cdot (3\mathbf{i} + 2\mathbf{j} - 5\mathbf{k})$

SOLUTION We use properties of the dot product to obtain

$$(\mathbf{i} + \mathbf{j} + \mathbf{k}) \cdot (3\mathbf{i} + 2\mathbf{j} - 5\mathbf{k}) = 3\mathbf{i} \cdot \mathbf{i} + 2\mathbf{i} \cdot \mathbf{j} - 5\mathbf{i} \cdot \mathbf{k} + 3\mathbf{j} \cdot \mathbf{i} + 2\mathbf{j} \cdot \mathbf{j} - 5\mathbf{j} \cdot \mathbf{k} + 3\mathbf{k} \cdot \mathbf{i} + 2\mathbf{k} \cdot \mathbf{j} - 5\mathbf{k} \cdot \mathbf{k}$$
$$= 3\|\mathbf{i}\|^2 + 2\|\mathbf{j}\|^2 - 5\|\mathbf{k}\|^2 = 3 \cdot 1 + 2 \cdot 1 - 5 \cdot 1 = 0$$

In Exercises 13–18, determine whether the two vectors are orthogonal and, if not, whether the angle between them is acute or obtuse.

13. $\langle 1, 1, 1 \rangle$, $\langle 1, -2, -2 \rangle$

SOLUTION We compute the dot product of the two vectors:

$$\langle 1, 1, 1 \rangle \cdot \langle 1, -2, -2 \rangle = 1 \cdot 1 + 1 \cdot (-2) + 1 \cdot (-2) = -3$$

Since the dot product is negative, the angle between the vectors is obtuse.

15. $\langle 1, 2, 1 \rangle$, $\langle 7, -3, -1 \rangle$

SOLUTION We compute the dot product:

$$\langle 1, 1, 1 \rangle \cdot \langle 3, -2, -1 \rangle = 1 \cdot 3 + 1 \cdot (-2) + 1 \cdot (-1) = 0$$

The dot product is zero, hence the vectors are orthogonal.

17. $\langle \frac{12}{5}, -\frac{4}{5} \rangle$, $\langle \frac{1}{2}, -\frac{7}{4} \rangle$

SOLUTION We find the dot product of the two vectors:

$$\left\langle \frac{12}{5}, -\frac{4}{5} \right\rangle \cdot \left\langle \frac{1}{2}, -\frac{7}{4} \right\rangle = \frac{12}{5} \cdot \frac{1}{2} + \left(-\frac{4}{5}\right) \cdot \left(-\frac{7}{4}\right) = \frac{12}{10} + \frac{28}{20} = \frac{13}{5}$$

The dot product is positive, hence the angle between the vectors is acute.

In Exercises 19–22, find the cosine of the angle between the vectors.

19. $\langle 0, 3, 1 \rangle$, $\langle 4, 0, 0 \rangle$

SOLUTION Since $\langle 0, 3, 1 \rangle \cdot \langle 4, 0, 0 \rangle = 0 \cdot 4 + 3 \cdot 0 + 1 \cdot 0 = 0$, the vectors are orthogonal, that is, the angle between them is $\theta = 90°$ and $\cos \theta = 0$.

21. $\mathbf{i} + \mathbf{j}$, $\mathbf{j} + 2\mathbf{k}$

SOLUTION We use the formula for the cosine of the angle between two vectors. Let $\mathbf{v} = \mathbf{i} + \mathbf{j}$ and $\mathbf{w} = \mathbf{j} + 2\mathbf{k}$. We compute the following values:

$$\|\mathbf{v}\| = \|\mathbf{i} + \mathbf{j}\| = \sqrt{1^2 + 1^2} = \sqrt{2}$$

$$\|\mathbf{w}\| = \|\mathbf{j} + 2\mathbf{k}\| = \sqrt{1^2 + 2^2} = \sqrt{5}$$

$$\mathbf{v} \cdot \mathbf{w} = (\mathbf{i} + \mathbf{j}) \cdot (\mathbf{j} + 2\mathbf{k}) = \mathbf{i} \cdot \mathbf{j} + 2\mathbf{i} \cdot \mathbf{k} + \mathbf{j} \cdot \mathbf{j} + 2\mathbf{j} \cdot \mathbf{k} = \|\mathbf{j}\|^2 = 1$$

Hence,

$$\cos \theta = \frac{\mathbf{v} \cdot \mathbf{w}}{\|\mathbf{v}\|\|\mathbf{w}\|} = \frac{1}{\sqrt{2}\sqrt{5}} = \frac{1}{\sqrt{10}}$$

In Exercises 23–30, find the angle between the vectors.

23. $\langle 2, \sqrt{2} \rangle$, $\langle 1 + \sqrt{2}, 1 - \sqrt{2} \rangle$

SOLUTION We write $\mathbf{v} = \langle 2, \sqrt{2} \rangle$ and $\mathbf{w} = \langle 2, \sqrt{2} \rangle$. To use the formula for the cosine of the angle θ between two vectors we need to compute the following values:

$$\|\mathbf{v}\| = \sqrt{4 + 2} = \sqrt{6}$$

$$\|\mathbf{w}\| = \sqrt{(1 + \sqrt{2})^2 + (1 - \sqrt{2})^2} = \sqrt{6}$$

$$\mathbf{v} \cdot \mathbf{w} = 2 + 2\sqrt{2} + \sqrt{2} - 2 = 3\sqrt{2}$$

Hence,

$$\cos \theta = \frac{\mathbf{v} \cdot \mathbf{w}}{\|\mathbf{v}\|\|\mathbf{w}\|} = \frac{3\sqrt{2}}{\sqrt{6}\sqrt{6}} = \frac{\sqrt{2}}{2}$$

and so,

$$\theta = \cos^{-1} \frac{\sqrt{2}}{2} = \pi/4$$

918 CHAPTER 12 | VECTOR GEOMETRY (LT CHAPTER 13)

25. $\langle 1,1,1 \rangle$, $\langle 1,0,1 \rangle$

SOLUTION We denote $\mathbf{v} = \langle 1, 1, 1 \rangle$ and $\mathbf{w} = \langle 1, 0, 1 \rangle$. To use the formula for the cosine of the angle θ between two vectors we need to compute the following values:

$$\|\mathbf{v}\| = \sqrt{1^2 + 1^2 + 1^2} = \sqrt{3}$$
$$\|\mathbf{w}\| = \sqrt{1^2 + 0^2 + 1^2} = \sqrt{2}$$
$$\mathbf{v} \cdot \mathbf{w} = 1 + 0 + 1 = 2$$

Hence,

$$\cos \theta = \frac{\mathbf{v} \cdot \mathbf{w}}{\|\mathbf{v}\|\|\mathbf{w}\|} = \frac{2}{\sqrt{3}\sqrt{2}} = \frac{\sqrt{6}}{3}$$

and so,

$$\theta = \cos^{-1} \frac{\sqrt{6}}{3} \approx 0.615$$

27. $\langle 0, 1, 1 \rangle$, $\langle 1, -1, 0 \rangle$

SOLUTION We denote $\mathbf{v} = \langle 0, 1, 1 \rangle$ and $\mathbf{w} = \langle 1, -1, 0 \rangle$. To use the formula for the cosine of the angle θ between two vectors we need to compute the following values:

$$\|\mathbf{v}\| = \sqrt{0^2 + 1^2 + 1^2} = \sqrt{2}$$
$$\|\mathbf{w}\| = \sqrt{1^2 + (-1)^2 + 0^2} = \sqrt{2}$$
$$\mathbf{v} \cdot \mathbf{w} = 0 + (-1) + 0 = -1$$

Hence,

$$\cos \theta = \frac{\mathbf{v} \cdot \mathbf{w}}{\|\mathbf{v}\|\|\mathbf{w}\|} = \frac{-1}{\sqrt{2}\sqrt{2}} = -\frac{1}{2}$$

and so,

$$\theta = \cos^{-1} -\frac{1}{2} = \frac{2\pi}{3}$$

29. $\mathbf{i}$, $3\mathbf{i} + 2\mathbf{j} + \mathbf{k}$

SOLUTION Writing these as vectors in three-dimensional space, we have

$$\mathbf{v} = \mathbf{i} = \langle 1, 0, 0 \rangle, \quad \mathbf{w} = 3\mathbf{i} + 2\mathbf{j} + \mathbf{k} = \langle 3, 2, 1 \rangle$$

To use the formula for the cosine of the angle θ between two vectors we need to compute the following values:

$$\|\mathbf{v}\| = \sqrt{1^2 + 0^2 + 0^2} = \sqrt{1} = 1$$
$$\|\mathbf{w}\| = \sqrt{3^2 + 2^2 + 1^2} = \sqrt{14}$$
$$\mathbf{v} \cdot \mathbf{w} = 1 \cdot 3 + 0 \cdot 2 + 0 \cdot 1 = 3$$

Hence,

$$\cos \theta = \frac{\mathbf{v} \cdot \mathbf{w}}{\|\mathbf{v}\|\|\mathbf{w}\|} = \frac{3}{\sqrt{14}}$$

and so

$$\theta = \cos^{-1} \frac{3}{\sqrt{14}} \approx 0.641$$

31. Find all values of b for which the vectors are orthogonal.

(a) $\langle b, 3, 2 \rangle$, $\langle 1, b, 1 \rangle$ **(b)** $\langle 4, -2, 7 \rangle$, $\langle b^2, b, 0 \rangle$

SECTION 12.3 | Dot Product and the Angle Between Two Vectors

SOLUTION

(a) The vectors are orthogonal if and only if the scalar product is zero. That is,

$$\langle b, 3, 2 \rangle \cdot \langle 1, b, 1 \rangle = 0$$
$$b \cdot 1 + 3 \cdot b + 2 \cdot 1 = 0$$
$$4b + 2 = 0 \quad \Rightarrow \quad b = -\frac{1}{2}$$

(b) We set the scalar product of the two vectors equal to zero and solve for b. This gives

$$\langle 4, -2, 7 \rangle \cdot \langle b^2, b, 0 \rangle = 0$$
$$4b^2 - 2b + 7 \cdot 0 = 0$$
$$2b(2b - 1) = 0 \quad \Rightarrow \quad b = 0 \text{ or } b = \frac{1}{2}$$

33. Find two vectors that are not multiples of each other and are both orthogonal to $\langle 2, 0, -3 \rangle$.

SOLUTION We denote by $\langle a, b, c \rangle$, a vector orthogonal to $\langle 2, 0, -3 \rangle$. Hence,

$$\langle a, b, c \rangle \cdot \langle 2, 0, -3 \rangle = 0$$
$$2a + 0 - 3c = 0$$
$$2a - 3c = 0 \quad \Rightarrow \quad a = \frac{3}{2}c$$

Thus, the vectors orthogonal to $\langle 2, 0, -3 \rangle$ are of the form

$$\left\langle \frac{3}{2}c, b, c \right\rangle$$

We may find two such vectors by setting $c = 0$, $b = 1$ and $c = 2$, $b = 2$. We obtain

$$\mathbf{v}_1 = \langle 0, 1, 0 \rangle, \quad \mathbf{v}_2 = \langle 3, 2, 2 \rangle$$

35. Find $\mathbf{v} \cdot \mathbf{e}$, where $\|\mathbf{v}\| = 3$, $\mathbf{e}$, is a unit vector, and the angle between $\mathbf{e}$ and $\mathbf{v}$ is $\frac{2\pi}{3}$.

SOLUTION Since $\mathbf{v} \cdot \mathbf{e} = \|\mathbf{v}\|\|\mathbf{e}\| \cos 2\pi/3$, and $\|\mathbf{v}\| = 3$ and $\|\mathbf{e}\| = 1$, we have $\mathbf{v} \cdot \mathbf{e} = 3 \cdot 1 \cdot (-1/2) = -3/2$.

In Exercises 37–40, simplify the expression.

37. $(\mathbf{v} - \mathbf{w}) \cdot \mathbf{v} + \mathbf{v} \cdot \mathbf{w}$

SOLUTION By properties of the dot product we obtain

$$(\mathbf{v} - \mathbf{w}) \cdot \mathbf{v} + \mathbf{v} \cdot \mathbf{w} = \mathbf{v} \cdot \mathbf{v} - \mathbf{w} \cdot \mathbf{v} + \mathbf{v} \cdot \mathbf{w} = \|\mathbf{v}\|^2 - \mathbf{v} \cdot \mathbf{w} + \mathbf{v} \cdot \mathbf{w} = \|\mathbf{v}\|^2$$

39. $(\mathbf{v} + \mathbf{w}) \cdot \mathbf{v} - (\mathbf{v} + \mathbf{w}) \cdot \mathbf{w}$

SOLUTION We use properties of the dot product to write

$$(\mathbf{v} + \mathbf{w}) \cdot \mathbf{v} - (\mathbf{v} + \mathbf{w}) \cdot \mathbf{w} = \mathbf{v} \cdot \mathbf{v} + \mathbf{w} \cdot \mathbf{v} - \mathbf{v} \cdot \mathbf{w} - \mathbf{w} \cdot \mathbf{w}$$
$$= \|\mathbf{v}\|^2 + \mathbf{w} \cdot \mathbf{v} - \mathbf{w} \cdot \mathbf{v} - \|\mathbf{w}\|^2 = \|\mathbf{v}\|^2 - \|\mathbf{w}\|^2$$

In Exercises 41–44, use the properties of the dot product to evaluate the expression, assuming that $\mathbf{u} \cdot \mathbf{v} = 2$, $\|\mathbf{u}\| = 1$, and $\|\mathbf{v}\| = 3$.

41. $\mathbf{u} \cdot (4\mathbf{v})$

SOLUTION Using properties of the dot product we get

$$\mathbf{u} \cdot (4\mathbf{v}) = 4(\mathbf{u} \cdot \mathbf{v}) = 4 \cdot 2 = 8$$

43. $2\mathbf{u} \cdot (3\mathbf{u} - \mathbf{v})$

SOLUTION By properties of the dot product we obtain

$$2\mathbf{u} \cdot (3\mathbf{u} - \mathbf{v}) = (2\mathbf{u}) \cdot (3\mathbf{u}) - (2\mathbf{u}) \cdot \mathbf{v} = 6(\mathbf{u} \cdot \mathbf{u}) - 2(\mathbf{u} \cdot \mathbf{v})$$
$$= 6\|\mathbf{u}\|^2 - 2(\mathbf{u} \cdot \mathbf{v}) = 6 \cdot 1^2 - 2 \cdot 2 = 2$$

45. Find the angle between $\mathbf{v}$ and $\mathbf{w}$ if $\mathbf{v} \cdot \mathbf{w} = -\|\mathbf{v}\| \|\mathbf{w}\|$.

SOLUTION Using the formula for dot product, and the given equation $\mathbf{v} \cdot \mathbf{w} = -\|\mathbf{v}\| \|\mathbf{w}\|$, we get:

$$\|\mathbf{v}\| \|\mathbf{w}\| \cos\theta = -\|\mathbf{v}\| \|\mathbf{w}\|$$

which implies $\cos\theta = -1$, and so the angle between the two vectors is $\theta = \pi$.

47. Assume that $\|\mathbf{v}\| = 3$, $\|\mathbf{w}\| = 5$, and the angle between $\mathbf{v}$ and $\mathbf{w}$ is $\theta = \frac{\pi}{3}$.
(a) Use the relation $\|\mathbf{v} + \mathbf{w}\|^2 = (\mathbf{v} + \mathbf{w}) \cdot (\mathbf{v} + \mathbf{w})$ to show that $\|\mathbf{v} + \mathbf{w}\|^2 = 3^2 + 5^2 + 2\mathbf{v} \cdot \mathbf{w}$.
(b) Find $\|\mathbf{v} + \mathbf{w}\|$.

SOLUTION For part (a), we use the distributive property to get:

$$\|\mathbf{v} + \mathbf{w}\|^2 = (\mathbf{v} + \mathbf{w}) \cdot (\mathbf{v} + \mathbf{w})$$
$$= \mathbf{v} \cdot \mathbf{v} + \mathbf{v} \cdot \mathbf{w} + \mathbf{w} \cdot \mathbf{v} + \mathbf{w} \cdot \mathbf{w}$$
$$= \|\mathbf{v}\|^2 + 2\mathbf{v} \cdot \mathbf{w} + \|\mathbf{w}\|^2$$
$$= 3^2 + 5^2 + 2\mathbf{v} \cdot \mathbf{w}$$

For part (b), we use the definition of dot product on the previous equation to get:

$$\|\mathbf{v} + \mathbf{w}\|^2 = 3^2 + 5^2 + 2\mathbf{v} \cdot \mathbf{w}$$
$$= 34 + 2 \cdot 3 \cdot 5 \cdot \cos\pi/3$$
$$= 34 + 15 = 49$$

Thus, $\|\mathbf{v} + \mathbf{w}\| = \sqrt{49} = 7$.

49. Show that if $\mathbf{e}$ and $\mathbf{f}$ are unit vectors such that $\|\mathbf{e} + \mathbf{f}\| = \frac{3}{2}$, then $\|\mathbf{e} - \mathbf{f}\| = \frac{\sqrt{7}}{2}$. *Hint:* Show that $\mathbf{e} \cdot \mathbf{f} = \frac{1}{8}$.

SOLUTION We use the relation of the dot product with length and properties of the dot product to write

$$9/4 = \|\mathbf{e} + \mathbf{f}\|^2 = (\mathbf{e} + \mathbf{f}) \cdot (\mathbf{e} + \mathbf{f}) = \mathbf{e} \cdot \mathbf{e} + \mathbf{e} \cdot \mathbf{f} + \mathbf{f} \cdot \mathbf{e} + \mathbf{f} \cdot \mathbf{f}$$
$$= \|\mathbf{e}\|^2 + 2\mathbf{e} \cdot \mathbf{f} + \|\mathbf{f}\|^2 = 1^2 + 2\mathbf{e} \cdot \mathbf{f} + 1^2 = 2 + 2\mathbf{e} \cdot \mathbf{f}$$

We now find $\mathbf{e} \cdot \mathbf{f}$:

$$9/4 = 2 + 2\mathbf{e} \cdot \mathbf{f} \quad \Rightarrow \quad \mathbf{e} \cdot \mathbf{f} = 1/8$$

Hence, using the same method as above, we have:

$$\|\mathbf{e} - \mathbf{f}\|^2 = (\mathbf{e} - \mathbf{f}) \cdot (\mathbf{e} - \mathbf{f}) = \mathbf{e} \cdot \mathbf{e} - \mathbf{e} \cdot \mathbf{f} - \mathbf{f} \cdot \mathbf{e} + \mathbf{f} \cdot \mathbf{f}$$
$$= \|\mathbf{e}\|^2 - 2\mathbf{e} \cdot \mathbf{f} + \|\mathbf{f}\|^2 = 1^2 - 2\mathbf{e} \cdot \mathbf{f} + 1^2 = 2 - 2\mathbf{e} \cdot \mathbf{f} = 2 - 2/8 = 7/4$$

Taking square roots, we get:

$$\|\mathbf{e} - \mathbf{f}\| = \frac{\sqrt{7}}{2}$$

51. Find the angle θ in the triangle in Figure 17.

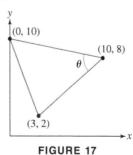

FIGURE 17

SOLUTION We denote by **u** and **v** the vectors in the figure.

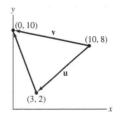

Hence,

$$\cos\theta = \frac{\mathbf{v}\cdot\mathbf{u}}{\|\mathbf{v}\|\|\mathbf{u}\|} \tag{1}$$

We find the vectors **v** and **u**, and then compute their length and the dot product $\mathbf{v}\cdot\mathbf{u}$. This gives

$$\mathbf{v} = \langle 0-10, 10-8\rangle = \langle -10, 2\rangle$$
$$\mathbf{u} = \langle 3-10, 2-8\rangle = \langle -7, -6\rangle$$
$$\|\mathbf{v}\| = \sqrt{(-10)^2+2^2} = \sqrt{104}$$
$$\|\mathbf{u}\| = \sqrt{(-7)^2+(-6)^2} = \sqrt{85}$$
$$\mathbf{v}\cdot\mathbf{u} = \langle -10, 2\rangle\cdot\langle -7, -6\rangle = (-10)\cdot(-7) + 2\cdot(-6) = 58$$

Substituting these values in (1) yields

$$\cos\theta = \frac{58}{\sqrt{104}\sqrt{85}} \approx 0.617$$

Hence the angle of the triangle is 0.906 radians (51.91°).

53. (a) Draw $\mathbf{u}_{\|\mathbf{v}}$ and $\mathbf{v}_{\|\mathbf{u}}$ for the vectors appearing as in Figure 19.
(b) Which of $\mathbf{u}_{\|\mathbf{v}}$ and $\mathbf{v}_{\|\mathbf{u}}$ has the greater magnitude?

FIGURE 19

SOLUTION
(a)

(b) The component of **u** parallel to **v**, $\mathbf{u}_{\|\mathbf{v}}$, has the greater magnitude.

In Exercises 55–62, find the projection of **u** *along* **v**.

55. $\mathbf{u} = \langle 2, 5\rangle$, $\mathbf{v} = \langle 1, 1\rangle$

SOLUTION We first compute the following dot products:

$$\mathbf{u}\cdot\mathbf{v} = \langle 2, 5\rangle\cdot\langle 1, 1\rangle = 7$$
$$\mathbf{v}\cdot\mathbf{v} = \|\mathbf{v}\|^2 = 1^2 + 1^2 = 2$$

The projection of **u** along **v** is the following vector:

$$\mathbf{u}_{\|\mathbf{v}} = \left(\frac{\mathbf{u}\cdot\mathbf{v}}{\mathbf{v}\cdot\mathbf{v}}\right)\mathbf{v} = \frac{7}{2}\mathbf{v} = \left\langle\frac{7}{2}, \frac{7}{2}\right\rangle$$

57. $\mathbf{u} = \langle -1, 2, 0 \rangle$, $\mathbf{v} = \langle 2, 0, 1 \rangle$

SOLUTION The projection of $\mathbf{u}$ along $\mathbf{v}$ is the following vector:

$$\mathbf{u}_{\|\mathbf{v}} = \left(\frac{\mathbf{u} \cdot \mathbf{v}}{\mathbf{v} \cdot \mathbf{v}}\right)\mathbf{v}$$

We compute the values in this expression:

$$\mathbf{u} \cdot \mathbf{v} = \langle -1, 2, 0 \rangle \cdot \langle 2, 0, 1 \rangle = -1 \cdot 2 + 2 \cdot 0 + 0 \cdot 1 = -2$$

$$\mathbf{v} \cdot \mathbf{v} = \|\mathbf{v}\|^2 = 2^2 + 0^2 + 1^2 = 5$$

Hence,

$$\mathbf{u}_{\|\mathbf{v}} = -\frac{2}{5}\langle 2, 0, 1 \rangle = \left\langle -\frac{4}{5}, 0, -\frac{2}{5} \right\rangle$$

59. $\mathbf{u} = 5\mathbf{i} + 7\mathbf{j} - 4\mathbf{k}$, $\mathbf{v} = \mathbf{k}$

SOLUTION The projection of $\mathbf{u}$ along $\mathbf{v}$ is the following vector:

$$\mathbf{u}_{\|\mathbf{v}} = \left(\frac{\mathbf{u} \cdot \mathbf{v}}{\mathbf{v} \cdot \mathbf{v}}\right)\mathbf{v}$$

We compute the dot products:

$$\mathbf{u} \cdot \mathbf{v} = (5\mathbf{i} + 7\mathbf{j} - 4\mathbf{k}) \cdot \mathbf{k} = -4\mathbf{k} \cdot \mathbf{k} = -4$$

$$\mathbf{v} \cdot \mathbf{v} = \|\mathbf{v}\|^2 = \|\mathbf{k}\|^2 = 1$$

Hence,

$$\mathbf{u}_{\|\mathbf{v}} = \frac{-4}{1}\mathbf{k} = -4\mathbf{k}$$

61. $\mathbf{u} = \langle a, b, c \rangle$, $\mathbf{v} = \mathbf{i}$

SOLUTION The component of $\mathbf{u}$ along $\mathbf{v}$ is a, since

$$\mathbf{u} \cdot \mathbf{e}_\mathbf{v} = (a\mathbf{i} + b\mathbf{j} + c\mathbf{k}) \cdot \mathbf{i} = a$$

Therefore, the projection of $\mathbf{u}$ along $\mathbf{v}$ is the vector

$$\mathbf{u}_{\|\mathbf{v}} = (\mathbf{u} \cdot \mathbf{e}_\mathbf{v})\mathbf{e}_\mathbf{v} = a\mathbf{i}$$

In Exercises 63 and 64, compute the component of $\mathbf{u}$ along $\mathbf{v}$.

63. $\mathbf{u} = \langle 3, 2, 1 \rangle$, $\mathbf{v} = \langle 1, 0, 1 \rangle$

SOLUTION We first compute the following dot products:

$$\mathbf{u} \cdot \mathbf{v} = \langle 3, 2, 1 \rangle \cdot \langle 1, 0, 1 \rangle = 4$$

$$\mathbf{v} \cdot \mathbf{v} = \|\mathbf{v}\|^2 = 1^2 + 1^2 = 2$$

The component of $\mathbf{u}$ along $\mathbf{v}$ is the length of the projection of $\mathbf{u}$ along $\mathbf{v}$

$$\left\|\left(\frac{\mathbf{u} \cdot \mathbf{v}}{\mathbf{v} \cdot \mathbf{v}}\right)\mathbf{v}\right\| = \frac{4}{2}\|\mathbf{v}\| = 2\|\mathbf{v}\| = 2\sqrt{2}$$

65. Find the length of $\overline{OP}$ in Figure 20.

SOLUTION This is just the component of $\mathbf{u} = \langle 3, 5 \rangle$ along $\mathbf{v} = \langle 8, 2 \rangle$. We first compute the following dot products:

$$\mathbf{u} \cdot \mathbf{v} = \langle 3, 5 \rangle \cdot \langle 8, 2 \rangle = 34$$

$$\mathbf{v} \cdot \mathbf{v} = \|\mathbf{v}\|^2 = 8^2 + 2^2 = 68$$

The component of $\mathbf{u}$ along $\mathbf{v}$ is the length of the projection of $\mathbf{u}$ along $\mathbf{v}$

$$\left\|\left(\frac{\mathbf{u} \cdot \mathbf{v}}{\mathbf{v} \cdot \mathbf{v}}\right)\mathbf{v}\right\| = \frac{34}{68}\|\mathbf{v}\| = \frac{34}{68}\sqrt{68}$$

SECTION 12.3 | Dot Product and the Angle Between Two Vectors (LT SECTION 13.3) 923

In Exercises 67–72, find the decomposition $\mathbf{a} = \mathbf{a}_{\|\mathbf{b}} + \mathbf{a}_{\perp\mathbf{b}}$ *with respect to* $\mathbf{b}$.

67. $\mathbf{a} = \langle 1, 0 \rangle$, $\mathbf{b} = \langle 1, 1 \rangle$

SOLUTION

Step 1. We compute $\mathbf{a} \cdot \mathbf{b}$ and $\mathbf{b} \cdot \mathbf{b}$

$$\mathbf{a} \cdot \mathbf{b} = \langle 1, 0 \rangle \cdot \langle 1, 1 \rangle = 1 \cdot 1 + 0 \cdot 1 = 1$$

$$\mathbf{b} \cdot \mathbf{b} = \|\mathbf{b}\|^2 = 1^2 + 1^2 = 2$$

Step 2. We find the projection of $\mathbf{a}$ along $\mathbf{b}$:

$$\mathbf{a}_{\|\mathbf{b}} = \left(\frac{\mathbf{a} \cdot \mathbf{b}}{\mathbf{b} \cdot \mathbf{b}} \right) \mathbf{b} = \frac{1}{2} \langle 1, 1 \rangle = \left\langle \frac{1}{2}, \frac{1}{2} \right\rangle$$

Step 3. We find the orthogonal part as the difference:

$$\mathbf{a}_{\perp\mathbf{b}} = \mathbf{a} - \mathbf{a}_{\|\mathbf{b}} = \langle 1, 0 \rangle - \left\langle \frac{1}{2}, \frac{1}{2} \right\rangle = \left\langle \frac{1}{2}, -\frac{1}{2} \right\rangle$$

Hence,

$$\mathbf{a} = \mathbf{a}_{\|\mathbf{b}} + \mathbf{a}_{\perp\mathbf{b}} = \left\langle \frac{1}{2}, \frac{1}{2} \right\rangle + \left\langle \frac{1}{2}, -\frac{1}{2} \right\rangle$$

69. $\mathbf{a} = \langle 4, -1, 0 \rangle$, $\mathbf{b} = \langle 0, 1, 1 \rangle$

SOLUTION We first compute $\mathbf{a} \cdot \mathbf{b}$ and $\mathbf{b} \cdot \mathbf{b}$ to find the projection of $\mathbf{a}$ along $\mathbf{b}$:

$$\mathbf{a} \cdot \mathbf{b} = \langle 4, -1, 0 \rangle \cdot \langle 0, 1, 1 \rangle = 4 \cdot 0 + (-1) \cdot 1 + 0 \cdot 1 = -1$$

$$\mathbf{b} \cdot \mathbf{b} = \|\mathbf{b}\|^2 = 0^2 + 1^2 + 1^2 = 2$$

Hence,

$$\mathbf{a}_{\|\mathbf{b}} = \left(\frac{\mathbf{a} \cdot \mathbf{b}}{\mathbf{b} \cdot \mathbf{b}} \right) \mathbf{b} = \frac{-1}{2} \langle 0, 1, 1 \rangle = \left\langle 0, -\frac{1}{2}, -\frac{1}{2} \right\rangle$$

We now find the vector $\mathbf{a}_{\perp\mathbf{b}}$ orthogonal to $\mathbf{b}$ by computing the difference:

$$\mathbf{a} - \mathbf{a}_{\|\mathbf{b}} = \langle 4, -1, 0 \rangle - \left\langle 0, -\frac{1}{2}, -\frac{1}{2} \right\rangle = \left\langle 4, -\frac{1}{2}, \frac{1}{2} \right\rangle$$

Thus, we have

$$\mathbf{a} = \mathbf{a}_{\|\mathbf{b}} + \mathbf{a}_{\perp\mathbf{b}} = \left\langle 0, -\frac{1}{2}, -\frac{1}{2} \right\rangle + \left\langle 4, -\frac{1}{2}, \frac{1}{2} \right\rangle$$

71. $\mathbf{a} = \langle x, y \rangle$, $\mathbf{b} = \langle 1, -1 \rangle$

SOLUTION We first compute $\mathbf{a} \cdot \mathbf{b}$ and $\mathbf{b} \cdot \mathbf{b}$ to find the projection of $\mathbf{a}$ along $\mathbf{b}$:

$$\mathbf{a} \cdot \mathbf{b} = \langle x, y \rangle \cdot \langle 1, -1 \rangle = x - y$$

$$\mathbf{b} \cdot \mathbf{b} = \|\mathbf{b}\|^2 = 1^2 + (-1)^2 = 2$$

Hence,

$$\mathbf{a}_{\|} = \left(\frac{\mathbf{a} \cdot \mathbf{b}}{\mathbf{b} \cdot \mathbf{b}} \right) \mathbf{b} = \frac{x-y}{2} \langle 1, -1 \rangle = \left\langle \frac{x-y}{2}, \frac{y-x}{2} \right\rangle$$

We now find the vector $\mathbf{a}_{\perp}$ orthogonal to $\mathbf{b}$ by computing the difference:

$$\mathbf{a} - \mathbf{a}_{\|} = \langle x, y \rangle - \left\langle \frac{x-y}{2}, \frac{y-x}{2} \right\rangle = \left\langle \frac{x+y}{2}, \frac{x+y}{2} \right\rangle$$

Thus, we have

$$\mathbf{a} = \mathbf{a}_{\|} + \mathbf{a}_{\perp} = \left\langle \frac{x-y}{2}, \frac{y-x}{2} \right\rangle + \left\langle \frac{x+y}{2}, \frac{x+y}{2} \right\rangle$$

73. Let $\mathbf{e}_\theta = \langle \cos\theta, \sin\theta \rangle$. Show that $\mathbf{e}_\theta \cdot \mathbf{e}_\psi = \cos(\theta - \psi)$ for any two angles θ and ψ.

SOLUTION First, $\mathbf{e}_\theta$ is a unit vector since by a trigonometric identity we have

$$\|\mathbf{e}_\theta\| = \sqrt{\cos^2\theta + \sin^2\theta} = \sqrt{1} = 1$$

The cosine of the angle α between $\mathbf{e}_\theta$ and the vector $\mathbf{i}$ in the direction of the positive x-axis is

$$\cos\alpha = \frac{\mathbf{e}_\theta \cdot \mathbf{i}}{\|\mathbf{e}_\theta\| \cdot \|\mathbf{i}\|} = \mathbf{e}_\theta \cdot \mathbf{i} = ((\cos\theta)\mathbf{i} + (\sin\theta)\mathbf{j}) \cdot \mathbf{i} = \cos\theta$$

The solution of $\cos\alpha = \cos\theta$ for angles between 0 and π is $\alpha = \theta$. That is, the vector $\mathbf{e}_\theta$ makes an angle θ with the x-axis. We now use the trigonometric identity

$$\cos\theta\cos\psi + \sin\theta\sin\psi = \cos(\theta - \psi)$$

to obtain the following equality:

$$\mathbf{e}_\theta \cdot \mathbf{e}_\psi = \langle \cos\theta, \sin\theta \rangle \cdot \langle \cos\psi, \sin\psi \rangle = \cos\theta\cos\psi + \sin\theta\sin\psi = \cos(\theta - \psi)$$

In Exercises 75–78, refer to Figure 21.

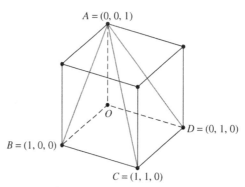

FIGURE 21 Unit cube in $\mathbf{R}^3$.

75. Find the angle between $\overrightarrow{AB}$ and $\overrightarrow{AC}$.

SOLUTION The cosine of the angle α between the vectors $\overrightarrow{AB}$ and $\overrightarrow{AC}$ is

$$\cos\alpha = \frac{\overrightarrow{AB} \cdot \overrightarrow{AC}}{\|\overrightarrow{AB}\|\|\overrightarrow{AC}\|} \tag{1}$$

We compute the vectors $\overrightarrow{AB}$ and $\overrightarrow{AC}$ and then calculate their dot product and lengths. We get

$$\overrightarrow{AB} = \langle 1-0, 0-0, 0-1 \rangle = \langle 1, 0, -1 \rangle$$

$$\overrightarrow{AC} = \langle 1-0, 1-0, 0-1 \rangle = \langle 1, 1, -1 \rangle$$

$$\overrightarrow{AB} \cdot \overrightarrow{AC} = \langle 1, 0, -1 \rangle \cdot \langle 1, 1, -1 \rangle = 1 \cdot 1 + 0 \cdot 1 + (-1) \cdot (-1) = 2$$

$$\|\overrightarrow{AB}\| = \sqrt{1^2 + 0^2 + (-1)^2} = \sqrt{2}$$

$$\|\overrightarrow{AC}\| = \sqrt{1^2 + 1^2 + (-1)^2} = \sqrt{3}$$

Substituting in (1) and solving for $0 \leq \alpha \leq 90°$ gives

$$\cos\alpha = \frac{2}{\sqrt{2} \cdot \sqrt{3}} \approx 0.816 \quad \Rightarrow \quad \alpha \approx 35.31°$$

77. Calculate the projection of $\overrightarrow{AC}$ along $\overrightarrow{AD}$.

SOLUTION $\overrightarrow{DC}$ is perpendicular to the face OAD of the cube. Hence, it is orthogonal to the segment $\overrightarrow{AD}$ on this face. Therefore, the projection of the vector $\overrightarrow{AC}$ along $\overrightarrow{AD}$ is the vector $\overrightarrow{AD}$ itself.

In Exercises 79–80, as in Example 9, assume that the carpenter's diagonal measurements are x and y, and compute the diagonal length that produces a rectangular frame. Compare the result with the corresponding split-difference approximation.

79. $x = 234\frac{1}{2}$ inches and $y = 223$ in.

SOLUTION The diagonal length that produces a rectangular frame is

$$\sqrt{\frac{x^2 + y^2}{2}} = \sqrt{\frac{234.5^2 + 223^2}{2}} \approx 228.82 \text{ in.}$$

Using the split-difference approximation, the diagonal length is computed as

$$\frac{x + y}{2} = \frac{234.5 + 223}{2} = 228.75 \text{ in.}$$

81. The methane molecule CH_4 consists of a carbon molecule bonded to four hydrogen molecules that are spaced as far apart from each other as possible. The hydrogen atoms then sit at the vertices of a tetrahedron, with the carbon atom at its center, as in Figure 22. We can model this with the carbon atom at the point $(\frac{1}{2}, \frac{1}{2}, \frac{1}{2})$ and the hydrogen atoms at $(0, 0, 0), (1, 1, 0), (1, 0, 1)$, and $(0, 1, 1)$. Use the dot product to find the bond angle α formed between any two of the line segments from the carbon atom to the hydrogen atoms.

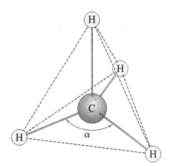

FIGURE 22 A methane molecule.

SOLUTION Use the atoms at $(0, 0, 0)$ and $(1, 1, 0)$. The vectors from C to these atoms are

$$\mathbf{u} = \langle 0, 0, 0 \rangle - \left\langle \frac{1}{2}, \frac{1}{2}, \frac{1}{2} \right\rangle = \left\langle -\frac{1}{2}, -\frac{1}{2}, -\frac{1}{2} \right\rangle$$

$$\mathbf{v} = \langle 1, 1, 0 \rangle - \left\langle \frac{1}{2}, \frac{1}{2}, \frac{1}{2} \right\rangle = \left\langle \frac{1}{2}, \frac{1}{2}, -\frac{1}{2} \right\rangle$$

The angle α between these two vectors is given by

$$\cos \alpha = \frac{\mathbf{u} \cdot \mathbf{v}}{\|\mathbf{u}\|\|\mathbf{v}\|} = \frac{-\frac{1}{2} \cdot \frac{1}{2} - \frac{1}{2} \cdot \frac{1}{2} - \frac{1}{2}\left(-\frac{1}{2}\right)}{\sqrt{\left(-\frac{1}{2}\right)^2 + \left(-\frac{1}{2}\right)^2 + \left(-\frac{1}{2}\right)^2} \sqrt{\left(\frac{1}{2}\right)^2 + \left(\frac{1}{2}\right)^2 + \left(-\frac{1}{2}\right)^2}} = \frac{-1/4}{3/4} = -\frac{1}{3}$$

It follows that

$$\alpha = \cos^{-1}\left(-\frac{1}{3}\right) \approx 1.9106 \approx 109.471°$$

83. Let $\mathbf{v}$ and $\mathbf{w}$ be nonzero vectors and set $\mathbf{u} = \mathbf{e}_\mathbf{v} + \mathbf{e}_\mathbf{w}$. Use the dot product to show that the angle between $\mathbf{u}$ and $\mathbf{v}$ is equal to the angle between $\mathbf{u}$ and $\mathbf{w}$. Explain this result geometrically with a diagram.

SOLUTION We denote by α the angle between $\mathbf{u}$ and $\mathbf{v}$ and by β the angle between $\mathbf{u}$ and $\mathbf{w}$. Since $\mathbf{e}_\mathbf{v}$ and $\mathbf{e}_\mathbf{w}$ are vectors in the directions of $\mathbf{v}$ and $\mathbf{w}$, respectively, α is the angle between $\mathbf{u}$ and $\mathbf{e}_\mathbf{v}$ and β is the angle between $\mathbf{u}$ and $\mathbf{e}_\mathbf{w}$. The cosines of these angles are thus

$$\cos \alpha = \frac{\mathbf{u} \cdot \mathbf{e}_\mathbf{v}}{\|\mathbf{u}\|\|\mathbf{e}_\mathbf{v}\|} = \frac{\mathbf{u} \cdot \mathbf{e}_\mathbf{v}}{\|\mathbf{u}\|}; \quad \cos \beta = \frac{\mathbf{u} \cdot \mathbf{e}_\mathbf{w}}{\|\mathbf{u}\|\|\mathbf{e}_\mathbf{w}\|} = \frac{\mathbf{u} \cdot \mathbf{e}_\mathbf{w}}{\|\mathbf{u}\|}$$

To show that $\cos \alpha = \cos \beta$ (which implies that $\alpha = \beta$) we must show that

$$\mathbf{u} \cdot \mathbf{e}_\mathbf{v} = \mathbf{u} \cdot \mathbf{e}_\mathbf{w}$$

We compute the two dot products:

$$\mathbf{u} \cdot \mathbf{e}_v = (\mathbf{e}_v + \mathbf{e}_w) \cdot \mathbf{e}_v = \mathbf{e}_v \cdot \mathbf{e}_v + \mathbf{e}_w \cdot \mathbf{e}_v = 1 + \mathbf{e}_w \cdot \mathbf{e}_v$$

$$\mathbf{u} \cdot \mathbf{e}_w = (\mathbf{e}_v + \mathbf{e}_w) \cdot \mathbf{e}_w = \mathbf{e}_v \cdot \mathbf{e}_w + \mathbf{e}_w \cdot \mathbf{e}_w = \mathbf{e}_v \cdot \mathbf{e}_w + 1$$

We see that $\mathbf{u} \cdot \mathbf{e}_v = \mathbf{u} \cdot \mathbf{e}_w$. We conclude that $\cos \alpha = \cos \beta$, hence $\alpha = \beta$. Geometrically, $\mathbf{u}$ is a diagonal in the rhombus $OABC$ (see figure), hence it bisects the angle $\sphericalangle AOC$ of the rhombus.

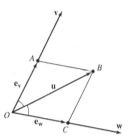

85. In Example 9, assume that the wind is out of the north at 45 km/h. Express the corresponding wind vector as a sum of vectors, one parallel to the bridge and one perpendicular to it. Also, compute the magnitude of the perpendicular term to determine the speed of the part of the wind blowing directly at the bridge.

SOLUTION As in Example 9, $\mathbf{u} = \langle \cos 58°, \sin 58° \rangle$ is a unit vector parallel to the bridge. The wind vector is $\mathbf{w} = \langle 0, -45 \rangle$, and we want to decompose $\mathbf{w}$ as the sum of $\mathbf{w}_{\|\mathbf{u}}$ and $\mathbf{w}_{\perp \mathbf{u}}$.

For $\mathbf{w}_{\|\mathbf{u}} = \left(\frac{\mathbf{w} \cdot \mathbf{u}}{\mathbf{u} \cdot \mathbf{u}} \right) \mathbf{u}$, $\mathbf{u} \cdot \mathbf{u} = 1$ since $\mathbf{u}$ is a unit vector. Thus

$$\mathbf{w}_{\|\mathbf{u}} = \left(\frac{\mathbf{w} \cdot \mathbf{u}}{\mathbf{u} \cdot \mathbf{u}} \right) \mathbf{u} = (\langle 0, -45 \rangle \cdot \langle \cos 58°, \sin 58° \rangle) \langle \cos 58°, \sin 58° \rangle$$

$$= -45 \cos 58° \langle \cos 58°, \sin 58° \rangle$$

$$\approx \langle -12.64, -20.22 \rangle$$

Then

$$\mathbf{w}_{\perp \mathbf{u}} = \mathbf{w} - \mathbf{w}_{\|\mathbf{u}} \approx \langle 12.64, -24.78 \rangle$$

The wind blowing at the bridge is the perpendicular part, $\mathbf{w}_{\perp \mathbf{u}}$. The magnitude (speed) of that wind is approximately

$$\|\mathbf{w}_{\perp \mathbf{u}}\| = \sqrt{12.64^2 + (-24.78)^2} \approx 27.81 \text{ km/h}$$

87. Calculate the force (in newtons) required to push a 40-kg wagon up a 10° incline (Figure 24).

FIGURE 24

SOLUTION Gravity exerts a force $\mathbf{F}_g$ of magnitude $40g$ newtons where $g = 9.8$. The magnitude of the force required to push the wagon equals the component of the force $\mathbf{F}_g$ along the ramp. Resolving $\mathbf{F}_g$ into a sum $\mathbf{F}_g = \mathbf{F}_\| + \mathbf{F}_\perp$, where $\mathbf{F}_\|$ is the force along the ramp and $\mathbf{F}_\perp$ is the force orthogonal to the ramp, we need to find the magnitude of $\mathbf{F}_\|$. The angle between $\mathbf{F}_g$ and the ramp is $90° - 10° = 80°$. Hence,

$$\mathbf{F}_\| = \|\mathbf{F}_g\| \cos 80° = 40 \cdot 9.8 \cdot \cos 80° \approx 68.07 N$$

Therefore the minimum force required to push the wagon is 68.07 N. (Actually, this is the force required to keep the wagon from sliding down the hill; any slight amount greater than this force will serve to push it up the hill.)

89. A light beam travels along the ray determined by a unit vector **L**, strikes a flat surface at point P, and is reflected along the ray determined by a unit vector **R**, where $\theta_1 = \theta_2$ (Figure 26). Show that if **N** is the unit vector orthogonal to the surface, then

$$\mathbf{R} = 2(\mathbf{L} \cdot \mathbf{N})\mathbf{N} - \mathbf{L}$$

FIGURE 26

SOLUTION We denote by **W** a unit vector orthogonal to **N** in the direction shown in the figure, and let $\theta_1 = \theta_2 = \theta$.

We resolve the unit vectors **R** and **L** into a sum of forces along **N** and **W**. This gives

$$\mathbf{R} = \cos(90 - \theta)\mathbf{W} + \cos\theta\mathbf{N} = \sin\theta\mathbf{W} + \cos\theta\mathbf{N}$$
$$\mathbf{L} = -\cos(90 - \theta)\mathbf{W} + \cos\theta\mathbf{N} = -\sin\theta\mathbf{W} + \cos\theta\mathbf{N} \quad (1)$$

Now, since

$$\mathbf{L} \cdot \mathbf{N} = \|\mathbf{L}\|\|\mathbf{N}\|\cos\theta = 1 \cdot 1 \cos\theta = \cos\theta$$

we have by (1):

$$2(\mathbf{L} \cdot \mathbf{N})\mathbf{N} - \mathbf{L} = (2\cos\theta)\mathbf{N} - \mathbf{L} = (2\cos\theta)\mathbf{N} - ((-\sin\theta)\mathbf{W} + (\cos\theta)\mathbf{N})$$
$$= (2\cos\theta)\mathbf{N} + (\sin\theta)\mathbf{W} - (\cos\theta)\mathbf{N} = (\sin\theta)\mathbf{W} + (\cos\theta)\mathbf{N} = \mathbf{R}$$

91. Prove that $\|\mathbf{v}+\mathbf{w}\|^2 - \|\mathbf{v}-\mathbf{w}\|^2 = 4\mathbf{v}\cdot\mathbf{w}$.

SOLUTION We compute the following values:

$$\|\mathbf{v}+\mathbf{w}\|^2 = (\mathbf{v}+\mathbf{w})\cdot(\mathbf{v}+\mathbf{w}) = \mathbf{v}\cdot\mathbf{v} + \mathbf{v}\cdot\mathbf{w} + \mathbf{w}\cdot\mathbf{v} + \mathbf{w}\cdot\mathbf{w} = \|\mathbf{v}\|^2 + 2\mathbf{v}\cdot\mathbf{w} + \|\mathbf{w}\|^2$$

$$\|\mathbf{v}-\mathbf{w}\|^2 = (\mathbf{v}-\mathbf{w})\cdot(\mathbf{v}-\mathbf{w}) = \mathbf{v}\cdot\mathbf{v} - \mathbf{v}\cdot\mathbf{w} - \mathbf{w}\cdot\mathbf{v} - \mathbf{w}\cdot\mathbf{w} = \|\mathbf{v}\|^2 - 2\mathbf{v}\cdot\mathbf{w} + \|\mathbf{w}\|^2$$

Hence,

$$\|\mathbf{v}+\mathbf{w}\|^2 - \|\mathbf{v}-\mathbf{w}\|^2 = (\|\mathbf{v}\|^2 + 2\mathbf{v}\cdot\mathbf{w} + \|\mathbf{w}\|^2) - (\|\mathbf{v}\|^2 - 2\mathbf{v}\cdot\mathbf{w} + \|\mathbf{w}\|^2) = 4\mathbf{v}\cdot\mathbf{w}$$

93. A rhombus is a parallelogram in which all four sides have equal length. Show that the diagonals of a parallelogram are perpendicular if and only if the parallelogram is a rhombus. *Hint:* Take an approach similar to the solution in Example 5, and consider $\mathbf{x}\cdot\mathbf{y}$.

SOLUTION Using the same notation as in Example 5, we have

$$\mathbf{x} = \mathbf{a} + \mathbf{b}, \quad \mathbf{y} = \mathbf{a} - \mathbf{b}$$

so that

$$\mathbf{x}\cdot\mathbf{y} = (\mathbf{a}+\mathbf{b})\cdot(\mathbf{a}-\mathbf{b}) = \mathbf{a}\cdot\mathbf{a} - \mathbf{a}\cdot\mathbf{b} + \mathbf{b}\cdot\mathbf{a} - \mathbf{b}\cdot\mathbf{b}$$

The two middle terms cancel, leaving

$$\mathbf{x}\cdot\mathbf{y} = \mathbf{a}\cdot\mathbf{a} - \mathbf{b}\cdot\mathbf{b} = \|\mathbf{a}\|^2 - \|\mathbf{b}\|^2$$

Note that the diagonals being perpendicular means that this dot product is zero. It follows that the diagonals are perpendicular if and only if the magnitudes of the sides are equal; that is, if and only if the parallelogram is a rhombus.

95. Verify that $(\lambda\mathbf{v})\cdot\mathbf{w} = \lambda(\mathbf{v}\cdot\mathbf{w})$ for any scalar λ.

SOLUTION We denote the components of the vectors $\mathbf{v}$ and $\mathbf{w}$ by

$$\mathbf{v} = \langle a_1, a_2, a_3\rangle \quad \mathbf{w} = \langle b_1, b_2, b_3\rangle$$

Thus,

$$(\lambda\mathbf{v})\cdot\mathbf{w} = (\lambda\langle a_1, a_2, a_3\rangle)\cdot\langle b_1, b_2, b_3\rangle = \langle \lambda a_1, \lambda a_2, \lambda a_3\rangle\cdot\langle b_1, b_2, b_3\rangle$$
$$= \lambda a_1 b_1 + \lambda a_2 b_2 + \lambda a_3 b_3$$

Recalling that λ, a_i, and b_i are scalars and using the definitions of scalar multiples of vectors and the dot product, we get

$$(\lambda\mathbf{v})\cdot\mathbf{w} = \lambda(a_1 b_1 + a_2 b_2 + a_3 b_3) = \lambda(\langle a_1, a_2, a_3\rangle\cdot\langle b_1, b_2, b_3\rangle) = \lambda(\mathbf{v}\cdot\mathbf{w})$$

Further Insights and Challenges

97. In this exercise, we prove the Cauchy–Schwarz inequality: If $\mathbf{v}$ and $\mathbf{w}$ are any two vectors, then

$$|\mathbf{v}\cdot\mathbf{w}| \leq \|\mathbf{v}\|\,\|\mathbf{w}\| \qquad \boxed{7}$$

(a) Let $f(x) = \|x\mathbf{v} + \mathbf{w}\|^2$ for x a scalar. Show that $f(x) = ax^2 + bx + c$, where $a = \|\mathbf{v}\|^2$, $b = 2\mathbf{v}\cdot\mathbf{w}$, and $c = \|\mathbf{w}\|^2$.

(b) Conclude that $b^2 - 4ac \leq 0$. *Hint:* Observe that $f(x) \geq 0$ for all x.

SOLUTION

(a) We express the norm as a dot product and compute it:

$$f(x) = \|x\mathbf{v}+\mathbf{w}\|^2 = (x\mathbf{v}+\mathbf{w})\cdot(x\mathbf{v}+\mathbf{w})$$
$$= x^2\mathbf{v}\cdot\mathbf{v} + x\mathbf{v}\cdot\mathbf{w} + x\mathbf{w}\cdot\mathbf{v} + \mathbf{w}\cdot\mathbf{w} = \|\mathbf{v}\|^2 x^2 + 2(\mathbf{v}\cdot\mathbf{w})x + \|\mathbf{w}\|^2$$

Hence, $f(x) = ax^2 + bx + c$, where $a = \|\mathbf{v}\|^2$, $b = 2\mathbf{v}\cdot\mathbf{w}$, and $c = \|\mathbf{w}\|^2$.

(b) If f has distinct real roots x_1 and x_2, then $f(x)$ is negative for x between x_1 and x_2, but this is impossible since f is the square of a length.

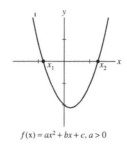

$f(x) = ax^2 + bx + c, \ a > 0$

Using properties of quadratic functions, it follows that f has a nonpositive discriminant. That is, $b^2 - 4ac \leq 0$. Substituting the values for a, b, and c, we get

$$4(\mathbf{v} \cdot \mathbf{w})^2 - 4\|\mathbf{v}\|^2 \|\mathbf{w}\|^2 \leq 0$$

$$(\mathbf{v} \cdot \mathbf{w})^2 \leq \|\mathbf{v}\|^2 \|\mathbf{w}\|^2$$

Taking the square root of both sides we obtain

$$|\mathbf{v} \cdot \mathbf{w}| \leq \|\mathbf{v}\| \|\mathbf{w}\|$$

99. This exercise gives another proof of the relation between the dot product and the angle θ between two vectors $\mathbf{v} = \langle a_1, b_1 \rangle$ and $\mathbf{w} = \langle a_2, b_2 \rangle$ in the plane. Observe that $\mathbf{v} = \|\mathbf{v}\| \langle \cos \theta_1, \sin \theta_1 \rangle$ and $\mathbf{w} = \|\mathbf{w}\| \langle \cos \theta_2, \sin \theta_2 \rangle$, with θ_1 and θ_2 as in Figure 29. Then use the addition formula for the cosine to show that

$$\mathbf{v} \cdot \mathbf{w} = \|\mathbf{v}\| \|\mathbf{w}\| \cos \theta$$

FIGURE 29

SOLUTION Using the trigonometric function for angles in right triangles, we have

$$a_2 = \|\mathbf{v}\| \sin \theta_1, \qquad a_1 = \|\mathbf{v}\| \cos \theta_1$$
$$b_2 = \|\mathbf{w}\| \sin \theta_2, \qquad b_1 = \|\mathbf{w}\| \cos \theta_2$$

Hence, using the given identity we obtain

$$\mathbf{v} \cdot \mathbf{w} = \langle a_1, a_2 \rangle \cdot \langle b_1, b_2 \rangle = a_1 b_1 + a_2 b_2 = \|\mathbf{v}\| \cos \theta_1 \|\mathbf{w}\| \cos \theta_2 + \|\mathbf{v}\| \sin \theta_1 \|\mathbf{w}\| \sin \theta_2$$
$$= \|\mathbf{v}\| \|\mathbf{w}\| (\cos \theta_1 \cos \theta_2 + \sin \theta_1 \sin \theta_2) = \|\mathbf{v}\| \|\mathbf{w}\| \cos(\theta_1 - \theta_2)$$

That is,

$$\mathbf{v} \cdot \mathbf{w} = \|\mathbf{v}\| \|\mathbf{w}\| \cos(\theta)$$

101. Let $\mathbf{v}$ be a nonzero vector. The angles α, β, γ between $\mathbf{v}$ and the unit vectors $\mathbf{i}, \mathbf{j}, \mathbf{k}$ are called the direction angles of $\mathbf{v}$ (Figure 30). The cosines of these angles are called the **direction cosines** of $\mathbf{v}$. Prove that

$$\cos^2 \alpha + \cos^2 \beta + \cos^2 \gamma = 1$$

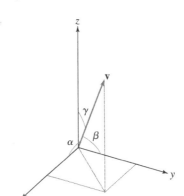

FIGURE 30 Direction angles of **v**.

SOLUTION We use the relation between the dot product and the angle between two vectors to write

$$\cos \alpha = \frac{\mathbf{v} \cdot \mathbf{i}}{\|\mathbf{v}\|\|\mathbf{i}\|} = \frac{\mathbf{v} \cdot \mathbf{i}}{\|\mathbf{v}\|}$$

$$\cos \beta = \frac{\mathbf{v} \cdot \mathbf{j}}{\|\mathbf{v}\|\|\mathbf{j}\|} = \frac{\mathbf{v} \cdot \mathbf{j}}{\|\mathbf{v}\|} \qquad (1)$$

$$\cos \gamma = \frac{\mathbf{v} \cdot \mathbf{k}}{\|\mathbf{v}\|\|\mathbf{k}\|} = \frac{\mathbf{v} \cdot \mathbf{k}}{\|\mathbf{v}\|}$$

We compute the values involved in (1). Letting $\mathbf{v} = \langle v_1, v_2, v_3 \rangle$ we get

$$\mathbf{v} \cdot \mathbf{i} = \langle v_1, v_2, v_3 \rangle \cdot \langle 1, 0, 0 \rangle = v_1$$

$$\mathbf{v} \cdot \mathbf{j} = \langle v_1, v_2, v_3 \rangle \cdot \langle 0, 1, 0 \rangle = v_2$$

$$\mathbf{v} \cdot \mathbf{k} = \langle v_1, v_2, v_3 \rangle \cdot \langle 0, 0, 1 \rangle = v_3$$

$$\|\mathbf{v}\| = \sqrt{v_1^2 + v_2^2 + v_3^2} \qquad (2)$$

We now substitute (2) into (1) to obtain

$$\cos \alpha = \frac{v_1}{\|\mathbf{v}\|}, \quad \cos \beta = \frac{v_2}{\|\mathbf{v}\|}, \quad \cos \gamma = \frac{v_3}{\|\mathbf{v}\|}$$

Finally, we compute the sum of squares of the direction cosines:

$$\cos^2 \alpha + \cos^2 \beta + \cos^2 \gamma = \left(\frac{v_1}{\|\mathbf{v}\|}\right)^2 + \left(\frac{v_2}{\|\mathbf{v}\|}\right)^2 + \left(\frac{v_3}{\|\mathbf{v}\|}\right)^2 = \frac{1}{\|\mathbf{v}\|^2}(v_1^2 + v_2^2 + v_3^2) = \frac{1}{\|\mathbf{v}\|^2} \cdot \|\mathbf{v}\|^2 = 1$$

103. The set of all points $X = (x, y, z)$ equidistant from two points P, Q in $\mathbf{R}^3$ is a plane (Figure 31). Show that X lies on this plane if

$$\overrightarrow{PQ} \cdot \overrightarrow{OX} = \frac{1}{2}\left(\|\overrightarrow{OQ}\|^2 - \|\overrightarrow{OP}\|^2\right) \qquad \boxed{8}$$

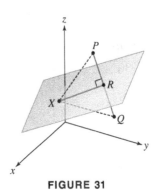

FIGURE 31

Hint: If R is the midpoint of $\overline{PQ}$, then X is equidistant from P and Q if and only if $\overrightarrow{XR}$ is orthogonal to $\overrightarrow{PQ}$.

SOLUTION Let R be the midpoint of the segment $\overline{PQ}$. The points $X = (x, y, z)$ that are equidistant from P and Q are the points for which the vector $\overrightarrow{XR}$ is orthogonal to $\overrightarrow{PQ}$. That is,

$$\overrightarrow{XR} \cdot \overrightarrow{PQ} = 0 \quad (1)$$

Since $\overrightarrow{XR} = \overrightarrow{XO} + \overrightarrow{OR}$ we have by (1):

$$0 = \left(\overrightarrow{XO} + \overrightarrow{OR}\right) \cdot \overrightarrow{PQ} = \overrightarrow{XO} \cdot \overrightarrow{PQ} + \overrightarrow{OR} \cdot \overrightarrow{PQ} = -\overrightarrow{OX} \cdot \overrightarrow{PQ} + \overrightarrow{OR} \cdot \overrightarrow{PQ}$$

Transferring sides we get

$$\overrightarrow{OX} \cdot \overrightarrow{PQ} = \overrightarrow{OR} \cdot \overrightarrow{PQ} \quad (2)$$

We now write $\overrightarrow{PQ} = \overrightarrow{PO} + \overrightarrow{OQ}$ on the right-hand-side of (2), and $\overrightarrow{OR} = \dfrac{\overrightarrow{OP} + \overrightarrow{OQ}}{2}$. We get

$$\overrightarrow{OX} \cdot \overrightarrow{PQ} = \frac{1}{2}\left(\overrightarrow{OP} + \overrightarrow{OQ}\right) \cdot \left(\overrightarrow{PO} + \overrightarrow{OQ}\right) = \frac{1}{2}\left(\overrightarrow{OP} + \overrightarrow{OQ}\right) \cdot \left(\overrightarrow{OQ} - \overrightarrow{OP}\right)$$

$$= \frac{1}{2}\left(\overrightarrow{OP} \cdot \overrightarrow{OQ} - \overrightarrow{OP} \cdot \overrightarrow{OP} + \overrightarrow{OQ} \cdot \overrightarrow{OQ} - \overrightarrow{OQ} \cdot \overrightarrow{OP}\right) = \frac{1}{2}\left(\|\overrightarrow{OQ}\|^2 - \|\overrightarrow{OP}\|^2\right)$$

Thus, we showed that the vector equation of the plane is

$$\overrightarrow{OX} \cdot \overrightarrow{PQ} = \frac{1}{2}\left(\|\overrightarrow{OQ}\|^2 - \|\overrightarrow{OP}\|^2\right)$$

105. Use Eq. (8) to find the equation of the plane consisting of all points $X = (x, y, z)$ equidistant from $P = (2, 1, 1)$ and $Q = (1, 0, 2)$.

SOLUTION Using Eq. (8) with $X = (x, y, z)$, $P = (2, 1, 1)$, and $Q = (1, 0, 2)$ gives

$$\langle x, y, z \rangle \cdot \langle -1, -1, 1 \rangle = \frac{1}{2}\left((\sqrt{5})^2 - (\sqrt{6})^2\right) = -\frac{1}{2}$$

This gives us $-1x - 1y + 1z = -\frac{1}{2}$, which leads to $2x + 2y - 2z = 1$.

12.4 The Cross Product   (LT Section 13.4)

Preliminary Questions

1. What is the $(1, 3)$ minor of the matrix $\begin{vmatrix} 3 & 4 & 2 \\ -5 & -1 & 1 \\ 4 & 0 & 3 \end{vmatrix}$?

SOLUTION The $(1, 3)$ minor is obtained by crossing out the first row and third column of the matrix. That is,

$$\begin{vmatrix} 3 & 4 & 2 \\ -5 & -1 & 1 \\ 4 & 0 & 3 \end{vmatrix} \quad \Rightarrow \quad \begin{vmatrix} -5 & -1 \\ 4 & 0 \end{vmatrix}$$

2. The angle between two unit vectors $\mathbf{e}$ and $\mathbf{f}$ is $\frac{\pi}{6}$. What is the length of $\mathbf{e} \times \mathbf{f}$?

SOLUTION We use the Formula for the Length of the Cross Product:

$$\|\mathbf{e} \times \mathbf{f}\| = \|\mathbf{e}\|\|\mathbf{f}\| \sin \theta$$

Since $\mathbf{e}$ and $\mathbf{f}$ are unit vectors, $\|\mathbf{e}\| = \|\mathbf{f}\| = 1$. Also $\theta = \frac{\pi}{6}$, therefore,

$$\|\mathbf{e} \times \mathbf{f}\| = 1 \cdot 1 \cdot \sin \frac{\pi}{6} = \frac{1}{2}$$

The length of $\mathbf{e} \times \mathbf{f}$ is $\frac{1}{2}$.

3. What is $\mathbf{u} \times \mathbf{w}$, assuming that $\mathbf{w} \times \mathbf{u} = \langle 2, 2, 1 \rangle$?

SOLUTION By anti-commutativity of the cross product, we have

$$\mathbf{u} \times \mathbf{w} = -\mathbf{w} \times \mathbf{u} = -\langle 2, 2, 1 \rangle = \langle -2, -2, -1 \rangle$$

4. Find the cross product without using the formula:

(a) $\langle 4, 8, 2 \rangle \times \langle 4, 8, 2 \rangle$ (b) $\langle 4, 8, 2 \rangle \times \langle 2, 4, 1 \rangle$

SOLUTION By properties of the cross product, the cross product of parallel vectors is the zero vector. In particular, the cross product of a vector with itself is the zero vector. Since $\langle 4, 8, 2 \rangle = 2 \langle 2, 4, 1 \rangle$, the vectors $\langle 4, 8, 2 \rangle$ and $\langle 2, 4, 1 \rangle$ are parallel. We conclude that

$$\langle 4, 8, 2 \rangle \times \langle 4, 8, 2 \rangle = \mathbf{0} \quad \text{and} \quad \langle 4, 8, 2 \rangle \times \langle 2, 4, 1 \rangle = \mathbf{0}$$

5. What are $\mathbf{i} \times \mathbf{j}$ and $\mathbf{i} \times \mathbf{k}$?

SOLUTION The cross product $\mathbf{i} \times \mathbf{j}$ and $\mathbf{i} \times \mathbf{k}$ are determined by the right-hand rule. We can also use the following figure to determine these cross products:

We get

$$\mathbf{i} \times \mathbf{j} = \mathbf{k} \text{ and } \mathbf{i} \times \mathbf{k} = -\mathbf{j}$$

6. When is the cross product $\mathbf{v} \times \mathbf{w}$ equal to zero?

SOLUTION The cross product $\mathbf{v} \times \mathbf{w}$ is equal to zero if one of the vectors $\mathbf{v}$ or $\mathbf{w}$ (or both) is the zero vector, or if $\mathbf{v}$ and $\mathbf{w}$ are parallel vectors.

7. Which of the following are meaningful and which are not? Explain.

(a) $(\mathbf{u} \cdot \mathbf{v}) \times \mathbf{w}$

(b) $(\mathbf{u} \times \mathbf{v}) \cdot \mathbf{w}$

(c) $\|\mathbf{w}\|(\mathbf{u} \cdot \mathbf{v})$

(d) $\|\mathbf{w}\|(\mathbf{u} \times \mathbf{v})$

SOLUTION

(a) Since $\mathbf{u} \cdot \mathbf{v}$ is a scalar, this product does not make sense: taking the cross product of a scalar with a vector is not defined.

(b) $\mathbf{u} \times \mathbf{v}$ is a vector, so the result of this expression is the dot product of two vectors, which is defined and is a scalar.

(c) Both $\|\mathbf{w}\|$ and $\mathbf{u} \cdot \mathbf{v}$ are scalars, so this expression is defined and is just the product of two real numbers.

(d) Since $\|\mathbf{w}\|$ is a scalar and $\mathbf{u} \times \mathbf{v}$ is a vector, this expression is defined. It is a multiple of the vector $\mathbf{u} \times \mathbf{v}$.

8. Which of the following vectors are equal to $\mathbf{j} \times \mathbf{i}$?

(a) $\mathbf{i} \times \mathbf{k}$

(b) $-\mathbf{k}$

(c) $\mathbf{i} \times \mathbf{j}$

SOLUTION Since $\mathbf{i} \times \mathbf{j} = \mathbf{k}$ (see Exercise 5 above), we see that $\mathbf{j} \times \mathbf{i} = -\mathbf{i} \times \mathbf{j} = -\mathbf{k}$. Thus **(b)** is equal to $\mathbf{j} \times \mathbf{i}$, but **(c)** is not. Finally, since $\mathbf{i} \times \mathbf{k} = -\mathbf{j}$, **(a)** is not equal to $\mathbf{j} \times \mathbf{i}$ either.

Exercises

In Exercises 1–4, calculate the 2×2 determinant.

1. $\begin{vmatrix} 1 & 2 \\ 4 & 3 \end{vmatrix}$

SOLUTION Using the definition of 2×2 determinant we get

$$\begin{vmatrix} 1 & 2 \\ 4 & 3 \end{vmatrix} = 1 \cdot 3 - 2 \cdot 4 = -5$$

3. $\begin{vmatrix} -6 & 9 \\ 1 & 1 \end{vmatrix}$

SOLUTION We evaluate the determinant to obtain

$$\begin{vmatrix} -6 & 9 \\ 1 & 1 \end{vmatrix} = -6 \cdot 1 - 9 \cdot 1 = -15$$

In Exercises 5–8, calculate the 3×3 determinant.

5. $\begin{vmatrix} 1 & 2 & 1 \\ 4 & -3 & 0 \\ 1 & 0 & 1 \end{vmatrix}$

SOLUTION Using the definition of 3×3 determinant we obtain

$$\begin{vmatrix} 1 & 2 & 1 \\ 4 & -3 & 0 \\ 1 & 0 & 1 \end{vmatrix} = 1 \begin{vmatrix} -3 & 0 \\ 0 & 1 \end{vmatrix} - 2 \begin{vmatrix} 4 & 0 \\ 1 & 1 \end{vmatrix} + 1 \begin{vmatrix} 4 & -3 \\ 1 & 0 \end{vmatrix}$$
$$= 1 \cdot (-3 \cdot 1 - 0 \cdot 0) - 2 \cdot (4 \cdot 1 - 0 \cdot 1) + 1 \cdot (4 \cdot 0 - (-3) \cdot 1)$$
$$= -3 - 8 + 3 = -8$$

7. $\begin{vmatrix} 1 & 2 & 3 \\ 2 & 4 & 6 \\ -3 & -4 & 2 \end{vmatrix}$

SOLUTION We have

$$\begin{vmatrix} 1 & 2 & 3 \\ 2 & 4 & 6 \\ -3 & -4 & 2 \end{vmatrix} = 1 \begin{vmatrix} 4 & 6 \\ -4 & 2 \end{vmatrix} - 2 \begin{vmatrix} 2 & 6 \\ -3 & 2 \end{vmatrix} + 3 \begin{vmatrix} 2 & 4 \\ -3 & -4 \end{vmatrix}$$
$$= 1(4 \cdot 2 - 6 \cdot (-4)) - 2(2 \cdot 2 - 6 \cdot (-3)) + 3(2 \cdot (-4) - 4 \cdot (-3))$$
$$= 32 - 44 + 12 = 0$$

In Exercises 9–14, calculate $\mathbf{v} \times \mathbf{w}$.

9. $\mathbf{v} = \langle 1, 2, 1 \rangle$, $\mathbf{w} = \langle 3, 1, 1 \rangle$

SOLUTION Using the definition of the cross product we get

$$\mathbf{v} \times \mathbf{w} = \begin{vmatrix} \mathbf{i} & \mathbf{j} & \mathbf{k} \\ 1 & 2 & 1 \\ 3 & 1 & 1 \end{vmatrix} = \begin{vmatrix} 2 & 1 \\ 1 & 1 \end{vmatrix} \mathbf{i} - \begin{vmatrix} 1 & 1 \\ 3 & 1 \end{vmatrix} \mathbf{j} + \begin{vmatrix} 1 & 2 \\ 3 & 1 \end{vmatrix} \mathbf{k}$$
$$= (2 - 1)\mathbf{i} - (1 - 3)\mathbf{j} + (1 - 6)\mathbf{k} = \mathbf{i} + 2\mathbf{j} - 5\mathbf{k}$$

934 CHAPTER 12 | VECTOR GEOMETRY (LT CHAPTER 13)

11. $\mathbf{v} = \langle \frac{2}{3}, 1, \frac{1}{2} \rangle$, $\mathbf{w} = \langle 4, -6, 3 \rangle$

SOLUTION We have

$$\mathbf{v} \times \mathbf{w} = \begin{vmatrix} \mathbf{i} & \mathbf{j} & \mathbf{k} \\ \frac{2}{3} & 1 & \frac{1}{2} \\ 4 & -6 & 3 \end{vmatrix} = \begin{vmatrix} 1 & \frac{1}{2} \\ -6 & 3 \end{vmatrix} \mathbf{i} - \begin{vmatrix} \frac{2}{3} & \frac{1}{2} \\ 4 & 3 \end{vmatrix} \mathbf{j} + \begin{vmatrix} \frac{2}{3} & 1 \\ 4 & -6 \end{vmatrix} \mathbf{k}$$

$$= (3 + 3)\mathbf{i} - (2 - 2)\mathbf{j} + (-4 - 4)\mathbf{k} = 6\mathbf{i} - 8\mathbf{k}$$

13. $\mathbf{v} = \langle 1, 2, 3 \rangle$, $\mathbf{w} = \langle 1, 2, 3.01 \rangle$

SOLUTION The cross product $\mathbf{v} \times \mathbf{w}$ is the following vector:

$$\mathbf{v} \times \mathbf{w} = \begin{vmatrix} \mathbf{i} & \mathbf{j} & \mathbf{k} \\ 1 & 2 & 3 \\ 1 & 2 & 3.01 \end{vmatrix} = \begin{vmatrix} 2 & 3 \\ 2 & 3.01 \end{vmatrix} \mathbf{i} - \begin{vmatrix} 1 & 3 \\ 1 & 3.01 \end{vmatrix} \mathbf{j} + \begin{vmatrix} 1 & 2 \\ 1 & 2 \end{vmatrix} \mathbf{k}$$

$$= (6.02 - 6)\mathbf{i} - (3.01 - 3)\mathbf{j} + (2 - 2)\mathbf{k} = 0.02\mathbf{i} - 0.01\mathbf{j}$$

In Exercises 15–18, use the relations in Eqs. (5) and (6) to calculate the cross product.

15. $(\mathbf{i} + \mathbf{j}) \times \mathbf{k}$

SOLUTION We use basic properties of the cross product to obtain

$$(\mathbf{i} + \mathbf{j}) \times \mathbf{k} = \mathbf{i} \times \mathbf{k} + \mathbf{j} \times \mathbf{k} = -\mathbf{j} + \mathbf{i}$$

$\mathbf{i} \times \mathbf{k} = -\mathbf{j}$

$\mathbf{j} \times \mathbf{k} = \mathbf{i}$

17. $(\mathbf{i} - 3\mathbf{j} + 2\mathbf{k}) \times (\mathbf{j} - \mathbf{k})$

SOLUTION Using the distributive law we obtain

$$(\mathbf{i} - 3\mathbf{j} + 2\mathbf{k}) \times (\mathbf{j} - \mathbf{k}) = (\mathbf{i} - 3\mathbf{j} + 2\mathbf{k}) \times \mathbf{j} - (\mathbf{i} - 3\mathbf{j} + 2\mathbf{k}) \times (\mathbf{k})$$

$$= \mathbf{i} \times \mathbf{j} + 2\mathbf{k} \times \mathbf{j} - \mathbf{i} \times \mathbf{k} - (-3\mathbf{j}) \times \mathbf{k}$$

$$= \mathbf{i} + \mathbf{j} + \mathbf{k}$$

In Exercises 19–24, calculate the cross product assuming that

$$\mathbf{u} \times \mathbf{v} = \langle 1, 1, 0 \rangle, \quad \mathbf{u} \times \mathbf{w} = \langle 0, 3, 1 \rangle, \quad \mathbf{v} \times \mathbf{w} = \langle 2, -1, 1 \rangle$$

19. $\mathbf{v} \times \mathbf{u}$

SOLUTION Using the properties of the cross product we obtain

$$\mathbf{v} \times \mathbf{u} = -\mathbf{u} \times \mathbf{v} = \langle -1, -1, 0 \rangle$$

21. $\mathbf{w} \times (\mathbf{u} + \mathbf{v})$

SOLUTION Using the properties of the cross product we obtain

$$\mathbf{w} \times (\mathbf{u} + \mathbf{v}) = \mathbf{w} \times \mathbf{u} + \mathbf{w} \times \mathbf{v} = -\mathbf{u} \times \mathbf{w} - \mathbf{v} \times \mathbf{w} = \langle -2, -2, -2 \rangle$$

23. $(\mathbf{u} - 2\mathbf{v}) \times (\mathbf{u} + 2\mathbf{v})$

SOLUTION Using the properties of the cross product we obtain

$$(\mathbf{u} - 2\mathbf{v}) \times (\mathbf{u} + 2\mathbf{v}) = (\mathbf{u} - 2\mathbf{v}) \times \mathbf{u} + (\mathbf{u} - 2\mathbf{v}) \times 2\mathbf{v} = \mathbf{u} \times \mathbf{u} - 2\mathbf{v} \times \mathbf{u} + \mathbf{u} \times 2\mathbf{v} - 4\mathbf{v} \times \mathbf{v}$$

$$= 0 + 2\mathbf{u} \times \mathbf{v} + 2\mathbf{u} \times \mathbf{v} - 0 = 0 + 4\mathbf{u} \times \mathbf{v} = \langle 4, 4, 0 \rangle$$

25. Let $\mathbf{v} = \langle a, b, c \rangle$. Calculate $\mathbf{v} \times \mathbf{i}$, $\mathbf{v} \times \mathbf{j}$, and $\mathbf{v} \times \mathbf{k}$.

SOLUTION We write $\mathbf{v} = a\mathbf{i} + b\mathbf{j} + c\mathbf{k}$ and use the distributive law:

$$\mathbf{v} \times \mathbf{i} = (a\mathbf{i} + b\mathbf{j} + c\mathbf{k}) \times \mathbf{i} = a\mathbf{i} \times \mathbf{i} + b\mathbf{j} \times \mathbf{i} + c\mathbf{k} \times \mathbf{i} = a \cdot \mathbf{0} - b\mathbf{k} + c\mathbf{j} = -b\mathbf{k} + c\mathbf{j} = \langle 0, c, -b \rangle$$

$$\mathbf{v} \times \mathbf{j} = (a\mathbf{i} + b\mathbf{j} + c\mathbf{k}) \times \mathbf{j} = a\mathbf{i} \times \mathbf{j} + b\mathbf{j} \times \mathbf{j} + c\mathbf{k} \times \mathbf{j} = a\mathbf{k} + b\mathbf{0} - c\mathbf{i} = a\mathbf{k} - c\mathbf{i} = \langle -c, 0, a \rangle$$

$$\mathbf{v} \times \mathbf{k} = (a\mathbf{i} + b\mathbf{j} + c\mathbf{k}) \times \mathbf{k} = a\mathbf{i} \times \mathbf{k} + b\mathbf{j} \times \mathbf{k} + c\mathbf{k} \times \mathbf{k} = -a\mathbf{j} + b\mathbf{i} + c\mathbf{0} = -a\mathbf{j} + b\mathbf{i} = \langle b, -a, 0 \rangle$$

In Exercises 27 and 28, refer to Figure 17.

FIGURE 17

27. Which of $\mathbf{u}$ and $-\mathbf{u}$ is equal to $\mathbf{v} \times \mathbf{w}$?

SOLUTION The direction of $\mathbf{v} \times \mathbf{w}$ is determined by the Right-Hand Rule, that is, our thumb points in the direction of $\mathbf{v} \times \mathbf{w}$ when the fingers of our right hand curl from $\mathbf{v}$ to $\mathbf{w}$. Therefore $\mathbf{v} \times \mathbf{w}$ equals $-\mathbf{u}$ rather than $\mathbf{u}$.

29. Let $\mathbf{v} = \langle 3, 0, 0 \rangle$ and $\mathbf{w} = \langle 0, 1, -1 \rangle$. Determine $\mathbf{u} = \mathbf{v} \times \mathbf{w}$ using the geometric properties of the cross product rather than the formula.

SOLUTION The cross product $\mathbf{u} = \mathbf{v} \times \mathbf{w}$ is orthogonal to $\mathbf{v}$.

Since $\mathbf{v}$ lies along the x-axis, $\mathbf{u}$ lies in the yz-plane, therefore $\mathbf{u} = \langle 0, b, c \rangle$. $\mathbf{u}$ is also orthogonal to $\mathbf{w}$, so $\mathbf{u} \cdot \mathbf{w} = 0$. This gives $\mathbf{u} \cdot \mathbf{w} = \langle 0, b, c \rangle \cdot \langle 0, 1, -1 \rangle = b - c = 0 \Rightarrow b = c$. Thus, $\mathbf{u} = \langle 0, b, b \rangle$. By the Right-Hand Rule, $\mathbf{u}$ points to the positive z-direction so $b > 0$. We compute the length of $\mathbf{u}$. Since $\mathbf{v} \cdot \mathbf{w} = \langle 3, 0, 0 \rangle \cdot \langle 0, 1, -1 \rangle = 0$, $\mathbf{v}$ and $\mathbf{w}$ are orthogonal. Hence,

$$\|\mathbf{v} \times \mathbf{w}\| = \|\mathbf{v}\|\|\mathbf{w}\| \sin \frac{\pi}{2} = \|\mathbf{v}\|\|\mathbf{w}\| = 3 \cdot \sqrt{2}$$

Also since $b > 0$, we have

$$\|\mathbf{u}\| = \|\langle 0, b, b \rangle\| = \sqrt{2b^2} = b\sqrt{2}$$

Equating the lengths gives

$$b\sqrt{2} = 3\sqrt{2} \quad \Rightarrow \quad b = 3$$

We conclude that $\mathbf{u} = \mathbf{v} \times \mathbf{w} = \langle 0, 3, 3 \rangle$.

31. Show that if **v** and **w** lie in the *yz*-plane, then **v** × **w** is a multiple of **i**.

SOLUTION **v** × **w** is orthogonal to **v** and **w**. Since **v** and **w** lie in the *yz*-plane, **v** × **w** must lie along the *x* axis, which is perpendicular to *yz*-plane. That is, **v** × **w** is a scalar multiple of the unit vector **i**.

33. Let **e** and **e'** be unit vectors in $\mathbf{R}^3$ such that $\mathbf{e} \perp \mathbf{e}'$. Use the geometric properties of the cross product to compute $\mathbf{e} \times (\mathbf{e}' \times \mathbf{e})$.

SOLUTION Let $\mathbf{u} = \mathbf{e} \times (\mathbf{e}' \times \mathbf{e})$ and $\mathbf{v} = \mathbf{e}' \times \mathbf{e}$. The vector **v** is orthogonal to **e'** and **e**, hence **v** is orthogonal to the plane π defined by **e'** and **e**. Now **u** is orthogonal to **v**, hence **u** lies in the plane π orthogonal to **v**. **u** is orthogonal to **e**, which is in this plane, hence **u** is a multiple of **e'**:

$$\mathbf{u} = \lambda \mathbf{e}' \qquad (1)$$

The Right-Hand Rule implies that **u** is in the direction of **e'**, hence $\lambda > 0$. To find λ, we compute the length of **u**:

$$\|\mathbf{v}\| = \|\mathbf{e}' \times \mathbf{e}\| = \|\mathbf{e}'\|\|\mathbf{e}\| \sin \frac{\pi}{2} = 1 \cdot 1 \cdot 1 = 1$$

$$\|\mathbf{u}\| = \|\mathbf{e} \times \mathbf{v}\| = \|\mathbf{e}\|\|\mathbf{v}\| \sin \frac{\pi}{2} = 1 \cdot 1 \cdot 1 = 1 \qquad (2)$$

Combining (1), (2), and $\lambda > 0$ we conclude that

$$\mathbf{u} = \mathbf{e} \times (\mathbf{e}' \times \mathbf{e}) = \mathbf{e}'$$

35. Determine the magnitude of each Coriolis force on a 2.3-kg parcel of air with wind **v**.

(a) **v** is a 20 m/s toward the west at the equator

(b) **v** is a 20 m/s toward the west at 60° N

(c) **v** is a 40 m/s toward the south at the equator

(d) **v** is a 40 m/s toward the south at 45° S

SOLUTION

(a) There is no poleward component of the velocity vector, so it is perpendicular to Ω and thus $\theta = 90°$. Then

$$\|\mathbf{F}_c\| = 2m\|\Omega\|\|\mathbf{v}\| \sin(90°) \approx 0.0060 \text{ N}$$

(b) There is no poleward component of the velocity vector, so it is perpendicular to Ω and thus $\theta = 90°$. Then

$$\|\mathbf{F}_c\| = 2m\|\Omega\|\|\mathbf{v}\| \sin(90°) \approx 0.0060 \text{ N}$$

Note that for east–west velocities, the latitude is irrelevant when computing the magnitude of the force [compare with part **(a)**].

(c) At the equator, $\mathbf{v} \| \Omega$, so their dot product is zero and the Coriolis force is zero.

(d) Following Example 5, the magnitude of the Coriolis force is

$$\|\mathbf{F}_c\| = 2m\|\Omega\|\|\mathbf{v}\| \sin(45°) \approx 0.0114 \text{ N}$$

*In Exercises 36 and 37, a force **F** (in newtons) on an electron moving at velocity **v** meters per second in a uniform magnetic field **B** (in teslas) is given by $\mathbf{F} = q(\mathbf{v} \times \mathbf{B})$, where $q = -1.6 \times 10^{-19}$ coulombs is the charge on the electron.*

37. Assume an electron moves with velocity **v** in the plane and **B** is a uniform magnetic field pointing directly out of the page. Which of the two vectors, $\mathbf{F}_1$ or $\mathbf{F}_2$, in Figure 18 represents the force on the electron? Remember that q is negative.

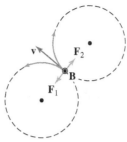

FIGURE 18 The magnetic field vector **B** points directly out of the page.

SOLUTION Since the magnetic field **B** points directly out of the page (toward us), the Right-Hand Rule implies that the cross product $\mathbf{v} \times \mathbf{B}$ is in the direction of $\mathbf{F}_2$ (see figure).

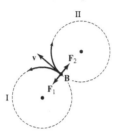

Since $\mathbf{F} = q\,(\mathbf{v} \times \mathbf{B})$ and $q < 0$, the force **F** on the electron is represented by the opposite vector $\mathbf{F}_1$.

39. Verify identity (12) for vectors $\mathbf{v} = \langle 3, -2, 2 \rangle$ and $\mathbf{w} = \langle 4, -1, 2 \rangle$.

SOLUTION We compute the cross product $\mathbf{v} \times \mathbf{w}$:

$$\mathbf{v} \times \mathbf{w} = \begin{vmatrix} \mathbf{i} & \mathbf{j} & \mathbf{k} \\ 3 & -2 & 2 \\ 4 & -1 & 2 \end{vmatrix} = \begin{vmatrix} -2 & 2 \\ -1 & 2 \end{vmatrix} \mathbf{i} - \begin{vmatrix} 3 & 2 \\ 4 & 2 \end{vmatrix} \mathbf{j} + \begin{vmatrix} 3 & -2 \\ 4 & -1 \end{vmatrix} \mathbf{k}$$

$$= (-4 + 2)\mathbf{i} - (6 - 8)\mathbf{j} + (-3 + 8)\mathbf{k} = -2\mathbf{i} + 2\mathbf{j} + 5\mathbf{k} = \langle -2, 2, 5 \rangle$$

We now find the dot product $\mathbf{v} \cdot \mathbf{w}$:

$$\mathbf{v} \cdot \mathbf{w} = \langle 3, -2, 2 \rangle \cdot \langle 4, -1, 2 \rangle = 3 \cdot 4 + (-2) \cdot (-1) + 2 \cdot 2 = 18$$

Finally we compute the squares of the lengths of **v**, **w** and $\mathbf{v} \times \mathbf{w}$:

$$\|\mathbf{v}\|^2 = 3^2 + (-2)^2 + 2^2 = 17$$
$$\|\mathbf{w}\|^2 = 4^2 + (-1)^2 + 2^2 = 21$$
$$\|\mathbf{v} \times \mathbf{w}\|^2 = (-2)^2 + 2^2 + 5^2 = 33$$

We now verify the equality:

$$\|\mathbf{v}\|^2 \|\mathbf{w}\|^2 - (\mathbf{v} \cdot \mathbf{w})^2 = 17 \cdot 21 - 18^2 = 33 = \|\mathbf{v} \times \mathbf{w}\|^2$$

41. Find the area of the parallelogram spanned by **v** and **w** in Figure 19.

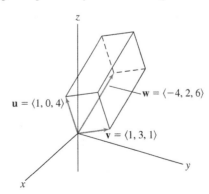

FIGURE 19

SOLUTION The area of the parallelogram equals the length of the cross product of the two vectors $\mathbf{v} = \langle 1, 3, 1 \rangle$ and $\mathbf{w} = \langle -4, 2, 6 \rangle$. We calculate the cross product as follows:

$$\mathbf{v} \times \mathbf{w} = \begin{vmatrix} \mathbf{i} & \mathbf{j} & \mathbf{k} \\ 1 & 3 & 1 \\ -4 & 2 & 6 \end{vmatrix} = (18 - 2)\mathbf{i} - (6 + 4)\mathbf{j} + (2 + 12)\mathbf{k} = 16\mathbf{i} - 10\mathbf{j} + 14\mathbf{k}$$

The length of this vector $16\mathbf{i} - 10\mathbf{j} + 14\mathbf{k}$ is $\sqrt{16^2 + 10^2 + 14^2} = 2\sqrt{138}$. Thus, the area of the parallelogram is $2\sqrt{138}$.

43. Sketch and compute the volume of the parallelepiped spanned by

$$\mathbf{u} = \langle 1, 0, 0 \rangle, \quad \mathbf{v} = \langle 0, 2, 0 \rangle, \quad \mathbf{w} = \langle 1, 1, 2 \rangle$$

SOLUTION Using $\mathbf{u} = \langle 1, 0, 0 \rangle$, $\mathbf{v} = \langle 0, 2, 0 \rangle$, and $\mathbf{w} = \langle 1, 1, 2 \rangle$, the volume is given by the following scalar triple product:

$$\mathbf{u} \cdot (\mathbf{v} \times \mathbf{w}) = \begin{vmatrix} 1 & 0 & 0 \\ 0 & 2 & 0 \\ 1 & 1 & 2 \end{vmatrix} = 1(4 - 0) - 0 + 0 = 4$$

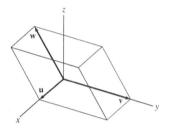

45. Calculate the area of the parallelogram spanned by $\mathbf{u} = \langle 1, 0, 3 \rangle$ and $\mathbf{v} = \langle 2, 1, 1 \rangle$.

SOLUTION The area of the parallelogram is the length of the vector $\mathbf{u} \times \mathbf{v}$. We first compute this vector:

$$\mathbf{u} \times \mathbf{v} = \begin{vmatrix} \mathbf{i} & \mathbf{j} & \mathbf{k} \\ 1 & 0 & 3 \\ 2 & 1 & 1 \end{vmatrix} = \begin{vmatrix} 0 & 3 \\ 1 & 1 \end{vmatrix} \mathbf{i} - \begin{vmatrix} 1 & 3 \\ 2 & 1 \end{vmatrix} \mathbf{j} + \begin{vmatrix} 1 & 0 \\ 2 & 1 \end{vmatrix} \mathbf{k} = -3\mathbf{i} - (1 - 6)\mathbf{j} + \mathbf{k} = -3\mathbf{i} + 5\mathbf{j} + \mathbf{k}$$

The area A is the length

$$A = \|\mathbf{u} \times \mathbf{v}\| = \sqrt{(-3)^2 + 5^2 + 1^2} = \sqrt{35} \approx 5.92$$

47. Sketch the triangle with vertices at the origin O, $P = (3, 3, 0)$, and $Q = (0, 3, 3)$, and compute its area using cross products.

SOLUTION The triangle OPQ is shown in the following figure.

The area S of the triangle is half of the area of the parallelogram determined by the vectors $\overrightarrow{OP} = \langle 3, 3, 0 \rangle$ and $\overrightarrow{OQ} = \langle 0, 3, 3 \rangle$. Thus,

$$S = \frac{1}{2} \|\overrightarrow{OP} \times \overrightarrow{OQ}\| \tag{1}$$

We compute the cross product:

$$\overrightarrow{OP} \times \overrightarrow{OQ} = \begin{vmatrix} \mathbf{i} & \mathbf{j} & \mathbf{k} \\ 3 & 3 & 0 \\ 0 & 3 & 3 \end{vmatrix} = \begin{vmatrix} 3 & 0 \\ 3 & 3 \end{vmatrix} \mathbf{i} - \begin{vmatrix} 3 & 0 \\ 0 & 3 \end{vmatrix} \mathbf{j} + \begin{vmatrix} 3 & 3 \\ 0 & 3 \end{vmatrix} \mathbf{k}$$

$$= 9\mathbf{i} - 9\mathbf{j} + 9\mathbf{k} = 9\langle 1, -1, 1 \rangle$$

Substituting into (1) gives

$$S = \frac{1}{2}\|9\langle 1,-1,1\rangle\| = \frac{9}{2}\|\langle 1,-1,1\rangle\| = \frac{9}{2}\sqrt{1^2+(-1)^2+1^2} = \frac{9\sqrt{3}}{2} \approx 7.8$$

The area of the triangle is $S = \frac{9\sqrt{3}}{2} \approx 7.8$.

49. Use the cross product to find the area of the triangle in the xy-plane defined by $(1,2)$, $(3,4)$, and $(-2,2)$.

SOLUTION Think of the triangle as lying in the xy-plane in a three-dimensional coordinate system, and let $P = (1,2,0)$, $Q = (3,4,0)$, and $R = (-2,2,0)$. Then the area T of the triangle is given by Eq. (7) in the text:

$$T = \frac{1}{2}\|\overrightarrow{PQ} \times \overrightarrow{PR}\|$$

We compute the vectors $\overrightarrow{PQ}$ and $\overrightarrow{PR}$:

$$\overrightarrow{PQ} = \langle 3-1, 4-2, 0-0\rangle = \langle 2,2,0\rangle$$
$$\overrightarrow{PR} = \langle -2-1, 2-2, 0-0\rangle = \langle -3,0,0\rangle$$

We now find the cross product $\overrightarrow{PQ} \times \overrightarrow{PR}$ by computing the following determinant:

$$\overrightarrow{PQ} \times \overrightarrow{PR} = \begin{vmatrix} \mathbf{i} & \mathbf{j} & \mathbf{k} \\ 2 & 2 & 0 \\ -3 & 0 & 0 \end{vmatrix} = \begin{vmatrix} 2 & 0 \\ 0 & 0 \end{vmatrix}\mathbf{i} - \begin{vmatrix} 2 & 0 \\ -3 & 0 \end{vmatrix}\mathbf{j} + \begin{vmatrix} 2 & 2 \\ -3 & 0 \end{vmatrix}\mathbf{k} = 6\mathbf{k}$$

Thus, we get

$$T = \frac{1}{2}\|6\mathbf{k}\| = \frac{1}{2} \cdot 6 = 3$$

51. Check that the four points $P(2,4,4)$, $Q(3,1,6)$, $R(2,8,0)$, and $S(7,2,1)$ all lie in a plane. Then use vectors to find the area of the quadrilateral they define.

SOLUTION The points P, Q, and R determine a plane with normal vector $\mathbf{n}$; S lies in that plane if $\overrightarrow{PS}$ is perpendicular to $\mathbf{n}$. To find $\mathbf{n}$, we compute

$$\overrightarrow{PQ} \times \overrightarrow{PR} = \langle 1,-3,2\rangle \times \langle 0,4,-4\rangle = \begin{vmatrix} \mathbf{i} & \mathbf{j} & \mathbf{k} \\ 1 & -3 & 2 \\ 0 & 4 & -4 \end{vmatrix} = \begin{vmatrix} -3 & 2 \\ 4 & -4 \end{vmatrix}\mathbf{i} - \begin{vmatrix} 1 & 2 \\ 0 & -4 \end{vmatrix}\mathbf{j} + \begin{vmatrix} 1 & -3 \\ 0 & 4 \end{vmatrix}\mathbf{k} = 4\mathbf{i} + 4\mathbf{j} + 4\mathbf{k}$$

Since

$$\langle 4,4,4\rangle \cdot \overrightarrow{PS} = \langle 4,4,4\rangle \cdot \langle 5,-2,-3\rangle = 0$$

the normal vector is also orthogonal to $\overrightarrow{PS}$, so that the vector $\overrightarrow{PS}$, and therefore the point S, also lies in the plane. So all four points lie in a plane.

To find the area of the quadrilateral of which they are the vertices, divide the quadrilateral into the two triangles $\triangle PQR$ and $\triangle SQR$. The area of each of these triangles is given by Eq. (7) in the text. First we must compute various vectors:

$$\overrightarrow{PQ} = \langle 3-2, 1-4, 6-4\rangle = \langle 1,-3,2\rangle$$
$$\overrightarrow{PR} = \langle 2-2, 8-4, 0-4\rangle = \langle 0,4,-4\rangle$$
$$\overrightarrow{SQ} = \langle 3-7, 1-2, 6-1\rangle = \langle -4,-1,5\rangle$$
$$\overrightarrow{SR} = \langle 2-7, 8-2, 0-1\rangle = \langle -5,6,-1\rangle$$

To find the area of $\triangle PQR$, we must compute $\overrightarrow{PQ} \times \overrightarrow{PR}$; to find the area of $\triangle SQR$ we must compute $\overrightarrow{SQ} \times \overrightarrow{SR}$:

$$\overrightarrow{PQ} \times \overrightarrow{PR} = \begin{vmatrix} \mathbf{i} & \mathbf{j} & \mathbf{k} \\ 1 & -3 & 2 \\ 0 & 4 & -4 \end{vmatrix} = \begin{vmatrix} -3 & 2 \\ 4 & -4 \end{vmatrix}\mathbf{i} - \begin{vmatrix} 1 & 2 \\ 0 & -4 \end{vmatrix}\mathbf{j} + \begin{vmatrix} 1 & -3 \\ 0 & 4 \end{vmatrix}\mathbf{k} = 4\mathbf{i} + 4\mathbf{j} + 4\mathbf{k}$$

$$\overrightarrow{SQ} \times \overrightarrow{SR} = \begin{vmatrix} \mathbf{i} & \mathbf{j} & \mathbf{k} \\ -4 & -1 & 5 \\ -5 & 6 & -1 \end{vmatrix} = \begin{vmatrix} -1 & 5 \\ 6 & -1 \end{vmatrix}\mathbf{i} - \begin{vmatrix} -4 & 5 \\ -5 & -1 \end{vmatrix}\mathbf{j} + \begin{vmatrix} -4 & -1 \\ -5 & 6 \end{vmatrix}\mathbf{k} = -29\mathbf{i} - 29\mathbf{j} - 29\mathbf{k}$$

The area of the quadrilateral, S, is given by

$$S = A(\triangle PQS) + A(\triangle RQS) = \frac{1}{2}\|4\mathbf{i} + 4\mathbf{j} + 4\mathbf{k}\| + \frac{1}{2}\|-29\mathbf{i} - 29\mathbf{j} - 29\mathbf{k}\| = \frac{1}{2}(4\sqrt{3} + 29\sqrt{3}) = \frac{33}{2}\sqrt{3}$$

In Exercises 53–55, verify the identity using the formula for the cross product.

53. $\mathbf{v} \times \mathbf{w} = -\mathbf{w} \times \mathbf{v}$

SOLUTION Let $\mathbf{v} = \langle a, b, c \rangle$ and $\mathbf{w} = \langle d, e, f \rangle$. By the definition of the cross product we have

$$\mathbf{v} \times \mathbf{w} = \begin{vmatrix} \mathbf{i} & \mathbf{j} & \mathbf{k} \\ a & b & c \\ d & e & f \end{vmatrix} = \begin{vmatrix} b & c \\ e & f \end{vmatrix}\mathbf{i} - \begin{vmatrix} a & c \\ d & f \end{vmatrix}\mathbf{j} + \begin{vmatrix} a & b \\ d & e \end{vmatrix}\mathbf{k} = (bf - ec)\mathbf{i} - (af - dc)\mathbf{j} + (ae - db)\mathbf{k}$$

We also have

$$-\mathbf{w} \times \mathbf{v} = \begin{vmatrix} \mathbf{i} & \mathbf{j} & \mathbf{k} \\ -d & -e & -f \\ a & b & c \end{vmatrix} = (-ec + bf)\mathbf{i} - (-dc + af)\mathbf{j} + (-db + ea)\mathbf{k}$$

Thus, $\mathbf{v} \times \mathbf{w} = -\mathbf{w} \times \mathbf{v}$, as desired.

55. $(\mathbf{u} + \mathbf{v}) \times \mathbf{w} = \mathbf{u} \times \mathbf{w} + \mathbf{v} \times \mathbf{w}$

SOLUTION We let $\mathbf{u} = \langle a_1, a_2, a_3 \rangle$, $\mathbf{v} = \langle b_1, b_2, b_3 \rangle$ and $\mathbf{w} = \langle c_1, c_2, c_3 \rangle$. Computing the left-hand side gives

$$(\mathbf{u} + \mathbf{v}) \times \mathbf{w} = \langle a_1 + b_1, a_2 + b_2, a_3 + b_3 \rangle \times \langle c_1, c_2, c_3 \rangle = \begin{vmatrix} \mathbf{i} & \mathbf{j} & \mathbf{k} \\ a_1 + b_1 & a_2 + b_2 & a_3 + b_3 \\ c_1 & c_2 & c_3 \end{vmatrix}$$

$$= \begin{vmatrix} a_2 + b_2 & a_3 + b_3 \\ c_2 & c_3 \end{vmatrix}\mathbf{i} - \begin{vmatrix} a_1 + b_1 & a_3 + b_3 \\ c_1 & c_3 \end{vmatrix}\mathbf{j} + \begin{vmatrix} a_1 + b_1 & a_2 + b_2 \\ c_1 & c_2 \end{vmatrix}\mathbf{k}$$

$$= (c_3(a_2 + b_2) - c_2(a_3 + b_3))\mathbf{i} - (c_3(a_1 + b_1) - c_1(a_3 + b_3))\mathbf{j} + (c_2(a_1 + b_1) - c_1(a_2 + b_2))\mathbf{k}$$

We now compute the right-hand-side of the equality:

$$\mathbf{u} \times \mathbf{w} + \mathbf{v} \times \mathbf{w} = \begin{vmatrix} \mathbf{i} & \mathbf{j} & \mathbf{k} \\ a_1 & a_2 & a_3 \\ c_1 & c_2 & c_3 \end{vmatrix} + \begin{vmatrix} \mathbf{i} & \mathbf{j} & \mathbf{k} \\ b_1 & b_2 & b_3 \\ c_1 & c_2 & c_3 \end{vmatrix}$$

$$= \begin{vmatrix} a_2 & a_3 \\ c_2 & c_3 \end{vmatrix}\mathbf{i} - \begin{vmatrix} a_1 & a_3 \\ c_1 & c_3 \end{vmatrix}\mathbf{j} + \begin{vmatrix} a_1 & a_2 \\ c_1 & c_2 \end{vmatrix}\mathbf{k} + \begin{vmatrix} b_2 & b_3 \\ c_2 & c_3 \end{vmatrix}\mathbf{i} - \begin{vmatrix} b_1 & b_3 \\ c_1 & c_3 \end{vmatrix}\mathbf{j} + \begin{vmatrix} b_1 & b_2 \\ c_1 & c_2 \end{vmatrix}\mathbf{k}$$

$$= (a_2c_3 - a_3c_2)\mathbf{i} - (a_1c_3 - a_3c_1)\mathbf{j} + (a_1c_2 - a_2c_1)\mathbf{k}$$
$$+ (b_2c_3 - b_3c_2)\mathbf{i} - (b_1c_3 - b_3c_1)\mathbf{j} + (b_1c_2 - b_2c_1)\mathbf{k}$$

$$= (a_2c_3 - a_3c_2 + b_2c_3 - b_3c_2)\mathbf{i} - (a_1c_3 - a_3c_1 + b_1c_3 - b_3c_1)\mathbf{j} + (a_1c_2 - a_2c_1 + b_1c_2 - b_2c_1)\mathbf{k}$$

$$= (c_3(a_2 + b_2) - c_2(a_3 + b_3))\mathbf{i} - (c_3(a_1 + b_1) - c_1(a_3 + b_3))\mathbf{j} + (c_2(a_1 + b_1) - c_1(a_2 + b_2))\mathbf{k}$$

The results are the same. Hence,

$$(\mathbf{u} + \mathbf{v}) \times \mathbf{w} = \mathbf{u} \times \mathbf{w} + \mathbf{v} \times \mathbf{w}$$

57. Prove $\mathbf{i} \times \mathbf{j} = \mathbf{k}$ and $\mathbf{k} \times \mathbf{j} = -\mathbf{i}$ by each of the following methods:
(a) Using the definition of cross product as a determinant
(b) Using the geometric description of the cross product in Theorem 1

SOLUTION
(a) Using the determinant definition, we have [with $\mathbf{i} = \langle 1, 0, 0 \rangle$, $\mathbf{j} = \langle 0, 1, 0 \rangle$, $\mathbf{k} = \langle 0, 0, 1 \rangle$]

$$\mathbf{i} \times \mathbf{j} = \begin{vmatrix} \mathbf{i} & \mathbf{j} & \mathbf{k} \\ 1 & 0 & 0 \\ 0 & 1 & 0 \end{vmatrix} = \begin{vmatrix} 0 & 0 \\ 1 & 0 \end{vmatrix}\mathbf{i} - \begin{vmatrix} 1 & 0 \\ 0 & 0 \end{vmatrix}\mathbf{j} + \begin{vmatrix} 1 & 0 \\ 0 & 1 \end{vmatrix}\mathbf{k} = \mathbf{k}$$

$$\mathbf{k} \times \mathbf{j} = \begin{vmatrix} \mathbf{i} & \mathbf{j} & \mathbf{k} \\ 0 & 0 & 1 \\ 0 & 1 & 0 \end{vmatrix} = \begin{vmatrix} 0 & 1 \\ 1 & 0 \end{vmatrix}\mathbf{i} - \begin{vmatrix} 0 & 1 \\ 0 & 0 \end{vmatrix}\mathbf{j} + \begin{vmatrix} 0 & 0 \\ 0 & 1 \end{vmatrix}\mathbf{k} = -\mathbf{i}$$

(b) Using the geometric description, note that $\mathbf{i}$, $\mathbf{j}$, and $\mathbf{k}$ are mutually pairwise orthogonal unit vectors, so that the cross product of any two of them has length 1 by part (ii) of Theorem 1. For $\mathbf{i} \times \mathbf{j}$, the possibilities are thus $\mathbf{k}$ or $-\mathbf{k}$, but $\{\mathbf{i}, \mathbf{j}, \mathbf{k}\}$ forms a right-handed system while $\{\mathbf{i}, \mathbf{j}, -\mathbf{k}\}$ forms a left-handed system. For $\mathbf{k} \times \mathbf{j}$, the possibilities are $\mathbf{i}$ or $-\mathbf{i}$. However, $\{\mathbf{k}, \mathbf{j}, \mathbf{i}\}$ forms a left-handed system (draw a picture!), while $\{\mathbf{k}, \mathbf{j}, -\mathbf{i}\}$ is right-handed.

59. The components of the cross product have a geometric interpretation. Show that the absolute value of the $\mathbf{k}$-component of $\mathbf{v} \times \mathbf{w}$ is equal to the area of the parallelogram spanned by the projections $\mathbf{v}_0$ and $\mathbf{w}_0$ onto the xy-plane (Figure 20).

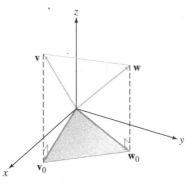

FIGURE 20

SOLUTION Let $\mathbf{v} = \langle a_1, a_2, a_3 \rangle$ and $\mathbf{w} = \langle b_1, b_2, b_3 \rangle$, hence, $\mathbf{v}_0 = \langle a_1, a_2, 0 \rangle$ and $\mathbf{w}_0 = \langle b_1, b_2, 0 \rangle$. The area S of the parallelogram spanned by $\mathbf{v}_0$ and $\mathbf{w}_0$ is the following value:

$$S = \|\mathbf{v}_0 \times \mathbf{w}_0\| \tag{1}$$

We compute the cross product:

$$\mathbf{v}_0 \times \mathbf{w}_0 = \begin{vmatrix} \mathbf{i} & \mathbf{j} & \mathbf{k} \\ a_1 & a_2 & 0 \\ b_1 & b_2 & 0 \end{vmatrix} = \begin{vmatrix} a_2 & 0 \\ b_2 & 0 \end{vmatrix} \mathbf{i} - \begin{vmatrix} a_1 & 0 \\ b_1 & 0 \end{vmatrix} \mathbf{j} + \begin{vmatrix} a_1 & a_2 \\ b_1 & b_2 \end{vmatrix} \mathbf{k}$$

$$= 0\mathbf{i} - 0\mathbf{j} + (a_1 b_2 - a_2 b_1)\mathbf{k} = \langle 0, 0, a_1 b_2 - a_2 b_1 \rangle$$

Using (1) we have

$$S = \sqrt{0^2 + 0^2 + (a_1 b_2 - a_2 b_1)^2} = |a_1 b_2 - a_2 b_1| \tag{2}$$

We now compute $\mathbf{v} \times \mathbf{w}$:

$$\mathbf{v} \times \mathbf{w} = \begin{vmatrix} \mathbf{i} & \mathbf{j} & \mathbf{k} \\ a_1 & a_2 & a_3 \\ b_1 & b_2 & b_3 \end{vmatrix} = \begin{vmatrix} a_2 & a_3 \\ b_2 & b_3 \end{vmatrix} \mathbf{i} - \begin{vmatrix} a_1 & a_3 \\ b_1 & b_3 \end{vmatrix} \mathbf{j} + \begin{vmatrix} a_1 & a_2 \\ b_1 & b_2 \end{vmatrix} \mathbf{k}$$

The $\mathbf{k}$-component of $\mathbf{v} \times \mathbf{w}$ is, thus,

$$\begin{vmatrix} a_1 & a_2 \\ b_1 & b_2 \end{vmatrix} = a_1 b_2 - a_2 b_1 \tag{3}$$

By (2) and (3) we obtain the desired result.

61. Show that three points P, Q, R are collinear (lie on a line) if and only if $\overrightarrow{PQ} \times \overrightarrow{PR} = \mathbf{0}$.

SOLUTION The points P, Q, and R lie on one line if and only if the vectors $\overrightarrow{PQ}$ and $\overrightarrow{PR}$ are parallel. By basic properties of the cross product this is equivalent to $\overrightarrow{PQ} \times \overrightarrow{PR} = \mathbf{0}$.

63. Solve the equation $\langle 1, 1, 1 \rangle \times \mathbf{X} = \langle 1, -1, 0 \rangle$, where $\mathbf{X} = \langle x, y, z \rangle$. *Note:* There are infinitely many solutions.

SOLUTION Let $\mathbf{X} = \langle a, b, c \rangle$. We compute the cross product:

$$\langle 1, 1, 1 \rangle \times \langle a, b, c \rangle = \begin{vmatrix} \mathbf{i} & \mathbf{j} & \mathbf{k} \\ 1 & 1 & 1 \\ a & b & c \end{vmatrix} = \begin{vmatrix} 1 & 1 \\ b & c \end{vmatrix} \mathbf{i} - \begin{vmatrix} 1 & 1 \\ a & c \end{vmatrix} \mathbf{j} + \begin{vmatrix} 1 & 1 \\ a & b \end{vmatrix} \mathbf{k}$$

$$= (c - b)\mathbf{i} - (c - a)\mathbf{j} + (b - a)\mathbf{k} = \langle c - b, a - c, b - a \rangle$$

The equation for $\mathbf{X}$ is, thus,

$$\langle c - b, a - c, b - a \rangle = \langle 1, -1, 0 \rangle$$

Equating corresponding components we get

$$c - b = 1$$
$$a - c = -1$$
$$b - a = 0$$

The third equation implies $a = b$. Substituting in the first and second equations gives

$$\begin{matrix} c - a = 1 \\ a - c = -1 \end{matrix} \Rightarrow c = a + 1$$

The solution is thus, $b = a$, $c = a + 1$. The corresponding solutions $\mathbf{X}$ are

$$\mathbf{X} = \langle a, b, c \rangle = \langle a, a, a + 1 \rangle$$

Therefore any vector of the form $\langle a, a, a + 1 \rangle$ where a is an arbitrary constant is a solution. For example, setting $a = 0$ gives $\langle 0, 0, 1 \rangle$.

65. Let $\mathbf{X} = \langle x, y, z \rangle$. Show that $\mathbf{i} \times \mathbf{X} = \mathbf{v}$ has a solution if and only if $\mathbf{v}$ is contained in the yz-plane (the $\mathbf{i}$-component is zero).

SOLUTION Let $\mathbf{X} = \langle a, b, c \rangle$. We compute the cross product:

$$\langle 1, 1, 1 \rangle \times \langle a, b, c \rangle = \begin{vmatrix} \mathbf{i} & \mathbf{j} & \mathbf{k} \\ 1 & 1 & 1 \\ a & b & c \end{vmatrix} = \begin{vmatrix} 1 & 1 \\ b & c \end{vmatrix} \mathbf{i} - \begin{vmatrix} 1 & 1 \\ a & c \end{vmatrix} \mathbf{j} + \begin{vmatrix} 1 & 1 \\ a & b \end{vmatrix} \mathbf{k}$$

$$= (c - b)\mathbf{i} - (c - a)\mathbf{j} + (b - a)\mathbf{k} = \langle c - b, a - c, b - a \rangle$$

The equation for $\mathbf{X}$ is, thus,

$$\langle c - b, a - c, b - a \rangle = \langle 1, -1, 0 \rangle$$

Equating corresponding components we get

$$c - b = 1$$
$$a - c = -1$$
$$b - a = 0$$

The third equation implies $a = b$. Substituting in the first and second equations gives

$$\begin{matrix} c - a = 1 \\ a - c = -1 \end{matrix} \Rightarrow c = a + 1$$

The solution is thus, $b = a$, $c = a + 1$. The corresponding solutions $\mathbf{X}$ are

$$\mathbf{X} = \langle a, b, c \rangle = \langle a, a, a + 1 \rangle$$

One possible solution is obtained for $a = 0$, that is, $\mathbf{X} = \langle 0, 0, 1 \rangle$.

In Exercises 67–70, the torque about the origin O due to a force **F** *acting on an object with position vector* **r** *is the vector quantity* $\tau = \mathbf{r} \times \mathbf{F}$. *If several forces* $\mathbf{F}_j$ *act at positions* $\mathbf{r}_j$, *then the net torque (units: N-m or lb-ft) is the sum*

$$\tau = \sum \mathbf{r}_j \times \mathbf{F}_j$$

67. Calculate the torque τ about O acting at the point P on the mechanical arm in Figure 21(A), assuming that a 25-newton force acts as indicated.

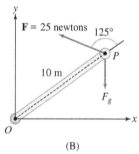

FIGURE 21

SOLUTION We denote by O and P the points shown in the figure and compute the position vector $\mathbf{r} = \overrightarrow{OP}$ and the force vector **F**.

Denoting by θ the angle between the arm and the x-axis we have

$$\mathbf{r} = \overrightarrow{OP} = 10(\cos\theta\,\mathbf{i} + \sin\theta\,\mathbf{j})$$

The angle between the force vector **F** and the x-axis is $(\theta + 125°)$, hence,

$$\mathbf{F} = 25(\cos(\theta + 125°)\,\mathbf{i} + \sin(\theta + 125°)\,\mathbf{j})$$

The torque τ about O acting at the point P is the cross product $\tau = \mathbf{r} \times \mathbf{F}$. We compute it using the cross products of the unit vectors **i** and **j**:

$$\tau = \mathbf{r} \times \mathbf{F} = 10(\cos\theta\,\mathbf{i} + \sin\theta\,\mathbf{j}) \times 25(\cos(\theta + 125°)\,\mathbf{i} + \sin(\theta + 125°)\,\mathbf{j})$$
$$= 250(\cos\theta\,\mathbf{i} + \sin\theta\,\mathbf{j}) \times (\cos(\theta + 125°)\,\mathbf{i} + \sin(\theta + 125°)\,\mathbf{j})$$
$$= 250(\cos\theta\sin(\theta + 125°)\,\mathbf{k} + \sin\theta\cos(\theta + 125°)(-\mathbf{k}))$$
$$= 250(\sin(\theta + 125°)\cos\theta - \sin\theta\cos(\theta + 125°))\,\mathbf{k}$$

We now use the identity $\sin\alpha\cos\beta - \sin\beta\cos\alpha = \sin(\alpha - \beta)$ to obtain

$$\tau = 250\sin(\theta + 125° - \theta)\,\mathbf{k} = 250\sin 125°\,\mathbf{k} \approx 204.79\mathbf{k}$$

69. Let τ be the net torque about O acting on the robotic arm of Figure 22, here taking into account the weight of the arms themselves. Assume that the arms have mass m_1 and m_2 (in kilograms) and that a weight of m_3 kg is located at the endpoint P. In calculating the torque, we may assume that the entire mass of each arm segment lies at the midpoint of the arm (its center of mass). Show that the position vectors of the masses m_1, m_2, and m_3 are

$$\mathbf{r}_1 = \frac{1}{2}L_1(\sin\theta_1 \mathbf{i} + \cos\theta_1 \mathbf{j})$$

$$\mathbf{r}_2 = L_1(\sin\theta_1 \mathbf{i} + \cos\theta_1 \mathbf{j}) + \frac{1}{2}L_2(\sin\theta_2 \mathbf{i} - \cos\theta_2 \mathbf{j})$$

$$\mathbf{r}_3 = L_1(\sin\theta_1 \mathbf{i} + \cos\theta_1 \mathbf{j}) + L_2(\sin\theta_2 \mathbf{i} - \cos\theta_2 \mathbf{j})$$

Then show that

$$\tau = -g\left(L_1\left(\frac{1}{2}m_1 + m_2 + m_3\right)\sin\theta_1 + L_2\left(\frac{1}{2}m_2 + m_3\right)\sin\theta_2\right)\mathbf{k}$$

where $g = 9.8$ m/s^2. To simplify the computation, note that all three gravitational forces act in the $-\mathbf{j}$ direction, so the $\mathbf{j}$-components of the position vectors $\mathbf{r}_i$ do not contribute to the torque.

FIGURE 22

SOLUTION We denote by O, P, and Q the points shown in the figure.

The coordinates of O and Q are

$$O = (0, 0), \quad Q = (L_1 \sin\theta_1, L_1 \cos\theta_1)$$

The midpoint of the segment OQ is, thus,

$$\left(\frac{0 + L_1 \sin\theta_1}{2}, \frac{0 + L_1 \cos\theta_1}{2}\right) = \left(\frac{L_1 \sin\theta_1}{2}, \frac{L_1 \cos\theta_1}{2}\right)$$

Since the mass m_1 is assumed to lie at the midpoint of the arm, the position vector of m_1 is

$$\mathbf{r}_1 = \frac{L_1}{2}(\sin\theta_1 \mathbf{i} + \cos\theta_1 \mathbf{j}) \tag{1}$$

We now find the position vector $\mathbf{r}_2$ of m_2. We have (see figure)

$$\mathbf{r}_2 = \overrightarrow{OQ} + \overrightarrow{QM} \tag{2}$$

$$\overrightarrow{OQ} = L_1 \sin\theta_1 \mathbf{i} + L_1 \cos\theta_1 \mathbf{j} = L_1(\sin\theta_1 \mathbf{i} + \cos\theta_1 \mathbf{j}) \tag{3}$$

The vector $\overrightarrow{QM}$ makes an angle of $-(90° - \theta_2)$ with the x axis and has length $\frac{L_2}{2}$, hence,

$$\overrightarrow{QM} = \frac{L_2}{2}(\cos(-(90° - \theta_2))\mathbf{i} + \sin(-(90° - \theta_2))\mathbf{j}) = \frac{L_2}{2}(\sin\theta_2 \mathbf{i} - \cos\theta_2 \mathbf{j}) \tag{4}$$

Combining (2), (3), and (4) we get

$$\mathbf{r}_2 = L_1(\sin\theta_1 \mathbf{i} + \cos\theta_1 \mathbf{j}) + \frac{L_2}{2}(\sin\theta_2 \mathbf{i} - \cos\theta_2 \mathbf{j}) \tag{5}$$

Finally, we find the position vector $\mathbf{r}_3$:

$$\mathbf{r}_3 = \overrightarrow{OQ} + \overrightarrow{QP} = \overrightarrow{OQ} + 2\overrightarrow{QM}$$

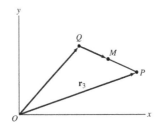

Substituting (3) and (4) we get

$$\mathbf{r}_3 = L_1(\sin\theta_1 \mathbf{i} + \cos\theta_1 \mathbf{j}) + L_2(\sin\theta_2 \mathbf{i} - \cos\theta_2 \mathbf{j}) \tag{6}$$

The net torque is the following vector:

$$\tau = \mathbf{r}_1 \times (-gm_1 \mathbf{j}) + \mathbf{r}_2 \times (-gm_2 \mathbf{j}) + \mathbf{r}_3 \times (-gm_3 \mathbf{j})$$

In computing the cross products, the $\mathbf{j}$ components of $\mathbf{r}_1$, $\mathbf{r}_2$, and $\mathbf{r}_3$ do not contribute to the torque since $\mathbf{j} \times \mathbf{j} = \mathbf{0}$. We thus consider only the $\mathbf{i}$ components of $\mathbf{r}_1$, $\mathbf{r}_2$, and $\mathbf{r}_3$ in (1), (5), and (6). This gives

$$\tau = \frac{L_1}{2}\sin\theta_1 \mathbf{i} \times (-gm_1\mathbf{j}) + \left(L_1\sin\theta_1 + \frac{L_2}{2}\sin\theta_2\right)\mathbf{i} \times (-gm_2\mathbf{j}) + (L_1\sin\theta_1 + L_2\sin\theta_2)\mathbf{i} \times (-gm_3\mathbf{j})$$

$$= -\frac{L_1 g m_1 \sin\theta_1}{2}\mathbf{k} - \left(L_1 gm_2\sin\theta_1 + \frac{L_2 gm_2}{2}\sin\theta_2\right)\mathbf{k} - (L_1 gm_3\sin\theta_1 + L_2 gm_3\sin\theta_2)\mathbf{k}$$

$$= -\left(L_1\left(\frac{1}{2}gm_1 + gm_2 + gm_3\right)\sin\theta_1 + L_2\left(\frac{1}{2}gm_2 + gm_3\right)\sin\theta_2\right)\mathbf{k}$$

$$= -g\left(L_1\left(\frac{1}{2}m_1 + m_2 + m_3\right)\sin\theta_1 + L_2\left(\frac{1}{2}m_2 + m_3\right)\sin\theta_2\right)\mathbf{k}$$

Further Insights and Challenges

71. Show that 3×3 determinants can be computed using the **diagonal rule**: Repeat the first two columns of the matrix and form the products of the numbers along the six diagonals indicated. Then add the products for the diagonals that slant from left to right and subtract the products for the diagonals that slant from right to left.

$$\det(A) = \begin{vmatrix} a_{11} & a_{12} & a_{13} \\ a_{21} & a_{22} & a_{23} \\ a_{31} & a_{32} & a_{33} \end{vmatrix} \begin{matrix} a_{11} & a_{12} \\ a_{21} & a_{22} \\ a_{31} & a_{32} \end{matrix}$$

$$= a_{11}a_{22}a_{33} + a_{12}a_{23}a_{31} + a_{13}a_{21}a_{32}$$

$$- a_{13}a_{22}a_{31} - a_{11}a_{23}a_{32} - a_{12}a_{21}a_{33}$$

SOLUTION Using the definition of 3×3 determinants given in Eq. (2) we get

$$\det(A) = a_{11}\begin{vmatrix} a_{22} & a_{23} \\ a_{32} & a_{33} \end{vmatrix} - a_{12}\begin{vmatrix} a_{21} & a_{23} \\ a_{31} & a_{33} \end{vmatrix} + a_{13}\begin{vmatrix} a_{21} & a_{22} \\ a_{31} & a_{32} \end{vmatrix}$$

Using the definition of 2×2 determinants given in Eq. (1) we get

$$\det(A) = a_{11}(a_{22}a_{33} - a_{23}a_{32}) - a_{12}(a_{21}a_{33} - a_{23}a_{31}) + a_{13}(a_{21}a_{32} - a_{22}a_{31})$$

$$= a_{11}a_{22}a_{33} - a_{11}a_{23}a_{32} - a_{12}a_{21}a_{33} + a_{12}a_{23}a_{31} + a_{13}a_{21}a_{32} - a_{13}a_{22}a_{31}$$

$$= a_{11}a_{22}a_{33} + a_{12}a_{23}a_{31} + a_{13}a_{21}a_{32} - a_{13}a_{22}a_{31} - a_{11}a_{23}a_{32} - a_{12}a_{21}a_{33}$$

73. Prove that $\mathbf{v} \times \mathbf{w} = \mathbf{v} \times \mathbf{u}$ if and only if $\mathbf{u} = \mathbf{w} + \lambda\mathbf{v}$ for some scalar λ. Assume that $\mathbf{v} \neq \mathbf{0}$.

SOLUTION Transferring sides and using the distributive law and the property of parallel vectors, we obtain the following equivalent equalities:

$$\mathbf{v} \times \mathbf{w} = \mathbf{v} \times \mathbf{u}$$

$$\mathbf{0} = \mathbf{v} \times \mathbf{u} - \mathbf{v} \times \mathbf{w}$$

$$\mathbf{0} = \mathbf{v} \times (\mathbf{u} - \mathbf{w})$$

This holds if and only if there exists a scalar λ such that

$$\mathbf{u} - \mathbf{w} = \lambda\mathbf{v}$$

$$\mathbf{u} = \mathbf{w} + \lambda\mathbf{v}$$

75. Show that if $\mathbf{u}$, $\mathbf{v}$, and $\mathbf{w}$ are nonzero vectors and $(\mathbf{u} \times \mathbf{v}) \times \mathbf{w} = \mathbf{0}$, then either (i) $\mathbf{u}$ and $\mathbf{v}$ are parallel, or (ii) $\mathbf{w}$ is orthogonal to $\mathbf{u}$ and $\mathbf{v}$.

SOLUTION By the theorem on basic properties of the cross product, part (c), it follows that $(\mathbf{u} \times \mathbf{v}) \times \mathbf{w} = \mathbf{0}$ if and only if

- $\mathbf{u} \times \mathbf{v} = \mathbf{0}$ or
- $\mathbf{w} = \lambda(\mathbf{u} \times \mathbf{v})$

We consider the two possibilities.

1. $\mathbf{u} \times \mathbf{v} = \mathbf{0}$ is equivalent to $\mathbf{u}$ and $\mathbf{v}$ being parallel vectors or one of them being the zero vector.
2. The cross product $\mathbf{u} \times \mathbf{v}$ is orthogonal to $\mathbf{u}$ and $\mathbf{v}$, hence $\mathbf{w} = \lambda(\mathbf{u} \times \mathbf{v})$ implies that $\mathbf{w}$ is also orthogonal to $\mathbf{u}$ and $\mathbf{v}$ (for $\lambda \neq 0$) or $\mathbf{w} = \mathbf{0}$ (for $\lambda = 0$).

Conclusions: $(\mathbf{u} \times \mathbf{v}) \times \mathbf{w} = \mathbf{0}$ implies that either $\mathbf{u}$ and $\mathbf{v}$ are parallel, or $\mathbf{w}$ is orthogonal to $\mathbf{u}$ and $\mathbf{v}$, or one of the vectors $\mathbf{u}$, $\mathbf{v}$, $\mathbf{w}$ is the zero vector.

77. Let $\mathbf{a}$, $\mathbf{b}$, $\mathbf{c}$ be nonzero vectors. Assume that $\mathbf{b}$ and $\mathbf{c}$ are not parallel, and set

$$\mathbf{v} = \mathbf{a} \times (\mathbf{b} \times \mathbf{c}), \qquad \mathbf{w} = (\mathbf{a} \cdot \mathbf{c})\mathbf{b} - (\mathbf{a} \cdot \mathbf{b})\mathbf{c}$$

(a) Prove that:
(i) $\mathbf{v}$ lies in the plane spanned by $\mathbf{b}$ and $\mathbf{c}$.
(ii) $\mathbf{v}$ is orthogonal to $\mathbf{a}$.
(b) Prove that $\mathbf{w}$ also satisfies (i) and (ii). Conclude that $\mathbf{v}$ and $\mathbf{w}$ are parallel.
(c) Show algebraically that $\mathbf{v} = \mathbf{w}$ (Figure 23).

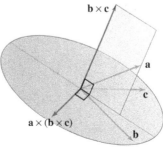

FIGURE 23

SOLUTION

(a) Since $\mathbf{v}$ is the cross product of $\mathbf{a}$ and another vector ($\mathbf{b} \times \mathbf{c}$), then $\mathbf{v}$ is orthogonal to $\mathbf{a}$. Furthermore, $\mathbf{v}$ is orthogonal to ($\mathbf{b} \times \mathbf{c}$), so it is orthogonal to a vector perpendicular to the plane containing $\mathbf{b}$ and $\mathbf{c}$, so $\mathbf{v}$ must be in that plane.

(b) $\mathbf{w} \cdot \mathbf{a} = ((\mathbf{a} \cdot \mathbf{c})\mathbf{b} - (\mathbf{a} \cdot \mathbf{b})\mathbf{c}) \cdot \mathbf{a} = (\mathbf{a} \cdot \mathbf{c})(\mathbf{b} \cdot \mathbf{a}) - (\mathbf{a} \cdot \mathbf{b})(\mathbf{c} \cdot \mathbf{a}) = 0$ (since $\mathbf{a} \cdot \mathbf{c} = \mathbf{c} \cdot \mathbf{a}$ and $\mathbf{b} \cdot \mathbf{a} = \mathbf{a} \cdot \mathbf{b}$). Thus, $\mathbf{w}$ is orthogonal to $\mathbf{a}$. Also, $\mathbf{w}$ is a multiple of $\mathbf{b}$ and $\mathbf{c}$, so $\mathbf{w}$ must be in the plane containing $\mathbf{b}$ and $\mathbf{c}$.

Now, if $\mathbf{a}$ is perpendicular to the plane spanned by $\mathbf{b}$ and $\mathbf{c}$, then $\mathbf{a}$ is parallel to $\mathbf{b} \times \mathbf{c}$ and so $\mathbf{a} \times (\mathbf{b} \times \mathbf{c}) = 0$, which means $\mathbf{v} = 0$, but also $\mathbf{a} \cdot \mathbf{b} = \mathbf{a} \cdot \mathbf{c} = 0$ which means $\mathbf{w} = 0$. Thus, $\mathbf{v}$ and $\mathbf{w}$ are parallel (in fact, equal).

Now, if $\mathbf{a}$ is not perpendicular to the plane spanned by $\mathbf{b}$ and $\mathbf{c}$, then the set of vectors on that plane that are also perpendicular to $\mathbf{a}$ form a line, and thus all such vectors are parallel. We conclude that $\mathbf{v}$ and $\mathbf{w}$, being on that plane and perpendicular to $\mathbf{a}$, are parallel.

(c) On the one hand,

$$\mathbf{v} = \mathbf{a} \times (\mathbf{b} \times \mathbf{c}) = \langle a_1, a_2, a_3 \rangle \times \begin{vmatrix} \mathbf{i} & \mathbf{j} & \mathbf{k} \\ b_1 & b_2 & b_3 \\ c_1 & c_2 & c_3 \end{vmatrix}$$

$$= \begin{vmatrix} \mathbf{i} & \mathbf{j} & \mathbf{k} \\ a_1 & a_2 & a_3 \\ (b_2 c_3 - b_3 c_2) & (b_3 c_1 - b_1 c_3) & (b_1 c_2 - b_2 c_1) \end{vmatrix}$$

$$= \langle a_2(b_1 c_2 - b_2 c_1) - a_3(b_3 c_1 - b_1 c_3), a_3(b_2 c_3 - b_3 c_2) - a_1(b_1 c_2 - b_2 c_1),$$
$$a_1(b_3 c_1 - b_1 c_3) - a_2(b_2 c_3 - b_3 c_2) \rangle$$

but on the other hand,

$$\mathbf{w} = (\mathbf{a} \cdot \mathbf{c})\mathbf{b} - (\mathbf{a} \cdot \mathbf{b})\mathbf{c}$$
$$= (a_1 c_1 + a_2 c_2 + a_3 c_3)\langle b_1, b_2, b_3 \rangle - (a_1 b_1 + a_2 b_2 + a_3 b_3)\langle c_1, c_2, c_3 \rangle$$
$$= \langle a_2 c_2 b_1 + a_3 c_3 b_1 - a_2 b_2 c_1 - a_3 b_3 c_1, a_1 c_1 b_2 + a_3 c_3 b_2 - a_1 b_1 c_2 - a_3 b_3 c_2,$$
$$a_1 c_1 b_3 + a_2 c_2 b_3 - a_1 b_1 c_3 - a_2 b_2 c_3 \rangle$$
$$= \langle a_2(b_1 c_2 - b_2 c_1) - a_3(b_3 c_1 - b_1 c_3), a_3(b_2 c_3 - b_3 c_2) - a_1(b_1 c_2 - b_2 c_1),$$
$$a_1(b_3 c_1 - b_1 c_3) - a_2(b_2 c_3 - b_3 c_2) \rangle$$

which is the same as $\mathbf{v}$.

79. Show that if $\mathbf{a}, \mathbf{b}$ are nonzero vectors such that $\mathbf{a} \perp \mathbf{b}$, then there exists a vector $\mathbf{X}$ such that

$$\mathbf{a} \times \mathbf{X} = \mathbf{b} \qquad \boxed{15}$$

Hint: Show that if $\mathbf{X}$ is orthogonal to $\mathbf{b}$ and is not a multiple of $\mathbf{a}$, then $\mathbf{a} \times \mathbf{X}$ is a multiple of $\mathbf{b}$.

SOLUTION We define the following vectors:

$$\mathbf{X} = \frac{\mathbf{a} \times \mathbf{b}}{\|\mathbf{a}\|^2}, \quad \mathbf{c} = \mathbf{a} \times \mathbf{X} \qquad (1)$$

We show that $\mathbf{c} = \mathbf{b}$. Since $\mathbf{X}$ is orthogonal to $\mathbf{a}$ and $\mathbf{b}$, $\mathbf{X}$ is orthogonal to the plane of $\mathbf{a}$ and $\mathbf{b}$. But $\mathbf{c}$ is orthogonal to $\mathbf{X}$, hence $\mathbf{c}$ is contained in the plane of $\mathbf{a}$ and $\mathbf{b}$, that is, $\mathbf{a}, \mathbf{b}$, and $\mathbf{c}$ are in the same plane. Now the vectors $\mathbf{a}, \mathbf{b}$, and $\mathbf{c}$ are in one plane, and the vectors $\mathbf{c}$ and $\mathbf{b}$ are orthogonal to $\mathbf{a}$. It follows that $\mathbf{c}$ and $\mathbf{b}$ are parallel.

We now show that $\|\mathbf{c}\| = \|\mathbf{b}\|$. We use the cross-product identity to obtain

$$\|\mathbf{c}\|^2 = \|\mathbf{a} \times \mathbf{X}\|^2 = \|\mathbf{a}\|^2 \|\mathbf{X}\|^2 - (\mathbf{a} \cdot \mathbf{X})^2$$

$\mathbf{X}$ is orthogonal to $\mathbf{a}$, hence $\mathbf{a} \cdot \mathbf{X} = 0$, and we obtain

$$\|\mathbf{c}\|^2 = \|\mathbf{a}\|^2 \|\mathbf{X}\|^2 = \|\mathbf{a}\|^2 \left\| \frac{\mathbf{a} \times \mathbf{b}}{\|\mathbf{a}\|^2} \right\|^2 = \frac{\|\mathbf{a}\|^2}{\|\mathbf{a}\|^4} \|\mathbf{a} \times \mathbf{b}\|^2 = \frac{1}{\|\mathbf{a}\|^2} \|\mathbf{a} \times \mathbf{b}\|^2$$

By the given data, $\mathbf{a}$ and $\mathbf{b}$ are orthogonal vectors, so $\|\mathbf{a} \times \mathbf{b}\|^2 = \|\mathbf{a}\|^2 \|\mathbf{b}\|^2$, and then

$$\|\mathbf{c}\|^2 = \frac{1}{\|\mathbf{a}\|^2} \|\mathbf{a} \times \mathbf{b}\|^2 = \frac{1}{\|\mathbf{a}\|^2} \left(\|\mathbf{a}\|^2 \|\mathbf{a}\|^2 \right) = \|\mathbf{b}\|^2 \Rightarrow \|\mathbf{c}\| = \|\mathbf{b}\| \qquad (2)$$

Since **c** and **b** are parallel, it follows that **c** = **b** or **c** = −**b**. We thus proved that the vector $\mathbf{X} = \dfrac{\mathbf{a} \times \mathbf{b}}{\|\mathbf{a}\|^2}$ satisfies $\mathbf{a} \times \mathbf{X} = \mathbf{b}$ or $\mathbf{a} \times \mathbf{X} = -\mathbf{b}$. If $\mathbf{a} \times \mathbf{X} = -\mathbf{b}$, then $\mathbf{a} \times (-\mathbf{X}) = \mathbf{b}$. Hence, there exists a vector $\mathbf{X}$ such that $\mathbf{a} \times \mathbf{X} = \mathbf{b}$.

81. Assume that **v** and **w** lie in the first quadrant in $\mathbf{R}^2$ as in Figure 24. Use geometry to prove that the area of the parallelogram is equal to $\det\begin{pmatrix}\mathbf{v}\\\mathbf{w}\end{pmatrix}$.

FIGURE 24

SOLUTION We denote the components of **u** and **v** by

$$\mathbf{u} = \langle c, d \rangle$$
$$\mathbf{v} = \langle a, b \rangle$$

We also denote by $O, A, B, C, D, E, F, G, H, K$ the points shown in the figure.

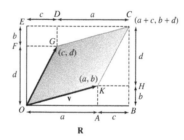

Since $OGCK$ is a parallelogram, it follows by geometrical properties that the triangles OFG and KHC and also the triangles DGC and AKO are congruent. It also follows that the rectangles $EFDG$ and $ABHK$ have equal areas. We use the following notation:

A: The area of the parallelogram
S: The area of the rectangle $OBCE$
S_1: The area of the rectangle $EFDG$
S_2: The area of the triangle OFG
S_3: The area of the triangle DGC

Hence,

$$A = S - 2(S_1 + S_2 + S_3) \tag{1}$$

Using the formulas for the areas of rectangles and triangles we have (see figure)

$$S = OB \cdot OE = (a+c)(d+b)$$
$$S_1 = bc, \quad S_2 = \frac{cd}{2}, \quad S_3 = \frac{ab}{2}$$

Substituting into (1) we get

$$A = (a+c)(d+b) - 2\left(bc + \frac{cd}{2} + \frac{ab}{2}\right)$$
$$= ad + ab + cd + cb - 2bc - cd - ab \tag{2}$$
$$= ad - bc$$

On the other hand,

$$\det\begin{pmatrix} \mathbf{v} \\ \mathbf{w} \end{pmatrix} = \begin{vmatrix} a & b \\ c & d \end{vmatrix} = ad - bc \tag{3}$$

By (2) and (3) we obtain the desired result.

83. In the notation of Exercise 82, suppose that $\mathbf{a}, \mathbf{b}, \mathbf{c}$ are mutually perpendicular as in Figure 25(B). Let S_F be the area of face F. Prove the following three-dimensional version of the Pythagorean Theorem:

$$S_A^2 + S_B^2 + S_C^2 = S_D^2$$

SOLUTION Since $\|\mathbf{v}_D\| = S_D$ then using Exercise 82 we obtain

$$S_D^2 = \|\mathbf{v}_D\|^2 = \mathbf{v}_D \cdot \mathbf{v}_D = (\mathbf{v}_A + \mathbf{v}_B + \mathbf{v}_C) \cdot (\mathbf{v}_A + \mathbf{v}_B + \mathbf{v}_C)$$

$$= \mathbf{v}_A \cdot \mathbf{v}_A + \mathbf{v}_A \cdot \mathbf{v}_B + \mathbf{v}_A \cdot \mathbf{v}_C + \mathbf{v}_B \cdot \mathbf{v}_A + \mathbf{v}_B \cdot \mathbf{v}_B + \mathbf{v}_B \cdot \mathbf{v}_C + \mathbf{v}_C \cdot \mathbf{v}_A + \mathbf{v}_C \cdot \mathbf{v}_B + \mathbf{v}_C \cdot \mathbf{v}_C$$

$$= \|\mathbf{v}_A\|^2 + \|\mathbf{v}_B\|^2 + \|\mathbf{v}_C\|^2 + 2(\mathbf{v}_A \cdot \mathbf{v}_B + \mathbf{v}_A \cdot \mathbf{v}_C + \mathbf{v}_B \cdot \mathbf{v}_C) \tag{1}$$

Now, the perpendicular vectors $\mathbf{v}_A$, $\mathbf{v}_B$, and $\mathbf{v}_C$ to the coordinate planes are mutually orthogonal, hence,

$$\mathbf{v}_A \cdot \mathbf{v}_B = \mathbf{v}_A \cdot \mathbf{v}_C = \mathbf{v}_B \cdot \mathbf{v}_C = 0 \tag{2}$$

Combining (1) and (2) and using the relations $\|\mathbf{v}_F\| = S_F$ we obtain

$$S_D^2 = S_A^2 + S_B^2 + S_C^2$$

12.5 Planes in 3-Space   (LT Section 13.5)

Preliminary Questions

1. What is the equation of the plane parallel to $3x + 4y - z = 5$ passing through the origin?

SOLUTION The two planes are parallel, therefore the vector $\mathbf{n} = \langle 3, 4, -1 \rangle$ that is normal to the given plane is also normal to the plane we need to find. This plane is passing through the origin, hence we may substitute $\langle x_0, y_0, z_0 \rangle = \langle 0, 0, 0 \rangle$ in the vector form of the equation of the plane. This gives

$$\mathbf{n} \cdot \langle x, y, z \rangle = \mathbf{n} \cdot \langle x_0, y_0, z_0 \rangle$$

$$\langle 3, 4, -1 \rangle \cdot \langle x, y, z \rangle = \langle 3, 4, -1 \rangle \cdot \langle 0, 0, 0 \rangle = 0$$

or in scalar form

$$3x + 4y - z = 0$$

2. The vector $\mathbf{k}$ is normal to which of the following planes?

(a) $x = 1$ **(b)** $y = 1$ **(c)** $z = 1$

SOLUTION The planes $x = 1$, $y = 1$, and $z = 1$ are orthogonal to the x, y, and z-axes respectively. Since the plane $z = 1$ is orthogonal to the z-axis, the vector $\mathbf{k}$ is normal to this plane.

3. Which of the following planes is not parallel to the plane $x + y + z = 1$?

(a) $2x + 2y + 2z = 1$ **(b)** $x + y + z = 3$

(c) $x - y + z = 0$

SOLUTION The two planes are parallel if vectors that are normal to the planes are parallel. The vector $\mathbf{n} = \langle 1, 1, 1 \rangle$ is normal to the plane $x + y + z = 1$. We identify the following normals:

- $\mathbf{v} = \langle 2, 2, 2 \rangle$ is normal to plane (a)
- $\mathbf{u} = \langle 1, 1, 1 \rangle$ is normal to plane (b)
- $\mathbf{w} = \langle 1, -1, 1 \rangle$ is normal to plane (c)

The vectors $\mathbf{v}$ and $\mathbf{u}$ are parallel to $\mathbf{n}$, whereas $\mathbf{w}$ is not. (These vectors are not constant multiples of each other.) Therefore, only plane (c) is not parallel to the plane $x + y + z = 1$.

4. To which coordinate plane is the plane $y = 1$ parallel?

SOLUTION The plane $y = 1$ is parallel to the xz-plane.

5. Which of the following planes contains the z-axis?

(a) $z = 1$ (b) $x + y = 1$ (c) $x + y = 0$

SOLUTION The points on the z-axis are the points with zero x and y coordinates. A plane contains the z-axis if and only if the points $(0, 0, c)$ satisfy the equation of the plane for all values of c.

(a) Plane (a) does not contain the z-axis, rather it is orthogonal to this axis. Only the point $(0, 0, 1)$ is on the plane.

(b) $x = 0$ and $y = 0$ do not satisfy the equation of the plane, since $0 + 0 \neq 1$. Therefore the plane does not contain the z-axis.

(c) The plane $x + y = 0$ contains the z-axis since $x = 0$ and $y = 0$ satisfy the equation of the plane.

6. Suppose that a plane $\mathcal{P}$ with normal vector $\mathbf{n}$ and a line $\mathcal{L}$ with direction vector $\mathbf{v}$ both pass through the origin and that $\mathbf{n} \cdot \mathbf{v} = 0$. Which of the following statements is correct?

(a) $\mathcal{L}$ is contained in $\mathcal{P}$.

(b) $\mathcal{L}$ is orthogonal to $\mathcal{P}$.

SOLUTION The direction vector of the line $\mathcal{L}$ is orthogonal to the vector $\mathbf{n}$ that is normal to the plane. Therefore, $\mathcal{L}$ is either parallel or contained in the plane. Since the origin is common to $\mathcal{L}$ and $\mathcal{P}$, the line is contained in the plane. That is, statement (a) is correct.

Exercises

In Exercises 1–8, write the equation of the plane with normal vector $\mathbf{n}$ passing through the given point in the scalar form $ax + by + cz = d$.

1. $\mathbf{n} = \langle 1, 3, 2 \rangle$, $(4, -1, 1)$

SOLUTION The vector equation is

$$\langle 1, 3, 2 \rangle \cdot \langle x, y, z \rangle = \langle 1, 3, 2 \rangle \cdot \langle 4, -1, 1 \rangle = 4 - 3 + 2 = 3$$

To obtain the scalar forms we compute the dot product on the left-hand side of the previous equation:

$$x + 3y + 2z = 3$$

or in the other scalar form:

$$(x - 4) + 3(y + 1) + 2(z - 1) + 4 - 3 + 2 = 3$$
$$(x - 4) + 3(y + 1) + 2(z - 1) = 0$$

3. $\mathbf{n} = \langle -1, 2, 1 \rangle$, $(4, 1, 5)$

SOLUTION The vector form is

$$\langle -1, 2, 1 \rangle \cdot \langle x, y, z \rangle = \langle -1, 2, 1 \rangle \cdot \langle 4, 1, 5 \rangle = -4 + 2 + 5 = 3$$

To obtain the scalar form we compute the dot product above:

$$-x + 2y + z = 3$$

or in the other scalar form:

$$-(x - 4) + 2(y - 1) + (z - 5) = 3 + 4 - 2 - 5 = 0$$
$$-(x - 4) + 2(y - 1) + (z - 5) = 0$$

5. $\mathbf{n} = \mathbf{i}$, $(3, 1, -9)$

SOLUTION We find the vector form of the equation of the plane. We write the vector $\mathbf{n} = \mathbf{i}$ as $\mathbf{n} = \langle 1, 0, 0 \rangle$ and obtain

$$\langle 1, 0, 0 \rangle \cdot \langle x, y, z \rangle = \langle 1, 0, 0 \rangle \cdot \langle 3, 1, -9 \rangle = 3 + 0 + 0 = 3$$

Computing the dot product above gives the scalar form:

$$x + 0 + 0 = 3$$
$$x = 3$$

Or in the other scalar form:

$$(x - 3) + 0 \cdot (y - 1) + 0 \cdot (z + 9) = 3 - 3 = 0$$

7. $\mathbf{n} = \mathbf{k}$, $(6, 7, 2)$

SOLUTION We write the normal $\mathbf{n} = \mathbf{k}$ in the form $\mathbf{n} = \langle 0, 0, 1 \rangle$ and obtain the following vector form of the equation of the plane:

$$\langle 0, 0, 1 \rangle \cdot \langle x, y, z \rangle = \langle 0, 0, 1 \rangle \cdot \langle 6, 7, 2 \rangle = 0 + 0 + 2 = 2$$

We compute the dot product to obtain the scalar form:

$$0x + 0y + 1z = 2$$
$$z = 2$$

or in the other scalar form:

$$0(x - 6) + 0(y - 7) + 1(z - 2) = 0$$

9. Write the equation of any plane through the origin.

SOLUTION We can use any equation $ax + by + cz = d$ which contains the point $(x, y, z) = (0, 0, 0)$. One solution (and there are many) is $x + y + z = 0$.

11. Which of the following statements are true of a plane that is parallel to the yz-plane?
(a) $\mathbf{n} = \langle 0, 0, 1 \rangle$ is a normal vector.
(b) $\mathbf{n} = \langle 1, 0, 0 \rangle$ is a normal vector.
(c) The equation has the form $ay + bz = d$.
(d) The equation has the form $x = d$.

SOLUTION
(a) For $\mathbf{n} = \langle 0, 0, 1 \rangle$ a normal vector, the plane would be parallel to the xy-plane, not the yz-plane. This statement is false.
(b) For $\mathbf{n} = \langle 1, 0, 0 \rangle$ a normal vector, the plane would be parallel to the yz-plane. This statement is true.
(c) For the equation $ay + bz = d$, this plane intersects the yz-plane at $y = 0, z = d/b$ or $y = d/a, z = 0$ depending on whether a or b is nonzero, but it is not equal to the yz-plane (which has equation $x = d$). Thus, it is not parallel to the yz-plane This statement is false.

(d) For the equation of the form $x = d$, this has $\langle 1, 0, 0 \rangle$ as a normal vector and is parallel to the yz-plane. This statement is true.

In Exercises 13–16, find a vector normal to the plane with the given equation.

13. $9x - 4y - 11z = 2$

SOLUTION Using the scalar form of the equation of the plane, a vector normal to the plane is the coefficients vector:

$$\mathbf{n} = \langle 9, -4, -11 \rangle$$

15. $3(x - 4) - 8(y - 1) + 11z = 0$

SOLUTION Using the scalar form of the equation of the plane, $3x - 8y + 11z = 4$ a vector normal to the plane is the coefficients vector:

$$\mathbf{n} = \langle 3, -8, 11 \rangle$$

In Exercises 17–20, find the equation of the plane with the given description.

17. Passes through O and is parallel to $4x - 9y + z = 3$

SOLUTION The vector $\mathbf{n} = \langle 4, -9, 1 \rangle$ is normal to the plane $4x - 9y + z = 3$, and so is also normal to the parallel plane. Setting $\mathbf{n} = \langle 4, -9, 1 \rangle$ and $(x_0, y_0, z_0) = (0, 0, 0)$ in the vector equation of the plane yields

$$\langle 4, -9, 1 \rangle \cdot \langle x, y, z \rangle = \langle 4, -9, 1 \rangle \cdot \langle 0, 0, 0 \rangle = 0$$

$$4x - 9y + z = 0$$

19. Passes through $(4, 1, 9)$ and is parallel to $x = 3$

SOLUTION The vector form of the plane $x = 3$ is

$$\langle 1, 0, 0 \rangle \cdot \langle x, y, z \rangle = 3$$

Hence, $\mathbf{n} = \langle 1, 0, 0 \rangle$ is normal to this plane. This vector is also normal to the parallel plane. Setting $(x_0, y_0, z_0) = (4, 1, 9)$ and $\mathbf{n} = \langle 1, 0, 0 \rangle$ in the vector equation of the plane yields

$$\langle 1, 0, 0 \rangle \cdot \langle x, y, z \rangle = \langle 1, 0, 0 \rangle \cdot \langle 4, 1, 9 \rangle = 4 + 0 + 0 = 4$$

or

$$x + 0 + 0 = 4 \quad \Rightarrow \quad x = 4$$

In Exercises 21–24, find an equation of the plane passing through the three points given.

21. $P = (2, -1, 4), \quad Q = (1, 1, 1), \quad R = (3, 1, -2)$

SOLUTION We go through the steps below:

Step 1. Find the normal vector $\mathbf{n}$. The vectors $\mathbf{a} = \overrightarrow{PQ}$ and $\mathbf{b} = \overrightarrow{PR}$ lie on the plane, hence the cross product $\mathbf{n} = \mathbf{a} \times \mathbf{b}$ is normal to the plane. We compute the cross product:

$$\mathbf{a} = \overrightarrow{PQ} = \langle 1 - 2, 1 - (-1), 1 - 4 \rangle = \langle -1, 2, -3 \rangle$$

$$\mathbf{b} = \overrightarrow{PR} = \langle 3 - 2, 1 - (-1), -2 - 4 \rangle = \langle 1, 2, -6 \rangle$$

$$\mathbf{n} = \mathbf{a} \times \mathbf{b} = \begin{vmatrix} \mathbf{i} & \mathbf{j} & \mathbf{k} \\ -1 & 2 & -3 \\ 1 & 2 & -6 \end{vmatrix} = \begin{vmatrix} 2 & -3 \\ 2 & -6 \end{vmatrix} \mathbf{i} - \begin{vmatrix} -1 & -3 \\ 1 & -6 \end{vmatrix} \mathbf{j} + \begin{vmatrix} -1 & 2 \\ 1 & 2 \end{vmatrix} \mathbf{k}$$

$$= -6\mathbf{i} - 9\mathbf{j} - 4\mathbf{k} = \langle -6, -9, -4 \rangle$$

Step 2. Choose a point on the plane. We choose any one of the three points on the plane, for instance $Q = (1, 1, 1)$. Using the vector form of the equation of the plane we get

$$\mathbf{n} \cdot \langle x, y, z \rangle = \mathbf{n} \cdot \langle x_0, y_0, z_0 \rangle$$
$$\langle -6, -9, -4 \rangle \cdot \langle x, y, z \rangle = \langle -6, -9, -4 \rangle \cdot \langle 1, 1, 1 \rangle$$

Computing the dot products we obtain the following equation:

$$-6x - 9y - 4z = -6 - 9 - 4 = -19$$
$$6x + 9y + 4z = 19$$

23. $P = (1, 0, 0)$, $Q = (0, 1, 1)$, $R = (2, 0, 1)$

SOLUTION We use the vector form of the equation of the plane:

$$\mathbf{n} \cdot \langle x, y, z \rangle = d \qquad (1)$$

To find the normal vector to the plane, $\mathbf{n}$, we first compute the vectors $\overrightarrow{PQ}$ and $\overrightarrow{PR}$ that lie in the plane, and then find the cross product of these vectors. This gives

$$\overrightarrow{PQ} = \langle 0, 1, 1 \rangle - \langle 1, 0, 0 \rangle = \langle -1, 1, 1 \rangle$$
$$\overrightarrow{PR} = \langle 2, 0, 1 \rangle - \langle 1, 0, 0 \rangle = \langle 1, 0, 1 \rangle$$
$$\mathbf{n} = \overrightarrow{PQ} \times \overrightarrow{PR} = \begin{vmatrix} \mathbf{i} & \mathbf{j} & \mathbf{k} \\ -1 & 1 & 1 \\ 1 & 0 & 1 \end{vmatrix} = \begin{vmatrix} 1 & 1 \\ 0 & 1 \end{vmatrix} \mathbf{i} - \begin{vmatrix} -1 & 1 \\ 1 & 1 \end{vmatrix} \mathbf{j} + \begin{vmatrix} -1 & 1 \\ 1 & 0 \end{vmatrix} \mathbf{k}$$
$$= \mathbf{i} + 2\mathbf{j} - \mathbf{k} = \langle 1, 2, -1 \rangle \qquad (2)$$

We now choose any one of the three points in the plane, say $P = (1, 0, 0)$, and compute d:

$$d = \mathbf{n} \cdot \overrightarrow{OP} = \langle 1, 2, -1 \rangle \cdot \langle 1, 0, 0 \rangle = 1 \cdot 1 + 2 \cdot 0 + (-1) \cdot 0 = 1 \qquad (3)$$

Finally we substitute (2) and (3) into (1) to obtain the following equation of the plane:

$$\langle 1, 2, -1 \rangle \cdot \langle x, y, z \rangle = 1$$
$$x + 2y - z = 1$$

25. In each case, describe how to find a normal vector to the plane:
(a) Three noncollinear points are given. The plane contains all three points.
(b) Two lines are given that intersect in a point. The plane contains the lines.

SOLUTION

(a) Let the three points be P, Q, and R, and consider $\overrightarrow{PQ} \times \overrightarrow{PR}$. Since P, Q, and R are noncollinear, these two vectors are not parallel and determine the plane, so their nonzero cross product is normal to both vectors and therefore to the plane.

(b) Choose a point on each line other than the point of intersection. These points together with the point of intersection are noncollinear; use part **(a)**.

27. In each case, determine whether or not the lines have a single point of intersection. If they do, give an equation of a plane containing them.
(a) $\mathbf{r}_1(t) = \langle t, 2t - 1, t - 3 \rangle$ and $\mathbf{r}_2(t) = \langle 4, 2t - 1, -1 \rangle$
(b) $\mathbf{r}_1(t) = \langle 3t, 2t + 1, t - 5 \rangle$ and $\mathbf{r}_2(t) = \langle 4t, 4t - 3, -1 \rangle$

SOLUTION
(a) If the two lines intersect, their first coordinates are equal at the point of intersection, so that $t = 4$ in $\mathbf{r}_1(t)$, and the point must be $(4, 7, 1)$. But $(4, 7, 1)$ does not lie on $\mathbf{r}_2$, so the lines do not intersect.

(b) If the two lines intersect, their third coordinates are equal at the point of intersection, so that $t = 4$ in $\mathbf{r}_1(t)$, and the point must be $(12, 9, -1)$. This point lies on $\mathbf{r}_2(t)$ as well, so there is a single point of intersection, $Q = (12, 9, -1)$. To determine a normal to the plane, we choose one other point on each of $\mathbf{r}_1$ and $\mathbf{r}_2$, say (for $t = 0$) $P_1 = (0, 1, -5)$ and $P_2 = (0, -3, -1)$. Then two vectors lying in the plane are

$$\overrightarrow{QP_1} = \langle -12, -8, -4 \rangle, \qquad \overrightarrow{QP_2} = \langle -12, -12, 0 \rangle$$

Their cross product is normal to the plane:

$$\overrightarrow{QP_1} \times \overrightarrow{QP_2} = \begin{vmatrix} \mathbf{i} & \mathbf{j} & \mathbf{k} \\ -12 & -8 & -4 \\ -12 & -12 & 0 \end{vmatrix} = \langle -48, 48, 48 \rangle$$

We may as well take the normal vector to be $\langle -1, 1, 1 \rangle$, a constant multiple of the cross product. Then the equation of the plane is $-x + y + z = c$; since Q lies on the plane, we know that $-12 + 9 + (-1) = -4 = c$, so that the equation is $-x + y + z = -4$, or $x - y - z = 4$.

29. In each case, determine whether or not the point lies on the line. If it does not, give an equation of a plane containing the point and the line.
(a) $(2, 2, -1)$ and $\mathbf{r}(t) = \langle 4t, 6t - 1, -1 \rangle$
(b) $(3, -3, 2)$ and $\mathbf{r}(t) = \langle 4t + 3, 4t - 3, t + 1 \rangle$

SOLUTION

(a) $\mathbf{r}(2) = \langle 2, 2, -1 \rangle$, so $(2, 2, -1)$ lies on the line.
(b) If $Q = (3, -3, 2)$ lies on $\mathbf{r}$, then looking at the third coordinate, we must have $t = 1$. But $\mathbf{r}(1) = \langle 7, 1, 2 \rangle$, so the point does not lie on the line. Choose two other points on the line, say (for $t = 0$) $P_1 = (3, -3, 1)$ and (for $t = -1$) $P_2 = (-1, -7, 0)$. Then two vectors lying in the plane are

$$\overrightarrow{QP_1} = \langle 0, 0, -1 \rangle, \qquad \overrightarrow{QP_2} = \langle -3, -9, 1 \rangle$$

Their cross product is normal to the plane:

$$\overrightarrow{QP_1} \times \overrightarrow{QP_2} = \begin{vmatrix} \mathbf{i} & \mathbf{j} & \mathbf{k} \\ 0 & 0 & -1 \\ -3 & -9 & 1 \end{vmatrix} = \langle -9, 3, 0 \rangle$$

We may as well take the normal vector to be $\langle -3, 1, 0 \rangle$, a constant multiple of the cross product. Then the equation of the plane is $-3x + y = c$; since Q lies on the plane, we have $-3 \cdot 3 + (-3) = -12 = c$, so the equation of the plane is $-3x + y = -12$, or $3x - y = 12$.

31. In each case, determine whether or not the lines are distinct parallel lines. If they are, give an equation of a plane containing them.
(a) $\mathbf{r}_1(t) = \langle t, 2t - 1, t - 3 \rangle$ and $\mathbf{r}_2(t) = \langle 3t - 3, 6t - 1, 3t - 1 \rangle$
(b) $\mathbf{r}_1(t) = \langle 3t, 2t + 1, t - 5 \rangle$ and $\mathbf{r}_2(t) = \langle -6t, 1 - 4t, 2t - 3 \rangle$

SOLUTION

(a) The direction vector of $\mathbf{r}_1$ is $\langle 1, 2, 1 \rangle$, and the direction vector of $\mathbf{r}_2$ is $\langle 3, 6, 3 \rangle$. Since these two vectors determine the same direction, the two lines are parallel. They are distinct since $\mathbf{r}_1(0) = \langle 0, -1, -3 \rangle$, but $\langle 0, -1, -3 \rangle$ is not on $\mathbf{r}_2(t)$ since we would need $t = 1$ to make the first coordinate zero, and $\mathbf{r}_2(1) = \langle 0, 5, 2 \rangle$. To find an equation of the plane containing both lines, we choose three points, two from $\mathbf{r}_1$ and one from $\mathbf{r}_2$:

$$\mathbf{r}_1(0) = \langle 0, -1, -3 \rangle, \quad \mathbf{r}_1(1) = \langle 1, 1, -2 \rangle, \quad \mathbf{r}_2(1) = \langle 0, 5, 2 \rangle.$$

Letting $P_1 = (0, -1, -3)$, $P_2 = (1, 1, -2)$, and $Q = (0, 5, 2)$, we see that two vectors lying in the plane are

$$\overrightarrow{QP_1} = \langle 0, -6, -5 \rangle, \quad \overrightarrow{QP_2} = \langle 1, -4, -4 \rangle$$

Their cross product is normal to the plane:

$$\overrightarrow{QP_1} \times \overrightarrow{QP_2} = \begin{vmatrix} \mathbf{i} & \mathbf{j} & \mathbf{k} \\ 0 & -6 & -5 \\ 1 & -4 & -4 \end{vmatrix} = \langle 4, -5, 6 \rangle$$

Then the equation of the plane is $4x - 5y + 6z = c$; since Q lies on the plane, we have $4 \cdot 0 - 5 \cdot 5 + 6 \cdot 2 = -13 = c$, so the equation of the plane is $4x - 5y + 6z = -13$.

(b) $\mathbf{r}_1$ has as direction vector $\langle 3, 2, 1 \rangle$, while $\mathbf{r}_2$ has $\langle -6, -4, 2 \rangle$. These two vectors do not point in the same direction, so the two lines are not parallel.

In Exercises 33–37, draw the plane given by the equation.

33. $x + y + z = 4$

SOLUTION

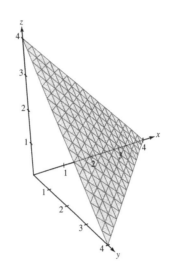

35. $12x - 6y + 4z = 6$

SOLUTION

37. $x + y + z = 0$

SOLUTION

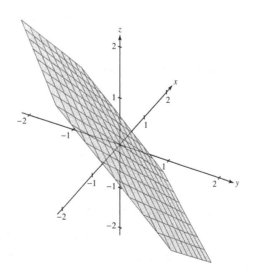

39. Find an equation of the plane $\mathcal{P}$ in Figure 9.

FIGURE 9

SOLUTION We must find the equation of the plane passing though the points $P = (3, 0, 0)$, $Q = (0, 2, 0)$, and $R = (0, 0, 5)$.

We use the following steps:

Step 1. Find a normal vector $\mathbf{n}$. The vectors $\mathbf{a} = \overrightarrow{PQ}$ and $\mathbf{b} = \overrightarrow{PR}$ lie in the plane, hence the cross product $\mathbf{n} = \mathbf{a} \times \mathbf{b}$ is normal to the plane. We compute the cross product:

$$\mathbf{a} = \overrightarrow{PQ} = \langle 0 - 3, 2 - 0, 0 - 0 \rangle = \langle -3, 2, 0 \rangle$$

$$\mathbf{b} = \overrightarrow{PR} = \langle 0 - 3, 0 - 0, 5 - 0 \rangle = \langle -3, 0, 5 \rangle$$

$$\mathbf{n} = \mathbf{a} \times \mathbf{b} = \begin{vmatrix} \mathbf{i} & \mathbf{j} & \mathbf{k} \\ -3 & 2 & 0 \\ -3 & 0 & 5 \end{vmatrix} = \begin{vmatrix} 2 & 0 \\ 0 & 5 \end{vmatrix} \mathbf{i} - \begin{vmatrix} -3 & 0 \\ -3 & 5 \end{vmatrix} \mathbf{j} + \begin{vmatrix} -3 & 2 \\ -3 & 0 \end{vmatrix} \mathbf{k}$$

$$= 10\mathbf{i} + 15\mathbf{j} + 6\mathbf{k} = \langle 10, 15, 6 \rangle$$

Step 2. Choose a point on the plane. We choose one of the points on the plane, say $P = (3, 0, 0)$. Substituting $\mathbf{n} = \langle 10, 15, 6 \rangle$ and $(x_0, y_0, z_0) = (3, 0, 0)$ in the vector form of the equation of the plane gives

$$\mathbf{n} \cdot \langle x, y, z \rangle = \mathbf{n} \cdot \langle x_0, y_0, z_0 \rangle$$

$$\langle 10, 15, 6 \rangle \cdot \langle x, y, z \rangle = \langle 10, 15, 6 \rangle \cdot \langle 3, 0, 0 \rangle$$

Computing the dot products we get the following scalar form of the equation of the plane:

$$10x + 15y + 6z = 10 \cdot 3 + 0 + 0 = 30$$

$$10x + 15y + 6z = 30$$

In Exercises 41–44, find the intersection of the line and the plane.

41. $x + y + z = 14$, $\mathbf{r}(t) = \langle 1, 1, 0 \rangle + t \langle 0, 2, 4 \rangle$

SOLUTION The line has parametric equations

$$x = 1, \quad y = 1 + 2t, \quad z = 4t$$

To find a value of t for which (x, y, z) lies on the plane, we substitute the parametric equations in the equation of the plane and solve for t:

$$x + y + z = 14$$
$$1 + (1 + 2t) + 4t = 14$$
$$6t = 12 \quad \Rightarrow \quad t = 2$$

The point P of intersection has coordinates

$$x = 1, \quad y = 1 + 2 \cdot 2 = 5, \quad z = 4 \cdot 2 = 8$$

That is, $P = (1, 5, 8)$.

43. $z = 12, \quad \mathbf{r}(t) = t\langle -6, 9, 36 \rangle$

SOLUTION The parametric equations of the line are

$$x = -6t, \quad y = 9t, \quad z = 36t \tag{1}$$

We substitute the parametric equations in the equation of the plane and solve for t:

$$z = 12$$
$$36t = 12 \quad \Rightarrow \quad t = \frac{1}{3}$$

The value of the parameter at the point of intersection is $t = \frac{1}{3}$. Substituting into (1) gives the coordinates of the point P of intersection:

$$x = -6 \cdot \frac{1}{3} = -2, \quad y = 9 \cdot \frac{1}{3} = 3, \quad z = 36 \cdot \frac{1}{3} = 12$$

That is,

$$P = (-2, 3, 12)$$

In Exercises 45–50, find the trace of the plane in the given coordinate plane.

45. $3x - 9y + 4z = 5, \quad yz$

SOLUTION The yz-plane has the equation $x = 0$, hence the intersection of the plane with the yz-plane must satisfy both $x = 0$ and the equation of the plane $3x - 9y + 4z = 5$. That is, this is the set of all points $(0, y, z)$ in the yz-plane such that $-9y + 4z = 5$.

47. $3x + 4z = -2, \quad xy$

SOLUTION The trace of the plane $3x + 4z = -2$ in the xy coordinate plane is the set of all points that satisfy the equation of the plane and the equation $z = 0$ of the xy coordinate plane. Thus, we substitute $z = 0$ in $3x + 4z = -2$ to obtain the line $3x = -2$ or $x = -\frac{2}{3}$ in the xy-plane.

49. $-x + y = 4, \quad xz$

SOLUTION The trace of the plane $-x + y = 4$ on the xz-plane is the set of all points that satisfy both the equation of the given plane and the equation $y = 0$ of the xz-plane. That is, the set of all points $(x, 0, z)$ such that $-x + 0 = 4$, or $x = -4$. This is a vertical line in the xz-plane.

51. Does the plane $x = 5$ have a trace in the yz-plane? Explain.

SOLUTION The yz-plane has the equation $x = 0$, hence the x-coordinates of the points in this plane are zero, whereas the x-coordinates of the points in the plane $x = 5$ are 5. Thus, the two planes have no common points.

53. Give equations for two distinct planes whose trace in the yz-plane has equation $y = 4z$.

SOLUTION The yz-plane has the equation $x = 0$, hence the trace of a plane $ax + by + cz = 0$ in the yz-plane is obtained by substituting $x = 0$ in the equation of the plane. Therefore, the following two planes have trace $y = 4z$ (that is, $y - 4z = 0$) in the yz-plane:

$$x + y - 4z = 0; \quad 2x + y - 4z = 0$$

55. Find all planes in $\mathbf{R}^3$ whose intersection with the xz-plane is the line with equation $3x + 2z = 5$.

SOLUTION The intersection of the plane $ax + by + cz = d$ with the xz-plane is obtained by substituting $y = 0$ in the equation of the plane. This gives the following line in the xz-plane:

$$ax + cz = d$$

This is the equation of the line $3x + 2z = 5$ if and only if for some $\lambda \neq 0$,

$$a = 3\lambda, \quad c = 2\lambda, \quad d = 5\lambda$$

Notice that b can have any value. The planes are thus

$$(3\lambda)x + by + (2\lambda)z = 5\lambda, \quad \lambda \neq 0$$

In Exercises 57–62, compute the angle between the two planes, defined as the angle θ (between 0 and π) between their normal vectors (Figure 10).

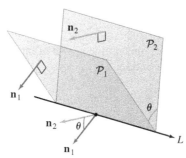

FIGURE 10 By definition, the angle between two planes is the angle between their normal vectors.

57. Planes with normals $\mathbf{n}_1 = \langle 1, 0, 1 \rangle$, $\mathbf{n}_2 = \langle -1, 1, 1 \rangle$

SOLUTION Using the formula for the angle between two vectors we get

$$\cos\theta = \frac{\mathbf{n}_1 \cdot \mathbf{n}_2}{\|\mathbf{n}_1\|\|\mathbf{n}_2\|} = \frac{\langle 1,0,1 \rangle \cdot \langle -1,1,1 \rangle}{\|\langle 1,0,1 \rangle\|\|\langle -1,1,1 \rangle\|} = \frac{-1+0+1}{\sqrt{1^2+0+1^2}\sqrt{(-1)^2+1^2+1^2}} = 0$$

The solution for $0 \leq \theta < \pi$ is $\theta = \frac{\pi}{2}$.

59. $2x + 3y + 7z = 2$ and $4x - 2y + 2z = 4$

SOLUTION The planes $2x + 3y + 7z = 2$ and $4x - 2y + 2z = 4$ have the normals $\mathbf{n}_1 = \langle 2, 3, 7 \rangle$ and $\mathbf{n}_2 = \langle 4, -2, 2 \rangle$, respectively. The cosine of the angle between $\mathbf{n}_1$ and $\mathbf{n}_2$ is

$$\cos\theta = \frac{\mathbf{n}_1 \cdot \mathbf{n}_2}{\|\mathbf{n}_1\|\|\mathbf{n}_2\|} = \frac{\langle 2,3,7 \rangle \cdot \langle 4,-2,2 \rangle}{\|\langle 2,3,7 \rangle\|\|\langle 4,-2,2 \rangle\|} = \frac{8-6+14}{\sqrt{2^2+3^2+7^2}\sqrt{4^2+(-2)^2+2^2}} = \frac{16}{\sqrt{62}\sqrt{24}} \approx 0.415$$

The solution for $0 \leq \theta < \pi$ is $\theta = 1.143$ rad or $\theta = 65.49°$.

61. $3(x - 1) - 5y + 2(z - 12) = 0$ and the plane with normal $\mathbf{n} = \langle 1, 0, 1 \rangle$

SOLUTION The plane $3(x - 1) - 5y + 2(z - 12) = 0$ has the normal $\mathbf{n}_1 = \langle 3, -5, 2 \rangle$, and our second plane has given normal $\mathbf{n}_2 = \langle 1, 0, 1 \rangle$. We use the formula for the angle between two vectors:

$$\cos\theta = \frac{\mathbf{n}_1 \cdot \mathbf{n}_2}{\|\mathbf{n}_1\|\|\mathbf{n}_2\|} = \frac{\langle 3,-5,2 \rangle \cdot \langle 1,0,1 \rangle}{\|\langle 3,-5,2 \rangle\|\|\langle 1,0,1 \rangle\|} = \frac{3+0+2}{\sqrt{3^2+(-5)^2+2^2}\sqrt{1^2+0+1^2}} = \frac{5}{\sqrt{38}\sqrt{2}} \approx 0.5735$$

The solution for $0 \leq \theta < \pi$ is $\theta = 0.96$ rad or $\theta = 55°$.

63. Find an equation of a plane making an angle of $\frac{\pi}{2}$ with the plane $3x + y - 4z = 2$.

SOLUTION The angle θ between two planes (chosen so that $0 \leq \theta < \pi$) is defined as the angle between their normal vectors. The following vector is normal to the plane $3x + y - 4z = 2$:

$$\mathbf{n}_1 = \langle 3, 1, -4 \rangle$$

Let $\mathbf{n} \cdot \langle x, y, z \rangle = d$ denote the equation of a plane making an angle of $\frac{\pi}{2}$ with the given plane, where $\mathbf{n} = \langle a, b, c \rangle$. Since the two planes are perpendicular, the dot product of their normal vectors is zero. That is,

$$\mathbf{n} \cdot \mathbf{n}_1 = \langle a, b, c \rangle \cdot \langle 3, 1, -4 \rangle = 3a + b - 4c = 0 \quad \Rightarrow \quad b = -3a + 4c$$

Thus, the required planes (there is more than one plane) have the following normal vector:

$$\mathbf{n} = \langle a, -3a + 4c, c \rangle$$

We obtain the following equation:

$$\mathbf{n} \cdot \langle x, y, c \rangle = d$$
$$\langle a, -3a + 4c, c \rangle \cdot \langle x, y, z \rangle = d$$
$$ax + (4c - 3a)y + cz = d$$

Every choice of the values of a, c, and d yields a plane with the desired property. For example, we set $a = c = d = 1$ to obtain

$$x + y + z = 1$$

65. Find a plane that is perpendicular to the two planes $x + y = 3$ and $x + 2y - z = 4$.

SOLUTION The vector forms of the equations of the planes are $\langle 1, 1, 0 \rangle \cdot \langle x, y, z \rangle = 3$ and $\langle 1, 2, -1 \rangle \cdot \langle x, y, z \rangle = 4$, hence the vectors $\mathbf{n}_1 = \langle 1, 1, 0 \rangle$ and $\mathbf{n}_2 = \langle 1, 2, -1 \rangle$ are normal to the planes. We denote the equation of the planes which are perpendicular to the two planes by

$$ax + by + cz = d \tag{1}$$

Then, the normal $\mathbf{n} = \langle a, b, c \rangle$ to the planes is orthogonal to the normals $\mathbf{n}_1$ and $\mathbf{n}_2$ of the given planes. Therefore, $\mathbf{n} \cdot \mathbf{n}_1 = 0$ and $\mathbf{n} \cdot \mathbf{n}_2 = 0$, which gives us

$$\langle a, b, c \rangle \cdot \langle 1, 1, 0 \rangle = 0, \quad \langle a, b, c \rangle \cdot \langle 1, 2, -1 \rangle = 0$$

We obtain the following equations:

$$\begin{cases} a + b = 0 \\ a + 2b - c = 0 \end{cases}$$

The first equation implies that $b = -a$. Substituting in the second equation we get $a - 2a - c = 0$, or $c = -a$. Substituting $b = -a$ and $c = -a$ in (1) gives (for $a \neq 0$):

$$ax - ay - az = d \quad \Rightarrow \quad x - y - z = \frac{d}{a}$$

$\frac{d}{a}$ is an arbitrary constant which we denote by f. The planes which are perpendicular to the given planes are, therefore,

$$x - y - z = f$$

67. Let $\mathcal{L}$ denote the intersection of the planes $x - y - z = 1$ and $2x + 3y + z = 2$. Find parametric equations for the line $\mathcal{L}$. *Hint:* To find a point on $\mathcal{L}$, substitute an arbitrary value for z (say, $z = 2$) and then solve the resulting pair of equations for x and y.

SOLUTION We use Exercise 64 to find a direction vector for the line of intersection $\mathcal{L}$ of the planes $x - y - z = 1$ and $2x + 3y + z = 2$. We identify the normals $\mathbf{n}_1 = \langle 1, -1, -1 \rangle$ and $\mathbf{n}_2 = \langle 2, 3, 1 \rangle$ to the two planes, respectively. Hence, a direction vector for $\mathcal{L}$ is the cross product $\mathbf{v} = \mathbf{n}_1 \times \mathbf{n}_2$. We find it here:

$$\mathbf{v} = \mathbf{n}_1 \times \mathbf{n}_2 = \begin{vmatrix} \mathbf{i} & \mathbf{j} & \mathbf{k} \\ 1 & -1 & -1 \\ 2 & 3 & 1 \end{vmatrix} = 2\mathbf{i} - 3\mathbf{j} + 5\mathbf{k} = \langle 2, -3, 5 \rangle$$

We now need to find a point on $\mathcal{L}$. We choose $z = 2$, substitute in the equations of the planes, and solve the resulting equations for x and y. This gives

$$\begin{array}{c} x - y - 2 = 1 \\ 2x + 3y + 2 = 2 \end{array} \quad \text{or} \quad \begin{array}{c} x - y = 3 \\ 2x + 3y = 0 \end{array}$$

The 1st equation implies that $y = x - 3$. Substituting in the 2nd equation and solving for x gives

$$2x + 3(x - 3) = 0$$

$$5x = 9 \quad \Rightarrow \quad x = \frac{9}{5}, \quad y = \frac{9}{5} - 3 = -\frac{6}{5}$$

We conclude that the point $(\frac{9}{5}, -\frac{6}{5}, 2)$ is on $\mathcal{L}$. We now use the vector parametrization of a line to obtain the following parametrization for $\mathcal{L}$:

$$\mathbf{r}(t) = \left\langle \frac{9}{5}, -\frac{6}{5}, 2 \right\rangle + t\langle 2, -3, 5 \rangle$$

This yields the parametric equations

$$x = \frac{9}{5} + 2t, \quad y = -\frac{6}{5} - 3t, \quad z = 2 + 5t$$

69. Two vectors $\mathbf{v}$ and $\mathbf{w}$, each of length 12, lie in the plane $x + 2y - 2z = 0$. The angle between $\mathbf{v}$ and $\mathbf{w}$ is $\pi/6$. This information determines $\mathbf{v} \times \mathbf{w}$ up to a sign ± 1. What are the two possible values of $\mathbf{v} \times \mathbf{w}$?

SOLUTION The length of $\mathbf{v} \times \mathbf{w}$ is $\|\mathbf{v}\|\|\mathbf{w}\| \sin \theta$, but since both vectors have length 12 and since the angle between them is $\pi/6$, then the length of $\mathbf{v} \times \mathbf{w}$ is $12 \cdot 12 \cdot 1/2 = 72$. The direction of $\mathbf{v} \times \mathbf{w}$ is perpendicular to the plane containing them, which is the plane $x + 2y - 2z = 0$, which has normal vector $\mathbf{n} = \langle 1, 2, -2 \rangle$. Since $\mathbf{v} \times \mathbf{w}$ must have length 72 and must be parallel to $\langle 1, 2, -2 \rangle$, then it must be ± 72 times the unit vector $\langle 1, 2, -2 \rangle / \sqrt{1^2 + 2^2 + (-2)^2} = \langle 1/3, 2/3, -2/3 \rangle$. Thus,

$$\mathbf{v} \times \mathbf{w} = \pm 72 \cdot \langle 1/3, 2/3, -2/3 \rangle = \pm 24 \cdot \langle 1, 2, -2 \rangle$$

71. In this exercise, we show that the orthogonal distance D from the plane $\mathcal{P}$ with equation $ax + by + cz = d$ to the origin O is equal to (Figure 11)

$$D = \frac{|d|}{\sqrt{a^2 + b^2 + c^2}}$$

Let $\mathbf{n} = \langle a, b, c \rangle$, and let P be the point where the line through the origin with direction vector $\mathbf{n}$ intersects $\mathcal{P}$. By definition, the orthogonal distance from $\mathcal{P}$ to O is the distance from P to O.

(a) Show that P is the terminal point of $\mathbf{v} = \left(\dfrac{d}{\mathbf{n} \cdot \mathbf{n}} \right) \mathbf{n}$.

(b) Show that the distance from P to O is D.

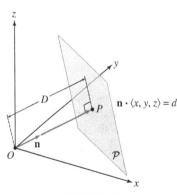

FIGURE 11

SOLUTION Let $\mathbf{v}$ be the vector $\mathbf{v} = \left(\dfrac{d}{\mathbf{n} \cdot \mathbf{n}} \right) \mathbf{n}$. Then $\mathbf{v}$ is parallel to $\mathbf{n}$ and the two vectors are on the same ray.

(a) First we must show that the terminal point of **v** lies on the plane $ax + by + cz = d$. Since the terminal point of **v** is the point

$$\left(\frac{d}{\mathbf{n} \cdot \mathbf{n}}\right)(a, b, c) = \left(\frac{da}{a^2 + b^2 + c^2}, \frac{db}{a^2 + b^2 + c^2}, \frac{dc}{a^2 + b^2 + c^2}\right)$$

then we need only show that this point satisfies $ax + by + cz = d$. Plugging in, we find:

$$ax + by + cz = a \cdot \frac{da}{a^2 + b^2 + c^2} + b \cdot \frac{db}{a^2 + b^2 + c^2} + c \cdot \frac{dc}{a^2 + b^2 + c^2} = \frac{a^2d + b^2d + c^2d}{a^2 + b^2 + c^2} = d$$

(b) We now show that the distance from P to O is D. This distance is just the length of the vector **v**, which is:

$$\|\mathbf{v}\| = \left(\frac{|d|}{\mathbf{n} \cdot \mathbf{n}}\right)\|\mathbf{n}\| = \frac{|d|}{\|\mathbf{n}\|} = \frac{|d|}{\sqrt{a^2 + b^2 + c^2}}$$

as desired.

Further Insights and Challenges

In Exercises 73 and 74, let $\mathcal{P}$ be a plane with equation

$$ax + by + cz = d$$

and normal vector $\mathbf{n} = \langle a, b, c \rangle$. *For any point Q, there is a unique point P on $\mathcal{P}$ that is closest to Q, and is such that $\overrightarrow{PQ}$ is orthogonal to $\mathcal{P}$ (Figure 12).*

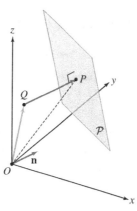

FIGURE 12

73. Show that the point P on $\mathcal{P}$ closest to Q is determined by the equation

$$\overrightarrow{OP} = \overrightarrow{OQ} + \left(\frac{d - \overrightarrow{OQ} \cdot \mathbf{n}}{\mathbf{n} \cdot \mathbf{n}}\right)\mathbf{n} \qquad \boxed{5}$$

SOLUTION Since $\overrightarrow{PQ}$ is orthogonal to the plane $\mathcal{P}$, it is parallel to the vector $\mathbf{n} = \langle a, b, c \rangle$, which is normal to the plane. Hence,

$$\overrightarrow{PQ} = \lambda \mathbf{n} \qquad (1)$$

Since $\overrightarrow{OP} + \overrightarrow{PQ} = \overrightarrow{OQ}$, we have $\overrightarrow{PQ} = \overrightarrow{OQ} - \overrightarrow{OP}$. Thus, by (1) we get

$$\overrightarrow{OQ} - \overrightarrow{OP} = \lambda \mathbf{n} \quad \Rightarrow \quad \overrightarrow{OP} = \overrightarrow{OQ} - \lambda \mathbf{n} \tag{2}$$

The point P is on the plane, hence $\overrightarrow{OP}$ satisfies the vector form of the equation of the plane, that is,

$$\mathbf{n} \cdot \overrightarrow{OP} = d \tag{3}$$

Substituting (2) into (3) and solving for λ yields

$$\mathbf{n} \cdot \left(\overrightarrow{OQ} - \lambda \mathbf{n}\right) = d$$

$$\mathbf{n} \cdot \overrightarrow{OQ} - \lambda \mathbf{n} \cdot \mathbf{n} = d$$

$$\lambda \mathbf{n} \cdot \mathbf{n} = \mathbf{n} \cdot \overrightarrow{OQ} - d \quad \Rightarrow \quad \lambda = \frac{\mathbf{n} \cdot \overrightarrow{OQ} - d}{\mathbf{n} \cdot \mathbf{n}} \tag{4}$$

Finally, we combine (2) and (4) to obtain

$$\overrightarrow{OP} = \overrightarrow{OQ} + \left(\frac{d - \mathbf{n} \cdot \overrightarrow{OQ}}{\mathbf{n} \cdot \mathbf{n}}\right)\mathbf{n}$$

75. Use Eq. (5) to find the point P nearest to $Q = (2, 1, 2)$ on the plane $x + y + z = 1$.

SOLUTION We identify $\mathbf{n} = \langle 1, 1, 1 \rangle$ as a vector normal to the plane. By Eq. (5) the nearest point P to Q is determined by

$$\overrightarrow{OP} = \overrightarrow{OQ} + \left(\frac{d - \overrightarrow{OQ} \cdot \mathbf{n}}{\mathbf{n} \cdot \mathbf{n}}\right)\mathbf{n}$$

We substitute $\mathbf{n} = \langle 1, 1, 1 \rangle$, $\overrightarrow{OQ} = \langle 2, 1, 2 \rangle$ and $d = 1$ in this equation to obtain

$$\overrightarrow{OP} = \langle 2, 1, 2 \rangle + \frac{1 - \langle 2, 1, 2 \rangle \cdot \langle 1, 1, 1 \rangle}{\langle 1, 1, 1 \rangle \cdot \langle 1, 1, 1 \rangle} \langle 1, 1, 1 \rangle = \langle 2, 1, 2 \rangle + \frac{1 - (2 + 1 + 2)}{1 + 1 + 1} \langle 1, 1, 1 \rangle$$

$$= \langle 2, 1, 2 \rangle - \frac{4}{3} \langle 1, 1, 1 \rangle = \left\langle \frac{2}{3}, -\frac{1}{3}, \frac{2}{3} \right\rangle$$

The terminal point $P = \left(\frac{2}{3}, -\frac{1}{3}, \frac{2}{3}\right)$ of $\overrightarrow{OP}$ is the nearest point to $Q = (2, 1, 2)$ on the plane.

77. Use Eq. (6) to find the distance from $Q = (1, 1, 1)$ to the plane $2x + y + 5z = 2$.

SOLUTION By Eq. (6), the distance from $Q = \langle x_1, y_1, z_1 \rangle$ to the plane $ax + by + cz = d$ is

$$\ell = \frac{|ax_1 + by_1 + cz_1 - d|}{\|\mathbf{n}\|} \tag{1}$$

We identify the vector $\mathbf{n} = \langle 2, 1, 5 \rangle$ as a normal to the plane $2x + y + 5z = 2$. Also $a = 2, b = 1, c = 5, d = 2$, and $(x_1, y_1, z_1) = (1, 1, 1)$. Substituting in (1) above we get

$$\ell = \frac{|2 \cdot 1 + 1 \cdot 1 + 5 \cdot 1 - 2|}{\|\langle 2, 1, 5 \rangle\|} = \frac{6}{\sqrt{2^2 + 1^2 + 5^2}} = \frac{6}{\sqrt{30}} \approx 1.095$$

79. What is the distance from $Q = (a, b, c)$ to the plane $x = 0$? Visualize your answer geometrically and explain without computation. Then verify that Eq. (6) yields the same answer.

SOLUTION The plane $x = 0$ is the yz-coordinate plane. The nearest point to Q on the plane is the projection of Q on the plane, which is the point $Q' = (0, b, c)$. Hence, the distance from Q to the plane is the length of the vector $\overrightarrow{Q'Q} = \langle a, 0, 0 \rangle$ which is $|a|$.

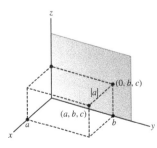

12.6 A Survey of Quadric Surfaces (LT Section 13.6)

Preliminary Questions

1. True or false? All traces of an ellipsoid are ellipses.

SOLUTION This statement is true, mostly. All traces of an ellipsoid $\left(\frac{x}{a}\right)^2 + \left(\frac{y}{b}\right)^2 + \left(\frac{z}{c}\right)^2 = 1$ are ellipses, except for the traces obtained by intersecting the ellipsoid with the planes $x = \pm a$, $y = \pm b$ and $z = \pm c$. These traces reduce to the single points $(\pm a, 0, 0)$, $(0, \pm b, 0)$, and $(0, 0, \pm c)$, respectively.

2. True or false? All traces of a hyperboloid are hyperbolas.

SOLUTION The statement is false. For a hyperbola in the standard orientation, the horizontal traces are ellipses (or perhaps empty for a hyperbola of two sheets), and the vertical traces are hyperbolas.

3. Which quadric surfaces have both hyperbolas and parabolas as traces?

SOLUTION The hyperbolic paraboloid $z = \left(\frac{x}{a}\right)^2 - \left(\frac{y}{b}\right)^2$ has vertical trace curves which are parabolas. If we set $x = x_0$ or $y = y_0$ we get

$$z = \left(\frac{x_0}{a}\right)^2 - \left(\frac{y}{b}\right)^2 \quad \Rightarrow \quad z = -\left(\frac{y}{b}\right)^2 + C$$

$$z = \left(\frac{x}{a}\right)^2 - \left(\frac{y_0}{b}\right)^2 \quad \Rightarrow \quad z = \left(\frac{x}{a}\right)^2 + C$$

The hyperbolic paraboloid has vertical traces which are hyperbolas, since for $z = z_0$, $(z_0 > 0)$, we get

$$z_0 = \left(\frac{x}{a}\right)^2 - \left(\frac{y}{b}\right)^2$$

4. Is there any quadric surface whose traces are all parabolas?

SOLUTION There is no quadric surface whose traces are all parabolas.

5. A surface is called **bounded** if there exists $M > 0$ such that every point on the surface lies at a distance of at most M from the origin. Which of the quadric surfaces are bounded?

SOLUTION The only quadric surface that is bounded is the ellipsoid

$$\left(\frac{x}{a}\right)^2 + \left(\frac{y}{b}\right)^2 + \left(\frac{z}{c}\right)^2 = 1$$

All other quadric surfaces are not bounded, since at least one of the coordinates can increase or decrease without bound.

6. What is the definition of a parabolic cylinder?

SOLUTION A parabolic cylinder consists of all vertical lines passing through a parabola C in the xy-plane.

Exercises

In Exercises 1–8, state whether the given equation defines an ellipsoid or hyperboloid, and if a hyperboloid, whether it is of one or two sheets.

1. $\left(\frac{x}{2}\right)^2 + \left(\frac{y}{3}\right)^2 + \left(\frac{z}{5}\right)^2 = 1$

SOLUTION This equation is the equation of an ellipsoid.

3. $x^2 + 3y^2 + 9z^2 = 1$

SOLUTION We rewrite the equation as follows:

$$x^2 + \left(\frac{y}{\frac{1}{\sqrt{3}}}\right)^2 + \left(\frac{z}{\frac{1}{3}}\right)^2 = 1$$

This equation defines an ellipsoid.

5. $x^2 - 3y^2 + 9z^2 = 1$

SOLUTION We rewrite the equation in the form

$$x^2 - \left(\frac{y}{\frac{1}{\sqrt{3}}}\right)^2 + \left(\frac{z}{\frac{1}{3}}\right)^2 = 1$$

This is the equation of a hyperboloid of one sheet.

7. $x^2 + y^2 = 4 - 4z^2$

SOLUTION Rewriting the equation in the form

$$x^2 + y^2 + 4z^2 = 4, \text{ or } \left(\frac{x}{2}\right)^2 + \left(\frac{y}{2}\right)^2 + z^2 = 1$$

we identify it as the equation of an ellipsoid.

In Exercises 9–16, state whether the given equation defines an elliptic paraboloid, a hyperbolic paraboloid, or an elliptic cone.

9. $z = \left(\frac{x}{4}\right)^2 + \left(\frac{y}{3}\right)^2$

SOLUTION This equation defines an elliptic paraboloid.

11. $z = \left(\frac{x}{9}\right)^2 - \left(\frac{y}{12}\right)^2$

SOLUTION This equation defines a hyperbolic paraboloid.

13. $3x^2 - 7y^2 = z$

SOLUTION Rewriting the equation as

$$z = \left(\frac{x}{\frac{1}{\sqrt{3}}}\right)^2 - \left(\frac{y}{\frac{1}{\sqrt{7}}}\right)^2$$

we identify it as the equation of a hyperbolic paraboloid.

15. $y^2 = 5x^2 - 4z^2$

SOLUTION We rewrite the equation first as $y^2 + 4z^2 = 5x^2$, and then as

$$y^2 + \left(\frac{z}{\frac{1}{2}}\right)^2 = \left(\frac{x}{\frac{1}{\sqrt{5}}}\right)^2$$

This is the equation of an elliptic cone.

SECTION 12.6 | **A Survey of Quadric Surfaces** (LT SECTION 13.6) **965**

In Exercises 17–24, state the type of the quadric surface and describe the trace obtained by intersecting with the given plane.

17. $x^2 + \left(\dfrac{y}{4}\right)^2 + z^2 = 1, \quad y = 0$

SOLUTION The equation $x^2 + \left(\dfrac{y}{4}\right)^2 + z^2 = 1$ defines an ellipsoid. The xz-trace is obtained by substituting $y = 0$ in the equation of the ellipsoid. This gives the equation $x^2 + z^2 = 1$, which defines a circle in the xz-plane.

19. $x^2 + \left(\dfrac{y}{4}\right)^2 + z^2 = 1, \quad z = \dfrac{1}{4}$

SOLUTION The quadric surface is an ellipsoid, since its equation has the form $\left(\dfrac{x}{a}\right)^2 + \left(\dfrac{y}{b}\right)^2 + \left(\dfrac{z}{c}\right)^2 = 1$ for $a = 1, b = 4, c = 1$. To find the trace obtained by intersecting the ellipsoid with the plane $z = \dfrac{1}{4}$, we set $z = \dfrac{1}{4}$ in the equation of the ellipsoid. This gives

$$1x^2 + \left(\dfrac{y}{4}\right)^2 + \left(\dfrac{1}{4}\right)^2 = 1$$

$$x^2 + \dfrac{y^2}{16} = \dfrac{15}{16}$$

To get the standard form we divide by $\dfrac{15}{16}$ to obtain

$$\dfrac{x^2}{\frac{15}{16}} + \dfrac{y^2}{\frac{16 \cdot 15}{16}} = 1 \quad \Rightarrow \quad \left(\dfrac{x}{\frac{\sqrt{15}}{4}}\right)^2 + \left(\dfrac{y}{\sqrt{15}}\right)^2 = 1 \tag{1}$$

We conclude that the trace is an ellipse on the xy-plane, whose equation is given in (1).

21. $\left(\dfrac{x}{3}\right)^2 + \left(\dfrac{y}{5}\right)^2 - 5z^2 = 1, \quad y = 1$

SOLUTION Rewriting the equation in the form

$$\left(\dfrac{x}{3}\right)^2 + \left(\dfrac{y}{5}\right)^2 - \left(\dfrac{z}{\frac{1}{\sqrt{5}}}\right)^2 = 1$$

we identify it as the equation of a hyperboloid of one sheet. Substituting $y = 1$ we get

$$\dfrac{x^2}{9} + \dfrac{1}{25} - 5z^2 = 1$$

$$\dfrac{x^2}{9} - 5z^2 = \dfrac{24}{25}$$

$$\dfrac{25}{24 \cdot 9}x^2 - \dfrac{25 \cdot 5}{24}z^2 = 1$$

$$\left(\dfrac{x}{\frac{6\sqrt{6}}{5}}\right)^2 - \left(\dfrac{z}{\frac{2}{5}\sqrt{\frac{6}{5}}}\right)^2 = 1$$

Thus, the trace on the plane $y = 1$ is a hyperbola.

23. $y = 3x^2, \quad z = 27$

SOLUTION This equation defines a parabolic cylinder, consisting of all vertical lines passing through the parabola $y = 3x^2$ in the xy-plane. Hence, the trace of the cylinder on the plane $z = 27$ is the parabola $y = 3x^2$ on this plane, that is, the following set:

$$\{(x, y, z) : y = 3x^2, \, z = 27\}$$

25. Match each of the ellipsoids in Figure 14 with the correct equation:
(a) $x^2 + 4y^2 + 4z^2 = 16$
(b) $4x^2 + y^2 + 4z^2 = 16$
(c) $4x^2 + 4y^2 + z^2 = 16$

966 CHAPTER 12 | VECTOR GEOMETRY (LT CHAPTER 13)

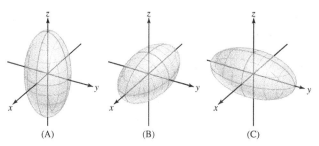

FIGURE 14

SOLUTION

(a) We rewrite the equation in the form

$$\left(\frac{x}{4}\right)^2 + \left(\frac{y}{2}\right)^2 + \left(\frac{z}{2}\right)^2 = 1$$

The ellipsoid intersects the x, y, and z axes at the points $(\pm 4, 0, 0)$, $(0, \pm 2, 0)$, and $(0, 0, \pm 2)$, hence (B) is the corresponding figure.

(b) We rewrite the equation in the form

$$\left(\frac{x}{2}\right)^2 + \left(\frac{y}{4}\right)^2 + \left(\frac{z}{2}\right)^2 = 1$$

The x, y, and z intercepts are $(\pm 2, 0, 0)$, $(0, \pm 4, 0)$, and $(0, 0, \pm 2)$, respectively, hence (C) is the correct figure.

(c) We write the equation in the form

$$\left(\frac{x}{2}\right)^2 + \left(\frac{y}{2}\right)^2 + \left(\frac{z}{4}\right)^2 = 1$$

The x, y, and z intercepts are $(\pm 2, 0, 0)$, $(0, \pm 2, 0)$, and $(0, 0, \pm 4)$, respectively, hence the corresponding figure is (A).

27. What is the equation of the surface obtained when the elliptic paraboloid $z = \left(\frac{x}{2}\right)^2 + \left(\frac{y}{4}\right)^2$ is rotated about the x-axis by $90°$? Refer to Figure 15.

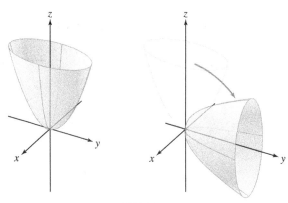

FIGURE 15

SOLUTION The axis of symmetry of the resulting surface is the y-axis rather than the z-axis. Interchanging y and z in the given equation gives the following equation of the rotated paraboloid:

$$y = \left(\frac{x}{2}\right)^2 + \left(\frac{z}{4}\right)^2$$

In Exercises 29–42, sketch the given surface.

29. $x^2 + y^2 - z^2 = 1$

SOLUTION This equation defines a hyperboloid of one sheet. The trace on the plane $z = z_0$ is the circle $x^2 + y^2 = 1 + z_0^2$. The trace on the plane $y = y_0$ is the hyperbola $x^2 - z^2 = 1 - y_0^2$ and the trace on the plane $x = x_0$ is the hyperbola $y^2 - z^2 = 1 - x_0^2$. We obtain the following surface:

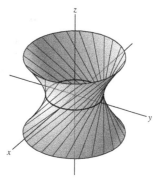

Graph of $x^2 + y^2 - z^2 = 1$

31. $z = \left(\dfrac{x}{4}\right)^2 + \left(\dfrac{y}{8}\right)^2$

SOLUTION This equation defines an elliptic paraboloid, as shown in the following figure:

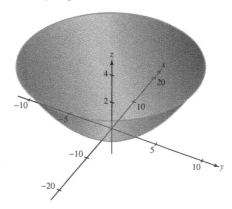

33. $z^2 = \left(\dfrac{x}{4}\right)^2 + \left(\dfrac{y}{8}\right)^2$

SOLUTION This equation defines the following elliptic cone:

35. $-x^2 - y^2 + 9z^2 = 9$

SOLUTION This is the equation of a hyperboloid of two sheets oriented along the y-axis. The graph of the surface is shown below:

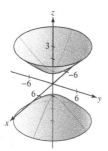

37. $xy = 1$

SOLUTION This is the equation of a hyperbolic cylinder oriented along the z-axis with base $xy = 1$ in the xy-plane. The graph of the cylinder is shown next:

39. $x = 1 + y^2 + z^2$

SOLUTION This is the equation of an elliptic paraboloid oriented along the x-axis. The graph of the surface is shown below:

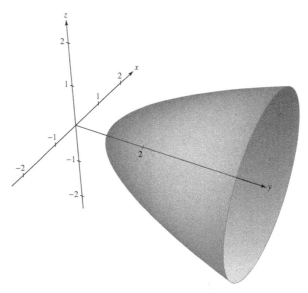

41. $x^2 + 9y^2 + 4z^2 = 36$

SOLUTION This is the equation of an ellipsoid. The graph of the surface is shown below:

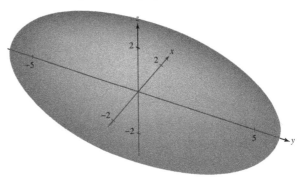

43. Find the equation of the ellipsoid passing through the points marked in Figure 16(A).

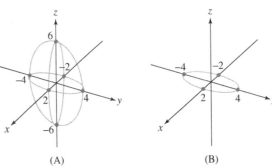

FIGURE 16

SOLUTION The equation of an ellipsoid is

$$\left(\frac{x}{a}\right)^2 + \left(\frac{y}{b}\right)^2 + \left(\frac{z}{c}\right)^2 = 1 \quad (1)$$

The x, y, and z intercepts are $(\pm a, 0, 0)$, $(0, \pm b, 0)$, and $(0, 0, \pm c)$, respectively. The x, y, and z intercepts of the desired ellipsoid are $(\pm 2, 0, 0)$, $(0, \pm 4, 0)$, and $(0, 0, \pm 6)$, respectively, hence $a = 2$, $b = 4$, and $c = 6$. Substituting into (1) we get

$$\left(\frac{x}{2}\right)^2 + \left(\frac{y}{4}\right)^2 + \left(\frac{z}{6}\right)^2 = 1$$

45. Find the equation of the hyperboloid shown in Figure 17(A).

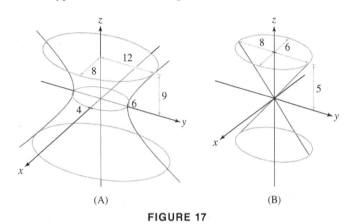

FIGURE 17

SOLUTION The hyperboloid in the figure is of one sheet and the intersections with the planes $z = z_0$ are ellipses. Hence, the equation of the hyperboloid has the form

$$\left(\frac{x}{a}\right)^2 + \left(\frac{y}{b}\right)^2 - \left(\frac{z}{c}\right)^2 = 1 \quad (1)$$

Substituting $z = 0$ we get

$$\left(\frac{x}{a}\right)^2 + \left(\frac{y}{b}\right)^2 = 1$$

By the given information this ellipse has x and y intercepts at the points $(\pm 4, 0)$ and $(0, \pm 6)$, hence $a = 4$, $b = 6$. Substituting in (1) we get

$$\left(\frac{x}{4}\right)^2 + \left(\frac{y}{6}\right)^2 - \left(\frac{z}{c}\right)^2 = 1 \quad (2)$$

Substituting $z = 9$ we get

$$\frac{x^2}{16} + \frac{y^2}{36} - \frac{9^2}{c^2} = 1$$

$$\frac{x^2}{16} + \frac{y^2}{36} = 1 + \frac{81}{c^2} = \frac{c^2 + 81}{c^2}$$

$$\frac{c^2 x^2}{16(81 + c^2)} + \frac{c^2 y^2}{36(81 + c^2)} = 1$$

$$\left(\frac{x}{\frac{4}{c}\sqrt{81 + c^2}}\right)^2 + \left(\frac{y}{\frac{6}{c}\sqrt{81 + c^2}}\right)^2 = 1$$

By the given information the following must hold:

$$\begin{aligned}\frac{4}{c}\sqrt{81 + c^2} &= 8 \\ \frac{6}{c}\sqrt{81 + c^2} &= 12\end{aligned} \quad \Rightarrow \quad \frac{\sqrt{81 + c^2}}{c} = 2 \quad \Rightarrow \quad 81 + c^2 = 4c^2 \quad \Rightarrow \quad 3c^2 = 81$$

Thus, $c = 3\sqrt{3}$, and by substituting in (2) we obtain the following equation:

$$\left(\frac{x}{4}\right)^2 + \left(\frac{y}{6}\right)^2 - \left(\frac{z}{3\sqrt{3}}\right)^2 = 1$$

47. Determine the vertical traces of elliptic and parabolic cylinders in standard form.

SOLUTION The vertical traces of elliptic or parabolic cylinders are one or two vertical lines, or an empty set.

49. Let C be an ellipse in a horizonal plane lying above the xy-plane. Which type of quadric surface is made up of all lines passing through the origin and a point on C?

SOLUTION The quadric surface is the upper part of an elliptic cone.

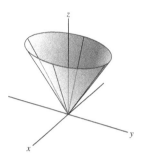

Further Insights and Challenges

51. Let S be the hyperboloid $x^2 + y^2 = z^2 + 1$ and let $P = (\alpha,\beta,0)$ be a point on S in the (x,y)-plane. Show that there are precisely two lines through P entirely contained in S (Figure 18). *Hint:* Consider the line $\mathbf{r}(t) = \langle \alpha + at, \beta + bt, t \rangle$ through P. Show that $\mathbf{r}(t)$ is contained in S if (a,b) is one of the two points on the unit circle obtained by rotating (α,β) through $\pm\frac{\pi}{2}$. This proves that a hyperboloid of one sheet is a **doubly ruled surface**, which means that it can be swept out by moving a line in space in two different ways.

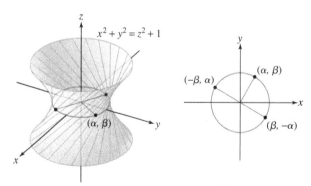

FIGURE 18

SOLUTION The parametric equations of the lines through $P = (\alpha,\beta,0)$ have the form

$$x = \alpha + ks, \quad y = \beta + \ell s, \quad z = ms$$

Setting the parameter $t = ms$ and replacing $\frac{k}{m}$ and $\frac{\ell}{m}$ by a and b, respectively, we obtain the following (normalized) form

$$x = \alpha + at, \quad y = \beta + bt, \quad z = t$$

The line is entirely contained in S if and only if for all values of the parameter t, the following equality holds:

$$(\alpha + at)^2 + (\beta + bt)^2 = t^2 + 1$$

That is, for all t,

$$\alpha^2 + 2\alpha at + a^2 t^2 + \beta^2 + 2\beta bt + b^2 t^2 = t^2 + 1$$

$$(a^2 + b^2 - 1)t^2 + 2(\alpha a + \beta b)t + (\alpha^2 + \beta^2 - 1) = 0$$

This equality holds for all t if and only if all the coefficients are zero. That is, if and only if

$$\begin{cases} a^2 + b^2 - 1 = 0 \\ \alpha a + \beta b = 0 \\ \alpha^2 + \beta^2 - 1 = 0 \end{cases}$$

The first and the third equations imply that (a, b) and (α, β) are points on the unit circle $x^2 + y^2 = 1$. The second equation implies that the vector $\mathbf{u} = \langle a, b \rangle$ is orthogonal to the vector $\mathbf{v} = \langle \alpha, \beta \rangle$ (since $\mathbf{u} \cdot \mathbf{v} = a\alpha + b\beta = 0$).

Conclusions: There are precisely two lines through P entirely contained in S. For the direction vectors $(a, b, 1)$ of these lines, (a, b) is obtained by rotating (α, β) through $\pm \frac{\pi}{2}$ about the origin.

In Exercises 52 and 53, let C be a curve in $\mathbf{R}^3$ not passing through the origin. The cone on C is the surface consisting of all lines passing through the origin and a point on C [Figure 19(A)].

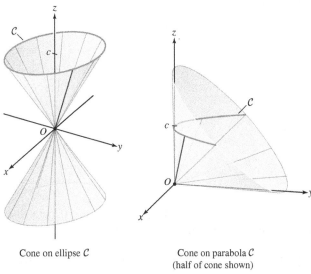

Cone on ellipse $\mathcal{C}$

Cone on parabola $\mathcal{C}$
(half of cone shown)

FIGURE 19

53. Let a and c be nonzero constants and let C be the parabola at height c consisting of all points (x, ax^2, c) [Figure 19(B)]. Let S be the cone consisting of all lines passing through the origin and a point on C. This exercise shows that S is also an elliptic cone.

(a) Show that S has equation $yz = acx^2$.

(b) Show that under the change of variables $y = u + v$ and $z = u - v$, this equation becomes $acx^2 = u^2 - v^2$ or $u^2 = acx^2 + v^2$ (the equation of an elliptic cone in the variables x, v, u).

SOLUTION

(a) A point P on the parabola C has the form $P = \left(x_0, ax_0^2, c\right)$, hence the parametric equations of the line through the origin and P are

$$x = tx_0, \quad y = tax_0^2, \quad z = tc$$

Then

$$yz = tax_0^2 ct = ac(tx_0)^2 = acx^2$$

(b) Define new variables $z = u - v$ and $y = u + v$. The equation in part (a) becomes

$$(u + v)(u - v) = acx^2$$
$$u^2 - v^2 = acx^2 \quad \Rightarrow \quad u^2 = acx^2 + v^2$$

This is the equation of an elliptic cone in the variables x, v, u. We, thus, showed that the cone on the parabola C is transformed to an elliptic cone by the transformation (change of variables) $y = u + v, z = u - v, x = x$.

12.7 Cylindrical and Spherical Coordinates (LT Section 13.7)

Preliminary Questions

1. Describe the surfaces $r = R$ in cylindrical coordinates and $\rho = R$ in spherical coordinates.

SOLUTION The surface $r = R$ consists of all points located at a distance R from the z-axis. This surface is the cylinder of radius R whose axis is the z-axis. The surface $\rho = R$ consists of all points located at a distance R from the origin. This is the sphere of radius R centered at the origin.

2. Which statement about cylindrical coordinates is correct?

(a) If $\theta = 0$, then P lies on the z-axis.

(b) If $\theta = 0$, then P lies in the xz-plane.

SOLUTION The equation $\theta = 0$ defines the half-plane of all points that project onto the ray $\theta = 0$, that is, onto the nonnegative x-axis. This half plane is part of the (x, z)-plane, therefore if $\theta = 0$, then P lies in the (x, z)-plane.

The half-plane $\theta = 0$

For instance, the point $P = (1, 0, 1)$ satisfies $\theta = 0$, but it does not lie on the z-axis. We conclude that statement (b) is correct and statement (a) is false.

3. Which statement about spherical coordinates is correct?

(a) If $\phi = 0$, then P lies on the z-axis.

(b) If $\phi = 0$, then P lies in the xy-plane.

SOLUTION The equation $\phi = 0$ describes the nonnegative z-axis. Therefore, if $\phi = 0$, P lies on the z-axis as stated in (a). Statement (b) is false, since the point $(0, 0, 1)$ satisfies $\phi = 0$, but it does not lie in the (x, y)-plane.

4. The level surface $\phi = \phi_0$ in spherical coordinates, usually a cone, reduces to a half-line for two values of ϕ_0. Which two values?

SOLUTION For $\phi_0 = 0$, the level surface $\phi = 0$ is the upper part of the z-axis. For $\phi_0 = \pi$, the level surface $\phi = \pi$ is the lower part of the z-axis. These are the two values of ϕ_0 where the level surface $\phi = \phi_0$ reduces to a half-line.

5. For which value of ϕ_0 is $\phi = \phi_0$ a plane? Which plane?

SOLUTION For $\phi_0 = \frac{\pi}{2}$, the level surface $\phi = \frac{\pi}{2}$ is the xy-plane.

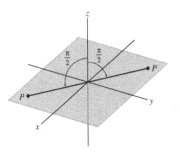

Exercises

In Exercises 1–4, convert from cylindrical to rectangular coordinates.

1. $(4, \pi, 4)$

SOLUTION By the given data $r = 4$, $\theta = \pi$ and $z = 4$. Hence,

$$x = r\cos\theta = 4\cos\pi = 4 \cdot (-1) = -4$$
$$y = r\sin\theta = 4\sin\pi = 4 \cdot 0 \quad \Rightarrow \quad (x, y, z) = (-4, 0, 4)$$
$$z = 4$$

3. $\left(0, \dfrac{\pi}{5}, \dfrac{1}{2}\right)$

SOLUTION We have $r = 0$, $\theta = \dfrac{\pi}{5}$, $z = \dfrac{1}{2}$. Thus,

$$x = r\cos\theta = 0 \cdot \cos\frac{\pi}{5} = 0$$
$$y = r\sin\theta = 0 \cdot \sin\frac{\pi}{5} = 0 \quad \Rightarrow \quad (x, y, z) = \left(0, 0, \frac{1}{2}\right)$$
$$z = \frac{1}{2}$$

In Exercises 5–10, convert from rectangular to cylindrical coordinates.

5. $(1, -1, 1)$

SOLUTION We are given that $x = 1$, $y = -1$, $z = 1$. We find r:

$$r = \sqrt{x^2 + y^2} = \sqrt{1^2 + (-1)^2} = \sqrt{2}$$

Next we find θ. The point $(x, y) = (1, -1)$ lies in the fourth quadrant, hence,

$$\tan\theta = \frac{y}{x} = \frac{-1}{1} = -1, \quad \frac{3\pi}{2} \le \theta \le 2\pi \quad \Rightarrow \quad \theta = \frac{7\pi}{4}$$

We conclude that the cylindrical coordinates of the point are

$$(r, \theta, z) = \left(\sqrt{2}, \frac{7\pi}{4}, 1\right)$$

7. $(1, \sqrt{3}, 7)$

SOLUTION We have $x = 1$, $y = \sqrt{3}$, $z = 7$. We first find r:

$$r = \sqrt{x^2 + y^2} = \sqrt{1^2 + \left(\sqrt{3}\right)^2} = 2$$

Since the point $(x, y) = \left(1, \sqrt{3}\right)$ lies in the first quadrant, $0 \le \theta \le \dfrac{\pi}{2}$. Hence,

$$\tan\theta = \frac{y}{x} = \frac{\sqrt{3}}{1} = \sqrt{3}, \quad 0 \le \theta \le \frac{\pi}{2} \quad \Rightarrow \quad \theta = \frac{\pi}{3}$$

The cylindrical coordinates are thus

$$(r, \theta, z) = \left(2, \frac{\pi}{3}, 7\right)$$

9. $\left(\dfrac{5}{\sqrt{2}}, \dfrac{5}{\sqrt{2}}, 2\right)$

SOLUTION We have $x = \dfrac{5}{\sqrt{2}}, y = \dfrac{5}{\sqrt{2}}, z = 2$. We find r:

$$r = \sqrt{x^2 + y^2} = \sqrt{\left(\dfrac{5}{\sqrt{2}}\right)^2 + \left(\dfrac{5}{\sqrt{2}}\right)^2} = \sqrt{25} = 5$$

Since the point $(x, y) = \left(\dfrac{5}{\sqrt{2}}, \dfrac{5}{\sqrt{2}}\right)$ is in the first quadrant, $0 \le \theta \le \dfrac{\pi}{2}$, therefore,

$$\tan\theta = \dfrac{y}{x} = \dfrac{5/\sqrt{2}}{5/\sqrt{2}} = 1, \quad 0 \le \theta \le \dfrac{\pi}{2} \quad \Rightarrow \quad \theta = \dfrac{\pi}{4}$$

The corresponding cylindrical coordinates are

$$(r, \theta, z) = \left(5, \dfrac{\pi}{4}, 2\right)$$

In Exercises 11–16, describe the set in cylindrical coordinates.

11. $x^2 + y^2 \le 3$

SOLUTION The inequality describes a solid cylinder of radius 3 centered on the z-axis. Since $x^2 + y^2 = r^2$, this inequality can be written as $r^2 \le 3$.

13. $y^2 + z^2 \le 4, \quad x = 0$

SOLUTION The projection of the points in this set onto the xy-plane are points on the y axis, thus $\theta = \dfrac{\pi}{2}$ or $\theta = \dfrac{3\pi}{2}$. Therefore, $y = r\sin\dfrac{\pi}{2} = r \cdot 1 = r$ or $y = r\sin\left(\dfrac{3\pi}{2}\right) = -r$. In both cases, $y^2 = r^2$, thus the inequality $y^2 + z^2 \le 4$ becomes $r^2 + z^2 \le 4$. In cylindrical coordinates, we obtain the following inequality

$$r^2 + z^2 \le 4, \quad \theta = \dfrac{\pi}{2} \quad \text{or} \quad \theta = \dfrac{3\pi}{2}$$

15. $x^2 + y^2 \le 9, \quad x \ge y$

SOLUTION The equation $x^2 + y^2 \le 9$ in cylindrical coordinates becomes $r^2 \le 9$, which becomes $r \le 3$. However, we also have the restriction that $x \ge y$. This means that the projection of our set onto the xy plane is below and to the right of the line $y = x$. In other words, our θ is restricted to $-3\pi/4 \le \theta \le \pi/4$. In conclusion, the answer is:

$$r \le 3, \quad -3\pi/4 \le \theta \le \pi/4$$

In Exercises 17–26, sketch the set (described in cylindrical coordinates).

17. $r = 4$

SOLUTION The surface $r = 4$ consists of all points located at a distance 4 from the z-axis. It is a cylinder of radius 4 whose axis is the z-axis. The cylinder is shown in the following figure:

19. $z = -2$

SOLUTION $z = -2$ is the horizontal plane at height -2, shown in the following figure:

21. $1 \leq r \leq 3, \quad 0 \leq z \leq 4$

SOLUTION The region $1 \leq r \leq 3, 0 \leq z \leq 4$ is shown in the following figure:

23. $r = \sin\theta$ (*Hint:* Convert to rectangular.)

SOLUTION To convert to rectangular coordinates, multiply both sides by r, giving $r^2 = r\sin\theta = y$, so that $x^2 + y^2 = y$. This simplifies to $x^2 + \left(y - \frac{1}{2}\right)^2 = \frac{1}{4}$, which is a cylinder oriented in the z direction whose base in the xy-plane is the circle of radius $\frac{1}{2}$ centered at $\left(0, \frac{1}{2}\right)$. A sketch of the surface is below:

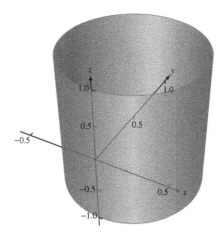

25. $z^2 + r^2 \leq 4$

SOLUTION The region $z^2 + r^2 \leq 4$ is shown in the following figure:

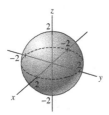

In rectangular coordinates the inequality is $z^2 + \left(x^2 + y^2\right) \leq 4$, or $x^2 + y^2 + z^2 \leq 4$, which is a ball of radius 2.

In Exercises 27–32, find an equation of the form $r = f(\theta, z)$ in cylindrical coordinates for the following surfaces.

27. $z = x + y$

SOLUTION We substitute $x = r\cos\theta$, $y = r\sin\theta$ to obtain the following equation in cylindrical coordinates:

$$z = r\cos\theta + r\sin\theta$$
$$z = r(\cos\theta + \sin\theta) \quad \Rightarrow \quad r = \frac{z}{\cos\theta + \sin\theta}$$

29. $\dfrac{x^2}{yz} = 1$

SOLUTION We rewrite the equation in the form

$$\frac{x}{\frac{y}{x}z} = 1$$

Substituting $x = r\cos\theta$ and $\frac{y}{x} = \tan\theta$ we get

$$\frac{r\cos\theta}{(\tan\theta)z} = 1$$
$$r = \frac{z\tan\theta}{\cos\theta}$$

31. $x^2 + y^2 = 4$

SOLUTION Since $x^2 + y^2 = r^2$, the equation in cylindrical coordinates is, $r^2 = 4$ or $r = 2$.

In Exercises 33–38, convert from spherical to rectangular coordinates.

33. $\left(3, 0, \dfrac{\pi}{2}\right)$

SOLUTION We are given that $\rho = 3$, $\theta = 0$, $\phi = \frac{\pi}{2}$. Using the relations between spherical and rectangular coordinates we have

$$x = \rho\sin\phi\cos\theta = 3\sin\frac{\pi}{2}\cos 0 = 3 \cdot 1 \cdot 1 = 3$$
$$y = \rho\sin\phi\sin\theta = 3\sin\frac{\pi}{2}\sin 0 = 3 \cdot 1 \cdot 0 = 0 \quad \Rightarrow \quad (x, y, z) = (3, 0, 0)$$
$$z = \rho\cos\phi = 3\cos\frac{\pi}{2} = 3 \cdot 0 = 0$$

35. $(3, \pi, 0)$

SOLUTION We have $\rho = 3$, $\theta = \pi$, $\phi = 0$. Hence,

$$x = \rho\sin\phi\cos\theta = 3\sin 0\cos\pi = 0$$
$$y = \rho\sin\phi\sin\theta = 3\sin 0\sin\pi = 0 \quad \Rightarrow \quad (x, y, z) = (0, 0, 3)$$
$$z = \rho\cos\phi = 3\cos 0 = 3$$

37. $\left(6, \dfrac{\pi}{6}, \dfrac{5\pi}{6}\right)$

SOLUTION Since $\rho = 6$, $\theta = \dfrac{\pi}{6}$, and $\phi = \dfrac{5\pi}{6}$ we get

$$x = \rho \sin\phi \cos\theta = 6\sin\dfrac{5\pi}{6}\cos\dfrac{\pi}{6} = 6 \cdot \dfrac{1}{2} \cdot \dfrac{\sqrt{3}}{2} = \dfrac{3\sqrt{3}}{2}$$

$$y = \rho \sin\phi \sin\theta = 6\sin\dfrac{5\pi}{6}\sin\dfrac{\pi}{6} = 6 \cdot \dfrac{1}{2} \cdot \dfrac{1}{2} = \dfrac{3}{2} \quad \Rightarrow \quad (x, y, z) = \left(\dfrac{3\sqrt{3}}{2}, \dfrac{3}{2}, -3\sqrt{3}\right)$$

$$z = \rho \cos\phi = 6\cos\dfrac{5\pi}{6} = 6 \cdot \left(-\dfrac{\sqrt{3}}{2}\right) = -3\sqrt{3}$$

In Exercises 39–44, convert from rectangular to spherical coordinates.

39. $(\sqrt{3}, 0, 1)$

SOLUTION By the given data $x = \sqrt{3}$, $y = 0$, and $z = 1$. We find the radial coordinate:

$$\rho = \sqrt{x^2 + y^2 + z^2} = \sqrt{\left(\sqrt{3}\right)^2 + 0^2 + 1^2} = 2$$

The angular coordinate θ satisfies

$$\tan\theta = \dfrac{y}{x} = \dfrac{0}{\sqrt{3}} = 0 \quad \Rightarrow \quad \theta = 0 \quad \text{or} \quad \theta = \pi$$

Since the point $(x, y) = \left(\sqrt{3}, 0\right)$ lies in the first quadrant, the correct choice is $\theta = 0$. The angle of declination ϕ satisfies

$$\cos\phi = \dfrac{z}{\rho} = \dfrac{1}{2}, \quad 0 \le \phi \le \pi \quad \Rightarrow \quad \phi = \dfrac{\pi}{3}$$

The spherical coordinates of the given points are thus

$$(\rho, \theta, \phi) = \left(2, 0, \dfrac{\pi}{3}\right)$$

41. $(1, 1, 1)$

SOLUTION We have $x = y = z = 1$. The radial coordinate is

$$\rho = \sqrt{x^2 + y^2 + z^2} = \sqrt{1^2 + 1^2 + 1^2} = \sqrt{3}$$

The angular coordinate θ is determined by $\tan\theta = \dfrac{y}{x} = \dfrac{1}{1} = 1$ and by the quadrant of the point $(x, y) = (1, 1)$, that is, $\theta = \dfrac{\pi}{4}$. The angle of declination ϕ satisfies

$$\cos\phi = \dfrac{z}{\rho} = \dfrac{1}{\sqrt{3}}, \quad 0 \le \phi \le \pi \quad \Rightarrow \quad \phi = 0.955$$

The spherical coordinates are thus

$$\left(\sqrt{3}, \dfrac{\pi}{4}, 0.955\right)$$

43. $\left(\frac{1}{2}, \frac{\sqrt{3}}{2}, \sqrt{3}\right)$

SOLUTION We have $x = \frac{1}{2}$, $y = \frac{\sqrt{3}}{2}$, and $z = \sqrt{3}$. Thus

$$\rho = \sqrt{x^2 + y^2 + z^2} = \sqrt{\left(\frac{1}{2}\right)^2 + \left(\frac{\sqrt{3}}{2}\right)^2 + \left(\sqrt{3}\right)^2} = 2$$

The angular coordinate θ satisfies $0 \le \theta \le \frac{\pi}{2}$, since the point $(x, y) = \left(\frac{1}{2}, \frac{\sqrt{3}}{2}\right)$ is in the first quadrant. Also $\tan \theta = \frac{y}{x} = \frac{\sqrt{3}/2}{1/2} = \sqrt{3}$, hence the angle is $\theta = \frac{\pi}{3}$. The angle of declination ϕ satisfies

$$\cos \phi = \frac{z}{\rho} = \frac{\sqrt{3}}{2}, \quad 0 \le \phi \le \pi \quad \Rightarrow \quad \phi = \frac{\pi}{6}$$

We conclude that

$$(\rho, \theta, \phi) = \left(2, \frac{\pi}{3}, \frac{\pi}{6}\right)$$

In Exercises 45 and 46, convert from cylindrical to spherical coordinates.

45. $(2, 0, 2)$

SOLUTION We are given that $r = 2, \theta = 0, z = 2$. Using the conversion formulas, we have

$$\rho = \sqrt{x^2 + y^2 + z^2} = \sqrt{r^2 + z^2} = \sqrt{2^2 + 2^2} = 2\sqrt{2}$$
$$\theta = \theta = 0$$
$$\phi = \cos^{-1}(z/\rho) = \cos^{-1}(2/(2\sqrt{2})) = \pi/4$$

In Exercises 47 and 48, convert from spherical to cylindrical coordinates.

47. $(4, 0, \frac{\pi}{4})$

SOLUTION We are given that $\rho = 4, \theta = 0$, and $\phi = \pi/4$. To find r, we use the formulas $x = r \cos \theta$ and $x = \rho \cos \theta \sin \phi$ to get $r \cos \theta = \rho \cos \theta \sin \phi$, and so

$$r = \rho \sin \phi = 4 \sin \pi/4 = 2\sqrt{2}$$

Clearly $\theta = 0$, and as for z,

$$z = \rho \cos \phi = 4 \cos \pi/4 = 2\sqrt{2}$$

So, in cylindrical coordinates, our point is $(2\sqrt{2}, 0, 2\sqrt{2})$

In Exercises 49–54, describe the given set in spherical coordinates.

49. $x^2 + y^2 + z^2 \le 100$

SOLUTION Substituting $\rho^2 = x^2 + y^2 + z^2$ we obtain $\rho^2 \le 100$ or $\rho \le 10$.

51. $x^2 + y^2 + z^2 = 10$, $x \ge 0$, $y \ge 0$, $z \ge 0$

SOLUTION Since $\rho^2 = x^2 + y^2 + z^2$, we get $\rho^2 = 10$ or $\rho = \sqrt{10}$. The inequalities $x \ge 0$, $y \ge 0$ determine the first quadrant, which is also determined by $0 \le \theta \le \frac{\pi}{2}$. Finally, $z \ge 0$ gives $\cos \phi = \frac{z}{\rho} \ge 0$. Also $0 \le \phi \le \pi$ by definition, hence $0 \le \phi \le \frac{\pi}{2}$. We obtain the following description:

$$\rho = \sqrt{10}, \quad 0 \le \theta \le \frac{\pi}{2}, \quad 0 \le \phi \le \frac{\pi}{2}$$

53. $y^2 + z^2 \leq 4$, $x = 0$

SOLUTION We substitute $y = \rho \sin \theta \sin \phi$ and $z = \rho \cos \phi$ in the given inequality. This gives

$$4 \geq \rho^2 \sin^2 \theta \sin^2 \phi + \rho^2 \cos^2 \phi \tag{1}$$

The equality $x = 0$ determines that $\theta = \frac{\pi}{2}$ or $\theta = \frac{3\pi}{2}$ (and the origin). In both cases, $\sin^2 \theta = 1$. Hence by (1) we get

$$\rho^2 \sin^2 \phi + \rho^2 \cos^2 \phi \leq 4$$
$$\rho^2(1) \leq 4$$
$$\rho \leq 2$$

We obtain the following description:

$$\left\{ (\rho, \theta, \phi) : 0 \leq \rho \leq 2, \theta = \frac{\pi}{2} \text{ or } \theta = \frac{3\pi}{2} \right\}$$

In Exercises 55–64, sketch the set of points (described in spherical coordinates).

55. $\rho = 4$

SOLUTION $\rho = 4$ describes the sphere of radius 4. This is shown in the following figure:

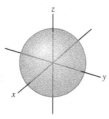

57. $\rho = 2$, $\theta = \frac{\pi}{4}$

SOLUTION The equation $\rho = 2$ is a sphere of radius 2, and the equation $\theta = \frac{\pi}{4}$ is the vertical plane $y = x$. These two surfaces intersect in a (vertical) circle of radius 2, as seen here.

59. $\rho = 2$, $0 \leq \phi \leq \frac{\pi}{2}$

SOLUTION The set

$$\rho = 2, \quad 0 \leq \phi \leq \frac{\pi}{2}$$

is shown in the following figure:

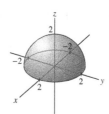

It is the upper half of the sphere with radius 2.

61. $\rho \leq 2, \quad 0 \leq \theta \leq \dfrac{\pi}{2}, \quad \dfrac{\pi}{2} \leq \phi \leq \pi$

SOLUTION This set is the part of the ball of radius 2 which is below the first quadrant of the xy-plane, as shown in the following figure:

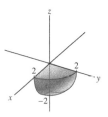

63. $\rho = \csc \phi$

SOLUTION Multiplying both sides by $\sin \phi$ gives $\rho \sin \phi = 1$; since $\rho^2 \sin^2 \phi = x^2 + y^2$, this is the equation of the surface $x^2 + y^2 = 1$, which is a cylinder oriented parallel to the z-axis having cross-section the circle of radius 1 centered at the origin. This set is shown in the following figure:

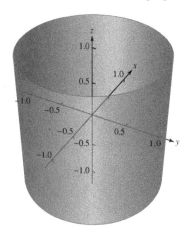

In Exercises 65–72, find an equation of the form $\rho = f(\theta, \phi)$ in spherical coordinates for the following surfaces.

65. $x^2 + y^2 = 9$

SOLUTION Since $x = \rho \cos \theta \sin \phi$, $y = \rho \sin \theta \sin \phi$, we get

$$x^2 + y^2 = \rho^2 \cos^2 \theta \sin^2 \phi + \rho^2 \sin^2 \theta \sin^2 \phi = \rho^2 \sin^2 \phi (\cos^2 \theta + \sin^2 \theta) = \rho^2 \sin^2 \phi$$

Thus the equation in spherical coordinates is $\rho^2 \sin^2 \phi = 9$. Since $\sin \phi$ is nonnegative for $0 \leq \phi \leq \pi$, this can be written $\rho \sin \phi = 3$, or $\rho = 3 \csc \phi$.

67. $z = 2$

SOLUTION Since $z = \rho \cos \phi$, we have $\rho \cos \phi = 2$, or $\rho = \dfrac{2}{\cos \phi}$.

69. $x = z^2$

SOLUTION Substituting $x = \rho \cos \theta \sin \phi$ and $z = \rho \cos \phi$ we obtain

$$\rho \cos \theta \sin \phi = \rho^2 \cos^2 \phi$$

$$\cos \theta \sin \phi = \rho \cos^2 \phi$$

$$\rho = \dfrac{\cos \theta \sin \phi}{\cos^2 \phi} = \dfrac{\cos \theta \tan \phi}{\cos \phi}$$

71. $x^2 - y^2 = 4$

SOLUTION We substitute $x = \rho \cos \theta \sin \phi$ and $y = \rho \sin \theta \sin \phi$ to obtain

$$4 = \rho^2 \cos^2 \theta \sin^2 \phi - \rho^2 \sin^2 \theta \sin^2 \phi = \rho^2 \sin^2 \phi (\cos^2 \theta - \sin^2 \theta)$$

Using the identity $\cos^2 \theta - \sin^2 \theta = \cos 2\theta$ we get

$$4 = \rho^2 \sin^2 \phi \cos 2\theta$$

$$\rho^2 = \frac{4}{\sin^2 \phi \cos 2\theta}$$

We take the square root of both sides. Since $0 < \phi < \pi$ we have $\sin \phi > 0$, hence,

$$\rho = \frac{2}{\sin \phi \sqrt{\cos 2\theta}}$$

73. Which of (a)–(c) is the equation of the cylinder of radius R in spherical coordinates? Refer to Figure 17.

(a) $R\rho = \sin \phi$ **(b)** $\rho \sin \phi = R$ **(c)** $\rho = R \sin \phi$

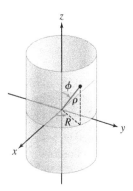

FIGURE 17

SOLUTION The equation of the cylinder of radius R in rectangular coordinates is $x^2 + y^2 = R^2$ (z is unlimited). Substituting the formulas for x and y in terms of ρ, θ, and ϕ yields

$$R^2 = \rho^2 \cos^2 \theta \sin^2 \phi + \rho^2 \sin^2 \theta \sin^2 \phi = \rho^2 \sin^2 \phi (\cos^2 \theta + \sin^2 \theta) = \rho^2 \sin^2 \phi$$

Hence,

$$R^2 = \rho^2 \sin^2 \phi$$

We take the square root of both sides. Since $0 \leq \phi \leq \pi$, we have $\sin \phi \geq 0$, therefore,

$$R = \rho \sin \phi$$

Equation (b) is the correct answer.

75. Find the spherical angles (θ, ϕ) for Helsinki, Finland (60.1° N, 25.0° E), and São Paulo, Brazil (23.52° S, 46.52° W).

SOLUTION For Helsinki, θ is 25° and ϕ is $90 - 60.1 = 29.9°$.
For São Paulo, θ is $360 - 46.52 = 313.48°$ and ϕ is $90 + 23.52 = 113.52°$.

77. Consider a rectangular coordinate system with its origin at the center of the earth, z-axis through the North Pole, and x-axis through the prime meridian. Find the rectangular coordinates of Sydney, Australia (34° S, 151° E), and Bogotá, Colombia (4° 32′ N, 74° 15′ W). A minute is 1/60°. Assume that the earth is a sphere of radius $R = 6370$ km.

SOLUTION We first find the angle (θ, ϕ) for the two towns. For Sydney $\theta = 151°$, since its longitude lies to the east of Greenwich, that is, in the positive θ direction. Sydney's latitude is south of the equator, hence $\phi = 90 + 34 = 124°$.

For Bogota, we have $\theta = 360° - 74°15′ = 285°45′$, since $74°15′W$ refers to $74°15′$ in the negative θ direction. The latitude is north of the equator hence $\phi = 90° - 4°32′ = 85°28′$.

We now use the formulas of x, y, and z in terms of ρ, θ, ϕ to find the rectangular coordinates of the two towns. (Notice that $285°45' = 285.75°$ and $85°28' = 85.47°$.)

Sydney:
$$x = \rho \cos\theta \sin\phi = 6367 \cos 151° \sin 124° = -4616.7$$
$$y = \rho \sin\theta \sin\phi = 6367 \sin 151° \sin 124° = 2559$$
$$z = \rho \cos\phi = 6367 \cos 124° = -3560.4$$

Bogota:
$$x = \rho \cos\theta \sin\phi = 6367 \cos 285.75° \sin 85.47° = 1722.9$$
$$y = \rho \sin\theta \sin\phi = 6367 \sin 285.75° \sin 85.47° = -6108.8$$
$$z = \rho \cos\phi = 6367 \cos 85.47° = 502.9$$

79. Find an equation of the form $z = f(r, \theta)$ in cylindrical coordinates for $z^2 = x^2 - y^2$.

SOLUTION In cylindrical coordinates, $x = r\cos\theta$ and $y = r\sin\theta$. Hence,
$$z^2 = x^2 - y^2 = r^2 \cos^2\theta - r^2 \sin^2\theta$$

We use the identity $\cos^2\theta - \sin^2\theta = \cos 2\theta$ to obtain
$$z^2 = r^2 \cos 2\theta \quad \Rightarrow \quad z = \pm r\sqrt{\cos 2\theta}$$

81. An apple modeled by taking all the points in and on a sphere of radius 2 inches is cored with a vertical cylinder of radius 1 inch. Use inequalities in cylindrical coordinates to describe the set of all points that remain in the apple once the core is removed.

SOLUTION The sphere together with its interior is, in rectangular coordinates, the set of points with $x^2 + y^2 + z^2 \le 4$. In cylindrical coordinates, this is the set of points with $r^2 + z^2 \le 4$. So we can parametrize the sphere and its interior as $-2 \le z \le 2$ and $0 \le r \le \sqrt{4 - z^2}$. The vertical cylinder together with its interior is parametrized by $r \le 1$. The cylinder intersects the sphere when $1^2 + z^2 = 4$, or $z = \pm\sqrt{3}$. When $|z| > \sqrt{3}$, all points in the sphere lie inside the cylinder, so are gone when the cylinder is removed. So removing the cylinder from the sphere gives a set of points parametrized as $-\sqrt{3} < z < \sqrt{3}$, $1 < r \le \sqrt{4 - z^2}$.

83. Explain the following statement: If the equation of a surface in cylindrical or spherical coordinates does not involve the coordinate θ, then the surface is rotationally symmetric with respect to the z-axis.

SOLUTION Since the equation of the surface does not involve the coordinate θ, then for every point P on the surface ($P = (\rho_0, \theta_0, \phi_0)$ in spherical coordinates or $P = (r_0, \theta_0, z_0)$ in cylindrical coordinates) so also all the points (ρ_0, θ, ϕ_0) or (r_0, θ, z_0) are on the surface. That is, all the points obtained by rotating P around the z-axis are on the surface. Hence, the surface is rotationally symmetric with respect to the z-axis.

85. Find equations $r = g(\theta, z)$ (cylindrical) and $\rho = f(\theta, \phi)$ (spherical) for the hyperboloid $x^2 + y^2 = z^2 + 1$ (Figure 18). Do there exist points on the hyperboloid with $\phi = 0$ or π? Which values of ϕ occur for points on the hyperboloid?

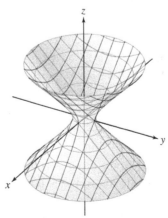

FIGURE 18 The hyperboloid $x^2 + y^2 = z^2 + 1$.

SOLUTION For the cylindrical coordinates (r, θ, z) we have $x^2 + y^2 = r^2$. Substituting into the equation $x^2 + y^2 = z^2 + 1$ gives

$$r^2 = z^2 + 1 \quad \Rightarrow \quad r = \sqrt{z^2 + 1}$$

For the spherical coordinates (ρ, θ, ϕ) we have $x = \rho \sin\phi \cos\theta$, $y = \rho \sin\phi \sin\theta$ and $z = \rho \cos\phi$. We substitute into the equation of the hyperboloid $x^2 + y^2 = z^2 + 1$ and simplify to obtain

$$\rho^2 \sin^2\phi \cos^2\theta + \rho^2 \sin^2\phi \sin^2\theta = \rho^2 \cos^2\phi + 1$$

$$\rho^2 \sin^2\phi(\cos^2\theta + \sin^2\theta) = \rho^2 \cos^2\phi + 1$$

$$\rho^2(\sin^2\phi - \cos^2\phi) = 1$$

Using the trigonometric identity $\cos 2\phi = \cos^2\phi - \sin^2\phi$ we get

$$\rho^2 \cdot (-\cos 2\phi) = 1 \quad \Rightarrow \quad \rho = \sqrt{-\frac{1}{\cos 2\phi}}$$

For $\phi = 0$ and $\phi = \pi$ we have $\cos 2 \cdot 0 = 1$ and $\cos 2\pi = 1$. In both cases $-\frac{1}{\cos 2\phi} = -1 < 0$, hence there is no real value of ρ satisfying $\rho = \sqrt{-\frac{1}{\cos 2\phi}}$. We conclude that there are no points on the hyperboloid with $\phi = 0$ or π.

To obtain a real ρ such that $\rho = \sqrt{-\frac{1}{\cos 2\phi}}$, we must have $-\frac{1}{\cos 2\phi} > 0$. That is, $\cos 2\phi < 0$ (and of course $0 \le \phi \le \pi$). The corresponding values of ϕ are

$$\frac{\pi}{2} < 2\phi \le \frac{3\pi}{2} \quad \Rightarrow \quad \frac{\pi}{4} < \phi \le \frac{3\pi}{4}$$

Further Insights and Challenges

*In Exercises 86–90, a **great circle** on a sphere S with center O and radius R is a circle obtained by intersecting S with a plane that passes through O (Figure 19). If P and Q are not antipodal (on opposite sides), there is a unique great circle through P and Q on S (intersect S with the plane through $O, P,$ and Q). The geodesic distance from P to Q is defined as the length of the smaller of the two circular arcs of this great circle.*

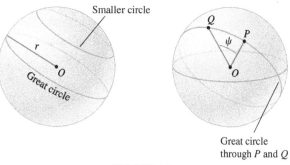

FIGURE 19

87. Show that the geodesic distance from $Q = (a, b, c)$ to the North Pole $P = (0, 0, R)$ is equal to $R \cos^{-1}\left(\frac{c}{R}\right)$.

SOLUTION Let ψ be the central angle between P and Q, that is, the angle between the vectors $\mathbf{v} = \overrightarrow{OP}$ and $\mathbf{u} = \overrightarrow{OQ}$. By Exercise 70 the geodesic distance from P to Q is $R\psi$. We find ψ. By the formula for the cosine of the angle between two vectors, we have

$$\cos\psi = \frac{\mathbf{u} \cdot \mathbf{v}}{\|\mathbf{u}\|\|\mathbf{v}\|} \tag{1}$$

We compute the values in this quotient:

$$\mathbf{u} \cdot \mathbf{v} = \langle 0, 0, R \rangle \cdot \langle a, b, c \rangle = 0 + 0 + Rc = Rc$$

$$\|\mathbf{v}\| = \|\overrightarrow{OP}\| = R$$

$$\|\mathbf{u}\| = \|\overrightarrow{OQ}\| = \sqrt{a^2 + b^2 + c^2} = R$$

Substituting in (1) we get

$$\cos\psi = \frac{Rc}{R^2} = \frac{c}{R} \quad \Rightarrow \quad \psi = \cos^{-1}\left(\frac{c}{R}\right)$$

The geodesic distance from Q to P is thus

$$R\psi = R\cos^{-1}\left(\frac{c}{R}\right)$$

89. Show that the central angle ψ between points P and Q on a sphere (of any radius) with angular coordinates (θ, ϕ) and (θ', ϕ') is equal to

$$\psi = \cos^{-1}(\sin\phi \sin\phi' \cos(\theta - \theta') + \cos\phi \cos\phi')$$

Hint: Compute the dot product of $\overrightarrow{OP}$ and $\overrightarrow{OQ}$. Check this formula by computing the geodesic distance between the North and South Poles.

SOLUTION We denote the vectors $\mathbf{u} = \overrightarrow{OP}$ and $\mathbf{v} = \overrightarrow{OQ}$. By the formula for the angle between two vectors we have

$$\psi = \cos^{-1}\left(\frac{\mathbf{u}\cdot\mathbf{v}}{\|\mathbf{u}\|\|\mathbf{v}\|}\right)$$

Denoting by R the radius of the sphere, we have $\|\mathbf{u}\| = \|\mathbf{v}\| = R$, hence,

$$\psi = \cos^{-1}\left(\frac{\mathbf{u}\cdot\mathbf{v}}{R^2}\right) \tag{1}$$

The rectangular coordinates of $\mathbf{u}$ and $\mathbf{v}$ are

u	v
$x = R\sin\phi\cos\theta$	$x' = R\sin\phi'\cos\theta'$
$y = R\sin\phi\sin\theta$	$y' = R\sin\phi'\sin\theta'$
$z = R\cos\phi$	$z' = R\cos\phi'$

Hence,

$$\mathbf{u}\cdot\mathbf{v} = R^2 \sin\phi\cos\theta\sin\phi'\cos\theta' + R^2 \sin\phi\sin\theta\sin\phi'\sin\theta' + R^2 \cos\phi\cos\phi'$$
$$= R^2 \left[\sin\phi\sin\phi'(\cos\theta\cos\theta' + \sin\theta\sin\theta') + \cos\phi\cos\phi'\right]$$

We use the identity $\cos(\alpha - \beta) = \cos\alpha\cos\beta + \sin\alpha\sin\beta$ to obtain

$$\mathbf{u}\cdot\mathbf{v} = R^2 \left(\sin\phi\sin\phi'\cos(\theta - \theta') + \cos\phi\cos\phi'\right)$$

Substituting in (1) we obtain

$$\psi = \cos^{-1}\left(\sin\phi\sin\phi'\cos(\theta - \theta') + \cos\phi\cos\phi'\right) \tag{2}$$

We now check this formula in the case where P and Q are the North and South Poles, respectively. In this case $\theta = \theta' = 0$, $\phi = 0$, $\phi' = \pi$. Substituting in (2) gives

$$\psi = \cos^{-1}(\sin 0 \sin\pi \cos 0 + \cos 0 \cos\pi) = \cos^{-1}(-1) = \pi$$

Using Exercise 70, the geodesic distance between the two poles is $R\psi = R\pi$, in accordance with the formula for the length of a semicircle.

CHAPTER REVIEW EXERCISES

In Exercises 1–6, let $\mathbf{v} = \langle -2, 5 \rangle$ *and* $\mathbf{w} = \langle 3, -2 \rangle$.

1. Calculate $5\mathbf{w} - 3\mathbf{v}$ and $5\mathbf{v} - 3\mathbf{w}$.

SOLUTION We use the definition of basic vector operations to compute the two linear combinations:

$$5\mathbf{w} - 3\mathbf{v} = 5\langle 3, -2 \rangle - 3\langle -2, 5 \rangle = \langle 15, -10 \rangle + \langle 6, -15 \rangle = \langle 21, -25 \rangle$$

$$5\mathbf{v} - 3\mathbf{w} = 5\langle -2, 5 \rangle - 3\langle 3, -2 \rangle = \langle -10, 25 \rangle + \langle -9, 6 \rangle = \langle -19, 31 \rangle$$

3. Find the unit vector in the direction of $\mathbf{v}$.

SOLUTION The unit vector in the direction of $\mathbf{v}$ is

$$\mathbf{e_v} = \frac{1}{\|\mathbf{v}\|}\mathbf{v}$$

We compute the length of $\mathbf{v}$:

$$\|\mathbf{v}\| = \sqrt{(-2)^2 + 5^2} = \sqrt{29}$$

Hence,

$$\mathbf{e_v} = \frac{\mathbf{v}}{\|\mathbf{v}\|} = \frac{\langle -2, 5 \rangle}{\sqrt{29}} = \left\langle \frac{-2}{\sqrt{29}}, \frac{5}{\sqrt{29}} \right\rangle$$

5. Express $\mathbf{i}$ as a linear combination $r\mathbf{v} + s\mathbf{w}$.

SOLUTION We use basic properties of vector algebra to write

$$\mathbf{i} = r\mathbf{v} + s\mathbf{w} \tag{1}$$

$$\langle 1, 0 \rangle = r\langle -2, 5 \rangle + s\langle 3, -2 \rangle = \langle -2r + 3s, 5r - 2s \rangle$$

The vector are equivalent, hence,

$$1 = -2r + 3s$$
$$0 = 5r - 2s$$

The second equation implies that $s = \frac{5}{2}r$. We substitute in the first equation and solve for r:

$$1 = -2r + 3 \cdot \frac{5}{2}r$$

$$1 = \frac{11}{2}r$$

$$r = \frac{2}{11} \quad \Rightarrow \quad s = \frac{5}{2} \cdot \frac{2}{11} = \frac{5}{11}$$

Substituting in (1) we obtain

$$\mathbf{i} = \frac{2}{11}\mathbf{v} + \frac{5}{11}\mathbf{w}$$

7. If $P = (1, 4)$ and $Q = (-3, 5)$, what are the components of $\overrightarrow{PQ}$? What is the length of $\overrightarrow{PQ}$?

SOLUTION By the Definition of Components of a Vector we have

$$\overrightarrow{PQ} = \langle -3 - 1, 5 - 4 \rangle = \langle -4, 1 \rangle$$

The length of $\overrightarrow{PQ}$ is

$$\|\overrightarrow{PQ}\| = \sqrt{(-4)^2 + 1^2} = \sqrt{17}$$

9. Find the vector with length 3 making an angle of $\frac{7\pi}{4}$ with the positive x-axis.

SOLUTION We denote the vector by $\mathbf{v} = \langle a, b \rangle$. $\mathbf{v}$ makes an angle $\theta = \frac{7\pi}{4}$ with the x-axis, and its length is 3, hence,

$$a = \|\mathbf{v}\| \cos \theta = 3 \cos \frac{7\pi}{4} = \frac{3}{\sqrt{2}}$$

$$b = \|\mathbf{v}\| \sin \theta = 3 \sin \frac{7\pi}{4} = -\frac{3}{\sqrt{2}}$$

That is,

$$\mathbf{v} = \langle a, b \rangle = \left\langle \frac{3}{\sqrt{2}}, -\frac{3}{\sqrt{2}} \right\rangle$$

11. Find the value of β for which $\mathbf{w} = \langle -2, \beta \rangle$ is parallel to $\mathbf{v} = \langle 4, -3 \rangle$.

SOLUTION If $\mathbf{v} = \langle 4, -3 \rangle$ and $\mathbf{w} = \langle -2, \beta \rangle$ are parallel, there exists a scalar λ such that $\mathbf{w} = \lambda \mathbf{v}$. That is,

$$\langle -2, \beta \rangle = \lambda \langle 4, -3 \rangle = \langle 4\lambda, -3\lambda \rangle$$

yielding

$$-2 = 4\lambda \quad \text{and} \quad \beta = -3\lambda$$

These equations imply that $\lambda = -\frac{1}{2}$ and $\lambda = -\frac{\beta}{3}$. Equating the two expressions for λ gives

$$-\frac{1}{2} = -\frac{\beta}{3} \quad \text{or} \quad \beta = \frac{3}{2}$$

13. Let $\mathbf{w} = \langle 2, -2, 1 \rangle$ and $\mathbf{v} = \langle 4, 5, -4 \rangle$. Solve for $\mathbf{u}$ if $\mathbf{v} + 5\mathbf{u} = 3\mathbf{w} - \mathbf{u}$.

SOLUTION Using vector algebra we have

$$\mathbf{v} + 5\mathbf{u} = 3\mathbf{w} - \mathbf{u}$$

$$6\mathbf{u} = 3\mathbf{w} - \mathbf{v}$$

$$\mathbf{u} = \frac{1}{2}\mathbf{w} - \frac{1}{6}\mathbf{v} = \left\langle 1, -1, \frac{1}{2} \right\rangle - \left\langle \frac{4}{6}, \frac{5}{6}, -\frac{4}{6} \right\rangle = \left\langle \frac{1}{3}, -\frac{11}{6}, \frac{7}{6} \right\rangle$$

15. Find a parametrization $\mathbf{r}_1(t)$ of the line passing through $(1, 4, 5)$ and $(-2, 3, -1)$. Then find a parametrization $\mathbf{r}_2(t)$ of the line parallel to $\mathbf{r}_1$ passing through $(1, 0, 0)$.

SOLUTION Since the points $P = (-2, 3, -1)$ and $Q = (1, 4, 5)$ are on the line l_1, the vector $\overrightarrow{PQ}$ is a direction vector for the line. We find this vector:

$$\overrightarrow{PQ} = \langle 1 - (-2), 4 - 3, 5 - (-1) \rangle = \langle 3, 1, 6 \rangle$$

Substituting $\mathbf{v} = \langle 3, 1, 6 \rangle$ and $P_0 = \langle 1, 4, 5 \rangle$ in the vector parametrization of the line we obtain the following equation for l_1:

$$\mathbf{r}_1(t) = \overrightarrow{OP_0} + t\mathbf{v}$$

$$\mathbf{r}_1(t) = \langle 1, 4, 5 \rangle + t \langle 3, 1, 6 \rangle = \langle 1 + 3t, 4 + t, 5 + 6t \rangle$$

The line l_2 is parallel to l_1, hence $\overrightarrow{PQ} = \langle 3, 1, 6 \rangle$ is also a direction vector for l_2. Substituting $\mathbf{v} = \langle 3, 1, 6 \rangle$ and $P_0 = (1, 0, 0)$ in the vector parametrization of the line we obtain the following equation for l_2:

$$\mathbf{r}_2(t) = \overrightarrow{OP_0} + t\mathbf{v}$$

$$\mathbf{r}_2(t) = \langle 1, 0, 0 \rangle + t \langle 3, 1, 6 \rangle = \langle 1 + 3t, t, 6t \rangle$$

17. Find a and b such that the lines $\mathbf{r}_1 = \langle 1, 2, 1 \rangle + t\langle 1, -1, 1 \rangle$ and $\mathbf{r}_2 = \langle 3, -1, 1 \rangle + t\langle a, b, -2 \rangle$ are parallel.

SOLUTION The lines are parallel if and only if the direction vectors $\mathbf{v}_1 = \langle 1, -1, 1 \rangle$ and $\mathbf{v}_2 = \langle a, b, -2 \rangle$ are parallel. That is, if and only if there exists a scalar λ such that:

$$\mathbf{v}_2 = \lambda \mathbf{v}_1$$

$$\langle a, b, -2 \rangle = \lambda \langle 1, -1, 1 \rangle = \langle \lambda, -\lambda, \lambda \rangle$$

We obtain the following equations:

$$\begin{aligned} a &= \lambda \\ b &= -\lambda \\ -2 &= \lambda \end{aligned} \quad \Rightarrow \quad a = -2, \quad b = 2$$

19. Sketch the vector sum $\mathbf{v} = \mathbf{v}_1 - \mathbf{v}_2 + \mathbf{v}_3$ for the vectors in Figure 1(A).

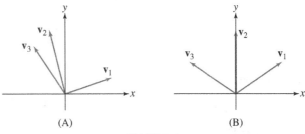

FIGURE 1

SOLUTION Using the Parallelogram Law, we obtain the vector sum shown in the figure.

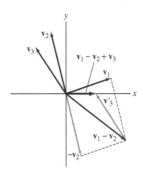

We first add $\mathbf{v}_1$ and $-\mathbf{v}_2$, then we add $\mathbf{v}_3$ to $\mathbf{v}_1 - \mathbf{v}_2$.

In Exercises 21–26, let $\mathbf{v} = \langle 1, 3, -2 \rangle$ and $\mathbf{w} = \langle 2, -1, 4 \rangle$.

21. Compute $\mathbf{v} \cdot \mathbf{w}$.

SOLUTION Using the definition of the dot product we have

$$\mathbf{v} \cdot \mathbf{w} = \langle 1, 3, -2 \rangle \cdot \langle 2, -1, 4 \rangle = 1 \cdot 2 + 3 \cdot (-1) + (-2) \cdot 4 = 2 - 3 - 8 = -9$$

23. Compute $\mathbf{v} \times \mathbf{w}$.

SOLUTION We use the definition of the cross product as a "determinant":

$$\mathbf{v} \times \mathbf{w} = \begin{vmatrix} \mathbf{i} & \mathbf{j} & \mathbf{k} \\ 1 & 3 & -2 \\ 2 & -1 & 4 \end{vmatrix} = \begin{vmatrix} 3 & -2 \\ -1 & 4 \end{vmatrix} \mathbf{i} - \begin{vmatrix} 1 & -2 \\ 2 & 4 \end{vmatrix} \mathbf{j} + \begin{vmatrix} 1 & 3 \\ 2 & -1 \end{vmatrix} \mathbf{k}$$

$$= (12 - 2)\mathbf{i} - (4 + 4)\mathbf{j} + (-1 - 6)\mathbf{k} = 10\mathbf{i} - 8\mathbf{j} - 7\mathbf{k} = \langle 10, -8, -7 \rangle$$

25. Find the volume of the parallelepiped spanned by **v**, **w**, and $\mathbf{u} = \langle 1, 2, 6 \rangle$.

SOLUTION The volume V of the parallelepiped spanned by **v**, **w** and **u** is the following determinant:

$$V = \left| \det \begin{pmatrix} \mathbf{v} \\ \mathbf{w} \\ \mathbf{u} \end{pmatrix} \right| = \begin{vmatrix} 1 & 3 & -2 \\ 2 & -1 & 4 \\ 1 & 2 & 6 \end{vmatrix} = \left| 1 \cdot \begin{vmatrix} -1 & 4 \\ 2 & 6 \end{vmatrix} - 3 \begin{vmatrix} 2 & 4 \\ 1 & 6 \end{vmatrix} - 2 \begin{vmatrix} 2 & -1 \\ 1 & 2 \end{vmatrix} \right|$$

$$= |1 \cdot (-6 - 8) - 3(12 - 4) - 2(4 + 1)| = 48$$

27. Use vectors to prove that the line connecting the midpoints of two sides of a triangle is parallel to the third side.

SOLUTION Let E and F be the midpoints of sides AC and BC in a triangle ABC (see figure).

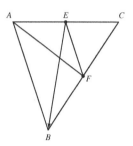

We must show that

$$\overrightarrow{EF} \parallel \overrightarrow{AB}$$

Using the Parallelogram Law we have

$$\overrightarrow{EF} = \overrightarrow{EA} + \overrightarrow{AB} + \overrightarrow{BF} \tag{1}$$

By the definition of the points E and F,

$$\overrightarrow{EA} = \frac{1}{2}\overrightarrow{CA}; \quad \overrightarrow{BF} = \frac{1}{2}\overrightarrow{BC}$$

We substitute (1) to obtain

$$\overrightarrow{EF} = \frac{1}{2}\overrightarrow{CA} + \overrightarrow{AB} + \frac{1}{2}\overrightarrow{BC} = \overrightarrow{AB} + \frac{1}{2}(\overrightarrow{CA} + \overrightarrow{BC})$$

$$= \overrightarrow{AB} + \frac{1}{2}(\overrightarrow{BC} + \overrightarrow{CA}) = \overrightarrow{AB} + \frac{1}{2}\overrightarrow{BA} = \overrightarrow{AB} - \frac{1}{2}\overrightarrow{AB} = \frac{1}{2}\overrightarrow{AB}$$

Therefore, $\overrightarrow{EF}$ is a constant multiple of $\overrightarrow{AB}$, which implies that $\overrightarrow{EF}$ and $\overrightarrow{AB}$ are parallel vectors.

29. Calculate the component of $\mathbf{v} = \langle -2, \frac{1}{2}, 3 \rangle$ along $\mathbf{w} = \langle 1, 2, 2 \rangle$.

SOLUTION We first compute the following dot products:

$$\mathbf{v} \cdot \mathbf{w} = \langle -2, \frac{1}{2}, 3 \rangle \cdot \langle 1, 2, 2 \rangle = 5$$

$$\mathbf{w} \cdot \mathbf{w} = \|\mathbf{w}\|^2 = 1^2 + 2^2 + 2^2 = 9$$

The component of **v** along **w** is the following number:

$$\left\| \left(\frac{\mathbf{v} \cdot \mathbf{w}}{\mathbf{w} \cdot \mathbf{w}} \right) \mathbf{w} \right\| = \frac{5}{9} \|\mathbf{w}\| = \frac{5}{9} \cdot 3 = \frac{5}{3}$$

31. A 50-kg wagon is pulled to the right by a force $\mathbf{F}_1$ making an angle of 30° with the ground. At the same time, the wagon is pulled to the left by a horizontal force $\mathbf{F}_2$.

(a) Find the magnitude of $\mathbf{F}_1$ in terms of the magnitude of $\mathbf{F}_2$ if the wagon does not move.

(b) What is the maximal magnitude of $\mathbf{F}_1$ that can be applied to the wagon without lifting it?

SOLUTION

(a) By Newton's Law, at equilibrium, the total force acting on the wagon is zero.

We resolve the force $\mathbf{F}_1$ into its components:

$$\mathbf{F}_1 = \mathbf{F}_\| + \mathbf{F}_\perp$$

where $\mathbf{F}_\|$ is the horizontal component and $\mathbf{F}_\perp$ is the vertical component. Since the wagon does not move, the magnitude of $\mathbf{F}_\|$ must be equal to the magnitude of $\mathbf{F}_2$. That is,

$$\|\mathbf{F}_\|\| = \|\mathbf{F}_1\| \cos 30° = \|\mathbf{F}_2\|$$

The above equation gives:

$$\|\mathbf{F}_1\| \frac{\sqrt{3}}{2} = \|\mathbf{F}_2\| \quad \Rightarrow \quad \|\mathbf{F}_1\| = \frac{2\|\mathbf{F}_2\|}{\sqrt{3}}$$

(b) The maximum magnitude of force $\mathbf{F}_1$ that can be applied to the wagon without lifting the wagon is found by comparing the vertical forces:

$$\|\mathbf{F}_1\| \sin 30° = 9.8 \cdot 50$$

$$\|\mathbf{F}_1\| \cdot \frac{1}{2} = 9.8 \cdot 50 \quad \Rightarrow \quad \|\mathbf{F}_1\| = 9.8 \cdot 100 = 980 \text{ N}$$

In Exercises 33–36, let $\mathbf{v} = \langle 1, 2, 4 \rangle$, $\mathbf{u} = \langle 6, -1, 2 \rangle$, *and* $\mathbf{w} = \langle 1, 0, -3 \rangle$. *Calculate the given quantity.*

33. $\mathbf{v} \times \mathbf{w}$

SOLUTION We use the definition of the cross product as a determinant to compute $\mathbf{v} \times \mathbf{w}$:

$$\mathbf{v} \times \mathbf{w} = \begin{vmatrix} \mathbf{i} & \mathbf{j} & \mathbf{k} \\ 1 & 2 & 4 \\ 1 & 0 & -3 \end{vmatrix} = \begin{vmatrix} 2 & 4 \\ 0 & -3 \end{vmatrix} \mathbf{i} - \begin{vmatrix} 1 & 4 \\ 1 & -3 \end{vmatrix} \mathbf{j} + \begin{vmatrix} 1 & 2 \\ 1 & 0 \end{vmatrix} \mathbf{k}$$

$$= (-6 - 0)\mathbf{i} - (-3 - 4)\mathbf{j} + (0 - 2)\mathbf{k} = -6\mathbf{i} + 7\mathbf{j} - 2\mathbf{k} = \langle -6, 7, -2 \rangle$$

35. $\det \begin{pmatrix} \mathbf{u} \\ \mathbf{v} \\ \mathbf{w} \end{pmatrix}$

SOLUTION We compute the determinant:

$$\det \begin{pmatrix} \mathbf{u} \\ \mathbf{v} \\ \mathbf{w} \end{pmatrix} = \begin{vmatrix} 6 & -1 & 2 \\ 1 & 2 & 4 \\ 1 & 0 & -3 \end{vmatrix} = 6 \cdot \begin{vmatrix} 2 & 4 \\ 0 & -3 \end{vmatrix} + 1 \cdot \begin{vmatrix} 1 & 4 \\ 1 & -3 \end{vmatrix} + 2 \begin{vmatrix} 1 & 2 \\ 1 & 0 \end{vmatrix}$$

$$= 6 \cdot (-6 - 0) + 1 \cdot (-3 - 4) + 2 \cdot (0 - 2) = -47$$

37. Use the cross product to find the area of the triangle whose vertices are $(1, 3, -1)$, $(2, -1, 3)$, and $(4, 1, 1)$.

SOLUTION Let $A = (1, 3, -1)$, $B = (2, -1, 3)$, and $C = (4, 1, 1)$.

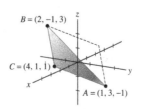

The area S of the triangle ABC is half the area of the parallelogram spanned by $\overrightarrow{AB}$ and $\overrightarrow{AC}$. Using the Formula for the Area of the Parallelogram, we conclude that the area of the triangle is:

$$S = \frac{1}{2} \left\| \overrightarrow{AB} \times \overrightarrow{AC} \right\| \tag{1}$$

We first compute the vectors $\overrightarrow{AB}$ and $\overrightarrow{AC}$:

$$\overrightarrow{AB} = \langle 2-1, -1-3, 3-(-1) \rangle = \langle 1, -4, 4 \rangle$$

$$\overrightarrow{AC} = \langle 4-1, 1-3, 1-(-1) \rangle = \langle 3, -2, 2 \rangle$$

We compute the cross product of the two vectors:

$$\overrightarrow{AB} \times \overrightarrow{AC} = \begin{vmatrix} \mathbf{i} & \mathbf{j} & \mathbf{k} \\ 1 & -4 & 4 \\ 3 & -2 & 2 \end{vmatrix} = \begin{vmatrix} -4 & 4 \\ -2 & 2 \end{vmatrix} \mathbf{i} - \begin{vmatrix} 1 & 4 \\ 3 & 2 \end{vmatrix} \mathbf{j} + \begin{vmatrix} 1 & -4 \\ 3 & -2 \end{vmatrix} \mathbf{k}$$

$$= (-8-(-8))\mathbf{i} - (2-12)\mathbf{j} + (-2-(-12))\mathbf{k}$$

$$= 10\mathbf{j} + 10\mathbf{k} = \langle 0, 10, 10 \rangle = 10\langle 0, 1, 1 \rangle$$

The length of $\overrightarrow{AB} \times \overrightarrow{AC}$ is, thus:

$$\left\| \overrightarrow{AB} \times \overrightarrow{AC} \right\| = \| 10\langle 0, 1, 1 \rangle \| = 10 \| \langle 0, 1, 1 \rangle \| = 10\sqrt{0^2 + 1^2 + 1^2} = 10\sqrt{2}$$

Substituting in (1) gives the following area:

$$S = \frac{1}{2} \cdot 10\sqrt{2} = 5\sqrt{2}$$

39. Show that if the vectors $\mathbf{v}, \mathbf{w}$ are orthogonal, then $\|\mathbf{v} + \mathbf{w}\|^2 = \|\mathbf{v}\|^2 + \|\mathbf{w}\|^2$.

SOLUTION The vectors $\mathbf{v}$ and $\mathbf{w}$ are orthogonal, hence:

$$\mathbf{v} \cdot \mathbf{w} = 0 \tag{1}$$

Using the relation of the dot product with length and properties of the dot product we obtain:

$$\|\mathbf{v} + \mathbf{w}\|^2 = (\mathbf{v} + \mathbf{w}) \cdot (\mathbf{v} + \mathbf{w}) = \mathbf{v} \cdot (\mathbf{v} + \mathbf{w}) + \mathbf{w} \cdot (\mathbf{v} + \mathbf{w})$$

$$= \mathbf{v} \cdot \mathbf{v} + \mathbf{v} \cdot \mathbf{w} + \mathbf{w} \cdot \mathbf{v} + \mathbf{w} \cdot \mathbf{w} = \|\mathbf{v}\|^2 + 2\mathbf{v} \cdot \mathbf{w} + \|\mathbf{w}\|^2 \tag{2}$$

Combining (1) and (2) we get:

$$\|\mathbf{v} + \mathbf{w}\|^2 = \|\mathbf{v}\|^2 + \|\mathbf{w}\|^2$$

41. Find $\|\mathbf{e} - 4\mathbf{f}\|$, assuming that $\mathbf{e}$ and $\mathbf{f}$ are unit vectors such that $\|\mathbf{e} + \mathbf{f}\| = \sqrt{3}$.

SOLUTION We use the relation of the dot product with length and properties of the dot product to write

$$3 = \|\mathbf{e} + \mathbf{f}\|^2 = (\mathbf{e} + \mathbf{f}) \cdot (\mathbf{e} + \mathbf{f}) = \mathbf{e} \cdot \mathbf{e} + \mathbf{e} \cdot \mathbf{f} + \mathbf{f} \cdot \mathbf{e} + \mathbf{f} \cdot \mathbf{f}$$

$$= \|\mathbf{e}\|^2 + 2\mathbf{e} \cdot \mathbf{f} + \|\mathbf{f}\|^2 = 1^2 + 2\mathbf{e} \cdot \mathbf{f} + 1^2 = 2 + 2\mathbf{e} \cdot \mathbf{f}$$

We now find $\mathbf{e} \cdot \mathbf{f}$:

$$3 = 2 + 2\mathbf{e} \cdot \mathbf{f} \quad \Rightarrow \quad \mathbf{e} \cdot \mathbf{f} = 1/2$$

Hence, using the same method as above, we have:

$$\|\mathbf{e} - 4\mathbf{f}\|^2 = (\mathbf{e} - 4\mathbf{f}) \cdot (\mathbf{e} - 4\mathbf{f})$$

$$= \|\mathbf{e}\|^2 - 2 \cdot \mathbf{e} \cdot 4\mathbf{f} + \|4\mathbf{f}\|^2 = 1^2 - 8\mathbf{e} \cdot \mathbf{f} + 4^2 = 17 - 4 = 13$$

Taking square roots, we get:

$$\|\mathbf{e} - 4\mathbf{f}\| = \sqrt{13}$$

43. Show that the equation $\langle 1, 2, 3\rangle \times \mathbf{v} = \langle -1, 2, a\rangle$ has no solution for $a \neq -1$.

SOLUTION By properties of the cross product, the vector $\langle -1, 2, a\rangle$ is orthogonal to $\langle 1, 2, 3\rangle$, hence the dot product of these vectors is zero. That is:

$$\langle -1, 2, a\rangle \cdot \langle 1, 2, 3\rangle = 0$$

We compute the dot product and solve for a:

$$-1 + 4 + 3a = 0$$

$$3a = -3 \quad \Rightarrow \quad a = -1$$

We conclude that if the given equation is solvable, then $a = -1$.

45. Use the identity

$$\mathbf{u} \times (\mathbf{v} \times \mathbf{w}) = (\mathbf{u} \cdot \mathbf{w})\mathbf{v} - (\mathbf{u} \cdot \mathbf{v})\mathbf{w}$$

to prove that

$$\mathbf{u} \times (\mathbf{v} \times \mathbf{w}) + \mathbf{v} \times (\mathbf{w} \times \mathbf{u}) + \mathbf{w} \times (\mathbf{u} \times \mathbf{v}) = \mathbf{0}$$

SOLUTION The given identity implies that:

$$\mathbf{u} \times (\mathbf{v} \times \mathbf{w}) = (\mathbf{u} \cdot \mathbf{w})\mathbf{v} - (\mathbf{u} \cdot \mathbf{v})\mathbf{w}$$

$$\mathbf{v} \times (\mathbf{w} \times \mathbf{u}) = (\mathbf{v} \cdot \mathbf{u})\mathbf{w} - (\mathbf{v} \cdot \mathbf{w})\mathbf{u}$$

$$\mathbf{w} \times (\mathbf{u} \times \mathbf{v}) = (\mathbf{w} \cdot \mathbf{v})\mathbf{u} - (\mathbf{w} \cdot \mathbf{u})\mathbf{v}$$

Adding the three equations and using the commutativity of the dot product we find that:

$$\mathbf{u} \times (\mathbf{v} \times \mathbf{w}) + \mathbf{v} \times (\mathbf{w} \times \mathbf{u}) + \mathbf{w} \times (\mathbf{u} \times \mathbf{v})$$

$$= (\mathbf{u} \cdot \mathbf{w} - \mathbf{w} \cdot \mathbf{u})\mathbf{v} + (\mathbf{v} \cdot \mathbf{u} - \mathbf{u} \cdot \mathbf{v})\mathbf{w} + (\mathbf{w} \cdot \mathbf{v} - \mathbf{v} \cdot \mathbf{w})\mathbf{u} = \mathbf{0}$$

47. Write the equation of the plane $\mathcal{P}$ with vector equation

$$\langle 1, 4, -3\rangle \cdot \langle x, y, z\rangle = 7$$

in the form

$$a(x - x_0) + b(y - y_0) + c(z - z_0) = 0$$

Hint: You must find a point $P = (x_0, y_0, z_0)$ on $\mathcal{P}$.

SOLUTION We identify the vector $\mathbf{n} = \langle a, b, c\rangle = \langle 1, 4, -3\rangle$ that is normal to the plane, hence we may choose,

$$a = 1, \quad b = 4, \quad c = -3$$

We now must find a point in the plane. The point $(x_0, y_0, z_0) = (0, 1, -1)$, for instance, satisfies the equation of the plane, therefore the equation may be written in the form:

$$1(x - 0) + 4(y - 1) - 3(z - (-1)) = 0$$

or

$$(x - 0) + 4(y - 1) - 3(z + 1) = 0$$

49. Find the plane through $P = (4, -1, 9)$ containing the line $\mathbf{r}(t) = \langle 1, 4, -3\rangle + t\langle 2, 1, 1\rangle$.

SOLUTION Since the plane contains the line, the direction vector of the line, $\mathbf{v} = \langle 2, 1, 1\rangle$, is in the plane. To find another vector in the plane, we use the points $A = (1, 4, -3)$ and $B = (4, -1, 9)$ that lie in the plane, and compute the vector $\mathbf{u} = \overrightarrow{AB}$:

$$\mathbf{u} = \overrightarrow{AB} = \langle 4 - 1, -1 - 4, 9 - (-3)\rangle = \langle 3, -5, 12\rangle$$

We now compute the cross product $\mathbf{n} = \mathbf{v} \times \mathbf{u}$ that is normal to the plane:

$$\mathbf{n} = \mathbf{v} \times \mathbf{u} = \begin{vmatrix} \mathbf{i} & \mathbf{j} & \mathbf{k} \\ 2 & 1 & 1 \\ 3 & -5 & 12 \end{vmatrix} = \begin{vmatrix} 1 & 1 \\ -5 & 12 \end{vmatrix}\mathbf{i} - \begin{vmatrix} 2 & 1 \\ 3 & 12 \end{vmatrix}\mathbf{j} + \begin{vmatrix} 2 & 1 \\ 3 & -5 \end{vmatrix}\mathbf{k}$$

$$= (12 + 5)\mathbf{i} - (24 - 3)\mathbf{j} + (-10 - 3)\mathbf{k} = 17\mathbf{i} - 21\mathbf{j} - 13\mathbf{k} = \langle 17, -21, -13\rangle$$

Finally, we use the vector form of the equation of the plane with $\mathbf{n} = \langle 17, -21, -13 \rangle$ and $P_0 = (4, -1, 9)$ to obtain the following equation:

$$\mathbf{n} \cdot \langle x, y, z \rangle = \mathbf{n} \cdot \langle x_0, y_0, z_0 \rangle$$
$$\langle 17, -21, -13 \rangle \cdot \langle x, y, z \rangle = \langle 17, -21, -13 \rangle \cdot \langle 4, -1, 9 \rangle$$
$$17x - 21y - 13z = 17 \cdot 4 + 21 - 13 \cdot 9 = -28$$

The equation of the plane is, thus,

$$17x - 21y - 13z = -28$$

51. Find the trace of the plane $3x - 2y + 5z = 4$ in the xy-plane.

SOLUTION The xy-plane has equation $z = 0$, therefore the intersection of the plane $3x - 2y + 5z = 4$ with the xy-plane must satisfy both $z = 0$ and the equation of the plane. Therefore the trace has the following equation:

$$3x - 2y + 5 \cdot 0 = 4 \quad \Rightarrow \quad 3x - 2y = 4$$

We conclude that the trace of the plane in the xy-plane is the line $3x - 2y = 4$ in the xy-plane.

In Exercises 53–58, determine the type of the quadric surface.

53. $\left(\dfrac{x}{3}\right)^2 + \left(\dfrac{y}{4}\right)^2 + 2z^2 = 1$

SOLUTION Writing the equation in the form:

$$\left(\frac{x}{3}\right)^2 + \left(\frac{y}{4}\right)^2 + \left(\frac{z}{\frac{1}{\sqrt{2}}}\right)^2 = 1$$

we identify the quadric surface as an ellipsoid.

55. $\left(\dfrac{x}{3}\right)^2 + \left(\dfrac{y}{4}\right)^2 - 2z = 0$

SOLUTION We rewrite this equation as:

$$2z = \left(\frac{x}{3}\right)^2 + \left(\frac{y}{4}\right)^2$$

or

$$z = \left(\frac{x}{3\sqrt{2}}\right)^2 + \left(\frac{y}{4\sqrt{2}}\right)^2$$

This is the equation of an elliptic paraboloid.

57. $\left(\dfrac{x}{3}\right)^2 - \left(\dfrac{y}{4}\right)^2 - 2z^2 = 0$

SOLUTION This equation may be rewritten in the form

$$\left(\frac{x}{3}\right)^2 - \left(\frac{y}{4}\right)^2 = \left(\frac{z}{\frac{1}{\sqrt{2}}}\right)^2$$

we identify the quadric surface as an elliptic cone.

59. Determine the type of the quadric surface $ax^2 + by^2 - z^2 = 1$ if:
  (a) $a < 0, \quad b < 0$
  (b) $a > 0, \quad b > 0$
  (c) $a > 0, \quad b < 0$

SOLUTION

(a) If $a < 0, b < 0$ then for all $x, y,$ and z we have $ax^2 + by^2 - z^2 < 0$, hence there are no points that satisfy $ax^2 + by^2 - z^2 = 1$. Therefore it is the empty set.

(b) For $a > 0$ and $b > 0$ we rewrite the equation as

$$\left(\frac{x}{\frac{1}{\sqrt{a}}}\right)^2 + \left(\frac{y}{\frac{1}{\sqrt{b}}}\right)^2 - z^2 = 1$$

which is the equation of a hyperboloid of one sheet.

(c) For $a > 0$, $b < 0$ we rewrite the equation in the form

$$\left(\frac{x}{\frac{1}{\sqrt{a}}}\right)^2 - \left(\frac{y}{\frac{1}{\sqrt{|b|}}}\right)^2 - z^2 = 1$$

which is the equation of a hyperboloid of two sheets.

61. Convert $(x, y, z) = (3, 4, -1)$ from rectangular to cylindrical and spherical coordinates.

SOLUTION In cylindrical coordinates (r, θ, z) we have

$$r = \sqrt{x^2 + y^2}, \quad \tan\theta = \frac{y}{x}$$

Therefore, $r = \sqrt{3^2 + 4^2} = 5$ and $\tan\theta = \frac{4}{3}$. The projection of the point $(3, 4, -1)$ onto the xy-plane is the point $(3, 4)$, in the first quadrant. Therefore, the corresponding value of θ is $\tan^{-1}\frac{4}{3} \approx 0.93$ rad. The cylindrical coordinates are, thus,

$$(r, \theta, z) = \left(5, \tan^{-1}\frac{4}{3}, -1\right)$$

The spherical coordinates (ρ, θ, ϕ) satisfy

$$\rho = \sqrt{x^2 + y^2 + z^2}, \quad \tan\theta = \frac{y}{x}, \quad \cos\phi = \frac{z}{\rho}$$

Therefore,

$$\rho = \sqrt{3^2 + 4^2 + (-1)^2} = \sqrt{26}$$

$$\tan\theta = \frac{4}{3}$$

$$\cos\phi = \frac{-1}{\sqrt{26}}$$

The angle θ is the same as in the cylindrical coordinates, that is, $\theta = \tan^{-1}\frac{4}{3}$. The angle ϕ is the solution of $\cos\phi = \frac{-1}{\sqrt{26}}$ that satisfies $0 \leq \phi \leq \pi$, that is, $\phi = \cos^{-1}\left(\frac{-1}{\sqrt{26}}\right) \approx 1.77$ rad. The spherical coordinates are, thus,

$$(\rho, \theta, \phi) = \left(\sqrt{26}, \tan^{-1}\frac{4}{3}, \cos^{-1}\left(\frac{-1}{\sqrt{26}}\right)\right)$$

63. Convert the point $(\rho, \theta, \phi) = \left(3, \frac{\pi}{6}, \frac{\pi}{3}\right)$ from spherical to cylindrical coordinates.

SOLUTION By the given information, $\rho = 3$, $\theta = \frac{\pi}{6}$, and $\phi = \frac{\pi}{3}$. We must determine the cylindrical coordinates (r, θ, z). The angle θ is the same as in spherical coordinates. We find z using the relation $\cos\phi = \frac{z}{\rho}$, or $z = \rho\cos\phi$. We obtain

$$z = \rho\cos\phi = 3\cos\frac{\pi}{3} = 3 \cdot \frac{1}{2} = \frac{3}{2}$$

We find r using the relation $\rho^2 = x^2 + y^2 + z^2 = r^2 + z^2$, or $r = \sqrt{\rho^2 - z^2}$, we get

$$r = \sqrt{3^2 - \left(\frac{3}{2}\right)^2} = \sqrt{\frac{27}{4}} = \frac{3\sqrt{3}}{2}$$

Hence, in cylindrical coordinates we obtain the following description:

$$(r, \theta, z) = \left(\frac{3\sqrt{3}}{2}, \frac{\pi}{6}, \frac{3}{2}\right)$$

65. Sketch the graph of the cylindrical equation $z = 2r \cos \theta$ and write the equation in rectangular coordinates.

SOLUTION To obtain the equation in rectangular coordinates, we substitute $x = r \cos \theta$ in the equation $z = 2r \cos \theta$:

$$z = 2r \cos \theta = 2x \quad \Rightarrow \quad z = 2x$$

This is the equation of a plane normal to the xz-plane, whose intersection with the xz-plane is the line $z = 2x$. The graph of the plane is shown in the following figure (the same plane drawn twice, using the cylindrical coordinates' equation and using the rectangular coordinates' equation):

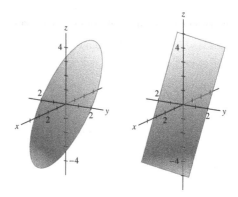

67. Show that the cylindrical equation

$$r^2(1 - 2\sin^2 \theta) + z^2 = 1$$

is a hyperboloid of one sheet.

SOLUTION We rewrite the equation in the form

$$r^2 - 2(r \sin \theta)^2 + z^2 = 1$$

To write this equation in rectangular coordinates, we substitute $r^2 = x^2 + y^2$ and $r \sin \theta = y$. This gives

$$x^2 + y^2 - 2y^2 + z^2 = 1$$
$$x^2 - y^2 + z^2 = 1$$

We now can identify the surface as a hyperboloid of one sheet.

69. Describe how the surface with spherical equation

$$\rho^2(1 + A \cos^2 \phi) = 1$$

depends on the constant A.

SOLUTION To identify the surface we convert the equation to rectangular coordinates. We write

$$\rho^2 + A\rho^2 \cos^2 \phi = 1$$

To obtain the following equation in terms of x, y, z only, we substitute $\rho^2 = x^2 + y^2 + z^2$ and $\rho \cos \phi = z$:

$$x^2 + y^2 + z^2 + Az^2 = 1$$
$$x^2 + y^2 + (1 + A)z^2 = 1 \tag{1}$$

Case 1: $A < -1$. Then $A + 1 < 0$ and the equation can be rewritten in the form

$$x^2 + y^2 - \left(\frac{z}{|1 + A|^{-1/2}}\right)^2 = 1$$

The corresponding surface is a hyperboloid of one sheet.

Case 2: $A = -1$. Equation (1) becomes:

$$x^2 + y^2 = 1$$

In R^3, this equation describes a cylinder with the z-axis as its central axis.

Case 3: $A > -1$. Then Eq. (1) can be rewritten as

$$x^2 + y^2 + \left(\frac{z}{(1+A)^{-1/2}}\right)^2 = 1$$

Then if $A = 0$ the equation $x^2 + y^2 + z^2 = 1$ describes the unit sphere in R^3. Otherwise, the surface is an ellipsoid.

71. Let c be a scalar, $\mathbf{a}$ and $\mathbf{b}$ be vectors, and $\mathbf{X} = \langle x, y, z \rangle$. Show that the equation $(\mathbf{X} - \mathbf{a}) \cdot (\mathbf{X} - \mathbf{b}) = c^2$ defines a sphere with center $\mathbf{m} = \frac{1}{2}(\mathbf{a} + \mathbf{b})$ and radius R, where $R^2 = c^2 + \left\|\frac{1}{2}(\mathbf{a} - \mathbf{b})\right\|^2$.

SOLUTION We evaluate the following length:

$$\|\mathbf{x} - \mathbf{m}\|^2 = \left\|\mathbf{x} - \frac{1}{2}(\mathbf{a} + \mathbf{b})\right\|^2 = \left((\mathbf{x} - \mathbf{a}) + \frac{1}{2}(\mathbf{a} - \mathbf{b})\right) \cdot \left((\mathbf{x} - \mathbf{b}) - \frac{1}{2}(\mathbf{a} - \mathbf{b})\right)$$

$$= (\mathbf{x} - \mathbf{a}) \cdot (\mathbf{x} - \mathbf{b}) - \frac{1}{2}(\mathbf{x} - \mathbf{a}) \cdot (\mathbf{a} - \mathbf{b}) + \frac{1}{2}(\mathbf{a} - \mathbf{b}) \cdot (\mathbf{x} - \mathbf{b}) - \frac{1}{4}(\mathbf{a} - \mathbf{b}) \cdot (\mathbf{a} - \mathbf{b})$$

$$= (\mathbf{x} - \mathbf{a}) \cdot (\mathbf{x} - \mathbf{b}) + \frac{1}{2}(\mathbf{a} - \mathbf{b}) \cdot (\mathbf{x} - \mathbf{b} - \mathbf{x} + \mathbf{a}) - \frac{1}{4}(\mathbf{a} - \mathbf{b}) \cdot (\mathbf{a} - \mathbf{b})$$

$$= (\mathbf{x} - \mathbf{a}) \cdot (\mathbf{x} - \mathbf{b}) + \frac{1}{2}(\mathbf{a} - \mathbf{b}) \cdot (\mathbf{a} - \mathbf{b}) - \frac{1}{4}(\mathbf{a} - \mathbf{b}) \cdot (\mathbf{a} - \mathbf{b})$$

$$= (\mathbf{x} - \mathbf{a}) \cdot (\mathbf{x} - \mathbf{b}) + \frac{1}{4}(\mathbf{a} - \mathbf{b}) \cdot (\mathbf{a} - \mathbf{b})$$

$$= (\mathbf{x} - \mathbf{a}) \cdot (\mathbf{x} - \mathbf{b}) + \left\|\frac{1}{2}(\mathbf{a} - \mathbf{b})\right\|^2$$

Since $R^2 = c^2 + \left\|\frac{1}{2}(\mathbf{a} - \mathbf{b})\right\|^2$ we get

$$\|\mathbf{x} - \mathbf{m}\|^2 = (\mathbf{x} - \mathbf{a}) \cdot (\mathbf{x} - \mathbf{b}) + R^2 - c^2$$

We conclude that if $(\mathbf{x} - \mathbf{a})(\mathbf{x} - \mathbf{b}) = c^2$ then $\|\mathbf{x} - \mathbf{m}\|^2 = R^2$. That is, the equation $(\mathbf{x} - \mathbf{a})(\mathbf{x} - \mathbf{b}) = c^2$ defines a sphere with center $\mathbf{m}$ and radius R.

13 CALCULUS OF VECTOR-VALUED FUNCTIONS

13.1 Vector-Valued Functions (LT Section 14.1)

Preliminary Questions

1. Which one of the following does *not* parametrize a line?

(a) $\mathbf{r}_1(t) = \langle 8 - t, 2t, 3t \rangle$

(b) $\mathbf{r}_2(t) = t^3\mathbf{i} - 7t^3\mathbf{j} + t^3\mathbf{k}$

(c) $\mathbf{r}_3(t) = \langle 8 - 4t^3, 2 + 5t^2, 9t^3 \rangle$

SOLUTION

(a) This is a parametrization of the line passing through the point $(8, 0, 0)$ in the direction parallel to the vector $\langle -1, 2, 3 \rangle$, since:

$$\langle 8 - t, 2t, 3t \rangle = \langle 8, 0, 0 \rangle + t\langle -1, 2, 3 \rangle$$

(b) Using the parameter $s = t^3$ we get:

$$\langle t^3\mathbf{i} - 7t^3\mathbf{j} + t^3\mathbf{k} \rangle = \langle s, -7s, s \rangle = s\langle 1, -7, 1 \rangle$$

This is a parametrization of the line through the origin, with the direction vector $\mathbf{v} = \langle -1, 7, 1 \rangle$.

(c) The parametrization $\langle 8 - 4t^3, 2 + 5t^2, 9t^3 \rangle$ does not parametrize a line. In particular, the points $(8, 2, 0)$ (at $t = 0$), $(4, 7, 9)$ (at $t = 1$), and $(-24, 22, 72)$ (at $t = 2$) are not collinear.

2. What is the projection of $\mathbf{r}(t) = t\mathbf{i} + t^4\mathbf{j} + e^t\mathbf{k}$ onto the xz-plane?

SOLUTION The projection of the path onto the xz-plane is the curve traced by $t\mathbf{i} + e^t\mathbf{k} = \langle t, 0, e^t \rangle$. This is the curve $z = e^x$ in the xz-plane.

3. Which projection of $\langle \cos t, \cos 2t, \sin t \rangle$ is a circle?

SOLUTION The parametric equations are

$$x = \cos t, \quad y = \cos 2t, \quad z = \sin t$$

The projection onto the xz-plane is $\langle \cos t, 0, \sin t \rangle$. Since $x^2 + z^2 = \cos^2 t + \sin^2 t = 1$, the projection is a circle in the xz-plane. The projection onto the xy-plane is traced by the curve $\langle \cos t, \cos 2t, 0 \rangle$. Therefore, $x = \cos t$ and $y = \cos 2t$. We express y in terms of x:

$$y = \cos 2t = 2\cos^2 t - 1 = 2x^2 - 1$$

The projection onto the xy-plane is a parabola. The projection onto the yz-plane is the curve $\langle 0, \cos 2t, \sin t \rangle$. Hence $y = \cos 2t$ and $z = \sin t$. We find y as a function of z:

$$y = \cos 2t = 1 - 2\sin^2 t = 1 - 2z^2$$

The projection onto the yz-plane is again a parabola.

4. What is the center of the circle with the following parametrization?

$$\mathbf{r}(t) = (-2 + \cos t)\mathbf{i} + 2\mathbf{j} + (3 - \sin t)\mathbf{k}$$

SOLUTION The parametric equations are

$$x = -2 + \cos t, \quad y = 2, \quad z = 3 - \sin t$$

Therefore, the curve is contained in the plane $y = 2$, and the following holds:

$$(x + 2)^2 + (z - 3)^2 = \cos^2 t + \sin^2 t = 1$$

We conclude that the curve $\mathbf{r}(t)$ is the circle of radius 1 in the plane $y = 2$ centered at the point $(-2, 2, 3)$.

5. How do the paths $\mathbf{r}_1(t) = \langle \cos t, \sin t \rangle$ and $\mathbf{r}_2(t) = \langle \sin t, \cos t \rangle$ around the unit circle differ?

SOLUTION The two paths describe the unit circle. However, as t increases from 0 to 2π, the point on the path $\sin t \mathbf{i} + \cos t \mathbf{j}$ moves in a clockwise direction, whereas the point on the path $\cos t \mathbf{i} + \sin t \mathbf{j}$ moves in a counterclockwise direction.

6. Which three of the following vector-valued functions parametrize the same space curve?
- **(a)** $(-2 + \cos t)\mathbf{i} + 9\mathbf{j} + (3 - \sin t)\mathbf{k}$
- **(b)** $(2 + \cos t)\mathbf{i} - 9\mathbf{j} + (-3 - \sin t)\mathbf{k}$
- **(c)** $(-2 + \cos 3t)\mathbf{i} + 9\mathbf{j} + (3 - \sin 3t)\mathbf{k}$
- **(d)** $(-2 - \cos t)\mathbf{i} + 9\mathbf{j} + (3 + \sin t)\mathbf{k}$
- **(e)** $(2 + \cos t)\mathbf{i} + 9\mathbf{j} + (3 + \sin t)\mathbf{k}$

SOLUTION All the curves except for (b) lie in the vertical plane $y = 9$. We identify each one of the curves (a), (c), (d) and (e).

(a) The parametric equations are:

$$x = -2 + \cos t, \quad y = 9, \quad z = 3 - \sin t$$

Hence,

$$(x + 2)^2 + (z - 3)^2 = (\cos t)^2 + (-\sin t)^2 = 1$$

This is the circle of radius 1 in the plane $y = 9$, centered at $(-2, 9, 3)$.

(c) The parametric equations are:

$$x = -2 + \cos 3t, \quad y = 9, \quad z = 3 - \sin 3t$$

Hence,

$$(x + 2)^2 + (z - 3)^2 = (\cos 3t)^2 + (-\sin 3t)^2 = 1$$

This is the circle of radius 1 in the plane $y = 9$, centered at $(-2, 9, 3)$.

(d) In this curve we have:

$$x = -2 - \cos t, \quad y = 9, \quad z = 3 + \sin t$$

Hence,

$$(x + 2)^2 + (z - 3)^2 = (-\cos t)^2 + (\sin t)^2 = 1$$

Again, the circle of radius 1 in the plane $y = 9$, centered at $(-2, 9, 3)$.

(e) In this parametrization we have:

$$x = 2 + \cos t, \quad y = 9, \quad z = 3 + \sin t$$

Hence,

$$(x - 2)^2 + (z - 3)^2 = (\cos t)^2 + (\sin t)^2 = 1$$

This is the circle of radius 1 in the plane $y = 9$, centered at $(2, 9, 3)$.

We conclude that (a), (c), and (d) parametrize the same circle whereas (b) and (e) are different curves.

Exercises

1. What is the domain of $\mathbf{r}(t) = e^t \mathbf{i} + \frac{1}{t}\mathbf{j} + (t + 1)^{-3}\mathbf{k}$?

SOLUTION $\mathbf{r}(t)$ is defined for $t \neq 0$ and $t \neq -1$, hence the domain of $\mathbf{r}(t)$ is:

$$D = \{t \in \mathbf{R} : t \neq 0, t \neq -1\}$$

3. Evaluate $\mathbf{r}(2)$ and $\mathbf{r}(-1)$ for $\mathbf{r}(t) = \left\langle \sin \frac{\pi}{2}t, t^2, (t^2+1)^{-1} \right\rangle$.

SOLUTION Since $\mathbf{r}(t) = \left\langle \sin \frac{\pi}{2}t, t^2, (t^2+1)^{-1} \right\rangle$, then

$$\mathbf{r}(2) = \left\langle \sin \pi, 4, 5^{-1} \right\rangle = \left\langle 0, 4, \frac{1}{5} \right\rangle$$

and

$$\mathbf{r}(-1) = \left\langle \sin \frac{-\pi}{2}, 1, 2^{-1} \right\rangle = \left\langle -1, 1, \frac{1}{2} \right\rangle$$

5. Find a vector parametrization of the line through $P = (3, -5, 7)$ in the direction $\mathbf{v} = \langle 3, 0, 1 \rangle$.

SOLUTION We use the vector parametrization of the line to obtain:

$$\mathbf{r}(t) = \overrightarrow{OP} + t\mathbf{v} = \langle 3, -5, 7 \rangle + t \langle 3, 0, 1 \rangle = \langle 3 + 3t, -5, 7 + t \rangle$$

or in the form:

$$\mathbf{r}(t) = (3 + 3t)\mathbf{i} - 5\mathbf{j} + (7 + t)\mathbf{k}$$

7. Determine whether the space curve given by $\mathbf{r}(t) = \langle \sin t, \cos t/2, t \rangle$ intersects the z-axis, and if it does, determine where.

SOLUTION The curve intersects the z-axis if there is some value of t such that both the x- and y-coordinates of $\mathbf{r}(t)$ are zero; that is, such that $\sin t = \cos \frac{t}{2} = 0$. Now, $\sin x = 0$ when $x = n\pi$, while $\cos x = 0$ when $x = \frac{\pi}{2} + n\pi = \frac{2n+1}{2}\pi$, where n is an integer. So if $t = k\pi$, then $\sin t = 0$, and if, further, $k = 2n+1$ is odd, then $\cos \frac{k}{2}\pi = 0$. So this curve intersects the z-axis whenever k is an odd integer multiple of π, at the points $(0, 0, (2n+1)\pi)$.

9. Determine whether the space curve given by $\mathbf{r}(t) = \left\langle t, t^3, t^2 + 1 \right\rangle$ intersects the xy-plane, and if it does, determine where.

SOLUTION This curve intersects the xy-plane whenever the z coordinate of $\mathbf{r}(t)$ is zero. But the z-coordinate is $t^2 + 1 > 0$, so this curve is always above the xy-plane and does not intersect it.

11. Show that the space curve given by $\mathbf{r}(t) = \left\langle 1 - \cos(2t), t + \sin t, t^2 \right\rangle$ intersects the yz-plane in infinitely many points but does not cross through it.

SOLUTION The given curve intersects the yz-plane whenever the x coordinate of $\mathbf{r}(t)$ is zero. $1 - \cos 2t = 0$ when $\cos 2t = 1$, which occurs when t is a multiple of π. Therefore there are infinitely many intersection points, at

$$\mathbf{r}(k\pi) = \left\langle 0, k\pi + \sin k\pi, k^2\pi^2 \right\rangle = \left\langle 0, 0, k^2\pi^2 \right\rangle$$

However, since the x-coordinate, $1 - \cos 2t$, is always nonnegative, the curve never passes through the yz-plane; it always lies on or above that plane.

13. Match the space curves in Figure 7 with their projections onto the xy-plane in Figure 8.

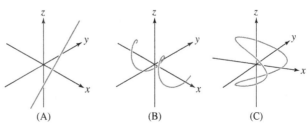

FIGURE 7

FIGURE 8

SOLUTION The projection of curve (C) onto the xy-plane is neither a segment nor a periodic wave. Hence, the correct projection is (iii), rather than the two other graphs. The projection of curve (A) onto the xy-plane is a vertical line, hence the corresponding projection is (ii). The projection of curve (B) onto the xy-plane is a periodic wave as illustrated in (i).

15. Match the vector-valued functions (a)–(f) with the space curves (i)–(vi) in Figure 9.

(a) $\mathbf{r}(t) = \langle t + 15, e^{0.08t} \cos t, e^{0.08t} \sin t \rangle$

(b) $\mathbf{r}(t) = \langle \cos t, \sin t, \sin 12t \rangle$

(c) $\mathbf{r}(t) = \left\langle t, t, \dfrac{25t}{1+t^2} \right\rangle$

(d) $\mathbf{r}(t) = \langle \cos^3 t, \sin^3 t, \sin 2t \rangle$

(e) $\mathbf{r}(t) = \langle t, t^2, 2t \rangle$

(f) $\mathbf{r}(t) = \langle \cos t, \sin t, \cos t \sin 12t \rangle$

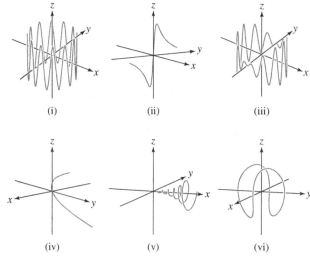

FIGURE 9

SOLUTION

(a) (v) (b) (i) (c) (ii)
(d) (vi) (e) (iv) (f) (iii)

17. Match the space curves (A)–(C) in Figure 10 with their projections (i)–(iii) onto the xy-plane.

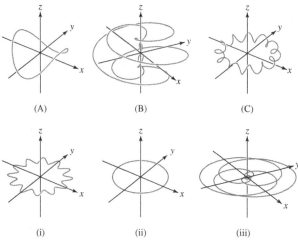

FIGURE 10

SOLUTION Curve (A) does not cross over itself, and appears to remain a constant distance from the z-axis, so it corresponds to projection (ii). Curve (C) oscillates, but does not cross over itself, so it must correspond to projection (i) and therefore curve (B) corresponds to projection (iii).

In Exercises 19–22, the function $\mathbf{r}(t)$ traces a circle. Determine the radius, center, and plane containing the circle.

19. $\mathbf{r}(t) = (9\cos t)\mathbf{i} + (9\sin t)\mathbf{j}$

SOLUTION Since $x(t) = 9\cos t$, $y(t) = 9\sin t$ we have:

$$x^2 + y^2 = 81\cos^2 t + 81\sin^2 t = 81(\cos^2 t + \sin^2 t) = 81$$

This is the equation of a circle with radius 9 centered at the origin. The circle lies in the xy-plane.

21. $\mathbf{r}(t) = \langle \sin t, 0, 4 + \cos t \rangle$

SOLUTION $x(t) = \sin t$, $z(t) = 4 + \cos t$, hence:

$$x^2 + (z-4)^2 = \sin^2 t + \cos^2 t = 1$$

$y = 0$ is the equation of the xz-plane. We conclude that the function traces the circle of radius 1, centered at the point $(0, 0, 4)$, and contained in the xz-plane.

23. Consider the curve C given by $\mathbf{r}(t) = \langle \cos(2t)\sin t, \sin(2t), \cos(2t)\cos t \rangle$.

(a) Show that C lies on the sphere of radius 1 centered at the origin.

(b) Show that C intersects the x-axis, the y-axis, and the z-axis.

SOLUTION

(a) It suffices to show that $\|\mathbf{r}(t)\| = 1$ everywhere, or equivalently that $\|\mathbf{r}(t)\|^2 = 1$ everywhere:

$$\|\mathbf{r}(t)\|^2 = (\cos(2t)\sin t)^2 + (\sin 2t)^2 + (\cos(2t)\cos t)^2$$
$$= \cos^2(2t)\sin^2 t + \sin^2(2t) + \cos^2(2t)\cos^2 t$$
$$= \cos^2(2t) + \sin^2(2t) = 1.$$

(b) The curve intersects the x-axis when the y- and z-coordinates of $\mathbf{r}(t)$ are both zero. Note that $\sin(2t) = 2\sin t\cos t$, so that, for example, $\sin(2t)$ and $\cos(2t)\cos t$ both vanish for $t = \frac{\pi}{2}$. Similarly, the curve intersects the y-axis when the x- and z-coordinates of $\mathbf{r}(t)$ are both zero. Clearly this happens whenever $\cos(2t) = 0$, so, for example, when $t = \frac{\pi}{4}$. Finally, the curve intersects the z-axis when the x- and y-coordinates of $\mathbf{r}(t)$ are both zero. Since $\sin(2t) = 2\sin t\cos t$, we see that the curve intersects the z-axis when $\sin t = 0$, so, for example, for $t = 0$.

25. Let C be the curve given by $\mathbf{r}(t) = \langle t\cos t, t\sin t, t\rangle$.

(a) Show that C lies on the cone $x^2 + y^2 = z^2$.

(b) Sketch the cone and make a rough sketch of C on the cone.

SOLUTION

(a) $x = t\cos t$, $y = t\sin t$ and $z = t$, hence for any point (x, y, z) on the curve we get

$$x^2 + y^2 = t^2\cos^2 t + t^2\sin^2 t = t^2(\cos^2 t + \sin^2 t) = t^2 = z^2$$

Thus the curve lies on the circular cone whose equation is $x^2 + y^2 = z^2$.

(b) As the height $z = t$ increases linearly with time, the x and y coordinates trace out points on the circles of increasing radius. We obtain the following curve:

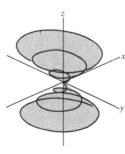

$\mathbf{r}(t) = \langle t\cos t, t\sin t, t\rangle$

In Exercises 27 and 28, let

$$\mathbf{r}(t) = \langle \sin t, \cos t, \sin t \cos 2t \rangle$$

be a parametrization of the curve shown in Figure 11.

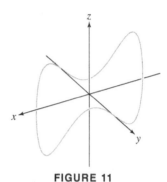

FIGURE 11

27. Find the points where $\mathbf{r}(t)$ intersects the xy-plane.

SOLUTION The curve intersects the xy-plane at the points where $z = 0$. That is, $\sin t \cos 2t = 0$ and so either $\sin t = 0$ or $\cos 2t = 0$. The solutions are, thus:

$$t = \pi k \text{ or } t = \frac{\pi}{4} + \frac{\pi k}{2}, \quad k = 0, \pm 1, \pm 2, \ldots$$

The values $t = \pi k$ yield the points: $(\sin \pi k, \cos \pi k, 0) = \left(0, (-1)^k, 0\right)$. The values $t = \frac{\pi}{4} + \frac{\pi k}{2}$ yield the points:

$$k = 0: \left(\sin \frac{\pi}{4}, \cos \frac{\pi}{4}, 0\right) = \left(\frac{1}{\sqrt{2}}, \frac{1}{\sqrt{2}}, 0\right)$$

$$k = 1: \left(\sin \frac{3\pi}{4}, \cos \frac{3\pi}{4}, 0\right) = \left(\frac{1}{\sqrt{2}}, -\frac{1}{\sqrt{2}}, 0\right)$$

$$k = 2: \left(\sin \frac{5\pi}{4}, \cos \frac{5\pi}{4}, 0\right) = \left(-\frac{1}{\sqrt{2}}, -\frac{1}{\sqrt{2}}, 0\right)$$

$$k = 3: \left(\sin \frac{7\pi}{4}, \cos \frac{7\pi}{4}, 0\right) = \left(-\frac{1}{\sqrt{2}}, \frac{1}{\sqrt{2}}, 0\right)$$

(Other values of k do not provide new points.) We conclude that the curve intersects the xy-plane at the following points: $(0, 1, 0), (0, -1, 0), \left(\frac{1}{\sqrt{2}}, \frac{1}{\sqrt{2}}, 0\right), \left(\frac{1}{\sqrt{2}}, -\frac{1}{\sqrt{2}}, 0\right), \left(-\frac{1}{\sqrt{2}}, -\frac{1}{\sqrt{2}}, 0\right), \left(-\frac{1}{\sqrt{2}}, \frac{1}{\sqrt{2}}, 0\right)$

29. Parametrize the part of the intersection of the surfaces

$$y^2 - z^2 = x - 2, \quad y^2 + z^2 = 9$$

where $z \geq 0$ using $t = y$ as the parameter.

SOLUTION We solve for z and x in terms of y. From the equation $y^2 + z^2 = 9$ we have $z^2 = 9 - y^2$ or $z = \pm \sqrt{9 - y^2}$. From the second equation we have:

$$x = y^2 - z^2 + 2 = y^2 - (9 - y^2) + 2 = 2y^2 - 7$$

Taking $t = y$ as a parameter, we have $z = \pm \sqrt{9 - t^2}$, $x = 2t^2 - 7$; since $z \geq 0$ we choose the positive root, so that $z = \sqrt{9 - t^2}$. This yields the following vector parametrization:

$$\mathbf{r}(t) = \left\langle 2t^2 - 7, t, \sqrt{9 - t^2} \right\rangle \text{ for } -3 \leq t \leq 3$$

31. Viviani's Curve C is the intersection of the surfaces (Figure 12)

$$x^2 + y^2 = z^2, \qquad y = z^2$$

(a) Separately parametrize each of the two parts of C corresponding to $x \geq 0$ and $x \leq 0$, taking $t = z$ as the parameter.

(b) Describe the projection of C onto the xy-plane.

(c) Show that C lies on the sphere of radius 1 with its center $(0, 1, 0)$. This curve looks like a figure eight lying on a sphere [Figure 12(B)].

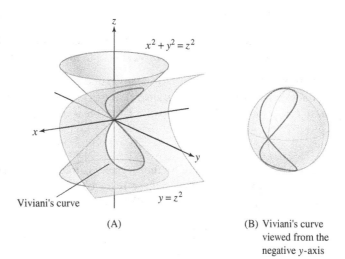

(A)

(B) Viviani's curve viewed from the negative y-axis

FIGURE 12 Viviani's curve is the intersection of the surfaces $x^2 + y^2 = z^2$ and $y = z^2$.

SOLUTION

(a) We must solve for y and x in terms of z (which is a parameter). We get:

$$y = z^2$$

$$x^2 = z^2 - y^2 \quad \Rightarrow \quad x = \pm\sqrt{z^2 - y^2} = \pm\sqrt{z^2 - z^4}$$

Here, the $\pm$ from $x = \pm\sqrt{z^2 - z^4}$ represents the two parts of the parametrization: $+$ for $x \geq 0$, and $-$ for $x \leq 0$. Substituting the parameter $z = t$ we get:

$$y = t^2, \quad x = \pm\sqrt{t^2 - t^4} = \pm|t|\sqrt{1 - t^2}$$

We obtain the following parametrization:

$$\mathbf{r}(t) = \left\langle \pm|t|\sqrt{1 - t^2}, t^2, t \right\rangle \text{ for } -1 \leq t \leq 1 \tag{1}$$

where the plus sign gives the parametrization for $x \geq 0$ and the minus sign for $x \leq 0$.

(b) The projection of the curve onto the xy-plane is the curve on the xy-plane obtained by setting the z-coordinate of $\mathbf{r}(t)$ equal to zero. We obtain the following curve:

$$\left\langle \pm|t|\sqrt{1 - t^2}, t^2, 0 \right\rangle, \quad -1 \leq t \leq 1$$

We also note that since $x = \pm|t|\sqrt{1 - t^2}$, then $x^2 = t^2(1 - t^2)$, but also $y = t^2$, so that gives us the equation $x^2 = y(1 - y)$ for the projection onto the xy plane. We rewrite this as follows.

$$x^2 = y(1 - y) \quad \Rightarrow \quad x^2 + y^2 - y = 0$$

$$x^2 + y^2 - y + 1/4 = 1/4$$

$$x^2 + (y - 1/2)^2 = (1/2)^2$$

We can now identify this projection as a circle in the xy plane, with radius $1/2$, centered at the xy point $(0, 1/2)$.

(c) The equation of the sphere of radius 1 with center $(0, 1, 0)$ is:

$$x^2 + (y-1)^2 + z^2 = 1 \qquad (2)$$

To show that C lies on this sphere, we show that the coordinates of the points on C (given in (1)) satisfy the equation of the sphere. Substituting the coordinates from (1) into the left side of (2) gives:

$$x^2 + (y-1)^2 + z^2 = \left(\pm|t|\sqrt{1-t^2}\right)^2 + (t^2-1)^2 + t^2 = t^2(1-t^2) + (t^2-1)^2 + t^2$$
$$= (t^2-1)(t^2-1-t^2) + t^2 = 1$$

We conclude that the curve C lies on the sphere of radius 1 with center $(0, 1, 0)$.

33. Use sine and cosine to parametrize the intersection of the cylinder $x^2 + y^2 = 1$ and the plane $x + y + z = 1$. Then describe the projections of this curve onto the three coordinate planes.

SOLUTION The circle $x^2 + y^2 = 1$ in the xy-plane is parametrized by $x = \cos t$, $y = \sin t$, so that points in the cylinder can be written as $(\cos t, \sin t, z)$. If this point is to lie on the plane $x + y + z = 1$ as well, then we must have $z = 1 - x - y = 1 - \cos t - \sin t$. Therefore a vector parametrization of the intersection is

$$\mathbf{r}(t) = \langle \cos t, \sin t, 1 - \cos t - \sin t \rangle$$

The projection of this curve on the xy-plane is traced by $\langle \cos t, \sin t, 0 \rangle$, which is the unit circle in the xy-plane. The projection on the xz-plane is traced by $\langle \cos t, 0, 1 - \cos t - \sin t \rangle$. Setting $x = \cos t$, we get the curve given by $\langle x, 0, 1 - x - \sqrt{1-x^2} \rangle$ for $-1 \le x \le 1$. This is the curve

$$z = 1 - x - \sqrt{1-x^2} \quad \Rightarrow \quad x + z - 1 = -\sqrt{1-x^2} \quad \Rightarrow \quad 2x^2 + z^2 + 2xz - 2x - 2z = 0$$

which is an ellipse. The projection on the yz-plane is traced by $\langle 0, \sin t, 1 - \cos t - \sin t \rangle$, which by similar calculations is also an ellipse.

35. Use sine and cosine to parametrize the intersection of the surfaces $x^2 + y^2 = 1$ and $z = 4x^2$ (Figure 13).

FIGURE 13 Intersection of the surfaces $x^2 + y^2 = 1$ and $z = 4x^2$.

SOLUTION The points on the cylinder $x^2 + y^2 = 1$ and on the parabolic cylinder $z = 4x^2$ can be written in the form:

$$x^2 + y^2 = 1: \quad (\cos t, \sin t, z)$$
$$z = 4x^2: \quad (x, y, 4x^2)$$

The points (x, y, z) on the intersection curve must satisfy the following equations:

$$\begin{aligned} x &= \cos t \\ y &= \sin t \\ z &= 4x^2 \end{aligned} \quad \Rightarrow \quad x = \cos t, \ y = \sin t, \ z = 4\cos^2 t$$

We obtain the vector parametrization:

$$\mathbf{r}(t) = \langle \cos t, \sin t, 4\cos^2 t \rangle$$

Using the CAS we obtain the following curve:

$$\mathbf{r}(t) = \langle \cos t, \sin t, 4\cos^2 t \rangle$$

In Exercises 36–38, two paths $\mathbf{r}_1(t)$ and $\mathbf{r}_2(t)$ intersect if there is a point P lying on both curves. We say that $\mathbf{r}_1(t)$ and $\mathbf{r}_2(t)$ collide if $\mathbf{r}_1(t_0) = \mathbf{r}_2(t_0)$ at some time t_0.

37. Determine whether $\mathbf{r}_1(t)$ and $\mathbf{r}_2(t)$ collide or intersect, giving the coordinates of the corresponding points if they exist:

$$\mathbf{r}_1(t) = \langle t^2 + 3, t + 1, 6t^{-1} \rangle$$

$$\mathbf{r}_2(t) = \langle 4t, 2t - 2, t^2 - 7 \rangle$$

SOLUTION To determine if the paths collide, we must examine whether the following equations have a solution:

$$\begin{cases} t^2 + 3 = 4t \\ t + 1 = 2t - 2 \\ \dfrac{6}{t} = t^2 - 7 \end{cases}$$

We simplify to obtain:

$$t^2 - 4t + 3 = (t-3)(t-1) = 0$$

$$t = 3$$

$$t^3 - 7t - 6 = 0$$

The solution of the second equation is $t = 3$. This is also a solution of the first and the third equations. It follows that $\mathbf{r}_1(3) = \mathbf{r}_2(3)$ so the curves collide. The curves also intersect at the point where they collide. We now check if there are other points of intersection by solving the following equation:

$$\mathbf{r}_1(t) = \mathbf{r}_2(s)$$

$$\left\langle t^2 + 3, t + 1, \frac{6}{t} \right\rangle = \langle 4s, 2s - 2, s^2 - 7 \rangle$$

Equating coordinates we get:

$$\begin{cases} t^2 + 3 = 4s \\ t + 1 = 2s - 2 \\ \dfrac{6}{t} = s^2 - 7 \end{cases}$$

By the second equation, $t = 2s - 3$. Substituting into the first equation yields:

$$(2s - 3)^2 + 3 = 4s$$

$$4s^2 - 12s + 9 + 3 = 4s$$

$$s^2 - 4s + 3 = 0 \quad \Rightarrow \quad s_1 = 1, \quad s_2 = 3$$

Substituting $s_1 = 1$ and $s_2 = 3$ into the second equation gives:

$$t_1 + 1 = 2 \cdot 1 - 2 \quad \Rightarrow \quad t_1 = -1$$
$$t_2 + 1 = 2 \cdot 3 - 2 \quad \Rightarrow \quad t_2 = 3$$

The solutions of the first two equations are:

$$t_1 = -1, \quad s_1 = 1; \qquad t_2 = 3, \quad s_2 = 3$$

We check if these solutions satisfy the third equation:

$$\frac{6}{t_1} = \frac{6}{-1} = -6, \quad s_1^2 - 7 = 1^2 - 7 = -6 \quad \Rightarrow \quad \frac{6}{t_1} = s_1^2 - 7$$

$$\frac{6}{t_2} = \frac{6}{3} = 2, \quad s_2^2 - 7 = 3^2 - 7 = 2 \quad \Rightarrow \quad \frac{6}{t_2} = s_2^2 - 7$$

We conclude that the paths intersect at the endpoints of the vectors $\mathbf{r}_1(-1)$ and $\mathbf{r}_1(3)$ (or equivalently $\mathbf{r}_2(1)$ and $\mathbf{r}_2(3)$). That is, at the points $(4, 0, -6)$ and $(12, 4, 2)$.

In Exercises 39–48, find a parametrization of the curve.

39. The vertical line passing through the point $(3, 2, 0)$

SOLUTION The points of the vertical line passing through the point $(3, 2, 0)$ can be written as $(3, 2, z)$. Using $z = t$ as parameter we get the following parametrization:

$$\mathbf{r}(t) = \langle 3, 2, t \rangle, \quad -\infty < t < \infty$$

41. The line through the origin whose projection on the xy-plane is a line of slope 3 and whose projection on the yz-plane is a line of slope 5 (i.e., $\Delta z / \Delta y = 5$)

SOLUTION We denote by (x, y, z) the points on the line. The projection of the line on the xy-plane is the line through the origin having slope 3, that is, the line $y = 3x$ in the xy-plane. The projection of the line on the yz-plane is the line through the origin with slope 5, that is the line $z = 5y$. Thus, the points on the desired line satisfy the following equalities:

$$\begin{matrix} y = 3x \\ z = 5y \end{matrix} \quad \Rightarrow \quad y = 3x, \ z = 5 \cdot 3x = 15x$$

We conclude that the points on the line are all the points in the form $(x, 3x, 15x)$. Using $x = t$ as parameter we obtain the following parametrization:

$$\mathbf{r}(t) = \langle t, 3t, 15t \rangle$$

43. The circle of radius 2 with center $(1, 2, 5)$ in a plane parallel to the yz-plane

SOLUTION The circle is parallel to the yz-plane and centered at $(1, 2, 5)$, hence the x-coordinates of the points on the circle are $x = 1$. The projection of the circle on the yz-plane is a circle of radius 2 centered at $(2, 5)$. This circle is parametrized by:

$$y = 2 + 2\cos t, \quad z = 5 + 2\sin t$$

We conclude that the points on the required circle can be written as $(1, 2 + 2\cos t, 5 + 2\sin t)$. This gives the following parametrization:

$$\mathbf{r}(t) = \langle 1, 2 + 2\cos t, 5 + 2\sin t \rangle$$

45. The intersection of the plane $y = \frac{1}{2}$ with the sphere $x^2 + y^2 + z^2 = 1$

SOLUTION Substituting $y = \frac{1}{2}$ in the equation of the sphere gives:

$$x^2 + \left(\frac{1}{2}\right)^2 + z^2 = 1 \quad \Rightarrow \quad x^2 + z^2 = \frac{3}{4}$$

This circle in the horizontal plane $y = \frac{1}{2}$ has the parametrization $x = \frac{\sqrt{3}}{2} \cos t$, $z = \frac{\sqrt{3}}{2} \sin t$. Therefore, the points on the intersection of the plane $y = \frac{1}{2}$ and the sphere $x^2 + y^2 + z^2 = 1$ can be written in the form $\left(\frac{\sqrt{3}}{2} \cos t, \frac{1}{2}, \frac{\sqrt{3}}{2} \sin t \right)$, yielding the following parametrization:

$$\mathbf{r}(t) = \left\langle \frac{\sqrt{3}}{2} \cos t, \frac{1}{2}, \frac{\sqrt{3}}{2} \sin t \right\rangle$$

47. The ellipse $\left(\frac{x}{2}\right)^2 + \left(\frac{z}{3}\right)^2 = 1$ in the xz-plane, translated to have center $(3, 1, 5)$ [Figure 14(A)]

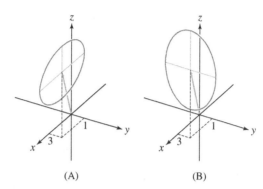

(A) (B)

FIGURE 14 The ellipses described in Exercises 47 and 48.

SOLUTION The translated ellipse is in the vertical plane $y = 1$, hence the y-coordinate of the points on this ellipse is $y = 1$. The x and z coordinates satisfy the equation of the ellipse:

$$\left(\frac{x-3}{2}\right)^2 + \left(\frac{z-5}{3}\right)^2 = 1$$

This ellipse is parametrized by the following equations:

$$x = 3 + 2 \cos t, \quad z = 5 + 3 \sin t$$

Therefore, the points on the translated ellipse can be written as $(3 + 2 \cos t, 1, 5 + 3 \sin t)$. This gives the following parametrization:

$$\mathbf{r}(t) = \langle 3 + 2 \cos t, 1, 5 + 3 \sin t \rangle$$

Further Insights and Challenges

49. Sketch the curve parametrized by $\mathbf{r}(t) = \langle |t| + t, |t| - t \rangle$.

SOLUTION We have:

$$|t| + t = \begin{cases} 0 & t \leq 0 \\ 2t & t > 0 \end{cases}; \quad |t| - t = \begin{cases} 2t & t \leq 0 \\ 0 & t > 0 \end{cases}$$

As t increases from $-\infty$ to 0, the x-coordinate is zero and the y-coordinate is positive and decreasing to zero. As t increases from 0 to $+\infty$, the y-coordinate is zero and the x-coordinate is positive and increasing to $+\infty$. We obtain the following curve:

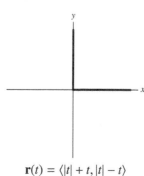

$\mathbf{r}(t) = \langle |t| + t, |t| - t \rangle$

51. Let C be the curve obtained by intersecting a cylinder of radius r and a plane. Insert two spheres of radius r into the cylinder above and below the plane, and let F_1 and F_2 be the points where the plane is tangent to the spheres [Figure 15(A)]. Let K be the vertical distance between the equators of the two spheres. Rediscover Archimedes's proof that C is an ellipse by showing that every point P on C satisfies

$$PF_1 + PF_2 = K \qquad \boxed{2}$$

Hint: If two lines through a point P are tangent to a sphere and intersect the sphere at Q_1 and Q_2 as in Figure 15(B), then the segments $\overline{PQ_1}$ and $\overline{PQ_2}$ have equal length. Use this to show that $PF_1 = PR_1$ and $PF_2 = PR_2$.

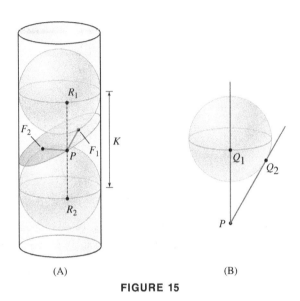

FIGURE 15

SOLUTION To show that C is an ellipse, we show that every point P on C satisfies:

$$\overline{F_1P} + \overline{F_2P} = K$$

We denote the points of intersection of the vertical line through P with the equators of the two spheres by R_1 and R_2 as in the figure, and denote the centers of the spheres by O_1 and O_2.

Since F_1 is the tangency point, the radius $\overline{O_1F_1}$ is perpendicular to the plane of the curve C, and therefore it is orthogonal to the segment $\overline{PF_1}$ on this plane. Hence, $\triangle O_1F_1P$ is a right triangle and by the Pythagorean Theorem we have:

$$\overline{O_1F_1}^2 + \overline{PF_1}^2 = \overline{O_1P}^2$$

$$r^2 + \overline{PF_1}^2 = \overline{O_1P}^2 \quad \Rightarrow \quad \overline{PF_1} = \sqrt{\overline{O_1P}^2 - r^2} \qquad (1)$$

$\triangle O_1 R_1 P$ is also a right triangle, hence by the Pythagorean Theorem we have:

$$\overline{O_1 R_1}^2 + \overline{R_1 P}^2 = \overline{O_1 P}^2$$

$$r^2 + \overline{R_1 P}^2 = \overline{O_1 P}^2 \quad \Rightarrow \quad \overline{PR_1} = \sqrt{\overline{O_1 P}^2 - r^2} \tag{2}$$

Combining (1) and (2) we get:

$$\overline{PF_1} = \overline{PR_1} \tag{3}$$

Similarly we have:

$$\overline{PF_2} = \overline{PR_2} \tag{4}$$

We now combine (3), (4) and the equality $\overline{PR_1} + \overline{PR_2} = K$ to obtain:

$$\overline{F_1 P} + \overline{F_2 P} = \overline{PR_1} + \overline{PR_2} = K$$

Thus, the sum of the distances of the points P on C to the two fixed points F_1 and F_2 is a constant $K > 0$, hence C is an ellipse.

53. *CAS* Now reprove the result of Exercise 51 using vector geometry. Assume that the cylinder has equation $x^2 + y^2 = r^2$ and the plane has equation $z = ax + by$.

(a) Show that the upper and lower spheres in Figure 15 have centers

$$C_1 = \left(0, 0, r\sqrt{a^2 + b^2 + 1}\right)$$

$$C_2 = \left(0, 0, -r\sqrt{a^2 + b^2 + 1}\right)$$

(b) Show that the points where the plane is tangent to the sphere are

$$F_1 = \frac{r}{\sqrt{a^2 + b^2 + 1}} (a, b, a^2 + b^2)$$

$$F_2 = \frac{-r}{\sqrt{a^2 + b^2 + 1}} (a, b, a^2 + b^2)$$

Hint: Show that $\overline{C_1 F_1}$ and $\overline{C_2 F_2}$ have length r and are orthogonal to the plane.

(c) Verify, with the aid of a computer algebra system, that Eq. (2) holds with

$$K = 2r\sqrt{a^2 + b^2 + 1}$$

To simplify the algebra, observe that since a and b are arbitrary, it suffices to verify Eq. (2) for the point $P = (r, 0, ar)$.

SOLUTION

(a) and (b) Since F_1 is the tangency point of the sphere and the plane, the radius to F_1 is orthogonal to the plane. Therefore to show that the center of the sphere is at C_1 and the tangency point is the given point we must show that:

$$\|\overrightarrow{C_1 F_1}\| = r \tag{1}$$

$$\overrightarrow{C_1 F_1} \text{ is orthogonal to the plane.} \tag{2}$$

We compute the vector $\overrightarrow{C_1 F_1}$:

$$\overrightarrow{C_1 F_1} = \left\langle \frac{ra}{\sqrt{a^2+b^2+1}}, \frac{rb}{\sqrt{a^2+b^2+1}}, \frac{r(a^2+b^2)}{\sqrt{a^2+b^2+1}} - r\sqrt{a^2+b^2+1} \right\rangle = \frac{r}{\sqrt{a^2+b^2+1}} \langle a, b, -1 \rangle$$

Hence,

$$\|\overrightarrow{C_1 F_1}\| = \frac{r}{\sqrt{a^2+b^2+1}} \|\langle a, b, -1 \rangle\| = \frac{r}{\sqrt{a^2+b^2+1}} \sqrt{a^2+b^2+(-1)^2} = r$$

We, thus, proved that (1) is satisfied. To show (2) we must show that $\overrightarrow{C_1 F_1}$ is parallel to the normal vector $\langle a, b, -1 \rangle$ to the plane $z = ax + by$ (i.e., $ax + by - z = 0$). The two vectors are parallel since by (1) $\overrightarrow{C_1 F_1}$ is a constant multiple of $\langle a, b, -1 \rangle$. In a similar manner one can show (1) and (2) for the vector $\overrightarrow{C_2 F_2}$.

(c) This is an extremely challenging problem. As suggested in the book, we use $P = (r, 0, ar)$, and we also use the expressions for F_1 and F_2 as given above. This gives us:

$$PF_1 = \sqrt{\left(1 + 2a^2 + b^2 - 2a\sqrt{1+a^2+b^2}\right)r^2}$$

$$PF_2 = \sqrt{\left(1 + 2a^2 + b^2 + 2a\sqrt{1+a^2+b^2}\right)r^2}$$

Their sum is not very inspiring:

$$PF_1 + PF_2 = \sqrt{\left(1 + 2a^2 + b^2 - 2a\sqrt{1+a^2+b^2}\right)r^2} + \sqrt{\left(1 + 2a^2 + b^2 + 2a\sqrt{1+a^2+b^2}\right)r^2}$$

Let us look, instead, at $(PF_1 + PF_2)^2$, and show that this is equal to K^2. Since everything is positive, this will imply that $PF_1 + PF_2 = K$, as desired.

$$(PF_1 + PF_2)^2 = 2r^2 + 4a^2r^2 + 2b^2r^2 + 2\sqrt{r^4 + 2b^2r^4 + b^4r^4}$$

$$= 2r^2 + 4a^2r^2 + 2b^2r^2 + 2(1 + b^2)r^2 = 4r^2(1 + a^2 + b^2) = K^2$$

13.2 Calculus of Vector-Valued Functions (LT Section 14.2)

Preliminary Questions

1. State the three forms of the Product Rule for vector-valued functions.

SOLUTION The Product Rule for scalar multiple $f(t)$ of a vector-valued function $\mathbf{r}(t)$ states that:

$$\frac{d}{dt} f(t)\mathbf{r}(t) = f(t)\mathbf{r}'(t) + f'(t)\mathbf{r}(t)$$

The Product Rule for dot products states that:

$$\frac{d}{dt} \mathbf{r}_1(t) \cdot \mathbf{r}_2(t) = \mathbf{r}_1(t) \cdot \mathbf{r}_2'(t) + \mathbf{r}_1'(t) \cdot \mathbf{r}_2(t)$$

Finally, the Product Rule for cross product is

$$\frac{d}{dt} \mathbf{r}_1(t) \times \mathbf{r}_2(t) = \mathbf{r}_1(t) \times \mathbf{r}_2'(t) + \mathbf{r}_1'(t) \times \mathbf{r}_2(t)$$

In Questions 2–6, indicate whether the statement is true or false, and if it is false, provide a correct statement.

2. The derivative of a vector-valued function is defined as the limit of a difference quotient, just as in the scalar-valued case.

SOLUTION The statement is true. The derivative of a vector-valued function $\mathbf{r}(t)$ is defined as a limit of the difference quotient:

$$\mathbf{r}'(t) = \lim_{t \to 0} \frac{\mathbf{r}(t+h) - \mathbf{r}(t)}{h}$$

in the same way as in the scalar-valued case.

3. The integral of a vector-valued function is obtained by integrating each component.

SOLUTION This statement is true. As discussed in the text, if $\mathbf{r}(t) = \langle x(t), y(t), z(t) \rangle$, then

$$\int_a^b \mathbf{r}(t)\, dt = \left\langle \int_a^b x(t)\, dt, \int_a^b y(t)\, dt, \int_a^b z(t)\, dt \right\rangle$$

4. The terms "velocity vector" and "tangent vector" for a path $\mathbf{r}(t)$ mean the same thing.

SOLUTION This statement is true.

5. The derivative of a vector-valued function is the slope of the tangent line, just as in the scalar case.

SOLUTION The statement is false. The derivative of a vector-valued function is again a vector-valued function, hence it cannot be the slope of the tangent line (which is a scalar). However, the derivative, $\mathbf{r}'(t_0)$ is the direction vector of the tangent line to the curve traced by $\mathbf{r}(t)$, at $\mathbf{r}(t_0)$.

6. The derivative of the cross product is the cross product of the derivatives.

SOLUTION The statement is false, since usually,

$$\frac{d}{dt}\mathbf{r}_1(t) \times \mathbf{r}_2(t) \neq \mathbf{r}_1'(t) \times \mathbf{r}_2'(t)$$

The correct statement is the Product Rule for Cross Products. That is,

$$\frac{d}{dt}\mathbf{r}_1(t) \times \mathbf{r}_2(t) = \mathbf{r}_1(t) \times \mathbf{r}_2'(t) + \mathbf{r}_1'(t) \times \mathbf{r}_2(t)$$

7. State whether the following derivatives of vector-valued functions $\mathbf{r}_1(t)$ and $\mathbf{r}_2(t)$ are scalars or vectors:

(a) $\dfrac{d}{dt}\mathbf{r}_1(t)$ **(b)** $\dfrac{d}{dt}(\mathbf{r}_1(t) \cdot \mathbf{r}_2(t))$ **(c)** $\dfrac{d}{dt}(\mathbf{r}_1(t) \times \mathbf{r}_2(t))$

SOLUTION (a) vector, (b) scalar, (c) vector.

Exercises

In Exercises 1–6, evaluate the limit.

1. $\lim\limits_{t \to 3}\left\langle t^2, 4t, \dfrac{1}{t}\right\rangle$

SOLUTION By the theorem on vector-valued limits we have:

$$\lim_{t \to 3}\left\langle t^2, 4t, \frac{1}{t}\right\rangle = \left\langle \lim_{t \to 3} t^2, \lim_{t \to 3} 4t, \lim_{t \to 3}\frac{1}{t}\right\rangle = \left\langle 9, 12, \frac{1}{3}\right\rangle$$

3. $\lim\limits_{t \to 0} e^{2t}\mathbf{i} + \ln(t+1)\mathbf{j} + 4\mathbf{k}$

SOLUTION Computing the limit of each component, we obtain:

$$\lim_{t \to 0}\left(e^{2t}\mathbf{i} + \ln(t+1)\mathbf{j} + 4\mathbf{k}\right) = \left(\lim_{t \to 0} e^{2t}\right)\mathbf{i} + \left(\lim_{t \to 0}\ln(t+1)\right)\mathbf{j} + \left(\lim_{t \to 0} 4\right)\mathbf{k} = e^0\mathbf{i} + (\ln 1)\mathbf{j} + 4\mathbf{k} = \mathbf{i} + 4\mathbf{k}$$

5. Evaluate $\lim\limits_{h \to 0}\dfrac{\mathbf{r}(t+h) - \mathbf{r}(t)}{h}$ for $\mathbf{r}(t) = \left\langle t^{-1}, \sin t, 4\right\rangle$.

SOLUTION This limit is the derivative $\frac{d\mathbf{r}}{dt}$. Using componentwise differentiation yields:

$$\lim_{h \to 0}\frac{\mathbf{r}(t+h) - \mathbf{r}(t)}{h} = \frac{d\mathbf{r}}{dt} = \left\langle \frac{d}{dt}\left(t^{-1}\right), \frac{d}{dt}(\sin t), \frac{d}{dt}(4)\right\rangle = \left\langle -\frac{1}{t^2}, \cos t, 0\right\rangle$$

In Exercises 7–12, compute the derivative.

7. $\mathbf{r}(t) = \langle t, t^2, t^3\rangle$

SOLUTION Using componentwise differentiation we get:

$$\frac{d\mathbf{r}}{dt} = \left\langle \frac{d}{dt}(t), \frac{d}{dt}(t^2), \frac{d}{dt}(t^3)\right\rangle = \langle 1, 2t, 3t^2\rangle$$

9. $\mathbf{r}(s) = \langle e^{3s}, e^{-s}, s^4\rangle$

SOLUTION Using componentwise differentiation we get:

$$\frac{d\mathbf{r}}{ds} = \left\langle \frac{d}{ds}e^{3s}, \frac{d}{ds}e^{-s}, \frac{d}{ds}s^4\right\rangle = \langle 3e^{3s}, -e^{-s}, 4s^3\rangle$$

11. $\mathbf{c}(t) = t^{-1}\mathbf{i} - e^{2t}\mathbf{k}$

SOLUTION Using componentwise differentiation we get:

$$\mathbf{c}'(t) = (t^{-1})'\mathbf{i} - (e^{2t})'\mathbf{k} = -t^{-2}\mathbf{i} - 2e^{2t}\mathbf{k}$$

13. Calculate $\mathbf{r}'(t)$ and $\mathbf{r}''(t)$ for $\mathbf{r}(t) = \langle t, t^2, t^3 \rangle$.

SOLUTION We perform the differentiation componentwise to obtain:

$$\mathbf{r}'(t) = \langle (t)', (t^2)', (t^3)' \rangle = \langle 1, 2t, 3t^2 \rangle$$

We now differentiate the derivative vector to find the second derivative:

$$\mathbf{r}''(t) = \frac{d}{dt}\langle 1, 2t, 3t^2 \rangle = \langle 0, 2, 6t \rangle$$

15. Sketch the curve parametrized by $\mathbf{r}_1(t) = \langle t, t^2 \rangle$ together with its tangent vector at $t = 1$. Then do the same for $\mathbf{r}_2(t) = \langle t^3, t^6 \rangle$.

SOLUTION Note that $\mathbf{r}_1'(t) = \langle 1, 2t \rangle$ and so $\mathbf{r}_1'(1) = \langle 1, 2 \rangle$. The graph of $\mathbf{r}_1(t)$ satisfies $y = x^2$. Likewise, $\mathbf{r}_2'(t) = \langle 3t^2, 6t^5 \rangle$ and so $\mathbf{r}_2'(1) = \langle 3, 6 \rangle$. The graph of $\mathbf{r}_2(t)$ also satisfies $y = x^2$. Both graphs and tangent vectors are given here.

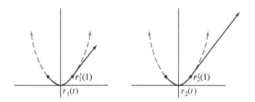

17. Determine the value of t between 0 and 2π such that the tangent vector to the cycloid $\mathbf{r}(t) = \langle t - \sin t, 1 - \cos t \rangle$ is parallel to $\langle \sqrt{3}, 1 \rangle$.

SOLUTION From Exercise 16, the tangent vector to the cycloid is

$$\mathbf{r}'(t) = \langle 1 - \cos t, \sin t \rangle$$

We want this vector to be parallel to $\langle \sqrt{3}, 1 \rangle$, so we want

$$\frac{1 - \cos t}{\sin t} = \sqrt{3} \quad \Rightarrow \quad (1 - \cos t)^2 = 3 \sin^2 t = 3(1 - \cos^2 t)$$

Expanding and simplifying gives $2\cos^2 t - \cos t - 1 = 0$. The roots of this quadratic are $\cos t = 1$ and $\cos t = -\frac{1}{2}$. We reject $\cos t = 1$, since this forces $\sin t = 0$. Thus $\cos t = -\frac{1}{2}$, so that $t = \frac{2\pi}{3}$ or $t = \frac{4\pi}{3}$. We get

$$\mathbf{r}'\left(\frac{2\pi}{3}\right) = \left\langle 1 - \cos\frac{2\pi}{3}, \sin\frac{2\pi}{3} \right\rangle = \left\langle \frac{3}{2}, \frac{\sqrt{3}}{2} \right\rangle = \frac{\sqrt{3}}{2}\langle \sqrt{3}, 1 \rangle$$

$$\mathbf{r}'\left(\frac{4\pi}{3}\right) = \left\langle 1 - \cos\frac{4\pi}{3}, \sin\frac{4\pi}{3} \right\rangle = \left\langle \frac{3}{2}, -\frac{\sqrt{3}}{2} \right\rangle = \frac{\sqrt{3}}{2}\langle \sqrt{3}, -1 \rangle$$

Only the first of these is a multiple of $\langle \sqrt{3}, 1 \rangle$, so the answer is $t = \frac{2\pi}{3}$.

In Exercises 19–22, evaluate the derivative by using the appropriate Product Rule, where

$$\mathbf{r}_1(t) = \langle t^2, t^3, t \rangle, \qquad \mathbf{r}_2(t) = \langle e^{3t}, e^{2t}, e^t \rangle$$

19. $\dfrac{d}{dt}(\mathbf{r}_1(t) \cdot \mathbf{r}_2(t))$

SOLUTION

$$\dfrac{d}{dt}(\mathbf{r}_1(t) \cdot \mathbf{r}_2(t)) = \mathbf{r}_1(t) \cdot \mathbf{r}_2'(t) + \mathbf{r}_1'(t) \cdot \mathbf{r}_2(t)$$

$$= \langle t^2, t^3, t \rangle \cdot \langle 3e^{3t}, 2e^{2t}, e^t \rangle + \langle 2t, 3t^2, 1 \rangle \cdot \langle e^{3t}, e^{2t}, e^t \rangle$$

$$= 3t^2 e^{3t} + 2t^3 e^{2t} + te^t + 2te^{3t} + 3t^2 e^{2t} + e^t$$

$$= (3t^2 + 2t)e^{3t} + (2t^3 + 3t^2)e^{2t} + (t+1)e^t$$

21. $\dfrac{d}{dt}(\mathbf{r}_1(t) \times \mathbf{r}_2(t))$

SOLUTION

$$\dfrac{d}{dt}(\mathbf{r}_1(t) \times \mathbf{r}_2(t)) = \mathbf{r}_1(t) \times \mathbf{r}_2'(t) + \mathbf{r}_1'(t) \times \mathbf{r}_2(t)$$

$$= \langle t^2, t^3, t \rangle \times \langle 3e^{3t}, 2e^{2t}, e^t \rangle + \langle 2t, 3t^2, 1 \rangle \times \langle e^{3t}, e^{2t}, e^t \rangle$$

$$= \begin{vmatrix} \mathbf{i} & \mathbf{j} & \mathbf{k} \\ t^2 & t^3 & t \\ 3e^{3t} & 2e^{2t} & e^t \end{vmatrix} + \begin{vmatrix} \mathbf{i} & \mathbf{j} & \mathbf{k} \\ 2t & 3t^2 & 1 \\ e^{3t} & e^{2t} & e^t \end{vmatrix}$$

$$= (t^3 e^t - 2te^{2t})\mathbf{i} + (3te^{3t} - t^2 e^t)\mathbf{j} + (2t^2 e^{2t} - 3t^3 e^{3t})\mathbf{k}$$

$$+ (3t^2 e^t - e^{2t})\mathbf{i} + (e^{3t} - 2te^t)\mathbf{j} + (2te^{2t} - 3t^2 e^{3t})\mathbf{k}$$

$$= [(t^3 + 3t^2)e^t - (2t+1)e^{2t}]\mathbf{i} + [(3t+1)e^{3t} - (t^2 + 2t)e^t]\mathbf{j}$$

$$+ [(2t^2 + 2t)e^{2t} - (3t^3 + 3t^2)e^{3t}]\mathbf{k}$$

In Exercises 23 and 24, let

$$\mathbf{r}_1(t) = \langle t^2, 1, 2t \rangle, \qquad \mathbf{r}_2(t) = \langle 1, 2, e^t \rangle$$

23. Compute $\dfrac{d}{dt}\mathbf{r}_1(t) \cdot \mathbf{r}_2(t)\bigg|_{t=1}$ in two ways:

(a) Calculate $\mathbf{r}_1(t) \cdot \mathbf{r}_2(t)$ and differentiate.

(b) Use the Dot Product Rule.

SOLUTION

(a) First we will calculate $\mathbf{r}_1(t) \cdot \mathbf{r}_2(t)$:

$$\mathbf{r}_1(t) \cdot \mathbf{r}_2(t) = \langle t^2, 1, 2t \rangle \cdot \langle 1, 2, e^t \rangle$$

$$= t^2 + 2 + 2te^t$$

And then differentiating we get:

$$\dfrac{d}{dt}(\mathbf{r}_1(t) \cdot \mathbf{r}_2(t)) = \dfrac{d}{dt}(t^2 + 2 + 2te^t) = 2t + 2te^t + 2e^t$$

$$\dfrac{d}{dt}(\mathbf{r}_1(t) \cdot \mathbf{r}_2(t))\bigg|_{t=1} = 2 + 2e + 2e = 2 + 4e$$

(b) First we differentiate:

$$\mathbf{r}_1(t) = \langle t^2, 1, 2t \rangle, \qquad \mathbf{r}_1'(t) = \langle 2t, 0, 2 \rangle$$

$$\mathbf{r}_2(t) = \langle 1, 2, e^t \rangle, \qquad \mathbf{r}_2'(t) = \langle 0, 0, e^t \rangle$$

Using the Dot Product Rule we see:

$$\frac{d}{dt}(\mathbf{r}_1(t) \cdot \mathbf{r}_2(t)) = \mathbf{r}_1(t) \cdot \mathbf{r}_2'(t) + \mathbf{r}_1'(t) \cdot \mathbf{r}_2(t)$$

$$= \langle t^2, 1, 2t \rangle \cdot \langle 0, 0, e^t \rangle + \langle 2t, 0, 2 \rangle \cdot \langle 1, 2, e^t \rangle$$

$$= 2te^t + 2t + 2e^t$$

$$\left.\frac{d}{dt}(\mathbf{r}_1(t) \cdot \mathbf{r}_2(t))\right|_{t=1} = 2e + 2 + 2e = 2 + 4e$$

In Exercises 25–28, evaluate $\frac{d}{dt}\mathbf{r}(g(t))$ using the Chain Rule.

25. $\mathbf{r}(t) = \langle t^2, 1-t \rangle, \quad g(t) = e^t$

SOLUTION We first differentiate the two functions:

$$\mathbf{r}'(t) = \frac{d}{dt}\langle t^2, 1-t \rangle = \langle 2t, -1 \rangle$$

$$g'(t) = \frac{d}{dt}(e^t) = e^t$$

Using the Chain Rule we get:

$$\frac{d}{dt}\mathbf{r}(g(t)) = g'(t)\mathbf{r}'(g(t)) = e^t \langle 2e^t, -1 \rangle = \langle 2e^{2t}, -e^t \rangle$$

27. $\mathbf{r}(t) = \langle e^t, e^{2t}, 4 \rangle, \quad g(t) = 4t + 9$

SOLUTION We first differentiate the two functions:

$$\mathbf{r}'(t) = \frac{d}{dt}\langle e^t, e^{2t}, 4 \rangle = \langle e^t, 2e^{2t}, 0 \rangle$$

$$g'(t) = \frac{d}{dt}(4t + 9) = 4$$

Using the Chain Rule we get:

$$\frac{d}{dt}\mathbf{r}(g(t)) = g'(t)\mathbf{r}'(g(t)) = 4\langle e^{4t+9}, 2e^{2(4t+9)}, 0 \rangle = \langle 4e^{4t+9}, 8e^{8t+18}, 0 \rangle$$

29. Let $\mathbf{r}(t) = \langle t^2, 1-t, 4t \rangle$. Calculate the derivative of $\mathbf{r}(t) \cdot \mathbf{a}(t)$ at $t = 2$, assuming that $\mathbf{a}(2) = \langle 1, 3, 3 \rangle$ and $\mathbf{a}'(2) = \langle -1, 4, 1 \rangle$.

SOLUTION By the Product Rule for dot products we have

$$\frac{d}{dt}\mathbf{r}(t) \cdot \mathbf{a}(t) = \mathbf{r}(t) \cdot \mathbf{a}'(t) + \mathbf{r}'(t) \cdot \mathbf{a}(t)$$

At $t = 2$ we have

$$\left.\frac{d}{dt}\mathbf{r}(t) \cdot \mathbf{a}(t)\right|_{t=2} = \mathbf{r}(2) \cdot \mathbf{a}'(2) + \mathbf{r}'(2) \cdot \mathbf{a}(2) \tag{1}$$

We compute the derivative $\mathbf{r}'(2)$:

$$\mathbf{r}'(t) = \frac{d}{dt}\langle t^2, 1-t, 4t \rangle = \langle 2t, -1, 4 \rangle \quad \Rightarrow \quad \mathbf{r}'(2) = \langle 4, -1, 4 \rangle \tag{2}$$

Also, $\mathbf{r}(2) = \langle 2^2, 1-2, 4 \cdot 2 \rangle = \langle 4, -1, 8 \rangle$. Substituting the vectors in the equation above, we obtain:

$$\left.\frac{d}{dt}\mathbf{r}(t) \cdot \mathbf{a}(t)\right|_{t=2} = \langle 4, -1, 8 \rangle \cdot \langle -1, 4, 1 \rangle + \langle 4, -1, 4 \rangle \cdot \langle 1, 3, 3 \rangle = (-4 - 4 + 8) + (4 - 3 + 12) = 13$$

The derivative of $\mathbf{r}(t) \cdot \mathbf{a}(t)$ at $t = 2$ is 13.

In Exercises 31–36, find a parametrization of the tangent line at the point indicated.

31. $\mathbf{r}(t) = \langle t^2, t^4 \rangle$, $t = -2$

SOLUTION The tangent line has the following parametrization:

$$\ell(t) = \mathbf{r}(1) + t\mathbf{r}'(1) \quad (1)$$

We compute the vectors $\mathbf{r}(1)$ and $\mathbf{r}'(1)$:

$$\mathbf{r}(1) = \langle 1^2, 1^4 \rangle = \langle 1, 1 \rangle$$

$$\mathbf{r}'(t) = \frac{d}{dt}\langle t^2, t^4 \rangle = \langle 2t, 4t^3 \rangle \quad \Rightarrow \quad \mathbf{r}'(1) = \langle 2, 4 \rangle$$

Substituting in (1) gives:

$$\ell(t) = \langle 1, 1 \rangle + t\langle 2, 4 \rangle = \langle 1 + 2t, 1 + 4t \rangle$$

The parametrization for the tangent line is, thus,

$$x = 1 + 2t, \quad y = 1 + 4t.$$

To find a direct relation between y and x, we express t in terms of x and substitute in $y = 1 + 4t$. This gives:

$$x = 1 + 2t \Rightarrow t = \frac{x-1}{2}$$

Hence,

$$y = 1 + 4t = 1 + 4 \cdot \frac{x-1}{2} = 1 + 2x - 2 = 2x - 1$$

The equation of the tangent line is $y = 2x - 1$.

33. $\mathbf{r}(t) = \langle 1 - t^2, 5t, 2t^3 \rangle$, $t = 2$

SOLUTION The tangent line is parametrized by:

$$\ell(t) = \mathbf{r}(2) + t\mathbf{r}'(2) \quad (1)$$

We compute the vectors in the above parametrization:

$$\mathbf{r}(2) = \langle 1 - 2^2, 5 \cdot 2, 2 \cdot 2^3 \rangle = \langle -3, 10, 16 \rangle$$

$$\mathbf{r}'(t) = \frac{d}{dt}\langle 1 - t^2, 5t, 2t^3 \rangle = \langle -2t, 5, 6t^2 \rangle \quad \Rightarrow \quad \mathbf{r}'(2) = \langle -4, 5, 24 \rangle$$

Substituting the vectors in (1) we obtain the following parametrization:

$$\ell(t) = \langle -3, 10, 16 \rangle + t\langle -4, 5, 24 \rangle = \langle -3 - 4t, 10 + 5t, 16 + 24t \rangle$$

35. $\mathbf{r}(s) = 4s^{-1}\mathbf{i} - \frac{8}{3}s^{-3}\mathbf{k}$, $s = 2$

SOLUTION The tangent line is parametrized by:

$$\ell(s) = \mathbf{r}(2) + s\mathbf{r}'(2) \quad (1)$$

We compute the vectors in the above parametrization:

$$\mathbf{r}(2) = 4(2)^{-1}\mathbf{i} - \frac{8}{3}(2)^{-3}\mathbf{k} = 2\mathbf{i} - \frac{1}{3}\mathbf{k}$$

$$\mathbf{r}'(s) = \frac{d}{ds}\left(4s^{-1}\mathbf{i} - \frac{8}{3}s^{-3}\mathbf{k}\right) = -4s^{-2}\mathbf{i} + 8s^{-4}\mathbf{k} \quad \Rightarrow \quad \mathbf{r}'(2) = -\mathbf{i} + \frac{1}{2}\mathbf{k}$$

Substituting the vectors in (1) we obtain the following parametrization:

$$\ell(t) = \left(2\mathbf{i} - \frac{1}{3}\mathbf{k}\right) + s\left(-\mathbf{i} + \frac{1}{2}\mathbf{k}\right) = (2 - s)\mathbf{i} + \left(\frac{1}{2}s - \frac{1}{3}\right)\mathbf{k}$$

37. Use Example 4 to calculate $\frac{d}{dt}(\mathbf{r} \times \mathbf{r}')$, where $\mathbf{r}(t) = \langle t, t^2, e^t \rangle$.

SOLUTION In Example 4 it is proved that:

$$\frac{d}{dt}\mathbf{r} \times \mathbf{r}' = \mathbf{r} \times \mathbf{r}'' \tag{1}$$

We compute the derivatives $\mathbf{r}'(t)$ and $\mathbf{r}''(t)$:

$$\mathbf{r}'(t) = \frac{d}{dt}\langle t^2, t^3, e^t \rangle = \langle 2t, 3t^2, e^t \rangle$$

$$\mathbf{r}''(t) = \frac{d}{dt}\langle 2t, 3t^2, e^t \rangle = \langle 2, 6t, e^t \rangle$$

Using (1) we get

$$\frac{d}{dt}\mathbf{r} \times \mathbf{r}' = \mathbf{r} \times \mathbf{r}'' = \langle t^2, t^3, e^t \rangle \times \langle 2, 6t, e^t \rangle = \begin{vmatrix} \mathbf{i} & \mathbf{j} & \mathbf{k} \\ t^2 & t^3 & e^t \\ 2 & 6t & e^t \end{vmatrix} = (t^3 e^t - 6te^t)\mathbf{i} - (t^2 e^t - 2e^t)\mathbf{j} + (6t^3 - 2t^3)\mathbf{k}$$

$$= (t^3 - 6t)e^t \mathbf{i} - (t^2 - 2)e^t \mathbf{j} + 4t^3 \mathbf{k} = \langle (t^3 - 6t)e^t, (2 - t^2)e^t, 4t^3 \rangle$$

39. Show that the *derivative of the norm* is not equal to the *norm of the derivative* by verifying that $\|\mathbf{r}(t)\|' \neq \|\mathbf{r}(t)'\|$ for $\mathbf{r}(t) = \langle t, 1, 1 \rangle$.

SOLUTION First let us compute $\|\mathbf{r}(t)\|'$ for $\mathbf{r}(t) = \langle t, 1, 1 \rangle$:

$$\|\mathbf{r}(t)\|' = \frac{d}{dt}(\sqrt{t^2 + 2}) = \frac{t}{\sqrt{t^2 + 2}}$$

Now, first let us compute the derivative, $\mathbf{r}'(t)$:

$$\mathbf{r}'(t) = \langle 1, 0, 0 \rangle$$

and then computing the norm:

$$\|\mathbf{r}'(t)\| = \|\langle 1, 0, 0 \rangle\| = \sqrt{1} = 1$$

It is clear in this example, that $\|\mathbf{r}(t)\|' \neq \|\mathbf{r}'(t)\|$.

In Exercises 41–48, evaluate the integrals.

41. $\int_{-2}^{2} \langle t^2 + 4t, 4t^3 - t \rangle \, dt$

SOLUTION Vector-valued integration is defined via componentwise integration. Thus, we first compute the integral of each component.

$$\int_{-2}^{2} t^2 + 4t \, dt = \frac{1}{3}t^3 + 2t^2 \Big|_{-2}^{2} = \left(\frac{8}{3} + 8\right) - \left(-\frac{8}{3} + 8\right) = \frac{16}{3}$$

$$\int_{-2}^{2} 4t^3 - t \, dt = t^4 - \frac{1}{2}t^2 \Big|_{-2}^{2} = (16 - 2) - (16 - 2) = 0$$

Therefore,

$$\int_{-2}^{2} \langle t^2 + 4t, 4t^3 - t \rangle \, dt = \left\langle \int_{-2}^{2} t^2 + 4t \, dt, \int_{-2}^{2} 4t^3 - t \, dt \right\rangle = \left\langle \frac{16}{3}, 0 \right\rangle$$

43. $\int_{-2}^{2} (u^3 \mathbf{i} + u^5 \mathbf{j})\, du$

SOLUTION The vector-valued integration is defined via componentwise integration. Thus, we first compute the integral of each component.

$$\int_{-2}^{2} u^3\, du = \frac{u^4}{4}\bigg|_{-2}^{2} = \frac{16}{4} - \frac{16}{4} = 0$$

$$\int_{-2}^{2} u^5\, du = \frac{u^6}{6}\bigg|_{-2}^{2} = \frac{64}{6} - \frac{64}{6} = 0$$

Therefore,

$$\int_{-2}^{2} (u^3\mathbf{i} + u^5\mathbf{j})\, du = \left(\int_{-2}^{2} u^3\, du\right)\mathbf{i} + \left(\int_{-2}^{2} u^5\, du\right)\mathbf{j} = 0\mathbf{i} + 0\mathbf{j}$$

45. $\int_{0}^{\pi} \langle -\sin t, 6t, 2t + \cos 2t \rangle\, dt$

SOLUTION Vector-valued integration is defined via componentwise integration. Thus, we first compute the integral of each component.

$$\int_{0}^{\pi} (-\sin t)\, dt = \cos t\bigg|_{0}^{\pi} = \cos \pi - \cos 0 = -2$$

$$\int_{0}^{\pi} 6t\, dt = 3t^2\bigg|_{0}^{\pi} = 3\pi^2 - 0 = 3\pi^2$$

$$\int_{0}^{\pi} (2t + \cos 2t)\, dt = t^2 + \frac{1}{2}\sin 2t\bigg|_{0}^{\pi} = (\pi^2 + 0) - (0^2 + 0) = \pi^2$$

Therefore,

$$\int_{0}^{\pi} \langle -\sin t, 6t, 2t + \cos 2t \rangle\, dt = \left\langle \int_{0}^{\pi}(-\sin t)\, dt,\ \int_{0}^{\pi} 6t\, dt,\ \int_{0}^{\pi}(2t + \cos 2t)\, dt\right\rangle = \langle -2, 3\pi^2, \pi^2 \rangle$$

47. $\int_{1}^{4} (t^{-1}\mathbf{i} + 4\sqrt{t}\,\mathbf{j} - 8t^{3/2}\mathbf{k})\, dt$

SOLUTION We perform the integration componentwise. Computing the integral of each component we get:

$$\int_{1}^{4} t^{-1}\, dt = \ln t\bigg|_{1}^{4} = \ln 4 - \ln 1 = \ln 4$$

$$\int_{1}^{4} 4\sqrt{t}\, dt = 4 \cdot \frac{2}{3} t^{3/2}\bigg|_{1}^{4} = \frac{8}{3}\left(4^{3/2} - 1\right) = \frac{56}{3}$$

$$\int_{1}^{4} -8t^{3/2}\, dt = -\frac{16}{5} t^{5/2}\bigg|_{1}^{4} = -\frac{16}{5}\left(4^{5/2} - 1\right) = -\frac{496}{5}$$

Hence,

$$\int_{1}^{4} \left(t^{-1}\mathbf{i} + 4\sqrt{t}\,\mathbf{j} - 8t^{3/2}\mathbf{k}\right) dt = (\ln 4)\mathbf{i} + \frac{56}{3}\mathbf{j} - \frac{496}{5}\mathbf{k}$$

In Exercises 49–56, find both the general solution of the differential equation and the solution with the given initial condition.

49. $\dfrac{d\mathbf{r}}{dt} = \langle 1 - 2t, 4t \rangle,\quad \mathbf{r}(0) = \langle 3, 1 \rangle$

SOLUTION We first find the general solution by integrating $\frac{d\mathbf{r}}{dt}$:

$$\mathbf{r}(t) = \int \langle 1 - 2t, 4t \rangle\, dt = \left\langle \int (1 - 2t)\, dt,\ \int 4t\, dt \right\rangle = \langle t - t^2, 2t^2 \rangle + \mathbf{c} \tag{1}$$

Since $\mathbf{r}(0) = \langle 3, 1 \rangle$, we have:

$$\mathbf{r}(0) = \langle 0 - 0^2, 2 \cdot 0^2 \rangle + \mathbf{c} = \langle 3, 1 \rangle \Rightarrow \mathbf{c} = \langle 3, 1 \rangle$$

Substituting in (1) gives the solution:

$$\mathbf{r}(t) = \langle t - t^2, 2t^2 \rangle + \langle 3, 1 \rangle = \langle -t^2 + t + 3, 2t^2 + 1 \rangle$$

51. $\mathbf{r}'(t) = t^2 \mathbf{i} + 5t\mathbf{j} + \mathbf{k}, \quad \mathbf{r}(1) = \mathbf{j} + 2\mathbf{k}$

SOLUTION We first find the general solution by integrating $\mathbf{r}'(t)$:

$$\mathbf{r}(t) = \int \left(t^2 \mathbf{i} + 5t\mathbf{j} + \mathbf{k} \right) dt = \left(\int t^2 \, dt \right) \mathbf{i} + \left(\int 5t \, dt \right) \mathbf{j} + \left(\int 1 \, dt \right) \mathbf{k} = \left(\frac{1}{3} t^3 \right) \mathbf{i} + \left(\frac{5}{2} t^2 \right) \mathbf{j} + t\mathbf{k} + \mathbf{c} \quad (1)$$

The solution which satisfies the initial condition must satisfy:

$$\mathbf{j} + 2\mathbf{k} = \mathbf{r}(0) = \left(\frac{1}{3} \cdot 1^3 \right) \mathbf{i} + \left(\frac{5}{2} \cdot 1^2 \right) \mathbf{j} + 1 \cdot \mathbf{k} + \mathbf{c} = \frac{1}{3} \mathbf{i} + \frac{5}{2} \mathbf{j} + \mathbf{k} + \mathbf{c}$$

That is,

$$\mathbf{c} = -\frac{1}{3} \mathbf{i} - \frac{3}{2} \mathbf{j} + \mathbf{k}$$

Substituting in (1) gives the following solution:

$$\mathbf{r}(t) = \left(\frac{1}{3} t^3 \right) \mathbf{i} + \left(\frac{5}{2} t^2 \right) \mathbf{j} + t\mathbf{k} - \frac{1}{3} \mathbf{i} - \frac{3}{2} \mathbf{j} + \mathbf{k} = \frac{t^3 - 1}{3} \mathbf{i} + \frac{5t^2 - 3}{2} \mathbf{j} + (t + 1)\mathbf{k}$$

53. $\mathbf{r}''(t) = 16\mathbf{k}, \quad \mathbf{r}(0) = \langle 1, 0, 0 \rangle, \quad \mathbf{r}'(0) = \langle 0, 1, 0 \rangle$

SOLUTION To find the general solution we first find $\mathbf{r}'(t)$ by integrating $\mathbf{r}''(t)$:

$$\mathbf{r}'(t) = \int \mathbf{r}''(t) \, dt = \int 16\mathbf{k} \, dt = (16t)\mathbf{k} + \mathbf{c}_1 \quad (1)$$

We now integrate $\mathbf{r}'(t)$ to find the general solution $\mathbf{r}(t)$:

$$\mathbf{r}(t) = \int \mathbf{r}'(t) \, dt = \int ((16t)\mathbf{k} + \mathbf{c}_1) \, dt = \left(\int 16(t) \, dt \right) \mathbf{k} + \mathbf{c}_1 t + \mathbf{c}_2 = (8t^2)\mathbf{k} + \mathbf{c}_1 t + \mathbf{c}_2 \quad (2)$$

We substitute the initial conditions in (1) and (2). This gives:

$$\mathbf{r}'(0) = \mathbf{c}_1 = \langle 0, 1, 0 \rangle = \mathbf{j}$$

$$\mathbf{r}(0) = 0\mathbf{k} + \mathbf{c}_1 \cdot 0 + \mathbf{c}_2 = \langle 1, 0, 0 \rangle \quad \Rightarrow \quad \mathbf{c}_2 = \langle 1, 0, 0 \rangle = \mathbf{i}$$

Combining with (2) we obtain the following solution:

$$\mathbf{r}(t) = (8t^2)\mathbf{k} + t\mathbf{j} + \mathbf{i} = \mathbf{i} + t\mathbf{j} + (8t^2)\mathbf{k}$$

55. $\mathbf{r}''(t) = \langle 0, 2, 0 \rangle, \quad \mathbf{r}(3) = \langle 1, 1, 0 \rangle, \mathbf{r}'(3) = \langle 0, 0, 1 \rangle$

SOLUTION To find the general solution we first find $\mathbf{r}'(t)$ by integrating $\mathbf{r}''(t)$:

$$\mathbf{r}'(t) = \int \mathbf{r}''(t) \, dt = \int \langle 0, 2, 0 \rangle \, dt = \langle 0, 2t, 0 \rangle + \mathbf{c}_1 \quad (1)$$

We now integrate $\mathbf{r}'(t)$ to find the general solution $\mathbf{r}(t)$:

$$\mathbf{r}(t) = \int \mathbf{r}'(t) \, dt = \int (\langle 0, 2t, 0 \rangle + \mathbf{c}_1) \, dt = \langle 0, t^2, 0 \rangle + \mathbf{c}_1 t + \mathbf{c}_2 \quad (2)$$

We substitute the initial conditions in (1) and (2). This gives:

$$\mathbf{r}'(3) = \langle 0, 6, 0 \rangle + \mathbf{c}_1 = \langle 0, 0, 1 \rangle \quad \Rightarrow \quad \mathbf{c}_1 = \langle 0, -6, 1 \rangle$$

$$\mathbf{r}(3) = \langle 0, 9, 0 \rangle + \mathbf{c}_1(3) + \mathbf{c}_2 = \langle 1, 1, 0 \rangle$$

$$\langle 0, 9, 0 \rangle + \langle 0, -18, 3 \rangle + \mathbf{c}_2 = \langle 1, 1, 0 \rangle$$

$$\Rightarrow \quad \mathbf{c}_2 = \langle 1, 10, -3 \rangle$$

Combining with (2) we obtain the following solution:

$$\mathbf{r}(t) = \langle 0, t^2, 0 \rangle + t \langle 0, -6, 1 \rangle + \langle 1, 10, -3 \rangle$$
$$= \langle 1, t^2 - 6t + 10, t - 3 \rangle$$

57. Find the location at $t = 3$ of a particle whose path (Figure 8) satisfies

$$\frac{d\mathbf{r}}{dt} = \left\langle 2t - \frac{1}{(t+1)^2}, 2t - 4 \right\rangle, \quad \mathbf{r}(0) = \langle 3, 8 \rangle$$

FIGURE 8 Particle path.

SOLUTION Integrate $\mathbf{r}'$ to obtain the general solution:

$$\mathbf{r}(t) = \int \left\langle 2t - \frac{1}{(t+1)^2}, 2t - 4 \right\rangle dt = \left\langle t^2 + \frac{1}{t+1}, t^2 - 4t \right\rangle + \mathbf{c}$$

Now substitute the initial condition $\mathbf{r}(0) = \langle 3, 8 \rangle$ to get

$$\left\langle 0^2 + \frac{1}{0+1}, 0^2 - 4 \cdot 0 \right\rangle + \mathbf{c} = \langle 3, 8 \rangle \quad \Rightarrow \quad \langle 1, 0 \rangle + \mathbf{c} = \langle 3, 8 \rangle \quad \Rightarrow \quad \mathbf{c} = \langle 2, 8 \rangle$$

Therefore

$$\mathbf{r}(t) = \left\langle t^2 + \frac{1}{t+1}, t^2 - 4t \right\rangle + \mathbf{c} = \left\langle t^2 + \frac{1}{t+1} + 2, t^2 - 4t + 8 \right\rangle$$

So at $t = 3$, the position of the particle is

$$\mathbf{r}(3) = \left\langle 3^2 + \frac{1}{3+1} + 2, 3^2 - 4 \cdot 3 + 8 \right\rangle = \left\langle \frac{45}{4}, 5 \right\rangle$$

59. A fighter plane, which can shoot a laser beam straight ahead, travels along the path $\mathbf{r}(t) = \langle 5 - t, 21 - t^2, 3 - t^3/27 \rangle$. Show that there is precisely one time t at which the pilot can hit a target located at the origin.

SOLUTION By the given information the laser beam travels in the direction of $\mathbf{r}'(t)$. The pilot hits a target located at the origin at the time t when $\mathbf{r}'(t)$ points toward the origin, that is, when $\mathbf{r}(t)$ and $\mathbf{r}'(t)$ are parallel and point to opposite directions.

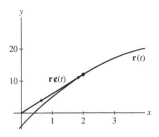

We find $\mathbf{r}'(t)$:

$$\mathbf{r}'(t) = \frac{d}{dt}\left\langle 5 - t, 21 - t^2, 3 - \frac{t^3}{27} \right\rangle = \left\langle -1, -2t, -\frac{t^2}{9} \right\rangle$$

We first find t such that $\mathbf{r}(t)$ and $\mathbf{r}'(t)$ are parallel, that is, we find t such that the cross product of the two vectors is zero. We obtain:

$$\mathbf{0} = \mathbf{r}'(t) \times \mathbf{r}(t) = \begin{vmatrix} \mathbf{i} & \mathbf{j} & \mathbf{k} \\ -1 & -2t & -\frac{t^2}{9} \\ 5-t & 21-t^2 & 3-\frac{t^3}{27} \end{vmatrix}$$

$$= \left(-2t\left(3 - \frac{t^3}{27}\right) + \frac{t^2}{9}(21 - t^2)\right)\mathbf{i} - \left(-\left(3 - \frac{t^3}{27}\right) + \frac{t^2}{9}(5-t)\right)\mathbf{j} + (-(21 - t^2) + 2t(5-t))\mathbf{k}$$

$$= \left(\frac{-t^4}{27} + \frac{7t^2}{3} - 6t\right)\mathbf{i} - \left(-\frac{2t^3}{27} + \frac{5t^2}{9} - 3\right)\mathbf{j} + (-t^2 + 10t - 21)\mathbf{k}$$

Equating each component to zero we obtain the following equations:

$$-\frac{t^4}{27} + \frac{7}{3}t^2 - 6t = 0$$

$$-\frac{2t^3}{27} + \frac{5t^2}{9} - 3 = 0$$

$$-t^2 + 10t - 21 = -(t-7)(t-3) = 0$$

The third equation implies that $t = 3$ or $t = 7$. Only $t = 3$ satisfies the other two equations as well. We now must verify that $\mathbf{r}(3)$ and $\mathbf{r}'(3)$ point in opposite directions. We find these vectors:

$$\mathbf{r}(3) = \left\langle 5-3, 21-3^2, 3 - \frac{3^3}{27}\right\rangle = \langle 2, 12, 2\rangle$$

$$\mathbf{r}'(3) = \left\langle -1, -2 \cdot 3, -\frac{3^2}{9}\right\rangle = \langle -1, -6, -1\rangle$$

Since $\mathbf{r}(3) = -2\mathbf{r}'(3)$, the vectors point in opposite direction. We conclude that only at time $t = 3$ can the pilot hit a target located at the origin.

61. Find all solutions to $\mathbf{r}'(t) = \mathbf{v}$ with initial condition $\mathbf{r}(1) = \mathbf{w}$, where $\mathbf{v}$ and $\mathbf{w}$ are constant vectors in $\mathbf{R}^3$.

SOLUTION We denote the components of the constant vector $\mathbf{v}$ by $\mathbf{v} = \langle v_1, v_2, v_3\rangle$ and integrate to find the general solution. This gives:

$$\mathbf{r}(t) = \int \mathbf{v}\, dt = \int \langle v_1, v_2, v_3\rangle\, dt = \left\langle \int v_1\, dt, \int v_2\, dt, \int v_3\, dt\right\rangle$$

$$= \langle v_1 t + c_1, v_2 t + c_2, v_3 t + c_3\rangle = t\langle v_1, v_2, v_3\rangle + \langle c_1, c_2, c_3\rangle$$

We let $\mathbf{c} = \langle c_1, c_2, c_3\rangle$ and obtain:

$$\mathbf{r}(t) = t\mathbf{v} + \mathbf{c} = \mathbf{c} + t\mathbf{v}$$

The initial condition gives

$$\mathbf{w} = \mathbf{r}(1) = \mathbf{c} + 1\mathbf{v} = \mathbf{c} + \mathbf{v} \quad \Rightarrow \quad \mathbf{c} = \mathbf{w} - \mathbf{v}$$

Therefore the solution is $\mathbf{r}(t) = \mathbf{w} - \mathbf{v} + t\mathbf{v}$.

63. Find all solutions to $\mathbf{r}'(t) = 2\mathbf{r}(t)$, where $\mathbf{r}(t)$ is a vector-valued function in 3-space.

SOLUTION We denote the components of $\mathbf{r}(t)$ by $\mathbf{r}(t) = \langle x(t), y(t), z(t)\rangle$. Then, $\mathbf{r}'(t) = \langle x'(t), y'(t), z'(t)\rangle$. Substituting in the differential equation we get:

$$\langle x'(t), y'(t), z'(t)\rangle = 2\langle x(t), y(t), z(t)\rangle$$

Equating corresponding components gives:

$$\begin{aligned} x'(t) &= 2x(t) & x(t) &= c_1 e^{2t} \\ y'(t) &= 2y(t) \quad \Rightarrow \quad y(t) &= c_2 e^{2t} \\ z'(t) &= 2z(t) & z(t) &= c_3 e^{2t} \end{aligned}$$

We denote the constant vector by $\mathbf{c} = \langle c_1, c_2, c_3\rangle$ and obtain the following solutions:

$$\mathbf{r}(t) = \langle c_1 e^{2t}, c_2 e^{2t}, c_3 e^{2t}\rangle = e^{2t}\langle c_1, c_2, c_3\rangle = e^{2t}\mathbf{c}$$

65. Prove that the **Bernoulli spiral** (Figure 9) with parametrization $\mathbf{r}(t) = \langle e^t \cos 4t, e^t \sin 4t \rangle$ has the property that the angle ψ between the position vector and the tangent vector is constant. Find the angle ψ in degrees.

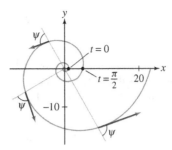

FIGURE 9 Bernoulli spiral.

SOLUTION First, let us compute the tangent vector, $\mathbf{r}'(t)$:

$$\mathbf{r}(t) = \langle e^t \cos 4t, e^t \sin 4t \rangle, \quad \Rightarrow \quad \mathbf{r}'(t) = \langle -4e^t \sin 4t + e^t \cos 4t, 4e^t \cos 4t + e^t \sin 4t \rangle$$

Then recall the identity that $\mathbf{a} \cdot \mathbf{b} = \|\mathbf{a}\| \cdot \|\mathbf{b}\| \cos\theta$, where θ is the angle between $\mathbf{a}$ and $\mathbf{b}$, so then,

$$\mathbf{r}(t) \cdot \mathbf{r}'(t) = \langle e^t \cos 4t, e^t \sin 4t \rangle \cdot \langle -4e^t \sin 4t + e^t \cos 4t, 4e^t \cos 4t + e^t \sin 4t \rangle$$

$$= -4e^{2t} \sin 4t \cos 4t + e^{2t} \cos^2 4t + 4e^{2t} \sin 4t \cos 4t + e^{2t} \sin^2 4t$$

$$= e^{2t}(\cos^2 4t + \sin^2 4t)$$

$$= e^{2t}$$

Then, computing norms, we get:

$$\|\mathbf{r}(t)\| = \sqrt{e^{2t} \cos^2 4t + e^{2t} \sin^2 4t} = \sqrt{e^{2t}(\cos^2 4t + \sin^2 4t)} = e^t$$

$$\|\mathbf{r}'(t)\| = \sqrt{(-4e^t \sin 4t + e^t \cos 4t)^2 + (4e^t \cos 4t + e^t \sin 4t)^2}$$

$$= \sqrt{16e^{2t} \sin^2 4t - 4e^{2t} \sin 4t \cos 4t + e^{2t} \cos^2 4t + 16e^{2t} \cos^2 4t + 4e^{2t} \sin 4t \cos 4t + e^{2t} \sin^2 4t}$$

$$= \sqrt{16e^{2t}(\sin^2 4t + \cos^2 4t) + e^{2t}(\cos^2 4t + \sin^2 4t)}$$

$$= \sqrt{16e^{2t} + e^{2t}}$$

$$= \sqrt{17} e^t$$

Then using the dot product relation listed above we get:

$$e^{2t} = e^t(\sqrt{17} e^t) \cos\theta = \sqrt{17} e^{2t} \cos\theta$$

Hence

$$\cos\theta = \frac{1}{\sqrt{17}}, \quad \Rightarrow \quad \theta \approx 75.96°$$

Therefore, the angle between the position vector and the tangent vector is constant.

67. Prove that if $\|\mathbf{r}(t)\|$ takes on a local minimum or maximum value at t_0, then $\mathbf{r}(t_0)$ is orthogonal to $\mathbf{r}'(t_0)$. Explain how this result is related to Figure 11. *Hint:* Observe that if $\|\mathbf{r}(t_0)\|$ is a minimum, then $\mathbf{r}(t)$ is tangent at t_0 to the sphere of radius $\|\mathbf{r}(t_0)\|$ centered at the origin.

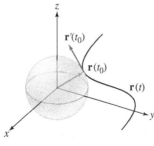

FIGURE 11

SOLUTION Suppose that $\|\mathbf{r}(t)\|$ takes on a minimum or maximum value at $t = t_0$. Hence, $\|\mathbf{r}(t)\|^2$ also takes on a minimum or maximum value at $t = t_0$, therefore $\frac{d}{dt}\|\mathbf{r}(t)\|^2\big|_{t=t_0} = 0$. Using the Product Rule for dot products we get

$$\frac{d}{dt}\|\mathbf{r}(t)\|^2\bigg|_{t=t_0} = \frac{d}{dt}\mathbf{r}(t) \cdot \mathbf{r}(t)\bigg|_{t=t_0} = \mathbf{r}(t_0) \cdot \mathbf{r}'(t_0) + \mathbf{r}'(t_0) \cdot \mathbf{r}(t_0) = 2\mathbf{r}(t_0) \cdot \mathbf{r}'(t_0) = 0$$

Thus $\mathbf{r}(t_0) \cdot \mathbf{r}'(t_0) = 0$, which implies the orthogonality of $\mathbf{r}(t_0)$ and $\mathbf{r}'(t_0)$. In Figure 11, $\|\mathbf{r}(t_0)\|$ is a minimum and the path intersects the sphere of radius $\|\mathbf{r}(t_0)\|$ at a single point. Therefore, the point of intersection is a tangency point which implies that $\mathbf{r}'(t_0)$ is tangent to the sphere at t_0. We conclude that $\mathbf{r}(t_0)$ and $\mathbf{r}'(t_0)$ are orthogonal.

69. Use FTC I from single-variable calculus to prove the first part of the Fundamental Theorem of Calculus for Vector-Valued Functions.

SOLUTION Suppose

$$\mathbf{r}(t) = \langle x_1(t), x_2(t), \ldots, x_n(t) \rangle$$

and let $X_i(t)$ be an antiderivative of $x_i(t)$. Since vector integration is performed componentwise, we have (using FTC I)

$$\int_a^b \mathbf{r}(t)\,dt = \left\langle \int_a^b x_1(t)\,dt, \int_a^b x_2(t)\,dt, \ldots, \int_a^b x_n(t)\,dt \right\rangle$$
$$= \langle X_1(b) - X_1(a), X_2(b) - X_2(a), \ldots, X_n(b) - X_n(a) \rangle$$
$$= \langle X_1(b), X_2(b), \ldots, X_n(b) \rangle - \langle X_1(a), X_2(a), \ldots, X_n(a) \rangle$$

Now, again since vector integration is performed componentwise, an antiderivative of $\mathbf{r}(t)$ is given by $\mathbf{R}(t) = \langle X_1, X_2, \ldots, X_n \rangle$, so that finally

$$\int_a^b \mathbf{r}(t) = \mathbf{R}(b) - \mathbf{R}(a)$$

Further Insights and Challenges

71. Let $\mathbf{r}(t) = \langle x(t), y(t) \rangle$ trace a plane curve C. Assume that $x'(t_0) \neq 0$. Show that the slope of the tangent vector $\mathbf{r}'(t_0)$ is equal to the slope dy/dx of the curve at $\mathbf{r}(t_0)$.

SOLUTION The slope of the tangent vector $\mathbf{r}'(t) = \langle x'(t), y'(t) \rangle$ at $t = t_0$ is (since $x'(t_0) \neq 0$)

$$\frac{y'(t_0)}{x'(t_0)}$$

On the other hand, by the Chain Rule we have

$$\frac{dy}{dt} = \frac{dy}{dx} \cdot \frac{dx}{dt}$$

Hence, at the points where $\frac{dx}{dt} \neq 0$ we have:

$$\frac{dy}{dx} = \frac{\frac{dy}{dt}}{\frac{dx}{dt}} = \frac{y'(t)}{x'(t)}$$

so that

$$\frac{dy}{dx}\bigg|_{x=x(t_0)} = \frac{y'(t_0)}{x'(t_0)}$$

Thus the two expressions are equal.

73. Prove the Sum and Constant Multiple Rules for derivatives of vector-valued functions.

SOLUTION We first verify the Sum Rule stating:

$$(\mathbf{r}_1(t) + \mathbf{r}_2(t))' = \mathbf{r}_1'(t) + \mathbf{r}_2'(t)$$

Let $\mathbf{r}_1(t) = \langle x_1(t), y_1(t), z_1(t) \rangle$ and $\mathbf{r}_2(t) = \langle x_2(t), y_2(t), z_2(t) \rangle$. Then,

$$(\mathbf{r}_1(t) + \mathbf{r}_2(t))' = \frac{d}{dt} \langle x_1(t) + x_2(t), y_1(t) + y_2(t), z_1(t) + z_2(t) \rangle$$

$$= \langle (x_1(t) + x_2(t))', (y_1(t) + y_2(t))', (z_1(t) + z_2(t))' \rangle$$

$$= \langle x_1'(t) + x_2'(t), y_1'(t) + y_2'(t), z_1'(t) + z_2'(t) \rangle$$

$$= \langle x_1'(t), y_1'(t), z_1'(t) \rangle + \langle x_2'(t), y_2'(t), z_2'(t) \rangle = \mathbf{r}_1'(t) + \mathbf{r}_2'(t)$$

The Constant Multiple Rule states that if $\mathbf{r}(t)$ is a differentiable scalar-valued function and c is any constant, then

$$(c\,\mathbf{r}(t))' = c\,\mathbf{r}'(t)$$

To verify this rule, let $\mathbf{r}(t) = \langle x_1(t), x_2(t), \ldots, x_n(t) \rangle$. Then

$$c\,\mathbf{r}(t) = \langle cx_1(t), cx_2(t), \ldots, cx_n(t) \rangle$$

Since differentiation is performed componentwise, we apply the constant multiple rule for scalar-valued functions, giving

$$(c\,\mathbf{r}(t))' = \frac{d}{dt}\langle cx_1(t), cx_2(t), \ldots, cx_n(t)\rangle = \langle cx_1'(t), cx_2'(t), \ldots, cx_n'(t)\rangle = c\langle x_1'(t), x_2'(t), \ldots, x_n'(t)\rangle = c\,\mathbf{r}'(t)$$

75. Prove the Cross Product Rule [Eq. (5)].

SOLUTION Let $\mathbf{r}_1(t) = \langle a_1(t), a_2(t), a_3(t) \rangle$ and $\mathbf{r}_2(t) = \langle b_1(t), b_2(t), b_3(t) \rangle$. Then (we omit the independent variable t for simplicity):

$$\mathbf{r}_1(t) \times \mathbf{r}_2(t) = \begin{vmatrix} \mathbf{i} & \mathbf{j} & \mathbf{k} \\ a_1 & a_2 & a_3 \\ b_1 & b_2 & b_3 \end{vmatrix} = (a_2 b_3 - a_3 b_2)\mathbf{i} - (a_1 b_3 - a_3 b_1)\mathbf{j} + (a_1 b_2 - a_2 b_1)\mathbf{k}$$

Differentiating this vector componentwise we get:

$$\frac{d}{dt}\mathbf{r}_1 \times \mathbf{r}_2 = \left(a_2 b_3' + a_2' b_3 - a_3 b_2' - a_3' b_2\right)\mathbf{i} - \left(a_1 b_3' + a_1' b_3 - a_3 b_1' - a_3' b_1\right)\mathbf{j} + \left(a_1 b_2' + a_1' b_2 - a_2 b_1' - a_2' b_1\right)\mathbf{k}$$

$$= \left((a_2 b_3' - a_3 b_2')\mathbf{i} - (a_1 b_3' - a_3 b_1')\mathbf{j} + (a_1 b_2' - a_2 b_1')\mathbf{k}\right)$$

$$+ \left((a_2' b_3 - a_3' b_2)\mathbf{i} - (a_1' b_3 - a_3' b_1)\mathbf{j} + (a_1' b_2 - a_2' b_1)\mathbf{k}\right)$$

Notice that the vectors in each of the two brackets can be written as the following formal determinants:

$$\frac{d}{dt}\mathbf{r}_1 \times \mathbf{r}_2 = \begin{vmatrix} \mathbf{i} & \mathbf{j} & \mathbf{k} \\ a_1 & a_2 & a_3 \\ b_1' & b_2' & b_3' \end{vmatrix} + \begin{vmatrix} \mathbf{i} & \mathbf{j} & \mathbf{k} \\ a_1' & a_2' & a_3' \\ b_1 & b_2 & b_3 \end{vmatrix} = \langle a_1, a_2, a_3 \rangle \times \langle b_1', b_2', b_3' \rangle + \langle a_1', a_2', a_3' \rangle \times \langle b_1, b_2, b_3 \rangle$$

$$= \mathbf{r}_1 \times \mathbf{r}_2' + \mathbf{r}_1' \times \mathbf{r}_2$$

77. Prove the Substitution Rule [where g is a differentiable scalar function with an inverse]:

$$\int_a^b \mathbf{r}(g(t))g'(t)\,dt = \int_{g^{-1}(a)}^{g^{-1}(b)} \mathbf{r}(u)\,du$$

SOLUTION (Note that an early edition of the textbook had the integral limits as $g(a)$ and $g(b)$; they should actually be $g^{-1}(a)$ and $g^{-1}(b)$.) We denote the components of the vector-valued function by $\mathbf{r}(t)\,dt = \langle x(t), y(t), z(t) \rangle$. Using componentwise integration we have:

$$\int_a^b \mathbf{r}(t)\,dt = \left\langle \int_a^b x(t)\,dt, \int_a^b y(t)\,dt, \int_a^b z(t)\,dt \right\rangle$$

Write $\int_a^b x(t)\,dt$ as $\int_a^b x(s)\,ds$. Let $s = g(t)$, so $ds = g'(t)\,dt$. The substitution gives us $\int_{g^{-1}(a)}^{g^{-1}(b)} x(g(t))g'(t)\,dt$. A similar procedure for the other two integrals gives us:

$$\int_a^b \mathbf{r}(t)\,dt = \left\langle \int_{g^{-1}(a)}^{g^{-1}(b)} x(g(t))\,g'(t)\,dt, \int_{g^{-1}(a)}^{g^{-1}(b)} y(g(t))\,g'(t)\,dt, \int_{g^{-1}(a)}^{g^{-1}(b)} z(g(t))\,g'(t)\,dt \right\rangle$$

$$= \int_{g^{-1}(a)}^{g^{-1}(b)} \langle x(g(t))\,g'(t), y(g(t))\,g'(t), z(g(t))\,g'(t) \rangle\,dt$$

$$= \int_{g^{-1}(a)}^{g^{-1}(b)} \langle x(g(t)), y(g(t)), z(g(t)) \rangle\,g'(t)\,dt = \int_{g^{-1}(a)}^{g^{-1}(b)} \mathbf{r}(g(t))\,g'(t)\,dt$$

13.3 Arc Length and Speed   (LT Section 14.3)

Preliminary Questions

1. At a given instant, a car on a roller coaster has velocity vector $\mathbf{v} = \langle 25, -35, 10 \rangle$ (in miles per hour). What would the velocity vector be if the speed were doubled? What would it be if the car's direction were reversed but its speed remained unchanged?

SOLUTION For any velocity vector $\mathbf{v}$, multiplying the speed by a factor λ is done by multiplying $\mathbf{v}$ by λ, giving $\lambda \mathbf{v}$. Thus if the speed is doubled but the direction is unchanged, the new velocity vector is $2\langle 25, -35, 10 \rangle = \langle 50, -70, 20 \rangle$. If the direction is reversed but the speed is unchanged, the new velocity vector is $-\mathbf{v} = \langle -25, 35, -10 \rangle$.

2. Two cars travel in the same direction along the same roller coaster (at different times). Which of the following statements about their velocity vectors at a given point P on the roller coaster are true?
(a) The velocity vectors are identical.
(b) The velocity vectors point in the same direction but may have different lengths.
(c) The velocity vectors may point in opposite directions.

SOLUTION

(a) The length of the velocity vector is the speed of the particle. Therefore, if the speeds of the cars are different the velocities are not identical. The statement is false.
(b) The velocity vector is tangent to the curve. Since the cars travel in the same direction, their velocity vectors point in the same direction. The statement is true.
(c) Since the cars travel in the same direction, the velocity vectors point in the same direction. The statement is false.

3. Starting at the origin, a mosquito flies along a parabola with speed $v(t) = t^2$. Let $L(t)$ be the total distance traveled at time t.
(a) How fast is $L(t)$ changing at $t = 2$?
(b) Is $L(t)$ equal to the mosquito's distance from the origin?

SOLUTION
(a) By the Arc Length Formula, we have:

$$L(t) = \int_0^t \|\mathbf{r}'(t)\|\,dt = \int_0^t v(t)\,dt$$

Therefore,

$$L'(t) = v(t)$$

To find the rate of change of $L(t)$ at $t = 2$ we compute the derivative of $L(t)$ at $t = 2$, that is,

$$L'(2) = v(2) = 2^2 = 4$$

(b) Since the mosquito travels along a parabola, the distance it travels will be longer than its distance from the origin (since the shortest distance between two points is a straight line).

4. What is the length of the path traced by $\mathbf{r}(t)$ for $4 \le t \le 10$ if $\mathbf{r}(t)$ is an arc length parametrization?

SOLUTION Since $\mathbf{r}(t)$ is an arc length parametrization, the length of the path for $4 \le t \le 10$ is equal to the length of the time interval $4 \le t \le 10$, which is 6.

Exercises

In Exercises 1–8, compute the length of the curve traced by $\mathbf{r}(t)$ over the given interval.

1. $\mathbf{r}(t) = \langle 3t, 4t - 3, 6t + 1 \rangle$, $\quad 0 \le t \le 3$

SOLUTION We have $x(t) = 3t$, $y(t) = 4t - 3$, $z(t) = 6t + 1$ hence

$$x'(t) = 3, \quad y'(t) = 4, \quad z'(t) = 6$$

We use the Arc Length Formula to obtain:

$$s = \int_0^3 \|\mathbf{r}'(t)\| \, dt = \int_0^3 \sqrt{x'(t)^2 + y'(t)^2 + z'(t)^2} \, dt = \int_0^3 \sqrt{3^2 + 4^2 + 6^2} \, dt = 3\sqrt{61}$$

3. $\mathbf{r}(t) = \langle 2t, \ln t, t^2 \rangle$, $\quad 1 \le t \le 4$

SOLUTION The derivative of $\mathbf{r}(t)$ is $\mathbf{r}'(t) = \langle 2, \frac{1}{t}, 2t \rangle$. We use the Arc Length Formula to obtain:

$$s = \int_1^4 \|\mathbf{r}'(t)\| \, dt = \int_1^4 \sqrt{2^2 + \left(\frac{1}{t}\right)^2 + (2t)^2} \, dt = \int_1^4 \sqrt{4t^2 + 4 + \frac{1}{t^2}} \, dt = \int_1^4 \sqrt{\left(2t + \frac{1}{t}\right)^2} \, dt$$

$$= \int_1^4 \left(2t + \frac{1}{t}\right) dt = t^2 + \ln t \Big|_1^4 = (16 + \ln 4) - (1 + \ln 1) = 15 + \ln 4$$

5. $\mathbf{r}(t) = \langle t, 4t^{3/2}, 2t^{3/2} \rangle$, $\quad 0 \le t \le 3$

SOLUTION Since $\mathbf{r}'(t) = \langle 1, \frac{3}{2} \cdot 4t^{1/2}, \frac{3}{2} \cdot 2t^{1/2} \rangle = \langle 1, 6t^{1/2}, 3t^{1/2} \rangle$, by Theorem 1 the arc length is

$$\int_0^3 \|\mathbf{r}'(t)\| \, dt = \int_0^3 \sqrt{1^2 + (6t^{1/2})^2 + (3t^{1/2})^2} \, dt$$

$$= \int_0^3 \sqrt{1 + 45t} \, dt$$

$$= \frac{2}{135}(1 + 45t)^{3/2} \Big|_0^3$$

$$= \frac{2}{135} \left(136^{3/2} - 1\right)$$

7. $\mathbf{r}(t) = \langle t \cos t, t \sin t, 3t \rangle$, $\quad 0 \le t \le 2\pi$

SOLUTION The derivative of $\mathbf{r}(t)$ is $\mathbf{r}'(t) = \langle \cos t - t \sin t, \sin t + t \cos t, 3 \rangle$. The length of $\mathbf{r}'(t)$ is, thus,

$$\|\mathbf{r}'(t)\| = \sqrt{(\cos t - t \sin t)^2 + (\sin t + t \cos t)^2 + 9}$$

$$= \sqrt{\cos^2 t - 2t \cos t \sin t + t^2 \sin^2 t + \sin^2 t + 2t \sin t \cos t + t^2 \cos^2 t + 9}$$

$$= \sqrt{\left(\cos^2 t + \sin^2 t\right) + t^2 \left(\sin^2 t + \cos^2 t\right) + 9} = \sqrt{t^2 + 10}$$

Using the Arc Length Formula and the integration formula given in Exercise 8, we obtain:

$$s = \int_0^{2\pi} \|\mathbf{r}'(t)\| \, dt = \int_0^{2\pi} \sqrt{t^2 + 10} \, dt = \frac{1}{2} t \sqrt{t^2 + 10} + \frac{1}{2} \cdot 10 \ln\left(t + \sqrt{t^2 + 10}\right) \Big|_0^{2\pi}$$

$$= \pi \sqrt{4\pi^2 + 10} + 5 \ln\left(2\pi + \sqrt{4\pi^2 + 10}\right) - 5 \ln \sqrt{10} = \pi \sqrt{4\pi^2 + 10} + 5 \ln \frac{2\pi + \sqrt{4\pi^2 + 10}}{\sqrt{10}} \approx 29.3$$

9. *CAS* The curve shown in Figure 5(A) is parametrized by $\mathbf{r}(t) = \langle \cos(7t), \sin(7t), 2\cos t \rangle$ for $0 \le t \le 2\pi$. Approximate its length.

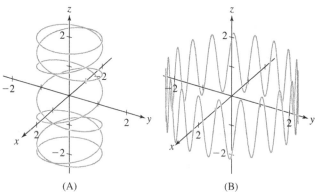

FIGURE 5

SOLUTION The derivative of $\mathbf{r}(t)$ is $\mathbf{r}'(t) = \langle -7\sin(7t), 7\cos(7t), -2\sin t \rangle$. Using the Arc Length Formula we get

$$s = \int_0^{2\pi} \|\mathbf{r}'(t)\|\, dt = \int_0^{2\pi} \sqrt{49\sin^2(7t) + 49\cos^2(7t) + 4\sin^2 t}\, dt = \int_0^{2\pi} \sqrt{49 + 4\sin^2 t}\, dt$$

Evaluating numerically gives $s \approx 44.8666$.

In Exercises 11 and 12, compute the arc length function $s(t) = \int_a^t \|\mathbf{r}'(u)\|\, du$ for the given value of a.

11. $\mathbf{r}(t) = \langle t^2, 2t^2, t^3 \rangle$, $a = 0$

SOLUTION The derivative of $\mathbf{r}(t)$ is $\mathbf{r}'(t) = \langle 2t, 4t, 3t^2 \rangle$. Hence,

$$\|\mathbf{r}'(t)\| = \sqrt{(2t)^2 + (4t)^2 + (3t^2)^2} = \sqrt{4t^2 + 16t^2 + 9t^4} = \sqrt{20 + 9t^2}\, t$$

Hence,

$$s(t) = \int_0^t \|\mathbf{r}'(u)\|\, du = \int_0^t \sqrt{20 + 9u^2}\, u\, du$$

We compute the integral using the substitution $v = 20 + 9u^2$, $dv = 18u\, du$. This gives:

$$s(t) = \frac{1}{18}\int_{20}^{20+9t^2} v^{1/2}\, dv = \frac{1}{18} \cdot \frac{2}{3} v^{3/2}\Big|_{20}^{20+9t^2} = \frac{1}{27}\left((20 + 9t^2)^{3/2} - 20^{3/2}\right)$$

In Exercises 13–18, find the speed at the given value of t.

13. $\mathbf{r}(t) = \langle 2t + 3, 4t - 3, 5 - t \rangle$, $t = 4$

SOLUTION The speed is the magnitude of the derivative $\mathbf{r}'(t) = \langle 2, 4, -1 \rangle$. That is,

$$v(t) = \|\mathbf{r}'(t)\| = \sqrt{2^2 + 4^2 + (-1)^2} = \sqrt{21} \approx 4.58$$

15. $\mathbf{r}(t) = \langle t, \ln t, (\ln t)^2 \rangle$, $t = 1$

SOLUTION The speed is the magnitude of the derivative $\mathbf{r}'(t) = \langle 1, \frac{1}{t}, \frac{2\ln t}{t} \rangle$. That is,

$$v(t) = \sqrt{1^2 + \left(\frac{1}{t}\right)^2 + \left(\frac{2\ln t}{t}\right)^2} = \frac{1}{|t|}\sqrt{t^2 + 1 + 4\ln^2 t}$$

At $t = 1$, we get $v(1) = \frac{1}{1}\sqrt{1 + 1 + 4\ln^2 1} = \sqrt{2}$.

17. $\mathbf{r}(t) = \langle \sin 3t, \cos 4t, \cos 5t \rangle, \quad t = \frac{\pi}{2}$

SOLUTION The velocity vector is $\mathbf{r}'(t) = \langle 3\cos 3t, -4\sin 4t, -5\sin 5t \rangle$. At $t = \frac{\pi}{2}$ the velocity vector is $\mathbf{r}'\left(\frac{\pi}{2}\right) = \langle 3\cos \frac{3\pi}{2}, -4\sin 2\pi, -5\sin \frac{5\pi}{2} \rangle = \langle 0, 0, -5 \rangle$. The speed is the magnitude of the velocity vector:

$$v\left(\frac{\pi}{2}\right) = \|\langle 0, 0, -5 \rangle\| = 5$$

19. At an air show, a jet has a trajectory following the curve $y = x^2$. If when the jet is at the point $(1, 1)$, it has a speed of 500 km/h, determine its velocity vector at this point.

SOLUTION The velocity vector points in the direction of motion, hence it is parallel to the tangent line to the curve $y = x^2$; assuming the jet is moving from left to right, it points to the right. We fine the slope of the tangent line at $x = 1$:

$$m = \left.\frac{dy}{dx}\right|_{x=1} = \left.\frac{d}{dx}(x^2)\right|_{x=1} = 2x|_{x=1} = 2$$

Therefore the vector $\langle 1, 2 \rangle$ is a direction vector of the tangent line at $x = 1$, and therefore the tangent vector of the jet at $x = 1$ is some multiple $\lambda \langle 1, 2 \rangle$. Since the speed is 500, we have

$$500 = \|\lambda \langle 1, 2 \rangle\| = |\lambda| \|\langle 1, 2 \rangle\| = \sqrt{5}\, |\lambda|$$

Therefore $\lambda = \pm 100\sqrt{5}$. Again assuming the jet is following the given trajectory from left to right, the correct value is the positive one, so the tangent vector at the given point is $100\sqrt{5}\,\langle 1, 2 \rangle$.

21. A bee with velocity vector $\mathbf{r}'(t)$ starts out at the origin at $t = 0$ and flies around for T seconds. Where is the bee located at time T if $\int_0^T \mathbf{r}'(u)\,du = \mathbf{0}$? What does the quantity $\int_0^T \|\mathbf{r}'(u)\|\,du$ represent?

SOLUTION By the Fundamental Theorem for vector-valued functions, $\int_0^T \mathbf{r}'(u)\,du = \mathbf{r}(T) - \mathbf{r}(0)$, hence by the given information $\mathbf{r}(T) = \mathbf{r}(0)$. It follows that at time T the bee is located at the starting point, which is at the origin. The integral $\int_0^T \|\mathbf{r}'(u)\|\,du$ is the length of the path traveled by the bee in the time interval $0 \le t \le T$. Notice that there is a difference between the displacement and the actual length traveled.

23. Let

$$\mathbf{r}(t) = \left\langle R\cos\left(\frac{2\pi Nt}{h}\right), R\sin\left(\frac{2\pi Nt}{h}\right), t \right\rangle, \qquad 0 \le t \le h$$

(a) Show that $\mathbf{r}(t)$ parametrizes a helix of radius R and height h making N complete turns.

(b) Guess which of the two springs in Figure 6 uses more wire.

(c) Compute the lengths of the two springs and compare.

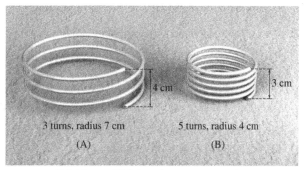

FIGURE 6 Which spring uses more wire?

SOLUTION Denote by $\mathbf{r}_1(t)$ and $\mathbf{r}_2(t)$ the parametrizations for the two springs using the formula given above, and denote by s_1 and s_2 the lengths of the two springs.

(a) We first verify that the projection $\mathbf{p}(t) = \left\langle R\cos\left(\frac{2\pi Nt}{h}\right), R\sin\left(\frac{2\pi Nt}{h}\right), 0 \right\rangle$ onto the xy-plane describes a point moving around the circle of radius R. We have:

$$x(t)^2 + y(t)^2 = R^2\cos^2\left(\frac{2\pi Nt}{h}\right) + R^2\sin^2\left(\frac{2\pi Nt}{h}\right) = R^2\left(\cos^2\left(\frac{2\pi Nt}{h}\right) + \sin^2\left(\frac{2\pi Nt}{h}\right)\right) = R^2$$

This is the equation of the circle of radius R in the xy-plane, which matches the projection of the spring onto the horizontal surface.

Now, as t changes in the interval $0 \le t \le h$ the argument $\frac{2\pi Nt}{h}$ changes from 0 to $2\pi N$, that is, it covers N periods of the cos and sin functions. It follows that the projection onto the xy-plane describes a point moving around the circle of radius R, making N complete turns. The height of the helix is the maximum value of the z-component, which is $t = h$.

(b) Answers may vary. It appears that the tighter separation and larger number of turns outweighs the larger radius and that the second spring is longer.

(c) Given the parametrization for $\mathbf{r}(t)$ above, we have

$$\mathbf{r}'(t) = \left\langle -\frac{2\pi NR}{h}\sin\left(\frac{2\pi Nt}{h}\right), \frac{2\pi NR}{h}\cos\left(\frac{2\pi Nt}{h}\right), 1 \right\rangle$$

Then

$$\|\mathbf{r}'(t)\| = \sqrt{\left(-\frac{2\pi NR}{h}\sin\left(\frac{2\pi Nt}{h}\right)\right)^2 + \left(\frac{2\pi NR}{h}\cos\left(\frac{2\pi Nt}{h}\right)\right)^2 + 1^2}$$

$$= \sqrt{\frac{4\pi^2 N^2 R^2}{h^2}\left(\sin^2\left(\frac{2\pi Nt}{h}\right) + \cos^2\left(\frac{2\pi Nt}{h}\right)\right) + 1}$$

$$= \sqrt{\frac{4\pi^2 N^2 R^2}{h^2} + 1}$$

$$= \frac{1}{h}\sqrt{4\pi^2 N^2 R^2 + h^2}$$

For the two springs, we get

$$R = 7,\, h = 4,\, N = 3 \quad \Rightarrow \quad \mathbf{r}'_1(t) = \frac{1}{4}\sqrt{4\pi^2 \cdot 9 \cdot 49 + 16} = \frac{1}{4}\sqrt{1764\pi^2 + 16}$$

$$R = 4,\, h = 3,\, N == 5 \quad \Rightarrow \quad \mathbf{r}'_2(t) = \frac{1}{3}\sqrt{4\pi^2 \cdot 25 \cdot 16 + 9} = \frac{1}{3}\sqrt{1600\pi^2 + 9}$$

Then by the Arc Length formula we obtain

$$s_1 = \int_0^4 \frac{1}{4}\sqrt{1764\pi^2 + 16}\, dt = \sqrt{1764\pi^2 + 16} \approx 132.008$$

$$s_2 = \int_0^3 \frac{1}{3}\sqrt{1600\pi^2 + 9}\, dt = \sqrt{1600\pi^2 + 9} \approx 125.6995$$

The lengths are quite close, but the first spring is longer.

25. The cycloid generated by the unit circle has parametrization

$$\mathbf{r}(t) = \langle t - \sin t, 1 - \cos t \rangle$$

(a) Find the value of t in $[0, 2\pi]$ where the speed is at a maximum.

(b) Show that one arch of the cycloid has length 8. Recall the identity $\sin^2(t/2) = (1 - \cos t)/2$.

SOLUTION Note that one arch of the cycloid is traced as $0 \le t \le 2\pi$.

(a) The speed is given by the function:

$$v(t) = \|\mathbf{r}'(t)\|$$

We compute the derivative and its length:

$$\mathbf{r}'(t) = \langle 1 - \cos t, \sin t \rangle$$

$$\|\mathbf{r}'(t)\| = \sqrt{(1-\cos t)^2 + (\sin t)^2} = \sqrt{1 - 2\cos t + \cos^2 t + \sin^2 t}$$

$$= \sqrt{2 - 2\cos t} = \sqrt{2(1-\cos t)} = \sqrt{2 \cdot 2\sin^2 \frac{t}{2}} = 2\left|\sin \frac{t}{2}\right|$$

For $0 \le t \le 2\pi$, we have $0 \le \frac{t}{2} \le \pi$, so $\sin \frac{t}{2} \ge 0$. Therefore we may omit the absolute value sign and write:

$$\|\mathbf{r}'(t)\| = 2\sin \frac{t}{2}$$

To find the value of t in $[0, 2\pi]$ where the speed is at maximum, we first find the critical point in this interval:

$$v'(t) = 2 \cdot \frac{1}{2}\cos \frac{t}{2} = \cos \frac{t}{2}$$

$$\cos \frac{t}{2} = 0 \quad\Rightarrow\quad \frac{t}{2} = \frac{\pi}{2} \text{ or } \frac{t}{2} = \frac{3\pi}{2} \quad\Rightarrow\quad t = \pi \text{ or } t = 3\pi$$

Since $t = 3\pi$ is outside the domain under consideration, we must have $t = \pi$. Since $v''(t) = -\frac{1}{2}\sin\frac{t}{2}$, we have $v''(\pi) = -\frac{1}{2}\sin\frac{\pi}{2} = -\frac{1}{2} < 0$, hence the speed $v(t)$ has a maximum value at $t = \pi$.

(b) By the Arc Length Formula we have, using the substitution $u = \frac{t}{2}$, $du = \frac{1}{2}dt$,

$$s = \int_0^{2\pi} \|\mathbf{r}'(t)\|\,dt = \int_0^{2\pi} 2\sin\frac{t}{2}\,dt = \int_0^{\pi} 2\sin u \cdot (2\,du) = 4\int_0^{\pi} \sin u\,du$$

$$= 4(-\cos u)\Big|_0^{\pi} = 4(-\cos\pi - (-\cos 0)) = 4(1+1) = 8$$

The length of one arch of the cycloid is $s = 8$.

27. Let $\mathbf{r}(t) = \langle 3t + 1, 4t - 5, 2t \rangle$.

(a) Evaluate the arc length integral $s(t) = \int_0^t \|\mathbf{r}'(u)\|\,du$.

(b) Find the inverse $g(s)$ of $s(t)$.

(c) Verify that $\mathbf{r}_1(s) = \mathbf{r}(g(s))$ is an arc length parametrization.

SOLUTION

(a) We differentiate $\mathbf{r}(t)$ componentwise and then compute the norm of the derivative vector. This gives:

$$\mathbf{r}'(t) = \langle 3, 4, 2 \rangle$$

$$\|\mathbf{r}'(t)\| = \sqrt{3^2 + 4^2 + 2^2} = \sqrt{29}$$

We compute $s(t)$:

$$s(t) = \int_0^t \|\mathbf{r}'(u)\|\,du = \int_0^t \sqrt{29}\,du = \sqrt{29}\,u\Big|_0^t = \sqrt{29}\,t$$

(b) We find the inverse $\varphi(s) = t(s)$ by solving $s = \sqrt{29}\,t$ for t. We obtain:

$$s = \sqrt{29}\,t \quad\Rightarrow\quad t = \varphi(s) = \frac{s}{\sqrt{29}}$$

We obtain the following arc length parametrization:

$$\mathbf{r}_1(s) = \mathbf{r}\left(\frac{s}{\sqrt{29}}\right) = \left\langle \frac{3s}{\sqrt{29}} + 1, \frac{4s}{\sqrt{29}} - 5, \frac{2s}{\sqrt{29}} \right\rangle$$

To verify that $\mathbf{r}_1(s)$ is an arc length parametrization we must show that $\|\mathbf{r}_1'(s)\| = 1$. We compute $\mathbf{r}_1'(s)$:

$$\mathbf{r}_1'(s) = \frac{d}{ds}\left\langle \frac{3s}{\sqrt{29}} + 1, \frac{4s}{\sqrt{29}} - 5, \frac{2s}{\sqrt{29}} \right\rangle = \left\langle \frac{3}{\sqrt{29}}, \frac{4}{\sqrt{29}}, \frac{2}{\sqrt{29}} \right\rangle = \frac{1}{\sqrt{29}}\langle 3, 4, 2 \rangle$$

Thus,

$$\|\mathbf{r}_1'(s)\| = \frac{1}{\sqrt{29}}\|\langle 3, 4, 2 \rangle\| = \frac{1}{\sqrt{29}}\sqrt{3^2 + 4^2 + 2^2} = \frac{1}{\sqrt{29}} \cdot \sqrt{29} = 1$$

29. Let $\mathbf{r}(t) = \mathbf{w} + t\mathbf{v}$ be a parametrization of a line.

(a) Show that the arc length function $s(t) = \int_0^t \|\mathbf{r}'(u)\| \, du$ is given by $s(t) = t\|\mathbf{v}\|$. This shows that $\mathbf{r}(t)$ is an arc length parametrizaton if and only if $\mathbf{v}$ is a unit vector.

(b) Find an arc length parametrization of the line with $\mathbf{w} = \langle 1, 2, 3 \rangle$ and $\mathbf{v} = \langle 3, 4, 5 \rangle$.

SOLUTION

(a) Since $\mathbf{r}(t) = \mathbf{w} + t\mathbf{v}$, then $\mathbf{r}'(t) = \mathbf{v}$ and $\|\mathbf{r}'(t)\| = \|\mathbf{v}\|$. Then computing $s(t)$ we get:

$$s(t) = \int_0^t \|\mathbf{r}'(u)\| \, du = \int_0^t \|\mathbf{v}\| \, du = t\|\mathbf{v}\|$$

If we consider $s(t)$,

$$s(t) = t \text{ if and only if } \|\mathbf{v}\| = 1$$

(b) Since $\mathbf{v} = \langle 3, 4, 5 \rangle$, then from part (a) we get:

$$s(t) = t\|\mathbf{v}\| = t\sqrt{3^2 + 4^2 + 5^2} = t\sqrt{50}, \quad \Rightarrow t = g(s) = \frac{s}{\sqrt{50}}$$

Therefore, since we are given $\mathbf{r}(t) = \mathbf{w} + t\mathbf{v}$, the arc length parametrization is:

$$\mathbf{r}_1(s) = \langle 1, 2, 3 \rangle + \frac{s}{\sqrt{50}} \langle 3, 4, 5 \rangle = \left\langle 1 + \frac{3s}{\sqrt{50}}, 2 + \frac{4s}{\sqrt{50}}, 3 + \frac{5s}{\sqrt{50}} \right\rangle$$

31. Find a path that traces the circle in the plane $y = 10$ with radius 4 and center $(2, 10, -3)$ with constant speed 8.

SOLUTION We start with the following parametrization of the circle:

$$\mathbf{r}(t) = \langle 2, 10, -3 \rangle + 4\langle \cos t, 0, \sin t \rangle = \langle 2 + 4\cos t, 10, -3 + 4\sin t \rangle$$

We need to reparametrize the curve by making a substitution $t = g(s)$, so that the new parametrization $\mathbf{r}_1(s) = \mathbf{r}(g(s))$ satisfies $\|\mathbf{r}_1'(s)\| = 8$ for all s. We find $\mathbf{r}_1'(s)$ using the Chain Rule:

$$\mathbf{r}_1'(s) = \frac{d}{ds}\mathbf{r}(g(s)) = g'(s)\mathbf{r}'(g(s)) \tag{1}$$

Next, we differentiate $\mathbf{r}(t)$ and then replace t by $g(s)$:

$$\mathbf{r}'(t) = \langle -4\sin t, 0, 4\cos t \rangle$$

$$\mathbf{r}'(g(s)) = \langle -4\sin g(s), 0, 4\cos g(s) \rangle$$

Substituting in (1) we get:

$$\mathbf{r}_1'(s) = g'(s)\langle -4\sin g(s), 0, 4\cos g(s) \rangle = -4g'(s)\langle \sin g(s), 0, -\cos g(s) \rangle$$

Hence,

$$\|\mathbf{r}_1'(s)\| = 4|g'(s)|\sqrt{(\sin g(s))^2 + (-\cos g(s))^2} = 4|g'(s)|$$

To satisfy $\|\mathbf{r}_1'(s)\| = 8$ for all s, we choose $g'(s) = 2$. We may take the antiderivative $g(s) = 2 \cdot s$, and obtain the following parametrization:

$$\mathbf{r}_1(s) = \mathbf{r}(g(s)) = \mathbf{r}(2s) = \langle 2 + 4\cos(2s), 10, -3 + 4\sin(2s) \rangle$$

This is a parametrization of the given circle, with constant speed 8.

33. Find an arc length parametrization of the curve parametrized by $\mathbf{r}(t) = \left\langle \cos t, \sin t, \frac{2}{3}t^{3/2} \right\rangle$, with the parameter s measuring from $(1, 0, 0)$.

SOLUTION We have $\mathbf{r}'(t) = \left\langle -\sin t, \cos t, t^{1/2} \right\rangle$, so that

$$\|\mathbf{r}'(t)\| = \sqrt{(-\sin t)^2 + (\cos t)^2 + (t^{1/2})^2} = \sqrt{1 + t}$$

The initial point $(1, 0, 0)$ corresponds to $t = 0$, so the arc length function is

$$s(t) = g(t) = \int_0^t \|\mathbf{r}'(u)\| \, du = \int_0^t \sqrt{1+u} \, du = \frac{2}{3}(1+u)^{3/2}\Big|_0^t = \frac{2}{3}\left((1+t)^{3/2} - 1\right)$$

Solving for t gives

$$t = g^{-1}(s) = \left(\frac{3}{2}s + 1\right)^{2/3} - 1$$

35. Find an arc length parametrization of the curve parametrized by $\mathbf{r}(t) = \langle t^2, t^3 \rangle$.

SOLUTION We follow two steps.

Step 1. Find the inverse of the arc length function. The arc length function is the following function:

$$s(t) = \int_0^t \|\mathbf{r}'(u)\| \, du \qquad (1)$$

In our case $\mathbf{r}'(t) = \langle 2t, 3t^2 \rangle$ hence $\|\mathbf{r}'(t)\| = \sqrt{4t^2 + 9t^4} = \sqrt{4 + 9t^2}\, t$. We substitute in (1) and compute the resulting integral using the substitution $v = 4 + 9u^2$, $dv = 18u \, du$. This gives:

$$s(t) = \int_0^t \sqrt{4 + 9u^2}\, u \, du = \frac{1}{18} \int_4^{4+9t^2} v^{1/2} \, dv = \frac{1}{18} \cdot \frac{2}{3} v^{3/2}\Big|_4^{4+9t^2} = \frac{1}{27}\left((4+9t^2)^{3/2} - 4^{3/2}\right)$$

$$= \frac{1}{27}\left((4+9t^2)^{3/2} - 8\right)$$

We find the inverse of $t = s(t)$ by solving for t in terms of s. This function is invertible for $t \geq 0$ and for $t \leq 0$.

$$s = \frac{1}{27}\left((4+9t^2)^{3/2} - 8\right)$$

$$27s + 8 = (4 + 9t^2)^{3/2}$$

$$(27s + 8)^{2/3} - 4 = 9t^2$$

$$t^2 = \frac{1}{9}\left((27s+8)^{2/3} - 4\right) = \frac{1}{9}(27s+8)^{2/3} - \frac{4}{9}$$

$$t = \pm \frac{1}{3}\sqrt{(27s+8)^{2/3} - 4} \qquad (2)$$

Step 2. Reparametrize the curve. The arc length parametrization is obtained by replacing t by (2) in $r(t)$:

$$\mathbf{r}_1(s) = \left\langle \frac{1}{9}(27s+8)^{2/3} - \frac{4}{9}, \pm \frac{1}{27}\left((27s+8)^{2/3} - 4\right)^{3/2} \right\rangle$$

37. Find an arc length parametrization of the line $y = mx$ for an arbitrary slope m.

SOLUTION

Step 1. Find the inverse of the arc length function. We are given the line $y = mx$ and a parametrization of this line is $\mathbf{r}(t) = \langle t, mt \rangle$, thus $\mathbf{r}'(t) = \langle 1, m \rangle$ and

$$\|\mathbf{r}'(t)\| = \sqrt{1 + m^2}$$

We then compute $s(t)$:

$$s(t) = \int_0^t \sqrt{1+m^2} \, du = t\sqrt{1+m^2}$$

Solving $s = t\sqrt{1+m^2}$ for t we get:

$$t = \frac{s}{\sqrt{1+m^2}}$$

Step 2. Reparametrize the curve using the t we just found.

$$\mathbf{r}_1(s) = \left\langle \frac{s}{\sqrt{1+m^2}}, \frac{sm}{\sqrt{1+m^2}} \right\rangle$$

39. The curve known as the **Bernoulli spiral** (Figure 7) has parametrization $\mathbf{r}(t) = \langle e^t \cos 4t, e^t \sin 4t \rangle$.

(a) Evaluate $s(t) = \int_{-\infty}^{t} \|\mathbf{r}'(u)\| \, du$. It is convenient to take lower limit $-\infty$ because $\lim_{t \to -\infty} \mathbf{r}(t) = \langle 0, 0 \rangle$.

(b) Use (a) to find an arc length parametrization of the spiral.

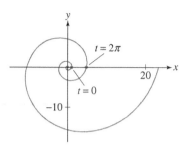

FIGURE 7 Bernoulli spiral.

SOLUTION

(a) We differentiate $\mathbf{r}(t)$ and compute the norm of the derivative vector. This gives:

$$\mathbf{r}'(t) = \langle e^t \cos 4t - 4e^t \sin 4t, e^t \sin 4t + 4e^t \cos 4t \rangle = e^t \langle \cos 4t - 4 \sin 4t, \sin 4t + 4 \cos 4t \rangle$$

$$\|\mathbf{r}'(t)\| = e^t \sqrt{(\cos 4t - 4 \sin 4t)^2 + (\sin 4t + 4 \cos 4t)^2}$$

$$= e^t (\cos^2 4t - 8 \cos 4t \sin 4t + 16 \sin^2 4t + \sin^2 4t + 8 \sin 4t \cos 4t + 16 \cos^2 4t)^{1/2}$$

$$= e^t \sqrt{\cos^2 4t + \sin^2 4t + 16(\sin^2 4t + \cos^2 4t)} = e^t \sqrt{1 + 16 \cdot 1} = \sqrt{17} e^t$$

We now evaluate the improper integral:

$$s(t) = \int_{-\infty}^{t} \|\mathbf{r}'(u)\| \, du = \lim_{R \to -\infty} \int_{R}^{t} \sqrt{17} e^u \, du = \lim_{R \to -\infty} \sqrt{17} e^u \Big|_{R}^{t} = \lim_{R \to -\infty} \sqrt{17}(e^t - e^R)$$

$$= \sqrt{17}(e^t - 0) = \sqrt{17} e^t$$

An arc length parametrization of $\mathbf{r}(t)$ is $\mathbf{r}_1(s) = \mathbf{r}(\varphi(s))$ where $t = \varphi(s)$ is the inverse function of $s(t)$. We find $t = \varphi(s)$ by solving $s = \sqrt{17} e^t$ for t:

$$s = \sqrt{17} e^t \quad \Rightarrow \quad e^t = \frac{s}{\sqrt{17}} \quad \Rightarrow \quad t = \varphi(s) = \ln \frac{s}{\sqrt{17}}$$

An arc length parametrization of $\mathbf{r}(t)$ is:

$$\mathbf{r}_1(s) = \mathbf{r}(\varphi(s)) = \left\langle e^{\ln(s/(\sqrt{17}))} \cos\left(4 \ln \frac{s}{\sqrt{17}}\right), e^{\ln(s/(\sqrt{17}))} \sin\left(4 \ln \frac{s}{\sqrt{17}}\right) \right\rangle$$

$$= \frac{s}{\sqrt{17}} \left\langle \cos\left(4 \ln \frac{s}{\sqrt{17}}\right), \sin\left(4 \ln \frac{s}{\sqrt{17}}\right) \right\rangle \quad (1)$$

(b) The cosine of the angle θ between the position vector $\mathbf{r}_1(s)$ and the tangent vector $\mathbf{r}_1'(s)$ is:

$$\cos \theta = \frac{\mathbf{r}_1(s) \cdot \mathbf{r}_1'(s)}{\|\mathbf{r}_1(s)\| \|\mathbf{r}_1'(s)\|}$$

Since for the arc length parametrization $\|\mathbf{r}_1'(s)\| = 1$, we obtain:

$$\cos \theta = \frac{\mathbf{r}_1(s) \cdot \mathbf{r}_1'(s)}{\|\mathbf{r}_1(s)\|} \quad (2)$$

We compute the dot product in (2). We first compute $\mathbf{r}_1'(s)$ from (1):

$$\mathbf{r}_1'(s) = \frac{1}{\sqrt{17}} \left\langle \cos\left(4 \ln \frac{s}{\sqrt{17}}\right), \sin\left(4 \ln \frac{s}{\sqrt{17}}\right) \right\rangle$$

$$+ \frac{s}{\sqrt{17}} \left\langle -\sin\left(4 \ln \frac{s}{\sqrt{17}}\right) \cdot \frac{4\sqrt{17}}{s} \cdot \frac{1}{\sqrt{17}}, \cos\left(4 \ln \frac{s}{\sqrt{17}}\right) \cdot \frac{4\sqrt{17}}{s} \cdot \frac{1}{\sqrt{17}} \right\rangle$$

$$= \frac{1}{\sqrt{17}}\left\langle \cos\left(4\ln\frac{s}{\sqrt{17}}\right), \sin\left(4\ln\frac{s}{\sqrt{17}}\right)\right\rangle + \frac{4}{\sqrt{17}}\left\langle -\sin\left(4\ln\frac{s}{\sqrt{17}}\right), \cos\left(4\ln\frac{s}{\sqrt{17}}\right)\right\rangle$$

$$= \frac{1}{\sqrt{17}}\left\langle \cos\left(4\ln\frac{s}{\sqrt{17}}\right) - 4\sin\left(4\ln\frac{s}{\sqrt{17}}\right), \sin\left(4\ln\frac{s}{\sqrt{17}}\right) + 4\cos\left(4\ln\frac{s}{\sqrt{17}}\right)\right\rangle$$

Thus,

$$\mathbf{r}_1(s) \cdot \mathbf{r}_1'(s) = \frac{s}{\sqrt{17}} \cdot \frac{1}{\sqrt{17}}\left\{\cos\left(4\ln\frac{s}{\sqrt{17}}\right)\left(\cos\left(4\ln\frac{s}{\sqrt{17}}\right) - 4\sin\left(4\ln\frac{s}{\sqrt{17}}\right)\right)\right.$$

$$\left. + \sin\left(4\ln\frac{s}{\sqrt{17}}\right)\left(\sin\left(4\ln\frac{s}{\sqrt{17}}\right) + 4\cos\left(4\ln\frac{s}{\sqrt{17}}\right)\right)\right\}$$

$$= \frac{s}{17}\left(\cos^2\left(4\ln\frac{s}{\sqrt{17}}\right) - 4\cos\left(4\ln\frac{s}{\sqrt{17}}\right)\sin\left(4\ln\frac{s}{\sqrt{17}}\right)\right.$$

$$\left. + \sin^2\left(4\ln\frac{s}{\sqrt{17}}\right) + 4\sin\left(4\ln\frac{s}{\sqrt{17}}\right)\cos\left(4\ln\frac{s}{\sqrt{17}}\right)\right)$$

$$= \frac{s}{17}\left(\cos^2\left(4\ln\frac{s}{\sqrt{17}}\right) + \sin^2\left(4\ln\frac{s}{\sqrt{17}}\right)\right) = \frac{s}{17} \cdot 1 = \frac{s}{17} \quad (3)$$

We now compute $\|\mathbf{r}_1(s)\|$ from (1) (Notice that $s(t) > 0$ for all t):

$$\|\mathbf{r}_1(s)\| = \frac{|s|}{\sqrt{17}}\sqrt{\cos^2\left(4\ln\frac{s}{\sqrt{17}}\right) + \sin^2\left(4\ln\frac{s}{\sqrt{17}}\right)} = \frac{s}{\sqrt{17}} \cdot 1 = \frac{s}{\sqrt{17}} \quad (4)$$

Combining (2), (3), and (4) yields:

$$\cos\theta = \frac{\frac{s}{17}}{\frac{s}{\sqrt{17}}} = \frac{1}{\sqrt{17}}$$

The solution for $0 \le \theta \le \pi$ is $\theta = 1.326$ rad. Thus, the angle between $\mathbf{r}_1(s)$ and $\mathbf{r}_1'(s)$ is constant.

Further Insights and Challenges

41. The unit circle with the point $(-1, 0)$ removed has parametrization (see Exercise 79 in Section 11.1)

$$\mathbf{r}(t) = \left\langle \frac{1-t^2}{1+t^2}, \frac{2t}{1+t^2} \right\rangle, \quad -\infty < t < \infty$$

Use this parametrization to compute the length of the unit circle as an improper integral. *Hint:* The expression for $\|\mathbf{r}'(t)\|$ simplifies.

SOLUTION Using the given parametrization, we have

$$\mathbf{r}'(t) = \left\langle \frac{-2t(1+t^2) - 2t(1-t^2)}{(1+t^2)^2}, \frac{2(1+t^2) - 2t \cdot 2t}{(1+t^2)^2} \right\rangle$$

$$= \left\langle -\frac{4t}{(1+t^2)^2}, \frac{2-2t^2}{(1+t^2)^2} \right\rangle = \frac{1}{(1+t^2)^2}\langle -4t, 2(1-t^2)\rangle$$

$$\|\mathbf{r}'(t)\| = \frac{1}{(1+t^2)^2}\sqrt{16t^2 + 4(1-t^2)^2} = \frac{2}{(1+t^2)^2}\sqrt{t^4 + 2t^2 + 1}$$

$$= \frac{2}{(1+t^2)^2}\sqrt{(t^2+1)^2} = \frac{2(t^2+1)}{(1+t^2)^2} = \frac{2}{1+t^2}$$

That is,

$$\|\mathbf{r}'(t)\| = \frac{2}{1+t^2}$$

We now use the Arc Length Formula to compute the length of the circle:

$$s = \int_{-\infty}^{\infty} \|\mathbf{r}'(t)\|\, dt = 2\int_{-\infty}^{\infty} \frac{dt}{1+t^2} = 2\left(\lim_{R\to\infty} \tan^{-1} R - \lim_{R\to -\infty} \tan^{-1} R\right) = 2\left(\frac{\pi}{2} - \left(-\frac{\pi}{2}\right)\right) = 2\pi$$

43. The curve $\mathbf{r}(t) = \langle t - \tanh t, \operatorname{sech} t \rangle$ is called a **tractrix** (see Exercise 100 in Section 11.1).

(a) Show that $s(t) = \int_0^t \|\mathbf{r}'(u)\| \, du$ is equal to $s(t) = \ln(\cosh t)$.

(b) Show that $t = g(s) = \ln(e^s + \sqrt{e^{2s} - 1})$ is an inverse of $s(t)$ and verify that

$$\mathbf{r}_1(s) = \left\langle \tanh^{-1}\left(\sqrt{1 - e^{-2s}}\right) - \sqrt{1 - e^{-2s}}, e^{-s} \right\rangle$$

is an arc length parametrization of the tractrix.

SOLUTION

(a) We compute the derivative vector and its length:

$$\mathbf{r}'(t) = \langle 1 - \operatorname{sech}^2 t, -\operatorname{sech} t \tanh t \rangle$$

$$\|\mathbf{r}'(t)\| = \sqrt{(1 - \operatorname{sech}^2 t)^2 + \operatorname{sech}^2 t \tanh^2 t} = \sqrt{1 - 2\operatorname{sech}^2 t + \operatorname{sech}^4 t + \operatorname{sech}^2 t \tanh^2 t}$$

$$= \sqrt{-\operatorname{sech}^2 t(2 - \tanh^2 t) + 1 + \operatorname{sech}^4 t}$$

We use the identity $1 - \tanh^2 t = \operatorname{sech}^2 t$ to write:

$$\|\mathbf{r}'(t)\| = \sqrt{-\operatorname{sech}^2 t(1 + \operatorname{sech}^2 t) + 1 + \operatorname{sech}^4 t} = \sqrt{-\operatorname{sech}^2 t - \operatorname{sech}^4 t + 1 + \operatorname{sech}^4 t}$$

$$= \sqrt{1 - \operatorname{sech}^2 t} = \sqrt{\tanh^2 t} = |\tanh t|$$

For $t \geq 0$, $\tanh t \geq 0$ hence, $\|\mathbf{r}'(t)\| = \tanh t$. We now apply the Arc Length Formula to obtain:

$$s(t) = \int_0^t \|\mathbf{r}'(u)\| \, du = \int_0^t (\tanh u) \, du = \ln(\cosh u) \Big|_0^t = \ln(\cosh t) - \ln(\cosh 0) = \ln(\cosh t) - \ln 1 = \ln(\cosh t)$$

That is:

$$s(t) = \ln(\cosh t)$$

(b) We show that the function $t = g(s) = \ln\left(e^s + \sqrt{e^{2s} - 1}\right)$ is an inverse of $s(t)$. First we note that $s'(t) = \tanh t$, hence $s'(t) > 0$ for $t > 0$, which implies that $s(t)$ has an inverse function for $t \geq 0$. Therefore, it suffices to verify that $g(s(t)) = t$. We have:

$$g(s(t)) = \ln\left(e^{\ln(\cosh t)} + \sqrt{e^{2\ln(\cosh t)} - 1}\right) = \ln\left(\cosh t + \sqrt{\cosh^2 t - 1}\right)$$

Since $\cosh^2 t - 1 = \sinh^2 t$ we obtain (for $t \geq 0$):

$$g(s(t)) = \ln\left(\cosh t + \sqrt{\sinh^2 t}\right) = \ln(\cosh t + \sinh t) = \ln\left(\frac{e^t + e^{-t}}{2} + \frac{e^t - e^{-t}}{2}\right) = \ln(e^t) = t$$

We thus proved that $t = g(s)$ is an inverse of $s(t)$. Therefore, the arc length parametrization is obtained by substituting $t = g(s)$ in $\mathbf{r}(t) = \langle t - \tanh t, \operatorname{sech} t \rangle$. We compute t, $\tanh t$, and $\operatorname{sech} t$ in terms of s. We have:

$$s = \ln(\cosh t) \quad \Rightarrow \quad e^s = \cosh t \quad \Rightarrow \quad \operatorname{sech} t = e^{-s}$$

Also:

$$\tanh^2 t = 1 - \operatorname{sech}^2 t = 1 - e^{-2s} \quad \Rightarrow \quad \tanh t = \sqrt{1 - e^{-2s}} \quad \Rightarrow \quad t = \tanh^{-1}\sqrt{1 - e^{-2s}}$$

Substituting in $\mathbf{r}(t)$ gives:

$$\mathbf{r}_1(s) = \langle t - \tanh t, \operatorname{sech} t \rangle = \left\langle \tanh^{-1}\sqrt{1 - e^{-2s}} - \sqrt{1 - e^{-2s}}, e^{-s} \right\rangle$$

13.4 Curvature (LT Section 14.4)

Preliminary Questions

1. What is the unit tangent vector of $\mathbf{r}(t)$ if the underlying curve is a line with direction vector $\mathbf{w} = \langle 2, 1, -2 \rangle$ and x is decreasing along $\mathbf{r}(t)$?

SOLUTION A line with direction vector $\mathbf{w}$ has the parametrization:

$$\mathbf{r}(t) = \overrightarrow{OP_0} + t\mathbf{w}$$

hence, since $\overrightarrow{OP_0}$ and $\mathbf{w}$ are constant vectors, we have:

$$\mathbf{r}'(t) = \mathbf{w}$$

Therefore, since $\|\mathbf{w}\| = 3$, a unit tangent vector is:

$$\mathbf{T}(t) = \frac{\mathbf{r}'(t)}{\|\mathbf{r}'(t)\|} = \frac{\mathbf{w}}{\|\mathbf{w}\|} = \left\langle \frac{2}{3}, \frac{1}{3}, -\frac{2}{3} \right\rangle$$

Since we assume x is decreasing along $\mathbf{r}(t)$, the tangent vector must point in the direction of decreasing x, so we choose the opposite of the unit tangent vector above, or

$$\left\langle -\frac{2}{3}, -\frac{1}{3}, \frac{2}{3} \right\rangle$$

2. What is the curvature of a circle of radius 4?

SOLUTION The curvature of a circle of radius R is $\frac{1}{R}$, hence the curvature of a circle of radius 4 is $\frac{1}{4}$.

3. Which has larger curvature, a circle of radius 2 or a circle of radius 4?

SOLUTION The curvature of a circle of radius 2 is $\frac{1}{2}$, and it is larger than the curvature of a circle of radius 4, which is $\frac{1}{4}$.

4. What is the curvature of $\mathbf{r}(t) = \langle 2 + 3t, 7t, 5 - t \rangle$?

SOLUTION $\mathbf{r}(t)$ parametrizes the line $\langle 2, 0, 5 \rangle + t \langle 3, 7, -1 \rangle$, and a line has zero curvature.

5. What is the curvature at a point where $\mathbf{T}'(s) = \langle 1, 2, 3 \rangle$ in an arc length parametrization $\mathbf{r}(s)$?

SOLUTION The curvature is given by the formula:

$$\kappa(t) = \frac{\|\mathbf{T}'(t)\|}{\|\mathbf{r}'(t)\|}$$

In an arc length parametrization, $\|\mathbf{r}'(t)\| = 1$ for all t, hence the curvature is $\kappa(t) = \|\mathbf{T}'(t)\|$. Using the given information we obtain the following curvature:

$$\kappa = \|\langle 1, 2, 3 \rangle\| = \sqrt{1^2 + 2^2 + 3^2} = \sqrt{14}$$

6. What is the radius of curvature of a circle of radius 4?

SOLUTION The definition of the osculating circle implies that the osculating circles at the points of a circle is the circle itself. Therefore, the radius of curvature is the radius of the circle, that is, 4.

7. What is the radius of curvature at P if $\kappa_P = 9$?

SOLUTION The radius of curvature is the reciprocal of the curvature, hence the radius of curvature at P is:

$$R = \frac{1}{\kappa_P} = \frac{1}{9}$$

Exercises

In Exercises 1–6, calculate $\mathbf{r}'(t)$ and $\mathbf{T}(t)$, and evaluate $\mathbf{T}(1)$.

1. $\mathbf{r}(t) = \langle 4t^2, 9t \rangle$

SOLUTION We differentiate $\mathbf{r}(t)$ to obtain:

$$\mathbf{r}'(t) = \langle 8t, 9 \rangle \quad \Rightarrow \quad \|\mathbf{r}'(t)\| = \sqrt{(8t)^2 + 9^2} = \sqrt{64t^2 + 81}$$

We now find the unit tangent vector:

$$\mathbf{T}(t) = \frac{\mathbf{r}'(t)}{\|\mathbf{r}'(t)\|} = \frac{1}{\sqrt{64t^2 + 81}} \langle 8t, 9 \rangle$$

For $t = 1$ we obtain the vector:

$$\mathbf{T}(t) = \frac{1}{\sqrt{64 + 81}} \langle 8, 9 \rangle = \left\langle \frac{8}{\sqrt{145}}, \frac{9}{\sqrt{145}} \right\rangle$$

3. $\mathbf{r}(t) = \langle 3 + 4t, 3 - 5t, 9t \rangle$

SOLUTION We first find the vector $\mathbf{r}'(t)$ and its length:

$$\mathbf{r}'(t) = \langle 4, -5, 9 \rangle \quad \Rightarrow \quad \|\mathbf{r}'(t)\| = \sqrt{4^2 + (-5)^2 + 9^2} = \sqrt{122}$$

The unit tangent vector is therefore:

$$\mathbf{T}(t) = \frac{\mathbf{r}'(t)}{\|\mathbf{r}'(t)\|} = \frac{1}{\sqrt{122}} \langle 4, -5, 9 \rangle = \left\langle \frac{4}{\sqrt{122}}, -\frac{5}{\sqrt{122}}, \frac{9}{\sqrt{122}} \right\rangle$$

We see that the unit tangent vector is constant, since the curve is a straight line.

5. $\mathbf{r}(t) = \langle \cos \pi t, \sin \pi t, t \rangle$

SOLUTION We compute the derivative vector and its length:

$$\mathbf{r}'(t) = \langle -\pi \sin \pi t, \pi \cos \pi t, 1 \rangle$$

$$\|\mathbf{r}'(t)\| = \sqrt{(-\pi \sin \pi t)^2 + (\pi \cos \pi t)^2 + 1^2} = \sqrt{\pi^2(\sin^2 \pi t + \cos^2 \pi t) + 1} = \sqrt{\pi^2 + 1}$$

The unit tangent vector is thus:

$$\mathbf{T}(t) = \frac{\mathbf{r}'(t)}{\|\mathbf{r}'(t)\|} = \frac{1}{\sqrt{\pi^2 + 1}} \langle -\pi \sin \pi t, \pi \cos \pi t, 1 \rangle$$

For $t = 1$ we get:

$$\mathbf{T}(1) = \frac{1}{\sqrt{\pi^2 + 1}} \langle -\pi \sin \pi, \pi \cos \pi, 1 \rangle = \frac{1}{\sqrt{\pi^2 + 1}} \langle 0, -\pi, 1 \rangle = \left\langle 0, -\frac{\pi}{\sqrt{\pi^2 + 1}}, \frac{1}{\sqrt{\pi^2 + 1}} \right\rangle$$

In Exercises 7–10, use Eq. (3) to calculate the curvature function $\kappa(t)$.

7. $\mathbf{r}(t) = \langle 1, e^t, t \rangle$

SOLUTION We compute the first and the second derivatives of $\mathbf{r}(t)$:

$$\mathbf{r}'(t) = \langle 0, e^t, 1 \rangle, \quad \mathbf{r}''(t) = \langle 0, e^t, 0 \rangle$$

Next, we find the cross product $\mathbf{r}'(t) \times \mathbf{r}''(t)$:

$$\mathbf{r}'(t) \times \mathbf{r}''(t) = \begin{vmatrix} \mathbf{i} & \mathbf{j} & \mathbf{k} \\ 0 & e^t & 1 \\ 0 & e^t & 0 \end{vmatrix} = \begin{vmatrix} e^t & 1 \\ e^t & 0 \end{vmatrix} \mathbf{i} - \begin{vmatrix} 0 & 1 \\ 0 & 0 \end{vmatrix} \mathbf{j} + \begin{vmatrix} 0 & e^t \\ 0 & e^t \end{vmatrix} \mathbf{k} = -e^t \mathbf{i} = \langle -e^t, 0, 0 \rangle$$

We need to find the lengths of the following vectors:

$$\|\mathbf{r}'(t) \times \mathbf{r}''(t)\| = \left| \langle -e^t, 0, 0 \rangle \right| = e^t$$

$$\|\mathbf{r}'(t)\| = \sqrt{0^2 + (e^t)^2 + 1^2} = \sqrt{1 + e^{2t}}$$

We now use the formula for curvature to calculate $\kappa(t)$:

$$\kappa(t) = \frac{\|\mathbf{r}'(t) \times \mathbf{r}''(t)\|}{\|\mathbf{r}'(t)\|^3} = \frac{e^t}{\left(\sqrt{1+e^{2t}}\right)^3} = \frac{e^t}{(1+e^{2t})^{3/2}}$$

9. $\mathbf{r}(t) = \langle 4t+1, 4t-3, 2t \rangle$

SOLUTION By Formula (3) we have:

$$\kappa(t) = \frac{\|\mathbf{r}'(t) \times \mathbf{r}''(t)\|}{\|\mathbf{r}'(t)\|^3}$$

We compute $\mathbf{r}'(t)$ and $\mathbf{r}''(t)$:

$$\mathbf{r}'(t) = \langle 4, 4, 2 \rangle, \quad \mathbf{r}''(t) = \langle 0, 0, 0 \rangle$$

Thus $\mathbf{r}'(t) \times \mathbf{r}''(t) = \langle 0, 0, 0 \rangle$, $\|\mathbf{r}'(t) \times \mathbf{r}''(t)\| = 0$, and $\kappa(t) = 0$, as expected.

In Exercises 11–14, use Eq. (3) to evaluate the curvature at the given point.

11. $\mathbf{r}(t) = \langle 1/t, 1/t^2, t^2 \rangle$, $t = -1$

SOLUTION By the formula for curvature we know:

$$\kappa(t) = \frac{\|\mathbf{r}'(t) \times \mathbf{r}''(t)\|}{\|\mathbf{r}'(t)\|^3}$$

We now find $\mathbf{r}'(t), \mathbf{r}''(t)$ and the cross product. These give:

$$\mathbf{r}'(t) = \langle -t^{-2}, -2t^{-3}, 2t \rangle, \quad \Rightarrow \mathbf{r}'(-1) = \langle -1, 2, -2 \rangle$$

$$\mathbf{r}''(t) = \langle 2t^{-3}, 6t^{-4}, 2 \rangle, \quad \Rightarrow \mathbf{r}''(-1) = \langle -2, 6, 2 \rangle$$

$$\mathbf{r}'(-1) \times \mathbf{r}''(-1) = \langle 16, 6, -2 \rangle$$

Now finding the norms, we get:

$$\|\mathbf{r}'(-1)\| = \sqrt{(-1)^2 + 2^2 + (-2)^2} = 3$$

$$\|\mathbf{r}'(-1) \times \mathbf{r}''(-1)\| = \sqrt{16^2 + 6^2 + (-2)^2} = \sqrt{296} = 2\sqrt{74}$$

Therefore,

$$\kappa(-1) = \frac{\|\mathbf{r}'(-1) \times \mathbf{r}''(-1)\|}{\|\mathbf{r}'(-1)\|^3} = \frac{2\sqrt{74}}{3^3} = \frac{2\sqrt{74}}{27}$$

13. $\mathbf{r}(t) = \langle \cos t, \sin t, t^2 \rangle$, $t = \frac{\pi}{2}$

SOLUTION By the formula for curvature we know:

$$\kappa(t) = \frac{\|\mathbf{r}'(t) \times \mathbf{r}''(t)\|}{\|\mathbf{r}'(t)\|^3}$$

We now find $\mathbf{r}'(t), \mathbf{r}''(t)$ and the cross product. These give:

$$\mathbf{r}'(t) = \langle -\sin t, \cos t, 2t \rangle \Rightarrow \mathbf{r}'(\pi/2) = \langle -1, 0, \pi \rangle$$

$$\mathbf{r}''(t) = \langle -\cos t, -\sin t, 2 \rangle \Rightarrow \mathbf{r}''(\pi/2) = \langle 0, -1, 2 \rangle$$

$$\mathbf{r}'(\pi/2) \times \mathbf{r}''(\pi/2) = \langle \pi, 2, 1 \rangle$$

Now finding norms we get:

$$\|\mathbf{r}'(\pi/2)\| = \sqrt{(-1)^2 + 0^2 + \pi^2} = \sqrt{1 + \pi^2}$$

1038 CHAPTER 13 | CALCULUS OF VECTOR-VALUED FUNCTIONS (LT CHAPTER 14)

$$\|\mathbf{r}'(\pi/2) \times \mathbf{r}''(\pi/2)\| = \sqrt{\pi^2 + (-1)^2 + 2^2} = \sqrt{\pi^2 + 5}$$

Therefore,

$$\kappa(\pi/2) = \frac{\|\mathbf{r}'(\pi/2) \times \mathbf{r}''(\pi/2)\|}{\|\mathbf{r}'(\pi/2)\|^3} = \frac{\sqrt{\pi^2 + 5}}{(\sqrt{1 + \pi^2})^3} = \frac{\sqrt{\pi^2 + 5}}{(1 + \pi^2)^{3/2}} \approx 0.108$$

In Exercises 15–18, find the curvature of the plane curve at the point indicated.

15. $y = e^t$, $t = 3$

SOLUTION We use the curvature of a graph in the plane:

$$\kappa(t) = \frac{|f''(t)|}{\left(1 + f'(t)^2\right)^{3/2}}$$

In our case $f(t) = e^t$, hence $f'(t) = f''(t) = e^t$ and we obtain:

$$\kappa(t) = \frac{e^t}{(1 + e^{2t})^{3/2}} \quad \Rightarrow \quad \kappa(3) = \frac{e^3}{(1 + e^6)^{3/2}} \approx 0.0025$$

17. $y = t^4$, $t = 2$

SOLUTION By the curvature of a graph in the plane, we have:

$$\kappa(t) = \frac{|f''(t)|}{\left(1 + f'(t)^2\right)^{3/2}}$$

In this case $f(t) = t^4$, $f'(t) = 4t^3$, $f''(t) = 12t^2$. Hence,

$$\kappa(t) = \frac{12t^2}{\left(1 + (4t^3)^2\right)^{3/2}} = \frac{12t^2}{(1 + 16t^6)^{3/2}}$$

At $t = 2$ we obtain the following curvature:

$$\kappa(2) = \frac{12 \cdot 2^2}{(1 + 16 \cdot 2^6)^{3/2}} = \frac{48}{(1{,}025)^{3/2}} \approx 0.0015$$

19. Find the curvature of $\mathbf{r}(t) = \langle 2 \sin t, \cos 3t, t \rangle$ at $t = \frac{\pi}{3}$ and $t = \frac{\pi}{2}$.

SOLUTION By the formula for curvature we have:

$$\kappa(t) = \frac{\|\mathbf{r}'(t) \times \mathbf{r}''(t)\|}{\|\mathbf{r}'(t)\|^3} \tag{1}$$

We compute the first and second derivatives:

$$\mathbf{r}'(t) = \langle 2 \cos t, -3 \sin 3t, 1 \rangle, \quad \mathbf{r}''(t) = \langle -2 \sin t, -9 \cos 3t, 0 \rangle$$

At the points $t = \frac{\pi}{3}$ and $t = \frac{\pi}{2}$ we have:

$$\mathbf{r}'\left(\frac{\pi}{3}\right) = \left\langle 2 \cos \frac{\pi}{3}, -3 \sin \frac{3\pi}{3}, 1 \right\rangle = \left\langle 2 \cos \frac{\pi}{3}, -3 \sin \pi, 1 \right\rangle = \langle 1, 0, 1 \rangle$$

$$\mathbf{r}''\left(\frac{\pi}{3}\right) = \left\langle -2 \sin \frac{\pi}{3}, -9 \cos \frac{3\pi}{3}, 0 \right\rangle = \langle -\sqrt{3}, 9, 0 \rangle$$

$$\mathbf{r}'\left(\frac{\pi}{2}\right) = \left\langle 2 \cos \frac{\pi}{2}, -3 \sin \frac{3\pi}{2}, 1 \right\rangle = \langle 0, 3, 1 \rangle$$

$$\mathbf{r}''\left(\frac{\pi}{2}\right) = \left\langle -2 \sin \frac{\pi}{2}, -9 \cos \frac{3\pi}{2}, 0 \right\rangle = \langle -2, 0, 0 \rangle$$

We compute the cross products required to use (1):

$$\mathbf{r}'\left(\frac{\pi}{3}\right) \times \mathbf{r}''\left(\frac{\pi}{3}\right) = \begin{vmatrix} \mathbf{i} & \mathbf{j} & \mathbf{k} \\ 1 & 0 & 1 \\ -\sqrt{3} & 9 & 0 \end{vmatrix} = \begin{vmatrix} 0 & 1 \\ 9 & 0 \end{vmatrix}\mathbf{i} - \begin{vmatrix} 1 & 1 \\ -\sqrt{3} & 0 \end{vmatrix}\mathbf{j} + \begin{vmatrix} 1 & 0 \\ -\sqrt{3} & 9 \end{vmatrix}\mathbf{k} = -9\mathbf{i} - \sqrt{3}\mathbf{j} + 9\mathbf{k}$$

$$\mathbf{r}'\left(\frac{\pi}{2}\right) \times \mathbf{r}''\left(\frac{\pi}{2}\right) = \begin{vmatrix} \mathbf{i} & \mathbf{j} & \mathbf{k} \\ 0 & 3 & 1 \\ -2 & 0 & 0 \end{vmatrix} = \begin{vmatrix} 3 & 1 \\ 0 & 0 \end{vmatrix}\mathbf{i} - \begin{vmatrix} 0 & 1 \\ -2 & 0 \end{vmatrix}\mathbf{j} + \begin{vmatrix} 0 & 3 \\ -2 & 0 \end{vmatrix}\mathbf{k} = -2\mathbf{j} + 6\mathbf{k}$$

Hence,

$$\left\| \mathbf{r}'\left(\frac{\pi}{3}\right) \times \mathbf{r}''\left(\frac{\pi}{3}\right) \right\| = \sqrt{(-9)^2 + \left(-\sqrt{3}\right)^2 + 9^2} = \sqrt{165}$$

$$\left\| \mathbf{r}'\left(\frac{\pi}{3}\right) \right\| = \sqrt{1^2 + 0^2 + 1^2} = \sqrt{2}$$

At $t = \frac{\pi}{2}$ we have:

$$\left\| \mathbf{r}'\left(\frac{\pi}{2}\right) \times \mathbf{r}''\left(\frac{\pi}{2}\right) \right\| = \sqrt{(-2)^2 + 6^2} = \sqrt{40} = 2\sqrt{10}$$

$$\left\| \mathbf{r}'\left(\frac{\pi}{2}\right) \right\| = \sqrt{0^2 + 3^2 + 1^2} = \sqrt{10}$$

Substituting the values for $t = \frac{\pi}{3}$ and $t = \frac{\pi}{2}$ in (1) we obtain the following curvatures:

$$\kappa\left(\frac{\pi}{3}\right) = \frac{\sqrt{165}}{\left(\sqrt{2}\right)^3} = \frac{\sqrt{165}}{2\sqrt{2}} \approx 4.54$$

$$\kappa\left(\frac{\pi}{2}\right) = \frac{2\sqrt{10}}{\left(\sqrt{10}\right)^3} = \frac{2\sqrt{10}}{10\sqrt{10}} = 0.2$$

21. Show that the tractrix $\mathbf{r}(t) = \langle t - \tanh t, \operatorname{sech} t \rangle$ has the curvature function $\kappa(t) = \operatorname{sech} t$.

SOLUTION Writing $\mathbf{r}(t) = \langle x(t), y(t) \rangle$, we have $x(t) = t - \tanh t$ and $y(t) = \operatorname{sech} t$. We compute the first and second derivatives of these functions. We use $\tanh^2 t = 1 - \operatorname{sech}^2 t$ to obtain:

$$x'(t) = 1 - \operatorname{sech}^2 t = \tanh^2 t$$

$$x''(t) = -2\operatorname{sech} t(-\operatorname{sech} t \tanh t) = 2\operatorname{sech}^2 t \tanh t$$

$$y'(t) = -\operatorname{sech} t \tanh t$$

$$y''(t) = -(-\operatorname{sech} t \tanh^2 t + \operatorname{sech}^3 t) = \operatorname{sech} t(\tanh^2 t - \operatorname{sech}^2 t) = \operatorname{sech} t(1 - 2\operatorname{sech}^2 t)$$

We compute the cross product $\|\mathbf{r}' \times \mathbf{r}''\|$:

$$x'(t)y''(t) - x''(t)y'(t) = \tanh^2 t \operatorname{sech} t(1 - 2\operatorname{sech}^2 t) + 2\operatorname{sech}^3 t \tanh^2 t$$

$$= \tanh^2 t \left[\operatorname{sech} t - 2\operatorname{sech}^3 t + 2\operatorname{sech}^3 t\right] = \tanh^2 t \operatorname{sech} t$$

We compute the length of $\mathbf{r}'$:

$$x'(t)^2 + y'(t)^2 = \tanh^4 t + \operatorname{sech}^2 t \tanh^2 t = \tanh^2 t(\tanh^2 t + \operatorname{sech}^2 t) = \tanh^2 t$$

Hence

$$\|\mathbf{r}'\|^3 = (\tanh^2 t)^{3/2} = \tanh^3 t$$

Substituting, we obtain

$$\kappa(t) = \frac{|\operatorname{sech} t \tanh^2 t|}{\tanh^3 t} = \frac{\operatorname{sech} t \tanh^2 t}{\tanh^3 t} = \frac{\operatorname{sech} t}{\tanh t}$$

23. Find the value(s) of α such that the curvature of $y = e^{\alpha x}$ at $x = 0$ is as large as possible.

SOLUTION Using the curvature of a graph in the plane we have:

$$\kappa(x) = \frac{|y''(x)|}{(1 + y'(x)^2)^{3/2}} \qquad (1)$$

In our case $y'(x) = \alpha e^{\alpha x}$, $y''(x) = \alpha^2 e^{\alpha x}$. Substituting in (1) we obtain

$$\kappa(x) = \frac{\alpha^2 e^{\alpha x}}{(1 + \alpha^2 e^{2\alpha x})^{3/2}}$$

The curvature at the origin is thus

$$\kappa(0) = \frac{\alpha^2 e^{\alpha \cdot 0}}{(1 + \alpha^2 e^{2\alpha \cdot 0})^{3/2}} = \frac{\alpha^2}{(1 + \alpha^2)^{3/2}}$$

Since $\kappa(0)$ and $\kappa^2(0)$ have their maximum values at the same values of α, we may maximize the function:

$$g(\alpha) = \kappa^2(0) = \frac{\alpha^4}{(1 + \alpha^2)^3}$$

We find the stationary points:

$$g'(\alpha) = \frac{4\alpha^3(1 + \alpha^2)^3 - \alpha^4(3)(1 + \alpha^2)^2 2\alpha}{(1 + \alpha^2)^6} = \frac{2\alpha^3(1 + \alpha^2)^2(2 - \alpha^2)}{(1 + \alpha^2)^6} = 0$$

The stationary points are the solutions of the following equation:

$$2\alpha^3(1 + \alpha^2)^2(2 - \alpha^2) = 0$$

$$\alpha^3 = 0 \qquad \text{or} \qquad 2 - \alpha^2 = 0$$
$$\alpha = 0 \qquad \qquad \alpha = \pm\sqrt{2}$$

Since $g(\alpha) \geq 0$ and $g(0) = 0$, $\alpha = 0$ is a minimum point. Also, $g'(\alpha)$ is positive immediately to the left of $\sqrt{2}$ and negative to the right. Hence, $\alpha = \sqrt{2}$ is a maximum point. Since $g(\alpha)$ is an even function, $\alpha = -\sqrt{2}$ is a maximum point as well. Conclusion: $\kappa(x)$ takes its maximum value at the origin when $\alpha = \pm\sqrt{2}$.

25. Show that the curvature function of the parametrization $\mathbf{r}(t) = \langle a \cos t, b \sin t \rangle$ of the ellipse $\left(\frac{x}{a}\right)^2 + \left(\frac{y}{b}\right)^2 = 1$ is

$$\kappa(t) = \frac{ab}{(b^2 \cos^2 t + a^2 \sin^2 t)^{3/2}} \qquad \boxed{11}$$

SOLUTION The curvature is the following function:

$$\kappa(t) = \frac{\|\mathbf{r}'(t) \times \mathbf{r}''(t)\|}{\|\mathbf{r}'(t)\|^3} \qquad (1)$$

We compute the derivatives and their cross product:

$$\mathbf{r}'(t) = \langle -a \sin t, b \cos t \rangle, \mathbf{r}''(t) = \langle -a \cos t, -b \sin t \rangle$$

$$\mathbf{r}'(t) \times \mathbf{r}''(t) = (-a \sin t \mathbf{i} + b \cos t \mathbf{j}) \times (-a \cos t \mathbf{i} - b \sin t \mathbf{j})$$

$$= ab \sin^2 t \mathbf{k} + ab \cos^2 t \mathbf{k} = ab(\sin^2 t + \cos^2 t)\mathbf{k} = ab\mathbf{k}$$

Thus,

$$\|\mathbf{r}'(t) \times \mathbf{r}''(t)\| = \|ab\mathbf{k}\| = ab$$

$$\|\mathbf{r}'(t)\| = \sqrt{(-a \sin t)^2 + (b \cos t)^2} = \sqrt{a^2 \sin^2 t + b^2 \cos^2 t}$$

Substituting in (1) we obtain the following curvature:

$$\kappa(t) = \frac{ab}{\left(\sqrt{a^2 \sin^2 t + b^2 \cos^2 t}\right)^3} = \frac{ab}{(a^2 \sin^2 t + b^2 \cos^2 t)^{3/2}}$$

27. In the notation of Exercise 25, assume that $a \geq b$. Show that $b/a^2 \leq \kappa(t) \leq a/b^2$ for all t.

SOLUTION In Exercise 25 we showed that the curvature of the ellipse $\mathbf{r}(t) = \langle a\cos t, b\sin t\rangle$ is the following function:

$$\kappa(t) = \frac{ab}{(b^2\cos^2 t + a^2\sin^2 t)^{3/2}}$$

Since $a \geq b > 0$ the quotient becomes greater if we replace a by b in the denominator, and it becomes smaller if we replace b by a in the denominator. We use the identity $\cos^2 t + \sin^2 t = 1$ to obtain:

$$\frac{ab}{(a^2\cos^2 t + a^2\sin^2 t)^{3/2}} \leq \kappa(t) \leq \frac{ab}{(b^2\cos^2 t + b^2\sin^2 t)^{3/2}}$$

$$\frac{ab}{(a^2(\cos^2 t + \sin^2 t))^{3/2}} \leq \kappa(t) \leq \frac{ab}{(b^2(\cos^2 t + \sin^2 t))^{3/2}}$$

$$\frac{ab}{a^3} = \frac{ab}{(a^2)^{3/2}} \leq \kappa(t) \leq \frac{ab}{(b^2)^{3/2}} = \frac{ab}{b^3}$$

$$\frac{b}{a^2} \leq \kappa(t) \leq \frac{a}{b^2}$$

In Exercises 29–32, use Eq. (5) to compute the curvature at the given point.

29. $\langle t^2, t^3\rangle$, $t = 2$

SOLUTION For the given parametrization, $x(t) = t^2$, $y(t) = t^3$, hence

$$x'(t) = 2t$$
$$x''(t) = 2$$
$$y'(t) = 3t^2$$
$$y''(t) = 6t$$

At the point $t = 2$ we have

$$x'(2) = 4, \quad x''(2) = 2, \quad y'(2) = 3 \cdot 2^2 = 12, \quad y''(2) = 12$$

Substituting in Eq. (5) we get

$$\kappa(2) = \frac{|x'(2)y''(2) - x''(2)y'(2)|}{(x'(2)^2 + y'(2)^2)^{3/2}} = \frac{|4 \cdot 12 - 2 \cdot 12|}{(4^2 + 12^2)^{3/2}} = \frac{24}{160^{3/2}} \approx 0.012$$

31. $\langle t\cos t, \sin t\rangle$, $t = \pi$

SOLUTION We have $x(t) = t\cos t$ and $y(t) = \sin t$, hence:

$$x'(t) = \cos t - t\sin t \quad \Rightarrow \quad x'(\pi) = \cos\pi - \pi\sin\pi = -1$$
$$x''(t) = -\sin t - (\sin t + t\cos t) = -2\sin t - t\cos t \quad \Rightarrow \quad x''(\pi) = -2\sin\pi - \pi\cos\pi = \pi$$
$$y'(t) = \cos t \quad \Rightarrow \quad y'(\pi) = \cos\pi = -1$$
$$y''(t) = -\sin t \quad \Rightarrow \quad y''(\pi) = -\sin\pi = 0$$

Substituting in Eq. (5) gives the following curvature:

$$\kappa(\pi) = \frac{|x'(\pi)y''(\pi) - x''(\pi)y'(\pi)|}{(x'(\pi)^2 + y'(\pi)^2)^{3/2}} = \frac{|-1 \cdot 0 - \pi \cdot (-1)|}{((-1)^2 + (-1)^2)^{3/2}} = \frac{\pi}{2\sqrt{2}} \approx 1.11$$

33. Let $s(t) = \int_{-\infty}^{t} \|\mathbf{r}'(u)\|\,du$ for the Bernoulli spiral $\mathbf{r}(t) = \langle e^t\cos 4t, e^t\sin 4t\rangle$ (see Exercise 39 in Section 13.3). Show that the radius of curvature is proportional to $s(t)$.

SOLUTION The radius of curvature is the reciprocal of the curvature:

$$R(t) = \frac{1}{\kappa(t)}$$

We compute the curvature using the equality given in Exercise 39 in Section 3:

$$\kappa(t) = \frac{|x'(t)y''(t) - x''(t)y'(t)|}{(x'(t)^2 + y'(t)^2)^{3/2}} \quad (1)$$

In our case, $x(t) = e^t \cos 4t$ and $y(t) = e^t \sin 4t$. Hence:

$$x'(t) = e^t \cos 4t - 4e^t \sin 4t = e^t(\cos 4t - 4\sin 4t)$$

$$x''(t) = e^t(\cos 4t - 4\sin 4t) + e^t(-4\sin 4t - 16\cos 4t) = -e^t(15\cos 4t + 8\sin 4t)$$

$$y'(t) = e^t \sin 4t + 4e^t \cos 4t = e^t(\sin 4t + 4\cos 4t)$$

$$y''(t) = e^t(\sin 4t + 4\cos 4t) + e^t(4\cos 4t - 16\sin 4t) = e^t(8\cos 4t - 15\sin 4t)$$

We compute the numerator in (1):

$$x'(t)y''(t) - x''(t)y'(t) = e^{2t}(\cos 4t - 4\sin 4t) \cdot (8\cos 4t - 15\sin 4t) + e^{2t}(15\cos 4t + 8\sin 4t) \cdot (\sin 4t + 4\cos 4t)$$

$$= e^{2t}(68\cos^2 4t + 68\sin^2 4t) = 68e^{2t}$$

We compute the denominator in (1):

$$x'(t)^2 + y'(t)^2 = e^{2t}(\cos 4t - 4\sin 4t)^2 + e^{2t}(\sin 4t + 4\cos 4t)^2$$

$$= e^{2t}(\cos^2 4t - 8\cos 4t \sin 4t + 16\sin^2 4t + \sin^2 4t + 8\sin 4t \cos 4t + 16\cos^2 4t)$$

$$= e^{2t}(\cos^2 4t + \sin^2 4t + 16(\sin^2 4t + \cos^2 4t))$$

$$= e^{2t}(1 + 16 \cdot 1) = 17e^{2t} \quad (2)$$

Hence

$$\left(x'(t)^2 + y'(t)^2\right)^{3/2} = 17^{3/2} e^{3t}$$

Substituting in (2) we have

$$\kappa(t) = \frac{68e^{2t}}{17^{3/2}e^{3t}} = \frac{4}{\sqrt{17}}e^{-t} \quad \Rightarrow \quad R = \frac{\sqrt{17}}{4}e^t \quad (3)$$

On the other hand, by the Fundamental Theorem and (2) we have

$$s'(t) = \|\mathbf{r}'(t)\| = \sqrt{x'(t)^2 + y'(t)^2} = \sqrt{17e^{2t}} = \sqrt{17}e^t$$

We integrate to obtain

$$s(t) = \int \sqrt{17}\, e^t \, dt = \sqrt{17}\, e^t + C \quad (4)$$

Since $s(t) = \int_{-\infty}^{t} \|\mathbf{r}'(u)\|\, du$, we have $\lim_{t \to -\infty} s(t) = 0$, hence by (4):

$$0 = \lim_{t \to -\infty}\left(\sqrt{17}e^t + C\right) = 0 + C = C$$

Substituting $C = 0$ in (4) we get:

$$s(t) = \sqrt{17}e^t \quad (5)$$

Combining (3) and (5) gives:

$$R(t) = \frac{1}{4}s(t)$$

which means that the radius of curvature is proportional to $s(t)$.

35. **CAS** Plot the **clothoid** $\mathbf{r}(t) = \langle x(t), y(t) \rangle$, and compute its curvature $\kappa(t)$ where

$$x(t) = \int_0^t \sin \frac{u^3}{3} \, du, \qquad y(t) = \int_0^t \cos \frac{u^3}{3} \, du$$

SOLUTION We use the following formula for the curvature (given earlier):

$$\kappa(t) = \frac{|x'(t)y''(t) - x''(t)y'(t)|}{(x'(t)^2 + y'(t)^2)^{3/2}} \qquad (1)$$

We compute the first and second derivatives of $x(t)$ and $y(t)$. Using the Fundamental Theorem and the Chain Rule we get:

$$x'(t) = \sin \frac{t^3}{3}$$

$$x''(t) = \frac{3t^2}{3} \cos \frac{t^3}{3} = t^2 \cos \frac{t^3}{3}$$

$$y'(t) = \cos \frac{t^3}{3}$$

$$y''(t) = \frac{3t^2}{3} \left(-\sin \frac{t^3}{3}\right) = -t^2 \sin \frac{t^3}{3}$$

Substituting in (1) gives the following curvature function:

$$\kappa(t) = \frac{\left|\sin \frac{t^3}{3} \left(-t^2 \sin \frac{t^3}{3}\right) - t^2 \cos \frac{t^3}{3} \cos \frac{t^3}{3}\right|}{\left(\left(\sin \frac{t^3}{3}\right)^2 + \left(\cos \frac{t^3}{3}\right)^2\right)^{3/2}} = \frac{t^2 \left(\sin^2 \frac{t^3}{3} + \cos^2 \frac{t^3}{3}\right)}{1^{3/2}} = t^2$$

That is, $\kappa(t) = t^2$. Here is a plot of the curvature as a function of t:

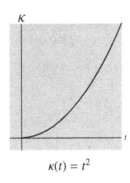

$\kappa(t) = t^2$

37. Find the normal vector $\mathbf{N}(t)$ to $\mathbf{r}(t) = \langle 4, \sin 2t, \cos 2t \rangle$.

SOLUTION We first find the unit tangent vector:

$$\mathbf{T}(t) = \frac{\mathbf{r}'(t)}{\|\mathbf{r}'(t)\|} \qquad (1)$$

We have

$$\mathbf{r}'(t) = \frac{d}{dt} \langle 4, \sin 2t, \cos 2t \rangle = \langle 0, 2\cos 2t, -2\sin 2t \rangle = 2\langle 0, \cos 2t, -\sin 2t \rangle$$

$$\|\mathbf{r}'(t)\| = 2\sqrt{0^2 + \cos^2 2t + (-\sin 2t)^2} = 2\sqrt{0+1} = 2$$

Substituting in (1) gives:

$$\mathbf{T}(t) = \frac{2\langle 0, \cos 2t, -\sin 2t \rangle}{2} = \langle 0, \cos 2t, -\sin 2t \rangle$$

The normal vector is the following vector:

$$\mathbf{N}(t) = \frac{\mathbf{T}'(t)}{\|\mathbf{T}'(t)\|} \qquad (2)$$

We compute the derivative of the unit tangent vector and its length:

$$\mathbf{T}'(t) = \frac{d}{dt}\langle 0, \cos 2t, -\sin 2t\rangle = \langle 0, -2\sin 2t, -2\cos 2t\rangle = -2\langle 0, \sin 2t, \cos 2t\rangle$$

$$\|\mathbf{T}'(t)\| = 2\sqrt{0^2 + \sin^2 2t + \cos^2 2t} = 2\sqrt{0+1} = 2$$

Substituting in (2) we obtain:

$$\mathbf{N}(t) = \frac{-2\langle 0, \sin 2t, \cos 2t\rangle}{2} = \langle 0, -\sin 2t, -\cos 2t\rangle$$

39. Find the normal vectors to $\mathbf{r}(t) = \langle t, \cos t\rangle$ at $t = \frac{\pi}{4}$ and $t = \frac{3\pi}{4}$.

SOLUTION The normal vector to $\mathbf{r}(t) = \langle t, \cos t\rangle$ is $\mathbf{T}'(t)$, where $\mathbf{T}(t) = \frac{\mathbf{r}'(t)}{\|\mathbf{r}'(t)\|}$ is the unit tangent vector. We have

$$\mathbf{r}'(t) = \langle 1, -\sin t\rangle \quad \Rightarrow \quad \|\mathbf{r}'(t)\| = \sqrt{1^2 + (\sin t)^2} = \sqrt{1 + \sin^2 t}$$

Hence,

$$\mathbf{T}(t) = \frac{1}{\sqrt{1 + \sin^2 t}}\langle 1, -\sin t\rangle$$

We compute the derivative of $\mathbf{T}(t)$ to find the normal vector. We use the Product Rule and the Chain Rule to obtain:

$$\mathbf{T}'(t) = \frac{1}{\sqrt{1 + \sin^2 t}}\frac{d}{dt}\langle 1, -\sin t\rangle + \left(\frac{1}{\sqrt{1 + \sin^2 t}}\right)'\langle 1, -\sin t\rangle$$

$$= \frac{1}{\sqrt{1 + \sin^2 t}}\langle 0, -\cos t\rangle - \frac{1}{1 + \sin^2 t} \cdot \frac{2\sin t \cos t}{2\sqrt{1 + \sin^2 t}}\langle 1, -\sin t\rangle$$

$$= \frac{1}{\sqrt{1 + \sin^2 t}}\langle 0, -\cos t\rangle - \frac{\sin 2t}{2(1 + \sin^2 t)^{3/2}}\langle 1, -\sin t\rangle$$

At $t = \frac{\pi}{4}$ we obtain the normal vector:

$$\mathbf{T}'\left(\frac{\pi}{4}\right) = \frac{1}{\sqrt{1+\frac{1}{2}}}\left\langle 0, -\frac{1}{\sqrt{2}}\right\rangle - \frac{1}{2\left(1+\frac{1}{2}\right)^{3/2}}\left\langle 1, -\frac{1}{\sqrt{2}}\right\rangle = \left\langle 0, -\frac{1}{\sqrt{3}}\right\rangle - \left\langle \frac{\sqrt{2}}{3\sqrt{3}}, \frac{-1}{3\sqrt{3}}\right\rangle = \left\langle \frac{-\sqrt{2}}{3\sqrt{3}}, \frac{-2}{3\sqrt{3}}\right\rangle$$

At $t = \frac{3\pi}{4}$ we obtain:

$$\mathbf{T}'\left(\frac{3\pi}{4}\right) = \frac{1}{\sqrt{1+\frac{1}{2}}}\left\langle 0, \frac{1}{\sqrt{2}}\right\rangle - \frac{-1}{2\left(1+\frac{1}{2}\right)^{3/2}}\left\langle 1, -\frac{1}{\sqrt{2}}\right\rangle = \left\langle 0, \frac{1}{\sqrt{3}}\right\rangle + \left\langle \frac{\sqrt{2}}{3\sqrt{3}}, \frac{-1}{3\sqrt{3}}\right\rangle = \left\langle \frac{\sqrt{2}}{3\sqrt{3}}, \frac{2}{3\sqrt{3}}\right\rangle$$

In Exercises 41–44, find $\mathbf{T}, \mathbf{N}$, and $\mathbf{B}$ for the curve at the indicated point. Hint: After finding $\mathbf{T}'$, plug in the specific value for t before computing $\mathbf{N}$ and $\mathbf{B}$.

41. $\mathbf{r}(t) = \langle 0, t, t^2\rangle$ at $(0, 1, 1)$. In this case, draw the curve and the three resultant vectors in 3-space.

SOLUTION Differentiating gives $\mathbf{r}'(t) = \langle 0, 1, 2t\rangle$, so that

$$\mathbf{T}(t) = \frac{\mathbf{r}'(t)}{\|\mathbf{r}'(t)\|} = \frac{1}{\sqrt{4t^2 + 1}}\langle 0, 1, 2t\rangle$$

The point $(0, 1, 1)$ corresponds to $t = 1$; substituting gives

$$\mathbf{T}(1) = \frac{1}{\sqrt{5}}\langle 0, 1, 2\rangle$$

Now,

$$\mathbf{T}'(t) = \frac{d}{dt}\left\langle 0, (1 + 4t^2)^{-1/2}, 2t(1 + 4t^2)^{-1/2}\right\rangle$$

$$= \left\langle 0, -\frac{1}{2} \cdot (1+4t^2)^{-3/2} \cdot 8t, 2(1+4t^2)^{-1/2} + 2t \cdot \left(-\frac{1}{2}(1+4t^2)^{-3/2} \cdot 8t\right)\right\rangle$$

$$= \left\langle 0, -4t(1+4t^2)^{-3/2}, 2(1+4t^2)^{-1/2} - 8t^2(1+4t^2)^{-3/2}\right\rangle$$

and at $t = 1$ we get

$$\mathbf{T}'(1) = \left\langle 0, -4 \cdot 1(1+4 \cdot 1^2)^{-3/2}, 2(1+4 \cdot 1^2)^{-1/2} - 8 \cdot 1^2(1+4 \cdot 1^2)^{-3/2}\right\rangle$$

$$= \left\langle 0, -4 \cdot 5^{-3/2}, 2 \cdot 5^{-1/2} - 8 \cdot 5^{-3/2}\right\rangle$$

$$= 5^{-3/2} \langle 0, -4, 2\rangle.$$

Then

$$\mathbf{N}(1) = \frac{\mathbf{T}'(1)}{\|\mathbf{T}'(1)\|} = \frac{5^{-3/2} \langle 0, -4, 2\rangle}{5^{-3/2}\sqrt{0^2 + (-4)^2 + 2^2}} = \frac{1}{2\sqrt{5}}\langle 0, -4, 2\rangle$$

$$\mathbf{B}(1) = \mathbf{T}(1) \times \mathbf{N}(1) = \frac{1}{\sqrt{5}} \cdot \frac{1}{2\sqrt{5}}\begin{vmatrix} \mathbf{i} & \mathbf{j} & \mathbf{k} \\ 0 & 1 & 2 \\ 0 & -4 & 2 \end{vmatrix} = \frac{1}{10}\langle 10, 0, 0\rangle = \langle 1, 0, 0\rangle.$$

A plot of the curve together with $\mathbf{T}(1)$, $\mathbf{N}(1)$, and $\mathbf{B}(1)$ is shown below:

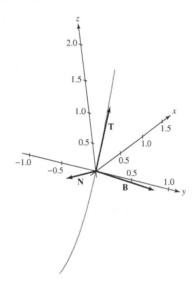

43. $\mathbf{r}(t) = \left\langle t, t^2, \frac{2}{3}t^3 \right\rangle$ at $\left(1, 1, \frac{2}{3}\right)$

SOLUTION Differentiating gives $\mathbf{r}'(t) = \left\langle 1, 2t, 2t^2 \right\rangle$, so that

$$\mathbf{T}(t) = \frac{\mathbf{r}'(t)}{\|\mathbf{r}'(t)\|} = \frac{1}{\sqrt{1+4t^2+4t^4}}\left\langle 1, 2t, 2t^2 \right\rangle = \frac{1}{2t^2+1}\left\langle 1, 2t, 2t^2 \right\rangle$$

The point $\left(1, 1, \frac{2}{3}\right)$ corresponds to $t = 1$; substituting gives

$$\mathbf{T}(1) = \frac{1}{3}\langle 1, 2, 2\rangle$$

Now,

$$\mathbf{T}'(t) = \frac{d}{dt}\left\langle \frac{1}{2t^2+1}, \frac{2t}{2t^2+1}, \frac{2t^2}{2t^2+1}\right\rangle$$

$$= \left\langle -\frac{4t}{(2t^2+1)^2}, \frac{(2t^2+1)(2) - 2t(4t)}{(2t^2+1)^2}, \frac{(2t^2+1)(4t) - 2t^2(4t)}{(2t^2+1)^2}\right\rangle$$

$$= \left\langle -\frac{4t}{(2t^2+1)^2}, \frac{2-4t^2}{(2t^2+1)^2}, \frac{4t}{(2t^2+1)^2}\right\rangle$$

and at $t=1$ we get

$$\mathbf{T}'(1) = \left\langle -\frac{4}{9}, -\frac{2}{9}, \frac{4}{9} \right\rangle = \frac{1}{9}\langle -4, -2, 4 \rangle$$

Then

$$\mathbf{N}(1) = \frac{\mathbf{T}'(1)}{\|\mathbf{T}'(1)\|} = \frac{\frac{1}{9}\langle -4, -2, 4 \rangle}{\frac{1}{9}\sqrt{(-4)^2 + (-2)^2 + 4^2}} = \frac{1}{6}\langle -4, -2, 4 \rangle$$

$$\mathbf{B}(1) = \mathbf{T}(1) \times \mathbf{N}(1) = \frac{1}{3} \cdot \frac{1}{6} \begin{vmatrix} \mathbf{i} & \mathbf{j} & \mathbf{k} \\ 1 & 2 & 2 \\ -4 & -2 & 4 \end{vmatrix} = \frac{1}{18}\langle 12, -12, 6 \rangle = \frac{1}{3}\langle 2, -2, 1 \rangle$$

45. Find the normal vector to the clothoid (Exercise 35) at $t = \pi^{1/3}$.

SOLUTION The Clothoid is the plane curve $\mathbf{r}(t) = \langle x(t), y(t) \rangle$ with

$$x(t) = \int_0^t \sin\frac{u^3}{3}\, du, \quad y(t) = \int_0^t \cos\frac{u^3}{3}\, du$$

The unit normal is the following vector:

$$\mathbf{N}(t) = \frac{\mathbf{T}'(t)}{\|\mathbf{T}'(t)\|} \tag{1}$$

We first find the unit tangent vector $\mathbf{T}(t) = \frac{\mathbf{r}'(t)}{\|\mathbf{r}'(t)\|}$. By the Fundamental Theorem we have

$$\mathbf{r}'(t) = \left\langle \sin\frac{t^3}{3}, \cos\frac{t^3}{3} \right\rangle \quad \Rightarrow \quad \|\mathbf{r}'(t)\| = \sqrt{\sin^2\frac{t^3}{3} + \cos^2\frac{t^3}{3}} = \sqrt{1} = 1$$

Hence,

$$\mathbf{T}(t) = \left\langle \sin\frac{t^3}{3}, \cos\frac{t^3}{3} \right\rangle$$

We now differentiate $\mathbf{T}(t)$ using the Chain Rule to obtain:

$$\mathbf{T}'(t) = \left\langle \frac{3t^2}{3}\cos\frac{t^3}{3}, \frac{-3t^2}{3}\sin\frac{t^3}{3} \right\rangle = t^2\left\langle \cos\frac{t^3}{3}, -\sin\frac{t^3}{3} \right\rangle$$

Hence,

$$\|\mathbf{T}'(t)\| = t^2\sqrt{\cos^2\frac{t^3}{3} + \left(-\sin\frac{t^3}{3}\right)^2} = t^2$$

Substituting in (1) we obtain the following unit normal:

$$\mathbf{N}(t) = \left\langle \cos\frac{t^3}{3}, -\sin\frac{t^3}{3} \right\rangle$$

At the point $T = \pi^{1/3}$ the unit normal is

$$\mathbf{N}(\pi^{1/3}) = \left\langle \cos\frac{(\pi^{1/3})^3}{3}, -\sin\frac{(\pi^{1/3})^3}{3} \right\rangle = \left\langle \cos\frac{\pi}{3}, -\sin\frac{\pi}{3} \right\rangle = \left\langle \frac{1}{2}, -\frac{\sqrt{3}}{2} \right\rangle$$

In Exercises 47–52, use Eq. (12) to find $\mathbf{N}$ at the point indicated.

47. $\langle t^2, t^3 \rangle$, $t = 1$

SOLUTION We use the equality

$$\mathbf{N}(t) = \frac{v(t)\mathbf{r}''(t) - v'(t)\mathbf{r}'(t)}{\|v(t)\mathbf{r}''(t) - v'(t)\mathbf{r}'(t)\|} \tag{1}$$

For $\mathbf{r}(t) = \langle t^2, t^3 \rangle$ we have

$$\mathbf{r}'(t) = \langle 2t, 3t^2 \rangle$$

$$\mathbf{r}''(t) = \langle 2, 6t \rangle$$

$$v(t) = \|\mathbf{r}'(t)\| = \sqrt{(2t)^2 + (3t^2)^2} = \sqrt{4t^2 + 9t^4}$$

$$v'(t) = \frac{8t + 36t^3}{2\sqrt{4t^2 + 9t^4}} = \frac{4t + 18t^3}{\sqrt{4t^2 + 9t^4}}$$

At the point $t = 1$ we get

$$\mathbf{r}'(1) = \langle 2, 3 \rangle, \quad \mathbf{r}''(1) = \langle 2, 6 \rangle, \quad v(1) = \sqrt{13}, \quad v'(1) = \frac{4 + 18}{\sqrt{4 + 9}} = \frac{22}{13}\sqrt{13}$$

Thus

$$v(1)\mathbf{r}''(1) - v'(1)\mathbf{r}'(1) = \sqrt{13}\langle 2, 6 \rangle - \frac{22}{13}\sqrt{13}\langle 2, 3 \rangle = \frac{\sqrt{13}}{13}\langle -18, 12 \rangle = \frac{1}{\sqrt{13}}\langle -18, 12 \rangle$$

Then

$$\mathbf{N}(1) = \frac{v(1)\mathbf{r}''(1) - v'(1)\mathbf{r}'(1)}{\|v(1)\mathbf{r}''(1) - v'(1)\mathbf{r}'(1)\|} = \frac{\langle -18, 12 \rangle}{\sqrt{(-18)^2 + 12^2}} = \frac{6\langle -3, 2 \rangle}{\sqrt{468}} = \frac{6\langle -3, 2 \rangle}{6\sqrt{13}} = \frac{1}{\sqrt{13}}\langle -3, 2 \rangle$$

49. $\langle t^2/2, t^3/3, t \rangle, \quad t = 1$

SOLUTION We use the equality

$$\mathbf{N}(t) = \frac{v(t)\mathbf{r}''(t) - v'(t)\mathbf{r}'(t)}{\|v(t)\mathbf{r}''(t) - v'(t)\mathbf{r}'(t)\|} \tag{1}$$

For $\mathbf{r}(t) = \langle \frac{t^2}{2}, \frac{t^3}{3}, t \rangle$ we have

$$\mathbf{r}'(t) = \langle t, t^2, 1 \rangle$$

$$\mathbf{r}''(t) = \langle 1, 2t, 0 \rangle$$

$$v(t) = \|\mathbf{r}'(t)\| = \sqrt{t^4 + t^2 + 1}$$

$$v'(t) = \frac{1}{2}(t^4 + t^2 + 1)^{-1/2}(4t^3 + 2t) = \frac{2t^3 + t}{\sqrt{t^4 + t^2 + 1}}$$

At the point $t = 1$ we get

$$\mathbf{r}'(1) = \langle 1, 1, 1 \rangle, \quad \mathbf{r}''(1) = \langle 1, 2, 0 \rangle, \quad v(1) = \sqrt{3}, \quad v'(1) = \sqrt{3}$$

Thus

$$v(1)\mathbf{r}''(1) - v'(1)\mathbf{r}'(1) = \sqrt{3}\langle 1, 2, 0 \rangle - \sqrt{3}\langle 1, 1, 1 \rangle = \sqrt{3}\langle 0, 1, -1 \rangle$$

Then

$$\mathbf{N}(1) = \frac{v(1)\mathbf{r}''(1) - v'(1)\mathbf{r}'(1)}{\|v(1)\mathbf{r}''(1) - v'(1)\mathbf{r}'(1)\|} = \frac{\langle 0, 1, -1 \rangle}{\sqrt{0^2 + 1^2 + (-1)^2}} = \frac{1}{\sqrt{2}}\langle 0, 1, -1 \rangle$$

51. $\langle t, e^t, t \rangle, \quad t = 0$

SOLUTION We use the equality

$$\mathbf{N}(t) = \frac{v(t)\mathbf{r}''(t) - v'(t)\mathbf{r}'(t)}{\|v(t)\mathbf{r}''(t) - v'(t)\mathbf{r}'(t)\|} \tag{1}$$

For $\mathbf{r}(t) = \langle t, e^t, t \rangle$ we have

$$\mathbf{r}'(t) = \langle 1, e^t, 1 \rangle$$

$$\mathbf{r}''(t) = \langle 0, e^t, 0 \rangle$$

$$v(t) = \|\mathbf{r}'(t)\| = \sqrt{1^2 + (e^t)^2 + 1^2} = \sqrt{e^{2t} + 2}$$

$$v'(t) = \frac{1}{2}(e^{2t} + 2)^{-1/2}(2e^{2t}) = \frac{e^{2t}}{\sqrt{e^{2t} + 2}}$$

At the point $t = 0$ we get

$$\mathbf{r}'(0) = \langle 1, 1, 1 \rangle, \quad \mathbf{r}''(0) = \langle 0, 1, 0 \rangle, \quad v(0) = \sqrt{3}, \quad v'(0) = \frac{1}{\sqrt{3}} = \frac{1}{3}\sqrt{3}$$

Thus

$$v(0)\mathbf{r}''(0) - v'(0)\mathbf{r}'(0) = \sqrt{3}\langle 0, 1, 0 \rangle - \frac{1}{3}\sqrt{3}\langle 1, 1, 1 \rangle = \frac{1}{3}\sqrt{3}\langle -1, 2, -1 \rangle$$

Then

$$\mathbf{N}(0) = \frac{v(0)\mathbf{r}''(0) - v'(0)\mathbf{r}'(0)}{\|v(0)\mathbf{r}''(0) - v'(0)\mathbf{r}'(0)\|} = \frac{\langle -1, 2, -1 \rangle}{\sqrt{(-1)^2 + 2^2 + (-1)^2}} = \frac{1}{\sqrt{6}}\langle -1, 2, -1 \rangle$$

53. Let $\mathbf{r}(t) = \langle t, \frac{4}{3}t^{3/2}, t^2 \rangle$.

(a) Find $\mathbf{T}, \mathbf{N}$, and $\mathbf{B}$ at the point corresponding to $t = 1$.

(b) Find the equation of the osculating plane at the point corresponding to $t = 1$.

SOLUTION

(a) Differentiating gives $\mathbf{r}'(t) = \langle 1, 2\sqrt{t}, 2t \rangle$, so that

$$\mathbf{T}(t) = \frac{\mathbf{r}'(t)}{\|\mathbf{r}'(t)\|} = \frac{1}{\sqrt{4t^2 + 4t + 1}}\langle 1, 2\sqrt{t}, 2t \rangle = \frac{1}{2t + 1}\langle 1, 2\sqrt{t}, 2t \rangle$$

Substituting $t = 1$ gives

$$\mathbf{T}(1) = \frac{1}{3}\langle 1, 2, 2 \rangle$$

Then

$$\mathbf{T}'(t) = \frac{d}{dt}\frac{1}{2t + 1}\langle 1, 2\sqrt{t}, 2t \rangle$$

$$= \frac{d}{dt}\left\langle \frac{1}{2t + 1}, \frac{2\sqrt{t}}{2t + 1}, 1 - \frac{1}{2t + 1} \right\rangle$$

$$= \left\langle -\frac{2}{(2t + 1)^2}, \frac{(2t + 1) \cdot t^{-1/2} - 2\sqrt{t} \cdot 2}{(2t + 1)^2}, \frac{2}{(2t + 1)^2} \right\rangle$$

$$= \frac{1}{(2t + 1)^2}\left\langle -2, \frac{2t + 1}{\sqrt{t}} - 4\sqrt{t}, 2 \right\rangle$$

$$= \frac{1}{(2t + 1)^2}\left\langle -2, \frac{1 - 2t}{\sqrt{t}}, 2 \right\rangle$$

and at $t = 1$ we get

$$\mathbf{T}'(1) = \frac{1}{9}\left\langle -2, \frac{1 - 2}{\sqrt{1}}, 2 \right\rangle = \frac{1}{9}\langle -2, -1, 2 \rangle$$

Then

$$\mathbf{N}(1) = \frac{\mathbf{T}'(1)}{\|\mathbf{T}'(1)\|} = \frac{\frac{1}{9}\langle -2, -1, 2 \rangle}{\frac{1}{9}\sqrt{(-2)^2 + 1^2 + 2^2}} = \frac{1}{3}\langle -2, -1, 2 \rangle$$

$$\mathbf{B}(1) = \mathbf{T}(1) \times \mathbf{N}(1) = \frac{1}{3} \cdot \frac{1}{3}\begin{vmatrix} \mathbf{i} & \mathbf{j} & \mathbf{k} \\ 1 & 2 & 2 \\ -2 & -1 & 2 \end{vmatrix} = \frac{1}{9}\langle 6, -6, 3 \rangle = \frac{1}{3}\langle 2, -2, 1 \rangle$$

(b) The binormal vector $\mathbf{B}(1)$ is normal to the osculating plane at $\mathbf{r}(1) = \left\langle 1, \frac{4}{3}, 1 \right\rangle$; for simplicity, we use instead the normal vector $2\mathbf{B}(1) = \langle 2, -2, 1 \rangle$. Then the equation of the osculating plane is

$$2(x-1) - 2\left(y - \frac{4}{3}\right) + 1(z-1) = 0$$

Simplifying gives $6(x-1) - 2(3y-4) + 3(z-1) = 0$, or $6x - 6y + 3z = 1$.

55. Let $\mathbf{r}(t) = \langle t, 1-t, t^2 \rangle$.

(a) Find the general formulas for $\mathbf{T}$ and $\mathbf{N}$ as functions of t.

(b) Find the general formula for $\mathbf{B}$ as a function of t.

(c) What can you conclude about the osculating planes of the curve based on your answer to b?

SOLUTION

(a) Differentiating gives $\mathbf{r}'(t) = \langle 1, -1, 2t \rangle$, so that

$$\mathbf{T}(t) = \frac{\mathbf{r}'(t)}{\|\mathbf{r}'(t)\|} = \frac{1}{\sqrt{1^2 + (-1)^2 + (2t)^2}} \langle 1, -1, 2t \rangle = \left\langle \frac{1}{\sqrt{4t^2+2}}, -\frac{1}{\sqrt{4t^2+2}}, \frac{2t}{\sqrt{4t^2+2}} \right\rangle$$

The derivatives of the components of $\mathbf{T}(t)$ are

$$\frac{d}{dt} \frac{1}{\sqrt{4t^2+2}} = -\frac{4t}{(4t^2+2)^{3/2}}$$

$$\frac{d}{dt}\left(-\frac{1}{\sqrt{4t^2+2}}\right) = \frac{4t}{(4t^2+2)^{3/2}}$$

$$\frac{d}{dt} \frac{2t}{\sqrt{4t^2+2}} = \frac{(4t^2+2)^{1/2}(2) - 2t \cdot \frac{1}{2}(4t^2+2)^{-1/2} \cdot 8t}{4t^2+2} = \frac{2(4t^2+2) - 8t^2}{(4t^2+2)^{3/2}} = \frac{4}{(4t^2+2)^{3/2}}$$

Thus

$$\mathbf{T}'(t) = \left\langle -\frac{4t}{(4t^2+2)^{3/2}}, \frac{4t}{(4t^2+2)^{3/2}}, \frac{4}{(4t^2+2)^{3/2}} \right\rangle = \frac{1}{(4t^2+2)^{3/2}} \langle -4t, 4t, 4 \rangle$$

To find $\mathbf{N}(t)$ we must find $\|\mathbf{T}'(t)\|$:

$$\|\mathbf{T}'(t)\| = \frac{1}{(4t^2+2)^{3/2}} \sqrt{(-4t)^2 + (4t)^2 + 4^2} = \frac{1}{(4t^2+2)^{3/2}} \sqrt{32t^2+16} = \frac{\sqrt{8} \cdot \sqrt{4t^2+2}}{(4t^2+2)^{3/2}} = \frac{\sqrt{8}}{4t^2+2}$$

So

$$\mathbf{N}(t) = \frac{\mathbf{T}'(t)}{\|\mathbf{T}'(t)\|} = \frac{\frac{1}{(4t^2+2)^{3/2}} \langle -4t, 4t, 4 \rangle}{\frac{\sqrt{8}}{4t^2+2}}$$

$$= \frac{(4t^2+2)}{(4t^2+2)^{3/2} \sqrt{8}} \langle -4t, 4t, 4 \rangle$$

$$= \frac{1}{\sqrt{4t^2+2}\sqrt{8}} \langle -4t, 4t, 4 \rangle$$

$$= \frac{1}{\sqrt{32t^2+16}} \langle -4t, 4t, 4 \rangle = \frac{1}{\sqrt{2t^2+1}} \langle -t, t, 1 \rangle$$

(b) The derivatives of the components of $\mathbf{N}(t)$ are

$$\frac{d}{dt}\left(-\frac{t}{\sqrt{2t^2+1}}\right) = -\frac{(2t^2+1)^{1/2}(1) - t \cdot \frac{1}{2}(2t^2+1)^{-1/2} \cdot 4t}{2t^2+1} = -\frac{(2t^2+1) - 2t^2}{(2t^2+1)^{3/2}} = -\frac{1}{(2t^2+1)^{3/2}}$$

$$\frac{d}{dt} \frac{t}{\sqrt{2t^2+1}} = \frac{1}{(2t^2+1)^{3/2}}$$

$$\frac{d}{dt} \frac{1}{\sqrt{2t^2+1}} = -\frac{2t}{(2t^2+1)^{3/2}}$$

so that

$$\mathbf{N}'(t) = \frac{1}{(2t^2+1)^{3/2}} \langle -1, 1, -2t \rangle$$

Finally,

$$\mathbf{B}(t) = \frac{\mathbf{N}'(t)}{\|\mathbf{N}'(t)\|} = \frac{\frac{1}{(2t^2+1)^{3/2}} \langle -1, 1, -2t \rangle}{\frac{1}{(2t^2+1)^{3/2}} \sqrt{(-1)^2 + 1^2 + (-2t)^2}} = \frac{1}{\sqrt{4t^2+2}} \langle -1, 1, -2t \rangle$$

(c) All the osculating planes are of the form $-x + y - 2tz = C$, or $x - y + Az = B$ for constants A and B depending on t.

57. Let $f(x) = x^2$. Show that the center of the osculating circle at (x_0, x_0^2) is given by $\left(-4x_0^3, \frac{1}{2} + 3x_0^2\right)$.

SOLUTION We parametrize the curve by $\mathbf{r}(x) = \langle x, x^2 \rangle$. The center Q of the osculating circle at $x = x_0$ has the position vector

$$\overrightarrow{OQ} = \mathbf{r}(x_0) + \kappa(x_0)^{-1} \mathbf{N}(x_0) \quad (1)$$

We first find the curvature, using the formula for the curvature of a graph in the plane. We have $f'(x) = 2x$ and $f''(x) = 2$, hence,

$$\kappa(x) = \frac{|f''(x)|}{(1 + f'(x)^2)^{3/2}} = \frac{2}{(1 + 4x^2)^{3/2}} \quad \Rightarrow \quad \kappa(x_0)^{-1} = \frac{1}{2}(1 + 4x_0^2)^{3/2}$$

To find the unit normal vector $\mathbf{N}(x_0)$ we use the following considerations:

- The tangent vector is $\mathbf{r}'(x_0) = \langle 1, 2x_0 \rangle$, hence the vector $\langle -2x_0, 1 \rangle$ is orthogonal to $\mathbf{r}'(x_0)$ (since their dot product is zero). Hence $\mathbf{N}(x_0)$ is one of the two unit vectors $\pm \frac{1}{\sqrt{1+4x_0^2}} \langle -2x_0, 1 \rangle$.

- The graph of $f(x) = x^2$ shows that the unit normal vector points in the positive y-direction, hence, the appropriate choice is:

$$\mathbf{N}(x_0) = \frac{1}{\sqrt{1 + 4x_0^2}} \langle -2x_0, 1 \rangle \quad (2)$$

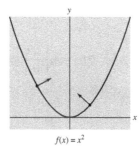

$f(x) = x^2$

We now substitute (2), (3), and $\mathbf{r}(x_0) = \langle x_0, x_0^2 \rangle$ in (1) to obtain

$$\overrightarrow{OQ} = \langle x_0, x_0^2 \rangle + \frac{1}{2}(1 + 4x_0^2)^{3/2} \cdot \frac{1}{\sqrt{1+4x_0^2}} \langle -2x_0, 1 \rangle = \langle x_0, x_0^2 \rangle + \frac{1}{2}(1 + 4x_0^2) \langle -2x_0, 1 \rangle$$

$$= \langle x_0, x_0^2 \rangle + \left\langle -x_0 - 4x_0^3, \frac{1}{2}(1 + 4x_0^2) \right\rangle = \left\langle -4x_0^3, \frac{1}{2} + 3x_0^2 \right\rangle$$

The center of the osculating circle is the terminal point of $\overrightarrow{OQ}$, that is,

$$Q = \left(-4x_0^3, \frac{1}{2} + 3x_0^2\right)$$

In Exercises 59–68, find an equation of the osculating circle at the point indicated or indicate that none exists.

59. $y = x^2$, $x = 1$

SOLUTION Let $f(x) = x^2$. We use the parametrization $\mathbf{r}(x) = \langle x, x^2 \rangle$ and proceed by the following steps.

Step 1. Find κ and $\mathbf{N}$. We compute κ using the curvature of a graph in the plane:

$$\kappa(x) = \frac{|f''(x)|}{(1 + f'(x)^2)^{3/2}}$$

We have $f'(x) = 2x$, $f''(x) = 2$, therefore,

$$\kappa(x) = \frac{2}{(1+(2x)^2)^{3/2}} = \frac{2}{(1+4x^2)^{3/2}} \quad \Rightarrow \quad \kappa(1) = \frac{2}{5^{3/2}} \tag{1}$$

To find $\mathbf{N}(x)$ we notice that the tangent vector is $\mathbf{r}'(x) = \langle 1, 2x \rangle$ hence $\langle -2x, 1 \rangle$ is orthogonal to $\mathbf{r}'(x)$ (their dot product is zero). Therefore, $\mathbf{N}(x)$ is the unit vector in the direction of $\langle -2x, 1 \rangle$ or $-\langle -2x, 1 \rangle$ that points to the "inside" of the curve.

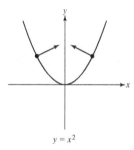

As shown in the figure, the unit normal vector points in the positive y-direction, hence:

$$\mathbf{N}(x) = \frac{\langle -2x, 1 \rangle}{\sqrt{4x^2 + 1}} \quad \Rightarrow \quad \mathbf{N}(1) = \frac{1}{\sqrt{5}} \langle -2, 1 \rangle \tag{2}$$

Step 2. Find the center of the osculating circle. The center Q at $\mathbf{r}(1)$ has the position vector

$$\overrightarrow{OQ} = \mathbf{r}(1) + \kappa(1)^{-1} \mathbf{N}(1)$$

Substituting (1), (2), and $\mathbf{r}(1) = \langle 1, 1 \rangle$ we get:

$$\overrightarrow{OQ} = \langle 1, 1 \rangle + \frac{5^{3/2}}{2} \cdot \frac{1}{5^{1/2}} \langle -2, 1 \rangle = \langle 1, 1 \rangle + \frac{5}{2} \langle -2, 1 \rangle = \left\langle -4, \frac{7}{2} \right\rangle$$

Step 3. Parametrize the osculating circle. The osculating circle has radius $R = \frac{1}{\kappa(1)} = \frac{5^{3/2}}{2}$ and is centered at the point $\left(-4, \frac{7}{2}\right)$, therefore it has the following parametrization:

$$\mathbf{c}(t) = \left\langle -4, \frac{7}{2} \right\rangle + \frac{5^{3/2}}{2} \langle \cos t, \sin t \rangle$$

61. $y = \sin x$, $x = \frac{\pi}{2}$

SOLUTION We use the parametrization $\mathbf{r}(x) = \langle x, \sin x \rangle$. The radius of the osculating circle is the radius of curvature $R = \frac{1}{\kappa\left(\frac{\pi}{2}\right)}$ and the center is the terminal point of the following vector:

$$\overrightarrow{OQ} = \mathbf{r}\left(\frac{\pi}{2}\right) + R\mathbf{N}\left(\frac{\pi}{2}\right) \tag{1}$$

We first compute the curvature. Since $y'(x) = \cos x$ and $y''(x) = -\sin x$, we have:

$$\kappa(x) = \frac{|y''(x)|}{(1+y'(x)^2)^{3/2}} = \frac{|-\sin x|}{(1+\cos^2 x)^{3/2}} \quad \Rightarrow \quad \kappa\left(\frac{\pi}{2}\right) = \frac{\sin \frac{\pi}{2}}{\left(1+\cos^2 \frac{\pi}{2}\right)^{3/2}} = \frac{1}{1} = 1$$

We compute the unit normal vector $\mathbf{N}(x)$. $\mathbf{N}(x)$ is a unit vector orthogonal to the tangent vector $\mathbf{r}'(x) = \langle 1, \cos x \rangle$. We observe that $\langle -\cos x, 1 \rangle$ is orthogonal to $\mathbf{r}'(x)$, since their dot product is zero. Therefore, $\mathbf{N}(x)$

is the unit vector in the direction of either $\langle -\cos x, 1\rangle$ or $-\langle -\cos x, 1\rangle$, depending on the graph. Considering the accompanying figure, we see that the unit normal vector at $x = \pi/2$ points to the negative y-direction. Thus,

$$\mathbf{N}(x) = \frac{\langle \cos x, -1\rangle}{\|\langle \cos x, -1\rangle\|} = \frac{\langle \cos x, -1\rangle}{\sqrt{\cos^2 x + (-1)^2}} \quad \Rightarrow \quad \mathbf{N}\left(\frac{\pi}{2}\right) = \langle 0, -1\rangle$$

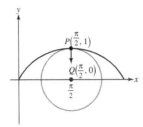

We now find the center of the osculating circle. We substitute $R = \frac{1}{\kappa(\frac{\pi}{2})} = 1$, $\mathbf{N}\left(\frac{\pi}{2}\right) = \langle 0, -1\rangle$, and $\mathbf{r}\left(\frac{\pi}{2}\right) = \left\langle \frac{\pi}{2}, \sin\frac{\pi}{2}\right\rangle = \left\langle \frac{\pi}{2}, 1\right\rangle$ into (1) to obtain

$$\overrightarrow{OQ} = \left\langle \frac{\pi}{2}, 1\right\rangle + 1 \cdot \langle 0, -1\rangle = \left\langle \frac{\pi}{2}, 0\right\rangle$$

The osculating circle is the circle with center at the point $\left(\frac{\pi}{2}, 0\right)$ and radius 1, so it has the following parametrization:

$$\mathbf{c}(t) = \left\langle \frac{\pi}{2}, 0\right\rangle + 1 \cdot \langle \cos t, \sin t\rangle = \left\langle \frac{\pi}{2}, 0\right\rangle + \langle \cos t, \sin t\rangle$$

63. $y = e^x$, $x = 0$

SOLUTION We use the parametrization $\mathbf{r}(x) = \langle x, e^x\rangle$. The radius of the osculating circle is the radius of curvature $R = \frac{1}{\kappa(0)}$ and the center is the terminal point of the vector

$$\overrightarrow{OQ} = \mathbf{r}(0) + R\mathbf{N}(0) \qquad (1)$$

We first compute the curvature. Since $y'(x) = y''(x) = e^x$, we have:

$$\kappa(x) = \frac{|y''(x)|}{(1 + y'(x)^2)^{3/2}} = \frac{|e^x|}{(1 + e^{2x})^{3/2}} \quad \Rightarrow \quad \kappa(0) = \frac{1}{(1+1)^{3/2}} = 2^{-3/2}$$

We compute the unit normal vector $\mathbf{N}(x)$. $\mathbf{N}(x)$ is a unit vector orthogonal to the tangent vector $\mathbf{r}'(x) = \langle 1, e^x\rangle$. We observe that $\langle -e^x, 1\rangle$ is orthogonal to $\mathbf{r}'(x)$, since their dot product is zero. Therefore, $\mathbf{N}(x)$ is the unit vector in the direction of either $\langle -e^x, 1\rangle$ or $-\langle -e^x, 1\rangle$, depending on the graph. Since the graph of e^x is concave upward, the unit normal points in the direction of increasing y. Thus,

$$\mathbf{N}(x) = \frac{\langle -e^x, 1\rangle}{\|\langle -e^x, 1\rangle\|} = \frac{\langle -e^x, 1\rangle}{\sqrt{e^{2x}+1}} \quad \Rightarrow \quad \mathbf{N}(0) = \frac{1}{\sqrt{2}}\langle -1, 1\rangle$$

We now find the center of the osculating circle. We substitute $R = \frac{1}{\kappa(0)} = 2^{3/2}$, $\mathbf{N}(0) = \frac{1}{\sqrt{2}}\langle -1, 1\rangle$, and $\mathbf{r}(0) = \langle 0, 1\rangle$ into (1) to obtain

$$\overrightarrow{OQ} = \langle 0, 1\rangle + 2^{3/2} \cdot \frac{1}{\sqrt{2}}\langle -1, 1\rangle = \langle -2, 3\rangle$$

The osculating circle is the circle with center at the point $(-2, 3)$ and radius $2^{3/2}$, so it has the following parametrization:

$$\mathbf{c}(t) = \langle -2, 3\rangle + 2^{3/2} \cdot \langle \cos t, \sin t\rangle$$

65. $\mathbf{r}(t) = \langle \cos t, \sin t\rangle$, $t = \frac{\pi}{4}$

SOLUTION The curve $\mathbf{r}(t) = \langle \cos t, \sin t\rangle$ is the unit circle. By the definition of the osculating circle, it follows that the osculating circle at each point of the circle is the circle itself. Therefore the osculating circle to the unit circle at $t = \frac{\pi}{4}$ is the unit circle itself.

67. $\mathbf{r}(t) = \langle 1 - \sin t, 1 - 2\cos t \rangle$, $t = \pi$

SOLUTION The radius of the osculating circle is the radius of curvature $R = \frac{1}{\kappa(\pi)}$ and the center is the terminal point of the vector

$$\overrightarrow{OQ} = \mathbf{r}(\pi) + R\mathbf{N}(\pi) \tag{1}$$

We first compute the curvature. We have $x(t) = 1 - \sin t$, $y(t) = 1 - 2\cos t$, so that

$$x'(t) = -\cos t, \quad x''(t) = \sin t$$

$$y'(t) = 2\sin t, \quad y''(t) = 2\cos t$$

Then

$$\kappa(t) = \frac{|x'(t)y''(t) - y'(t)x''(t)|}{(x'(t)^2 + y'(t)^2)^{3/2}} = \frac{|-2\cos^2 t - 2\sin^2 t|}{(\cos^2 t + 4\sin^2 t)^{3/2}} = \frac{2}{(\cos^2 t + 4\sin^2 t)^{3/2}}$$

$$\Rightarrow \quad \kappa(\pi) = \frac{2}{(1 + 4 \cdot 0)^{3/2}} = 2$$

We compute the unit normal vector $\mathbf{N}(t)$. $\mathbf{N}(t)$ is a unit vector orthogonal to the tangent vector $\mathbf{r}'(t) = \langle -\cos t, 2\sin t \rangle$. We observe that $\langle 2\sin t, \cos t \rangle$ is orthogonal to $\mathbf{r}'(t)$, since their dot product is zero. Therefore, $\mathbf{N}(t)$ is the unit vector in the direction of either $\langle 2\sin t, \cos t \rangle$ or $-\langle 2\sin t, \cos t \rangle$, depending on the graph. From the accompanying figure, we see that the unit normal points in the direction of decreasing y at $t = \pi$, so the unit normal is $\langle 2\sin t, \cos t \rangle$.

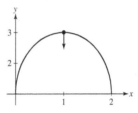

Thus,

$$\mathbf{N}(t) = \frac{\langle 2\sin t, \cos t \rangle}{\|\langle 2\sin t, \cos t \rangle\|} = \frac{\langle 2\sin t, \cos t \rangle}{\sqrt{4\sin^2 t + \cos^2 t}} \quad \Rightarrow \quad \mathbf{N}(\pi) = \langle 0, -1 \rangle$$

We now find the center of the osculating circle. We substitute $R = \frac{1}{\kappa(\pi)} = \frac{1}{2}$, $\mathbf{N}(\pi) = \langle 0, -1 \rangle$, and $\mathbf{r}(\pi) = \langle 1, 3 \rangle$ into (1) to obtain

$$\overrightarrow{OQ} = \langle 1, 3 \rangle + \frac{1}{2} \cdot \langle 0, -1 \rangle = \left\langle 1, \frac{5}{2} \right\rangle$$

The osculating circle is the circle with center at the point $\left(1, \frac{11}{4}\right)$ and radius $\frac{1}{2}$, so it has the following parametrization:

$$\mathbf{c}(t) = \left\langle 1, \frac{5}{2} \right\rangle + \frac{1}{2} \cdot \langle \cos t, \sin t \rangle$$

69. Figure 19 shows the graph of the half-ellipse $y = \pm\sqrt{2rx - px^2}$, where r and p are positive constants. Show that the radius of curvature at the origin is equal to r. *Hint:* One way of proceeding is to write the ellipse in the form of Exercise 25 and apply Eq. (11).

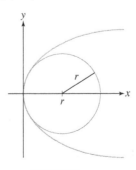

FIGURE 19 The curve $y = \pm\sqrt{2rx - px^2}$ and the osculating circle at the origin.

SOLUTION The radius of curvature is the reciprocal of the curvature. We thus must find the curvature at the origin. We use the following simple variant of the formula for the curvature of a graph in the plane:

$$\kappa(y) = \frac{|x''(y)|}{(1 + x'(y)^2)^{3/2}} \quad (1)$$

(The traditional formula of $\kappa(x) = \frac{|y''(x)|}{(1+y'(x)^2)^{3/2}}$ is inappropriate for this problem, as $y'(x)$ is undefined at $x = 0$.)
We find x in terms of y:

$$y = \sqrt{2rx - px^2}$$
$$y^2 = 2rx - px^2$$
$$px^2 - 2rx + y^2 = 0$$

We solve for x and obtain:

$$x = \pm \frac{1}{p}\sqrt{r^2 - py^2} + \frac{r}{p}, \quad y \geq 0$$

We find x' and x'':

$$x' = \pm \frac{-2py}{2p\sqrt{r^2 - py^2}} = \pm \frac{y}{\sqrt{r^2 - py^2}}$$

$$x'' = \pm \frac{1 \cdot \sqrt{r^2 - py^2} - y \cdot \frac{-py}{\sqrt{r^2-py^2}}}{r^2 - py^2} = \pm \frac{r^2 - py^2 + py^2}{(r^2 - py^2)^{3/2}} = \pm \frac{r^2}{(r^2 - py^2)^{3/2}}$$

At the origin we get:

$$x'(0) = 0, \quad x''(0) = \frac{\pm r^2}{(r^2)^{3/2}} = \frac{\pm 1}{r}$$

Substituting in (1) gives the following curvature at the origin:

$$\kappa(0) = \frac{|x''(0)|}{(1 + x'(0)^2)^{3/2}} = \frac{|\pm \frac{1}{r}|}{(1 + 0)^{3/2}} = \frac{1}{|r|} = \frac{1}{r}$$

We conclude that the radius of curvature at the origin is

$$R = \frac{1}{\kappa(0)} = r$$

71. The **angle of inclination** at a point P on a plane curve is the angle θ between the unit tangent vector **T** and the x-axis (Figure 21). Assume that $\mathbf{r}(s)$ is a arc length parametrization, and let $\theta = \theta(s)$ be the angle of inclination at $\mathbf{r}(s)$. Prove that

$$\kappa(s) = \left|\frac{d\theta}{ds}\right| \quad \boxed{13}$$

Hint: Observe that $\mathbf{T}(s) = \langle \cos\theta(s), \sin\theta(s) \rangle$.

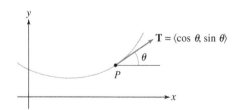

FIGURE 21 The curvature at P is the quantity $|d\theta/ds|$.

SOLUTION Since $\mathbf{T}(t)$ is a unit vector that makes an angle $\theta(t)$ with the positive x-axis, we have

$$\mathbf{T}(t) = \langle \cos\theta(t), \sin\theta(t) \rangle$$

Differentiating this vector using the Chain Rule gives:

$$\mathbf{T}'(t) = \langle -\theta'(t)\sin\theta(t), \theta'(t)\cos\theta(t)\rangle = \theta'(t)\langle -\sin\theta(t), \cos\theta(t)\rangle$$

We compute the norm of the vector $\mathbf{T}'(t)$:

$$\|\mathbf{T}'(t)\| = \|\theta'(t)\langle -\sin\theta(t), \cos\theta(t)\rangle\| = |\theta'(t)|\sqrt{(-\sin\theta(t))^2 + (\cos\theta(t))^2} = |\theta'(t)|\cdot 1 = |\theta'(t)|$$

When $\mathbf{r}(s)$ is a parametrization by arc length we have:

$$\kappa(s) = \left\|\frac{d\mathbf{T}}{ds}\right\| = \left\|\frac{d\mathbf{T}}{dt}\right\|\left|\frac{dt}{d\theta}\frac{d\theta}{ds}\right| = |\theta'(t)|\frac{1}{|\theta'(t)|}\left|\frac{d\theta}{ds}\right| = \left|\frac{d\theta}{ds}\right|$$

as desired.

73. Let $\theta(x)$ be the angle of inclination at a point on the graph $y = f(x)$ (see Exercise 71).

(a) Use the relation $f'(x) = \tan\theta$ to prove that $\dfrac{d\theta}{dx} = \dfrac{f''(x)}{(1 + f'(x)^2)}$.

(b) Use the arc length integral to show that $\dfrac{ds}{dx} = \sqrt{1 + f'(x)^2}$.

(c) Now give a proof of Eq. (6) using Eq. (13).

SOLUTION

(a) By the relation $f'(x) = \tan\theta$ we have $\theta = \tan^{-1}f'(x)$. Differentiating using the Chain Rule we get:

$$\frac{d\theta}{dx} = \frac{d}{dx}(\tan^{-1}f'(x)) = \frac{1}{1 + f'(x)^2}\frac{d}{dx}(f'(x)) = \frac{f''(x)}{1 + f'(x)^2}$$

(b) We use the parametrization $\mathbf{r}(x) = \langle x, f(x)\rangle$. Hence, $\mathbf{r}'(x) = \langle 1, f'(x)\rangle$ and we obtain the following arc length function:

$$S(x) = \int_0^x \|\mathbf{r}'(u)\|\,du = \int_0^x \|\langle 1, f'(u)\rangle\|\,du = \int_0^x \sqrt{1 + f'(u)^2}\,du$$

Differentiating using the Fundamental Theorem gives:

$$\frac{ds}{dx} = \frac{d}{dx}\left(\int_0^x \sqrt{1 + f'(u)^2}\,du\right) = \sqrt{1 + f'(x)^2}$$

(c) By Eq. (13),

$$\kappa(s) = \left|\frac{d\theta}{ds}\right| \tag{1}$$

Using the Chain Rule and the equalities in part (a) and part (b), we obtain:

$$\frac{d\theta}{ds} = \frac{d\theta}{dx}\cdot\frac{dx}{ds} = \frac{d\theta}{dx}\cdot\frac{1}{\frac{ds}{dx}} = \frac{f''(x)}{1 + f'(x)^2}\cdot\frac{1}{\sqrt{1 + f'(x)^2}} = \frac{f''(x)}{(1 + f'(x)^2)^{3/2}}$$

Combining with (1) we obtain the curvature as the following function of x:

$$\kappa(x) = \frac{|f''(x)|}{(1 + f'(x)^2)^{3/2}}$$

which proves Eq. (6).

In Exercises 75–77, use Eq. (14) to find the curvature of the curve given in polar form.

75. $f(\theta) = 2\cos\theta$

SOLUTION By Eq. (14):,

$$\kappa(\theta) = \frac{|f(\theta)^2 + 2f'(\theta)^2 - f(\theta)f''(\theta)|}{(f(\theta)^2 + f'^2(\theta))^{3/2}}$$

We compute the derivatives $f'(\theta)$ and $f''(\theta)$ and evaluate the numerator of $\kappa(\theta)$. This gives:

$$f'(\theta) = -2\sin\theta$$

$$f''(\theta) = -2\cos\theta$$

$$f(\theta)^2 + 2f'(\theta)^2 - f(\theta)f''(\theta) = 4\cos^2\theta + 2\cdot 4\sin^2\theta - 2\cos\theta(-2\cos\theta)$$

$$= 8\cos^2\theta + 8\sin^2\theta = 8$$

We compute the denominator of $\kappa(\theta)$:

$$\left(f(\theta)^2 + f'(\theta)^2\right)^{3/2} = \left(4\cos^2\theta + 4\sin^2\theta\right)^{3/2} = 4^{3/2} = 8$$

Hence,

$$\kappa(\theta) = \frac{8}{8} = 1$$

77. $f(\theta) = e^\theta$

SOLUTION By Eq. (14) we have the following curvature:

$$\kappa(\theta) = \frac{|f(\theta)^2 + 2f'(\theta)^2 - f(\theta)f''(\theta)|}{\left(f(\theta)^2 + f'^2(\theta)\right)^{3/2}}$$

Since $f(\theta) = e^\theta$ also $f'(\theta) = f''(\theta) = e^\theta$. We compute the numerator and denominator of $\kappa(\theta)$:

$$f(\theta)^2 + 2f'(\theta)^2 - f(\theta)f''(\theta) = e^{2\theta} + 2e^{2\theta} - e^\theta \cdot e^\theta = 2e^{2\theta}$$

$$\left(f(\theta)^2 + f'(\theta)^2\right)^{3/2} = \left(e^{2\theta} + e^{2\theta}\right)^{3/2} = \left(2e^{2\theta}\right)^{3/2} = 2\sqrt{2}e^{3\theta}$$

Substituting in the formula for $\kappa(\theta)$ we obtain:

$$\kappa(\theta) = \frac{2e^{2\theta}}{2\sqrt{2}e^{3\theta}} = \frac{1}{\sqrt{2}}e^{-\theta}$$

79. Show that both $\mathbf{r}'(t)$ and $\mathbf{r}''(t)$ lie in the osculating plane for a vector function $\mathbf{r}(t)$. *Hint:* Differentiate $\mathbf{r}'(t) = v(t)\mathbf{T}(t)$.

SOLUTION The osculating plane at P is the plane through P determined by the unit tangent $\mathbf{T}$ and the unit normal $\mathbf{N}$ at P. Since $\mathbf{T}(t) = \frac{\mathbf{r}'(t)}{\|\mathbf{r}'(t)\|}$ we have $\mathbf{r}'(t) = v(t)\mathbf{T}(t)$ where $v(t) = \|\mathbf{r}'(t)\|$. That is, $\mathbf{r}'(t)$ is a scalar multiple of $\mathbf{T}(t)$, hence it lies in every plane containing $\mathbf{T}(t)$, in particular in the osculating plane. We now differentiate $\mathbf{r}'(t) = v(t)\mathbf{T}(t)$ using the Product Rule:

$$\mathbf{r}''(t) = v'(t)\mathbf{T}(t) + v(t)\mathbf{T}'(t) \tag{1}$$

By $\mathbf{N}(t) = \frac{\mathbf{T}'(t)}{\|\mathbf{T}'(t)\|}$ we have $\mathbf{T}'(t) = b(t)\mathbf{N}(t)$ for $b(t) = \|\mathbf{T}'(t)\|$. Substituting in (1) gives:

$$\mathbf{r}''(t) = v'(t)\mathbf{T}(t) + v(t)b(t)\mathbf{N}(t)$$

We see that $\mathbf{r}''(t)$ is a linear combination of $\mathbf{T}(t)$ and $\mathbf{N}(t)$, hence $\mathbf{r}''(t)$ lies in the plane determined by these two vectors, that is, $\mathbf{r}''(t)$ lies in the osculating plane.

81. Two vector-valued functions $\mathbf{r}_1(s)$ and $\mathbf{r}_2(s)$ are said to *agree to order 2* at s_0 if

$$\mathbf{r}_1(s_0) = \mathbf{r}_2(s_0), \quad \mathbf{r}_1'(s_0) = \mathbf{r}_2'(s_0), \quad \mathbf{r}_1''(s_0) = \mathbf{r}_2''(s_0)$$

Let $\mathbf{r}(s)$ be an arc length parametrization of a curve C, and let P be the terminal point of $\mathbf{r}(0)$. Let $\gamma(s)$ be the arc length parametrization of the osculating circle given in Exercise 80. Show that $\mathbf{r}(s)$ and $\gamma(s)$ agree to order 2 at $s = 0$ (in fact, the osculating circle is the unique circle that approximates C to order 2 at P).

SOLUTION The arc length parametrization of the osculating circle at P, described in the xy-coordinate system with P at the origin and the x and y axes in the directions of $\mathbf{T}$ and $\mathbf{N}$ respectively, is given in Exercise 80 by:

$$\gamma(s) = \frac{1}{\kappa}\mathbf{N} + \frac{1}{\kappa}((\sin\kappa s)\mathbf{T} - (\cos\kappa s)\mathbf{N})$$

Hence

$$\gamma(0) = \frac{1}{\kappa}\mathbf{N} + \frac{1}{\kappa}((\sin 0)\mathbf{T} - (\cos 0)\mathbf{N}) = \frac{1}{\kappa}\mathbf{N} + \frac{1}{\kappa}(0 - 1 \cdot \mathbf{N}) = \frac{1}{\kappa}\mathbf{N} - \frac{1}{\kappa}\mathbf{N} = \mathbf{0}$$

$$\mathbf{r}(0) = \overrightarrow{OP} = \mathbf{0}$$

We get:

$$\gamma(0) = \mathbf{r}(0) \tag{1}$$

Differentiating $\gamma(s)$ gives (notice that $\mathbf{N}$, $\mathbf{T}$, and κ are fixed):

$$\gamma'(s) = \frac{1}{\kappa}((\kappa \cos \kappa s)\mathbf{T} + (\kappa \sin \kappa s)\mathbf{N}) = (\cos \kappa s)\mathbf{T} + (\sin \kappa s)\mathbf{N}$$

Hence:

$$\gamma'(0) = (\cos \kappa \cdot 0)\mathbf{T} + (\sin \kappa \cdot 0)\mathbf{N} = 1 \cdot \mathbf{T} + 0 \cdot \mathbf{N} = \mathbf{T}$$

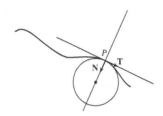

Also, since $\mathbf{r}(s)$ is the arc length parametrization, $\|\mathbf{r}'(s)\| = 1$, hence:

$$\mathbf{T} = \mathbf{T}(0) = \frac{\mathbf{r}'(0)}{\|\mathbf{r}'(0)\|} = \mathbf{r}'(0)$$

We conclude that:

$$\gamma'(0) = \mathbf{r}'(0) \tag{2}$$

We differentiate $\gamma'(s)$ to obtain:

$$\gamma''(s) = (-\kappa \sin \kappa s)\mathbf{T} + (\kappa \cos \kappa s)\mathbf{N}$$

Hence:

$$\gamma''(0) = (-\kappa \sin 0)\mathbf{T} + (\kappa \cos 0)\mathbf{N} = 0\mathbf{T} + \kappa\mathbf{N} = \kappa\mathbf{N}$$

For the arc length parametrization $\mathbf{r}(s)$ we have:

$$\mathbf{r}''(s) = \mathbf{T}'(s) = \|\mathbf{T}'(s)\|\mathbf{N}(s) = \|\mathbf{r}'(s)\|\kappa(s)\mathbf{N}(s) = 1 \cdot \kappa(s)\mathbf{N}(s)$$

Hence:

$$\mathbf{r}''(0) = \kappa(0)\mathbf{N}(0) = \kappa\mathbf{N}$$

We conclude that:

$$\gamma''(0) = \mathbf{r}''(0) \tag{3}$$

(1), (2), and (3) show that $\mathbf{r}(s)$ and $\gamma(s)$ agree to order two at $s = 0$.

Further Insights and Challenges

83. Show that the curvature of Viviani's curve, given by $\mathbf{r}(t) = \langle 1 + \cos t, \sin t, 2\sin(t/2)\rangle$, is

$$\kappa(t) = \frac{\sqrt{13 + 3\cos t}}{(3 + \cos t)^{3/2}}$$

SOLUTION We use the formula for curvature:

$$\kappa(t) = \frac{\|\mathbf{r}'(t) \times \mathbf{r}''(t)\|}{\|\mathbf{r}'(t)\|^3} \quad (1)$$

Differentiating $\mathbf{r}(t)$ gives

$$\mathbf{r}'(t) = \left\langle -\sin t, \cos t, 2 \cdot \frac{1}{2} \cos \frac{t}{2} \right\rangle = \left\langle -\sin t, \cos t, \cos \frac{t}{2} \right\rangle$$

$$\mathbf{r}''(t) = \left\langle -\cos t, -\sin t, -\frac{1}{2} \sin \frac{t}{2} \right\rangle$$

We compute the cross product in (1):

$$\mathbf{r}'(t) \times \mathbf{r}''(t) = \begin{vmatrix} \mathbf{i} & \mathbf{j} & \mathbf{k} \\ -\sin t & \cos t & \cos \frac{t}{2} \\ -\cos t & -\sin t & -\frac{1}{2}\sin \frac{t}{2} \end{vmatrix}$$

$$= \left(-\frac{1}{2} \cos t \sin \frac{t}{2} + \sin t \cos \frac{t}{2} \right) \mathbf{i} - \left(\frac{1}{2} \sin t \sin \frac{t}{2} + \cos t \cos \frac{t}{2} \right) \mathbf{j} + \mathbf{k}$$

We find the length of the cross product:

$$\|\mathbf{r}'(t) \times \mathbf{r}''(t)\|^2 = \left(-\frac{1}{2} \cos t \sin \frac{t}{2} + \sin t \cos \frac{t}{2} \right)^2 + \left(\frac{1}{2} \sin t \sin \frac{t}{2} + \cos t \cos \frac{t}{2} \right)^2 + 1$$

$$= \frac{1}{4} \sin^2 \frac{t}{2} \left(\cos^2 t + \sin^2 t \right) + \cos^2 \frac{t}{2} \left(\sin^2 t + \cos^2 t \right) + 1$$

$$= \frac{1}{4} \sin^2 \frac{t}{2} + \cos^2 \frac{t}{2} + 1$$

We use the identities $\sin^2 \frac{t}{2} + \cos^2 \frac{t}{2} = 1$ and $\cos^2 \frac{t}{2} = \frac{1}{2} + \frac{1}{2} \cos t$ to write:

$$\|\mathbf{r}'(t) \times \mathbf{r}''(t)\|^2 = \frac{1}{4} \sin^2 \frac{t}{2} + \cos^2 \frac{t}{2} + 1 = \frac{1}{4} \left(\sin^2 \frac{t}{2} + \cos^2 \frac{t}{2} \right) + \frac{3}{4} \cos^2 \frac{t}{2} + 1$$

$$= \frac{1}{4} + \frac{3}{4} \left(\frac{1}{2} + \frac{1}{2} \cos t \right) + 1 = \frac{3}{8} \cos t + \frac{13}{8}$$

Hence:

$$\|\mathbf{r}'(t) \times \mathbf{r}''(t)\| = \frac{1}{\sqrt{8}} \sqrt{13 + 3 \cos t} \quad (2)$$

We compute the length of $\mathbf{r}'(t)$:

$$\|\mathbf{r}'(t)\|^2 = (-\sin t)^2 + \cos^2 t + \cos^2 \frac{t}{2} = 1 + \cos^2 \frac{t}{2} = 1 + \left(\frac{1}{2} + \frac{1}{2} \cos t \right) = \frac{3}{2} + \frac{1}{2} \cos t$$

Hence,

$$\|\mathbf{r}'(t)\| = \frac{1}{\sqrt{2}} \sqrt{3 + \cos t} \quad (3)$$

Substituting (2) and (3) in (1) gives:

$$\kappa(t) = \frac{\frac{1}{\sqrt{8}} \sqrt{13 + 3 \cos t}}{\left(\frac{1}{\sqrt{2}} \sqrt{3 + \cos t} \right)^3} = \frac{\frac{1}{\sqrt{8}} \sqrt{13 + 3 \cos t}}{\frac{1}{2} \frac{1}{\sqrt{2}} (3 + \cos t)^{3/2}} = \frac{\sqrt{13 + 3 \cos t}}{(3 + \cos t)^{3/2}}$$

In Exercises 85–93, we investigate the binormal vector further.

85. Let $\mathbf{r}(t) = \langle x(t), y(t), 0 \rangle$. Assuming that $\mathbf{T}(t) \times \mathbf{N}(t)$ is nonzero, there are two possibilities for the vector $\mathbf{B}(t)$. What are they? Explain.

SOLUTION Since the curve defined by $\mathbf{r}(t)$ lies in the xy-plane, $\mathbf{T}(t)$ and $\mathbf{N}(t)$ do as well. Since $\mathbf{B}(t) = \mathbf{T}(t) \times \mathbf{N}(t)$, it is a unit vector perpendicular to both $\mathbf{T}(t)$ and $\mathbf{N}(t)$, so it is in either the positive or negative z-direction. Since it is a unit vector, it follows that $\mathbf{B}(t) = \mathbf{k}$ or $\mathbf{B}(t) = -\mathbf{k}$.

87. Show that if C is contained in a plane $\mathcal{P}$, then $\mathbf{B}$ is a unit vector normal to $\mathcal{P}$. Conclude that $\tau = 0$ for a plane curve.

SOLUTION If C is contained in a plane $\mathcal{P}$, then the unit normal $\mathbf{N}$ and the unit tangent $\mathbf{T}$ are in $\mathcal{P}$. The cross product $\mathbf{B} = \mathbf{T} \times \mathbf{N}$ is orthogonal to $\mathbf{T}$ and $\mathbf{N}$ which are in the plane, hence $\mathbf{B}$ is normal to the plane. Thus, $\mathbf{B}$ is a unit vector normal to the plane. There are only two different unit normal vectors to a plane, one pointing "up" and the other pointing "down." Thus, we can assume (due to continuity) that $\mathbf{B}$ is a constant vector, therefore

$$\frac{d\mathbf{B}}{ds} = 0 \quad \text{or} \quad \tau = 0$$

89. Use the identity

$$\mathbf{a} \times (\mathbf{b} \times \mathbf{c}) = (\mathbf{a} \cdot \mathbf{c})\mathbf{b} - (\mathbf{a} \cdot \mathbf{b})\mathbf{c}$$

to prove

$$\mathbf{N} \times \mathbf{B} = \mathbf{T}, \qquad \mathbf{B} \times \mathbf{T} = \mathbf{N} \qquad \boxed{16}$$

SOLUTION We use the given equality and the definition $\mathbf{B} = \mathbf{T} \times \mathbf{N}$ to write:

$$\mathbf{N} \times \mathbf{B} = \mathbf{N} \times (\mathbf{T} \times \mathbf{N}) = (\mathbf{N} \cdot \mathbf{N})\mathbf{T} - (\mathbf{N} \cdot \mathbf{T})\mathbf{N} \tag{1}$$

The unit normal $\mathbf{N}$ and the unit tangent $\mathbf{T}$ are orthogonal unit vectors, hence $\mathbf{N} \cdot \mathbf{N} = \|\mathbf{N}\|^2 = 1$ and $\mathbf{N} \cdot \mathbf{T} = 0$. Therefore, (1) gives:

$$\mathbf{N} \times \mathbf{B} = 1 \cdot \mathbf{T} - 0\mathbf{N} = \mathbf{T}$$

To prove the second equality, we substitute $\mathbf{T} = \mathbf{N} \times \mathbf{B}$ and then use the given equality. We obtain:

$$\mathbf{B} \times \mathbf{T} = \mathbf{B} \times (\mathbf{N} \times \mathbf{B}) = (\mathbf{B} \cdot \mathbf{B})\mathbf{N} - (\mathbf{B} \cdot \mathbf{N})\mathbf{B} \tag{2}$$

Now, $\mathbf{B}$ is a unit vector, hence $\mathbf{B} \cdot \mathbf{B} = \|\mathbf{B}\|^2 = 1$. Also, since $\mathbf{B} = \mathbf{T} \times \mathbf{N}$, $\mathbf{B}$ is orthogonal to $\mathbf{N}$ which implies that $\mathbf{B} \cdot \mathbf{N} = 0$. Substituting in (2) we get:

$$\mathbf{B} \times \mathbf{T} = 1\mathbf{N} - 0\mathbf{B} = \mathbf{N}.$$

91. Show that $\mathbf{r}' \times \mathbf{r}''$ is a multiple of $\mathbf{B}$. Conclude that

$$\mathbf{B} = \frac{\mathbf{r}' \times \mathbf{r}''}{\|\mathbf{r}' \times \mathbf{r}''\|} \qquad \boxed{18}$$

SOLUTION By the definition of the binormal vector, $\mathbf{B} = \mathbf{T} \times \mathbf{N}$. We denote $a(t) = \frac{1}{\|\mathbf{r}'(t)\|}$ and write:

$$\mathbf{T}(t) = \frac{\mathbf{r}'(t)}{\|\mathbf{r}'(t)\|} = a(t)\mathbf{r}'(t) \tag{1}$$

We differentiate $\mathbf{T}(t)$ using the Product Rule:

$$\mathbf{T}'(t) = a(t)\mathbf{r}''(t) + a'(t)\mathbf{r}'(t)$$

We denote $b(t) = \|\mathbf{T}'(t)\|$ and obtain:

$$\mathbf{N}(t) = \frac{\mathbf{T}'(t)}{\|\mathbf{T}'(t)\|} = \frac{a(t)}{b(t)}\mathbf{r}''(t) + \frac{a'(t)}{b(t)}\mathbf{r}'(t)$$

For $c_1 = \frac{a(t)}{b(t)}$ and $c_2 = \frac{a'(t)}{b(t)}$ we have:

$$\mathbf{N}(t) = c_1(t)\mathbf{r}''(t) + c_2(t)\mathbf{r}'(t) \tag{2}$$

We now find **B** as the cross product of **T**(t) in (1) and **N**(t) in (2). This gives:

$$\mathbf{B}(t) = a(t)\mathbf{r}'(t) \times (c_1(t)\mathbf{r}''(t) + c_2(t)\mathbf{r}'(t)) = a(t)c_1(t)\mathbf{r}'(t) \times \mathbf{r}''(t) + a(t)c_2(t)\mathbf{r}'(t) \times \mathbf{r}'(t)$$
$$= a(t)c_1(t)\mathbf{r}'(t) \times \mathbf{r}''(t) + \mathbf{0} = a(t)c_1(t)\mathbf{r}'(t) \times \mathbf{r}''(t)$$

We see that **B** is parallel to $\mathbf{r}' \times \mathbf{r}''$. Since **B** is a unit vector we have:

$$\mathbf{B} = \frac{\mathbf{r}' \times \mathbf{r}''}{\|\mathbf{r}' \times \mathbf{r}''\|}.$$

93. The vector **N** can be computed using $\mathbf{N} = \mathbf{B} \times \mathbf{T}$ [Eq. (16)] with **B**, as in Eq. (18). Use this method to find **N** in the following cases:

(a) $\mathbf{r}(t) = \langle \cos t, t, t^2 \rangle$ at $t = 0$

(b) $\mathbf{r}(t) = \langle t^2, t^{-1}, t \rangle$ at $t = 1$

SOLUTION

(a) We first compute the vector **B** using Eq. (18):

$$\mathbf{B} = \frac{\mathbf{r}' \times \mathbf{r}''}{\|\mathbf{r}' \times \mathbf{r}''\|} \quad (1)$$

Differentiating $\mathbf{r}(t) = \langle \cos t, t, t^2 \rangle$ gives

$$\begin{aligned}\mathbf{r}'(t) &= \langle -\sin t, 1, 2t \rangle \\ \mathbf{r}''(t) &= \langle -\cos t, 0, 2 \rangle\end{aligned} \Rightarrow \begin{aligned}\mathbf{r}'(0) &= \langle 0, 1, 0 \rangle \\ \mathbf{r}''(0) &= \langle -1, 0, 2 \rangle\end{aligned}$$

We compute the cross product:

$$\mathbf{r}'(0) \times \mathbf{r}''(0) = \mathbf{j} \times (-\mathbf{i} + 2\mathbf{k}) = -\mathbf{j} \times \mathbf{i} + 2\mathbf{j} \times \mathbf{k} = \mathbf{k} + 2\mathbf{i} = \langle 2, 0, 1 \rangle$$
$$\|\mathbf{r}'(0) \times \mathbf{r}''(0)\| = \sqrt{2^2 + 0^2 + 1^2} = \sqrt{5}$$

Substituting in (1) we obtain:

$$\mathbf{B}(0) = \frac{\langle 2, 0, 1 \rangle}{\sqrt{5}} = \frac{1}{\sqrt{5}}\langle 2, 0, 1 \rangle$$

We now compute **T**(0):

$$\mathbf{T}(0) = \frac{\mathbf{r}'(0)}{\|\mathbf{r}'(0)\|} = \frac{\langle 0, 1, 0 \rangle}{\|\langle 0, 1, 0 \rangle\|} = \langle 0, 1, 0 \rangle$$

Finally we find $\mathbf{N} = \mathbf{B} \times \mathbf{T}$:

$$\mathbf{N}(0) = \frac{1}{\sqrt{5}}\langle 2, 0, 1 \rangle \times \langle 0, 1, 0 \rangle = \frac{1}{\sqrt{5}}(2\mathbf{i} + \mathbf{k}) \times \mathbf{j} = \frac{1}{\sqrt{5}}(2\mathbf{i} \times \mathbf{j} + \mathbf{k} \times \mathbf{j}) = \frac{1}{\sqrt{5}}(2\mathbf{k} - \mathbf{i}) = \frac{1}{\sqrt{5}}\langle -1, 0, 2 \rangle$$

(b) Differentiating $\mathbf{r}(t) = \langle t^2, t^{-1}, t \rangle$ gives

$$\begin{aligned}\mathbf{r}'(t) &= \langle 2t, -t^{-2}, 1 \rangle \\ \mathbf{r}''(t) &= \langle 2, 2t^{-3}, 0 \rangle\end{aligned} \Rightarrow \begin{aligned}\mathbf{r}'(1) &= \langle 2, -1, 1 \rangle \\ \mathbf{r}''(1) &= \langle 2, 2, 0 \rangle\end{aligned}$$

We compute the cross product:

$$\mathbf{r}'(1) \times \mathbf{r}''(1) = \begin{vmatrix} \mathbf{i} & \mathbf{j} & \mathbf{k} \\ 2 & -1 & 1 \\ 2 & 2 & 0 \end{vmatrix} = \begin{vmatrix} -1 & 1 \\ 2 & 0 \end{vmatrix}\mathbf{i} - \begin{vmatrix} 2 & 1 \\ 2 & 0 \end{vmatrix}\mathbf{j} + \begin{vmatrix} 2 & -1 \\ 2 & 2 \end{vmatrix}\mathbf{k} = -2\mathbf{i} + 2\mathbf{j} + 6\mathbf{k} = \langle -2, 2, 6 \rangle$$

$$\|\mathbf{r}'(1) \times \mathbf{r}''(1)\| = \sqrt{(-2)^2 + 2^2 + 6^2} = \sqrt{44} = 2\sqrt{11}$$

Substituting in (1) gives:

$$\mathbf{B}(1) = \frac{\langle -2, 2, 6 \rangle}{2\sqrt{11}} = \frac{1}{\sqrt{11}}\langle -1, 1, 3 \rangle$$

We now find $\mathbf{T}(1)$:

$$\mathbf{T}(1) = \frac{\mathbf{r}'(1)}{\|\mathbf{r}'(1)\|} = \frac{\langle 2, -1, 1 \rangle}{\sqrt{4+1+1}} = \frac{1}{\sqrt{6}} \langle 2, -1, 1 \rangle$$

Finally we find $\mathbf{N}(1)$ by computing the following cross product:

$$\mathbf{N}(1) = \mathbf{B}(1) \times \mathbf{T}(1) = \frac{1}{\sqrt{11}} \langle -1, 1, 3 \rangle \times \frac{1}{\sqrt{6}} \langle 2, -1, 1 \rangle = \frac{1}{\sqrt{66}} \begin{vmatrix} \mathbf{i} & \mathbf{j} & \mathbf{k} \\ -1 & 1 & 3 \\ 2 & -1 & 1 \end{vmatrix}$$

$$= \frac{1}{\sqrt{66}} \left\{ \begin{vmatrix} 1 & 3 \\ -1 & 1 \end{vmatrix} \mathbf{i} - \begin{vmatrix} -1 & 3 \\ 2 & 1 \end{vmatrix} \mathbf{j} + \begin{vmatrix} -1 & 1 \\ 2 & -1 \end{vmatrix} \mathbf{k} \right\} = \frac{1}{\sqrt{66}} (4\mathbf{i} + 7\mathbf{j} - \mathbf{k}) = \frac{1}{\sqrt{66}} \langle 4, 7, -1 \rangle$$

13.5 Motion in 3-Space (LT Section 14.5)

Preliminary Questions

1. If a particle travels with constant speed, must its acceleration vector be zero? Explain.

SOLUTION If the speed of the particle is constant, the tangential component, $a_T(t) = v'(t)$, of the acceleration is zero. However, the normal component, $a_N(t) = \kappa(t)v(t)^2$ is not necessarily zero, since the particle may change its direction.

2. For a particle in uniform circular motion around a circle, which of the vectors $\mathbf{v}(t)$ or $\mathbf{a}(t)$ always points toward the center of the circle?

SOLUTION For a particle in uniform circular motion around a circle, the acceleration vector $\mathbf{a}(t)$ points towards the center of the circle, whereas $\mathbf{v}(t)$ is tangent to the circle.

3. Two objects travel to the right along the parabola $y = x^2$ with nonzero speed. Which of the following statements must be true?
(a) Their velocity vectors point in the same direction.
(b) Their velocity vectors have the same length.
(c) Their acceleration vectors point in the same direction.

SOLUTION

(a) The velocity vector points in the direction of motion, hence the velocities of the two objects point in the same direction.

(b) The length of the velocity vector is the speed. Since the speeds are not necessarily equal, the velocity vectors may have different lengths.

(c) The acceleration is determined by the tangential component $v'(t)$ and the normal component $\kappa(t)v(t)^2$. Since v and v' may be different for the two objects, the acceleration vectors may have different directions.

4. Use the decomposition of acceleration into tangential and normal components to explain the following statement: If the speed is constant, then the acceleration and velocity vectors are orthogonal.

SOLUTION If the speed is constant, $v'(t) = 0$. Therefore, the acceleration vector has only the normal component:

$$\mathbf{a}(t) = a_N(t)\mathbf{N}(t)$$

The velocity vector always points in the direction of motion. Since the vector $\mathbf{N}(t)$ is orthogonal to the direction of motion, the vectors $\mathbf{a}(t)$ and $\mathbf{v}(t)$ are orthogonal.

5. If a particle travels along a straight line, then the acceleration and velocity vectors are (choose the correct description):
(a) orthogonal **(b)** parallel

SOLUTION Since a line has zero curvature, the normal component of the acceleration is zero, hence $\mathbf{a}(t)$ has only the tangential component. The velocity vector is always in the direction of motion, hence the acceleration and the velocity vectors are parallel to the line. We conclude that (b) is the correct statement.

6. What is the length of the acceleration vector of a particle traveling around a circle of radius 2 cm with constant velocity 4 cm/s?

SOLUTION The acceleration vector is given by the following decomposition:

$$\mathbf{a}(t) = v'(t)\mathbf{T}(t) + \kappa(t)v(t)^2\mathbf{N}(t) \quad (1)$$

In our case $v(t) = 4$ is constant hence $v'(t) = 0$. In addition, the curvature of a circle of radius 2 is $\kappa(t) = \frac{1}{2}$. Substituting $v(t) = 4$, $v'(t) = 0$ and $\kappa(t) = \frac{1}{2}$ in (1) gives:

$$\mathbf{a}(t) = \frac{1}{2} \cdot 4^2 N(t) = 8N(t)$$

The length of the acceleration vector is, thus,

$$\|\mathbf{a}(t)\| = 8 \text{ cm/s}^2$$

7. Two cars are racing around a circular track. If, at a certain moment, both of their speedometers read 110 mph. then the two cars have the same (choose one):

(a) a_T **(b)** a_N

SOLUTION The tangential acceleration a_T and the normal acceleration a_N are the following values:

$$a_T(t) = v'(t); \quad a_N(t) = \kappa(t)v(t)^2$$

At the moment where both speedometers read 110 mph, the speeds of the two cars are $v = 110$ mph. Since the track is circular, the curvature $\kappa(t)$ is constant, hence the normal accelerations of the two cars are equal at this moment. Statement (b) is correct.

Exercises

1. Use the table to calculate the difference quotients $\dfrac{\mathbf{r}(1+h) - \mathbf{r}(1)}{h}$ for $h = -0.2, -0.1, 0.1, 0.2$. Then estimate the velocity and speed at $t = 1$.

$\mathbf{r}(0.8)$	$\langle 1.557, 2.459, -1.970 \rangle$
$\mathbf{r}(0.9)$	$\langle 1.559, 2.634, -1.740 \rangle$
$\mathbf{r}(1)$	$\langle 1.540, 2.841, -1.443 \rangle$
$\mathbf{r}(1.1)$	$\langle 1.499, 3.078, -1.035 \rangle$
$\mathbf{r}(1.2)$	$\langle 1.435, 3.342, -0.428 \rangle$

SOLUTION

$$(h = -0.2)$$

$$\frac{\mathbf{r}(1 - 0.2) - \mathbf{r}(1)}{-0.2} = \frac{\mathbf{r}(0.8) - \mathbf{r}(1)}{-0.2} = \frac{\langle 1.557, 2.459, -1.970 \rangle - \langle 1.540, 2.841, -1.443 \rangle}{-0.2}$$

$$= \frac{\langle 0.017, -0.382, -0.527 \rangle}{-0.2} = \langle -0.085, 1.91, 2.635 \rangle$$

$$(h = -0.1)$$

$$\frac{\mathbf{r}(1 - 0.1) - \mathbf{r}(1)}{-0.1} = \frac{\mathbf{r}(0.9) - \mathbf{r}(1)}{-0.1} = \frac{\langle 1.559, 2.634, -1.740 \rangle - \langle 1.540, 2.841, -1.443 \rangle}{-0.1}$$

$$= \frac{\langle 0.019, -0.207, -0.297 \rangle}{-0.1} = \langle -0.19, 2.07, 2.97 \rangle$$

$$(h = 0.1)$$

$$\frac{\mathbf{r}(1 + 0.1) - \mathbf{r}(1)}{0.1} = \frac{\mathbf{r}(1.1) - \mathbf{r}(1)}{0.1} = \frac{\langle 1.499, 3.078, -1.035 \rangle - \langle 1.540, 2.841, -1.443 \rangle}{0.1}$$

$$= \frac{\langle -0.041, 0.237, 0.408 \rangle}{0.1} = \langle -0.41, 2.37, 4.08 \rangle$$

$$\frac{\mathbf{r}(1+0.2)-\mathbf{r}(1)}{0.2} = \frac{\mathbf{r}(1.2)-\mathbf{r}(1)}{0.2} = \frac{\langle 1.435, 3.342, -0.428\rangle - \langle 1.540, 2.841, -1.443\rangle}{0.2}$$

$$(h = 0.2)$$

$$= \frac{\langle -0.105, 0.501, 1.015\rangle}{0.2} = \langle -0.525, 2.505, 5.075\rangle$$

The velocity vector is defined by:

$$\mathbf{v}(t) = \mathbf{r}'(t) = \lim_{h\to 0} \frac{\mathbf{r}(t+h) - \mathbf{r}(t)}{h}$$

We may estimate the velocity at $t = 1$ by:

$$\mathbf{v}(1) \approx \langle -0.3, 2.2, 3.5\rangle$$

and the speed by:

$$v(1) = \|\mathbf{v}(1)\| \approx \sqrt{0.3^2 + 2.2^2 + 3.5^2} \cong 4.1$$

In Exercises 3–6, calculate the velocity and acceleration vectors and the speed at the time indicated.

3. $\mathbf{r}(t) = \langle t^3, 1-t, 4t^2\rangle$, $t = 1$

SOLUTION In this case $\mathbf{r}(t) = \langle t^3, 1-t, 4t^2\rangle$ hence:

$$\mathbf{v}(t) = \mathbf{r}'(t) = \langle 3t^2, -1, 8t\rangle \quad \Rightarrow \quad \mathbf{v}(1) = \langle 3, -1, 8\rangle$$

$$\mathbf{a}(t) = \mathbf{r}''(t) = \langle 6t, 0, 8\rangle \quad \Rightarrow \quad \mathbf{a}(1) = \langle 6, 0, 8\rangle$$

The speed is the magnitude of the velocity vector, that is,

$$v(1) = \|\mathbf{v}(1)\| = \sqrt{3^2 + (-1)^2 + 8^2} = \sqrt{74}$$

5. $\mathbf{r}(\theta) = \langle \sin\theta, \cos\theta, \cos 3\theta\rangle$, $\theta = \frac{\pi}{3}$

SOLUTION Differentiating $\mathbf{r}(\theta) = \langle \sin\theta, \cos\theta, \cos 3\theta\rangle$ gives:

$$\mathbf{v}(\theta) = \mathbf{r}'(\theta) = \langle \cos\theta, -\sin\theta, -3\sin 3\theta\rangle$$

$$\Rightarrow \mathbf{v}\left(\frac{\pi}{3}\right) = \left\langle \cos\frac{\pi}{3}, -\sin\frac{\pi}{3}, -3\sin\pi\right\rangle = \left\langle \frac{1}{2}, -\frac{\sqrt{3}}{2}, 0\right\rangle$$

$$\mathbf{a}(\theta) = \mathbf{r}''(\theta) = \langle -\sin\theta, -\cos\theta, -9\cos 3\theta\rangle$$

$$\Rightarrow \mathbf{a}\left(\frac{\pi}{3}\right) = \left\langle -\sin\frac{\pi}{3}, -\cos\frac{\pi}{3}, -9\cos\pi\right\rangle = \left\langle -\frac{\sqrt{3}}{2}, -\frac{1}{2}, 9\right\rangle$$

The speed is the magnitude of the velocity vector, that is:

$$v\left(\frac{\pi}{3}\right) = \left\|\mathbf{v}\left(\frac{\pi}{3}\right)\right\| = \sqrt{\left(\frac{1}{2}\right)^2 + \left(-\frac{\sqrt{3}}{2}\right)^2 + 0^2} = 1$$

7. Find $\mathbf{a}(t)$ for a particle moving around a circle of radius 8 cm at a constant speed of $v = 4$ cm/s (see Example 4). Draw the path, and on it, draw the acceleration vector at $t = \frac{\pi}{4}$.

SOLUTION The position vector is:

$$\mathbf{r}(t) = 8\langle \cos\omega t, \sin\omega t\rangle$$

Hence,

$$\mathbf{v}(t) = \mathbf{r}'(t) = 8\langle -\omega\sin\omega t, \omega\cos\omega t\rangle = 8\omega\langle -\sin\omega t, \cos\omega t\rangle \tag{1}$$

We are given that the speed of the particle is $v = 4$cm/s. The speed is the magnitude of the velocity vector, hence:

$$v = 8\omega\sqrt{(-\sin\omega t)^2 + \cos^2\omega t} = 8\omega = 4 \quad \Rightarrow \quad \omega = \frac{1}{2} \text{ rad/s}$$

Substituting in (1) we get:
$$\mathbf{v}(t) = 4\left\langle -\sin\frac{t}{2}, \cos\frac{t}{2} \right\rangle$$

We now find $\mathbf{a}(t)$ by differentiating the velocity vector. This gives
$$\mathbf{a}(t) = \mathbf{v}'(t) = 4\left\langle -\frac{1}{2}\cos\frac{t}{2}, -\frac{1}{2}\sin\frac{t}{2} \right\rangle = -2\left\langle \cos\frac{t}{2}, \sin\frac{t}{2} \right\rangle$$

The path of the particle is $\mathbf{r}(t) = 8\left\langle \cos\frac{t}{2}, \sin\frac{t}{2} \right\rangle$ and the acceleration vector at $t = \frac{\pi}{4}$ is:
$$\mathbf{a}\left(\frac{\pi}{4}\right) = -2\left\langle \cos\frac{\pi}{8}, \sin\frac{\pi}{8} \right\rangle \approx \langle -1.85, -0.77 \rangle$$

The path $\mathbf{r}(t)$ and the acceleration vector at $t = \frac{\pi}{4}$ are shown in the following figure:

$\mathbf{r}(t) = 8\left\langle \cos\frac{t}{2}, \sin\frac{t}{2} \right\rangle$

9. Sketch the path $\mathbf{r}(t) = \langle t^2, t^3 \rangle$ together with the velocity and acceleration vectors at $t = 1$.

SOLUTION We compute the velocity and acceleration vectors at $t = 1$:
$$\mathbf{v}(t) = \mathbf{r}'(t) = \langle 2t, 3t^2 \rangle \quad \Rightarrow \quad \mathbf{v}(1) = \langle 2, 3 \rangle$$
$$\mathbf{a}(t) = \mathbf{v}'(t) = \langle 2, 6t \rangle \quad \Rightarrow \quad \mathbf{a}(1) = \langle 2, 6 \rangle$$

The following figure shows the path $\mathbf{r}(t) = \langle t^2, t^3 \rangle$ and the vectors $\mathbf{v}(1)$ and $\mathbf{a}(1)$:

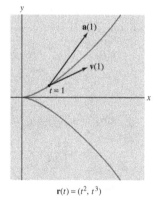

$\mathbf{r}(t) = \langle t^2, t^3 \rangle$

In Exercises 11–14, find $\mathbf{v}(t)$ given $\mathbf{a}(t)$ and the initial velocity.

11. $\mathbf{a}(t) = \langle t, 4 \rangle$, $\quad \mathbf{v}(0) = \langle \frac{1}{3}, -2 \rangle$

SOLUTION We find $\mathbf{v}(t)$ by integrating $\mathbf{a}(t)$:
$$\mathbf{v}(t) = \int_0^t \mathbf{a}(u)\,du = \int_0^t \langle u, 4 \rangle\,du = \left\langle \frac{1}{2}u^2, 4u \right\rangle\bigg|_0^t + \mathbf{v}_0 = \left\langle \frac{t^2}{2}, 4t \right\rangle + \mathbf{v}_0$$

The initial condition gives:
$$\mathbf{v}(0) = \langle 0, 0 \rangle + \mathbf{v}_0 = \left\langle \frac{1}{3}, -2 \right\rangle \quad \Rightarrow \quad \mathbf{v}_0 = \left\langle \frac{1}{3}, -2 \right\rangle$$

Hence,
$$\mathbf{v}(t) = \left\langle \frac{t^2}{2}, 4t \right\rangle + \left\langle \frac{1}{3}, -2 \right\rangle = \left\langle \frac{3t^2 + 2}{6}, 4t - 2 \right\rangle$$

13. $\mathbf{a}(t) = \mathbf{k}$, $\mathbf{v}(0) = \mathbf{i}$

SOLUTION We compute $\mathbf{v}(t)$ by integrating the acceleration vector:

$$\mathbf{v}(t) = \int_0^t \mathbf{a}(u)\, du = \int_0^t \mathbf{k}\, du = \mathbf{k} u \Big|_0^t + \mathbf{v}_0 = t\mathbf{k} + \mathbf{v}_0 \tag{1}$$

Substituting the initial condition gives:

$$\mathbf{v}(0) = 0\mathbf{k} + \mathbf{v}_0 = \mathbf{i} \quad \Rightarrow \quad \mathbf{v}_0 = \mathbf{i}$$

Combining with (1) we obtain:

$$\mathbf{v}(t) = \mathbf{i} + t\mathbf{k}$$

In Exercises 15–18, find $\mathbf{r}(t)$ and $\mathbf{v}(t)$ given $\mathbf{a}(t)$ and the initial velocity and position.

15. $\mathbf{a}(t) = \langle t, 4 \rangle$, $\mathbf{v}(0) = \langle 3, -2 \rangle$, $\mathbf{r}(0) = \langle 0, 0 \rangle$

SOLUTION We first integrate $\mathbf{a}(t)$ to find the velocity vector:

$$\mathbf{v}(t) = \int_0^t \langle u, 4 \rangle\, du = \left\langle \frac{u^2}{2}, 4u \right\rangle \Big|_0^t + \mathbf{v}_0 = \left\langle \frac{t^2}{2}, 4t \right\rangle + \mathbf{v}_0 \tag{1}$$

The initial condition $\mathbf{v}(0) = \langle 3, -2 \rangle$ gives:

$$\mathbf{v}(0) = \langle 0, 0 \rangle + \mathbf{v}_0 = \langle 3, -2 \rangle \quad \Rightarrow \quad \mathbf{v}_0 = \langle 3, -2 \rangle$$

Substituting in (1) we get:

$$\mathbf{v}(t) = \left\langle \frac{t^2}{2}, 4t \right\rangle + \langle 3, -2 \rangle = \left\langle \frac{t^2}{2} + 3, 4t - 2 \right\rangle$$

We now integrate the velocity vector to find $\mathbf{r}(t)$:

$$\mathbf{r}(t) = \int_0^t \left\langle \frac{u^2}{2} + 3, 4u - 2 \right\rangle du = \left\langle \frac{u^3}{6} + 3u, 2u^2 - 2u \right\rangle \Big|_0^t + \mathbf{r}_0 = \left\langle \frac{t^3}{6} + 3t, 2t^2 - 2t \right\rangle + \mathbf{r}_0$$

The initial condition $\mathbf{r}(0) = \langle 0, 0 \rangle$ gives:

$$\mathbf{r}(0) = \langle 0, 0 \rangle + \mathbf{r}_0 = \langle 0, 0 \rangle \quad \Rightarrow \quad \mathbf{r}_0 = \langle 0, 0 \rangle$$

Hence,

$$\mathbf{r}(t) = \left\langle \frac{t^3}{6} + 3t, 2t^2 - 2t \right\rangle$$

17. $\mathbf{a}(t) = t\mathbf{k}$, $\mathbf{v}(0) = \mathbf{i}$, $\mathbf{r}(0) = \mathbf{j}$

SOLUTION Integrating the acceleration vector gives:

$$\mathbf{v}(t) = \int_0^t u\mathbf{k}\, du = \frac{u^2}{2}\mathbf{k} \Big|_0^t + \mathbf{v}_0 = \frac{t^2}{2}\mathbf{k} + \mathbf{v}_0 \tag{1}$$

The initial condition for $\mathbf{v}(t)$ gives:

$$\mathbf{v}(0) = \frac{0^2}{2}\mathbf{k} + \mathbf{v}_0 = \mathbf{i} \quad \Rightarrow \quad \mathbf{v}_0 = \mathbf{i}$$

We substitute in (1):

$$\mathbf{v}(t) = \frac{t^2}{2}\mathbf{k} + \mathbf{i} = \mathbf{i} + \frac{t^2}{2}\mathbf{k}$$

We now integrate $\mathbf{v}(t)$ to find $\mathbf{r}(t)$:

$$\mathbf{r}(t) = \int_0^t \left(\mathbf{i} + \frac{u^2}{2}\mathbf{k} \right) du = u\mathbf{i} + \frac{u^3}{6}\mathbf{k} \Big|_0^t + \mathbf{r}_0 = t\mathbf{i} + \frac{t^3}{6}\mathbf{k} + \mathbf{r}_0 \tag{2}$$

The initial condition for $\mathbf{r}(t)$ gives:

$$\mathbf{r}(0) = 0\mathbf{i} + 0\mathbf{k} + \mathbf{r}_0 = \mathbf{j} \quad \Rightarrow \quad \mathbf{r}_0 = \mathbf{j}$$

Combining with (2) gives the position vector:

$$\mathbf{r}(t) = t\mathbf{i} + \mathbf{j} + \frac{t^3}{6}\mathbf{k}$$

19. A projectile is launched from the ground at an angle of 45°. What initial speed must the projectile have in order to hit the top of a 120-m tower located 180 m away?

SOLUTION We put the launch point at the origin and let $\mathbf{r}(t)$ be the projectile's position vector.

Step 1. Use Newton's Law. The net force vector acting on the projectile is the force of gravity $\mathbf{F} = \langle 0, -gm \rangle = m\langle 0, -g \rangle$. By Newton's Second Law, $\mathbf{F} = m\mathbf{r}''(t)$, hence:

$$m\langle 0, -g \rangle = m\mathbf{r}''(t) \quad \Rightarrow \quad \mathbf{r}''(t) = \langle 0, -g \rangle$$

We compute the position vector by integrating twice:

$$\mathbf{r}'(t) = \int_0^t \mathbf{r}''(u)\,du = \int_0^t \langle 0, -g \rangle\,du = \langle 0, -gt \rangle + \mathbf{v}_0$$

$$\mathbf{r}(t) = \int_0^t \mathbf{r}'(u)\,du = \int_0^t (\langle 0, -gu \rangle + \mathbf{v}_0)\,du = \left\langle 0, -g\frac{t^2}{2} \right\rangle + \mathbf{v}_0 t + \mathbf{r}_0$$

That is,

$$\mathbf{r}(t) = \left\langle 0, \frac{-g}{2} t^2 \right\rangle + \mathbf{v}_0 t + \mathbf{r}_0 \tag{1}$$

Since the launch point is at the origin, $\mathbf{r}_0 = \mathbf{0}$. The projectile is launched at an angle of 45°, hence the initial velocity $\mathbf{v}_0$ points in the direction of the unit vector $\langle \cos 45°, \sin 45° \rangle = \left\langle \frac{\sqrt{2}}{2}, \frac{\sqrt{2}}{2} \right\rangle$ therefore, $\mathbf{v}_0 = v_0 \left\langle \frac{\sqrt{2}}{2}, \frac{\sqrt{2}}{2} \right\rangle$. Substituting these initial values in (1) gives:

$$\mathbf{r}(t) = \left\langle 0, \frac{-g}{2} t^2 \right\rangle + tv_0 \left\langle \frac{\sqrt{2}}{2}, \frac{\sqrt{2}}{2} \right\rangle$$

Step 2. Solve for v_0. The position vector of the top of the tower is $\langle 600, 400 \rangle$, hence at the moment of hitting the tower we have,

$$\mathbf{r}(t) = \left\langle 0, \frac{-g}{2} t^2 \right\rangle + tv_0 \left\langle \frac{\sqrt{2}}{2}, \frac{\sqrt{2}}{2} \right\rangle = \langle 600, 400 \rangle$$

$$\left\langle tv_0 \frac{\sqrt{2}}{2}, \frac{-g}{2} t^2 + \frac{\sqrt{2}}{2} tv_0 \right\rangle = \langle 600, 400 \rangle$$

Equating components, we get the equations:

$$\begin{cases} tv_0 \dfrac{\sqrt{2}}{2} = 600 \\ -\dfrac{g}{2} t^2 + \dfrac{\sqrt{2}}{2} tv_0 = 400 \end{cases}$$

The first equation implies that $t = \frac{1{,}200}{\sqrt{2}v_0}$. We substitute in the second equation and solve for v_0 (we use $g = 32$ ft/s²):

$$-16 \left(\frac{1{,}200}{\sqrt{2}v_0} \right)^2 + \frac{\sqrt{2}}{2} \left(\frac{1{,}200}{\sqrt{2}v_0} \right) v_0 = 400$$

$$-8 \left(\frac{1{,}200}{v_0} \right)^2 + 600 = 400$$

$$\left(\frac{1{,}200}{v_0} \right)^2 = 25 \quad \Rightarrow \quad \frac{1{,}200}{v_0} = 5 \quad \Rightarrow \quad v_0 = 240 \text{ ft/s}$$

The initial speed of the projectile must be $v_0 = 240$ ft/s.

21. Assume that astronaut Alan Shepard hit his golf shot on the moon (acceleration due to gravity = 1.6 m/s²) with a modest initial speed of 35 m/s at an angle of 30°. How far did the ball travel?

SOLUTION Place the launch point at the origin, and let $\mathbf{r}(t)$ be the position vector, with $\mathbf{i}$ the horizontal axis and $\mathbf{j}$ the vertical. Then $\mathbf{r}(0) = \mathbf{0}$. Since the launch speed is 35 m/s at an angle of 30°, we have

$$\mathbf{v}(0) = 35\cos(30°)\mathbf{i} + 35\sin(30°)\mathbf{j} = \frac{35\sqrt{3}}{2}\mathbf{i} + \frac{35}{2}\mathbf{j}$$

Since the acceleration due to gravity is 1.6 m/s², the acceleration vector is $\mathbf{a}(t) = -1.6\mathbf{j}$. Then

$$\mathbf{v}(t) = \int \mathbf{a}(t)\,dt = -1.6t\mathbf{j} + \mathbf{C}_0$$

Since $\mathbf{v}(0) = -1.6 \cdot 0\mathbf{j} + \mathbf{C}_0 = \frac{35\sqrt{3}}{2}\mathbf{i} + \frac{35}{2}\mathbf{j}$, we get

$$\mathbf{v}(t) = \frac{35\sqrt{3}}{2}\mathbf{i} + \left(\frac{35}{2} - 1.6t\right)\mathbf{j}$$

Integrate again to get

$$\mathbf{r}(t) = \int \mathbf{v}(t)\,dt = \frac{35\sqrt{3}}{2}t\mathbf{i} + \left(\frac{35}{2}t - 0.8t^2\right)\mathbf{j} + \mathbf{C}_1$$

Since $\mathbf{r}(0) = 0\mathbf{i} + 0\mathbf{j} + \mathbf{C}_1 = \mathbf{0}$, the constant of integration is zero, so that

$$\mathbf{r}(t) = \frac{35\sqrt{3}}{2}t\mathbf{i} + \left(\frac{35}{2}t - 0.8t^2\right)\mathbf{j}$$

Finally, we want to know when the golf ball hits the ground; that happens when the vertical coordinate is zero:

$$\left(\frac{35}{2}t - 0.8t^2\right)\mathbf{j} = 0\mathbf{j} \quad\Rightarrow\quad t = 0 \text{ or } t = \frac{35}{1.6} = 21.875$$

The ball lands on the moon at time $t = 21.875$; it travels

$$\frac{35\sqrt{3}}{2} \cdot 21.875 \approx 663 \text{ m}$$

23. Show that a projectile launched at an angle θ with initial speed v_0 travels a distance $(v_0^2/g)\sin 2\theta$ before hitting the ground. Conclude that the maximum distance (for a given v_0) is attained at $\theta = \frac{\pi}{4}$.

SOLUTION We place the launch point at the origin and let $\mathbf{r}(t)$ be the projectile's position vector. The net force acting on the projectile is $\mathbf{F} = \langle 0, -mg \rangle = m\langle 0, -g \rangle$. By Newton's Second Law, $\mathbf{F} = m\mathbf{r}''(t)$, hence:

$$m\langle 0, -g \rangle = m\mathbf{r}''(t) \quad\Rightarrow\quad \mathbf{r}''(t) = \langle 0, -g \rangle$$

Integrating twice we get:

$$\mathbf{r}'(t) = \int_0^t \mathbf{r}''(u)\,du = \int_0^t \langle 0, -g \rangle\,du = \langle 0, -gt \rangle + \mathbf{v}_0$$

$$\mathbf{r}(t) = \int_0^t \mathbf{r}'(u)\,du = \int_0^t (\langle 0, -g \cdot u \rangle + \mathbf{v}_0)\,du = \left\langle 0, -\frac{g}{2}t^2 \right\rangle + \mathbf{v}_0 t + \mathbf{r}_0 \qquad (1)$$

Since the launch point is at the origin, $\mathbf{r}_0 = \mathbf{0}$. The firing was at an angle θ, hence the initial velocity points in the direction of the unit vector $\langle \cos\theta, \sin\theta \rangle$. Hence, $\mathbf{v}_0 = v_0 \langle \cos\theta, \sin\theta \rangle$. We substitute the initial vectors in (1) to obtain:

$$\mathbf{r}(t) = \left\langle 0, -\frac{g}{2}t^2 \right\rangle + v_0 t \langle \cos\theta, \sin\theta \rangle \qquad (2)$$

The total distance is obtained when the y-component of $\mathbf{r}(t)$ is zero (besides the original moment), that is,

$$-\frac{g}{2}t^2 + (v_0 \sin\theta)t = 0$$

$$t\left(-\frac{g}{2}t + v_0 \sin\theta\right) = 0 \quad\Rightarrow\quad t = 0 \text{ or } t = \frac{2v_0 \sin\theta}{g}$$

The appropriate choice is $t = \frac{2v_0 \sin \theta}{g}$. We now find the total distance x_T by substituting this value of t in the x-component of $\mathbf{r}(t)$ in (2). We obtain:

$$x(t) = v_0 t \cos \theta$$

$$x_T = v_0 \cos \theta \cdot \frac{2v_0 \sin \theta}{g} = \frac{2v_0^2 \cos \theta \sin \theta}{g} = \frac{v_0^2 \sin 2\theta}{g}$$

The maximum distance is attained when $\sin 2\theta = 1$, that is $2\theta = \frac{\pi}{2}$ or $\theta = \frac{\pi}{4}$.

25. A quarterback throws a football while standing at the very center of the field on the 50-yard line. The ball leaves his hand at a height of 5 ft and has initial velocity $\mathbf{v}_0 = 40\mathbf{i} + 35\mathbf{j} + 32\mathbf{k}$ ft/s. Assume an acceleration of 32 ft/s² due to gravity and that the $\mathbf{i}$ vector points down the field toward the endzone and the $\mathbf{j}$ vector points to the sideline. The field is 150 ft in width and 300 ft in length.

(a) Determine the position function that gives the position of the ball t seconds after it is thrown.

(b) The ball is caught by a player 5 ft above the ground. Is the player in bounds or out of bounds when he receives the ball? Assume the player is standing vertically with both toes on the ground at the time of reception.

SOLUTION Place the origin at the center of the field, with the z-axis ($\mathbf{k}$) pointing up.

(a) The only acceleration is due to gravity, and it is $\mathbf{a}(t) = -32\mathbf{k}$, so that the velocity vector is

$$\mathbf{v}(t) = \int \mathbf{a}(t)\, dt = -32t\mathbf{k} + \mathbf{C}_0$$

We are given $\mathbf{v}(0)$; use that to determine $\mathbf{C}_0$:

$$40\mathbf{i} + 35\mathbf{j} + 32\mathbf{k} = \mathbf{v}(0) = -32 \cdot 0\mathbf{k} + \mathbf{C}_0$$

so that $\mathbf{C}_0 = 40\mathbf{i} + 35\mathbf{j} + 32\mathbf{k}$ and

$$\mathbf{v}(t) = -32t\mathbf{k} + 40\mathbf{i} + 35\mathbf{j} + 32\mathbf{k} = 40\mathbf{i} + 35\mathbf{j} + (32 - 32t)\mathbf{k}$$

To find $\mathbf{r}(t)$, integrate again:

$$\mathbf{r}(t) = \int \mathbf{v}(t)\, dt = 40t\mathbf{i} + 35t\mathbf{j} + (32t - 16t^2)\mathbf{k} + \mathbf{R}_0$$

where $\mathbf{R}_0$ is a constant. Since the ball is thrown at an initial height of 5 feet, we have $\mathbf{r}(0) = \langle 0, 0, 5 \rangle = \mathbf{R}_0$. Then the position function is

$$\mathbf{r}(t) = 40t\mathbf{i} + 35t\mathbf{j} + (32t - 16t^2 + 5)\mathbf{k}$$

(b) The ball is 5 feet above the ground when $32t - 16t^2 + 5 = 5$, or $32t - 16t^2 = 16t(2 - t) = 0$. Thus $t = 0$ or $t = 2$. $t = 0$ is when the quarterback releases the ball, so we are interested in the position of the ball at $t = 2$:

$$\mathbf{r}(2) = 80\mathbf{i} + 70\mathbf{j} + 5\mathbf{k}$$

Since the field is 150 feet wide, it is 75 feet from the center of the field to the sideline, so the receiver is 5 feet inside the sideline, in bounds, when he catches the ball.

27. A constant force $\mathbf{F} = \langle 5, 2 \rangle$ (in newtons) acts on a 10-kg mass. Find the position of the mass at $t = 10$ seconds if it is located at the origin at $t = 0$ and has initial velocity $\mathbf{v}_0 = \langle 2, -3 \rangle$ (in meters per second).

SOLUTION $\mathbf{F} = m\mathbf{a}$ and thus $\mathbf{a} = \langle 0.5, 0.2 \rangle$, and

$$\mathbf{v} = \langle 0.5t + 2, 0.2t - 3 \rangle$$

and

$$\mathbf{r} = \langle 0.25t^2 + 2t, 0.1t^2 - 3t \rangle$$

and

$$\mathbf{r}(10) = \langle 45, -20 \rangle$$

29. A particle follows a path $\mathbf{r}(t)$ for $0 \le t \le T$, beginning at the origin O. The vector $\bar{\mathbf{v}} = \dfrac{1}{T}\displaystyle\int_0^T \mathbf{r}'(t)\,dt$ is called the **average velocity** vector. Suppose that $\bar{\mathbf{v}} = \mathbf{0}$. Answer and explain the following:

(a) Where is the particle located at time T if $\bar{\mathbf{v}} = \mathbf{0}$?

(b) Is the particle's average speed necessarily equal to zero?

SOLUTION

(a) If the average velocity is 0, then the particle must be back at its original position at time $t = T$. This is perhaps best seen by noting that $\bar{\mathbf{v}} = \dfrac{1}{T}\displaystyle\int_0^T \mathbf{r}'(t)\,dt = \mathbf{r}(t)\Big|_0^T$.

(b) The average speed need not be zero! Consider a particle moving at constant speed around a circle, with position vector $\mathbf{r}(t) = \langle \cos t, \sin t \rangle$. From 0 to 2π, this has average velocity of 0, but constant average speed of 1.

31. At a certain moment, a particle moving along a path has velocity $\mathbf{v} = \langle 12, 20, 20 \rangle$ and acceleration $\mathbf{a} = \langle 2, 1, -3 \rangle$. Is the particle speeding up or slowing down?

SOLUTION We are asked if the particle is speeding up or slowing down, that is if $\|\mathbf{v}\|$ or $\|\mathbf{v}\|^2$ is increasing or decreasing. We check $(\|\mathbf{v}\|^2)'$:

$$(\|\mathbf{v}\|^2)' = (\mathbf{v} \cdot \mathbf{v})' = 2\mathbf{v}' \cdot \mathbf{v} = 2 \cdot \mathbf{a} \cdot \mathbf{v} = 2\langle 2, 1, -3\rangle \cdot \langle 12, 20, 20\rangle = 2 \cdot (24 + 20 - 60) = -32 < 0$$

So the speed is decreasing.

In Exercises 32–35, use Eq. (6) to find the coefficients a_T and a_N as a function of t (or at the specified value of t).

33. $\mathbf{r}(t) = \langle t, \cos t, \sin t \rangle$

SOLUTION We find a_T and a_N using the following equalities:

$$a_T = \mathbf{a} \cdot \mathbf{T}, \quad a_N = \frac{\|\mathbf{a} \times \mathbf{v}\|}{\|\mathbf{v}\|}$$

We compute $\mathbf{v}$ and $\mathbf{a}$ by differentiating $\mathbf{r}$ twice:

$$\mathbf{v}(t) = \mathbf{r}'(t) = \langle 1, -\sin t, \cos t \rangle \quad \Rightarrow \quad \|\mathbf{v}(t)\| = \sqrt{1 + (-\sin t)^2 + \cos^2 t} = \sqrt{2}$$

$$\mathbf{a}(t) = \mathbf{r}''(t) = \langle 0, -\cos t, -\sin t \rangle$$

The unit tangent vector $\mathbf{T}$ is, thus:

$$\mathbf{T}(t) = \frac{\mathbf{v}(t)}{\|\mathbf{v}(t)\|} = \frac{1}{\sqrt{2}} \langle 1, -\sin t, \cos t \rangle$$

Since the speed is constant ($v = \|\mathbf{v}(t)\| = \sqrt{2}$), the tangential component of the acceleration is zero, that is:

$$a_T = 0$$

To find a_N we first compute the following cross product:

$$\mathbf{a} \times \mathbf{v} = \begin{vmatrix} \mathbf{i} & \mathbf{j} & \mathbf{k} \\ 0 & -\cos t & -\sin t \\ 1 & -\sin t & \cos t \end{vmatrix} = \begin{vmatrix} -\cos t & -\sin t \\ -\sin t & \cos t \end{vmatrix}\mathbf{i} - \begin{vmatrix} 0 & -\sin t \\ 1 & \cos t \end{vmatrix}\mathbf{j} + \begin{vmatrix} 0 & -\cos t \\ 1 & -\sin t \end{vmatrix}\mathbf{k}$$

$$= -\left(\cos^2 t + \sin^2 t\right)\mathbf{i} - \sin t\,\mathbf{j} + \cos t\,\mathbf{k} = -\mathbf{i} - \sin t\,\mathbf{j} + \cos t\,\mathbf{k} = \langle -1, -\sin t, \cos t \rangle$$

Hence,

$$a_N = \frac{\|\mathbf{a} \times \mathbf{v}\|}{\|\mathbf{v}\|} = \frac{\sqrt{(-1)^2 + (-\sin t)^2 + \cos^2 t}}{\sqrt{2}} = \frac{\sqrt{2}}{\sqrt{2}} = 1$$

35. $\mathbf{r}(t) = \langle e^{2t}, t, e^{-t}\rangle$, $\quad t = 0$

SOLUTION We will use the following equalities:

$$a_T = \mathbf{a} \cdot \mathbf{T}, \quad a_N = \frac{\|\mathbf{a} \times \mathbf{v}\|}{\|\mathbf{v}\|}$$

We first find $\mathbf{a}$ and $\mathbf{v}$ by twice differentiating $\mathbf{r}$. We get:

$$\mathbf{v}(t) = \mathbf{r}'(t) = \langle 2e^{2t}, 1, -e^{-t}\rangle$$

$$\mathbf{a}(t) = \mathbf{r}''(t) = \langle 4e^{2t}, 0, e^{-t}\rangle$$

Then evaluating at $t = 0$ we get:

$$\mathbf{v}(0) = \langle 2, 1, -1\rangle, \quad \Rightarrow \|\mathbf{v}(0)\| = \sqrt{2^2 + 1^2 + (-1)^2} = \sqrt{6}$$

$$\mathbf{a}(0) = \langle 4, 0, 1\rangle$$

Hence, $\mathbf{T} = \frac{\mathbf{v}}{\|\mathbf{v}\|} = \frac{1}{\sqrt{6}}\langle 2, 1, -1\rangle$ and we obtain:

$$a_T = \mathbf{a} \cdot \mathbf{T} = \langle 4, 0, 1\rangle \cdot \frac{1}{\sqrt{6}}\langle 2, 1, -1\rangle = \frac{1}{\sqrt{6}}(8 + 0 - 1) = \frac{7}{\sqrt{6}}$$

To find a_N we first compute the following cross product:

$$\mathbf{a} \times \mathbf{v} = \begin{vmatrix} \mathbf{i} & \mathbf{j} & \mathbf{k} \\ 4 & 0 & 1 \\ 2 & 1 & -1 \end{vmatrix} = \langle -1, 6, 4\rangle$$

Therefore,

$$a_N = \frac{\|\mathbf{a} \times \mathbf{v}\|}{\|\mathbf{v}\|} = \frac{\sqrt{(-1)^2 + 6^2 + 4^2}}{\sqrt{6}} = \sqrt{\frac{53}{6}}$$

In Exercise 36–43, find the decomposition of $\mathbf{a}(t)$ into tangential and normal components at the point indicated, as in Example 6.

37. $\mathbf{r}(t) = \langle \frac{1}{3}t^3, 1 - 3t\rangle$, $\quad t = -1$

SOLUTION We have

$$\mathbf{v}(t) = \mathbf{r}'(t) = \langle t^2, -3\rangle, \quad \mathbf{a}(t) = \mathbf{v}'(t) = \mathbf{r}''(t) = \langle 2t, 0\rangle$$

At $t = -1$, this gives

$$\mathbf{v} = \langle 1, -3\rangle, \quad \mathbf{a} = \langle -2, 0\rangle$$

Therefore

$$\mathbf{T} = \frac{\mathbf{v}}{\|\mathbf{v}\|} = \frac{\langle 1, -3\rangle}{\sqrt{1^2 + (-3)^2}} = \frac{1}{\sqrt{10}}\langle 1, -3\rangle$$

and by Eq. (2) in the text,

$$a_T = \mathbf{a} \cdot \mathbf{T} = \frac{1}{\sqrt{10}}\langle -2, 0\rangle \cdot \langle 1, -3\rangle = -\frac{2}{\sqrt{10}}$$

Next, by Eq. (3) in the text,

$$a_N \mathbf{N} = \mathbf{a} - a_T \mathbf{T} = \langle -2, 0\rangle + \frac{2}{\sqrt{10}}\left(\frac{1}{\sqrt{10}}\right)\langle 1, -3\rangle = \left\langle -\frac{9}{5}, -\frac{3}{5}\right\rangle$$

This vector has length

$$a_N = \|a_N \mathbf{N}\| = \sqrt{\frac{81}{25} + \frac{9}{25}} = \frac{\sqrt{90}}{5} = \frac{3}{5}\sqrt{10}$$

so that
$$\mathbf{N} = \frac{a_{\mathbf{N}}\mathbf{N}}{a_{\mathbf{N}}} = \frac{\langle -\frac{9}{5}, -\frac{3}{5}\rangle}{3\sqrt{10}/5} = \left\langle -\frac{3}{\sqrt{10}}, -\frac{1}{\sqrt{10}}\right\rangle$$

This gives the decomposition
$$\mathbf{a} = \langle -2, 0\rangle = a_{\mathbf{T}}\mathbf{T} + a_{\mathbf{N}}\mathbf{N} = -\frac{2}{\sqrt{10}}\mathbf{T} + \frac{3}{5}\sqrt{10}\mathbf{N} = -\frac{2}{\sqrt{10}}\mathbf{T} + \frac{6}{\sqrt{10}}\mathbf{N}$$

39. $\mathbf{r}(t) = \langle t, \frac{1}{2}t^2, \frac{1}{6}t^3\rangle$, $t = 4$

SOLUTION We have
$$\mathbf{v}(t) = \mathbf{r}'(t) = \left\langle 1, t, \frac{1}{2}t^2\right\rangle, \quad \mathbf{a}(t) = \mathbf{v}'(t) = \mathbf{r}''(t) = \langle 0, 1, t\rangle$$

At $t = 4$, this gives
$$\mathbf{v} = \langle 1, 4, 8\rangle, \quad \mathbf{a} = \langle 0, 1, 4\rangle$$

Therefore
$$\mathbf{T} = \frac{\mathbf{v}}{\|\mathbf{v}\|} = \frac{\langle 1, 4, 8\rangle}{\sqrt{1^2 + 4^2 + 8^2}} = \left\langle \frac{1}{9}, \frac{4}{9}, \frac{8}{9}\right\rangle$$

and by Eq. (2) in the text,
$$a_{\mathbf{T}} = \mathbf{a} \cdot \mathbf{T} = \langle 0, 1, 4\rangle \cdot \left\langle \frac{1}{9}, \frac{4}{9}, \frac{8}{9}\right\rangle = 4$$

Next, by Eq. (3) in the text,
$$a_{\mathbf{N}}\mathbf{N} = \mathbf{a} - a_{\mathbf{T}}\mathbf{T} = \langle 0, 1, 4\rangle - 4\left\langle \frac{1}{9}, \frac{4}{9}, \frac{8}{9}\right\rangle = \left\langle -\frac{4}{9}, -\frac{7}{9}, \frac{4}{9}\right\rangle$$

This vector has length
$$a_{\mathbf{N}} = \|a_{\mathbf{N}}\mathbf{N}\| = \sqrt{\frac{16}{9} + \frac{49}{9} + \frac{16}{9}} = 1$$

so that
$$\mathbf{N} = \frac{a_{\mathbf{N}}\mathbf{N}}{a_{\mathbf{N}}} = \left\langle -\frac{4}{9}, -\frac{7}{9}, \frac{4}{9}\right\rangle$$

This gives the decomposition
$$\mathbf{a} = \langle 0, 1, 4\rangle = a_{\mathbf{T}}\mathbf{T} + a_{\mathbf{N}}\mathbf{N} = 4\mathbf{T} + 1\mathbf{N}$$

41. $\mathbf{r}(t) = \langle t, e^t, te^t\rangle$, $t = 0$

SOLUTION We have
$$\mathbf{v}(t) = \mathbf{r}'(t) = \left\langle 1, e^t, (t+1)e^t\right\rangle, \quad \mathbf{a}(t) = \mathbf{v}'(t) = \mathbf{r}''(t) = \left\langle 0, e^t, (t+2)e^t\right\rangle$$

At $t = 0$, this gives
$$\mathbf{v} = \langle 1, 1, 1\rangle, \quad \mathbf{a} = \langle 0, 1, 2\rangle$$

Therefore
$$\mathbf{T} = \frac{\mathbf{v}}{\|\mathbf{v}\|} = \frac{\langle 1, 1, 1\rangle}{\sqrt{1^2 + 1^2 + 1^2}} = \frac{1}{\sqrt{3}}\langle 1, 1, 1\rangle$$

and by Eq. (2) in the text,
$$a_{\mathbf{T}} = \mathbf{a} \cdot \mathbf{T} = \frac{1}{\sqrt{3}}\langle 0, 1, 2\rangle \cdot \langle 1, 1, 1\rangle = \sqrt{3}$$

Next, by Eq. (3) in the text,

$$a_N \mathbf{N} = \mathbf{a} - a_T \mathbf{T} = \langle 0, 1, 2 \rangle - \sqrt{3}\left(\frac{1}{\sqrt{3}}\langle 1, 1, 1\rangle\right) = \langle -1, 0, 1\rangle$$

This vector has length $a_N = \sqrt{2}$, so that

$$\mathbf{N} = \frac{a_N \mathbf{N}}{a_N} = \frac{\langle -1, 0, 1\rangle}{\sqrt{2}} = \frac{1}{\sqrt{2}}\langle -1, 0, 1\rangle$$

This gives the decomposition

$$\mathbf{a} = \langle 0, 1, 2\rangle = a_T \mathbf{T} + a_N \mathbf{N} = \sqrt{3}\,\mathbf{T} + \sqrt{2}\,\mathbf{N}$$

43. $\mathbf{r}(t) = \langle t, \cos t, t \sin t\rangle$, $t = \frac{\pi}{2}$

SOLUTION We have

$$\mathbf{v}(t) = \mathbf{r}'(t) = \langle 1, -\sin t, \sin t + t\cos t\rangle, \qquad \mathbf{a}(t) = \mathbf{v}'(t) = \mathbf{r}''(t) = \langle 0, -\cos t, 2\cos t - t\sin t\rangle$$

At $t = \frac{\pi}{2}$, this gives

$$\mathbf{v} = \langle 1, -1, 1\rangle, \qquad \mathbf{a} = \left\langle 0, 0, -\frac{\pi}{2}\right\rangle$$

Therefore

$$\mathbf{T} = \frac{\mathbf{v}}{\|\mathbf{v}\|} = \frac{\langle 1, -1, 1\rangle}{\sqrt{1^2 + (-1)^2 + 1^2}} = \frac{1}{\sqrt{3}}\langle 1, -1, 1\rangle$$

and by Eq. (2) in the text,

$$a_T = \mathbf{a} \cdot \mathbf{T} = \frac{1}{\sqrt{3}}\left\langle 0, 0, -\frac{\pi}{2}\right\rangle \cdot \langle 1, -1, 1\rangle = -\frac{\pi}{2\sqrt{3}}$$

Next, by Eq. (3) in the text,

$$a_N \mathbf{N} = \mathbf{a} - a_T \mathbf{T} = \left\langle 0, 0, -\frac{\pi}{2}\right\rangle + \frac{\pi}{2\sqrt{3}}\left(\frac{1}{\sqrt{3}}\langle 1, -1, 1\rangle\right) = \left\langle \frac{\pi}{6}, -\frac{\pi}{6}, -\frac{\pi}{3}\right\rangle = \frac{\pi}{6}\langle 1, -1, -2\rangle$$

This vector has length

$$a_N = \|a_N \mathbf{N}\| = \frac{\pi}{6}\sqrt{1^2 + (-1)^2 + (-2)^2} = \frac{\pi\sqrt{6}}{6}$$

so that

$$\mathbf{N} = \frac{a_N \mathbf{N}}{a_N} = \frac{\frac{\pi}{6}\langle 1, -1, -2\rangle}{\pi\sqrt{6}/6} = \frac{1}{\sqrt{6}}\langle 1, -1, -2\rangle$$

This gives the decomposition

$$\mathbf{a} = \left\langle 0, 0, -\frac{\pi}{2}\right\rangle = a_T \mathbf{T} + a_N \mathbf{N} = -\frac{\pi}{2\sqrt{3}}\mathbf{T} + \frac{\pi\sqrt{6}}{6}\mathbf{N}$$

45. Find the components a_T and a_N of the acceleration vector of a particle moving along a circular path of radius $R = 100$ cm with constant velocity $v_0 = 5$ cm/s.

SOLUTION Since the particle moves with constant speed, we have $v'(t) = 0$, hence:

$$a_T = v'(t) = 0$$

The normal component of the acceleration is $a_N = \kappa(t)v(t)^2$. The curvature of a circular path of radius $R = 100$ is $\kappa(t) = \frac{1}{R} = \frac{1}{100}$, and the velocity is the constant value $v(t) = v_0 = 5$. Hence,

$$a_N = \frac{1}{R}v_0^2 = \frac{25}{100} = 0.25 \text{ cm/s}^2$$

47. Suppose that the Ferris wheel in Example 5 is rotating clockwise and that the point P at angle $45°$ has acceleration vector $\mathbf{a} = \langle 0, -50 \rangle$ m/min^2 pointing down, as in Figure 12. Determine the speed and tangential component of the acceleration of the Ferris wheel.

FIGURE 12

SOLUTION Since the tangential acceleration points tangent to the circle, pointing in the direction $\langle 1, -1 \rangle$, and the normal points inwards, in the direction $\langle -1, -1 \rangle$, by symmetry (since the net acceleration points downwards) it is clear that the tangential and normal accelerations are equal in magnitude, say to a. Therefore

$$\langle 0, -50 \rangle = a_\mathbf{T}\mathbf{T} + a_\mathbf{N}\mathbf{N} = a\frac{1}{\sqrt{2}}\langle 1, -1 \rangle + a\frac{1}{\sqrt{2}}\langle -1, -1 \rangle = a\langle 0, -\sqrt{2} \rangle$$

so that $a = \frac{50}{\sqrt{2}}$. The tangential acceleration is thus $\frac{50}{\sqrt{2}} \approx 35.36$ m/s^2. The speed can be computed from

$$\frac{50}{\sqrt{2}} = a_\mathbf{N} = \kappa v^2 = \frac{1}{30}v^2$$

solving for v gives $v^2 = 750\sqrt{2}$, so that $v \approx 32.56$ m/s.

49. A satellite orbits the earth at an altitude 400 km above the earth's surface, with constant speed $v = 28{,}000$ km/h. Find the magnitude of the satellite's acceleration (in kilometers per square hour), assuming that the radius of the earth is 6378 km (Figure 13).

FIGURE 13 Satellite orbit.

SOLUTION The shuttle is in a uniform circular motion, therefore the tangential component of its acceleration is zero, and the acceleration can be written as:

$$\mathbf{a} = \kappa v^2 \mathbf{N} \quad (1)$$

The radius of motion is $6378 + 400 = 6778$ km; hence the curvature is $\kappa = \frac{1}{6778}$. Also by the given information the constant speed is $v = 28000$ km/h. Substituting these values in (1) we get:

$$\mathbf{a} = \frac{1}{6778} \cdot 28000^2 \mathbf{N} \text{ km/h}$$

The magnitude of the shuttle's acceleration is thus:

$$\mathbf{a} = \left\|\frac{1}{6778} \cdot 28000^2 \mathbf{N}\right\| = \frac{1}{6778} \cdot 28000^2 \|\mathbf{N}\| = \frac{1}{6778} \cdot 28000^2 \approx 115668 \text{ km/h}^2$$

51. A particle follows a path $\mathbf{r}_1(t)$ on the helical curve with parametrization $\mathbf{r}(\theta) = \langle \cos\theta, \sin\theta, \theta \rangle$. When it is at position $\mathbf{r}(\frac{\pi}{2})$, its speed is 3 m/s and it is accelerating at a rate of $\frac{1}{2}$ m/s^2. Find its acceleration vector $\mathbf{a}$ at this moment. *Note:* The particle's acceleration vector does not coincide with $\mathbf{r}''(\theta)$.

SOLUTION Since $\mathbf{r}(t) = \langle \cos t, \sin t, t \rangle$, we have

$$\mathbf{r}'(t) = \langle -\sin t, \cos t, 1 \rangle, \quad \|\mathbf{r}'(t)\| = \sqrt{2}, \quad \Rightarrow \quad \mathbf{T} = \frac{\mathbf{r}'(t)}{\|\mathbf{r}'(t)\|} = \frac{1}{\sqrt{2}} \langle -\sin t, \cos t, 1 \rangle$$

By definition, $\mathbf{N}$ is the unit vector in the direction of $\mathbf{T}'$:

$$\mathbf{T}'(t) = \frac{1}{\sqrt{2}} \langle -\cos t, -\sin t, 0 \rangle \quad \Rightarrow \quad \mathbf{N} = \langle -\cos t, -\sin t, 0 \rangle$$

Therefore at $t = \pi/2$, we have

$$\mathbf{T} = \frac{1}{\sqrt{2}} \left\langle -\sin\frac{\pi}{2}, \cos\frac{\pi}{2}, 1 \right\rangle = \frac{1}{\sqrt{2}} \langle -1, 0, 1 \rangle, \quad \mathbf{N} = \left\langle -\cos\frac{\pi}{2}, -\sin\frac{\pi}{2}, 0 \right\rangle = \langle 0, -1, 0 \rangle.$$

The acceleration vector is

$$\mathbf{a} = a_T \mathbf{T} + a_N \mathbf{N} = v' \mathbf{T} + \kappa v^2 \mathbf{N}$$

Computing the curvature gives

$$\kappa = \left\| \frac{d\mathbf{T}}{ds} \right\| = \frac{\left\| \frac{d\mathbf{T}}{dt} \right\|}{\|\mathbf{r}'\|} = \frac{\left\| \frac{1}{\sqrt{2}} \langle -\cos t, -\sin t, 0 \rangle \right\|}{\sqrt{2}} = \frac{1}{2}$$

which gives

$$\mathbf{a} = v' \mathbf{T} + \kappa v^2 \mathbf{N} = \left(\frac{1}{2}\right) \mathbf{T} + \left(\frac{1}{2}\right)(3^2) \mathbf{N} = \left(\frac{1}{2}\right)\left(\frac{1}{\sqrt{2}}\right) \langle -1, 0, 1 \rangle + \frac{9}{2} \langle 0, -1, 0 \rangle = \left\langle -\frac{1}{2\sqrt{2}}, -\frac{9}{2}, \frac{1}{2\sqrt{2}} \right\rangle$$

53. Figure 15 shows the acceleration vectors of a particle moving clockwise around a circle. In each case, state whether the particle is speeding up, slowing down, or momentarily at constant speed. Explain.

(A)   (B)   (C)

FIGURE 15

SOLUTION In (A) and (B) the acceleration vector has a nonzero tangential and normal components; these are both possible acceleration vectors. In (C) the normal component of the acceleration toward the inside of the curve is zero, that is, $\mathbf{a}$ is parallel to $\mathbf{T}$, so $\kappa \cdot v(t)^2 = 0$, so either $\kappa = 0$ (meaning our curve is not a circle) or $v(t) = 0$ (meaning our particle isn't moving). Either way, (C) is not a possible acceleration vector.

55. Suppose that $\mathbf{r}(t)$ lies on a sphere of radius R for all t. Let $\mathbf{J} = |, \mathbf{r} \times \mathbf{r}'$. Show that $\mathbf{r}' = (\mathbf{J} \times \mathbf{r})/\|\mathbf{r}\|^2$. *Hint:* Observe that $\mathbf{r}$ and $\mathbf{r}'$ are perpendicular.

SOLUTION

(a) Solution 1. Since $\mathbf{r} = \mathbf{r}(t)$ lies on the sphere, the vectors $\mathbf{r} = \mathbf{r}(t)$ and $\mathbf{r}' = \mathbf{r}'(t)$ are orthogonal, therefore:

$$\mathbf{r} \cdot \mathbf{r}' = 0 \tag{1}$$

We use the following well-known equality:

$$\mathbf{a} \times (\mathbf{b} \times \mathbf{c}) = (\mathbf{a} \cdot \mathbf{c}) \mathbf{b} - (\mathbf{a} \cdot \mathbf{b}) \cdot \mathbf{c}$$

Using this equality and (1) we obtain:

$$\mathbf{J} \times \mathbf{r} = (\mathbf{r} \times \mathbf{r}') \times \mathbf{r} = -\mathbf{r} \times (\mathbf{r} \times \mathbf{r}') = -((\mathbf{r} \cdot \mathbf{r}') \mathbf{r} - (\mathbf{r} \cdot \mathbf{r}) \mathbf{r}')$$
$$= -(\mathbf{r} \cdot \mathbf{r}') \mathbf{r} + \|\mathbf{r}\|^2 \mathbf{r}' = 0\mathbf{r} + \|\mathbf{r}\|^2 \mathbf{r}' = \|\mathbf{r}\|^2 \mathbf{r}'$$

Divided by the scalar $\|\mathbf{r}\|^2$ we obtain:

$$\mathbf{r}' = \frac{\mathbf{J} \times \mathbf{r}}{\|\mathbf{r}\|^2}$$

(b) Solution 2. The cross product $\mathbf{J} = \mathbf{r} \times \mathbf{r}'$ is orthogonal to $\mathbf{r}$ and $\mathbf{r}'$. Also, $\mathbf{r}$ and $\mathbf{r}'$ are orthogonal, hence the vectors $\mathbf{r}$, $\mathbf{r}''$ and $\mathbf{J}$ are mutually orthogonal. Now, since $\mathbf{r}'$ is orthogonal to $\mathbf{r}$ and $\mathbf{J}$, the right-hand rule implies that $\mathbf{r}'$ points in the direction of $\mathbf{J} \times \mathbf{r}$. Therefore, for some $\alpha > 0$ we have:

$$\mathbf{r}' = \alpha \mathbf{J} \times \mathbf{r} = \|\mathbf{r}'\| \cdot \frac{\mathbf{J} \times \mathbf{r}}{\|\mathbf{J} \times \mathbf{r}\|} \tag{2}$$

By properties of the cross product and since $\mathbf{J}$, $\mathbf{r}$, and $\mathbf{r}'$ are mutually orthogonal we have:

$$\|\mathbf{J} \times \mathbf{r}\| = \|\mathbf{J}\|\|\mathbf{r}\| = \|\mathbf{r} \times \mathbf{r}'\|\|\mathbf{r}\| = \|\mathbf{r}\|\|\mathbf{r}'\|\|\mathbf{r}\| = \|\mathbf{r}\|^2 \|\mathbf{r}'\|$$

Substituting in (2) we get:

$$\mathbf{r}' = \|\mathbf{r}'\| \frac{\mathbf{J} \times \mathbf{r}}{\|\mathbf{r}\|^2 \|\mathbf{r}'\|} = \frac{\mathbf{J} \times \mathbf{r}}{\|\mathbf{r}\|^2}$$

Further Insights and Challenges

In Exercises 57–61, we consider an automobile of mass m traveling along a curved but level road. To avoid skidding, the road must supply a frictional force $\mathbf{F} = m\mathbf{a}$, where $\mathbf{a}$ is the car's acceleration vector. The maximum magnitude of the frictional force is $\mu m g$, where μ is the coefficient of friction and $g = 9.8$ m/s^2. Let v be the car's speed in meters per second.

57. Show that the car will not skid if the curvature κ of the road is such that (with $R = 1/\kappa$)

$$(v')^2 + \left(\frac{v^2}{R}\right)^2 \leq (\mu g)^2 \quad \boxed{8}$$

Note that braking ($v' < 0$) and speeding up ($v' > 0$) contribute equally to skidding.

SOLUTION To avoid skidding, the frictional force the road must supply is:

$$\mathbf{F} = m\mathbf{a}$$

where $\mathbf{a}$ is the acceleration of the car. We consider the decomposition of the acceleration $\mathbf{a}$ into normal and tangential directions:

$$\mathbf{a}(t) = v'(t)\mathbf{T}(t) + \kappa v^2(t)\mathbf{N}(t)$$

Since $\mathbf{N}$ and $\mathbf{T}$ are orthogonal unit vectors, $\mathbf{T} \cdot \mathbf{N} = 0$ and $\mathbf{T} \cdot \mathbf{T} = \mathbf{N} \cdot \mathbf{N} = 1$. Thus:

$$\|\mathbf{a}\|^2 = (v'\mathbf{T} + \kappa v^2 \mathbf{N}) \cdot (v'\mathbf{T} + \kappa v^2 \mathbf{N}) = v'^2 \mathbf{T} \cdot \mathbf{T} + 2\kappa v^2 v' \mathbf{N} \cdot \mathbf{T} + \kappa^2 v^4 \mathbf{N} \cdot \mathbf{N}$$

$$= v'^2 + \kappa^2 v^4 = v'^2 + \frac{v^4}{R^2}$$

Therefore:

$$\|\mathbf{a}\| = \sqrt{(v')^2 + \frac{v^4}{R^2}}$$

Since the maximal fractional force is $\mu m g$ we obtain that to avoid skidding the curvature must satisfy:

$$m\sqrt{(v')^2 + \frac{v^4}{R^2}} \leq m\mu g$$

Hence,

$$(v')^2 + \frac{v^4}{R^2} \leq (\mu g)^2$$

which becomes:

$$(v')^2 + \left(\frac{v^2}{R}\right)^2 \leq (\mu g)^2$$

59. Beginning at rest, an automobile drives around a circular track of radius $R = 300$ m, accelerating at a rate of 0.3 m/s^2. After how many seconds will the car begin to skid if the coefficient of friction is $\mu = 0.6$?

SOLUTION By Exercise 57 the car will begin to skid when:

$$(v')^2 + \frac{v^4}{R^2} = \mu^2 g^2 \quad (1)$$

We are given that $v' = 0.3$ and $v_0 = 0$. Integrating gives:

$$v = \int_0^t v' \, dt = \int_0^t 0.3 \, dt = 0.3t + v_0 = 0.3t$$

We substitute $v = 0.3t$, $v' = 0.3$, $R = 300$, $\mu = 0.6$ and $g = 9.8$ in (1) and solve for t. This gives:

$$0.3^2 + \frac{(0.3t)^4}{300^2} = 0.6^2 \cdot 9.8^2$$

$$0.09 + \frac{9}{10^8} t^4 \approx 34.57$$

$$t^4 \approx 3.83 \times 10^8 \quad \Rightarrow \quad t \approx 139.90$$

After about 139.90 s the car will begin to skid.

61. What is the smallest radius R about which an automobile can turn without skidding at 100 km/h if $\mu = 0.75$ (a typical value)?

SOLUTION The solution to Exercise 58 showed that the maximum constant speed achievable on a circle of radius R without skidding is $v = \sqrt{\mu g R}$. Converting 100 km/h to m/s gives

$$100 \text{ km/h} \cdot 1000 \text{m/km} \cdot \frac{1}{3600 \text{ s/h}} \approx 27.78 \text{ m/s}$$

Then with the given values, we want to find R if $27.78 = \sqrt{0.75 \cdot 9.8 \cdot R}$. Squaring and simplifying gives $R \approx 104.98$. So the smallest possible turning radius at 100 km/h with the given μ is about 105 m.

13.6 Planetary Motion According to Kepler and Newton (LT Section 14.6)

Preliminary Questions

1. Describe the relation between the vector $\mathbf{J} = \mathbf{r} \times \mathbf{r}'$ and the rate at which the radial vector sweeps out area.

SOLUTION The rate at which the radial vector sweeps out area equals half the magnitude of the vector $\mathbf{J}$. This relation is expressed in the formula:

$$\frac{dA}{dt} = \frac{1}{2} \|\mathbf{J}\|$$

2. Eq. (1) shows that $\mathbf{r}''$ is proportional to $\mathbf{r}$. Explain how this fact is used to prove Kepler's Second Law.

SOLUTION In the proof of Kepler's Second Law it is shown that the rate at which area is swept out is

$$\frac{dA}{dt} = \frac{1}{2} \|\mathbf{J}\|, \quad \text{where} \quad \mathbf{J} = \mathbf{r}(t) \times \mathbf{r}'(t)$$

To show that $\|\mathbf{J}\|$ is constant, show that $\mathbf{J}$ is constant. This is done using the proportionality of $\mathbf{r}''$ and $\mathbf{r}$ which implies that $\mathbf{r}(t) \times \mathbf{r}''(t) = 0$. Using this we get:

$$\frac{d\mathbf{J}}{dt} = \frac{d}{dt}(\mathbf{r} \times \mathbf{r}') = \mathbf{r} \times \mathbf{r}'' + \mathbf{r}' \times \mathbf{r}' = 0 + 0 = 0 \Rightarrow \mathbf{J} = \text{const}$$

SECTION 13.6 | Planetary Motion According to Kepler and Newton (LT SECTION 14.6)

3. How is the period T affected if the semimajor axis a is increased 4-fold?

SOLUTION Kepler's Third Law states that the period T of the orbit is given by:

$$T^2 = \left(\frac{4\pi^2}{GM}\right) a^3$$

or

$$T = \frac{2\pi}{\sqrt{GM}} a^{3/2}$$

If a is increased 4-fold the period becomes:

$$\frac{2\pi}{\sqrt{GM}} (4a)^{3/2} = 8 \cdot \frac{2\pi}{\sqrt{GM}} a^{3/2}$$

That is, the period is increased 8-fold.

Exercises

1. Kepler's Third Law states that T^2/a^3 has the same value for each planetary orbit. Do the data in the following table support this conclusion? Estimate the length of Jupiter's period, assuming that $a = 77.8 \times 10^{10}$ m.

Planet	Mercury	Venus	Earth	Mars
a (10^{10} m)	5.79	10.8	15.0	22.8
T (years)	0.241	0.615	1.00	1.88

SOLUTION Using the given data we obtain the following values of T^2/a^3, where a, as always, is measured not in meters but in 10^{10} m:

Planet	Mercury	Venus	Earth	Mars
T^2/a^3	$2.99 \cdot 10^{-4}$	$3 \cdot 10^{-4}$	$2.96 \cdot 10^{-4}$	$2.98 \cdot 10^{-4}$

The data on the planets supports Kepler's prediction. We estimate Jupiter's period (using the given a) as $T \approx \sqrt{a^3 \cdot 3 \cdot 10^{-4}} \approx 11.9$ years.

3. Ganymede, one of Jupiter's moons discovered by Galileo, has an orbital period of 7.154 days and a semimajor axis of 1.07×10^9 m. Use Exercise 2 to estimate the mass of Jupiter.

SOLUTION By Exercise 2, the mass of Jupiter can be computed using the following equality:

$$M = \frac{4\pi^2}{G} \frac{a^3}{T^2}$$

We substitute the given data $T = 7.154 \cdot 24 \cdot 60^2 = 618{,}105.6$ $a = 1.07 \times 10^9$ m and $G = 6.67300 \times 10^{-11} \text{m}^3\text{kg}^{-1}\text{s}^{-1}$, to obtain:

$$M = \frac{4\pi^2 \cdot \left(1.07 \times 10^9\right)^3}{6.67300 \times 10^{-11} \cdot (618{,}105.6)^2} \approx 1.897 \times 10^{27} \text{ kg}.$$

5. Mass of the Milky Way The sun revolves around the center of mass of the Milky Way galaxy in an orbit that is approximately circular, of radius $a \approx 2.8 \times 10^{17}$ km and velocity $v \approx 250$ km/s. Use the result of Exercise 2 to estimate the mass of the portion of the Milky Way inside the sun's orbit (place all of this mass at the center of the orbit).

SOLUTION Write $a = 2.8 \times 10^{20}$ m and $v = 250 \times 10^3$ m/s. The circumference of the sun's orbit (which is assumed circular) is $2\pi a$ m; since the sun's speed is a constant v m/s, its period is $T = \frac{2\pi a}{v}$ s. By Exercise 2, the mass of the portion of the Milky Way inside the sun's orbit is

$$M = \left(\frac{4\pi^2}{G}\right)\left(\frac{a^3}{T^2}\right)$$

Substituting the values of a and T from above, $G = 6.673 \times 10^{-11}$ m^{3}kg^{-1}s^{-2} gives

$$M = \frac{4\pi^2 a^3}{G\left(\frac{4\pi^2 a^2}{v^3}\right)} = \frac{av^2}{G} = \frac{2.8 \cdot 10^{20} \cdot (250 \times 10^3)^2}{6.673 \times 10^{-11}} = 2.6225 \times 10^{41} \text{ kg}$$

The mass of the sun is 1.989×10^{30} kg, hence M is 1.32×10^{11} times the mass of the sun (132 billions times the mass of the sun).

7. Show that a planet in a circular orbit travels at constant speed. *Hint:* Use the facts that **J** is constant and that $\mathbf{r}(t)$ is orthogonal to $\mathbf{r}'(t)$ for a circular orbit.

SOLUTION It is shown in the proof of Kepler's Second Law that the vector $\mathbf{J} = \mathbf{r}(t) \times \mathbf{r}'(t)$ is constant, hence its length is constant:

$$\|\mathbf{J}\| = \|\mathbf{r}(t) \times \mathbf{r}'(t)\| = \text{const} \tag{1}$$

We consider the orbit as a circle of radius R, therefore, $\mathbf{r}(t)$ and $\mathbf{r}'(t)$ are orthogonal and $\|\mathbf{r}(t)\| = R$. By (1) and using properties of the cross product we obtain:

$$\|\mathbf{r}(t) \times \mathbf{r}'(t)\| = \|\mathbf{r}(t)\|\|\mathbf{r}'(t)\| \sin\frac{\pi}{2} = R \cdot \|\mathbf{r}'(t)\| = \text{const}$$

We conclude that $\|\mathbf{r}'(t)\|$ is constant, that is the speed $v = \|\mathbf{r}'(t)\|$ of the planet is constant.

9. Prove that if a planetary orbit is circular of radius R, then $vT = 2\pi R$, where v is the planet's speed (constant by Exercise 7) and T is the period. Then use Kepler's Third Law to prove that $v = \sqrt{\frac{k}{R}}$.

SOLUTION By the Arc Length Formula and since the speed $v = \|\mathbf{r}'(t)\|$ is constant, the length L of the circular orbit can be computed by the following integral:

$$L = \int_0^T \|\mathbf{r}'(t)\|\, dt = \int_0^T v\, dt = vt \Big|_0^T = vT$$

On the other hand, the length of a circular orbit of radius R is $2\pi R$, so we obtain:

$$vT = 2\pi R \Rightarrow T = \frac{2\pi R}{v} \tag{1}$$

In a circular orbit of radius R, $a = R$, hence by Kepler's Third Law we have:

$$T^2 = \frac{4\pi^2}{GM} R^3 \tag{2}$$

We now substitute (1) in (2) and solve for v. This gives:

$$\left(\frac{2\pi R}{v}\right)^2 = \frac{4\pi^2 R^3}{GM}$$

$$\frac{4\pi^2 R^2}{v^2} = \frac{4\pi^2 R^3}{GM}$$

$$\frac{1}{v^2} = \frac{R}{GM} \Rightarrow v = \sqrt{\frac{GM}{R}}$$

11. A communications satellite orbiting the earth has initial position $\mathbf{r}(0) = \langle 29{,}000, 20{,}000, 0\rangle$ (in kilometers) and initial velocity $\mathbf{r}'(0) = \langle 1, 1, 1\rangle$ (in kilometers per second), where the origin is the earth's center. Find the equation of the plane containing the satellite's orbit. *Hint:* This plane is orthogonal to $\mathbf{J}$.

SOLUTION The vectors $\mathbf{r}(t)$ and $\mathbf{r}'(t)$ lie in the plane containing the satellite's orbit, in particular the initial position $\mathbf{r} = \langle 1000, 2000, 0\rangle$ and the initial velocity $\mathbf{r}' = \langle 1, 1, 1\rangle$. Therefore, the cross product $\mathbf{J} = \mathbf{r} \times \mathbf{r}'$ is perpendicular to the plane. We compute $\mathbf{J}$:

$$\mathbf{J} = \mathbf{r} \times \mathbf{r}' = \begin{vmatrix} \mathbf{i} & \mathbf{j} & \mathbf{k} \\ 29000 & 20000 & 0 \\ 1 & 1 & 1 \end{vmatrix} = \begin{vmatrix} 20000 & 0 \\ 1 & 1 \end{vmatrix}\mathbf{i} - \begin{vmatrix} 29000 & 0 \\ 1 & 1 \end{vmatrix}\mathbf{j} + \begin{vmatrix} 29000 & 20000 \\ 1 & 1 \end{vmatrix}\mathbf{k}$$

$$= 20000\mathbf{i} - 29000\mathbf{j} + 9000\mathbf{k} = \langle 20000, -29000, 9000\rangle$$

We now use the vector form of the equation of the plane with $\mathbf{n} = \mathbf{J} = \langle 20000, -29000, 9000\rangle$ and $\langle x_0, y_0, z_0\rangle = \mathbf{r} = \langle 29000, 20000, 0\rangle$, to obtain the following equation:

$$\langle 20000, -29000, 9000\rangle \cdot \langle x, y, z\rangle = \langle 20000, -29000, 9000\rangle \cdot \langle 29000, 20000, 0\rangle$$

$$20000x - 29000y + 9000z = 0$$

$$20x - 29y + 9z = 0$$

The plane containing the satellite's orbit is, thus:

$$\mathcal{P} = \{(x, y, z) : 20x - 29y + 9z = 0\}$$

In Exercises 13–19, the perihelion and aphelion are the points on the orbit closest to and farthest from the sun, respectively (Figure 8). The distance from the sun at the perihelion is denoted r_{per} and the speed at this point is denoted v_{per}. Similarly, we write r_{ap} and v_{ap} for the distance and speed at the aphelion. The semimajor axis is denoted a.

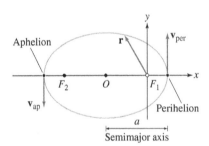

FIGURE 8 $\mathbf{r}$ and $\mathbf{v} = \mathbf{r}'$ are perpendicular at the perihelion and aphelion.

13. Use the polar equation of an ellipse

$$r = \frac{p}{1 + e\cos\theta}$$

to show that $r_{\text{per}} = a(1 - e)$ and $r_{\text{ap}} = a(1 + e)$. *Hint:* Use the fact that $r_{\text{per}} + r_{\text{ap}} = 2a$.

SOLUTION We use the polar equation of the elliptic orbit:

$$r = \frac{p}{1 + e\cos\theta} \qquad (1)$$

At the perigee, $\theta = 0$ and at the apogee $\theta = \pi$. Substituting these values in (1) gives the distances r_{per} and r_{ap} respectively. That is,

$$r_{\text{per}} = \frac{p}{1 + e\cos\theta} = \frac{p}{1 + e} \qquad (2)$$

$$r_{\text{ap}} = \frac{p}{1 + e\cos\pi} = \frac{p}{1 - e} \qquad (3)$$

To obtain the solutions in terms of a rather than p, we notice that:

$$r_{\text{per}} + r_{\text{ap}} = 2a$$

Hence:

$$2a = \frac{p}{1+e} + \frac{p}{1-e} = \frac{p(1-e) + p(1+e)}{(1+e)(1-e)} = \frac{2p}{(1+e)(1-e)}$$

yielding

$$p = a(1+e)(1-e)$$

Substituting in (2) and (3) we obtain:

$$r_{\text{per}} = \frac{a(1+e)(1-e)}{1+e} = a(1-e)$$

$$r_{\text{ap}} = \frac{a(1+e)(1-e)}{1-e} = a(1+e)$$

15. Use the fact that $\mathbf{J} = \mathbf{r} \times \mathbf{r}'$ is constant to prove

$$v_{\text{per}}(1-e) = v_{\text{ap}}(1+e)$$

Hint: $\mathbf{r}$ is perpendicular to $\mathbf{r}'$ at the perihelion and aphelion.

SOLUTION Since the vector $\mathbf{J}(t) = \mathbf{r}(t) \times \mathbf{r}'(t)$ is constant, it is the same vector at the perigee and at the apogee, hence we may equate the length of $\mathbf{J}(t)$ at these two points. Since at the perigee and at the apogee $\mathbf{r}(t)$ and $\mathbf{r}'(t)$ are orthogonal we have by properties of the cross product:

$$\|\mathbf{r}_{\text{ap}} \times \mathbf{r}'_{\text{ap}}\| = \|\mathbf{r}_{\text{ap}}\|\|\mathbf{r}'_{\text{ap}}\| = r_{\text{ap}} v_{\text{ap}}$$

$$\|\mathbf{r}_{\text{per}} \times \mathbf{r}'_{\text{per}}\| = \|\mathbf{r}_{\text{per}}\|\|\mathbf{r}'_{\text{per}}\| = r_{\text{per}} v_{\text{per}}$$

Equating the two values gives:

$$r_{\text{ap}} v_{\text{ap}} = r_{\text{per}} v_{\text{per}} \tag{1}$$

In Exercise 13 we showed that $r_{\text{per}} = a(1-e)$ and $r_{\text{ap}} = a(1+e)$. Substituting in (1) we obtain:

$$a(1+e)v_{\text{ap}} = a(1-e)v_{\text{per}}$$

$$(1+e)v_{\text{ap}} = (1-e)v_{\text{per}}$$

17. Conservation of Energy The total mechanical energy (kinetic energy plus potential energy) of a planet of mass m orbiting a sun of mass M with position $\mathbf{r}$ and speed $v = \|\mathbf{r}'\|$ is

$$E = \frac{1}{2}mv^2 - \frac{GMm}{\|\mathbf{r}\|} \qquad \boxed{8}$$

(a) Prove the equations

$$\frac{d}{dt} \frac{1}{2} mv^2 = \mathbf{v} \cdot (m\mathbf{a}), \qquad \frac{d}{dt} \frac{GMm}{\|\mathbf{r}\|} = \mathbf{v} \cdot \left(-\frac{GMm}{\|\mathbf{r}\|^3} \mathbf{r} \right)$$

(b) Then use Newton's Law $\mathbf{F} = m\mathbf{a}$ and Eq. (1) to prove that energy is conserved—that is, $\dfrac{dE}{dt} = 0$.

SOLUTION

(a) We will need to know $\frac{d}{dt}\|\mathbf{r}'\|$ and $\frac{d}{dt}\|\mathbf{r}\|$. More generally, suppose that $\mathbf{v}(t)$ is a vector-valued function. Then

$$\frac{d}{dt}\|\mathbf{v}(t)\| = \frac{d}{dt}\sqrt{\mathbf{v}(t) \cdot \mathbf{v}(t)}$$

$$= \frac{1}{2}(\mathbf{v}(t) \cdot \mathbf{v}(t))^{-1/2} \frac{d}{dt}(\mathbf{v}(t) \cdot \mathbf{v}(t))$$

$$= \frac{1}{2}(\mathbf{v}(t) \cdot \mathbf{v}(t))^{-1/2}(\mathbf{v}'(t) \cdot \mathbf{v}(t) + \mathbf{v}(t) \cdot \mathbf{v}'(t))$$

$$= \frac{\mathbf{v}(t) \cdot \mathbf{v}'(t)}{\|\mathbf{v}(t)\|}$$

Then using the chain rule,

$$\frac{d}{dt}\frac{1}{2}mv^2 = \frac{d}{dt}\frac{1}{2}m\|\mathbf{r}\|^2 = \frac{1}{2}\cdot 2m\|\mathbf{r}\|\cdot\|\mathbf{r}\|' = m\|\mathbf{r}\|\cdot\frac{\mathbf{r}\cdot\mathbf{r}'}{\|\mathbf{r}\|} = m\mathbf{r}\cdot\mathbf{r}' = \mathbf{v}\cdot(m\mathbf{a})$$

For the second equation,

$$\frac{d}{dt}\frac{GMm}{\|\mathbf{r}\|} = \frac{d}{dt}GMm\|\mathbf{r}\|^{-1} = -GMm\|\mathbf{r}\|^{-2}\|\mathbf{r}\|' = -GMm\|\mathbf{r}\|^{-2}\cdot\frac{\mathbf{r}\cdot\mathbf{r}'}{\|\mathbf{r}\|} = \mathbf{r}'\cdot\left(-\frac{GMm}{\|\mathbf{r}\|^3}\mathbf{r}\right) = \mathbf{v}\cdot\left(-\frac{GMm}{\|\mathbf{r}\|^3}\mathbf{r}\right)$$

(b) Compute $\frac{dE}{dt}$:

$$\frac{dE}{dt} = \frac{d}{dt}\frac{1}{2}mv^2 - \frac{d}{dt}\frac{GMm}{\|\mathbf{r}\|}$$

$$= \mathbf{v}\cdot(m\mathbf{a}) - \mathbf{v}\cdot\left(-\frac{GMm}{\|\mathbf{r}\|^3}\mathbf{r}\right)$$

$$= \mathbf{v}\cdot\left(\mathbf{F} + \frac{GMm}{\|\mathbf{r}\|^3}\mathbf{r}\right)$$

But by Newton's Law of Universal Gravitation,

$$\mathbf{F} = \mathbf{F}(\mathbf{r}(t)) = -\left(\frac{GMm}{\|\mathbf{r}\|^2}\right)\mathbf{e}_r = -\left(\frac{GMm}{\|\mathbf{r}\|^2}\right)\frac{\mathbf{r}}{\|\mathbf{r}\|} = -\left(\frac{GMm}{\|\mathbf{r}\|^3}\right)\mathbf{r}$$

and the two terms in the computation of $\frac{dE}{dt}$ cancel, so that $\frac{dE}{dt} = 0$. Since the derivative is zero, energy $E = E(t)$ is a constant, so that energy is conserved.

19. Prove that $v_{\text{per}} = \sqrt{\left(\frac{GM}{a}\right)\frac{1+e}{1-e}}$ as follows:

(a) Use Conservation of Energy (Exercise 17) to show that

$$v_{\text{per}}^2 - v_{\text{ap}}^2 = 2GM(r_{\text{per}}^{-1} - r_{\text{ap}}^{-1})$$

(b) Show that $r_{\text{per}}^{-1} - r_{\text{ap}}^{-1} = \frac{2e}{a(1-e^2)}$ using Exercise 13.

(c) Show that $v_{\text{per}}^2 - v_{\text{ap}}^2 = 4\frac{e}{(1+e)^2}v_{\text{per}}^2$ using Exercise 15. Then solve for v_{per} using (a) and (b).

SOLUTION

(a) The total mechanical energy of a planet is constant. That is,

$$E = \frac{1}{2}mv^2 - \frac{GMm}{\|\mathbf{r}\|} = \text{const}$$

Therefore, E has equal values at the perigee and apogee. Hence,

$$\frac{1}{2}mv_{\text{per}}^2 - \frac{GMm}{r_{\text{per}}} = \frac{1}{2}mv_{\text{ap}}^2 - \frac{GMm}{r_{\text{ap}}}$$

$$\frac{1}{2}m\left(v_{\text{per}}^2 - v_{\text{ap}}^2\right) = GMm\left(\frac{1}{r_{\text{per}}} - \frac{1}{r_{\text{ap}}}\right)$$

$$v_{\text{per}}^2 - v_{\text{ap}}^2 = 2GM\left(r_{\text{per}}^{-1} - r_{\text{ap}}^{-1}\right)$$

(b) In Exercise 13 we showed that $r_{\text{per}} = a(1-e)$ and $r_{\text{ap}} = a(1+e)$. Therefore,

$$r_{\text{per}}^{-1} - r_{\text{ap}}^{-1} = \frac{1}{a(1-e)} - \frac{1}{a(1+e)} = \frac{1+e-(1-e)}{a(1-e)(1+e)} = \frac{2e}{a(1-e^2)}$$

(c) In Exercise 15 we showed that

$$v_{\text{per}}(1-e) = v_{\text{ap}}(1+e)$$

Hence,

$$v_{\text{ap}} = \frac{1-e}{1+e}v_{\text{per}}$$

We compute the following difference,

$$v_{\text{per}}^2 - v_{\text{ap}}^2 = v_{\text{per}}^2 - \left(\frac{1-e}{1+e} v_{\text{per}}\right)^2 = v_{\text{per}}^2 \left(1 - \left(\frac{1-e}{1+e}\right)^2\right)$$

$$= v_{\text{per}}^2 \frac{(1+e)^2 - (1-e)^2}{(1+e)^2} = v_{\text{per}}^2 \frac{1 + 2e + e^2 - (1 - 2e + e^2)}{(1+e)^2} = 4\frac{e}{(1+e)^2} v_{\text{per}}^2$$

We combine this equality with the equality in part (a) to write

$$\frac{4e}{(1+e)^2} v_{\text{per}}^2 = 2GM \left(r_{\text{per}}^{-1} - r_{\text{ap}}^{-1}\right)$$

Replacing the difference in the right-hand side by $\frac{2e}{a(1-e^2)}$ (from part (b)) and solving for v_{per} we obtain:

$$\frac{4e}{(1+e)^2} v_{\text{per}}^2 = 2GM \cdot \frac{2e}{a(1-e^2)}$$

$$v_{\text{per}}^2 = \frac{4GMe}{a(1-e)(1+e)} \cdot \frac{(1+e)^2}{4e} = \frac{GM(1+e)}{a(1-e)}$$

or,

$$v_{\text{per}} = \sqrt{\frac{GM}{a} \frac{1+e}{1-e}}$$

21. Prove that $v^2 = GM\left(\frac{2}{r} - \frac{1}{a}\right)$ at any point on an elliptical orbit, where $r = \|\mathbf{r}\|$, v is the velocity, and a is the semimajor axis of the orbit.

SOLUTION The total energy $E = \frac{1}{2}mv^2 - \frac{GMm}{\|\mathbf{r}\|}$ is conserved, and in Exercise 20 we showed that its constant value is $-\frac{GMm}{2a}$. We obtain the following equality:

$$\frac{1}{2}mv^2 - \frac{GMm}{r} = -\frac{GMm}{2a}$$

Algebraic manipulations yield:

$$v^2 = \frac{2GM}{r} - \frac{GM}{a} = GM\left(\frac{2}{r} - \frac{1}{a}\right)$$

Further Insights and Challenges

Exercises 23 and 24 prove Kepler's Third Law. Figure 10 shows an elliptical orbit with polar equation

$$r = \frac{p}{1 + e\cos\theta}$$

where $p = J^2/k$. The origin of the polar coordinates occurs at F_1. Let a and b be the semimajor and semiminor axes, respectively.

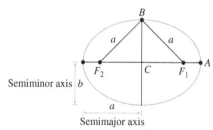

FIGURE 10

23. This exercise shows that $b = \sqrt{pa}$.
(a) Show that $CF_1 = ae$. Hint: $r_{\text{per}} = a(1-e)$ by Exercise 13.
(b) Show that $a = \frac{p}{1-e^2}$.

(c) Show that $F_1A + F_2A = 2a$. Conclude that $F_1B + F_2B = 2a$ and hence $F_1B = F_2B = a$.

(d) Use the Pythagorean Theorem to prove that $b = \sqrt{pa}$.

SOLUTION

(a) From the hint, $r_{\text{per}} = AF_1 = a(1-e)$, and $AC = a$, so $CF_1 = AC - AF_1 = a - (a(1-e)) = ae$.

(b) The polar equation of the ellipse, where the focus F_1 is at the origin is

$$r = \frac{p}{1 + e \cos \theta}$$

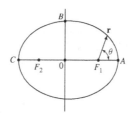

The point A corresponds to $\theta = 0$, hence,

$$F_1A = \frac{p}{1 + e \cos 0} = \frac{p}{1 + e}$$

The point C corresponds to $\theta = \pi$ hence,

$$F_1C = \frac{p}{1 + e \cos \pi} = \frac{p}{1 - e}$$

We now find F_2A. Using the equality $CF_2 = AF_1$ we get:

$$F_2A = F_2F_1 + F_1A = F_2F_1 + F_2C = F_1C = \frac{p}{1-e}$$

That is,

$$F_2A = \frac{p}{1 - e}$$

Then we obtain

$$2a = F_1A + F_2A = \frac{p}{1+e} + \frac{p}{1-e}$$

Hence,

$$a = \frac{1}{2}\left(\frac{p}{1+e} + \frac{p}{1-e}\right) = \frac{p(1-e) + p(1+e)}{2(1+e)(1-e)} = \frac{2p}{2(1-e^2)} = \frac{p}{1-e^2}$$

(c) Denote by D the left end of the major axis in Figure 10, the point on the ellipse to the left of F_2. Since $DF_2 = F_1A$, we have:

$$F_2A = DA - DF_2 = 2a - F_1A$$

Therefore,

$$F_1A + F_2A = 2a$$

The ellipse is the set of all points such that the sum of the distances to the two foci F_1 and F_2 is constant. Therefore,

$$2a = F_1A + F_2A = F_1B + F_2B$$

Since $F_2C = F_1C$, $CB = CB$, and $\angle F_2CB = \angle F_1CB$ (they are both right angles), the triangles $\triangle F_2CB$ and $\triangle F_1CB$ are congruent, so that $F_2B = F_1B$, and we conclude that $F_2B = F_1B = a$.

(d) We use the Pythagorean Theorem for the triangle F_1CB in Figure 10:

$$CF_1^2 + CB^2 = BF_1^2$$

We know that $CF_1 = ae$ from part (a), and $BF_1 = a$ from part (c), so that

$$b = CB = \sqrt{BF_1^2 - CF_1^2} = \sqrt{a^2 - a^2e^2} = a\sqrt{1-e^2}$$

But from part (b), $1 - e^2 = \frac{p}{a}$, so we get

$$b = a\sqrt{1-e^2} = a\sqrt{\frac{p}{a}} = \sqrt{ap}$$

25. According to Eq. (7), the velocity vector of a planet as a function of the angle θ is

$$\mathbf{v}(\theta) = \frac{k}{J}\mathbf{e}_\theta + \mathbf{c}$$

Use this to explain the following statement: As a planet revolves around the sun, its velocity vector traces out a circle of radius k/J with its center at the terminal point of $\mathbf{c}$ (Figure 11). This beautiful but hidden property of orbits was discovered by William Rowan Hamilton in 1847.

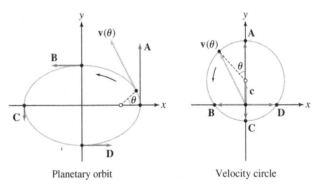

FIGURE 11 The velocity vector traces out a circle as the planet travels along its orbit.

SOLUTION By Eq. (1)

$$\mathbf{r}''(t) = \frac{-k}{\|\mathbf{r}\|^3}\mathbf{r}(t) = \frac{-GM}{\|\mathbf{r}\|^3}\mathbf{r}(t)$$

Writing $\mathbf{r} = \|\mathbf{r}\|\mathbf{e}_\mathbf{r}$ and $\mathbf{r}''(t) = \frac{d}{dt}\mathbf{r}'(t) = \frac{d\mathbf{v}}{dt}$ we get:

$$\frac{d\mathbf{v}}{dt} = \frac{-GM}{\|\mathbf{r}\|^3}\|\mathbf{r}\|\mathbf{e}_\mathbf{r} = \frac{-GM}{\|\mathbf{r}\|^2}\mathbf{e}_\mathbf{r} = \frac{-GM}{r^2}\mathbf{e}_\mathbf{r}$$

That is,

$$\frac{d\mathbf{v}}{dt} = -\frac{GM}{r^2}\mathbf{e}_\mathbf{r}$$

By the Chain Rule, $\frac{d\mathbf{v}}{dt} = \frac{d\mathbf{v}}{d\theta} \cdot \frac{d\theta}{dt}$. Further, differentiating $\mathbf{e}_\theta$ gives

$$\frac{d\mathbf{e}_\theta}{d\theta} = \frac{d}{d\theta}\langle -\sin\theta, \cos\theta\rangle = \langle -\cos\theta, -\sin\theta\rangle = -\langle \cos\theta, \sin\theta\rangle = -\mathbf{e}_\mathbf{r}$$

Substituting in the equation above, we get:

$$\frac{d\mathbf{v}}{d\theta}\frac{d\theta}{dt} = \frac{GM}{r^2}\frac{d\mathbf{e}_\theta}{d\theta}$$

By Eq. 2,

$$r(t)^2\theta'(t) = \|\mathbf{J}\| \quad\Rightarrow\quad \frac{d\theta}{dt} = \frac{\|\mathbf{J}\|}{r^2} = \frac{J}{r^2}$$

Therefore

$$\frac{d\mathbf{v}}{d\theta}\frac{d\theta}{dt} = \frac{d\mathbf{v}}{d\theta}\frac{J}{r^2} = \frac{GM}{r^2}\frac{d\mathbf{e}_\theta}{d\theta} = \frac{k}{r^2}\frac{d\mathbf{e}_\theta}{d\theta} \quad\Rightarrow\quad \frac{d\mathbf{v}}{d\theta} = \frac{k}{J}\frac{d\mathbf{e}_\theta}{d\theta}$$

Integrating both sides of this equation gives

$$\mathbf{v}(\theta) = \int \frac{d\mathbf{v}}{d\theta}d\theta = \frac{k}{J}\int \frac{d\mathbf{e}_\theta}{d\theta}d\theta = \frac{k}{J}\mathbf{e}_\theta + \mathbf{u}$$

CHAPTER REVIEW EXERCISES

1. Determine the domains of the vector-valued functions.
(a) $\mathbf{r}_1(t) = \langle t^{-1}, (t+1)^{-1}, \sin^{-1} t \rangle$
(b) $\mathbf{r}_2(t) = \langle \sqrt{8-t^3}, \ln t, e^{\sqrt{t}} \rangle$

SOLUTION

(a) We find the domain of $\mathbf{r}_1(t) = \langle t^{-1}, (t+1)^{-1}, \sin^{-1} t \rangle$. The function t^{-1} is defined for $t \neq 0$. $(t+1)^{-1}$ is defined for $t \neq -1$ and $\sin^{-1} t$ is defined for $-1 \leq t \leq 1$. Hence, the domain of $\mathbf{r}_1(t)$ is defined by the following inequalities:

$$t \neq 0$$
$$t \neq -1 \quad \Rightarrow \quad -1 < t < 0 \quad \text{or} \quad 0 < t \leq 1$$
$$-1 \leq t \leq 1$$

(b) We find the domain of $\mathbf{r}_2(t) = \langle \sqrt{8-t^3}, \ln t, e^{\sqrt{t}} \rangle$. The domain of $\sqrt{8-t^3}$ is $8-t^3 \geq 0$. The domain of $\ln t$ is $t > 0$ and $e^{\sqrt{t}}$ is defined for $t \geq 0$. Hence, the domain of $\mathbf{r}_2(t)$ is defined by the following inequalities:

$$\begin{array}{c} 8 - t^3 \geq 0 \\ t > 0 \\ t \geq 0 \end{array} \Rightarrow \begin{array}{c} t^3 \leq 8 \\ t > 0 \end{array} \Rightarrow 0 < t \leq 2$$

3. Find a vector parametrization of the intersection of the surfaces $x^2 + y^4 + 2z^3 = 6$ and $x = y^2$ in $\mathbf{R}^3$.

SOLUTION We need to find a vector parametrization $\mathbf{r}(t) = \langle x(t), y(t), z(t) \rangle$ for the intersection curve. Using $t = y$ as a parameter, we have $x = t^2$ and $y = t$. We substitute in the equation of the surface $x^2 + y^4 + 2z^3 = 6$ and solve for z in terms of t. This gives:

$$t^4 + t^4 + 2z^3 = 6$$
$$2t^4 + 2z^3 = 6$$
$$z^3 = 3 - t^4 \quad \Rightarrow \quad z = \sqrt[3]{3-t^4}$$

We obtain the following parametrization of the intersection curve:

$$\mathbf{r}(t) = \langle t^2, t, \sqrt[3]{3-t^4} \rangle$$

In Exercises 5–10, calculate the derivative indicated.

5. $\mathbf{r}'(t), \quad \mathbf{r}(t) = \langle 1-t, t^{-2}, \ln t \rangle$

SOLUTION We use the Theorem on Componentwise Differentiation to compute the derivative $\mathbf{r}'(t)$. We get

$$\mathbf{r}'(t) = \langle (1-t)', (t^{-2})', (\ln t)' \rangle = \left\langle -1, -2t^{-3}, \frac{1}{t} \right\rangle$$

7. $\mathbf{r}'(0), \quad \mathbf{r}(t) = \langle e^{2t}, e^{-4t^2}, e^{6t} \rangle$

SOLUTION We differentiate $\mathbf{r}(t)$ componentwise to find $\mathbf{r}'(t)$:

$$\mathbf{r}'(t) = \langle (e^{2t})', (e^{-4t^2})', (e^{6t})' \rangle = \langle 2e^{2t}, -8te^{-4t^2}, 6e^{6t} \rangle$$

The derivative $\mathbf{r}'(0)$ is obtained by setting $t = 0$ in $\mathbf{r}'(t)$. This gives

$$\mathbf{r}'(0) = \langle 2e^{2 \cdot 0}, -8 \cdot 0 e^{-4 \cdot 0^2}, 6e^{6 \cdot 0} \rangle = \langle 2, 0, 6 \rangle$$

9. $\dfrac{d}{dt} e^t \langle 1, t, t^2 \rangle$

SOLUTION Using the Product Rule for differentiation gives

$$\dfrac{d}{dt} e^t \langle 1, t, t^2 \rangle = e^t \dfrac{d}{dt} \langle 1, t, t^2 \rangle + \left(e^t\right)' \langle 1, t, t^2 \rangle = e^t \langle 0, 1, 2t \rangle + e^t \langle 1, t, t^2 \rangle$$

$$= e^t \left(\langle 0, 1, 2t \rangle + \langle 1, t, t^2 \rangle \right) = e^t \langle 1, 1+t, 2t+t^2 \rangle$$

In Exercises 11–14, calculate the derivative at t = 3, assuming that

$$\mathbf{r}_1(3) = \langle 1, 1, 0 \rangle, \qquad \mathbf{r}_2(3) = \langle 1, 1, 0 \rangle$$
$$\mathbf{r}_1'(3) = \langle 0, 0, 1 \rangle, \qquad \mathbf{r}_2'(3) = \langle 0, 2, 4 \rangle$$

11. $\dfrac{d}{dt} (6\mathbf{r}_1(t) - 4 \cdot \mathbf{r}_2(t))$

SOLUTION Using Differentiation Rules we obtain:

$$\left. \dfrac{d}{dt} (6\mathbf{r}_1(t) - 4\mathbf{r}_2(t)) \right|_{t=3} = 6\mathbf{r}_1'(3) - 4\mathbf{r}_2'(3) = 6 \cdot \langle 0, 0, 1 \rangle - 4 \cdot \langle 0, 2, 4 \rangle$$

$$= \langle 0, 0, 6 \rangle - \langle 0, 8, 16 \rangle = \langle 0, -8, -10 \rangle$$

13. $\dfrac{d}{dt} (\mathbf{r}_1(t) \cdot \mathbf{r}_2(t))$

SOLUTION Using Product Rule for Dot Products we obtain:

$$\dfrac{d}{dt} \mathbf{r}_1(t) \cdot \mathbf{r}_2(t) = \mathbf{r}_1(t) \cdot \mathbf{r}_2'(t) + \mathbf{r}_1'(t) \cdot \mathbf{r}_2(t)$$

Setting $t = 3$ gives:

$$\left. \dfrac{d}{dt} \mathbf{r}_1(t) \cdot \mathbf{r}_2(t) \right|_{t=3} = \mathbf{r}_1(3) \cdot \mathbf{r}_2'(3) + \mathbf{r}_1'(3) \cdot \mathbf{r}_2(3) = \langle 1, 1, 0 \rangle \cdot \langle 0, 2, 4 \rangle + \langle 0, 0, 1 \rangle \cdot \langle 1, 1, 0 \rangle = 2 + 0 = 2$$

15. Calculate $\displaystyle\int_0^3 \langle 4t + 3, t^2, -4t^3 \rangle \, dt$.

SOLUTION By the definition of vector-valued integration, we have

$$\int_0^3 \langle 4t + 3, t^2, -4t^3 \rangle \, dt = \left\langle \int_0^3 (4t + 3) \, dt, \int_0^3 t^2 \, dt, \int_0^3 -4t^3 \, dt \right\rangle \tag{1}$$

We compute the integrals on the right-hand side:

$$\int_0^3 (4t + 3) \, dt = 2t^2 + 3t \bigg|_0^3 = 2 \cdot 9 + 3 \cdot 3 - 0 = 27$$

$$\int_0^3 t^2 \, dt = \dfrac{t^3}{3} \bigg|_0^3 = \dfrac{3^3}{3} = 9$$

$$\int_0^3 -4t^3 \, dt = -t^4 \bigg|_0^3 = -3^4 = -81$$

Substituting in (1) gives the following integral:

$$\int_0^3 \langle 4t + 3, t^2, -4t^3 \rangle \, dt = \langle 27, 9, -81 \rangle$$

17. A particle located at $(1, 1, 0)$ at time $t = 0$ follows a path whose velocity vector is $\mathbf{v}(t) = \langle 1, t, 2t^2 \rangle$. Find the particle's location at $t = 2$.

SOLUTION We first find the path $\mathbf{r}(t)$ by integrating the velocity vector $\mathbf{v}(t)$:

$$\mathbf{r}(t) = \int \langle 1, t, 2t^2 \rangle \, dt = \left\langle \int 1 \, dt, \int t \, dt, \int 2t^2 \, dt \right\rangle = \left\langle t + c_1, \frac{1}{2}t^2 + c_2, \frac{2}{3}t^3 + c_3 \right\rangle$$

Denoting by $\mathbf{c} = \langle c_1, c_2, c_3 \rangle$ the constant vector, we obtain:

$$\mathbf{r}(t) = \left\langle t, \frac{1}{2}t^2, \frac{2}{3}t^3 \right\rangle + \mathbf{c} \tag{1}$$

To find the constant vector $\mathbf{c}$, we use the given information on the initial position of the particle. At time $t = 0$ it is at the point $(1, 1, 0)$. That is, by (1):

$$\mathbf{r}(0) = \langle 0, 0, 0 \rangle + \mathbf{c} = \langle 1, 1, 0 \rangle$$

or,

$$\mathbf{c} = \langle 1, 1, 0 \rangle$$

We substitute in (1) to obtain:

$$\mathbf{r}(t) = \left\langle t, \frac{1}{2}t^2, \frac{2}{3}t^3 \right\rangle + \langle 1, 1, 0 \rangle = \left\langle t + 1, \frac{1}{2}t^2 + 1, \frac{2}{3}t^3 \right\rangle$$

Finally, we substitute $t = 2$ to obtain the particle's location at $t = 2$:

$$\mathbf{r}(2) = \left\langle 2 + 1, \frac{1}{2} \cdot 2^2 + 1, \frac{2}{3} \cdot 2^3 \right\rangle = \left\langle 3, 3, \frac{16}{3} \right\rangle$$

At time $t = 2$ the particle is located at the point

$$\left(3, 3, \frac{16}{3} \right)$$

19. Calculate $\mathbf{r}(t)$, assuming that

$$\mathbf{r}''(t) = \langle 4 - 16t, 12t^2 - t \rangle, \qquad \mathbf{r}'(0) = \langle 1, 0 \rangle, \qquad \mathbf{r}(0) = \langle 0, 1 \rangle$$

SOLUTION Using componentwise integration we get:

$$\mathbf{r}'(t) = \int \langle 4 - 16t, 12t^2 - t \rangle \, dt$$

$$= \left\langle \int 4 - 16t \, dt, \int 12t^2 - t \, dt \right\rangle$$

$$= \left\langle 4t - 8t^2, 4t^3 - \frac{t^2}{2} \right\rangle + \mathbf{c}_1$$

Then using the initial condition $\mathbf{r}'(0) = \langle 1, 0 \rangle$ we get:

$$\mathbf{r}'(0) = \langle 1, 0 \rangle = \mathbf{c}_1$$

so then

$$\mathbf{r}'(t) = \left\langle 4t - 8t^2, 4t^3 - \frac{t^2}{2} \right\rangle + \langle 1, 0 \rangle = \left\langle 4t - 8t^2 + 1, 4t^3 - \frac{t^2}{2} \right\rangle$$

Then integrating componentwise once more we get:

$$\mathbf{r}(t) = \int \left\langle 4t - 8t^2 + 1, 4t^3 - \frac{t^2}{2} \right\rangle dt$$

$$= \left\langle \int 4t - 8t^2 + 1 \, dt, \int 4t^3 - \frac{t^2}{2} \, dt \right\rangle$$

$$= \left\langle 2t^2 - \frac{8}{3}t^3 + t, t^4 - \frac{t^3}{6} \right\rangle + \mathbf{c}_2$$

Using the initial condition $\mathbf{r}(0) = \langle 0, 1 \rangle$ we have:

$$\mathbf{r}(0) = \langle 0, 1 \rangle = \mathbf{c}_2$$

Therefore,

$$\mathbf{r}(t) = \left\langle 2t^2 - \frac{8}{3}t^3 + t, t^4 - \frac{t^3}{6} \right\rangle + \langle 0, 1 \rangle = \left\langle 2t^2 - \frac{8}{3}t^3 + t, t^4 - \frac{t^3}{6} + 1 \right\rangle$$

21. Compute the length of the path

$$\mathbf{r}(t) = \langle \sin 2t, \cos 2t, 3t - 1 \rangle \quad \text{for } 1 \leq t \leq 3$$

SOLUTION We use the formula for the arc length:

$$L = \int_1^3 \|\mathbf{r}'(t)\| \, dt \tag{1}$$

We compute the derivative vector $\mathbf{r}'(t)$ and its length:

$$\mathbf{r}'(t) = \langle 2 \cos 2t, -2 \sin 2t, 3 \rangle$$

$$\|\mathbf{r}'(t)\| = \sqrt{(2 \cos 2t)^2 + (-2 \sin 2t)^2 + 3^2} = \sqrt{4 \cos^2 2t + 4 \sin^2 2t + 9}$$

$$= \sqrt{4\left(\cos^2 2t + \sin^2 2t\right) + 9} = \sqrt{4 \cdot 1 + 9} = \sqrt{13}$$

We substitute in (1) and compute the integral to obtain the following length:

$$L = \int_1^3 \sqrt{13} \, dt = \sqrt{13}\, t \Big|_1^3 = 2\sqrt{13}.$$

23. Find an arc length parametrization of a helix of height 20 cm that makes four full rotations over a circle of radius 5 cm.

SOLUTION Since the radius is 5 cm and the height is 20 cm, the helix is traced by a parametrization of the form:

$$\mathbf{r}(t) = \langle 5 \cos at, 5 \sin at, t \rangle, \quad 0 \leq t \leq 20$$

Since the helix makes exactly 4 full rotations, we have:

$$a \cdot 20 = 4 \cdot 2\pi \quad \Rightarrow \quad a = \frac{2\pi}{5}$$

The parametrization of the helix is, thus:

$$\mathbf{r}(t) = \left\langle 5 \cos \frac{2\pi t}{5}, 5 \sin \frac{2\pi t}{5}, t \right\rangle, \quad 0 \leq t \leq 20$$

The helix is shown in the following figure:

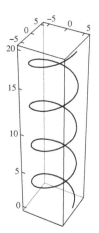

To find the length of the helix, we use the arc length formula:

$$L = \int_0^{20} \|\mathbf{r}'(t)\| \, dt \tag{1}$$

We find $\mathbf{r}'(t)$ and its length:

$$\mathbf{r}'(t) = \left\langle -5 \cdot \frac{2\pi}{5} \sin \frac{2\pi t}{5}, 5 \cdot \frac{2\pi}{5} \cos \frac{2\pi t}{5}, 1 \right\rangle = \left\langle -2\pi \sin \frac{2\pi t}{5}, 2\pi \cos \frac{2\pi t}{5}, 1 \right\rangle$$

$$\|\mathbf{r}'(t)\| = \sqrt{4\pi^2 \sin^2 \frac{2\pi t}{5} + 4\pi^2 \cos^2 \frac{2\pi t}{5} + 1} = \sqrt{4\pi^2 \left(\sin^2 \frac{2\pi t}{5} + \cos^2 \frac{2\pi t}{5} \right) + 1} = \sqrt{1 + 4\pi^2}$$

Substituting in (1) we get:

$$L = \int_0^{20} \sqrt{1 + 4\pi^2} \, dt = 20 \sqrt{1 + 4\pi^2} \approx 127.2$$

25. A projectile fired at an angle of 60° lands 400 m away. What was its initial speed?

SOLUTION Place the projectile at the origin, and let $\mathbf{r}(t)$ be the position vector of the projectile.

Step 1. Use Newton's Law

Gravity exerts a downward force of magnitude mg, where m is the mass of the bullet and $g = 9.8$ m/s^2. In vector form,

$$\mathbf{F} = \langle 0, -mg \rangle = m \langle 0, -g \rangle$$

Newton's Second Law $\mathbf{F} = m\mathbf{r}''(t)$ yields $m \langle 0, -g \rangle = m\mathbf{r}''(t)$ or $\mathbf{r}''(t) = \langle 0, -g \rangle$. We determine $\mathbf{r}(t)$ by integrating twice:

$$\mathbf{r}'(t) = \int_0^t \mathbf{r}''(u) \, du = \int_0^t \langle 0, -g \rangle \, du = \langle 0, -gt \rangle + \mathbf{v}_0$$

$$\mathbf{r}(t) = \int_0^t \mathbf{r}'(u) \, du = \int_0^t (\langle 0, -gu \rangle + \mathbf{v}_0) \, du = \left\langle 0, -\frac{1}{2} g t^2 \right\rangle + t\mathbf{v}_0 + \mathbf{r}_0$$

Step 2. Use the initial conditions

By our choice of coordinates, $\mathbf{r}_0 = \mathbf{0}$. The initial velocity $\mathbf{v}_0$ has unknown magnitude v_0, but we know that it points in the direction of the unit vector $\langle \cos 60°, \sin 60° \rangle$. Therefore,

$$\mathbf{v}_0 = v_0 \langle \cos 60°, \sin 60° \rangle = v_0 \left\langle \frac{1}{2}, \frac{\sqrt{3}}{2} \right\rangle$$

$$\mathbf{r}(t) = \left\langle 0, -\frac{1}{2} g t^2 \right\rangle + t v_0 \left\langle \frac{1}{2}, \frac{\sqrt{3}}{2} \right\rangle$$

Step 3. Solve for v_0.

The projectile hits the point $\langle 400, 0 \rangle$ on the ground if there exists a time t such that $\mathbf{r}(t) = \langle 400, 0 \rangle$; that is,

$$\left\langle 0, -\frac{1}{2} g t^2 \right\rangle + t v_0 \left\langle \frac{1}{2}, \frac{\sqrt{3}}{2} \right\rangle = \langle 400, 0 \rangle$$

Equating components, we obtain

$$\frac{1}{2} t v_0 = 400, \quad -\frac{1}{2} g t^2 + \frac{\sqrt{3}}{2} t v_0 = 0$$

The first equation yields $t = \frac{800}{v_0}$. Now substitute in the second equation and solve, using $g = 9.8$ m/s^2:

$$-4.9 \left(\frac{800}{v_0} \right)^2 + \frac{\sqrt{3}}{2} \left(\frac{800}{v_0} \right) v_0 = 0$$

$$\left(\frac{800}{v_0} \right)^2 = \frac{400 \sqrt{3}}{4.9}$$

$$\left(\frac{v_0}{800}\right)^2 = \frac{4.9}{400\sqrt{3}} \approx 0.00707$$

$$v_0^2 = 4526.42611, \quad v_0 \approx 67.279 \text{ m/s}$$

We obtain $v_0 \approx 67.279$ m/s.

27. During a short time interval $[0.5, 1.5]$, the path of an unmanned spy plane is described by

$$\mathbf{r}(t) = \left\langle -\frac{100}{t^2}, 7 - t, 40 - t^2 \right\rangle$$

A laser is fired (in the tangential direction) toward the yz-plane at time $t = 1$. Which point in the yz-plane does the laser beam hit?

SOLUTION Notice first that by differentiating we get the tangent vector:

$$\mathbf{r}'(t) = \left\langle \frac{200}{t^3}, -1, -2t \right\rangle, \quad \Rightarrow \quad \mathbf{r}'(1) = \langle 200, -1, -2 \rangle$$

and the tangent line to the path would be:

$$\ell(s) = \mathbf{r}(1) + s\mathbf{r}'(1) = \langle -100, 6, 39 \rangle + s\langle 200, -1, -2 \rangle = \langle -100 + 200s, 6 - s, 39 - 2s \rangle$$

If the laser is fired in the tangential direction toward the yz-plane means that the x-coordinate will be zero—this is when $s = 1/2$. Therefore,

$$\ell(1/2) = \langle 0, 11/2, 38 \rangle$$

Hence, the laser beam will hit the point $(0, 11/2, 38)$.

29. Find the unit tangent vector to $\mathbf{r}(t) = \langle \sin t, t, \cos t \rangle$ at $t = \pi$.

SOLUTION The unit tangent vector at $t = \pi$ is

$$\mathbf{T}(\pi) = \frac{\mathbf{r}'(\pi)}{\|\mathbf{r}'(\pi)\|} \tag{1}$$

We differentiate $\mathbf{r}(t)$ componentwise to obtain:

$$\mathbf{r}'(t) = \langle \cos t, 1, -\sin t \rangle$$

Therefore,

$$\mathbf{r}'(\pi) = \langle \cos \pi, 1, -\sin \pi \rangle = \langle -1, 1, 0 \rangle$$

We compute the length of $\mathbf{r}'(\pi)$:

$$\|\mathbf{r}'(\pi)\| = \sqrt{(-1)^2 + 1^2 + 0^2} = \sqrt{2}$$

Substituting in (1) gives:

$$\mathbf{T}(\pi) = \left\langle \frac{-1}{\sqrt{2}}, \frac{1}{\sqrt{2}}, 0 \right\rangle$$

31. Calculate $\kappa(1)$ for $\mathbf{r}(t) = \langle \ln t, t \rangle$.

SOLUTION Recall,

$$\kappa(t) = \frac{\|\mathbf{r}'(t) \times \mathbf{r}''(t)\|}{\|\mathbf{r}'(t)\|^3}$$

Computing derivatives we get:

$$\mathbf{r}'(t) = \left\langle \frac{1}{t}, 1 \right\rangle, \quad \Rightarrow \quad \mathbf{r}'(1) = \langle 1, 1 \rangle, \quad \Rightarrow \quad \|\mathbf{r}'(1)\| = \sqrt{2}$$

$$\mathbf{r}''(t) = \left\langle -\frac{1}{t^2}, 0 \right\rangle, \quad \Rightarrow \quad \mathbf{r}''(1) = \langle -1, 0 \rangle$$

Computing the cross product we get:

$$\mathbf{r}'(1) \times \mathbf{r}''(1) = \begin{vmatrix} \mathbf{i} & \mathbf{j} & \mathbf{k} \\ 1 & 1 & 0 \\ -1 & 0 & 0 \end{vmatrix} = \langle 0, 0, 1 \rangle$$

and $\|\mathbf{r}'(1) \times \mathbf{r}''(1)\| = 1$. Therefore,

$$\kappa(1) = \frac{\|\mathbf{r}'(1) \times \mathbf{r}''(1)\|}{\|\mathbf{r}'(1)\|^3} = \frac{1}{(\sqrt{2})^3} = \frac{1}{2^{3/2}}$$

In Exercises 33 and 34, write the acceleration vector **a** *at the point indicated as a sum of tangential and normal components.*

33. $\mathbf{r}(\theta) = \langle \cos \theta, \sin 2\theta \rangle$, $\theta = \frac{\pi}{4}$

SOLUTION

Step 1. Compute **T** and a_T. Since $\mathbf{r}(\theta) = \langle \cos \theta, \sin 2\theta \rangle$, we have $\mathbf{r}'(\theta) = \langle -\sin \theta, 2 \cos 2\theta \rangle$ and $\mathbf{r}''(\theta) = \langle -\cos \theta, -4 \sin 2\theta \rangle$. Thus, at $\theta = \frac{\pi}{4}$,

$$\mathbf{v} = \mathbf{r}'\left(\frac{\pi}{4}\right) = \left\langle -\sin \frac{\pi}{4}, 2 \cos \frac{\pi}{2} \right\rangle = \left\langle -\frac{1}{\sqrt{2}}, 0 \right\rangle$$

$$\mathbf{T} = \frac{\mathbf{v}}{\|\mathbf{v}\|} = \frac{\left\langle -\frac{1}{\sqrt{2}}, 0 \right\rangle}{\sqrt{\left(\frac{1}{\sqrt{2}}\right)^2 + 0}} = \sqrt{2}\left\langle -\frac{1}{\sqrt{2}}, 0 \right\rangle = \langle -1, 0 \rangle$$

$$\mathbf{a} = \mathbf{r}''\left(\frac{\pi}{4}\right) = \left\langle -\cos \frac{\pi}{4}, -4 \sin \frac{\pi}{2} \right\rangle = \left\langle -\frac{1}{\sqrt{2}}, -4 \right\rangle$$

$$a_\mathrm{T} = \mathbf{a} \cdot \mathbf{T} = \left\langle -\frac{1}{\sqrt{2}}, -4 \right\rangle \cdot \langle -1, 0 \rangle = \frac{1}{\sqrt{2}}$$

Step 2. Compute a_N and **N**. We have

$$a_\mathrm{N}\mathbf{N} = \mathbf{a} - a_\mathrm{T}\mathbf{T} = \left\langle -\frac{1}{\sqrt{2}}, -4 \right\rangle - \frac{1}{\sqrt{2}}\langle -1, 0 \rangle = \langle 0, -4 \rangle$$

Since **N** is a unit vector, we can find a_N by:

$$a_\mathrm{N} = \|a_\mathrm{N}\mathbf{N}\| = \|\langle 0, -4 \rangle\| = 4$$

Therefore,

$$\mathbf{N} = \frac{a_\mathrm{N}\mathbf{N}}{a_\mathrm{N}} = \frac{\langle 0, -4 \rangle}{4} = \langle 0, -1 \rangle$$

Step 3. Write the decomposition. We found that $a_\mathrm{T} = \frac{1}{\sqrt{2}}$ and $a_\mathrm{N} = 4$, hence the decomposition of **a** is

$$\mathbf{a} = a_\mathrm{T}\mathbf{T} + a_\mathrm{N}\mathbf{N} = \frac{1}{\sqrt{2}}\mathbf{T} + 4\mathbf{N}$$

where $\mathbf{T} = \langle -1, 0 \rangle$ and $\mathbf{N} = \langle 0, -1 \rangle$.

35. At a certain time t_0, the path of a moving particle is tangent to the *y*-axis in the positive direction. The particle's speed at time t_0 is 4 m/s, and its acceleration vector is $\mathbf{a} = \langle 5, 4, 12 \rangle$. Determine the curvature of the path at t_0.

SOLUTION We are given that the particle is moving tangent to the *y*-axis with speed 4 m/s, so then:

$$\mathbf{r}' = \langle 0, 4, 0 \rangle$$

and $\mathbf{a} = \mathbf{r}'' = \langle 5, 4, 12 \rangle$. Recall the formula for curvature:

$$\kappa = \frac{\|\mathbf{r}' \times \mathbf{r}''\|}{\|\mathbf{r}'\|^3}$$

First calculate the cross product:

$$\mathbf{r}' \times \mathbf{r}'' = \begin{vmatrix} \mathbf{i} & \mathbf{j} & \mathbf{k} \\ 0 & 4 & 0 \\ 5 & 4 & 12 \end{vmatrix} = \langle 48, 0, -20 \rangle$$

Then the length of $\mathbf{r}'$ and $\mathbf{r}' \times \mathbf{r}''$:

$$\|\mathbf{r}'\| = 4, \quad \|\mathbf{r}' \times \mathbf{r}''\| = \sqrt{48^2 + 20^2} = \sqrt{2704} = 52$$

so then for curvature we get:

$$\kappa = \frac{\|\mathbf{r}' \times \mathbf{r}''\|}{\|\mathbf{r}'\|^3} = \frac{52}{4^3} = \frac{13}{16}$$

37. Give an equation for the osculating circle to $y = \sqrt{x}$ at $x = 4$.

SOLUTION Let $f(x) = \sqrt{x}$. Applying equation (5) in the text to $f(x)$ gives for the curvature

$$\kappa(x) = \frac{|f''(x)|}{(1 + f'(x)^2)^{3/2}} = \frac{\left|-\frac{1}{4}x^{-3/2}\right|}{\left(1 + \left(\frac{1}{2}x^{-1/2}\right)^2\right)^{3/2}}$$

$$\kappa(1) = \frac{\left|-\frac{1}{4 \cdot 4^{3/2}}\right|}{\left(1 + \left(\frac{1}{2 \cdot 4^{1/2}}\right)^2\right)^{3/2}} = \frac{1/32}{(17/16)^{3/2}} = \frac{2}{17^{3/2}}$$

The osculating circle has radius $R = \frac{1}{\kappa(1)} = \frac{1}{2}17^{3/2}$.

Parametrize the curve by

$$\mathbf{r}(y) = \langle f^{-1}(y), y \rangle = \langle y^2, y \rangle$$

To find a normal to the curve at the given point, note that the tangent vector is $\mathbf{r}'(y) = \langle 2y, 1 \rangle$, so two normal vectors are $\langle 1, -2y \rangle$ and $\langle -1, 2y \rangle$. At $x = 4$, $y = 2$, these are $\langle 1, -4 \rangle$ and $\langle -1, 4 \rangle$. Since the curve opens down, we want the normal vector that points downwards and to the right, so we choose $\mathbf{N}(1)$ to be the unit vector in the direction $\langle 1, -4 \rangle$: $\mathbf{N}(1) = \frac{1}{\sqrt{17}}\langle 1, -4 \rangle$. To find the center, we apply equation (9) in the text:

$$\mathbf{r}(4) + \frac{1}{\kappa(1)}\mathbf{N}(1) = \langle 4, 2 \rangle + \frac{17^{3/2}}{2} \cdot \frac{1}{\sqrt{17}}\langle 1, -4 \rangle = \left\langle \frac{25}{2}, -32 \right\rangle$$

The osculating circle has radius $\frac{1}{2}17^{3/2}$, so it can be parametrized as

$$\mathbf{c}(t) = \left\langle \frac{25}{2}, -32 \right\rangle + \frac{1}{2}17^{3/2}\langle \cos t, \sin t \rangle$$

39. Let $\mathbf{r}(t) = \left\langle \ln t, t, \frac{t^2}{2} \right\rangle$. Find the equation of the osculating plane corresponding to $t = 1$.

SOLUTION Differentiating gives $\mathbf{r}'(t) = \left\langle \frac{1}{t}, 1, t \right\rangle$, so that

$$\mathbf{T}(t) = \frac{\mathbf{r}'(t)}{\|\mathbf{r}'(t)\|} = \frac{1}{\sqrt{\frac{1}{t^2} + 1 + t^2}}\left\langle \frac{1}{t}, 1, t \right\rangle = \frac{t}{\sqrt{t^4 + t^2 + 1}}\left\langle \frac{1}{t}, 1, t \right\rangle = \frac{1}{\sqrt{t^4 + t^2 + 1}}\langle 1, t, t^2 \rangle$$

Substituting $t = 1$ in the above gives

$$\mathbf{T}(1) = \frac{1}{\sqrt{3}}\langle 1, 1, 1 \rangle$$

Differentiating the components of $\mathbf{T}(t)$ gives

$$\frac{d}{dt}\frac{1}{\sqrt{t^4 + t^2 + 1}} = -\frac{2t^3 + t}{(t^4 + t^2 + 1)^{3/2}}$$

$$\frac{d}{dt}\frac{t}{\sqrt{t^4 + t^2 + 1}} = \frac{(t^4 + t^2 + 1)^{1/2}(1) - t \cdot \frac{1}{2}(t^4 + t^2 + 1)^{-1/2} \cdot (4t^3 + 2t)}{t^4 + t^2 + 1}$$

$$= \frac{(t^4+t^2+1)-(2t^4+t^2)}{(t^4+t^2+1)^{3/2}} = \frac{1-t^4}{(t^4+t^2+1)^{3/2}}$$

$$\frac{d}{dt}\frac{t^2}{\sqrt{t^4+t^2+1}} = \frac{(t^4+t^2+1)^{1/2}(2t)-t^2\cdot\frac{1}{2}(t^4+t^2+1)^{-1/2}\cdot(4t^3+2t)}{t^4+t^2+1}$$

$$= \frac{(t^4+t^2+1)(2t)-(2t^5+t^3)}{(t^4+t^2+1)^{3/2}}$$

$$= \frac{t^3+2t}{(t^4+t^2+1)^{3/2}}$$

so that

$$\mathbf{T}'(t)\frac{1}{(t^4+t^2+1)^{3/2}}\left\langle -2t^3-t, 1-t^4, t^3+2t\right\rangle$$

At $t = 1$ we get

$$\mathbf{T}'(1) = \frac{1}{3\sqrt{3}}\langle -3,0,3\rangle = \frac{1}{\sqrt{3}}\langle -1,0,1\rangle$$

Then

$$\mathbf{N}(1) = \frac{\mathbf{T}'(1)}{\|\mathbf{T}'(1)\|} = \frac{\frac{1}{\sqrt{3}}\langle -1,0,1\rangle}{\frac{1}{\sqrt{3}}\sqrt{(-1)^2+0^2+1^2}} = \frac{1}{\sqrt{2}}\langle -1,0,1\rangle$$

$$\mathbf{B}(1) = \mathbf{T}(1)\times\mathbf{N}(1) = \frac{1}{\sqrt{3}}\frac{1}{\sqrt{2}}\begin{vmatrix} \mathbf{i} & \mathbf{j} & \mathbf{k} \\ 1 & 1 & 1 \\ -1 & 0 & 1 \end{vmatrix} = \frac{1}{\sqrt{6}}\langle 1,-2,1\rangle$$

The binormal vector $\mathbf{B}(1)$ is normal to the osculating plane at $\mathbf{r}(1) = \left(0,1,\frac{1}{2}\right)$; for simplicity, we use instead the normal vector $\sqrt{6}\mathbf{B}(1) = \langle 1,-2,1\rangle$. Then the equation of the osculating plane is

$$1(x-0)-2(y-1)+1\left(z-\frac{1}{2}\right) = 0, \text{ or } 2x-4y+2z = -3$$

41. Suppose the orbit of a planet is an ellipse of eccentricity $e = c/a$ and period T (Figure 2). Use Kepler's Second Law to show that the time required to travel from A' to B' is equal to

$$\left(\frac{1}{4}+\frac{e}{2\pi}\right)T$$

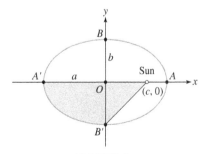

FIGURE 2

SOLUTION By the Law of Equal Areas, the position vector pointing from the sun to the planet sweeps out equal areas in equal times. We denote by S_1 the area swept by the position vector when the planet moves from A' to B', and t is the desired time. Since the position vector sweeps out the whole area of the ellipse (πab) in time T, the Law of Equal Areas implies that:

$$\frac{S_1}{\pi ab} = \frac{t}{T} \quad\Rightarrow\quad t = \frac{TS_1}{\pi ab} \tag{1}$$

We now find the area S_1 as the sum of the area of a quarter of the ellipse and the area of the triangle ODB. That is,

$$S_1 = \frac{\pi ab}{4} + \frac{\overline{OD} \cdot \overline{OB'}}{2} = \frac{\pi ab}{4} + \frac{cb}{2} = \frac{b}{4}(\pi a + 2c)$$

Substituting in (1) we get:

$$t = \frac{Tb(\pi a + 2c)}{4\pi ab} = \frac{T(\pi a + 2c)}{4\pi a} = T\left(\frac{1}{4} + \frac{1}{2\pi}\frac{c}{a}\right) = T\left(\frac{1}{4} + \frac{e}{2\pi}\right)$$

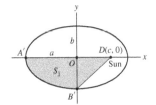

14 DIFFERENTIATION IN SEVERAL VARIABLES

14.1 Functions of Two or More Variables (LT Section 15.1)

Preliminary Questions

1. What is the difference between a horizontal trace and a level curve? How are they related?

SOLUTION A horizontal trace at height c consists of all points (x, y, c) such that $f(x, y) = c$. A level curve is the curve $f(x, y) = c$ in the xy-plane. The horizontal trace is in the $z = c$ plane. The two curves are related in the sense that the level curve is the projection of the horizontal trace on the xy-plane. The two curves have the same shape but they are located in parallel planes.

2. Describe the trace of $f(x, y) = x^2 - \sin(x^3 y)$ in the xz-plane.

SOLUTION The intersection of the graph of $f(x, y) = x^2 - \sin(x^3 y)$ with the xz-plane is obtained by setting $y = 0$ in the equation $z = x^2 - \sin(x^3 y)$. We get the equation $z = x^2 - \sin 0 = x^2$. This is the parabola $z = x^2$ in the xz-plane.

3. Is it possible for two different level curves of a function to intersect? Explain.

SOLUTION Two different level curves of $f(x, y)$ are the curves in the xy-plane defined by equations $f(x, y) = c_1$ and $f(x, y) = c_2$ for $c_1 \neq c_2$. If the curves intersect at a point (x_0, y_0), then $f(x_0, y_0) = c_1$ and $f(x_0, y_0) = c_2$, which implies that $c_1 = c_2$. Therefore, two different level curves of a function do not intersect.

4. Describe the contour map of $f(x, y) = x$ with contour interval 1.

SOLUTION The level curves of the function $f(x, y) = x$ are the vertical lines $x = c$. Therefore, the contour map of f with contour interval 1 consists of vertical lines so that every two adjacent lines are distanced one unit from another.

5. How will the contour maps of

$$f(x, y) = x \quad \text{and} \quad g(x, y) = 2x$$

with contour interval 1 look different?

SOLUTION The level curves of $f(x, y) = x$ are the vertical lines $x = c$, and the level curves of $g(x, y) = 2x$ are the vertical lines $2x = c$ or $x = \frac{c}{2}$. Therefore, the contour map of $f(x, y) = x$ with contour interval 1 consists of vertical lines with distance 1 unit between adjacent lines, whereas in the contour map of $g(x, y) = 2x$ (with contour interval 1) the distance between two adjacent vertical lines is $\frac{1}{2}$.

Exercises

In Exercises 1–6, at each point evaluate the function or indicate that the function is undefined there.

1. $f(x, y) = x + yx^3$, $(1, 2), (-1, 6), (e, \pi)$

SOLUTION We substitute the values for x and y in $f(x, y)$ and compute the values of f at the given points. This gives

$$f(1, 2) = 1 + 2 \cdot 1^3 = 3$$
$$f(-1, 6) = -1 + 6 \cdot (-1)^3 = -7$$
$$f(e, \pi) = e + \pi e^3$$

3. $h(x, y) = \dfrac{\sqrt{x - y^2}}{x - y}$, $(20, 2), (1, -2), (1, 1)$

SOLUTION We substitute the values for x and y in $h(x,y)$ and compute the values of h at the given points. This gives

$$h(20, 2) = \frac{\sqrt{20 - 2^2}}{20 - 2} = \frac{4}{18} = \frac{2}{9}$$

$$h(1, -2) = \frac{\sqrt{1 - (-2)^2}}{1 - (-2)} \quad \text{(Numerator is } \sqrt{-3}, \text{ so } h(1, -2) \text{ is undefined in real numbers)}$$

$$h(1, 1) = \frac{\sqrt{1 - 1^2}}{1 - 1} \quad \text{(Numerator and denominator are both zero, so } h(1, 1) \text{ is undefined)}$$

5. $h(x, y, z) = xyz^{-2}$, $(3, 7, -2), (3, 2, \frac{1}{4}), (4, -4, 0)$

SOLUTION We substitute the values for x, y, and z in $h(x, y, z)$ and compute the values of h at the given points. This gives

$$h(3, 7, 2) = 3 \cdot 7 \cdot (-2)^{-2} = 3 \cdot 7 \cdot \frac{1}{4} = \frac{21}{4}$$

$$h\left(3, 2, \frac{1}{4}\right) = 3 \cdot 2 \cdot \left(\frac{1}{4}\right)^{-2} = 6 \cdot 16 = 96$$

$$h(4, -4, 0) = 4 \cdot (-4) \cdot 0^{-2} \quad (0^{-2} \text{ is undefined, so } h(4, -4, 0) \text{ is undefined})$$

In Exercises 7–14, sketch the domain of the function.

7. $f(x, y) = 12x - 5y$

SOLUTION The function is defined for all x and y, hence the domain is the entire xy-plane.

9. $f(x, y) = \ln(4x^2 - y)$

SOLUTION The function is defined if $4x^2 - y > 0$, that is, $y < 4x^2$. The domain is the region in the xy-plane that is below the parabola $y = 4x^2$.

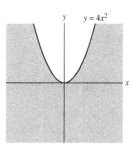

11. $g(y, z) = \dfrac{1}{z + y^2}$

SOLUTION The function is defined if $z + y^2 \neq 0$, that is, $z \neq -y^2$. The domain is the (y, z) plane with the parabola $z = -y^2$ excluded.

$$D = \{(y, z) : z \neq -y^2\}$$

$z + y^2 \neq 0$

13. $F(I, R) = \sqrt{IR}$

SOLUTION The function is defined if $IR \geq 0$. Therefore the domain is the first and the third quadrants of the IR-plane including both axes.

$IR \geq 0$

In Exercises 15–18, describe the domain and range of the function.

15. $f(x, y, z) = xz + e^y$

SOLUTION The domain of f is the entire (x, y, z)-space. Since f takes all the real values, the range is the entire real line.

17. $P(r, s, t) = \sqrt{16 - r^2 s^2 t^2}$

SOLUTION The domain is the subset of $\mathbf{R}^3$ where $r^2 s^2 t^2 \leq 16$, or $|rst| \leq 4$. The range is $\{w : 0 \leq w \leq 4\}$ since the values of P clearly range from $\sqrt{16 - 16} = 0$ to $\sqrt{16 - 0} = 4$.

19. Match graphs (A) and (B) in Figure 18 with the functions:

(i) $f(x, y) = -x + y^2$

(ii) $g(x, y) = x + y^2$

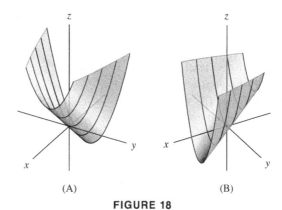

FIGURE 18

SOLUTION

(i) The vertical trace for $f(x, y) = -x + y^2$ in the xz-plane ($y = 0$) is $z = -x$. This matches the graph shown in (B).

(ii) The vertical trace for $f(x, y) = x + y^2$ in the xz-plane ($y = 0$) is $z = x$. This matches the graph show in (A).

21. Match the functions (a)–(f) with their graphs (A)–(F) in Figure 20.

(a) $f(x, y) = |x| + |y|$

(b) $f(x, y) = \cos(x - y)$

(c) $f(x, y) = \dfrac{-1}{1 + 9x^2 + y^2}$

(d) $f(x, y) = \cos(y^2) e^{-0.1(x^2 + y^2)}$

(e) $f(x, y) = \dfrac{-1}{1 + 9x^2 + 9y^2}$

(f) $f(x, y) = \cos(x^2 + y^2) e^{-0.1(x^2 + y^2)}$

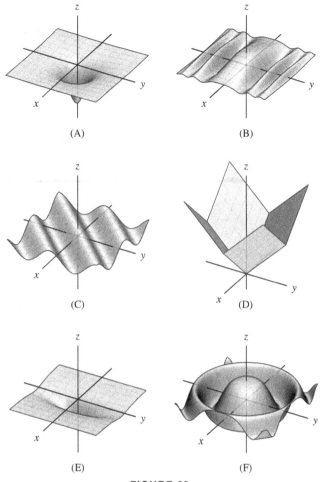

FIGURE 20

SOLUTION

(a) $|x| + |y|$. The level curves are $|x| + |y| = c$, $y = c - |x|$, or $y = -c + |x|$. The graph (D) corresponds to the function with these level curves.

(b) $\cos(x - y)$. The vertical trace in the plane $x = c$ is the curve $z = \cos(c - y)$ in the plane $x = c$. These traces correspond to the graph (C).

(c) $\dfrac{-1}{1 + 9x^2 + y^2}$ **(e)** $\dfrac{-1}{1 + 9x^2 + 9y^2}$. The level curves of the two functions are:

$$\dfrac{-1}{1 + 9x^2 + y^2} = c \qquad \dfrac{-1}{1 + 9x^2 + 9y^2} = c$$

$$1 + 9x^2 + y^2 = -\dfrac{1}{c} \qquad 1 + 9x^2 + 9y^2 = -\dfrac{1}{c}$$

$$9x^2 + y^2 = -1 - \dfrac{1}{c} \qquad 9x^2 + 9y^2 = -1 - \dfrac{1}{c}$$

$$x^2 + y^2 = -\dfrac{1 + c}{9c}$$

For suitable values of c, the level curves of the function in (c) are ellipses as in (E), and the level curves of the function (e) are circles as in (A).

(d) $\cos(x^2)e^{-1/(x^2+y^2)}$

(f) $\cos(x^2 + y^2)e^{-1/(x^2+y^2)}$. The value of $|z|$ is decreasing to zero as x or y are decreasing, hence the possible graphs are (B) and (F).

In (f), z is constant whenever $x^2 + y^2$ is constant, that is, z is constant whenever (x, y) varies on a circle. Hence (f) corresponds to the graph (F) and (d) corresponds to (B). To summarize, we have the following matching:

(a) ↔ (D)   (b) ↔ (C)   (c) ↔ (E)

(d) ↔ (B)   (e) ↔ (A)   (f) ↔ (F)

In Exercises 23–28, sketch the graph and draw several vertical and horizontal traces.

23. $f(x, y) = 12 - 3x - 4y$

SOLUTION The graph of $f(x, y) = 12 - 3x - 4y$ is shown in the figure:

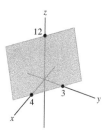

The horizontal trace at height c is the line $12 - 3x - 4y = c$ or $3x + 4y = 12 - c$ in the plane $z = c$.

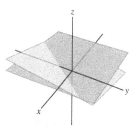

The vertical traces obtained by setting $x = a$ or $y = a$ are the lines $z = (12 - 3a) - 4y$ and $z = -3x + (12 - 4a)$ in the planes $x = a$ and $y = a$, respectively.

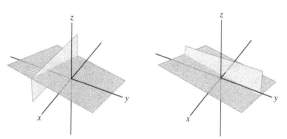

25. $f(x, y) = x^2 + 4y^2$

SOLUTION The graph of the function is shown in the figure:

The horizontal trace at height c is the curve $x^2 + 4y^2 = c$, where $c \geq 0$ (if $c = 0$, it is the origin). The horizontal traces are ellipses for $c > 0$.

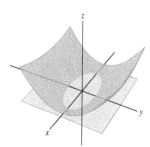

The vertical trace in the plane $x = a$ is the parabola $z = a^2 + 4y^2$, and the vertical trace in the plane $y = a$ is the parabola $z = x^2 + 4a^2$.

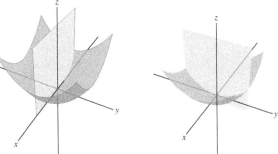

27. $f(x, y) = \sin(x - y)$

SOLUTION The graph of $f(x, y) = \sin(x - y)$ is shown in the figure:

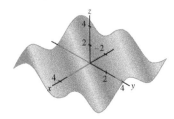

The horizontal trace at the height $z = c$ is $\sin(x - y) = c$ (we could also write $x - y = \sin^{-1}(c)$ or $y = x - \sin^{-1}(c)$). The trace consists of multiple lines all having slope 1, with y-intercepts separated by multiples of 2π.

The vertical trace in the plane $x = a$ is $\sin(a - y) = -\sin(y - a) = z$. This curve is a shifted sine curve reflected through the z-axis.

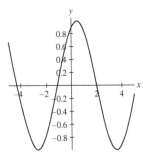

The vertical trace in the plane $y = a$ is $\sin(x - a) = z$. This curve is a shifted sine curve as well.

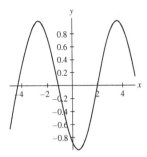

29. Sketch contour maps of $f(x, y) = x + y$ with contour intervals $m = 1$ and 2.

SOLUTION The level curves are $x + y = c$ or $y = c - x$. Using contour interval $m = 1$, we plot $y = c - x$ for various values of c.

Using contour interval $m = 2$, we plot $y = c - x$ for various values of c.

In Exercises 31–38, draw a contour map of $f(x, y)$ with an appropriate contour interval, showing at least six level curves.

31. $f(x, y) = x^2 - y$

SOLUTION The level curves are the parabolas $y = x^2 + c$. We draw a contour plot with contour interval $m = 1$, for $c = 0, 1, 2, 3, 4, 5$:

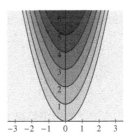

33. $f(x, y) = \dfrac{y}{x}$

SOLUTION The level curves are $\dfrac{y}{x} = c$ or $y = cx$. We plot $y = cx$ for $c = -2, -1, 0, 1, 2, 3$ using contour interval $m = 1$:

35. $f(x, y) = x^2 + 4y^2$

SOLUTION The level curves are $x^2 + 4y^2 = c$. These are ellipses centered at the origin in the xy-plane.

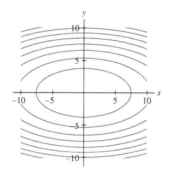

37. $f(x, y) = x^2$

SOLUTION The level curves are $x^2 = c$. For $c > 0$ these are the two vertical lines $x = \sqrt{c}$ and $x = -\sqrt{c}$ and for $c = 0$ it is the y-axis. We draw a contour map using contour interval $m = 4$ and $c = 0, 4, 8, 12, 16, 20$:

39. Find the linear function whose contour map (with contour interval $m = 6$) is shown in Figure 22. What is the linear function if $m = 3$ (and the curve labeled $c = 6$ is relabeled $c = 3$)?

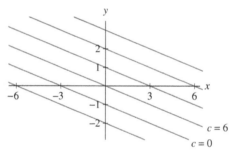

FIGURE 22 Contour map with contour interval $m = 6$.

SOLUTION

A linear function has the form $f(x, y) = Ax + By + C$.

According to the contour map, the level curve through the origin $(0, 0)$ has equation $f(x, y) = 6$. Therefore

$$f(0, 0) = A(0) + B(0) + C = 6 \quad \Rightarrow \quad C = 6$$

Next, we see from the contour map that the points $(-3, 0)$ and $f(0, -1)$ lie on the level curve $f(x, y) = 0$. Hence

$$f(-3, 0) = A(-3) + B(0) + 6 = 0 \quad \Rightarrow \quad A = 2$$
$$f(0, -1) = A(0) + B(-1) + 6 = 0 \quad \Rightarrow \quad B = 6$$

Therefore $f(x, y) = 2x + 6y + 6$.

Next, if we relabel $c = 6$ as $c = 3$ and consider a contour interval of $m = 3$, then $(0, 0)$ lies on the level curve $f(x, y) = 3$, and we proceed as before

$$f(0, 0) = A(0) + B(0) + C = 3 \quad \Rightarrow \quad C = 3$$
$$f(-3, 0) = A(-3) + B(0) + 3 = 0 \quad \Rightarrow \quad A = 1$$
$$f(0, -1) = A(0) + B(-1) + 3 = 0 \quad \Rightarrow \quad B = 2$$

Therefore $f(x, y) = x + 3y + 3$.

Exercises 41–43 refer to the map in Figure 24.

FIGURE 24 Atmospheric pressure (in millibars) over North America on March 26, 2009.

41. (a) At which of A–C is pressure increasing in the northern direction?

(b) At which of A–C is pressure increasing in the westerly direction?

SOLUTION

(a) At B and C the level curves go north-south, so the direction of pressure change is primarily east-west. At A, since the level curves go east-west, the pressure change occurs north-south. The level curves below A have a lower pressure than above, so the pressure increases in the northern direction.

(b) At C, level curves to the west have a lower pressure than those to the east, so the pressure *decreases* in the westerly direction. However, at B, the level curves to the west have a higher pressure than those to the east, so the pressure at B increases in the westerly direction.

43. Rank the following states in order from greatest change in pressure across the state to least: Arkansas, Colorado, North Dakota, Wisconsin.

SOLUTION For Arkansas, the lowest pressure (at the west edge of the state) is about 1004, and the highest (at the northeast corner) is about 1009, so the pressure increase is about 5 millibars. For Colorado, the lowest pressure (at the southeast corner) is about 1004, and the highest (at the northwest corner) is about 1016, so the pressure change is about 12 millibars. For North Dakota, the lowest pressure (at the southeast corner) is about 1020, while the greatest pressure (at the northwest corner) is about 1029, so the pressure change is about 9 millibars. Finally, for Wisconsin, the low pressure (at the northeast corner) is about 1012, while the high pressure (at the northwest corner) is about 1020, so the pressure change is about 8 millibars.

Thus ordered from most to least, the states are Colorado (≈ 12), North Dakota (≈ 9), Wisconsin (≈ 8), Arkansas (≈ 5).

In Exercises 44–47, let $T(x, y, z)$ denote temperature at each point in space. Draw level surfaces (also called isotherms) corresponding to the fixed temperatures given.

45. $T(x, y, z) = x - y + 2z$, $T = 0, 1, 2$

SOLUTION Each level surface of T is of the form $x - y + 2z = a$ for some constant a, so the level surfaces are planes. In each plot below, the axis intercepts are marked to make the graph clearer.

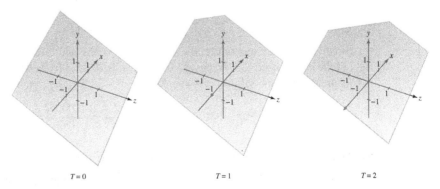

$T = 0$ $T = 1$ $T = 2$

47. $T(x, y, z) = x^2 - y^2 + z^2$, $T = 0, 1, 2, -1, -2$

SOLUTION Each level surface of T is of the form $x^2 - y^2 + z^2 = a$ for some constant a, so the level surfaces for $a > 0$ are hyperboloids of one sheet; for $a < 0$ they are hyperboloids of two sheets, and for $a = 0$ it is a cone. Note that all of these surfaces are oriented along the y-axis.

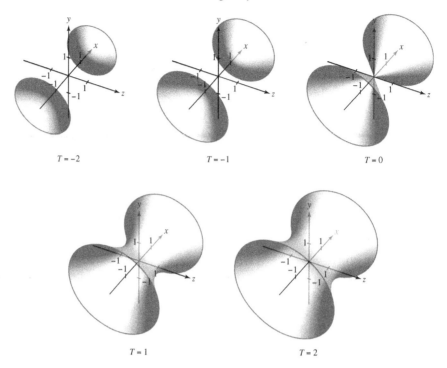

In Exercises 48–51, $\rho(S, T)$ is seawater density (kilograms per cubic meter) as a function of salinity S (parts per thousand) and temperature T (degrees Celsius). Refer to the contour map in Figure 25.

FIGURE 25 Contour map of seawater density $\rho(S, T)$ (kilograms per cubic meter).

49. Calculate the average rate of change of ρ with respect to S from B to C.

SOLUTION For fixed temperature, the segment $\overline{BC}$ spans one level curve, and the level curve of C is to the right of the level curve of B. Therefore, the change in density from B to C is $\Delta d = 0.0005$ kg/m^3. The salinity at C is greater than the salinity at B and $\Delta s = 33.6 - 32.7 = 0.9$ ppt. Therefore,

$$\text{Average ROC from } B \text{ to } C = \frac{\Delta d}{\Delta S} = \frac{0.0005}{0.9} = 0.000556 \text{ kg/m}^3 \cdot \text{ppt}$$

51. Does water density appear to be more sensitive to a change in temperature at point A or point B?

SOLUTION The two adjacent level curves are closer to the level curve of A than the corresponding two adjacent level curves are to the level curve of B. This suggests that water density is more sensitive to a change in temperature at A than at B.

In Exercises 52–55, refer to Figure 26.

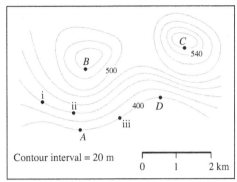

FIGURE 26

53. Estimate the average rate of change from A to B and from A to C.

SOLUTION The change in elevation from A to B is 140 m. The scale shows that $\overline{AB}$ is approximately 1800 m. Therefore,

$$\text{Average ROC from } A \text{ to } B = \frac{140}{1800} \approx 0.078$$

The change in elevation from A to C is obtained by multiplying the number of level curves between A and C, which is 8, by the contour interval 20 meters, giving $8 \cdot 20 = 160$ m. Using the scale, we approximate the distance $\overline{AC}$ by 3500 m. Therefore,

$$\text{Average ROC from } A \text{ to } C = \frac{160}{3500} \approx 0.046$$

55. Sketch the path of steepest ascent beginning at D.

SOLUTION Starting at D we draw a path that everywhere along the way points on the steepest direction, that is, moves as straight as possible from one level curve to the next. We obtain the following paths:

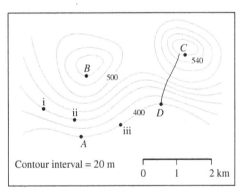

57. Let temperature in 3-space be given by $T(x, y, z) = \frac{x^2}{4} + \frac{y^2}{9} + z^2$. Draw isotherms corresponding to temperatures $T = 0, 1, 4$.

SOLUTION Each level surface of T is of the form $\frac{x^2}{4} + \frac{y^2}{9} + z^2 = a$ for some constant a, so the level surfaces are ellipsoids for positive values of a, empty for negative values of a, and consist of the origin if $a = 0$. Plots of the various surfaces are below:

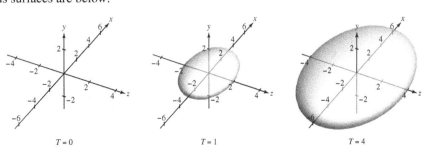

59. Let temperature in 3-space be given by $T(x, y, z) = x^2 - y^2 - z^2$. Draw isotherms corresponding to temperatures $T = -2, -1, 0, 1, 2$.

SOLUTION Each level surface of T is of the form $x^2 - y^2 - z^2 = a$ for some constant a. So for $a > 0$ the surfaces are hyperboloids of two sheets, while for $a < 0$ they are hyperboloids of one sheet. For $a = 0$ the level surface is a cone. Plots are below; in all cases the surfaces are oriented along the x-axis.

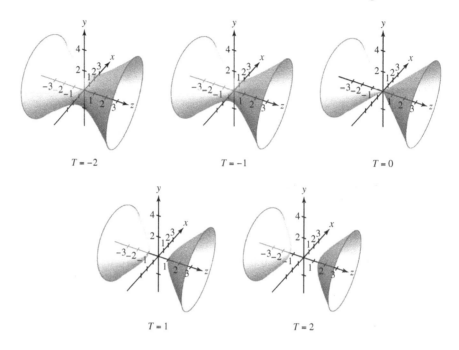

Further Insights and Challenges

61. Let

$$f(x, y) = \frac{x}{\sqrt{x^2 + y^2}} \text{ for} (x, y) \neq (0, 0)$$

Write f as a function $f(r, \theta)$ in polar coordinates, and use this to find the level curves of f.

SOLUTION In polar coordinates $x = r \cos \theta$ and $r = \sqrt{x^2 + y^2}$. Hence,

$$f(r, \theta) = \frac{r \cos \theta}{r} = \cos \theta$$

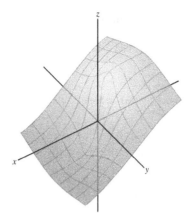

The level curves are the curves $\cos\theta = c$ in the $r\theta$-plane, for $|c| \le 1$. For $-1 < c < 1$, $c \ne 0$, the level curves $\cos\theta = c$ are the two rays $\theta = \cos^{-1} c$ and $\theta = -\cos^{-1} c$.

For $c = 0$, the level curve $\cos\theta = 0$ is the y-axis; for $c = 1$ the level curve $\cos\theta = 1$ is the nonnegative x-axis.

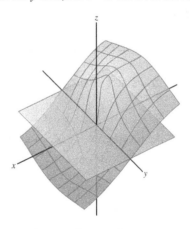

For $c = -1$, the level curve $\cos\theta = -1$ is the negative x-axis.

14.2 Limits and Continuity in Several Variables (LT Section 15.2)

Preliminary Questions

1. What is the difference between $D(P, r)$ and $D^*(P, r)$?

SOLUTION $D(P, r)$ is the open disk of radius r and center (a, b). It consists of all points distanced less than r from P, hence $D(P, r)$ includes the point P. $D^*(P, r)$ consists of all points in $D(P, r)$ other than P itself.

2. Suppose that $f(x, y)$ is continuous at $(2, 3)$ and that $f(2, y) = y^3$ for $y \ne 3$. What is the value $f(2, 3)$?

SOLUTION $f(x, y)$ is continuous at $(2, 3)$, hence the following holds:

$$f(2, 3) = \lim_{(x,y) \to (2,3)} f(x, y)$$

Since the limit exists, we may compute it by approaching $(2, 3)$ along the vertical line $x = 2$. This gives

$$f(2, 3) = \lim_{(x,y) \to (2,3)} f(x, y) = \lim_{y \to 3} f(2, y) = \lim_{y \to 3} y^3 = 3^3 = 27$$

We conclude that $f(2, 3) = 27$.

3. Suppose that $Q(x, y)$ is a function such that $1/Q(x, y)$ is continuous for all (x, y). Which of the following statements are true?

(a) $Q(x, y)$ is continuous for all (x, y).

(b) $Q(x, y)$ is continuous for $(x, y) \neq (0, 0)$.

(c) $Q(x, y) \neq 0$ for all (x, y).

SOLUTION All three statements are true. Let $f(x, y) = \frac{1}{Q(x,y)}$. Hence $Q(x, y) = \frac{1}{f(x,y)}$.

(a) Since f is continuous, Q is continuous whenever $f(x, y) \neq 0$. But by the definition of f it is never zero, therefore Q is continuous at all (x, y).

(b) Q is continuous everywhere including at $(0, 0)$.

(c) Since $f(x, y) = \frac{1}{Q(x,y)}$ is continuous, the denominator is never zero, that is, $Q(x, y) \neq 0$ for all (x, y). Moreover, there are no points where $Q(x, y) = 0$. (The equality $Q(x, y) = (0, 0)$ is meaningless since the range of Q consists of real numbers.)

4. Suppose that $f(x, 0) = 3$ for all $x \neq 0$ and $f(0, y) = 5$ for all $y \neq 0$. What can you conclude about $\lim_{(x,y) \to (0,0)} f(x, y)$?

SOLUTION We show that the limit $\lim_{(x,y) \to (0,0)} f(x, y)$ does not exist. Indeed, if the limit exists, it may be computed by approaching $(0, 0)$ along the x-axis or along the y-axis. We compute these two limits:

$$\lim_{\substack{(x,y) \to (0,0) \\ \text{along } y=0}} f(x, y) = \lim_{x \to 0} f(x, 0) = \lim_{x \to 0} 3 = 3$$

$$\lim_{\substack{(x,y) \to (0,0) \\ \text{along } x=0}} f(x, y) = \lim_{y \to 0} f(0, y) = \lim_{y \to 0} 5 = 5$$

Since the limits are different, $f(x, y)$ does not approach one limit as $(x, y) \to (0, 0)$, hence the limit $\lim_{(x,y) \to (0,0)} f(x, y)$ does not exist.

Exercises

In Exercises 1–8, evaluate the limit using continuity.

1. $\lim_{(x,y) \to (1,2)} (x^2 + y)$

SOLUTION Since the function $x^2 + y$ is continuous, we evaluate the limit by substitution:

$$\lim_{(x,y) \to (1,2)} (x^2 + y) = 1^2 + 2 = 3$$

3. $\lim_{(x,y) \to (-2,1)} (x^2 y - 3x^4 y^3)$

SOLUTION The function $x^2 y - 3x^4 y^3$ is a polynomial, so it is continuous everywhere. Thus we compute the limit by substitution:

$$\lim_{(x,y) \to (-2,1)} (x^2 y - 3x^4 y^3) = (-2)^2 \cdot 1 - 3 \cdot (-2)^4 \cdot 1^3 = -44$$

5. $\lim_{(x,y) \to (\frac{\pi}{4}, 0)} \tan x \cos y$

SOLUTION We use the continuity of $\tan x \cos y$ at the point $\left(\frac{\pi}{4}, 0\right)$ to evaluate the limit by substitution:

$$\lim_{(x,y) \to (\frac{\pi}{4}, 0)} \tan x \cos y = \tan \frac{\pi}{4} \cos 0 = 1 \cdot 1 = 1$$

7. $\lim_{(x,y) \to (1,1)} \frac{e^{x^2} - e^{-y^2}}{x + y}$

SOLUTION The function is the quotient of two continuous functions, and the denominator is not zero at the point $(1, 1)$. Therefore, the function is continuous at this point, and we may compute the limit by substitution:

$$\lim_{(x,y)\to(1,1)} \frac{e^{x^2} - e^{-y^2}}{x+y} = \frac{e^{1^2} - e^{-1^2}}{1+1} = \frac{e - \frac{1}{e}}{2} = \frac{1}{2}(e - e^{-1})$$

In Exercises 9–12, assume that

$$\lim_{(x,y)\to(2,5)} f(x,y) = 3, \qquad \lim_{(x,y)\to(2,5)} g(x,y) = 7$$

to find the limit.

9. $\lim_{(x,y)\to(2,5)} (g(x,y) - 2f(x,y))$

SOLUTION

$$\lim_{(x,y)\to(2,5)} (g(x,y) - 2f(x,y)) = 7 - 2(3) = 1$$

11. $\lim_{(x,y)\to(2,5)} e^{f(x,y)^2 - g(x,y)}$

SOLUTION

$$\lim_{(x,y)\to(2,5)} e^{f(x,y)^2 - g(x,y)} = e^{3^2 - 7} = e^2$$

13. Does $\lim_{(x,y)\to(0,0)} \frac{y^2}{x^2+y^2}$ exist? Explain.

SOLUTION This limit does not exist. Consider the following approaches to the point $(x, y) = (0, 0)$, first along the line $x = 0$ and second, along the line $y = x$.

First along the line $x = 0$ we calculate:

$$\lim_{(x,y)\to(0,0)} \frac{y^2}{x^2+y^2} = \lim_{y\to 0} \frac{y^2}{0^2+y^2} = \lim_{y\to 0} 1 = 1$$

Second, along the line $y = x$ we calculate:

$$\lim_{(x,y)\to(0,0)} \frac{y^2}{x^2+y^2} = \lim_{x\to 0} \frac{x^2}{x^2+x^2} = \lim_{x\to 0} \frac{1}{2} = \frac{1}{2}$$

Since these two limits are not equal, the limit in question, $\lim_{(x,y)\to(0,0)} \frac{y^2}{x^2+y^2}$ does not exist.

15. Let $f(x, y) = \frac{x^3 + y^3}{xy^2}$. Set $y = mx$ and show that the resulting limit depends on m, and therefore the limit $\lim_{(x,y)\to(0,0)} f(x,y)$ does not exist.

SOLUTION We have

$$f(x, mx) = \frac{x^3 + (mx)^3}{x(mx)^2} = \frac{(m^3 + 1)x^3}{m^2 x^3} = \frac{m^3 + 1}{m^2}$$

So f is constant along any line through the origin. Since $\frac{m^3+1}{m^2}$ is not a constant function of m, the limit of $f(x, mx)$ depends on m. For example,

$$m = 2: \quad \lim_{x\to 0} f(x, mx) = \lim_{x\to 0} \frac{2^3 + 1}{2^2} = \frac{9}{4}$$

$$m = 1: \quad \lim_{x\to 0} f(x, mx) = \lim_{x\to 0} \frac{1^3 + 1}{1^2} = 2$$

Therefore $\lim_{(x,y)\to(0,0)} f(x,y)$ does not exist.

17. Prove that
$$\lim_{(x,y)\to(0,0)} \frac{x}{x^2+y^2}$$
does not exist by considering the limit along the x-axis.

SOLUTION Compute this limit approaching $(x,y)=(0,0)$ along the x-axis ($y=0$):
$$\lim_{(x,y)\to(0,0)} \frac{x}{x^2+y^2} = \lim_{x\to 0} \frac{x}{x^2+0^2} = \lim_{x\to 0} \frac{1}{x}$$

This limit is known not to exist (it gets arbitrarily large from the right and arbitrarily small from the left), therefore the limit in question, $\lim_{(x,y)\to(0,0)} \frac{x}{x^2+y^2}$, also does not exist.

In Exercises 19–22, use any method to evaluate the limit or show that it does not exist.

19. $\displaystyle\lim_{(x,y)\to(0,0)} \frac{x^2-y^2}{\sqrt{x^2+y^2}}$

SOLUTION This limit exists and is equal to zero. Convert the function to polar coordinates:
$$\frac{x^2-y^2}{\sqrt{x^2+y^2}} = \frac{(r\cos\theta)^2 - (r\sin\theta)^2}{r} = r(\cos^2\theta - \sin^2\theta) = r\cos 2\theta$$

Then
$$0 \le \left|\frac{x^2-y^2}{\sqrt{x^2+y^2}}\right| = |r\cos 2\theta| = r$$

As $(x,y)\to(0,0)$, we also must have $r\to 0$, so that by the Squeeze Theorem
$$0 \le \lim_{(x,y)\to(0,0)} \left|\frac{x^2-y^2}{\sqrt{x^2+y^2}}\right| \le \lim_{r\to 0} r = 0$$

and therefore $\displaystyle\lim_{(x,y)\to(0,0)} \frac{x^2-y^2}{\sqrt{x^2+y^2}} = 0$.

21. $\displaystyle\lim_{(x,y)\to(0,0)} \frac{xy}{3x^2+2y^2}$

SOLUTION This limit does not exist. Consider the limit as $(x,y)\to(0,0)$ along the line $y=mx$:
$$\lim_{x\to 0} \frac{x(mx)}{3x^2+2(mx)^2} = \lim_{x\to 0} \frac{mx^2}{x^2(3+2m^2)} = \lim_{x\to 0} \frac{m}{3+2m^2}$$

This value is different for different values of m, so the limit depends on the path taken and $\displaystyle\lim_{(x,y)\to(0,0)} \frac{xy}{3x^2+2y^2}$ does not exist.

In Exercises 23–24, show that the limit does not exist by approaching the origin along one or more of the coordinate axes.

23. $\displaystyle\lim_{(x,y,z)\to(0,0,0)} \frac{x+y+z}{x^2+y^2+z^2}$

SOLUTION Consider the limit as $(x,y,z)\to(0,0,0)$ along the x-axis. Then $y=z=0$ and we get
$$\lim_{x\to 0} \frac{x+0+0}{x^2+0^2+0^2} = \lim_{x\to 0} \frac{x}{x^2} = \lim_{x\to 0} \frac{1}{x} = \infty$$

Since the limit does not exist along this path, $\displaystyle\lim_{(x,y,z)\to(0,0,0)} \frac{x+y+z}{x^2+y^2+z^2}$ does not exist.

25. Use the Squeeze Theorem to evaluate
$$\lim_{(x,y)\to(4,0)} (x^2-16)\cos\left(\frac{1}{(x-4)^2+y^2}\right)$$

SOLUTION Consider the following inequalities:

$$-1 \le \cos\left(\frac{1}{(x-4)^2 + y^2}\right) \le 1$$

Then for x such that $x \ge 4$ then $x^2 - 16 \ge 0$ and we have:

$$(-1)(x^2 - 16) \le (x^2 - 16)\cos\left(\frac{1}{(x-4)^2 + y^2}\right) \le (x^2 - 16)$$

$$\lim_{(x,y)\to(4,0)} (-1)(x^2 - 16) \le \lim_{(x,y)\to(4,0)} (x^2 - 16)\cos\left(\frac{1}{(x-4)^2 + y^2}\right) \le \lim_{(x,y)\to(4,0)} (x^2 - 16)$$

Then the two limits at the ends of the inequality are clearly equal to 0, by the Squeeze Theorem.

Now, if $x < 4$, then $x^2 - 16 < 0$ and we have:

$$(x^2 - 16) \le (x^2 - 16)\cos\left(\frac{1}{(x-4)^2 + y^2}\right) \le (-1)(x^2 - 16)$$

$$\lim_{(x,y)\to(4,0)} (x^2 - 16) \le \lim_{(x,y)\to(4,0)} (x^2 - 16)\cos\left(\frac{1}{(x-4)^2 + y^2}\right) \le \lim_{(x,y)\to(4,0)} (-1)(x^2 - 16)$$

Then the two limits at the ends of the inequality are clearly equal to 0, by the Squeeze Theorem.

Thus we can conclude

$$\lim_{(x,y)\to(4,0)} (x^2 - 16)\cos\left(\frac{1}{(x-4)^2 + y^2}\right) = 0$$

In Exercises 27–42, evaluate the limit or determine that it does not exist.

27. $\displaystyle\lim_{(z,w)\to(-2,1)} \frac{z^4 \cos(\pi w)}{e^{z+w}}$

SOLUTION This function is continuous everywhere since the denominator is never equal to 0, therefore, we will evaluate the limit by substitution:

$$\lim_{(z,w)\to(-2,1)} \frac{z^4 \cos(\pi w)}{e^{z+w}} = \frac{(-2)^4 \cos(\pi)}{e^{-2+1}} = \frac{16(-1)}{e^{-1}} = -16e$$

29. $\displaystyle\lim_{(x,y)\to(4,2)} \frac{y-2}{\sqrt{x^2 - 4}}$

SOLUTION The function is continuous at the point $(4, 2)$, since it is the quotient of two continuous functions and the denominator is not zero at $(4, 2)$. We compute the limit by substitution:

$$\lim_{(x,y)\to(4,2)} \frac{y-2}{\sqrt{x^2-4}} = \frac{2-2}{\sqrt{4^2-4}} = \frac{0}{\sqrt{12}} = 0$$

31. $\displaystyle\lim_{(x,y)\to(3,4)} \frac{1}{\sqrt{x^2 + y^2}}$

SOLUTION The function $\dfrac{1}{\sqrt{x^2+y^2}}$ is continuous at the point $(3, 4)$ since it is the quotient of two continuous functions and the denominator is not zero at $(3, 4)$. We compute the limit by substitution:

$$\lim_{(x,y)\to(3,4)} \frac{1}{\sqrt{x^2+y^2}} = \frac{1}{\sqrt{9+16}} = \frac{1}{5}$$

33. $\displaystyle\lim_{(x,y)\to(\pi,0)} \frac{\cos x}{\sin y}$

SOLUTION Since $\sin 0 = 0$ but $\cos \pi = -1$, this limit approaches $\frac{-1}{0}$, so it does not exist.

35. $\displaystyle\lim_{(x,y)\to(1,-3)} e^{x-y} \ln(x-y)$

SOLUTION This function $e^{x-y} \ln(x - y)$ is continuous at the point $(1, -3)$ since it is the product of two continuous functions. We can compute the limit by substitution:

$$\lim_{(x,y) \to (1,-3)} e^{x-y} \ln(x - y) = e^{1+3} \ln(1 + 3) = e^4 \ln 4$$

37. $\lim_{(x,y) \to (-3,-2)} (x^2 y^3 + 4xy)$

SOLUTION The function $x^2 y^3 + 4xy$ is continuous everywhere because it is a polynomial. We can compute this limit by substitution:

$$\lim_{(x,y) \to (-3,-2)} (x^2 y^3 + 4xy) = 9(-8) + 4(-3)(-2) = -72 + 24 = -48$$

39. $\lim_{(x,y) \to (0,0)} \tan(x^2 + y^2) \tan^{-1}\left(\frac{1}{x^2 + y^2}\right)$

SOLUTION Consider the following inequalities:

$$-\frac{\pi}{2} \leq \tan^{-1}\left(\frac{1}{x^2 + y^2}\right) \leq \frac{\pi}{2}$$

$$-\frac{\pi}{2} \cdot \tan(x^2 + y^2) \leq \tan(x^2 + y^2) \cdot \left(\frac{1}{x^2 + y^2}\right) \leq \frac{\pi}{2} \tan(x^2 + y^2)$$

and then taking limits:

$$\lim_{(x,y) \to (0,0)} -\frac{\pi}{2} \cdot \tan(x^2 + y^2) \leq \lim_{(x,y) \to (0,0)} \tan(x^2 + y^2) \cdot \left(\frac{1}{x^2 + y^2}\right) \leq \lim_{(x,y) \to (0,0)} \frac{\pi}{2} \tan(x^2 + y^2)$$

Each of the limits on the endpoints of this inequality is equal to 0, thus we can conclude:

$$\lim_{(x,y) \to (0,0)} \tan(x^2 + y^2) \cdot \left(\frac{1}{x^2 + y^2}\right) = 0$$

41. $\lim_{(x,y) \to (0,0)} \frac{x^2 + y^2}{\sqrt{x^2 + y^2 + 1} - 1}$

SOLUTION We rewrite the function by dividing and multiplying it by the conjugate of $\sqrt{x^2 + y^2 + 1} - 1$ and using the identity $(a - b)(a + b) = a^2 - b^2$. This gives

$$\frac{x^2 + y^2}{\sqrt{x^2 + y^2 + 1} - 1} = \frac{(x^2 + y^2)\left(\sqrt{x^2 + y^2 + 1} + 1\right)}{\left(\sqrt{x^2 + y^2 + 1} - 1\right)\left(\sqrt{x^2 + y^2 + 1} + 1\right)} = \frac{(x^2 + y^2)\left(\sqrt{x^2 + y^2 + 1} + 1\right)}{(x^2 + y^2 + 1) - 1}$$

$$= \frac{(x^2 + y^2)\left(\sqrt{x^2 + y^2 + 1} + 1\right)}{x^2 + y^2} = \sqrt{x^2 + y^2 + 1} + 1$$

The resulting function is continuous, hence we may compute the limit by substitution. This gives

$$\lim_{(x,y) \to (0,0)} \frac{x^2 + y^2}{\sqrt{x^2 + y^2 + 1} - 1} = \lim_{(x,y) \to (0,0)} \left(\sqrt{x^2 + y^2 + 1} + 1\right) = \sqrt{0^2 + 0^2 + 1} + 1 = 2$$

43. Let $f(x, y) = \frac{x^3 + y^3}{x^2 + y^2}$.

(a) Show that

$$|x^3| \leq |x|(x^2 + y^2), \quad |y^3| \leq |y|(x^2 + y^2)$$

(b) Show that $|f(x, y)| \leq |x| + |y|$.

(c) Use the Squeeze Theorem to prove that $\lim_{(x,y) \to (0,0)} f(x, y) = 0$.

SOLUTION

(a) Since $|x|y^2 \geq 0$, we have

$$|x^3| \leq |x^3| + |x|y^2 = |x|^3 + |x|y^2 = |x|(x^2 + y^2)$$

Similarly, since $|y|x^2 \geq 0$, we have

$$|y^3| \leq |y^3| + |y|x^2 = |y|^3 + |y|x^2 = |y|(x^2 + y^2)$$

(b) We use the triangle inequality to write

$$|f(x,y)| = \frac{|x^3 + y^3|}{x^2 + y^2} \leq \frac{|x^3| + |y^3|}{x^2 + y^2}$$

We continue using the inequality in part (a):

$$|f(x,y)| \leq \frac{|x|(x^2 + y^2) + |y|(x^2 + y^2)}{x^2 + y^2} = \frac{(|x| + |y|)(x^2 + y^2)}{x^2 + y^2} = |x| + |y|$$

That is,

$$|f(x,y)| \leq |x| + |y|$$

(c) In part (b) we showed that

$$|f(x,y)| \leq |x| + |y| \quad (1)$$

Let $\epsilon > 0$. Then if $|x| < \frac{\epsilon}{2}$ and $|y| < \frac{\epsilon}{2}$, we have by (1)

$$|f(x,y) - 0| \leq |x| + |y| < \frac{\epsilon}{2} + \frac{\epsilon}{2} = \epsilon \quad (2)$$

Notice that if $x^2 + y^2 < \frac{\epsilon^2}{4}$, then $x^2 < \frac{\epsilon^2}{4}$ and $y^2 < \frac{\epsilon^2}{4}$. Hence $|x| < \frac{\epsilon}{2}$ and $|y| < \frac{\epsilon}{2}$, so (1) holds. In other words, using $D^\star\left(\frac{\epsilon}{2}\right)$ to represent the punctured disc of radius $\epsilon/2$ centered at the origin, we have

$$(x,y) \in D^\star\left(\frac{\epsilon}{2}\right) \Rightarrow |x| < \frac{\epsilon}{2}$$

and

$$|y| < \frac{\epsilon}{2} \Rightarrow |f(x,y) - 0| < \epsilon$$

We conclude by the limit definition that

$$\lim_{(x,y) \to (0,0)} f(x,y) = 0$$

45. Figure 7 shows the contour maps of two functions. Explain why the limit $\lim_{(x,y) \to P} f(x,y)$ in (A) does not exist. Does $\lim_{(x,y) \to Q} g(x,y)$ appear to exist in (B)? If so, what is its limit?

(A) Contour map of $f(x,y)$ (B) Contour map of $g(x,y)$

FIGURE 7

SOLUTION As (x,y) approaches arbitrarily close to P, the function $f(x,y)$ takes the values ± 1, ± 3, and ± 5 (among others). Therefore $f(x,y)$ does not approach one limit as $(x,y) \to P$. Rather, the limit depends on the contour along which (x,y) is approaching P. This implies that the limit $\lim_{(x,y) \to P} f(x,y)$ does not exist. In (B) the limit $\lim_{(x,y) \to Q} g(x,y)$ appears to exist. If it exists, it must be 4, which is the level curve of Q.

Further Insights and Challenges

47. Is the following function continuous?

$$f(x,y) = \begin{cases} x^2 + y^2 & \text{if } x^2 + y^2 < 1 \\ 1 & \text{if } x^2 + y^2 \geq 1 \end{cases}$$

SOLUTION $f(x,y)$ is defined by a polynomial in the domain $x^2 + y^2 < 1$, hence f is continuous in this domain. In the domain $x^2 + y^2 > 1$, f is a constant function, hence f is continuous in this domain also. Thus, we must examine continuity at the points on the circle $x^2 + y^2 = 1$.

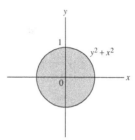

We express $f(x,y)$ using polar coordinates:

$$f(r,\theta) = \begin{cases} r^2 & 0 \leq r < 1 \\ 1 & r \geq 1 \end{cases}$$

Since $\lim_{r \to 1-} f(r,\theta) = \lim_{r \to 1-} r^2 = 1$ and $\lim_{r \to 1+} f(r,\theta) = \lim_{r \to 1+} 1 = 1$, we have $\lim_{r \to 1} f(r,\theta) = 1$. Therefore $f(r,\theta)$ is continuous at $r = 1$, or $f(x,y)$ is continuous on $x^2 + y^2 = 1$. We conclude that f is continuous everywhere on R^2.

49. Prove that the function

$$f(x,y) = \begin{cases} \dfrac{(2^x - 1)(\sin y)}{xy} & \text{if } xy \neq 0 \\ \ln 2 & \text{if } xy = 0 \end{cases}$$

is continuous at $(0,0)$.

SOLUTION To solve this problem it is necessary to show that $\lim_{(x,y)\to(0,0)} f(x,y) = f(0,0) = \ln 2$. Consider the following:

$$\lim_{(x,y)\to(0,0)} \frac{(2^x - 1)\sin y}{xy} = \lim_{(x,y)\to(0,0)} \frac{2^x - 1}{x} \cdot \frac{\sin y}{y}$$

$$= \left(\lim_{x \to 0} \frac{2^x - 1}{x}\right)\left(\lim_{y \to 0} \frac{\sin y}{y}\right)$$

$$= \lim_{x \to 0} \frac{(\ln 2)2^x}{1} \cdot (1) = \ln 2$$

(Using L'Hopital's Rule on the limit in terms of x.) Thus since $\lim_{(x,y)\to(0,0)} f(x,y) = f(0,0)$, we see that $f(x,y)$ is continuous at $(0,0)$.

51. Consider the function $f(x,y) = \dfrac{x^3 y}{x^6 + 2y^2}$.

(a) Show that as $(x,y) \to (0,0)$ along any line $y = mx$, the limit equals 0.

(b) Show that as $(x,y) \to (0,0)$ along the curve $y = x^3$, the limit does not equal 0, and hence, $\lim_{(x,y)\to(0,0)} f(x,y)$ does not exist.

SOLUTION

(a) Along the line $y = mx$, the limit is

$$\lim_{x \to 0} \frac{x^3(mx)}{x^6 + 2(mx)^2} = \lim_{x \to 0} \frac{mx^2}{x^4 + 2m^2}$$

If $m = 0$, this simplifies to

$$\lim_{x \to 0} \frac{0}{x^4} = 0$$

while if $m \neq 0$, the numerator approaches zero and the denominator approaches $2m^2 \neq 0$, so again the limit is zero. Thus the limit along any line $y = mx$ is zero.

(b) Along the curve $y = x^3$ we have

$$\lim_{x \to 0} \frac{x^3(x^3)}{x^6 + 2(x^3)^2} = \lim_{x \to 0} \frac{x^6}{3x^6} = \lim_{x \to 0} \frac{1}{3} = \frac{1}{3}$$

So the limit along this curve is nonzero; since it does not agree with the limit along the lines $y = mx$, it follows that $\lim_{(x,y) \to (0,0)} f(x,y)$ does not exist.

14.3 Partial Derivatives (LT Section 15.3)

Preliminary Questions

1. Patricia derived the following *incorrect* formula by misapplying the Product Rule:

$$\frac{\partial}{\partial x}(x^2 y^2) = x^2(2y) + y^2(2x)$$

What was her mistake and what is the correct calculation?

SOLUTION To compute the partial derivative with respect to x, we treat y as a constant. Therefore the Constant Multiple Rule must be used rather than the Product Rule. The correct calculation is:

$$\frac{\partial}{\partial x}(x^2 y^2) = y^2 \frac{\partial}{\partial x}(x^2) = y^2 \cdot 2x = 2xy^2$$

2. Explain why it is not necessary to use the Quotient Rule to compute $\frac{\partial}{\partial x}\left(\frac{x+y}{y+1}\right)$. Should the Quotient Rule be used to compute $\frac{\partial}{\partial y}\left(\frac{x+y}{y+1}\right)$?

SOLUTION In differentiating with respect to x, y is considered a constant. Therefore in this case the Constant Multiple Rule can be used to obtain

$$\frac{\partial}{\partial x}\left(\frac{x+y}{y+1}\right) = \frac{1}{y+1} \frac{\partial}{\partial x}(x+y) = \frac{1}{y+1} \cdot 1 = \frac{1}{y+1}$$

As for the second part, since y appears in both the numerator and the denominator, the Quotient Rule is indeed needed.

3. Which of the following partial derivatives should be evaluated without using the Quotient Rule?

(a) $\frac{\partial}{\partial x} \frac{xy}{y^2 + 1}$ **(b)** $\frac{\partial}{\partial y} \frac{xy}{y^2 + 1}$ **(c)** $\frac{\partial}{\partial x} \frac{y^2}{y^2 + 1}$

SOLUTION

(a) This partial derivative does not require use of the Quotient Rule, since the Constant Multiple Rule gives

$$\frac{\partial}{\partial x}\left(\frac{xy}{y^2 + 1}\right) = \frac{y}{y^2 + 1} \frac{\partial}{\partial x}(x) = \frac{y}{y^2 + 1} \cdot 1 = \frac{y}{y^2 + 1}$$

(b) This partial derivative requires use of the Quotient Rule.

(c) Since y is considered a constant in differentiating with respect to x, we do not need the Quotient Rule to state that $\frac{\partial}{\partial x}\left(\frac{y^2}{y^2 + 1}\right) = 0$.

4. What is f_x, where $f(x, y, z) = (\sin yz)e^{z^3 - z^{-1}\sqrt{y}}$?

SOLUTION In differentiating with respect to x, we treat y and z as constants. Therefore, the whole expression for $f(x, y, z)$ is treated as constant, so the derivative is zero:

$$\frac{\partial}{\partial x}(\sin yze^{z^3-z^{-1}}\sqrt{y}) = 0$$

5. Assuming the hypotheses of Clairaut's Theorem are satisfied, which of the following partial derivatives are equal to f_{xxy}?

(a) f_{xyx} **(b)** f_{yyx} **(c)** f_{xyy} **(d)** f_{yxx}

SOLUTION f_{xxy} involves two differentiations with respect to x and one differentiation with respect to y. Therefore, if f satisfies the assumptions of Clairaut's Theorem, then

$$f_{xxy} = f_{xyx} = f_{yxx}$$

Exercises

1. Use the limit definition of the partial derivative to verify the formulas

$$\frac{\partial}{\partial x}xy^2 = y^2, \quad \frac{\partial}{\partial y}xy^2 = 2xy$$

SOLUTION Using the limit definition of the partial derivative, we have

$$\frac{\partial}{\partial x}xy^2 = \lim_{h \to 0}\frac{(x+h)y^2 - xy^2}{h} = \lim_{h \to 0}\frac{xy^2 + hy^2 - xy^2}{h} = \lim_{h \to 0}\frac{hy^2}{h} = \lim_{h \to 0}y^2 = y^2$$

$$\frac{\partial}{\partial y}xy^2 = \lim_{k \to 0}\frac{x(y+k)^2 - xy^2}{k} = \lim_{k \to 0}\frac{x(y^2 + 2yk + k^2) - xy^2}{k} = \lim_{k \to 0}\frac{xy^2 + 2xyk + xk^2 - xy^2}{k}$$

$$= \lim_{k \to 0}\frac{k(2xy + xk)}{k} = \lim_{k \to 0}(2xy + k) = 2xy + 0 = 2xy$$

3. Use the Product Rule to compute f_x and f_y for $f(x, y) = (x^2 - y)(x - y^2)$.

SOLUTION Using the Product Rule we obtain

$$f_x(x, y) = (x^2 - y)\frac{\partial}{\partial x}(x - y^2) + (x - y^2)\frac{\partial}{\partial x}(x^2 - y)$$

$$= (x^2 - y)(1) + (x - y^2)(2x) = 3x^2 - 2xy^2 - y$$

$$f_y(x, y) = (x^2 - y)\frac{\partial}{\partial y}(x - y^2) + (x - y^2)\frac{\partial}{\partial y}(x^2 - y)$$

$$= (x^2 - y)(-2y) + (x - y^2)(-1) = 3y^2 - 2x^2y - x$$

5. Use the Quotient Rule to compute $\dfrac{\partial}{\partial y}\dfrac{y}{x+y}$.

SOLUTION Using the Quotient Rule we obtain

$$\frac{\partial}{\partial y}\frac{y}{x+y} = \frac{(x+y)\frac{\partial}{\partial y}(y) - y\frac{\partial}{\partial y}(x+y)}{(x+y)^2} = \frac{(x+y) \cdot 1 - y \cdot 1}{(x+y)^2} = \frac{x}{(x+y)^2}$$

7. Calculate $f_z(2, 3, 1)$, where $f(x, y, z) = xyz$.

SOLUTION We first find the partial derivative $f_z(x, y, z)$:

$$f_z(x, y, z) = \frac{\partial}{\partial z}(xyz) = xy$$

Substituting the given point we get

$$f_z(2, 3, 1) = 2 \cdot 3 = 6$$

9. The plane $y = 1$ intersects the surface $z = x^4 + 6xy - y^4$ in a certain curve. Find the slope of the tangent line to this curve at the point $P = (1, 1, 6)$.

SOLUTION The slope of the tangent line to the curve $z = z(x, 1) = x^4 + 6x - 1$, obtained by intersecting the surface $z = x^4 + 6xy - y^4$ with the plane $y = 1$, is the partial derivative $\frac{\partial z}{\partial x}(1, 1)$.

$$\frac{\partial z}{\partial x} = \frac{\partial}{\partial x}(x^4 + 6xy - y^4) = 4x^3 + 6y$$

$$m = \frac{\partial z}{\partial x}(1, 1) = 4 \cdot 1^3 + 6 \cdot 1 = 10$$

In Exercises 11–14, refer to Figure 7.

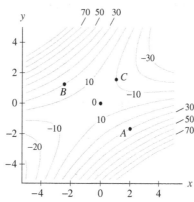

FIGURE 7 Contour map of $f(x, y)$.

11. Estimate f_x and f_y at point A.

SOLUTION To estimate f_x we move horizontally to the next level curve in the direction of growing x, to a point A'. The change in f from A to A' is the contour interval, $\Delta f = 40 - 30 = 10$.

The distance between A and A' is approximately $\Delta x \approx 1.25$. Hence,

$$f_x(A) \approx \frac{\Delta f}{\Delta x} = \frac{10}{1.25} = 8$$

To estimate f_y we move vertically from A to a point A'' on the next level curve in the direction of growing y. The change in f from A to A'' is $\Delta f = 20 - 30 = -10$.

The distance between A and A'' is $\Delta y \approx 0.6$. Hence,

$$f_y(A) \approx \frac{\Delta f}{\Delta y} = \frac{-10}{0.6} \approx -16.7$$

13. Starting at point B, in which compass direction (N, NE, SW, etc.) does f increase most rapidly?

SOLUTION The distances between adjacent level curves starting at B are the smallest along the line with slope -1, upward.

Therefore, f is increasing most rapidly in the NW direction.

In Exercises 15–42, compute the first-order partial derivatives.

15. $z = x^2 + y^2$

SOLUTION We compute $z_x(x, y)$ by treating y as a constant, and we compute $z_y(x, y)$ by treating x as a constant:

$$\frac{\partial}{\partial x}(x^2 + y^2) = 2x; \quad \frac{\partial}{\partial y}(x^2 + y^2) = 2y$$

17. $z = x^4 y + xy^{-2}$

SOLUTION We obtain the following partial derivatives:

$$\frac{\partial}{\partial x}(x^4 y + xy^{-2}) = 4x^3 y + y^{-2}$$

$$\frac{\partial}{\partial y}(x^4 y + xy^{-2}) = x^4 + x \cdot (-2y^{-3}) = x^4 - 2xy^{-3}$$

19. $z = \dfrac{x}{y}$

SOLUTION Treating y as a constant we have

$$\frac{\partial}{\partial x}\left(\frac{x}{y}\right) = \frac{1}{y}\frac{\partial}{\partial x}(x) = \frac{1}{y} \cdot 1 = \frac{1}{y}$$

We now find the derivative $z_y(x, y)$, treating x as a constant:

$$\frac{\partial}{\partial y}\left(\frac{x}{y}\right) = x \cdot \frac{\partial}{\partial y}\left(\frac{1}{y}\right) = x \cdot \frac{-1}{y^2} = \frac{-x}{y^2}$$

21. $z = \sqrt{9 - x^2 - y^2}$

SOLUTION Differentiating with respect to x, treating y as a constant, and using the Chain Rule, we obtain

$$\frac{\partial}{\partial x}\left(\sqrt{9 - x^2 - y^2}\right) = \frac{1}{2\sqrt{9 - x^2 - y^2}}\frac{\partial}{\partial x}(9 - x^2 - y^2) = \frac{-2x}{2\sqrt{9 - x^2 - y^2}} = \frac{-x}{\sqrt{9 - x^2 - y^2}}$$

We now differentiate with respect to y, treating x as a constant:

$$\frac{\partial}{\partial y}\left(\sqrt{9 - x^2 - y^2}\right) = \frac{1}{2\sqrt{9 - x^2 - y^2}}\frac{\partial}{\partial y}(9 - x^2 - y^2) = \frac{-2y}{2\sqrt{9 - x^2 - y^2}} = \frac{-y}{\sqrt{9 - x^2 - y^2}}$$

23. $z = (\sin x)(\cos y)$

SOLUTION We obtain the following partial derivatives:

$$\frac{\partial}{\partial x}(\sin x \cos y) = \cos y \frac{\partial}{\partial x} \sin x = (\cos x)(\cos y)$$

$$\frac{\partial}{\partial y}(\sin x \cos y) = \sin x \frac{\partial}{\partial y} \cos y = -(\sin x)(\sin y)$$

25. $z = \cos \dfrac{1 - x}{y}$

SOLUTION By the Chain Rule,

$$\frac{d}{dx}\cos u = -\sin u \cdot \frac{du}{dx} \quad \text{and} \quad \frac{d}{dy}\cos u = -\sin u \cdot \frac{du}{dy}$$

Using these equations with $u = \frac{1-x}{y}$ gives

$$\frac{\partial z}{\partial x} = -\sin\left(\frac{1-x}{y}\right) \cdot \frac{d}{dx}\frac{1-x}{y} = \frac{1}{y}\sin\left(\frac{1-x}{y}\right)$$

$$\frac{\partial z}{\partial u} = -\sin\left(\frac{1-x}{y}\right) \cdot \frac{d}{dy}\frac{1-x}{y} = -\frac{1-x}{y^2} \cdot \left(-\sin\left(\frac{1-x}{y}\right)\right) = \frac{1-x}{y^2}\sin\left(\frac{1-x}{y}\right)$$

27. $w = \ln(x^2 - y^2)$

SOLUTION Using the Chain Rule we have

$$\frac{\partial w}{\partial x} = \frac{1}{x^2 - y^2}\frac{\partial}{\partial x}(x^2 - y^2) = \frac{1}{x^2 - y^2} \cdot 2x = \frac{2x}{x^2 - y^2}$$

$$\frac{\partial w}{\partial y} = \frac{1}{x^2 - y^2}\frac{\partial}{\partial y}(x^2 - y^2) = \frac{1}{x^2 - y^2} \cdot (-2y) = -\frac{2y}{x^2 - y^2}$$

29. $W = e^{r+s}$

SOLUTION We use the Chain Rule to compute $\frac{\partial W}{\partial r}$ and $\frac{\partial W}{\partial s}$:

$$\frac{\partial W}{\partial r} = e^{r+s} \cdot \frac{\partial}{\partial r}(r+s) = e^{r+s} \cdot 1 = e^{r+s}$$

$$\frac{\partial W}{\partial s} = e^{r+s} \cdot \frac{\partial}{\partial s}(r+s) = e^{r+s} \cdot 1 = e^{r+s}$$

31. $z = e^{xy}$

SOLUTION We use the Chain Rule, $\frac{d}{dx}e^u = e^u\frac{du}{dx}$, $\frac{d}{dy}e^u = e^u\frac{du}{dy}$ with $u = xy$ to obtain

$$\frac{\partial}{\partial x}e^{xy} = e^{xy}\frac{\partial}{\partial x}(xy) = e^{xy}y = ye^{xy}$$

$$\frac{\partial}{\partial y}e^{xy} = e^{xy}\frac{\partial}{\partial y}(xy) = e^{xy}x = xe^{xy}$$

33. $z = e^{-x^2-y^2}$

SOLUTION We use the Chain Rule to find $\frac{\partial z}{\partial x}$ and $\frac{\partial z}{\partial y}$:

$$\frac{\partial z}{\partial x} = e^{-x^2-y^2}\frac{\partial}{\partial x}(-x^2 - y^2) = e^{-x^2-y^2} \cdot (-2x) = -2xe^{-x^2-y^2}$$

$$\frac{\partial z}{\partial y} = e^{-x^2-y^2}\frac{\partial}{\partial y}(-x^2 - y^2) = e^{-x^2-y^2} \cdot (-2y) = -2ye^{-x^2-y^2}$$

35. $U = \frac{e^{-rt}}{r}$

SOLUTION We have

$$\frac{\partial U}{\partial r} = \frac{-te^{-rt} \cdot r - e^{-rt} \cdot 1}{r^2} = \frac{-(1+rt)e^{-rt}}{r^2}$$

and also

$$\frac{\partial U}{\partial t} = \frac{-re^{-rt}}{r} = -e^{-rt}$$

37. $z = \sinh(x^2 y)$

SOLUTION By the Chain Rule, $\frac{d}{dx}\sinh u = \cosh u \frac{du}{dx}$ and $\frac{d}{dy}\sinh u = \cosh u \frac{du}{dy}$. We use the Chain Rule with $u = x^2 y$ to obtain

$$\frac{\partial}{\partial x}\sinh(x^2 y) = \cosh(x^2 y)\frac{\partial}{\partial x}(x^2 y) = 2xy\cosh(x^2 y)$$

$$\frac{\partial}{\partial y}\sinh(x^2 y) = \cosh(x^2 y)\frac{\partial}{\partial y}(x^2 y) = x^2 \cosh(x^2 y)$$

39. $w = xy^2 z^3$

SOLUTION The partial derivatives of w are

$$\frac{\partial w}{\partial x} = y^2 z^3$$

$$\frac{\partial w}{\partial y} = xz^3 \frac{\partial}{\partial y}(y^2) = xz^3 \cdot 2y = 2xz^3 y$$

$$\frac{\partial w}{\partial z} = xy^2 \frac{\partial}{\partial z}(z^3) = xy^2 \cdot 3z^2 = 3xy^2 z^2$$

41. $Q = \frac{L}{M} e^{-Lt/M}$

SOLUTION

$$\frac{\partial Q}{\partial L} = \frac{\partial}{\partial L}\left(\frac{L}{M}e^{-Lt/M}\right)$$

$$= \frac{L}{M} \cdot e^{-Lt/M} \cdot (-t/M) + e^{-Lt/M} \cdot \frac{1}{M}$$

$$= -\frac{Lt}{M^2}e^{-Lt/M} + \frac{e^{-Lt/M}}{M}$$

$$\frac{\partial Q}{\partial M} = \frac{\partial}{\partial M}\left(\frac{L}{M}e^{-Lt/M}\right)$$

$$= \frac{L}{M} \cdot e^{-Lt/M} \cdot \frac{Lt}{M^2} + e^{-Lt/M} \cdot \frac{-L}{M^2}$$

$$= \frac{L^2 t}{M^3}e^{-Lt/M} - \frac{L}{M^2}e^{-Lt/M}$$

$$\frac{\partial Q}{\partial t} = \frac{\partial}{\partial t}\left(\frac{L}{M}e^{-Lt/M}\right)$$

$$= -\frac{L^2}{M^2}e^{-Lt/M}$$

In Exercises 43–46, compute the given partial derivatives.

43. $f(x, y) = 3x^2 y + 4x^3 y^2 - 7xy^5$, $f_x(1, 2)$

SOLUTION Differentiating with respect to x gives

$$f_x(x, y) = 6xy + 12x^2 y^2 - 7y^5$$

Evaluating at $(1, 2)$ gives

$$f_x(1, 2) = 6 \cdot 1 \cdot 2 + 12 \cdot 1^2 \cdot 2^2 - 7 \cdot 2^5 = -164$$

45. $g(u, v) = u \ln(u + v)$, $g_u(1, 2)$

SOLUTION Using the Product Rule and the Chain Rule we get

$$g_u(u, v) = \frac{\partial}{\partial u}(u \ln(u + v)) = 1 \cdot \ln(u + v) + u \cdot \frac{1}{u + v} = \ln(u + v) + \frac{u}{u + v}$$

At the point $(1, 2)$ we have

$$g_u(1, 2) = \ln(1 + 2) + \frac{1}{1 + 2} = \ln 3 + \frac{1}{3}$$

47. The **heat index** I is a measure of how hot it feels when the relative humidity is H (as a percentage) and the actual air temperature is T (in degrees Fahrenheit). An approximate formula for the heat index that is valid for (T, H) near $(90, 40)$ is

$$I(T, H) = 45.33 + 0.6845T + 5.758H - 0.00365T^2 \\ - 0.1565HT + 0.001HT^2$$

(a) Calculate I at $(T, H) = (95, 50)$.

(b) Which partial derivative tells us the increase in I per degree increase in T when $(T, H) = (95, 50)$? Calculate this partial derivative.

SOLUTION

(a) Let us compute I when $T = 95$ and $H = 50$:

$$I(95, 50) = 45.33 + 0.6845(95) + 5.758(50) - 0.00365(95)^2 - 0.1565(50)(95) + 0.001(50)(95)^2$$
$$= 73.19125$$

(b) The partial derivative we are looking for here is $\partial I/\partial T$:

$$\frac{\partial I}{\partial T} = 0.6845 - 0.00730T - 0.1565H + 0.002HT$$

and evaluating we have:

$$\frac{\partial I}{\partial T}(95, 50) = 0.6845 - 0.00730(95) - 0.1565(50) + 0.002(50)(95) = 1.666$$

49. Use the contour map of $f(x, y)$ in Figure 8 to explain the following statements:

(a) f_y is larger at P than at Q, and f_x is more negative at P than at Q.

(b) $f_x(x, y)$ is decreasing as a function of y; that is, for any fixed value $x = a$, $f_x(a, y)$ is decreasing in y.

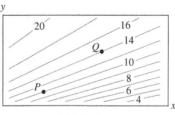

FIGURE 8

SOLUTION

(a) A vertical segment through P meets more level curves than a vertical segment of the same size through Q, so f is increasing more rapidly in the y at P than at Q. Therefore, f_y are both larger at P than at Q.

Similarly, a horizontal segment through P meets more level curves at P than at Q, but f is *decreasing* in the positive x-direction, so f is decreasing more rapidly in the x-direction at P than at Q. Therefore, f_x is more negative at P than at Q.

(b) For any fixed value $x = a$, a horizontal segment meets fewer level curves as we move it vertically upward. This indicates that $f_x(a, y)$ in a decreasing function of y.

51. Over most of the earth, a magnetic compass does not point to true (geographic) north; instead, it points at some angle east or west of true north. The angle D between magnetic north and true north is called the **magnetic declination**. Use Figure 10 to determine which of the following statements is true:

(a) $\left.\dfrac{\partial D}{\partial y}\right|_A > \left.\dfrac{\partial D}{\partial y}\right|_B$ **(b)** $\left.\dfrac{\partial D}{\partial x}\right|_C > 0$ **(c)** $\left.\dfrac{\partial D}{\partial y}\right|_C > 0$

Note that the horizontal axis increases from right to left because of the way longitude is measured.

FIGURE 10 Contour interval 1°.

SOLUTION

(a) To estimate $\frac{\partial D}{\partial y}\big|_A$ and $\frac{\partial D}{\partial y}\big|_B$, we move vertically from A and B to the points on the next level curve in the direction of increasing y (upward). From A, we quickly come to a level curve corresponding to higher value of D; but from B, moving vertically, there is hardly any change as we move along the curve. The statement is thus true.

(b) The derivative $\frac{\partial D}{\partial x}\big|_C$ is estimated by $\frac{\Delta D}{\Delta x}$. Since x varies in the horizontal direction, we move horizontally from C to a point on the next level curve in the direction of increasing x (leftwards). Since the value of D on this level curve is greater than on the level curve of C, $\Delta D = 1$. Also $\Delta x > 0$, hence

$$\frac{\partial D}{\partial x}\bigg|_C \approx \frac{\Delta D}{\Delta x} = \frac{1}{\Delta x} > 0$$

The statement is correct.

(c) Moving from C vertically upward (in the direction of increasing y), we come to a point on a level curve with a smaller value of D. Therefore, $\Delta D = -1$ and $\Delta y > 0$, so we obtain

$$\frac{\partial D}{\partial y}\bigg|_C \approx \frac{\Delta D}{\Delta y} = \frac{-1}{\Delta y} < 0$$

Hence, the statement is false.

In Exercises 53–58, compute the derivatives indicated.

53. $f(x, y) = 3x^2 y - 6xy^4$, $\frac{\partial^2 f}{\partial x^2}$ and $\frac{\partial^2 f}{\partial y^2}$

SOLUTION We first compute the partial derivatives $\frac{\partial f}{\partial x}$ and $\frac{\partial f}{\partial y}$:

$$\frac{\partial f}{\partial x} = 6xy - 6y^4; \quad \frac{\partial f}{\partial y} = 3x^2 - 6x \cdot 4y^3 = 3x^2 - 24xy^3$$

We now differentiate $\frac{\partial f}{\partial x}$ with respect to x and $\frac{\partial f}{\partial y}$ with respect to y. We get

$$\frac{\partial^2 f}{\partial x^2} = \frac{\partial}{\partial x} f_x = 6y; \quad \frac{\partial^2 f}{\partial y^2} = \frac{\partial}{\partial y} f_y = -24x \cdot 3y^2 = -72xy^2$$

55. $h(u, v) = \dfrac{u}{u + 4v}$, $h_{vv}(u, v)$

SOLUTION We first note

$$\frac{\partial h}{\partial v} = \frac{-4u}{(u + 4v)^2}$$

so thus

$$\frac{\partial h^2}{\partial v^2} = \frac{\partial}{\partial v}\left(\frac{-4u}{(u + 4v)^2}\right) = \frac{32u}{(u + 4v)^3}$$

57. $f(x, y) = x \ln(y^2)$, $\quad f_{yy}(2, 3)$

SOLUTION We find f_y using the Chain Rule:

$$f_y = \frac{\partial}{\partial y} x \ln y^2 = x \frac{\partial}{\partial y} \ln y^2 = x \frac{1}{y^2} \cdot 2y = \frac{2x}{y}$$

We now differentiate f_y with respect to y, obtaining

$$f_{yy}(x, y) = \frac{\partial}{\partial y} f_y = 2x \frac{\partial}{\partial y} \cdot \frac{1}{y} = \frac{-2x}{y^2}$$

The derivative at $(2, 3)$ is thus

$$f_{yy}(2, 3) = \frac{-2 \cdot 2}{3^2} = -\frac{4}{9}$$

59. Compute f_{xyxzy} for

$$f(x, y, z) = y \sin(xz) \sin(x + z) + (x + z^2) \tan y + x \tan\left(\frac{z + z^{-1}}{y - y^{-1}}\right)$$

Hint: Use a well-chosen order of differentiation on each term.

SOLUTION At the points where the derivatives are continuous, the partial derivative f_{xyz} may be performed in any order on each term. Let

$$g(x, y, z) = y \sin(xz) \sin(x + z), \quad h(x, y, z) = (x + z^2) \tan y, \quad k(x, y, z) = x \tan\left(\frac{z + z^{-1}}{y - y^{-1}}\right)$$

Then $f(x, y, z) = g(x, y, z) + h(x, y, z) + k(x, y, z)$.

Compute g_{xyxzy} by first differentiating twice with respect to y:

$$g_y(x, y, z) = \sin(xz) \sin(x + z) \quad \Rightarrow \quad g_{yy}(x, y, z) = 0 \quad \Rightarrow \quad g_{xyxzy}(x, y, z) = 0$$

Next, compute h_{xyxzy} by first differentiating twice with respect to x:

$$h_x(x, y, z) = \tan y, \quad \Rightarrow \quad h_{xx}(x, y, z) = 0 \quad \Rightarrow \quad h_{xyxzy}(x, y, z) = 0$$

Finally, compute k_{xyxzy} by first differentiating twice with respect to x:

$$k_x(x, y, z) = \tan\left(\frac{z + z^{-1}}{y - y^{-1}}\right) \quad \Rightarrow \quad k_{xx}(x, y, z) = 0 \quad \Rightarrow \quad k_{xyxzy}(x, y, z) = 0$$

therefore

$$f_{xyxzy}(x, y, z) = g_{xyxzy}(x, y, z) + h_{xyxzy}(x, y, z) + k_{xyxzy}(x, y, z) = 0$$

In Exercises 61–68, compute the derivative indicated.

61. $f(u, v) = \cos(u + v^2)$, $\quad f_{uuv}$

SOLUTION Using the Chain Rule, we have

$$f_u = \frac{\partial}{\partial u} \cos(u + v^2) = -\sin(u + v^2) \cdot \frac{\partial}{\partial u}(u + v^2) = -\sin(u + v^2)$$

$$f_{uu} = \frac{\partial}{\partial u}(-\sin(u + v^2)) = -\cos(u + v^2)$$

$$f_{uuv} = \frac{\partial}{\partial v}(-\cos(u + v^2)) = \sin(u + v^2) \cdot \frac{\partial}{\partial v}(u + v^2) = 2v \sin(u + v^2)$$

63. $F(r, s, t) = r(s^2 + t^2)$, $\quad F_{rst}$

SOLUTION For $F(r, s, t) = r(s^2 + t^2)$, we have

$$F_r = s^2 + t^2$$

$$F_{rs} = 2s$$

$$F_{rst} = 0$$

65. $F(\theta, u, v) = \sinh(uv + \theta^2)$, $F_{uu\theta}$

SOLUTION We can compute:

$$F_u = v \cdot \cosh(uv + \theta^2)$$

$$F_{uu} = v^2 \cdot \sinh(uv + \theta^2)$$

$$F_{uu\theta} = 2\theta v^2 \cosh(uv + \theta^2)$$

67. $g(x, y, z) = \sqrt{x^2 + y^2 + z^2}$, g_{xyz}

SOLUTION Differentiating with respect to x, using the Chain Rule, we get

$$g_x = \frac{\partial}{\partial x}\sqrt{x^2 + y^2 + z^2} = \frac{1}{2\sqrt{x^2 + y^2 + z^2}} \frac{\partial}{\partial x}(x^2 + y^2 + z^2) = \frac{1}{2\sqrt{x^2 + y^2 + z^2}} \cdot 2x = \frac{x}{\sqrt{x^2 + y^2 + z^2}}$$

We now differentiate g_x with respect to y, using the Chain Rule. This gives

$$g_{xy} = x\frac{\partial}{\partial y}(x^2 + y^2 + z^2)^{-1/2} = x \cdot \left(-\frac{1}{2}\right)(x^2 + y^2 + z^2)^{-3/2} \cdot 2y = \frac{-xy}{(x^2 + y^2 + z^2)^{3/2}}$$

Finally, we differentiate g_{xy} with respect to z, obtaining

$$g_{xyz} = -xy\frac{\partial}{\partial z}(x^2 + y^2 + z^2)^{-3/2} = -xy \cdot \left(-\frac{3}{2}\right)(x^2 + y^2 + z^2)^{-5/2} \cdot 2z = \frac{3xyz}{(x^2 + y^2 + z^2)^{5/2}}$$

69. Find a function such that $\frac{\partial f}{\partial x} = 2xy$ and $\frac{\partial f}{\partial y} = x^2$.

SOLUTION The function $f(x, y) = x^2 y$ satisfies $\frac{\partial f}{\partial y} = x^2$ and $\frac{\partial f}{\partial x} = 2xy$.

71. Assume that f_{xy} and f_{yx} are continuous and that f_{yxx} exists. Show that f_{xyx} also exists and that $f_{yxx} = f_{xyx}$.

SOLUTION Since f_{xy} and f_{yx} are continuous, Clairaut's Theorem implies that

$$f_{xy} = f_{yx} \tag{1}$$

We are given that f_{yxx} exists. Using (1) we get

$$f_{yxx} = \frac{\partial}{\partial x}\frac{\partial}{\partial x}f_y = \frac{\partial}{\partial x}f_{yx} = \frac{\partial}{\partial x}f_{xy} = f_{xyx}$$

Therefore, f_{xyx} also exists and $f_{yxx} = f_{xyx}$.

73. Find all values of A and B such that $f(x, t) = e^{Ax+Bt}$ satisfies Eq. (3).

SOLUTION We compute the following partials, using the Chain Rule:

$$\frac{\partial f}{\partial t} = \frac{\partial}{\partial t}(e^{Ax+Bt}) = e^{Ax+Bt}\frac{\partial}{\partial t}(Ax + Bt) = Be^{Ax+Bt}$$

$$\frac{\partial f}{\partial x} = \frac{\partial}{\partial x}(e^{Ax+Bt}) = e^{Ax+Bt}\frac{\partial}{\partial x}(Ax + Bt) = Ae^{Ax+Bt}$$

$$\frac{\partial^2 f}{\partial x^2} = \frac{\partial}{\partial x}(Ae^{Ax+Bt}) = A\frac{\partial}{\partial x}(e^{Ax+Bt}) = Ae^{Ax+Bt}\frac{\partial}{\partial x}(Ax + Bt) = A^2 e^{Ax+Bt}$$

Substituting these partials in the differential equation (3), we get

$$Be^{Ax+Bt} = A^2 e^{Ax+Bt}$$

We divide by the nonzero e^{Ax+Bt} to obtain

$$B = A^2$$

We conclude that $f(x, t) = e^{Ax+Bt}$ satisfies equation (5) if and only if $B = A^2$, where A is arbitrary.

*In Exercises 75–78, the **Laplace operator** Δ is defined by $\Delta f = f_{xx} + f_{yy}$. A function $u(x, y)$ satisfying the Laplace equation $\Delta u = 0$ is called **harmonic**.*

75. Show that the following functions are harmonic:

(a) $u(x, y) = x$

(b) $u(x, y) = e^x \cos y$

(c) $u(x, y) = \tan^{-1} \dfrac{y}{x}$

(d) $u(x, y) = \ln(x^2 + y^2)$

SOLUTION

(a) We compute u_{xx} and u_{yy} for $u(x, y) = x$:

$$u_x = \frac{\partial}{\partial x}(x) = 1; \quad u_{xx} = \frac{\partial}{\partial x}(1) = 0$$

$$u_y = \frac{\partial}{\partial y}(x) = 0; \quad u_{yy} = \frac{\partial}{\partial y}(0) = 0$$

Since $u_{xx} + u_{yy} = 0$, u is harmonic.

(b) We compute the partial derivatives of $u(x, y) = e^x \cos y$:

$$u_x = \frac{\partial}{\partial x}(e^x \cos y) = \cos y \frac{\partial}{\partial x} e^x = (\cos y) e^x$$

$$u_y = \frac{\partial}{\partial y}(e^x \cos y) = e^x \frac{\partial}{\partial y} \cos y = -e^x \sin y$$

$$u_{xx} = \frac{\partial}{\partial x}((\cos y) e^x) = \cos y \frac{\partial}{\partial x} e^x = (\cos y) e^x$$

$$u_{yy} = \frac{\partial}{\partial y}(-e^x \sin y) = -e^x \frac{\partial}{\partial y} \sin y = -e^x \cos y$$

Thus,

$$u_{xx} + u_{yy} = (\cos y) e^x - e^x \cos y = 0$$

Hence $u(x, y) = e^x \cos y$ is harmonic.

(c) We compute the partial derivatives of $u(x, y) = \tan^{-1} \dfrac{y}{x}$ using the Chain Rule and the formula

$$\frac{d}{dt} \tan^{-1} t = \frac{1}{1 + t^2}$$

We have

$$u_x = \frac{\partial}{\partial x} \tan^{-1} \frac{y}{x} = \frac{1}{1 + (y/x)^2} \frac{\partial}{\partial x} \frac{y}{x} = \frac{1}{1 + (y/x)^2} \left(\frac{-y}{x^2}\right) = -\frac{y}{x^2 + y^2}$$

$$u_y = \frac{\partial}{\partial y} \tan^{-1} \frac{y}{x} = \frac{1}{1 + (y/x)^2} \frac{\partial}{\partial y} \frac{y}{x} = \frac{1}{1 + (y/x)^2} \left(\frac{1}{x}\right) = \frac{x}{x^2 + y^2}$$

$$u_{xx} = \frac{\partial}{\partial x} \left(-\frac{y}{x^2 + y^2}\right) = \frac{2xy}{(x^2 + y^2)^2}$$

$$u_{yy} = \frac{\partial}{\partial y} \frac{x}{x^2 + y^2} = -\frac{2xy}{(x^2 + y^2)^2}$$

Therefore $u_{xx} + u_{xx} = 0$. This shows that $u(x, y) = \tan^{-1} \dfrac{y}{x}$ is harmonic.

(d) We compute the partial derivatives of $u(x, y) = \ln(x^2 + y^2)$ using the Chain Rule:

$$u_x = \frac{\partial}{\partial x} \ln(x^2 + y^2) = \frac{1}{x^2 + y^2} \cdot 2x = \frac{2x}{x^2 + y^2}$$

$$u_y = \frac{\partial}{\partial y} \ln(x^2 + y^2) = \frac{1}{x^2 + y^2} \cdot 2y = \frac{2y}{x^2 + y^2}$$

We now find u_{xx} and u_{yy} using the Quotient Rule:

$$u_{xx} = \frac{\partial}{\partial x} \frac{2x}{x^2 + y^2} = \frac{2(x^2 + y^2) - 2x \cdot 2x}{(x^2 + y^2)^2} = \frac{2(y^2 - x^2)}{(x^2 + y^2)^2}$$

$$u_{yy} = \frac{\partial}{\partial y} \frac{2y}{x^2 + y^2} = \frac{2(x^2 + y^2) - 2y \cdot 2y}{(x^2 + y^2)^2} = \frac{2(x^2 - y^2)}{(x^2 + y^2)^2}$$

Thus,

$$u_{xx} + u_{yy} = \frac{2(y^2 - x^2)}{(x^2 + y^2)^2} + \frac{2(x^2 - y^2)}{(x^2 + y^2)^2} = 0$$

Therefore, $u(x, y) = \ln(x^2 + y^2)$ is harmonic.

77. Show that if $u(x, y)$ is harmonic, then the partial derivatives $\partial u/\partial x$ and $\partial u/\partial y$ are harmonic.

SOLUTION We assume that the second-order partials are continuous, hence the partial differentiation may be performed in any order. By the given data, we have

$$u_{xx} + u_{yy} = 0 \qquad (1)$$

We must show that

$$(u_x)_{xx} + (u_x)_{yy} = 0 \quad \text{and} \quad (u_y)_{xx} + (u_y)_{yy} = 0$$

We differentiate (1) with respect to x, obtaining

$$0 = (u_{xx})_x + (u_{yy})_x = u_{xxx} + u_{xyy} = (u_x)_{xx} + (u_x)_{yy} \qquad (2)$$

We differentiate (1) with respect to y:

$$0 = (u_{xx})_y + (u_{yy})_y = u_{xxy} + u_{yyy} = u_{yxx} + u_{yyy} = (u_y)_{xx} + (u_y)_{yy} \qquad (3)$$

Equalities (2) and (3) prove that u_x and u_y are harmonic.

79. Show that $u(x, t) = \text{sech}^2(x - t)$ satisfies the **Korteweg–deVries equation** (which arises in the study of water waves):

$$4u_t + u_{xxx} + 12uu_x = 0$$

SOLUTION In Exercise 68 we found the following derivatives:

$$u_x = -2\,\text{sech}^2(x - t)\tanh(x - t)$$

$$u_{xxx} = 16\,\text{sech}^4(x - t)\tanh(x - t) - 8\,\text{sech}^2(x - t)\tanh^3(x - t)$$

Hence,

$$\begin{aligned}4u_t + u_{xxx} + 12uu_x &= 8\,\text{sech}^2(x - t)\tanh(x - t) + 16\,\text{sech}^4(x - t)\tanh(x - t) \\ &\quad - 8\,\text{sech}^2(x - t)\tanh^3(x - t) - 24\,\text{sech}^4(x - t)\tanh(x - t) \\ &= 8\,\text{sech}^2(x - t)\{\tanh(x - t) - \tanh^3(x - t)\} - 8\,\text{sech}^4(x - t)\tanh(x - t) \\ &= 8\,\text{sech}^2(x - t)\tanh(x - t)\{1 - \tanh^2(x - t)\} - 8\,\text{sech}^4(x - t)\tanh(x - t) \\ &= 8\,\text{sech}^2(x - t)\tanh(x - t)\{\text{sech}^2(x - t)\} - 8\,\text{sech}^4(x - t)\tanh(x - t) \\ &= 0\end{aligned}$$

14.4 Differentiability, Tangent Planes, and Linear Approximation (LT Section 15.4)

Preliminary Questions

1. How is the linearization of $f(x, y)$ centered at (a, b) defined?

SOLUTION The linearization of $f(x, y)$ centered at (a, b) is the linear function

$$L(x, y) = f(a, b) + f_x(a, b)(x - a) + f_y(a, b)(y - b)$$

This function is the equation of the tangent plane to the surface $z = f(x, y)$ at $(a, b, f(a, b))$.

2. If f is differentiable at (a, b) and $f_x(a, b) = f_y(a, b) = 0$, what can we conclude about the tangent plane at (a, b)?

SECTION 14.4 | Differentiability, Tangent Planes, and Linear Approximation

SOLUTION $df = f_x(x, y)\,dx + f_y(x, y)\,dy$ represents the change in the height of the tangent plane for given changes in x and y. If $f_x(a, b) = f_y(a, b) = 0$, then $df = 0$ at (a, b), so the tangent plane is horizontal (its height does not change).

In Exercises 3–5, assume that

$$f(2, 3) = 8, \quad f_x(2, 3) = 5, \quad f_y(2, 3) = 7$$

3. Which of (a)–(b) is the linearization of f centered at $(2, 3)$?

(a) $L(x, y) = 8 + 5x + 7y$

(b) $L(x, y) = 8 + 5(x - 2) + 7(y - 3)$

SOLUTION The linearization of f at $(2, 3)$ is the following linear function:

$$L(x, y) = f(2, 3) + f_x(2, 3)(x - 2) + f_y(2, 3)(y - 3)$$

That is,

$$L(x, y) = 8 + 5(x - 2) + 7(y - 3) = -23 + 5x + 7y$$

The function in (b) is the correct answer.

4. Estimate $f(2, 3.1)$.

SOLUTION We use the linear approximation

$$f(a + h, b + k) \approx f(a, b) + f_x(a, b)h + f_y(a, b)k$$

We let $(a, b) = (2, 3)$, $h = 0$, $k = 3.1 - 3 = 0.1$. Then,

$$f(2, 3.1) \approx f(2, 3) + f_x(2, 3) \cdot 0 + f_y(2, 3) \cdot 0.1 = 8 + 0 + 7 \cdot 0.1 = 8.7$$

We get the estimation $f(2, 3.1) \approx 8.7$.

5. Estimate Δf at $(2, 3)$ if $\Delta x = -0.3$ and $\Delta y = 0.2$.

SOLUTION The change in f can be estimated by the linear approximation as follows:

$$\Delta f \approx f_x(a, b)\Delta x + f_y(a, b)\Delta y$$

$$\Delta f \approx f_x(2, 3) \cdot (-0.3) + f_y(2, 3) \cdot 0.2$$

or

$$\Delta f \approx 5 \cdot (-0.3) + 7 \cdot 0.2 = -0.1$$

The estimated change is $\Delta f \approx -0.1$.

6. In the derivation of the equation for the plane determined by f_x and f_y, we used $\mathbf{w} \times \mathbf{v}$ for a normal vector to the plane. How would the choice of $\mathbf{v} \times \mathbf{w}$ for a normal vector have affected the resultant equation?

SOLUTION It would not have affected the result. Reversing the order of the vectors would result in negating the cross product, giving $\langle -f_x(a, b), -f_y(a, b), 1 \rangle$ as a normal, so the equation of the plane would be

$$-f_x(a, b)(x - a) - f_y(a, b)(y - b) + (z - f(a, b)) = 0$$

which also simplifies to

$$z = f(a, b) + f_x(a, b)(x - a) + f_y(a, b)(y - b)$$

Exercises

1. Find an equation of the tangent plane to the graph of $f(x, y) = 2x^2 - 4xy^2$ at $(-1, 2)$.

SOLUTION The equation of the tangent plane at the point $(-1, 2, 18)$ is

$$z = f(-1, 2) + f_x(-1, 2)(x + 1) + f_y(-1, 2)(y - 2) \tag{1}$$

We compute the function and its partial derivatives at the point $(-1, 2)$:

$$\begin{aligned} f(x, y) &= 2x^2 - 4xy^2 & f(-1, 2) &= 18 \\ f_x(x, y) &= 4x - 4y^2 &\Rightarrow\quad f_x(-1, 2) &= -20 \\ f_y(x, y) &= -8xy & f_y(-1, 2) &= 16 \end{aligned}$$

Substituting in (1) we obtain the following equation of the tangent plane:

$$z = 18 - 20(x + 1) + 16(y - 2) = -34 - 20x + 16y$$

That is,

$$z = -34 - 20x + 16y$$

In Exercises 3–10, find an equation of the tangent plane at the given point.

3. $f(x, y) = xy^2 + x^3y^2$, $(-1, 2)$

SOLUTION The equation of the tangent plane at $(-1, 2)$ is

$$z = f(-1, 2) + f_x(-1, 2)(x - (-1)) + f_y(-1, 2)(y - 2) \tag{1}$$

Computing the values of f and its partial derivatives, we get

$$\begin{aligned} f(x, y) &= xy^2 + x^3y^2 & f(-1, 2) &= -8 \\ f_x(x, y) &= y^2 + 3x^2y^2 &\Rightarrow\quad f_x(-1, 2) &= 16 \\ f_y(x, y) &= 2xy + 2x^3y & f_y(-1, 2) &= -8 \end{aligned}$$

Substituting in (1) we obtain the following equation of the tangent plane:

$$z = -8 + 16(x + 1) - 8(y - 2) = 16x - 8y + 24$$

5. $f(x, y) = x^2 + y^{-2}$, $(4, 1)$

SOLUTION The equation of the tangent plane at $(4, 1)$ is

$$z = f(4, 1) + f_x(4, 1)(x - 4) + f_y(4, 1)(y - 1) \tag{1}$$

We compute the values of f and its partial derivatives at $(4, 1)$:

$$\begin{aligned} f(x, y) &= x^2 + y^{-2} & f(4, 1) &= 17 \\ f_x(x, y) &= 2x &\Rightarrow\quad f_x(4, 1) &= 8 \\ f_y(x, y) &= -2y^{-3} & f_y(4, 1) &= -2 \end{aligned}$$

Substituting in (1) we obtain the following equation of the tangent plane:

$$z = 17 + 8(x - 4) - 2(y - 1) = 8x - 2y - 13$$

7. $F(r, s) = r^2 s^{-1/2} + s^{-3}$, $(2, 1)$

SOLUTION The equation of the tangent plane at $(2, 1)$ is

$$z = f(2, 1) + f_r(2, 1)(r - 2) + f_s(2, 1)(s - 1) \tag{1}$$

We compute f and its partial derivatives at $(2, 1)$:

$$f(r, s) = r^2 s^{-1/2} + s^{-3} \qquad f(2, 1) = 5$$

$$f_r(r, s) = 2rs^{-1/2} \qquad \Rightarrow \qquad f_r(2, 1) = 4$$

$$f_s(r, s) = -\frac{1}{2}r^2 s^{-3/2} - 3s^{-4} \qquad f_s(2, 1) = -5$$

We substitute these values in (1) to obtain the following equation of the tangent plane:

$$z = 5 + 4(r - 2) - 5(s - 1) = 4r - 5s + 2$$

9. $f(x, y) = \operatorname{sech}(x - y)$, $(\ln 4, \ln 2)$

SOLUTION The equation of the tangent plane at $(\ln 4, \ln 2)$ is:

$$z = f(\ln 4, \ln 2) + f_x(\ln 4, \ln 2)(x - \ln 4) + f_y(\ln 4, \ln 2)(y - \ln 2)$$

We compute f and its partial derivatives at $(\ln 4, \ln 2)$:

$$f(x, y) = \operatorname{sech}(x - y), \quad f(\ln 4, \ln 2) = \operatorname{sech}(\ln 2) = \frac{4}{5}$$

$$f_x(x, y) = -\tanh(x - y)\operatorname{sech}(x - y), \quad f_x(\ln 4, \ln 2) = -\tanh(\ln 2)\operatorname{sech}(\ln 2) = -\frac{12}{25}$$

$$f_y(x, y) = \tanh(x - y)\operatorname{sech}(x - y), \quad f_y(\ln 4, \ln 2) = \tanh(\ln 2)\operatorname{sech}(\ln 2) = \frac{12}{25}$$

We substitute these values in the tangent plane equation to obtain:

$$z = \frac{4}{5} - \frac{12}{25}(x - \ln 4) + \frac{12}{25}(x - \ln 2) = -\frac{4}{25}(3x - 3y - 5 - \ln 8)$$

11. Find the points on the graph of $z = 3x^2 - 4y^2$ at which the vector $\mathbf{n} = \langle 3, 2, 2 \rangle$ is normal to the tangent plane.

SOLUTION The equation of the tangent plane at the point $(a, b, f(a, b))$ on the graph of $z = f(x, y)$ is

$$z = f(a, b) + f_x(a, b)(x - a) + f_y(a, b)(y - b)$$

or

$$f_x(a, b)(x - a) + f_y(a, b)(y - b) - z + f(a, b) = 0$$

Therefore, the following vector is normal to the plane:

$$\mathbf{v} = \langle f_x(a, b), f_y(a, b), -1 \rangle$$

We compute the partial derivatives of the function $f(x, y) = 3x^2 - 4y^2$:

$$f_x(x, y) = 6x \quad \Rightarrow \quad f_x(a, b) = 6a$$

$$f_y(x, y) = -8y \quad \Rightarrow \quad f_y(a, b) = -8b$$

Therefore, the vector $\mathbf{v} = \langle 6a, -8b, -1 \rangle$ is normal to the tangent plane at (a, b). Since we want $\mathbf{n} = \langle 3, 2, 2 \rangle$ to be normal to the plane, the vectors $\mathbf{v}$ and $\mathbf{n}$ must be parallel. That is, the following must hold:

$$\frac{6a}{3} = \frac{-8b}{2} = -\frac{1}{2}$$

which implies that $a = -\frac{1}{4}$ and $b = \frac{1}{8}$. We compute the z-coordinate of the point:

$$z = 3 \cdot \left(-\frac{1}{4}\right)^2 - 4\left(\frac{1}{8}\right)^2 = \frac{1}{8}$$

The point on the graph at which the vector $\mathbf{n} = \langle 3, 2, 2 \rangle$ is normal to the tangent plane is $\left(-\frac{1}{4}, \frac{1}{8}, \frac{1}{8}\right)$.

13. Find the points on the graph of $f(x, y) = 3x^2 - xy - y^2$ at which the tangent plane is horizontal.

SOLUTION The tangent plane is horizontal at the points (x, y) where $f_x(x, y) = f_y(x, y) = 0$. We have

$$f_x(x, y) = 6x - y, \quad f_y(x, y) = -x - 2y$$

Solving the system

$$6x - y = 0$$
$$-x - 2y = 0$$

gives $x = y = 0$, so the only such point is $(0, 0)$.

15. Find the linearization $L(x, y)$ of $f(x, y) = x^2 y^3$ at $(a, b) = (2, 1)$. Use it to estimate $f(2.01, 1.02)$ and $f(1.97, 1.01)$, and compare with values obtained using a calculator.

SOLUTION

(a) We compute the value of the function and its partial derivatives at $(a, b) = (2, 1)$:

$$f(x, y) = x^2 y^3 \qquad f(2, 1) = 4$$
$$f_x(x, y) = 2xy^3 \quad \Rightarrow \quad f_x(2, 1) = 4$$
$$f_y(x, y) = 3x^2 y^2 \qquad f_y(2, 1) = 12$$

The linear approximation is therefore

$$f(x, y) \approx f(2, 1) + f_x(2, 1)(x - 2) + f_y(2, 1)(y - 1)$$
$$f(x, y) \approx 4 + 4(x - 2) + 12(y - 1) = -16 + 4x + 12y$$

(b) For $h = x - 2$ and $k = y - 1$ we have the following form of the linear approximation at $(a, b) = (2, 1)$:

$$f(x, y) \approx f(2, 1) + f_x(2, 1)h + f_y(2, 1)k = 4 + 4h + 12k$$

To compute $f(2.01, 1.02)$ we set $h = 2.01 - 2 = 0.01$, $k = 1.02 - 1 = 0.02$ to obtain

$$f(2.01, 1.02) \approx 4 + 4 \cdot 0.01 + 12 \cdot 0.02 = 4.28$$

The actual value is

$$f(2.01, 1.02) = 2.01^2 \cdot 1.02^3 = 4.2874$$

To compute $f(1.97, 1.01)$ we set $h = 1.97 - 2 = -0.03$, $k = 1.01 - 1 = 0.01$ to obtain

$$f(1.97, 1.01) \approx 4 + 4 \cdot (-0.03) + 12 \cdot 0.01 = 4$$

The actual value is

$$f(1.97, 1.01) = 1.97^2 \cdot 1.01^3 = 3.998$$

17. Let $f(x, y) = x^3 y^{-4}$. Use Eq. (4) to estimate the change

$$\Delta f = f(2.03, 0.9) - f(2, 1)$$

SOLUTION We compute the function and its partial derivatives at $(a, b) = (2, 1)$:

$$f(x, y) = x^3 y^{-4} \qquad f(2, 1) = 8$$
$$f_x(x, y) = 3x^2 y^{-4} \quad \Rightarrow \quad f_x(2, 1) = 12$$
$$f_y(x, y) = -4x^3 y^{-5} \qquad f_y(2, 1) = -32$$

Also, $\Delta x = 2.03 - 2 = 0.03$ and $\Delta y = 0.9 - 1 = -0.1$. Therefore,

$$\Delta f = f(2.03, 0.9) - f(2, 1) \approx f_x(2, 1)\Delta x + f_y \Delta y = 12 \cdot 0.03 + (-32) \cdot (-0.1) = 3.56$$
$$\Delta f \approx 3.56$$

19. Use the Linear Approximation of $f(x, y) = e^{x^2 + y}$ at $(0, 0)$ to estimate $f(0.01, -0.02)$. Compare with the value obtained using a calculator.

SOLUTION The linear approximation of f at the point $(0, 0)$ is

$$f(h, k) \approx f(0, 0) + f_x(0, 0)h + f_y(0, 0)k \quad (1)$$

We first must compute f and its partial derivative at the point $(0, 0)$. Using the Chain Rule we obtain

$$f(x, y) = e^{x^2+y} \quad\quad f(0, 0) = e^0 = 1$$
$$f_x(x, y) = 2xe^{x^2+y} \quad\Rightarrow\quad f_x(0, 0) = 2 \cdot 0 \cdot e^0 = 0$$
$$f_y(x, y) = e^{x^2+y} \quad\quad f_y(0, 0) = e^0 = 1$$

We substitute these values and $h = 0.01$, $k = -0.02$ in (1) to obtain

$$f(0.01, -0.02) \approx 1 + 0 \cdot 0.01 + 1 \cdot (-0.02) = 0.98$$

The actual value is $f(0.01, -0.02) = e^{0.01^2 - 0.02} \approx 0.9803$.

21. Find the linearization of $f(x, y, z) = z\sqrt{x+y}$ centered at $(8, 4, 5)$.

SOLUTION The linear approximation of f at the point $(8, 4, 5)$ is:

$$f(x, y, z) \approx f(8, 4, 5) + f_x(8, 4, 5)(x - 8) + f_y(8, 4, 5)(y - 4) + f_z(8, 4, 5)(z - 5)$$

We compute the values of f and its partial derivatives at $(8, 4, 5)$:

$$f(x, y, z) = z\sqrt{x+y}, \quad\quad f(8, 4, 5) = 5\sqrt{12} = 10\sqrt{3}$$

$$f_x(x, y, z) = \frac{z}{2\sqrt{x+y}}, \quad\quad f_x(8, 4, 5) = \frac{5}{2\sqrt{12}} = \frac{5}{4\sqrt{3}}$$

$$f_y(x, y, z) = \frac{z}{2\sqrt{x+y}}, \quad\quad f_y(8, 4, 5) = \frac{5}{2\sqrt{12}} = \frac{5}{4\sqrt{3}}$$

$$f_z(x, y, z) = \sqrt{x+y}, \quad\quad f_z(8, 4, 5) = \sqrt{12} = 4\sqrt{3}$$

Substituting these values we obtain the linearization:

$$f(x, y, z) \approx 10\sqrt{3} + \frac{5}{4\sqrt{3}}(x - 8) + \frac{5}{4\sqrt{3}}(y - 4) + 4\sqrt{3}(z - 5)$$

$$= \frac{5}{4\sqrt{3}}(x - 8) + \frac{5}{4\sqrt{3}}(y - 4) + 4\sqrt{3}z - 15\sqrt{3}$$

23. Estimate $f(2.1, 3.8)$ assuming that

$$f(2, 4) = 5, \quad f_x(2, 4) = 0.3, \quad f_y(2, 4) = -0.2$$

SOLUTION We use the linear approximation of f at the point $(2, 4)$, which is

$$f(2 + h, 4 + k) \approx f(2, 4) + f_x(2, 4)h + f_y(2, 4)k$$

Substituting the given values and $h = 0.1$, $k = -0.2$ we obtain the following approximation:

$$f(2.1, 3.8) \approx 5 + 0.3 \cdot 0.1 + 0.2 \cdot 0.2 = 5.07$$

In Exercises 25–30, use the Linear Approximation to estimate the value. Compare with the value given by a calculator.

25. $(2.01)^3(1.02)^2$

SOLUTION The number $(2.01)^3(1.02)^2$ is a value of the function $f(x, y) = x^3y^2$. We use the linear approximation at $(2, 1)$, which is

$$f(2 + h, 1 + k) \approx f(2, 1) + f_x(2, 1)h + f_y(2, 1)k \quad (1)$$

We compute the value of the function and its partial derivatives at $(2, 1)$:

$$f(x, y) = x^3y^2 \quad\quad f(2, 1) = 8$$
$$f_x(x, y) = 3x^2y^2 \quad\Rightarrow\quad f_x(2, 1) = 12$$
$$f_y(x, y) = 2x^3y \quad\quad f_y(2, 1) = 16$$

Substituting these values and $h = 0.01$, $k = 0.02$ in (1) gives the approximation

$$(2.01)^3(1.02)^2 \approx 8 + 12 \cdot 0.01 + 16 \cdot 0.02 = 8.44$$

The value given by a calculator is 8.4487. The error is 0.0087 and the percentage error is

$$\text{Percentage error} \approx \frac{0.0087 \cdot 100}{8.4487} = 0.103\%$$

27. $\sqrt{3.01^2 + 3.99^2}$

SOLUTION This is a value of the function $f(x, y) = \sqrt{x^2 + y^2}$. We use the linear approximation at the point $(3, 4)$, which is

$$f(3 + h, 4 + k) \approx f(3, 4) + f_x(3, 4)h + f_y(3, 4)k \tag{1}$$

Using the chain Rule gives the following partial derivatives:

$$f(x, y) = \sqrt{x^2 + y^2} \qquad\qquad f(3, 4) = 5$$

$$f_x(x, y) = \frac{2x}{2\sqrt{x^2 + y^2}} = \frac{x}{\sqrt{x^2 + y^2}} \quad\Rightarrow\quad f_x(3, 4) = \frac{3}{5}$$

$$f_y(x, y) = \frac{2y}{2\sqrt{x^2 + y^2}} = \frac{y}{\sqrt{x^2 + y^2}} \qquad f_y(3, 4) = \frac{4}{5}$$

Substituting these values and $h = 0.01$, $k = -0.01$ in (1) gives the following approximation:

$$\sqrt{3.01^2 + 3.99^2} \approx 5 + \frac{3}{5} \cdot 0.01 + \frac{4}{5} \cdot (-0.01) = 4.998$$

The value given by a calculator is $\sqrt{3.01^2 + 3.99^2} \approx 4.99802$. The error is 0.00002 and the percentage error is at most

$$\text{Percentage error} \approx \frac{0.00002 \cdot 100}{4.99802} = 0.0004002\%$$

29. $\sqrt{(1.9)(2.02)(4.05)}$

SOLUTION We use the linear approximation of the function $f(x, y, z) = \sqrt{xyz}$ at the point $(2, 2, 4)$, which is

$$f(2 + h, 2 + k, 4 + l) \approx f(2, 2, 4) + f_x(2, 2, 4)h + f_y(2, 2, 4)k + f_z(2, 2, 4)l \tag{1}$$

We compute the values of the function and its partial derivatives at $(2, 2, 4)$:

$$f(x, y, z) = \sqrt{xyz} \qquad\qquad f(2, 2, 4) = 4$$

$$f_x(x, y, z) = \frac{yz}{2\sqrt{xyz}} = \frac{1}{2}\sqrt{\frac{yz}{x}} \quad\Rightarrow\quad f_x(2, 2, 4) = 1$$

$$f_y(x, y, z) = \frac{xz}{2\sqrt{xyz}} = \frac{1}{2}\sqrt{\frac{xz}{y}} \qquad f_y(2, 2, 4) = 1$$

$$f_z(x, y, z) = \frac{xy}{2\sqrt{xyz}} = \frac{1}{2}\sqrt{\frac{xy}{z}} \qquad f_z(2, 2, 4) = \frac{1}{2}$$

Substituting these values and $h = -0.1$, $k = 0.02$, $l = 0.05$ in (1) gives the following approximation:

$$\sqrt{(1.9)(2.02)(4.05)} = 4 + 1 \cdot (-0.1) + 1 \cdot 0.02 + \frac{1}{2}(0.05) = 3.945$$

The value given by a calculator is:

$$\sqrt{(1.9)(2.02)(4.05)} \approx 3.9426$$

31. Suppose that the plane tangent to $z = f(x, y)$ at $(-2, 3, 4)$ has equation $4x + 2y + z = 2$. Estimate $f(-2.1, 3.1)$.

SOLUTION The tangent plane $z = 2 - 4x - 2y$ is also a linear approximation for f near $(-2, 3)$, so we can thus calculate the following:

$$f(-2.1, 3.1) \approx 2 - 4(-2.1) - 2(3.1) = 4.2$$

In Exercises 33–36, let $I = W/H^2$ denote the BMI described in Example 6.

33. A child has weight $W = 34$ kg and height $H = 1.3$ m. Use the linear approximation to estimate the change in I if (W, H) changes to $(36, 1.32)$.

SOLUTION Let $\Delta I = I(36, 1.32) - I(34, 1.3)$ denote the change in I. Using the linear approximation of I at the point $(34, 1.3)$ we have

$$I(34 + h, 1.3 + k) - I(34, 1.3) \approx \frac{\partial I}{\partial W}(34, 1.3)h + \frac{\partial I}{\partial H}(34, 1.3)k$$

For $h = 2, k = 0.02$ we obtain

$$\Delta I \approx \frac{\partial I}{\partial W}(34, 1.3) \cdot 2 + \frac{\partial I}{\partial H}(34, 1.3) \cdot 0.02 \tag{1}$$

We compute the partial derivatives in (1):

$$\frac{\partial I}{\partial W} = \frac{\partial}{\partial W}\frac{W}{H^2} = \frac{1}{H^2} \quad\Rightarrow\quad \frac{\partial I}{\partial W}(34, 1.3) = 0.5917$$

$$\frac{\partial I}{\partial H} = W\frac{\partial}{\partial H}H^{-2} = W \cdot (-2H^{-3}) = \frac{-2W}{H^3} \quad\Rightarrow\quad \frac{\partial I}{\partial H}(34, 1.3) = -30.9513$$

Substituting the partial derivatives in (1) gives the following estimation of ΔI:

$$\Delta I \approx 0.5917 \cdot 2 - 30.9513 \cdot 0.02 = 0.5644$$

35. (a) Show that $\Delta I \approx 0$ if $\Delta H / \Delta W \approx H/2W$.

(b) Suppose that $(W, H) = (25, 1.1)$. What increase in H will leave I (approximately) constant if W is increased by 1 kg?

SOLUTION

(a) The linear approximation implies that

$$\Delta I \approx \frac{\partial I}{\partial W}\Delta W + \frac{\partial I}{\partial H}\Delta H$$

Hence, $\Delta I \approx 0$ if

$$\frac{\partial I}{\partial W}\Delta W + \frac{\partial I}{\partial H}\Delta H = 0 \tag{1}$$

We compute the partial derivatives of $I = \frac{W}{H^2}$:

$$\frac{\partial I}{\partial W} = \frac{\partial}{\partial W}\left(\frac{W}{H^2}\right) = \frac{1}{H^2}$$

$$\frac{\partial I}{\partial H} = W\frac{\partial}{\partial H}(H^{-2}) = -2WH^{-3} = \frac{-2W}{H^3}$$

We substitute the partial derivatives in (1) to obtain

$$\frac{1}{H^2}\Delta W - \frac{2W}{H^3}\Delta H = 0$$

Hence,

$$\frac{1}{H^2}\Delta W = \frac{2W}{H^3}\Delta H$$

or

$$\frac{\Delta H}{\Delta W} = \frac{1}{H^2} \cdot \frac{H^3}{2W} = \frac{H}{2W}$$

(b) In part (a) we showed that if $\frac{\Delta H}{\Delta W} = \frac{H}{2W}$, then I remains approximately constant. We thus substitute $W = 25$, $H = 1.1$, $\Delta W = 1$, and solve for ΔH. This gives

$$\frac{\Delta H}{1} = \frac{1.1}{50} \quad \Rightarrow \quad \Delta H \approx 0.022 \text{ meters}$$

That is, an increase of 0.022 meters in H will leave I approximately constant.

37. A cylinder of radius r and height h has volume $V = \pi r^2 h$.

(a) Use the Linear Approximation to show that

$$\frac{\Delta V}{V} \approx \frac{2\Delta r}{r} + \frac{\Delta h}{h}$$

(b) Estimate the percentage increase in V if r and h are each increased by 2%.

(c) The volume of a certain cylinder V is determined by measuring r and h. Which will lead to a greater error in V: a 1% error in r or a 1% error in h?

SOLUTION

(a) The linear approximation is

$$\Delta V \approx V_r \Delta r + V_h \Delta h \tag{1}$$

We compute the partial derivatives of $V = \pi r^2 h$:

$$V_r = \pi h \frac{\partial}{\partial r} r^2 = 2\pi h r$$

$$V_h = \pi r^2 \frac{\partial}{\partial h} h = \pi r^2$$

Substituting in (1) gives

$$\Delta V \approx 2\pi h r \Delta r + \pi r^2 \Delta h$$

We divide by $V = \pi r^2 h$ to obtain

$$\frac{\Delta V}{V} \approx \frac{2\pi h r \Delta r}{V} + \frac{\pi r^2 \Delta h}{V} = \frac{2\pi h r \Delta r}{\pi r^2 h} + \frac{\pi r^2 \Delta h}{\pi r^2 h} = \frac{2\Delta r}{r} + \frac{\Delta h}{h}$$

That is,

$$\frac{\Delta V}{V} \approx \frac{2\Delta r}{r} + \frac{\Delta h}{h}$$

(b) The percentage increase in V is, by part (a),

$$\frac{\Delta V}{V} \cdot 100 \approx 2\frac{\Delta r}{r} \cdot 100 + \frac{\Delta h}{h} \cdot 100$$

We are given that $\frac{\Delta r}{r} \cdot 100 = 2$ and $\frac{\Delta h}{h} \cdot 100 = 2$, hence the percentage increase in V is

$$\frac{\Delta V}{V} \cdot 100 = 2 \cdot 2 + 2 = 6\%$$

(c) The percentage error in V is

$$\frac{\Delta V}{V} \cdot 100 = 2\frac{\Delta r}{r} \cdot 100 + \frac{\Delta h}{h} \cdot 100$$

A 1% error in r implies that $\frac{\Delta r}{r} \cdot 100 = 1$. Assuming that there is no error in h, we get

$$\frac{\Delta V}{V} \cdot 100 = 2 \cdot 1 + 0 = 2\%$$

A 1% in h implies that $\frac{\Delta h}{h} \cdot 100 = 1$. Assuming that there is no error in r, we get

$$\frac{\Delta V}{V} \cdot 100 = 0 + 1 = 1\%$$

We conclude that a 1% error in r leads to a greater error in V than a 1% error in h.

39. The monthly payment for a home loan is given by a function $f(P, r, N)$, where P is the principal (initial size of the loan), r the interest rate, and N the length of the loan in months. Interest rates are expressed as a decimal: A 6% interest rate is denoted by $r = 0.06$. If $P = \$100{,}000$, $r = 0.06$, and $N = 240$ (a 20-year loan), then the monthly payment is $f(100{,}000, 0.06, 240) = 716.43$. Furthermore, at these values, we have

$$\frac{\partial f}{\partial P} = 0.0071, \qquad \frac{\partial f}{\partial r} = 5769, \qquad \frac{\partial f}{\partial N} = -1.5467$$

Estimate:

(a) The change in monthly payment per $1000 increase in loan principal
(b) The change in monthly payment if the interest rate increases to $r = 6.5\%$ and $r = 7\%$
(c) The change in monthly payment if the length of the loan increases to 24 years

SOLUTION

(a) The linear approximation to $f(P, r, N)$ is

$$\Delta f \approx \frac{\partial f}{\partial P}\Delta P + \frac{\partial f}{\partial r}\Delta r + \frac{\partial f}{\partial N}\Delta N$$

We are given that $\frac{\partial f}{\partial P} = 0.0071$, $\frac{\partial f}{\partial r} = 5769$, $\frac{\partial f}{\partial N} = -1.5467$, and $\Delta P = 1000$. Assuming that $\Delta r = 0$ and $\Delta N = 0$, we get

$$\Delta f \approx 0.0071 \cdot 1000 = 7.1$$

The change in monthly payment per thousand dollar increase in loan principal is $7.1.

(b) By the given data, we have

$$\Delta f \approx 0.0071\Delta P + 5769\Delta r - 1.5467\Delta N \qquad (1)$$

The interest rate 6.5% corresponds to $r = 0.065$, and the interest rate 7% corresponds to $r = 0.07$. In the first case $\Delta r = 0.065 - 0.06 = 0.005$ and in the second case $\Delta r = 0.07 - 0.06 = 0.01$. Substituting in (1), assuming that $\Delta P = 0$ and $\Delta N = 0$, gives

$$\Delta f = 5769 \cdot 0.005 = \$28.845$$

$$\Delta f = 5769 \cdot 0.01 = \$57.69$$

(c) We substitute $\Delta N = (24 - 20) \cdot 12 = 48$ months and $\Delta r = \Delta N = 0$ in (1) to obtain

$$\Delta f \approx -1.5467 \cdot 48 = -74.2416$$

The monthly payment will be reduced by $74.2416.

41. The volume V of a right-circular cylinder is computed using the values 3.5 m for diameter and 6.2 m for height. Use the Linear Approximation to estimate the maximum error in V if each of these values has a possible error of at most 5%. Recall that $V = \pi r^2 h$.

SOLUTION We denote by d and h the diameter and height of the cylinder, respectively. By the Formula for the Volume of a Cylinder we have

$$V = \pi \left(\frac{d}{2}\right)^2 h = \frac{\pi}{4}d^2 h$$

The linear approximation is

$$\Delta V \approx \frac{\partial V}{\partial d}\Delta d + \frac{\partial V}{\partial h}\Delta h \qquad (1)$$

We compute the partial derivatives at $(d, h) = (3.5, 6.2)$:

$$\frac{\partial V}{\partial d}(d, h) = \frac{\pi}{4}h \cdot 2d = \frac{\pi}{2}hd \qquad \frac{\partial V}{\partial d}(3.5, 6.2) \approx 34.086$$

$$\frac{\partial V}{\partial h}(d, h) = \frac{\pi}{4}d^2 \qquad \Rightarrow \qquad \frac{\partial V}{\partial h}(3.5, 6.2) = 9.621$$

Substituting these derivatives in (1) gives

$$\Delta V \approx 34.086\Delta d + 9.621\Delta h \qquad (2)$$

We are given that the errors in the measurements of d and h are at most 5%. Hence,

$$\frac{\Delta d}{3.5} = 0.05 \quad \Rightarrow \quad \Delta d = 0.175$$

$$\frac{\Delta h}{6.2} = 0.05 \quad \Rightarrow \quad \Delta h = 0.31$$

Substituting in (2) we obtain

$$\Delta V \approx 34.086 \cdot 0.175 + 9.621 \cdot 0.31 \approx 8.948$$

The error in V is approximately 8.948 meters. The percentage error is at most

$$\frac{\Delta V \cdot 100}{V} = \frac{8.948 \cdot 100}{\frac{\pi}{4} \cdot 3.5^2 \cdot 6.2} = 15\%$$

Further Insights and Challenges

43. This exercise shows directly (without using Theorem 1) that the function $f(x, y) = 5x + 4y^2$ from Example 1 is differentiable at $(a, b) = (2, 1)$.
(a) Show that $f(x, y) = L(x, y) + e(x, y)$ with $e(x, y) = 4(y - 1)^2$.
(b) Show that

$$0 \le \frac{e(x, y)}{\sqrt{(x - 2)^2 + (y - 1)^2}} \le 4|y - 1|$$

(c) Verify that $f(x, y)$ is differentiable.

SOLUTION According to Example 1,

$$L(x, y) = -4 + 5x + 8y$$

(a) We compute the difference:

$$f(x, y) - L(x, y) = (5x + 4y^2) - (-4 + 5x + 8y)$$
$$= 4y^2 - 8y + 4 = 4(y - 1)^2$$

Therefore, $f(x, y) = L(x, y) + e(x, y) = L(x, y) + 4(y - 1)^2$, so that $e(x, y) = 4(y - 1)^2$.
(b)

$$\frac{e(x, y)}{\sqrt{(x - 2)^2 + (y - 1)^2}} = \frac{4(y - 1)^2}{\sqrt{(x - 2)^2 + (y - 1)^2}} \le \frac{4(y - 1)^2}{\sqrt{(y - 1)^2}} = 4|y - 1|$$

because we have made the denominator smaller.
(c) We have

$$\frac{f(x, y) - L(x, y)}{\sqrt{(x - 2)^2 + (y - 1)^2}} = \frac{e(x, y)}{\sqrt{(x - 2)^2 + (y - 1)^2}}$$

and

$$0 \le \frac{e(x, y)}{\sqrt{(x - 2)^2 + (y - 1)^2}} \le 4|y - 1|$$

But $\lim_{(x,y) \to (2,1)} 4|y - 1| = 0$, and therefore

$$\lim_{(x,y) \to (2,1)} (f(x, y) - L(x, y)) = \lim_{(x,y) \to (2,1)} \frac{e(x, y)}{\sqrt{(x - 2)^2 + (y - 1)^2}} = 0$$

by the Squeeze Theorem. This proves that $f(x, y)$ is differentiable at $(2, 1)$ since $f_x(2, 1)$ and $f_y(2, 1)$ are both defined.

45. Differentiability Implies Continuity Use the definition of differentiability to prove that if f is differentiable at (a, b), then f is continuous at (a, b).

SECTION 14.4 | **Differentiability, Tangent Planes, and Linear Approximation** (LT SECTION 15.4)

SOLUTION Suppose that f is differentiable at (a,b), then we know f is differentiable at (a,b), that is

$$f(x,y) = L(x,y) + e(x,y)$$

where $e(x,y)$ satisfies

$$\lim_{(x,y)\to(a,b)} \frac{e(x,y)}{\sqrt{(x-a)^2+(y-b)^2}} = \lim_{(x,y)\to(a,b)} E(x,y) = 0$$

and

$$L(x,y) = f(a,b) + f_x(a,b)(x-a) + f_y(a,b)(y-b)$$

We would like to show $\lim_{(x,y)\to(a,b)} f(x,y) = f(a,b)$, then f would be continuous at (a,b). Consider the following computation:

$$\lim_{(x,y)\to(a,b)} f(x,y) = \lim_{(x,y)\to(a,b)} L(x,y) + e(x,y)$$

$$= \lim_{(x,y)\to(a,b)} L(x,y) + E(x,y)\sqrt{(x-a)^2+(y-b)^2}$$

$$= \lim_{(x,y)\to(a,b)} f(a,b) + f_x(a,b)(x-a) + f_y(a,b)(y-b) + E(x,y)\sqrt{(x-a)^2+(y-b)^2}$$

$$= f(a,b) + 0 + 0 + 0 = f(a,b)$$

Therefore we have shown that if f is differentiable at (a,b) then f is continuous at (a,b).

47. Assumptions Matter Define $g(x,y) = 2xy(x+y)/(x^2+y^2)$ for $(x,y) \neq 0$ and $g(0,0) = 0$. In this exercise, we show that $g(x,y)$ is continuous at $(0,0)$ and that $g_x(0,0)$ and $g_y(0,0)$ exist, but $g(x,y)$ is not differentiable at $(0,0)$.

(a) Show using polar coordinates that $g(x,y)$ is continuous at $(0,0)$.
(b) Use the limit definitions to show that $g_x(0,0)$ and $g_y(0,0)$ exist and that both are equal to zero.
(c) Show that the linearization of $g(x,y)$ at $(0,0)$ is $L(x,y) = 0$.
(d) Show that if $g(x,y)$ were differentiable at $(0,0)$, we would have $\lim_{h\to 0} \frac{g(h,h)}{h} = 0$. Then observe that this is not the case because $g(h,h) = 2h$. This shows that $g(x,y)$ is not differentiable at $(0,0)$ and hence not differentiable at $(0,0)$.

SOLUTION
(a) We would like to show $\lim_{(x,y)\to(0,0)} g(x,y) = g(0,0)$. Consider the following, using polar coordinates, $x = r\cos\theta$ and $y = r\sin\theta$:

$$\lim_{(x,y)\to(0,0)} \frac{2xy(x+y)}{x^2+y^2} = \lim_{(r,\theta)\to(0,0)} \frac{2r^2\cos\theta\sin\theta(r\cos\theta + r\sin\theta)}{r^2\cos^2\theta + r^2\sin^2\theta}$$

$$= \lim_{(r,\theta)\to(0,0)} \frac{2r^3\cos\theta\sin\theta(\cos\theta + \sin\theta)}{r^2}$$

$$= \lim_{(r,\theta)\to(0,0)} 2r\cos\theta\sin\theta(\cos\theta + \sin\theta) = 0 = g(0,0)$$

Therefore $g(x,y)$ is continuous at $(0,0)$.
(b) Taking partial derivatives we have:

$$g_x(x,y) = \frac{2y^2(y-x)^2}{(x^2+y^2)^2}, \quad g_y(x,y) = \frac{2x^2(x-y)^2}{(x^2+y^2)^2}$$

But we need to use limit definitions for the partial derivatives. Consider the following:

$$g_x(0,0) = \lim_{h\to 0} \frac{g(h,0) - g(0,0)}{h}$$

$$= \lim_{h\to 0} \frac{1}{h}(0-0) = 0$$

$$g_y(0,0) = \lim_{h\to 0} \frac{g(0,h) - g(0,0)}{h}$$

$$= \lim_{h\to 0} \frac{1}{h}(0-0) = 0$$

Thus both partial derivatives exist and $g_x(0,0) = 0$ and $g_y(0,0) = 0$.

(c) We know that the linearization of g will be:

$$g(x, y) \approx g(0,0) + g_x(0,0)(x-0) + g_y(0,0)(y-0)$$

We are given that $g(0,0) = 0$. In part (b) we know $g_x(0,0) = 0$ and $g_y(0,0) = 0$. Substituting in these values in the linearization we have:

$$g(x, y) \approx 0 + 0 + 0 = 0$$

(d) We know if g were differentiable at $(0,0)$, we would have:

$$\lim_{h \to 0} \frac{g(h,h)}{h} = 0$$

However, we know:

$$g(h, h) = \frac{2h^2(2h)}{2h^2} = 2h, \quad \frac{g(h,h)}{h} = \frac{2h}{h} = 2$$

This is a contradiction, $g(x, y)$ is not differentiable at $(0,0)$ and hence, is not differentiable at $(0,0)$.

14.5 The Gradient and Directional Derivatives   (LT Section 15.5)

Preliminary Questions

1. Which of the following is a possible value of the gradient ∇f of a function $f(x, y)$ of two variables?

(a) 5 (b) $\langle 3, 4 \rangle$ (c) $\langle 3, 4, 5 \rangle$

SOLUTION The gradient of $f(x, y)$ is a vector with two components, hence the possible value of the gradient $\nabla f = \left\langle \frac{\partial f}{\partial x}, \frac{\partial f}{\partial y} \right\rangle$ is (b).

2. True or false? A differentiable function increases at the rate $\|\nabla f_P\|$ in the direction of ∇f_P.

SOLUTION The statement is true. The value $\|\nabla f_P\|$ is the rate of increase of f in the direction ∇f_P.

3. Describe the two main geometric properties of the gradient ∇f.

SOLUTION The gradient of f points in the direction of maximum rate of increase of f and is normal to the level curve (or surface) of f.

4. You are standing at a point where the temperature gradient vector is pointing in the northeast (NE) direction. In which direction(s) should you walk to avoid a change in temperature?

(a) NE (b) NW (c) SE (d) SW

SOLUTION The rate of change of the temperature T at a point P in the direction of a unit vector $\mathbf{u}$ is the directional derivative $D_{\mathbf{u}}T(P)$, which is given by the formula

$$D_{\mathbf{u}}T(P) = \|\nabla f_P\| \cos \theta$$

To avoid a change in temperature, we must choose the direction $\mathbf{u}$ so that $D_{\mathbf{u}}T(P) = 0$, that is, $\cos \theta = 0$, so $\theta = \frac{\pi}{2}$ or $\theta = \frac{3\pi}{2}$. Since the gradient at P is pointing NE, we should walk NW or SE to avoid a change in temperature. Thus, the answer is (b) and (c).

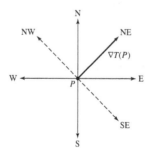

5. What is the rate of change of $f(x, y)$ at $(0,0)$ in the direction making an angle of $45°$ with the x-axis if $\nabla f(0,0) = \langle 2, 4 \rangle$?

SOLUTION By the formula for directional derivatives, and using the unit vector $\langle 1/\sqrt{2}, 1/\sqrt{2} \rangle$, we get $\langle 2, 4 \rangle \cdot \langle 1/\sqrt{2}, 1/\sqrt{2} \rangle = 6/\sqrt{2} = 3\sqrt{2}$.

Exercises

1. Let $f(x,y) = xy^2$ and $\mathbf{r}(t) = \left\langle \frac{1}{2}t^2, t^3 \right\rangle$.

(a) Calculate ∇f and $\mathbf{r}'(t)$.

(b) Use the Chain Rule for Paths to evaluate $\frac{d}{dt} f(\mathbf{r}(t))$ at $t = 1$ and $t = -1$.

SOLUTION

(a) We compute the partial derivatives of $f(x,y) = xy^2$:

$$\frac{\partial f}{\partial x} = y^2, \quad \frac{\partial f}{\partial y} = 2xy$$

The gradient vector is thus

$$\nabla f = \left\langle y^2, 2xy \right\rangle$$

Also,

$$\mathbf{r}'(t) = \left\langle \left(\frac{1}{2}t^2\right)', \left(t^3\right)' \right\rangle = \left\langle t, 3t^2 \right\rangle$$

(b) Using the Chain Rule gives

$$\frac{d}{dt} f(\mathbf{r}(t)) = \frac{\partial f}{\partial x} \frac{dx}{dt} + \frac{\partial f}{\partial y} \frac{dy}{dt} = y^2 \cdot t + 2xy \cdot 3t^2$$

To express this purely in terms of t, substitute the parametrization for x and y:

$$y^2 \cdot t + 2xy \cdot 3t^2 = \left(t^3\right)^2 \cdot t + 2\left(\frac{1}{2}t^2\right) \cdot t^3 \cdot 3t^2 = 4t^7$$

At the points $t = 1$ and $t = -1$, we get

$$\left.\frac{d}{dt}(f(\mathbf{r}(t)))\right|_{t=1} = 4 \cdot 1^7 = 4, \quad \left.\frac{d}{dt}(f(\mathbf{r}(t)))\right|_{t=-1} = 4 \cdot (-1)^7 = -4$$

3. Figure 14 shows the level curves of a function $f(x,y)$ and a path $\mathbf{r}(t)$, traversed in the direction indicated. State whether the derivative $\frac{d}{dt} f(\mathbf{r}(t))$ is positive, negative, or zero at points A–D.

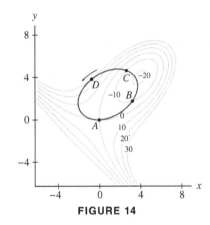

FIGURE 14

SOLUTION At points A and D, the path is (temporarily) tangent to one of the contour lines, which means that along the path $\mathbf{r}(t)$ the function $f(x,y)$ is (temporarily) constant, and so the derivative $\frac{d}{dt} f(\mathbf{r}(t))$ is zero. At point B, the path is moving from a higher contour (of -10) to a lower one (of -20), so the derivative is negative. At the point C, where the path moves from the contour of -10 toward the contour of value 0, the derivative is positive.

In Exercises 5–8, calculate the gradient.

5. $f(x,y) = \cos(x^2 + y)$

SOLUTION We find the partial derivatives using the Chain Rule:

$$\frac{\partial f}{\partial x} = -\sin\left(x^2 + y\right) \frac{\partial}{\partial x}\left(x^2 + y\right) = -2x \sin\left(x^2 + y\right)$$

$$\frac{\partial f}{\partial y} = -\sin\left(x^2 + y\right) \frac{\partial}{\partial y}\left(x^2 + y\right) = -\sin\left(x^2 + y\right)$$

The gradient vector is thus

$$\nabla f = \left\langle \frac{\partial f}{\partial x}, \frac{\partial f}{\partial y} \right\rangle = \left\langle -2x\sin\left(x^2 + y\right), -\sin\left(x^2 + y\right) \right\rangle = -\sin\left(x^2 + y\right)\langle 2x, 1 \rangle$$

7. $h(x, y, z) = xyz^{-3}$

SOLUTION We compute the partial derivatives of $h(x, y, z) = xyz^{-3}$, obtaining

$$\frac{\partial h}{\partial x} = yz^{-3}, \quad \frac{\partial h}{\partial y} = xz^{-3}, \quad \frac{\partial h}{\partial z} = xy \cdot \left(-3z^{-4}\right) = -3xyz^{-4}$$

The gradient vector is thus

$$\nabla h = \left\langle \frac{\partial h}{\partial x}, \frac{\partial h}{\partial y}, \frac{\partial h}{\partial z} \right\rangle = \left\langle yz^{-3}, xz^{-3}, -3xyz^{-4} \right\rangle$$

In Exercises 9–20, use the Chain Rule to calculate $\frac{d}{dt}f(\mathbf{r}(t))$ at the value of t given.

9. $f(x, y) = 3x - 7y$, $\quad \mathbf{r}(t) = \langle \cos t, \sin t \rangle$, $\quad t = 0$

SOLUTION By the Chain Rule for paths, we have

$$\frac{d}{dt}f(\mathbf{r}(t)) = \nabla f_{\mathbf{r}(t)} \cdot \mathbf{r}'(t) \tag{1}$$

We compute the gradient and the derivative $\mathbf{r}'(t)$:

$$\nabla f = \left\langle \frac{\partial f}{\partial x}, \frac{\partial f}{\partial y} \right\rangle = \langle 3, -7 \rangle, \quad \mathbf{r}'(t) = \langle -\sin t, \cos t \rangle$$

We determine these vectors at $t = 0$:

$$\mathbf{r}'(0) = \langle -\sin 0, \cos 0 \rangle = \langle 0, 1 \rangle$$

and since the gradient is a constant vector, we have

$$\nabla f_{\mathbf{r}(0)} = \nabla f_{(1,0)} = \langle 3, -7 \rangle$$

Substituting these vectors in (1) gives

$$\left.\frac{d}{dt}f(\mathbf{r}(t))\right|_{t=0} = \langle 3, -7 \rangle \cdot \langle 0, 1 \rangle = 0 - 7 = -7$$

11. $f(x, y) = x^2 - 3xy$, $\quad \mathbf{r}(t) = \langle \cos t, \sin t \rangle$, $\quad t = 0$

SOLUTION By the Chain Rule For Paths we have

$$\frac{d}{dt}f(\mathbf{r}(t)) = \nabla f_{\mathbf{r}(t)} \cdot \mathbf{r}'(t) \tag{1}$$

We compute the gradient and $\mathbf{r}'(t)$:

$$\nabla f = \left\langle \frac{\partial f}{\partial x}, \frac{\partial f}{\partial y} \right\rangle = \langle 2x - 3y, -3x \rangle$$

$$\mathbf{r}'(t) = \langle -\sin t, \cos t \rangle$$

At the point $t = 0$ we have

$$\mathbf{r}(0) = \langle \cos 0, \sin 0 \rangle = \langle 1, 0 \rangle$$

$$\mathbf{r}'(0) = \langle -\sin 0, \cos 0 \rangle = \langle 0, 1 \rangle$$

$$\left.\nabla f\right|_{\mathbf{r}(0)} = \nabla f_{(1,0)} = \langle 2 \cdot 1 - 3 \cdot 0, -3 \cdot 1 \rangle = \langle 2, -3 \rangle$$

Substituting in (1) we obtain

$$\left.\frac{d}{dt}f(\mathbf{r}(t))\right|_{t=0} = \langle 2, -3 \rangle \cdot \langle 0, 1 \rangle = -3$$

13. $f(x, y) = \sin(xy)$, $\mathbf{r}(t) = \langle e^{2t}, e^{3t} \rangle$, $t = 0$

SOLUTION By the Chain Rule for Paths we have

$$\frac{d}{dt} f(\mathbf{r}(t)) = \nabla f_{\mathbf{r}(t)} \cdot \mathbf{r}'(t) \qquad (1)$$

We compute the gradient and $\mathbf{r}'(t)$:

$$\nabla f = \left\langle \frac{\partial f}{\partial x}, \frac{\partial f}{\partial y} \right\rangle = \langle y\cos(xy), x\cos(xy) \rangle$$

$$\mathbf{r}'(t) = \langle 2e^{2t}, 3e^{3t} \rangle$$

At the point $t = 0$ we have

$$\mathbf{r}(0) = \left(e^0, e^0 \right) = (1, 1)$$

$$\mathbf{r}'(0) = \langle 2e^0, 3e^0 \rangle = \langle 2, 3 \rangle$$

$$\nabla f_{\mathbf{r}(0)} = \nabla f_{(1,1)} = \langle \cos 1, \cos 1 \rangle$$

Substituting the vectors in (1) we get

$$\left. \frac{d}{dt} f(\mathbf{r}(t)) \right|_{t=0} = \langle \cos 1, \cos 1 \rangle \cdot \langle 2, 3 \rangle = 5 \cos 1$$

15. $f(x, y) = x - xy$, $\mathbf{r}(t) = \langle t^2, t^2 - 4t \rangle$, $t = 4$

SOLUTION We compute the gradient and $\mathbf{r}'(t)$:

$$\nabla f = \left\langle \frac{\partial f}{\partial x}, \frac{\partial f}{\partial y} \right\rangle = \langle 1 - y, -x \rangle$$

$$\mathbf{r}'(t) = \langle 2t, 2t - 4 \rangle$$

At the point $t = 4$ we have

$$\mathbf{r}(4) = \left(4^2, 4^2 - 4 \cdot 4 \right) = (16, 0)$$

$$\mathbf{r}'(4) = \langle 2 \cdot 4, 2 \cdot 4 - 4 \rangle = \langle 8, 4 \rangle$$

$$\nabla f_{\mathbf{r}(4)} = \nabla f_{(16,0)} = \langle 1 - 0, -16 \rangle = \langle 1, -16 \rangle$$

We now use the Chain Rule for Paths to compute the following derivative:

$$\left. \frac{d}{dt} f(\mathbf{r}(t)) \right|_{t=4} = \nabla f_{\mathbf{r}(4)} \cdot \mathbf{r}'(4) = \langle 1, -16 \rangle \cdot \langle 8, 4 \rangle = 8 - 64 = -56$$

17. $f(x, y) = \ln x + \ln y$, $\mathbf{r}(t) = \langle \cos t, t^2 \rangle$, $t = \frac{\pi}{4}$

SOLUTION We compute the gradient and $\mathbf{r}'(t)$:

$$\nabla f = \left\langle \frac{\partial f}{\partial x}, \frac{\partial f}{\partial y} \right\rangle = \left\langle \frac{1}{x}, \frac{1}{y} \right\rangle$$

$$\mathbf{r}'(t) = \langle -\sin t, 2t \rangle$$

At the point $t = \frac{\pi}{4}$ we have

$$\mathbf{r}\left(\frac{\pi}{4}\right) = \left(\cos \frac{\pi}{4}, \left(\frac{\pi}{4}\right)^2 \right) = \left(\frac{\sqrt{2}}{2}, \frac{\pi^2}{16} \right)$$

$$\mathbf{r}'\left(\frac{\pi}{4}\right) = \left\langle -\sin \frac{\pi}{4}, \frac{2\pi}{4} \right\rangle = \left\langle -\frac{\sqrt{2}}{2}, \frac{\pi}{2} \right\rangle$$

$$\nabla f_{\mathbf{r}(\frac{\pi}{4})} = \nabla f_{\left(\frac{\sqrt{2}}{2}, \frac{\pi^2}{16}\right)} = \left\langle \sqrt{2}, \frac{16}{\pi^2} \right\rangle$$

Using the Chain Rule for Paths we obtain the following derivative:

$$\frac{d}{dt}f(\mathbf{r}(t))\bigg|_{t=\frac{\pi}{4}} = \nabla f_{\mathbf{r}(\frac{\pi}{4})} \cdot \mathbf{r}'\left(\frac{\pi}{4}\right) = \left\langle\sqrt{2}, \frac{16}{\pi^2}\right\rangle \cdot \left\langle-\frac{\sqrt{2}}{2}, \frac{\pi}{2}\right\rangle = -1 + \frac{8}{\pi} \approx 1.546$$

19. $g(x,y,z) = xyz^{-1}$, $\mathbf{r}(t) = \langle e^t, t, t^2 \rangle$, $t = 1$

SOLUTION By the Chain Rule for Paths we have

$$\frac{d}{dt}g(\mathbf{r}(t)) = \nabla g_{\mathbf{r}(t)} \cdot \mathbf{r}'(t) \qquad (1)$$

We compute the gradient and $\mathbf{r}'(t)$:

$$\nabla g = \left\langle\frac{\partial g}{\partial x}, \frac{\partial g}{\partial y}, \frac{\partial g}{\partial z}\right\rangle = \langle yz^{-1}, xz^{-1}, -xyz^{-2}\rangle$$

$$\mathbf{r}'(t) = \langle e^t, 1, 2t\rangle$$

At the point $t = 1$ we have

$$\mathbf{r}(1) = \langle e, 1, 1\rangle$$
$$\mathbf{r}'(1) = \langle e, 1, 2\rangle$$
$$\nabla g_{\mathbf{r}(1)} = \nabla g_{(e,1,1)} = \langle 1, e, -e\rangle$$

Substituting the vectors in (1) gives the following derivative:

$$\frac{d}{dt}g(\mathbf{r}(t))\bigg|_{t=1} = \langle 1, e, -e\rangle \cdot \langle e, 1, 2\rangle = e + e - 2e = 0$$

In Exercises 21–30, calculate the directional derivative in the direction of $\mathbf{v}$ *at the given point. Remember to use a unit vector in your directional derivative computation.*

21. $f(x,y) = x^2 + y^3$, $\mathbf{v} = \langle 4, 3\rangle$, $P = (1, 2)$

SOLUTION We first normalize the direction vector $\mathbf{v}$:

$$\mathbf{u} = \frac{\mathbf{v}}{\|\mathbf{v}\|} = \frac{\langle 4, 3\rangle}{\sqrt{4^2 + 3^2}} = \left\langle\frac{4}{5}, \frac{3}{5}\right\rangle$$

We compute the gradient of $f(x,y) = x^2 + y^3$ at the given point:

$$\nabla f = \left\langle\frac{\partial f}{\partial x}, \frac{\partial f}{\partial y}\right\rangle = \langle 2x, 3y^2\rangle \quad \Rightarrow \quad \nabla f_{(1,2)} = \langle 2, 12\rangle$$

Using the Theorem on Evaluating Directional Derivatives, we get

$$D_{\mathbf{u}}f(1,2) = \nabla f_{(1,2)} \cdot \mathbf{u} = \langle 2, 12\rangle \cdot \left\langle\frac{4}{5}, \frac{3}{5}\right\rangle = \frac{8}{5} + \frac{36}{5} = \frac{44}{5} = 8.8$$

23. $f(x,y) = x^2y^3$, $\mathbf{v} = \mathbf{i} + \mathbf{j}$, $P = (\frac{1}{6}, 3)$

SOLUTION We normalize $\mathbf{v}$ to obtain a unit vector $\mathbf{u}$ in the direction of $\mathbf{v}$:

$$\mathbf{u} = \frac{\mathbf{v}}{\|\mathbf{v}\|} = \frac{1}{\sqrt{2}}(\mathbf{i} + \mathbf{j}) = \frac{1}{\sqrt{2}}\mathbf{i} + \frac{1}{\sqrt{2}}\mathbf{j}$$

We compute the gradient of $f(x,y) = x^2y^3$ at the point P:

$$\nabla f = \left\langle\frac{\partial f}{\partial x}, \frac{\partial f}{\partial y}\right\rangle = \langle 2xy^3, 3x^2y^2\rangle \quad \Rightarrow \quad \nabla f_{(\frac{1}{6}, 3)} = \left\langle 2 \cdot \frac{1}{6} \cdot 3^3, 3 \cdot \frac{1}{6^2} \cdot 3^2\right\rangle = \left\langle 9, \frac{3}{4}\right\rangle = 9\mathbf{i} + \frac{3}{4}\mathbf{j}$$

The directional derivative in the direction $\mathbf{v}$ is thus

$$D_{\mathbf{u}}f\left(\frac{1}{6}, 3\right) = \nabla f_{(\frac{1}{6},3)} \cdot \mathbf{u} = \left(9\mathbf{i} + \frac{3}{4}\mathbf{j}\right) \cdot \left(\frac{1}{\sqrt{2}}\mathbf{i} + \frac{1}{\sqrt{2}}\mathbf{j}\right) = \frac{9}{\sqrt{2}} + \frac{3}{4\sqrt{2}} = \frac{39}{4\sqrt{2}}$$

25. $f(x,y) = \tan^{-1}(xy)$, $\mathbf{v} = \langle 1,1 \rangle$, $P = (3,4)$

SOLUTION We first normalize $\mathbf{v}$ to obtain a unit vector $\mathbf{u}$ in the direction $\mathbf{v}$:

$$\mathbf{u} = \frac{\mathbf{v}}{\|\mathbf{v}\|} = \frac{1}{\sqrt{2}} \langle 1,1 \rangle$$

We compute the gradient of $f(x,y) = \tan^{-1}(xy)$ at the point $P = (3,4)$:

$$\nabla f = \left\langle \frac{\partial f}{\partial x}, \frac{\partial f}{\partial y} \right\rangle = \left\langle \frac{y}{1+(xy)^2}, \frac{x}{1+(xy)^2} \right\rangle = \frac{1}{1+x^2 y^2} \langle y,x \rangle$$

$$\nabla f_{(3,4)} = \frac{1}{1+3^2 \cdot 4^2} \langle 4,3 \rangle = \frac{1}{145} \langle 4,3 \rangle$$

Therefore, the directional derivative in the direction $\mathbf{v}$ is

$$D_{\mathbf{u}} f(3,4) = \nabla f_{(3,4)} \cdot \mathbf{u} = \frac{1}{145} \langle 4,3 \rangle \cdot \frac{1}{\sqrt{2}} \langle 1,1 \rangle = \frac{1}{145\sqrt{2}} (4+3) = \frac{7}{145\sqrt{2}} = \frac{7\sqrt{2}}{290}$$

27. $f(x,y) = \ln(x^2 + y^2)$, $\mathbf{v} = 3\mathbf{i} - 2\mathbf{j}$, $P = (1,0)$

SOLUTION We normalize $\mathbf{v}$ to obtain a unit vector $\mathbf{u}$ in the direction $\mathbf{v}$:

$$\mathbf{u} = \frac{\mathbf{v}}{\|\mathbf{v}\|} = \frac{1}{\sqrt{3^2 + (-2)^2}} (3\mathbf{i} - 2\mathbf{j}) = \frac{1}{\sqrt{13}} (3\mathbf{i} - 2\mathbf{j})$$

We compute the gradient of $f(x,y) = \ln(x^2 + y^2)$ at the point $P = (1,0)$:

$$\nabla f = \left\langle \frac{\partial f}{\partial x}, \frac{\partial f}{\partial y} \right\rangle = \left\langle \frac{2x}{x^2 + y^2}, \frac{2y}{x^2 + y^2} \right\rangle = \frac{2}{x^2 + y^2} \langle x,y \rangle$$

$$\nabla f_{(1,0)} = \frac{2}{1^2 + 0^2} \langle 1,0 \rangle = \langle 2,0 \rangle = 2\mathbf{i}$$

The directional derivative in the direction $\mathbf{v}$ is thus

$$D_{\mathbf{u}} f(1,0) = \nabla f_{(1,0)} \cdot \mathbf{u} = 2\mathbf{i} \cdot \frac{1}{\sqrt{13}} (3\mathbf{i} - 2\mathbf{j}) = \frac{6}{\sqrt{13}}$$

29. $g(x,y,z) = xe^{-yz}$, $\mathbf{v} = \langle 1,1,1 \rangle$, $P = (1,2,0)$

SOLUTION We first compute a unit vector $\mathbf{u}$ in the direction $\mathbf{v}$:

$$\mathbf{u} = \frac{\mathbf{v}}{\|\mathbf{v}\|} = \frac{\langle 1,1,1 \rangle}{\sqrt{1^2 + 1^2 + 1^2}} = \frac{1}{\sqrt{3}} \langle 1,1,1 \rangle$$

We find the gradient of $f(x,y,z) = xe^{-yz}$ at the point $P = (1,2,0)$:

$$\nabla f = \left\langle \frac{\partial f}{\partial x}, \frac{\partial f}{\partial y}, \frac{\partial f}{\partial z} \right\rangle = \langle e^{-yz}, -xze^{-yz}, -xye^{-yz} \rangle = e^{-yz} \langle 1, -xz, -xy \rangle$$

$$\nabla f_{(1,2,0)} = e^0 \langle 1,0,-2 \rangle = \langle 1,0,-2 \rangle$$

The directional derivative in the direction $\mathbf{v}$ is thus

$$D_{\mathbf{u}} f(1,2,0) = \nabla f_{(1,2,0)} \cdot \mathbf{u} = \langle 1,0,-2 \rangle \cdot \frac{1}{\sqrt{3}} \langle 1,1,1 \rangle = \frac{1}{\sqrt{3}}(1+0-2) = -\frac{1}{\sqrt{3}}$$

31. Find the directional derivative of $f(x,y) = x^2 + 4y^2$ at the point $P = (3,2)$ in the direction pointing to the origin.

SOLUTION The direction vector is $\mathbf{v} = \overrightarrow{PO} = \langle -3,-2 \rangle$. A unit vector $\mathbf{u}$ in the direction $\mathbf{v}$ is obtained by normalizing $\mathbf{v}$. That is,

$$\mathbf{u} = \frac{\mathbf{v}}{\|\mathbf{v}\|} = \frac{\langle -3,-2 \rangle}{\sqrt{3^2 + 2^2}} = \frac{-1}{\sqrt{13}} \langle 3,2 \rangle$$

We compute the gradient of $f(x, y) = x^2 + 4y^2$ at the point $P = (3, 2)$:

$$\nabla f = \left\langle \frac{\partial f}{\partial x}, \frac{\partial f}{\partial y} \right\rangle = \langle 2x, 8y \rangle \quad \Rightarrow \quad \nabla f_{(3,2)} = \langle 6, 16 \rangle$$

The directional derivative is thus

$$D_{\mathbf{u}} f(3, 2) = \nabla f_{(3,2)} \cdot \mathbf{u} = \langle 6, 16 \rangle \cdot \frac{-1}{\sqrt{13}} \langle 3, 2 \rangle = \frac{-50}{\sqrt{13}}$$

In Exercises 33–36, determine the direction in which f has maximum rate of increase from P, and give the rate of change in that direction.

33. $f(x, y) = xe^{-y}$, $P = (2, 0)$

SOLUTION The gradient points in the direction of maximum rate increase, so we compute the gradient at P:

$$\nabla f = \langle e^{-y}, -xe^{-y} \rangle, \quad \nabla f_{(2,0)} = \langle 1, -2 \rangle$$

The direction of greatest increase is the direction $\langle 1, -2 \rangle$, and the rate of change in this direction is

$$\|\nabla f_{(2,0)}\| = \sqrt{1^2 + (-2)^2} = \sqrt{5}$$

35. $f(x, y, z) = \dfrac{xy}{z}$, $P = (1, -1, 3)$

SOLUTION The gradient points in the direction of maximum rate increase, so we compute the gradient at P:

$$\nabla f = \left\langle \frac{y}{z}, \frac{x}{z}, -\frac{xy}{z^2} \right\rangle, \quad \nabla f_{(1,-1,3)} = \left\langle -\frac{1}{3}, \frac{1}{3}, \frac{1}{9} \right\rangle$$

The direction of greatest increase is the direction $\left\langle -\frac{1}{3}, \frac{1}{3}, \frac{1}{9} \right\rangle$ (or $\langle -3, 3, 1 \rangle$), and the rate of change in this direction is

$$\|\nabla f_{(1,-1,3)}\| = \sqrt{\left(-\frac{1}{3}\right)^2 + \left(\frac{1}{3}\right)^2 + \left(\frac{1}{9}\right)^2} = \frac{1}{9}\sqrt{19}$$

37. Suppose that $\nabla f_P = \langle 2, -4, 4 \rangle$. Is f increasing or decreasing at P in the direction $\mathbf{v} = \langle 2, 1, 3 \rangle$?

SOLUTION We compute the derivative of f at P with respect to $\mathbf{v}$:

$$D_{\mathbf{v}} f(P) = \nabla f_P \cdot \mathbf{v} = \langle 2, -4, 4 \rangle \cdot \langle 2, 1, 3 \rangle = 4 - 4 + 12 = 12 > 0$$

Since the derivative is positive, f is increasing at P in the direction of $\mathbf{v}$.

39. Let $f(x, y, z) = \sin(xy + z)$ and $P = (0, -1, \pi)$. Calculate $D_{\mathbf{u}} f(P)$, where $\mathbf{u}$ is a unit vector making an angle $\theta = 30°$ with ∇f_P.

SOLUTION The directional derivative $D_{\mathbf{u}} f(P)$ is the following dot product:

$$D_{\mathbf{u}} f(P) = \nabla f_P \cdot \mathbf{u}$$

Since $\mathbf{u}$ is a unit vector making an angle $\theta = 30°$ with ∇f_P, we have by the properties of the dot product

$$D_{\mathbf{u}} f(P) = \|\nabla f_P\| \cdot \|\mathbf{u}\| \cos 30° = \frac{\sqrt{3}}{2} \|\nabla f_P\| \tag{1}$$

We now must find the gradient at P and its length:

$$\nabla f = \left\langle \frac{\partial f}{\partial x}, \frac{\partial f}{\partial y}, \frac{\partial f}{\partial z} \right\rangle = \langle y \cos(xy + z), x \cos(xy + z), \cos(xy + z) \rangle = \cos(xy + z) \langle y, x, 1 \rangle$$

$$\nabla f_{(0,-1,\pi)} = \cos \pi \langle -1, 0, 1 \rangle = -1 \langle -1, 0, 1 \rangle = \langle 1, 0, -1 \rangle$$

Hence,

$$\|\nabla f_{(0,-1,\pi)}\| = \sqrt{1^2 + 0^2 + (-1)^2} = \sqrt{2}$$

Substituting in (1) we get

$$D_{\mathbf{u}} f(P) = \frac{\sqrt{3}}{2} \sqrt{2} = \frac{\sqrt{6}}{2}$$

41. Find a vector normal to the surface $x^2 + y^2 - z^2 = 6$ at $P = (3, 1, 2)$.

SOLUTION The gradient ∇f_P is normal to the level curve $f(x, y, z) = x^2 + y^2 - z^2 = 6$ at P. We compute this vector:

$$f_x(x, y, z) = 2x$$
$$f_y(x, y, z) = 2y \quad \Rightarrow \quad \nabla f_P = \nabla f_{(3,1,2)} = \langle 6, 2, -4 \rangle$$
$$f_z(x, y, z) = -2z$$

The vector $\langle 6, 2, -4 \rangle$ is normal to the surface $x^2 + y^2 - z^2 = 6$ at P.

43. Find the two points on the ellipsoid

$$\frac{x^2}{4} + \frac{y^2}{9} + z^2 = 1$$

where the tangent plane is normal to $\mathbf{v} = \langle 1, 1, -2 \rangle$.

SOLUTION The gradient ∇f_P is normal to the level surface $f(x, y, z) = \frac{x^2}{4} + \frac{y^2}{9} + z^2 = 1$. If $\mathbf{v} = \langle 1, 1, -2 \rangle$ is also normal, then ∇f_P and $\mathbf{v}$ are parallel, that is, $\nabla f_P = k\mathbf{v}$ for some constant k. This yields the equation

$$\nabla f_P = \langle \frac{x}{2}, \frac{2y}{9}, 2z \rangle = k \langle 1, 1, -2 \rangle$$

Thus $x = 2k$, $y = 9k/2$, and $z = -k$. To determine k, substitute in the equation of the ellipsoid:

$$\frac{x^2}{4} + \frac{y^2}{9} + z^2 = \frac{(2k)^2}{4} + \frac{(9k/2)^2}{9} + (-k)^2 = 1$$

This yields $k^2 + \frac{9}{4}k^2 + k^2 = 1$ or $k = \pm 2/\sqrt{17}$. The two points are

$$(x, y, z) = (2k, \frac{9}{2}k, -k) = \pm \left(\frac{4}{\sqrt{17}}, \frac{9}{\sqrt{17}}, -\frac{2}{\sqrt{17}} \right)$$

45. Use the geostrophic flow model to explain the following: In the Southern Hemisphere, winds blow with low pressure to the right, and the closer together the isobars, the stronger the winds. In particular, winds blow clockwise around low pressure systems and counterclockwise around high pressure systems.

SOLUTION As in Example 9, from the previous exercise (or from Eq. 5),

$$\left\langle V \frac{\partial p}{\partial x}, V \frac{\partial p}{\partial y} \right\rangle = \langle 2mw_2 \omega \sin L, -2mw_1 \omega \sin L \rangle = 2m\omega \sin L \langle w_2, -w_1 \rangle$$

so that

$$\langle w_1, w_2 \rangle = \left\langle -\frac{V}{2mw_1 \omega \sin L} \frac{\partial p}{\partial y}, \frac{V}{2mw_1 \omega \sin L} \frac{\partial p}{\partial x} \right\rangle = C \left\langle -\frac{\partial p}{\partial y}, \frac{\partial p}{\partial x} \right\rangle$$

where $C = \frac{V}{2mw_1 \omega \sin L}$.

In the Southern Hemisphere, $\sin L$ is negative since $-\frac{\pi}{2} \leq L < 0$, and m, ω, and V are all positive, so $C < 0$. Set $D = -C$; then

$$\langle w_1, w_2 \rangle = D \left\langle \frac{\partial p}{\partial y}, -\frac{\partial p}{\partial x} \right\rangle$$

where D is a positive constant. The vector $\left\langle \frac{\partial p}{\partial y}, -\frac{\partial p}{\partial x} \right\rangle$ is ∇p rotated clockwise by $90°$, so the wind vector is proportional in magnitude to the pressure gradient, and points in the direction $90°$ clockwise from it. Since the pressure gradient points outward from a low pressure system, the wind vectors create a clockwise rotation in the Southern Hemisphere. For high pressure systems, the pressure gradient points inward, so the wind vectors create a counterclockwise rotation. Additionally, from the proportionality, the wind speed is proportional to the pressure gradient.

In Exercises 46–49, find an equation of the tangent plane to the surface at the given point.

47. $xz + 2x^2y + y^2z^3 = 11$, $\quad P = (2, 1, 1)$

SOLUTION The equation of the tangent plane at P is

$$\nabla f_P \cdot \langle x - 2, y - 1, z - 1 \rangle = 0 \tag{1}$$

We compute the gradient of $f(x, y, z) = xz + 2x^2y + y^2z^3$ at the point $P = (2, 1, 1)$:

$$\nabla f = \left\langle \frac{\partial f}{\partial x}, \frac{\partial f}{\partial y}, \frac{\partial f}{\partial z} \right\rangle = \langle z + 4xy, 2x^2 + 2yz^3, x + 3y^2z^2 \rangle$$

At the point P we have

$$\nabla f_P = \langle 9, 10, 5 \rangle$$

Substituting in (1) we obtain the following equation of the tangent plane:

$$\langle 9, 10, 5 \rangle \cdot \langle x - 2, y - 1, z - 1 \rangle = 0$$
$$9(x - 2) + 10(y - 1) + 5(z - 1) = 0$$

or

$$9x + 10y + 5z = 33$$

49. $\ln(1 + 4x^2 + 9y^4) - 0.1z^2 = 0$, $\quad P = (3, 1, 6.1876)$

SOLUTION The equation of the tangent plane at P is

$$\nabla f_P \cdot (x - 3, y - 1, z - 6.1876) = 0 \tag{1}$$

We compute the gradient of $f(x, y, z) = \ln(1 + 4x^2 + 9y^4) - 0.1z^2$ at the point P:

$$\nabla f = \left\langle \frac{\partial f}{\partial x}, \frac{\partial f}{\partial y}, \frac{\partial f}{\partial z} \right\rangle = \left\langle \frac{8x}{1 + 4x^2 + 9y^4}, \frac{36y^3}{1 + 4x^2 + 9y^4}, -0.2z \right\rangle$$

At the point $P = (3, 1, 6.1876)$ we have

$$\nabla f_P = \left\langle \frac{24}{1 + 36 + 9}, \frac{36}{46}, -1.2375 \right\rangle = \langle 0.5217, 0.7826, -1.2375 \rangle$$

We substitute in (1) to obtain the following equation of the tangent plane:

$$0.5217(x - 3) + 0.7826(y - 1) - 1.2375(z - 6.1876) = 0$$

or

$$0.5217x + 0.7826y - 1.2375z = -5.309$$

51. *CAS* Use a computer algebra system to produce a contour plot of $f(x, y) = x^2 - 3xy + y - y^2$ together with its gradient vector field on the domain $[-4, 4] \times [-4, 4]$.

SOLUTION

53. Find a function $f(x, y, z)$ such that $\nabla f = \langle 2x, 1, 2 \rangle$.

SOLUTION The following equality must hold:

$$\nabla f = \left\langle \frac{\partial f}{\partial x}, \frac{\partial f}{\partial y}, \frac{\partial f}{\partial z} \right\rangle = \langle 2x, 1, 2 \rangle$$

Equating corresponding components gives

$$\frac{\partial f}{\partial x} = 2x$$

$$\frac{\partial f}{\partial y} = 1$$

$$\frac{\partial f}{\partial z} = 2$$

One of the functions that satisfies these equalities is $f(x, y, z) = x^2 + y + 2z$.

55. Find a function $f(x, y, z)$ such that $\nabla f = \langle z, 2y, x \rangle$.

SOLUTION $f(x, y, z) = xz + y^2$ is a good choice.

57. Show that there does not exist a function $f(x, y)$ such that $\nabla f = \langle y^2, x \rangle$. *Hint:* Use Clairaut's Theorem $f_{xy} = f_{yx}$.

SOLUTION Suppose that for some differentiable function $f(x, y)$,

$$\nabla f = \langle f_x, f_y \rangle = \langle y^2, x \rangle$$

That is, $f_x = y^2$ and $f_y = x$. Therefore,

$$f_{xy} = \frac{\partial}{\partial y} f_x = \frac{\partial}{\partial y} y^2 = 2y \quad \text{and} \quad f_{yx} = \frac{\partial}{\partial x} f_y = \frac{\partial}{\partial x} x = 1$$

Since f_{xy} and f_{yx} are both continuous, they must be equal by Clairaut's Theorem. Since $f_{xy} \neq f_{yx}$ we conclude that such a function f does not exist.

59. Use Eq. (6) to estimate

$$\Delta f = f(3.53, 8.98) - f(3.5, 9)$$

assuming that $\nabla f_{(3.5,9)} = \langle 2, -1 \rangle$.

SOLUTION By Eq. (6),

$$\Delta f \approx \nabla f_P \cdot \Delta \mathbf{v}$$

The vector $\Delta \mathbf{v}$ is the following vector:

$$\Delta \mathbf{v} = \langle 3.53 - 3.5, 8.98 - 9 \rangle = \langle 0.03, -0.02 \rangle$$

Hence,

$$\Delta f \approx \nabla f_{(3.5,9)} \cdot \Delta \mathbf{v} = \langle 2, -1 \rangle \cdot \langle 0.03, -0.02 \rangle = 0.08$$

61. Suppose, in the previous exercise, that a particle located at the point $P = (2, 2, 8)$ travels toward the xy-plane in the direction normal to the surface.

(a) Through which point Q on the xy-plane will the particle pass?

(b) Suppose the axes are calibrated in centimeters. Determine the path $\mathbf{r}(t)$ of the particle if it travels at a constant speed of 8 cm/s. How long will it take the particle to reach Q?

SOLUTION

(a) The particle travels along the line through $P = (2, 2, 8)$ in the direction $(4, 2, -1)$. The vector parametrization of this line is

$$\mathbf{r}(t) = \langle 2, 2, 8 \rangle + t \langle 4, 2, -1 \rangle = \langle 2 + 4t, 2 + 2t, 8 - t \rangle \tag{1}$$

We must find the point where this line intersects the xy-plane. At this point the z-component is zero. Hence,

$$8 - t = 0 \quad \Rightarrow \quad t = 8$$

Substituting $t = 8$ in (1) we obtain

$$\mathbf{r}(8) = \langle 2 + 4 \cdot 8, 2 + 2 \cdot 8, 0 \rangle = \langle 34, 18, 0 \rangle$$

The particle will pass through the point $Q = (34, 18, 0)$ on the xy-plane.

(b) If $\mathbf{v}$ is a direction vector of the line PQ, so that $\|\mathbf{v}\| = 8$, the following parametrization of the line has constant speed 8:

$$\mathbf{r}(t) = \langle 2, 2, 8 \rangle + t\mathbf{v}$$

(This has speed 8 because $\|\mathbf{r}'(t)\| = \|\mathbf{v}\| = 8$.) In the previous exercise, we found the unit vector $\mathbf{n} = \frac{1}{\sqrt{21}} \langle 4, 2, -1 \rangle$, therefore we use the direction vector $\mathbf{v} = 8\mathbf{n} = \frac{8}{\sqrt{21}} \langle 4, 2, -1 \rangle$, obtaining the following parametrization of the line:

$$\mathbf{r}(t) = \langle 2, 2, 8 \rangle + t \cdot \frac{8}{\sqrt{21}} \langle 4, 2, -1 \rangle = \left\langle 2 + \frac{32}{\sqrt{21}} t, 2 + \frac{16}{\sqrt{21}} t, 8 - \frac{8t}{\sqrt{21}} \right\rangle$$

To find the time needed for the particle to reach Q if it travels along $\mathbf{r}(t)$, we first compute the distance $\overline{PQ}$:

$$\overline{PQ} = \sqrt{(34 - 2)^2 + (18 - 2)^2 + (0 - 8)^2} = \sqrt{1344} = 8\sqrt{21}$$

The time needed is thus

$$T = \frac{\overline{PQ}}{8} = \frac{8\sqrt{21}}{8} = \sqrt{21} \approx 4.58 \text{ s}$$

63. Suppose that the intersection of two surfaces $F(x, y, z) = 0$ and $G(x, y, z) = 0$ is a curve C, and let P be a point on C. Explain why the vector $\mathbf{v} = \nabla F_P \times \nabla G_P$ is a direction vector for the tangent line to C at P.

SOLUTION The gradient ∇F_P is orthogonal to all the curves in the level surface $F(x, y, z) = 0$ passing through P. Similarly, ∇G_P is orthogonal to all the curves in the level surface $G(x, y, z) = 0$ passing through P. Therefore, both ∇F_P and ∇G_P are orthogonal to the intersection curve C at P, hence the cross product $\nabla F_P \times \nabla G_P$ is parallel to the tangent line to C at P.

65. Let C be the curve obtained as the intersection of the two surfaces $x^3 + 2xy + yz = 7$ and $3x^2 - yz = 1$. Find the parametric equations of the tangent line to C at $P = (1, 2, 1)$.

SOLUTION The parametric equations of the tangent line to C at $P = (1, 2, 1)$ are

$$x = 1 + at, \quad y = 2 + bt, \quad z = 1 + ct \tag{1}$$

where $\mathbf{v} = \langle a, b, c \rangle$ is a direction vector for the line. By Exercise 63, $\mathbf{v}$ may be chosen as the cross product:

$$\mathbf{v} = \nabla F_P \times \nabla G_P \tag{2}$$

where $F(x, y, z) = x^3 + 2xy + yz$ and $G(x, y, z) = 3x^2 - yz$. We compute the gradient vectors:

$$F_x(x, y, z) = 3x^2 + 2y \qquad F_x(1, 2, 1) = 7$$
$$F_y(x, y, z) = 2x + z \quad \Rightarrow \quad F_y(1, 2, 1) = 3 \quad \Rightarrow \quad \nabla F_P = \langle 7, 3, 2 \rangle$$
$$F_z(x, y, z) = y \qquad F_z(1, 2, 1) = 2$$

$$G_x(x, y, z) = 6x \qquad G_x(1, 2, 1) = 6$$
$$G_y(x, y, z) = -z \quad \Rightarrow \quad G_y(1, 2, 1) = -1 \quad \Rightarrow \quad \nabla G_P = \langle 6, -1, -2 \rangle$$
$$G_z(x, y, z) = -y \qquad G_z(1, 2, 1) = -2$$

Hence,

$$\mathbf{v} = \langle 7, 3, 2 \rangle \times \langle 6, -1, -2 \rangle = \begin{vmatrix} \mathbf{i} & \mathbf{j} & \mathbf{k} \\ 7 & 3 & 2 \\ 6 & -1 & -2 \end{vmatrix} = -4\mathbf{i} + 26\mathbf{j} - 25\mathbf{k} = \langle -4, 26, -25 \rangle$$

Therefore, $\mathbf{v} = \langle a, b, c \rangle = \langle -4, 26, -25 \rangle$, so we obtain

$$a = -4, \quad b = 26, \quad c = -25$$

Substituting in (1) gives the following parametric equations of the tangent line:

$$x = 1 - 4t, \quad y = 2 + 26t, \quad z = 1 - 25t$$

67. Prove the Chain Rule for Gradients in Theorem 1.

SOLUTION We must show that if $F(t)$ is a differentiable function of t and $f(x, y, z)$ is differentiable, then

$$\nabla F(f(x,y,z)) = F'(f(x,y,z)) \nabla f$$

Using the Chain Rule for partial derivatives we get

$$\nabla F(f(x,y,z)) = \left\langle \frac{\partial}{\partial x} F(f(x,y,z)), \frac{\partial}{\partial y} F(f(x,y,z)), \frac{\partial}{\partial z} F(f(x,y,z)) \right\rangle$$

$$= \left\langle \frac{dF}{dt} \cdot \frac{\partial f}{\partial x}, \frac{dF}{dt} \cdot \frac{\partial f}{\partial y}, \frac{dF}{dt} \cdot \frac{\partial f}{\partial z} \right\rangle = \frac{dF}{dt} \left\langle \frac{\partial f}{\partial x}, \frac{\partial f}{\partial y}, \frac{\partial f}{\partial z} \right\rangle = F'(f(x,y,z)) \nabla F$$

Further Insights and Challenges

69. Let $\mathbf{u}$ be a unit vector. Show that the directional derivative $D_{\mathbf{u}} f$ is equal to the component of ∇f along $\mathbf{u}$.

SOLUTION The component of ∇f along $\mathbf{u}$ is $\nabla f \cdot \mathbf{u}$. By the Theorem on Evaluating Directional Derivatives, $D_{\mathbf{u}} f = \nabla f \cdot \mathbf{u}$, which is the component of ∇f along $\mathbf{u}$.

71. Use the definition of differentiability to show that if $f(x, y)$ is differentiable at $(0, 0)$ and

$$f(0,0) = f_x(0,0) = f_y(0,0) = 0$$

then

$$\lim_{(x,y) \to (0,0)} \frac{f(x,y)}{\sqrt{x^2 + y^2}} = 0 \qquad \boxed{7}$$

SOLUTION If $f(x, y)$ is differentiable at $(0, 0)$, then there exists a function $\epsilon(x, y)$ satisfying $\lim_{(x,y) \to (0,0)} \epsilon(x, y) = 0$ such that

$$f(x, y) = L(x, y) + \epsilon(x, y) \sqrt{x^2 + y^2} \qquad (1)$$

Since $f(0, 0) = 0$, the linear function $L(x, y)$ is

$$L(x, y) = f(0,0) + f_x(0,0)x + f_y(0,0)y = f_x(0,0)x + f_y(0,0)y$$

Substituting in (1) gives

$$f(x, y) = f_x(0,0)x + f_y(0,0)y + \epsilon(x, y) \sqrt{x^2 + y^2}$$

Therefore,

$$\lim_{(x,y) \to (0,0)} \frac{f(x,y) - f_x(0,0)x - f_y(0,0)y}{\sqrt{x^2 + y^2}} = \lim_{(x,y) \to (0,0)} \epsilon(x, y) = 0$$

73. Prove that if $f(x, y)$ is differentiable and $\nabla f_{(x,y)} = \mathbf{0}$ for all (x, y), then f is constant.

SOLUTION Since $\nabla f = \langle f_x, f_y \rangle = \langle 0, 0 \rangle$ for all (x, y), we have

$$f_x(x, y) = f_y(x, y) = 0 \text{ for all } (x, y) \qquad (1)$$

Let $Q_0 = (x_0, y_0)$ be a fixed point and let $P = (x_1, y_1)$ be any other point. Let $\mathbf{r}(t) = \langle x(t), y(t) \rangle$ be a parametric equation of the line joining Q_0 and P, with $P = \mathbf{r}(t_1)$ and $Q_0 = \mathbf{r}(t_0)$. We define the following function:

$$F(t) = f(x(t), y(t))$$

$F(t)$ is defined for all t, since $f(x, y)$ is defined for all (x, y). By the Chain Rule we have

$$F'(t) = f_x(x(t), y(t)) \frac{dx}{dt} + f_y(x(t), y(t)) \frac{dy}{dt}$$

Combining with (1) we get $F'(t) = 0$ for all t. We conclude that $F(t) = $ const. That is, f is constant on the line $\mathbf{r}(t)$. In particular, $f(P) = f(Q_0)$. Since P is any point, it follows that $f(x, y)$ is a constant function.

In Exercises 75–77, a path $\mathbf{r}(t) = \langle x(t), y(t) \rangle$ follows the gradient of a function $f(x, y)$ if the tangent vector $\mathbf{r}'(t)$ points in the direction of ∇f for all t. In other words, $\mathbf{r}'(t) = k(t)\nabla f_{\mathbf{r}(t)}$ for some positive function $k(t)$. Note that in this case, $\mathbf{r}(t)$ crosses each level curve of $f(x, y)$ at a right angle.

75. Show that if the path $\mathbf{r}(t) = \langle x(t), y(t) \rangle$ follows the gradient of $f(x, y)$, then

$$\frac{y'(t)}{x'(t)} = \frac{f_y}{f_x}$$

SOLUTION Since $\mathbf{r}(t)$ follows the gradient of $f(x, y)$, we have

$$\mathbf{r}'(t) = k(t) \nabla f_{\mathbf{r}(t)} = k(t) \left\langle f_x(\mathbf{r}(t)), f_y(\mathbf{r}(t)) \right\rangle$$

which implies that

$$x'(t) = k(t) f_x(\mathbf{r}(t)) \quad \text{and} \quad y'(t) = k(t) f_y(\mathbf{r}(t))$$

Hence,

$$\frac{y'(t)}{x'(t)} = \frac{k(t) f_y(\mathbf{r}(t))}{k(t) f_x(\mathbf{r}(t))} = \frac{f_y(\mathbf{r}(t))}{f_x(\mathbf{r}(t))}$$

or in short notation,

$$\frac{y'(t)}{x'(t)} = \frac{f_y}{f_x}$$

77. *CAS* Find the curve $y = g(x)$ passing through $(0, 1)$ that crosses each level curve of $f(x, y) = y \sin x$ at a right angle. Using a computer algebra system, graph $y = g(x)$ together with the level curves of f.

SOLUTION Using $f_x = y \cos x$, $f_y = \sin x$, and $y(0) = 1$, we get

$$\frac{dy}{dx} = \frac{\tan x}{y} \quad \Rightarrow \quad y(0) = 1$$

We solve the differential equation using separation of variables:

$$y \, dy = \tan x \, dx$$

$$\int y \, dy = \int \tan x \, dx$$

$$\frac{1}{2} y^2 = -\ln|\cos x| + k$$

$$y^2 = -2 \ln|\cos x| + k = -\ln\left(\cos^2 x\right) + k$$

$$y = \pm \sqrt{-\ln\left(\cos^2 x\right) + k}$$

Since $y(0) = 1 > 0$, the appropriate sign is the positive sign. That is,

$$y = \sqrt{-\ln\left(\cos^2 x\right) + k} \tag{1}$$

We find the constant k by substituting $x = 0$, $y = 1$ and solve for k. This gives

$$1 = \sqrt{-\ln(\cos^2 0) + k} = \sqrt{-\ln 1 + k} = \sqrt{k}$$

Hence,

$$k = 1$$

Substituting in (2) gives the following solution:

$$y = \sqrt{1 - \ln\left(\cos^2 x\right)} \tag{2}$$

The following figure shows the graph of the curve (3) together with some level curves of f.

14.6 Multivariable Calculus Chain Rules (LT Section 15.6)

Preliminary Questions

1. Let $f(x, y) = xy$, where $x = uv$ and $y = u + v$.
   (a) What are the primary derivatives of f?
   (b) What are the independent variables?

SOLUTION

(a) The primary derivatives of f are $\frac{\partial f}{\partial x}$ and $\frac{\partial f}{\partial y}$.

(b) The independent variables are u and v, on which x and y depend.

In Exercises 2 and 3, suppose that $f(u, v) = ue^v$, where $u = rs$ and $v = r + s$.

2. The composite function $f(u, v)$ is equal to:
   (a) rse^{r+s} (b) re^s (c) rse^{rs}

SOLUTION The composite function $f(u, v)$ is obtained by replacing u and v in the formula for $f(u, v)$ by the corresponding functions $u = rs$ and $v = r + s$. This gives

$$f(u(r, s), v(r, s)) = u(r, s)e^{v(r,s)} = rse^{r+s}$$

Answer (a) is the correct answer.

3. What is the value of $f(u, v)$ at $(r, s) = (1, 1)$?

SOLUTION We compute $u = rs$ and $v = r + s$ at the point $(r, s) = (1, 1)$:

$$u(1, 1) = 1 \cdot 1 = 1; \quad v(1, 1) = 1 + 1 = 2$$

Substituting in $f(u, v) = ue^v$, we get

$$f(u, v)\Big|_{(r,s)=(1,1)} = 1 \cdot e^2 = e^2$$

4. According to the Chain Rule, $\partial f / \partial r$ is equal to (choose the correct answer):

(a) $\dfrac{\partial f}{\partial x}\dfrac{\partial x}{\partial r} + \dfrac{\partial f}{\partial x}\dfrac{\partial x}{\partial s}$

(b) $\dfrac{\partial f}{\partial x}\dfrac{\partial x}{\partial r} + \dfrac{\partial f}{\partial y}\dfrac{\partial y}{\partial r}$

(c) $\dfrac{\partial f}{\partial r}\dfrac{\partial r}{\partial x} + \dfrac{\partial f}{\partial s}\dfrac{\partial s}{\partial x}$

SOLUTION For a function $f(x, y)$ where $x = x(r, s)$ and $y = y(r, s)$, the Chain Rule states that the partial derivative $\frac{\partial f}{\partial r}$ is as given in (b). That is,

$$\frac{\partial f}{\partial x}\frac{\partial x}{\partial r} + \frac{\partial f}{\partial y}\frac{\partial y}{\partial r}$$

5. Suppose that x, y, z are functions of the independent variables u, v, w. Which of the following terms appear in the Chain Rule expression for $\partial f/\partial w$?

(a) $\dfrac{\partial f}{\partial v}\dfrac{\partial x}{\partial v}$ 　　(b) $\dfrac{\partial f}{\partial w}\dfrac{\partial w}{\partial x}$ 　　(c) $\dfrac{\partial f}{\partial z}\dfrac{\partial z}{\partial w}$

SOLUTION By the Chain Rule, the derivative $\dfrac{\partial f}{\partial w}$ is

$$\frac{\partial f}{\partial w} = \frac{\partial f}{\partial x}\frac{\partial x}{\partial w} + \frac{\partial f}{\partial y}\frac{\partial y}{\partial w} + \frac{\partial f}{\partial z}\frac{\partial z}{\partial w}$$

Therefore (c) is the only correct answer.

6. With notation as in the previous exercise, does $\partial x/\partial v$ appear in the Chain Rule expression for $\partial f/\partial u$?

SOLUTION The Chain Rule expression for $\dfrac{\partial f}{\partial u}$ is

$$\frac{\partial f}{\partial u} = \frac{\partial f}{\partial x}\frac{\partial x}{\partial u} + \frac{\partial f}{\partial y}\frac{\partial y}{\partial u} + \frac{\partial f}{\partial z}\frac{\partial z}{\partial u}$$

The derivative $\dfrac{\partial x}{\partial v}$ does not appear in differentiating f with respect to the independent variable u.

Exercises

1. Let $f(x, y, z) = x^2y^3 + z^4$ and $x = s^2$, $y = st^2$, and $z = s^2t$.

(a) Calculate the primary derivatives $\dfrac{\partial f}{\partial x}, \dfrac{\partial f}{\partial y}, \dfrac{\partial f}{\partial z}$.

(b) Calculate $\dfrac{\partial x}{\partial s}, \dfrac{\partial y}{\partial s}, \dfrac{\partial z}{\partial s}$.

(c) Compute $\dfrac{\partial f}{\partial s}$ using the Chain Rule:

$$\frac{\partial f}{\partial s} = \frac{\partial f}{\partial x}\frac{\partial x}{\partial s} + \frac{\partial f}{\partial y}\frac{\partial y}{\partial s} + \frac{\partial f}{\partial z}\frac{\partial z}{\partial s}$$

Express the answer in terms of the independent variables s, t.

SOLUTION

(a) The primary derivatives of $f(x, y, z) = x^2y^3 + z^4$ are

$$\frac{\partial f}{\partial x} = 2xy^3, \quad \frac{\partial f}{\partial y} = 3x^2y^2, \quad \frac{\partial f}{\partial z} = 4z^3$$

(b) The partial derivatives of x, y, and z with respect to s are

$$\frac{\partial x}{\partial s} = 2s, \quad \frac{\partial y}{\partial s} = t^2, \quad \frac{\partial z}{\partial s} = 2st$$

(c) We use the Chain Rule and the partial derivatives computed in parts (a) and (b) to find the following derivative:

$$\frac{\partial f}{\partial s} = \frac{\partial f}{\partial x}\frac{\partial x}{\partial s} + \frac{\partial f}{\partial y}\frac{\partial y}{\partial s} + \frac{\partial f}{\partial z}\frac{\partial z}{\partial s} = 2xy^3 \cdot 2s + 3x^2y^2t^2 + 4z^3 \cdot 2st = 4xy^3s + 3x^2y^2t^2 + 8z^3st$$

To express the answer in terms of the independent variables s, t we substitute $x = s^2$, $y = st^2$, $z = s^2t$. This gives

$$\frac{\partial f}{\partial s} = 4s^2(st^2)^3s + 3(s^2)^2(st^2)^2t^2 + 8(s^2t)^3st = 4s^6t^6 + 3s^6t^6 + 8s^7t^4 = 7s^6t^6 + 8s^7t^4$$

In Exercises 3–10, use the Chain Rule to calculate the partial derivatives. Express the answer in terms of the independent variables.

3. $\dfrac{\partial f}{\partial s}, \dfrac{\partial f}{\partial r}$; $f(x, y, z) = xy + z^2$, $x = s^2$, $y = 2rs$, $z = r^2$

SOLUTION We perform the following steps:

Step 1. Compute the primary derivatives. The primary derivatives of $f(x, y, z) = xy + z^2$ are

$$\frac{\partial f}{\partial x} = y, \quad \frac{\partial f}{\partial y} = x, \quad \frac{\partial f}{\partial z} = 2z$$

Step 2. Apply the Chain Rule. By the Chain Rule,

$$\frac{\partial f}{\partial s} = \frac{\partial f}{\partial x} \cdot \frac{\partial x}{\partial s} + \frac{\partial f}{\partial y} \cdot \frac{\partial y}{\partial s} + \frac{\partial f}{\partial z} \cdot \frac{\partial z}{\partial s} \tag{1}$$

$$\frac{\partial f}{\partial r} = \frac{\partial f}{\partial x} \cdot \frac{\partial x}{\partial r} + \frac{\partial f}{\partial y} \cdot \frac{\partial y}{\partial r} + \frac{\partial f}{\partial z} \cdot \frac{\partial z}{\partial r} \tag{2}$$

We compute the partial derivatives of x, y, z with respect to s and r:

$$\frac{\partial x}{\partial s} = 2s, \quad \frac{\partial y}{\partial s} = 2r, \quad \frac{\partial z}{\partial s} = 0.$$

$$\frac{\partial x}{\partial r} = 0, \quad \frac{\partial y}{\partial r} = 2s, \quad \frac{\partial z}{\partial r} = 2r$$

Substituting these derivatives and the primary derivatives computed in step 1 in (1) and (2), we get

$$\frac{\partial f}{\partial s} = y \cdot 2s + x \cdot 2r + 2z \cdot 0 = 2ys + 2xr$$

$$\frac{\partial f}{\partial r} = y \cdot 0 + x \cdot 2s + 2z \cdot 2r = 2xs + 4zr$$

Step 3. Express the answer in terms of r and s. We substitute $x = s^2$, $y = 2rs$, and $z = r^2$ in $\frac{\partial f}{\partial s}$ and $\frac{\partial f}{\partial r}$ in step 2, to obtain

$$\frac{\partial f}{\partial s} = 2rs \cdot 2s + s^2 \cdot 2r = 4rs^2 + 2rs^2 = 6rs^2$$

$$\frac{\partial f}{\partial r} = 2s^2 \cdot s + 4r^2 \cdot r = 2s^3 + 4r^3$$

5. $\frac{\partial g}{\partial x}, \frac{\partial g}{\partial y}; g(\theta, \phi) = \tan(\theta + \phi), \theta = xy, \phi = x + y$

SOLUTION We perform the following steps:

Step 1. Compute the primary derivatives. The primary derivatives of $g(\theta, \phi) = \tan(\theta + \phi)$ are

$$\frac{\partial g}{\partial \theta} = \sec^2(\theta + \phi), \quad \frac{\partial g}{\partial \phi} = \sec^2(\theta + \phi)$$

Step 2. Apply the Chain Rule. By the Chain Rule,

$$\frac{\partial g}{\partial x} = \frac{\partial g}{\partial \theta} \cdot \frac{\partial \theta}{\partial x} + \frac{\partial g}{\partial \phi} \cdot \frac{\partial \phi}{\partial x} \tag{1}$$

$$\frac{\partial g}{\partial y} = \frac{\partial g}{\partial \theta} \cdot \frac{\partial \theta}{\partial y} + \frac{\partial g}{\partial \phi} \cdot \frac{\partial \phi}{\partial y} \tag{2}$$

We compute the partial derivatives of θ and ϕ with respect to x and y:

$$\frac{\partial \theta}{\partial x} = y, \quad \frac{\partial \theta}{\partial y} = x, \quad \frac{\partial \phi}{\partial x} = 1, \quad \frac{\partial \phi}{\partial y} = 1$$

Substituting these derivatives and the primary derivatives computed in step 1 in (1) and (2), we get

$$\frac{\partial g}{\partial x} = \sec^2(\theta + \phi) \cdot y + \sec^2(\theta + \phi) \cdot 1 = (y + 1)\sec^2(\theta + \phi)$$

$$\frac{\partial g}{\partial y} = \sec^2(\theta + \phi) \cdot x + \sec^2(\theta + \phi) \cdot 1 = (x + 1)\sec^2(\theta + \phi)$$

Step 3. Express the answer in terms of x and y. We substitute $\theta = xy$ and $\phi = x + y$ in $\frac{\partial g}{\partial x}$ and $\frac{\partial g}{\partial y}$ in step 2, to obtain

$$\frac{\partial g}{\partial x} = (y + 1)\sec^2(xy + x + y)$$

$$\frac{\partial g}{\partial y} = (x + 1)\sec^2(xy + x + y)$$

7. $\dfrac{\partial F}{\partial y}$; $F(u, v) = e^{u+v}$, $u = x^2$, $v = xy$

SOLUTION We use the following steps:

Step 1. Compute the primary derivatives. The primary derivatives of $F(u, v) = e^{u+v}$ are

$$\frac{\partial f}{\partial u} = e^{u+v}, \quad \frac{\partial f}{\partial v} = e^{u+v}$$

Step 2. Apply the Chain Rule. By the Chain Rule,

$$\frac{\partial F}{\partial y} = \frac{\partial F}{\partial u}\frac{\partial u}{\partial y} + \frac{\partial F}{\partial v}\frac{\partial v}{\partial y} = e^{u+v}\frac{\partial u}{\partial y} + e^{u+v}\frac{\partial v}{\partial y} = e^{u+v}\left(\frac{\partial u}{\partial y} + \frac{\partial v}{\partial y}\right)$$

We compute the partial derivatives of u and v with respect to y:

$$\frac{\partial u}{\partial y} = 0, \quad \frac{\partial v}{\partial y} = x$$

We substitute to obtain

$$\frac{\partial F}{\partial y} = xe^{u+v} \tag{1}$$

Step 3. Express the answer in terms of x and y. We substitute $u = x^2$, $v = xy$ in (1), obtaining

$$\frac{\partial F}{\partial y} = xe^{x^2+xy}$$

9. $\dfrac{\partial h}{\partial t_2}$; $h(x, y) = \dfrac{x}{y}$, $x = t_1 t_2$, $y = t_1^2 t_2$

SOLUTION We use the following steps:

Step 1. Compute the primary derivatives. The primary derivatives of $h(x, y) = \frac{x}{y}$ are

$$\frac{\partial h}{\partial x} = \frac{1}{y}, \quad \frac{\partial h}{\partial y} = -\frac{x}{y^2}$$

Step 2. Apply the Chain Rule. By the Chain Rule,

$$\frac{\partial h}{\partial t_2} = \frac{\partial h}{\partial x}\frac{\partial x}{\partial t_2} + \frac{\partial h}{\partial y}\frac{\partial y}{\partial t_2} = \frac{1}{y}\frac{\partial x}{\partial t_2} - \frac{x}{y^2}\frac{\partial y}{\partial t_2}$$

We compute the partial derivatives of x and y with respect to t_2:

$$\frac{\partial x}{\partial t_2} = t_1, \quad \frac{\partial y}{\partial t_2} = t_1^2$$

Hence,

$$\frac{\partial h}{\partial t_2} = \frac{t_1}{y} - \frac{x}{y^2}t_1^2$$

Step 3. Express the answer in terms of t_1 and t_2. We substitute $x = t_1 t_2$, $y = t_1^2 t_2$ in $\frac{\partial h}{\partial t_2}$ computed in step 2, to obtain

$$\frac{\partial h}{\partial t_2} = \frac{t_1}{t_1^2 t_2} - \frac{t_1 t_2 \cdot t_1^2}{(t_1^2 t_2)^2} = \frac{1}{t_1 t_2} - \frac{1}{t_1 t_2} = 0$$

Remark: Notice that $h(x(t_1, t_2), y(t_1, t_2)) = h(t_1, t_2) = \frac{t_1 t_2}{t_1^2 t_2} = \frac{1}{t_1}$. $h(t_1, t_2)$ is independent of t_2, hence $\frac{\partial h}{\partial t_2} = 0$ (as obtained in our computations).

In Exercises 11–16, use the Chain Rule to evaluate the partial derivative at the point specified.

11. $\partial f/\partial u$ and $\partial f/\partial v$ at $(u, v) = (-1, -1)$, where $f(x, y, z) = x^3 + yz^2$, $x = u^2 + v$, $y = u + v^2$, $z = uv$

SOLUTION The primary derivatives of $f(x, y, z) = x^3 + yz^2$ are

$$\frac{\partial f}{\partial x} = 3x^2, \quad \frac{\partial f}{\partial y} = z^2, \quad \frac{\partial f}{\partial z} = 2yz$$

By the Chain Rule we have

$$\frac{\partial f}{\partial u} = \frac{\partial f}{\partial x}\frac{\partial x}{\partial u} + \frac{\partial f}{\partial y}\frac{\partial y}{\partial u} + \frac{\partial f}{\partial z}\frac{\partial z}{\partial u} = 3x^2\frac{\partial x}{\partial u} + z^2\frac{\partial y}{\partial u} + 2yz\frac{\partial z}{\partial u} \tag{1}$$

$$\frac{\partial f}{\partial v} = \frac{\partial f}{\partial x}\frac{\partial x}{\partial v} + \frac{\partial f}{\partial y}\frac{\partial y}{\partial v} + \frac{\partial f}{\partial z}\frac{\partial z}{\partial v} = 3x^2\frac{\partial x}{\partial v} + z^2\frac{\partial y}{\partial v} + 2yz\frac{\partial z}{\partial v} \tag{2}$$

We compute the partial derivatives of x, y, and z with respect to u and v:

$$\frac{\partial x}{\partial u} = 2u, \quad \frac{\partial y}{\partial u} = 1, \quad \frac{\partial z}{\partial u} = v$$

$$\frac{\partial x}{\partial v} = 1, \quad \frac{\partial y}{\partial v} = 2v, \quad \frac{\partial z}{\partial v} = u$$

Substituting in (1) and (2) we get

$$\frac{\partial f}{\partial u} = 6x^2 u + z^2 + 2yzv \tag{3}$$

$$\frac{\partial f}{\partial v} = 3x^2 + 2vz^2 + 2yzu \tag{4}$$

We determine (x, y, z) for $(u, v) = (-1, -1)$:

$$x = (-1)^2 - 1 = 0, \quad y = -1 + (-1)^2 = 0, \quad z = (-1) \cdot (-1) = 1$$

Finally, we substitute $(x, y, z) = (0, 0, 1)$ and $(u, v) = (-1, -1)$ in (3), (4) to obtain the following derivatives:

$$\left.\frac{\partial f}{\partial u}\right|_{(u,v)=(-1,-1)} = 6 \cdot 0^2 \cdot (-1) + 1^2 + 2 \cdot 0 \cdot 1 \cdot (-1) = 1$$

$$\left.\frac{\partial f}{\partial v}\right|_{(u,v)=(-1,-1)} = 3 \cdot 0^2 + 2 \cdot (-1) \cdot 1^2 + 2 \cdot 0 \cdot 1 \cdot (-1) = -2$$

13. $\partial g/\partial \theta$ at $(r, \theta) = (2\sqrt{2}, \frac{\pi}{4})$, where $g(x, y) = 1/(x + y^2)$, $x = r\cos\theta$, $y = r\sin\theta$

SOLUTION We compute the primary derivatives of $g(x, y) = \frac{1}{x+y^2}$:

$$\frac{\partial g}{\partial x} = -\frac{1}{(x+y^2)^2}, \quad \frac{\partial g}{\partial y} = -\frac{2y}{(x+y^2)^2}$$

By the Chain Rule we have

$$\frac{\partial g}{\partial \theta} = \frac{\partial g}{\partial x}\frac{\partial x}{\partial \theta} + \frac{\partial g}{\partial y}\frac{\partial y}{\partial \theta} = -\frac{1}{(x+y^2)^2}\frac{\partial x}{\partial \theta} - \frac{2y}{(x+y^2)^2}\frac{\partial y}{\partial \theta} = -\frac{1}{(x+y^2)^2}\left(\frac{\partial x}{\partial \theta} + 2y\frac{\partial y}{\partial \theta}\right)$$

We find the partial derivatives $\frac{\partial x}{\partial \theta}$, $\frac{\partial y}{\partial \theta}$:

$$\frac{\partial x}{\partial \theta} = -r\sin\theta, \quad \frac{\partial y}{\partial \theta} = r\cos\theta$$

Hence,

$$\frac{\partial g}{\partial \theta} = -\frac{r}{(x+y^2)^2}(-\sin\theta + 2y\cos\theta) \tag{1}$$

At the point $(r, \theta) = (2\sqrt{2}, \frac{\pi}{4})$, we have $x = 2\sqrt{2}\cos\frac{\pi}{4} = 2$ and $y = 2\sqrt{2}\sin\frac{\pi}{4} = 2$. Substituting $(r, \theta) = (2\sqrt{2}, \frac{\pi}{4})$ and $(x, y) = (2, 2)$ in (1) gives the following derivative:

$$\left.\frac{\partial g}{\partial \theta}\right|_{(r,\theta)=(2\sqrt{2}, \frac{\pi}{4})} = \frac{-2\sqrt{2}}{(2+2^2)^2}\left(\cos\frac{\pi}{4} - 4\sin\frac{\pi}{4}\right) = \frac{-\sqrt{2}}{18}\left(\frac{1}{\sqrt{2}} - \frac{4}{\sqrt{2}}\right) = \frac{1}{6}$$

15. $\partial g/\partial u$ at $(u, v) = (0, 1)$, where $g(x, y) = x^2 - y^2$, $x = e^u \cos v$, $y = e^u \sin v$

SOLUTION The primary derivatives of $g(x, y) = x^2 - y^2$ are

$$\frac{\partial g}{\partial x} = 2x, \quad \frac{\partial g}{\partial y} = -2y$$

By the Chain Rule we have

$$\frac{\partial g}{\partial u} = \frac{\partial g}{\partial x} \cdot \frac{\partial x}{\partial u} + \frac{\partial g}{\partial y} \cdot \frac{\partial y}{\partial u} = 2x\frac{\partial x}{\partial u} - 2y\frac{\partial y}{\partial u} \tag{1}$$

We find $\frac{\partial x}{\partial u}$ and $\frac{\partial y}{\partial u}$:

$$\frac{\partial x}{\partial u} = e^u \cos v, \quad \frac{\partial y}{\partial u} = e^u \sin v$$

Substituting in (1) gives

$$\frac{\partial g}{\partial u} = 2xe^u \cos v - 2ye^u \sin v = 2e^u(x \cos v - y \sin v) \tag{2}$$

We determine (x, y) for $(u, v) = (0, 1)$:

$$x = e^0 \cos 1 = \cos 1, \quad y = e^0 \sin 1 = \sin 1$$

Finally, we substitute $(u, v) = (0, 1)$ and $(x, y) = (\cos 1, \sin 1)$ in (2) and use the identity $\cos^2 \alpha - \sin^2 \alpha = \cos 2\alpha$, to obtain the following derivative:

$$\left.\frac{\partial g}{\partial u}\right|_{(u,v)=(0,1)} = 2e^0 \left(\cos^2 1 - \sin^2 1\right) = 2 \cdot \cos 2 \cdot 1 = 2 \cos 2$$

17. Given $f(x, y)$ and $y = y(x)$, we can define a composite function $g(x) = f(x, y(x))$.

(a) Show that $g'(x) = \frac{\partial f}{\partial x} + \frac{\partial f}{\partial y}y'(x)$.

(b) Let $f(x, y) = x^3 - xy^2$ and $y(x) = 1 - x$. With $g(x) = f(x, y(x))$, use the formula in (a) to determine $g'(x)$, expressing the result in terms of x only.

(c) With $f(x, y)$ and $y(x)$ as in (b), give an expression for $g(x)$ in terms of x. Then compute $g'(x)$ from $g(x)$, and show that the result coincides with the one from (b).

SOLUTION
(a) By Theorem 2,

$$g'(x) = \frac{\partial g}{\partial x} = \frac{\partial f}{\partial x}\frac{\partial x}{\partial x} + \frac{\partial f}{\partial y}\frac{\partial y}{\partial x} = \frac{\partial f}{\partial x} + \frac{\partial f}{\partial y}y'(x)$$

(b) With the given functions, we have

$$\frac{\partial f}{\partial x} = 3x^2 - y^2, \quad \frac{\partial f}{\partial y} = -2xy, \quad y'(x) = -1$$

so that

$$g'(x) = \frac{\partial f}{\partial x} + \frac{\partial f}{\partial y}y'(x) = 3x^2 - y^2 - 2xy(-1) = 3x^2 + 2xy - y^2$$

Substituting $1 - x$ for y, this becomes

$$g'(x) = 3x^2 + 2x(1 - x) - (1 - x)^2 = 4x - 1$$

(c) We have

$$g(x) = f(x, y(x)) = f(x, 1 - x) = x^3 - x(1 - x)^2 = 2x^2 - x$$

Differentiating gives $g'(x) = 4x - 1$, which agrees with the answer from (b).

19. A baseball player hits the ball and then runs down the first base line at 20 ft/s. The first baseman fields the ball and then runs toward first base along the second base line at 18 ft/s as in Figure 7.

FIGURE 7

Determine how fast the distance between the two players is changing at a moment when the hitter is 8 ft from first base and the first baseman is 6 ft from first base.

SOLUTION As in the figure, let $x(t)$ be the distance of the runner from first base at time t, and $y(t)$ the distance of the fielder from first base at time t. With first base at the origin, fix signs so that both speeds are positive and the distances from the base are negative. Then the distance between the two players is $D(t) = \sqrt{x(t)^2 + y(t)^2}$. Using the Chain Rule, the rate of change of distance with respect to time is

$$\frac{dD}{dt} = \frac{dD}{dx} \cdot \frac{dx}{dt} + \frac{dD}{dy} \cdot \frac{dy}{dt}$$

We are given that $\frac{dx}{dt} = 20$ and $\frac{dy}{dt} = 18$. Further,

$$\frac{dD}{dx} = \frac{1}{2}(x^2 + y^2)^{-1/2} \cdot 2x = \frac{x}{\sqrt{x^2 + y^2}}$$

$$\frac{dD}{dy} = \frac{1}{2}(x^2 + y^2)^{-1/2} \cdot 2y = \frac{y}{\sqrt{x^2 + y^2}}$$

so that for the given values of x and y

$$\left.\frac{dD}{dx}\right|_{(x,y)=(-8,-6)} = -\frac{8}{\sqrt{8^2 + 6^2}} = -\frac{4}{5}$$

$$\left.\frac{dD}{dy}\right|_{(x,y)=(-8,-6)} = -\frac{6}{\sqrt{8^2 + 6^2}} = -\frac{3}{5}$$

Thus

$$\frac{dD}{dt} = -\frac{4}{5} \cdot 20 - \frac{3}{5} \cdot 18 = -\frac{134}{5} = -26.8 \text{ ft/sec}$$

The distance between the players is decreasing at a rate of 26.8 ft/sec.

21. Two spacecraft are following paths in space given by $\mathbf{r}_1 = \langle \sin t, t, t^2 \rangle$ and $\mathbf{r}_2 = \langle \cos t, 1 - t, t^3 \rangle$. If the temperature for points in space is given by $T(x, y, z) = x^2 y(1 - z)$, use the Chain Rule to determine the rate of change of the difference D in the temperatures the two spacecraft experience at time $t = \pi$.

SOLUTION Write $T_1(x_1, y_1, z_1)$ for the temperatures seen by the first spacecraft and $T_2(x_2, y_2, z_2)$ for the second, just to keep notation straight below. Then by the chain rule

$$\frac{dD}{dt} = \frac{dT_1}{dt} - \frac{dT_2}{dt}$$

$$= \frac{dT_1}{dx_1} \cdot \frac{dx_1}{dt} + \frac{dT_1}{dy_1} \cdot \frac{dy_1}{dt} + \frac{dT_1}{dz_1} \cdot \frac{dz_1}{dt} - \left(\frac{dT_2}{dx_2} \cdot \frac{dx_2}{dt} + \frac{dT_2}{dy_2} \cdot \frac{dy_2}{dt} + \frac{dT_2}{dz_2} \cdot \frac{dz_2}{dt}\right)$$

$$= 2x_1 y_1 (1 - z_1) \cos t + x_1^2 (1 - z_1) \cdot 1 - x_1^2 y_1 \cdot 2t - (2x_2 y_2 (1 - z_2)(-\sin t) + x_2^2 (1 - z_2) \cdot (-1) - x_2^2 y_2 \cdot 3t^2)$$

$$= 2x_1 y_1 (1 - z_1) \cos t + x_1^2 (1 - z_1) - 2t x_1^2 y_1 + 2x_2 y_2 (1 - z_2) \sin t + x_2^2 (1 - z_2) + 3t^2 x_2^2 y_2$$

When $t = \pi$, we have

$$x_1(\pi) = \sin \pi = 0, \quad y_1(\pi) = \pi, \quad z_1(\pi) = \pi^2$$
$$x_2(\pi) = \cos \pi = -1, \quad y_2(\pi) = 1 - \pi, \quad z_2(\pi) = \pi^3$$

Substituting gives

$$\frac{dD}{dt} = 2 \cdot 0 \cdot \pi \cdot (1 - \pi^2) \cos \pi + 0^2(1 - \pi^2) - 2\pi 0^2 \cdot \pi$$
$$+ 2(-1)(1 - \pi)(1 - \pi^3) \sin \pi + (-1)^2(1 - \pi^3) + 3\pi^2(-1)^2(1 - \pi)$$
$$= 1 - \pi^3 + 3\pi^2(1 - \pi) = -4\pi^3 + 3\pi^2 + 1$$

23. Let $u = u(x, y)$, and let (r, θ) be polar coordinates. Verify the relation

$$\|\nabla u\|^2 = u_r^2 + \frac{1}{r^2} u_\theta^2 \qquad \boxed{8}$$

Hint: Compute the right-hand side by expressing u_θ and u_r in terms of u_x and u_y.

SOLUTION By the Chain Rule we have

$$u_\theta = u_x x_\theta + u_y y_\theta \qquad (1)$$
$$u_r = u_x x_r + u_y y_r \qquad (2)$$

Since $x = r \cos \theta$ and $y = r \sin \theta$, the partial derivatives of x and y with respect to r and θ are

$$x_\theta = -r \sin \theta, \quad y_\theta = r \cos \theta$$
$$x_r = \cos \theta, \quad y_r = \sin \theta$$

Substituting in (1) and (2) gives

$$u_\theta = (-r \sin \theta) u_x + (r \cos \theta) u_y \qquad (3)$$
$$u_r = (\cos \theta) u_x + (\sin \theta) u_y \qquad (4)$$

We now solve these equations for u_x and u_y in terms of u_θ and u_r. Multiplying (3) by $(-\sin \theta)$ and (4) by $r \cos \theta$ and adding the resulting equations gives

$$(-\sin \theta) u_\theta = (r \sin^2 \theta) u_x - (r \cos \theta \sin \theta) u_y$$
$$+ \quad r \cos \theta u_r = (r \cos^2 \theta) u_x + (r \cos \theta \sin \theta) u_y$$
$$\overline{(r \cos \theta) u_r - (\sin \theta) u_\theta = r u_x}$$

or

$$u_x = (\cos \theta) u_r - \frac{\sin \theta}{r} u_\theta \qquad (5)$$

Similarly, we multiply (3) by $\cos \theta$ and (4) by $r \sin \theta$ and add the resulting equations. We get

$$(\cos \theta) u_\theta = (-r \sin \theta \cos \theta) u_x + \left(r \cos^2 \theta\right) u_y$$
$$+ \quad r \sin \theta u_r = (r \sin \theta \cos \theta) u_x + (r \sin^2 \theta) u_y$$
$$\overline{(\cos \theta) u_\theta + (r \sin \theta) u_r = r u_y}$$

or

$$u_y = (\sin \theta) u_r + \frac{\cos \theta}{r} u_\theta \qquad (6)$$

We now use (5) and (6) to compute $\|\nabla u\|^2$ in terms of u_r and u_θ. We get

$$\|\nabla u\|^2 = u_x^2 + u_y^2 = \left((\cos \theta) u_r - \frac{\sin \theta}{r} u_\theta\right)^2 + \left((\sin \theta) u_r + \frac{\cos \theta}{r} u_\theta\right)^2$$

$$= \left(\cos^2 \theta\right) u_r^2 - \frac{2 \cos \theta \sin \theta}{r} u_r u_\theta + \frac{\sin^2 \theta}{r^2} u_\theta^2 + \left(\sin^2 \theta\right) u_r^2 + \frac{2 \sin \theta \cos \theta}{r} u_r u_\theta + \frac{\cos^2 \theta}{r^2} u_\theta^2$$

$$= \left(\cos^2\theta + \sin^2\theta\right)u_r^2 + \frac{1}{r^2}\left(\sin^2\theta + \cos^2\theta\right)u_\theta^2 = u_r^2 + \frac{1}{r^2}u_\theta^2$$

That is,

$$\|\nabla u\|^2 = u_r^2 + \frac{1}{r^2}u_\theta^2$$

25. Let $x = s + t$ and $y = s - t$. Show that for any differentiable function $f(x, y)$,

$$\left(\frac{\partial f}{\partial x}\right)^2 - \left(\frac{\partial f}{\partial y}\right)^2 = \frac{\partial f}{\partial s}\frac{\partial f}{\partial t}$$

SOLUTION By the Chain Rule we have

$$\frac{\partial f}{\partial s} = \frac{\partial f}{\partial x}\frac{\partial x}{\partial s} + \frac{\partial f}{\partial y}\frac{\partial y}{\partial s} = \frac{\partial f}{\partial x}\cdot 1 + \frac{\partial f}{\partial y}\cdot 1 = \frac{\partial f}{\partial x} + \frac{\partial f}{\partial y}$$

$$\frac{\partial f}{\partial t} = \frac{\partial f}{\partial x}\frac{\partial x}{\partial t} + \frac{\partial f}{\partial y}\frac{\partial y}{\partial t} = \frac{\partial f}{\partial x}\cdot 1 + \frac{\partial f}{\partial y}\cdot(-1) = \frac{\partial f}{\partial x} - \frac{\partial f}{\partial y}$$

Hence, using the algebraic identity $(a + b)(a - b) = a^2 - b^2$, we get

$$\frac{\partial f}{\partial s}\cdot\frac{\partial f}{\partial t} = \left(\frac{\partial f}{\partial x} + \frac{\partial f}{\partial y}\right)\cdot\left(\frac{\partial f}{\partial x} - \frac{\partial f}{\partial y}\right) = \left(\frac{\partial f}{\partial x}\right)^2 - \left(\frac{\partial f}{\partial y}\right)^2$$

27. Suppose that z is defined implicitly as a function of x and y by the equation $F(x, y, z) = xz^2 + y^2z + xy - 1 = 0$.
(a) Calculate F_x, F_y, F_z.
(b) Use Eq. (7) to calculate $\dfrac{\partial z}{\partial x}$ and $\dfrac{\partial z}{\partial y}$.

SOLUTION
(a) The partial derivatives of F are

$$F_x = z^2 + y, \quad F_y = 2yz + x, \quad F_z = 2xz + y^2$$

(b) By Eq. (7) we have

$$\frac{\partial z}{\partial x} = -\frac{F_x}{F_z} = -\frac{z^2 + y}{2xz + y^2}$$

$$\frac{\partial z}{\partial y} = -\frac{F_y}{F_z} = -\frac{2yz + x}{2xz + y^2}$$

In Exercises 29–34, calculate the partial derivative using implicit differentiation.

29. $\dfrac{\partial z}{\partial x}$, $x^2y + y^2z + xz^2 = 10$

SOLUTION For $F(x, y, z) = x^2y + y^2z + xz^2 = 10$ we have

$$\frac{\partial z}{\partial x} = -\frac{F_x}{F_z} \tag{1}$$

We compute the partial derivatives of F:

$$F_x = 2xy + z^2, \quad F_z = y^2 + 2xz$$

Substituting in (1) gives the following derivative:

$$\frac{\partial z}{\partial x} = -\frac{2xy + z^2}{2xz + y^2}$$

31. $\dfrac{\partial z}{\partial y}$, $e^{xy} + \sin(xz) + y = 0$

SOLUTION We use Eq. (7):

$$\frac{\partial z}{\partial y} = -\frac{F_y}{F_z} \quad (1)$$

The partial derivatives of $F(x, y, z) = e^{xy} + \sin(xz) + y$ are

$$F_y = xe^{xy} + 1, \quad F_z = x\cos(xz)$$

Substituting in (1), we get

$$\frac{\partial z}{\partial y} = -\frac{xe^{xy} + 1}{x\cos(xz)}$$

33. $\dfrac{\partial w}{\partial y}$, $\dfrac{1}{w^2 + x^2} + \dfrac{1}{w^2 + y^2} = 1$ at $(x, y, w) = (1, 1, 1)$

SOLUTION Using the formula obtained by implicit differentiation (Eq. (7)), we have

$$\frac{\partial w}{\partial y} = -\frac{F_y}{F_w} \quad (1)$$

We find the partial derivatives of $F(x, y, w) = \dfrac{1}{w^2+x^2} + \dfrac{1}{w^2+y^2} - 1$:

$$F_y = -\frac{2y}{(w^2+y^2)^2}, \quad F_w = \frac{-2w}{(w^2+x^2)^2} - \frac{2w}{(w^2+y^2)^2}$$

We substitute in (1) to obtain

$$\frac{\partial w}{\partial y} = -\frac{\frac{-2y}{(w^2+y^2)^2}}{\frac{-2w}{(w^2+x^2)^2} - \frac{2w}{(w^2+y^2)^2}} = -\frac{y(w^2+x^2)^2}{w(w^2+y^2)^2 + w(w^2+x^2)^2} = \frac{-y(w^2+x^2)^2}{w((w^2+y^2)^2 + (w^2+x^2)^2)}$$

35. Let $\mathbf{r} = \langle x, y, z \rangle$ and $\mathbf{e_r} = \mathbf{r}/\|\mathbf{r}\|$. Show that if a function $f(x, y, z) = F(r)$ depends only on the distance from the origin $r = \|\mathbf{r}\| = \sqrt{x^2 + y^2 + z^2}$, then

$$\nabla f = F'(r)\mathbf{e_r} \quad \boxed{9}$$

SOLUTION The gradient of f is the following vector:

$$\nabla f = \left\langle \frac{\partial f}{\partial x}, \frac{\partial f}{\partial y}, \frac{\partial f}{\partial z} \right\rangle$$

We must express this vector in terms of $\mathbf{r}$ and r. Using the Chain Rule, we have

$$\frac{\partial f}{\partial x} = F'(r)\frac{\partial r}{\partial x} = F'(r) \cdot \frac{2x}{2\sqrt{x^2+y^2+z^2}} = F'(r) \cdot \frac{x}{r}$$

$$\frac{\partial f}{\partial y} = F'(r)\frac{\partial r}{\partial y} = F'(r) \cdot \frac{2y}{2\sqrt{x^2+y^2+z^2}} = F'(r) \cdot \frac{y}{r}$$

$$\frac{\partial f}{\partial z} = F'(r)\frac{\partial r}{\partial z} = F'(r) \cdot \frac{2z}{2\sqrt{x^2+y^2+z^2}} = F'(r) \cdot \frac{z}{r}$$

Hence,

$$\nabla f = \left\langle F'(r)\frac{x}{r}, F'(r)\frac{y}{r}, F'(r)\frac{z}{r} \right\rangle = \frac{F'(r)}{r}\langle x, y, z \rangle = F'(r)\frac{\mathbf{r}}{\|\mathbf{r}\|} = F'(r)\mathbf{e_r}$$

37. Use Eq. (9) to compute $\nabla\left(\dfrac{1}{r}\right)$.

SOLUTION To compute $\nabla(\frac{1}{r})$ using Eq. (9), we let $F(r) = \frac{1}{r}$:

$$F'(r) = -\frac{1}{r^2}$$

We obtain

$$\nabla\left(\frac{1}{r}\right) = F'(r)\mathbf{e_r} = -\frac{1}{r^2} \cdot \frac{\mathbf{r}}{\|\mathbf{r}\|} = -\frac{1}{r^3}\mathbf{r}$$

39. Figure 9 shows the graph of the equation

$$F(x, y, z) = x^2 + y^2 - z^2 - 12x - 8z - 4 = 0$$

(a) Use the quadratic formula to solve for z as a function of x and y. This gives two formulas, depending on the choice of sign.

(b) Which formula defines the portion of the surface satisfying $z \geq -4$? Which formula defines the portion satisfying $z \leq -4$?

(c) Calculate $\partial z/\partial x$ using the formula $z = f(x, y)$ (for both choices of sign) and again via implicit differentiation. Verify that the two answers agree.

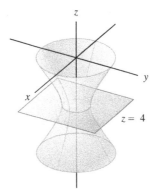

FIGURE 9 Graph of $x^2 + y^2 - z^2 - 12x - 8z - 4 = 0$.

SOLUTION

(a) We rewrite $F(x, y, z) = 0$ as a quadratic equation in the variable z:

$$z^2 + 8z + \left(4 + 12x - x^2 - y^2\right) = 0$$

We solve for z. The discriminant is

$$8^2 - 4\left(4 + 12x - x^2 - y^2\right) = 4x^2 + 4y^2 - 48x + 48 = 4\left(x^2 + y^2 - 12x + 12\right)$$

Hence,

$$z_{1,2} = \frac{-8 \pm \sqrt{4\left(x^2 + y^2 - 12x + 12\right)}}{2} = -4 \pm \sqrt{x^2 + y^2 - 12x + 12}$$

We obtain two functions:

$$z = -4 + \sqrt{x^2 + y^2 - 12x + 12}, \quad z = -4 - \sqrt{x^2 + y^2 - 12x + 12}$$

(b) The formula with the positive root defines the portion of the surface satisfying $z \geq -4$, and the formula with the negative root defines the portion satisfying $z \leq -4$.

(c) Differentiating $z = -4 + \sqrt{x^2 + y^2 - 12x + 12}$ with respect to x, using the Chain Rule, gives

$$\frac{\partial z}{\partial x} = \frac{2x - 12}{2\sqrt{x^2 + y^2 - 12x + 12}} = \frac{x - 6}{\sqrt{x^2 + y^2 - 12x + 12}} \quad (1)$$

Alternatively, using the formula for $\frac{\partial z}{\partial x}$ obtained by implicit differentiation gives

$$\frac{\partial z}{\partial x} = -\frac{F_x}{F_z} \quad (2)$$

We find the partial derivatives of $F(x, y, z) = x^2 + y^2 - z^2 - 12x - 8z - 4$:

$$F_x = 2x - 12, \quad F_z = -2z - 8$$

Substituting in (2) gives

$$\frac{\partial z}{\partial x} = -\frac{2x - 12}{-2z - 8} = \frac{x - 6}{z + 4}$$

This result is the same as the result in (1), since $z = -4 + \sqrt{x^2 + y^2 - 12x + 12}$ implies that

$$\sqrt{x^2 + y^2 - 12x + 12} = z + 4$$

For $z = -4 - \sqrt{x^2 + y^2 - 12x + 12}$, differentiating with respect to x gives

$$\frac{\partial z}{\partial x} = -\frac{2x - 12}{2\sqrt{x^2 + y^2 - 12x + 12}} = \frac{x - 6}{-\sqrt{x^2 + y^2 - 12x + 12}} = \frac{x - 6}{z + 4}$$

which is equal to $-\frac{F_x}{F_z}$ computed above.

41. The pressure P, volume V, and temperature T of a van der Waals gas with n molecules (n constant) are related by the equation

$$\left(P + \frac{an^2}{V^2}\right)(V - nb) = nRT$$

where a, b, and R are constant. Calculate $\partial P/\partial T$ and $\partial V/\partial P$.

SOLUTION Let F be the following function:

$$F(P, V, T) = \left(P + \frac{an^2}{V^2}\right)(V - nb) - nRT$$

By Eq. (7),

$$\frac{\partial P}{\partial T} = -\frac{\frac{\partial F}{\partial T}}{\frac{\partial F}{\partial P}}, \quad \frac{\partial V}{\partial P} = -\frac{\frac{\partial F}{\partial P}}{\frac{\partial F}{\partial V}} \quad (1)$$

We compute the partial derivatives of F:

$$\frac{\partial F}{\partial P} = V - nb$$

$$\frac{\partial F}{\partial T} = -nR$$

$$\frac{\partial F}{\partial V} = -2an^2 V^{-3}(V - nb) + \left(P + \frac{an^2}{V^2}\right) = P + \frac{2an^3 b}{V^3} - \frac{an^2}{V^2}$$

Substituting in (1) gives

$$\frac{\partial P}{\partial T} = -\frac{-nR}{V - nb} = \frac{nR}{V - nb}$$

$$\frac{\partial V}{\partial P} = -\frac{V - nb}{P + \frac{2an^3 b}{V^3} - \frac{an^2}{V^2}} = \frac{nbV^3 - V^4}{PV^3 + 2an^3 b - an^2 V}$$

43. Show that if $f(x)$ is differentiable and $c \neq 0$ is a constant, then $u(x, t) = f(x - ct)$ satisfies the so-called **advection equation**

$$\frac{\partial u}{\partial t} + c\frac{\partial u}{\partial x} = 0$$

SOLUTION For $s = x - ct$, we have $u(x, t) = f(s)$. We use the Chain Rule to compute $\frac{\partial u}{\partial t}$ and $\frac{\partial u}{\partial x}$:

$$\frac{\partial u}{\partial t} = f'(s)\frac{\partial s}{\partial t} = f'(s) \cdot (-c) = -cf'(s) \quad (1)$$

$$\frac{\partial u}{\partial x} = f'(s)\frac{\partial s}{\partial x} = f'(s) \cdot 1 = f'(s) \quad (2)$$

Equalities (1) and (2) imply that:

$$\frac{\partial u}{\partial t} = -c\frac{\partial u}{\partial x} \quad \text{or} \quad \frac{\partial u}{\partial t} + c\frac{\partial u}{\partial x} = 0$$

Further Insights and Challenges

In Exercises 44–47, a function $f(x, y, z)$ is called **homogeneous of degree n** if $f(\lambda x, \lambda y, \lambda z) = \lambda^n f(x, y, z)$ for all $\lambda \in \mathbf{R}$.

45. Prove that if $f(x, y, z)$ is homogeneous of degree n, then $f_x(x, y, z)$ is homogeneous of degree $n - 1$. *Hint:* Either use the limit definition or apply the Chain Rule to $f(\lambda x, \lambda y, \lambda z)$.

SOLUTION We are given that $f(\lambda x, \lambda y, \lambda z) = \lambda^n f(x, y, z)$ for all λ, and we must show that $f_x(\lambda x, \lambda y, \lambda z) = \lambda^{n-1} f_x(x, y, z)$. We use the limit definition of f_x. Since for all $\lambda \ne 0$, $\lambda h \to 0$ if and only if $h \to 0$, we get

$$f_x(\lambda x, \lambda y, \lambda z) = \lim_{h \to 0} \frac{f(\lambda x + \lambda h, \lambda y, \lambda z) - f(\lambda x, \lambda y, \lambda z)}{\lambda h} = \lim_{h \to 0} \frac{f(\lambda(x + h), \lambda y, \lambda z) - f(\lambda x, \lambda y, \lambda z)}{\lambda h}$$

$$= \lim_{h \to 0} \frac{\lambda^n f(x + h, y, z) - \lambda^n f(x, y, z)}{\lambda h} = \lim_{h \to 0} \frac{\lambda^{n-1} f(x + h, y, z) - \lambda^{n-1} f(x, y, z)}{h}$$

$$= \lambda^{n-1} \lim_{h \to 0} \frac{f(x + h, y, z) - f(x, y, z)}{h} = \lambda^{n-1} f_x(x, y, z)$$

Alternatively, we prove this property using the Chain Rule. We use the Chain Rule to differentiate the following equality with respect to x:

$$f(\lambda x, \lambda y, \lambda z) = \lambda^n f(x, y, z)$$

We get

$$f_x(\lambda x, \lambda y, \lambda z) \cdot \frac{\partial(\lambda x)}{\partial x} + f_y(\lambda x, \lambda y, \lambda z) \cdot \frac{\partial(\lambda y)}{\partial x} + f_z(\lambda x, \lambda y, \lambda z) \cdot \frac{\partial(\lambda z)}{\partial x} = \lambda^n f_x(x, y, z)$$

Since $\frac{\partial(\lambda y)}{\partial x} = \frac{\partial(\lambda z)}{\partial x} = 0$ and $\frac{\partial(\lambda x)}{\partial x} = \lambda$, we obtain for $\lambda \ne 0$,

$$\lambda f_x(\lambda x, \lambda y, \lambda z) = \lambda^n f_x(x, y, z) \quad \text{or} \quad f_x(\lambda x, \lambda y, \lambda z) = \lambda^{n-1} f_x(x, y, z)$$

47. Verify Eq. (11) for the functions in Exercise 44.

SOLUTION Eq. (11) states that if f is homogeneous of degree n, then

$$x \frac{\partial f}{\partial x} + y \frac{\partial f}{\partial y} + z \frac{\partial f}{\partial z} = nf$$

(a) $f(x, y, z) = x^2 y + xyz$. f is homogeneous of degree $n = 3$. The partial derivatives of f are

$$\frac{\partial f}{\partial x} = 2xy + yz, \quad \frac{\partial f}{\partial y} = x^2 + xz, \quad \frac{\partial f}{\partial z} = xy$$

Hence,

$$x \frac{\partial f}{\partial x} + y \frac{\partial f}{\partial y} + z \frac{\partial f}{\partial z} = x(2xy + yz) + y(x^2 + xz) + zxy = 3x^2 y + 3xyz = 3(x^2 y + xyz) = 3f(x, y, z)$$

(b) $f(x, y, z) = 3x + 2y - 8z$. f is homogeneous of degree $n = 1$. We have

$$x \frac{\partial f}{\partial x} + y \frac{\partial f}{\partial y} + z \frac{\partial f}{\partial z} = x \cdot 3 + y \cdot 2 + z \cdot (-8) = 3x + 2y - 8z = 1 \cdot f(x, y, z)$$

(c) $f(x, y, z) = \ln\left(\frac{xy}{z^2}\right)$. f is homogeneous of degree $n = 0$. The partial derivatives of f are

$$\frac{\partial f}{\partial x} = \frac{\frac{y}{z^2}}{\frac{xy}{z^2}} = \frac{1}{x}, \quad \frac{\partial f}{\partial y} = \frac{\frac{x}{z^2}}{\frac{xy}{z^2}} = \frac{1}{y}, \quad \frac{\partial f}{\partial z} = \frac{-2z^{-3}xy}{xyz^{-2}} = -\frac{2}{z}$$

Hence,

$$x \frac{\partial f}{\partial x} + y \frac{\partial f}{\partial y} + z \frac{\partial f}{\partial z} = x \cdot \frac{1}{x} + y \cdot \frac{1}{y} + z \cdot \left(-\frac{2}{z}\right) = 0 = 0 \cdot f(x, y, z)$$

(d) $f(x, y, z) = z^4$. f is homogeneous of degree $n = 4$. We have

$$x \frac{\partial f}{\partial x} + y \frac{\partial f}{\partial y} + z \frac{\partial f}{\partial z} = x \cdot 0 + y \cdot 0 + z \cdot 4z^3 = 4z^4 = 4f(x, y, z)$$

49. Let $r = \sqrt{x_1^2 + \cdots + x_n^2}$ and let $g(r)$ be a function of r. Prove the formulas

$$\frac{\partial g}{\partial x_i} = \frac{x_i}{r} g_r, \qquad \frac{\partial^2 g}{\partial x_i^2} = \frac{x_i^2}{r^2} g_{rr} + \frac{r^2 - x_i^2}{r^3} g_r$$

SOLUTION By the Chain Rule, we have

$$\frac{\partial g}{\partial x_i} = g'(r)\frac{\partial r}{\partial x_i} = g_r \cdot \frac{2x_i}{2\sqrt{x_1^2 + \cdots + x_n^2}} = g_r \frac{x_i}{r}$$

We differentiate $\frac{\partial g}{\partial x_i}$ with respect to x_i. Using the Product Rule we get

$$\frac{\partial^2 g}{\partial x_i^2} = \frac{\partial}{\partial x_i}(g_r) \cdot \frac{x_i}{r} + g_r \frac{\partial}{\partial x_i}\left(\frac{x_i}{r}\right) \tag{1}$$

We use the Chain Rule to compute $\frac{\partial}{\partial x_i}(g_r)$:

$$\frac{\partial}{\partial x_i}(g_r) = \frac{d}{dr}(g_r) \cdot \frac{\partial r}{\partial x_i} = g_{rr} \cdot \frac{2x_i}{2\sqrt{x_1^2 + \cdots + x_n^2}} = g_{rr} \cdot \frac{x_i}{r} \tag{2}$$

We compute $\frac{\partial}{\partial x_i} \cdot \left(\frac{x_i}{r}\right)$ using the Quotient Rule and the Chain Rule:

$$\frac{\partial}{\partial x_i} \cdot \left(\frac{x_i}{r}\right) = \frac{1 \cdot r - x_i \cdot \frac{\partial r}{\partial x_i}}{r^2} = \frac{r - x_i \cdot \frac{x_i}{r}}{r^2} = \frac{r^2 - x_i^2}{r^3} \tag{3}$$

Substituting (2) and (3) in (1), we obtain

$$\frac{\partial^2 g}{\partial x_i^2} = g_{rr} \cdot \frac{x_i}{r} \cdot \frac{x_i}{r} + g_r \frac{r^2 - x_i^2}{r^3} = \frac{x_i^2}{r^2} g_{rr} + \frac{r^2 - x_i^2}{r^3} g_r$$

*In Exercises 51–55, the **Laplace operator** is defined by $\Delta f = f_{xx} + f_{yy}$. A function $f(x, y)$ satisfying the Laplace equation $\Delta f = 0$ is called **harmonic**. A function $f(x, y)$ is called **radial** if $f(x, y) = g(r)$, where $r = \sqrt{x^2 + y^2}$.*

51. Use Eq. (12) to prove that in polar coordinates (r, θ),

$$\Delta f = f_{rr} + \frac{1}{r^2} f_{\theta\theta} + \frac{1}{r} f_r \qquad \boxed{13}$$

SOLUTION The polar coordinates are $x = r\cos\theta$, $y = r\sin\theta$. Hence,

$$\frac{\partial x}{\partial \theta} = -r\sin\theta, \quad \frac{\partial y}{\partial \theta} = r\cos\theta, \quad \frac{\partial x}{\partial r} = \cos\theta, \quad \frac{\partial y}{\partial r} = \sin\theta,$$

$$\frac{\partial^2 x}{\partial \theta^2} = -r\cos\theta, \quad \frac{\partial^2 y}{\partial \theta^2} = -r\sin\theta, \quad \frac{\partial^2 x}{\partial r^2} = \frac{\partial^2 y}{\partial r^2} = 0$$

By Eq. (12) we have

$$f_{\theta\theta} = f_{xx}\left(\frac{\partial x}{\partial \theta}\right)^2 + f_{yy}\left(\frac{\partial y}{\partial \theta}\right)^2 + 2f_{xy}\left(\frac{\partial x}{\partial \theta}\right)\left(\frac{\partial y}{\partial \theta}\right) + f_x \frac{\partial^2 x}{\partial \theta^2} + f_y \frac{\partial^2 y}{\partial \theta^2}$$

$$= f_{xx}\left(r^2 \sin^2\theta\right) + f_{yy}\left(r^2 \cos^2\theta\right) - \left(2r^2 \sin\theta\cos\theta\right) f_{xy} - (r\cos\theta)f_x - (r\sin\theta)f_y \tag{1}$$

and

$$f_{rr} = f_{xx}\left(\frac{\partial x}{\partial r}\right)^2 + f_{yy}\left(\frac{\partial y}{\partial r}\right)^2 + 2f_{xy}\left(\frac{\partial x}{\partial r}\right)\left(\frac{\partial y}{\partial r}\right) + f_x \frac{\partial^2 x}{\partial r^2} + f_y \frac{\partial^2 y}{\partial r^2}$$

$$= f_{xx}\left(\cos^2\theta\right) + f_{yy}\left(\sin^2\theta\right) + (2\cos\theta\sin\theta)f_{xy} \tag{2}$$

$$f_r = f_x \frac{\partial x}{\partial r} + f_y \frac{\partial y}{\partial r} = f_x(\cos\theta) + f_y(\sin\theta) \tag{3}$$

We now compute the right-hand side of the equality we need to prove. Using (1), (2), and (3), we obtain

$$f_{rr} + \frac{1}{r^2}f_{\theta\theta} + \frac{1}{r}f_r = f_{xx}\left(\cos^2\theta\right) + f_{yy}\left(\sin^2\theta\right) + (2\cos\theta\sin\theta)f_{xy} + f_{xx}\left(\sin^2\theta\right)$$

$$+ f_{yy}\left(\cos^2\theta\right) - (2\sin\theta\cos\theta)f_{xy} - \frac{\cos\theta}{r}f_x - \frac{\sin\theta}{r}f_y + f_x\frac{\cos\theta}{r} + f_y\frac{\sin\theta}{r}$$

$$= f_{xx}\left(\cos^2\theta + \sin^2\theta\right) + f_{yy}\left(\sin^2\theta + \cos^2\theta\right)$$

$$= f_{xx} + f_{yy} = \Delta f$$

We thus showed that

$$\Delta f = f_{rr} + \frac{1}{r^2}f_{\theta\theta} + \frac{1}{r}f_r$$

53. Verify that $f(x,y) = x$ and $f(x,y) = y$ are harmonic using both the rectangular and polar expressions for Δf.

SOLUTION We must show that $\Delta f = 0$.

(a) Using the rectangular expression for Δf:

$$\Delta f = f_{xx} + f_{yy}$$

For $f(x,y) = x$ we have $f_x = 1$, $f_y = 0$, hence, $f_{xx} = 0$, $f_{yy} = 0$. Therefore $\Delta f = f_{xx} + f_{yy} = 0 + 0 = 0$. For $f(x,y) = y$ we have $f_y = 1$, $f_x = 0$, hence, $f_{xx} = 0$, $f_{yy} = 0$, and again, $\Delta f = f_{xx} + f_{yy} = 0 + 0 = 0$.

(b) Using the polar expression for Δf,

$$\Delta f = f_{rr} + \frac{1}{r^2}f_{\theta\theta} + \frac{1}{r}f_r \tag{1}$$

Since $x = r\cos\theta$, we have $f(r,\theta) = x = r\cos\theta$. Hence,

$$f_r = \cos\theta, \quad f_\theta = -r\sin\theta, \quad f_{rr} = 0, \quad f_{\theta\theta} = -r\cos\theta$$

We now show that $\Delta f = 0$:

$$\Delta f = f_{rr} + \frac{1}{r^2}f_{\theta\theta} + \frac{1}{r}f_r = 0 + \frac{1}{r^2}\cdot(-r\cos\theta) + \frac{1}{r}\cos\theta = 0$$

Similarly, since $y = r\sin\theta$, we have $f(r,\theta) = y = r\sin\theta$. Hence,

$$f_r = \sin\theta, \quad f_\theta = r\cos\theta, \quad f_{rr} = 0, \quad f_{\theta\theta} = -r\sin\theta$$

Substituting in (1) gives

$$\Delta f = 0 + \frac{1}{r^2}(-r\sin\theta) + \frac{1}{r}\sin\theta = 0$$

55. Use the Product Rule to show that

$$f_{rr} + \frac{1}{r}f_r = r^{-1}\frac{\partial}{\partial r}\left(r\frac{\partial f}{\partial r}\right)$$

Use this formula to show that if f is a radial harmonic function, then $rf_r = C$ for some constant C. Conclude that $f(x,y) = C\ln r + b$ for some constant b.

SOLUTION We show that $f_{rr} + \frac{1}{r}f_r = r^{-1}\frac{\partial}{\partial r}\left(r\frac{\partial f}{\partial r}\right)$. We use the Product Rule to compute the following derivative:

$$\frac{\partial}{\partial r}\left(r\frac{\partial f}{\partial r}\right) = 1\cdot\frac{\partial f}{\partial r} + r\frac{\partial}{\partial r}\left(\frac{\partial f}{\partial r}\right) = \frac{\partial f}{\partial r} + r\frac{\partial^2 f}{\partial r^2} = f_r + rf_{rr} = r\left(f_{rr} + \frac{1}{r}f_r\right)$$

Hence,

$$f_{rr} + \frac{1}{r}f_r = r^{-1}\frac{\partial}{\partial r}\left(r\frac{\partial f}{\partial r}\right) \tag{1}$$

Now, suppose that $f(x,y)$ is a radial harmonic function. Since f is radial, $f(x,y) = g(r)$, therefore $f_{\theta\theta} = 0$. Substituting in the polar expressions for Δf gives

$$\Delta f = f_{rr} + \frac{1}{r^2}f_{\theta\theta} + \frac{1}{r}f_r = f_{rr} + \frac{1}{r}f_r = 0$$

Combining with (1), we get

$$r^{-1}\frac{\partial}{\partial r}\left(r\frac{\partial f}{\partial r}\right) = 0 \quad \text{or} \quad \frac{\partial}{\partial r}\left(r\frac{\partial f}{\partial r}\right) = 0$$

yielding

$$r\frac{\partial f}{\partial r} = C \quad \Rightarrow \quad f_r = \frac{C}{r}$$

We now integrate the two sides to obtain

$$\int f_r \, dr = \int \frac{C}{r} \, dr \quad \text{or} \quad f(r) = C \ln r + b$$

14.7 Optimization in Several Variables (LT Section 15.7)

Preliminary Questions

1. The functions $f(x, y) = x^2 + y^2$ and $g(x, y) = x^2 - y^2$ both have a critical point at $(0, 0)$. How is the behavior of the two functions at the critical point different?

SOLUTION Let $f(x, y) = x^2 + y^2$ and $g(x, y) = x^2 - y^2$. In the domain $\mathbf{R}^2$, the partial derivatives of f and g are

$$f_x = 2x, \quad f_{xx} = 2, \quad f_y = 2y, \quad f_{yy} = 2, \quad f_{xy} = 0$$

$$g_x = 2x, \quad g_{xx} = 2, \quad g_y = -2y, \quad g_{yy} = -2, \quad g_{xy} = 0$$

Therefore, $f_x = f_y = 0$ at $(0, 0)$ and $g_x = g_y = 0$ at $(0, 0)$. That is, the two functions have one critical point, which is the origin. Since the discriminant of f is $D = 4 > 0$, $f_{xx} > 0$, and the discriminant of g is $D = -4 < 0$, f has a local minimum (which is also a global minimum) at the origin, whereas g has a saddle point there. Moreover, since $\lim_{y \to \infty} g(0, y) = -\infty$ and $\lim_{x \to \infty} g(x, 0) = \infty$, g does not have global extrema on the plane. Similarly, f does not have a global maximum but does have a global minimum, which is $f(0, 0) = 0$.

2. Identify the points indicated in the contour maps as local minima, local maxima, saddle points, or neither (Figure 16).

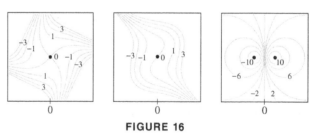

FIGURE 16

SOLUTION If $f(P)$ is a local minimum or maximum, then the nearby level curves are closed curves encircling P. In Figure (C), f increases in all directions emanating from P and decreases in all directions emanating from Q. Hence, f has a local minimum at P and local maximum at Q.

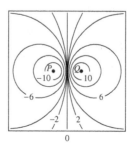

In Figure (A), the level curves through the point R consist of two intersecting lines that divide the neighborhood near R into four regions. f is decreasing in some directions and increasing in other directions. Therefore, R is a saddle point.

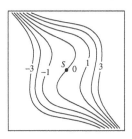

Figure (A)

Point S in Figure (B) is neither a local extremum nor a saddle point of f.

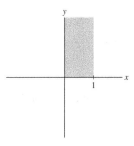

Figure (B)

3. Let $f(x, y)$ be a continuous function on a domain $\mathcal{D}$ in $\mathbf{R}^2$. Determine which of the following statements are true:

(a) If $\mathcal{D}$ is closed and bounded, then f takes on a maximum value on $\mathcal{D}$.

(b) If $\mathcal{D}$ is neither closed nor bounded, then f does not take on a maximum value of $\mathcal{D}$.

(c) $f(x, y)$ need not have a maximum value on the domain $\mathcal{D}$ defined by $0 \leq x \leq 1, 0 \leq y \leq 1$.

(d) A continuous function takes on neither a minimum nor a maximum value on the open quadrant

$$\{(x, y) : x > 0, y > 0\}$$

SOLUTION

(a) This statement is true. It follows by the Theorem on Existence of Global Extrema.

(b) The statement is false. Consider the constant function $f(x, y) = 2$ in the following domain:

$$D = \{(x, y) : 0 < x \leq 1, \ 0 \leq y < \infty\}$$

Obviously f is continuous and D is neither closed nor bounded. However, f takes on a maximum value (which is 2) on D.

(c) The domain $D = \{(x, y) : 0 \leq x, y \leq 1\}$ is the following rectangle:

$$D = \{(x, y) : 0 \leq x, y \leq 1\}$$

D is closed and bounded, hence f takes on a maximum value on D. Thus the statement is false.

(d) The statement is false. The constant function $f(x, y) = c$ takes on minimum and maximum values on the open quadrant.

Exercises

1. Let $P = (a, b)$ be a critical point of $f(x, y) = x^2 + y^4 - 4xy$.

(a) First use $f_x(x, y) = 0$ to show that $a = 2b$. Then use $f_y(x, y) = 0$ to show that $P = (0, 0)$, $(2\sqrt{2}, \sqrt{2})$, or $(-2\sqrt{2}, -\sqrt{2})$.

(b) Referring to Figure 17, determine the local minima and saddle points of $f(x, y)$ and find the absolute minimum value of $f(x, y)$.

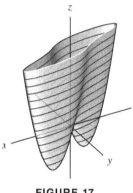

FIGURE 17

SOLUTION

(a) We find the partial derivatives:

$$f_x(x, y) = \frac{\partial}{\partial x}\left(x^2 + y^4 - 4xy\right) = 2x - 4y$$

$$f_y(x, y) = \frac{\partial}{\partial y}\left(x^2 + y^4 - 4xy\right) = 4y^3 - 4x$$

Since $P = (a, b)$ is a critical point, $f_x(a, b) = 0$. That is,

$$2a - 4b = 0 \quad \Rightarrow \quad a = 2b$$

Also $f_y(a, b) = 0$, hence,

$$4b^3 - 4a = 0 \quad \Rightarrow \quad a = b^3$$

We obtain the following equations for the critical points (a, b):

$$\begin{cases} a = 2b \\ a = b^3 \end{cases}$$

Equating the two equations, we get

$$2b = b^3$$

$$b^3 - 2b = b(b^2 - 2) = 0 \quad \Rightarrow \quad \begin{cases} b_1 = 0 \\ b_2 = \sqrt{2} \\ b_3 = -\sqrt{2} \end{cases}$$

Since $a = 2b$, we have $a_1 = 0$, $a_2 = 2\sqrt{2}$, $a_3 = -2\sqrt{2}$. The critical points are thus

$$P_1 = (0, 0), \quad P_2 = \left(2\sqrt{2}, \sqrt{2}\right), \quad P_3 = \left(-2\sqrt{2}, -\sqrt{2}\right)$$

(b) Referring to Figure 14, we see that $P_1 = (0, 0)$ is a saddle point and $P_2 = \left(2\sqrt{2}, \sqrt{2}\right)$, $P_3 = \left(-2\sqrt{2}, -\sqrt{2}\right)$ are local minima. The absolute minimum value of f is -4.

3. Find the critical points of

$$f(x, y) = 8y^4 + x^2 + xy - 3y^2 - y^3$$

Use the contour map in Figure 19 to determine their nature (local minimum, local maximum, or saddle point).

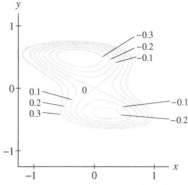

FIGURE 19 Contour map of $f(x, y) = 8y^4 + x^2 + xy - 3y^2 - y^3$.

SOLUTION The critical points are the solutions of $f_x = 0$ and $f_y = 0$. That is,

$$f_x(x, y) = 2x + y = 0$$
$$f_y(x, y) = 32y^3 + x - 6y - 3y^2 = 0$$

The first equation gives $y = -2x$. We substitute in the second equation and solve for x. This gives

$$32(-2x)^3 + x - 6(-2x) - 3(-2x)^2 = 0$$
$$-256x^3 + 13x - 12x^2 = 0$$
$$-x(256x^2 + 12x - 13) = 0$$

Hence $x = 0$ or $256x^2 + 12x - 13 = 0$. Solving the quadratic,

$$x_{1,2} = \frac{-12 \pm \sqrt{12^2 - 4 \cdot 256 \cdot (-13)}}{512} = \frac{-12 \pm 116}{512} \quad \Rightarrow \quad x = \frac{13}{64} \quad \text{or} \quad -\frac{1}{4}$$

Substituting in $y = -2x$ gives the y-coordinates of the critical points. The critical points are thus

$$(0, 0), \quad \left(\frac{13}{64}, -\frac{13}{32}\right), \quad \left(-\frac{1}{4}, \frac{1}{2}\right)$$

We now use the contour map to determine the type of each critical point. The level curves through $(0, 0)$ consist of two intersecting lines that divide the neighborhood near $(0, 0)$ into four regions. The function is decreasing in the y-direction and increasing in the x-direction. Therefore, $(0, 0)$ is a saddle point. The level curves near the critical points $\left(\frac{13}{64}, -\frac{13}{32}\right)$ and $\left(-\frac{1}{4}, \frac{1}{2}\right)$ are closed curves encircling the points, hence these are local minima or maxima. The graph shows that both $\left(\frac{13}{64}, -\frac{13}{32}\right)$ and $\left(-\frac{1}{4}, \frac{1}{2}\right)$ are local minima.

5. Let $f(x, y) = y^2 x - yx^2 + xy$.

(a) Show that the critical points (x, y) satisfy the equations

$$y(y - 2x + 1) = 0, \qquad x(2y - x + 1) = 0$$

(b) Show that f has three critical points where $x = 0$ or $y = 0$ (or both) and one critical point where x and y are nonzero.

(c) Use the Second Derivative Test to determine the nature of the critical points.

SOLUTION

(a) The critical points are the solutions of the two equations $f_x(x, y) = 0$ and $f_y(x, y) = 0$. That is,

$$\begin{aligned} f_x(x, y) &= y^2 - 2yx + y = 0 \\ f_y(x, y) &= 2yx - x^2 + x = 0 \end{aligned} \quad \Rightarrow \quad \begin{aligned} y(y - 2x + 1) &= 0 \\ x(2y - x + 1) &= 0 \end{aligned}$$

(b) We find the critical points by solving the equations obtained in part (a):

$$y(y - 2x + 1) = 0 \quad (1)$$
$$x(2y - x + 1) = 0 \quad (2)$$

Equation (1) implies that $y = 0$ or $y = 2x - 1$. Substituting $y = 0$ in (2) and solving for x gives

$$x(-x + 1) = 0 \quad \Rightarrow \quad x = 0 \quad \text{or} \quad x = 1$$

We obtain the solutions $(0, 0)$ and $(1, 0)$. We now substitute $y = 2x - 1$ in (2) and solve for x. We get

$$x(4x - 2 - x + 1) = 0$$
$$x(3x - 1) = 0 \quad \Rightarrow \quad x = 0 \quad \text{or} \quad x = \frac{1}{3}$$

We compute the y-coordinate, using $y = 2x - 1$:

$$y = 2 \cdot 0 - 1 = -1$$
$$y = 2 \cdot \frac{1}{3} - 1 = -\frac{1}{3}$$

We obtain the solutions $(0, -1)$ and $\left(\frac{1}{3}, -\frac{1}{3}\right)$. To summarize, the critical points are $(0, 0)$, $(1, 0)$, $(0, -1)$, and $\left(\frac{1}{3}, -\frac{1}{3}\right)$. Three of the critical points have at least one zero coordinate, and one has two nonzero coordinates.

(c) We compute the second-order partial derivatives:

$$f_{xx}(x, y) = \frac{\partial}{\partial x}(y^2 - 2yx + y) = -2y$$

$$f_{yy}(x, y) = \frac{\partial}{\partial y}(2yx - x^2 + x) = 2x$$

$$f_{xy}(x, y) = \frac{\partial}{\partial y}(y^2 - 2yx + y) = 2y - 2x + 1$$

The discriminant is

$$D(x, y) = f_{xx}f_{yy} - f_{xy}^2 = -2y \cdot 2x - (2y - 2x + 1)^2 = -4xy - (2y - 2x + 1)^2$$

We now apply the Second Derivative Test. We first compute the discriminants at the critical points:

$$D(0, 0) = -1 < 0$$
$$D(1, 0) = -1 < 0$$
$$D(0, -1) = -1 < 0$$
$$D\left(\frac{1}{3}, -\frac{1}{3}\right) = -4 \cdot \frac{1}{3}\left(-\frac{1}{3}\right) - \left(-\frac{2}{3} - \frac{2}{3} + 1\right)^2 = \frac{1}{3} > 0,$$
$$f_{xx}\left(\frac{1}{3}, -\frac{1}{3}\right) = -2 \cdot \left(-\frac{1}{3}\right) = \frac{2}{3} > 0$$

The Second Derivative Test implies that the points $(0, 0)$, $(1, 0)$, and $(0, -1)$ are saddle points, and $f\left(\frac{1}{3}, -\frac{1}{3}\right)$ is a local minimum.

In Exercises 7–23, find the critical points of the function. Then use the Second Derivative Test to determine whether they are local minima, local maxima, or saddle points (or state that the test fails).

7. $f(x, y) = x^2 + y^2 - xy + x$

SOLUTION

Step 1. Find the critical points. We set the first-order partial derivatives of $f(x, y) = x^2 + y^2 - xy + x$ equal to zero and solve:

$$f_x(x, y) = 2x - y + 1 = 0 \quad (1)$$
$$f_y(x, y) = 2y - x = 0 \quad (2)$$

Equation (2) implies that $x = 2y$. Substituting in (1) and solving for y gives

$$2 \cdot 2y - y + 1 = 0 \implies 3y = -1 \implies y = -\frac{1}{3}$$

The corresponding value of x is $x = 2 \cdot \left(-\frac{1}{3}\right) = -\frac{2}{3}$. The critical point is $\left(-\frac{2}{3}, -\frac{1}{3}\right)$.

Step 2. Compute the Discriminant. We find the second-order partials:

$$f_{xx}(x, y) = 2, \quad f_{yy}(x, y) = 2, \quad f_{xy}(x, y) = -1$$

The discriminant is

$$D(x, y) = f_{xx}f_{yy} - f_{xy}^2 = 2 \cdot 2 - (-1)^2 = 3$$

Step 3. Applying the Second Derivative Test. We have

$$D\left(-\frac{2}{3}, -\frac{1}{3}\right) = 3 > 0 \quad \text{and} \quad f_{xx}\left(-\frac{2}{3}, -\frac{1}{3}\right) = 2 > 0$$

The Second Derivative Test implies that $f\left(-\frac{2}{3}, -\frac{1}{3}\right)$ is a local minimum.

9. $f(x, y) = x^3 + 2xy - 2y^2 - 10x$

SOLUTION

Step 1. Find the critical points. We set the first-order partial derivatives of $f(x, y) = x^3 + 2xy - 2y^2 - 10x$ equal to zero and solve:

$$f_x(x, y) = 3x^2 + 2y - 10 = 0 \tag{1}$$

$$f_y(x, y) = 2x - 4y = 0 \tag{2}$$

Equation (2) implies that $x = 2y$. We substitute in (1) and solve for y. This gives

$$3 \cdot (2y)^2 + 2y - 10 = 0$$

$$12y^2 + 2y - 10 = 0$$

$$6y^2 + y - 5 = 0$$

$$y_{1,2} = \frac{-1 \pm \sqrt{1 - 4 \cdot 6 \cdot (-5)}}{12} = \frac{-1 \pm 11}{12} \implies y_1 = -1 \quad \text{and} \quad y_2 = \frac{5}{6}$$

We find the x-coordinates using $x = 2y$:

$$x_1 = 2 \cdot (-1) = -2, \quad x_2 = 2 \cdot \frac{5}{6} = \frac{5}{3}$$

The critical points are thus $(-2, -1)$ and $\left(\frac{5}{3}, \frac{5}{6}\right)$.

Step 2. Compute the Discriminant. We find the second-order partials:

$$f_{xx}(x, y) = 6x, \quad f_{yy}(x, y) = -4, \quad f_{xy}(x, y) = 2$$

The discriminant is

$$D(x, y) = f_{xx}f_{yy} - f_{xy}^2 = 6x \cdot (-4) - 2^2 = -24x - 4$$

Step 3. Apply the Second Derivative Test. We have

$$D(-2, -1) = -24 \cdot (-2) - 4 = 44 > 0,$$

$$f_{xx}(-2, -1) = 6 \cdot (-2) = -12 < 0$$

$$D\left(\frac{5}{3}, \frac{5}{6}\right) = -24 \cdot \frac{5}{3} - 4 = -44 < 0$$

We conclude that $f(-2, -1)$ is a local maximum and $\left(\frac{5}{3}, \frac{5}{6}\right)$ is a saddle point.

11. $f(x, y) = 4x - 3x^3 - 2xy^2$

SOLUTION

Step 1. Find the critical points. We set the first-order derivatives of $f(x,y) = 4x - 3x^3 - 2xy^2$ equal to zero and solve:

$$f_x(x,y) = 4 - 9x^2 - 2y^2 = 0 \quad (1)$$

$$f_y(x,y) = -4xy = 0 \quad (2)$$

Equation (2) implies that $x = 0$ or $y = 0$. If $x = 0$, then equation (1) gives

$$4 - 2y^2 = 0 \quad \Rightarrow \quad y^2 = 2 \quad \Rightarrow \quad y = \sqrt{2}, \quad y = -\sqrt{2}$$

If $y = 0$, then equation (1) gives

$$4 - 9x^2 = 0 \quad \Rightarrow \quad 9x^2 = 4 \quad \Rightarrow \quad x = \frac{2}{3}, \quad x = -\frac{2}{3}$$

The critical points are therefore

$$\left(0, \sqrt{2}\right), \quad \left(0, -\sqrt{2}\right), \quad \left(\frac{2}{3}, 0\right), \quad \left(-\frac{2}{3}, 0\right)$$

Step 2. Compute the discriminant. The second-order partials are

$$f_{xx}(x,y) = -18x, \quad f_{yy}(x,y) = -4x, \quad f_{xy} = -4y$$

The discriminant is thus

$$D(x,y) = f_{xx}f_{yy} - f_{xy}^2 = -18x \cdot (-4x) - (-4y)^2 = 72x^2 - 16y^2$$

Step 3. Apply the Second Derivative Test. We have

$$D\left(0, \sqrt{2}\right) = -32 < 0$$

$$D\left(0, -\sqrt{2}\right) = -32 < 0$$

$$D\left(\frac{2}{3}, 0\right) = 72 \cdot \frac{4}{9} = 32 > 0,$$

$$f_{xx}\left(\frac{2}{3}, 0\right) = -18 \cdot \frac{2}{3} = -12 < 0$$

$$D\left(-\frac{2}{3}, 0\right) = 72 \cdot \frac{4}{9} = 32 > 0,$$

$$f_{xx}\left(-\frac{2}{3}, 0\right) = -18 \cdot \left(-\frac{2}{3}\right) = 12 > 0$$

The Second Derivative Test implies that the points $\left(0, \pm\sqrt{2}\right)$ are the saddle points, $f\left(\frac{2}{3}, 0\right)$ is a local maximum, and $f\left(-\frac{2}{3}, 0\right)$ is a local minimum.

13. $f(x,y) = x^4 + y^4 - 4xy$

SOLUTION

Step 1. Find the critical points. We set the first-order derivatives of $f(x,y) = x^4 + y^4 - 4xy$ equal to zero and solve:

$$f_x(x,y) = 4x^3 - 4y = 0, \quad f_y(x,y) = 4y^3 - 4x = 0 \quad (1)$$

Equation (1) implies that $y = x^3$. Substituting in (2) and solving for x, we obtain

$$(x^3)^3 - x = x^9 - x = x(x^8 - 1) = 0 \quad \Rightarrow \quad x = 0, \quad x = 1, \quad x = -1$$

The corresponding y coordinates are

$$y = 0^3 = 0, \quad y = 1^3 = 1, \quad y = (-1)^3 = -1$$

The critical points are therefore

$$(0,0), \quad (1,1), \quad (-1,-1)$$

Step 2. Compute the discriminant. We find the second-order partials:

$$f_{xx}(x,y) = 12x^2, \quad f_{yy}(x,y) = 12y^2, \quad f_{xy}(x,y) = -4$$

The discriminant is thus

$$D(x,y) = f_{xx}f_{yy} - f_{xy}^2 = 12x^2 \cdot 12y^2 - (-4)^2 = 144x^2y^2 - 16$$

Step 3. Apply the Second Derivative Test. We have

$$D(0,0) = -16 < 0$$
$$D(1,1) = 144 - 16 = 128 > 0, \quad f_{xx}(1,1) = 12 > 0$$
$$D(-1,-1) = 144 - 16 = 128 > 0, \quad f_{xx}(-1,-1) = 12 > 0$$

We conclude that $(0,0)$ is a saddle point, whereas $f(1,1)$ and $f(-1,-1)$ are local minima.

15. $f(x,y) = xye^{-x^2-y^2}$

SOLUTION

Step 1. Find the critical points. We compute the partial derivatives of $f(x,y) = xye^{-x^2-y^2}$, using the Product Rule and the Chain Rule:

$$f_x(x,y,z) = y\left(1 \cdot e^{-x^2-y^2} + xe^{-x^2-y^2} \cdot (-2x)\right) = ye^{-x^2-y^2}\left(1 - 2x^2\right)$$

$$f_y(x,y,z) = x\left(1 \cdot e^{-x^2-y^2} + ye^{-x^2-y^2} \cdot (-2y)\right) = xe^{-x^2-y^2}\left(1 - 2y^2\right)$$

We set the partial derivatives equal to zero and solve to find the critical points. This gives

$$ye^{-x^2-y^2}\left(1 - 2x^2\right) = 0$$
$$xe^{-x^2-y^2}\left(1 - 2y^2\right) = 0$$

Since $e^{-x^2-y^2} \neq 0$, the first equation gives $y = 0$ or $1 - 2x^2 = 0$, that is, $y = 0$, $x = \frac{1}{\sqrt{2}}$, $x = -\frac{1}{\sqrt{2}}$. We substitute each of these values in the second equation and solve to obtain

$$y = 0: \quad xe^{-x^2} = 0 \quad \Rightarrow \quad x = 0$$

$$x = \frac{1}{\sqrt{2}}: \quad \frac{1}{\sqrt{2}}e^{-\frac{1}{2}-y^2}\left(1 - 2y^2\right) = 0 \quad \Rightarrow \quad 1 - 2y^2 = 0 \quad \Rightarrow \quad y = \pm\frac{1}{\sqrt{2}}$$

$$x = -\frac{1}{\sqrt{2}}: \quad -\frac{1}{\sqrt{2}}e^{-\frac{1}{2}-y^2}\left(1 - 2y^2\right) = 0 \quad \Rightarrow \quad 1 - 2y^2 = 0 \quad \Rightarrow \quad y = \pm\frac{1}{\sqrt{2}}$$

We obtain the following critical points: $(0,0)$,

$$\left(\frac{1}{\sqrt{2}}, \frac{1}{\sqrt{2}}\right), \quad \left(\frac{1}{\sqrt{2}}, -\frac{1}{\sqrt{2}}\right), \quad \left(-\frac{1}{\sqrt{2}}, \frac{1}{\sqrt{2}}\right), \quad \left(-\frac{1}{\sqrt{2}}, -\frac{1}{\sqrt{2}}\right)$$

Step 2. Compute the second-order partials.

$$f_{xx}(x,y) = y\frac{\partial}{\partial x}\left(e^{-x^2-y^2}\left(1 - 2x^2\right)\right) = y\left(e^{-x^2-y^2}(-2x)\left(1 - 2x^2\right) + e^{-x^2-y^2}(-4x)\right)$$
$$= -2xye^{-x^2-y^2}\left(3 - 2x^2\right)$$

$$f_{yy}(x,y) = x\frac{\partial}{\partial y}\left(e^{-x^2-y^2}\left(1 - 2y^2\right)\right) = x\left(e^{-x^2-y^2}(-2y)\left(1 - 2y^2\right) + e^{-x^2-y^2}(-4y)\right)$$
$$= -2yxe^{-x^2-y^2}\left(3 - 2y^2\right)$$

$$f_{xy}(x,y) = \frac{\partial}{\partial y}f_x = \left(1 - 2x^2\right)\frac{\partial}{\partial y}\left(ye^{-x^2-y^2}\right) = \left(1 - 2x^2\right)\left(1 \cdot e^{-x^2-y^2} + ye^{-x^2-y^2}(-2y)\right)$$
$$= e^{-x^2-y^2}\left(1 - 2x^2\right)\left(1 - 2y^2\right)$$

The discriminant is

$$D(x,y) = f_{xx}f_{yy} - f_{xy}^2$$

Step 3. Apply the Second Derivative Test. We construct the following table:

Critical Point	f_{xx}	f_{yy}	f_{xy}	D	Type
$(0,0)$	0	0	1	-1	$D<0$, saddle point
$\left(\frac{1}{\sqrt{2}}, \frac{1}{\sqrt{2}}\right)$	$-\frac{2}{e}$	$-\frac{2}{e}$	0	$\frac{4}{e^2}$	$D>0$, $f_{xx}<0$ local maximum
$\left(\frac{1}{\sqrt{2}}, -\frac{1}{\sqrt{2}}\right)$	$\frac{2}{e}$	$\frac{2}{e}$	0	$\frac{4}{e^2}$	$D>0$, $f_{xx}>0$ local minimum
$\left(-\frac{1}{\sqrt{2}}, \frac{1}{\sqrt{2}}\right)$	$\frac{2}{e}$	$\frac{2}{e}$	0	$\frac{4}{e^2}$	$D>0$, $f_{xx}>0$ local minimum
$\left(-\frac{1}{\sqrt{2}}, -\frac{1}{\sqrt{2}}\right)$	$-\frac{2}{e}$	$-\frac{2}{e}$	0	$\frac{4}{e^2}$	$D>0$, $f_{xx}<0$ local maximum

17. $f(x,y) = \sin(x+y) - \cos x$

SOLUTION

Step 1. Find the critical points. We set the first-order derivatives of $f(x,y) = \sin(x+y) - \cos x$ equal to zero and solve:

$$f_x(x,y) = \cos(x+y) + \sin x = 0$$

$$f_y(x,y) = \cos(x+y) = 0$$

First consider the second equation, $\cos(x+y) = 0$ this is when

$$x+y = \frac{(2k+1)\pi}{2} \rightarrow y = \frac{(2k+1)\pi}{2} - x \text{ where } k \text{ is an integer}$$

Then setting the two equations equal to one another we gain $\sin x = 0$ which are the values:

$$x = 0, \pm\pi, \pm 2\pi, \cdots = \pm k\pi \text{ where } k \text{ is an integer}$$

Thus we have:

$$x = k\pi \text{ and } y = \frac{(2n+1)\pi}{2} \text{ where } n, k \text{ are integers}$$

Step 2. Compute the discriminant. We find the second-order partial derivatives:

$$f_{xx}(x,y) = -\sin(x+y) + \cos x, \quad f_{yy}(x,y) = -\sin(x+y), \quad f_{xy}(x,y) = -\sin(x+y)$$

The discriminant is:

$$D(x,y) = f_{xx}f_{yy} - f_{xy}^2 = (-\sin(x+y) + \cos x)(-\sin(x+y)) - \sin^2(x+y) = -\cos(x)\sin(x+y)$$

Step 3. Apply the Second Derivative Test. We have

$$D = \begin{cases} +1, & \text{if } y = \frac{4n+3}{2}\pi \\ -1, & y = \frac{4n+1}{2}\pi \end{cases}$$

Therefore, the points $\left(k\pi, \frac{4n+1}{2}\pi\right)$ are saddle points since $D < 0$.

Since $D > 0$ for the points $\left(k\pi, \frac{4n+3}{2}\pi\right)$, we need to examine f_{xx}. The results show:

$$f_{xx} > 0 \text{ if } k \text{ is even and } f_{xx} < 0 \text{ if } k \text{ is odd}$$

Thus:

$$\left(k\pi, \frac{4n+3}{2}\pi\right) \text{ are local minima if } k \text{ is even}$$

while

$$\left(k\pi, \frac{4n+3}{2}\pi\right) \text{ are local maxima if } k \text{ is odd}$$

19. $f(x, y) = \ln x + 2 \ln y - x - 4y$

SOLUTION

Step 1. Find the critical points. We set the first-order partials of $f(x, y) = \ln x + 2 \ln y - x - 4y$ equal to zero and solve:

$$f_x(x, y) = \frac{1}{x} - 1 = 0, \quad f_y(x, y) = \frac{2}{y} - 4 = 0$$

The first equation gives $x = 1$, and the second equation gives $y = \frac{1}{2}$. We obtain the critical point $\left(1, \frac{1}{2}\right)$. Notice that f_x and f_y do not exist if $x = 0$ or $y = 0$, respectively, but these are not critical points since they are not in the domain of f. The critical point is thus $\left(1, \frac{1}{2}\right)$.

Step 2. Compute the discriminant. We find the second-order partials:

$$f_{xx}(x, y) = -\frac{1}{x^2}, \quad f_{yy}(x, y) = -\frac{2}{y^2}, \quad f_{xy}(x, y) = 0$$

The discriminant is

$$D(x, y) = f_{xx} f_{yy} - f_{xy}^2 = \frac{2}{x^2 y^2}$$

Step 3. Apply the Second Derivative Test. We have
$f(x, y) = x - y^2 - \ln(x + y)$

SOLUTION

Step 1. Find the critical points. We set the partial derivatives of $f(x, y) = x - y^2 - \ln(x + y)$ equal to zero and solve.

$$f_x(x, y) = 1 - \frac{1}{x + y} = 0, \quad f_y(x, y) = -2y - \frac{1}{x + y} = 0$$

The first equation implies that $\frac{1}{x+y} = 1$. Substituting in the second equation gives

$$-2y - 1 = 0 \quad \Rightarrow \quad 2y = -1 \quad \Rightarrow \quad y = -\frac{1}{2}$$

We substitute $y = -\frac{1}{2}$ in the first equation and solve for x:

$$1 - \frac{1}{x - \frac{1}{2}} = 0 \quad \Rightarrow \quad x - \frac{1}{2} = 1 \quad \Rightarrow \quad x = \frac{3}{2}$$

We obtain the critical point $\left(\frac{3}{2}, -\frac{1}{2}\right)$. Notice that although f_x and f_y do not exist where $x + y = 0$, these are not critical points since f is not defined at these points.

Step 2. Compute the discriminant. We compute the second-order partial derivatives:

$$f_{xx}(x, y) = \frac{\partial}{\partial x}\left(1 - \frac{1}{x + y}\right) = \frac{1}{(x + y)^2}$$

$$f_{yy}(x, y) = \frac{\partial}{\partial y}\left(-2y - \frac{1}{x + y}\right) = -2 + \frac{1}{(x + y)^2}$$

$$f_{xy}(x, y) = \frac{\partial}{\partial y}\left(1 - \frac{1}{x + y}\right) = \frac{1}{(x + y)^2}$$

The discriminant is

$$D(x, y) = f_{xx} f_{yy} - f_{xy}^2 = \frac{1}{(x + y)^2}\left(-2 + \frac{1}{(x + y)^2}\right) - \frac{1}{(x + y)^4} = \frac{-2}{(x + y)^2}$$

Step 3. Apply the Second Derivative Test. We have

$$D\left(\frac{3}{2}, -\frac{1}{2}\right) = \frac{-2}{\left(\frac{3}{2} - \frac{1}{2}\right)^2} = -2 < 0$$

We conclude that $\left(\frac{3}{2}, -\frac{1}{2}\right)$ is a saddle point.

$$D\left(1, \frac{1}{2}\right) = \frac{2}{1^2 \cdot \left(\frac{1}{2}\right)^2} = 8 > 0, \quad f_{xx}\left(1, \frac{1}{2}\right) = -\frac{1}{1^2} = -1 < 0$$

We conclude that $f\left(1, \frac{1}{2}\right)$ is a local maximum.

21. $f(x, y) = x - y^2 - \ln(x + y)$

SOLUTION

Step 1. Find the critical points. We set the partial derivatives of $f(x, y) = x - y^2 - \ln(x + y)$ equal to zero and solve.

$$f_x(x, y) = 1 - \frac{1}{x+y} = 0, \quad f_y(x, y) = -2y - \frac{1}{x+y} = 0$$

The first equation implies that $\frac{1}{x+y} = 1$. Substituting in the second equation gives

$$-2y - 1 = 0 \quad \Rightarrow \quad 2y = -1 \quad \Rightarrow \quad y = -\frac{1}{2}$$

We substitute $y = -\frac{1}{2}$ in the first equation and solve for x:

$$1 - \frac{1}{x - \frac{1}{2}} = 0 \quad \Rightarrow \quad x - \frac{1}{2} = 1 \quad \Rightarrow \quad x = \frac{3}{2}$$

We obtain the critical point $\left(\frac{3}{2}, -\frac{1}{2}\right)$. Notice that although f_x and f_y do not exist where $x + y = 0$, these are not critical points since f is not defined at these points.

Step 2. Compute the discriminant. We compute the second-order partial derivatives:

$$f_{xx}(x, y) = \frac{\partial}{\partial x}\left(1 - \frac{1}{x+y}\right) = \frac{1}{(x+y)^2}$$

$$f_{yy}(x, y) = \frac{\partial}{\partial y}\left(-2y - \frac{1}{x+y}\right) = -2 + \frac{1}{(x+y)^2}$$

$$f_{xy}(x, y) = \frac{\partial}{\partial y}\left(1 - \frac{1}{x+y}\right) = \frac{1}{(x+y)^2}$$

The discriminant is

$$D(x, y) = f_{xx} f_{yy} - f_{xy}^2 = \frac{1}{(x+y)^2}\left(-2 + \frac{1}{(x+y)^2}\right) - \frac{1}{(x+y)^4} = \frac{-2}{(x+y)^2}$$

Step 3. Apply the Second Derivative Test. We have

$$D\left(\frac{3}{2}, -\frac{1}{2}\right) = \frac{-2}{\left(\frac{3}{2} - \frac{1}{2}\right)^2} = -2 < 0$$

We conclude that $\left(\frac{3}{2}, -\frac{1}{2}\right)$ is a saddle point.

23. $f(x, y) = (x + 3y)e^{y - x^2}$

SOLUTION

Step 1. Find the critical points. We compute the partial derivatives of $f(x, y) = (x + 3y)e^{y - x^2}$, using the Product Rule and the Chain Rule:

$$f_x(x, y) = 1 \cdot e^{y - x^2} + (x + 3y)e^{y - x^2} \cdot (-2x) = e^{y - x^2}\left(1 - 2x^2 - 6xy\right)$$

$$f_y(x, y) = 3e^{y - x^2} + (x + 3y)e^{y - x^2} \cdot 1 = e^{y - x^2}(3 + x + 3y)$$

We set the partial derivatives equal to zero and solve to find the critical points:

$$e^{y - x^2}\left(1 - 2x^2 - 6xy\right) = 0$$

$$e^{y - x^2}(3 + x + 3y) = 0$$

Since $e^{y-x^2} \neq 0$, we obtain the following equations:

$$1 - 2x^2 - 6xy = 0$$
$$3 + x + 3y = 0$$

The second equation gives $x = -3(1 + y)$. We substitute for x in the first equation and solve for y:

$$1 - 2 \cdot 9(1 + y)^2 + 18(1 + y)y = 0$$
$$1 - 18(1 + 2y + y^2) + 18(y + y^2) = 0$$
$$-17 - 18y = 0 \quad \Rightarrow \quad y = -\frac{17}{18}, \quad x = -3\left(1 - \frac{17}{18}\right) = -\frac{1}{6}$$

The critical point is $\left(-\frac{1}{6}, -\frac{17}{18}\right)$.

Step 2. Compute the second-order partials.

$$f_{xx}(x, y) = \frac{\partial}{\partial x} f_x = e^{y-x^2}(-2x)\left(1 - 2x^2 - 6xy\right) + e^{y-x^2}(-4x - 6y) = 2e^{y-x^2}\left(2x^3 + 6x^2y - 3x - 3y\right)$$

$$f_{yy}(x, y) = \frac{\partial}{\partial y} f_y = e^{y-x^2}(3 + x + 3y) + e^{y-x^2} \cdot 3 = e^{y-x^2}(6 + x + 3y)$$

$$f_{xy}(x, y) = \frac{\partial}{\partial x} f_y = e^{y-x^2}(-2x)(3 + x + 3y) + e^{y-x^2} \cdot 1 = e^{y-x^2}\left(1 - 6xy - 2x^2 - 6x\right)$$

The discriminant is

$$D(x, y) = f_{xx} f_{yy} - f_{xy}^2$$

Step 3. Apply the Second Derivative Test. We obtain the following table:

Critical Point	f_{xx}	f_{yy}	f_{xy}	D	Type
$\left(-\frac{1}{6}, -\frac{17}{18}\right)$	2.4	1.13	0.38	2.57	$D > 0$, $f_{xx} > 0$, local minimum

25. Prove that the function $f(x, y) = \frac{1}{3}x^3 + \frac{2}{3}y^{3/2} - xy$ satisfies $f(x, y) \geq 0$ for $x \geq 0$ and $y \geq 0$.
(a) First, verify that the set of critical points of f is the parabola $y = x^2$ and that the Second Derivative Test fails for these points.
(b) Show that for fixed b, the function $g(x) = f(x, b)$ is concave up for $x > 0$ with a critical point at $x = b^{1/2}$.
(c) Conclude that $f(a, b) \geq f(b^{1/2}, b) = 0$ for all $a, b \geq 0$.

SOLUTION
(a) To find the critical points, we need the first-order partial derivatives, set them equal to zero and solve:

$$f_x(x, y) = x^2 - y = 0, \quad f_y(x, y) = y^{1/2} - x = 0$$

This gives us:

$$y = x^2$$

as the solution set for the critical points.
Now to compute the discriminant, we need the second-order partials

$$f_{xx}(x, y) = 2x, \quad f_{yy}(x, y) = \frac{1}{2} y^{-1/2}, \quad f_{xy}(x, y) = -1$$

Thus the discriminant is

$$D(x, y) = \frac{x}{\sqrt{y}} - 1$$

Since $y = x^2$ is the solution set for the critical points we see:

$$D(x, y) = 1 - 1 = 0$$

Therefore the Second Derivative Test is inconclusive and fails us.

(b) If we fix a value b and consider $g(x) = f(x,b) = \frac{1}{3}x^3 + \frac{2}{3}b^{3/2} - bx$ to find the concavity, we see

$$g'(x) = x^2 - b, \quad g''(x) = 2x$$

Then certainly, for $x > 0$, this function is concave up. The critical point will occur at the point when $x^2 - b = 0$ or $x = b^{1/2}$.

(c) Now, since for fixed b, we know that $g(x) = f(x,b)$ is concave up if $x > 0$, and the critical point is $x = b^{1/2}$. Therefore

$$f(a,b) \geq f(b^{1/2},b) = 0 \text{ for all } b \geq 0$$

27. **CAS** Use a computer algebra system to find a numerical approximation to the critical point of

$$f(x,y) = (1 - x + x^2)e^{y^2} + (1 - y + y^2)e^{x^2}$$

Apply the Second Derivative Test to confirm that it corresponds to a local minimum as in Figure 21.

FIGURE 21 Plot of $f(x,y) = (1 - x + x^2)e^{y^2} + (1 - y + y^2)e^{x^2}$.

SOLUTION The critical points are the solutions of $f_x(x,y) = 0$ and $f_y(x,y) = 0$. We compute the partial derivatives:

$$f_x(x,y) = (-1 + 2x)e^{y^2} + \left(1 - y + y^2\right)e^{x^2} \cdot 2x$$

$$f_y(x,y) = \left(1 - x + x^2\right)e^{y^2} \cdot 2y + (-1 + 2y)e^{x^2}$$

Hence, the critical points are the solutions of the following equations:

$$(2x - 1)e^{y^2} + 2x\left(1 - y + y^2\right)e^{x^2} = 0$$

$$(2y - 1)e^{x^2} + 2y\left(1 - x + x^2\right)e^{y^2} = 0$$

Using a CAS we obtain the following solution: $x = y \approx 0.27788$, which from the figure is a local minimum. To verify this using the Second Derivative Test, we compute

$$f_{xx}(x,y) = 2e^{y^2} + 2(1 - y + y^2)e^{x^2} + 4x^2(1 - y + y^2)e^{x^2} = 2e^{y^2} + (4x^2 + 2)(1 - y + y^2)e^{x^2}$$

$$f_{yy}(x,y) = 2(1 - x + x^2)e^{y^2} + 4y^2(1 - x + x^2)e^{y^2} + 2e^{x^2} = 2e^{x^2} + (4y^2 + 2)(1 - x + x^2)e^{y^2}$$

$$f_{xy}(x,y) = 2y(-1 + 2x)e^{y^2} + (-1 + 2y)e^{x^2} \cdot 2x = 2y(-1 + 2x)e^{y^2} + 2x(-1 + 2y)e^{y^2}$$

At the critical point $x = y \approx 0.27788$,

$$f_{xx}(x,y) = f_{yy}(x,y) \approx 4.15427, \quad f_{xy}(x,y) \approx -0.53342$$

so that $f_{xx}f_{yy} - f_{xy}^2 \approx 16.9735 > 0$. Since $f_{xx} > 0$, we conclude that the critical point is indeed a local minimum.

In Exercises 29–32, determine the global extreme values of the function on the given set without using calculus.

29. $f(x,y) = x + y, \quad 0 \leq x \leq 1, \quad 0 \leq y \leq 1$

SOLUTION The sum $x + y$ is maximum when $x = 1$ and $y = 1$, and it is minimum when $x = 0$ and $y = 0$. Therefore, the global maximum of f on the given set is $f(1, 1) = 1 + 1 = 2$ and the global minimum is $f(0, 0) = 0 + 0 = 0$.

31. $f(x, y) = (x^2 + y^2 + 1)^{-1}$, $0 \le x \le 3$, $0 \le y \le 5$

SOLUTION $f(x, y) = \frac{1}{x^2+y^2+1}$ is maximum when x^2 and y^2 are minimum, that is, when $x = y = 0$. f is minimum when x^2 and y^2 are maximum, that is, when $x = 3$ and $y = 5$. Therefore, the global maximum of f on the given set is $f(0, 0) = (0^2 + 0^2 + 1)^{-1} = 1$, and the global minimum is $f(3, 5) = (3^2 + 5^2 + 1)^{-1} = \frac{1}{35}$.

A linear function $f(x, y) = ax + by + c$ has no critical points. Therefore, the global minimum and maximum values of $f(x, y)$ on a closed and bounded domain must occur on the boundary of the domain. Furthermore, it is not difficult to see that if the domain is a polygon, as in Figure 22, then the global minimum and maximum values of f must occur at a vertex of the polygon. In Exercises 33–36, find the global minimum and maximum values of $f(x, y)$ on the specified polygon, and indicate where on the polygon they occur.

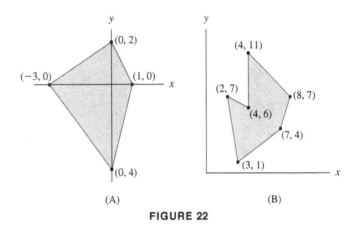

FIGURE 22

33. $f(x, y) = 2x - 6y + 4$ on the polygon in Figure 22(A).

SOLUTION We check the value of f at each vertex:

$$f(0, 2) = 2 \cdot 0 - 6 \cdot 2 + 4 = -8$$
$$f(1, 0) = 2 \cdot 1 - 6 \cdot 0 + 4 = 6$$
$$f(0, -4) = 2 \cdot 0 - 6 \cdot (-4) + 4 = 28$$
$$f(-3, 0) = 2 \cdot (-3) - 6 \cdot 0 + 4 = -2$$

Then the global maximum of f on the polygon is 28, occurring at $(0, -4)$, and the global minimum is -8, occurring at $(0, 2)$.

35. $f(x, y) = 12 + 5y - 20x$ on the polygon in Figure 22(A).

SOLUTION We check the value of f at each vertex:

$$f(0, 2) = 12 + 5 \cdot 2 - 20 \cdot 0 = 22$$
$$f(1, 0) = 12 + 5 \cdot 0 - 20 \cdot 1 = -8$$
$$f(0, -4) = 12 + 5 \cdot (-4) - 20 \cdot 0 = -8$$
$$f(-3, 0) = 12 + 5 \cdot 0 - 20 \cdot (-3) = 72$$

Then the global maximum of f on the polygon is 72, occurring at $(-3, 0)$, and the global minimum is -8, occurring at $(1, 0)$ and at $(0, -4)$ (as well as at any point on the line segment from $(1, 0)$ to $(0, -4)$).

37. Assumptions Matter Show that $f(x, y) = xy$ does not have a global minimum or a global maximum on the domain

$$\mathcal{D} = \{(x, y) : 0 < x < 1, 0 < y < 1\}$$

Explain why this does not contradict Theorem 3.

SOLUTION The largest and smallest values of f on the closed square $0 \le x, y \le 1$ are $f(1,1) = 2$ and $f(-1,-1) = -2$. However, on the open square $0 < x, y < 1$, f can never attain these maximum and minimum values, since the boundary (and in particular the points $(1,1)$ and $(-1,-1)$) is not included in the domain. This does not contradict Theorem 3 since the domain is open.

39. Find the maximum of

$$f(x,y) = x + y - x^2 - y^2 - xy$$

on the square, $0 \le x \le 2, 0 \le y \le 2$ (Figure 23).

(a) First, locate the critical point of f in the square, and evaluate f at this point.
(b) On the bottom edge of the square, $y = 0$ and $f(x, 0) = x - x^2$. Find the extreme values of f on the bottom edge.
(c) Find the extreme values of f on the remaining edges.
(d) Find the greatest among the values computed in (a), (b), and (c).

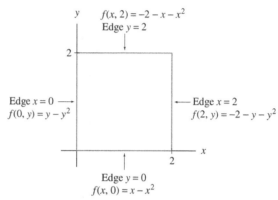

FIGURE 23 The function $f(x, y) = x + y - x^2 - y^2 - xy$ on the boundary segments of the square $0 \le x \le 2, 0 \le y \le 2$.

SOLUTION
(a) Since

$$f_x(x,y) = 1 - 2x - y, \qquad f_y(x,y) = 1 - 2y - x$$

setting both equal to zero gives $y = 1 - 2x$ and $2y = 1 - x$, so that $2(1 - 2x) = 1 - x$, and then $x = y = \frac{1}{3}$. We have

$$f\left(\frac{1}{3}, \frac{1}{3}\right) = \frac{1}{3} + \frac{1}{3} - \frac{1}{9} - \frac{1}{9} - \frac{1}{9} = \frac{1}{3}$$

(b) Along the bottom edge, where $f(x, 0) = x - x^2$, the derivative is $f'(x, 0) = 1 - 2x$, which is zero at $x = \frac{1}{2}$. The values of f here and at the endpoints of the edge are

$$f\left(\frac{1}{2}\right) = \frac{1}{4}, \qquad f(0,0) = 0, \qquad f(2,0) = 0$$

so that the maximum value of f on the bottom edge is $\frac{1}{4}$ and its minimum value is 0.

(c) Along the left edge, $f(0, y) = y - y^2$; this is identical to part (b) except that the variable is y instead of x, so that the maximum value along this edge is $\frac{1}{4}$ and the minimum value is 0. Along the top edge, $f(x, 2) = -2 - x - x^2$. Then $f'(x, 2) = -1 - 2x$, and $f'(x, 2) = 0$ when $x = -\frac{1}{2}$. Since this is outside the interval $[0, 2]$, we see that f has no critical points along the top edge. At the two endpoints, we have $f(0, 2) = -2$ and $f(2, 2) = -8$, so the maximum value of f along the top edge is -2 and the minimum is -8. Finally, along the right edge, we have $f(2, y) = -2 - y - y^2$. This is identical to the top edge except that the variable is y instead of x, and again we get a minimum value of -8 and a maximum value of -2.

(d) The largest value is $\frac{1}{3}$, achieved at $(x, y) = \left(\frac{1}{3}, \frac{1}{3}\right)$. So this is the maximum value of f on the given square.

In Exercises 41–49, determine the global extreme values of the function on the given domain.

41. $f(x, y) = x^3 - 2y$, $0 \le x \le 1$, $0 \le y \le 1$

SOLUTION We use the following steps.

Step 1. Find the critical points. We set the first derivative equal to zero and solve:
$$f_x(x,y) = 3x^2 = 0, \quad f_y(x,y) = -2$$

The two equations have no solutions, hence there are no critical points.

Step 2. Check the boundary. The extreme values occur either at the critical points or at a point on the boundary of the domain. Since there are no critical points, the extreme values occur at boundary points. We consider each edge of the square $0 \le x, y \le 1$ separately.

The segment $\overline{OA}$: On this segment $y = 0$, $0 \le x \le 1$, and f takes the values $f(x, 0) = x^3$. The minimum value is $f(0, 0) = 0$ and the maximum value is $f(1, 0) = 1$.

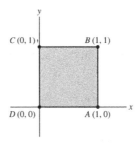

The segment $\overline{AB}$: On this segment $x = 1$, $0 \le y \le 1$, and f takes the values $f(1, y) = 1 - 2y$. The minimum value is $f(1, 1) = 1 - 2 \cdot 1 = -1$ and the maximum value is $f(1, 0) = 1 - 2 \cdot 0 = 1$.
The segment $\overline{BC}$: On this segment $y = 1$, $0 \le x \le 1$, and f takes the values $f(x, 1) = x^3 - 2$. The minimum value is $f(0, 1) = 0^3 - 2 = -2$ and the maximum value is $f(1, 1) = 1^3 - 2 = -1$.
The segment $\overline{OC}$: On this segment $x = 0$, $0 \le y \le 1$, and f takes the values $f(0, y) = -2y$. The minimum value is $f(0, 1) = -2 \cdot 1 = -2$ and the maximum value is $f(0, 0) = -2 \cdot 0 = 0$.

Step 3. Conclusions. The values obtained in the previous steps are
$$f(0,0) = 0, \quad f(1,0) = 1, \quad f(1,1) = -1, \quad f(0,1) = -2$$

The smallest value is $f(0, 1) = -2$ and it is the global minimum of f on the square. The global maximum is the largest value $f(1, 0) = 1$.

43. $f(x,y) = x^2 + 2y^2, \quad 0 \le x \le 1, \quad 0 \le y \le 1$

SOLUTION The sum $x^2 + 2y^2$ is maximum at the point $(1, 1)$, where x^2 and y^2 are maximum. It is minimum if $x = y = 0$, that is, at the point $(0, 0)$. Hence,

$$\text{Global maximum} = f(1,1) = 1^2 + 2 \cdot 1^2 = 3$$
$$\text{Global minimum} = f(0,0) = 0^2 + 2 \cdot 0^2 = 0$$

45. $f(x,y) = x^2 + xy^2 + y^2, \quad x, y \ge 0, \quad x + y \le 1$

SOLUTION We use the following steps:

Step 1. Examine the critical points in the interior of the domain. We set the partial derivatives equal to zero and solve:
$$f_x(x,y) = 2x + y^2 = 0$$
$$f_y(x,y) = 2xy + 2y = 0$$

The first of these gives $x = -\frac{y^2}{2}$. We substitute in the second and solve for y:

$$2\left(-\frac{y^2}{2}\right)y + 2y = 0 \quad \Rightarrow \quad -y^3 + 2y = 0 \quad \Rightarrow \quad y = 0 \text{ or } y = 2$$

When $y = 0$, we get $x = 0$, and $(0, 0)$ is not in the interior of the domain. If $y = 2$, then $x = -2$, and this point is also not in the domain. So there are no critical points in the interior of the domain.

Step 2. Find the extreme values on the boundary. We consider each part of the boundary separately.

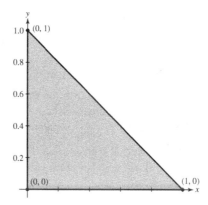

The edge from $(0,0)$ to $(1,0)$: On this edge $y = 0$, $0 \le x \le 1$, and $f(x,0) = x^2$. The maximum value of 1 is obtained at $(1,0)$ and the minimum value of 0 at $(0,0)$.

The edge from $(1,0)$ to $(0,1)$: On this edge $x = 1 - y$, $0 \le y \le 1$, and

$$g(y) = f(1-y, y) = (1-y)^2 + (1-y)y^2 + y^2 = -y^3 + 3y^2 - 2y + 1$$

Then $g'(y) = -3y^2 + 6y - 2$, which has roots $y = \frac{1}{3}(3 \pm \sqrt{3})$. The root $\frac{1}{3}(3 - \sqrt{3}) \approx 0.42265$ lies along that edge, but the other root does not.

The edge from $(0,0)$ to $(0,1)$: On this edge $x = 0$, $0 \le y \le 1$, and $f(0, y) = y^2$. The maximum value of 1 is obtained at $(0,1)$ and the minimum value of 0 at $(0,0)$.

By Theorem 3, the extreme values occur either at a critical point in the interior of the square or at a point on the boundary of the square. Since there are no critical points in the interior of the square, the candidates for extreme values are the points above on the boundary:

$$(0,0), \quad (1,0), \quad (0,1), \quad \approx (0.42265, 0.57735)$$

We compute $f(x, y) = x^2 + xy^2 + y^2$ at these points:

$$f(0,0) = 0$$
$$f(1,0) = 1$$
$$f(0,1) = 1$$
$$f(0.42265, 0.57735) \approx 0.6151$$

We conclude that in the given domain, the global maximum is $f(1,0) = f(0,1) = 1$ and the global minimum is $f(0,0) = 0$.

47. $f(x, y) = x^2 + y^2 - 2x - 4y, \quad x \ge 0, \quad 0 \le y \le 3, \quad y \ge x$

SOLUTION We use the following steps:

Step 1. Examine the critical points in the interior of the domain. We set the partial derivatives equal to zero and solve:

$$f_x(x,y) = 2x - 2, \quad f_y(x,y) = 2y - 4$$

Setting each equal to zero and solving we get: $x = 1$ and $y = 2$. Evaluating at the point $(1, 2)$ we see:

$$f(1,2) = -5$$

Step 2. Find the extreme values on the boundary. We consider each part of the boundary separately. The region that is described is the triangle bounded by the lines $x = 0$, $y = 3$, and $y = x$ with vertices $(0,0), (3,3), (0,3)$.

First consider the line $x = 0$:

$$f(0, y) = y^2 - 4y \quad \Rightarrow \quad f' = 2y - 4$$

Setting f' equal to zero and solving we get $y = 2$. So we must consider the point $(0, 2)$:

$$f(0, 2) = -4$$

We must also consider the endpoints of this line segment, $(0, 0)$ and $(0, 3)$:

$$f(0,0) = 0, \quad f(0,3) = -3$$

Next, consider the line $y = 3$:

$$f(x, 3) = x^2 + 9 - 2x - 12 = x^2 - 2x - 3 \quad \Rightarrow \quad f' = 2x - 2$$

Setting f' equal to zero and solving we get $x = 1$. So we must also consider the point $(1, 3)$:

$$f(1, 3) = -4$$

We must also consider the endpoints of this line segment, $(0, 3)$ and $(3, 3)$:

$$f(0, 3) = -3, \quad f(3, 3) = 0$$

Finally, consider the line $y = x$:

$$f(x, x) = x^2 + x^2 - 2x - 4x = 2x^2 - 6x \quad \Rightarrow \quad f' = 4x - 6$$

Setting f' equal to zero and solving, we get $x = 3/2$. So we must also consider the point $(3/2, 3/2)$:

$$f(3/2, 3/2) = -\frac{9}{2}$$

We have already examined the endpoints of this line segment in the steps above.

Step 3. Conclusions. The points that we have considered in this problem are

$$f(1, 2) = -5, \quad f(0, 2) = -4, \quad f(1, 3) = -4, \quad f(3/2, 3/2) = -\frac{9}{2}$$

$$f(0, 0) = 0, \quad f(0, 3) = -3, \quad f(3, 3) = 0$$

Therefore the minimum value is -5 and occurs at the point $(1, 2)$ and the maximum value is 0 and occurs in two places, at the points $(0, 0)$ and $(3, 3)$.

49. $f(x, y) = x^2 + 2xy^2, \quad x^2 + y^2 \leq 1$

SOLUTION We use the following steps:

Step 1. Examine the critical points in the interior of the domain. We set the partial derivatives equal to zero and solve:

$$f_x(x, y) = 2x + 2y^2 = 0$$

$$f_y(x, y) = 4xy = 0$$

The first equation gives $x = -y^2$; substituting this in the second equation gives $-4y^3 = 0$, so that $(0, 0)$ is the only critical point on the interior, and $f(0, 0) = 0$.

Step 2. Find the extreme values on the boundary. Along the boundary, $y^2 = 1 - x^2$; substituting this into the equation for f gives $g(x) = f(x, \sqrt{1 - x^2}) = x^2 + 2x(1 - x^2) = -2x^3 + x^2 + 2x$. Then $g'(x) = -6x^2 + 2x + 2$; setting this to zero and solving gives $x = \frac{1}{6}(1 \pm \sqrt{13})$. Each of these values of x gives two possible values for y:

$$x = \frac{1}{6}(1 + \sqrt{13}) \Rightarrow y^2 = 1 - x^2 = 1 - \frac{1}{36}(14 + 2\sqrt{13}) = \frac{1}{18}(11 - \sqrt{13}) \Rightarrow y = \pm\sqrt{\frac{1}{18}(11 - \sqrt{13})}$$

$$x = \frac{1}{6}(1 - \sqrt{13}) \Rightarrow y^2 = 1 - x^2 = 1 - \frac{1}{36}(14 - 2\sqrt{13}) = \frac{1}{18}(11 + \sqrt{13}) \Rightarrow y = \pm\sqrt{\frac{1}{18}(11 + \sqrt{13})}$$

So there are four critical points on the boundary, and

$$f\left(\frac{1}{6}(1 + \sqrt{13}), \pm\sqrt{\frac{1}{18}(11 - \sqrt{13})}\right) = \frac{1}{54}(19 - 13\sqrt{13}) \approx -0.5162$$

$$f\left(\frac{1}{6}(1 - \sqrt{13}), \pm\sqrt{\frac{1}{18}(11 + \sqrt{13})}\right) = \frac{1}{54}(19 + 13\sqrt{13}) \approx 1.2199$$

By Theorem 3, the extreme values occur either at a critical point in the interior of the square or at a point on the boundary of the square. Comparing the extreme values of the critical point in the interior and those along the boundary, we conclude that the global maximum of f on this region is approximately 1.2199, achieved twice along the boundary, and the global minimum is approximately -0.5162, also achieved twice along the boundary.

51. Find the volume of the largest box of the type shown in Figure 24, with one corner at the origin and the opposite corner at a point $P = (x, y, z)$ on the paraboloid

$$z = 1 - \frac{x^2}{4} - \frac{y^2}{9} \quad \text{with } x, y, z \geq 0$$

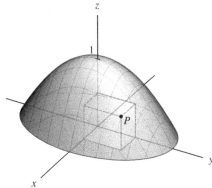

FIGURE 24

SOLUTION To maximize the volume of a rectangular box, start with the relation $V = xyz$ and using the paraboloid equation we see

$$z = 1 - \frac{x^2}{4} - \frac{y^2}{9} \quad \Rightarrow \quad V(x, y) = xy\left(1 - \frac{x^2}{4} - \frac{y^2}{9}\right)$$

Therefore we will consider

$$V(x, y) = xy - \frac{1}{4}x^3 y - \frac{1}{9}xy^3$$

First to find the critical points, we take the first-order partial derivatives and set them equal to zero, and solve:

$$V_x(x, y) = y - \frac{3}{4}x^2 y - \frac{1}{9}y^3, \quad V_y(x, y) = x - \frac{1}{4}x^3 - \frac{1}{3}xy^2$$

Using the equation $V_y = 0$ we see

$$x - \frac{1}{4}x^3 - \frac{1}{3}xy^2 = 0 \quad \Rightarrow \quad x = 0, \quad y^2 = 3 - \frac{3}{4}x^2 \quad \Rightarrow \quad y = \sqrt{3 - \frac{3}{4}x^2}$$

(Note here, we can ignore the value $x = 0$, since it produces a box having zero volume.)

Using this relation in the first equation, $V_x = 0$, we see:

$$\sqrt{3 - \frac{3}{4}x^2} - \frac{3}{4}x^2 \sqrt{3 - \frac{3}{4}x^2} - \frac{1}{9}\left(3 - \frac{3}{4}x^2\right)^{3/2} = 0$$

Factoring we see:

$$\sqrt{3 - \frac{3}{4}x^2} \left[1 - \frac{3}{4}x^2 - \frac{1}{9}\left(3 - \frac{3}{4}x^2\right)\right] = 0$$

and thus

$$3 - \frac{3}{4}x^2 = 0 \quad \Rightarrow \quad x^2 = 4 \quad \Rightarrow \quad x = \pm 2$$

or

$$1 - \frac{3}{4}x^2 - \frac{1}{3} + \frac{1}{12}x^2 = 0 \quad \Rightarrow \quad \frac{2}{3} - \frac{2}{3}x^2 = 0 \quad \Rightarrow \quad x = \pm 1$$

Since the governing equation $f(x, y)$ is a paraboloid, that is symmetric about the z-axis, we need only consider the point when $x = 2$ or $x = 1$.

Therefore, since $y = \sqrt{3 - \frac{3}{4}x^2}$ and $z = 1 - \frac{1}{4}x^2 - \frac{1}{9}y^2$, we have, if $x = 2$

$$y = \sqrt{3 - \frac{3}{4} \cdot 4} = 0 \quad \Rightarrow \quad z = 1 - \frac{1}{4} \cdot 4 - \frac{1}{9} \cdot 0 = 0$$

This will give a box having zero volume—not a maximum volume at all.

Using $x = 1$, and $y = \sqrt{3 - \frac{3}{4}x^2}$, $z = 1 - \frac{1}{4}x^2 - \frac{1}{9}y^2$, we have

$$y = \sqrt{3 - \frac{3}{4}} = \frac{3}{2}, \quad z = 1 - \frac{1}{4} \cdot 1^2 - \frac{1}{9} \cdot \frac{9}{4} = \frac{1}{2}$$

Therefore, the box having maximum volume has dimensions, $x = 1$, $y = 3/2$, and $z = 1/2$ and maximum value for the volume:

$$V = xyz = 1 \cdot \frac{3}{2} \cdot \frac{1}{2} = \frac{3}{4}$$

53. Show that the sum of the squares of the distances from a point $P = (c, d)$ to n fixed points $(a_1, b_1), \ldots, (a_n, b_n)$ is minimized when c is the average of the x-coordinates a_i and d is the average of the y-coordinates b_i.

SOLUTION First we must form the sum of the squares of the distances from a point $P(c, d)$ to n fixed points. For instance, the square of the distance from (c, d) to (a_1, b_1) would be:

$$(c - a_1)^2 + (d - b_1)^2$$

using this pattern, the sum in question would be

$$S = \sum_{i=1}^{n} [(c - a_i)^2 + (d - b_i)^2]$$

Using the methods discussed in this section of the text, we want to minimize the sum S. We will examine the first-order partial derivatives with respect to c and d and set them equal to zero and solve:

$$S_c = \sum_{i=1}^{n} 2(c - a_i) = 0, \quad S_d = \sum_{i=1}^{n} 2(d - b_i) = 0$$

Consider first the following:

$$\sum_{i=1}^{n} 2(c - a_i) = 0 \quad \Rightarrow \quad \sum_{i=1}^{n} (c - a_i) = 0 \quad \Rightarrow \quad \sum_{i=1}^{n} c - \sum_{i=1}^{n} a_i = 0$$

Therefore

$$\sum_{i=1}^{n} c = \sum_{i=1}^{n} a_i \quad \Rightarrow \quad n \cdot c = \sum_{i=1}^{n} a_i \quad \Rightarrow \quad c = \frac{1}{n} \sum_{i=1}^{n} a_i$$

Similarly we can examine $S_d = 0$ to see

$$\sum_{i=1}^{n} 2(d - b_i) = 0 \quad \Rightarrow \quad \sum_{i=1}^{n} (d - b_i) = 0 \quad \Rightarrow \quad \sum_{i=1}^{n} d - \sum_{i=1}^{n} b_i = 0$$

and

$$\sum_{i=1}^{n} d = \sum_{i=1}^{n} b_i \quad \Rightarrow \quad n \cdot d = \sum_{i=1}^{n} b_i \quad \Rightarrow \quad d = \frac{1}{n} \sum_{i=1}^{n} b_i$$

Therefore, the sum is minimized when c is the average of the x-coordinates a_i and d is the average of the y-coordinates b_i.

55. Consider a rectangular box B that has a bottom and sides but no top and has minimal surface area among all boxes with fixed volume $V = 64$ m^3.

(a) Do you think B is a cube as in the solution to Exercise 54? If not, how would its shape differ from a cube?

(b) Find the dimensions of B and compare with your response to (a).

SOLUTION

(a) Each of the variables x and y is the length of a side of three faces (for example, x is the length of the front, back, and bottom sides), whereas z is the length of a side of four faces.

Therefore, the variables x, y, and z do not have equal influence on the surface area. We expect that in the box B with minimal surface area, z is smaller than $\sqrt[3]{64} = 4$, which is the side of a cube with volume 64.

(b) We must find the dimensions of the box B, with fixed volume 64 m^3 and with smallest possible surface area, when the top is not included.

Step 1. Find a function to be minimized. The surface area of the box with sides lengths x, y, z when the top is not included is

$$S = 2xz + 2yz + xy \tag{1}$$

To express the surface in terms of x and y only, we use the formula for the volume of the box, $64 = xyz$, giving $z = \frac{64}{xy}$. We substitute in (1) to obtain

$$S = 2x \cdot \frac{64}{xy} + 2y \cdot \frac{64}{xy} + xy = \frac{128}{y} + \frac{128}{x} + xy$$

That is,

$$S = \frac{128}{y} + \frac{128}{x} + xy$$

Step 2. Determine the domain. The variables x, y denote lengths, hence they must be nonnegative. Moreover, S is not defined for $x = 0$ or $y = 0$. Since there are no other limitations on the variables, the domain is

$$D = \{(x, y) : x > 0, y > 0\}$$

We must find the minimum value of S on D. Because this domain is neither closed nor bounded, we are not sure that a minimum value exists. However, it can be proved (in like manner as in Exercise 54) that S does have a minimum value on D. This value occurs at a critical point in D, hence we set the partial derivatives equal to zero and solve. This gives

$$S_x(x, y) = -\frac{128}{x^2} + y = 0$$

$$S_y(x, y) = -\frac{128}{y^2} + x = 0$$

The first equation gives $y = \frac{128}{x^2}$. Substituting in the second equation yields

$$x - \frac{128}{\frac{128^2}{x^4}} = x - \frac{x^4}{128} = x\left(1 - \frac{x^3}{128}\right) = 0$$

The solutions are $x = 0$ and $x = 128^{1/3}$. The solution $x = 0$ is not included in D, so the only solution is $x = 128^{1/3}$.

We find the value of y using $y = \frac{128}{x^2}$:

$$y = \frac{128}{128^{2/3}} = 128^{1/3}$$

We conclude that the critical point, which is the point where the minimum value of S in D occurs, is $\left(128^{1/3}, 128^{1/3}\right)$. We find the corresponding value of z using $z = \frac{64}{xy}$. We get

$$z = \frac{64}{128^{1/3} 128^{1/3}} = \frac{64}{2^{2/3} 64^{2/3}} = \frac{64^{1/3}}{2^{2/3}} = 16^{1/3}$$

We conclude that the sizes of the box with minimum surface area are

width: $x = 128^{1/3}$
length: $y = 128^{1/3}$
height: $z = 16^{1/3}$

We see that z is smaller than x and y as predicted.

57. A 120-m long fence is to be cut into pieces to make three enclosures, each of which is square. How should the fence be cut up in order to minimize the total area enclosed by the fence?

SOLUTION Suppose the sides of the three squares are x, y, and z. Then $4x + 4y + 4z = 120$, so that $x + y + z = 30$. We want to minimize $x^2 + y^2 + z^2$. We can substitute $30 - x - y$ for z, giving

$$f(x, y) = x^2 + y^2 + (30 - x - y)^2$$

The domain is $x > 0$ and $y > 0$. Taking derivatives and setting them equal to zero gives

$$f_x(x, y) = 2x + 2(30 - x - y)(-1) = 4x + 2y - 60 = 0$$
$$f_y(x, y) = 2y + 2(30 - x - y)(-1) = 4y + 2x - 60 = 0$$

These two equations in two unknowns have the unique solution $x = y = 10$ with $f(x, y) = 300$. It is a local minimum since

$$f_{xx}(10, 10) = 4, \quad f_{yy}(10, 10) = 4, \quad f_{xy}(10, 10) = 2$$

and $f_{xx}(10, 10)f_{yy}(10, 10) - f_{xy}(10, 10)^2 = 12 > 0$ while $f_{xx}(10, 10) > 0$.

The domain, which is $\mathcal{D} = \{(x, y) \mid x, y > 0\}$, is open, so we must justify the statement that this is also a global minimum. To do this, we will show that $(10, 10)$ is the location of a global minimum on the closed domain $C = \{(x, y) \mid 0 \le x, y \le 20\}$ and that $f(10, 10)$ is less than the value of $f(x, y)$ at any point $(x, y) \in \mathcal{D} - C$. Note that $f(x, y)$ is a sum of three squares, all of which must be nonnegative, so if $(x, y) \in \mathcal{D} - C$, then either $x > 20$ or $y > 20$, so that $f(x, y) > 20^2 > 300 = f(10, 10)$. Now consider $f(x, y)$ on the domain C. Along the boundary where $x = 0$ the function is $g(y) = f(0, y) = y^2 + (30 - y)^2$. Then $g'(y) = 2y - 2(30 - y) = 4y - 60$, so g has a critical point at $y = 15$. Checking this critical point and the endpoints of the interval $[0, 20]$, we get

$$g(0) = 900, \quad g(15) = 450, \quad g(20) = 500$$

Similarly, along the boundary where $y = 0$ the function is $g(x) = f(x, 0) = x^2 + (30 - x)^2$ and we get the same critical values. Finally, we know that along the two remaining boundaries, $f(x, y) > 100$. So $f(10, 10) = 300$ is indeed a global minimum in the domain C. We have shown that $f(10, 10)$ is a global minimum on C and also that $f(10, 10)$ exceeds the value of f at any point in $\mathcal{D} - C$, so it is a global minimum on the given domain $\mathcal{D}$.

Finally, $x = y = 10$ implies that $z = 30 - x - y = 10$, so the sides of the three enclosures should each be 10 m and the total area is 300 m^2.

59. Find the maximum volume of a cylindrical can such that the sum of its height and its circumference is 120 cm.

SOLUTION Suppose the radius of the base is r and the height is h. Then the volume is $v(r, h) = \pi r^2 h$. We wish to minimize $v(r, h)$ subject to the condition that the sum of its height and circumference is $h + 2\pi r = 120$. Substituting $120 - 2\pi r$ for h in $v(r, h)$ gives

$$f(r) = v(r, 120 - 2\pi r) = \pi r^2 (120 - 2\pi r) = 120\pi r^2 - 2\pi^2 r^3$$

Setting $f'(r) = 0$ gives

$$f'(r) = 240\pi r - 6\pi^2 r^2 = 0$$

so that $r = 0$ or $r = \frac{40}{\pi}$. $r = 0$ is impossible, and $r = \frac{40}{\pi}$ is a local maximum since

$$f''(r) = 240\pi - 12\pi^2 r \quad \Rightarrow \quad f''\left(\frac{40}{\pi}\right) = 240\pi - 12\pi^2 \frac{40}{\pi} = 240\pi - 480\pi < 0$$

Further, $r = 0$ is a local minimum, since $f''(0) = 240\pi > 0$. It follows that f is increasing on $\left(0, \frac{40}{\pi}\right)$ and decreasing on $\left(\frac{40}{\pi}, \infty\right)$, so that on the domain $\{r > 0\}$ the point $r = \frac{40}{\pi}$ represents a global maximum. There,

$$f\left(\frac{40}{\pi}\right) = 120\pi \left(\frac{40}{\pi}\right)^2 - 2\pi^2 \left(\frac{40}{\pi}\right)^3 = \frac{64000}{\pi} \text{ cm}^3$$

The maximum volume is $\frac{64000}{\pi}$ cm³ with a radius of $\frac{40}{\pi}$ cm and a height of $120 - 2\pi \cdot \frac{40}{\pi} = 40$ cm.

61. The power (in microwatts) of a laser is measured as a function of current (in milliamps). Find the linear least-squares fit (Exercise 60) for the data points.

Current (milliamps)	1.0	1.1	1.2	1.3	1.4	1.5
Laser power (microwatts)	0.52	0.56	0.82	0.78	1.23	1.50

SOLUTION By Exercise 60, the coefficients of the linear least-square fit $f(x) = mx + b$ are determined by the following equations:

$$m \sum_{j=1}^{n} x_j + bn = \sum_{j=1}^{n} y_j$$

$$m \sum_{j=1}^{n} x_j^2 + b \sum_{j=1}^{n} x_j = \sum_{j=1}^{n} x_j \cdot y_j \quad (1)$$

In our case there are $n = 6$ data points:

$$(x_1, y_1) = (1, 0.52), (x_2, y_2) = (1.1, 0.56)$$
$$(x_3, y_3) = (1.2, 0.82), (x_4, y_4) = (1.3, 0.78)$$
$$(x_5, y_5) = (1.4, 1.23), (x_6, y_6) = (1.5, 1.50)$$

We compute the sums in (1):

$$\sum_{j=1}^{6} x_j = 1 + 1.1 + 1.2 + 1.3 + 1.4 + 1.5 = 7.5$$

$$\sum_{j=1}^{6} y_j = 0.52 + 0.56 + 0.82 + 0.78 + 1.23 + 1.50 = 5.41$$

$$\sum_{j=1}^{6} x_j^2 = 1^2 + 1.1^2 + 1.2^2 + 1.3^2 + 1.4^2 + 1.5^2 = 9.55$$

$$\sum_{j=1}^{6} x_j \cdot y_j = 1 \cdot 0.52 + 1.1 \cdot 0.56 + 1.2 \cdot 0.82 + 1.3 \cdot 0.78 + 1.4 \cdot 1.23 + 1.5 \cdot 1.50 = 7.106$$

Substituting in (1) gives the following equations:

$$7.5m + 6b = 5.41$$
$$9.55m + 7.5b = 7.106 \quad (2)$$

We multiply the first equation by 9.55 and the second by (-7.5), then add the resulting equations. This gives

$$\begin{aligned} 71.625m + 57.3b &= 51.6655 \\ + \; -71.625m - 56.25b &= -53.295 \\ \hline 1.05b &= -1.6295 \end{aligned} \quad \Rightarrow \quad b = -1.5519$$

We now substitute $b = -1.5519$ in the first equation in (2) and solve for m:

$$7.5m + 6 \cdot (-1.5519) = 5.41$$
$$7.5m = 14.7214$$
$$\Rightarrow \quad m = 1.9629$$

The linear least squares fit $f(x) = mx + b$ is thus

$$f(x) = 1.9629x - 1.5519$$

Further Insights and Challenges

63. In this exercise, we prove that for all $x, y \geq 0$:

$$\frac{1}{\alpha}x^{\alpha} + \frac{1}{\beta}x^{\beta} \geq xy$$

where $\alpha \geq 1$ and $\beta \geq 1$ are numbers such that $\alpha^{-1} + \beta^{-1} = 1$. To do this, we prove that the function

$$f(x, y) = \alpha^{-1}x^{\alpha} + \beta^{-1}y^{\beta} - xy$$

satisfies $f(x, y) \geq 0$ for all $x, y \geq 0$.

(a) Show that the set of critical points of $f(x, y)$ is the curve $y = x^{\alpha-1}$ (Figure 28). Note that this curve can also be described as $x = y^{\beta-1}$. What is the value of $f(x, y)$ at points on this curve?

(b) Verify that the Second Derivative Test fails. Show, however, that for fixed $b > 0$, the function $g(x) = f(x, b)$ is concave up with a critical point at $x = b^{\beta-1}$.

(c) Conclude that for all $x > 0$, $f(x, b) \geq f(b^{\beta-1}, b) = 0$.

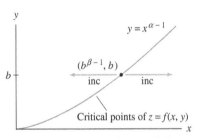

FIGURE 28 The critical points of $f(x, y) = \alpha^{-1}x^{\alpha} + \beta^{-1}y^{\beta} - xy$ form a curve $y = x^{\alpha-1}$.

SOLUTION Notice that $f(0, 0) = 0$, so it suffices to show that 0 is the minimum value of f in the domain $x \geq 0, y \geq 0$.

(a) For $R > 1$, we denote the square $\mathcal{D}_R = \{(x, y) : 0 \leq x \leq R, 0 \leq y \leq R\}$. f is continuous on the closed and bounded set $\mathcal{D}_R$, therefore f has a minimum value on $\mathcal{D}$. We show that the minimum value is zero.

First examine the critical points in the interior of $\mathcal{D}_R$. We set the partial derivatives equal to zero and solve:

$$f_x(x, y) = \frac{\partial}{\partial x}\left(\frac{1}{\alpha}x^{\alpha} + \frac{1}{\beta}y^{\beta} - xy\right) = x^{\alpha-1} - y = 0$$

$$f_y(x, y) = \frac{\partial}{\partial y}\left(\frac{1}{\alpha}x^{\alpha} + \frac{1}{\beta}y^{\beta} - xy\right) = y^{\beta-1} - x = 0$$

The first equation gives $y = x^{\alpha-1}$ and the second gives $x = y^{\beta-1}$. Therefore, the critical points in the interior are the points on the curve $y = x^{\alpha-1}$ (or, equivalently, $x = y^{\beta-1}$).

Now find the minimum on the boundary of $\mathcal{D}_R$. We consider each part of the boundary separately.

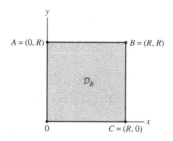

- The segment OC: On the x-axis, $y = 0$, hence,

$$f(x,0) = \frac{1}{\alpha}x^\alpha, \quad 0 \leq x \leq R$$

$f(x,0)$ is minimized at $x = 0$ corresponding to the point $(0,0)$, and $f(0,0) = 0$. Note that $(0,0)$ lies on the curve $y = x^{\alpha-1}$.

- The segment OA: On the y-axis, $x = 0$, hence $f(0,y) = \frac{1}{\beta}y^\beta$, $0 \leq y \leq R$. This function is minimized at $y = 0$ corresponding to the point $(0,0)$, and $f(0,0) = 0$. Note that $(0,0)$ lies on the curve $y = x^{\alpha-1}$.
- The segment AB: On this segment, $f(x,R) = \frac{1}{\alpha}x^\alpha - Rx + \frac{1}{\beta}R^\beta$, $0 \leq x \leq R$. The critical point is

$$\frac{d}{dx}f(x,R) = x^{\alpha-1} - R = 0, \text{ or } R = x^{\alpha-1}$$

Note that this point also lies on the curve $y = x^{\alpha-1}$.

- The segment BC: On this segment $f(R,y) = \frac{1}{\beta}y^\beta - Ry + \frac{1}{\alpha}R^\alpha$. The critical point is

$$\frac{d}{dy}f(R,y) = y^{\beta-1} - R = 0, \text{ or } R = y^{\beta-1}$$

This point also lies on $x = y^{\beta-1}$ (or, equivalently, $y = x^{\alpha-1}$).

As R increases, we see that the critical values of f consist exactly of points on the curve $y = x^{\alpha-1}$. Choose a point (a,b) on that curve with $a \geq 0$. Then $b = a^{\alpha-1}$ and $a = b^{\beta-1}$, and

$$f(a,b) = \frac{1}{\alpha}a^\alpha + \frac{1}{\beta}b^\beta - ab$$

$$= \frac{a}{\alpha}a^{\alpha-1} + \frac{b}{\beta}b^{\beta-1} - ab$$

$$= \frac{ab}{\alpha} + \frac{ab}{\beta} - ab$$

$$= ab\left(\frac{1}{\alpha} + \frac{1}{\beta}\right) - ab$$

$$= ab - ab = 0$$

So the value of f at any point on this curve (that is, at any critical point) is 0.

(b) Computing derivatives we get

$$f_x(x,y) = x^{\alpha-1} - y$$

$$f_y(x,y) = y^{\beta-1} - x$$

$$f_{xx}(x,y) = (\alpha - 1)x^{\alpha-2}$$

$$f_{yy}(x,y) = (\beta - 1)y^{\beta-2}$$

$$f_{xy}(x,y) = -1$$

Try applying the second derivative test to a critical point (x,y), so that $y = x^{\alpha-1}$:

$$f_{xx}(x,y)f_{yy}(x,y) - f_{xy}(x,y)^2 = (\alpha-1)(\beta-1)x^{\alpha-2}y^{\beta-2} - 1$$

$$= (\alpha-1)(\beta-1)\frac{x^{\alpha-1}}{x} \cdot \frac{y^{\beta-1}}{y} - 1$$

$$= (\alpha - 1)(\beta - 1)\frac{y}{x} \cdot \frac{x}{y} - 1$$
$$= (\alpha - 1)(\beta - 1) - 1$$

Now use the fact that $\frac{1}{\alpha} + \frac{1}{\beta} = 1$ to conclude that $\beta = \frac{\alpha}{\alpha-1}$. Substituting in the above, we get

$$f_{xx}(x,y)f_{yy}(x,y) - f_{xy}(x,y)^2 = (\alpha-1)(\beta-1) - 1 = (\alpha-1)\left(\frac{\alpha}{\alpha-1} - 1\right) - 1 = \alpha - (\alpha-1) - 1 = 0$$

Therefore the second derivative test fails, since the value of the determinant is zero.

Now fix $b > 0$, and consider the function

$$g(x) = f(x,b) = \frac{1}{\alpha}x^\alpha + \frac{1}{\beta}b^\beta - bx$$

Taking derivatives we get

$$g'(x) = x^{\alpha-1} - b, \qquad g''(x) = (\alpha-1)x^{\alpha-2}$$

It follows that $g'(x) = 0$ when $b = x^{\alpha-1}$, or $x = b^{\beta-1}$, so this is a critical point. Further, since $g''(x) > 0$ for all $x > 0$, we see that g is concave up everywhere.

(c) From the analysis in part (b), it follows that $b^{\beta-1}$ is a global minimum for g; that is, that for every $b > 0$,

$$f(x,b) \geq f\left(b^{\beta-1}, b\right) = 0$$

Since 0 is also the minimum value of f along $y = 0$ and along $x = 0$, it follows that 0 is the minimum value of f on $\mathcal{D}$, so that $f(x,y) \geq 0$ for all $x, y \geq 0$.

14.8 Lagrange Multipliers: Optimizing with a Constraint (LT Section 15.8)

Preliminary Questions

1. Suppose that the maximum of $f(x,y)$ subject to the constraint $g(x,y) = 0$ occurs at a point $P = (a,b)$ such that $\nabla f_P \neq 0$. Which of the following statements is true?

(a) ∇f_P is tangent to $g(x,y) = 0$ at P.

(b) ∇f_P is orthogonal to $g(x,y) = 0$ at P.

SOLUTION

(a) Since the maximum of f subject to the constraint occurs at P, it follows by Theorem 1 that ∇f_P and ∇g_P are parallel vectors. The gradient ∇g_P is orthogonal to $g(x,y) = 0$ at P, hence ∇f_P is also orthogonal to this curve at P. We conclude that statement (b) is false (yet the statement can be true if $\nabla f_P = (0,0)$).

(b) This statement is true by the reasoning given in the previous part.

2. Figure 10 shows a constraint $g(x,y) = 0$ and the level curves of a function f. In each case, determine whether f has a local minimum, a local maximum, or neither at the labeled point.

FIGURE 10

SOLUTION The level curve $f(x,y) = 2$ is tangent to the constraint curve at the point A. A close level curve that intersects the constraint curve is $f(x,y) = 1$, hence we may assume that f has a local maximum 2 under the constraint at A. The level curve $f(x,y) = 3$ is tangent to the constraint curve. However, in approaching B under the constraint, from one side f is increasing and from the other side f is decreasing. Therefore, $f(B)$ is neither local minimum nor local maximum of f under the constraint.

3. On the contour map in Figure 11:
(a) Identify the points where $\nabla f = \lambda \nabla g$ for some scalar λ.
(b) Identify the minimum and maximum values of $f(x,y)$ subject to $g(x,y) = 0$.

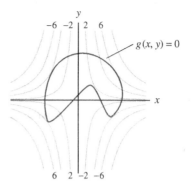

Contour plot of $f(x, y)$
(contour interval 2)
FIGURE 11

SOLUTION

(a) The gradient ∇g is orthogonal to the constraint curve $g(x,y) = 0$, and ∇f is orthogonal to the level curves of f. These two vectors are parallel at the points where the level curve of f is tangent to the constraint curve. These are the points A, B, C, D, E in the figure:

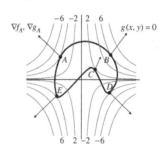

(b) The minimum and maximum occur where the level curve of f is tangent to the constraint curve. The level curves tangent to the constraint curve are

$$f(A) = -4, \quad f(C) = 2, \quad f(B) = 6, \quad f(D) = -4, \quad f(E) = 4$$

Therefore the global minimum of f under the constraint is -4 and the global maximum is 6.

Exercises

In this exercise set, use the method of Lagrange multipliers unless otherwise stated.

1. Find the extreme values of the function $f(x, y) = 2x + 4y$ subject to the constraint $g(x, y) = x^2 + y^2 - 5 = 0$.
(a) Show that the Lagrange equation $\nabla f = \lambda \nabla g$ gives $\lambda x = 1$ and $\lambda y = 2$.
(b) Show that these equations imply $\lambda \neq 0$ and $y = 2x$.
(c) Use the constraint equation to determine the possible critical points (x, y).
(d) Evaluate $f(x, y)$ at the critical points and determine the minimum and maximum values.

SOLUTION

(a) The Lagrange equations are determined by the equality $\nabla f = \lambda \nabla g$. We find them:

$$\nabla f = \langle f_x, f_y \rangle = \langle 2, 4 \rangle, \quad \nabla g = \langle g_x, g_y \rangle = \langle 2x, 2y \rangle$$

Hence,

$$\langle 2, 4 \rangle = \lambda \langle 2x, 2y \rangle$$

or

$$\lambda(2x) = 2 \quad\Rightarrow\quad \lambda x = 1$$
$$\lambda(2y) = 4 \quad\quad\quad \lambda y = 2$$

(b) The Lagrange equations in part (a) imply that $\lambda \neq 0$. The first equation implies that $x = \frac{1}{\lambda}$ and the second equation gives $y = \frac{2}{\lambda}$. Therefore $y = 2x$.

(c) We substitute $y = 2x$ in the constraint equation $x^2 + y^2 - 5 = 0$ and solve for x and y. This gives

$$x^2 + (2x)^2 - 5 = 0$$
$$5x^2 = 5$$
$$x^2 = 1 \quad\Rightarrow\quad x_1 = -1, \quad x_2 = 1$$

Since $y = 2x$, we have $y_1 = 2x_1 = -2$, $y_2 = 2x_2 = 2$. The critical points are thus

$$(-1, -2) \quad \text{and} \quad (1, 2)$$

Extreme values can also occur at the points where $\nabla g = \langle 2x, 2y \rangle = \langle 0, 0 \rangle$. However, $(0, 0)$ is not on the constraint.

(d) We evaluate $f(x, y) = 2x + 4y$ at the critical points, obtaining

$$f(-1, -2) = 2 \cdot (-1) + 4 \cdot (-2) = -10$$
$$f(1, 2) = 2 \cdot 1 + 4 \cdot 2 = 10$$

Since f is continuous and the graph of $g = 0$ is closed and bounded, global minimum and maximum points exist. So according to Theorem 1, we conclude that the maximum of $f(x, y)$ on the constraint is 10 and the minimum is -10.

3. Apply the method of Lagrange multipliers to the function $f(x, y) = (x^2 + 1)y$ subject to the constraint $x^2 + y^2 = 5$. *Hint:* First show that $y \neq 0$; then treat the cases $x = 0$ and $x \neq 0$ separately.

SOLUTION We first write out the Lagrange equations. We have $\nabla f = \langle 2xy, x^2 + 1 \rangle$ and $\nabla g = \langle 2x, 2y \rangle$. Hence, the Lagrange condition for $\nabla g \neq 0$ is

$$\nabla f = \lambda \nabla g$$
$$\langle 2xy, x^2 + 1 \rangle = \lambda \langle 2x, 2y \rangle$$

We obtain the following equations:

$$\begin{aligned} 2xy &= \lambda(2x) \\ x^2 + 1 &= \lambda(2y) \end{aligned} \quad\Rightarrow\quad \begin{aligned} 2x(y - \lambda) &= 0 \\ x^2 + 1 &= 2\lambda y \end{aligned} \tag{1}$$

The second equation implies that $y \neq 0$, since there is no real value of x such that $x^2 + 1 = 0$. Likewise, $\lambda \neq 0$. The solutions of the first equation are $x = 0$ and $y = \lambda$.

Case 1: $x = 0$. Substituting $x = 0$ in the second equation gives $2\lambda y = 1$, or $y = \frac{1}{2\lambda}$. We substitute $x = 0$, $y = \frac{1}{2\lambda}$ (recall that $\lambda \neq 0$) in the constraint to obtain

$$0^2 + \frac{1}{4\lambda^2} = 5 \quad\Rightarrow\quad 4\lambda^2 = \frac{1}{5} \quad\Rightarrow\quad \lambda = \pm\frac{1}{\sqrt{20}} = \pm\frac{1}{2\sqrt{5}}$$

The corresponding values of y are

$$y = \frac{1}{2 \cdot \frac{1}{2\sqrt{5}}} = \sqrt{5} \quad \text{and} \quad y = \frac{1}{2 \cdot \left(-\frac{1}{2\sqrt{5}}\right)} = -\sqrt{5}$$

We obtain the critical points:

$$\left(0, \sqrt{5}\right) \quad \text{and} \quad \left(0, -\sqrt{5}\right)$$

Case 2: $x \neq 0$. Then the first equation in (1) implies $y = \lambda$. Substituting in the second equation gives

$$x^2 + 1 = 2\lambda^2 \quad \Rightarrow \quad x^2 = 2\lambda^2 - 1$$

We now substitute $y = \lambda$ and $x^2 = 2\lambda^2 - 1$ in the constraint $x^2 + y^2 = 5$ to obtain

$$2\lambda^2 - 1 + \lambda^2 = 5$$
$$3\lambda^2 = 6$$
$$\lambda^2 = 2 \quad \Rightarrow \quad \lambda = \pm\sqrt{2}$$

The solution (x, y) are thus

$$\lambda = \sqrt{2}: \quad y = \sqrt{2}, \quad x = \pm\sqrt{2 \cdot 2 - 1} = \pm\sqrt{3}$$
$$\lambda = -\sqrt{2}: \quad y = -\sqrt{2}, \quad x = \pm\sqrt{2 \cdot 2 - 1} = \pm\sqrt{3}$$

We obtain the critical points:

$$\left(\sqrt{3},\, \sqrt{2}\right), \quad \left(-\sqrt{3},\, \sqrt{2}\right), \quad \left(\sqrt{3},\, -\sqrt{2}\right), \quad \left(-\sqrt{3},\, -\sqrt{2}\right)$$

We conclude that the critical points are

$$\left(0,\, \sqrt{5}\right), \quad \left(0,\, -\sqrt{5}\right), \quad \left(\sqrt{3},\, \sqrt{2}\right), \quad \left(-\sqrt{3},\, \sqrt{2}\right), \quad \left(\sqrt{3},\, -\sqrt{2}\right), \quad \left(-\sqrt{3},\, -\sqrt{2}\right)$$

We now calculate $f(x, y) = (x^2 + 1)y$ at the critical points:

$$f\left(0,\, \sqrt{5}\right) = \sqrt{5} \approx 2.24$$
$$f\left(0,\, -\sqrt{5}\right) = -\sqrt{5} \approx -2.24$$
$$f\left(\sqrt{3},\, \sqrt{2}\right) = f\left(-\sqrt{3},\, \sqrt{2}\right) = 4\sqrt{2} \approx 5.66$$
$$f\left(\sqrt{3},\, -\sqrt{2}\right) = f\left(-\sqrt{3},\, -\sqrt{2}\right) = -4\sqrt{2} \approx -5.66$$

Since the constraint gives a closed and bounded curve, f achieves a minimum and a maximum under it. We conclude that the maximum of $f(x, y)$ on the constraint is $4\sqrt{2}$ and the minimum is $-4\sqrt{2}$.

In Exercises 4–15, find the minimum and maximum values of the function subject to the given constraint.

5. $f(x, y) = x^2 + y^2, \quad 2x + 3y = 6$

SOLUTION We find the extreme values of $f(x, y) = x^2 + y^2$ under the constraint $g(x, y) = 2x + 3y - 6 = 0$.

Step 1. Write out the Lagrange equations. The gradients of f and g are $\nabla f = \langle 2x, 2y \rangle$ and $\nabla g = \langle 2, 3 \rangle$. The Lagrange condition is

$$\nabla f = \lambda \nabla g$$
$$\langle 2x, 2y \rangle = \lambda \langle 2, 3 \rangle$$

We obtain the following equations:

$$2x = \lambda \cdot 2$$
$$2y = \lambda \cdot 3$$

Step 2. Solve for λ in terms of x and y. Notice that if $x = 0$, then the first equation gives $\lambda = 0$, therefore by the second equation also $y = 0$. The point $(0, 0)$ does not satisfy the constraint. Similarly, if $y = 0$ also $x = 0$. We therefore may assume that $x \neq 0$ and $y \neq 0$ and obtain by the two equations:

$$\lambda = x \quad \text{and} \quad \lambda = \frac{2}{3}y$$

Step 3. Solve for x and y using the constraint. Equating the two expressions for λ gives

$$x = \frac{2}{3}y \quad \Rightarrow \quad y = \frac{3}{2}x$$

We substitute $y = \frac{3}{2}x$ in the constraint $2x + 3y = 6$ and solve for x and y:

$$2x + 3 \cdot \frac{3}{2}x = 6$$

$$13x = 12 \quad \Rightarrow \quad x = \frac{12}{13}, \quad y = \frac{3}{2} \cdot \frac{12}{13} = \frac{18}{13}$$

We obtain the critical point $\left(\frac{12}{13}, \frac{18}{13}\right)$.

Step 4. Calculate f at the critical point. We evaluate $f(x, y) = x^2 + y^2$ at the critical point:

$$f\left(\frac{12}{13}, \frac{18}{13}\right) = \left(\frac{12}{13}\right)^2 + \left(\frac{18}{13}\right)^2 = \frac{468}{169} \approx 2.77$$

Rewriting the constraint as $y = -\frac{2}{3}x + 2$, we see that as $|x| \to +\infty$ then so does $|y|$, and hence $x^2 + y^2$ is increasing without bound on the constraint as $|x| \to \infty$. We conclude that the value $468/169$ is the minimum value of f under the constraint, rather than the maximum value.

7. $f(x, y) = xy, \quad 4x^2 + 9y^2 = 32$

SOLUTION We find the extreme values of $f(x, y) = xy$ under the constraint $g(x, y) = 4x^2 + 9y^2 - 32 = 0$.

Step 1. Write out the Lagrange equation. The gradient vectors are $\nabla f = \langle y, x \rangle$ and $\nabla g = \langle 8x, 18y \rangle$, hence the Lagrange condition is

$$\nabla f = \lambda \nabla g$$

$$\langle y, x \rangle = \lambda \langle 8x, 18y \rangle$$

We obtain the following equations:

$$y = \lambda(8x)$$

$$x = \lambda(18y)$$

Step 2. Solve for λ in terms of x and y. If $x = 0$, then the Lagrange equations also imply that $y = 0$ and vice versa. Since the point $(0, 0)$ does not satisfy the equation of the constraint, we may assume that $x \neq 0$ and $y \neq 0$. The two equations give

$$\lambda = \frac{y}{8x} \quad \text{and} \quad \lambda = \frac{x}{18y}$$

Step 3. Solve for x and y using the constraint. We equate the two expressions for λ to obtain

$$\frac{y}{8x} = \frac{x}{18y} \quad \Rightarrow \quad 18y^2 = 8x^2 \quad \Rightarrow \quad y = \pm\frac{2}{3}x$$

We now substitute $y = \pm\frac{2}{3}x$ in the equation of the constraint and solve for x and y:

$$4x^2 + 9 \cdot \left(\pm\frac{2}{3}x\right)^2 = 32$$

$$4x^2 + 9 \cdot \frac{4x^2}{9} = 32$$

$$8x^2 = 32 \quad \Rightarrow \quad x = -2, \quad x = 2$$

We find y by the relation $y = \pm\frac{2}{3}x$:

$$y = \frac{2}{3} \cdot (-2) = -\frac{4}{3}, \quad y = -\frac{2}{3} \cdot (-2) = \frac{4}{3}, \quad y = \frac{2}{3} \cdot 2 = \frac{4}{3}, \quad y = -\frac{2}{3} \cdot 2 = -\frac{4}{3}$$

We obtain the following critical points:

$$\left(-2, -\frac{4}{3}\right), \quad \left(-2, \frac{4}{3}\right), \quad \left(2, \frac{4}{3}\right), \quad \left(2, -\frac{4}{3}\right)$$

Extreme values can also occur at the point where $\nabla g = \langle 8x, 18y \rangle = \langle 0, 0 \rangle$, that is, at the point $(0, 0)$. However, the point does not lie on the constraint.

Step 4. Calculate f at the critical points. We evaluate $f(x, y) = xy$ at the critical points:

$$f\left(-2, -\frac{4}{3}\right) = f\left(2, \frac{4}{3}\right) = \frac{8}{3}$$

$$f\left(-2, \frac{4}{3}\right) = f\left(2, -\frac{4}{3}\right) = -\frac{8}{3}$$

Since f is continuous and the constraint is a closed and bounded set in R^2 (an ellipse), f attains global extrema on the constraint. We conclude that $\frac{8}{3}$ is the maximum value and $-\frac{8}{3}$ is the minimum value.

9. $f(x, y) = x^2 + y^2$, $\quad x^4 + y^4 = 1$

SOLUTION We find the extreme values of $f(x, y) = x^2 + y^2$ under the constraint $g(x, y) = x^4 + y^4 - 1 = 0$.

Step 1. Write out the Lagrange equations. We have $\nabla f = \langle 2x, 2y \rangle$ and $\nabla g = \langle 4x^3, 4y^3 \rangle$, hence the Lagrange condition $\nabla f = \lambda \nabla g$ gives

$$\langle 2x, 2y \rangle = \lambda \langle 4x^3, 4y^3 \rangle$$

or

$$\begin{aligned} 2x &= \lambda\left(4x^3\right) \\ 2y &= \lambda\left(4y^3\right) \end{aligned} \quad \Rightarrow \quad \begin{aligned} x &= 2\lambda x^3 \\ y &= 2\lambda y^3 \end{aligned} \tag{1}$$

Step 2. Solve for λ in terms of x and y. We first assume that $x \neq 0$ and $y \neq 0$. Then the Lagrange equations give

$$\lambda = \frac{1}{2x^2} \quad \text{and} \quad \lambda = \frac{1}{2y^2}$$

Step 3. Solve for x and y using the constraint. Equating the two expressions for λ gives

$$\frac{1}{2x^2} = \frac{1}{2y^2} \quad \Rightarrow \quad y^2 = x^2 \quad \Rightarrow \quad y = \pm x$$

We now substitute $y = \pm x$ in the equation of the constraint $x^4 + y^4 = 1$ and solve for x and y:

$$x^4 + (\pm x)^4 = 1$$

$$2x^4 = 1$$

$$x^4 = \frac{1}{2} \quad \Rightarrow \quad x = \frac{1}{2^{1/4}}, \quad x = -\frac{1}{2^{1/4}}$$

The corresponding values of y are obtained by the relation $y = \pm x$. The critical points are thus

$$\left(\frac{1}{2^{1/4}}, \frac{1}{2^{1/4}}\right), \quad \left(\frac{1}{2^{1/4}}, -\frac{1}{2^{1/4}}\right), \quad \left(-\frac{1}{2^{1/4}}, \frac{1}{2^{1/4}}\right), \quad \left(-\frac{1}{2^{1/4}}, -\frac{1}{2^{1/4}}\right) \tag{2}$$

We examine the case $x = 0$ or $y = 0$. Notice that the point $(0, 0)$ does not satisfy the equation of the constraint, hence either $x = 0$ or $y = 0$ can hold, but not both at the same time.

Case 1: $x = 0$. Substituting $x = 0$ in the constraint $x^4 + y^4 = 1$ gives $y = \pm 1$. We thus obtain the critical points

$$(0, -1), \quad (0, 1) \tag{3}$$

Case 2: $y = 0$. We may interchange x and y in the discussion in case 1, and obtain the critical points:

$$(-1, 0), \quad (1, 0) \tag{4}$$

Combining (2), (3), and (4) we conclude that the critical points are

$$A_1 = \left(\frac{1}{2^{1/4}}, \frac{1}{2^{1/4}}\right), \quad A_2 = \left(\frac{1}{2^{1/4}}, -\frac{1}{2^{1/4}}\right), \quad A_3 = \left(-\frac{1}{2^{1/4}}, \frac{1}{2^{1/4}}\right)$$

$$A_4 = \left(-\frac{1}{2^{1/4}}, -\frac{1}{2^{1/4}}\right), \quad A_5 = (0, -1), \quad A_6 = (0, 1), \quad A_7 = (-1, 0), \quad A_8 = (1, 0)$$

The point where $\nabla g = \langle 4x^3, 4y^3 \rangle = \langle 0, 0 \rangle$, that is, $(0, 0)$, does not lie on the constraint.

Step 4. Compute f at the critical points. We evaluate $f(x, y) = x^2 + y^2$ at the critical points:

$$f(A_1) = f(A_2) = f(A_3) = f(A_4) = \left(\frac{1}{2^{1/4}}\right)^2 + \left(\frac{1}{2^{1/4}}\right)^2 = \frac{2}{2^{1/2}} = \sqrt{2}$$

$$f(A_5) = f(A_6) = f(A_7) = f(A_8) = 1$$

The constraint $x^4 + y^4 = 1$ is a closed and bounded set in R^2 and f is continuous on this set, hence f has global extrema on the constraint. We conclude that $\sqrt{2}$ is the maximum value and 1 is the minimum value.

11. $f(x, y, z) = 3x + 2y + 4z, \quad x^2 + 2y^2 + 6z^2 = 1$

SOLUTION We find the extreme values of $f(x, y, z) = 3x + 2y + 4z$ under the constraint $g(x, y, z) = x^2 + 2y^2 + 6z^2 - 1 = 0$.

Step 1. Write out the Lagrange equations. The gradient vectors are $\nabla f = \langle 3, 2, 4 \rangle$ and $\nabla g = \langle 2x, 4y, 12z \rangle$, therefore the Lagrange condition $\nabla f = \lambda \nabla g$ is:

$$\langle 3, 2, 4 \rangle = \lambda \langle 2x, 4y, 12z \rangle$$

The Lagrange equations are, thus:

$$3 = \lambda(2x) \qquad \frac{3}{2} = \lambda x$$

$$2 = \lambda(4y) \quad \Rightarrow \quad \frac{1}{2} = \lambda y$$

$$4 = \lambda(12z) \qquad \frac{1}{3} = \lambda z$$

Step 2. Solve for λ in terms of x, y, and z. The Lagrange equations imply that $x \ne 0$, $y \ne 0$, and $z \ne 0$. Solving for λ we get

$$\lambda = \frac{3}{2x}, \quad \lambda = \frac{1}{2y}, \quad \lambda = \frac{1}{3z}$$

Step 3. Solve for x, y, and z using the constraint. Equating the expressions for λ gives

$$\frac{3}{2x} = \frac{1}{2y} = \frac{1}{3z} \quad \Rightarrow \quad x = \frac{9}{2}z, \quad y = \frac{3}{2}z$$

Substituting $x = \frac{9}{2}z$ and $y = \frac{3}{2}z$ in the equation of the constraint $x^2 + 2y^2 + 6z^2 = 1$ and solving for z we get

$$\left(\frac{9}{2}z\right)^2 + 2\left(\frac{3}{2}z\right)^2 + 6z^2 = 1$$

$$\frac{123}{4}z^2 = 1 \quad \Rightarrow \quad z_1 = \frac{2}{\sqrt{123}}, z_2 = -\frac{2}{\sqrt{123}}$$

Using the relations $x = \frac{9}{2}z, y = \frac{3}{2}z$ we get

$$x_1 = \frac{9}{2} \cdot \frac{2}{\sqrt{123}} = \frac{9}{\sqrt{123}}, \quad y_1 = \frac{3}{2} \cdot \frac{2}{\sqrt{123}} = \frac{3}{\sqrt{123}}, \quad z_1 = \frac{2}{\sqrt{123}}$$

$$x_2 = \frac{9}{2} \cdot \frac{-2}{\sqrt{123}} = -\frac{9}{\sqrt{123}}, \quad y_2 = \frac{3}{2} \cdot \frac{-2}{\sqrt{123}} = -\frac{3}{\sqrt{123}}, \quad z_2 = -\frac{2}{\sqrt{123}}$$

We obtain the following critical points:

$$p_1 = \left(\frac{9}{\sqrt{123}}, \frac{3}{\sqrt{123}}, \frac{2}{\sqrt{123}}\right) \quad \text{and} \quad p_2 = \left(-\frac{9}{\sqrt{123}}, -\frac{3}{\sqrt{123}}, -\frac{2}{\sqrt{123}}\right)$$

Critical points are also the points on the constraint where $\nabla g = 0$. However, $\nabla g = \langle 2x, 4y, 12z \rangle = \langle 0, 0, 0 \rangle$ only at the origin, and this point does not lie on the constraint.

Step 4. Computing f at the critical points. We evaluate $f(x, y, z) = 3x + 2y + 4z$ at the critical points:

$$f(p_1) = \frac{27}{\sqrt{123}} + \frac{6}{\sqrt{123}} + \frac{8}{\sqrt{123}} = \frac{41}{\sqrt{123}} = \sqrt{\frac{41}{3}} \approx 3.7$$

$$f(p_2) = -\frac{27}{\sqrt{123}} - \frac{6}{\sqrt{123}} - \frac{8}{\sqrt{123}} = -\frac{41}{\sqrt{123}} = -\sqrt{\frac{41}{3}} \approx -3.7$$

Since f is continuous and the constraint is closed and bounded in R^3, f has global extrema under the constraint. We conclude that the minimum value of f under the constraint is about -3.7 and the maximum value is about 3.7.

13. $f(x, y, z) = xy + 2z$, $x^2 + y^2 + z^2 = 36$

SOLUTION We find the extreme values of $f(x, y, z) = xy + 2z$ under the constraint $g(x, y, z) = x^2 + y^2 + z^2 - 36 = 0$.

Step 1. Write out the Lagrange equations. The gradient vectors are $\nabla f = \langle y, x, 2 \rangle$ and $\nabla g = \langle 2x, 2y, 2z \rangle$; therefore, the Lagrange condition $\nabla f = \lambda \nabla g$ is:

$$\langle y, x, 2 \rangle = \lambda \langle 2x, 2y, 2z \rangle.$$

It follows that the Lagrange equations are

$$y = 2\lambda x, \qquad x = 2\lambda y, \qquad 2 = 2\lambda z$$

Step 2. Solve for λ in terms of x, y, and z. The Lagrange equations imply that $z \neq 0$, and that $x = 0$ if and only if $y = 0$. If $x = y = 0$, then $z = \pm 6$. This gives the points $(0, 0, -6)$ and $(0, 0, 6)$. Assuming that x and y are nonzero and solving for λ gives

$$\lambda = \frac{y}{2x}, \qquad \lambda = \frac{x}{2y}, \qquad \lambda = \frac{1}{z}$$

Step 3. Solve for x, y, and z using the constraint. Equating the expressions for λ gives

$$\frac{y}{2x} = \frac{x}{2y} = \frac{1}{z} \quad \Rightarrow \quad x^2 = y^2, \ z = \frac{2x}{y} \quad x = \pm y, \ z = \frac{2x}{y}$$

Substituting $x = \pm y$ for x in the equation for z gives $z = \pm 2$. Substituting $z = \pm 2$ in the constraint equation, and setting $x^2 = y^2$, gives

$$x^2 + y^2 + z^2 - 36 = 0 \quad \Rightarrow \quad 2x^2 - 32 = 0 \quad \Rightarrow \quad x = \pm 4 \quad \Rightarrow \quad y = \pm 4$$

This gives us eight critical points: $(\pm 4, \pm 4, \pm 2)$.

Step 4. Computing f at the critical points. We evaluate $f(x, y, z) = xy + 2z$ at the critical points and compare it to the value of f at the two points from part (a):

$$f(0, 0, 6) = 12 \quad f(0, 0, -6) = -12$$
$$f(4, 4, 2) = 20 \quad f(4, 4, -2) = 12$$
$$f(-4, 4, 2) = -12 \quad f(-4, 4, -2) = -20$$
$$f(4, -4, 2) = -12 \quad f(4, -4, -2) = -20$$
$$f(-4, -4, 2) = 20 \quad f(-4, -4, -2) = 12$$

Since f is continuous and the constraint is closed and bounded in R^3, f has global extrema under the constraint. We conclude that the minimum value of f under the constraint is -20, achieved at two points, and the maximum is 20, also achieved at two points.

15. $f(x, y, z) = xy + xz$, $x^2 + y^2 + z^2 = 4$

SOLUTION We find the extreme values of $f(x, y, z) = xy + xz$ under the constraint $g(x, y, z) = x^2 + y^2 + z^2 - 4 = 0$.

Step 1. Write out the Lagrange equations. The gradient vectors are $\nabla f = \langle y + z, x, x \rangle$ and $\nabla g = \langle 2x, 2y, 2z \rangle$; therefore, the Lagrange condition $\nabla f = \lambda \nabla g$ is:

$$\langle y + z, x, x \rangle = \lambda \langle 2x, 2y, 2z \rangle$$

It follows that the Lagrange equations are

$$y + z = 2\lambda x, \qquad x = 2\lambda y, \qquad x = 2\lambda z$$

Step 2. Solve for λ in terms of x, y, and z. The Lagrange equations imply that if any of x, y, or z is zero then the others are as well. However, $(0,0,0)$ does not lie on the constraint curve, so we reject it. Thus all three are nonzero; solving for λ gives

$$\lambda = \frac{y+z}{2x}, \qquad \lambda = \frac{x}{2y}, \qquad \lambda = \frac{x}{2z}$$

Step 3. Solve for x, y, and z using the constraint. Equating the expressions for λ gives

$$\frac{y+z}{2x} = \frac{x}{2y} = \frac{x}{2z}$$

The last equality implies that $y = z$; substituting y for z in the first expression gives $\frac{2y}{2x} = \frac{x}{2y}$, so that $2y^2 = x^2$ and therefore $x = y\sqrt{2} = z\sqrt{2}$. Substituting $\frac{x}{\sqrt{2}}$ for y and z in the constraint equation gives

$$x^2 + \left(\frac{x}{\sqrt{2}}\right)^2 + \left(\frac{x}{\sqrt{2}}\right)^2 - 4 = 0 \quad \Rightarrow \quad 2x^2 = 4 \quad \Rightarrow \quad x = \pm\sqrt{2} \quad \Rightarrow \quad y = z = \pm 1$$

This gives us four critical points: $(\sqrt{2}, 1, 1)$, $(\sqrt{2}, -1, -1)$, $(-\sqrt{2}, 1, 1)$, $(-\sqrt{2}, -1, -1)$.

Step 4. Computing f at the critical points. We evaluate $f(x, y, z) = xy + xz$ at the critical points:

$$f(\sqrt{2}, 1, 1) = 2\sqrt{2} \qquad f(\sqrt{2}, -1, -1) = -2\sqrt{2}$$
$$f(-\sqrt{2}, 1, 1) = -2\sqrt{2} \qquad f(-\sqrt{2}, -1, -1) = 2\sqrt{2}$$

Since f is continuous and the constraint is closed and bounded in R^3, f has global extrema under the constraint. We conclude that the minimum value of f under the constraint is $-2\sqrt{2}$, achieved at two points, and the maximum is $2\sqrt{2}$, also achieved at two points.

17. Find the point (a, b) on the graph of $y = e^x$ where the value ab is the least.

SOLUTION We must find the point where $f(x, y) = xy$ has a minimum value subject to the constraint $g(x, y) = e^x - y = 0$.

Step 1. Write out the Lagrange equations. Since $\nabla f = \langle y, x \rangle$ and $\nabla g = \langle e^x, -1 \rangle$, the Lagrange condition $\nabla f = \lambda \nabla g$ is

$$\langle y, x \rangle = \lambda \langle e^x, -1 \rangle$$

The Lagrange equations are thus

$$y = \lambda e^x$$
$$x = -\lambda$$

Step 2. Solve for λ in terms of x and y. The Lagrange equations imply that

$$\lambda = ye^{-x} \quad \text{and} \quad \lambda = -x$$

Step 3. Solve for x and y using the constraint. We equate the two expressions for λ to obtain

$$ye^{-x} = -x \quad \Rightarrow \quad y = -xe^x$$

We now substitute $y = -xe^x$ in the equation of the constraint and solve for x:

$$e^x - (-xe^x) = 0$$
$$e^x(1+x) = 0$$

Since $e^x \neq 0$ for all x, we have $x = -1$. The corresponding value of y is determined by the relation $y = -xe^x$. That is,

$$y = -(-1)e^{-1} = e^{-1}$$

We obtain the critical point

$$(-1, e^{-1})$$

Step 4. Calculate f at the critical point. We evaluate $f(x, y) = xy$ at the critical point.

$$f(-1, e^{-1}) = (-1) \cdot e^{-1} = -e^{-1}$$

We conclude (see Remark) that the minimum value of xy on the graph of $y = e^x$ is $-e^{-1}$, and it is obtained for $x = -1$ and $y = e^{-1}$.

Remark: Since the constraint is not bounded, we need to justify the existence of a minimum value. The values $f(x, y) = xy$ on the constraint $y = e^x$ are $f(x, e^x) = h(x) = xe^x$. Since $h(x) > 0$ for $x > 0$, the minimum value (if it exists) occurs at a point $x < 0$. Since

$$\lim_{x \to -\infty} xe^x = \lim_{x \to -\infty} \frac{x}{e^{-x}} = \lim_{x \to -\infty} \frac{1}{-e^{-x}} = \lim_{x \to -\infty} -e^x = 0$$

then for $x <$ some negative number $-R$, we have $|f(x) - 0| < 0.1$, say. Thus, on the bounded region $-R \le x \le 0$, f has a minimum value of $-e^{-1} \approx -0.37$, and this is thus a global minimum (for all x).

19. The surface area of a right-circular cone of radius r and height h is $S = \pi r \sqrt{r^2 + h^2}$, and its volume is $V = \frac{1}{3}\pi r^2 h$.

(a) Determine the ratio h/r for the cone with given surface area S and maximum volume V.

(b) What is the ratio h/r for a cone with given volume V and minimum surface area S?

(c) Does a cone with given volume V and maximum surface area exist?

SOLUTION

(a) Let S_0 denote a given surface area. We must find the ratio $\frac{h}{r}$ for which the function $V(r, h) = \frac{1}{3}\pi r^2 h$ has maximum value under the constraint $S(r, h) = \pi r \sqrt{r^2 + h^2} = \pi \sqrt{r^4 + h^2 r^2} = S_0$.

Step 1. Write out the Lagrange equation. We have

$$\nabla V = \pi \left\langle \frac{2rh}{3}, \frac{r^2}{3} \right\rangle \quad \text{and} \quad \nabla S = \pi \left\langle \frac{2r^3 + h^2 r}{\sqrt{r^4 + h^2 r^2}}, \frac{hr^2}{\sqrt{r^4 + h^2 r^2}} \right\rangle$$

The Lagrange condition $\nabla V = \lambda \nabla S$ gives the following equations:

$$\frac{2rh}{3} = \frac{2r^3 + h^2 r}{\sqrt{r^4 + h^2 r^2}} \lambda \quad \Rightarrow \quad \frac{2h}{3} = \frac{2r^2 + h^2}{\sqrt{r^4 + h^2 r^2}} \lambda$$

$$\frac{r^2}{3} = \frac{hr^2}{\sqrt{r^4 + h^2 r^2}} \lambda \quad \Rightarrow \quad \frac{1}{3} = \frac{h}{\sqrt{r^4 + h^2 r^2}} \lambda$$

Step 2. Solve for λ in terms of r and h. These equations yield two expressions for λ that must be equal:

$$\lambda = \frac{2h}{3} \frac{\sqrt{r^4 + h^2 r^2}}{2r^2 + h^2} = \frac{1}{3h} \sqrt{r^4 + h^2 r^2}$$

Step 3. Solve for r and h using the constraint. We have

$$\frac{2h}{3} \frac{\sqrt{r^4 + h^2 r^2}}{2r^2 + h^2} = \frac{1}{3h} \sqrt{r^4 + h^2 r^2}$$

$$2h \frac{1}{2r^2 + h^2} = \frac{1}{h}$$

$$2h^2 = 2r^2 + h^2 \quad \Rightarrow \quad h^2 = 2r^2 \quad \Rightarrow \quad \frac{h}{r} = \sqrt{2}$$

We substitute $h^2 = 2r^2$ in the constraint $\pi r \sqrt{r^2 + h^2} = S_0$ and solve for r. This gives

$$\pi r \sqrt{r^2 + 2r^2} = S_0$$

$$\pi r \sqrt{3r^2} = S_0$$

$$\sqrt{3}\pi r^2 = S_0 \quad \Rightarrow \quad r^2 = \frac{S_0}{\sqrt{3}\pi}, \quad h^2 = 2r^2 = \frac{2S_0}{\sqrt{3}\pi}$$

Extreme values can occur also at points on the constraint where $\nabla S = \left\langle \frac{2r^2 + h^2 r}{\sqrt{r^4 + h^2 r^2}}, \frac{hr^2}{\sqrt{r^4 + h^2 r^2}} \right\rangle = \langle 0, 0 \rangle$, that is, at $(r, h) = (0, h)$, $h \ne 0$. However, since the radius of the cone is positive ($r > 0$), these points are irrelevant. We conclude that for the cone with surface area S_0 and maximum volume, the following holds:

$$\frac{h}{r} = \sqrt{2}, \quad h = \sqrt{\frac{2S_0}{\sqrt{3}\pi}}, \quad r = \sqrt{\frac{S_0}{\sqrt{3}\pi}}$$

For the surface area $S_0 = 1$ we get

$$h = \sqrt{\frac{2}{\sqrt{3}\pi}} \approx 0.6, \quad r = \sqrt{\frac{1}{\sqrt{3}\pi}} = 0.43$$

(b) We now must find the ratio $\frac{h}{r}$ that minimizes the function $S(r,h) = \pi r \sqrt{r^2 + h^2}$ under the constraint

$$V(r,h) = \frac{1}{3}\pi r^2 h = V_0$$

Using the gradients computed in part (a), the Lagrange condition $\nabla S = \lambda \nabla V$ gives the following equations:

$$\frac{2r^3 + h^2 r}{\sqrt{r^4 + h^2 r^2}} = \lambda \frac{2rh}{3} \qquad \frac{2r^2 + h^2}{\sqrt{r^4 + h^2 r^2}} = \lambda \frac{2h}{3}$$

$$\Rightarrow$$

$$\frac{hr^2}{\sqrt{r^4 + h^2 r^2}} = \lambda \frac{r^2}{3} \qquad \frac{h}{\sqrt{r^4 + h^2 r^2}} = \frac{\lambda}{3}$$

These equations give

$$\frac{\lambda}{3} = \frac{1}{2h} \frac{2r^2 + h^2}{\sqrt{r^4 + h^2 r^2}} = \frac{h}{\sqrt{r^4 + h^2 r^2}}$$

We simplify and solve for $\frac{h}{r}$:

$$\frac{2r^2 + h^2}{2h} = h$$

$$2r^2 + h^2 = 2h^2$$

$$2r^2 = h^2 \quad \Rightarrow \quad \frac{h}{r} = \sqrt{2}$$

We conclude that the ratio $\frac{h}{r}$ for a cone with a given volume and minimal surface area is

$$\frac{h}{r} = \sqrt{2}$$

(c) The constant $V = 1$ gives $\frac{1}{3}\pi r^2 h = 1$ or $h = \frac{3}{\pi r^2}$. As $r \to \infty$, we have $h \to 0$, therefore

$$\lim_{\substack{r \to \infty \\ h \to 0}} S(r,h) = \lim_{\substack{r \to \infty \\ h \to 0}} \pi r \sqrt{r^2 + h^2} = \infty$$

That is, S does not have maximum value on the constraint, hence there is no cone of volume 1 and maximal surface area.

21. Find the point on the ellipse

$$x^2 + 6y^2 + 3xy = 40$$

with the greatest x-coordinate (Figure 14).

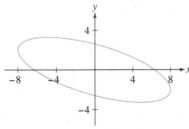

FIGURE 14 Graph of $x^2 + 6y^2 + 3xy = 40$.

SOLUTION We need to maximize $f(x, y) = x$ subject to the constraint

$$g(x, y) = x^2 + 6y^2 + 3xy = 40$$

Step 1. Write out the Lagrange equations. The gradient vectors are $\nabla f = \langle 1, 0 \rangle$ and $\nabla g = \langle 2x + 3y, 12y + 3x \rangle$, hence the Lagrange condition $\nabla f = \lambda \nabla g$ gives:

$$\langle 1, 0 \rangle = \lambda \langle 2x + 3y, 12y + 3x \rangle$$

or

$$1 = \lambda(2x + 3y), \qquad 0 = \lambda(12y + 3x)$$

this yields

$$x = -4y$$

Step 2. Solve for x and y using the constraint.

$$x^2 + 6y^2 + 3xy = (-4y)^2 + 6y^2 + 3(-4y)y = (16 + 6 - 12)y^2 = 10y^2 = 40$$

so $y = \pm 2$. If $y = 2$ then $x = -8$ and if $y = -2$ then $x = 8$. The extreme points are $(-8, 2)$ and $(8, -2)$. We conclude that the point with largest x-coordinate is $P = (8, -2)$.

23. Find the point (x_0, y_0) on the line $4x + 9y = 12$ that is closest to the origin.

SOLUTION Since we are minimizing distance, we can minimize the square of the distance function without loss of generality:

$$f(x, y) = (x - 0)^2 + (y - 0)^2 = x^2 + y^2$$

subject to the constraint $g(x, y) = 4x + 9y - 12$.

Step 1. Write out the Lagrange equations. The gradient vectors are $\nabla f = \langle 2x, 2y \rangle$ and $\nabla g = \langle 4, 9 \rangle$, hence the Lagrange condition $\nabla f = \lambda \nabla g$ gives

$$\langle 2x, 2y \rangle = \lambda \langle 4, 9 \rangle$$

or

$$2x = 4\lambda \quad \Rightarrow \quad x = 2\lambda, \quad 2y = 9\lambda$$

Step 2. Solve for λ in terms of x and y. The Lagrange equations give the following two expressions for λ:

$$\lambda = \frac{x}{2}, \quad \lambda = \frac{9}{2}y$$

Equating these two

$$\frac{x}{2} = \frac{9}{2}y \quad \Rightarrow \quad x = 9y$$

Step 3. Solve for x and y using the constraint. We are given $4x + 9y = 12$, therefore we can write:

$$4(9y) + 9y = 12 \quad \Rightarrow \quad 45y = 12 \quad \Rightarrow \quad y = \frac{12}{45} = \frac{4}{15}$$

Since $x = 9y$, then we conclude:

$$y = \frac{4}{15}, \quad x = 9 \cdot \frac{4}{15} = \frac{12}{5}$$

Step 4. Conclusions. Therefore the point closest to the origin lying on the plane $4x + 9y = 12$ is the point $(12/5, 4/15)$.

25. Find the maximum value of $f(x, y) = x^a y^b$ for $x \geq 0, y \geq 0$ on the line $x + y = 1$, where $a, b > 0$ are constants.

SECTION 14.8 | Lagrange Multipliers: Optimizing with a Constraint

SOLUTION

Step 1. Write the Lagrange equations. We must find the maximum value of $f(x,y) = x^a y^b$ under the constraints $g(x,y) = x + y - 1$, $x > 0$, $y > 0$. The gradient vectors are $\nabla f = \langle ax^{a-1}y^b, bx^a y^{b-1}\rangle$ and $\nabla g = \lambda\langle 1, 1\rangle$, hence the Lagrange condition $\nabla f = \lambda \nabla g$ is

$$\langle ax^{a-1}y^b, bx^a y^{b-1}\rangle = \lambda\langle 1, 1\rangle$$

We obtain the following equations:

$$\begin{aligned} ax^{a-1}y^b &= \lambda \\ bx^a y^{b-1} &= \lambda \end{aligned} \quad \Rightarrow \quad ax^{a-1}y^b = bx^a y^{b-1}$$

Step 2. Solve for x and y using the constraint. We solve the equation in step 1 for y in terms of x. This gives

$$ax^{a-1}y^b = bx^a y^{b-1}$$
$$ay = bx \quad \Rightarrow \quad y = \frac{b}{a}x$$

We now substitute $y = \frac{b}{a}x$ in the constraint $x + y = 1$ and solve for x:

$$x + \frac{b}{a}x = 1$$
$$(a+b)x = a \quad \Rightarrow \quad x = \frac{a}{a+b}$$

We find y using the relation $y = \frac{b}{a}x$:

$$y = \frac{b}{a} \cdot \frac{a}{a+b} = \frac{b}{a+b}$$

The critical point is thus

$$\left(\frac{a}{a+b}, \frac{b}{a+b}\right) \tag{1}$$

Step 3. Conclusions. We compute $f(x,y) = x^a y^b$ at the critical point:

$$f\left(\frac{a}{a+b}, \frac{b}{a+b}\right) = \left(\frac{a}{a+b}\right)^a \left(\frac{b}{a+b}\right)^b = \frac{a^a b^b}{(a+b)^{a+b}}$$

Now, since f is continuous on the segment $x + y = 1$, $x \geq 0$, $y \geq 0$, which is a closed and bounded set in R^2, then f has minimum and maximum values on this segment. The minimum value is 0 (obtained at $(0, 1)$ and $(1, 0)$), therefore the critical point (1) corresponds to the maximum value. We conclude that the maximum value of $x^a y^b$ on $x + y = 1$, $x > 0$, $y > 0$ is

$$\frac{a^a b^b}{(a+b)^{a+b}}$$

27. Find the maximum value of $f(x,y) = x^a y^b$ for $x \geq 0, y \geq 0$ on the unit circle, where $a, b > 0$ are constants.

SOLUTION We must find the maximum value of $f(x,y) = x^a y^b$ ($a, b > 0$) subject to the constraint $g(x,y) = x^2 + y^2 = 1$.

Step 1. Write out the Lagrange equations. We have $\nabla f = \langle ax^{a-1}y^b, bx^a y^{b-1}\rangle$ and $\nabla g = \langle 2x, 2y\rangle$. Therefore the Lagrange condition $\nabla f = \lambda \nabla g$ is

$$\langle ax^{a-1}y^b, bx^a y^{b-1}\rangle = \lambda\langle 2x, 2y\rangle$$

or

$$\begin{aligned} ax^{a-1}y^b &= 2\lambda x \\ bx^a y^{b-1} &= 2\lambda y \end{aligned} \tag{1}$$

Step 2. Solve for λ in terms of x and y. If $x = 0$ or $y = 0$, f has the minimum value 0. We thus may assume that $x > 0$ and $y > 0$. The equations (1) imply that

$$\lambda = \frac{ax^{a-2}y^b}{2}, \quad \lambda = \frac{bx^a y^{b-2}}{2}$$

Step 3. Solve for x and y using the constraint. Equating the two expressions for λ and solving for y in terms of x gives

$$\frac{ax^{a-2}y^b}{2} = \frac{bx^a y^{b-2}}{2}$$

$$ax^{a-2}y^b = bx^a y^{b-2}$$

$$ay^2 = bx^2$$

$$y^2 = \frac{b}{a}x^2 \quad \Rightarrow \quad y = \sqrt{\frac{b}{a}}x$$

We now substitute $y = \sqrt{\frac{b}{a}}x$ in the constraint $x^2 + y^2 = 1$ and solve for $x > 0$. We obtain

$$x^2 + \frac{b}{a}x^2 = 1$$

$$(a+b)x^2 = a$$

$$x^2 = \frac{a}{a+b} \quad \Rightarrow \quad x = \sqrt{\frac{a}{a+b}}$$

We find y using the relation $y = \sqrt{\frac{b}{a}}x$:

$$y = \sqrt{\frac{b}{a}}\sqrt{\frac{a}{a+b}} = \sqrt{\frac{ab}{a(a+b)}} = \sqrt{\frac{b}{a+b}}$$

We obtain the critical point:

$$\left(\sqrt{\frac{a}{a+b}}, \sqrt{\frac{b}{a+b}}\right)$$

Extreme points can also occur where $\nabla g = \mathbf{0}$, that is, $\langle 2x, 2y \rangle = \langle 0, 0 \rangle$ or $(x, y) = (0, 0)$. However, this point is not on the constraint.

Step 4. Conclusions. We compute $f(x, y) = x^a y^b$ at the critical point:

$$f\left(\sqrt{\frac{a}{a+b}}, \sqrt{\frac{b}{a+b}}\right) = \left(\frac{a}{a+b}\right)^{a/2}\left(\frac{b}{a+b}\right)^{b/2} = \frac{a^{a/2}b^{b/2}}{(a+b)^{(a+b)/2}} = \sqrt{\frac{a^a b^b}{(a+b)^{a+b}}}$$

The function $f(x, y) = x^a y^b$ is continuous on the set $x^2 + y^2 = 1$, $x \geq 0$, $y \geq 0$, which is a closed and bounded set in R^2, hence f has minimum and maximum values on the set. The minimum value is 0 (obtained at $(0, 1)$ and $(1, 0)$), hence the critical point that we found corresponds to the maximum value. We conclude that the maximum value of $x^a y^b$ on $x^2 + y^2 = 1$, $x > 0$, $y > 0$ is

$$\sqrt{\frac{a^a b^b}{(a+b)^{a+b}}}$$

29. Show that the minimum distance from the origin to a point on the plane $ax + by + cz = d$ is

$$\frac{|d|}{\sqrt{a^2 + b^2 + c^2}}$$

SOLUTION We want to minimize the distance $P = \sqrt{x^2 + y^2 + z^2}$ subject to $ax + by + cz = d$. Since the square function u^2 is increasing for $u \geq 0$, the square P^2 attains its minimum at the same point where the distance P attains its minimum. Thus, we may minimize the function $f(x, y, z) = x^2 + y^2 + z^2$ subject to the constraint $g(x, y, z) = ax + by + cz = d$.

Step 1. Write out the Lagrange equations. We have $\nabla f = \langle 2x, 2y, 2z \rangle$ and $\nabla g = \langle a, b, c \rangle$, hence the Lagrange condition $\nabla f = \lambda \nabla g$ gives the following equations:

$$2x = \lambda a$$
$$2y = \lambda b$$
$$2z = \lambda c$$

Assume for now that $a \neq 0, b \neq 0, c \neq 0$.

Step 2. Solve for λ in terms of x, y, and z. The Lagrange equations imply that

$$\lambda = \frac{2x}{a}, \quad \lambda = \frac{2y}{b}, \quad \lambda = \frac{2z}{c}$$

Step 3. Solve for x, y, and z using the constraint. Equating the expressions for λ give the following equations:

$$\begin{aligned} \frac{2x}{a} = \frac{2z}{c} \\ \frac{2y}{b} = \frac{2z}{c} \end{aligned} \quad \Rightarrow \quad \begin{aligned} x = \frac{a}{c}z \\ y = \frac{b}{c}z \end{aligned} \tag{1}$$

We now substitute $x = \frac{a}{c}z$ and $y = \frac{b}{c}z$ in the equation of the constraint $ax + by + cz = d$ and solve for z. This gives

$$a\left(\frac{a}{c}z\right) + b\left(\frac{b}{c}z\right) + cz = d$$

$$\frac{a^2}{c}z + \frac{b^2}{c}z + cz = d$$

$$\left(a^2 + b^2 + c^2\right)z = dc$$

Since $a^2 + b^2 + c^2 \neq 0$, we get $z = \frac{dc}{a^2+b^2+c^2}$. We now use (1) to compute y and x:

$$x = \frac{a}{c} \cdot \frac{dc}{a^2+b^2+c^2} = \frac{ad}{a^2+b^2+c^2}, \quad y = \frac{b}{c} \cdot \frac{dc}{a^2+b^2+c^2} = \frac{bd}{a^2+b^2+c^2}$$

We obtain the critical point:

$$P = \left(\frac{ad}{a^2+b^2+c^2}, \frac{bd}{a^2+b^2+c^2}, \frac{dc}{a^2+b^2+c^2}\right) \tag{2}$$

Step 4. Conclusions. It is clear geometrically that f has a minimum value subject to the constraint, hence the minimum value occurs at the point P. We conclude that the point P is the point on the plane closest to the origin. We now consider the case where $a = 0$. We consider the planes $ax + by + cz = d$, where $a \neq 0$ and $a \to 0$. A continuous change in a causes a continuous change in the closest point P. Therefore, the point P closest to the origin in case of $a = 0$ can be obtained by computing the limit of P in (2) as $a \to 0$, that is, by substituting $a = 0$. Similar considerations hold for $b = 0$ or $c = 0$. We conclude that the closest point P in (2) holds also for the planes with $a = 0$, $b = 0$, or $c = 0$ (but not all of them 0). The distance P of that point to the origin is

$$P = \sqrt{\frac{(ad)^2 + (bd)^2 + (dc)^2}{(a^2+b^2+c^2)^2}} = |d|\sqrt{\frac{a^2+b^2+c^2}{(a^2+b^2+c^2)^2}} = \frac{|d|}{\sqrt{a^2+b^2+c^2}}$$

31. Let Q be the point on an ellipse closest to a given point P outside the ellipse. It was known to the Greek mathematician Apollonius (third century BCE) that $\overline{PQ}$ is perpendicular to the tangent to the ellipse at Q (Figure 16). Explain in words why this conclusion is a consequence of the method of Lagrange multipliers. *Hint:* The circles centered at P are level curves of the function to be minimized.

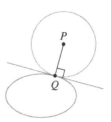

FIGURE 16

SOLUTION Let $P = (x_0, y_0)$. The distance d between the point P and a point $Q = (x, y)$ on the ellipse is minimum where the square d^2 is minimum (since the square function u^2 is increasing for $u \geq 0$). Therefore, we want to minimize the function

$$f(x, y) = (x - x_0)^2 + (y - y_0)^2$$

subject to the constraint

$$g(x, y) = \frac{x^2}{a^2} + \frac{y^2}{b^2} = 1$$

The method of Lagrange indicates that the solution Q is the point on the ellipse where $\nabla f = \lambda \nabla g$, that is, the point on the ellipse where the gradients ∇f and ∇g are parallel. Since the gradient is orthogonal to the level curves of the function, ∇g is orthogonal to the ellipse $g(x, y) = 1$, and ∇f is orthogonal to the level curve of f passing through Q. But this level curve is a circle through Q centered at P, hence the parallel vectors ∇g and ∇f are orthogonal to the ellipse and to the circle centered at P respectively. We conclude that the point Q is the point at which the tangent to the ellipse is also the tangent to the circle through Q centered at P. That is, the tangent to the ellipse at Q is perpendicular to the radius $\overline{PQ}$ of the circle.

In Exercises 33 and 34, let V be the volume of a can of radius r and height h, and let S be its surface area (including the top and bottom).

33. Find r and h that minimize S subject to the constraint $V = 54\pi$.

SOLUTION The volume of the can is $V(r, h) = \pi r^2 h$, and its surface area is $S(r, h) = 2\pi r^2 + 2\pi rh$. We want to minimize $S(r, h)$ subject to the constraint $V(r, h) = 54\pi$, or $\pi r^2 h - 54\pi = 0$. Divide through by π to simplify, giving the constraint equation $g(r, h) = r^2 h - 54 = 0$. We have

$$\nabla S(r, h) = \langle 4\pi r + 2\pi h, 2\pi r \rangle, \qquad \nabla g(r, h) = \langle 2rh, r^2 \rangle$$

so that the Lagrange condition $\nabla f(r, h) = \lambda \nabla g(r, h)$ gives

$$\langle 4\pi r + 2\pi h, 2\pi r \rangle = \lambda \langle 2rh, r^2 \rangle \quad \Rightarrow \quad 2\pi(2r + h) = \lambda(2rh), \ 2\pi r = \lambda(r^2)$$

Both r and h must be positive; solving for λ gives

$$\lambda = \frac{\pi(2r + h)}{rh}, \qquad \lambda = \frac{2\pi}{r}$$

Equating these two expressions for λ gives $\frac{2r+h}{rh} = \frac{2}{r}$, or $2r^2 + rh = 2rh$. Solving for h gives

$$h = \frac{2r^2}{2r - r} = 2r$$

Substitute $2r$ for h in the constraint equation, giving $\pi r^2(2r) - 54\pi = 2\pi(r^3 - 27) = 0$; this yields $r = 3$, so that $h = 6$. So the only critical point is $(3, 6)$, and $S(3, 6) = 18\pi + 36\pi = 54\pi$. This is clearly a minimum, since if r is very small, then h must be very large to give a volume of 54π, so the surface area will be very large. Therefore the minimum suface area is 54π achieved with a radius of 3 and a height of 6.

35. Figure 19 depicts a tetrahedron whose faces lie in the coordinate planes and in the plane with equation $\frac{x}{a} + \frac{y}{b} + \frac{z}{c} = 1$ $(a, b, c > 0)$. The volume of the tetrahedron is given by $V = \frac{1}{6}abc$. Find the minimum value of V among all planes passing through the point $P = (1, 1, 1)$.

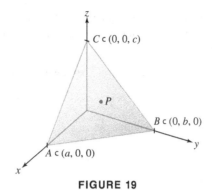

FIGURE 19

SOLUTION The plane is constrained to pass through the point $P = (1, 1, 1)$, hence this point must satisfy the equation of the plane. That is,

$$\frac{1}{a} + \frac{1}{b} + \frac{1}{c} = 1$$

We thus must minimize the function $V(a, b, c) = \frac{1}{6}abc$ subject to the constraint $g(a, b, c) = \frac{1}{a} + \frac{1}{b} + \frac{1}{c} = 1$, $a > 0, b > 0, c > 0$.

Step 1. Write out the Lagrange equations. We have $\nabla V = \left\langle \frac{1}{6}bc, \frac{1}{6}ac, \frac{1}{6}ab \right\rangle$ and $\nabla g = \left\langle -\frac{1}{a^2}, -\frac{1}{b^2}, -\frac{1}{c^2} \right\rangle$, hence the Lagrange condition $\nabla V = \lambda \nabla g$ yields the following equations:

$$\frac{1}{6}bc = -\frac{1}{a^2}\lambda$$

$$\frac{1}{6}ac = -\frac{1}{b^2}\lambda$$

$$\frac{1}{6}ab = -\frac{1}{c^2}\lambda$$

Step 2. Solve for λ in terms of a, b, and c. The Lagrange equations imply that

$$\lambda = -\frac{bca^2}{6}, \quad \lambda = -\frac{acb^2}{6}, \quad \lambda = -\frac{abc^2}{6}$$

Step 3. Solve for a, b, and c using the constraint. Equating the expressions for λ, we obtain the following equations:

$$\begin{aligned} bca^2 &= acb^2 \\ abc^2 &= acb^2 \end{aligned} \quad \Rightarrow \quad \begin{aligned} abc(a - b) &= 0 \\ abc(c - b) &= 0 \end{aligned}$$

Since a, b, c are positive numbers, we conclude that $a = b$ and $c = b$. We now substitute $a = b$ and $c = b$ in the equation of the constraint $\frac{1}{a} + \frac{1}{b} + \frac{1}{c} = 1$ and solve for b. This gives

$$\frac{1}{b} + \frac{1}{b} + \frac{1}{b} = 1$$

$$\frac{3}{b} = 1 \quad \Rightarrow \quad b = 3$$

Therefore also $a = b = 3$ and $c = b = 3$. We obtain the critical point $(3, 3, 3)$.

Step 4. Conclusions. If V has a minimum value subject to the constraint then it occurs at the point $(3, 3, 3)$. That is, the plane that minimizes V is

$$\frac{x}{3} + \frac{y}{3} + \frac{z}{3} = 1 \quad \text{or} \quad x + y + z = 3$$

Remark: Since the constraint is not bounded, we need to justify the existence of a minimum value of $V = \frac{1}{6}abc$ under the constraint $\frac{1}{a} + \frac{1}{b} + \frac{1}{c} = 1$. First notice that since a, b, c are nonnegative and the sum of their reciprocals is 1, none of them can tend to zero. In fact, none of a, b, c can be less than 1. Therefore, if $a \to \infty, b \to \infty$, or $c \to \infty$, then $V \to \infty$. This means that we can find a cube that includes the point $\left(\frac{1}{3}, \frac{1}{3}, \frac{1}{3}\right)$ such that, on the part of the constraint that is outside the cube, it holds that $V > V\left(\frac{1}{3}, \frac{1}{3}, \frac{1}{3}\right) = \frac{1}{162}$. On the part of the constraint inside the cube, V has a minimum value m, since it is a closed and bounded set. Clearly m is the minimum of V on the whole constraint.

37. Show that the Lagrange equations for $f(x,y) = x + y$ subject to the constraint $g(x,y) = x + 2y = 0$ have no solution. What can you conclude about the minimum and maximum values of f subject to $g = 0$? Show this directly.

SOLUTION Using the methods of Lagrange we can write $\nabla f = \lambda \nabla g$ and see

$$\langle 1, 1 \rangle = \lambda \langle 1, 2 \rangle$$

Which gives us the equations:

$$1 = \lambda, \quad 1 = 2\lambda$$

hence, $\lambda = 1$ or $\lambda = 1/2$. This is an inconsistent set of equations, thus the Lagrange method has no solution. What we can conclude from this is that the maximum and minimum values of f subject to $g = 0$ does not exist. This means that $f(x,y)$ increases without an upper bound and decreases without a lower bound.

To show this directly, we can write $y = -1/2x$ from the constraint equation and substitute it into $f(x,y) = f(x, -1/2x) = x - 1/2x = 1/2x$. We know that $y = 1/2x$ is a straight line having slope $1/2$, increasing, with no maximum nor minimum values.

39. Let L be the minimum length of a ladder that can reach over a fence of height h to a wall located a distance b behind the wall.

(a) Use Lagrange multipliers to show that $L = (h^{2/3} + b^{2/3})^{3/2}$ (Figure 20). *Hint:* Show that the problem amounts to minimizing $f(x,y) = (x+b)^2 + (y+h)^2$ subject to $y/b = h/x$ or $xy = bh$.

(b) Show that the value of L is also equal to the radius of the circle with center $(-b, -h)$ that is tangent to the graph of $xy = bh$.

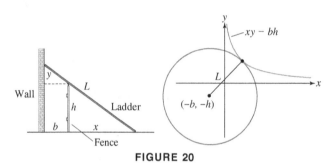

FIGURE 20

SOLUTION

(a) We denote by x and y the lengths shown in the figure, and express the length l of the ladder in terms of x and y.

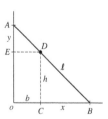

Using the Pythagorean Theorem, we have

$$l = \sqrt{\overline{OA}^2 + \overline{OB}^2} = \sqrt{(y+h)^2 + (x+b)^2} \tag{1}$$

Since the function u^2 is increasing for $u \geq 0$, l and l^2 have their minimum values at the same point. Therefore, we may minimize the function $f(x,y) = l^2(x,y)$, which is

$$f(x,y) = (x+b)^2 + (y+h)^2$$

We now identify the constraint on the variables x and y. (Notice that h, b are constants while x and y are free to change.) Using proportional lengths in the similar triangles $\triangle AED$ and $\triangle DCB$, we have

$$\frac{\overline{AE}}{\overline{DC}} = \frac{\overline{ED}}{\overline{CB}}$$

SECTION 14.8 | Lagrange Multipliers: Optimizing with a Constraint

That is,

$$\frac{y}{h} = \frac{b}{x} \quad \Rightarrow \quad xy = bh$$

We thus must minimize $f(x, y) = (x + b)^2 + (y + h)^2$ subject to the constraint $g(x, y) = xy = bh$, $x > 0$, $y > 0$.

Step 1. Write out the Lagrange equations. We have $\nabla f = \langle 2(x+b), 2(y+h) \rangle$ and $\nabla g = \langle y, x \rangle$, hence the Lagrange condition $\nabla f = \lambda \nabla g$ gives the following equations:

$$2(x + b) = \lambda y$$

$$2(y + h) = \lambda x$$

Step 2. Solve for λ in terms of x and y. The equation of the constraint implies that $y \neq 0$ and $x \neq 0$. Therefore, the Lagrange equations yield

$$\lambda = \frac{2(x+b)}{y}, \quad \lambda = \frac{2(y+h)}{x}$$

Step 3. Solve for x and y using the constraint. Equating the two expressions for λ gives

$$\frac{2(x+b)}{y} = \frac{2(y+h)}{x}$$

We simplify:

$$x(x + b) = y(y + h)$$

$$x^2 + xb = y^2 + yh$$

The equation of the constraint implies that $y = \frac{bh}{x}$. We substitute and solve for $x > 0$. This gives

$$x^2 + xb = \left(\frac{bh}{x}\right)^2 + \frac{bh}{x} \cdot h$$

$$x^2 + xb = \frac{b^2 h^2}{x^2} + \frac{bh^2}{x}$$

$$x^4 + x^3 b = b^2 h^2 + bh^2 x$$

$$x^4 + bx^3 - bh^2 x - b^2 h^2 = 0$$

$$x^3(x + b) - bh^2(x + b) = 0$$

$$\left(x^3 - bh^2\right)(x + b) = 0$$

Since $x > 0$ and $b > 0$, also $x + b > 0$ and the solution is

$$x^3 - bh^2 = 0 \quad \Rightarrow \quad x = (bh^2)^{1/3}$$

We compute y. Using the relation $y = \frac{bh}{x}$,

$$y = \frac{bh}{(bh^2)^{1/3}} = \frac{bh}{b^{1/3} h^{2/3}} = b^{2/3} h^{1/3} = (b^2 h)^{1/3}$$

We obtain the solution

$$x = \left(bh^2\right)^{1/3}, \quad y = \left(b^2 h\right)^{1/3} \tag{2}$$

Extreme values may also occur at the point on the constraint where $\nabla g = \mathbf{0}$. However, $\nabla g = \langle y, x \rangle = \langle 0, 0 \rangle$ only at the point $(0, 0)$, which is not on the constraint.

Step 4. Conclusions. Notice that on the constraint $y = \frac{bh}{x}$ or $x = \frac{bh}{y}$, as $x \to 0+$ then $y \to \infty$, and as $x \to \infty$, then $y \to 0+$. Also, as $y \to 0+$, $x \to \infty$ and as $y \to \infty$, $x \to 0+$. In either case, $f(x, y)$ is increasing without bound. Using this property and the theorem on the existence of extreme values for a continuous function on a closed and bounded set (for a certain part of the constraint), one can show that f has a minimum value on the constraint. This minimum value occurs at the point (2). We substitute this point in (1) to obtain the following minimum length L:

$$L = \sqrt{\left((b^2 h)^{1/3} + h\right)^2 + \left((bh^2)^{1/3} + b\right)^2}$$

$$= \sqrt{(b^2h)^{2/3} + 2h(b^2h)^{1/3} + h^2 + (bh^2)^{2/3} + 2b(bh^2)^{1/3} + b^2}$$

$$= \sqrt{b^{\frac{4}{3}}h^{2/3} + 2h^{\frac{4}{3}}b^{2/3} + h^2 + b^{2/3}h^{\frac{4}{3}} + 2b^{\frac{4}{3}}h^{2/3} + b^2}$$

$$= \sqrt{3b^{\frac{4}{3}}h^{2/3} + 3h^{\frac{4}{3}}b^{2/3} + h^2 + b^2}$$

$$= \sqrt{(h^{2/3})^3 + 3(h^{2/3})^2 b^{2/3} + 3h^{2/3}(b^{2/3})^2 + (b^{2/3})^3}$$

Using the identity $(\alpha + \beta)^3 = \alpha^3 + 3\alpha^2\beta + 3\alpha\beta^2 + \beta^3$, we conclude that

$$L = \sqrt{\left(h^{2/3} + b^{2/3}\right)^3} = \left(h^{2/3} + b^{2/3}\right)^{3/2}$$

(b) The Lagrange condition states that the gradient vectors ∇f_P and ∇g_P are parallel (where P is the minimizing point). The gradient ∇f_P is orthogonal to the level curve of f passing through P, which is a circle through P centered at $(-b, -h)$. ∇g_P is orthogonal to the level curve of g passing through P, which is the curve of the constraint $xy = bh$. We conclude that the circle and the curve $xy = bh$, both being perpendicular to parallel vectors, are tangent at P. The radius of the circle is the minimum value L, of $f(x, y)$.

41. Find the minimum of $f(x, y, z) = x^2 + y^2 + z^2$ subject to the two constraints $x + y + z = 1$ and $x + 2y + 3z = 6$.

SOLUTION The constraint equations are:

$$g(x, y, z) = x + y + z - 1 = 0, \quad h(x, y, z) = x + 2y + 3z - 6 = 0$$

We have

$$\nabla f = \langle 2x, 2y, 2z \rangle, \quad \nabla g = \langle 1, 1, 1 \rangle, \quad \nabla h = \langle 1, 2, 3 \rangle$$

so the Lagrange condition is

$$\nabla f = \lambda \nabla g + \mu \nabla h \quad \Rightarrow \quad \langle 2x, 2y, 2z \rangle = \lambda \langle 1, 1, 1 \rangle + \mu \langle 1, 2, 3 \rangle$$

This gives three equations

$$2x = \lambda + \mu, \quad 2y = \lambda + 2\mu, \quad 2z = \lambda + 3\mu$$

Subtracting each equation from the next gives $\mu = 2y - 2x = 2z - 2y$, so that $y - x = z - y$, so that $x = 2y - z$. Plugging this into the two constraint equations gives

$$(2y - z) + y + z = 1, \; (2y - z) + 2y + 3z - 6 = 0 \quad \Rightarrow \quad 3y = 1, \; 2y + z = 3 \quad \Rightarrow \quad y = \frac{1}{3}, \; z = \frac{7}{3}, \; x = -\frac{5}{3}$$

So there is only one critical point, and

$$f\left(-\frac{5}{3}, \frac{1}{3}, \frac{7}{3}\right) = \frac{25}{3}$$

This is a minimum, since for example $x = -4$, $y = 5$, $z = 0$ satisfy the constraints, and $f(-4, 5, 0) > \frac{25}{3}$. There is no maximum value, since for any (large) value of x we can find values of y and z satisfying both constraints, and then $f(x, y, z)$ is large as well.

43. Find the point lying on the intersection of the plane $x + \frac{1}{2}y + \frac{1}{4}z = 0$ and the sphere $x^2 + y^2 + z^2 = 9$ with the greatest z-coordinate.

SOLUTION We will use the method of Lagrange multipliers with two constraints here. We want to maximize $f(x, y, z) = z$ subject to the two surfaces. Set the first constraint as $g(x, y, z) = x + \frac{1}{2}y + \frac{1}{4}z = 0$ and the second as $h(x, y, z) = x^2 + y^2 + z^2 - 9 = 0$.

Write out the Lagrange equations. We have $\nabla f = \langle 0, 0, 1 \rangle$, $\nabla g = \left\langle 1, \frac{1}{2}, \frac{1}{4} \right\rangle$ and $\nabla g = \langle 2x, 2y, 2z \rangle$, hence the Lagrange condition, $\nabla f = \lambda \nabla g + \mu \nabla h$ yields the following equations:

$$\langle 0, 0, 1 \rangle = \lambda \left\langle 1, \frac{1}{2}, \frac{1}{4} \right\rangle + \mu \langle 2x, 2y, 2z \rangle$$

and

$$0 = \lambda + 2\mu x, \quad 0 = \frac{1}{2}\lambda + 2\mu y, \quad 1 = \frac{1}{4}\lambda + 2\mu z$$

Hence, from the first two equations we see

$$\lambda = -2\mu x, \quad \lambda = -4\mu y$$

Therefore

$$-2\mu x = -4\mu y \quad \Rightarrow \quad x = 2y$$

since $\mu \neq 0$. Using the first constraint equation $x + \frac{1}{2}y + \frac{1}{4}z = 0$ we have

$$2y + \frac{1}{2}y + \frac{1}{4}z = 0 \quad \Rightarrow \quad \frac{5}{2}y + \frac{1}{4}z = 0 \quad \Rightarrow \quad y = -\frac{1}{10}z$$

Finally, we can substitute $y = -1/10z$ and $x = 2y = -1/5z$ into the second constraint equation $x^2 + y^2 + z^2 = 9$ to see

$$\left(-\frac{1}{5}z\right)^2 + \left(-\frac{1}{10}z\right)^2 + z^2 = 9 \quad \Rightarrow \quad \frac{1}{25}z^2 + \frac{1}{100}z^2 + z^2 = 9 \quad \Rightarrow \quad 4z^2 + z^2 + 100z^2 = 900$$

Hence

$$105z^2 = 900 \quad \Rightarrow \quad z^2 = \frac{900}{105} = \frac{60}{7}$$

Therefore $z = \pm\sqrt{\frac{60}{7}} = \pm 2\sqrt{\frac{15}{7}}$. The two critical points are:

$$P\left(-\frac{2}{5}\sqrt{\frac{15}{7}}, -\frac{1}{5}\sqrt{\frac{15}{7}}, 2\sqrt{\frac{15}{7}}\right), \quad Q\left(\frac{2}{5}\sqrt{\frac{15}{7}}, \frac{1}{5}\sqrt{\frac{15}{7}}, -2\sqrt{\frac{15}{7}}\right)$$

The critical point with the largest z-coordinate (the maximum of $f(x, y, z)$) is P with z-coordinate $2\sqrt{\frac{15}{7}} \approx 2.928$.

45. The cylinder $x^2 + y^2 = 1$ intersects the plane $x + z = 1$ in an ellipse. Find the point on such an ellipse that is farthest from the origin.

SOLUTION Maximizing the distance of a point (x, y, z) on the ellipse from the origin is the same as maximizing the square of its distance from the origin, so we want to maximize $f(x, y, z) = x^2 + y^2 + z^2$ subject to the two constraints

$$g(x, y, z) = x^2 + y^2 - 1 = 0$$
$$h(x, y, z) = x + z - 1 = 0.$$

We have

$$\nabla f = \langle 2x, 2y, 2z \rangle, \quad \nabla g = \langle 2x, 2y, 0 \rangle, \quad \nabla h = \langle 1, 0, 1 \rangle$$

so the Lagrange condition is

$$\nabla f = \lambda \nabla g + \mu \nabla h \quad \Rightarrow \quad \langle 2x, 2y, 2z \rangle = \lambda \langle 2x, 2y, 0 \rangle + \mu \langle 1, 0, 1 \rangle$$

This gives the three equations

$$2x = 2x\lambda + \mu$$
$$2y = 2y\lambda$$
$$2z = \mu$$

so that from the second equation either $y = 0$ or $\lambda = 1$. If $y = 0$, then $g(x, y, z) = 0$ implies that $x = \pm 1$. If $x = 1$, then $h(1, 0, z) = 0$ forces $z = 0$, while if $x = -1$, then $h(-1, 0, z) = 0$ implies $z = 2$. So we get two points in this case: $(1, 0, 0)$ and $(-1, 0, 2)$. If $\lambda = 1$, then the first equation gives $2x = 2x + \mu$, so that $\mu = 0$; the third equation then implies $z = 0$. $h(x, y, z) = 0$ forces $x = 1$, and then $g(x, y, z) = 0$ means that $y = 0$. So we get point $(1, 0, 0)$ again. Evaluating f at these two critical points gives

$$f(1, 0, 0) = 1^2 + 0^2 + 0^2 = 1, \quad f(-1, 0, 2) = (-1)^2 + 0^2 + 2^2 = 5$$

So the required point is $(-1, 0, 2)$, and it is at distance $\sqrt{5}$ from the origin.

47. Find the minimum value of $f(x,y,z) = x^2 + y^2 + z^2$ subject to two constraints, $x + 2y + z = 3$ and $x - y = 4$.

SOLUTION The constraint equations are

$$g(x,y,z) = x + 2y + z - 3 = 0, \quad h(x,y) = x - y - 4 = 0$$

Step 1. Write out the Lagrange equations. We have $\nabla f = \langle 2x, 2y, 2z \rangle$, $\nabla g = \langle 1, 2, 1 \rangle$, and $\nabla h = \langle 1, -1, 0 \rangle$, hence the Lagrange condition is

$$\nabla f = \lambda \nabla g + \mu \nabla h$$
$$\langle 2x, 2y, 2z \rangle = \lambda \langle 1, 2, 1 \rangle + \mu \langle 1, -1, 0 \rangle$$
$$= \langle \lambda + \mu, 2\lambda - \mu, \lambda \rangle$$

We obtain the following equations:

$$2x = \lambda + \mu$$
$$2y = 2\lambda - \mu$$
$$2z = \lambda$$

Step 2. Solve for λ and μ. The first equation gives $\lambda = 2x - \mu$. Combining with the third equation we get

$$2z = 2x - \mu \tag{1}$$

The second equation gives $\mu = 2\lambda - 2y$, combining with the third equation we get $\mu = 4z - 2y$. Substituting in (1) we obtain

$$2z = 2x - (4z - 2y) = 2x - 4z + 2y$$
$$6z = 2x + 2y \quad \Rightarrow \quad z = \frac{x+y}{3} \tag{2}$$

Step 3. Solve for x, y, and z using the constraints. The constraints give x and y as functions of z:

$$x - y = 4 \quad \Rightarrow \quad y = x - 4$$
$$x + 2y + z = 3 \quad \Rightarrow \quad y = \frac{3 - x - z}{2}$$

Combining the two equations we get

$$x - 4 = \frac{3 - x - z}{2}$$
$$2x - 8 = 3 - x - z$$
$$3x = 11 - z \quad \Rightarrow \quad x = \frac{11 - z}{3}$$

We find y using $y = x - 4$:

$$y = \frac{11 - z}{3} - 4 = \frac{-1 - z}{3}$$

We substitute x and y in (2) and solve for z:

$$z = \frac{\frac{11-z}{3} + \frac{-1-z}{3}}{3} = \frac{11 - z - 1 - z}{9} = \frac{10 - 2z}{9}$$
$$9z = 10 - 2z$$
$$11z = 10 \quad \Rightarrow \quad z = \frac{10}{11}$$

We find x and y:

$$y = \frac{-1 - z}{3} = \frac{-1 - \frac{10}{11}}{3} = -\frac{21}{33} = -\frac{7}{11}$$

$$x = \frac{11 - z}{3} = \frac{11 - \frac{10}{11}}{3} = \frac{111}{33} = \frac{37}{11}$$

We obtain the solution

$$P = \left(\frac{37}{11}, -\frac{7}{11}, \frac{10}{11}\right)$$

Step 4. Calculate the critical values. We compute $f(x, y, z) = z^2 + y^2 + z^2$ at the critical point:

$$f(P) = \left(\frac{37}{11}\right)^2 + \left(-\frac{7}{11}\right)^2 + \left(\frac{10}{11}\right)^2 = \frac{1518}{121} = \frac{138}{11} \approx 12.545$$

As x tends to infinity, so also does $f(x, y, z)$ tend to ∞. Therefore f has no maximum value and the given critical point P must produce a minimum. We conclude that the minimum value of f subject to the two constraints is $f(P) = \frac{138}{11} \approx 12.545$.

Further Insights and Challenges

49. Assumptions Matter Consider the problem of minimizing $f(x, y) = x$ subject to $g(x, y) = (x - 1)^3 - y^2 = 0$.
(a) Show, without using calculus, that the minimum occurs at $P = (1, 0)$.
(b) Show that the Lagrange condition $\nabla f_P = \lambda \nabla g_P$ is not satisfied for any value of λ.
(c) Does this contradict Theorem 1?

SOLUTION
(a) The equation of the constraint can be rewritten as

$$(x - 1)^3 = y^2 \quad \text{or} \quad x = y^{2/3} + 1$$

Therefore, at the points under the constraint, $x \geq 1$, hence $f(x, y) \geq 1$. Also at the point $P = (1, 0)$ we have $f(1, 0) = 1$, hence $f(1, 0) = 1$ is the minimum value of f under the constraint.
(b) We have $\nabla f = \langle 1, 0 \rangle$ and $\nabla g = \langle 3(x - 1)^2, -2y \rangle$, hence the Lagrange condition $\nabla f = \lambda \nabla g$ yields the following equations:

$$1 = \lambda \cdot 3(x - 1)^2$$
$$0 = -2\lambda y$$

The first equation implies that $\lambda \neq 0$ and $x - 1 = \pm \frac{1}{\sqrt{3\lambda}}$. The second equation gives $y = 0$. Substituting in the equation of the constraint gives

$$(x - 1)^3 - y^2 = \left(\frac{\pm 1}{\sqrt{3\lambda}}\right)^3 - 0^2 = \frac{\pm 1}{(3\lambda)^{3/2}} \neq 0$$

We conclude that the Lagrange condition is not satisfied by any point under the constraint.
(c) Theorem 1 is not violated since at the point $P = (1, 0)$, $\nabla g = \mathbf{0}$, whereas the theorem is valid for points where $\nabla g_P \neq \mathbf{0}$.

51. Consider the utility function $U(x_1, x_2) = x_1 x_2$ with budget constraint $p_1 x_1 + p_2 x_2 = c$.
(a) Show that the maximum of $U(x_1, x_2)$ subject to the budget constraint is equal to $c^2/(4 p_1 p_2)$.
(b) Calculate the value of the Lagrange multiplier λ occurring in (a).
(c) Prove the following interpretation: λ is the rate of increase in utility per unit increase in total budget c.

SOLUTION
(a) By the earlier exercise, the utility is maximized at a point where the following equality holds:

$$\frac{U_{x_1}}{U_{x_2}} = \frac{p_1}{p_2}$$

Since $U_{x_1} = x_2$ and $U_{x_2} = x_1$, we get

$$\frac{x_2}{x_1} = \frac{p_1}{p_2} \quad \Rightarrow \quad x_2 = \frac{p_1}{p_2} x_1$$

We now substitute x_2 in terms of x_1 in the constraint $p_1 x_1 + p_2 x_2 = c$ and solve for x_1. This gives

$$p_1 x_1 + p_2 \cdot \frac{p_1}{p_2} x_1 = c$$

$$2p_1x_1 = c \quad \Rightarrow \quad x_1 = \frac{c}{2p_1}$$

The corresponding value of x_2 is computed by $x_2 = \frac{p_1}{p_2}x_1$:

$$x_2 = \frac{p_1}{p_2} \cdot \frac{c}{2p_1} = \frac{c}{2p_2}$$

That is, $U(x_1, x_2)$ is maximized at the consumption level $x_1 = \frac{c}{2p_1}$, $x_2 = \frac{c}{2p_2}$. The maximum value is

$$U\left(\frac{c}{2p_1}, \frac{c}{2p_2}\right) = \frac{c}{2p_1} \cdot \frac{c}{2p_2} = \frac{c^2}{4p_1p_2}$$

(b) The Lagrange condition $\nabla U = \lambda \nabla g$ for $U(x_1, x_2) = x_1 x_2$ and $g(x_1, x_2) = p_1 x_1 + p_2 x_2 - c = 0$ is

$$\langle x_2, x_1 \rangle = \lambda \langle p_1, p_2 \rangle \tag{1}$$

or

$$\begin{aligned} x_2 &= \lambda p_1 \\ x_1 &= \lambda p_2 \end{aligned} \quad \Rightarrow \quad \lambda = \frac{x_2}{p_1} = \frac{x_1}{p_2}$$

In part (a) we showed that at the maximizing point $x_1 = \frac{c}{2p_1}$, therefore the value of λ is

$$\lambda = \frac{x_1}{p_2} = \frac{c}{2p_1 p_2}$$

(c) We compute $\frac{dU}{dc}$ using the Chain Rule:

$$\frac{dU}{dc} = \frac{\partial U}{\partial x_1} x_1'(c) + \frac{\partial U}{\partial x_2} x_2'(c) = x_2 x_1'(c) + x_1 x_2'(c) = \langle x_2, x_1 \rangle \cdot \langle x_1'(c), x_2'(c) \rangle$$

Substituting in (1) we get

$$\frac{dU}{dc} = \lambda \langle p_1, p_2 \rangle \cdot \langle x_1'(c), x_2'(c) \rangle = \lambda (p_1 x_1'(c) + p_2 x_2'(c)) \tag{2}$$

We now use the Chain Rule to differentiate the equation of the constraint $p_1 x_1 + p_2 x_2 = c$ with respect to c:

$$p_1 x_1'(c) + p_2 x_2'(c) = 1$$

Substituting in (2), we get

$$\frac{dU}{dc} = \lambda \cdot 1 = \lambda$$

Using the approximation $\Delta U \approx \frac{dU}{dc} \Delta c$, we conclude that λ is the rate of increase in utility per unit increase of total budget L.

53. Let $B > 0$. Show that the maximum of

$$f(x_1, \ldots, x_n) = x_1 x_2 \cdots x_n$$

subject to the constraints $x_1 + \cdots + x_n = B$ and $x_j \geq 0$ for $j = 1, \ldots, n$ occurs for $x_1 = \cdots = x_n = B/n$. Use this to conclude that

$$(a_1 a_2 \cdots a_n)^{1/n} \leq \frac{a_1 + \cdots + a_n}{n}$$

for all positive numbers $a_1, \ldots, a_n$.

SOLUTION We first notice that the constraints $x_1 + \cdots + x_n = B$ and $x_j \geq 0$ for $j = 1, \ldots, n$ define a closed and bounded set in the nth dimensional space, hence f (continuous, as a polynomial) has extreme values on this set. The minimum value zero occurs where one of the coordinates is zero (for example, for $n = 2$ the constraint $x_1 + x_2 = B$, $x_1 \geq 0$, $x_2 \geq 0$ is a triangle in the first quadrant). We need to maximize the function $f(x_1, \ldots, x_n) = x_1 x_2 \cdots x_n$ subject to the constraints $g(x_1, \ldots, x_n) = x_1 + \cdots + x_n - B = 0$, $x_j \geq 0$, $j = 1, \ldots, n$.

Step 1. Write out the Lagrange equations. The gradient vectors are

$$\nabla f = \langle x_2 x_3 \cdots x_n, x_1 x_3 \cdots x_n, \ldots, x_1 x_2 \cdots x_{n-1} \rangle$$
$$\nabla g = \langle 1, 1, \ldots, 1 \rangle$$

The Lagrange condition $\nabla f = \lambda \nabla g$ yields the following equations:

$$x_2 x_3 \cdots x_n = \lambda$$
$$x_1 x_3 \cdots x_n = \lambda$$
$$x_1 x_2 \cdots x_{n-1} = \lambda$$

Step 2. Solving for $x_1, x_2, \ldots, x_n$ using the constraint. The Lagrange equations imply the following equations:

$$x_2 x_3 \cdots x_n = x_1 x_2 \cdots x_{n-1}$$
$$x_1 x_3 \cdots x_n = x_1 x_2 \cdots x_{n-1}$$
$$x_1 x_2 x_4 \cdots x_n = x_1 x_2 \cdots x_{n-1}$$
$$\vdots$$
$$x_1 x_2 \cdots x_{n-2} x_n = x_1 x_2 \cdots x_{n-1}$$

We may assume that $x_j \neq 0$ for $j = 1, \ldots, n$, since if one of the coordinates is zero, f has the minimum value zero. We divide each equation by its right-hand side to obtain

$$\begin{array}{ll} \dfrac{x_n}{x_1} = 1 & \\ \dfrac{x_n}{x_2} = 1 & x_1 = x_n \\ \dfrac{x_n}{x_3} = 1 \Rightarrow & x_2 = x_n \\ \vdots & x_3 = x_n \\ \dfrac{x_n}{x_{n-1}} = 1 & \vdots \\ & x_{n-1} = x_n \end{array}$$

Substituting in the constraint $x_1 + \cdots + x_n = B$ and solving for x_n gives

$$\underbrace{x_n + x_n + \cdots + x_n}_{n} = B$$

$$n x_n = B \quad \Rightarrow \quad x_n = \frac{B}{n}$$

Hence $x_1 = \cdots = x_n = \frac{B}{n}$.

Step 3. Conclusions. The maximum value of $f(x_1, \ldots, x_n) = x_1 x_2 \cdots x_n$ on the constraint $x_1 + \cdots + x_n = B$, $x_j \geq 0$, $j = 1, \ldots, n$ occurs at the point at which all coordinates are equal to $\frac{B}{n}$. The value of f at this point is

$$f\left(\frac{B}{n}, \frac{B}{n}, \ldots, \frac{B}{n}\right) = \left(\frac{B}{n}\right)^n$$

It follows that for any point $(x_1, \ldots, x_n)$ on the constraint, that is, for any point satisfying $x_1 + \cdots + x_n = B$ with x_j positive, the following holds:

$$f(x_1, \ldots, x_n) \leq \left(\frac{B}{n}\right)^n$$

That is,

$$x_1 \cdots x_n \leq \left(\frac{x_1 + \cdots + x_n}{n}\right)^n$$

or

$$(x_1 \cdots x_n)^{1/n} \leq \frac{x_1 + \cdots + x_n}{n}$$

55. Given constants E, E_1, E_2, E_3, consider the maximum of

$$S(x_1, x_2, x_3) = x_1 \ln x_1 + x_2 \ln x_2 + x_3 \ln x_3$$

subject to two constraints:

$$x_1 + x_2 + x_3 = N, \qquad E_1 x_1 + E_2 x_2 + E_3 x_3 = E$$

Show that there is a constant μ such that $x_i = A^{-1} e^{\mu E_i}$ for $i = 1, 2, 3$, where $A = N^{-1}(e^{\mu E_1} + e^{\mu E_2} + e^{\mu E_3})$.

SOLUTION The constraints equations are

$$g(x_1, x_2, x_3) = x_1 + x_2 + x_3 - N = 0$$

$$h(x_1, x_2, x_3) = E_1 x_1 + E_2 x_2 + E_3 x_3 - E = 0$$

We first find the Lagrange equations. The gradient vectors are

$$\nabla S = \left\langle \ln x_1 + x_1 \cdot \frac{1}{x_1}, \ln x_2 + x_2 \cdot \frac{1}{x_2}, \ln x_3 + x_3 \cdot \frac{1}{x_3} \right\rangle = \langle 1 + \ln x_1, 1 + \ln x_2, 1 + \ln x_3 \rangle$$

$$\nabla g = \langle 1, 1, 1 \rangle, \quad \nabla h = \langle E_1, E_2, E_3 \rangle$$

The Lagrange condition $\nabla f = \lambda \nabla g + \mu \nabla h$ gives the following equation:

$$\langle 1 + \ln x_1, 1 + \ln x_2, 1 + \ln x_3 \rangle = \lambda \langle 1, 1, 1 \rangle + \mu \langle E_1, E_2, E_3 \rangle = \langle \lambda + \mu E_1, \lambda + \mu E_2, \lambda + \mu E_3 \rangle$$

We obtain the Lagrange equations:

$$1 + \ln x_1 = \lambda + \mu E_1$$

$$1 + \ln x_2 = \lambda + \mu E_2$$

$$1 + \ln x_3 = \lambda + \mu E_3$$

We subtract the third equation from the other equations to obtain

$$\ln x_1 - \ln x_3 = \mu (E_1 - E_3)$$

$$\ln x_2 - \ln x_3 = \mu (E_2 - E_3)$$

or

$$\ln \frac{x_1}{x_3} = \mu (E_1 - E_3) \qquad x_1 = x_3 e^{\mu (E_1 - E_3)}$$
$$\Rightarrow \qquad \qquad \qquad \qquad \qquad \qquad (1)$$
$$\ln \frac{x_2}{x_3} = \mu (E_2 - E_3) \qquad x_2 = x_3 e^{\mu (E_2 - E_3)}$$

Substituting x_1 and x_2 in the equation of the constraint $g(x_1, x_2, x_3) = 0$ and solving for x_3 gives

$$x_3 e^{\mu (E_1 - E_3)} + x_3 e^{\mu (E_2 - E_3)} + x_3 = N$$

We multiply by $e^{\mu E_3}$:

$$x_3 (e^{\mu E_1} + e^{\mu E_2} + e^{\mu E_3}) = N e^{\mu E_3}$$

$$x_3 = \frac{N e^{\mu E_3}}{e^{\mu E_1} + e^{\mu E_2} + e^{\mu E_3}}$$

Substituting in (1) we get

$$x_1 = \frac{N e^{\mu E_3}}{e^{\mu E_1} + e^{\mu E_2} + e^{\mu E_3}} \cdot e^{\mu (E_1 - E_3)} = \frac{N e^{\mu E_1}}{e^{\mu E_1} + e^{\mu E_2} + e^{\mu E_3}}$$

$$x_2 = \frac{N e^{\mu E_3}}{e^{\mu E_1} + e^{\mu E_2} + e^{\mu E_3}} \cdot e^{\mu (E_2 - E_3)} = \frac{N e^{\mu E_2}}{e^{\mu E_1} + e^{\mu E_2} + e^{\mu E_3}}$$

Letting $A = \frac{e^{\mu E_1} + e^{\mu E_2} + e^{\mu E_3}}{N}$, we obtain

$$x_1 = A^{-1} e^{\mu E_1}, \quad x_2 = A^{-1} e^{\mu E_2}, \quad x_3 = A^{-1} e^{\mu E_3}$$

The value of μ is determined by the second constraint $h(x_1, x_2, x_3) = 0$.

CHAPTER REVIEW EXERCISES

1. Given $f(x, y) = \dfrac{\sqrt{x^2 - y^2}}{x + 3}$:

(a) Sketch the domain of f.

(b) Calculate $f(3, 1)$ and $f(-5, -3)$.

(c) Find a point satisfying $f(x, y) = 1$.

SOLUTION

(a) f is defined where $x^2 - y^2 \geq 0$ and $x + 3 \neq 0$. We solve these two inequalities:

$$x^2 - y^2 \geq 0 \quad \Rightarrow \quad x^2 \geq y^2 \quad \Rightarrow \quad |x| \geq |y|$$

$$x + 3 \neq 0 \quad \Rightarrow \quad x \neq -3$$

Therefore, the domain of f is the following set:

$$D = \{(x, y) : |x| \geq |y|, x \neq -3\}$$

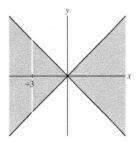

(b) To find $f(3, 1)$ we substitute $x = 3$, $y = 1$ in $f(x, y)$. We get

$$f(3, 1) = \frac{\sqrt{3^2 - 1^2}}{3 + 3} = \frac{\sqrt{8}}{6} = \frac{\sqrt{2}}{3}$$

Similarly, setting $x = -5$, $y = -3$, we get

$$f(-5, -3) = \frac{\sqrt{(-5)^2 - (-3)^2}}{-5 + 3} = \frac{\sqrt{16}}{-2} = -2$$

(c) We must find a point (x, y) such that

$$f(x, y) = \frac{\sqrt{x^2 - y^2}}{x + 3} = 1$$

We choose, for instance, $y = 1$, substitute and solve for x. This gives

$$\frac{\sqrt{x^2 - 1^2}}{x + 3} = 1$$

$$\sqrt{x^2 - 1} = x + 3$$

$$x^2 - 1 = (x + 3)^2 = x^2 + 6x + 9$$

$$6x = -10 \quad \Rightarrow \quad x = -\frac{5}{3}$$

Thus, the point $\left(-\frac{5}{3}, 1\right)$ satisfies $f\left(-\frac{5}{3}, 1\right) = 1$.

3. Sketch the graph $f(x, y) = x^2 - y + 1$ and describe its vertical and horizontal traces.

SOLUTION The graph is shown in the following figure.

The trace obtained by setting $x = c$ is the line $z = c^2 - y + 1$ or $z = (c^2 + 1) - y$ in the plane $x = c$. The trace obtained by setting $y = c$ is the parabola $z = x^2 - c + 1$ in the plane $y = c$. The trace obtained by setting $z = c$ is the parabola $y = x^2 + 1 - c$ in the plane $z = c$.

5. Match the functions (a)–(d) with their graphs in Figure 1.

(a) $f(x, y) = x^2 + y$
(b) $f(x, y) = x^2 + 4y^2$
(c) $f(x, y) = \sin(4xy)e^{-x^2-y^2}$
(d) $f(x, y) = \sin(4x)e^{-x^2-y^2}$

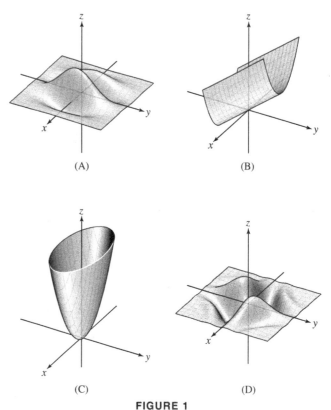

FIGURE 1

SOLUTION The function $f = x^2 + y$ matches picture (b), as can be seen by taking the $x = 0$ slice. The function $f = x^2 + 4y^2$ matches picture (c), as can be seen by taking $z = c$ slices (giving ellipses). Since $\sin(4xy)e^{-x^2-y^2}$ is symmetric with respect to x and y, and so also is picture (d), we match $\sin(4xy)e^{-x^2-y^2}$ with (d). That leaves the third function, $\sin(4x)e^{-x^2-y^2}$, to match with picture (a).

7. Describe the level curves of:

(a) $f(x, y) = e^{4x-y}$
(b) $f(x, y) = \ln(4x - y)$
(c) $f(x, y) = 3x^2 - 4y^2$
(d) $f(x, y) = x + y^2$

SOLUTION

(a) The level curves of $f(x, y) = e^{4x-y}$ are the curves $e^{4x-y} = c$ in the xy-plane, where $c > 0$. Taking ln from both sides we get $4x - y = \ln c$. Therefore, the level curves are the parallel lines of slope 4, $4x - y = \ln c$, $c > 0$, in the xy-plane.
(b) The level curves of $f(x, y) = \ln(4x - y)$ are the curves $\ln(4x - y) = c$ in the xy-plane. We rewrite it as $4x - y = e^c$ to obtain the parallel lines of slope 4, with negative y-intercepts.
(c) The level curves of $f(x, y) = 3x^2 - 4y^2$ are the hyperbolas $3x^2 - 4y^2 = c$ in the xy plane.

(d) The level curves of $f(x, y) = x + y^2$ are the curves $x + y^2 = c$ or $x = c - y^2$ in the xy-plane. These are parabolas whose axis is the x-axis.

In Exercises 9–14, evaluate the limit or state that it does not exist.

9. $\lim\limits_{(x,y) \to (1,-3)} (xy + y^2)$

SOLUTION The function $f(x, y) = xy + y^2$ is continuous everywhere because it is a polynomial, therefore we evaluate the limit using substitution:

$$\lim_{(x,y) \to (1,-3)} (xy + y^2) = 1 \cdot (-3) + (-3)^2 = 6$$

11. $\lim\limits_{(x,y) \to (0,0)} \dfrac{xy + xy^2}{x^2 + y^2}$

SOLUTION We evaluate the limits as (x, y) approaches the origin along the lines $y = x$ and $y = 2x$:

$$\lim_{\substack{(x,y) \to (0,0) \\ \text{along } y=x}} \frac{xy + xy^2}{x^2 + y^2} = \lim_{x \to 0} \frac{x \cdot x + x \cdot x^2}{x^2 + x^2} = \lim_{x \to 0} \frac{x^2 + x^3}{2x^2} = \lim_{x \to 0} \frac{1 + x}{2} = \frac{1}{2}$$

$$\lim_{\substack{x \to (0,0) \\ \text{along } y=2x}} \frac{xy + xy^2}{x^2 + y^2} = \lim_{x \to 0} \frac{x \cdot 2x + x \cdot (2x)^2}{x^2 + (2x)^2} = \lim_{x \to 0} \frac{2x^2 + 4x^3}{5x^2} = \lim_{x \to 0} \frac{2 + 4x}{5} = \frac{2}{5}$$

Since the two limits are different, $f(x, y)$ does not approach one limit as $(x, y) \to (0, 0)$, therefore the limit does not exist.

13. $\lim\limits_{(x,y) \to (1,-3)} (2x + y)e^{-x+y}$

SOLUTION The function $f(x, y) = (2x + y)e^{-x+y}$ is continuous, hence we evaluate the limit using substitution:

$$\lim_{(x,y) \to (1,-3)} (2x + y)e^{-x+y} = (2 \cdot 1 - 3)e^{-1-3} = -e^{-4}$$

15. Let

$$f(x, y) = \begin{cases} \dfrac{(xy)^p}{x^4 + y^4} & (x, y) \neq (0, 0) \\ 0 & (x, y) = (0, 0) \end{cases}$$

Use polar coordinates to show that $f(x, y)$ is continuous at all (x, y) if $p > 2$ but is discontinuous at $(0, 0)$ if $p \leq 2$.

SOLUTION We show using the polar coordinates $x = r\cos\theta$, $y = r\sin\theta$, that the limit of $f(x, y)$ as $(x, y) \to (0, 0)$ is zero for $p > 2$. This will prove that f is continuous at the origin. Since f is a rational function with nonzero denominator for $(x, y) \neq (0, 0)$, f is continuous there. We have

$$\lim_{(x,y) \to (0,0)} f(x, y) = \lim_{r \to 0+} \frac{(r\cos\theta)^p (r\sin\theta)^p}{(r\cos\theta)^4 + (r\sin\theta)^4} = \lim_{r \to 0+} \frac{r^{2p}(\cos\theta \sin\theta)^p}{r^4(\cos^4\theta + \sin^4\theta)} \quad (1)$$

$$= \lim_{r \to 0+} \frac{r^{2(p-2)}(\cos\theta \sin\theta)^p}{\cos^4\theta + \sin^4\theta}$$

We use the following inequalities:

$$|\cos^4\theta \sin^4\theta| \leq 1$$

$$\cos^4\theta + \sin^4\theta = (\cos^2\theta + \sin^2\theta)^2 - 2\cos^2\theta \sin^2\theta = 1 - \frac{1}{2} \cdot (2\cos\theta \sin\theta)^2$$

$$= 1 - \frac{1}{2}\sin^2 2\theta \geq 1 - \frac{1}{2} = \frac{1}{2}$$

Therefore,

$$0 \leq \left| \frac{r^{2(p-2)}(\cos\theta \sin\theta)^p}{\cos^4\theta + \sin^4\theta} \right| \leq \frac{r^{2(p-2)} \cdot 1}{\frac{1}{2}} = 2r^{2(p-2)}$$

Since $p - 2 > 0$, $\lim_{r \to 0+} 2r^{2(p-2)} = 0$, hence by the Squeeze Theorem the limit in (1) is also zero. We conclude that f is continuous for $p > 2$.

We now show that for $p < 2$ the limit of $f(x, y)$ as $(x, y) \to (0, 0)$ does not exist. We compute the limit as (x, y) approaches the origin along the line $y = x$.

$$\lim_{\substack{(x,y) \to (0,0) \\ \text{along } y=x}} f(x, y) = \lim_{x \to 0} \frac{(x^2)^p}{x^4 + x^4} = \lim_{x \to 0} \frac{x^{2p}}{2x^4} = \lim_{x \to 0} \frac{x^{2(p-2)}}{2} = \infty$$

Therefore the limit of $f(x, y)$ as $(x, y) \to (0, 0)$ does not exist for $p < 2$. We now show that the limit $\lim_{(x,y) \to (0,0)} \frac{x^2 y^2}{x^4 + y^4}$ does not exist for $p = 2$ as well. We compute the limits along the line $y = 0$ and $y = x$:

$$\lim_{\substack{(x,y) \to (0,0) \\ \text{along } y=0}} \frac{x^2 y^2}{x^4 + y^4} = \lim_{x \to 0} \frac{x^2 \cdot 0^2}{x^4 + 0^4} = \lim_{x \to 0} \frac{0}{x^4} = 0$$

$$\lim_{\substack{(x,y) \to (0,0) \\ \text{along } y=x}} \frac{x^2 y^2}{x^4 + y^4} = \lim_{x \to 0} \frac{x^2 \cdot x^2}{x^4 + x^4} = \lim_{x \to 0} \frac{x^4}{2x^4} = \frac{1}{2}$$

Since the limits along two paths are different, $f(x, y)$ does not approach one limit as $(x, y) \to (0, 0)$. We thus showed that if $p \le 2$, the limit $\lim_{(x,y) \to (0,0)} f(x, y)$ does not exist, and f is not continuous at the origin for $p \le 2$.

In Exercises 17–20, compute f_x and f_y.

17. $f(x, y) = 2x + y^2$

SOLUTION To find f_x we treat y as a constant, and to find f_y we treat x as a constant. We get

$$f_x = \frac{\partial}{\partial x}\left(2x + y^2\right) = \frac{\partial}{\partial x}(2x) + \frac{\partial}{\partial x}\left(y^2\right) = 2 + 0 = 2$$

$$f_y = \frac{\partial}{\partial y}\left(2x + y^2\right) = \frac{\partial}{\partial y}(2x) + \frac{\partial}{\partial y}\left(y^2\right) = 0 + 2y = 2y$$

19. $f(x, y) = \sin(xy)e^{-x-y}$

SOLUTION We compute f_x, treating y as a constant and using the Product Rule and the Chain Rule. We get

$$f_x = \frac{\partial}{\partial x}\left(\sin(xy)e^{-x-y}\right) = \frac{\partial}{\partial x}(\sin(xy))\, e^{-x-y} + \sin(xy)\frac{\partial}{\partial x}e^{-x-y}$$
$$= \cos(xy) \cdot ye^{-x-y} + \sin(xy) \cdot (-1)e^{-x-y} = e^{-x-y}(y\cos(xy) - \sin(xy))$$

We compute f_y similarly, treating x as a constant. Notice that since $f(y, x) = f(x, y)$, the partial derivative f_y can be obtained from f_x by interchanging x and y. That is,

$$f_y = e^{-x-y}(x\cos(yx) - \sin(yx))$$

21. Calculate f_{xxyz} for $f(x, y, z) = y\sin(x + z)$.

SOLUTION We differentiate f twice with respect to x, once with respect to y, and finally with respect to z. This gives

$$f_x = \frac{\partial}{\partial x}(y\sin(x+z)) = y\cos(x+z)$$

$$f_{xx} = \frac{\partial}{\partial x}(y\cos(x+z)) = -y\sin(x+z)$$

$$f_{xxy} = \frac{\partial}{\partial y}(-y\sin(x+z)) = -\sin(x+z)$$

$$f_{xxyz} = \frac{\partial}{\partial z}(-\sin(x+z)) = -\cos(x+z)$$

23. Find an equation of the tangent plane to the graph of $f(x, y) = xy^2 - xy + 3x^3y$ at $P = (1, 3)$.

SOLUTION The tangent plane has the equation

$$z = f(1,3) + f_x(1,3)(x-1) + f_y(1,3)(y-3) \tag{1}$$

We compute the partial derivatives of $f(x,y) = xy^2 - xy + 3x^3y$:

$$f_x(x,y) = y^2 - y + 9x^2y \qquad f_x(1,3) = 3^2 - 3 + 9 \cdot 1^2 \cdot 3 = 33$$
$$f_y(x,y) = 2xy - x + 3x^3 \qquad f_y(1,3) = 2 \cdot 1 \cdot 3 - 1 + 3 \cdot 1^3 = 8$$

Also, $f(1,3) = 1 \cdot 3^2 - 1 \cdot 3 + 3 \cdot 1^3 \cdot 3 = 15$. Substituting these values in (1), we obtain the following equation:

$$z = 15 + 33(x-1) + 8(y-3)$$

or

$$z = 33x + 8y - 42$$

25. Use a Linear Approximation of $f(x,y,z) = \sqrt{x^2 + y^2 + z}$ to estimate $\sqrt{7.1^2 + 4.9^2 + 69.5}$. Compare with a calculator value.

SOLUTION The function whose value we want to approximate is

$$f(x,y,z) = \sqrt{x^2 + y^2 + z}$$

We will use the linear approximation at the point $(7,5,70)$. Recall that the linear approximation to a surface will be:

$$L(x,y,z) = f(7,5,70) + f_x(7,5,70)(x-7) + f_y(7,5,70)(y-5) + f_z(7,5,70)(z-70)$$

We compute the partial derivatives of f:

$$f_x(x,y,z) = \frac{2x}{2\sqrt{x^2+y^2+z}} = \frac{x}{\sqrt{x^2+y^2+z}} \quad \Rightarrow \quad f_x(7,5,70) = \frac{7}{\sqrt{7^2+5^2+70}} = \frac{7}{12}$$

$$f_y(x,y,z) = \frac{2y}{2\sqrt{x^2+y^2+z}} = \frac{y}{\sqrt{x^2+y^2+z}} \quad \Rightarrow \quad f_y(7,5,70) = \frac{5}{\sqrt{7^2+5^2+70}} = \frac{5}{12}$$

$$f_z(x,y,z) = \frac{1}{2\sqrt{x^2+y^2+z}} \quad \Rightarrow \quad f_z(7,5,70) = \frac{1}{2\sqrt{7^2+5^2+70}} = \frac{1}{24}$$

Also, $f(7,5,70) = \sqrt{7^2 + 5^2 + 70} = 12$. Substituting the values in the linear approximation equation we obtain the following approximation:

$$L(x,y,z) = 12 + \frac{7}{12}(x-7) + \frac{5}{12}(y-5) + \frac{1}{24}(z-70)$$

Now we are ready to approximate $\sqrt{7.1^2 + 4.9^2 + 69.5}$. That is, using the linear approximation,

$$L(7.1, 4.9, 69.5) = 12 + \frac{7}{12}(7.1-7) + \frac{5}{12}(4.9-5) + \frac{1}{24}(69.5-70)$$

$$= 12 + \frac{7}{12} \cdot \frac{1}{10} + \frac{5}{12} \cdot -\frac{1}{10} + \frac{1}{24} \cdot -\frac{1}{2}$$

$$= 12 + \frac{7}{120} - \frac{5}{120} - \frac{1}{48}$$

$$= \frac{2879}{240} = 11.9958333$$

The value obtained using a calculator is 11.996667.

27. Figure 4 shows the contour map of a function $f(x,y)$ together with a path $\mathbf{r}(t)$ in the counterclockwise direction. The points $\mathbf{r}(1)$, $\mathbf{r}(2)$, and $\mathbf{r}(3)$ are indicated on the path. Let $g(t) = f(\mathbf{r}(t))$. Which of statements (i)–(iv) are true? Explain.

(i) $g'(1) > 0$.

(ii) $g(t)$ has a local minimum for some $1 \leq t \leq 2$.
(iii) $g'(2) = 0$.
(iv) $g'(3) = 0$.

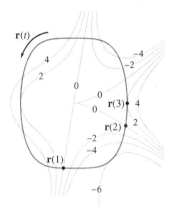

FIGURE 4

SOLUTION (ii) and (iv) are true.

In Exercises 29–32, compute $\dfrac{d}{dt} f(\mathbf{r}(t))$ at the given value of t.

29. $f(x, y) = x + e^y$, $\mathbf{r}(t) = \left\langle 3t - 1, t^2 \right\rangle$ at $t = 2$

SOLUTION By the Chain Rule for Paths we have

$$\frac{d}{dt} f(\mathbf{c}(t)) = \nabla f \cdot \mathbf{c}'(t) \tag{1}$$

We evaluate the gradient ∇f and $\mathbf{c}'(t)$:

$$\mathbf{c}'(t) = \langle 3, 2t \rangle$$

$$\nabla f = \left\langle f_x, f_y \right\rangle = \langle 1, e^y \rangle \quad \Rightarrow \quad \nabla f_{\mathbf{c}(t)} = \left\langle 1, e^{t^2} \right\rangle$$

Substituting in (1) we get

$$\frac{d}{dt} f(\mathbf{c}(t)) = \left\langle 1, e^{t^2} \right\rangle \cdot \langle 3, 2t \rangle = 3 + 2te^{t^2}$$

At $t = 2$ we have

$$\left. \frac{d}{dt} f(\mathbf{c}(t)) \right|_{t=2} = 3 + 2 \cdot 2 \cdot e^{2^2} = 3 + 4e^4 \approx 221.4$$

31. $f(x, y) = xe^{3y} - ye^{3x}$, $\mathbf{r}(t) = \langle e^t, \ln t \rangle$ at $t = 1$

SOLUTION We use the Chain Rule for Paths:

$$\frac{d}{dt} f(\mathbf{c}(t)) = \nabla f_{\mathbf{c}(t)} \cdot \mathbf{c}'(t) \tag{1}$$

We find the ∇f at the point $\mathbf{c}(1)$ and compute $\mathbf{c}'(1)$. We get

$$\nabla f = \left\langle f_x, f_y \right\rangle = \left\langle e^{3y} - 3ye^{3x}, 3xe^{3y} - e^{3x} \right\rangle$$

$$\mathbf{c}(1) = \left\langle e^1, \ln 1 \right\rangle = \langle e, 0 \rangle$$

$$\nabla f_{\mathbf{c}(1)} = \left\langle e^{3 \cdot 0} - 3 \cdot 0 e^{3e}, 3ee^{3 \cdot 0} - e^{3e} \right\rangle = \left\langle 1, 3e - e^{3e} \right\rangle \tag{2}$$

$$\mathbf{c}'(t) = \frac{d}{dt} \left\langle e^t, \ln t \right\rangle = \left\langle e^t, t^{-1} \right\rangle \quad \Rightarrow \quad \mathbf{c}'(1) = \langle e, 1 \rangle \tag{3}$$

Substituting (2) and (3) in (1) gives

$$\left. \frac{d}{dt} f(\mathbf{c}(t)) \right|_{t=1} = \nabla f_{c(1)} \cdot \mathbf{c}'(1) = \left\langle 1, 3e - e^{3e} \right\rangle \cdot \langle e, 1 \rangle = e + 3e - e^{3e} = 4e - e^{3e}$$

In Exercises 33–36, compute the directional derivative at P in the direction of **v**.

33. $f(x, y) = x^3 y^4$, $\quad P = (3, -1)$, $\quad \mathbf{v} = 2\mathbf{i} + \mathbf{j}$

SOLUTION We first normalize **v** to find a unit vector **u** in the direction of **v**:

$$\mathbf{u} = \frac{\mathbf{v}}{\|\mathbf{v}\|} = \frac{2\mathbf{i} + \mathbf{j}}{\sqrt{2^2 + 1^2}} = \frac{2}{\sqrt{5}}\mathbf{i} + \frac{1}{\sqrt{5}}\mathbf{j}$$

We compute the directional derivative using the following equality:

$$D_\mathbf{u} f(3, -1) = \nabla f_{(3,-1)} \cdot \mathbf{u}$$

The gradient vector at the given point is the following vector:

$$\nabla f = \langle f_x, f_y \rangle = \langle 3x^2 y^4, 4x^3 y^3 \rangle \quad \Rightarrow \quad \nabla f_{(3,-1)} = \langle 27, -108 \rangle$$

Hence,

$$D_\mathbf{u} f(3, -1) = \langle 27, -108 \rangle \cdot \left\langle \frac{2}{\sqrt{5}}, \frac{1}{\sqrt{5}} \right\rangle = \frac{54}{\sqrt{5}} - \frac{108}{\sqrt{5}} = -\frac{54}{\sqrt{5}}$$

35. $f(x, y) = e^{x^2 + y^2}$, $\quad P = \left(\frac{\sqrt{2}}{2}, \frac{\sqrt{2}}{2}\right)$, $\quad \mathbf{v} = \langle 3, -4 \rangle$

SOLUTION We normalize **v** to obtain a vector **u** in the direction of **v**:

$$\mathbf{u} = \frac{\langle 3, -4 \rangle}{\sqrt{3^2 + (-4)^2}} = \left\langle \frac{3}{5}, -\frac{4}{5} \right\rangle$$

We use the following theorem:

$$D_\mathbf{u} f(P) = \nabla f_P \cdot \mathbf{u} \tag{1}$$

We find the gradient of f at the given point:

$$\nabla f = \langle f_x, f_y \rangle = \langle 2x e^{x^2 + y^2}, 2y e^{x^2 + y^2} \rangle = 2 e^{x^2 + y^2} \langle x, y \rangle$$

Hence,

$$\nabla f_P = 2 e^{\left(\frac{\sqrt{2}}{2}\right)^2 + \left(\frac{\sqrt{2}}{2}\right)^2} \left\langle \frac{\sqrt{2}}{2}, \frac{\sqrt{2}}{2} \right\rangle = e\sqrt{2} \langle 1, 1 \rangle$$

Substituting in (1) we get

$$D_\mathbf{u} f(P) = \sqrt{2} e \langle 1, 1 \rangle \cdot \left\langle \frac{3}{5}, -\frac{4}{5} \right\rangle = \sqrt{2} e \left(\frac{3}{5} - \frac{4}{5}\right) = -\frac{\sqrt{2} e}{5}$$

37. Find the unit vector **e** at $P = (0, 0, 1)$ pointing in the direction along which $f(x, y, z) = xz + e^{-x^2 + y}$ increases most rapidly.

SOLUTION The gradient vector ∇f_P points in the direction of maximum rate of increase of f. Therefore we need to find a unit vector in the direction of ∇f_P. We first find the gradient of $f(x, y, z) = xz + e^{-x^2 + y}$:

$$\nabla f = \left\langle \frac{\partial f}{\partial x}, \frac{\partial f}{\partial y}, \frac{\partial f}{\partial z} \right\rangle = \left\langle z - 2x e^{-x^2 + y}, e^{-x^2 + y}, x \right\rangle$$

At the point $P = (0, 0, 1)$ we have

$$\nabla f_P = \langle 1, 1, 0 \rangle$$

We normalize ∇f_P to obtain the unit vector **e** at P pointing in the direction of maximum increase of f:

$$\mathbf{e} = \frac{\nabla f_P}{\|\nabla f_P\|} = \left\langle \frac{1}{\sqrt{2}}, \frac{1}{\sqrt{2}}, 0 \right\rangle$$

39. Let $n \neq 0$ be an integer and r an arbitrary constant. Show that the tangent plane to the surface $x^n + y^n + z^n = r$ at $P = (a, b, c)$ has equation

$$a^{n-1}x + b^{n-1}y + c^{n-1}z = r$$

SOLUTION The tangent plane to the surface, defined implicitly by $F(x, y, z) = r$ at a point (a, b, c) on the surface, has the following equation:

$$0 = F_x(a, b, c)(x - a) + F_y(a, b, c)(y - b) + F_z(a, b, c)(z - c) \tag{1}$$

The given surface is defined by the function $F(x, y, z) = x^n + y^n + z^n$. We find the partial derivative of F at a point $P = (a, b, c)$ on the surface:

$$F_x(x, y, z) = nx^{n-1} \qquad F_x(a, b, c) = na^{n-1}$$
$$F_y(x, y, z) = ny^{n-1} \quad \Rightarrow \quad F_y(a, b, c) = nb^{n-1}$$
$$F_z(x, y, z) = nz^{n-1} \qquad F_z(a, b, c) = nc^{n-1}$$

Substituting in (1) we get

$$na^{n-1}(x - a) + nb^{n-1}(y - b) + nc^{n-1}(z - c) = 0$$

We divide by n and simplify:

$$a^{n-1}x - a^n + b^{n-1}y - b^n + c^{n-1}z - c^n = 0$$
$$a^{n-1}x + b^{n-1}y + c^{n-1}z = a^n + b^n + c^n \tag{2}$$

The point $P = (a, b, c)$ lies on the surface, hence it satisfies the equation of the surface. That is,

$$a^n + b^n + c^n = r$$

Substituting in (2) we obtain the following equation of the tangent plane:

$$a^{n-1}x + b^{n-1}y + c^{n-1}z = r$$

41. Let $f(x, y, z) = x^2y + y^2z$. Use the Chain Rule to calculate $\partial f/\partial s$ and $\partial f/\partial t$ (in terms of s and t), where

$$x = s + t, \quad y = st, \quad z = 2s - t$$

SOLUTION We compute the Primary Derivatives:

$$\frac{\partial f}{\partial x} = 2xy, \quad \frac{\partial f}{\partial y} = x^2 + 2yz, \quad \frac{\partial f}{\partial z} = y^2$$

Since $\frac{\partial x}{\partial s} = 1$, $\frac{\partial y}{\partial s} = t$, $\frac{\partial z}{\partial s} = 2$, $\frac{\partial x}{\partial t} = 1$, $\frac{\partial y}{\partial t} = s$, and $\frac{\partial z}{\partial t} = -1$, the Chain Rule gives

$$\frac{\partial f}{\partial s} = \frac{\partial f}{\partial x}\frac{\partial x}{\partial s} + \frac{\partial f}{\partial y}\frac{\partial y}{\partial s} + \frac{\partial f}{\partial z}\frac{\partial z}{\partial s} = 2xy \cdot 1 + \left(x^2 + 2yz\right)t + y^2 \cdot 2$$
$$= 2xy + \left(x^2 + 2yz\right)t + 2y^2$$
$$\frac{\partial f}{\partial t} = \frac{\partial f}{\partial x}\frac{\partial x}{\partial t} + \frac{\partial f}{\partial y}\frac{\partial y}{\partial t} + \frac{\partial f}{\partial z}\frac{\partial z}{\partial t} = 2xy \cdot 1 + \left(x^2 + 2yz\right)s + y^2 \cdot (-1)$$
$$= 2xy + \left(x^2 + 2yz\right)s - y^2$$

We now substitute $x = s + t$, $y = st$, and $z = 2s - t$ to express the answer in terms of the independent variables s, t. We get

$$\frac{\partial f}{\partial s} = 2(s + t)st + \left((s + t)^2 + 2st(2s - t)\right)t + 2s^2t^2$$
$$= 2s^2t + 2st^2 + \left(s^2 + 2st + t^2 + 4s^2t - 2st^2\right)t + 2s^2t^2$$
$$= 3s^2t + 4st^2 + t^3 - 2st^3 + 6s^2t^2$$
$$\frac{\partial f}{\partial t} = 2(s + t)st + \left((s + t)^2 + 2st(2s - t)\right)s - s^2t^2$$

$$= 2s^2t + 2st^2 + \left(s^2 + 2st + t^2 + 4s^2t - 2st^2\right)s - s^2t^2$$
$$= 4s^2t + 3st^2 + s^3 + 4s^3t - 3s^2t^2$$

43. Let $g(u, v) = f(u^3 - v^3, v^3 - u^3)$. Prove that

$$v^2 \frac{\partial g}{\partial u} + u^2 \frac{\partial g}{\partial v} = 0$$

SOLUTION We are given the function $f(x, y)$, where $x = u^3 - v^3$ and $y = v^3 - u^3$. Using the Chain Rule we have the following derivatives:

$$\frac{\partial g}{\partial u} = \frac{\partial f}{\partial x} \frac{\partial x}{\partial u} + \frac{\partial f}{\partial y} \frac{\partial y}{\partial u}$$

$$\frac{\partial g}{\partial v} = \frac{\partial f}{\partial x} \frac{\partial x}{\partial v} + \frac{\partial f}{\partial y} \frac{\partial y}{\partial v} \qquad (1)$$

We compute the following partial derivatives:

$$\frac{\partial x}{\partial u} = 3u^2, \quad \frac{\partial y}{\partial u} = -3u^2$$

$$\frac{\partial x}{\partial v} = -3v^2, \quad \frac{\partial y}{\partial v} = 3v^2$$

Substituting in (1) we obtain

$$\frac{\partial g}{\partial u} = \frac{\partial f}{\partial x} \cdot 3u^2 + \frac{\partial f}{\partial y}\left(-3u^2\right) = 3u^2\left(\frac{\partial f}{\partial x} - \frac{\partial f}{\partial y}\right)$$

$$\frac{\partial g}{\partial v} = \frac{\partial f}{\partial x}\left(-3v^2\right) + \frac{\partial f}{\partial y}\left(3v^2\right) = -3v^2\left(\frac{\partial f}{\partial x} - \frac{\partial f}{\partial y}\right)$$

Therefore,

$$v^2 \frac{\partial g}{\partial u} + u^2 \frac{\partial g}{\partial v} = 3u^2v^2\left(\frac{\partial f}{\partial x} - \frac{\partial f}{\partial y}\right) - 3u^2v^2\left(\frac{\partial f}{\partial x} - \frac{\partial f}{\partial y}\right) = 0$$

45. Calculate $\partial z/\partial x$, where $xe^z + ze^y = x + y$.

SOLUTION The function $F(x, y, z) = xe^z + ze^y - x - y = 0$ defines z implicitly as a function of x and y. Using implicit differentiation, the partial derivative of z with respect to x is

$$\frac{\partial z}{\partial x} = -\frac{F_x}{F_z} \qquad (1)$$

We compute the partial derivatives F_x and F_z:

$$F_x = e^z - 1$$
$$F_z = xe^z + e^y$$

Substituting in (1) gives

$$\frac{\partial z}{\partial x} = -\frac{e^z - 1}{xe^z + e^y}.$$

In Exercises 47–50, find the critical points of the function and analyze them using the Second Derivative Test.

47. $f(x, y) = x^4 - 4xy + 2y^2$

SOLUTION To find the critical points, we need the first-order partial derivatives and set them equal to zero to solve for x and y:

$$f_x(x, y) = 4x^3 - 4y = 0, \quad f_y(x, y) = -4x + 4y = 0$$

Looking at the second equation we see $x = y$. Using this in the first equation, then

$$4x^3 - 4x = 0 \quad \Rightarrow \quad 4x(x^2 - 1) = 0 \quad \Rightarrow \quad x = 0, \pm 1$$

Therefore, our critical points are:

$$(0, 0), (1, 1), (-1, -1)$$

Now to find the discriminant, D, we need the second-order partial derivatives:

$$f_{xx}(x, y) = 12x^2, \quad f_{yy}(x, y) = 4, \quad f_{xy}(x, y) = -4$$

Hence,

$$D(x, y) = f_{xx}f_{yy} - f_{xy}^2 = 48x^2 - 16 = 16(3x^2 - 1)$$

Analyzing our three critical points we see:

$$D(0, 0) = -16 < 0, \quad D(1, 1) = 32 > 0, \quad D(-1, -1) = 32 > 0$$

Since the discriminant for $(0, 0)$ is negative, $(0, 0)$ is a saddle point.
   Looking at $f_{xx}(1, 1) = 12 > 0$ and $f_{xx}(-1, -1) = 12 > 0$ hence, the points $(1, 1)$ and $(-1, -1)$ are both local minima.

49. $f(x, y) = e^{x+y} - xe^{2y}$

SOLUTION We find the critical point by setting the partial derivatives of $f(x, y) = e^{x+y} - xe^{2y}$ equal to zero and solve. This gives

$$f_x(x, y) = e^{x+y} - e^{2y} = 0$$
$$f_y(x, y) = e^{x+y} - 2xe^{2y} = 0$$

The first equation gives $e^{x+y} = e^{2y}$ and the second equation gives $e^{x+y} = 2xe^{2y}$. Equating the two expressions, dividing by the nonzero function e^{2y}, and solving for x, we obtain

$$e^{2y} = 2xe^{2y} \quad \Rightarrow \quad 1 = 2x \quad \Rightarrow \quad x = \frac{1}{2}$$

We now substitute $x = \frac{1}{2}$ in the first equation and solve for y, to obtain

$$e^{\frac{1}{2}+y} - e^{2y} = 0 \quad \Rightarrow \quad e^{\frac{1}{2}+y} = e^{2y} \quad \Rightarrow \quad \frac{1}{2} + y = 2y \quad \Rightarrow \quad y = \frac{1}{2}$$

There is one critical point, $\left(\frac{1}{2}, \frac{1}{2}\right)$. We examine the critical point using the Second Derivative Test. We compute the second derivatives at this point:

$$f_{xx}(x, y) = e^{x+y} \quad \Rightarrow \quad f_{xx}\left(\frac{1}{2}, \frac{1}{2}\right) = e^{\frac{1}{2}+\frac{1}{2}} = e$$

$$f_{yy}(x, y) = e^{x+y} - 4xe^{2y} \quad \Rightarrow \quad f_{yy}\left(\frac{1}{2}, \frac{1}{2}\right) = e^{\frac{1}{2}+\frac{1}{2}} - 4 \cdot \frac{1}{2}e^{2 \cdot \frac{1}{2}} = -e$$

$$f_{xy}(x, y) = e^{x+y} - 2e^{2y} \quad \Rightarrow \quad f_{xy}\left(\frac{1}{2}, \frac{1}{2}\right) = e^{\frac{1}{2}+\frac{1}{2}} - 2e^{2 \cdot \frac{1}{2}} = -e$$

Therefore the discriminant at the critical point is

$$D\left(\frac{1}{2}, \frac{1}{2}\right) = f_{xx}f_{yy} - f_{xy}^2 = e \cdot (-e) - (-e)^2 = -2e^2 < 0$$

We conclude that $\left(\frac{1}{2}, \frac{1}{2}\right)$ is a saddle point.

51. Prove that $f(x, y) = (x + 2y)e^{xy}$ has no critical points.

SOLUTION We find the critical points by setting the partial derivatives of $f(x, y) = (x + 2y)e^{xy}$ equal to zero and solving. We get

$$f_x(x, y) = e^{xy} + (x + 2y)ye^{xy} = e^{xy}\left(1 + xy + 2y^2\right) = 0$$

$$f_y(x,y) = 2e^{xy} + (x+2y)xe^{xy} = e^{xy}\left(2 + x^2 + 2xy\right) = 0$$

We divide the two equations by the nonzero expression e^{xy} to obtain the following equations:

$$1 + xy + 2y^2 = 0$$
$$2 + 2xy + x^2 = 0$$

The first equation implies that $xy = -1 - 2y^2$. Substituting in the second equation gives

$$2 + 2\left(-1 - 2y^2\right) + x^2 = 0$$
$$2 - 2 - 4y^2 + x^2 = 0$$
$$x^2 = 4y^2 \quad \Rightarrow \quad x = 2y \quad \text{or} \quad x = -2y$$

We substitute in the first equation and solve for y:

$x = 2y$	$x = -2y$
$1 + 2y^2 + 2y^2 = 0$	$1 - 2y^2 + 2y^2 = 0$
$1 + 4y^2 = 0$	$1 = 0$
$y^2 = -\frac{1}{4}$	

In both cases there is no solution. We conclude that there are no solutions for $f_x = 0$ and $f_y = 0$, that is, there are no critical points.

53. Find the global extrema of $f(x, y) = 2xy - x - y$ on the domain $\{y \le 4, y \ge x^2\}$.

SOLUTION The region is shown in the figure.

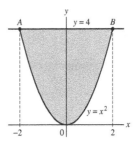

Step 1. Finding the critical points. We find the critical points in the interior of the domain by setting the partial derivatives equal to zero and solving. We get

$$f_x = 2y - 1 = 0$$
$$f_y = 2x - 1 = 0 \quad \Rightarrow \quad x = \frac{1}{2}, \quad y = \frac{1}{2}$$

The critical point is $\left(\frac{1}{2}, \frac{1}{2}\right)$. (It lies in the interior of the domain since $\frac{1}{2} < 4$ and $\frac{1}{2} > \left(\frac{1}{2}\right)^2$.)

Step 2. Finding the global extrema on the boundary. We consider the two parts of the boundary separately.

The parabola $y = x^2$, $-2 \le x \le 2$:

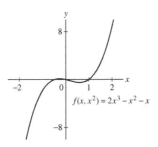

On this curve, $f(x, x^2) = 2 \cdot x \cdot x^2 - x - x^2 = 2x^3 - x^2 - x$. Using calculus in one variable or the graph of the function, we see that the minimum of $f(x, x^2)$ on the interval occurs at $x = -2$ and the maximum at $x = 2$. The corresponding points are $(-2, 4)$ and $(2, 4)$.

The segment $\overline{AB}$: On this segment $y = 4$, $-2 \leq x \leq 2$, hence $f(x, 4) = 2 \cdot x \cdot 4 - x - 4 = 7x - 4$. The maximum value occurs at $x = 2$ and the minimum value at $x = -2$. The corresponding points on the segment $\overline{AB}$ are $(-2, 4)$ and $(2, 4)$

Step 3. Conclusions. Since the global extrema occur either at critical points in the interior of the domain or on the boundary of the domain, the candidates for global extrema are the following points:

$$\left(\frac{1}{2}, \frac{1}{2}\right), \quad (-2, 4), \quad (2, 4)$$

We compute the values of $f = 2xy - x - y$ at these points:

$$f\left(\frac{1}{2}, \frac{1}{2}\right) = 2 \cdot \frac{1}{2} \cdot \frac{1}{2} - \frac{1}{2} - \frac{1}{2} = -\frac{1}{2}$$

$$f(-2, 4) = 2 \cdot (-2) \cdot 4 + 2 - 4 = -18$$

$$f(2, 4) = 2 \cdot 2 \cdot 4 - 2 - 4 = 10$$

We conclude that the global maximum is $f(2, 4) = 10$ and the global minimum is $f(-2, 4) = -18$.

55. Use Lagrange multipliers to find the minimum and maximum values of $f(x, y) = 3x - 2y$ on the circle $x^2 + y^2 = 4$.

SOLUTION

Step 1. Write out the Lagrange equations. The constraint curve is $g(x, y) = x^2 + y^2 - 4 = 0$, hence $\nabla g = \langle 2x, 2y \rangle$ and $\nabla f = \langle 3, -2 \rangle$. The Lagrange condition $\nabla f = \lambda \nabla g$ is thus $\langle 3, -2 \rangle = \lambda \langle 2x, 2y \rangle$. That is,

$$3 = \lambda \cdot 2x$$
$$-2 = \lambda \cdot 2y$$

Note that $\lambda \neq 0$.

Step 2. Solve for x and y using the constraint. The Lagrange equations gives

$$\begin{aligned} 3 &= \lambda \cdot 2x \\ -2 &= \lambda \cdot 2y \end{aligned} \quad \Rightarrow \quad \begin{aligned} x &= \frac{3}{2\lambda} \\ y &= -\frac{1}{\lambda} \end{aligned} \tag{1}$$

We substitute x and y in the equation of the constraint and solve for λ. We get

$$\left(\frac{3}{2\lambda}\right)^2 + \left(-\frac{1}{\lambda}\right)^2 = 4$$

$$\frac{9}{4\lambda^2} + \frac{1}{\lambda^2} = 4$$

$$\frac{1}{\lambda^2} \cdot \frac{13}{4} = 4 \quad \Rightarrow \quad \lambda = \frac{\sqrt{13}}{4} \quad \text{or} \quad \lambda = -\frac{\sqrt{13}}{4}$$

Substituting in (1), we obtain the points

$$x = \frac{6}{\sqrt{13}}, \quad y = -\frac{4}{\sqrt{13}}$$

$$x = -\frac{6}{\sqrt{13}}, \quad y = \frac{4}{\sqrt{13}}$$

The critical points are thus

$$P_1 = \left(\frac{6}{\sqrt{13}}, -\frac{4}{\sqrt{13}}\right)$$

$$P_2 = \left(-\frac{6}{\sqrt{13}}, \frac{4}{\sqrt{13}}\right)$$

Step 3. Calculate the value at the critical points. We find the value of $f(x,y) = 3x - 2y$ at the critical points:

$$f(P_1) = 3 \cdot \frac{6}{\sqrt{13}} - 2 \cdot \frac{-4}{\sqrt{13}} = \frac{26}{\sqrt{13}}$$

$$f(P_2) = 3 \cdot \frac{-6}{\sqrt{13}} - 2 \cdot \frac{4}{\sqrt{13}} = \frac{-26}{\sqrt{13}}$$

Thus, the maximum value of f on the circle is $\frac{26}{\sqrt{13}}$, and the minimum is $-\frac{26}{\sqrt{13}}$.

57. Find the minimum and maximum values of $f(x,y) = x^2y$ on the ellipse $4x^2 + 9y^2 = 36$.

SOLUTION We must find the minimum and maximum values of $f(x,y) = x^2y$ subject to the constraint $g(x,y) = 4x^2 + 9y^2 - 36 = 0$.

Step 1. Write out the Lagrange equations. The gradient vectors are $\nabla f = \langle 2xy, x^2 \rangle$ and $\nabla g = \langle 8x, 18y \rangle$, hence the Lagrange condition $\nabla f = \lambda \nabla g$ gives

$$\langle 2xy, x^2 \rangle = \lambda \langle 8x, 18y \rangle = \langle 8\lambda x, 18\lambda y \rangle$$

We obtain the following Lagrange equations:

$$2xy = 8\lambda x$$
$$x^2 = 18\lambda y$$

Step 2. Solve for λ in terms of x and y. If $x = 0$, the equation of the constraint implies that $y = \pm 2$. The points $(0, 2)$ and $(0, -2)$ satisfy the Lagrange equations for $\lambda = 0$. If $x \neq 0$, the second Lagrange equation implies that $y \neq 0$. Therefore the Lagrange equations give

$$2xy = 8\lambda x \quad \Rightarrow \quad \lambda = \frac{y}{4}$$

$$x^2 = 18\lambda y \quad \Rightarrow \quad \lambda = \frac{x^2}{18y}$$

Step 3. Solve for x and y using the constraint. We equate the two expressions for λ to obtain

$$\frac{y}{4} = \frac{x^2}{18y}$$

$$18y^2 = 4x^2$$

We now substitute $4x^2 = 18y^2$ in the equation of the constraint $4x^2 + 9y^2 = 36$ and solve for y. This gives

$$18y^2 + 9y^2 = 36 \quad \Rightarrow \quad y^2 = \frac{36}{27} \quad \Rightarrow \quad y_1 = \frac{2}{\sqrt{3}}, \quad y_2 = -\frac{2}{\sqrt{3}}$$

$$27y^2 = 36$$

We find the x-coordinates using $x^2 = \frac{9y^2}{2}$:

$$x^2 = \frac{9y^2}{2}$$

$$x^2 = \frac{9}{2} \cdot \frac{4}{3} = 6 \quad \Rightarrow \quad x_1 = \sqrt{6}, \quad x_2 = -\sqrt{6}$$

We obtain the following critical points:

$$P_1 = (0, 2), \quad P_2 = (0, -2), \quad P_3 = \left(\sqrt{6}, \frac{2}{\sqrt{3}}\right)$$

$$P_4 = \left(\sqrt{6}, -\frac{2}{\sqrt{3}}\right), \quad P_5 = \left(-\sqrt{6}, \frac{2}{\sqrt{3}}\right), \quad P_6 = \left(-\sqrt{6}, -\frac{2}{\sqrt{3}}\right)$$

Step 4. Conclusions. We evaluate the function $f(x,y) = x^2y$ at the critical points:

$$f(P_1) = 0^2 \cdot 2 = 0$$

$$f(P_2) = 0^2 \cdot (-2) = 0$$

$$f(P_3) = f(P_5) = 6 \cdot \frac{2}{\sqrt{3}} = \frac{12}{\sqrt{3}}$$

$$f(P_4) = f(P_5) = 6 \cdot \left(-\frac{2}{\sqrt{3}}\right) = -\frac{12}{\sqrt{3}}$$

Since the min and max of f occur on the ellipse, it must occur at critical points. Thus, we conclude that the maximum and minimum of f subject to the constraint are $\frac{12}{\sqrt{3}}$ and $-\frac{12}{\sqrt{3}}$, respectively.

59. Find the extreme values of $f(x, y, z) = x + 2y + 3z$ subject to the two constraints $x + y + z = 1$ and $x^2 + y^2 + z^2 = 1$.

SOLUTION We must find the extreme values of $f(x, y, z) = x + 2y + 3z$ subject to the constraints $g(x, y, z) = x + y + z - 1 = 0$ and $h(x, y, z) = x^2 + y^2 + z^2 - 1 = 0$.

Step 1. Write out the Lagrange equations. We have $\nabla f = <1, 2, 3>$, $\nabla g = <1, 1, 1>$, $\nabla h = <2x, 2y, 2z>$, hence the Lagrange condition $\nabla f = \lambda \nabla g + \mu \nabla h$ gives

$$<1, 2, 3> = \lambda <1, 1, 1> + \mu <2x, 2y, 2z> = <\lambda + 2\mu x, \lambda + 2\mu y, \lambda + 2\mu z>$$

or

$$1 = \lambda + 2\mu x$$
$$2 = \lambda + 2\mu y$$
$$3 = \lambda + 2\mu z$$

Step 2. Solve for λ and μ. The Lagrange equations give

$$\begin{array}{ll} 1 = \lambda + 2\mu x & \lambda = 1 - 2\mu x \\ 2 = \lambda + 2\mu y \Rightarrow & \lambda = 2 - 2\mu y \\ 3 = \lambda + 2\mu z & \lambda = 3 - 2\mu z \end{array}$$

Equating the three expressions for λ, we get the following equations:

$$\begin{array}{l} 1 - 2\mu x = 2 - 2\mu y \\ 1 - 2\mu x = 3 - 2\mu z \end{array} \Rightarrow \begin{array}{l} 2\mu(y - x) = 1 \\ \mu(z - x) = 2 \end{array}$$

The first equation implies that $\mu = \frac{1}{2(y-x)}$, and the second implies that $\mu = \frac{2}{z-x}$. Equating the two expressions for μ, we get

$$\frac{1}{2(y-x)} = \frac{2}{z-x}$$

$$z - x = 4y - 4x \quad \Rightarrow \quad z = 4y - 3x$$

Step 3. Solve for x, y, and z using the constraints. We substitute $z = 4y - 3x$ in the equations of the constraints and solve to find x and y. This gives

$$\begin{array}{l} x + y + (4y - 3x) = 1 \\ x^2 + y^2 + (4y - 3x)^2 = 1 \end{array} \Rightarrow \begin{array}{l} y = \frac{1 + 2x}{5} \\ 10x^2 + 17y^2 - 24xy = 1 \end{array}$$

Substituting in the second equation and solving for x, we get

$$y = \frac{1 + 2x}{5}$$

$$10x^2 + 17\left(\frac{1 + 2x}{5}\right)^2 - 24x \cdot \frac{1 + 2x}{5} = 1$$

$$250x^2 + 17(1 + 2x)^2 - 120x(1 + 2x) = 25$$

$$39x^2 - 26x - 4 = 0$$

$$x_{1,2} = \frac{26 \pm \sqrt{1300}}{78}$$

$$\Rightarrow x_1 = \frac{1}{3} + \frac{5\sqrt{13}}{39} \approx 0.8, \quad x_2 = \frac{1}{3} - \frac{5\sqrt{13}}{39} \approx -0.13$$

We find the y-coordinates using $y = \frac{1+2x}{5}$.

$$y_1 = \frac{1 + 2 \cdot 0.8}{5} = 0.52, \quad y_2 = \frac{1 - 2 \cdot 0.13}{5} = 0.15$$

Finally, we find the z-coordinate using $z = 4y - 3x$:

$$z_1 = 4 \cdot 0.52 - 3 \cdot 0.8 = -0.32, \quad z_2 = 4 \cdot 0.15 + 3 \cdot 0.13 = 0.99$$

We obtain the critical points:

$$P_1 = (0.8, 0.52, -0.32), \quad P_2 = (-0.13, 0.15, 0.99)$$

Step 4. Conclusions. We evaluate the function $f(x, y, z) = x + 2y + 3z$ at the critical points:

$$f(P_1) = 0.8 + 2 \cdot 0.52 - 3 \cdot 0.32 = 0.88$$
$$f(P_2) = -0.13 + 2 \cdot 0.15 + 3 \cdot 0.99 = 3.14 \tag{1}$$

The two constraints determine the common points of the unit sphere $x^2 + y^2 + z^2 = 1$ and the plane $x + y + z = 1$. This set is a circle that is a closed and bounded set in R^3. Therefore, f has a minimum and maximum values on this set. These extrema are given in (1).

61. Use Lagrange multipliers to find the dimensions of a cylindrical can with a bottom but no top, of fixed volume V with minimum surface area.

SOLUTION We denote the radius of the cylinder by r and the height by h.

The volume of the cylinder is $g = \pi r^2 h$ and the surface area is

$$f = 2\pi r h + 2\pi r^2$$

We need to minimize $f(r, h) = 2\pi r h + 2\pi r^2$ subject to the constraint $g(r, h) = \pi r^2 h - V = 0$.

Step 1. Write out the Lagrange equations. We have $\nabla f = \langle 2\pi h + 4\pi r, 2\pi r \rangle = 2\pi \langle h + 2r, r \rangle$ and $\nabla g = \langle 2\pi h r, \pi r^2 \rangle = \pi \langle 2hr, r^2 \rangle$, hence the Lagrange condition $\nabla f = \lambda \nabla g$ is

$$2\pi \langle h + 2r, r \rangle = \pi \lambda \langle 2hr, r^2 \rangle$$

or

$$2 \langle h + 2r, r \rangle = \lambda \langle 2hr, r^2 \rangle$$

We obtain the following equations:

$$\begin{aligned} 2(h + 2r) &= 2hr\lambda \\ 2r &= \lambda r^2 \end{aligned} \quad \Rightarrow \quad \begin{aligned} h + 2r &= hr\lambda \\ 2r &= \lambda r^2 \end{aligned}$$

Step 2. Solve for λ in terms of r and h. The equation of the constraint implies that $r \neq 0$ and $h \neq 0$ (we assume that $V > 0$). Therefore, the Lagrange equations give

$$\lambda = \frac{h + 2r}{hr} = \frac{1}{r} + \frac{2}{h}, \quad \lambda = \frac{2}{r}$$

Step 3. Solve for r and h using the constraint. Equating the two expressions for λ gives

$$\frac{1}{r} + \frac{2}{h} = \frac{2}{r}$$

$$\frac{2}{h} = \frac{1}{r} \quad \Rightarrow \quad h = 2r$$

We substitute $h = 2r$ in the equation of the constraint $\pi r^2 h = V$ and solve for r. We obtain

$$\pi r^2 \cdot 2r = V$$

$$2\pi r^3 = V \quad \Rightarrow \quad r = \left(\frac{V}{2\pi}\right)^{1/3}$$

We find h using the relation $h = 2r$:

$$h = 2\left(\frac{V}{2\pi}\right)^{1/3}$$

The critical point is $h = 2\left(\frac{V}{2\pi}\right)^{1/3}$, $r = \left(\frac{V}{2\pi}\right)^{1/3}$.

Step 4. Conclusions. On the constraint $\pi r^2 h = V$ we have $h = \frac{V}{\pi r^2}$ and $r = \sqrt{\frac{V}{\pi h}}$, hence

$$f\left(r, \frac{V}{\pi r^2}\right) = 2\pi r \cdot \frac{V}{\pi r^2} + 2\pi r^2 = \frac{2V}{r} + 2\pi r^2$$

$$f\left(\sqrt{\frac{V}{\pi h}}, h\right) = 2\pi \sqrt{\frac{V}{\pi h}} h + 2\pi \cdot \frac{V}{\pi h} = 2\sqrt{\pi V}\sqrt{h} + \frac{2V}{h}$$

We see that as $h \to 0+$ or $h \to \infty$, we have $f(r, h) \to \infty$, and as $r \to 0+$ or $r \to \infty$, we have $f(r, h) \to \infty$. Therefore, f has a minimum value on the constraint, which occurs at the critical point. We evaluate $f(r, h) = 2\pi rh + 2\pi r^2 = 2\pi(rh + r^2)$ at the critical point P:

$$f(P) = 2\pi\left(\left(\frac{V}{2\pi}\right)^{1/3} \cdot 2\left(\frac{V}{2\pi}\right)^{1/3} + \left(\frac{V}{2\pi}\right)^{2/3}\right) = 2\pi\left(2\left(\frac{V}{2\pi}\right)^{2/3} + \left(\frac{V}{2\pi}\right)^{2/3}\right) = 6\pi\left(\frac{V}{2\pi}\right)^{2/3}$$

We conclude that the minimum surface area is $6\pi\left(\frac{V}{2\pi}\right)^{2/3}$, and the dimensions of the corresponding cylinder are $r = \left(\frac{V}{2\pi}\right)^{1/3}$, $h = 2\left(\frac{V}{2\pi}\right)^{1/3}$.

63. Given n nonzero numbers $\sigma_1, \ldots, \sigma_n$, show that the minimum value of

$$f(x_1, \ldots, x_n) = x_1^2 \sigma_1^2 + \cdots + x_n^2 \sigma_n^2$$

subject to $x_1 + \cdots + x_n = 1$ is c, where $c = \left(\sum_{j=1}^{n} \sigma_j^{-2}\right)^{-1}$.

SOLUTION We must minimize the function $f(x_1, \ldots, x_n) = x_1^2 \sigma_1^2 + \cdots + x_n^2 \sigma_n^2$ subject to the constraint $g(x_1, \ldots, x_n) = x_1 + \cdots + x_n - 1 = 0$.

First write out the Lagrange equations. We have $\nabla f = \langle 2\sigma_1^2 x_1, \ldots, 2\sigma_n^2 x_n \rangle$ and $\nabla g = \langle 1, \ldots, 1 \rangle$, hence the Lagrange condition $\nabla f = \lambda \nabla g$ gives the following equations:

$$2\sigma_1^2 x_1 = \lambda$$

$$2\sigma_2^2 x_2 = \lambda$$

$$\vdots$$

$$2\sigma_n^2 x_n = \lambda$$

It follows from these equations that $\sigma_i^2 x_i$ is independent of i; call their common value c. Then from the constraint,

$$1 = x_1 + \cdots + x_n = c\sigma_1^{-2} + c\sigma_2^{-2} + \cdots + c\sigma_n^{-2} = c\sum_{j=1}^{n}\sigma_j^{-2}.$$

It follows that $c = \left(\sum_{j=1}^{n} \sigma_j^{-2}\right)^{-1}$ and that

$$(x_1, \ldots, x_n) = \left(c\sigma_1^{-2}, \ldots, c\sigma_n^{-2}\right).$$

We must show that these values for the x_i produce a minimum for f. As $x_i \to \infty$ or $x_i \to -\infty$, for one or more i's the function $f(x_1, \ldots, x_n)$ tends to ∞. f is continuous since it is a polynomial, hence f has a minimum value on the constraint. This minimum occurs at the critical point. We find it:

$$f(P) = \sum_{j=1}^{n} \sigma_j^2 \left(\frac{c}{\sigma_j^2} \right)^2 = \sum_{j=1}^{n} \frac{\sigma_j^2 c^2}{\sigma_j^4} = c^2 \sum_{j=1}^{n} \sigma_j^{-2} = c^2 \cdot c^{-1} = c$$

15 MULTIPLE INTEGRATION

15.1 Integration in Two Variables (LT Section 16.1)

Preliminary Questions

1. If $S_{8,4}$ is a Riemann sum for a double integral over a rectangle $\mathcal{R} = [1,5] \times [2,10]$ using a regular partition, what is the area of each subrectangle? How many subrectangles are there?

SOLUTION Since the partition is regular, all subrectangles have sides of length

$$\Delta x = \frac{5-1}{8} = \frac{1}{2}, \quad \Delta y = \frac{10-2}{4} = 2$$

Therefore the area of each subrectangle is $\Delta A = \Delta x \Delta y = \frac{1}{2} \cdot 2 = 1$, and the number of subrectangles is $8 \cdot 4 = 32$.

2. Estimate the double integral of a continuous function f over the small rectangle $\mathcal{R} = [0.9, 1.1] \times [1.9, 2.1]$ if $f(1,2) = 4$.

SOLUTION Since we are given the value of f at one point in $\mathcal{R}$ only, we can only use the approximation S_{11} for the integral of f over $\mathcal{R}$. For S_{11} we have one rectangle with sides

$$\Delta x = 1.1 - 0.9 = 0.2, \quad \Delta y = 2.1 - 1.9 = 0.2$$

Hence, the area of the rectangle is $\Delta A = \Delta x \Delta y = 0.2 \cdot 0.2 = 0.04$. We obtain the following approximation:

$$\iint_\mathcal{R} f\,dA \approx S_{1,1} = f(1,2)\Delta A = 4 \cdot 0.04 = 0.16$$

3. What is the integral of the constant function $f(x,y) = 5$ over the rectangle $[-2,3] \times [2,4]$?

SOLUTION The integral of f over the unit square $\mathcal{R} = [-2,3] \times [2,4]$ is the volume of the box of base $\mathcal{R}$ and height 5. That is,

$$\iint_\mathcal{R} 5\,dA = 5 \cdot \text{Area}(\mathcal{R}) = 5 \cdot 5 \cdot 2 = 50$$

4. What is the interpretation of $\iint_\mathcal{R} f(x,y)\,dA$ if $f(x,y)$ takes on both positive and negative values on $\mathcal{R}$?

SOLUTION The double integral $\iint_\mathcal{R} f(x,y)\,dA$ is the signed volume between the graph $z = f(x,y)$ for $(x,y) \in \mathcal{R}$, and the xy-plane. The region below the xy-plane is treated as negative volume.

5. Which of (a) or (b) is equal to $\int_1^2 \int_4^5 f(x,y)\,dy\,dx$?

(a) $\int_1^2 \int_4^5 f(x,y)\,dx\,dy$ **(b)** $\int_4^5 \int_1^2 f(x,y)\,dx\,dy$

SOLUTION The integral $\int_1^2 \int_4^5 f(x,y)\,dy\,dx$ is written with dy preceding dx, therefore the integration is first with respect to y over the interval $4 \le y \le 5$, and then with respect to x over the interval $1 \le x \le 2$. By Fubini's Theorem, we may replace the order of integration over the corresponding intervals. Therefore the given integral is equal to (b) rather than to (a).

6. For which of the following functions is the double integral over the rectangle in Figure 15 equal to zero? Explain your reasoning.

(a) $f(x,y) = x^2 y$ **(b)** $f(x,y) = xy^2$
(c) $f(x,y) = \sin x$ **(d)** $f(x,y) = e^x$

FIGURE 15

SOLUTION The double integral is the signed volume of the region between the graph of $f(x,y)$ and the xy-plane over $\mathcal{R}$. In (b) and (c) the function satisfies $f(-x,y) = -f(x,y)$, hence the region below the xy-plane, where $-1 \leq x \leq 0$ cancels with the region above the xy-plane, where $0 \leq x \leq 1$. Therefore, the double integral is zero. In (a) and (d), the function $f(x,y)$ is always positive on the rectangle, so the double integral is greater than zero.

Exercises

1. Compute the Riemann sum $S_{4,3}$ to estimate the double integral of $f(x,y) = xy$ over $\mathcal{R} = [1,3] \times [1, 2.5]$. Use the regular partition and upper-right vertices of the subrectangles as sample points.

SOLUTION The rectangle $\mathcal{R}$ and the subrectangles are shown in the following figure:

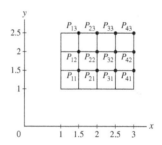

The subrectangles have sides of length

$$\Delta x = \frac{3-1}{4} = 0.5, \quad \Delta y = \frac{2.5-1}{3} = 0.5 \quad \Rightarrow \quad \Delta A = 0.5 \cdot 0.5 = 0.25$$

The upper right vertices are the following points:

$$\begin{array}{llll} P_{11} = (1.5, 1.5) & P_{21} = (2, 1.5) & P_{31} = (2.5, 1.5) & P_{41} = (3, 1.5) \\ P_{12} = (1.5, 2) & P_{22} = (2, 2) & P_{32} = (2.5, 2) & P_{42} = (3, 2) \\ P_{13} = (1.5, 2.5) & P_{23} = (2, 2.5) & P_{33} = (2.5, 2.5) & P_{43} = (3, 2.5) \end{array}$$

We compute $f(x,y) = xy$ at these points:

$$\begin{array}{lll} f(P_{11}) = 1.5 \cdot 1.5 = 2.25 & f(P_{12}) = 1.5 \cdot 2 = 3 & f(P_{13}) = 3.75 \\ f(P_{21}) = 2 \cdot 1.5 = 3 & f(P_{22}) = 2 \cdot 2 = 4 & f(P_{23}) = 5 \\ f(P_{31}) = 2.5 \cdot 1.5 = 3.75 & f(P_{32}) = 2.5 \cdot 2 = 5 & f(P_{33}) = 6.25 \\ f(P_{41}) = 3 \cdot 1.5 = 4.5 & f(P_{42}) = 3 \cdot 2 = 6 & f(P_{43}) = 7.5 \end{array}$$

Hence, $S_{4,3}$ is the following sum:

$$S_{4,3} = \sum_{i=1}^{4} \sum_{j=1}^{3} f(P_{ij}) \Delta A = 0.25(2.25 + 3 + 3.75 + 4.5 + 3 + 4 + 5 + 6 + 3.75 + 5 + 6.25 + 7.5) = 13.5$$

In Exercises 3–6, compute the Riemann sums for the double integral $\iint_{\mathcal{R}} f(x, y) \, dA$, where $\mathcal{R} = [1, 4] \times [1, 3]$, for the grid and two choices of sample points shown in Figure 16.

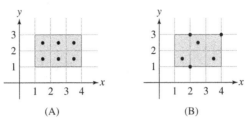

FIGURE 16

3. $f(x, y) = 2x + y$

SOLUTION The subrectangles have sides of length $\Delta x = \frac{4-1}{3} = 1$ and $\Delta y = \frac{3-1}{2} = 1$, and area $\Delta A = \Delta x \Delta y = 1$. We find the sample points in (A) and (B):

(A)

$$P_{11} = (1.5, 1.5) \; P_{21} = (2.5, 1.5) \; P_{31} = (3.5, 1.5)$$
$$P_{12} = (1.5, 2.5) \; P_{22} = (2.5, 2.5) \; P_{32} = (3.5, 2.5)$$

(A)

(B)

$$P_{11} = (1.5, 1.5) \; P_{21} = (2, 1) \; P_{31} = (3.5, 1.5)$$
$$P_{21} = (2, 3) \; P_{22} = (2.5, 2.5) \; P_{23} = (4, 3)$$

(B)

The Riemann Sum S_{32} is the following estimation of the double integral:

$$\iint_{\mathcal{R}} f(x, y) \, dA \approx S_{32} = \sum_{i=1}^{3} \sum_{j=1}^{2} f\left(P_{ij}\right) \Delta A = \sum_{i=1}^{3} \sum_{j=1}^{2} f\left(P_{ij}\right)$$

We compute S_{32} for the two choices of sample points (A) and (B), and the following function:

$$f(x, y) = 2x + y$$

We compute $f\left(P_{ij}\right)$ for the sample points computed above:

(A)

$$f(P_{11}) = f(1.5, 1.5) = 2 \cdot 1.5 + 1.5 = 4.5$$
$$f(P_{21}) = f(2.5, 1.5) = 2 \cdot 2.5 + 1.5 = 6.5$$

$$f(P_{31}) = f(3.5, 1.5) = 2 \cdot 3.5 + 1.5 = 8.5$$
$$f(P_{12}) = f(1.5, 2.5) = 2 \cdot 1.5 + 2.5 = 5.5$$
$$f(P_{22}) = f(2.5, 2.5) = 2 \cdot 2.5 + 2.5 = 7.5$$
$$f(P_{32}) = f(3.5, 2.5) = 2 \cdot 3.5 + 2.5 = 9.5$$

Hence,

$$S_{32} = \sum_{i=1}^{3} \sum_{j=1}^{2} f(P_{ij}) \Delta A = 4.5 + 6.5 + 8.5 + 5.5 + 7.5 + 9.5 = 42$$

(B)

$$f(P_{11}) = f(1.5, 1.5) = 2 \cdot 1.5 + 1.5 = 4.5$$
$$f(P_{21}) = f(2, 1) = 2 \cdot 2 + 1 = 5$$
$$f(P_{31}) = f(3.5, 1.5) = 2 \cdot 3.5 + 1.5 = 8.5$$
$$f(P_{21}) = f(2, 3) = 2 \cdot 2 + 3 = 7$$
$$f(P_{22}) = f(2.5, 2.5) = 2 \cdot 2.5 + 2.5 = 7.5$$
$$f(P_{23}) = f(4, 3) = 2 \cdot 4 + 3 = 11$$

Hence,

$$S_{32} = \sum_{i=1}^{3} \sum_{j=1}^{2} f(P_{ij}) \Delta A = 4.5 + 5 + 8.5 + 7 + 7.5 + 11 = 43.5$$

5. $f(x, y) = 4x$

SOLUTION We compute the values of f at the sample points:

(A)

$$f(P_{11}) = f(1.5, 1.5) = 4 \cdot 1.5 = 6$$
$$f(P_{21}) = f(2.5, 1.5) = 4 \cdot 2.5 = 10$$
$$f(P_{31}) = f(3.5, 1.5) = 4 \cdot 3.5 = 14$$
$$f(P_{12}) = f(1.5, 2.5) = 4 \cdot 1.5 = 6$$
$$f(P_{22}) = f(2.5, 2.5) = 4 \cdot 2.5 = 10$$
$$f(P_{32}) = f(3.5, 2.5) = 4 \cdot 3.5 = 14$$
$$\Delta x = \frac{4-1}{3} = 1, \quad \Delta y = \frac{3-1}{2} = 1$$

Hence $\Delta A = \Delta x \cdot \Delta y = 1$ and we get

$$S_{32} = \sum_{i=1}^{3} \sum_{j=1}^{2} f(P_{ij}) \Delta A = 6 + 10 + 14 + 6 + 10 + 14 = 60$$

(B)

$$f(P_{11}) = f(1.5, 1.5) = 4 \cdot 1.5 = 6$$
$$f(P_{21}) = f(2, 1) = 4 \cdot 2 = 8$$
$$f(P_{31}) = f(3.5, 1.5) = 4 \cdot 3.5 = 14$$
$$f(P_{12}) = f(2, 3) = 4 \cdot 2 = 8$$
$$f(P_{22}) = f(2.5, 2.5) = 4 \cdot 2.5 = 10$$
$$f(P_{32}) = f(4, 3) = 4 \cdot 4 = 16$$

$\Delta A = 1$. Hence,

$$S_{32} = \sum_{i=1}^{3}\sum_{j=1}^{2} f(P_{ij})\Delta A = 6 + 8 + 14 + 8 + 10 + 16 = 62$$

7. Let $\mathcal{R} = [0,1] \times [0,1]$. Estimate $\iint_{\mathcal{R}} (x+y)\,dA$ by computing two different Riemann sums, each with at least six rectangles.

SOLUTION We define the following subrectangles and sample points:

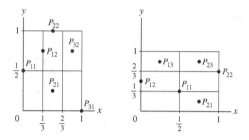

The sample points defined in the two figures are:

(A)
$$P_{11} = \left(0, \tfrac{1}{2}\right) \quad P_{21} = \left(\tfrac{1}{2}, \tfrac{1}{4}\right) \quad P_{31} = (1, 0)$$
$$P_{12} = \left(\tfrac{1}{3}, \tfrac{3}{4}\right) \quad P_{22} = \left(\tfrac{1}{2}, 1\right) \quad P_{32} = \left(\tfrac{5}{6}, \tfrac{3}{4}\right)$$

(B)
$$P_{11} = \left(\tfrac{1}{2}, \tfrac{1}{3}\right) \quad P_{21} = \left(\tfrac{3}{4}, \tfrac{1}{6}\right) \quad P_{12} = \left(0, \tfrac{1}{2}\right)$$
$$P_{22} = \left(1, \tfrac{2}{3}\right) \quad P_{13} = \left(\tfrac{1}{4}, \tfrac{5}{6}\right) \quad P_{23} = \left(\tfrac{3}{4}, \tfrac{5}{6}\right)$$

We compute the values of $f(x,y) = x + y$ at the sample points:

(A)
$$f(P_{11}) = f\left(0, \tfrac{1}{2}\right) = 0 + \tfrac{1}{2} = \tfrac{1}{2}$$
$$f(P_{21}) = f\left(\tfrac{1}{2}, \tfrac{1}{4}\right) = \tfrac{1}{2} + \tfrac{1}{4} = \tfrac{3}{4}$$
$$f(P_{31}) = f(1, 0) = 1 + 0 = 1$$
$$f(P_{12}) = f\left(\tfrac{1}{3}, \tfrac{3}{4}\right) = \tfrac{1}{3} + \tfrac{3}{4} = \tfrac{13}{12}$$
$$f(P_{22}) = f\left(\tfrac{1}{2}, 1\right) = \tfrac{1}{2} + 1 = \tfrac{3}{2}$$
$$f(P_{32}) = f\left(\tfrac{5}{6}, \tfrac{3}{4}\right) = \tfrac{5}{6} + \tfrac{3}{4} = \tfrac{19}{12}$$

Each subrectangle has sides of length $\Delta x = \tfrac{1}{3}, \Delta y = \tfrac{1}{2}$ and area $\Delta A = \Delta x \Delta y = \tfrac{1}{3} \cdot \tfrac{1}{2} = \tfrac{1}{6}$. We obtain the following Riemann sum:

$$S_{32} = \sum_{i=1}^{3}\sum_{j=1}^{2} f(P_{ij})\Delta A = \tfrac{1}{6}\left(\tfrac{1}{2} + \tfrac{3}{4} + 1 + \tfrac{13}{12} + \tfrac{3}{2} + \tfrac{19}{12}\right) = \tfrac{77}{72} \approx 1.069$$

(B)
$$f(P_{11}) = f\left(\tfrac{1}{2}, \tfrac{1}{3}\right) = \tfrac{1}{2} + \tfrac{1}{3} = \tfrac{5}{6}$$
$$f(P_{21}) = f\left(\tfrac{3}{4}, \tfrac{1}{6}\right) = \tfrac{3}{4} + \tfrac{1}{6} = \tfrac{11}{12}$$

$$f(P_{12}) = f\left(0, \frac{1}{2}\right) = 0 + \frac{1}{2} = \frac{1}{2}$$

$$f(P_{22}) = f\left(1, \frac{2}{3}\right) = 1 + \frac{2}{3} = \frac{5}{3}$$

$$f(P_{13}) = f\left(\frac{1}{4}, \frac{5}{6}\right) = \frac{1}{4} + \frac{5}{6} = \frac{13}{12}$$

$$f(P_{23}) = f\left(\frac{3}{4}, \frac{5}{6}\right) = \frac{3}{4} + \frac{5}{6} = \frac{19}{12}$$

Each subrectangle has sides of length $\Delta x = \frac{1}{2}, \Delta y = \frac{1}{3}$ and area $\Delta A = \Delta x \Delta y = \frac{1}{2} \cdot \frac{1}{3} = \frac{1}{6}$. We obtain the following Riemann sum:

$$S_{23} = \sum_{i=1}^{3} \sum_{j=1}^{2} f(P_{ij}) \Delta A = \frac{1}{6}\left(\frac{5}{6} + \frac{11}{12} + \frac{1}{2} + \frac{5}{3} + \frac{13}{12} + \frac{19}{12}\right) = \frac{79}{72} \approx 1.097$$

9. Evaluate $\iint_{\mathcal{R}} (15 - 3x)\, dA$, where $\mathcal{R} = [0, 5] \times [0, 3]$, and sketch the corresponding solid region (see Example 2).

SOLUTION This double integral is the volume V of the solid wedge underneath the graph of $f(x, y) = 15 - 3x$. The triangular face of the wedge has area

$$A = \frac{1}{2} \cdot 5 \cdot 15 = \frac{75}{2}$$

The volume of the wedge is equal to the area A times the length $\ell = 3$; that is

$$V = \ell A = 3\left(\frac{75}{2}\right) = \frac{225}{2}$$

11. The following table gives the approximate height at quarter-meter intervals of a mound of gravel. Estimate the volume of the mound by computing the average of the two Riemann sums $S_{4,3}$ with lower-left and upper-right vertices of the subrectangles as sample points.

0.75	0.1	0.2	0.2	0.15	0.1
0.5	0.2	0.3	0.5	0.4	0.2
0.25	0.15	0.2	0.4	0.3	0.2
0	0.1	0.15	0.2	0.15	0.1
y \ x	0	0.25	0.5	0.75	1

SOLUTION Each subrectangle is a square of side 0.25, hence the area of each subrectangle is $\Delta A = 0.25^2 = 0.0625$. By the given data, the lower-left vertex sample points are:

$$\begin{array}{lll}
f(P_{11}) = f(0, 0) & f(P_{12}) = f(0, 0.25) & f(P_{13}) = f(0, 0.50) \\
f(P_{21}) = f(0.25, 0) & f(P_{22}) = f(0.25, 0.25) & f(P_{23}) = f(0.25, 0.50) \\
f(P_{31}) = f(0.50, 0) & f(P_{32}) = f(0.50, 0.25) & f(P_{33}) = f(0.50, 0.50) \\
f(P_{41}) = f(0.75, 0) & f(P_{42}) = f(0.75, 0.25) & f(P_{43}) = f(0.75, 0.50)
\end{array}$$

The Riemann sum $S_{4,3}$ that corresponds to these lower-left vertex sample points is the following sum:

$$S_{4,3} = \sum_{i=1}^{4} \sum_{j=1}^{3} f(P_{ij}) \Delta A$$

$$= 0.0625(0.1 + 0.15 + 0.2 + 0.15 + 0.2 + 0.3 + 0.2 + 0.4 + 0.5 + 0.15 + 0.3 + 0.4) \approx 0.190625$$

Now by the given data, the upper-right vertex sample points are:

$$\begin{array}{lll}
f(P_{11}) = f(0.25, 0.25) & f(P_{12}) = f(0.25, 0.50) & f(P_{13}) = f(0.25, 0.75) \\
f(P_{21}) = f(0.50, 0.25) & f(P_{22}) = f(0.50, 0.50) & f(P_{23}) = f(0.50, 0.75) \\
f(P_{31}) = f(0.75, 0.25) & f(P_{32}) = f(0.75, 0.50) & f(P_{33}) = f(0.75, 0.75) \\
f(P_{41}) = f(1, 0.25) & f(P_{42}) = f(1, 0.50) & f(P_{43}) = f(1, 0.75)
\end{array}$$

The Riemann sum $S'_{4,3}$ that corresponds to these upper-right vertex sample points is the following sum:

$$S'_{4,3} = \sum_{i=1}^{4} \sum_{j=1}^{3} f(P_{ij}) \Delta A$$

$$= 0.0625(0.2 + 0.3 + 0.2 + 0.4 + 0.5 + 0.2 + 0.3 + 0.4 + 0.15 + 0.2 + 0.2 + 0.1) \approx 0.196875$$

Taking the average of the two Riemann sums we have:

$$\text{volume} \approx \frac{S_{4,3} + S'_{4,3}}{2} = \frac{0.190625 + 0.196875}{2} = 0.19375$$

13. CAS Let $S_{N,N}$ be the Riemann sum for $\int_0^1 \int_0^1 e^{x^3-y^3} \, dy \, dx$ using the regular partition and the lower-left vertex of each subrectangle as sample points. Use a computer algebra system to calculate $S_{N,N}$ for $N = 25$, 50, 100.

SOLUTION Using a computer algebra system, we compute $S_{N,N}$ to be 1.0731, 1.0783, and 1.0809.

In Exercises 15–18, use symmetry to evaluate the double integral.

15. $\iint_{\mathcal{R}} x^3 \, dA, \quad \mathcal{R} = [-4, 4] \times [0, 5]$

SOLUTION The double integral is the signed volume of the region between the graph of $f(x, y) = x^3$ and the xy-plane. However, $f(x, y)$ takes opposite values at (x, y) and $(-x, y)$:

$$f(-x, y) = (-x)^3 = -x^3 = -f(x, y)$$

Because of symmetry, the (negative) signed volume of the region below the xy-plane where $-4 \leq x \leq 0$ cancels with the (positive) signed volume of the region above the xy-plane where $0 \leq x \leq 4$. The net result is

$$\iint_{\mathcal{R}} x^3 \, dA = 0$$

17. $\iint_{\mathcal{R}} \sin x \, dA, \quad \mathcal{R} = [0, 2\pi] \times [0, 2\pi]$

SOLUTION Since $\sin(\pi + x) = -\sin x$, the region between the graph and the xy-plane where $\pi \leq x \leq 2\pi$, is below the xy-plane, and it cancels with the region above the xy-plane where $0 \leq x \leq \pi$. Hence,

$$\iint_{\mathcal{R}} \sin x \, dA = 0$$

In Exercises 19–36, evaluate the iterated integral.

19. $\int_1^3 \int_0^2 x^3 y \, dy \, dx$

SOLUTION We first compute the inner integral, treating x as a constant, then integrate the result with respect to x:

$$\int_1^3 \int_0^2 x^3 y \, dy \, dx = \int_1^3 x^3 \frac{y^2}{2} \bigg|_{y=0}^{2} dx = \int_1^3 x^3 \left(\frac{2^2}{2} - 0 \right) dx = \int_1^3 2x^3 \, dx = \frac{x^4}{2} \bigg|_1^3 = 40$$

21. $\int_4^9 \int_{-3}^8 1 \, dx \, dy$

SOLUTION

$$\int_4^9 \int_{-3}^8 1 \, dx \, dy = \int_4^9 1 \left(\int_{-3}^8 1 \, dy \right) dx$$

$$= \int_4^9 1 \left(y \bigg|_{-3}^8 \right) dx$$

$$= \int_4^9 11\,dx$$

$$= 11x\Big|_4^9$$

$$= 99 - 44 = 55$$

23. $\displaystyle\int_{-1}^{1}\int_{0}^{\pi} x^2 \sin y\,dy\,dx$

SOLUTION We first evaluate the inner integral, treating x as a constant, then integrate the result with respect to x. This gives

$$\int_{-1}^{1}\int_{0}^{\pi} x^2 \sin y\,dy\,dx = \int_{-1}^{1} x^2(-\cos y)\Big|_{y=0}^{\pi} dx = \int_{-1}^{1} x^2(-\cos\pi + \cos 0)\,dx$$

$$= \int_{-1}^{1} x^2(1+1)\,dx = \int_{-1}^{1} 2x^2\,dx = \frac{2}{3}x^3\Big|_{-1}^{1} = \frac{2}{3}(1^3-(-1)^3) = \frac{4}{3}$$

25. $\displaystyle\int_{2}^{6}\int_{1}^{4} x^2\,dx\,dy$

SOLUTION We use Iterated Integral of a Product Function to compute the integral as follows:

$$\int_{2}^{6}\int_{1}^{4} x^2\,dx\,dy = \int_{2}^{6}\int_{1}^{4} x^2\cdot 1\,dx\,dy = \left(\int_{1}^{4} x^2\,dx\right)\left(\int_{2}^{6} 1\,dy\right) = \left(\frac{x^3}{3}\Big|_{1}^{4}\right)\left(y\Big|_{2}^{6}\right)$$

$$= \left(\frac{4^3}{3} - \frac{1^3}{3}\right)(6-2) = 21\cdot 4 = 84$$

27. $\displaystyle\int_{0}^{1}\int_{0}^{2} (x+4y^3)\,dx\,dy$

SOLUTION We use additivity of the double integral to write

$$\int_{0}^{1}\int_{0}^{2} (x+4y^3)\,dx\,dy = \int_{0}^{1}\int_{0}^{2} x\,dx\,dy + \int_{0}^{1}\int_{0}^{2} 4y^3\,dx\,dy \qquad (1)$$

We now compute each of the double integrals using product of iterated integrals:

$$\int_{0}^{1}\int_{0}^{2} x\,dx\,dy = \left(\int_{0}^{2} x\,dx\right)\left(\int_{0}^{1} 1\,dy\right) = \left(\frac{1}{2}x^2\Big|_{0}^{2}\right)\left(y\Big|_{0}^{1}\right) = 2\cdot 1 = 2$$

$$\int_{0}^{1}\int_{0}^{2} 4y^3\,dx\,dy = \left(\int_{0}^{1} 4y^3\,dy\right)\left(\int_{0}^{2} 1\,dx\right) = \left(y^4\Big|_{0}^{1}\right)\left(x\Big|_{0}^{2}\right) = 1\cdot 2 = 2$$

Substituting in (1) gives

$$\int_{0}^{1}\int_{0}^{2} (x+4y^3)\,dx\,dy = 2+2 = 4$$

29. $\displaystyle\int_{0}^{4}\int_{0}^{9} \sqrt{x+4y}\,dx\,dy$

SOLUTION We compute the inner integral, treating y as a constant. Then we evaluate the resulting integral with respect to y:

$$\int_{0}^{4}\int_{0}^{9} \sqrt{x+4y}\,dx\,dy = \int_{0}^{4} \frac{2}{3}(x+4y)^{3/2}\Big|_{x=0}^{9} dy = \int_{0}^{4} \frac{2}{3}\left((9+4y)^{3/2} - (4y)^{3/2}\right) dy$$

$$= \frac{2}{3}\left(\frac{2}{5\cdot 4}(9+4y)^{5/2} - \frac{2}{5\cdot 4}(4y)^{5/2}\right)\Big|_{0}^{4}$$

$$= \frac{1}{15}(5^5 - 4^5) - \frac{1}{15}(3^5 - 0) \approx 123.8667$$

31. $\displaystyle\int_1^2 \int_0^4 \frac{dy\,dx}{x+y}$

SOLUTION The inner integral is an iterated integral with respect to y. We evaluate it first and then compute the resulting integral with respect to x. This gives

$$\int_1^2 \int_0^4 \frac{dy\,dx}{x+y} = \int_1^2 \left(\int_0^4 \frac{dy}{x+y}\right) dx = \int_1^2 \ln(x+y)\Big|_{y=0}^4 dx = \int_1^2 (\ln(x+4) - \ln x)\,dx$$

We use the integral formula:

$$\int \ln(x+a)\,dx = (x+a)(\ln(x+a) - 1) + C$$

We get

$$\int_1^2 \int_0^4 \frac{dy\,dx}{x+y} = (x+4)(\ln(x+4) - 1) - x(\ln x - 1)\Big|_1^2 = 6(\ln 6 - 1) - 2(\ln 2 - 1) - (5(\ln 5 - 1) - (\ln 1 - 1))$$

$$= 6\ln 6 - 2\ln 2 - 5\ln 5 \approx 1.31$$

33. $\displaystyle\int_0^4 \int_0^5 \frac{dy\,dx}{\sqrt{x+y}}$

SOLUTION

$$\int_0^4 \int_0^5 \frac{dy\,dx}{\sqrt{x+y}} = \int_0^4 \left(\int_0^5 \frac{dy}{\sqrt{x+y}}\right) dx$$

$$= \int_0^4 \left(2\sqrt{x+y}\Big|_{y=0}^5\right) dx$$

$$= 2\int_0^4 (\sqrt{x+5} - \sqrt{x})\,dx$$

$$= 2\left(\frac{2}{3}(x+5)^{3/2} - \frac{2}{3}x^{3/2}\right)\Big|_0^4$$

$$= 2\left(\frac{2}{3}\cdot 27 - \frac{2}{3}\cdot 8\right) - 2\left(\frac{2}{3}\cdot 5^{3/2} - 0\right)$$

$$= 36 - \frac{32}{3} - \frac{20}{3}\sqrt{5} = \frac{76}{3} - \frac{20}{3}\sqrt{5} \approx 10.426$$

35. $\displaystyle\int_1^2 \int_1^3 \frac{\ln(xy)\,dy\,dx}{y}$

SOLUTION

$$\int_1^2 \int_1^3 \frac{\ln(xy)\,dy\,dx}{y} = \int_1^2 \left(\frac{1}{2}[\ln(xy)]^2\Big|_1^3\right) dx$$

$$= \frac{1}{2}\int_1^2 [\ln(3x)]^2 - [\ln(x)]^2\,dx$$

$$= \frac{1}{2}\int_1^2 [\ln(3x)]^2\,dx - \frac{1}{2}\int_1^2 [\ln(x)]^2\,dx$$

$$= \frac{1}{2}\left[x(\ln 3x)^2\Big|_1^2 - 2\int_1^2 \ln(3x)\,dx\right] - \frac{1}{2}\left[x(\ln x)^2\Big|_1^2 - 2\int_1^2 \ln x\,dx\right]$$

$$= \frac{1}{2}\left[2(\ln 6)^2 - (\ln 3)^2\right] - \int_1^2 \ln(3x)\,dx - \frac{1}{2}\left[2(\ln 2)^2 - 0\right] + \int_1^2 \ln x\,dx$$

$$= (\ln 6)^2 - \frac{1}{2}(\ln 3)^2 - \left[x\ln(3x) - x\Big|_1^2\right] - (\ln 2)^2 + \left[x\ln x - x\Big|_1^2\right]$$

In Exercises 37–44, evaluate the integral.

37. $\iint_{\mathcal{R}} \dfrac{x}{y}\, dA$, $\mathcal{R} = [-2, 4] \times [1, 3]$

SOLUTION We compute the double integral as the product of two single integrals:

$$\iint_{\mathcal{R}} \frac{x}{y}\, dA = \int_{-2}^{4} \int_{1}^{3} \frac{x}{y}\, dy\, dx = \int_{-2}^{4} x\, dx \cdot \int_{1}^{3} \frac{1}{y}\, dy$$

$$= \left(\frac{1}{2}x^2\bigg|_{-2}^{4}\right)\left(\ln y \bigg|_{1}^{3}\right) = \frac{1}{2}(16 - 4) \cdot (\ln 3 - \ln 1)$$

$$= 6 \ln 3$$

39. $\iint_{\mathcal{R}} \cos x \sin 2y\, dA$, $\mathcal{R} = [0, \tfrac{\pi}{2}] \times [0, \tfrac{\pi}{2}]$

SOLUTION Since the integrand has the form $f(x, y) = g(x)h(y)$, we may compute the double integral as the product of two single integrals. That is,

$$\iint_{\mathcal{R}} \cos x \sin 2y\, dA = \int_{0}^{\pi/2} \int_{0}^{\pi/2} \cos x \sin 2y\, dx\, dy = \left(\int_{0}^{\pi/2} \cos x\, dx\right)\left(\int_{0}^{\pi/2} \sin 2y\, dy\right)$$

$$= \left(\sin x \bigg|_{0}^{\pi/2}\right)\left(-\frac{1}{2}\cos 2y \bigg|_{0}^{\pi/2}\right) = \left(\sin \frac{\pi}{2} - \sin 0\right)\left(-\frac{1}{2}\cos \pi + \frac{1}{2}\cos 0\right)$$

$$= (1 - 0)\left(\frac{1}{2} + \frac{1}{2}\right) = 1$$

SOLUTION Using additivity of the double integral and then computing the integrals as products of single integrals, we obtain

$$\int_{0}^{2}\int_{0}^{3}(2x + 6y)\, dx\, dy = \int_{0}^{2}\int_{0}^{3} 2x\, dx\, dy + \int_{0}^{2}\int_{0}^{3} 6y\, dx\, dy$$

$$= \left(\int_{0}^{3} 2x\, dx\right)\left(\int_{0}^{2} 1\, dy\right) + \left(\int_{0}^{3} 1\, dx\right)\left(\int_{0}^{2} 6y\, dy\right)$$

$$= \left(x^2\bigg|_{0}^{3}\right)\left(y\bigg|_{0}^{2}\right) + \left(x\bigg|_{0}^{3}\right)\left(3y^2\bigg|_{0}^{2}\right) = 9 \cdot 2 + 3 \cdot 12 = 54$$

41. $\iint_{\mathcal{R}} e^x \sin y\, dA$, $\mathcal{R} = [0, 2] \times [0, \tfrac{\pi}{4}]$

SOLUTION We compute the double integral as the product of two single integrals. This can be done since the integrand has the form $f(x, y) = g(x)h(y)$. We get

$$\iint_{\mathcal{R}} e^x \sin y\, dA = \int_{0}^{\pi/4} \int_{0}^{2} e^x \sin y\, dx\, dy = \left(\int_{0}^{2} e^x\, dx\right)\left(\int_{0}^{\pi/4} \sin y\, dy\right)$$

$$= \left(e^x \bigg|_{0}^{2}\right)\left(-\cos y \bigg|_{0}^{\pi/4}\right) = (e^2 - e^0)\left(-\cos \frac{\pi}{4} + \cos 0\right) = (e^2 - 1)\left(1 - \frac{\sqrt{2}}{2}\right) \approx 1.87$$

43. $\iint_{\mathcal{R}} x \ln y \, dA$, $\mathcal{R} = [0, 3] \times [1, e]$

SOLUTION We can compute this integral as the product of two single integrals:

$$\iint_{\mathcal{R}} x \ln y \, dA = \int_0^3 \int_1^e x \ln y \, dy \, dx = \int_0^3 x \, dx \cdot \int_1^e \ln y \, dy = \frac{1}{2} x^2 \Big|_0^3 \cdot (y \ln y - y) \Big|_1^e = \frac{9}{2} \cdot 1 = \frac{9}{2}$$

45. Let $f(x, y) = mxy^2$, where m is a constant. Find a value of m such that $\iint_{\mathcal{R}} f(x, y) \, dA = 1$, where $\mathcal{R} = [0, 1] \times [0, 2]$.

SOLUTION Since $f(x, y) = mxy^2$ is a product of a function of x and a function of y, we may compute the double integral as the product of two single integrals. That is,

$$\int_0^2 \int_0^1 mxy^2 \, dx \, dy = m \left(\int_0^1 x \, dx \right) \left(\int_0^2 y^2 \, dy \right) \quad (1)$$

We compute each integral:

$$\int_0^1 x \, dx = \frac{1}{2} x^2 \Big|_0^1 = \frac{1}{2}(1^2 - 0^2) = \frac{1}{2}$$

$$\int_0^2 y^2 \, dy = \frac{1}{3} y^3 \Big|_0^2 = \frac{1}{3}(2^3 - 0^3) = \frac{8}{3}$$

We substitute in (1), equate to 1, and solve the resulting equation for m. This gives

$$m \cdot \frac{1}{2} \cdot \frac{8}{3} = 1 \Rightarrow m = \frac{3}{4}$$

47. (a) Which is easier, antidifferentiating $y\sqrt{1 + xy}$ with respect to x or with respect to y? Explain.

(b) Evaluate $\iint_{\mathcal{R}} y\sqrt{1 + xy} \, dA$, where $\mathcal{R} = [0, 1] \times [0, 1]$.

SOLUTION

(a) Since $\frac{\partial}{\partial x}(1 + xy) = y$, making the substitution $u = 1 + xy$, $du = y \, dx$ reduces the integrand (as a function of x) to $\sqrt{u} \, du$. Thus it is easier to integrate with respect to x.

(b) Making the substitution from part (a), with new bounds $u = 1$ to $u = 1 + y$, gives

$$\iint_{\mathcal{R}} y\sqrt{1 + xy} \, dA = \int_0^1 \int_0^1 y\sqrt{1 + xy} \, dy \, dx = \int_0^1 \int_0^1 y\sqrt{1 + xy} \, dx \, dy$$

$$= \int_0^1 \int_1^{1+y} \sqrt{u} \, du \, dy$$

$$= \int_0^1 \frac{2}{3} u^{3/2} \Big|_1^{1+y} dy$$

$$= \frac{2}{3} \int_0^1 \left((1 + y)^{3/2} - 1 \right) dy$$

$$= \frac{2}{3} \left(\frac{2}{5}(1 + y)^{5/2} - y \right) \Big|_0^1$$

$$= \frac{2}{3} \left(\frac{2}{5} \cdot 2^{5/2} - 1 - \frac{2}{5} \cdot 1^{5/2} + 0 \right)$$

$$= \frac{4}{15} \cdot 2^{5/2} - \frac{14}{15} = \frac{16}{15} \sqrt{2} - \frac{14}{15}.$$

49. (a) Which is easier, antidifferentiating $\frac{y}{1+xy}$ with respect to x or with respect to y? Explain.

(b) Evaluate $\iint_{\mathcal{R}} \frac{y}{1 + xy} \, dA$, where $\mathcal{R} = [0, 1] \times [0, 1]$.

SOLUTION

(a) Since $\frac{\partial}{\partial x}(1+xy) = y$, making the substitution $u = 1+xy$, $du = y\,dx$ reduces the integrand (as a function of x) to $\frac{du}{u}$. Thus it is easier to integrate with respect to x.

(b) Making the substitution from part (a), with new bounds $u=1$ to $u=1+y$, gives

$$\iint_{\mathcal{R}} \frac{y}{1+xy}\,dA = \int_0^1 \int_0^1 \frac{y}{1+xy}\,dx\,dy = \int_0^1 \int_1^{1+y} \frac{du}{u}\,dy$$

$$= \int_0^1 \ln u \Big|_1^{1+y}\,dy$$

$$= \int_0^1 \ln(1+y)\,dy$$

$$= ((1+y)\ln(1+y) - (1+y))\Big|_0^1$$

$$= -1 + 2\ln 2$$

51. Using Fubini's Theorem, argue that the solid in Figure 18 has volume AL, where A is the area of the front face of the solid.

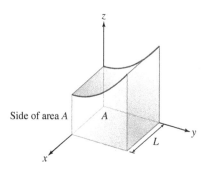

FIGURE 18

SOLUTION We denote by M the length of the other side of the rectangle in the basis of the solid. The volume of the solid is the double integral of the function $f(x,y) = g(x)$ over the rectangle

$$\mathcal{R} = [0, M] \times [0, L]$$

$$V = \iint_{\mathcal{R}} g(x)\,dA$$

We use Fubini's Theorem to write the double integral as iterated integral, and then compute the resulting integral as the product of two single integrals. This gives

$$V = \iint_{\mathcal{R}} g(x)\,dA = \int_0^L \int_0^M g(x)\,dx\,dy = \left(\int_0^M g(x)\,dx\right)\left(\int_0^L 1\,dy\right) = \left(\int_0^M g(x)\,dx\right) \cdot L \quad (1)$$

The integral $\int_0^M g(x)\,dx$ is the area A of the region under the graph of $z = g(x)$ over the interval $0 \le x \le M$. Substituting in (1) gives the following volume of the solid:

$$V = \left(\int_0^M g(x)\,dx\right) \cdot L = AL$$

Further Insights and Challenges

53. Let $F(x,y) = x^{-1}e^{xy}$. Show that $\dfrac{\partial^2 F}{\partial x\, \partial y} = ye^{xy}$ and use the result of Exercise 52 to evaluate $\iint_{\mathcal{R}} ye^{xy}\, dA$ for $\mathcal{R} = [1, 3] \times [0, 1]$.

SOLUTION Differentiating $F(x, y) = x^{-1}e^{xy}$ with respect to y gives

$$\frac{\partial F}{\partial y} = \frac{\partial}{\partial y}\left(x^{-1}e^{xy}\right) = x^{-1}xe^{xy} = e^{xy}$$

We now differentiate $\dfrac{\partial F}{\partial y}$ with respect to x:

$$\frac{\partial^2 F}{\partial x\, \partial y} = \frac{\partial}{\partial x}(e^{xy}) = ye^{xy}$$

In Exercise 52 we showed that

$$\int_c^d \int_a^b \frac{\partial^2 F}{\partial x\, \partial y}\, dx\, dy = F(b,d) - F(b,c) - F(a,d) + F(a,c)$$

Therefore, for $F(x,y) = x^{-1}e^{xy} = \dfrac{e^{xy}}{x}$ we obtain

$$\iint_{\mathcal{R}} ye^{xy}\, dA = \int_0^1 \int_1^3 ye^{xy}\, dx\, dy = F(3,1) - F(3,0) - F(1,1) + F(1,0)$$

$$= \frac{e^3}{3} - \frac{e^0}{3} - \frac{e^1}{1} + \frac{e^0}{1} = \frac{e^3}{3} - \frac{1}{3} - e + 1 = 4.644$$

55. In this exercise, we use double integration to evaluate the following improper integral for $a > 0$ a positive constant:

$$I(a) = \int_0^{\infty} \frac{e^{-x} - e^{-ax}}{x}\, dx$$

(a) Use L'Hôpital's Rule to show that $f(x) = \dfrac{e^{-x} - e^{-ax}}{x}$, though not defined at $x = 0$, can be defined and made continuous at $x = 0$ by assigning the value $f(0) = a - 1$.

(b) Prove that $|f(x)| \le e^{-x} + e^{-ax}$ for $x > 1$ (use the Triangle Inequality), and apply the Comparison Theorem to show that $I(a)$ converges.

(c) Show that $I(a) = \displaystyle\int_0^{\infty}\int_1^a e^{-xy}\, dy\, dx$.

(d) Prove, by interchanging the order of integration, that

$$I(a) = \ln a - \lim_{T \to \infty} \int_1^a \frac{e^{-Ty}}{y}\, dy \qquad \boxed{1}$$

(e) Use the Comparison Theorem to show that the limit in Eq. (1) is zero and therefore that $I(a) = \ln a$ (see Figure 19, for example). *Hint:* If $a \ge 1$, show that $e^{-Ty}/y \le e^{-T}$ for $y \ge 1$, and if $a < 1$, show that $e^{-Ty}/y \le e^{-aT}/a$ for $a \le y \le 1$.

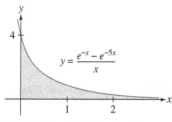

FIGURE 19 The shaded region has area $\ln 5$.

SOLUTION

(a) The function $f(x) = \frac{e^{-x}-e^{-ax}}{x}$, $f(0) = a - 1$ is continuous if $\lim_{x \to 0} f(x) = f(0) = a - 1$. We verify this limit using L'Hôpital's Rule:

$$\lim_{x \to 0} \frac{e^{-x} - e^{-ax}}{x} = \lim_{x \to 0} \frac{-e^{-x} + ae^{-ax}}{1} = -1 + a = a - 1$$

Therefore, f is continuous. We now show that the following integral converges:

$$I(a) = \int_0^\infty \frac{e^{-x} - e^{-ax}}{x} dx \quad (a > 0)$$

Since $e^{-x} - e^{-ax} < e^{-x} + e^{-ax}$ then also $\frac{e^{-x}-e^{-ax}}{x} < \frac{e^{-x}+e^{-ax}}{x}$ for $x > 0$. Therefore, if $x > 1$ we have

$$\frac{e^{-x} - e^{-ax}}{x} < \frac{e^{-x} + e^{-ax}}{x} < e^{-x} + e^{-ax}$$

That is, for $x > 1$,

$$f(x) < e^{-x} + e^{-ax} \tag{1}$$

Also, since $e^{-ax} - e^{-x} < e^{-ax} + e^{-x}$ we have for $x > 1$,

$$\frac{e^{-ax} - e^{-x}}{x} < \frac{e^{-ax} + e^{-x}}{x} < e^{-ax} + e^{-x}$$

Thus, we get

$$-f(x) < e^{-x} + e^{-ax} \tag{2}$$

Together with (1) we have

$$0 \leq |f(x)| < e^{-x} + e^{-ax} \tag{3}$$

We now show that the integral of the right-hand side converges:

$$\int_0^\infty (e^{-x} + e^{-ax}) \, dx = \lim_{R \to \infty} \int_0^R (e^{-x} + e^{-ax}) \, dx = \lim_{R \to \infty} \left(-e^{-x} - \frac{e^{-ax}}{a} \Big|_{x=0}^R \right)$$

$$= \lim_{R \to \infty} \left(-e^{-R} - \frac{e^{-aR}}{a} + e^0 + \frac{e^0}{a} \right) = \lim_{R \to \infty} \left(-e^{-R} - \frac{e^{-aR}}{a} + 1 + \frac{1}{a} \right) = 1 + \frac{1}{a}$$

Since the integral converges, we conclude by (3) and the Comparison Test for Improper Integrals that

$$\int_0^\infty \frac{e^{-x} - e^{-ax}}{x} dx$$

also converges for $a > 0$.

(b) We compute the inner integral with respect to y:

$$\int_1^a e^{-xy} \, dy = -\frac{1}{x} e^{-xy} \Big|_{y=1}^a = -\frac{1}{x}\left(e^{-xa} - e^{-x \cdot 1}\right) = \frac{e^{-x} - e^{-xa}}{x}$$

Therefore,

$$\int_0^\infty \int_1^a e^{-xy} \, dy \, dx = \int_0^\infty \left(\int_1^a e^{-xy} \, dy \right) dx = \int_0^\infty \frac{e^{-x} - e^{-xa}}{x} dx = I(a)$$

(c) By the definition of the improper integral,

$$I(a) = \lim_{T \to \infty} \int_0^T \int_1^a e^{-xy} \, dy \, dx \tag{4}$$

We compute the double integral. Using Fubini's Theorem we may compute the iterated integral using reversed order of integration. That is,

$$\int_0^T \int_1^a e^{-xy}\,dy\,dx = \int_1^a \int_0^T e^{-xy}\,dx\,dy = \int_1^a \left(\int_0^T e^{-xy}\,dx\right) dy = \int_1^a \left(-\frac{1}{y}e^{-xy}\bigg|_{x=0}^T\right) dy$$

$$= \int_1^a \left(-\frac{1}{y}\left(e^{-Ty} - e^{-0\cdot y}\right)\right) dy = \int_1^a \frac{1 - e^{-Ty}}{y}\,dy = \int_1^a \frac{dy}{y} - \int_1^a \frac{e^{-Ty}}{y}\,dy$$

$$= \ln y \bigg|_1^a - \int_1^a \frac{e^{-Ty}}{y}\,dy = \ln a - \ln 1 - \int_1^a \frac{e^{-Ty}}{y}\,dy = \ln a - \int_1^a \frac{e^{-Ty}}{y}\,dy$$

Combining with (4) we get

$$I(a) = \ln a - \lim_{T\to\infty} \int_1^a \frac{e^{-Ty}}{y}\,dy \tag{5}$$

(d) We now show, using the Comparison Theorem, that

$$\lim_{T\to\infty} \int_1^a \frac{e^{-Ty}}{y}\,dy = 0$$

We consider the following possible cases:

Case 1: $a \geq 1$. Then in the interval of integration $y \geq 1$. Also since $T \to \infty$, we may assume that $T > 0$. Thus,

$$\frac{e^{-Ty}}{y} \leq \frac{e^{-T\cdot 1}}{1} = e^{-T}$$

Hence,

$$0 \leq \int_1^a \frac{e^{-Ty}}{y}\,dy \leq \int_1^a e^{-T}\,dy = e^{-T}(a - 1)$$

By the limit $\lim_{T\to\infty} e^{-T}(a - 1) = 0$ and the Squeeze Theorem we conclude that,

$$\lim_{T\to\infty} \int_1^a \frac{e^{-Ty}}{y}\,dy = 0$$

Case 2: $0 < a < 1$. Then,

$$\int_1^a \frac{e^{-Ty}}{y}\,dy = -\int_a^1 \frac{e^{-Ty}}{y}\,dy$$

and in the interval of integration $a \leq y \leq 1$, therefore

$$\frac{e^{-Ty}}{y} \leq \frac{e^{-Ta}}{a}$$

(the function $\frac{e^{-Ty}}{y}$ is decreasing). Hence,

$$0 \leq \int_a^1 \frac{e^{-Ty}}{y}\,dy \leq \int_a^1 \frac{e^{-Ta}}{a}\,dy = \frac{(1-a)}{a}e^{-Ta}$$

By the limit $\lim_{T\to\infty} \frac{1-a}{a}e^{-Ta} = 0$ and the Squeeze Theorem we conclude also that

$$\lim_{T\to\infty} \int_1^a \frac{e^{-Ty}}{y}\,dy = -\lim_{T\to\infty} \int_a^1 \frac{e^{-Ty}}{y}\,dy = 0$$

We thus showed that for all $a > 0$, $\lim_{T\to\infty} \int_1^a \frac{e^{-Ty}}{y} = 0$. Combining with Eq. (5) obtained in part (c), we find that $I(a) = \ln a$.

15.2 Double Integrals over More General Regions
(LT Section 16.2)

Preliminary Questions

1. Which of the following expressions do not make sense?

(a) $\displaystyle\int_0^1 \int_1^x f(x,y)\,dy\,dx$
(b) $\displaystyle\int_0^1 \int_1^y f(x,y)\,dy\,dx$
(c) $\displaystyle\int_0^1 \int_x^y f(x,y)\,dy\,dx$
(d) $\displaystyle\int_0^1 \int_x^1 f(x,y)\,dy\,dx$

SOLUTION
(a) This integral is the following iterated integral:
$$\int_0^1 \int_1^x f(x,y)\,dy\,dx = \int_0^1 \left(\int_1^x f(x,y)\,dy\right)dx$$
The inner integral is a function of x and it is integrated with respect to x over the interval $0 \le x \le 1$. The result is a number. This integral makes sense.

(b) This integral is the same as
$$\int_0^1 \int_1^y f(x,y)\,dy\,dx = \int_0^1 \left(\int_1^y f(x,y)\,dy\right)dx$$
The inner integral is an integral with respect to y, over the interval $[1, y]$. This does not make sense.

(c) This integral is the following iterated integral:
$$\int_0^1 \left(\int_x^y f(x,y)\right)dy\,dx$$
The inner integral is a function of x and y and it is integrated with respect to y over the interval $x \le y \le y$. This does not make sense.

(d) This integral is the following iterated integral:
$$\int_0^1 \left(\int_x^1 f(x,y)\,dy\right)dx$$
The inner integral is a function of x and it is integrated with respect to x. This makes sense.

2. Draw a domain in the plane that is neither vertically nor horizontally simple.

SOLUTION The following region cannot be described in the form $\{a \le x \le b, \alpha(x) \le y \le \beta(x)\}$ nor in the form $\{c \le y \le d, \alpha(y) \le x \le \beta(y)\}$, hence it is neither vertically nor horizontally simple.

3. Which of the four regions in Figure 21 is the domain of integration for $\displaystyle\int_{-\sqrt{2}/2}^{0} \int_{-x}^{\sqrt{1-x^2}} f(x,y)\,dy\,dx$?

FIGURE 21

SOLUTION The region B is defined by the inequalities

$$-x \le y \le \sqrt{1-x^2}, \qquad -\frac{\sqrt{2}}{2} \le x \le 0$$

To compute $\int_{-\sqrt{2}/2}^{0} \int_{-x}^{\sqrt{1-x^2}} f(x,y)\,dy\,dx$, we first integrate with respect to y over the interval $-x \le y \le \sqrt{1-x^2}$, and then with respect to x over $-\frac{\sqrt{2}}{2} \le x \le 0$. That is, the domain of integration is B.

4. Let $\mathcal{D}$ be the unit disk. If the maximum value of $f(x,y)$ on $\mathcal{D}$ is 4, then the largest possible value of $\iint_{\mathcal{D}} f(x,y)\,dA$ is (choose the correct answer):

(a) 4   (b) 4π (c) $\dfrac{4}{\pi}$

SOLUTION The area of the unit disk is π and the maximum value of $f(x,y)$ on this region is $M = 4$, therefore we have,

$$\iint_{\mathcal{D}} f(x,y)\,dx\,dy \le 4\pi$$

The correct answer is (b).

Exercises

1. Calculate the Riemann sum for $f(x,y) = x - y$ and the shaded domain $\mathcal{D}$ in Figure 22 with two choices of sample points, • and ∘. Which do you think is a better approximation to the integral of f over $\mathcal{D}$? Why?

FIGURE 22

SOLUTION The subrectangles in Figure 22 have sides of length $\Delta x = \Delta y = 1$ and area $\Delta A = 1 \cdot 1 = 1$.

- Sample points •. There are six sample points that lie in the domain $\mathcal{D}$. We compute the values of $f(x,y) = x - y$ at these points:

$$f(1,1) = 0, \quad f(1,2) = -1, \quad f(1,3) = -2,$$
$$f(2,1) = 1, \quad f(2,2) = 0, \quad f(2,3) = -1$$

The Riemann sum is

$$S_{3,4} = (0 - 1 - 2 + 1 + 0 - 1) \cdot 1 = -3$$

- Sample points ∘. We compute the values of $f(x,y) = x - y$ at the eight sample points that lie in $\mathcal{D}$:

$$f(1.5, 0.5) = 1, \quad f(0.5, 1.5) = -1, \quad f(0.5, 2.5) = -2,$$
$$f(1.5, 3.5) = -2, \quad f(1.5, 1.5) = 0, \quad f(1.5, 2.5) = -1,$$
$$\qquad\qquad\qquad f(2.5, 1.5) = 1, \quad f(2.5, 2.5) = 0$$

The corresponding Riemann sum is thus

$$S_{34} = (1 - 1 - 2 + 0 - 1 - 2 + 1 + 0) \cdot 1 = -4$$

Since there are eight sample points ○ but only six ●, we might expect that the choice of ○ gives a better approximation.

3. Express the domain $\mathcal{D}$ in Figure 24 as both a vertically simple region and a horizontally simple region, and evaluate the integral of $f(x, y) = xy$ over $\mathcal{D}$ as an iterated integral in two ways.

FIGURE 24

SOLUTION The domain $\mathcal{D}$ can be described as a vertically simple region as follows:

$$0 \le x \le 1, \quad 0 \le y \le 1 - x^2 \tag{1}$$

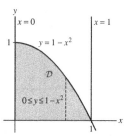

Vertically simple region

The domain $\mathcal{D}$ can also be described as a horizontally simple region. To do this, we must express x in terms of y, for nonnegative values of x. This gives

$$y = 1 - x^2 \quad \Rightarrow \quad x^2 = 1 - y \quad \Rightarrow \quad x = \sqrt{1 - y}$$

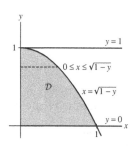

Horizontally simple region

Therefore, we can describe $\mathcal{D}$ by the following inequalities:

$$0 \le y \le 1, \quad 0 \le x \le \sqrt{1 - y} \tag{2}$$

We now compute the integral of $f(x, y) = xy$ over $\mathcal{D}$ first using definition (1) and then using definition (2). We obtain

$$\iint_{\mathcal{D}} xy \, dA = \int_0^1 \int_0^{1-x^2} xy \, dy \, dx = \int_0^1 \left.\frac{xy^2}{2}\right|_{y=0}^{1-x^2} dx = \int_0^1 \frac{x}{2}\left((1-x^2)^2 - 0^2\right) dx = \int_0^1 \frac{x(1-x^2)^2}{2} dx$$

$$= \frac{1}{2} \int_0^1 (x - 2x^3 + x^5) \, dx = \frac{1}{2}\left.\left(\frac{x^2}{2} - \frac{x^4}{2} + \frac{x^6}{6}\right)\right|_0^1 = \frac{1}{2}\left(\frac{1}{2} - \frac{1}{2} + \frac{1}{6}\right) = \frac{1}{12}$$

Using definition (2) gives

$$\iint_D xy\, dA = \int_0^1 \int_0^{\sqrt{1-y}} xy\, dx\, dy = \int_0^1 \frac{yx^2}{2}\bigg|_{x=0}^{\sqrt{1-y}} dy = \int_0^1 \frac{y}{2}\left(\left(\sqrt{1-y}\right)^2 - 0^2\right) dy$$
$$= \int_0^1 \frac{y}{2}(1-y)\, dy = \int_0^1 \left(\frac{y}{2} - \frac{y^2}{2}\right) dy = \frac{y^2}{4} - \frac{y^3}{6}\bigg|_0^1 = \frac{1}{4} - \frac{1}{6} = \frac{1}{12}$$

The answers agree as expected.

In Exercises 5–7, compute the double integral of $f(x,y) = x^2 y$ over the given shaded domain in Figure 25.

(A) (B) (C)

FIGURE 25

5. (A)

SOLUTION

We describe the domain $\mathcal{D}$ as a vertically simple region. We find the equation of the line connecting the points $(0, 2)$ and $(4, 0)$.

$$y - 0 = \frac{2-0}{0-4}(x-4) \quad \Rightarrow \quad y = -\frac{1}{2}x + 2$$

Therefore the domain is described as a vertically simple region by the inequalities

$$0 \le x \le 4, \quad -\frac{1}{2}x + 2 \le y \le 2$$

We use Theorem 2 to evaluate the double integral:

$$\iint_D x^2 y\, dA = \int_0^4 \int_{-\frac{x}{2}+2}^{2} x^2 y\, dy\, dx = \int_0^4 \frac{x^2 y^2}{2}\bigg|_{y=-\frac{x}{2}+2}^{2} dx = \int_0^4 \frac{x^2}{2}\left(2^2 - \left(-\frac{x}{2}+2\right)^2\right) dx$$
$$= \int_0^4 \left(x^3 - \frac{x^4}{8}\right) dx = \frac{x^4}{4} - \frac{x^5}{40}\bigg|_0^4 = \frac{4^4}{4} - \frac{4^5}{40} = \frac{192}{5} = 38.4$$

7. (C)

SOLUTION The domain in (C) is a horizontally simple region, described by the inequalities

$$0 \le y \le 2, \quad y \le x \le 4$$

Using Theorem 2 we obtain the following integral:

$$\iint_D x^2 y\, dA = \int_0^2 \int_y^4 x^2 y\, dx\, dy = \int_0^2 \frac{x^3 y}{3}\bigg|_{x=y}^{x=4} dy = \int_0^2 \frac{y}{3}(4^3 - y^3)\, dy = \int_0^2 \left(\frac{64y}{3} - \frac{y^4}{3}\right) dy$$
$$= \frac{32}{3}y^2 - \frac{y^5}{15}\bigg|_0^2 = \frac{32 \cdot 2^2}{3} - \frac{2^5}{15} = \frac{608}{15} \approx 40.53$$

9. Integrate $f(x, y) = x$ over the region bounded by $y = x^2$ and $y = x + 2$.

SOLUTION The domain of integration is shown in the following figure:

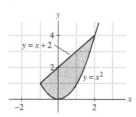

To find the inequalities defining the domain as a vertically simple region we first must find the x-coordinates of the two points where the line $y = x + 2$ and the parabola $y = x^2$ intersect. That is,

$$x + 2 = x^2 \quad \Rightarrow \quad x^2 - x - 2 = (x - 2)(x + 1) = 0$$
$$\Rightarrow \quad x_1 = -1, \quad x_2 = 2$$

We describe the domain by the following inequalities:

$$-1 \le x \le 2, \quad x^2 \le y \le x + 2$$

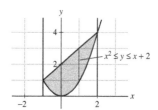

We now evaluate the integral of $f(x, y) = x$ over the vertically simple region $\mathcal{D}$:

$$\iint_{\mathcal{D}} x\, dA = \int_{-1}^{2} \int_{x^2}^{x+2} x\, dy\, dx = \int_{-1}^{2} xy \Big|_{y=x^2}^{x+2} dx = \int_{-1}^{2} x(x + 2 - x^2)\, dx$$
$$= \int_{-1}^{2} (x^2 + 2x - x^3)\, dx = \frac{x^3}{3} + x^2 - \frac{x^4}{4} \Big|_{-1}^{2} = \left(\frac{8}{3} + 4 - 4\right) - \left(-\frac{1}{3} + 1 - \frac{1}{4}\right) = 2\frac{1}{4}$$

11. Evaluate $\iint_{\mathcal{D}} \frac{y}{x}\, dA$, where $\mathcal{D}$ is the shaded part inside the semicircle of radius 2 in Figure 26.

FIGURE 26 $y = \sqrt{4 - x^2}$.

SOLUTION The region $\mathcal{D}$ is defined by the following inequalities:

$$1 \le x \le 2, \quad 0 \le y \le \sqrt{4 - x^2}$$

The double integral of f over $\mathcal{D}$ is computed using Theorem 2:

$$\iint_{\mathcal{D}} \frac{y}{x}\, dA = \int_{1}^{2} \int_{0}^{\sqrt{4-x^2}} \frac{y}{x}\, dy\, dx$$
$$= \int_{1}^{2} \frac{y^2}{2x} \Big|_{0}^{\sqrt{4-x^2}} dx$$
$$= \int_{1}^{2} \left(\frac{4 - x^2}{2x} - 0\right) dx$$

$$= \int_1^2 \left(\frac{2}{x} - \frac{1}{2}x\right) dx$$

$$= \left(2\ln|x| - \frac{1}{4}x^2\right)\Big|_1^2$$

$$= 2\ln 2 - \frac{1}{4}\cdot 2^2 - 0 + \frac{1}{4}\cdot 1^2 = 2\ln 2 - \frac{3}{4}$$

13. Calculate the double integral of $f(x,y) = x + y$ over the domain $\mathcal{D} = \{(x,y): x^2 + y^2 \leq 4, y \geq 0\}$.

SOLUTION

The domain $\mathcal{D}$

The semicircle can be described as a vertically simple region, by the following inequalities:

$$-2 \leq x \leq 2, \quad 0 \leq y \leq \sqrt{4-x^2}$$

We evaluate the double integral by the following iterated integral:

$$\iint_{\mathcal{D}} (x+y)\, dA = \int_{-2}^{2} \int_0^{\sqrt{4-x^2}} (x+y)\, dy\, dx = \int_{-2}^{2} xy + \frac{1}{2}y^2 \Big|_{y=0}^{\sqrt{4-x^2}} dx = \int_{-2}^{2} \left(x\sqrt{4-x^2} + \frac{1}{2}\left(\sqrt{4-x^2}\right)^2\right) dx$$

$$= \int_{-2}^{2} x\sqrt{4-x^2}\, dx + \frac{1}{2}\int_{-2}^{2} (4-x^2)\, dx = \int_{-2}^{2} x\sqrt{4-x^2}\, dx + 2x - \frac{x^3}{6}\Big|_{x=-2}^{2}$$

$$= \int_{-2}^{2} x\sqrt{4-x^2}\, dx + 4 - \frac{8}{6} - \left(-4 - \frac{-8}{6}\right) = \int_{-2}^{2} x\sqrt{4-x^2}\, dx + \frac{16}{3}$$

The integral of an odd function over an interval that is symmetric with respect to the origin is zero. Hence $\int_{-2}^{2} x\sqrt{4-x^2}\, dx = 0$, so we get

$$\iint_{\mathcal{D}} (x+y)\, dA = 0 + \frac{16}{3} = \frac{16}{3} \approx 5.33$$

15. Calculate the integral of $f(x,y) = x$ over the region $\mathcal{D}$ bounded above by $y = x(2-x)$ and below by $x = y(2-y)$. *Hint:* Apply the quadratic formula to the lower boundary curve to solve for y as a function of x.

SOLUTION The two graphs are symmetric with respect to the line $y = x$, thus their point of intersection is $(1, 1)$. The region $\mathcal{D}$ is shown in the following figure:

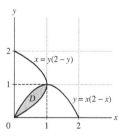

To find the inequalities defining the region $\mathcal{D}$ as a vertically simple region, we first must solve the lower boundary curve for y in terms of x. We get

$$x = y(2-y) = 2y - y^2$$
$$y^2 - 2y + x = 0$$

We solve the quadratic equation in y:

$$y = 1 \pm \sqrt{1-x}$$

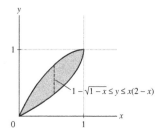

The domain $\mathcal{D}$ lies below the line $y = 1$, hence the appropriate solution is $y = 1 - \sqrt{1-x}$. We obtain the following inequalities for $\mathcal{D}$:

$$0 \le x \le 1, \quad 1 - \sqrt{1-x} \le y \le x(2-x)$$

We now evaluate the double integral of $f(x, y) = x$ over $\mathcal{D}$:

$$\iint_{\mathcal{D}} x\, dA = \int_0^1 \int_{1-\sqrt{1-x}}^{x(2-x)} x\, dy\, dx = \int_0^1 xy \Big|_{y=1-\sqrt{1-x}}^{x(2-x)} dx = \int_0^1 \left(x^2(2-x) - \left(x - x\sqrt{1-x}\right)\right) dx$$

$$= \int_0^1 \left(2x^2 - x^3 - x + x\sqrt{1-x}\right) dx = \frac{2x^3}{3} - \frac{x^4}{4} - \frac{x^2}{2}\Big|_0^1 + \int_0^1 x\sqrt{1-x}\, dx$$

$$= -\frac{1}{12} + \int_0^1 x\sqrt{1-x}\, dx$$

Using the substitution $u = \sqrt{1-x}$ it can be shown that $\int_0^1 x\sqrt{1-x}\, dx = \frac{4}{15}$. Therefore we get

$$\iint_{\mathcal{D}} x\, dA = -\frac{1}{12} + \frac{4}{15} = \frac{11}{60}$$

In Exercises 17–24, compute the double integral of $f(x, y)$ over the domain $\mathcal{D}$ indicated.

17. $f(x, y) = x^3 y; \quad 0 \le x \le 5, \quad x \le y \le 2x + 3$

SOLUTION The vertically simple region $\mathcal{D}$ defined by the given inequalities is shown in the figure below:

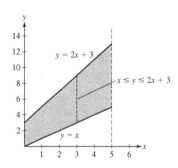

We compute the double integral as the following iterated integral:

$$\iint_{\mathcal{D}} x^3 y\, dA = \int_0^5 \int_x^{2x+3} x^3 y\, dy\, dx$$

$$= \int_0^5 \frac{1}{2} x^3 y^2 \Big|_x^{2x+3} dx$$

$$= \frac{1}{2} \int_0^5 x^3 ((2x+3)^2 - x^2)\, dx$$

$$= \frac{1}{2} \int_0^5 (3x^5 + 12x^4 + 9x^3)\, dx$$

$$= \frac{1}{2} \left(\frac{1}{2} x^6 + \frac{12}{5} x^5 + \frac{9}{4} x^4\right) \Big|_0^5$$

$$= \frac{1}{2}\left(\frac{15625}{2} + 7500 + \frac{5625}{4}\right)$$

$$= \frac{66875}{8}$$

19. $f(x, y) = x;\quad 0 \le x \le 1,\quad 1 \le y \le e^{x^2}$

SOLUTION We compute the double integral of $f(x, y) = x$ over the vertically simple region $\mathcal{D}$, as the following iterated integral:

$$\iint_{\mathcal{D}} x\, dA = \int_0^1 \int_1^{e^{x^2}} x\, dy\, dx = \int_0^1 xy\Big|_{y=1}^{e^{x^2}} dx = \int_0^1 \left(xe^{x^2} - x \cdot 1\right) dx$$

$$= \int_0^1 xe^{x^2}\, dx - \int_0^1 x\, dx = \int_0^1 xe^{x^2}\, dx - \frac{x^2}{2}\Big|_0^1 = \int_0^1 xe^{x^2}\, dx - \frac{1}{2} \tag{1}$$

The resulting integral can be computed using the substitution $u = x^2$. The value of this integral is

$$\int_0^1 xe^{x^2}\, dx = \frac{e-1}{2}$$

Combining with (1) we get

$$\iint_{\mathcal{D}} x\, dA = \frac{e-1}{2} - \frac{1}{2} = \frac{e-2}{2} \approx 0.359$$

21. $f(x, y) = 6xy - x^2$; bounded below by $y = x^2$, above by $y = \sqrt{x}$

SOLUTION The curves $y = x^2$ and $y = \sqrt{x}$ meet at $(0, 0)$ and $(1, 1)$, so the given area can be described by the inequalities

$$0 \le x \le 1,\quad x^2 \le y \le \sqrt{x}$$

These inequalities describe $\mathcal{D}$ as a vertically simple region, shown in the figure below:

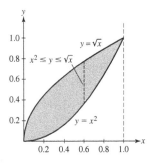

The double integral can then be computed as an iterated integral:

$$\iint_{\mathcal{D}} (6xy - x^2)\, dA = \int_0^1 \int_{x^2}^{\sqrt{x}} (6xy - x^2)\, dy\, dx$$

$$= \int_0^1 (3xy^2 - x^2 y)\Big|_{x^2}^{\sqrt{x}} dx$$

$$= \int_0^1 ((3x^2 - x^{5/2}) - (3x^5 - x^4))\, dx$$

$$= \left(x^3 - \frac{2}{7}x^{7/2} - \frac{1}{2}x^6 + \frac{1}{5}x^5\right)\Big|_0^1$$
$$= \frac{29}{70}$$

23. $f(x, y) = e^{x+y}$; bounded by $y = x - 1$, $y = 12 - x$ for $2 \le y \le 4$

SOLUTION The horizontally simple region $\mathcal{D}$ is shown in the figure:

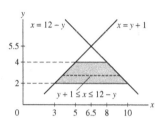

We compute the double integral of $f(x, y) = e^{x+y}$ over $\mathcal{D}$ by evaluating the following iterated integral:

$$\iint_{\mathcal{D}} e^{x+y}\, dA = \int_2^4 \int_{y+1}^{12-y} e^{x+y}\, dx\, dy = \int_2^4 e^{x+y}\Big|_{x=y+1}^{12-y}\, dy = \int_2^4 \left(e^{12-y+y} - e^{y+1+y}\right) dy$$

$$= \int_2^4 \left(e^{12} - e^{2y+1}\right) dy = e^{12} \cdot y - \frac{1}{2}e^{2y+1}\Big|_2^4 = \left(e^{12} \cdot 4 - \frac{1}{2}e^{2\cdot 4+1}\right) - \left(e^{12} \cdot 2 - \frac{1}{2}e^{2\cdot 2+1}\right)$$

$$= 4e^{12} - \frac{1}{2}e^9 - 2e^{12} + \frac{1}{2}e^5 = 2e^{12} - \frac{1}{2}e^9 + \frac{1}{2}e^5 \approx 321532.2$$

In Exercises 25–28, sketch the domain of integration and express as an iterated integral in the opposite order.

25. $\displaystyle\int_0^4 \int_x^4 f(x, y)\, dy\, dx$

SOLUTION The limits of integration correspond to the inequalities describing the following domain $\mathcal{D}$:

$$0 \le x \le 4, \quad x \le y \le 4$$

From the sketch of $\mathcal{D}$ we see that $\mathcal{D}$ can also be expressed as a horizontally simple region as follows:

$$0 \le y \le 4, \quad 0 \le x \le y$$

Therefore we can reverse the order of integration as follows:

$$\int_0^4 \int_x^4 f(x, y)\, dy\, dx = \int_0^4 \int_0^y f(x, y)\, dx\, dy$$

27. $\displaystyle\int_4^9 \int_2^{\sqrt{y}} f(x,y)\,dx\,dy$

SOLUTION The limits of integration correspond to the following inequalities defining the horizontally simple region $\mathcal{D}$:

$$4 \le y \le 9, \quad 2 \le x \le \sqrt{y}$$

The region $\mathcal{D}$ can also be expressed as a vertically simple region. We first need to write the equation of the curve $x = \sqrt{y}$ in the form $y = x^2$. The corresponding inequalities are

$$2 \le x \le 3, \quad x^2 \le y \le 9$$

We now can write the iterated integral with reversed order of integration:

$$\int_4^9 \int_2^{\sqrt{y}} f(x,y)\,dx\,dy = \int_2^3 \int_{x^2}^9 f(x,y)\,dy\,dx$$

29. Sketch the domain $\mathcal{D}$ corresponding to

$$\int_0^4 \int_{\sqrt{y}}^2 \sqrt{4x^2 + 5y}\,dx\,dy$$

Then change the order of integration and evaluate.

SOLUTION The limits of integration correspond to the following inequalities describing the domain $\mathcal{D}$:

$$0 \le y \le 4, \quad \sqrt{y} \le x \le 2$$

The horizontally simple region $\mathcal{D}$ is shown in the figure:

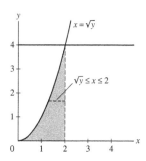

The domain $\mathcal{D}$ can also be described as a vertically simple region. Rewriting the equation $x = \sqrt{y}$ in the form $y = x^2$, we define $\mathcal{D}$ by the following inequalities (see figure):

$$0 \le x \le 2, \quad 0 \le y \le x^2$$

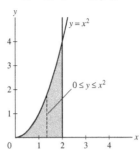

The corresponding iterated integral is

$$\int_0^2 \int_0^{x^2} \sqrt{4x^2 + 5y}\, dy\, dx$$

We evaluate this integral:

$$\int_0^2 \int_0^{x^2} \sqrt{4x^2 + 5y}\, dy\, dx = \int_0^2 \frac{2}{15}(4x^2 + 5y)^{3/2}\bigg|_0^{x^2} dx$$

$$= \frac{2}{15}\int_0^2 \left((4x^2 + 5x^2)^{3/2} - (4x^2)^{3/2}\right) dx$$

$$= \frac{2}{15}\int_0^2 (27x^3 - 8x^3)\, dx = \frac{2}{15}\int_0^2 19x^3\, dx$$

$$= \frac{2}{15} \cdot \frac{1}{4} \cdot 19x^4\bigg|_0^2 = \frac{152}{15}$$

31. Compute the integral of $f(x,y) = (\ln y)^{-1}$ over the domain $\mathcal{D}$ bounded by $y = e^x$ and $y = e^{\sqrt{x}}$. *Hint:* Choose the order of integration that enables you to evaluate the integral.

SOLUTION The region $\mathcal{D}$ bounded by the curves $y = e^x$, $y = e^{\sqrt{x}}$, $y = 1$, and $y = 2$ is shown in the following figure:

To express $\mathcal{D}$ as a horizontally simple region, we first must rewrite the equations of the curves $y = e^x$ and $y = e^{\sqrt{x}}$ with x as a function of y. That is,

$$y = e^x \quad \Rightarrow \quad x = \ln y$$

$$y = e^{\sqrt{x}} \quad \Rightarrow \quad \sqrt{x} = \ln y \quad \Rightarrow \quad x = \ln^2 y$$

We obtain the following inequalities:

$$1 \le y \le 2, \quad \ln^2 y \le x \le \ln y$$

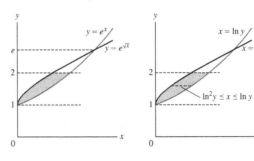

Using Theorem 2, we compute the double integral of $f(x,y) = (\ln y)^{-1}$ over $\mathcal{D}$ as the following iterated integral:

$$\iint_{\mathcal{D}} (\ln y)^{-1} \, dA = \int_1^2 \int_{\ln^2 y}^{\ln y} (\ln y)^{-1} \, dx \, dy = \int_1^2 (\ln y)^{-1} x \Big|_{x=\ln^2 y}^{\ln y} dy = \int_1^2 (\ln y)^{-1} \left(\ln y - \ln^2 y \right) dy$$

$$= \int_1^2 (1 - \ln y) \, dy = \int_1^2 1 \, dy - \int_1^2 \ln y \, dy = y \Big|_1^2 - y(\ln y - 1) \Big|_1^2$$

$$= 1 - (2(\ln 2 - 1) - 1(\ln 1 - 1)) = 2 - 2\ln 2 \approx 0.614$$

In Exercises 33–36, sketch the domain of integration. Then change the order of integration and evaluate. Explain the simplification achieved by changing the order.

33. $\int_0^1 \int_y^1 \frac{\sin x}{x} \, dx \, dy$

SOLUTION The limits of integration correspond to the following inequalities:

$$0 \leq y \leq 1, \quad y \leq x \leq 1$$

The horizontally simple region $\mathcal{D}$ is shown in the figure.

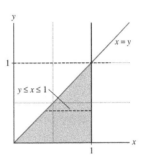

We see that $\mathcal{D}$ can also be described as a vertically simple region, by the following inequalities:

$$0 \leq x \leq 1, \quad 0 \leq y \leq x$$

We evaluate the corresponding iterated integral:

$$\int_0^1 \int_0^x \frac{\sin x}{x} \, dy \, dx = \int_0^1 \frac{\sin x}{x} y \Big|_{y=0}^x dx = \int_0^1 \frac{\sin x}{x} (x - 0) \, dx = \int_0^1 \sin x \, dx = -\cos x \Big|_0^1 = 1 - \cos 1 \approx 0.46$$

Trying to integrate in reversed order we obtain a complicated integral in the inner integral. That is,

$$\int_0^1 \int_y^1 \frac{\sin x}{x} \, dx \, dy = \int_0^1 \left(\int_y^1 \frac{\sin x}{x} \, dx \right) dy$$

Remark: $f(x,y) = \frac{\sin x}{x}$ is not continuous at the point $(0,0)$ in $\mathcal{D}$. To make it continuous we need to define $f(0,0) = 1$.

35. $\displaystyle\int_0^1 \int_{y=x}^1 xe^{y^3}\, dy\, dx$

SOLUTION The limits of integration define a vertically simple region $\mathcal{D}$ by the following inequalities:

$$0 \le x \le 1, \quad x \le y \le 1$$

This region can also be described as a horizontally simple region by the following inequalities (see figure):

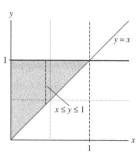

$$0 \le y \le 1, \quad 0 \le x \le y$$

We thus can rewrite the given integral in reversed order of integration as follows:

$$\int_0^1 \int_0^y xe^{y^3}\, dx\, dy = \int_0^1 \frac{x^2}{2}e^{y^3}\bigg|_{x=0}^y dy = \int_0^1 e^{y^3}\left(\frac{y^2}{2} - 0\right) dy = \int_0^1 \frac{1}{2}e^{y^3} y^2\, dy$$

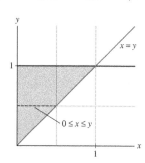

We compute this integral using the substitution $u = y^3$, $du = 3y^2\, dy$. This gives

$$\int_0^1 \int_0^y xe^{y^3}\, dx\, dy = \int_0^1 \frac{1}{2}e^{y^3} y^2\, dy = \int_0^1 e^u \cdot \frac{1}{6}\, du = \frac{e^u}{6}\bigg|_0^1 = \frac{e-1}{6} \approx 0.286$$

Trying to evaluate the double integral in the original order of integration, we find that the inner integral is impossible to compute:

$$\int_0^1 \int_x^1 xe^{y^3}\, dy\, dx = \int_0^1 \left(\int_x^1 xe^{y^3}\, dy\right) dx$$

37. Sketch the domain $\mathcal{D}$ where $0 \le x \le 2$, $0 \le y \le 2$, and x or y is greater than 1. Then compute $\displaystyle\iint_\mathcal{D} e^{x+y}\, dA$.

SOLUTION The domain $\mathcal{D}$ within the square $0 \le x, y \le 2$ is shown in the figure.

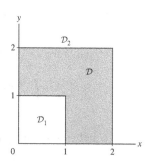

We denote the unit square $0 \leq x, y \leq 1$ and the square $0 \leq x, y \leq 2$ by $\mathcal{D}_1$ and $\mathcal{D}_2$ respectively. Then $\mathcal{D}_2$ is the union of $\mathcal{D}_1$ and $\mathcal{D}$, and these two domains do not overlap except on the boundary of $\mathcal{D}_1$. Therefore, by properties of the double integral, we have

$$\iint_{\mathcal{D}_2} e^{x+y} dA = \iint_{\mathcal{D}_1} e^{x+y} dA + \iint_{\mathcal{D}} e^{x+y} dA$$

Hence,

$$\iint_{\mathcal{D}} e^{x+y} dA = \iint_{\mathcal{D}_2} e^{x+y} dA - \iint_{\mathcal{D}_1} e^{x+y} dA = \int_0^2 \int_0^2 e^{x+y} \, dx\, dy - \int_0^1 \int_0^1 e^{x+y} \, dx\, dy$$

$$= \int_0^2 e^{x+y} \Big|_{x=0}^2 dy - \int_0^1 e^{x+y} \Big|_{x=0}^1 dy = \int_0^2 (e^{2+y} - e^y) \, dy - \int_0^1 (e^{1+y} - e^y) \, dy$$

$$= e^{2+y} - e^y \Big|_{y=0}^2 - (e^{1+y} - e^y) \Big|_{y=0}^1 = e^4 - e^2 - (e^2 - e^0) - \left(e^2 - e - (e - e^0)\right)$$

$$= e^4 - 3e^2 + 2e \approx 37.87$$

In Exercises 39–42, calculate the double integral of $f(x, y)$ over the triangle indicated in Figure 28.

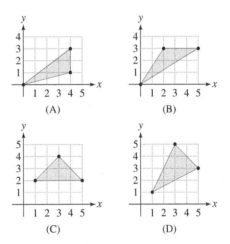

FIGURE 28

39. $f(x, y) = e^{x^2}$, (A)

SOLUTION The equations of the lines OA and OB are $y = \frac{3}{4}x$ and $y = \frac{1}{4}x$, respectively. Therefore, the triangle may be expressed as a vertically simple region by the following inequalities:

$$0 \leq x \leq 4, \quad \frac{x}{4} \leq y \leq \frac{3x}{4}$$

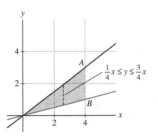

The double integral of $f(x,y) = e^{x^2}$ over the triangle is the following iterated integral:

$$\int_0^4 \int_{x/4}^{3x/4} e^{x^2} \, dy \, dx = \int_0^4 \left(y e^{x^2}\right)\Big|_{y=x/4}^{y=3x/4} dx$$

$$= \frac{1}{2} \int_0^4 x e^{x^2} \, dx$$

$$= \frac{1}{4}\left(e^{x^2}\right)\Big|_0^4 = \frac{1}{4}(e^{16} - 1)$$

41. $f(x,y) = \dfrac{x}{y^2}$, (C)

SOLUTION To find the inequalities defining the triangle as a horizontally simple region, we first find the inequalities of the lines AB and BC:

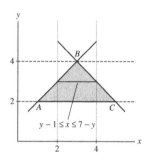

AB: $y - 2 = \dfrac{4-2}{3-1}(x-1) \;\Rightarrow\; y - 2 = x - 1 \;\Rightarrow\; x = y - 1$

BC: $y - 2 = \dfrac{4-2}{3-5}(x-5) \;\Rightarrow\; y - 2 = 5 - x \;\Rightarrow\; x = 7 - y$

We obtain the following inequalities for the triangle:

$$2 \le y \le 4, \quad y - 1 \le x \le 7 - y$$

The double integral of $f(x,y) = \dfrac{x}{y^2}$ over the triangle is the following iterated integral:

$$\int_2^4 \int_{y-1}^{7-y} \frac{x}{y^2} \, dx \, dy = \int_2^4 \frac{x^2}{2y^2}\bigg|_{x=y-1}^{7-y} dy = \int_2^4 \frac{(7-y)^2 - (y-1)^2}{2y^2} \, dy = \int_2^4 \left(\frac{24}{y^2} - \frac{6}{y}\right) dy$$

$$= -\frac{24}{y} - 6\ln y \bigg|_2^4 = -\frac{24}{4} - 6\ln 4 - \left(-\frac{24}{2} - 6\ln 2\right) = 6 - 6\ln 2 = 1.84$$

43. Calculate the double integral of $f(x,y) = \dfrac{\sin y}{y}$ over the region $\mathcal{D}$ in Figure 29.

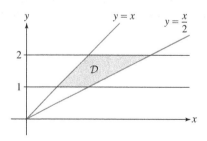

FIGURE 29

SOLUTION To describe $\mathcal{D}$ as a horizontally simple region, we rewrite the equations of the lines with x as a function of y, that is, $x = y$ and $x = 2y$. The inequalities for $\mathcal{D}$ are

$$1 \le y \le 2, \quad y \le x \le 2y$$

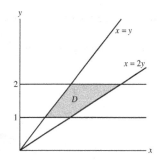

We now compute the double integral of $f(x, y) = \frac{\sin y}{y}$ over $\mathcal{D}$ by the following iterated integral:

$$\iint_{\mathcal{D}} \frac{\sin y}{y} \, dA = \int_1^2 \int_y^{2y} \frac{\sin y}{y} \, dx \, dy = \int_1^2 \frac{\sin y}{y} x \Big|_{x=y}^{2y} \, dy = \int_1^2 \frac{\sin y}{y} (2y - y) \, dy$$

$$= \int_1^2 \frac{\sin y}{y} \cdot y \, dy = \int_1^2 \sin y \, dy = -\cos y \Big|_1^2 = \cos 1 - \cos 2 \approx 0.956$$

45. Find the volume of the region bounded by $z = 40 - 10y$, $z = 0$, $y = 0$, and $y = 4 - x^2$.

SOLUTION The region lies over the following domain $\mathcal{D}$ in the xy-plane:

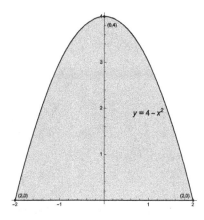

This is a vertically simple region described by

$$-2 \leq x \leq 2, \quad 0 \leq y \leq 4 - x^2$$

Note that $40 - 10y \geq 0$ everywhere in this region, since $y \leq 4$, so that the volume of the solid in question is

$$\iint_{\mathcal{D}} (40 - 10y) \, dA = \int_{-2}^{2} \int_{0}^{4-x^2} (40 - 10y) \, dy \, dx$$

$$= \int_{-2}^{2} (40y - 5y^2) \Big|_0^{4-x^2} \, dx$$

$$= \int_{-2}^{2} (40(4 - x^2) - 5(4 - x^2)^2) \, dx$$

$$= \int_{-2}^{2} (-5x^4 + 80) \, dx$$

$$= (-x^5 + 80x) \Big|_{-2}^{2} = 256$$

47. Find the volume of the region bounded by $z = 16 - y$, $z = y$, $y = x^2$, and $y = 8 - x^2$.

SOLUTION A diagram of the region is below:

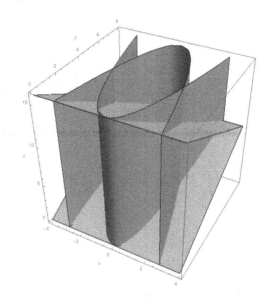

and the region lies over the domain $\mathcal{D}$ shown below:

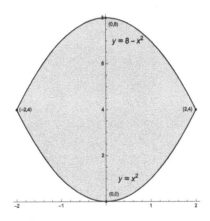

This is a vertically simple domain described by

$$-2 \leq x \leq 2, \quad x^2 \leq y \leq 8 - x^2$$

At each point in $\mathcal{D}$, the first diagram shows that $z = y$ lies below $z = 16 - y$ (this can also be seen clearly since $0 \leq y \leq 8$, so that $y \leq 8 \leq 16 - y$). Thus the volume of the solid is

$$\iint_{\mathcal{D}} ((16 - y) - y)\, dA = \int_{-2}^{2} \int_{x^2}^{8-x^2} (16 - 2y)\, dy\, dx$$

$$= \int_{-2}^{2} \left(16y - y^2\right)\Big|_{x^2}^{8-x^2} dx$$

$$= \int_{-2}^{2} \left(16(8 - x^2) - (8 - x^2)^2 - 16x^2 + (x^2)^2\right) dx$$

$$= \int_{-2}^{2} (64 - 16x^2)\, dx$$

$$= \left(64x - \frac{16}{3}x^3\right)\Big|_{-2}^{2} = \frac{512}{3}$$

49. Set up a double integral that gives the volume of the region bounded by the two paraboloids $z = x^2 + y^2$ and $z = 8 - x^2 - y^2$. (Do not evaluate the double integral.)

SOLUTION A diagram of the region is below:

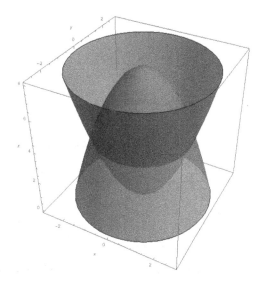

The two paraboloids intersect when $x^2 + y^2 = 8 - x^2 - y^2$, so when $x^2 + y^2 = 4$. Thus the region lies over the domain $\mathcal{D}$ in the xy-plane given by $x^2 + y^2 = 4$, which is a circle of radius 2. Since $x^2 + y^2 \le 8 - x^2 - y^2$ everywhere in $\mathcal{D}$, the double integral giving the volume of the region is

$$\iint_{\mathcal{D}} ((8 - x^2 - y^2) - (x^2 + y^2)) \, dA = \int_{-2}^{2} \int_{-\sqrt{4-x^2}}^{\sqrt{4-x^2}} (8 - 2x^2 - 2y^2) \, dy \, dx$$

51. On April 15, the snow depth in Jocoro Provincial Park (Example 6) was given by $d(x, y) = -0.01x + 0.004y + 1.3$ m. Estimate the volume of the snowpack in the park in cubic meters.

SOLUTION Using the description of the park given in Example 6, the volume of the snowpack is

$$V = \int_{0}^{80} \int_{0}^{120-\frac{1}{2}y} (-0.01x + 0.004y + 1.3) \, dx \, dy$$

$$= \int_{0}^{80} (-0.005x^2 + 0.004xy + 1.3x)\Big|_{x=0}^{120-\frac{1}{2}y} dy$$

$$= \int_{0}^{80} \left(-0.005\left(120 - \frac{1}{2}y\right)^2 + 0.004y\left(120 - \frac{1}{2}y\right) + 1.3\left(120 - \frac{1}{2}y\right) \right) dy$$

$$= \int_{0}^{80} (-0.00325y^2 + 0.43y + 84) \, dy$$

$$= -0.00108y^3 + 0.215y^2 + 84y \Big|_{0}^{80}$$

$$= 7541.33$$

Since x and y are in kilometers, while the depth function is in meters, we multiply this result by 10^6 to get an estimate of 7.541 billion cubic meters.

53. Calculate the average value of $f(x, y) = e^{x+y}$ on the square domain $[0, 1] \times [0, 1]$.

SOLUTION

Since the area of the square $\mathcal{D}$ is 1, the average value of $f(x, y) = e^{x+y}$ on $\mathcal{D}$ is the following value:

$$\overline{f} = \frac{1}{\text{Area}(\mathcal{D})} \iint_{\mathcal{D}} f(x, y) \, dA = \frac{1}{1} \int_0^1 \int_0^1 e^{x+y} \, dx \, dy = \int_0^1 e^{x+y} \Big|_{x=0}^1 dy = \int_0^1 \left(e^{1+y} - e^{0+y} \right) dy$$

$$= \int_0^1 \left(e^{1+y} - e^y \right) dy = e^{1+y} - e^y \Big|_0^1 = \left(e^2 - e \right) - \left(e^1 - e^0 \right) = e^2 - 2e + 1 \approx 2.95$$

55. Find the average height of the "ceiling" in Figure 31 defined by $z = y^2 \sin x$ for $0 \leq x \leq \pi$, $0 \leq y \leq 1$.

FIGURE 31

SOLUTION

The average height is

$$\overline{H} = \frac{1}{\text{Area}(\mathcal{D})} \iint_{\mathcal{D}} y^2 \sin x \, dA = \frac{1}{\pi \cdot 1} \int_0^1 \int_0^\pi y^2 \sin x \, dx \, dy = \frac{1}{\pi} \int_0^1 y^2 (-\cos x) \Big|_{x=0}^\pi dy$$

$$= \frac{1}{\pi} \int_0^1 y^2 (-\cos \pi + \cos 0) \, dy = \frac{1}{\pi} \int_0^1 2y^2 \, dy = \frac{1}{\pi} \cdot \frac{2}{3} y^3 \Big|_0^1 = \frac{2}{3\pi}$$

57. What is the average value of the linear function

$$f(x, y) = mx + ny + p$$

on the ellipse $\left(\dfrac{x}{a}\right)^2 + \left(\dfrac{y}{b}\right)^2 \leq 1$? Argue by symmetry rather than calculation.

SOLUTION The average value of the linear function $f(x,y) = mx + ny + p$ over the ellipse $\mathcal{D}$ is

$$\overline{f} = \frac{1}{\text{Area}(\mathcal{D})} \iint_{\mathcal{D}} f(x,y)\, dA = \frac{1}{\text{Area}(\mathcal{D})} \iint_{\mathcal{D}} (mx + ny + p)\, dA$$

$$= m \cdot \underbrace{\frac{1}{\text{Area}(\mathcal{D})} \iint_{\mathcal{D}} x\, dA}_{I_1} + n \cdot \underbrace{\frac{1}{\text{Area}(\mathcal{D})} \iint_{\mathcal{D}} y\, dA}_{I_2} + \frac{1}{\text{Area}(\mathcal{D})} \iint_{\mathcal{D}} p\, dA \qquad (1)$$

I_1 and I_2 are the average values of the x and y coordinates of a point in the region enclosed by the ellipse. This region is symmetric with respect to the y-axis, hence $I_1 = 0$. It is also symmetric with respect to the x-axis, hence $I_2 = 0$. We use the formula

$$\iint_{\mathcal{D}} p\, dA = p \cdot \text{Area}(\mathcal{D})$$

to conclude by (1) that

$$\overline{f} = m \cdot 0 + n \cdot 0 + \frac{1}{\text{Area}(\mathcal{D})} \cdot p \cdot \text{Area}(\mathcal{D}) = p$$

59. Let $\mathcal{D}$ be the rectangle $0 \le x \le 2$, $-\frac{1}{8} \le y \le \frac{1}{8}$, and let $f(x,y) = \sqrt{x^3 + 1}$. Prove that

$$\iint_{\mathcal{D}} f(x,y)\, dA \le \frac{3}{2}$$

SOLUTION Recall that we can write

$$\iint_{\mathcal{D}} f(x,y)\, dA \le M \cdot \text{Area}(\mathcal{D})$$

where M is a constant such that $f(x,y) \le M$. We can see that $\text{Area}(\mathcal{D}) = 2(1/4) = 1/2$. So it remains to show that there is some constant M so that $f(x,y) \le M$. Consider the following:

$$x \le 2 \quad \Rightarrow \quad x^3 + 1 \le 0 \quad \Rightarrow \quad \sqrt{x^3 + 1} \le 3$$

Thus we can let $M = 3$. So then we have

$$\iint_{\mathcal{D}} f(x,y)\, dA \le M \cdot \text{Area}(\mathcal{D}) \quad \Rightarrow \quad \iint_{\mathcal{D}} \sqrt{x^3 + 1}\, dA \le 3 \cdot \frac{1}{2} = \frac{3}{2}$$

61. Prove the inequality $\iint_{\mathcal{D}} \frac{dA}{4 + x^2 + y^2} \le \pi$, where $\mathcal{D}$ is the disk $x^2 + y^2 \le 4$.

SOLUTION The function $f(x,y) = \frac{1}{4+x^2+y^2}$ satisfies

$$f(x,y) = \frac{1}{4 + x^2 + y^2} \le \frac{1}{4}$$

Also, the area of the disk is

$$\text{Area}(\mathcal{D}) = \pi \cdot 2^2 = 4\pi$$

Therefore, by Theorem 3, we have

$$\iint_{\mathcal{D}} \frac{dA}{4 + x^2 + y^2} \le \frac{1}{4} \cdot 4\pi = \pi$$

63. Let $\overline{f}$ be the average of $f(x,y) = xy^2$ on $\mathcal{D} = [0,1] \times [0,4]$. Find a point $P \in \mathcal{D}$ such that $f(P) = \overline{f}$ (the existence of such a point is guaranteed by the Mean Value Theorem for Double Integrals).

SOLUTION We first compute the average $\overline{f}$ of $f(x,y) = xy^2$ on $\mathcal{D}$.

$\overline{f}$ is

$$\overline{f} = \frac{1}{\text{Area}(\mathcal{D})} \iint_{\mathcal{D}} xy^2 \, dA = \frac{1}{4 \cdot 1} \int_0^1 \int_0^4 xy^2 \, dy \, dx = \frac{1}{4} \int_0^1 \left. \frac{xy^3}{3} \right|_{y=0}^4 dx$$

$$= \frac{1}{4} \int_0^1 \left(\frac{x \cdot 4^3}{3} - \frac{x \cdot 0^3}{3} \right) dx = \int_0^1 \frac{16x}{3} dx = \left. \frac{8x^2}{3} \right|_0^1 = \frac{8}{3}$$

We now must find a point $P = (a, b)$ in $\mathcal{D}$ such that

$$f(P) = ab^2 = \frac{8}{3}$$

We choose $b = 2$, obtaining

$$a \cdot 2^2 = \frac{8}{3} \quad \Rightarrow \quad a = \frac{2}{3}$$

The point $P = \left(\frac{2}{3}, 2\right)$ in the rectangle $\mathcal{D}$ satisfies

$$f(P) = \overline{f} = \frac{8}{3}$$

In Exercises 65 and 66, use the approximation in (12) to estimate the double integral.

65. The following table lists the areas of the subdomains $\mathcal{D}_j$ of the domain $\mathcal{D}$ in Figure 33 and the values of a function $f(x, y)$ at sample points $P_j \in \mathcal{D}_j$. Estimate $\iint_{\mathcal{D}} f(x, y) \, dA$.

j	1	2	3	4	5	6
Area($\mathcal{D}_j$)	1.2	1.1	1.4	0.6	1.2	0.8
$f(P_j)$	9	9.1	9.3	9.1	8.9	8.8

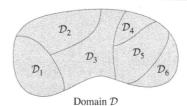

Domain $\mathcal{D}$

FIGURE 33

SOLUTION By Eq. (12) we have

$$\iint_{\mathcal{D}} f(x, y) \, dA \approx \sum_{j=1}^{6} f(P_j) \, \text{Area}(\mathcal{D}_j)$$

Substituting the data given in the table, we obtain

$$\iint_{\mathcal{D}} f(x, y) \, dA \approx 9 \cdot 1.2 + 9.1 \cdot 1.1 + 9.3 \cdot 1.4 + 9.1 \cdot 0.6 + 8.9 \cdot 1.2 + 8.8 \cdot 0.8 = 57.01$$

Thus,

$$\iint_{\mathcal{D}} f(x, y) \, dA \approx 57.01$$

67. According to Eq. (3), the area of a domain $\mathcal{D}$ is equal to $\iint_{\mathcal{D}} 1\, dA$. Prove that if $\mathcal{D}$ is the region between two curves $y = g_1(x)$ and $y = g_2(x)$ with $g_2(x) \le g_1(x)$ for $a \le x \le b$, then

$$\iint_{\mathcal{D}} 1\, dA = \int_a^b (g_1(x) - g_2(x))\, dx$$

SOLUTION The region $\mathcal{D}$ is defined by the inequalities

$$a \le x \le b, \quad g_2(x) \le y \le g_1(x)$$

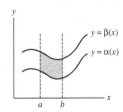

We compute the double integral of $f(x, y) = 1$ on $\mathcal{D}$, using Theorem 2, by evaluating the following iterated integral:

$$\int_{\mathcal{D}} 1\, dA = \int_a^b \int_{g_2(x)}^{g_1(x)} 1\, dy\, dx = \int_a^b \left(\int_{g_2(x)}^{g_1(x)} 1\, dy \right) dx = \int_a^b y \Big|_{y=g_2(x)}^{g_1(x)} dx = \int_a^b (g_1(x) - g_2(x))\, dx$$

Further Insights and Challenges

69. Use the fact that a continuous function on a closed bounded domain $\mathcal{D}$ attains both a minimum value m and a maximum value M, together with Theorem 3, to prove that the average value $\overline{f}$ lies between m and M. Then use the IVT in Exercise 68 to prove the Mean Value Theorem for Double Integrals.

SOLUTION Suppose that $f(x, y)$ is continuous and $\mathcal{D}$ is closed, bounded, and connected. By Theorem 3 in Chapter 15.7 ("Existence of Global Extrema"), $f(x, y)$ takes on a minimum value (call it m) at some point (x_m, y_m) and a maximum value (call it M) at some point (x_M, y_M) in the domain $\mathcal{D}$. Now, by Theorem 3,

$$m\, \text{Area}(\mathcal{D}) \le \iint_{\mathcal{D}} f(x, y)\, dA \le M\, \text{Area}(\mathcal{D})$$

which can be restated as

$$m \le \frac{1}{\text{Area}(\mathcal{D})} \iint_{\mathcal{D}} f(x, y)\, dA \le M$$

By the IVT in two variables (stated and proved in the previous problem), $f(x, y)$ takes on every value between m and M at some point in $\mathcal{D}$. In particular, f must take on the value $\frac{1}{\text{Area}(\mathcal{D})} \iint_{\mathcal{D}} f(x, y)\, dA$ at some point P. So, $f(P) = \frac{1}{\text{Area}(\mathcal{D})} \iint_{\mathcal{D}} f(x, y)\, dA$, which is rewritten as

$$f(P)\, \text{Area}(\mathcal{D}) = \iint_{\mathcal{D}} f(x, y)\, dA$$

15.3 Triple Integrals (LT Section 16.3)

Preliminary Questions

1. Which of (a)–(c) is *not* equal to $\displaystyle\int_0^1 \int_3^4 \int_6^7 f(x, y, z)\, dz\, dy\, dx$?

(a) $\displaystyle\int_6^7 \int_0^1 \int_3^4 f(x, y, z)\, dy\, dx\, dz$

(b) $\displaystyle\int_3^4 \int_0^1 \int_6^7 f(x, y, z)\, dz\, dx\, dy$

(c) $\displaystyle\int_0^1 \int_3^4 \int_6^7 f(x, y, z)\, dx\, dz\, dy$

SOLUTION The given integral, I, is a triple integral of f over the box $B = [0, 1] \times [3, 4] \times [6, 7]$. In (a) the limits of integration are $0 \le x \le 1$, $3 \le y \le 4$, $6 \le z \le 7$, hence this integral is equal to I. In (b) the limits of integration are $0 \le x \le 1$, $3 \le y \le 4$, $6 \le z \le 7$, hence it is also equal to I. In (c) the limits of integration are $6 \le x \le 7$, $0 \le y \le 1$, $3 \le z \le 4$. This is the triple integral of f over the box $[6, 7] \times [0, 1] \times [3, 4]$, which is different from B. Therefore, the triple integral is usually unequal to I.

2. Which of the following is *not* a meaningful triple integral?

(a) $\displaystyle\int_0^1 \int_0^x \int_{x+y}^{2x+y} e^{x+y+z} \, dz \, dy \, dx$

(b) $\displaystyle\int_0^1 \int_0^z \int_{x+y}^{2x+y} e^{x+y+z} \, dz \, dy \, dx$

SOLUTION

(a) The limits of integration determine the following inequalities:

$$0 \le x \le 1, \quad 0 \le y \le x, \quad x + y \le z \le 2x + y$$

The integration is over the simple region W, which lies between the planes $z = x + y$ and $z = 2x + y$ over the domain $\mathcal{D}_1 = \{(x, y) : 0 \le x \le 1, \ 0 \le y \le x\}$ in the xy-plane.

Thus, the integral represents a meaningful triple integral.

(b) Note that the inner integral is with respect to z, but then the middle integral has limits from 0 to z! This makes no sense.

3. Describe the projection of the region of integration W onto the xy-plane:

(a) $\displaystyle\int_0^1 \int_0^x \int_0^{x^2+y^2} f(x, y, z) \, dz \, dy \, dx$

(b) $\displaystyle\int_0^1 \int_0^{\sqrt{1-x^2}} \int_2^4 f(x, y, z) \, dz \, dy \, dx$

SOLUTION

(a) The region of integration is defined by the limits of integration, yielding the following inequalities:

$$0 \le x \le 1, \quad 0 \le y \le x, \quad 0 \le z \le x^2 + y^2$$

W is the region between the paraboloid $z = x^2 + y^2$ and the xy-plane which is above the triangle $\mathcal{D} = \{(x, y) : 0 \le x \le 1, 0 \le y \le x\}$ in the xy-plane. This triangle is the projection of W onto the xy-plane.

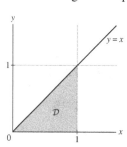

(b) The inequalities determined by the limits of integration are

$$0 \le x \le 1, \quad 0 \le y \le \sqrt{1 - x^2}, \quad 2 \le z \le 4$$

This is the region between the planes $z = 2$ and $z = 4$, which is above the region $\mathcal{D} = \{(x, y) : 0 \le x \le 1, \ 0 \le y \le \sqrt{1 - x^2}\}$ in the xy-plane. The projection $\mathcal{D}$ of W onto the xy-plane is the part of the unit disk in the first quadrant.

Exercises

In Exercises 1–8, evaluate $\iiint_{\mathcal{B}} f(x, y, z)\, dV$ for the specified function f and box $\mathcal{B}$.

1. $f(x, y, z) = xz + yz^2$; $0 \le x \le 2$, $2 \le y \le 4$, $0 \le z \le 4$

SOLUTION We write the integral as an iterated integral in any order we choose, and evaluate the inner, middle, and outer integral one after the other. This gives

$$\iiint_{\mathcal{B}} (xz + yz^2)\, dV = \int_0^4 \int_2^4 \int_0^2 (xz + yz^2)\, dx\, dy\, dz$$

$$= \int_0^4 \int_2^4 \left(\frac{1}{2}x^2 z + xyz^2\right)\bigg|_0^2 dy\, dz$$

$$= \int_0^4 \int_2^4 (2z + 2yz^2)\, dy\, dz$$

$$= \int_0^4 (2yz + y^2 z^2)\big|_2^4\, dz$$

$$= \int_0^4 (4z + 12z^2)\, dz$$

$$= (2z^2 + 4z^3)\big|_0^4$$

$$= 288$$

3. $f(x, y, z) = xe^{y-2z}$; $0 \le x \le 2$, $0 \le y \le 1$, $0 \le z \le 1$

SOLUTION We write the triple integral as an iterated integral. Since $f(x, y, z) = xe^y \cdot e^{-2z}$, we may evaluate the iterated integral as the product of three single integrals. We get

$$\iiint_{\mathcal{B}} xe^{y-2z}\, dV = \int_0^2 \int_0^1 \int_0^1 xe^{y-2z}\, dz\, dy\, dx = \left(\int_0^2 x\, dx\right)\left(\int_0^1 e^y\, dy\right)\left(\int_0^1 e^{-2z}\, dz\right)$$

$$= \left(\frac{1}{2}x^2\bigg|_0^2\right)\left(e^y\big|_0^1\right)\left(-\frac{1}{2}e^{-2z}\bigg|_0^1\right) = 2(e - 1) \cdot -\frac{1}{2}(e^{-2} - 1) = (e - 1)(1 - e^{-2})$$

5. $f(x, y, z) = (x - y)(y - z)$; $[0, 1] \times [0, 3] \times [0, 3]$

SOLUTION We write the triple integral as an iterated integral and evaluate the inner, middle, and outer integrals successively. This gives

$$\iiint_{\mathcal{B}} (x - y)(y - z)\, dV = \int_0^1 \int_0^3 \int_0^3 (x - y)(y - z)\, dz\, dy\, dx = \int_0^1 \int_0^3 \left(\int_0^3 (x - y)(y - z)\, dz\right) dy\, dx$$

$$= \int_0^1 \int_0^3 (x - y)\left(yz - \frac{1}{2}z^2\right)\bigg|_{z=0}^3 dy\, dx = \int_0^1 \int_0^3 (x - y)\left(3y - \frac{9}{2}\right) dy\, dx$$

$$= \int_0^1 \int_0^3 \left(\left(3x + \frac{9}{2}\right)y - \frac{9}{2}x - 3y^2\right) dy\, dx = \int_0^1 \left(\frac{3}{2}x + \frac{9}{4}\right)y^2 - \frac{9}{2}xy - y^3 \bigg|_{y=0}^3 dx$$

$$= \int_0^1 \left(\left(\frac{3}{2}x + \frac{9}{4}\right)\cdot 9 - \frac{9}{2}x \cdot 3 - 27\right) dx = \int_0^1 -\frac{27}{4}\, dx = -\frac{27}{4} = -6.75$$

7. $f(x, y, z) = (x + z)^3$; $[0, a] \times [0, b] \times [0, c]$

SOLUTION We write the triple integral as an iterated integral and evaluate it to obtain

$$\iiint_B f(x, y, z) \, dV = \int_0^a \int_0^b \int_0^c (x + z)^3 \, dz \, dy \, dx = \int_0^a \int_0^b \frac{(x + z)^4}{4} \bigg|_{z=0}^c dy \, dx$$

$$= \int_0^a \int_0^b \left(\frac{(x + c)^4}{4} - \frac{x^4}{4} \right) dy \, dx = \int_0^a \frac{(x + c)^4 - x^4}{4} y \bigg|_{y=0}^b dx$$

$$= \int_0^a \frac{b}{4} \left[(x + c)^4 - x^4 \right] dx = \frac{b}{4} \left[\frac{(x + c)^5}{5} - \frac{x^5}{5} \right] \bigg|_{x=0}^a$$

$$= \frac{b}{4} \cdot \frac{(a + c)^5 - a^5 - c^5}{5} = \frac{b}{20} \left[(a + c)^5 - a^5 - c^5 \right]$$

In Exercises 9–14, evaluate $\iiint_{\mathcal{W}} f(x, y, z) \, dV$ *for the function f and region $\mathcal{W}$ specified.*

9. $f(x, y, z) = x + y$; $\mathcal{W} : y \le z \le x$, $0 \le y \le x$, $0 \le x \le 1$

SOLUTION $\mathcal{W}$ is the region between the planes $z = y$ and $z = x$ lying over the triangle $\mathcal{D}$ in the xy-plane defined by the inequalities $0 \le y \le x$, $0 \le x \le 1$.

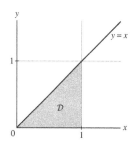

We compute the integral, using Theorem 2, by evaluating the following iterated integral:

$$\iiint_{\mathcal{W}} (x + y) \, dV = \iint_{\mathcal{D}} \left(\int_y^x (x + y) \, dz \right) dA = \iint_{\mathcal{D}} (x + y) z \bigg|_{z=y}^x dA = \iint_{\mathcal{D}} (x + y)(x - y) \, dA$$

$$= \iint_{\mathcal{D}} \left(x^2 - y^2 \right) dA = \int_0^1 \int_0^x \left(x^2 - y^2 \right) dy \, dx = \int_0^1 \left(\int_0^x \left(x^2 - y^2 \right) dy \right) dx$$

$$= \int_0^1 x^2 y - \frac{y^3}{3} \bigg|_{y=0}^x dx = \int_0^1 \frac{2x^3}{3} \, dx = \frac{2}{12} x^4 \bigg|_0^1 = \frac{1}{6}$$

11. $f(x, y, z) = xyz$; $\mathcal{W} : 0 \le z \le 1$, $0 \le y \le \sqrt{1 - x^2}$, $0 \le x \le 1$

SOLUTION $\mathcal{W}$ is the region between the planes $z = 0$ and $z = 1$, lying over the part $\mathcal{D}$ of the disk in the first quadrant.

Using Theorem 2, we compute the triple integral as the following iterated integral:

$$\iiint_{\mathcal{W}} xyz\, dV = \iint_{\mathcal{D}} \left(\int_0^1 xyz\, dz \right) dA = \iint_{\mathcal{D}} \left. \frac{xyz^2}{2} \right|_{z=0}^{1} dA = \iint_{\mathcal{D}} \frac{xy}{2}\, dA$$

$$= \int_0^1 \left(\int_0^{\sqrt{1-x^2}} \frac{xy}{2}\, dy \right) dx = \int_0^1 \left. \frac{xy^2}{4} \right|_{y=0}^{\sqrt{1-x^2}} dx = \int_0^1 \frac{x(1-x^2)}{4}\, dx$$

$$= \int_0^1 \frac{x - x^3}{4}\, dx = \left. \frac{x^2}{8} - \frac{x^4}{16} \right|_0^1 = \frac{1}{8} - \frac{1}{16} = \frac{1}{16}$$

13. $f(x, y, z) = e^z$; $\mathcal{W} : x + y + z \le 1$, $x \ge 0$, $y \ge 0$, $z \ge 0$

SOLUTION Notice that $\mathcal{W}$ is the tetrahedron under the plane $x + y + z = 1$ above the first quadrant.

First, we must determine the projection $\mathcal{D}$ of $\mathcal{W}$ onto the xy-plane. The intersection of the plane $x + y + z = 1$ with the xy-plane is obtained by solving

$$\begin{matrix} x + y + z = 1 \\ z = 0 \end{matrix} \quad \Rightarrow \quad x + y = 1$$

Therefore, the projection $\mathcal{D}$ of $\mathcal{W}$ onto the xy-plane is the triangle enclosed by the line $x + y = 1$ and the positive axes.

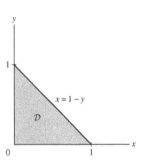

The region $\mathcal{W}$ is the region between the planes $z = 1 - x - y$ and $z = 0$, lying above the triangle $\mathcal{D}$ in the xy-plane. The triple integral can be written as the following iterated integral:

$$\iiint_{\mathcal{W}} e^z\, dV = \iint_{\mathcal{D}} \left(\int_0^{1-x-y} e^z\, dz \right) dA$$

$$= \iint_{\mathcal{D}} \left. e^z \right|_{z=0}^{1-x-y} dA$$

$$= \iint_{\mathcal{D}} \left(e^{1-x-y} - 1 \right) dA$$

$$= \int_0^1 \left(\int_0^{1-y} \left(e^{1-x-y} - 1 \right) dx \right) dy$$

$$= \int_0^1 \left. \left(-e^{1-x-y} - x \right) \right|_{x=0}^{1-y} dy$$

$$= \int_0^1 \left(-e^{1-(1-y)-y} - (1-y) + e^{1-0-y} + 0 \right) dy$$

$$= \int_0^1 \left(y - 2 + e^{1-y}\right) dy$$

$$= \left(\frac{1}{2}y^2 - 2y - e^{1-y}\right)\Big|_0^1$$

$$= e - \frac{5}{2}$$

15. Calculate the integral of $f(x, y, z) = z$ over the region $\mathcal{W}$ in Figure 11, below the hemisphere of radius 3 and lying over the triangle $\mathcal{D}$ in the xy-plane bounded by $x = 1$, $y = 0$, and $x = y$.

FIGURE 11

SOLUTION

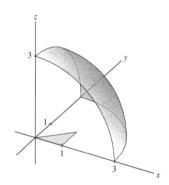

The upper surface is the hemisphere $z = \sqrt{9 - x^2 - y^2}$ and the lower surface is the xy-plane $z = 0$. The projection of $\mathcal{V}$ onto the xy-plane is the triangle $\mathcal{D}$ shown in the figure.

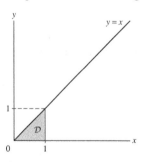

We compute the triple integral as the following iterated integral:

$$\iiint_\mathcal{V} z\, dV = \iint_\mathcal{D} \left(\int_0^{\sqrt{9-x^2-y^2}} z\, dz\right) dA = \iint_\mathcal{D} \frac{z^2}{2}\Big|_0^{\sqrt{9-x^2-y^2}} dA = \iint_\mathcal{D} \frac{9 - x^2 - y^2}{2}\, dA$$

$$= \int_0^1 \left(\int_0^x \frac{9 - x^2 - y^2}{2}\, dy\right) dx = \int_0^1 \frac{9y - x^2 y - \frac{y^3}{3}}{2}\Big|_{y=0}^{x} dx = \int_0^1 \left(\frac{9x}{2} - \frac{2x^3}{3}\right) dx$$

$$= \frac{9x^2}{4} - \frac{x^4}{6}\Big|_0^1 = 2\frac{1}{12}$$

17. Integrate $f(x, y, z) = x$ over the region in the first octant bounded above by $z = 8 - 2x^2 - y^2$ and below by $z = y^2$.

SOLUTION We first find the projection of the region $\mathcal{W}$ onto the xy-plane. We find the curve of intersection between the upper and lower surfaces, by solving the following equation for $x, y \geq 0$:

$$8 - 2x^2 - y^2 = y^2 \quad \Rightarrow \quad y^2 = 4 - x^2 \quad \Rightarrow \quad y = \sqrt{4 - x^2}, x \geq 0$$

The projection $\mathcal{D}$ of $\mathcal{W}$ onto the xy-plane is the region bounded by the circle $x^2 + y^2 = 4$ and the positive axes.

We now compute the triple integral over $\mathcal{W}$ by evaluating the following iterated integral:

$$\iiint_{\mathcal{W}} x \, dV = \iint_{\mathcal{D}} \left(\int_{y^2}^{8 - 2x^2 - y^2} x \, dz \right) dA$$

$$= \iint_{\mathcal{D}} xz \Big|_{y^2}^{8 - 2x^2 - y^2} dA$$

$$= \iint_{\mathcal{D}} x(8 - 2x^2 - y^2 - y^2) \, dA$$

$$= \iint_{\mathcal{D}} (8x - 2x^3 - 2xy^2) \, dA = \int_0^2 \int_0^{\sqrt{4 - x^2}} (8x - 2x^3 - 2xy^2) \, dy \, dx$$

$$= \int_0^2 \left(8xy - 2x^3 y - \frac{2}{3} xy^3 \right) \Big|_0^{\sqrt{4 - x^2}} dx$$

$$= \int_0^2 \left(8x \sqrt{4 - x^2} - 2x^3 \sqrt{4 - x^2} - \frac{2}{3} x(4 - x^2)^{3/2} \right) dx$$

$$= \int_0^2 \left(8x \sqrt{4 - x^2} - 2x^3 \sqrt{4 - x^2} - \frac{8}{3} x \sqrt{4 - x^2} + \frac{2}{3} x^3 \sqrt{4 - x^2} \right) dx$$

$$= \int_0^2 \left(\frac{16}{3} x \sqrt{4 - x^2} - \frac{4}{3} x^3 \sqrt{4 - x^2} \right) dx$$

This integral is evaluated using the substitution $u = 4 - x^2$; then $du = -2x \, dx$ and $x^2 = 4 - u$, so we get

$$\int_0^2 \left(\frac{16}{3} x \sqrt{4 - x^2} - \frac{4}{3} x^3 \sqrt{4 - x^2} \right) dx = \int_4^0 \left(-\frac{8}{3} \sqrt{u} + \frac{2}{3} (4 - u) \sqrt{u} \right) du$$

$$= -\int_4^0 \frac{2}{3} u^{3/2} \, du$$

$$= -\left(\frac{4}{15} u^{5/2} \right) \Big|_4^0 \, du = \frac{128}{15}$$

19. Find the triple integral of the function $F(x, y, z) = z$ over the region in Figure 13.

FIGURE 13

1278 CHAPTER 15 | MULTIPLE INTEGRATION (LT CHAPTER 16)

SOLUTION We place the coordinate axes as shown in the figure:

The upper surface is the plane passing through the points $O = (0, 0, 0)$, $A = (3, 0, 0)$, and $B = (3, 4, 1)$. We find a normal to this plane and then determine the equation of the plane. We get

$$\overrightarrow{OA} \times \overrightarrow{AB} = \langle 3, 0, 0 \rangle \times \langle 0, 4, 1 \rangle = \begin{vmatrix} \mathbf{i} & \mathbf{j} & \mathbf{k} \\ 3 & 0 & 0 \\ 0 & 4 & 1 \end{vmatrix} = -3\mathbf{j} + 12\mathbf{k} = 3(-\mathbf{j} + 4\mathbf{k})$$

The plane is orthogonal to the vector $\langle 0, -1, 4 \rangle$ and passes through the origin, hence the equation of the plane is

$$0 \cdot x - y + 4z = 0 \quad \Rightarrow \quad z = \frac{y}{4}$$

The projection of the region of integration $\mathcal{W}$ onto the xy-plane is the rectangle $\mathcal{D}$ defined by

$$0 \leq x \leq 3, \quad 0 \leq y \leq 4$$

We now compute the triple integral of f over $\mathcal{W}$, as the following iterated integral:

$$\iiint_{\mathcal{W}} z \, dV = \iint_{\mathcal{D}} \left(\int_0^{y/4} z \, dz \right) dA = \iint_{\mathcal{D}} \frac{z^2}{2} \bigg|_{z=0}^{y/4} dA = \iint_{\mathcal{D}} \frac{y^2}{32} dA = \int_0^4 \left(\int_0^3 \frac{y^2}{32} dx \right) dy$$

$$= \int_0^4 \frac{y^2 x}{32} \bigg|_{x=0}^3 dy = \int_0^4 \frac{3y^2}{32} dy = \frac{y^3}{32} \bigg|_0^4 = \frac{4^3}{32} = 2$$

21. Find the volume of the solid in the first octant bounded between the planes $x + y + z = 1$ and $x + y + 2z = 1$.

SOLUTION The solid $\mathcal{W}$ is shown in the figure:

The upper and lower surfaces are the planes $x + y + z = 1$ (or $z = 1 - x - y$) and $x + y + 2z = 1$ (or $z = \frac{1-x-y}{2}$), respectively. The projection of $\mathcal{W}$ onto the xy-plane is the triangle enclosed by the line $AB: y = 1 - x$ and the positive x and y-axes.

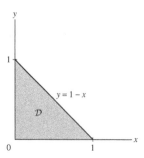

Using the volume of a solid as a triple integral, we have

$$\text{Volume}(\mathcal{W}) = \iiint_{\mathcal{W}} 1 \, dV = \iint_{\mathcal{D}} \left(\int_{(1-x-y)/2}^{1-x-y} 1 \, dz \right) dA = \iint_{\mathcal{D}} z \Big|_{z=(1-x-y)/2}^{1-x-y} dA$$

$$= \iint_{\mathcal{D}} \left((1-x-y) - \frac{1-x-y}{2} \right) dA = \iint_{\mathcal{D}} \frac{1-x-y}{2} \, dA$$

$$= \int_0^1 \left(\int_0^{1-x} \frac{1-x-y}{2} \, dy \right) dx = \int_0^1 \frac{y - xy - \frac{y^2}{2}}{2} \Big|_{y=0}^{1-x} dx$$

$$= \int_0^1 \frac{1 - x - x(1-x) - \frac{(1-x)^2}{2}}{2} \, dx = \frac{1}{2} \int_0^1 \left(\frac{x^2}{2} - x + \frac{1}{2} \right) dx$$

$$= \frac{1}{2} \left(\frac{x^3}{6} - \frac{x^2}{2} + \frac{1}{2}x \right) \Big|_0^1 = \frac{1}{2} \left(\frac{1}{6} - \frac{1}{2} + \frac{1}{2} \right) = \frac{1}{12}$$

23. Evaluate $\iiint_{\mathcal{W}} xz \, dV$, where $\mathcal{W}$ is the domain bounded by the elliptic cylinder $\frac{x^2}{4} + \frac{y^2}{9} = 1$ and the sphere $x^2 + y^2 + z^2 = 16$ in the first octant (Figure 14).

FIGURE 14

SOLUTION

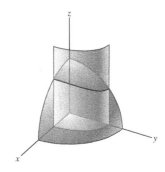

The upper surface is the sphere $x^2 + y^2 + z^2 = 16$, or $z = \sqrt{16 - x^2 - y^2}$, and the lower surface is the xy-plane, $z = 0$. The projection of $\mathcal{W}$ onto the xy-plane is the region in the first quadrant bounded by the ellipse $(x/2)^2 + (y/3)^2 = 1$, or $y = \frac{3}{2}\sqrt{4 - x^2}$.

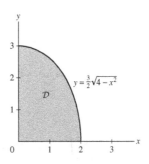

Therefore, the triple integral over $\mathcal{W}$ is equal to the following iterated integral:

$$\iiint_{\mathcal{W}} xz\, dV = \iint_{\mathcal{D}} \left(\int_0^{\sqrt{16-x^2-y^2}} xz\, dz \right) dA$$

$$= \iint_{\mathcal{D}} \frac{1}{2} xz^2 \bigg|_{z=0}^{\sqrt{16-x^2-y^2}} dA$$

$$= \frac{1}{2} \iint_{\mathcal{D}} x(16 - x^2 - y^2)\, dA$$

$$= \frac{1}{2} \int_0^2 \left(\int_0^{\frac{3}{2}\sqrt{4-x^2}} (16x - x^3 - xy^2)\, dy \right) dx$$

$$= \frac{1}{2} \int_0^2 \left(16xy - x^3 y - \frac{1}{3} xy^3 \right) \bigg|_{y=0}^{\frac{3}{2}\sqrt{4-x^2}} dx$$

$$= \frac{1}{2} \int_0^2 \left(24x\sqrt{4-x^2} - \frac{3}{2} x^3 \sqrt{4-x^2} - \frac{9}{8} x(4-x^2)\sqrt{4-x^2} \right) dx$$

$$= \frac{1}{2} \int_0^2 \left(\frac{39}{2} x\sqrt{4-x^2} - \frac{3}{8} x^3 \sqrt{4-x^2} \right) dx$$

$$= \frac{1}{16} \int_0^2 (156 x\sqrt{4-x^2} - 3x^3 \sqrt{4-x^2})\, dx$$

To integrate, use the substitution $u = 4 - x^2$, so that $du = -2x\, dx$, and $x^2 = 4 - u$. Then we get

$$\frac{1}{16} \int_0^2 (156x \sqrt{4-x^2} - 3x^3 \sqrt{4-x^2})\, dx = \frac{1}{16} \int_4^0 \left(-78\sqrt{u} + \frac{3}{2}(4-u)\sqrt{u} \right) du$$

$$= \frac{1}{16} \int_4^0 \left(-72 u^{1/2} - \frac{3}{2} u^{3/2} \right) du$$

$$= \frac{1}{16} \left(-48 u^{3/2} - \frac{3}{5} u^{5/2} \right) \bigg|_4^0$$

$$= \frac{1}{16} \left(48 \cdot 4^{3/2} + \frac{3}{5} \cdot 4^{5/2} \right)$$

$$= 24 + \frac{6}{5} = \frac{126}{5}$$

25. Describe the domain of integration of the following integral:

$$\int_{-2}^2 \int_{-\sqrt{4-z^2}}^{\sqrt{4-z^2}} \int_1^{\sqrt{5-x^2-z^2}} f(x, y, z)\, dy\, dx\, dz$$

SOLUTION The domain of integration of $\mathcal{W}$ is defined by the following inequalities:

$$-2 \le z \le 2, \quad -\sqrt{4-z^2} \le x \le \sqrt{4-z^2}, \quad 1 \le y \le \sqrt{5 - x^2 - z^2}$$

This region is bounded by the plane $y = 1$ and the sphere $y^2 = 5 - x^2 - z^2$ or $x^2 + y^2 + z^2 = 5$, lying over the disk $x^2 + z^2 \le 4$ in the xz-plane. This is the central cylinder oriented along the y-axis of radius 2 inside a sphere of radius $\sqrt{5}$.

27. Assume $f(x, y, z)$ can be expressed as a product, $f(x, y, z) = g(x)h(y)k(z)$. Show that the integral of f over a box $\mathcal{B} = [a, b] \times [c, d] \times [p, q]$ can be expressed as a product of integrals as follows:

$$\iiint_{\mathcal{B}} f(x, y, z)\, dV = \left(\int_a^b g(x)\, dx \right) \left(\int_c^d h(y)\, dy \right) \left(\int_p^q k(z)\, dz \right)$$

SOLUTION The triple integral is

$$\iiint_{\mathcal{B}} f(x, y, z)\, dV = \int_a^b \int_c^d \int_e^f g(x) h(y) k(z)\, dz\, dy\, dx$$

Considering the inner integral, since we are integrating with respect to z, $g(x)h(y)$ may be considered as a constant and moved outside the integral sign, so that

$$\int_a^b \int_c^d \int_e^f g(x)h(y)k(z)\,dz\,dy\,dx = \int_a^b \int_c^d \left(g(x)h(y) \int_e^f k(z)\,dz\right) dy\,dx$$

But $\int_e^f k(z)\,dz$ is a constant, so may be moved outside the remaining integrals, giving

$$\int_a^b \int_c^d \left(g(x)h(y) \int_e^f k(z)\,dz\right) dy\,dx = \left(\int_a^b \int_c^d g(x)h(y)\,dy\,dx\right)\left(\int_e^f k(z)\,dz\right)$$

Considering now the integral with respect to y, the factor $g(x)$ is effectively constant and may be moved outside the integral sign, giving

$$\left(\int_a^b \int_c^d g(x)h(y)\,dy\,dx\right)\left(\int_e^f k(z)\,dz\right) = \left(\int_a^b g(x) \int_c^d h(y)\,dy\,dx\right)\left(\int_e^f k(z)\,dz\right)$$

Finally, $\int_c^d h(y)\,dy$ is a constant; moving it outside of the integral with respect to x gives

$$\iiint_{\mathcal{B}} f(x,y,z)\,dV = \left(\int_a^b g(x) \int_c^d h(y)\,dy\,dx\right)\left(\int_e^f k(z)\,dz\right) = \left(\int_a^b g(x)\,dx\right)\left(\int_c^d h(y)\,dy\right)\left(\int_e^f k(z)\,dz\right)$$

29. In Example 5, we expressed a triple integral as an iterated integral in the three orders

$$dz\,dy\,dx, \quad dx\,dz\,dy, \quad \text{and} \quad dy\,dz\,dx$$

Write this integral in the three other orders:

$$dz\,dx\,dy, \quad dx\,dy\,dz, \quad \text{and} \quad dy\,dx\,dz$$

SOLUTION In Example 5 we considered the triple integral $\iiint_{\mathcal{W}} xyz^2\,dV$, where $\mathcal{W}$ is the region bounded by

$$z = 4 - y^2, \quad z = 0, \quad y = 2x, \quad x = 0$$

We now write the triple integral as an iterated integral in the orders $dz\,dx\,dy$, $dx\,dy\,dz$, and $dy\,dx\,dz$.

- $dz\,dx\,dy$: The upper surface $z = 4 - y^2$ projects onto the xy-plane on the triangle defined by the lines $y = 2$, $y = 2x$, and $x = 0$.

We express the line $y = 2x$ as $x = \frac{y}{2}$ and write the triple integral as

$$\iiint_{\mathcal{W}} xyz^2\,dV = \iint_{\mathcal{D}} \left(\int_0^{4-y^2} xyz^2\,dz\right) dA = \int_0^2 \int_0^{y/2} \int_0^{4-y^2} xyz^2\,dz\,dx\,dy$$

- $dx\,dy\,dz$: The projection of $\mathcal{W}$ onto the yz-plane is the domain T (see Example 5) defined by the inequalities

$$T: 0 \le y \le 2, \ 0 \le z \le 4 - y^2$$

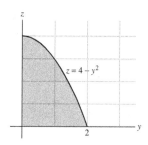

This region can also be expressed as

$$0 \le z \le 4, \quad 0 \le y \le \sqrt{4-z}$$

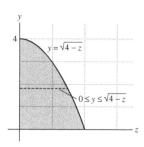

As explained in Example 5, the region $\mathcal{W}$ consists of all points lying between T and the "left-face" $y = 2x$, or $x = \frac{y}{2}$. Therefore, we obtain the following inequalities for $\mathcal{W}$:

$$0 \le z \le 4, \quad 0 \le y \le \sqrt{4-z}, \quad 0 \le x \le \frac{1}{2}y$$

This yields the following iterated integral:

$$\iiint_{\mathcal{W}} xyz^2 \, dV = \iint_T \left(\int_0^{y/2} xyz^2 \, dx \right) dA = \int_0^4 \int_0^{\sqrt{4-z}} \int_0^{y/2} xyz^2 \, dx \, dy \, dz$$

- $dy \, dx \, dz$: As explained in Example 5, the projection of $\mathcal{W}$ onto the xz-plane is determined by the inequalities

$$S : 0 \le x \le 1, \, 0 \le z \le 4 - 4x^2$$

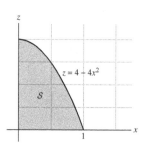

This region can also be described if we write x as a function of z:

$$z = 4 - 4x^2 \quad \Rightarrow \quad 4x^2 = 4 - z \quad \Rightarrow \quad x = \sqrt{1 - \frac{z}{4}}$$

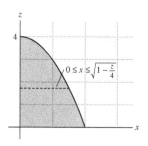

This gives the following inequalities of S:

$$S : 0 \le z \le 4, \ 0 \le x \le \sqrt{1 - \frac{z}{4}}$$

The upper surface $z = 4 - y^2$ can be described by $y = \sqrt{4 - z}$, hence the limits of y are $2x \le y \le \sqrt{4 - z}$. We obtain the following iterated integral:

$$\iiint_W xyz^2 \, dV = \iint_S \left(\int_{2x}^{\sqrt{4-z}} xyz^2 \, dy \right) dA = \int_0^4 \int_0^{\sqrt{1-\frac{z}{4}}} \int_{2x}^{\sqrt{4-z}} xyz^2 \, dy \, dx \, dz$$

31. Let

$$W = \{(x, y, z) : \sqrt{x^2 + y^2} \le z \le 1\}$$

(see Figure 16). Express $\iiint_W f(x, y, z) \, dV$ as an iterated integral in the order $dz \, dy \, dx$ (for an arbitrary function f).

FIGURE 16

SOLUTION To express the triple integral as an iterated integral in order $dz \, dy \, dx$, we must find the projection of W onto the xy-plane. The upper circle is $\sqrt{x^2 + y^2} = 1$ or $x^2 + y^2 = 1$, hence the projection of W onto the xy plane is the disk

$$\mathcal{D} : x^2 + y^2 \le 1$$

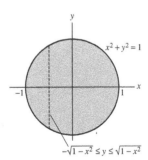

The upper surface is the plane $z = 1$ and the lower surface is $z = \sqrt{x^2 + y^2}$, therefore the triple integral over W is equal to the following iterated integral:

$$\iiint_W f(x, y, z) \, dV = \iint_\mathcal{D} \left(\int_{\sqrt{x^2+y^2}}^1 f(x, y, z) \, dz \right) dA = \int_{-1}^1 \int_{-\sqrt{1-x^2}}^{\sqrt{1-x^2}} \int_{\sqrt{x^2+y^2}}^1 f(x, y, z) \, dz \, dy \, dx$$

To express the triple integral as an iterated integral in order $dx \, dy \, dz$, we must find the projection of W onto the yz-plane. The points on the surface $z = \sqrt{x^2 + y^2}$ are $\left(x, y, \sqrt{x^2 + y^2}\right)$, hence the projection on the yz-plane consists of the points $(0, y, |y|)$, that is, $z = |y|$.

Therefore, the projection of $\mathcal{W}$ onto the yz-plane is the triangle defined by the lines $z = |y|$ and $z = 1$. The region $\mathcal{W}$ consists of all points lying between T and the surface $z = \sqrt{x^2 + y^2}$, or $x = \pm\sqrt{z^2 - y^2}$. $\mathcal{W}$ can be described by the inequalities

$$0 \le z \le 1$$
$$-z \le y \le z$$
$$-\sqrt{z^2 - y^2} \le x \le \sqrt{z^2 - y^2}$$

The triple integral is equal to the following iterated integral:

$$\iiint_{\mathcal{W}} f(x,y,z)\,dV = \int_0^1 \int_{-z}^{z} \int_{-\sqrt{z^2-y^2}}^{\sqrt{z^2-y^2}} f(x,y,z)\,dx\,dy\,dz$$

33. Let $\mathcal{W}$ be the region bounded by $z = 1 - y^2$, $y = x^2$, and the plane $z = 0$. Calculate the volume of $\mathcal{W}$ as a triple integral in the order $dz\,dy\,dx$.

SOLUTION

The projection of $\mathcal{W}$ onto the xy-plane is the region $\mathcal{D}$ bounded by the curve $y = x^2$ and the line $y = 1$. The region $\mathcal{W}$ consists of all points lying between $\mathcal{D}$ and the cylinder $z = 1 - y^2$.

Therefore, $\mathcal{W}$ can be described by the following inequalities:

$$-1 \le x \le 1, \quad x^2 \le y \le 1, \quad 0 \le z \le 1 - y^2$$

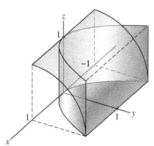

We use the formula for the volume as a triple integral, write the triple integral as an iterated integral, and compute it. We obtain

$$\text{Volume}(\mathcal{W}) = \iiint_{\mathcal{W}} 1\,dV = \int_{-1}^{1}\int_{x^2}^{1}\int_{0}^{1-y^2} 1\,dz\,dy\,dx = \int_{-1}^{1}\int_{x^2}^{1} z\Big|_{z=0}^{1-y^2} dy\,dx$$

$$= \int_{-1}^{1}\int_{x^2}^{1} (1 - y^2)\,dy\,dx = \int_{-1}^{1}\left(y - \frac{y^3}{3}\right)\Big|_{y=x^2}^{1} dx = \int_{-1}^{1}\left(1 - \frac{1}{3} - \left(x^2 - \frac{x^6}{3}\right)\right)dx$$

$$= 2\int_{0}^{1}\left(\frac{x^6}{3} - x^2 + \frac{2}{3}\right)dx = 2\left(\frac{x^7}{21} - \frac{x^3}{3} + \frac{2x}{3}\right)\Big|_{0}^{1} = 2\left(\frac{1}{21} - \frac{1}{3} + \frac{2}{3}\right) = \frac{16}{21}$$

In Exercises 35–38, draw the region $\mathcal{W}$ and then set up but do not compute a single triple integral that yields the volume of $\mathcal{W}$.

35. The region $\mathcal{W}$ is bounded by the surfaces given by $z = 1 - y^2$, $x = 0$ and $z = 0$, $z + x = 3$.

SOLUTION The region in question is the volume bounded above by the surface $z = 1 - y^2$ and $z + x = 3$, below by the plane $z = 0$, and on the left by the plane $x = 0$.

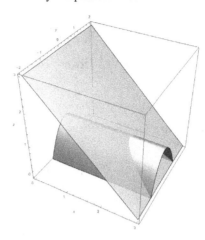

Since $z = 1 - y^2$ meets the plane $z = 0$ at $y = \pm 1$, the outer integration will be for $-1 \leq y \leq 1$. For each y, we integrate for z from 0 to $1 - y^2$, and for x from 0 to $3 - z$. So the integral is:

$$\int_{-1}^{1} \int_{0}^{1-y^2} \int_{0}^{3-z} 1 \, dx \, dz \, dy$$

37. The region $\mathcal{W}$ is bounded by the surfaces given by $z = y^2$, $y = z^2$ and $x = 0$, $x + y + z = 4$.

SOLUTION The region is the propeller-shaped region contained between $z = y^2$ and $y = z^2$; the planes $x = 0$ and $x = 4 - y - z$ provide a boundary to that region:

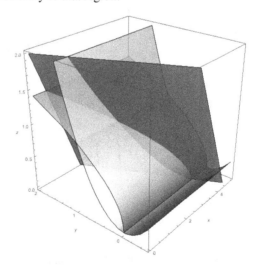

The two parabolic cylinders intersect when $y = z = 0$ and when $y = z = 1$; for $0 \leq y \leq 1$, $z = y^2$ lies below $y = z^2$ since $y^2 \leq y$. So the integral is

$$\int_{0}^{1} \int_{y^2}^{\sqrt{y}} \int_{0}^{4-y-z} 1 \, dx \, dz \, dy$$

In Exercises 39–42, compute the average value of $f(x, y, z)$ over the region $\mathcal{W}$.

39. $f(x, y, z) = xy \sin(\pi z); \quad \mathcal{W} = [0, 1] \times [0, 1] \times [0, 1]$

SOLUTION The volume of the cube is $V = 1$, hence the average of f over the cube is the following value:

$$\bar{f} = \iiint_W xy \sin(\pi z)\, dV = \int_0^1 \int_0^1 \int_0^1 xy \sin(\pi z)\, dx\, dy\, dz$$

$$= \int_0^1 \int_0^1 \frac{1}{2}x^2 y \sin(\pi z)\Big|_{x=0}^1 dy\, dz = \int_0^1 \int_0^1 \frac{1}{2} y \sin(\pi z)\, dy\, dz$$

$$= \int_0^1 \frac{y^2}{4} \sin(\pi z)\Big|_{y=0}^1 dz = \int_0^1 \frac{1}{4} \sin(\pi z)\, dz = -\frac{1}{4\pi} \cos(\pi z)\Big|_0^1$$

$$= -\frac{1}{4\pi}(\cos \pi - \cos 0) = -\frac{1}{4\pi}(-1 - 1) = \frac{1}{2\pi}$$

41. $f(x, y, z) = e^y;\ \ W: 0 \le y \le 1 - x^2,\ \ 0 \le z \le x$

SOLUTION First we must calculate the volume of W. We will use the symmetry of $y = 1 - x^2$ to write:

$$V = \iiint_W 1\, dV = 2 \int_0^1 \int_0^{1-x^2} \int_0^x 1\, dz\, dy\, dx$$

$$= 2 \int_0^1 \int_0^{1-x^2} z\Big|_{z=0}^x dy\, dx = 2 \int_0^1 \int_0^{1-x^2} x\, dy\, dx$$

$$= 2 \int_0^1 xy\Big|_{y=0}^{1-x^2} dx = 2 \int_0^1 x(1-x^2)\, dx$$

$$= 2 \int_0^1 x - x^3\, dx = x^2 - \frac{1}{2}x^4\Big|_0^1 = \frac{1}{2}$$

Now we can compute the average value for $f(x, y, z) = e^y$:

$$\bar{f} = \frac{1}{V}\iiint_W e^y\, dV = 2 \cdot \frac{1}{1/2} \int_0^1 \int_0^{1-x^2} \int_0^x e^y\, dz\, dy\, dx$$

$$= 4 \int_0^1 \int_0^{1-x^2} ze^y\Big|_{z=0}^x dy\, dx = 4 \int_0^1 \int_0^{1-x^2} xe^y\, dy\, dx$$

$$= 4 \int_0^1 xe^y\Big|_{y=0}^{1-x^2} dx = 4 \int_0^1 xe^{1-x^2} - x\, dx$$

$$= 4\left(-\frac{1}{2}e^{1-x^2} - \frac{1}{2}x^2\right)\Big|_0^1 = (-2e^0 - 2) - (-2e^1 - 0) = 2e - 4$$

In Exercises 43 and 44, let $I = \int_0^1 \int_0^1 \int_0^1 f(x, y, z)\, dV$ and let $S_{N,N,N}$ be the Riemann sum approximation

$$S_{N,N,N} = \frac{1}{N^3} \sum_{i=1}^N \sum_{j=1}^N \sum_{k=1}^N f\left(\frac{i}{N}, \frac{j}{N}, \frac{k}{N}\right)$$

43. *CAS* Calculate $S_{N,N,N}$ for $f(x, y, z) = e^{x^2 - y - z}$ for $N = 10, 20, 30$. Then evaluate I and find an N such that $S_{N,N,N}$ approximates I to two decimal places.

SOLUTION Using a CAS, we get $S_{N,N,N} \approx 0.561, 0.572$, and 0.576 for $N = 10, 20,$ and 30, respectively. We get $I \approx 0.584$, and using $N = 100$ we get $S_{N,N,N} \approx 0.582$, accurate to two decimal places.

Further Insights and Challenges

45. Use Integration by Parts to verify Eq. (9).

SOLUTION If $C_n = \int_{-\pi/2}^{\pi/2} \cos^n \theta\, d\theta$, we use integration by parts to show that

$$C_n = \left(\frac{n-1}{n}\right) C_{n-2}$$

We use integration by parts with $u = \cos^{n-1}\theta$ and $V' = \cos\theta$. Hence, $u' = (n-1)\cos^{n-2}\theta(-\sin\theta)$ and $v = \sin\theta$. Thus,

$$C_n = \int_{-\pi/2}^{\pi/2} \cos^n\theta\, d\theta = \int_{-\pi/2}^{\pi/2} \cos^{n-1}\theta \cos\theta\, d\theta = \cos^{n-1}\theta \sin\theta \Big|_{\theta=-\pi/2}^{\pi/2} + \int_{-\pi/2}^{\pi/2} (n-1)\cos^{n-2}\theta \sin^2\theta\, d\theta$$

$$= \cos^{n-1}\frac{\pi}{2}\sin\frac{\pi}{2} - \cos^{n+1}\left(-\frac{\pi}{2}\right)\sin\left(-\frac{\pi}{2}\right) + (n-1)\int_{-\pi/2}^{\pi/2} \cos^{n-2}\theta \sin^2\theta\, d\theta$$

$$= 0 + (n-1)\int_{-\pi/2}^{\pi/2} \cos^{n-2}\theta\left(1 - \cos^2\theta\right) d\theta = (n-1)\int_{-\pi/2}^{\pi/2} \cos^{n-2}\theta\, d\theta - (n-1)\int_{-\pi/2}^{\pi/2} \cos^n\theta\, d\theta$$

$$= (n-1)C_{n-2} - (n-1)C_n$$

We obtain the following equality:

$$C_n = (n-1)C_{n-2} - (n-1)C_n$$

or

$$C_n + (n-1)C_n = (n-1)C_{n-2}$$
$$nC_n = (n-1)C_{n-2}$$
$$C_n = \frac{n-1}{n}C_{n-2}$$

15.4 Integration in Polar, Cylindrical, and Spherical Coordinates (LT Section 16.4)

Preliminary Questions

1. Which of the following represent the integral of $f(x,y) = x^2 + y^2$ over the unit circle?

(a) $\int_0^1 \int_0^{2\pi} r^2\, dr\, d\theta$

(b) $\int_0^{2\pi} \int_0^1 r^2\, dr\, d\theta$

(c) $\int_0^1 \int_0^{2\pi} r^3\, dr\, d\theta$

(d) $\int_0^{2\pi} \int_0^1 r^3\, dr\, d\theta$

SOLUTION The unit circle is described in polar coordinates by the inequalities

$$0 \le \theta \le 2\pi, \quad 0 \le r \le 1$$

Using double integral in polar coordinates, we have

$$\iint_D f(x,y)\, dA = \int_0^{2\pi}\int_0^1 \left((r\cos\theta)^2 + (r\sin\theta)^2\right) r\, dr\, d\theta = \int_0^{2\pi}\int_0^1 r^2\left(\cos^2\theta + \sin^2\theta\right) r\, dr\, d\theta$$

$$= \int_0^{2\pi}\int_0^1 r^3\, dr\, d\theta$$

Therefore (d) is the correct answer.

2. What are the limits of integration in $\iiint f(r,\theta,z) r\, dr\, d\theta\, dz$ if the integration extends over the following regions?

(a) $x^2 + y^2 \le 4$, $-1 \le z \le 2$

(b) Lower hemisphere of the sphere of radius 2, center at origin

SOLUTION

(a) This is a cylinder of radius 2. In the given region the z coordinate is changing between the values -1 and 2, and the angle θ is changing between the values $\theta = 0$ and 2π. Therefore the region is described by the inequalities

$$-1 \le z \le 2, \quad 0 \le \theta < 2\pi, \quad 0 \le r \le 2$$

Using triple integral in cylindrical coordinates gives

$$\int_{-1}^{2}\int_{0}^{2\pi}\int_{0}^{2} f(P)\, r\, dr\, d\theta\, dz$$

(b) The sphere of radius 2 is $x^2 + y^2 + z^2 = r^2 + z^2 = 4$, or $r = \sqrt{4 - z^2}$.

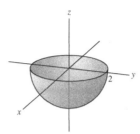

In the lower hemisphere we have $-2 \le z \le 0$ and $0 \le \theta < 2\pi$. Therefore, it has the description

$$-2 \le z \le 0, \quad 0 \le \theta < 2\pi, \quad 0 \le r \le \sqrt{4 - z^2}$$

We obtain the following integral in cylindrical coordinates:

$$\int_{-2}^{0}\int_{0}^{2\pi}\int_{0}^{\sqrt{4-z^2}} r\, dr\, d\theta\, dz$$

3. What are the limits of integration in

$$\iiint f(\rho, \phi, \theta)\, \rho^2 \sin\phi\, d\rho\, d\phi\, d\theta$$

if the integration extends over the following spherical regions centered at the origin?
(a) Sphere of radius 4
(b) Region between the spheres of radii 4 and 5
(c) Lower hemisphere of the sphere of radius 2

SOLUTION

(a) In the sphere of radius 4, θ varies from 0 to 2π, ϕ varies from 0 to π, and ρ varies from 0 to 4. Using triple integral in spherical coordinates, we obtain the following integral:

$$\int_{0}^{2\pi}\int_{0}^{\pi}\int_{0}^{4} f(P)\rho^2 \sin\phi\, d\rho\, d\phi\, d\theta$$

(b) In the region between the spheres of radii 4 and 5, ρ varies from 4 to 5, ϕ varies from 0 to π, and θ varies from 0 to 2π. We obtain the following integral:

$$\int_{0}^{2\pi}\int_{0}^{\pi}\int_{4}^{5} f(P)\rho^2 \sin\phi\, d\rho\, d\phi\, d\theta$$

(c) The inequalities in spherical coordinates for the lower hemisphere of radius 2 are

$$0 \le \theta \le 2\pi, \quad \frac{\pi}{2} \le \phi \le \pi, \quad 0 \le \rho \le 2$$

Therefore we obtain the following integral:

$$\int_{0}^{2\pi}\int_{\pi/2}^{\pi}\int_{0}^{2} f(P)\rho^2 \sin\phi\, d\rho\, d\phi\, d\theta$$

4. An ordinary rectangle of sides Δx and Δy has area $\Delta x\, \Delta y$, no matter where it is located in the plane. However, the area of a polar rectangle of sides Δr and $\Delta \theta$ depends on its distance from the origin. How is this difference reflected in the Change of Variables Formula for polar coordinates?

SOLUTION The area ΔA of a small polar rectangle is

$$\Delta A = \frac{1}{2}(r + \Delta r)^2 \Delta\theta - \frac{1}{2}r^2 \Delta\theta = r(\Delta r\, \Delta\theta) + \frac{1}{2}(\Delta r)^2 \Delta\theta \approx r(\Delta r\, \Delta\theta)$$

SECTION 15.4 | Integration in Polar, Cylindrical, and Spherical Coordinates (LT SECTION 16.4) 1289

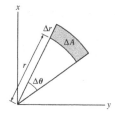

The factor r, due to the distance of the polar rectangle from the origin, appears in $dA = r\, dr\, d\theta$, in the Change of Variables formula.

5. The volume of a sphere of radius 3 is 36π. What is the value of each of the following integrals?

(a) $\displaystyle\int_0^\pi \int_0^\pi \int_0^3 \rho^2 \sin\phi\, d\rho\, d\phi\, d\theta$

(b) $\displaystyle\int_0^{2\pi} \int_0^{\pi/2} \int_0^3 \rho^2 \sin\phi\, d\rho\, d\phi\, d\theta$

(c) $\displaystyle\int_0^{\pi/4} \int_0^{\pi/2} \int_0^3 \rho^2 \sin\phi\, d\rho\, d\phi\, d\theta$

(d) $\displaystyle\int_{-\pi/4}^{\pi/4} \int_0^\pi \int_0^3 \rho^2 \sin\phi\, d\rho\, d\phi\, d\theta$

SOLUTION

(a) θ goes halfway around the sphere (from 0 to π), while ϕ traverses all its values, so this is the hemisphere lying above the xz-plane (positive y-values); its volume is 18π.

(b) θ traverses all its values, while ϕ goes from the z-axis at $\phi = 0$ to the xz-plane at $\phi = \frac{\pi}{2}$. This is the hemisphere lying above the xy-plane; its volume is 18π.

(c) θ traverses one-eighth of the circle $x^2 + y^2 = 9$, since it goes from 0 to $\frac{\pi}{4}$. ϕ goes from the z-axis at $\phi = 0$ to the xz-plane at $\phi = \frac{\pi}{2}$. So of the hemisphere lying above the xy-plane (traversed by ϕ), we take one-eighth of it. Thus the volume of this solid is $\frac{1}{8} \cdot 18\pi = \frac{9}{4}\pi$.

(d) θ traverses one-quarter of the circle $x^2 + y^2 = 9$, since it goes from $-\frac{\pi}{4}$ to $\frac{\pi}{4}$. ϕ traverses all its values, so it traces out the entire sphere. Of the total sphere, then, we take the quarter of it traversed by θ, so the volume is $\frac{1}{4} \cdot 36\pi = 9\pi$.

Exercises

In Exercises 1–6, sketch the region $\mathcal{D}$ indicated and integrate $f(x, y)$ over $\mathcal{D}$ using polar coordinates.

1. $f(x, y) = \sqrt{x^2 + y^2}, \quad x^2 + y^2 \leq 2$

SOLUTION The domain $\mathcal{D}$ is the disk of radius $\sqrt{2}$ shown in the figure:

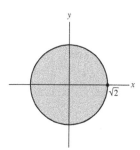

The inequalities defining $\mathcal{D}$ in polar coordinates are

$$0 \leq \theta \leq 2\pi, \quad 0 \leq r \leq \sqrt{2}$$

We describe $f(x, y) = \sqrt{x^2 + y^2}$ in polar coordinates:

$$f(x, y) = \sqrt{x^2 + y^2} = \sqrt{r^2} = r$$

Using change of variables in polar coordinates, we get

$$\iint_{\mathcal{D}} \sqrt{x^2+y^2}\,dA = \int_0^{2\pi}\int_0^{\sqrt{2}} r\cdot r\,dr\,d\theta = \int_0^{2\pi}\int_0^{\sqrt{2}} r^2\,dr\,d\theta = \int_0^{2\pi} \frac{r^3}{3}\bigg|_{r=0}^{\sqrt{2}} d\theta$$

$$= \int_0^{2\pi} \frac{(\sqrt{2})^3}{3}\,d\theta = \frac{2\sqrt{2}}{3}\theta\bigg|_0^{2\pi} = \frac{4\sqrt{2}\pi}{3}$$

3. $f(x,y) = xy;\quad x \geq 0,\quad y \geq 0,\quad x^2 + y^2 \leq 4$

SOLUTION The domain $\mathcal{D}$ is the quarter circle of radius 2 in the first quadrant.

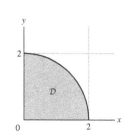

It is described by the inequalities

$$0 \leq \theta \leq \frac{\pi}{2},\quad 0 \leq r \leq 2$$

We write f in polar coordinates:

$$f(x,y) = xy = (r\cos\theta)(r\sin\theta) = r^2\cos\theta\sin\theta = \frac{1}{2}r^2\sin 2\theta$$

Using change of variables in polar coordinates gives

$$\iint_{\mathcal{D}} xy\,dA = \int_0^{\pi/2}\int_0^2 \left(\frac{1}{2}r^2\sin 2\theta\right)r\,dr\,d\theta = \int_0^{\pi/2}\int_0^2 \frac{1}{2}r^3\sin 2\theta\,dr\,d\theta = \int_0^{\pi/2} \frac{1}{2}\cdot\frac{r^4}{4}\sin 2\theta\bigg|_{r=0}^2 d\theta$$

$$= \int_0^{\pi/2} 2\sin 2\theta\,d\theta = -\cos 2\theta\bigg|_0^{\pi/2} = -(\cos\pi - \cos 0) = 2$$

5. $f(x,y) = y(x^2+y^2)^{-1};\quad y \geq \frac{1}{2},\quad x^2 + y^2 \leq 1$

SOLUTION The region $\mathcal{D}$ is the part of the unit circle lying above the line $y = \frac{1}{2}$.

The angle α in the figure is

$$\alpha = \tan^{-1}\frac{\frac{1}{2}}{\frac{\sqrt{3}}{2}} = \tan^{-1}\frac{1}{\sqrt{3}} = \frac{\pi}{6}$$

Therefore, θ varies between $\frac{\pi}{6}$ and $\pi - \frac{\pi}{6} = \frac{5\pi}{6}$. The horizontal line $y = \frac{1}{2}$ has polar equation $r\sin\theta = \frac{1}{2}$ or $r = \frac{1}{2}\csc\theta$. The circle of radius 1 centered at the origin has polar equation $r = 1$. Therefore, r varies between $\frac{1}{2}\csc\theta$ and 1. The inequalities describing $\mathcal{D}$ in polar coordinates are thus

$$\frac{\pi}{6} \leq \theta \leq \frac{5\pi}{6},\quad \frac{1}{2}\csc\theta \leq r \leq 1$$

We write f in polar coordinates:

$$f(x,y) = y(x^2 + y^2)^{-1} = (r\sin\theta)(r^2)^{-1} = r^{-1}\sin\theta$$

Using change of variables in polar coordinates, we obtain

$$\iint_\mathcal{D} y(x^2+y^2)^{-1}\,dA = \int_{\pi/6}^{5\pi/6}\int_{\frac{1}{2}\csc\theta}^{1} r^{-1}\sin\theta\, r\,dr\,d\theta = \int_{\pi/6}^{5\pi/6}\int_{\frac{1}{2}\csc\theta}^{1} \sin\theta\,dr\,d\theta$$

$$= \int_{\pi/6}^{5\pi/6} r\sin\theta\Big|_{r=\frac{1}{2}\csc\theta}^{1} d\theta = \int_{\pi/6}^{5\pi/6}\left(\sin\theta - \frac{1}{2}\sin\theta\csc\theta\right)d\theta$$

$$= \int_{\pi/6}^{5\pi/6}\left(\sin\theta - \frac{1}{2}\right)d\theta = -\cos\theta - \frac{\theta}{2}\Big|_{\pi/6}^{5\pi/6} = -\cos\frac{5\pi}{6} - \frac{5\pi}{12} - \left(-\cos\frac{\pi}{6} - \frac{\pi}{12}\right)$$

$$= \frac{\sqrt{3}}{2} - \frac{\pi}{3} + \frac{\sqrt{3}}{2} = \sqrt{3} - \frac{\pi}{3} \approx 0.685$$

In Exercises 7–14, sketch the region of integration and evaluate by changing to polar coordinates.

7. $\displaystyle\int_{-2}^{2}\int_{0}^{\sqrt{4-x^2}} (x^2+y^2)\,dy\,dx$

SOLUTION The domain $\mathcal{D}$ is described by the inequalities

$$\mathcal{D}: -2 \le x \le 2,\ 0 \le y \le \sqrt{4-x^2}$$

That is, $\mathcal{D}$ is the semicircle $x^2 + y^2 \le 4,\ 0 \le y$.

We describe $\mathcal{D}$ in polar coordinates:

$$\mathcal{D}: 0 \le \theta \le \pi,\ 0 \le r \le 2$$

The function f in polar coordinates is $f(x,y) = x^2 + y^2 = r^2$. We use the Change of Variables Formula to write

$$\int_{-2}^{2}\int_{0}^{\sqrt{4-x^2}}(x^2+y^2)\,dy\,dx = \int_0^\pi\int_0^2 r^2\cdot r\,dr\,d\theta = \int_0^\pi\int_0^2 r^3\,dr\,d\theta = \int_0^\pi \frac{r^4}{4}\Big|_{r=0}^{2} d\theta = \int_0^\pi \frac{2^4}{4}\,d\theta = 4\pi$$

9. $\displaystyle\int_0^{1/2}\int_{\sqrt{3}x}^{\sqrt{1-x^2}} x\,dy\,dx$

SOLUTION The region of integration is described by the inequalities

$$0 \le x \le \frac{1}{2},\quad \sqrt{3}x \le y \le \sqrt{1-x^2}$$

$\mathcal{D}$ is the circular sector shown in the figure.

1292 CHAPTER 15 | MULTIPLE INTEGRATION (LT CHAPTER 16)

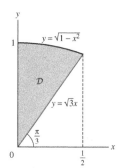

The ray $y = \sqrt{3}x$ in the first quadrant has the polar equation

$$r \sin\theta = \sqrt{3} r \cos\theta \quad \Rightarrow \quad \tan\theta = \sqrt{3} \quad \Rightarrow \quad \theta = \frac{\pi}{3}$$

Therefore, $\mathcal{D}$ lies in the angular sector $\frac{\pi}{3} \leq \theta \leq \frac{\pi}{2}$. Also, the circle $y = \sqrt{1-x^2}$ has the polar equation $r = 1$, hence $\mathcal{D}$ can be described by the inequalities

$$\frac{\pi}{3} \leq \theta \leq \frac{\pi}{2}, \quad 0 \leq r \leq 1$$

We use change of variables to obtain

$$\int_0^{1/2} \int_{\sqrt{3}x}^{\sqrt{1-x^2}} x \, dy \, dx = \int_{\pi/3}^{\pi/2} \int_0^1 r(\cos\theta) r \, dr \, d\theta = \int_{\pi/3}^{\pi/2} \int_0^1 r^2 \cos\theta \, dr \, d\theta = \int_{\pi/3}^{\pi/2} \frac{r^3 \cos\theta}{3}\bigg|_{r=0}^1 d\theta$$

$$= \int_{\pi/3}^{\pi/2} \frac{\cos\theta}{3} d\theta = \frac{\sin\theta}{3}\bigg|_{\pi/3}^{\pi/2} = \frac{1}{3}\left(\sin\frac{\pi}{2} - \sin\frac{\pi}{3}\right) = \frac{1}{3}\left(1 - \frac{\sqrt{3}}{2}\right) \approx 0.045$$

11. $\int_0^5 \int_0^y x \, dx \, dy$

SOLUTION

$$\int_0^5 \int_0^y x \, dx \, dy = \int_{\pi/4}^{\pi/2} \int_{r=0}^{5/\sin\theta} r^2 \cos\theta \, dr \, d\theta = \int_{\pi/4}^{\pi/2} \frac{1}{3} r^3 \cos\theta \bigg|_{r=0}^{5/\sin\theta} d\theta$$

$$= \frac{1}{3} \int_{\pi/4}^{\pi/2} \frac{125}{\sin^3\theta} \cos\theta \, d\theta = \frac{125}{3} \int_{\pi/4}^{\pi/2} \frac{\cos\theta}{\sin^3\theta} d\theta$$

$$= -\frac{125}{6} \frac{1}{\sin^2\theta}\bigg|_{\pi/4}^{\pi/2} = -\frac{125}{6}(1 - 2) = \frac{125}{6}$$

13. $\int_{-1}^2 \int_0^{\sqrt{4-x^2}} (x^2 + y^2) \, dy \, dx$

SOLUTION The domain $\mathcal{D}$, shown in the figure, is described by the inequalities

$$-1 \leq x \leq 2, \quad 0 \leq y \leq \sqrt{4-x^2}$$

We denote by $\mathcal{D}_1$ and $\mathcal{D}_2$ the triangle and the circular sections, respectively, shown in the figure.

By properties of integrals we have

$$\iint_D (x^2 + y^2) \, dA = \iint_{D_1} (x^2 + y^2) \, dA + \iint_{D_2} (x^2 + y^2) \, dA \quad (1)$$

We compute each integral separately, starting with D_1. The vertical line $x = -1$ has polar equation $r \cos \theta = -1$ or $r = -\sec \theta$. The ray OA has polar equation $\theta = \frac{2\pi}{3}$. Therefore D_1 is described by

$$\frac{2\pi}{3} \leq \theta \leq \pi, \quad 0 \leq r \leq -\sec \theta$$

Using change of variables gives

$$\iint_{D_1} (x^2 + y^2) \, dA = \int_{2\pi/3}^{\pi} \int_0^{-\sec \theta} r^2 \cdot r \, dr \, d\theta = \int_{2\pi/3}^{\pi} \int_0^{-\sec \theta} r^3 \, dr \, d\theta$$

$$= \int_{2\pi/3}^{\pi} \frac{r^4}{4} \bigg|_{r=0}^{-\sec \theta} d\theta = \int_{2\pi/3}^{\pi} \frac{\sec^4 \theta}{4} \, d\theta = \frac{1}{4} \int_{2\pi/3}^{\pi} \sec^4 \theta \, d\theta \quad (2)$$

We compute the integral (we use substitution $u = \tan \theta$ for the second integral):

$$\int_{2\pi/3}^{\pi} \frac{1}{\cos^4 \theta} \, d\theta = \int_{2\pi/3}^{\pi} \frac{\sin^2 \theta + \cos^2 \theta}{\cos^4 \theta} \, d\theta = \int_{2\pi/3}^{\pi} \frac{d\theta}{\cos^2 \theta} + \int_{2\pi/3}^{\pi} \tan^2 \theta \cdot \frac{1}{\cos^2 \theta} \, d\theta$$

$$= \tan \theta \bigg|_{\theta = 2\pi/3}^{\pi} + \int_{-\sqrt{3}}^{0} u^2 \, du = \tan \pi - \tan \frac{2\pi}{3} + \frac{u^3}{3} \bigg|_{-\sqrt{3}}^{0} = \sqrt{3} + \frac{3\sqrt{3}}{3} = 2\sqrt{3}$$

Hence, by (2) we get

$$\iint_{D_1} (x^2 + y^2) \, dA = \frac{\sqrt{3}}{2} \quad (3)$$

D_2 is described by the inequalities

$$0 \leq \theta \leq \frac{2\pi}{3}, \quad 0 \leq r \leq 2$$

Hence,

$$\iint_{D_2} (x^2 + y^2) \, dA = \int_0^{2\pi/3} \int_0^2 r^2 \cdot r \, dr \, d\theta = \int_0^{2\pi/3} \int_0^2 r^3 \, dr \, d\theta$$

$$= \int_0^{2\pi/3} \frac{r^4}{4} \bigg|_{r=0}^{2} d\theta = \int_0^{2\pi/3} 4 \, d\theta = 4 \cdot \theta \bigg|_0^{2\pi/3} = \frac{8\pi}{3} \quad (4)$$

Combining (1), (3), and (4) we obtain the following solution:

$$\iint_D (x^2 + y^2) \, dA = \frac{\sqrt{3}}{2} + \frac{8\pi}{3} \approx 9.24$$

In Exercises 15–20, calculate the integral over the given region by changing to polar coordinates.

15. $f(x, y) = (x^2 + y^2)^{-2}; \quad x^2 + y^2 \leq 2, \quad x \geq 1$

SOLUTION The region D lies in the angular sector

$$-\frac{\pi}{4} \leq \theta \leq \frac{\pi}{4}$$

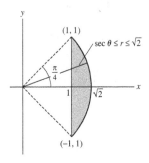

The vertical line $x = 1$ has polar equation $r\cos\theta = 1$ or $r = \sec\theta$. The circle $x^2 + y^2 = 2$ has polar equation $r = \sqrt{2}$. Therefore, $\mathcal{D}$ has the following description:

$$-\frac{\pi}{4} \leq \theta \leq \frac{\pi}{4}, \quad \sec\theta \leq r \leq \sqrt{2}$$

The function in polar coordinates is

$$f(x,y) = (x^2 + y^2)^{-2} = (r^2)^{-2} = r^{-4}$$

Using change of variables we obtain

$$\iint_{\mathcal{D}} (x^2 + y^2)^{-2} dA = \int_{-\pi/4}^{\pi/4} \int_{\sec\theta}^{\sqrt{2}} r^{-4} r\, dr\, d\theta = \int_{-\pi/4}^{\pi/4} \int_{\sec\theta}^{\sqrt{2}} r^{-3}\, dr\, d\theta = \int_{-\pi/4}^{\pi/4} \left.\frac{r^{-2}}{-2}\right|_{\sec\theta}^{\sqrt{2}} d\theta$$

$$= \int_{-\pi/4}^{\pi/4} \left(\frac{1}{2\sec^2\theta} - \frac{1}{4}\right) d\theta = 2\int_{0}^{\pi/4} \left(\frac{1}{2}\cos^2\theta - \frac{1}{4}\right) d\theta$$

$$= \left.\left(\frac{\theta}{2} + \frac{\sin 2\theta}{4}\right)\right|_{0}^{\pi/4} - \left.\frac{\theta}{2}\right|_{0}^{\pi/4} = \frac{\pi}{8} + \frac{1}{4} - \frac{\pi}{8} = \frac{1}{4}$$

17. $f(x,y) = |xy|; \quad x^2 + y^2 \leq 1$

SOLUTION The unit disk is described in polar coordinates by

$$\mathcal{D}: 0 \leq \theta \leq 2\pi, \; 0 \leq r \leq 1$$

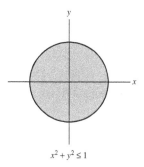

The function is $f(x,y) = |xy| = |r\cos\theta \cdot r\sin\theta| = \frac{1}{2}r^2|\sin 2\theta|$. Using change of variables we obtain

$$\iint_{\mathcal{D}} |xy|\, dA = \int_{0}^{2\pi} \int_{0}^{1} \frac{1}{2}r^2 |\sin 2\theta| \cdot r\, dr\, d\theta = \int_{0}^{2\pi} \int_{0}^{1} \frac{1}{2}r^3 |\sin 2\theta|\, dr\, d\theta$$

$$= \int_{0}^{2\pi} \left.\frac{r^4}{8}|\sin 2\theta|\right|_{r=0}^{1} d\theta = \int_{0}^{2\pi} \frac{1}{8}|\sin 2\theta|\, d\theta \qquad (1)$$

The signs of $\sin 2\theta$ in the interval of integration are

For $0 \leq \theta \leq \frac{\pi}{2}$ or $\pi \leq \theta \leq \frac{3\pi}{2}$, $\sin 2\theta \geq 0$, hence $|\sin 2\theta| = \sin 2\theta$.
For $\frac{\pi}{2} \leq \theta \leq \pi$ or $\frac{3\pi}{2} \leq \theta \leq 2\pi$, $\sin 2\theta \leq 0$, hence $|\sin 2\theta| = -\sin 2\theta$.

Therefore, by (1) we get

$$\iint_{\mathcal{D}} |xy|\, dA = \int_{0}^{\pi/2} \frac{1}{8}\sin 2\theta\, d\theta - \int_{\pi/2}^{\pi} \frac{1}{8}\sin 2\theta\, d\theta + \int_{\pi}^{3\pi/2} \frac{1}{8}\sin 2\theta\, d\theta - \int_{3\pi/2}^{2\pi} \frac{1}{8}\sin 2\theta\, d\theta$$

$$= \left.-\frac{1}{16}\cos 2\theta\right|_{0}^{\pi/2} + \left.\frac{1}{16}\cos 2\theta\right|_{\pi/2}^{\pi} - \left.\frac{1}{16}\cos 2\theta\right|_{\pi}^{3\pi/2} + \left.\frac{1}{16}\cos 2\theta\right|_{3\pi/2}^{2\pi}$$

$$= -\frac{1}{16}(\cos\pi - 1) + \frac{1}{16}(\cos 2\pi - \cos\pi) - \frac{1}{16}(\cos 3\pi - \cos 2\pi) + \frac{1}{16}(\cos 4\pi - \cos 3\pi)$$

$$= \frac{2}{16} + \frac{2}{16} + \frac{2}{16} + \frac{2}{16} = \frac{1}{2}$$

That is,

$$\iint_{\mathcal{D}} |xy|\, dA = \frac{1}{2}$$

19. $f(x,y) = x - y;\quad x^2 + y^2 \le 1,\quad x + y \ge 1$

SOLUTION As shown in Exercise 18, the region $\mathcal{D}$ is described by the following inequalities:

$$\mathcal{D}: 0 \le \theta \le \frac{\pi}{2},\ \frac{1}{\cos\theta + \sin\theta} \le r \le 1$$

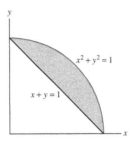

The function in polar coordinates is

$$f(x, y) = x - y = r\cos\theta - r\sin\theta = r(\cos\theta - \sin\theta)$$

Using the Change of Variables Formula we get

$$\iint_{\mathcal{D}} (x-y)\, dA = \int_0^{\pi/2} \int_{\frac{1}{\cos\theta+\sin\theta}}^{1} r(\cos\theta - \sin\theta) r\, dr\, d\theta$$

$$= \int_0^{\pi/2} \int_{\frac{1}{\cos\theta+\sin\theta}}^{1} r^2(\cos\theta - \sin\theta)\, dr\, d\theta$$

$$= \frac{1}{3}\int_0^{\pi/2} r^3(\cos\theta - \sin\theta)\Big|_{r=\frac{1}{\cos\theta+\sin\theta}}^{1}\, d\theta$$

$$= \frac{1}{3}\int_0^{\pi/2} (\cos\theta - \sin\theta)\left(1 - \frac{1}{(\cos\theta + \sin\theta)^3}\right) d\theta$$

$$= \frac{1}{3}\left(\int_0^{\pi/2} (\cos\theta - \sin\theta)\, d\theta - \int_0^{\pi/2} \frac{\cos\theta - \sin\theta}{(\cos\theta + \sin\theta)^3}\, d\theta\right)$$

The first integral is straightforward:

$$\int_0^{\pi/2} (\cos\theta - \sin\theta)\, d\theta = (\sin\theta + \cos\theta)\Big|_0^{\pi/2} = (1 + 0) - (0 + 1) = 0 \tag{1}$$

To compute the second integral, use the substitution $u = \sin\theta + \cos\theta$; then $du = (\cos\theta - \sin\theta)\, d\theta$:

$$\int_0^{\pi/2} \frac{\cos\theta - \sin\theta}{(\cos\theta + \sin\theta)^3}\, d\theta = \int_1^1 \frac{1}{u^3}\, du = 0$$

since the upper and lower bounds are equal. Since both integrals are zero we conclude that

$$\iint_{\mathcal{D}} (x - y)\, dA = 0$$

21. Find the volume of the wedge-shaped region (Figure 18) contained in the cylinder $x^2 + y^2 = 9$, bounded above by the plane $z = x$ and below by the xy-plane.

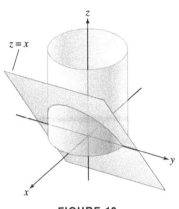

FIGURE 18

SOLUTION

Step 1. Express $\mathcal{W}$ in cylindrical coordinates. $\mathcal{W}$ is bounded above by the plane $z = x$ and below by $z = 0$, therefore $0 \leq z \leq x$, in particular $x \geq 0$. Hence, $\mathcal{W}$ projects onto the semicircle $\mathcal{D}$ in the xy-plane of radius 3, where $x \geq 0$.

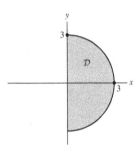

In polar coordinates,

$$\mathcal{D}: -\frac{\pi}{2} \leq \theta \leq \frac{\pi}{2}, \ 0 \leq r \leq 3$$

The upper surface is $z = x = r\cos\theta$ and the lower surface is $z = 0$. Therefore,

$$\mathcal{W}: -\frac{\pi}{2} \leq \theta \leq \frac{\pi}{2}, \ 0 \leq r \leq 3, \ 0 \leq z \leq r\cos\theta$$

Step 2. Set up an integral in cylindrical coordinates and evaluate. The volume of $\mathcal{W}$ is the triple integral $\iiint_{\mathcal{W}} 1 \, dV$. Using change of variables in cylindrical coordinates gives

$$\iiint_{\mathcal{W}} 1 \, dV = \int_{-\pi/2}^{\pi/2} \int_0^3 \int_0^{r\cos\theta} r \, dz \, dr \, d\theta = \int_{-\pi/2}^{\pi/2} \int_0^3 rz \Big|_{z=0}^{r\cos\theta} dr \, d\theta = \int_{-\pi/2}^{\pi/2} \int_0^3 r^2 \cos\theta \, dr \, d\theta$$

$$= \int_{-\pi/2}^{\pi/2} \frac{r^3}{3} \cos\theta \Big|_{r=0}^3 d\theta = \int_{-\pi/2}^{\pi/2} 9\cos\theta \, d\theta = 9\sin\theta \Big|_{-\pi/2}^{\pi/2} = 9\left(\sin\frac{\pi}{2} - \sin\left(-\frac{\pi}{2}\right)\right) = 18$$

23. Evaluate $\iint_{\mathcal{D}} \sqrt{x^2 + y^2} \, dA$, where $\mathcal{D}$ is the domain in Figure 20. *Hint:* Find the equation of the inner circle in polar coordinates and treat the right and left parts of the region separately.

SECTION 15.4 | Integration in Polar, Cylindrical, and Spherical Coordinates

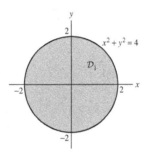

FIGURE 20

SOLUTION We denote by $\mathcal{D}_1$ and $\mathcal{D}_2$ the regions enclosed by the circles $x^2 + y^2 = 4$ and $(x-1)^2 + y^2 = 1$. Therefore,

$$\iint_{\mathcal{D}} \sqrt{x^2 + y^2}\, dx\, dy = \iint_{\mathcal{D}_1} \sqrt{x^2 + y^2}\, dx\, dy - \iint_{\mathcal{D}_2} \sqrt{x^2 + y^2}\, dx\, dy \tag{1}$$

We compute the integrals on the right hand-side.

$\mathcal{D}_1$:

The circle $x^2 + y^2 = 4$ has polar equation $r = 2$, therefore $\mathcal{D}_1$ is determined by the following inequalities:

$$\mathcal{D}_1 : 0 \le \theta \le 2\pi,\ 0 \le r \le 2$$

The function in polar coordinates is $f(x, y) = \sqrt{x^2 + y^2} = r$. Using change of variables in the integral gives

$$\iint_{\mathcal{D}_1} \sqrt{x^2 + y^2}\, dx\, dy = \int_0^{2\pi} \int_0^2 r \cdot r\, dr\, d\theta = \int_0^{2\pi} \int_0^2 r^2\, dr\, d\theta = \int_0^{2\pi} \frac{r^3}{3}\bigg|_{r=0}^2 d\theta = \int_0^{2\pi} \frac{8}{3}\, d\theta = \frac{16\pi}{3} \tag{2}$$

$\mathcal{D}_2$:

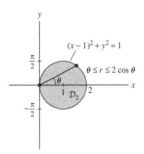

$\mathcal{D}_2$ lies in the angular sector $-\frac{\pi}{2} \le \theta \le \frac{\pi}{2}$. We find the polar equation of the circle $(x-1)^2 + y^2 = 1$:

$$(x - 1)^2 + y^2 = x^2 - 2x + 1 + y^2 = x^2 + y^2 - 2x + 1 = 1 \quad \Rightarrow \quad x^2 + y^2 = 2x$$

$$\Rightarrow \quad r^2 = 2r \cos \theta$$

$$\Rightarrow \quad r = 2 \cos \theta$$

Thus, the domain $\mathcal{D}_2$ is defined by the following inequalities:

$$\mathcal{D}_2 : -\frac{\pi}{2} \le \theta \le \frac{\pi}{2},\ 0 \le r \le 2\cos\theta$$

We use the change of variables in the integral and integration table to obtain

$$\iint_{D_2} \sqrt{x^2+y^2}\,dx\,dy = \int_{-\pi/2}^{\pi/2}\int_0^{2\cos\theta} r\cdot r\,dr\,d\theta = \int_{-\pi/2}^{\pi/2}\int_0^{2\cos\theta} r^2\,dr\,d\theta = \int_{-\pi/2}^{\pi/2}\int_0^{2\cos\theta} \frac{r^3}{3}\bigg|_{r=0}^{2\cos\theta}\,d\theta$$

$$= \int_{-\pi/2}^{\pi/2} \frac{8\cos^3\theta}{3}\,d\theta = 2\int_0^{\pi/2}\frac{8\cos^3\theta}{3}\,d\theta = \frac{16}{3}\left(\frac{\cos^2\theta\sin\theta}{3} + \frac{2}{3}\sin\theta\right)\bigg|_{\theta=0}^{\pi/2}$$

$$= \frac{16}{3}\cdot\frac{2}{3}\sin\frac{\pi}{2} = \frac{32}{9} \quad (3)$$

Substituting (2) and (3) in (1), we obtain the following solution:

$$\iint_D \sqrt{x^2+y^2}\,dx\,dy = \frac{16\pi}{3} - \frac{32}{9} = \frac{48\pi - 32}{9} \approx 13.2$$

Remark: The integral can also be evaluated using the hint as the sum of

$$\int_{D^*}\!\!\iint \sqrt{x^2+y^2}\,dA \quad \text{and} \quad \int_{D^{**}}\!\!\iint \sqrt{x^2+y^2}\,dA$$

where D^* is the left semicircle $x^2+y^2=4$ and D^* is the right part of D. Since

$$D^*: \frac{\pi}{2} \le \theta \le \frac{3\pi}{2},\ 0 \le r \le 2$$

$$D^{**}: -\frac{\pi}{2} \le \theta \le \frac{\pi}{2},\ 2\cos\theta \le r \le 2$$

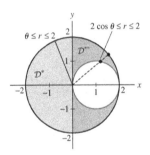

we get

$$\iint_D \sqrt{x^2+y^2}\,dA = \int_{\pi/2}^{3\pi/2}\int_0^2 r^2\,dr\,d\theta + \int_{-\pi/2}^{\pi/2}\int_{2\cos\theta}^2 r^2\,dr\,d\theta$$

Obviously, computing the integrals leads to the same result.

25. Let W be the region above the plane $z=2$ and below the paraboloid $z=6-(x^2+y^2)$.

(a) Describe W in cylindrical coordinates.

(b) Use cylindrical coordinates to compute the volume of W.

SOLUTION

(a) A diagram of the region is below:

SECTION 15.4 Integration in Polar, Cylindrical, and Spherical Coordinates (LT SECTION 16.4)

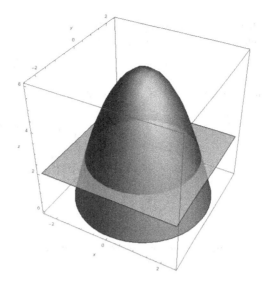

The region is question is above the plane $z = 2$. The plane $z = 2$ has cylindrical equation $z = 2$. Since $x^2 + y^2 = r^2$ in cylindrical coordinates, the equation of the paraboloid is $z = 6 - r^2$. The plane and the paraboloid intersect where

$$z = 2 \text{ and } z = 6 - (x^2 + y^2) \quad \Rightarrow \quad 2 = 6 - (x^2 + y^2) \quad \Rightarrow \quad x^2 + y^2 = 4$$

so that the projection $\mathcal{D}$ of $\mathcal{W}$ onto the xy-plane is the circle of radius 2 centered at the origin. This circle is described in polar coordinates by

$$\mathcal{D}: \ 0 \le \theta \le 2\pi, \ 0 \le r \le 2$$

Therefore the inequalities for $\mathcal{W}$ in cylindrical coordinates are

$$\mathcal{W}: \ 0 \le \theta \le 2\pi, \ 0 \le r \le 2, \ 2 \le z \le 6 - r^2$$

(b) Using the inequalities above, the volume of $\mathcal{W}$ is

$$\text{Volume}(\mathcal{W}) = \iiint_{\mathcal{W}} 1 \, dV = \int_0^{2\pi} \int_0^2 \int_2^{6-r^2} 1 \cdot r \, dz \, dr \, d\theta$$

$$= \int_0^{2\pi} \int_0^2 rz \Big|_{z=2}^{6-r^2} dr \, d\theta$$

$$= \int_0^{2\pi} \int_0^2 r(6 - r^2 - 2) \, dr \, d\theta$$

$$= \int_0^{2\pi} \int_0^2 (4r - r^3) \, dr \, d\theta$$

$$= \int_0^{2\pi} \left(2r^2 - \frac{1}{4}r^4 \right) \Big|_0^2 d\theta$$

$$= \int_0^{2\pi} 4 \, d\theta = 8\pi$$

In Exercises 27–32, use cylindrical coordinates to calculate $\iiint_{\mathcal{W}} f(x, y, z) \, dV$ *for the given function and region.*

27. $f(x, y, z) = x^2 + y^2; \quad x^2 + y^2 \le 9, \ 0 \le z \le 5$

SOLUTION The projection of $\mathcal{W}$ onto the xy-plane is the region inside the circle $x^2 + y^2 = 9$. In polar coordinates,

$$\mathcal{D}: \ 0 \le \theta \le 2\pi, \ 0 \le r \le 3$$

The upper and lower boundaries are the planes $z = 5$ and $z = 0$, respectively. Therefore, $\mathcal{W}$ has the following description in cylindrical coordinates:

$$\mathcal{W}: 0 \leq \theta \leq 2\pi, \; 0 \leq r \leq 3, \; 0 \leq z \leq 5$$

The integral in cylindrical coordinates is thus

$$\iiint_{\mathcal{W}} (x^2 + y^2)\, dV = \int_0^{2\pi} \int_0^3 \int_0^5 r^2 \cdot r \, dz \, dr \, d\theta = \int_0^{2\pi} \int_0^3 \int_0^5 r^3 \, dz \, dr \, d\theta$$

$$= \left(\int_0^{2\pi} 1\, d\theta\right)\left(\int_0^3 r^3 \, dr\right)\left(\int_0^5 1\, dz\right) = 2\pi \cdot 5 \cdot \left.\frac{r^4}{4}\right|_0^3 = \frac{5 \cdot 3^4 \pi}{2} \approx 636.17$$

29. $f(x, y, z) = x;$ $\;x^2 + y^2 \leq 16,$ $\;x \geq 0,$ $\;y \geq 0,$ $\;-3 \leq z \leq 3$

SOLUTION $\mathcal{W}$ is a portion of a cylinder that projects onto the quarter-disk $\mathcal{D}$ of radius $\sqrt{16} = 4$ centered at the origin, with $x \geq 0$ and $y \geq 0$. This domain is shown in the figure below:

The upper and lower boundaries of $\mathcal{W}$ are the planes $z = 3$ and $z = -3$, so $\mathcal{W}$ has the following description in cylindrical coordinates:

$$\mathcal{W}: 0 \leq \theta \leq \frac{\pi}{2}, \; 0 \leq r \leq 4, \; -3 \leq z \leq 3$$

The function in cylindrical coordinates is $x = r\cos\theta$. We obtain the following integral:

$$\iiint_{\mathcal{W}} x \, dV = \int_{-3}^3 \int_0^{\pi/2} \int_0^4 r \cos\theta \cdot r \, dr \, d\theta \, dz = \int_{-3}^3 \int_0^{\pi/2} \int_0^4 r^2 \cos\theta \, dr \, d\theta \, dz$$

$$= \left(\int_{-3}^3 1\, dx\right)\left(\int_0^{\pi/2} \cos\theta\, d\theta\right)\left(\int_0^4 r^2 \, dr\right)$$

$$= \left. x\right|_{-3}^3 \cdot \left.\sin\theta\right|_0^{\pi/2} \cdot \left.\frac{1}{3}r^3\right|_0^4$$

$$= 6 \cdot 1 \cdot \frac{64}{3} = 128$$

31. $f(x, y, z) = z;$ $\;x^2 + y^2 \leq z \leq 9$

SOLUTION

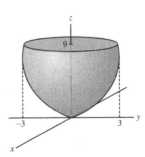

The upper boundary of $\mathcal{W}$ is the plane $z = 9$, and the lower boundary is $z = x^2 + y^2 = r^2$. Therefore, $r^2 \leq z \leq 9$. The projection $\mathcal{D}$ onto the xy-plane is the circle $x^2 + y^2 = 9$ or $r = 3$. That is,

$$\mathcal{D}: 0 \leq \theta \leq 2\pi, \; 0 \leq r \leq 3$$

SECTION 15.4 | Integration in Polar, Cylindrical, and Spherical Coordinates (LT SECTION 16.4) 1301

The inequalities defining $\mathcal{W}$ in cylindrical coordinates are thus

$$\mathcal{W}: 0 \le \theta \le 2\pi,\ 0 \le r \le 3,\ r^2 \le z \le 9$$

Therefore, we obtain the following integral:

$$\iiint_{\mathcal{W}} z\, dV = \int_0^{2\pi}\int_0^3\int_{r^2}^9 zr\, dz\, dr\, d\theta = \int_0^{2\pi}\int_0^3 \left.\frac{z^2 r}{2}\right|_{z=r^2}^9 dr\, d\theta = \int_0^{2\pi}\int_0^3 \frac{r(81 - r^4)}{2}\, dr\, d\theta$$

$$= \int_0^{2\pi}\int_0^3 \frac{81r - r^5}{2}\, dr\, d\theta = \int_0^{2\pi} \left.\frac{81r^2}{4} - \frac{r^6}{12}\right|_0^3 d\theta = \int_0^{2\pi} 121.5\, d\theta = 243\pi$$

In Exercises 33–36, express the triple integral in cylindrical coordinates.

33. $\displaystyle\int_{-1}^{1}\int_{y=-\sqrt{1-x^2}}^{y=\sqrt{1-x^2}}\int_{z=0}^{4} f(x, y, z)\, dz\, dy\, dx$

SOLUTION The region of integration is determined by the limits of integration. That is,

$$\mathcal{W}: -1 \le x \le 1,\ -\sqrt{1 - x^2} \le y \le \sqrt{1 - x^2},\ 0 \le z \le 4$$

Therefore the projection of $\mathcal{W}$ onto the xy-plane is the disk $x^2 + y^2 \le 1$. This region has the following definition in polar coordinates:

$$\mathcal{D}: 0 \le \theta \le 2\pi,\ 0 \le r \le 1$$

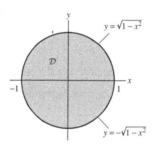

The upper and lower boundaries of $\mathcal{W}$ are the planes $z = 4$ and $z = 0$, respectively. Hence,

$$\mathcal{W}: 0 \le \theta \le 2\pi,\ 0 \le r \le 1,\ 0 \le z \le 4$$

Using change of variables in cylindrical coordinates, we get the integral

$$\int_0^{2\pi}\int_0^1\int_0^4 f(r\cos\theta, r\sin\theta, z) r\, dz\, dr\, d\theta$$

35. $\displaystyle\int_{-1}^{1}\int_{y=0}^{y=\sqrt{1-x^2}}\int_{z=0}^{x^2+y^2} f(x, y, z)\, dz\, dy\, dx$

SOLUTION The inequalities defining the region of integration are

$$\mathcal{W}: -1 \le x \le 1,\ 0 \le y \le \sqrt{1 - x^2},\ 0 \le z \le x^2 + y^2$$

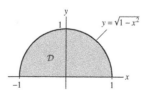

The projection of $\mathcal{W}$ onto the xy-plane is the semicircle $x^2 + y^2 = 1$, where $-1 \le x \le 1$. This domain is defined by the polar inequalities

$$\mathcal{D}: 0 \le \theta \le \pi,\ 0 \le r \le 1$$

The lower surface is $z = 0$ and upper surface is $z = x^2 + y^2 = r^2$, hence $\mathcal{W}$ has the following description in cylindrical coordinates:

$$\mathcal{W}: 0 \le \theta \le \pi, \, 0 \le r \le 1, \, 0 \le z \le r^2$$

We obtain the following integral:

$$\int_0^\pi \int_0^1 \int_0^{r^2} f(r\cos\theta, r\sin\theta, z) r \, dz \, dr \, d\theta$$

37. Find the equation of the right-circular cone in Figure 22 in cylindrical coordinates and compute its volume.

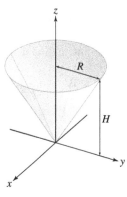

FIGURE 22

SOLUTION To find the equation of the surface we use proportion in similar triangles.

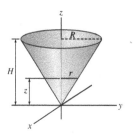

This gives

$$\frac{z}{H} = \frac{r}{R} \quad \Rightarrow \quad z = \frac{H}{R} r$$

The volume of the right circular cone is

$$V = \iiint_{\mathcal{W}} 1 \, dV$$

The projection of $\mathcal{W}$ onto the xy-plane is the region $x^2 + y^2 \le R^2$, or in polar coordinates,

$$\mathcal{D}: 0 \le \theta \le 2\pi, \, 0 \le r \le R$$

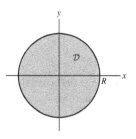

The upper and lower boundaries are the surfaces $z = H$ and $z = \frac{H}{R}r$, respectively. Hence $\mathcal{W}$ is determined by the following cylindrical inequalities:

$$\mathcal{W}: 0 \leq \theta \leq 2\pi,\ 0 \leq r \leq R,\ \frac{H}{R}r \leq z \leq H$$

We compute the volume using the following integral:

$$V = \iiint_{\mathcal{W}} 1\,dv = \int_0^{2\pi}\int_0^R \int_{\frac{H}{R}r}^H r\,dz\,dr\,d\theta = \int_0^{2\pi}\int_0^R rz\Big|_{z=\frac{Hr}{R}}^H dr\,d\theta = \int_0^{2\pi}\int_0^R r\left(H - \frac{Hr}{R}\right)dr\,d\theta$$

$$= \int_0^{2\pi}\int_0^R \left(rH - \frac{r^2 H}{R}\right)dr\,d\theta = \int_0^{2\pi} \frac{r^2 H}{2} - \frac{r^3 H}{3R}\Big|_{r=0}^R d\theta = \int_0^{2\pi} \frac{R^2 H}{6}\,d\theta = \frac{R^2 H}{6}\cdot 2\pi = \frac{\pi R^2 H}{3}$$

39. Find the volume of the region appearing between the two surfaces in Figure 23.

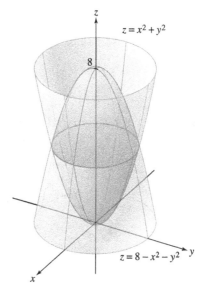

FIGURE 23

SOLUTION We first must identify the domain of integration in the xy-plane. To do this we equate the equations of the two surfaces. That is,

$$x^2 + y^2 = 8 - x^2 - y^2 \quad \Rightarrow \quad 2(x^2 + y^2) = 8 \quad \Rightarrow \quad x^2 + y^2 = 4$$

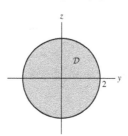

Therefore, the domain of integration is

$$\mathcal{D} = \left\{(x, y);\ x^2 + y^2 \leq 4\right\}$$

The double integrals $\iint_{\mathcal{D}} (x^2 + y^2)\,dA$ and $\iint_{\mathcal{D}} (8 - x^2 - y^2)\,dA$ give the volumes of the regions bounded by the graphs of $z = x^2 + y^2$ and $z = 8 - x^2 - y^2$, respectively, and the xy-plane over the region $\mathcal{D}$. Therefore, the volume of the required region is the difference between these two integrals (see figure). That is,

$$V = \iint_{\mathcal{D}} (8 - x^2 - y^2)\,dA - \iint_{\mathcal{D}} (x^2 + y^2)\,dA = \iint_{\mathcal{D}} \left((8 - x^2 - y^2) - (x^2 + y^2)\right)dA$$

or

$$V = \iint_{\mathcal{D}} (8 - 2x^2 - 2y^2)\,dA$$

To compute the double integral we first notice that, since the function $f(x,y) = 8 - 2x^2 - 2y^2$ satisfies $f(x,-y) = f(x,y)$ and $f(-x,y) = f(x,y)$, and the circle $\mathcal{D}$ is symmetric with respect to the x and y axes, the double integral over $\mathcal{D}$ is four times the integral over the part $\mathcal{D}_1$ of the circle in the first quadrant. That is,

$$V = 4 \iint_{\mathcal{D}_1} (8 - 2x^2 - 2y^2)\, dA$$

$\mathcal{D}_1$ is the vertically simple region described by the inequalities

$$0 \le x \le 2, \quad 0 \le y \le \sqrt{4 - x^2}$$

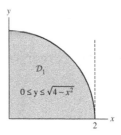

Therefore,

$$V = 4 \iint_{\mathcal{D}_1} (8 - 2x^2 - 2y^2)\, dA = 4 \int_0^2 \int_0^{\sqrt{4-x^2}} (8 - 2x^2 - 2y^2)\, dy\, dx$$

$$= 4 \int_0^2 \left(8y - 2x^2 y - \frac{2}{3} y^3 \right)\bigg|_{y=0}^{\sqrt{4-x^2}} dx$$

$$= 4 \int_0^2 \left(8\sqrt{4-x^2} - 2x^2 \sqrt{4-x^2} - \frac{2}{3}(4-x^2)^{2/3} \right) dx$$

$$= 4 \int_0^2 \left(2\sqrt{4-x^2}(4-x^2) - \frac{2}{3}(4-x^2)^{3/2} \right) dx$$

$$= 4 \int_0^2 \frac{4}{3}(4-x^2)^{3/2}\, dx = \frac{16}{3} \int_0^2 (4-x^2)^{3/2}\, dx.$$

If we let $x = 2\sin\theta$, then $dx = 2\cos\theta\, d\theta$, and so

$$V = \frac{32}{3} \int_0^{\pi/2} (4 - 4\sin^2\theta)^{3/2} \cdot \cos\theta\, d\theta = \frac{32}{3} \int_0^{\pi/2} 8\cos^3\theta \cdot \cos\theta\, d\theta = \frac{256}{3} \int_0^{\pi/2} \cos^4\theta\, d\theta$$

To perform the remaining integration, we expand $\cos^4\theta$:

$$\cos^4\theta = (\cos^2\theta)^2 = \left(\frac{1}{2} + \frac{1}{2}\cos 2\theta\right)^2 = \frac{1}{4}(1 + 2\cos 2\theta + \cos^2 2\theta)$$

$$= \frac{1}{4}\left(1 + 2\cos 2\theta + \frac{1}{2} + \frac{1}{2}\cos 4\theta\right) = \frac{1}{4}\left(\frac{3}{2} + 2\cos 2\theta + \frac{1}{2}\cos 4\theta\right).$$

Then

$$\frac{256}{3} \int_0^{\pi/2} \cos^4\theta\, d\theta = \frac{64}{3} \int_0^{\pi/2} \left(\frac{3}{2} + 2\cos 2\theta + \frac{1}{2}\cos 4\theta\right) d\theta$$

$$= \frac{64}{3}\left(\frac{3}{2}\theta + \sin 2\theta + 2\sin 4\theta\right)\bigg|_0^{\pi/2}$$

$$= 16\pi$$

Thus $V = 16\pi$.

41. Use cylindrical coordinates to show that the volume of a sphere of radius a from which a central cylinder of radius b has been removed, where $0 < b < a$, only depends on the height of the band that results. In particular, this implies that such a band of radius 2 m and height 1 m has the same volume as such a band of radius 6,400 km (the radius of the earth) and height 1 m.

SOLUTION A cross-sectional view of the diagram of $\mathcal{W}$ from the previous exercise shows the following:

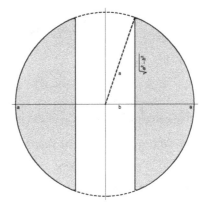

Using the Pythagorean Theorem shows that the height of the band (that is, the distance from the x-axis to the top) is $h = \sqrt{a^2 - b^2}$. Substituting in the volume formula from the previous exercise gives

$$V = \frac{4\pi}{3}(a^2 - b^2)^{3/2} = \frac{4\pi}{3}h^3$$

This shows that the volume depends neither on a nor on b, but only on the height remaining.

43. Use spherical coordinates to find the volume of the region bounded below by the plane $z = 1$ and above by the sphere $x^2 + y^2 + z^2 = 4$.

SOLUTION Use the diagram from the previous exercise, since the region in question is identical. The circle of intersection of the two is the same. Consider the following cross-sectional diagram:

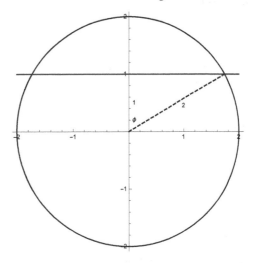

The angle ϕ determined by the intersection of the sphere and the plane satisfies $\cos\phi = \frac{1}{2}$, so that $\phi = \frac{\pi}{3}$. The plane $z = 1$ in spherical coordinates is $\rho\cos\phi = 1$, or $\rho = \sec\phi$, and the sphere is $\rho = 2$. Thus the region $\mathcal{W}$ is described by

$$\mathcal{W}: 0 \le \theta \le 2\pi,\ 0 \le \phi \le \frac{\pi}{3},\ \sec\phi \le \rho \le 2$$

The volume of the region is then

$$\text{Volume}(\mathcal{W}) = \iiint_{\mathcal{W}} 1\, dV = \int_0^{2\pi}\int_0^{\pi/3}\int_{\sec\phi}^2 \rho^2 \sin\phi\, d\rho\, d\phi\, d\theta$$

$$= \int_0^{2\pi}\int_0^{\pi/3} \left(\frac{1}{3}\rho^3 \sin\phi\right)\bigg|_{\rho=\sec\phi}^2 d\phi\, d\theta$$

$$= \frac{1}{3}\int_0^{2\pi}\int_0^{\pi/3} (8\sin\phi - \sec^3\phi \sin\phi)\, d\phi\, d\theta$$

$$= \frac{1}{3}\int_0^{2\pi}\int_0^{\pi/3} \left(8\sin\phi - \frac{\sin\phi}{\cos^3\phi}\right) d\phi\, d\theta$$

$$= \frac{1}{3} \int_0^{2\pi} \left(-8\cos\phi - \frac{1}{2\cos^2\phi} \right) \Big|_0^{\pi/3} d\theta$$

$$= \frac{1}{3} \int_0^{2\pi} \left(-4 - \frac{1}{2 \cdot \frac{1}{4}} + 8 + \frac{1}{2} \right) d\theta$$

$$= \frac{1}{3} \int_0^{2\pi} \frac{5}{2} d\theta = \frac{5}{3}\pi$$

As expected, this agrees with the result of the previous exercise.

In Exercises 45–50, use spherical coordinates to calculate the triple integral of $f(x,y,z)$ over the given region.

45. $f(x,y,z) = y$; $\quad x^2 + y^2 + z^2 \leq 1$, $\quad x, y, z \leq 0$

SOLUTION

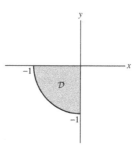

The region inside the unit sphere in the octant $x, y, z \leq 0$ is defined by the inequalities

$$\mathcal{W}: \pi \leq \theta \leq \frac{3\pi}{2}, \frac{\pi}{2} \leq \phi \leq \pi, 0 \leq \rho \leq 1$$

The function in spherical coordinates is $f(x,y,z) = y = \rho \sin\theta \sin\phi$. Using a triple integral in spherical coordinates, we obtain

$$\iiint_{\mathcal{W}} y \, dV = \int_\pi^{3\pi/2} \int_{\pi/2}^\pi \int_0^1 (\rho \sin\theta \sin\phi)\rho^2 \sin\phi \, d\rho \, d\phi \, d\theta = \int_\pi^{3\pi/2} \int_{\pi/2}^\pi \int_0^1 \rho^3 \sin\theta \sin^2\phi \, d\rho \, d\phi \, d\theta$$

$$= \left(\int_\pi^{3\pi/2} \sin\theta \, d\theta \right) \left(\int_{\pi/2}^\pi \sin^2\phi \, d\phi \right) \left(\int_0^1 \rho^3 \, d\rho \right) = \left(-\cos\theta \Big|_\pi^{3\pi/2} \right) \left(\frac{\theta}{2} - \frac{\sin 2\theta}{4} \Big|_{\pi/2}^\pi \right) \left(\frac{\rho^4}{4} \Big|_0^1 \right)$$

$$= (-1) \cdot \left(\frac{\pi}{2} - \frac{\pi}{4} \right) \cdot \frac{1}{4} = -\frac{\pi}{16}$$

47. $f(x,y,z) = x^2 + y^2$; $\quad \rho \leq 1$

SOLUTION $\mathcal{W}$ is the region inside the unit sphere, therefore it is described by the following inequalities:

$$\mathcal{W}: 0 \leq \theta \leq 2\pi, 0 \leq \phi \leq \pi, 0 \leq \rho \leq 1$$

The function in spherical coordinates is

$$f(x,y,z) = x^2 + y^2 = (\rho \cos\theta \sin\phi)^2 + (\rho \sin\theta \sin\phi)^2$$
$$= \rho^2 \sin^2\phi \left(\cos^2\theta + \sin^2\theta \right) = \rho^2 \sin^2\phi$$

Using triple integrals in spherical coordinates we get

$$\iiint_{\mathcal{W}} (x^2 + y^2) \, dV = \int_0^{2\pi} \int_0^\pi \int_0^1 (\rho^2 \sin^2\phi)\rho^2 \sin\phi \, d\rho \, d\phi \, d\theta$$

$$= \int_0^{2\pi} \int_0^\pi \int_0^1 \rho^4 \sin^3\phi \, d\rho \, d\phi \, d\theta = \left(\int_0^{2\pi} 1 \, d\theta \right) \left(\int_0^\pi \sin^3\phi \, d\phi \right) \left(\int_0^1 \rho^4 \, d\rho \right)$$

$$= \left(\theta \Big|_0^{2\pi} \right) \left(-\frac{\sin^2\theta \cos\theta}{3} - \frac{2}{3} \cos\theta \Big|_0^\pi \right) \left(\frac{\rho^5}{5} \Big|_0^1 \right) = 2\pi \cdot \left(\frac{2}{3} + \frac{2}{3} \right) \cdot \frac{1}{5} = \frac{8\pi}{15}$$

49. $f(x, y, z) = \sqrt{x^2 + y^2 + z^2}$; $x^2 + y^2 + z^2 \leq 2z$

SOLUTION We rewrite the inequality for the region using spherical coordinates:

$$\rho^2 \leq 2\rho \cos\phi \quad \Rightarrow \quad \rho \leq 2\cos\phi$$

Completing the square in $x^2 + y^2 + z^2 = 2z$, we see that this is the equation of the sphere of radius 1 centered at $(0, 0, 1)$. That is,

$$x^2 + y^2 + z - 2z = 0$$
$$x^2 + y^2 + (z - 1)^2 = 1$$

$\mathcal{W}$ is the region inside the sphere, hence θ varies from 0 to 2π, and ϕ varies from 0 to $\frac{\pi}{2}$. The inequalities describing $\mathcal{W}$ in spherical coordinates are thus

$$\mathcal{W}: 0 \leq \theta \leq 2\pi,\ 0 \leq \phi \leq \frac{\pi}{2},\ 0 \leq \rho \leq 2\cos\phi$$

The function in spherical coordinates is

$$f(x, y, z) = \sqrt{x^2 + y^2 + z^2} = \rho$$

We obtain the following integral:

$$I = \int_0^{2\pi} \int_0^{\pi/2} \int_0^{2\cos\phi} \rho \cdot \rho^2 \sin\phi\, d\rho\, d\phi\, d\theta = \int_0^{2\pi} \int_0^{\pi/2} \int_0^{2\cos\phi} \rho^3 \sin\phi\, d\rho\, d\phi\, d\theta$$

$$= \int_0^{2\pi} \int_0^{\pi/2} \frac{\rho^4 \sin\phi}{4}\Big|_{\rho=0}^{2\cos\phi} d\phi\, d\theta = \int_0^{2\pi} \int_0^{\pi/2} \frac{16\cos^4\phi \sin\phi}{4}\, d\phi\, d\theta$$

$$= \left(\int_0^{2\pi} 4\, d\theta\right)\left(\int_0^{\pi/2} \cos^4\phi \sin\phi\, d\phi\right) = 8\pi \int_0^{\pi/2} \cos^4\phi \sin\phi\, d\phi$$

We compute the integral using the substitution $u = \cos\phi$, $du = -\sin\phi\, d\phi$. We obtain

$$I = 8\pi \int_1^0 u^4(-du) = 8\pi \int_0^1 u^4\, du = 8\pi \frac{u^5}{5}\Big|_0^1 = \frac{8\pi}{5}$$

51. Use spherical coordinates to evaluate the triple integral of $f(x, y, z) = z$ over the region

$$0 \leq \theta \leq \frac{\pi}{3}, \qquad 0 \leq \phi \leq \frac{\pi}{2}, \qquad 1 \leq \rho \leq 2$$

SOLUTION The function in spherical coordinates is $f(x, y, z) = z = \rho \cos\phi$. Using triple integral in spherical coordinates gives

$$\iiint_\mathcal{W} z\, dV = \int_0^{\pi/3} \int_0^{\pi/2} \int_1^2 (\rho\cos\phi)\rho^2 \sin\phi\, d\rho\, d\phi\, d\theta = \int_0^{\pi/3} \int_0^{\pi/2} \int_1^2 \rho^3 \cos\phi \sin\phi\, d\rho\, d\phi\, d\theta$$

$$= \left(\int_0^{\pi/3} 1\, d\theta\right)\left(\int_0^{\pi/2} \frac{1}{2}\sin 2\phi\, d\phi\right)\left(\int_1^2 \rho^3\, d\rho\right)$$

$$= \frac{\pi}{3} \cdot \left(-\frac{1}{4}\cos 2\phi\right)\Big|_0^{\pi/2} \cdot \left(\frac{\rho^4}{4}\Big|_1^2\right) = \frac{\pi}{3} \cdot \frac{1}{2} \cdot \left(4 - \frac{1}{4}\right) = \frac{5}{8}\pi$$

53. Calculate the integral of

$$f(x, y, z) = z(x^2 + y^2 + z^2)^{-3/2}$$

over the part of the ball $x^2 + y^2 + z^2 \leq 16$ defined by $z \geq 2$.

SOLUTION

The equation of the sphere in spherical coordinates is $\rho^2 = 16$ or $\rho = 4$.

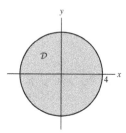

We write the equation of the plane $z = 2$ in spherical coordinates:

$$\rho \cos \phi = 2 \quad \Rightarrow \quad \rho = \frac{2}{\cos \phi}$$

To compute the interval of ϕ, we must find the value of ϕ corresponding to $\rho = 4$ on the plane $z = 2$. We get

$$4 = \frac{2}{\cos \phi} \quad \Rightarrow \quad \cos \phi = \frac{1}{2} \quad \Rightarrow \quad \phi = \frac{\pi}{3}$$

Therefore, ϕ is changing from 0 to $\frac{\pi}{3}$, θ is changing from 0 to 2π, and ρ is changing from $\frac{2}{\cos \phi}$ to 4. We obtain the following description for $\mathcal{W}$:

$$\mathcal{W}: 0 \leq \theta \leq 2\pi,\ 0 \leq \phi \leq \frac{\pi}{3},\ \frac{2}{\cos \phi} \leq \rho \leq 4$$

The function is

$$f(x, y, z) = z(x^2 + y^2 + z^2)^{-3/2} = \rho \cos \phi \cdot (\rho^2)^{-3/2} = \rho^{-2} \cos \phi$$

We use triple integrals in spherical coordinates to write

$$\iiint_{\mathcal{W}} f(x, y, z)\, dV = \int_0^{2\pi} \int_0^{\pi/3} \int_{2/\cos\phi}^4 (\rho^{-2} \cos \phi) \rho^2 \sin \phi\, d\rho\, d\phi\, d\theta = \int_0^{2\pi} \int_0^{\pi/3} \int_{2/\cos\phi}^4 \frac{\sin 2\phi}{2}\, d\rho\, d\phi\, d\theta$$

$$= \int_0^{2\pi} \int_0^{\pi/3} \frac{\sin 2\phi}{2} \rho \Big|_{\rho = \frac{2}{\cos\phi}}^4 d\phi\, d\theta = \int_0^{2\pi} \int_0^{\pi/3} \left(2 \sin 2\phi - \frac{\sin 2\phi}{2} \cdot \frac{2}{\cos \phi}\right) d\phi\, d\theta$$

$$= \int_0^{2\pi} \int_0^{\pi/3} (2 \sin 2\phi - 2 \sin \phi)\, d\phi\, d\theta = 2\pi \cdot \left(-\cos 2\phi + 2 \cos \phi \Big|_{\phi=0}^{\pi/3}\right)$$

$$= 2\pi \cdot \left(-\cos \frac{2\pi}{3} + 2 \cos \frac{\pi}{3} + 1 - 2\right) = \pi$$

55. Calculate the volume of the sphere $x^2 + y^2 + z^2 = a^2$, using both spherical and cylindrical coordinates.

SOLUTION Spherical coordinates: In the entire sphere of radius a, we have

$$\mathcal{W}: 0 \leq \theta \leq 2\pi,\ 0 \leq \phi \leq \pi,\ 0 \leq \rho \leq a$$

Using triple integral in spherical coordinates we get

$$V = \iiint_W 1\,dV = \int_0^{2\pi} \int_0^{\pi} \int_0^a \rho^2 \sin\phi\,d\rho\,d\phi\,d\theta = \left(\int_0^a \rho^2 d\rho\right)\left(\int_0^{\pi} \sin\phi\,d\phi\right)\left(\int_0^{2\pi} 1\,d\theta\right)$$

$$= \left(\frac{\rho^3}{3}\bigg|_0^a\right)\left(-\cos\phi\bigg|_0^{\pi}\right)\left(\theta\bigg|_0^{2\pi}\right) = \frac{a^3}{3}\cdot 2 \cdot 2\pi = \frac{4\pi a^3}{3}$$

Cylindrical coordinates: The projection of $\mathcal{W}$ onto the xy-plane is the circle of radius a, that is,

$$\mathcal{D}: 0 \le \theta \le 2\pi,\ 0 \le r \le a$$

The upper surface is $z = \sqrt{a^2 - (x^2+y^2)} = \sqrt{a^2 - r^2}$ and the lower surface is $z = -\sqrt{a^2-r^2}$. Therefore, $\mathcal{W}$ has the following description in cylindrical coordinates:

$$\mathcal{W}: 0 \le \theta \le 2\pi,\ 0 \le r \le a,\ -\sqrt{a^2-r^2} \le z \le \sqrt{a^2-r^2}$$

We obtain the following integral:

$$V = \int_0^{2\pi}\int_0^a \int_{-\sqrt{a^2-r^2}}^{\sqrt{a^2-r^2}} r\,dz\,dr\,d\theta = \int_0^{2\pi}\int_0^a rz\bigg|_{z=-\sqrt{a^2-r^2}}^{\sqrt{a^2-r^2}} dr\,d\theta = \int_0^{2\pi}\int_0^a 2r\sqrt{a^2-r^2}\,dr\,d\theta \quad (1)$$

We compute the inner integral using the substitution $u = \sqrt{a^2-r^2}$, $du = -\frac{r}{u}dr$. We get

$$\int_0^a 2r\sqrt{a^2-r^2}\,dr = \int_a^0 -2u^2\,du = \int_0^a 2u^2\,du = \frac{2u^3}{3}\bigg|_0^a = \frac{2a^3}{3}$$

Substituting in (1) gives

$$V = \int_0^{2\pi} \frac{2a^3}{3}\,d\theta = \frac{2a^3}{3}\theta\bigg|_0^{2\pi} = \frac{2a^3}{3}\cdot 2\pi = \frac{4\pi a^3}{3}$$

57. Bell-Shaped Curve One of the key results in calculus is the computation of the area under the bell-shaped curve (Figure 24):

$$I = \int_{-\infty}^{\infty} e^{-x^2}\,dx$$

This integral appears throughout engineering, physics, and statistics, and although e^{-x^2} does not have an elementary antiderivative, we can compute I using multiple integration.

(a) Show that $I^2 = J$, where J is the improper double integral

$$J = \int_{-\infty}^{\infty}\int_{-\infty}^{\infty} e^{-x^2-y^2}\,dx\,dy$$

Hint: Use Fubini's Theorem and $e^{-x^2-y^2} = e^{-x^2}e^{-y^2}$.

(b) Evaluate J in polar coordinates.

(c) Prove that $I = \sqrt{\pi}$.

FIGURE 24 The bell-shaped curve $y = e^{-x^2}$.

SOLUTION

(a) Since $e^{-x^2-y^2} = e^{-x^2}e^{-y^2}$, we have

$$J = \int_{-\infty}^{\infty}\int_{-\infty}^{\infty} e^{-x^2-y^2}\,dx\,dy = \int_{-\infty}^{\infty}\int_{-\infty}^{\infty}\left(e^{-x^2}e^{-y^2}\right)dx\,dy = \left(\int_{-\infty}^{\infty} e^{-y^2}\,dy\right)\left(\int_{-\infty}^{\infty} e^{-x^2}\,dx\right) = I^2$$

(b) To evaluate J, convert to polar coordinates. Define $\mathcal{D}_R$ as

$$\mathcal{D}_R : 0 \le \theta \le 2\pi,\ 0 \le r \le \mathcal{R}$$

A diagram of $\mathcal{D}_R$ is below:

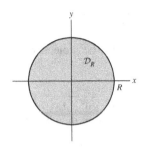

The improper integral over the xy-plane can be computed as the limit as $\mathcal{R} \to \infty$ of the double integrals over $\mathcal{D}_R$. That is,

$$J = \lim_{\mathcal{R} \to \infty} \iint_{\mathcal{D}_R} e^{-x^2 - y^2}\, dx\, dy \tag{1}$$

Since $-x^2 - y^2 = -r^2$ in polar coordinates, we have

$$J = \lim_{\mathcal{R} \to \infty} \iint_{\mathcal{D}_R} e^{-x^2 - y^2}\, dA = \lim_{\mathcal{R} \to \infty} \iint_{\mathcal{D}_R} e^{-r^2} \cdot r\, dA$$

$$= \lim_{\mathcal{R} \to \infty} \int_0^{2\pi} \int_0^{\mathcal{R}} r e^{-r^2}\, dr\, d\theta$$

$$= \lim_{\mathcal{R} \to \infty} \int_0^{2\pi} \left(-\frac{1}{2} e^{-r^2} \right)\Big|_{r=0}^{\mathcal{R}} d\theta$$

$$= \frac{1}{2} \lim_{\mathcal{R} \to \infty} \int_0^{2\pi} \left(-e^{-\mathcal{R}^2} + 1 \right) d\theta$$

$$= \pi \lim_{\mathcal{R} \to \infty} \left(-e^{-\mathcal{R}^2} + 1 \right)$$

$$= \pi$$

(c) Since $J = I^2 = \pi$, we must have either $I = \sqrt{\pi}$ or $I = -\sqrt{\pi}$. But since I is the integral of a function that is positive everywhere, its integral must be positive, so we choose $I = \sqrt{\pi}$*.

Further Insights and Challenges

59. Prove the formula

$$\iint_{\mathcal{D}} \ln r\, dA = -\frac{\pi}{2}$$

where $r = \sqrt{x^2 + y^2}$ and $\mathcal{D}$ is the unit disk $x^2 + y^2 \le 1$. This is an improper integral since $\ln r$ is not defined at $(0,0)$, so integrate first over the annulus $a \le r \le 1$, where $0 < a < 1$, and let $a \to 0$.

SOLUTION

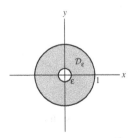

---

*"A mathematician is one to whom that is as obvious as that twice two makes four is to you." – Lord Kelvin

The improper integral I is computed by the limit as $a \to 0^+$ of the integrals over the annulus $\mathcal{D}_a$ defined by

$$\mathcal{D}_a: 0 \le \theta \le 2\pi, \ a \le r \le 1$$

Using double integrals in polar coordinates and integration by parts, we get

$$I_a = \int_0^{2\pi} \int_a^1 (\ln r) \cdot r \, dr \, d\theta = 2\pi \int_a^1 r \ln r \, dr = 2\pi \left(\frac{r^2 \ln r}{2} - \frac{r^2}{4} \bigg|_a^1 \right)$$

$$= 2\pi \left(\frac{\ln 1}{2} - \frac{1}{4} - \frac{a^2 \ln a}{2} + \frac{a^2}{4} \right) = \frac{\pi}{2} \left(a^2 - 2a^2 \ln a - 1 \right)$$

We now compute the limit of I_a as $a \to 0^+$. We use L'Hôpital's rule to obtain

$$I = \lim_{a \to 0^+} \frac{\pi}{2}(a^2 - 2a^2 \ln a - 1) = -\frac{\pi}{2} - \pi \lim_{a \to 0^+} a^2 \ln a = -\frac{\pi}{2} - \pi \lim_{a \to 0^+} \frac{\ln a}{a^{-2}}$$

$$= -\frac{\pi}{2} - \pi \lim_{a \to 0^+} \frac{a^{-1}}{-2a^{-3}} = -\frac{\pi}{2} + \frac{\pi}{2} \lim_{a \to 0^+} a^2 = -\frac{\pi}{2}$$

15.5 Applications of Multiple Integrals (LT Section 16.5)

Preliminary Questions

1. What is the mass density $\delta(x, y, z)$ of a solid of volume 5 m³ with uniform mass density and total mass 25 kg?

SOLUTION Here, recall that

$$\text{total mass} = \iiint_{\mathcal{W}} \rho(x, y, z) \, dV$$

Since we are told that the solid has volume 5, and $\rho(x, y, z)$ is uniform (i.e., constant, let $\rho(x, y, z) = \rho$), we can write:

$$25 = \iiint_{\mathcal{W}} \rho(x, y, z) \, dV = \rho \cdot V(\mathcal{W}) = 5\rho, \quad \Rightarrow \quad \rho = 5 \text{ kg/m}^3$$

2. A domain $\mathcal{D}$ in $\mathbf{R}^2$ with uniform mass density is symmetric with respect to the y-axis. Which of the following are true?

(a) $x_{\text{CM}} = 0$ **(b)** $y_{\text{CM}} = 0$ **(c)** $I_x = 0$ **(d)** $I_y = 0$

SOLUTION Here, the x-coordinate of the center of mass, $x_{\text{CM}} = 0$ (a) since $x_{\text{CM}} = \frac{M_y}{M}$ and $M_y = \iint_\mathcal{D} x\rho(x, y) \, dA$. Since $\rho(x, y) = \rho(-x, y)$, then we see that $(-x)\rho(-x, y) = -x\rho(x, y)$ and M_y is an integral of an odd function over a symmetric region, hence $M_y = 0$.

3. If $p(x, y)$ is the joint probability density function of random variables X and Y, what does the double integral of $p(x, y)$ over $[0, 1] \times [0, 1]$ represent? What does the integral of $p(x, y)$ over the triangle bounded by $x = 0$, $y = 0$, and $x + y = 1$ represent?

SOLUTION The double integral of $p(x, y)$ over $[0, 1] \times [0, 1]$ represents the probability that both X and Y are between 0 and 1. The integral of $p(x, y)$ over the triangle bounded by $x = 0$, $y = 0$, and $x + y = 1$ represents the probability that both X and Y are nonnegative and $X + Y \le 1$.

Exercises

1. Find the total mass of the rectangle $0 \le x \le 1$, $0 \le y \le 2$ assuming a mass density of

$$\delta(x, y) = 2x^2 + y^2$$

SOLUTION The total mass M is obtained by integrating the mass density $\delta(x, y) = 2x^2 + y^2$ over the rectangle

$$\mathcal{D}: 0 \le x \le 1, \ 0 \le y \le 2$$

in the xy-plane. This gives

$$M = \iint_D \delta(x,y)\, dA = \int_0^1 \int_0^2 (2x^2 + y^2)\, dy\, dx = \int_0^1 \left(2x^2 y + \frac{1}{3}y^3\right)\Big|_0^2 dx$$

$$= \int_0^1 \left(4x^2 + \frac{8}{3}\right) dx = \frac{4}{3}x^3 + \frac{8}{3}x \Big|_0^1 = 4$$

3. Find the total charge in the region under the graph of $y = 4e^{-x^2/2}$ for $0 \le x \le 10$ (in centimeters) assuming a charge density of $\delta(x,y) = 10^{-6} xy$ coulombs per square centimeter (C/cm²).

SOLUTION The total charge C of the region is obtained by computing the double integral of charge density $\rho(x,y) = 10^{-6} xy$ over the region defined by the inequalities

$$0 \le x \le 10, \quad 0 \le y \le 4e^{-x^2/2}$$

Therefore, we compute the double integral

$$C = \iint_D \rho(x,y)\, dA = \int_0^{10} \int_0^{4e^{-x^2/2}} 10^{-6} xy\, dy\, dx = 10^{-6} \int_0^{10} \left(\frac{1}{2}xy^2 \Big|_{y=0}^{4e^{-x^2/2}}\right) dx$$

$$= \frac{1}{2} \cdot 10^{-6} \int_0^{10} 16xe^{-x^2}\, dx = -4 \cdot 10^{-6} \int_0^{10} e^{-x^2}(-2x\, dx) = -4 \cdot 10^{-6} \left(e^{-x^2}\Big|_0^{10}\right)$$

$$= -4 \cdot 10^{-6}\left(e^{-10^2} - 1\right) = \frac{1}{250{,}000}\left(1 - e^{-100}\right)$$

5. Find the total population within the sector $2|x| \le y \le 8$ assuming a population density of $\delta(x,y) = 100e^{-0.1y}$ people per square kilometer.

SOLUTION The total population P of the region is obtained by computing the double integral of population density $\rho(x,y) = 100e^{-0.1y}$ over the region defined by the inequality $2|x| \le y \le 8$. This means the region can be split into two vertically simple regions described by the inequalities:

$$0 \le x \le 4, \quad 2x \le y \le 8$$

and

$$-4 \le x \le 4, \quad -2x \le y \le 8$$

Now to compute the double integral:

$$\iint_D \rho(x,y)\, dA = \iint_{D_1} \rho(x,y)\, dA + \iint_{D_2} \rho(x,y)\, dA$$

$$\iint_{D_1} \rho(x,y)\, dA + \iint_{D_2} \rho(x,y)\, dA = \int_0^4 \int_{2x}^8 100e^{-0.1y}\, dy\, dx + \int_{-4}^0 \int_{-2x}^8 100e^{-0.1y}\, dy\, dx$$

$$= \int_0^4 \frac{100}{-0.1} e^{-0.1y}\Big|_{y=2x}^8 dx + \int_{-4}^0 \frac{100}{-0.1} e^{-0.1y}\Big|_{y=-2x}^8 dx$$

$$= -1000\int_0^4 e^{-0.8} - e^{-0.2x}\, dx - 1000\int_{-4}^0 e^{-0.8} - e^{0.2x}\, dx$$

$$= -1000\left(e^{-0.8}x + 5e^{-0.2x}\Big|_0^4\right) - 1000\left(e^{-0.8}x - 5e^{0.2x}\Big|_{-4}^0\right)$$

$$= -1000\left(4e^{-0.8} + 5e^{-0.8} - 5\right) - 1000\left(-5 + 4e^{-0.8} + 5e^{-0.8}\right)$$

$$= -1000\left(18e^{-0.8} - 10\right) \approx 1912$$

7. Calculate the total charge of the solid ball $x^2 + y^2 + z^2 \le 5$ (in centimeters) assuming a charge density (in coulombs per cubic centimeter) of

$$\delta(x,y,z) = (3 \cdot 10^{-8})(x^2 + y^2 + z^2)^{1/2}$$

SOLUTION To calculate total charge, first we consider the solid ball in spherical coordinates:

$$x^2 + y^2 + z^2 \le 5 \quad \Rightarrow \quad 0 \le \rho \le \sqrt{5}, \quad 0 \le \theta \le 2\pi, \quad 0 \le \phi \le \pi$$

And the charge density function too, let us rename it $R(x, y, z)$:

$$R(x, y, z) = (3 \cdot 10^{-8})(x^2 + y^2 + z^2)^{1/2} \quad \Rightarrow \quad R(\rho \sin\phi \cos\theta, \rho \sin\phi \sin\theta, \rho \cos\phi) = (3 \cdot 10^{-8})\rho$$

Then integrating to compute the total charge we have:

$$\int_0^{2\pi} \int_0^{\pi} \int_0^{\sqrt{5}} (3 \cdot 10^{-8})\rho \cdot \rho^2 \sin\phi \, d\rho \, d\phi \, d\theta = 3 \cdot 10^{-8} \int_0^{2\pi} \int_0^{\pi} \int_0^{\sqrt{5}} \rho^3 \sin\phi \, d\rho \, d\phi \, d\theta$$

$$= 3 \cdot 10^{-8} \int_0^{2\pi} \int_0^{\pi} \sin\phi \left(\frac{1}{4}\rho^4 \Big|_0^{\sqrt{5}}\right) d\phi \, d\theta = 3 \cdot 10^{-8} \cdot \frac{25}{4} \int_0^{2\pi} \int_0^{\pi} \sin\phi \, d\phi \, d\theta$$

$$= 3 \cdot 10^{-8} \cdot \frac{25}{4} \int_0^{2\pi} -\cos\phi \Big|_0^{\pi} d\theta = 3 \cdot 10^{-8} \cdot \frac{25}{4} \int_0^{2\pi} 2 \, d\theta = 3 \cdot 10^{-8} \cdot 25\pi$$

$$\approx 2.356 \cdot 10^{-6}$$

9. Assume that the density of the atmosphere as a function of altitude h (in kilometers) above sea level is $\delta(h) = ae^{-bh}$ kg/km^3, where $a = 1.225 \times 10^9$ and $b = 0.13$. Calculate the total mass of the atmosphere contained in the cone-shaped region $\sqrt{x^2 + y^2} \leq h \leq 3$.

SOLUTION First we must consider the given cone in cylindrical coordinates:

$$\sqrt{x^2 + y^2} \leq z \leq 3 \quad \Rightarrow \quad r \leq z \leq 3$$

while

$$0 \leq r \leq 3, \quad 0 \leq \theta \leq 2\pi$$

And the density function as well:

$$\rho(x, y, z) = ae^{-bz} \quad \Rightarrow \quad \rho(r\cos\theta, r\sin\theta, z) = ae^{-bz}$$

Now to compute the total mass of the atmosphere in question:

$$\int_0^{\theta=2\pi} \int_{r=0}^{3} \int_{z=r}^{3} ae^{-bz} \cdot r \, dz \, dr \, d\theta = \int_0^{2\pi} \int_0^{3} \int_r^{3} r(ae^{-bz}) \, dz \, dr \, d\theta$$

$$= \int_0^{2\pi} \int_0^{3} r\left(-\frac{1}{b}ae^{-bz}\Big|_{z=r}^{3}\right) dr \, d\theta$$

$$= -\frac{a}{b} \int_0^{2\pi} \int_0^{3} re^{-3b} - re^{-br} \, dr \, d\theta$$

$$= -\frac{a}{b} \int_0^{2\pi} \frac{1}{2}r^2 e^{-3b}\Big|_0^{3} - \left(-\frac{1}{b}re^{-br} - \frac{1}{b^2}e^{-br}\Big|_0^{3}\right) d\theta$$

$$= -\frac{a}{b} \cdot 2\pi \left(\frac{9}{2}e^{-3b} + \frac{3}{b}e^{-3b} + \frac{1}{b^2}e^{-3b} - \frac{1}{b^2}\right)$$

Now, since $a = 1.225 \times 10^9$ and $b = 0.13$ we have that the total mass is

$$-\frac{a}{b} \cdot 2\pi \left(\frac{9}{2}e^{-3b} + \frac{3}{b}e^{-3b} + \frac{1}{b^2}e^{-3b} - \frac{1}{b^2}\right) \approx 2.593 \times 10^{10}$$

In Exercises 11–16, find the centroid of the given region assuming the density $\delta(x, y) = 1$.

11. Region bounded by $y = 1 - x^2$ and $y = 0$

SOLUTION First we will compute the area of the region:

$$\text{Area}(\mathcal{D}) = \int_{-1}^{1} \int_0^{1-x^2} dy \, dx = \int_{-1}^{1} 1 - x^2 \, dx = x - \frac{1}{3}x^3 \Big|_{-1}^{1} = 1 - \frac{1}{3} - \left(-1 + \frac{1}{3}\right) = \frac{4}{3}$$

It is clear from symmetry that $\bar{x} = 0$, and

$$\bar{y} = \frac{1}{\text{Area}(\mathcal{D})} \iint_{\mathcal{D}} y \, dA = \frac{3}{4} \int_{-1}^{1} \int_0^{1-x^2} y \, dy \, dx$$

$$= \frac{3}{4} \int_{-1}^{1} \frac{1}{2}y^2 \Big|_0^{1-x^2} dx = \frac{3}{8} \int_{-1}^{1} (1 - x^2)^2 \, dx$$

$$= \frac{3}{8} \int_{-1}^{1} 1 - 2x^2 + x^4 \, dx = \frac{3}{8} \left(x - \frac{2}{3}x^3 + \frac{1}{5}x^5 \right) \Big|_{-1}^{1}$$

$$= \frac{3}{8}\left(1 - \frac{2}{3} + \frac{1}{5}\right) - \frac{3}{8}\left(-1 + \frac{2}{3} - \frac{1}{5}\right) = \frac{2}{5}$$

The centroid has coordinates $(\bar{x}, \bar{y}) = \left(0, \frac{2}{5}\right)$.

13. Quarter circle $x^2 + y^2 \le R^2$, $x \ge 0$, $y \ge 0$

SOLUTION

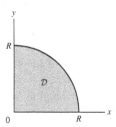

The centroid $P = (\bar{x}, \bar{y})$ has the following coordinates:

$$\bar{x} = \frac{1}{\text{Area}(\mathcal{D})} \iint_{\mathcal{D}} x \, dA = \frac{4}{\pi R^2} \iint_{\mathcal{D}} x \, dA$$

$$\bar{y} = \frac{1}{\text{Area}(\mathcal{D})} \iint_{\mathcal{D}} y \, dA = \frac{4}{\pi R^2} \iint_{\mathcal{D}} y \, dA$$

We compute the integrals using polar coordinates. The domain $\mathcal{D}$ is described in polar coordinates by the inequalities

$$\mathcal{D}: 0 \le \theta \le \frac{\pi}{2},\ 0 \le r \le R$$

The functions are $x = r\cos\theta$ and $y = r\sin\theta$, respectively. Using the Change of Variables Formula gives

$$\bar{x} = \frac{4}{\pi R^2} \int_0^{\pi/2} \int_0^R r\cos\theta \cdot r \, dr \, d\theta = \frac{4}{\pi R^2} \int_0^{\pi/2} \int_0^R r^2 \cos\theta \, dr \, d\theta = \frac{4}{\pi R^2} \int_0^{\pi/2} \frac{r^3 \cos\theta}{3} \Big|_{r=0}^R d\theta$$

$$= \frac{4}{\pi R^2} \int_0^{\pi/2} \frac{R^3 \cos\theta}{3} d\theta = \frac{4R}{3\pi} \sin\theta \Big|_0^{\pi/2} = \frac{4R}{3\pi}\left(\sin\frac{\pi}{2} - \sin 0\right) = \frac{4R}{3\pi}$$

And,

$$\bar{y} = \frac{4}{\pi R^2} \int_0^{\pi/2} \int_0^R r\sin\theta \cdot r \, dr \, d\theta = \frac{4}{\pi R^2} \int_0^{\pi/2} \int_0^R r^2 \sin\theta \, dr \, d\theta = \frac{4}{\pi R^2} \int_0^{\pi/2} \frac{r^3 \sin\theta}{3} \Big|_{r=0}^R d\theta$$

$$= \frac{4}{\pi R^2} \int_0^{\pi/2} \frac{R^3 \sin\theta}{3} d\theta = \frac{4R}{3\pi}(-\cos\theta) \Big|_0^{\pi/2} = \frac{4R}{3\pi}\left(-\cos\frac{\pi}{2} + \cos 0\right) = \frac{4R}{3\pi}$$

Notice that we can use the symmetry of $\mathcal{D}$ with respect to x and y to conclude that $\bar{y} = \bar{x}$, and save the computation of $\bar{y}$. We obtain the centroid $P = \left(\frac{4R}{3\pi}, \frac{4R}{3\pi}\right)$.

15. Lamina bounded by the x- and y-axes, the line $x = M$, and the graph of $y = e^{-x}$

SOLUTION The domain $\mathcal{D}$ is shown below:

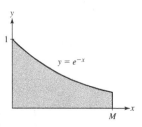

The area of this region (with density $\delta(x, y) = 1$) is

$$A = \iint_D 1\, dA = \int_0^M \int_0^{e^{-x}} 1\, dy\, dx = \int_0^M e^{-x}\, dx = (-e^{-x})\Big|_0^M = 1 - e^{-M}$$

The coordinates of the centroid are given by

$$\bar{x} = \frac{1}{\text{Area}(D)} \iint_D x\, dA = \frac{1}{1 - e^{-M}} \int_0^M \int_0^{e^{-x}} x\, dy\, dx$$

$$\bar{y} = \frac{1}{\text{Area}(D)} \iint_D y\, dA = \frac{1}{1 - e^{-M}} \int_0^M \int_0^{e^{-x}} y\, dy\, dx.$$

$\bar{x}$ is computed using integration by parts for the outer integral:

$$\bar{x} = \frac{1}{1 - e^{-M}} \int_0^M \int_0^{e^{-x}} x\, dy\, dx$$

$$= \frac{1}{1 - e^{-M}} \int_0^M xe^{-x}\, dx$$

$$= \frac{1}{1 - e^{-M}} \left((-xe^{-x})\Big|_0^M + \int_0^M e^{-x}\, dx \right)$$

$$= \frac{1}{1 - e^{-M}} \left(-Me^{-M} + (-e^{-x})\Big|_0^M \right)$$

$$= \frac{1}{1 - e^{-M}} \left(-Me^{-M} - e^{-M} + 1 \right)$$

$$= \frac{1 - e^{-M} - Me^{-M}}{1 - e^{-M}}$$

The computation of $\bar{y}$ is straightforward:

$$\bar{y} = \frac{1}{1 - e^{-M}} \int_0^M \int_0^{e^{-x}} y\, dy\, dx$$

$$= \frac{1}{1 - e^{-M}} \int_0^M \frac{1}{2} e^{-2x}\, dx$$

$$= \frac{1}{1 - e^{-M}} \left(-\frac{1}{4} e^{-2x} \right)\Big|_0^M$$

$$= \frac{1}{1 - e^{-M}} \left(\frac{1}{4} - \frac{1}{4} e^{-2M} \right)$$

$$= \frac{1 - e^{-2M}}{4(1 - e^{-M})}$$

The centroid is

$$(\bar{x}, \bar{y}) = \left(\frac{1 - e^{-M} - Me^{-M}}{1 - e^{-M}}, \frac{1 - e^{-2M}}{4(1 - e^{-M})} \right) = \left(1 + \frac{M}{1 - e^M}, \frac{1 + e^{-M}}{4} \right)$$

17. *CAS* Use a computer algebra system to compute numerically the centroid of the shaded region in Figure 12 bounded by $r^2 = \cos 2\theta$ for $x \geq 0$.

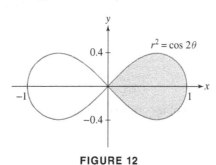

FIGURE 12

SOLUTION Using symmetry, it is easy to see $\bar{y} = 0$. Also, computing the area of the region,

$$\text{Area} = 2 \cdot \frac{1}{2} \int_{-\pi/4}^{\pi/4} r^2 \, d\theta = \int_{-\pi/4}^{\pi/4} \cos 2\theta \, d\theta = \frac{1}{2} \sin 2\theta \Big|_{-\pi/4}^{\pi/4} = 1$$

and we will compute $\bar{x}$ as

$$\bar{x} = \frac{1}{A} \iint_{\mathcal{D}} x \, dA = \int_{\theta=-\pi/4}^{\pi/4} \int_{r=0}^{\sqrt{\cos 2\theta}} r \cos\theta \cdot r \, dr \, d\theta = \frac{\sqrt{2}}{16} \pi \approx 0.278$$

Therefore, we have that the centroid is $(\bar{x}, \bar{y}) = (\sqrt{2}\pi/16, 0)$.

In Exercises 19–21, find the centroid of the given solid region assuming a density of $\delta(x, y) = 1$.

19. Hemisphere $x^2 + y^2 + z^2 \leq R^2$, $z \geq 0$

SOLUTION First we need to find the volume of the solid in question. It is a hemisphere, so using geometry, we have

$$\text{Volume} = \frac{1}{2} \cdot \frac{4}{3}\pi R^3 = \frac{2\pi R^3}{3}$$

The centroid is the point P with the following coordinates:

$$\bar{x} = \frac{1}{V} \iiint_{\mathcal{W}} x \, dV, \quad \bar{y} = \frac{1}{V} \iiint_{\mathcal{W}} y \, dV, \quad \bar{z} = \frac{1}{V} \iiint_{\mathcal{W}} z \, dV$$

By symmetry, it is clear that $\bar{x} = \bar{y} = 0$, and using spherical coordinates,

$$\bar{z} = \frac{1}{V} \iiint_{\text{region}} z \, dV = \frac{3}{2\pi R^3} \int_0^{2\pi} \int_0^{\pi/2} \int_0^R \rho \cos\phi \cdot \rho^2 \sin\phi \, d\rho \, d\phi \, d\theta$$

$$= \frac{3}{2\pi R^3} \int_0^{2\pi} \int_0^{\pi/2} \int_0^R \rho^3 \cos\phi \sin\phi \, d\rho \, d\phi \, d\theta$$

$$= \frac{3}{2\pi R^3} \int_0^{2\pi} 1 \, d\theta \cdot \int_0^{\pi/2} \cos\phi \sin\phi \, d\phi \cdot \int_0^R \rho^3 \, d\rho$$

$$= \frac{3}{2\pi R^3} \cdot 2\pi \left(\frac{1}{2} \sin^2 \phi \Big|_0^{\pi/2} \right) \left(\frac{1}{4} \rho^4 \Big|_0^R \right) = \frac{3}{2\pi R^3} \cdot 2\pi \cdot \frac{1}{2} \cdot \frac{1}{4} R^4 = \frac{3R}{8}$$

Therefore, the coordinates of the centroid of a hemisphere having radius R are $(0, 0, 3R/8)$.

21. The "ice cream cone" region $\mathcal{W}$ bounded, in spherical coordinates, by the cone $\phi = \pi/3$ and the sphere $\rho = 2$

SOLUTION First we must find the volume of this solid:

$$V = \int_{\theta=0}^{2\pi} \int_{\phi=0}^{\pi/3} \int_{\rho=0}^{2} \rho^2 \sin\phi \, d\rho \, d\phi \, d\theta$$

$$= 2\pi \left(\int_0^{\pi/3} \sin\phi \, d\phi \right) \left(\int_0^2 \rho^2 \, d\rho \right) = 2\pi \cdot \frac{8}{3} \left(-\cos\phi \Big|_0^{\pi/3} \right)$$

$$= \frac{16\pi}{3} \cdot \frac{1}{2} = \frac{8\pi}{3}$$

And now compute the coordinates of the centroid. By symmetry, it is clear that $\bar{x} = \bar{y} = 0$.

$$\bar{z} = \frac{1}{V} \iiint_{\mathcal{W}} z \, dV = \frac{3}{8\pi} \int_{\theta=0}^{2\pi} \int_{\phi=0}^{\pi/3} \int_{\rho=0}^{2} \rho \cos\phi \cdot \rho^2 \sin\phi \, d\rho \, d\phi \, d\theta$$

$$= \frac{3}{8\pi} \int_0^{2\pi} d\theta \cdot \int_0^{\pi/3} \cos\phi \sin\phi \, d\phi \cdot \int_0^2 \rho^3 \, d\rho$$

$$= \frac{3}{8\pi} \cdot 2\pi \cdot \left(\frac{1}{2} \sin^2 \phi \Big|_0^{\pi/3} \right) \left(\frac{1}{4} \rho^4 \Big|_0^2 \right) = \frac{3}{4} \left(\frac{1}{2} \cdot \frac{3}{4} \right)(4) = \frac{9}{8}$$

The coordinates of the centroid are $(0, 0, 9/8)$.

23. Find the centroid of the region $\mathcal{W}$ in Figure 15, lying above the sphere $x^2 + y^2 + z^2 = 6$ and below the paraboloid $z = 4 - x^2 - y^2$.

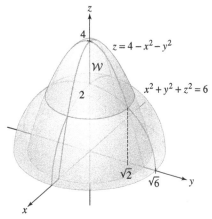

FIGURE 15

SOLUTION The centroid is the point P with the following coordinates:

$$\overline{x} = \frac{1}{V} \iiint_{\mathcal{W}} x\, dV, \quad \overline{y} = \frac{1}{V} \iiint_{\mathcal{W}} y\, dV, \quad \overline{z} = \frac{1}{V} \iiint_{\mathcal{W}} z\, dV$$

where V is the volume of $\mathcal{W}$. To compute V, first we must find the intersection of the sphere $x^2 + y^2 + z^2 = 6$ and the paraboloid $z = 4 - x^2 - y^2$. The equation $z = 4 - x^2 - y^2$ implies $x^2 + y^2 = 4 - z$. Substituting in $x^2 + y^2 + z^2 = 6$ and solving for z gives

$$4 - z + z^2 = 6 \quad \Rightarrow \quad z^2 - z - 2 = (z-2)(z+1) = 0$$
$$\Rightarrow \quad z = 2, \quad z = -1$$

We omit the negative solution $z = -1$, since in $\mathcal{W}$, $z \geq 0$. Substituting $z = 2$ in $z = 4 - x^2 - y^2$, we get

$$2 = 4 - x^2 - y^2 \quad \Rightarrow \quad x^2 + y^2 = 2$$

That is, the projection of $\mathcal{W}$ on the xy-plane is the disk $x^2 + y^2 \leq 2$.

We therefore obtain the following inequalities for $\mathcal{D}$:

$$\mathcal{D}: 0 \leq \theta \leq 2\pi, \quad 0 \leq r \leq \sqrt{2}$$

The upper boundary is the surface $z = 4 - x^2 - y^2$ or $z = 4 - r^2$. The lower boundary is the sphere $x^2 + y^2 + z^2 = 6$, $r^2 + z^2 = 6$, or $z = \sqrt{6 - r^2}$. Therefore, the inequalities describing $\mathcal{W}$ in cylindrical coordinates are

$$0 \leq \theta \leq 2\pi, \quad 0 \leq r \leq \sqrt{2}, \quad \sqrt{6 - r^2} \leq z \leq 4 - r^2$$

We compute the volume as a triple integral using the change of variables in cylindrical coordinates to obtain

$$V = \iiint_{\mathcal{W}} 1\, dV = \int_0^{2\pi} \int_0^{\sqrt{2}} \int_{\sqrt{6-r^2}}^{4-r^2} r\, dz\, dr\, d\theta$$

$$= \int_0^{2\pi} \int_0^{\sqrt{2}} r \cdot z \Big|_{z=\sqrt{6-r^2}}^{4-r^2} dr\, d\theta$$

$$= \int_0^{2\pi} \int_0^{\sqrt{2}} \left(4 - r^2 - \sqrt{6-r^2}\right) r\, dr\, d\theta$$

$$= \int_0^{2\pi} \int_0^{\sqrt{2}} (4r - r^3)\, - \int_0^{2\pi} \int_0^{\sqrt{2}} \left(r\sqrt{6-r^2}\right) dr\, d\theta$$

Compute the second integral using the substitution $u = \sqrt{6 - r^2}$, $du = -\frac{r}{u}\, dr$:

$$\int_0^{2\pi} \int_0^{\sqrt{2}} (4r - r^3)\, dr\, d\theta - \int_0^{2\pi} \int_0^{\sqrt{2}} \left(r\sqrt{6-r^2}\right) dr\, d\theta$$

$$= \int_0^{2\pi} \left(2r^2 - \frac{1}{4}r^4\right)\Big|_0^{\sqrt{2}} d\theta - \int_0^{2\pi} \int_{\sqrt{6}}^{2} (-u^2)\, du\, d\theta$$

1318 CHAPTER 15 | MULTIPLE INTEGRATION (LT CHAPTER 16)

$$= \int_0^{2\pi} (4-1)\, d\theta + \int_0^{2\pi} \frac{1}{3} u^3 \Big|_{\sqrt{6}}^{2} d\theta$$

$$= 6\pi + \int_0^{2\pi} \frac{1}{3}(8 - 6\sqrt{6})\, d\theta$$

$$= 6\pi + \frac{16}{3}\pi - 4\pi\sqrt{6} = \left(\frac{34}{3} - 4\sqrt{6}\right)\pi$$

Since the region is symmetric with respect to the *yz*-plane, we know that $\bar{x} = 0$. Similarly, since the region is symmetric with respect to the *xz*-plane, we have $\bar{y} = 0$. It remains to compute $\bar{z}$:

$$\bar{z} = \frac{1}{V} \int_0^{2\pi} \int_0^{\sqrt{2}} \int_{\sqrt{6-r^2}}^{4-r^2} zr\, dz\, dr\, d\theta$$

$$= \frac{1}{V} \int_0^{2\pi} \int_0^{\sqrt{2}} \frac{1}{2} r z^2 \Big|_{z=\sqrt{6-r^2}}^{4-r^2} dr\, d\theta$$

$$= \frac{1}{2V} \int_0^{2\pi} \int_0^{\sqrt{2}} r((4-r^2)^2 - (6-r^2))\, dr\, d\theta$$

$$= \frac{1}{2V} \int_0^{2\pi} \int_0^{\sqrt{2}} (10r - 7r^3 + r^5)\, dr\, d\theta$$

$$= \frac{1}{2V} \int_0^{2\pi} \left(5r^2 - \frac{7}{4}r^4 + \frac{1}{6}r^6\right)\Big|_0^{\sqrt{2}} d\theta$$

$$= \frac{1}{2V} \int_0^{2\pi} \frac{13}{3}\, d\theta$$

$$= \frac{1}{2V} \cdot \frac{26\pi}{3}$$

$$= \frac{13}{34 - 12\sqrt{6}} \approx 2.82$$

Therefore the centroid of $\mathcal{W}$ is

$$P \approx (0, 0, 2.82)$$

In Exercises 25–28, find the center of mass of the region with the given mass density δ.

25. Region bounded by $y = 6 - x$, $x = 0$, $y = 0$; $\delta(x, y) = x^2$

SOLUTION A diagram of the region $\mathcal{W}$ is below:

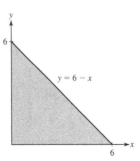

The mass is

$$M = \int_0^6 \int_0^{6-x} x^2\, dy\, dx = \int_0^6 x^2(6 - x)\, dx = \int_0^6 6x^2 - x^3\, dx = \left(2x^3 - \frac{1}{4}x^4\right)\Big|_0^6 = 432 - 324 = 108$$

The coordinates of the center of mass are then

$$x_{CM} = \frac{1}{M} \iint_D x\delta(x,y)\, dA$$

$$= \frac{1}{108} \int_0^6 \int_0^{6-x} x^3\, dy\, dx$$

$$= \frac{1}{108} \int_0^6 x^3(6-x)\, dx$$

$$= \frac{1}{108} \int_0^6 (6x^3 - x^4)\, dx$$

$$= \frac{1}{108} \left(\frac{3}{2}x^4 - \frac{1}{5}x^5\right)\Big|_0^6$$

$$= \frac{1}{108}\left(1944 - \frac{7776}{5}\right) = \frac{18}{5}$$

$$y_{CM} = \frac{1}{M} \iint_D y\delta(x,y)\, dA$$

$$= \frac{1}{108} \int_0^6 \int_0^{6-x} x^2 y\, dy\, dx$$

$$= \frac{1}{108} \int_0^6 \left(\frac{1}{2}x^2 y^2\right)\Big|_0^{6-x} dx$$

$$= \frac{1}{216} \int_0^6 x^2(6-x)^2\, dx$$

$$= \frac{1}{216} \int_0^6 (36x^2 - 12x^3 + x^4)\, dx$$

$$= \frac{1}{216}\left(12x^3 - 3x^4 + \frac{1}{5}x^5\right)\Big|_0^6$$

$$= \frac{1}{216}\left(2592 - 3888 + \frac{7776}{5}\right) = \frac{6}{5}$$

The centroid is at $\left(\frac{18}{5}, \frac{6}{5}\right)$.

27. Region $|x| + |y| \leq 1$; $\delta(x,y) = (x+1)(y+1)$

SOLUTION A diagram of the region $\mathcal{W}$ is below:

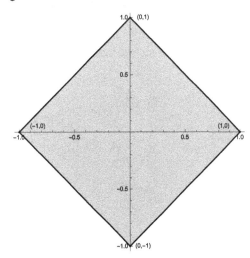

The equations of the four lines bounding the region are $y = x + 1$, $y = 1 - x$, $y = x - 1$, and $y = -1 - x$. Then computing the moment about the y axis gives

$$M_y = \int_{-1}^{0} \int_{-x-1}^{x+1} x(x+1)(y+1)\,dy\,dx + \int_{0}^{1} \int_{x-1}^{1-x} x(x+1)(y+1)\,dy\,dx$$

$$= \int_{-1}^{0} \frac{1}{2}(y+1)^2 x(x+1) \bigg|_{y=-x-1}^{x+1} dx + \int_{0}^{1} \frac{1}{2}(y+1)^2 x(x+1) \bigg|_{y=x-1}^{1-x} dx$$

$$= \frac{1}{2}\left(\int_{-1}^{0} \left((x(x+1)(x+2)^2 - x^3(x+1)\right) dx + \int_{0}^{1} \left((2-x)^2 x(x+1) - x^3(x+1)\right) dx\right)$$

$$= \frac{1}{2}\left(\int_{-1}^{0} (4x^3 + 8x^2 + 4x)\,dx + \int_{0}^{1} (4x - 4x^3)\,dx\right)$$

$$= \frac{1}{2}\left(\left(x^4 + \frac{8}{3}x^3 + 2x^2\right)\bigg|_{-1}^{0} + \left(2x^2 - x^4\right)\bigg|_{0}^{1}\right)$$

$$= \frac{1}{2}\left(\left(-1 + \frac{8}{3} - 2\right) + (2 - 1)\right) = \frac{1}{3}$$

Since both the region and the density function are symmetric in x and y, the moment about the x-axis will also be $\frac{1}{3}$.

It remains to compute the mass.

$$M = \int_{-1}^{0} \int_{-x-1}^{x+1} (x+1)(y+1)\,dy\,dx + \int_{0}^{1} \int_{x-1}^{1-x} (x+1)(y+1)\,dy\,dx$$

$$= \int_{-1}^{0} \frac{1}{2}(x+1)(y+1)^2 \bigg|_{-x-1}^{x+1} dx + \int_{0}^{1} \frac{1}{2}(x+1)(y+1)^2 \bigg|_{x-1}^{1-x} dx$$

$$= \frac{1}{2}\left(\int_{-1}^{0} \left((x+1)(x+2)^2 - (x+1)x^2\right) dx + \int_{0}^{1} \left((x+1)(2-x)^2 - (x+1)x^2\right) dx\right)$$

$$= \frac{1}{2}\left(\int_{-1}^{0} (4x^2 + 8x + 4)\,dx + \int_{0}^{1} (4 - 4x^2)\,dx\right)$$

$$= \frac{1}{2}\left(\left(\frac{4}{3}x^3 + 4x^2 + 4x\right)\bigg|_{-1}^{0} + \left(4x - \frac{4}{3}x^3\right)\bigg|_{0}^{1}\right)$$

$$= \frac{1}{2}\left(\left(\frac{4}{3} - 4 + 4\right) + \left(4 - \frac{4}{3}\right)\right) = 2$$

Therefore the center of mass lies at

$$(\overline{x}, \overline{y}) = \left(\frac{M_y}{M}, \frac{M_x}{M}\right) = \left(\frac{1}{6}, \frac{1}{6}\right)$$

29. Find the z-coordinate of the center of mass of the first octant of the unit sphere with mass density $\delta(x, y, z) = y$ (Figure 17).

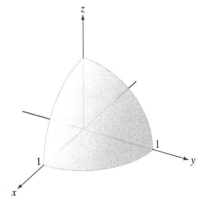

FIGURE 17

SOLUTION We use spherical coordinates:

$$x = \rho \cos\theta \sin\phi, \quad y = \rho \sin\theta \sin\phi, \quad z = \rho \cos\phi, \quad dV = \rho^2 \sin\phi \, d\rho \, d\phi \, d\theta$$

The octant $\mathcal{W}$ is defined by $0 \le \theta \le 2\pi$, $0 \le \phi \le \frac{\pi}{2}$, $0 \le \rho \le 1$, so we have

$$M_{xy} = \iiint_{\mathcal{W}} z\rho(x,y,z) \, dV = \int_{\theta=0}^{\pi/2} \int_{\phi=0}^{\pi/2} \int_{\rho=0}^{1} (\rho\cos\phi)(\rho\sin\theta\sin\phi)\rho^2 \sin\phi \, d\rho \, d\phi \, d\theta$$

$$= \left(\int_{\theta=0}^{\pi/2} \sin\theta \, d\theta\right)\left(\int_{\phi=0}^{\pi/2} \cos\phi \sin^2\phi \, d\phi\right)\left(\int_{\rho=0}^{1} \rho^4 \, d\rho\right)$$

$$= (1)\left(\frac{1}{3} \sin^3\phi \Big|_0^{\pi/2}\right)\left(\frac{1}{5}\right) = \frac{1}{15}$$

The total mass M of the $\mathcal{W}$ is equal to the integral of the mass density $\rho(x,y,z)$:

$$M = \iiint_{\mathcal{W}} \rho(x,y,z) \, dV = \int_{\theta=0}^{\pi/2} \int_{\phi=0}^{\pi/2} \int_{\rho=0}^{1} (\rho\sin\theta\sin\phi)\rho^2 \sin\phi \, d\rho \, d\phi \, d\theta$$

$$= \left(\int_{\theta=0}^{\pi/2} \sin\theta \, d\theta\right)\left(\int_{\phi=0}^{\pi/2} \sin^2\phi \, d\phi\right)\left(\int_{\rho=0}^{1} \rho^3 \, d\rho\right) = (1)\left(\frac{\pi}{4}\right)\left(\frac{1}{4}\right) = \frac{\pi}{16}$$

We conclude that

$$z_{CM} = \frac{1}{M}\iiint_{\mathcal{W}} z\rho(x,y,z) \, dV = \frac{1/15}{\pi/16} = \frac{16}{15\pi} \approx 0.34$$

31. Let $\mathcal{R}$ be the rectangle $[-a, a] \times [b, -b]$ with uniform density and total mass M. Calculate:
(a) The mass density δ of $\mathcal{R}$
(b) I_x and I_0
(c) The radius of gyration about the x-axis

SOLUTION
(a) Since the area of the region is $2a \cdot 2b = 4ab$, and the total mass is M distributed uniformly, the mass density is $\frac{M}{4ab}$.
(b) By formula 3 in the text, we have

$$I_x = \iint_{\mathcal{R}} y^2 \delta(x,y) \, dA = \int_{-a}^{a} \int_{-b}^{b} y^2 \cdot \frac{M}{4ab} \, dy \, dx$$

$$= \frac{M}{4ab} \int_{-a}^{a} \left(\frac{1}{3}y^3\right)\Big|_{-b}^{b} dx$$

$$= \frac{M}{12ab} \int_{-a}^{a} 2b^3 \, dx$$

$$= \frac{M}{12ab} \cdot 2a \cdot 2b^3 = \frac{M}{3}b^2$$

$$I_0 = \iint_{\mathcal{R}} (x^2 + y^2)\delta(x,y) \, dA = \int_{-a}^{a} \int_{-b}^{b} (x^2 + y^2) \cdot \frac{M}{4ab} \, dy \, dx$$

$$= \frac{M}{4ab} \int_{-a}^{a} \left(x^2 y + \frac{1}{3}y^3\right)\Big|_{-b}^{b} dx$$

$$= \frac{M}{4ab} \int_{-a}^{a} \left(2bx^2 + \frac{2}{3}b^3\right) dx$$

$$= \frac{M}{4ab} \left(\frac{2}{3}bx^3 + \frac{2}{3}b^3 x\right)\Big|_{-a}^{a}$$

$$= \frac{M}{4ab} \left(\frac{4}{3}ba^3 + \frac{4}{3}ab^3\right)$$

$$= \frac{M}{3}(a^2 + b^2)$$

1322 CHAPTER 15 | MULTIPLE INTEGRATION (LT CHAPTER 16)

(c) The radius of gyration about the x-axis is

$$r_g = \left(\frac{I_x}{M}\right)^{1/2} = \left(\frac{Mb^2}{3M}\right)^{1/2} = \frac{b}{\sqrt{3}}$$

33. Calculate I_0 and I_x for the disk $\mathcal{D}$ defined by $x^2 + y^2 \leq 16$ (in meters), with total mass 1000 kg and uniform mass density. *Hint:* Calculate I_0 first and observe that $I_0 = 2I_x$. Express your answer in the correct units.

SOLUTION Note that the area of the disk is $\pi r^2 = 16\pi$ so that the mass density is

$$\rho(x, y) = \frac{1000}{16\pi} = \frac{125}{2\pi}$$

Then using polar coordinates we have

$$I_0 = \iint_\mathcal{D} (x^2 + y^2) \frac{125}{2\pi} \, dA = \frac{125}{2\pi} \int_0^4 \int_0^{2\pi} r^2 \cdot r \, d\theta \, dr = 125 \cdot \frac{1}{4} r^4 \Big|_0^4 = 125 \cdot 64 = 8000 \text{ kg-m}^2$$

Since both the region and the mass density are symmetric in x and y, we have $I_x = I_y$. But then $I_0 = I_x + I_y = 2I_x$ so that

$$I_x = 4000 \text{ kg-m}^2$$

In Exercises 35–38, let $\mathcal{D}$ be the triangular domain bounded by the coordinate axes and the line $y = 3 - x$, with mass density $\delta(x, y) = y$. Compute the given quantities.

35. Total mass

SOLUTION The total mass is simply

$$\iint_\mathcal{D} \rho(x, y) \, dA = \int_0^3 \int_0^{3-x} y \, dy \, dx = \frac{1}{2} \int_0^3 y^2 \Big|_0^{3-x} dx = \frac{1}{2} \int_0^3 (3 - x)^2 \, dx = -\frac{1}{6}(3 - x)^3 \Big|_0^3 = \frac{27}{6} = \frac{9}{2}$$

37. I_x

SOLUTION

$$I_x = \iint_\mathcal{D} y^2 \rho(x, y) \, dA = \int_0^3 \int_0^{3-x} y^3 \, dy \, dx = \frac{1}{4} \int_0^3 (3 - x)^4 \, dx$$

$$= -\frac{1}{20}(3 - x)^5 \Big|_0^3 = \frac{1}{20} 3^5 = \frac{243}{20}$$

In Exercises 39–42, let $\mathcal{D}$ be the domain between the line $y = bx/a$ and the parabola $y = bx^2/a^2$, where $a, b > 0$. Assume the mass density is $\delta(x, y) = 1$ for Exercise 37 and $\delta(x, y) = xy$ for Exercises 38–40. Compute the given quantities.

39. Centroid

SOLUTION The curves intersect at $x = 0$ and at $x = a$. The area is

$$A = \iint_\mathcal{D} 1 \, dA = \int_0^a \int_{bx^2/a^2}^{bx/a} 1 \, dy \, dx = \int_0^a \frac{bx}{a} - \frac{bx^2}{a^2} \, dx$$

$$= \left(\frac{bx^2}{2a} - \frac{bx^3}{3a^2}\right)\Big|_0^a = \frac{ab}{2} - \frac{ab}{3} = \frac{ab}{6}$$

Then

$$\bar{x} = \frac{1}{A} \iint_\mathcal{D} x \, dA = \frac{6}{ab} \int_0^a \int_{bx^2/a^2}^{bx/a} x \, dy \, dx = \frac{6}{ab} \int_0^a x\left(\frac{bx}{a} - \frac{bx^2}{a^2}\right) dx$$

$$= \frac{6}{ab} \int_0^a \frac{b}{a} x^2 - \frac{b}{a^2} x^3 \, dx = \frac{6}{ab}\left(\frac{b}{3a} a^3 - \frac{b}{4a^2} a^4\right)$$

$$= \frac{6}{ab}\left(\frac{a^2b}{3} - \frac{a^2b}{4}\right) = \frac{a}{2}$$

and

$$\bar{y} = \frac{1}{A}\iint_\mathcal{D} y\, dA = \frac{6}{ab}\int_0^a \int_{bx^2/a^2}^{bx/a} y\, dy\, dx = \frac{6}{2ab}\int_0^a y^2\Big|_{bx^2/a^2}^{bx/a} dx$$

$$= \frac{3}{ab}\int_0^a \frac{b^2}{a^2}x^2 - \frac{b^2}{a^4}x^4\, dx = \frac{3}{ab}\left(\frac{b^2}{3a^2}x^3 - \frac{b^2}{5a^4}x^5\right)\Big|_0^a$$

$$= \frac{3}{ab}\left(\frac{ab^2}{3} - \frac{ab^2}{5}\right) = \frac{2b}{5}$$

41. I_x

SOLUTION The curves intersect at $x = 0$ and at $x = a$, so

$$I_x = \iint_\mathcal{D} y^2 \rho(x,y)\, dA = \int_0^a \int_{bx^2/a^2}^{bx/a} xy^3\, dy\, dx = \frac{1}{4}\int_0^a xy^4\Big|_{bx^2/a^2}^{bx/a} dx$$

$$= \frac{1}{4}\int_0^a x\left(\frac{b^4}{a^4}x^4 - \frac{b^4}{a^8}x^8\right) dx = \frac{b^4}{4a^8}\int_0^a a^4 x^5 - x^9\, dx$$

$$= \frac{b^4}{4a^8}\left(\frac{a^4}{6}x^6 - \frac{1}{10}x^{10}\right)\Big|_0^a = \frac{b^4}{4a^8}\left(\frac{a^{10}}{6} - \frac{a^{10}}{10}\right) = \frac{a^2 b^4}{60}$$

43. Calculate the moment of inertia I_x of the disk $\mathcal{D}$ defined by $x^2 + y^2 \le R^2$ (in meters), with total mass M kilograms. How much kinetic energy (in joules) is required to rotate the disk about the x-axis with angular velocity 10 radians per second?

SOLUTION The disk has area πR^2 m² and mass density $\delta = \frac{M}{\pi R^2}$ kg/m². We compute I_x using polar coordinates:

$$I_x = \int\int_\mathcal{D} y^2 \delta\, dA$$

$$= \int_0^{2\pi}\int_{r=0}^R (r\sin\theta)^2 \delta r\, dr\, d\theta$$

$$= \delta\int_0^{2\pi}\int_{r=0}^R r^3 \sin^2\theta\, dr\, d\theta$$

$$= \frac{M}{\pi R^2}\left(\int_0^{2\pi} \sin^2\theta\, d\theta\right)\left(\int_{r=0}^R r^3\, dr\right)$$

$$= \frac{M}{\pi R^2}(\pi)\left(\frac{R^4}{4}\right) = \frac{1}{4}MR^2$$

45. Show that the moment of inertia of a sphere of radius R of total mass M with uniform mass density about any axis passing through the center of the sphere is $\frac{2}{5}MR^2$. Note that the mass density of the sphere is $\delta = M/(\frac{4}{3}\pi R^3)$.

SOLUTION Since the sphere is symmetric under an arbitrary rotation, and since the mass density is uniform, it follows that the moments of inertia of the sphere about all axes passing through its center are equal. Thus it suffices to prove the result for an arbitrary axis; we choose the z-axis. Then, using spherical coordinates, we have

$$I_z = \iiint_S (x^2 + y^2)\rho(x,y,z)\, dA$$

$$= \frac{3M}{4\pi R^3}\int_0^{2\pi}\int_0^\pi \int_0^R ((r\cos\theta\sin\phi)^2 + (r\sin\theta\sin\phi)^2) r^2 \sin\phi\, dr\, d\theta\, d\phi$$

$$= \frac{3M}{4\pi R^3}\int_0^{2\pi}\int_0^\pi\int_0^R r^4 \sin^3\phi\, dr\, d\theta\, d\phi = \frac{3M}{4\pi R^3}\cdot 2\pi\left(\int_0^\pi \sin^3\phi\, d\phi\right)\left(\int_0^R r^4\, dr\right)$$

$$= \frac{3M}{2R^3}\cdot\frac{4}{3}\cdot\frac{1}{5}R^5 = \frac{2}{5}MR^2$$

In Exercises 47 and 48, prove the formula for the right circular cylinder in Figure 18.

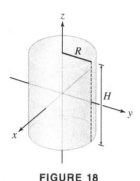

FIGURE 18

47. $I_z = \frac{1}{2}MR^2$

SOLUTION Assuming the cylinder has uniform mass density 1, and using cylindrical coordinates, we have

$$I_z = \iiint_C (x^2 + y^2)\rho(x,y,z)\,dA = \int_0^R \int_0^{2\pi} \int_{-H/2}^{H/2} r^2 \cdot r\,dz\,d\theta\,dr$$

$$= 2\pi H \int_0^R r^3\,dr = \frac{1}{2}\pi R^4 H$$

But the volume of the cylinder, which is equal to its mass, is $\pi R^2 H$, so that

$$I_z = \frac{1}{2}\pi R^4 H = \frac{1}{2}MR^2$$

49. The yo-yo in Figure 19 is made up of two disks of radius $r = 3$ cm and an axle of radius $b = 1$ cm. Each disk has mass $M_1 = 20$ g, and the axle has mass $M_2 = 5$ g.

(a) Use the result of Exercise 47 to calculate the moment of inertia I of the yo-yo with respect to the axis of symmetry. Note that I is the sum of the moments of the three components of the yo-yo.

(b) The yo-yo is released and falls to the end of a 100-cm string, where it spins with angular velocity ω. The total mass of the yo-yo is $m = 45$ g, so the potential energy lost is PE $= mgh = (45)(980)100$ g-cm^2/s^2. Find ω using the fact that the potential energy is the sum of the rotational kinetic energy and the translational kinetic energy and that the velocity $v = b\omega$ since the string unravels at this rate.

FIGURE 19

SOLUTION

i. Since the yo-yo is composed of three components, the moment of inertia is the sum of the moments of each of them:

$$I = 2M_1 r^2 + M_2 b^2 = 2(20)(3^2) + 5(1^2) = 365 \text{ g-cm}^2$$

ii. If half of the potential energy is converted to kinetic energy, then

$$\frac{1}{3}(45)(980)100 = \frac{1}{2}(365)\omega^2$$

and $\omega = \sqrt{(2/3)45(980)(100)/365} \approx 90$ radians per second or $90/(2\pi) \approx 14$ rotations per second.

51. Calculate $P(0 \leq X \leq 2; 1 \leq Y \leq 2)$, where X and Y have joint probability density function

$$p(x,y) = \begin{cases} \frac{1}{72}(2xy + 2x + y) & \text{if } 0 \leq x \leq 4 \text{ and } 0 \leq y \leq 2 \\ 0 & \text{otherwise} \end{cases}$$

SOLUTION The region $0 \leq X \leq 2; 1 \leq Y \leq 2$ falls into the region where $p(x,y)$ is defined by the first line of the given formula, so that

$$P(0 \leq X \leq 2; 1 \leq Y \leq 2) = \int_0^2 \int_1^2 \frac{1}{72}(2xy + 2x + y)\,dy\,dx = \frac{1}{72}\int_0^2 \left(xy^2 + 2xy + \frac{1}{2}y^2\right)\Big|_1^2 dx$$

$$= \frac{1}{72}\int_0^2 5x + \frac{3}{2}\,dx = \frac{1}{72}\left(\frac{5}{2}x^2 + \frac{3}{2}x\right)\Big|_0^2 = \frac{1}{72} \cdot 13 = \frac{13}{72} \approx 0.18$$

53. The lifetime (in months) of two components in a certain device are random variables X and Y that have joint probability distribution function

$$p(x,y) = \begin{cases} \frac{1}{9216}(48 - 2x - y) & \text{if } x \geq 0, y \geq 0,\ 2x + y \leq 48 \\ 0 & \text{otherwise} \end{cases}$$

Calculate the probability that both components function for at least 12 months without failing. Note that $p(x,y)$ is nonzero only within the triangle bounded by the coordinate axes and the line $2x + y = 48$ shown in Figure 20.

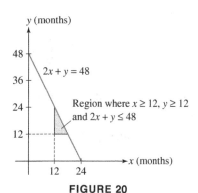

FIGURE 20

SOLUTION Both components function for at least 12 months without failing if $X + Y \geq 12$; however, we must also have $2X + Y \leq 48$. Then the region of integration is the shaded triangle in the figure; the lower-left corner of that triangle is $(12, 12)$. One of the remaining vertices is the intersection of $x = 12$ and $2x + y = 48$; solving for y we have $y = 24$, so the point is $(12, 24)$. The other vertex is the intersection of $y = 12$ and $2x + y = 48$; solving for x gives $x = 18$, so the point is $(18, 12)$. The region of integration is then $12 \leq x \leq 18$ and $12 \leq y \leq 48 - 2x$. Thus the probability is

$$P(X \geq 12, Y \geq 12) = \int_{12}^{18} \int_{12}^{48-2x} \frac{1}{9216}(48 - 2x - y)\,dy\,dx$$

$$= \frac{1}{9216} \int_{12}^{18} \left(48y - 2xy - \frac{1}{2}y^2\right)\Big|_{12}^{48-2x} dx$$

$$= \frac{1}{9216} \int_{12}^{18} 1800 - 96x - 2x(36 - 2x) - \frac{1}{2}(48 - 2x)^2\,dx$$

$$= \frac{1}{9216}\left(1800x - 48x^2 - 36x^2 + \frac{4}{3}x^3 + \frac{1}{12}(48 - 2x)^3\right)\Big|_{12}^{18}$$

$$= \frac{144}{9216} = \frac{1}{64} = .015625$$

55. Find a constant C such that

$$p(x,y) = \begin{cases} Cy & \text{if } 0 \leq x \leq 1 \text{ and } x^2 \leq y \leq x \\ 0 & \text{otherwise} \end{cases}$$

is a joint probability density function. Then calculate the probability that $Y \geq X^{3/2}$.

SOLUTION $p(x, y)$ is a joint probability density function if

$$1 = \int_0^1 \int_{x^2}^x p(x,y)\,dy\,dx = \int_0^1 \int_{x^2}^x Cy\,dy\,dx$$

$$= \frac{C}{2} \int_0^1 y^2 \Big|_{x^2}^x dx = \frac{C}{2} \int_0^1 x^2 - x^4\,dx = \frac{C}{15}$$

so that we must have $C = 15$. Now, for $0 \leq x \leq 1$ we have $x^2 \leq x^{3/2} \leq x$, so that

$$P(Y \geq X^{3/2}) = \int_0^1 \int_{x^{3/2}}^x 15y\,dy\,dx = \frac{15}{2} \int_0^1 y^2 \Big|_{x^{3/2}}^x dx$$

$$= \frac{15}{2} \int_0^1 x^2 - x^3\,dx = \frac{15}{2} \left(\frac{1}{3} - \frac{1}{4} \right) = \frac{15}{24} = \frac{5}{8} = 0.375$$

57. According to quantum mechanics, the x- and y-coordinates of a particle confined to the region $\mathcal{R} = [0, 1] \times [0, 1]$ are random variables with joint probability density function

$$p(x, y) = \begin{cases} C \sin^2(2\pi\ell x) \sin^2(2\pi n y) & \text{if } (x, y) \in \mathcal{R} \\ 0 & \text{otherwise} \end{cases}$$

The integers ℓ and n determine the energy of the particle, and C is a constant.

(a) Find the constant C.

SOLUTION We have

$$\int_0^1 \int_0^1 \sin^2(2\pi\ell x) \sin^2(2\pi n y)\,dx\,dy = \frac{1}{4}$$

so $c = 4$.

(b) Calculate the probability that a particle with $\ell = 2$, $n = 3$ lies in the region $[0, \frac{1}{4}] \times [0, \frac{1}{8}]$.

SOLUTION We compute

$$\int_0^{1/4} \int_0^{1/8} 4\sin^2(2\pi\ell x)\sin^2(2\pi n y)\,dx\,dy = \frac{1}{48}(3 + \frac{2}{\pi}) \approx 0.076$$

59. According to Coulomb's Law, the attractive force between two electric charges of magnitude q_1 and q_2 separated by a distance r is kq_1q_2/r^2 (k is a constant). Let F be the net force on a charged particle P of charge Q coulombs located d centimeters above the center of a circular disk of radius R, with a uniform charge distribution of density ρ coulombs per square meter (Figure 21). By symmetry, F acts in the vertical direction.

(a) Let $\mathcal{R}$ be a small polar rectangle of size $\Delta r \times \Delta\theta$ located at distance r. Show that $\mathcal{R}$ exerts a force on P whose vertical component is

$$\left(\frac{k\rho Qd}{(r^2 + d^2)^{3/2}} \right) r\,\Delta r\,\Delta\theta$$

(b) Explain why F is equal to the following double integral, and evaluate:

$$F = k\rho Qd \int_0^{2\pi} \int_0^R \frac{r\,dr\,d\theta}{(r^2 + d^2)^{3/2}}$$

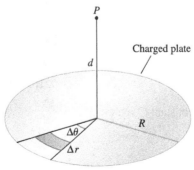

FIGURE 21

SOLUTION

(a) The area of the small polar rectangle $\mathcal{R}$ is

$$\Delta A = \frac{1}{2}(r + \Delta r)^2 \Delta\theta - \frac{1}{2}r^2 \Delta\theta = r(\Delta r \Delta\theta) + \frac{1}{2}\Delta r^2 \Delta\theta \approx r(\Delta r \Delta\theta)$$

Therefore, the charge on $\mathcal{R}$ is $q_1 = r(\Delta r \Delta\theta)k$. The distance between P and $\mathcal{R}$ is, by the Pythagorean Law, $\sqrt{r^2 + d^2}$.

Therefore, the magnitude of the force between P and $\mathcal{R}$ is

$$\frac{Cq_1q_2}{\left(\sqrt{r^2+d^2}\right)^2} = \frac{C(r\Delta r \Delta\theta k)Q}{r^2+d^2} = \frac{kCQ}{r^2+d^2}r\Delta r \Delta\theta$$

The vertical component of this force is obtained by multiplying the force by $\cos\alpha = \frac{d}{\sqrt{r^2+d^2}}$. That is,

$$\Delta F = \frac{kCQ}{r^2+d^2} \cdot \frac{d}{\sqrt{r^2+d^2}} r\Delta r \Delta\theta = \frac{kCQd}{(r^2+d^2)^{3/2}} r\Delta r \Delta\theta$$

(b) Since F acts in the vertical direction, it is approximated by the Riemann sum of the force ΔF in part (a), over the polar rectangles. This Riemann sum approximates F in higher precision if we let $\Delta\theta \to 0$ and $\Delta r \to 0$. The result is the double integral of $\frac{kCQd}{(r^2+d^2)^{3/2}}$ over the disk. The disk is determined by $0 \le \theta \le 2\pi$ and $0 \le r \le R$. Therefore we get

$$F = \int_0^{2\pi}\int_0^R \frac{kCQd}{(r^2+d^2)^{3/2}} r\, dr\, d\theta = kCQd\int_0^{2\pi}\int_0^R \frac{r\, dr\, d\theta}{(r^2+d^2)^{3/2}}$$

Using a simple u-substitution of $u = r^2 + d^2$, and integrating the $d\theta$ integral, we get

$$F = kCQd \cdot 2\pi \cdot \int_{r^2}^{r^2+R^2} \frac{1}{2}u^{-3/2}\, du = -2kCQd\pi \cdot u^{-1/2}\Big|_{u=r^2}^{r^2+R^2} = 2\pi kCQd\left[\frac{1}{r} - \frac{1}{\sqrt{r^2+R^2}}\right]$$

Further Insights and Challenges

61. Let $\mathcal{D}$ be the domain in Figure 22. Assume that $\mathcal{D}$ is symmetric with respect to the y-axis; that is, both $g_1(x)$ and $g_2(x)$ are even functions.

(a) Prove that the centroid lies on the y-axis—that is, that $\overline{x} = 0$.

(b) Show that if the mass density satisfies $\delta(-x, y) = \delta(x, y)$, then $M_y = 0$ and $x_{\text{CM}} = 0$.

FIGURE 22

SOLUTION

(a) Assume $\mathcal{D}$ has area A. Then the x coordinate of the centroid is

$$\bar{x} = \frac{1}{A} \iint_{\mathcal{D}} x\, dA = \frac{1}{A} \int_{-a}^{a} \int_{g_1(x)}^{g_2(x)} x\, dy\, dx = \frac{1}{A} \int_{-a}^{a} x(g_2(x) - g_1(x))\, dx$$

Since g_1 and g_2 are both even functions, $x(g_2(x) - g_1(x))$ is an odd function, so its integral over a region symmetric about the x-axis is zero. Thus $\bar{x} = 0$.

(b) Let $R(x, y)$ be an antiderivative of $\rho(x, y)$ with respect to y (i.e. $R(x, y) = \int \rho(x, y)\, dy$). Note that

$$R(-x, y) = \int \rho(-x, y)\, dy = \int \rho(x, y)\, dy = R(x, y)$$

so that R is an even function with respect to x. Now, we have

$$M_y \iint_{\mathcal{D}} x\rho(x, y)\, dA = \int_{-a}^{a} \int_{g_1(x)}^{g_2(x)} x\rho(x, y)\, dy\, dx = \int_{-a}^{a} x(R(x, g_2(x)) - R(x, g_1(x)))\, dx$$

Since g_1, g_2, and R are all even functions of x, we have

$$R(-x, g_2(-x)) - R(-x, g_1(-x)) = R(-x, g_2(x)) - R(-x, g_1(x)) = R(x, g_2(x)) - R(x, g_1(x))$$

so that the second factor in the integrand is an even function of x. But x is an odd function of x, so their product is odd. It follows that the integral over the range $-a \leq x \leq a$ is zero. Thus $M_y = x_{CM} = 0$.

63. Use Pappus's Theorem in Exercise 62 to show that the torus obtained by revolving a circle of radius b centered at $(0, a)$ about the x-axis (where $b < a$) has volume $V = 2\pi^2 ab^2$.

SOLUTION The centroid of the circle is obviously at the center, $(0, a)$, and the area of the circle is πb^2, so that by Pappus' Theorem,

$$V = 2\pi \pi b^2 a = 2\pi^2 ab^2$$

65. Parallel-Axis Theorem Let $\mathcal{W}$ be a region in $\mathbf{R}^3$ with center of mass at the origin. Let I_z be the moment of inertia of $\mathcal{W}$ about the z-axis, and let I_h be the moment of inertia about the vertical axis through a point $P = (a, b, 0)$, where $h = \sqrt{a^2 + b^2}$. By definition,

$$I_h = \iiint_{\mathcal{W}} ((x-a)^2 + (y-b)^2)\delta(x, y, z)\, dV$$

Prove the Parallel-Axis Theorem: $I_h = I_z + Mh^2$.

SOLUTION We have

$$I_h - I_z = \iiint_{\mathcal{W}} ((x-a)^2 + (y-b)^2)\rho(x, y, z)\, dV - \iiint_{\mathcal{W}} (x^2 + y^2)\rho(x, y, z)\, dV$$

$$= \iiint_{\mathcal{W}} (-2ax + a^2 - 2by + b^2)\rho(x, y, z)\, dV$$

$$= (a^2 + b^2) \iiint_{\mathcal{W}} \rho(x, y, z)\, dV - 2a \iiint_{\mathcal{W}} x\rho(x, y, z)\, dV - 2b \iiint_{\mathcal{W}} y\rho(x, y, z)\, dV$$

$$= (a^2 + b^2)M - 2aM_{yz} - 2bM_{xz} = Mh^2$$

since the last two terms are zero because the center of mass of $\mathcal{W}$ is at the origin.

15.6 Change of Variables (LT Section 16.6)

Preliminary Questions

1. Which of the following maps is linear?

(a) (uv, v) (b) $(u + v, u)$ (c) $(3, e^u)$

SOLUTION

(a) This map is not linear since it does not satisfy the linearity property:

$$\Phi(2u, 2v) = (2u \cdot 2v, 2v) = (4uv, 2v) = 2(2uv, v)$$

$$2\Phi(u, v) = 2(uv, v) \quad \Rightarrow \quad \Phi(2u, 2v) \neq 2\Phi(u, v)$$

(b) This map is linear since it has the form $\Phi(u, v) = (Au + Cv, Bu + Dv)$ where $A = C = 1, B = 1, D = 0$.

(c) This map is not linear since it does not satisfy the linearity properties. For example,

$$\Phi(2u, 2v) = (3, e^{2u})$$
$$2\Phi(u, v) = 2(3, e^u)$$
$$\Rightarrow \quad \Phi(2u, 2v) \neq 2\Phi(u, v)$$

2. Suppose that Φ is a linear map such that $\Phi(2, 0) = (4, 0)$ and $\Phi(0, 3) = (-3, 9)$. Find the images of:
(a) $\Phi(1, 0)$ **(b)** $\Phi(1, 1)$ **(c)** $\Phi(2, 1)$

SOLUTION We denote the linear map by $\Phi(u, v) = (Au + Cv, Bu + Dv)$. By the given information we have

$$\Phi(2, 0) = (A \cdot 2 + C \cdot 0, B \cdot 2 + D \cdot 0) = (2A, 2B) = (4, 0)$$

$$\Phi(0, 3) = (A \cdot 0 + C \cdot 3, B \cdot 0 + D \cdot 3) = (3C, 3D) = (-3, 9)$$

Therefore,

$$\begin{matrix} 2A = 4 \\ 2B = 0 \\ 3C = -3 \\ 3D = 9 \end{matrix} \quad \Rightarrow \quad A = 2, \quad B = 0, \quad C = -1, \quad D = 3$$

The linear map is thus

$$\Phi(u, v) = (2u - v, 3v)$$

We now compute the images:
(a) $\Phi(1, 0) = (2 \cdot 1 - 0, 3 \cdot 0) = (2, 0)$
(b) $\Phi(1, 1) = (2 \cdot 1 - 1, 3 \cdot 1) = (1, 3)$
(c) $\Phi(2, 1) = (2 \cdot 2 - 1, 3 \cdot 1) = (3, 3)$

3. What is the area of $\Phi(\mathcal{R})$ if $\mathcal{R}$ is a rectangle of area 9 and Φ is a mapping whose Jacobian has constant value 4?

SOLUTION

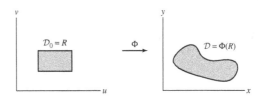

The areas of $\mathcal{D}_0 = \Phi(\mathcal{R})$ and $\mathcal{D} = \mathcal{R}$ are the following integrals:

$$\text{Area}(\mathcal{R}) = 9 = \iint_{\mathcal{D}_0} 1 \, du \, dv$$

$$\text{Area}(\Phi(\mathcal{R})) = \iint_{\mathcal{D}} 1 \, dx \, dy$$

Using the Change of Variables Formula, we have

$$\text{Area}(\Phi(\mathcal{R})) = \iint_{\mathcal{D}} 1 \, dx \, dy = \iint_{\mathcal{D}_0} 1 |\text{Jac}\Phi| \, du \, dv = \iint_{\mathcal{D}_0} 4 \, du \, dv = 4 \iint_{\mathcal{D}_0} 1 \, du \, dv = 4 \cdot 9 = 36$$

The area of $\Phi(\mathcal{R})$ is 36.

4. Estimate the area of $\Phi(\mathcal{R})$, where $\mathcal{R} = [1, 1.2] \times [3, 3.1]$ and Φ is a mapping such that $\text{Jac}(\Phi)(1, 3) = 3$.

SOLUTION

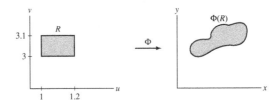

We use the following estimation:

$$\text{Area}(\Phi(\mathcal{R})) \approx |\text{Jac}(\Phi)(P)|\text{Area}(\mathcal{R})$$

The area of the rectangle $\mathcal{R}$ is

$$\text{Area}(\mathcal{R}) = 0.2 \cdot 0.1 = 0.02$$

We choose the sample point $P = (1, 3)$ in $\mathcal{R}$ to obtain the following estimation:

$$\text{Area}(\Phi(\mathcal{R})) \approx |\text{Jac}(\Phi)(1, 3)|\text{Area}(\mathcal{R}) = 3 \cdot 0.02 = 0.06$$

Exercises

1. Determine the image under $\Phi(u, v) = (2u, u + v)$ of the following sets:

(a) The u- and v-axes

(b) The rectangle $\mathcal{R} = [0, 5] \times [0, 7]$

(c) The line segment joining $(1, 2)$ and $(5, 3)$

(d) The triangle with vertices $(0, 1)$, $(1, 0)$, and $(1, 1)$

SOLUTION

(a) The image of the u-axis is obtained by substituting $v = 0$ in $\Phi(u, v) = (2u, u + v)$. That is,

$$\Phi(u, 0) = (2u, u + 0) = (2u, u)$$

The image of the u-axis is the set of points $(x, y) = (2u, u)$, which is the line $y = \frac{1}{2}x$ in the xy-plane. The image of the v-axis is obtained by substituting $u = 0$ in $\Phi(u, v) = (2u, u + v)$. That is,

$$\Phi(0, v) = (0, 0 + v) = (0, v)$$

Therefore, the image of the v-axis is the set $(x, y) = (0, v)$, which is the vertical line $x = 0$ (the y-axis).

(b) Since Φ is a linear map, the segment through points P and Q is mapped to the segment through $\Phi(P)$ and $\Phi(Q)$. We thus must find the images of the vertices of $\mathcal{R}$:

$$\Phi(0, 0) = (2 \cdot 0, 0 + 0) = (0, 0)$$
$$\Phi(5, 0) = (2 \cdot 5, 5 + 0) = (10, 5)$$
$$\Phi(5, 7) = (2 \cdot 5, 5 + 7) = (10, 12)$$
$$\Phi(0, 7) = (2 \cdot 0, 0 + 7) = (0, 7)$$

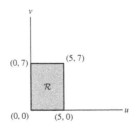

The image of $\mathcal{R}$ is the parallelogram with vertices $(0, 0)$, $(10, 5)$, $(10, 12)$, and $(0, 7)$ in the xy-plane.

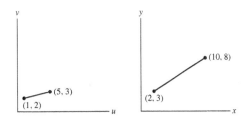

(c) We compute the images of the endpoints of the segment:

$$\Phi(1,2) = (2 \cdot 1, 1+2) = (2,3)$$

$$\Phi(5,3) = (2 \cdot 5, 5+3) = (10,8)$$

The image is the segment in the xy-plane joining the points $(2,3)$ and $(10,8)$.

(d) Since Φ is linear, the image of the triangle is the triangle whose vertices are the images of the vertices of the triangle. We compute these images:

$$\Phi(0,1) = (2 \cdot 0, 0+1) = (0,1)$$

$$\Phi(1,0) = (2 \cdot 1, 1+0) = (2,1)$$

$$\Phi(1,1) = (2 \cdot 1, 1+1) = (2,2)$$

Therefore the image is the triangle in the xy-plane whose vertices are at the points $(0,1)$, $(2,1)$, and $(2,2)$.

3. Let $\Phi(u,v) = (u^2, v)$. Is Φ one-to-one? If not, determine a domain on which Φ is one-to-one. Find the image under Φ of:

(a) The u- and v-axes
(b) The rectangle $\mathcal{R} = [-1,1] \times [-1,1]$
(c) The line segment joining $(0,0)$ and $(1,1)$
(d) The triangle with vertices $(0,0)$, $(0,1)$, and $(1,1)$

SOLUTION Φ is not one-to-one since for any $u \neq 0$, (u,v) and $(-u,v)$ are two different points with the same image. However, Φ is one-to-one on the domain $\{(u,v) : u \geq 0\}$ and on the domain $\{(u,v) : u \leq 0\}$.

(a) The image of the u-axis is the set of the points

$$(x,y) = \Phi(u,0) = (u^2, 0) \quad \Rightarrow \quad x = u^2, \quad y = 0$$

That is, the positive x-axis, including the origin. The image of the v-axis is the set of the following points:

$$(x,y) = \Phi(0,v) = (0^2, v) = (0,v) \quad \Rightarrow \quad x = 0, \quad y = v$$

That is, the line $x = 0$, which is the y-axis.

(b) The rectangle $\mathcal{R}$ is defined by

$$|u| \leq 1, \quad |v| \leq 1$$

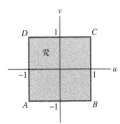

Since $x = u^2$ and $y = v$, we have $u = \pm\sqrt{x}$ and $v = y$ (depending on our choice of domain). Therefore, the inequalities for x and y are

$$|\pm\sqrt{x}| \leq 1, \quad |y| \leq 1$$

or

$$0 \leq x \leq 1 \quad \text{and} \quad -1 \leq y \leq 1$$

We conclude that the image of $\mathcal{R}$ in the xy-plane is the rectangle $[0, 1] \times [-1, 1]$.

(c) The line segment joining the points $(0, 0)$ and $(1, 1)$ in the uv-plane is defined by

$$0 \leq u \leq 1, \quad v = u$$

Substituting $u = \sqrt{x}$ and $v = y$, we get

$$0 \leq \sqrt{x} \leq 1, \quad y = \sqrt{x}$$

or

$$0 \leq x \leq 1, \quad y = \sqrt{x}$$

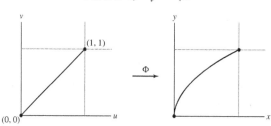

The image is the curve $y = \sqrt{x}$ for $0 \leq x \leq 1$.

(d) We identify the image of the sides of the triangle OAB.

The image of $\overline{OA}$: This segment is defined by $u = 0$ and $0 \leq v \leq 1$. That is,

$$\pm\sqrt{x} = 0 \quad \text{and} \quad 0 \leq y \leq 1$$

or

$$x = 0, \quad 0 \leq y \leq 1$$

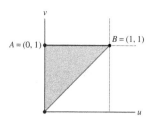

This is the segment joining the points $(0, 0)$ and $(0, 1)$ in the xy-plane.
The image of $\overline{AB}$: This segment is defined by $0 \leq u \leq 1$ and $v = 1$. That is,

$$0 \leq \sqrt{x} \leq 1, \quad y = 1$$

or

$$0 \leq x \leq 1, \quad y = 1$$

This is the segment joining the points $(0, 1)$ and $(1, 1)$ in the xy-plane.

The image of $\overline{OB}$: In part (c) we showed that the image of the segment is the curve $y = \sqrt{x}$, $0 \le x \le 1$. Therefore, the image of the triangle is the region shown in the figure:

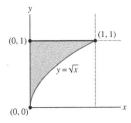

In Exercises 5–12, let $\Phi(u, v) = (2u + v, 5u + 3v)$ be a map from the uv-plane to the xy-plane.

5. Show that the image of the horizontal line $v = c$ is the line with equation $y = \frac{5}{2}x + \frac{1}{2}c$. What is the image (in slope-intercept form) of the vertical line $u = c$?

SOLUTION The image of the horizontal line $v = c$ is the set of the following points:

$$(x, y) = \Phi(u, c) = (2u + c, 5u + 3c) \quad \Rightarrow \quad x = 2u + c, \quad y = 5u + 3c$$

The first equation implies $u = \frac{x-c}{2}$. Substituting in the second equation gives

$$y = 5\frac{(x-c)}{2} + 3c = \frac{5x}{2} + \frac{c}{2}$$

Therefore, the image of the line $v = c$ is the line $y = \frac{5x}{2} + \frac{c}{2}$ in the xy-plane.

The image of the vertical line $u = c$ is the set of the following points:

$$(x, y) = \Phi(c, v) = (2c + v, 5c + 3v) \quad \Rightarrow \quad x = 2c + v, \quad y = 5c + 3v$$

By the first equation, $v = x - 2c$. Substituting in the second equation gives

$$y = 5c + 3(x - 2c) = 5c + 3x - 6c = 3x - c$$

Therefore, the image of the line $u = c$ is the line $y = 3x - c$ in the xy-plane.

7. Describe the image of the line $v = 4u$ under Φ in slope-intercept form.

SOLUTION We choose any two points on the line $v = 4u$, for example $(u, v) = (1, 4)$ and $(u, v) = (0, 0)$. By a property of linear maps, the image of the line $v = 4u$ under the linear map $\Phi(u, v) = (2u + v, 5u + 3v)$ is the line in the xy-plane through the points $\Phi(1, 4)$ and $\Phi(0, 0)$. We find these points:

$$\Phi(0, 0) = (2 \cdot 0 + 0, 5 \cdot 0 + 3 \cdot 0) = (0, 0)$$
$$\Phi(1, 4) = (2 \cdot 1 + 4, 5 \cdot 1 + 3 \cdot 4) = (6, 17)$$

We now find the slope-intercept equation of the line in the xy-plane through the points $(0, 0)$ and $(6, 17)$:

$$y - 0 = \frac{17 - 0}{6 - 0}(x - 0) \quad \Rightarrow \quad y = \frac{17}{6}x$$

9. Show that the inverse of Φ is

$$\Phi^{-1}(x, y) = (3x - y, -5x + 2y)$$

Hint: Show that $\Phi(\Phi^{-1}(x, y)) = (x, y)$ and $\Phi^{-1}(\Phi(u, v)) = (u, v)$.

SOLUTION By the definition of the inverse map, we must show that the given maps $\Phi^{-1}(x, y) = (3x - y, -5x + 2y)$ and $\Phi(u, v) = (2u + v, 5u + 3v)$ satisfy $\Phi\left(\Phi^{-1}(x, y)\right) = (x, y)$ and $\Phi^{-1}(\Phi(u, v)) = (u, v)$. We have

$$\Phi\left(\Phi^{-1}(x, y)\right) = \Phi(3x - y, -5x + 2y) = (2(3x - y) + (-5x + 2y), 5(3x - y) + 3(-5x + 2y)) = (x, y)$$

$$\Phi^{-1}(\Phi(u, v)) = \Phi^{-1}(2u + v, 5u + 3v) = (3(2u + v) - (5u + 3v), -5(2u + v) + 2(5u + 3v)) = (u, v)$$

We conclude that Φ^{-1} is the inverse of Φ.

11. Calculate $\text{Jac}(\Phi) = \dfrac{\partial(x,y)}{\partial(u,v)}$.

SOLUTION The Jacobian of the linear mapping $\Phi(u,v) = (2u + v, 5u + 3v)$ is the following determinant:

$$\text{Jac}(\Phi) = \dfrac{\partial(x,y)}{\partial(u,v)} = \begin{vmatrix} 2 & 1 \\ 5 & 3 \end{vmatrix} = 2 \cdot 3 - 5 \cdot 1 = 1$$

In Exercises 13–18, compute the Jacobian (at the point, if indicated).

13. $\Phi(u,v) = (3u + 4v, u - 2v)$

SOLUTION Using the Jacobian of linear mappings we get

$$\text{Jac}(\Phi) = \dfrac{\partial(x,y)}{\partial(u,v)} = \begin{vmatrix} 3 & 4 \\ 1 & -2 \end{vmatrix} = 3 \cdot (-2) - 1 \cdot 4 = -10$$

15. $\Phi(r,t) = (r \sin t, r - \cos t)$, $(r,t) = (1, \pi)$

SOLUTION We have $x = r \sin t$ and $y = r - \cos t$. Therefore,

$$\text{Jac}(\Phi) = \dfrac{\partial(x,y)}{\partial(r,t)} = \begin{vmatrix} \dfrac{\partial x}{\partial r} & \dfrac{\partial x}{\partial t} \\ \dfrac{\partial y}{\partial r} & \dfrac{\partial y}{\partial t} \end{vmatrix} = \begin{vmatrix} \sin t & r \cos t \\ 1 & \sin t \end{vmatrix} = \sin^2 t - r \cos t$$

At the point $(r,t) = (1, \pi)$ we get

$$\text{Jac}(\Phi)(1, \pi) = \sin^2 \pi - 1 \cdot \cos \pi = 0 - 1 \cdot (-1) = 1$$

17. $\Phi(r, \theta) = (r \cos \theta, r \sin \theta)$, $(r, \theta) = \left(4, \dfrac{\pi}{6}\right)$

SOLUTION Since $x = r \cos \theta$ and $y = r \sin \theta$, the Jacobian of Φ is the following determinant:

$$\text{Jac}(\Phi) = \dfrac{\partial(x,y)}{\partial(r,\theta)} = \begin{vmatrix} \dfrac{\partial x}{\partial r} & \dfrac{\partial x}{\partial \theta} \\ \dfrac{\partial y}{\partial r} & \dfrac{\partial y}{\partial \theta} \end{vmatrix} = \begin{vmatrix} \cos \theta & -r \sin \theta \\ \sin \theta & r \cos \theta \end{vmatrix} = r \cos^2 \theta + r \sin^2 \theta = r(\cos^2 \theta + \sin^2 \theta) = r \cdot 1 = r.$$

At the point $(r, \theta) = \left(4, \dfrac{\pi}{6}\right)$ we get

$$\text{Jac}(\Phi)\left(4, \dfrac{\pi}{6}\right) = 4$$

19. Find a linear mapping Φ that maps $[0, 1] \times [0, 1]$ to the parallelogram in the xy-plane spanned by the vectors $\langle 2, 3 \rangle$ and $\langle 4, 1 \rangle$.

SOLUTION

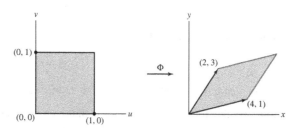

We denote the linear map by

$$\Phi(u,v) = (Au + Cv, Bu + Dv) \tag{1}$$

The image of the unit square $\mathcal{R} = [0, 1] \times [0, 1]$ under the linear map is the parallelogram whose vertices are the images of the vertices of $\mathcal{R}$. Two of vertices of the given parallelogram are $(2, 3)$ and $(4, 1)$. To find

A, B, C, and D it suffices to determine four equations. Therefore, we ask that (notice that for linear maps $\Phi(0,0) = (0,0)$)

$$\Phi(0, 1) = (2, 3), \quad \Phi(1, 0) = (4, 1)$$

We substitute in (1) and solve for A, B, C, and D:

$$(A \cdot 0 + C \cdot 1, B \cdot 0 + D \cdot 1) = (C, D) = (2, 3) \quad\quad C = 2, \quad D = 3$$
$$(A \cdot 1 + C \cdot 0, B \cdot 1 + D \cdot 0) = (A, B) = (4, 1) \quad\Rightarrow\quad A = 4, \quad B = 1$$

Substituting in (1) we obtain the following map:

$$\Phi(u, v) = (4u + 2v, u + 3v)$$

21. Let $\mathcal{D}$ be the parallelogram in Figure 13. Apply the Change of Variables Formula to the map $\Phi(u, v) = (5u + 3v, u + 4v)$ to evaluate $\iint_{\mathcal{D}} xy\,dx\,dy$ as an integral over $\mathcal{D}_0 = [0, 1] \times [0, 1]$.

FIGURE 13

SOLUTION

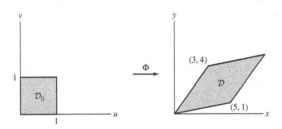

We express $f(x, y) = xy$ in terms of u and v. Since $x = 5u + 3v$ and $y = u + 4v$, we have

$$f(x, y) = xy = (5u + 3v)(u + 4v) = 5u^2 + 12v^2 + 23uv$$

The Jacobian of the linear map $\Phi(u, v) = (5u + 3v, u + 4v)$ is

$$\operatorname{Jac}(\Phi) = \frac{\partial(x, y)}{\partial(u, v)} = \begin{vmatrix} 5 & 3 \\ 1 & 4 \end{vmatrix} = 20 - 3 = 17$$

Applying the Change of Variables Formula we get

$$\iint_{\mathcal{D}} xy\,dA = \iint_{\mathcal{D}_0} f(x, y) \left|\frac{\partial(x, y)}{\partial(u, v)}\right| du\,dv = \int_0^1 \int_0^1 (5u^2 + 12v^2 + 23uv) \cdot 17\,du\,dv$$

$$= 17 \int_0^1 \left.\frac{5u^3}{3} + 12v^2 u + \frac{23u^2 v}{2}\right|_{u=0}^1 dv = 17 \int_0^1 \left(\frac{5}{3} + 12v^2 + \frac{23v}{2}\right) dv$$

$$= 17 \left.\left(\frac{5v}{3} + 4v^3 + \frac{23v^2}{4}\right)\right|_0^1 = 17\left(\frac{5}{3} + 4 + \frac{23}{4}\right) = \frac{2329}{12} \approx 194.08$$

23. Let $\Phi(u, v) = (3u + v, u - 2v)$. Use the Jacobian to determine the area of $\Phi(\mathcal{R})$ for:
(a) $\mathcal{R} = [0, 3] \times [0, 5]$
(b) $\mathcal{R} = [2, 5] \times [1, 7]$

SOLUTION The Jacobian of the linear map $\Phi(u, v) = (3u + v, u - 2v)$ is the following determinant:

$$\text{Jac}\,\Phi = \frac{\partial(x, y)}{\partial(u, v)} = \begin{vmatrix} 3 & 1 \\ 1 & -2 \end{vmatrix} = -6 - 1 = -7$$

By properties of linear maps, we have

$$\text{Area}\,(\Phi(\mathcal{R})) = |\text{Jac}\,\Phi|\text{Area}(\mathcal{R}) = 7 \cdot \text{Area}(\mathcal{R})$$

(a) The area of the rectangle $R = [0, 3] \times [0, 5]$ is $3 \cdot 5 = 15$, therefore the area of $\Phi(\mathcal{R})$ is

$$\text{Area}\,(\Phi(\mathcal{R})) = 7 \cdot 15 = 105$$

(b) The area of the rectangle $\mathcal{R} = [2, 5] \times [1, 7]$ is $3 \cdot 6 = 18$, hence the area of $\Phi(\mathcal{R})$ is

$$\text{Area}\,(\Phi(\mathcal{R})) = 7 \cdot 18 = 126$$

25. With Φ as in Example 3, use the Change of Variables Formula to compute the area of the image of $[1, 4] \times [1, 4]$.

SOLUTION Let $\mathcal{R}$ represent the rectangle $[1, 4] \times [1, 4]$. We proceed as follows. $\text{Jac}(\Phi)$ is easily calculated as

$$\text{Jac}(T) = \frac{\partial(x, y)}{\partial(u, v)} = \begin{vmatrix} 1/v & -u/v^2 \\ v & u \end{vmatrix} = 2u/v$$

Now, the area is given by the Change of Variables Formula as

$$\iint_{\Phi(\mathcal{R})} 1 \, dA = \iint_{\mathcal{R}} 1 |\text{Jac}(\Phi)| \, du \, dv = \iint_{\mathcal{R}} 1 |2u/v| \, du \, dv = \int_1^4 \int_1^4 2u/v \, du \, dv$$

$$= \int_1^4 2u \, du \cdot \int_1^4 \frac{1}{v} \, dv = (16 - 1)(\ln 4 - \ln 1) = 15 \ln 4$$

In Exercises 26–28, let $\mathcal{R}_0 = [0, 1] \times [0, 1]$ be the unit square. The translate of a map $\Phi_0(u, v) = (\phi(u, v), \psi(u, v))$ is a map

$$\Phi(u, v) = (a + \phi(u, v), b + \psi(u, v))$$

where a, b are constants. Observe that the map Φ_0 in Figure 15 maps $\mathcal{R}_0$ to the parallelogram $\mathcal{P}_0$ and that the translate

$$\Phi_1(u, v) = (2 + 4u + 2v, 1 + u + 3v)$$

maps $\mathcal{R}_0$ to $\mathcal{P}_1$.

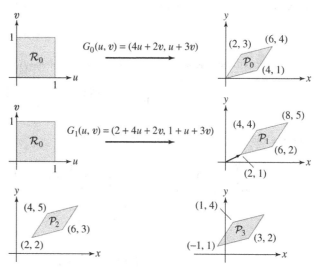

FIGURE 15

27. Sketch the parallelogram $\mathcal{P}$ with vertices $(1, 1), (2, 4), (3, 6), (4, 9)$ and find the translate of a linear mapping that maps $\mathcal{R}_0$ to $\mathcal{P}$.

SOLUTION The parallelogram $\mathcal{P}$ is shown in the figure:

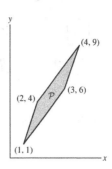

We first translate the parallelogram $\mathcal{P}$ one unit to the left and one unit downward to obtain a parallelogram $\mathcal{P}_0$ with a vertex at the origin.

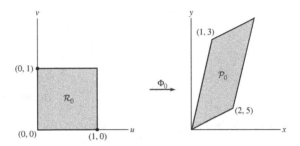

We find a linear map $\Phi_0(u, v) = (Au + Cv, Bu + Dv)$ that maps $\mathcal{R}_0$ to $\mathcal{P}_0$:

$$\Phi_0(0, 1) = (1, 3) \quad \Rightarrow \quad (C, D) = (1, 3) \quad \Rightarrow \quad C = 1, \quad D = 3$$
$$\Phi_0(1, 0) = (2, 5) \quad \Rightarrow \quad (A, B) = (2, 5) \quad \Rightarrow \quad A = 2, \quad B = 5$$

Therefore,

$$\Phi_0(u, v) = (2u + v, 5u + 3v)$$

Now we can determine the translate Φ of Φ_0 that maps $\mathcal{R}_0$ to $\mathcal{P}$. Since $\mathcal{P}$ is obtained by translating $\mathcal{P}_0$ one unit upward and one unit to the right, the map Φ is the following translate of Φ_0:

$$\Phi(u, v) = (1 + 2u + v, 1 + 5u + 3v)$$

29. Let $\mathcal{D} = \Phi(\mathcal{R})$, where $\Phi(u, v) = (u^2, u + v)$ and $\mathcal{R} = [1, 2] \times [0, 6]$. Calculate $\iint_{\mathcal{D}} y\, dx\, dy$. *Note:* It is not necessary to describe $\mathcal{D}$.

SOLUTION

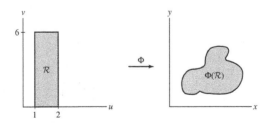

Changing variables, we have

$$\iint_{\mathcal{D}} y\, dA = \iint_{\mathcal{R}} (u + v) \left| \frac{\partial(x, y)}{\partial(u, v)} \right| du\, dv \qquad (1)$$

We compute the Jacobian of Φ. Since $x = u^2$ and $y = u + v$, we have

$$\frac{\partial(x, y)}{\partial(u, v)} = \begin{vmatrix} \dfrac{\partial x}{\partial u} & \dfrac{\partial x}{\partial v} \\ \dfrac{\partial y}{\partial u} & \dfrac{\partial y}{\partial v} \end{vmatrix} = \begin{vmatrix} 2u & 0 \\ 1 & 1 \end{vmatrix} = 2u$$

We substitute in (1) and compute the resulting integral:

$$\iint_{\mathcal{D}} y\,dA = \int_0^6 \int_1^2 (u+v)\cdot 2u\,du\,dv = \int_0^6 \int_1^2 (2u^2 + 2uv)\,du\,dv = \int_0^6 \frac{2u^3}{3} + u^2 v \Big|_{u=1}^2 dv$$

$$= \int_0^6 \left(\left(\frac{16}{3} + 4v\right) - \left(\frac{2}{3} + v\right)\right) dv = \int_0^6 \left(3v + \frac{14}{3}\right) dv = \frac{3}{2}v^2 + \frac{14}{3}v \Big|_0^6 = 82$$

31. Compute $\iint_{\mathcal{D}} (x+3y)\,dx\,dy$, where $\mathcal{D}$ is the shaded region in Figure 16. *Hint:* Use the map $\Phi(u,v) = (u-2v, v)$.

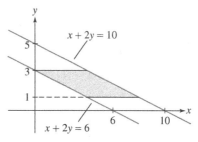

FIGURE 16

SOLUTION The boundary of $\mathcal{D}$ is defined by the lines $x+2y=6$, $x+2y=10$, $y=1$, and $y=3$.

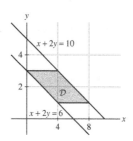

Therefore, $\mathcal{D}$ is mapped to a rectangle $\mathcal{D}_0$ in the uv-plane under the map

$$u = x+2y, \quad v = y \tag{1}$$

or

$$(u,v) = \Phi^{-1}(x,y) = (x+2y, y)$$

Since $\mathcal{D}$ is defined by the inequalities $6 \le x+2y \le 10$ and $1 \le y \le 3$, the corresponding domain in the uv-plane is the rectangle

$$\mathcal{D}_0: 6 \le u \le 10,\ 1 \le v \le 3$$

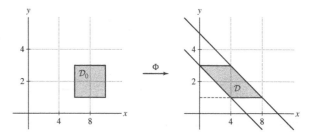

To find $\Phi(u,v)$ we must solve the equations for x and y in terms of u and v. We obtain

$$\begin{matrix} u = x+2y \\ v = y \end{matrix} \Rightarrow \begin{matrix} x = u-2v \\ y = v \end{matrix} \Rightarrow \Phi(u,v) = (u-2v, v)$$

We compute the Jacobian of the linear mapping Φ:

$$\text{Jac}(\Phi) = \frac{\partial(x,y)}{\partial(u,v)} = \begin{vmatrix} 1 & -2 \\ 0 & 1 \end{vmatrix} = 1\cdot 1 + 2\cdot 0 = 1$$

The function $f(x, y) = x + 3y$ expressed in terms of the new variables u and v is

$$f(x, y) = u - 2v + 3v = u + v$$

We now use the Change of Variables Formula to compute the required integral. We get

$$\iint_D f(x,y)\,dx\,dy = \iint_{D_0} (u+v)\left|\frac{\partial(x,y)}{\partial(u,v)}\right|\,du\,dv = \int_1^3 \int_6^{10} (u+v) \cdot 1\,du\,dv$$

$$= \int_1^3 \frac{u^2}{2} + vu\bigg|_{u=6}^{10}\,dv = \int_1^3 ((50 + 10v) - (18 + 6v))\,dv$$

$$= \int_1^3 (32 + 4v)\,dv = 32v + 2v^2\bigg|_1^3 = (96 + 18) - (32 + 2) = 80$$

33. Show that $T(u, v) = (u^2 - v^2, 2uv)$ maps the triangle $D_0 = \{(u, v) : 0 \le v \le u \le 1\}$ to the domain D bounded by $x = 0$, $y = 0$, and $y^2 = 4 - 4x$. Use T to evaluate

$$\iint_D \sqrt{x^2 + y^2}\,dx\,dy$$

SOLUTION We show that the boundary of D_0 is mapped to the boundary of D.

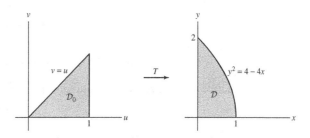

We have

$$x = u^2 - v^2 \quad \text{and} \quad y = 2uv$$

The line $v = u$ is mapped to the following set:

$$(x, y) = (u^2 - u^2, 2u^2) = (0, 2u^2) \quad \Rightarrow \quad x = 0, \quad y \ge 0$$

That is, the image of the line $u = v$ is the positive y-axis. The line $v = 0$ is mapped to the following set:

$$(x, y) = (u^2, 0) \quad \Rightarrow \quad x = u^2, \quad y = 0 \quad \Rightarrow \quad y = 0, \quad x \ge 0$$

Thus, the line $v = 0$ is mapped to the positive x-axis. We now show that the vertical line $u = 1$ is mapped to the curve $y^2 + 4x = 4$. The image of the line $u = 1$ is the following set:

$$(x, y) = (1 - v^2, 2v) \quad \Rightarrow \quad x = 1 - v^2, \quad y = 2v$$

We substitute $v = \frac{y}{2}$ in the equation $x = 1 - v^2$ to obtain

$$x = 1 - \left(\frac{y}{2}\right)^2 = 1 - \frac{y^2}{4} \quad \Rightarrow \quad 4x = 4 - y^2 \quad \Rightarrow \quad y^2 + 4x = 4$$

Since the boundary of D_0 is mapped to the boundary of D, we conclude that the domain D_0 is mapped by T to the domain D in the xy-plane. We now compute the integral $\iint_D \sqrt{x^2 + y^2}\,dx\,dy$. We express the function $f(x, y) = \sqrt{x^2 + y^2}$ in terms of the new variables u and v:

$$f(x, y) = \sqrt{(u^2 - v^2)^2 + (2uv)^2} = \sqrt{u^4 - 2u^2v^2 + v^4 + 4u^2v^2}$$

$$= \sqrt{u^4 + 2u^2v^2 + v^4} = \sqrt{(u^2 + v^2)^2} = u^2 + v^2$$

We compute the Jacobian of T:

$$\text{Jac}(T) = \frac{\partial(x,y)}{\partial(u,v)} = \begin{vmatrix} \frac{\partial x}{\partial u} & \frac{\partial x}{\partial v} \\ \frac{\partial y}{\partial u} & \frac{\partial y}{\partial v} \end{vmatrix} = \begin{vmatrix} 2u & -2v \\ 2v & 2u \end{vmatrix} = 4u^2 + 4v^2 = 4(u^2 + v^2)$$

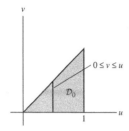

Using the Change of Variables Formula gives

$$\iint_D \sqrt{x^2 + y^2}\, dx\, dy = \iint_{D_0} (u^2 + v^2) \cdot 4(u^2 + v^2)\, du\, dv = 4 \int_0^1 \int_0^u (u^4 + 2u^2v^2 + v^4)\, dv\, du$$

$$= 4 \int_0^1 \left. u^4 v + \frac{2}{3} u^2 v^3 + \frac{v^5}{5} \right|_{v=0}^u du = 4 \int_0^1 \left(u^5 + \frac{2}{3} u^5 + \frac{u^5}{5} \right) du$$

$$= 4 \int_0^1 \frac{28}{15} u^5\, du = \frac{112}{15} \cdot \left. \frac{u^6}{6} \right|_0^1 = \frac{56}{45}$$

35. Calculate $\iint_D e^{9x^2 + 4y^2}\, dx\, dy$, where D is the interior of the ellipse $\left(\frac{x}{2}\right)^2 + \left(\frac{y}{3}\right)^2 \leq 1$.

SOLUTION We define a map that maps the unit disk $u^2 + v^2 \leq 1$ onto the interior of the ellipse. That is,

$$x = 2u, \quad y = 3v \quad \Rightarrow \quad \Phi(u,v) = (2u, 3v)$$

Since $\left(\frac{x}{2}\right)^2 + \left(\frac{y}{3}\right)^2 \leq 1$ if and only if $u^2 + v^2 \leq 1$, Φ is the map we need.

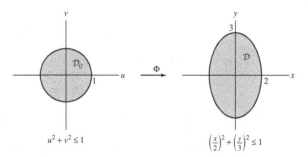

We express the function $f(x,y) = e^{9x^2 + 4y^2}$ in terms of u and v:

$$f(x,y) = e^{9(2u)^2 + 4(3v)^2} = e^{36u^2 + 36v^2} = e^{36(u^2 + v^2)}$$

We compute the Jacobian of Φ:

$$\text{Jac}(\Phi) = \begin{vmatrix} \frac{\partial x}{\partial u} & \frac{\partial x}{\partial v} \\ \frac{\partial y}{\partial u} & \frac{\partial y}{\partial v} \end{vmatrix} = \begin{vmatrix} 2 & 0 \\ 0 & 3 \end{vmatrix} = 6$$

Using the Change of Variables Formula gives

$$\iint_D e^{9x^2 + 4y^2}\, dA = \iint_{D_0} e^{36(u^2 + v^2)} \cdot 6\, du\, dv$$

We compute the integral using polar coordinates $u = r \cos\theta$, $v = r \sin\theta$:

$$\iint_D e^{9x^2 + 4y^2}\, dA = \int_0^{2\pi} \int_0^1 6 e^{36 r^2} \cdot r\, dr\, d\theta = \left(6 \int_0^{2\pi} d\theta \right) \left(\int_0^1 e^{36 r^2} r\, dr \right)$$

$$= 12\pi \left. \frac{e^{36 r^2}}{72} \right|_{r=0}^1 = \frac{12\pi (e^{36} - 1)}{72} = \frac{\pi(e^{36} - 1)}{6}$$

37. Sketch the domain $\mathcal{D}$ bounded by $y = x^2$, $y = \frac{1}{2}x^2$, and $y = x$. Use a change of variables with the map $x = uv$, $y = u^2$ to calculate

$$\iint_{\mathcal{D}} y^{-1}\, dx\, dy$$

This is an improper integral since $f(x, y) = y^{-1}$ is undefined at $(0, 0)$, but it becomes proper after changing variables.

SOLUTION The domain $\mathcal{D}$ is shown in the figure.

We must identify the domain $\mathcal{D}_0$ in the uv-plane. Notice that Φ is one-to-one, where $u \geq 0$ (or $u \leq 0$), since in $\mathcal{D}$, $x \geq 0$, so it also follows by $x = uv$ that $v \geq 0$. Therefore, we search the domain $\mathcal{D}_0$ in the first quadrant of the uv-plane. To do this, we examine the curves that are mapped to the curves defining the boundary of $\mathcal{D}$. We examine each curve separately.

$y = x^2$: Since $x = uv$ and $y = u^2$ we get

$$u^2 = (uv)^2 \quad \Rightarrow \quad 1 = v^2 \quad \Rightarrow \quad v = 1$$

$y = \frac{1}{2}x^2$:

$$u^2 = \frac{1}{2}(uv)^2 \quad \Rightarrow \quad 1 = \frac{1}{2}v^2 \quad \Rightarrow \quad v^2 = 2 \quad \Rightarrow \quad v = \sqrt{2}$$

$y = x$: $u^2 = uv \quad \Rightarrow \quad v = u$. The region $\mathcal{D}_0$ is the region in the first quadrant of the uv-plane enclosed by the curves $v = 1$, $v = \sqrt{2}$, and $v = u$.

We now use change of variables to compute the integral $\iint_{\mathcal{D}} y^{-1}\, dx\, dy$. The function in terms of the new variables is $f(x, y) = u^{-2}$. We compute the Jacobian of $\Phi(u, v) = (x, y) = (uv, u^2)$:

$$\text{Jac}(\Phi) = \begin{vmatrix} \dfrac{\partial x}{\partial u} & \dfrac{\partial x}{\partial v} \\ \dfrac{\partial y}{\partial u} & \dfrac{\partial y}{\partial v} \end{vmatrix} = \begin{vmatrix} v & u \\ 2u & 0 \end{vmatrix} = -2u^2$$

Using the Change of Variables Formula gives

$$\iint_{\mathcal{D}} y^{-1}\, dx\, dy = \iint_{\mathcal{D}_0} u^{-2} \cdot 2u^2\, du\, dv = \int_1^{\sqrt{2}} \int_0^v 2\, du\, dv$$

$$= \int_1^{\sqrt{2}} 2u \bigg|_{u=0}^{v} dv = \int_1^{\sqrt{2}} 2v\, dv = v^2 \bigg|_1^{\sqrt{2}} = 2 - 1 = 1$$

39. Let Φ be the inverse of the map $F(x, y) = (xy, x^2 y)$ from the xy-plane to the uv-plane. Let $\mathcal{D}$ be the domain in Figure 18. Show, by applying the Change of Variables Formula to the inverse $\Phi = F^{-1}$, that

$$\iint_{\mathcal{D}} e^{xy} \, dx \, dy = \int_{10}^{20} \int_{20}^{40} e^u v^{-1} \, dv \, du$$

and evaluate this result. *Hint:* See Example 8.

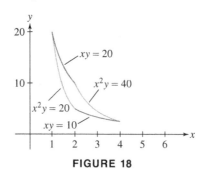

FIGURE 18

SOLUTION The domain $\mathcal{D}$ is defined by the inequalities

$$\mathcal{D}: 10 \le xy \le 20, \ 20 \le x^2 y \le 40$$

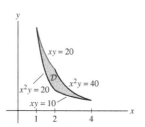

Since $u = xy$ and $v = x^2 y$, the image $\mathcal{D}_0$ of $\mathcal{D}$ (in the uv-plane) under F is the rectangle

$$\mathcal{D}_0: 10 \le u \le 20, \ 20 \le v \le 40$$

The function expressed in the new variables is

$$f(x, y) = e^{xy} = e^u$$

To find the Jacobian of the inverse Φ of F, we use the formula for the Jacobian of the inverse mapping. That is,

$$\frac{\partial(x, y)}{\partial(u, v)} = \left(\frac{\partial(u, v)}{\partial(x, y)} \right)^{-1}$$

We find the Jacobian of F. Since $u = xy$ and $v = x^2 y$, we have

$$\mathrm{Jac}(F) = \begin{vmatrix} \dfrac{\partial u}{\partial x} & \dfrac{\partial u}{\partial y} \\ \dfrac{\partial v}{\partial x} & \dfrac{\partial v}{\partial y} \end{vmatrix} = \begin{vmatrix} y & x \\ 2xy & x^2 \end{vmatrix} = yx^2 - 2x^2 y = -x^2 y$$

Hence,

$$\mathrm{Jac}(\Phi) = -\frac{1}{x^2 y}$$

We now compute the double integral $\iint_{\mathcal{D}} e^{xy}\, dA$ using the Change of Variables Formula. Since $y > 0$ in $\mathcal{D}$, we have $|\text{Jac}(\Phi)| = \left|-\frac{1}{x^2 y}\right| = \frac{1}{x^2 y} = v^{-1}$. Therefore,

$$\iint_{\mathcal{D}} e^{xy}\, dA = \iint_{\mathcal{D}_0} e^u v^{-1}\, dv\, du = \int_{10}^{20}\int_{20}^{40} e^u v^{-1}\, dv\, du = \left(\int_{10}^{20} e^u\, du\right)\left(\int_{20}^{40} v^{-1}\, dv\right)$$

$$= e^u\Big|_{10}^{20} \cdot \ln v\Big|_{20}^{40} = (e^{20} - e^{10})(\ln(40) - \ln(20)) = (e^{20} - e^{10})\ln 2$$

41. Let $I = \iint_{\mathcal{D}} (x^2 - y^2)\, dx\, dy$, where

$$\mathcal{D} = \{(x, y) : 2 \leq xy \leq 4,\, 0 \leq x - y \leq 3,\, x \geq 0, y \geq 0\}$$

(a) Show that the mapping $u = xy$, $v = x - y$ maps $\mathcal{D}$ to the rectangle $\mathcal{R} = [2, 4] \times [0, 3]$.

(b) Compute $\partial(x, y)/\partial(u, v)$ by first computing $\partial(u, v)/\partial(x, y)$.

(c) Use the Change of Variables Formula to show that I is equal to the integral of $f(u, v) = v$ over $\mathcal{R}$ and evaluate.

SOLUTION

(a) The domain $\mathcal{D}$ is defined by the inequalities

$$\mathcal{D}: 2 \leq xy \leq 4,\ 0 \leq x - y \leq 3,\ x \geq 0,\ y \geq 0$$

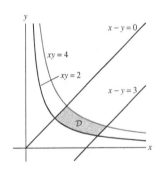

Since $u = xy$ and $v = x - y$, the image of $\mathcal{D}$ under this mapping is the rectangle defined by

$$\mathcal{D}_0: 2 \leq u \leq 4,\ 0 \leq v \leq 3$$

That is, $\mathcal{D}_0 = [2, 4] \times [0, 3]$.

(b) We compute the Jacobian $\frac{\partial(u,v)}{\partial(x,y)}$ and then use the formula for the Jacobian of the inverse mapping to compute $\frac{\partial(x,y)}{\partial(u,v)}$. Since $u = xy$ and $v = x - y$, we have

$$\frac{\partial(u, v)}{\partial(x, y)} = \begin{vmatrix} \dfrac{\partial u}{\partial x} & \dfrac{\partial u}{\partial y} \\ \dfrac{\partial v}{\partial x} & \dfrac{\partial v}{\partial y} \end{vmatrix} = \begin{vmatrix} y & x \\ 1 & -1 \end{vmatrix} = -y - x = -(x + y)$$

Therefore,

$$\frac{\partial(x, y)}{\partial(u, v)} = \left(\frac{\partial(u, v)}{\partial(x, y)}\right)^{-1} = -\frac{1}{x + y}$$

(c) In $\mathcal{D}$, $x \geq 0$ and $y \geq 0$, hence $\left|\frac{\partial(x,y)}{\partial(u,v)}\right| = \frac{1}{x+y}$. Using the change of variable formula gives:

$$I = \iint_{\mathcal{D}_0} (x^2 - y^2) \cdot \frac{1}{x + y}\, du\, dv = \iint_{\mathcal{D}_0} (x - y)\, du\, dv = \int_0^3\int_2^4 v\, du\, dv$$

$$= \left(\int_0^3 v\, dv\right)\left(\int_2^4 du\right) = \left(\frac{v^2}{2}\Big|_0^3\right)\left(u\Big|_2^4\right) = \frac{9}{2} \cdot 2 = 9$$

43. Derive formula (8) in Section 15.4 for integration in spherical coordinates from the general Change of Variables Formula.

SOLUTION The spherical coordinates are

$$x = \rho \cos\theta \sin\phi, \quad y = \rho \sin\theta \sin\phi, \quad z = \rho \cos\phi$$

Suppose that a region $\mathcal{W}$ in the (x, y, z)-plane is the image of a region $\mathcal{W}_0$ in the (θ, ϕ, ρ)-space defined by:

$$\mathcal{W}_0: \theta_1 \le \theta \le \theta_2, \quad \phi_1 \le \phi \le \phi_2, \quad \rho_1(\theta, \phi) \le \rho \le \rho_2(\theta, \phi) \tag{1}$$

Then, by the Change of Variables Formula, we have

$$\iiint_{\mathcal{W}} f(x, y, z) = \iiint_{\mathcal{W}_0} f(\rho \cos\theta \sin\phi, \rho \sin\theta \sin\phi, \rho \cos\phi) = \left|\frac{\partial(x, y, z)}{\partial(\theta, \phi, \rho)}\right| d\rho\, d\phi\, d\theta \tag{2}$$

We compute the Jacobian:

$$\frac{\partial(x, y, z)}{\partial(\theta, \phi, \rho)} = \begin{vmatrix} \frac{\partial x}{\partial \theta} & \frac{\partial x}{\partial \phi} & \frac{\partial x}{\partial \rho} \\ \frac{\partial y}{\partial \theta} & \frac{\partial y}{\partial \phi} & \frac{\partial y}{\partial \rho} \\ \frac{\partial z}{\partial \theta} & \frac{\partial z}{\partial \phi} & \frac{\partial z}{\partial \rho} \end{vmatrix} = \begin{vmatrix} -\rho \sin\theta \sin\phi & \rho \cos\theta \cos\phi & \cos\theta \sin\phi \\ \rho \cos\theta \sin\phi & \rho \sin\theta \cos\phi & \sin\theta \sin\phi \\ 0 & -\rho \sin\phi & \cos\phi \end{vmatrix}$$

$$= -\rho \sin\theta \sin\phi \begin{vmatrix} \rho \sin\theta \cos\phi & \sin\theta \sin\phi \\ -\rho \sin\phi & \cos\phi \end{vmatrix} - \rho \cos\theta \cos\phi \begin{vmatrix} \rho \cos\theta \sin\phi & \sin\theta \sin\phi \\ 0 & \cos\phi \end{vmatrix}$$

$$+ \cos\theta \sin\phi \begin{vmatrix} \rho \cos\theta \sin\phi & \rho \sin\theta \cos\phi \\ 0 & -\rho \sin\phi \end{vmatrix}$$

$$= -\rho \sin\theta \sin\phi(\rho \sin\theta \cos^2\phi + \rho \sin\theta \sin^2\phi) - \rho \cos\theta \cos\phi(\rho \cos\theta \cos\phi \sin\phi - 0)$$

$$+ \cos\theta \sin\phi(-\rho^2 \cos\theta \sin^2\phi - 0)$$

$$= -\rho^2 \sin^2\theta \sin\phi(\cos^2\phi + \sin^2\phi) - \rho^2 \cos^2\theta \cos^2\phi \sin\phi - \rho^2 \cos^2\theta \sin^3\phi$$

$$= -\rho^2 \sin^2\theta \sin\phi - \rho^2 \cos^2\theta \cos^2\phi \sin\phi - \rho^2 \cos^2\theta \sin^3\phi$$

$$= -\rho^2 \sin\phi(\sin^2\theta + \cos^2\theta \cos^2\phi + \cos^2\theta \sin^2\phi)$$

$$= -\rho^2 \sin\phi \left(\sin^2\theta + \cos^2\theta(\cos^2\phi + \sin^2\phi)\right)$$

$$= -\rho^2 \sin\phi(\sin^2\theta + \cos^2\theta) = -\rho^2 \sin\phi$$

Since $0 \le \phi \le \pi$, we have $\sin\phi \ge 0$. Therefore,

$$\left|\frac{\partial(x, y, z)}{\partial(\theta, \phi, \rho)}\right| = \rho^2 \sin\phi \tag{3}$$

Combining (1), (2), and (3) gives

$$\iiint_{\mathcal{W}} f(x, y, z)\, dv = \int_{\theta_1}^{\theta_2} \int_{\phi_1}^{\phi_2} \int_{\rho_1(\theta,\phi)}^{\rho_2(\theta,\phi)} f(\rho \cos\theta \sin\phi, \rho \sin\theta \sin\phi, \rho \cos\phi) \rho^2 \sin\phi\, d\rho\, d\phi\, d\theta$$

Further Insights and Challenges

45. Use the map

$$x = \frac{\sin u}{\cos v}, \quad y = \frac{\sin v}{\cos u}$$

to evaluate the integral

$$\int_0^1 \int_0^1 \frac{dx\, dy}{1 - x^2 y^2}$$

This integral is an improper integral since the integrand is infinite if $x = \pm 1$ and $y = \pm 1$, but applying the Change of Variables Formula shows that the result is finite.

SOLUTION We express the function $f(x, y) = \frac{1}{1-x^2y^2}$ in terms of the new variables u and v:

$$1 - x^2y^2 = 1 - \frac{\sin^2 u}{\cos^2 v} \cdot \frac{\sin^2 v}{\cos^2 u} = 1 - \left(\frac{\sin u}{\cos u}\right)^2 \cdot \left(\frac{\sin v}{\cos v}\right)^2 = 1 - \tan^2 u \tan^2 v$$

Hence,

$$f(x, y) = \frac{1}{1 - \tan^2 u \tan^2 v}$$

We compute the Jacobian of the mapping:

$$\frac{\partial(x, y)}{\partial(u, v)} = \begin{vmatrix} \frac{\partial x}{\partial u} & \frac{\partial x}{\partial v} \\ \frac{\partial y}{\partial u} & \frac{\partial y}{\partial v} \end{vmatrix} = \begin{vmatrix} \frac{\cos u}{\cos v} & \frac{\sin u \sin v}{\cos^2 v} \\ \frac{\sin v \sin u}{\cos^2 u} & \frac{\cos v}{\cos u} \end{vmatrix} = \frac{\cos u}{\cos v} \cdot \frac{\cos v}{\cos u} - \frac{\sin u \sin v}{\cos^2 v} \cdot \frac{\sin v \sin u}{\cos^2 u}$$

$$= 1 - \frac{\sin^2 u}{\cos^2 u} \cdot \frac{\sin^2 v}{\cos^2 v} = 1 - \tan^2 u \tan^2 v$$

Now, since $0 \le x \le 1$ and $0 \le y \le 1$, we have $0 \le \frac{\sin u}{\cos v} \cdot \frac{\sin v}{\cos u} \le 1$ or $0 \le \tan u \tan v \le 1$. Therefore, $0 \le \tan^2 u \tan^2 v \le 1$, hence

$$\left|\frac{\partial(x, y)}{\partial(u, v)}\right| = 1 - \tan^2 u \tan^2 v$$

We now identify a domain $\mathcal{D}_0$ in the uv-plane that is mapped by Φ onto $\mathcal{D}$ and Φ is one-to-one on $\mathcal{D}_0$.

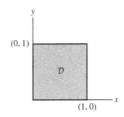

We examine each segment on the boundary of $\mathcal{D}$ separately.

$y = 0$:

$$\frac{\sin v}{\cos u} = 0 \quad \Rightarrow \quad \sin v = 0 \quad \Rightarrow \quad v = \pi k$$

$x = 0$:

$$\frac{\sin u}{\cos v} = 0 \quad \Rightarrow \quad \sin u = 0 \quad \Rightarrow \quad u = \pi k$$

$y = 1$:

$$\frac{\sin v}{\cos u} = 1 \quad \Rightarrow \quad \sin v = \cos u \quad \Rightarrow \quad v + u = \frac{\pi}{2} + 2\pi k \quad \text{or} \quad v - u = \frac{\pi}{2} + 2\pi k \quad \quad (1)$$

$x = 1$:

$$\frac{\sin u}{\cos v} = 1 \quad \Rightarrow \quad \sin u = \cos v \quad \Rightarrow \quad v + u = \frac{\pi}{2} + 2\pi k \quad \text{or} \quad u - v = \frac{\pi}{2} + 2\pi k \quad \quad (2)$$

One of the possible regions $\mathcal{D}_0$ is obtained by choosing $k = 0$ in all solutions. We get

$$v = 0, \quad u = 0, \quad \left(v + u = \frac{\pi}{2} \text{ or } v - u = \frac{\pi}{2}\right), \quad \left(u + v = \frac{\pi}{2} \text{ or } u - v = \frac{\pi}{2}\right)$$

The corresponding regions are:

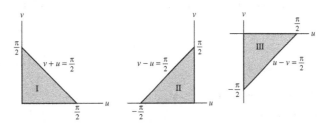

In II, $x = \frac{\sin u}{\cos v} < 0$ and in III $y = \frac{\sin v}{\cos u} < 0$, therefore these regions are not mapped to the unit square in the xy-plane. The appropriate region is I.

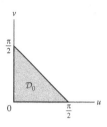

We now use the Change of Variables Formula and the result obtained previously to obtain the following integral:

$$\int_0^1 \int_0^1 \frac{dx\,dy}{1-x^2y^2} = \iint_{\mathcal{D}_0} \frac{1}{1-\tan^2 u \tan^2 v} \cdot (1 - \tan^2 u \tan^2 v)\,du\,dv$$

$$= \iint_{\mathcal{D}_0} 1\,du\,dv = \text{Area}(\mathcal{D}_0) = \frac{\frac{\pi}{2} \cdot \frac{\pi}{2}}{2} = \frac{\pi^2}{8}$$

47. Let P and Q be points in $\mathbf{R}^2$. Show that a linear map $\Phi(u, v) = (Au + Cv, Bu + Dv)$ maps the segment joining P and Q to the segment joining $\Phi(P)$ to $\Phi(Q)$. *Hint:* The segment joining P and Q has parametrization

$$(1-t)\overrightarrow{OP} + t\overrightarrow{OQ} \quad \text{for} \quad 0 \leq t \leq 1$$

SOLUTION First let $P(x_0, y_0)$ and $Q(x_1, y_1)$ so that we see if

$$\mathbf{r}(0) = \overrightarrow{OP} = (x_0, y_0), \quad \mathbf{r}(1) = \overrightarrow{OQ} = (x_1, y_1)$$

Then, using the linear map we see:

$$\Phi(x_0, y_0) = (Ax_0 + Cy_0, Bx_0 + Dy_0) = \Phi(P)$$

and

$$\Phi(x_1, y_1) = (Ax_1 + Cy_1, Bx_1 + Dy_1) = \Phi(Q)$$

Hence this linear map take the endpoints P and Q to the new endpoints $\Phi(P)$ and $\Phi(Q)$. Now to determine the line segment mapping, consider the following:

$$\Phi(\mathbf{r}(t)) = \Phi((1-t)\overrightarrow{OP} + t\overrightarrow{OQ}) = \Phi((1-t)x_0 + tx_1, (1-t)y_0 + ty_1)$$

$$= (A((1-t)x_0 + tx_1) + C((1-t)y_0 + ty_1), B((1-t)x_0 + tx_1 + D((1-t)y_0 + ty_1))$$

$$= ((1-t)Ax_0 + t(Ax_1) + (1-t)Cy_0 + t(Cy_1), (1-t)Bx_0 + t(Bx_1) + (1-t)Dy_0 + t(Dy_1))$$

$$= ((1-t)(Ax_0 + Cy_0) + t(Ax_1 + Cy_1), (1-t)(Bx_0 + Dy_0) + t(Bx_1 + Dy_1))$$

$$= (1-t)(Ax_0 + Cy_0, Bx_0 + Dy_0) + t(Ax_1 + Cy_1, Bx_1 + Dy_1)$$

$$= (1-t)\Phi(P) + t\Phi(Q)$$

This is a parameterization for the line segment joining $\Phi(P)$ and $\Phi(Q)$. Therefore, the linear map maps the line segment joining P and Q to the line segment joining $\Phi(P)$ and $\Phi(Q)$.

49. The product of 2×2 matrices A and B is the matrix AB defined by

$$\underbrace{\begin{pmatrix} a & b \\ c & d \end{pmatrix}}_{A} \underbrace{\begin{pmatrix} a' & b' \\ c' & d' \end{pmatrix}}_{B} = \underbrace{\begin{pmatrix} aa' + bc' & ab' + bd' \\ ca' + dc' & cb' + dd' \end{pmatrix}}_{AB}$$

The (i, j)-entry of A is the **dot product** of the ith row of A and the jth column of B. Prove that $\det(AB) = \det(A)\det(B)$.

SOLUTION The determinants of A and B are

$$\det(A) = \begin{vmatrix} a & b \\ c & d \end{vmatrix} = ad - bc,$$

$$\det(B) = \begin{vmatrix} a' & b' \\ c' & d' \end{vmatrix} = a'd' - b'c' \quad (1)$$

We now compute the determinant of AB:

$$\det(AB) = \begin{vmatrix} aa' + bc' & ab' + bd' \\ ca' + dc' & cb' + dd' \end{vmatrix} = (aa' + bc')(cb' + dd') - (ab' + bd')(ca' + dc')$$

$$= aa'cb' + aa'dd' + bc'cb' + bc'dd' - ab'ca' - ab'dc' - bd'ca' - bd'dc'$$

$$= (aa'dd' - bd'ca') + (bc'cb' - ab'dc') = a'd'(ad - bc) - b'c'(ad - bc)$$

$$= (ad - bc)(a'd' - b'c') \quad (2)$$

We combine (1) and (2) to conclude

$$\det(AB) = \det(A)\det(B)$$

51. Use Exercise 50 to prove that

$$\text{Jac}(\Phi^{-1}) = \text{Jac}(\Phi)^{-1}$$

Hint: Verify that $\text{Jac}(I) = 1$, where I is the identity map $I(u, v) = (u, v)$.

SOLUTION Since $\Phi^{-1}(\Phi(u, v)) = (u, v)$, we have $(\Phi^{-1} \circ \Phi)(u, v) = (u, v)$. Therefore, $\Phi^{-1} \circ \Phi = I$. Using Exercise 50, we have

$$\text{Jac}(I) = \text{Jac}(\Phi^{-1} \circ \Phi) = \text{Jac}(\Phi^{-1})\text{Jac}(\Phi) \quad (1)$$

The Jacobian of the linear map $I(u, v) = (u, v) = (1 \cdot u + 0 \cdot v, 0 \cdot u + 1 \cdot v)$ is

$$\text{Jac}(I) = \begin{vmatrix} 1 & 0 \\ 0 & 1 \end{vmatrix} = 1 \cdot 1 - 0 \cdot 0 = 1$$

Substituting in (1) gives

$$1 = \text{Jac}(\Phi^{-1})\text{Jac}(\Phi)$$

or

$$\text{Jac}(\Phi^{-1}) = (\text{Jac}(\Phi))^{-1}.$$

CHAPTER REVIEW EXERCISES

1. Calculate the Riemann sum $S_{2,3}$ for $\int_1^4 \int_2^6 x^2 y \, dx \, dy$ using two choices of sample points:

(a) Lower-left vertex

(b) Midpoint of rectangle

Then calculate the exact value of the double integral.

SOLUTION

(a) The rectangle $[2, 6] \times [1, 3]$ is divided into 2×3 subrectangles. The lower-left vertices of the subrectangles are

$$P_{11} = (2, 1) \quad P_{21} = (2, 2) \quad P_{31} = (2, 3)$$
$$P_{12} = (3, 1) \quad P_{22} = (3, 2) \quad P_{32} = (3, 3)$$

Also $\Delta x = \frac{6-2}{2} = 2$, $\Delta y = \frac{4-1}{3} = 1$, hence $\Delta A = 2 \cdot 1 = 2$.

The Riemann sum $S_{3,4}$ is the following sum:

$$S_{2,3} = 2\left(2^2 \cdot 1 + 2^2 \cdot 2 + 2^2 \cdot 3 + 3^2 \cdot 1 + 3^2 \cdot 2 + 3^2 \cdot 3\right)$$

$$= 2(4 + 8 + 12 + 9 + 18 + 27) = 156$$

(b) The midpoints of the subrectangles are

$$P_{11} = (3, 3/2) \quad P_{21} = (3, 5/2) \quad P_{31} = (3, 7/2)$$
$$P_{12} = (5, 3/2) \quad P_{22} = (5, 5/2) \quad P_{32} = (5, 7/2)$$

Also $\Delta x = 2$, $\Delta y = 1$, hence $\Delta A = 2 \cdot 1 = 2$. The Riemann sum $S_{2,3}$ is

$$S_{2,3} = 2\left(3^2 \cdot \frac{3}{2} + 3^2 \cdot \frac{5}{2} + 3^2 \cdot \frac{7}{2} + 5^2 \cdot \frac{3}{2} + 5^2 \cdot \frac{5}{2} + 5^2 \cdot \frac{7}{2}\right)$$

$$= 2\left(\frac{27}{2} + \frac{45}{2} + \frac{63}{2} + \frac{75}{2} + \frac{125}{2} + \frac{175}{2}\right)$$

$$= 510$$

We compute the exact value of the double integral, using an iterated integral of a product function. We get

$$\int_1^4 \int_2^6 x^2 y \, dx \, dy = \left(\int_1^4 y \, dy\right)\left(\int_2^6 x^2 \, dx\right) = \left(\frac{y^2}{2}\bigg|_1^4\right)\left(\frac{x^3}{3}\bigg|_2^6\right)$$

$$= \frac{16-1}{2} \cdot \frac{216-8}{3} = \frac{3120}{6} = 520$$

3. Let $\mathcal{D}$ be the shaded domain in Figure 1.

FIGURE 1

Estimate $\iint_{\mathcal{D}} xy \, dA$ by the Riemann sum whose sample points are the midpoints of the squares in the grid.

SOLUTION The subrectangles have sides of length $\Delta x = \Delta y = 0.5$ and area $\Delta A = 0.5^2 = 0.25$. Of 16 sample points only 10 lie in $\mathcal{D}$. The sample points that lie in $\mathcal{D}$ are

$$(0.75, 0.75), \ (0.75, 1.25), \ (0.75, 1.75), \ (1.25, 0.25), \ (1.25, 0.75),$$
$$(1.25, 1.25), \ (1.25, 1.75), \ (1.75, 0.25), \ (1.75, 0.75), \ (1.75, 1.25)$$

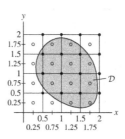

The Riemann sum S_{44} is thus

$$S_{44} = 0.25 \, (f(0.75, 0.75) + f(0.75, 1.25) + f(0.75, 1.75) + f(1.25, 0.25) + f(1.25, 0.75)$$
$$+ f(1.25, 1.25) + f(1.25, 1.75) + f(1.75, 0.25) + f(1.75, 0.75) + f(1.75, 1.25))$$

$$= 0.25 \left(0.75^2 + 0.75 \cdot 1.25 + 0.75 \cdot 1.75 + 1.25 \cdot 0.25 + 1.25 \cdot 0.75 + 1.25^2\right.$$
$$\left. + 1.25 \cdot 1.75 + 1.75 \cdot 0.25 + 1.75 \cdot 0.75 + 1.75 \cdot 1.25\right)$$

$$= 0.25 \cdot 11.75 = 2.9375$$

In Exercises 5–8, evaluate the iterated integral.

5. $\int_0^2 \int_3^5 y(x-y)\,dx\,dy$

SOLUTION First we evaluate the inner integral treating y as a constant:

$$\int_3^5 y(x-y)\,dx = y\left(\frac{x^2}{2} - yx\right)\bigg|_{x=3}^5 = y\left(\left(\frac{25}{2} - 5y\right) - \left(\frac{9}{2} - 3y\right)\right) = y(8 - 2y) = 8y - 2y^2$$

Now we integrate this result with respect to y:

$$\int_0^2 (8y - 2y^2)\,dy = 4y^2 - \frac{2}{3}y^3\bigg|_0^2 = 16 - \frac{16}{3} = \frac{32}{3}$$

Therefore,

$$\int_0^2 \int_3^5 y(x-y)\,dx\,dy = \frac{32}{3}$$

7. $\int_0^{\pi/3} \int_0^{\pi/6} \sin(x+y)\,dx\,dy$

SOLUTION We compute the inner integral treating y as a constant:

$$\int_0^{\pi/6} \sin(x+y)\,dx = -\cos(x+y)\bigg|_{x=0}^{\pi/6} = -\cos\left(\frac{\pi}{6} + y\right) + \cos y = \cos y - \cos\left(y + \frac{\pi}{6}\right)$$

We now integrate the result with respect to y:

$$\int_0^{\pi/3} \int_0^{\pi/6} \sin(x+y)\,dx\,dy = \int_0^{\pi/3} \left(\cos y - \cos\left(y + \frac{\pi}{6}\right)\right) dy = \sin y - \sin\left(y + \frac{\pi}{6}\right)\bigg|_0^{\pi/3}$$

$$= \sin\frac{\pi}{3} - \sin\left(\frac{\pi}{3} + \frac{\pi}{6}\right) - \left(\sin 0 - \sin\frac{\pi}{6}\right) = \frac{\sqrt{3}}{2} - 1 + \frac{1}{2} = \frac{\sqrt{3} - 1}{2}$$

In Exercises 9–14, sketch the domain $\mathcal{D}$ and calculate $\iint_{\mathcal{D}} f(x,y)\,dA$.

9. $\mathcal{D} = \{0 \le x \le 4,\ 0 \le y \le x\}$, $f(x,y) = \cos y$

SOLUTION The domain $\mathcal{D}$ is shown in the figure:

We compute the double integral, considering $\mathcal{D}$ as a vertically simple region. We describe $\mathcal{D}$ by the inequalities

$$0 \le x \le 4, \quad 0 \le y \le x$$

We now write the double integral as an iterated integral and compute:

$$\iint_{\mathcal{D}} \cos y\,dA = \int_0^4 \int_0^x \cos y\,dy\,dx = \int_0^4 \sin y\bigg|_{y=0}^x dx$$

$$= \int_0^4 (\sin x - \sin 0)\,dx = \int_0^4 \sin x\,dx = -\cos x\bigg|_0^4 = 1 - \cos 4$$

11. $\mathcal{D} = \{0 \le x \le 1,\ 1 - x \le y \le 2 - x\}$, $f(x, y) = e^{x+2y}$

SOLUTION

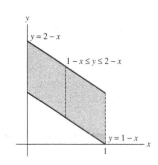

$\mathcal{D}$ is a vertically simple region, hence the double integral over $\mathcal{D}$ is the following iterated integral:

$$\iint_{\mathcal{D}} e^{x+2y}\, dA = \int_0^1 \int_{1-x}^{2-x} e^{x+2y}\, dy\, dx = \int_0^1 \frac{1}{2} e^{x+2y}\Big|_{y=1-x}^{2-x} dx = \int_0^1 \left(\frac{1}{2} e^{x+2(2-x)} - \frac{1}{2} e^{x+2(1-x)}\right) dx$$

$$= \int_0^1 \left(\frac{1}{2} e^{4-x} - \frac{1}{2} e^{2-x}\right) dx = -\frac{1}{2} e^{4-x} + \frac{1}{2} e^{2-x}\Big|_0^1 = -\frac{1}{2} e^3 + \frac{1}{2} e + \frac{1}{2} e^4 - \frac{1}{2} e^2$$

$$= \frac{1}{2} e(e^3 - e^2 - e + 1) = \frac{1}{2} e(e+1)(e-1)^2$$

13. $\mathcal{D} = \{0 \le y \le 1,\ 0.5 y^2 \le x \le y^2\}$, $f(x,y) = ye^{1+x}$

SOLUTION

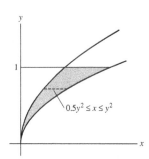

The region is horizontally simple, hence the double integral is equal to the following iterated integral:

$$\iint_{\mathcal{D}} ye^{1+x}\, dA = \int_0^1 \int_{0.5y^2}^{y^2} ye^{1+x}\, dx\, dy = \int_0^1 ye^{1+x}\Big|_{x=0.5y^2}^{y^2} dy$$

$$= \int_0^1 y\left(e^{1+y^2} - e^{1+0.5y^2}\right) dy = \int_0^1 ye^{1+y^2}\, dy - \int_0^1 ye^{1+0.5y^2}\, dy$$

We compute the integrals using the substitutions $u = 1 + y^2$, $du = 2y\, dy$, and $v = 1 + 0.5y^2$, $dv = y\, dy$, respectively. We get

$$\iint_{\mathcal{D}} ye^{1+x}\, dA = \frac{1}{2} \int_1^2 e^u\, du - \int_1^{1.5} e^v\, dv = \frac{1}{2} e^u\Big|_1^2 - e^v\Big|_1^{1.5} = \frac{1}{2}(e^2 - e) - (e^{3/2} - e)$$

$$= \frac{1}{2} e^2 + \frac{1}{2} e - e^{3/2} = 0.5(e^2 - 2e^{1.5} + e)$$

15. Express $\int_{-3}^{3} \int_0^{9-x^2} f(x, y)\, dy\, dx$ as an iterated integral in the order $dx\, dy$.

SOLUTION The limits of integration correspond to the inequalities describing the domain $\mathcal{D}$:

$$-3 \le x \le 3, \quad 0 \le y \le 9 - x^2$$

A quick sketch verifies that this is the region under the upper part of the parabola $y = 9 - x^2$, that is, the part that is above the x-axis. Therefore, the double integral can be rewritten as the following sum:

$$\int_{-3}^{3} \int_{0}^{9-x^2} f(x, y)\, dy\, dx = \int_{0}^{9} \int_{-\sqrt{9-y}}^{\sqrt{9-y}} f(x, y)\, dx\, dy$$

17. Let $\mathcal{D}$ be the domain between $y = x$ and $y = \sqrt{x}$. Calculate $\iint_{\mathcal{D}} xy\, dA$ as an iterated integral in the order $dx\, dy$ and $dy\, dx$.

SOLUTION In the order $dx\, dy$: The inequalities describing $\mathcal{D}$ as a horizontally simple region are obtained by first rewriting the equations of the curves with x as a function of y, that is, $x = y$ and $x = y^2$, respectively. The points of intersection are found solving the equation

$$y = y^2 \quad \Rightarrow \quad y(1 - y) = 0 \quad \Rightarrow \quad y = 0, \quad y = 1$$

We obtain the following inequalities for $\mathcal{D}$ (see figure):

$$\mathcal{D}: 0 \leq y \leq 1,\; y^2 \leq x \leq y$$

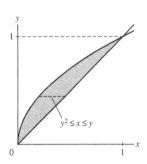

We now compute the double integral as the following iterated integral:

$$\iint_{\mathcal{D}} xy\, dA = \int_{0}^{1} \int_{y^2}^{y} xy\, dx\, dy = \int_{0}^{1} \left. \frac{x^2 y}{2} \right|_{x=y^2}^{x=y} dy = \int_{0}^{1} \left(\frac{y \cdot y^2}{2} - \frac{y^4 \cdot y}{2} \right) dy$$

$$= \int_{0}^{1} \left(\frac{y^3}{2} - \frac{y^5}{2} \right) dy = \left. \frac{y^4}{8} - \frac{y^6}{12} \right|_{0}^{1} = \frac{1}{8} - \frac{1}{12} = \frac{1}{24}$$

In the order $dy\, dx$: $\mathcal{D}$ is described as a vertically simple region by the following inequalities (see figure):

$$\mathcal{D}: 0 \leq x \leq 1,\; x \leq y \leq \sqrt{x}$$

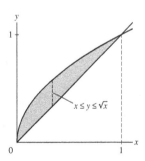

The corresponding iterated integral is

$$\iint_{\mathcal{D}} xy\, dA = \int_{0}^{1} \int_{x}^{\sqrt{x}} xy\, dy\, dx = \int_{0}^{1} \left. \frac{xy^2}{2} \right|_{y=x}^{\sqrt{x}} dx = \int_{0}^{1} \left(\frac{x \cdot x}{2} - \frac{x \cdot x^2}{2} \right) dx$$

$$= \int_{0}^{1} \left(\frac{x^2}{2} - \frac{x^3}{2} \right) dx = \left. \frac{x^3}{6} - \frac{x^4}{8} \right|_{0}^{1} = \frac{1}{6} - \frac{1}{8} = \frac{1}{24}$$

19. Change the order of integration and evaluate $\int_0^9 \int_0^{\sqrt{y}} \frac{x\,dx\,dy}{(x^2+y)^{1/2}}$.

SOLUTION The region here is described by the inequalities:

$$0 \le x \le \sqrt{y}, \quad 0 \le y \le 9$$

This region can also be described by writing these inequalities:

$$0 \le x \le 3, \quad x^2 \le y \le 9$$

Hence, changing the order of integration and evaluating we get:

$$\int_0^9 \int_0^{\sqrt{y}} \frac{x}{\sqrt{x^2+y}}\,dx\,dy = \int_0^3 \int_{x^2}^9 \frac{x}{\sqrt{x^2+y}}\,dy\,dx = \int_0^3 x\left(2\sqrt{x^2+y}\,\Big|_{x^2}^9\right) dx$$

$$= 2\int_0^3 x\sqrt{x^2+9} - x\sqrt{x^2+x^2}\,dx = 2\int_0^3 x\sqrt{x^2+9} - x^2\sqrt{2}\,dx$$

$$= 2\left(\frac{1}{3}(x^2+9)^{3/2} - \frac{\sqrt{2}}{3}x^3\,\Big|_0^3\right)$$

$$= \frac{2}{3} \cdot 18^{3/2} - \frac{2\sqrt{2}}{3} \cdot 27 - 2 \cdot \frac{1}{3} \cdot 9^{3/2}$$

$$= 36\sqrt{2} - 18\sqrt{2} - 18 = 18\sqrt{2} - 18$$

21. Prove the formula

$$\int_0^1 \int_0^y f(x)\,dx\,dy = \int_0^1 (1-x)f(x)\,dx$$

Then use it to calculate $\int_0^1 \int_0^y \frac{\sin x}{1-x}\,dx\,dy$.

SOLUTION The region of integration of the double integral $\int_0^1 \int_0^y f(x)\,dx\,dy$ is described as horizontally simple by the inequalities

$$0 \le y \le 1, \quad 0 \le x \le y$$

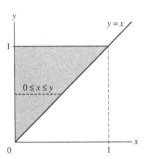

The region can also be described as a vertically simple region, by the inequalities

$$0 \le x \le 1, \quad x \le y \le 1$$

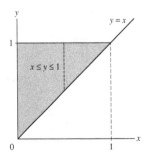

Therefore,
$$\int_0^1 \int_0^y f(x)\,dx\,dy = \int_0^1 \int_x^1 f(x)\,dy\,dx = \int_0^1 f(x)y\Big|_{y=x}^1 dx = \int_0^1 f(x)(1-x)\,dx$$

We use the formula with $f(x) = \frac{\sin x}{1-x}$. We get
$$\int_0^1 \int_0^y \frac{\sin x}{1-x}\,dx\,dy = \int_0^1 (1-x)\cdot\frac{\sin x}{1-x}\,dx = \int_0^1 \sin x\,dx = -\cos x\Big|_0^1 = 1-\cos 1$$

23. Use cylindrical coordinates to compute the volume of the region defined by $4 - x^2 - y^2 \le z \le 10 - 4x^2 - 4y^2$.

SOLUTION

We first find the projection of $\mathcal{W}$ onto the xy-plane. The intersection curve of the upper and lower boundaries of $\mathcal{W}$ is obtained by solving
$$10 - 4x^2 - 4y^2 = 4 - x^2 - y^2$$
$$6 = 3(x^2 + y^2) \quad\Rightarrow\quad x^2 + y^2 = 2$$

Therefore, the projection of $\mathcal{W}$ onto the xy-plane is the circle $x^2 + y^2 \le 2$. The upper surface is $z = 10 - 4(x^2 + y^2)$ or $z = 10 - 4r^2$ and the lower surface is $z = 4 - (x^2 + y^2) = 4 - r^2$. Therefore, the inequalities for $\mathcal{W}$ in cylindrical coordinates are
$$0 \le \theta \le 2\pi, \quad 0 \le r \le \sqrt{2}, \quad 4 - r^2 \le z \le 10 - 4r^2$$

We use the volume as a triple integral and change of variables in cylindrical coordinates to write
$$V = \text{Volume}(\mathcal{W}) = \iiint_{\mathcal{W}} 1\,dV = \int_0^{2\pi}\int_0^{\sqrt{2}}\int_{4-r^2}^{10-4r^2} r\,dz\,dr\,d\theta = \int_0^{2\pi}\int_0^{\sqrt{2}} rz\Big|_{z=4-r^2}^{10-4r^2} dr\,d\theta$$
$$= \int_0^{2\pi}\int_0^{\sqrt{2}} r\left(10 - 4r^2 - (4-r^2)\right)dr\,d\theta = \int_0^{2\pi}\int_0^{\sqrt{2}} (6r - 3r^3)\,dr\,d\theta$$
$$= \int_0^{2\pi} 3r^2 - \frac{3}{4}r^4\Big|_{r=0}^{\sqrt{2}} d\theta = \int_0^{2\pi} (6 - 3)\,d\theta = 6\pi$$

25. Find the volume of the region between the graph of the function $f(x,y) = 1 - (x^2 + y^2)$ and the xy-plane.

SOLUTION

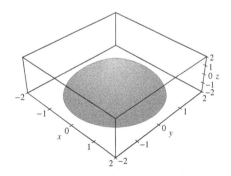

The intersection of the surface $z = 1 - (x^2 + y^2)$ with the xy-plane is obtained by setting $z = 0$. That is, $1 - (x^2 + y^2) = 0$ or $x^2 + y^2 = 1$. Therefore, the projection of the solid onto the xy-plane is the disk $x^2 + y^2 \le 1$. We describe the disk as a vertically simple region:
$$\mathcal{D}: -1 \le x \le 1,\ -\sqrt{1-x^2} \le y \le \sqrt{1-x^2}$$

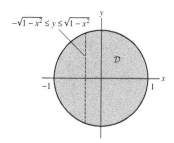

The volume V is the double integral of $z = 1 - (x^2 + y^2)$ over $\mathcal{D}$. That is,

$$V = \iint_{\mathcal{D}} \left(1 - (x^2 + y^2)\right) dA = \int_0^{2\pi} \int_0^1 (1 - r^2) r\, dr\, d\theta = 2\pi \int_0^1 (r - r^3)\, dr = \pi/2$$

27. Calculate $\iiint_{\mathcal{B}} (xy + z)\, dV$, where

$$\mathcal{B} = \{0 \le x \le 2,\ 0 \le y \le 1,\ 1 \le z \le 3\}$$

as an iterated integral in two different ways.

SOLUTION The triple integral over the box may be evaluated in any order. For instance,

$$\iiint_{\mathcal{B}} (xy + z)\, dV = \int_0^2 \int_0^1 \int_1^3 (xy + z)\, dz\, dy\, dx$$

$$= \int_0^1 \int_0^2 \int_1^3 (xy + z)\, dz\, dx\, dy = \int_1^3 \int_0^2 \int_0^1 (xy + z)\, dy\, dx\, dz$$

We compute the integral in two of the possible orders:

$$\iiint_{\mathcal{B}} (xy + z)\, dV = \int_0^2 \int_0^1 \int_1^3 (xy + z)\, dz\, dy\, dx = \int_0^2 \int_0^1 \left. xyz + \frac{z^2}{2} \right|_{z=1}^3 dy\, dx$$

$$= \int_0^2 \int_0^1 \left(\left(3xy + \frac{9}{2}\right) - \left(xy + \frac{1}{2}\right) \right) dy\, dx = \int_0^2 \int_0^1 (2xy + 4)\, dy\, dx$$

$$= \int_0^2 \left. xy^2 + 4y \right|_{y=0}^1 dx = \int_0^2 (x + 4)\, dx = \left. \frac{x^2}{2} + 4x \right|_0^2 = \frac{4}{2} + 8 = 10$$

$$\iiint_{\mathcal{B}} (xy + z)\, dV = \int_0^1 \int_0^2 \int_1^3 (xy + z)\, dz\, dx\, dy = \int_0^1 \int_0^2 \left. xyz + \frac{z^2}{2} \right|_{z=1}^3 dx\, dy$$

$$= \int_0^1 \int_0^2 \left(\left(3xy + \frac{9}{2}\right) - \left(xy + \frac{1}{2}\right) \right) dx\, dy = \int_0^1 \int_0^2 (2xy + 4)\, dx\, dy$$

$$= \int_0^1 \left. x^2 y + 4x \right|_{x=0}^2 dy = \int_0^1 (4y + 8)\, dy = \left. 2y^2 + 8y \right|_0^1 = 2 + 8 = 10$$

29. Evaluate $I = \int_{-1}^1 \int_0^{\sqrt{1-x^2}} \int_0^1 (x + y + z)\, dz\, dy\, dx$.

SOLUTION We compute the triple integral:

$$I = \int_{-1}^1 \int_0^{\sqrt{1-x^2}} \int_0^1 (x + y + z)\, dz\, dy\, dx = \int_{-1}^1 \int_0^{\sqrt{1-x^2}} \left. \left((x+y)z + \frac{z^2}{2}\right) \right|_{y=0}^1 dy\, dx$$

$$= \int_{-1}^1 \int_0^{\sqrt{1-x^2}} \left(x + y + \frac{1}{2}\right) dy\, dx = \int_{-1}^1 \left. \left(\left(x + \frac{1}{2}\right) y + \frac{y^2}{2}\right) \right|_{y=0}^{\sqrt{1-x^2}} dx$$

$$= \int_{-1}^1 \left(\left(x + \frac{1}{2}\right) \sqrt{1 - x^2} + \frac{1 - x^2}{2} \right) dx = \int_{-1}^1 x\sqrt{1 - x^2}\, dx + \int_{-1}^1 \frac{1}{2} \sqrt{1 - x^2}\, dx + \int_{-1}^1 \frac{1 - x^2}{2}\, dx$$

The first integral is zero since the integrand is an odd function. Therefore, using Integration Formulas we get

$$I = \int_0^1 \sqrt{1-x^2}\,dx + \int_0^1 (1-x^2)\,dx = \left(\frac{x}{2}\sqrt{1-x^2} + \frac{1}{2}\sin^{-1}x\right)\Big|_0^1 + \left(x - \frac{x^3}{3}\right)\Big|_0^1$$

$$= \frac{1}{2}\sin^{-1} 1 + \frac{2}{3} = \frac{\pi}{4} + \frac{2}{3}$$

31. Find the volume of the solid contained in the cylinder $x^2 + y^2 = 1$ below the surface $z = (x+y)^2$ and above the surface $z = -(x-y)^2$.

SOLUTION

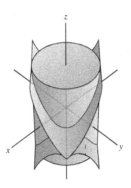

We rewrite the equations of the surfaces using cylindrical coordinates:

$$z = (x+y)^2 = x^2 + y^2 + 2xy = r^2 + 2(r\cos\theta)(r\sin\theta) = r^2(1 + \sin 2\theta)$$

$$z = -(x-y)^2 = -(x^2 + y^2 - 2xy) = -(r^2 - 2r^2\cos\theta\sin\theta) = -r^2(1 - \sin 2\theta)$$

The projection of the solid onto the xy-plane is the unit disk. Therefore, the solid is described by the following inequalities:

$$\mathcal{W}: 0 \leq \theta \leq 2\pi,\ 0 \leq r \leq 1,\ -r^2(1 - \sin 2\theta) \leq z \leq r^2(1 + \sin 2\theta)$$

Expressing the volume as a triple integral and converting the triple integral to cylindrical coordinates, we get

$$V = \text{Volume}(\mathcal{W}) = \iiint_{\mathcal{W}} 1\,dv = \int_0^{2\pi}\int_0^1 \int_{-r^2(1-\sin 2\theta)}^{r^2(1+\sin 2\theta)} r\,dz\,dr\,d\theta$$

$$= \int_0^{2\pi}\int_0^1 rz\Big|_{z=-r^2(1-\sin 2\theta)}^{r^2(1+\sin 2\theta)}\,dr\,d\theta = \int_0^{2\pi}\int_0^1 r\left(r^2(1+\sin 2\theta) + r^2(1-\sin 2\theta)\right)dr\,d\theta$$

$$= \int_0^{2\pi}\int_0^1 r^3 \cdot 2\,dr\,d\theta = \left(\int_0^{2\pi} 2\,d\theta\right)\left(\int_0^1 r^3\,dr\right) = 4\pi \cdot \frac{r^4}{4}\Big|_0^1 = \pi$$

33. Use polar coordinates to calculate $\iint_{\mathcal{D}} \sqrt{x^2 + y^2}\,dA$, where $\mathcal{D}$ is the region in the first quadrant bounded by the spiral $r = \theta$, the circle $r = 1$, and the x-axis.

SOLUTION The region of integration, shown in the figure, has the following description in polar coordinates:

$$\mathcal{D}: 0 \leq \theta \leq 1,\ \theta \leq r \leq 1$$

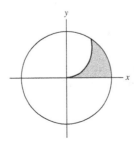

The function is $f(x,y) = \sqrt{x^2+y^2} = r$. We convert the double integral to polar coordinates and compute to obtain

$$\iint_{\mathcal{D}} \sqrt{x^2+y^2}\, dA = \int_0^1 \int_\theta^1 r \cdot r\, dr\, d\theta = \int_0^1 \int_\theta^1 r^2\, dr\, d\theta = \int_0^1 \frac{r^3}{3}\bigg|_{r=\theta}^1 d\theta$$

$$= \int_0^1 \left(\frac{1}{3} - \frac{\theta^3}{3}\right) d\theta = \frac{\theta}{3} - \frac{\theta^4}{12}\bigg|_0^1 = \frac{1}{3} - \frac{1}{12} = \frac{1}{4}$$

35. Express in cylindrical coordinates and evaluate:

$$\int_0^1 \int_0^{\sqrt{1-x^2}} \int_0^{\sqrt{x^2+y^2}} z\, dz\, dy\, dx$$

SOLUTION We evaluate the integral by converting it to cylindrical coordinates. The projection of the region of integration onto the xy-plane, as defined by the limits of integration, is

$$\mathcal{D}: 0 \leq x \leq 1,\ 0 \leq y \leq \sqrt{1-x^2}$$

That is, $\mathcal{D}$ is the part of the disk $x^2 + y^2 \leq 1$ in the first quadrant. The inequalities defining $\mathcal{D}$ in polar coordinates are

$$\mathcal{D}: 0 \leq \theta \leq \frac{\pi}{2},\ 0 \leq r \leq 1$$

The upper surface is $z = \sqrt{x^2+y^2} = r$ and the lower surface is $z = 0$. Therefore, the inequalities defining the region of integration in cylindrical coordinates are

$$\mathcal{W}: 0 \leq \theta \leq \frac{\pi}{2},\ 0 \leq r \leq 1,\ 0 \leq z \leq r$$

Converting the double integral to cylindrical coordinates gives

$$I = \int_0^{\pi/2} \int_0^1 \int_0^r zr\, dz\, dr\, d\theta = \int_0^{\pi/2} \int_0^1 \frac{z^2 r}{2}\bigg|_{z=0}^r dr\, d\theta = \int_0^{\pi/2} \int_0^1 \frac{r^3}{2}\, dr\, d\theta$$

$$= \left(\int_0^{\pi/2} d\theta\right)\left(\int_0^1 \frac{r^3}{2}\, dr\right) = \frac{\pi}{2} \cdot \frac{r^4}{8}\bigg|_0^1 = \frac{\pi}{16}$$

37. Convert to spherical coordinates and evaluate:

$$\int_{-2}^2 \int_{-\sqrt{4-x^2}}^{\sqrt{4-x^2}} \int_0^{\sqrt{4-x^2-y^2}} e^{-(x^2+y^2+z^2)^{3/2}}\, dz\, dy\, dx$$

SOLUTION The region of integration as defined by the limits of integration is

$$\mathcal{W}: -2 \leq x \leq 2,\ -\sqrt{4-x^2} \leq y \leq \sqrt{4-x^2},\ 0 \leq z \leq \sqrt{4-x^2-y^2}$$

That is, $\mathcal{W}$ is the region enclosed by the sphere $x^2 + y^2 + z^2 = 4$ and the xy-plane. We see that the region of integration is the upper half-ball $x^2 + y^2 + z^2 \leq 4$, hence the inequalities defining $\mathcal{W}$ in spherical coordinates are

$$\mathcal{W}: 0 \leq \theta \leq 2\pi,\ 0 \leq \phi \leq \frac{\pi}{2},\ 0 \leq \rho \leq 2$$

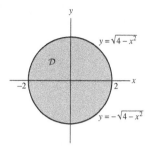

The function is $f(x,y,z) = e^{-(x^2+y^2+z^2)^{3/2}} = e^{-(\rho^2)^{3/2}} = e^{-\rho^3}$, therefore the integral in spherical coordinates is

$$I = \int_0^{2\pi}\int_0^{\pi/2}\int_0^2 e^{-\rho^3}\rho^2\sin\phi\,d\rho\,d\phi\,d\theta = \left(\int_0^{2\pi}d\theta\right)\left(\int_0^{\pi/2}\sin\phi\,d\phi\right)\left(\int_0^2 e^{-\rho^3}\rho^2\,d\rho\right)$$

$$= 2\pi\left(-\cos\phi\Big|_0^{\pi/2}\right)\int_0^2 e^{-\rho^3}\rho^2\,d\rho = 2\pi\int_0^2 e^{-\rho^3}\rho^2\,d\rho$$

We compute the integral using the substitution $u = \rho^3$, $du = 3\rho^2 d\rho$. We get

$$I = 2\pi\int_0^8 e^{-u}\frac{du}{3} = \frac{2\pi}{3}(-e^{-u})\Big|_0^8 = \frac{2\pi}{3}(-e^{-8}+1) = \frac{2\pi(-1+e^8)}{3e^8}$$

39. Let $\mathcal{W}$ be the ball of radius R in $\mathbf{R}^3$ centered at the origin, and let $P = (0,0,R)$ be the North Pole. Let $d_P(x,y,z)$ be the distance from P to (x,y,z). Show that the average value of d_P over the ball $\mathcal{W}$ is equal to $\overline{d} = 6R/5$. *Hint:* Show that

$$\overline{d} = \frac{1}{\frac{4}{3}\pi R^3}\int_{\theta=0}^{2\pi}\int_{\rho=0}^R\int_{\phi=0}^\pi \rho^2\sin\phi\sqrt{R^2+\rho^2-2\rho R\cos\phi}\,d\phi\,d\rho\,d\theta$$

and evaluate.

SOLUTION We know that the volume of the ball is $\frac{4}{3}\pi R^3$.

In spherical coordinates, the distance from P to a point on the ball is

$$\sqrt{(x-0)^2+(y-0)^2+(z-R)^2} = \sqrt{\rho^2\sin^2\phi\cos^2\theta+\rho^2\sin^2\phi\sin^2\theta+(\rho\cos\phi-R)^2}$$

$$= \sqrt{\rho^2\sin^2\phi(\cos^2\theta+\sin^2\theta)+\rho^2\cos^2\phi-2\rho R\cos\phi+R^2}$$

$$= \sqrt{\rho^2\sin^2\phi+\rho^2\cos^2\phi-2\rho R\cos\phi+R^2}$$

$$= \sqrt{\rho^2(\sin^2\phi+\cos^2\phi)-2\rho R\cos\phi+R^2}$$

$$= \sqrt{R^2+\rho^2-2\rho R\cos\phi}$$

Now, to write the average value of d_P we have:

$$d_P = \frac{1}{\frac{4}{3}\pi R^3}\int_{\theta=0}^{2\pi}\int_{\rho=0}^R\int_{\phi=0}^\pi \rho^2\sin\phi\sqrt{R^2+\rho^2-2\rho R\cos\phi}\,d\phi\,d\rho\,d\theta$$

Using substitution, and the fact that $0 \le \rho \le R$,

$$\int_{\phi=0}^\pi \sin\phi\sqrt{R^2+\rho^2-2\rho R\cos\phi}\,d\phi = \frac{2\rho}{3R}(R^2+\rho^2-2\rho R\cos\phi)^{3/2}\Big|_0^\pi$$

$$= \frac{2\rho}{3R}\big((R+\rho)^3-(R-\rho)^3\big) = \frac{2\rho}{3R}(\rho^3+3R^2\rho)$$

Now integrate with respect to θ and ρ:

$$d_P = \frac{1}{\frac{4}{3}\pi R^3}\int_0^{2\pi}\int_0^R \rho^2\cdot\frac{2\rho}{3R}(\rho^3+3R^2\rho)\,d\rho\,d\theta$$

$$= \frac{3}{4\pi R^3}\cdot\frac{2}{3R}\int_0^{2\pi}\int_0^R \rho^3(\rho^3+3R^2\rho)\,d\rho\,d\theta$$

$$= \frac{1}{2\pi R^4} \int_0^{2\pi} \int_0^R \rho^6 + 3R^2 \rho^4 \, d\rho \, d\theta$$

$$= \frac{1}{2\pi R^4} \int_0^{2\pi} \frac{1}{7}\rho^7 + \frac{3}{5}R^2 \rho^5 \Big|_0^R \, d\theta$$

$$= \frac{1}{2\pi R^4} \int_0^{2\pi} \frac{1}{7}R^7 + \frac{3}{5}R^7 \, d\theta$$

$$= \frac{1}{2\pi R^4} \left(\frac{26}{35}R^7\right) \cdot 2\pi = \frac{26}{35}R^3$$

to get $8\pi R^4/5$. Dividing by the volume of the sphere gives us $6R/5$

41. Use cylindrical coordinates to find the mass of the solid bounded by $z = 8 - x^2 - y^2$ and $z = x^2 + y^2$, assuming a mass density of $f(x, y, z) = (x^2 + y^2)^{1/2}$.

SOLUTION The mass of the solid $\mathcal{W}$ is the following integral:

$$M = \iiint_{\mathcal{W}} (x^2 + y^2)^{1/2} \, dV$$

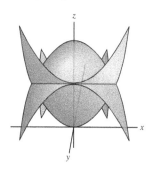

The projection of $\mathcal{W}$ on the xy-plane is obtained by equating the equations of the two surfaces:

$$8 - x^2 - y^2 = x^2 + y^2$$
$$2(x^2 + y^2) = 8 \quad \Rightarrow \quad x^2 + y^2 = 4$$

We conclude that the projection is the disk $\mathcal{D}: x^2 + y^2 \leq 4$.

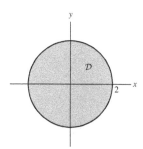

Therefore, $\mathcal{W}$ is described by

$$\mathcal{W}: x^2 + y^2 \leq z \leq 8 - (x^2 + y^2), \ (x, y) \in \mathcal{D}$$

Thus,

$$M = \iint_{\mathcal{D}} \int_{x^2+y^2}^{8-(x^2+y^2)} (x^2 + y^2)^{1/2} \, dz \, dx \, dy$$

We convert the integral to cylindrical coordinates. The inequalities for $\mathcal{W}$ are

$$0 \leq r \leq 2, \quad 0 \leq \theta \leq 2\pi, \quad r^2 \leq z \leq 8 - r^2$$

Also, $(x^2 + y^2)^{1/2} = r$, hence we obtain the following integral:

$$M = \int_0^2 \int_0^{2\pi} \int_{r^2}^{8-r^2} r \cdot r \, dz \, d\theta \, dr = \int_0^2 \int_0^{2\pi} \int_{r^2}^{8-r^2} r^2 \, dz \, d\theta \, dr = \int_0^2 \int_0^{2\pi} r^2 z \Big|_{z=r^2}^{8-r^2} \, d\theta \, dr$$

$$= \int_0^2 \int_0^{2\pi} r^2(8 - r^2 - r^2)\,d\theta\,dr = \int_0^2 \int_0^{2\pi} (8r^2 - 2r^4)\,d\theta\,dr = \left(\int_0^{2\pi} 1\,d\theta\right)\left(\int_0^2 (8r^2 - 2r^4)\,dr\right)$$

$$= 2\pi \left(\frac{8r^3}{3} - \frac{2}{5}r^5\right)\Big|_0^2 = \frac{256}{15}\pi \approx 53.62$$

43. Use cylindrical coordinates to find the mass of a cylinder of radius 4 and height 10 if the mass density at a point is equal to the square of the distance from the cylinder's central axis.

SOLUTION

The mass density is $\rho(x, y, z) = x^2 + y^2 = r^2$, hence the mass of the cylinder is

$$M = \iiint_W (x^2 + y^2)\,dV$$

The region W is described using cylindrical coordinates by the following inequalities:

$$W:\ 0 \le \theta \le 2\pi,\ 0 \le r \le 4,\ 0 \le z \le 10$$

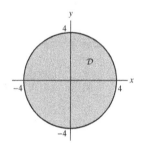

Thus,

$$M = \iiint_W (x^2 + y^2)\,dV = \int_0^{2\pi} \int_0^4 \int_0^{10} r^2 \cdot r\,dz\,dr\,d\theta = \int_0^{2\pi} \int_0^4 \int_0^{10} r^3\,dz\,dr\,d\theta = \int_0^{2\pi} \int_0^4 r^3 z \Big|_{z=0}^{10}\,dr\,d\theta$$

$$= \int_0^{2\pi} \int_0^4 r^3 \cdot 10\,dr\,d\theta = \int_0^{2\pi} \frac{10r^4}{4}\Big|_{r=0}^4 d\theta = \int_0^{2\pi} 640\,d\theta = 640 \cdot 2\pi = 1280\pi$$

45. Find the centroid of the solid bounded by the xy-plane, the cylinder $x^2 + y^2 = R^2$, and the plane $x/R + z/H = 1$.

SOLUTION First to find the volume of this solid. The first equation lends itself well to cylindrical coordinates:

$$x^2 + y^2 = R^2 \quad \Rightarrow \quad r = R, 0 \le \theta \le 2\pi$$

and

$$\frac{x}{R} + \frac{z}{H} = 1 \quad \Rightarrow \quad z = H\left(1 - \frac{x}{R}\right) = H\left(1 - \frac{r\cos\theta}{R}\right)$$

The volume is:

$$\text{Volume} = \int_0^{2\pi} \int_0^R \int_0^{H(1 - r\cos\theta/R)} 1\,dz\,dr\,d\theta$$

$$= \int_0^{2\pi} \int_0^R H\left(1 - \frac{r\cos\theta}{R}\right)\,dr\,d\theta$$

$$= H\int_0^{2\pi} r - \frac{1}{2}\cdot\frac{r^2\cos\theta}{R}\bigg|_{r=0}^R d\theta$$

$$= H\int_0^{2\pi} R - \frac{1}{2}R\cos\theta\, d\theta$$

$$= H\left(R\theta - \frac{1}{2}R\sin\theta\bigg|_0^{2\pi}\right)$$

$$= 2\pi HR$$

Now to compute the coordinates of the centroid:

$$\bar{x} = \frac{1}{V}\iiint_W x\, dV = \frac{1}{2\pi HR}\int_0^{2\pi}\int_0^R\int_0^{H(1-r\cos\theta/R)} r\cos\theta\, dz\, dr\, d\theta$$

$$= \frac{1}{2\pi HR}\int_0^{2\pi}\int_0^R r\cos\theta\cdot z\bigg|_0^{H(1-r\cos\theta/R)} dr\, d\theta$$

$$= \frac{H}{2\pi HR}\int_0^{2\pi}\int_0^R r\cos\theta\left(1 - \frac{r\cos\theta}{R}\right) dr\, d\theta$$

$$= \frac{1}{2\pi R}\int_0^{2\pi}\int_0^R r\cos\theta - \frac{1}{R}r^2\cos^2\theta\, dr\, d\theta$$

$$= \frac{1}{2\pi R}\int_0^{2\pi} \frac{1}{2}r^2\cos\theta - \frac{1}{3R}r^3\cos^2\theta\bigg|_0^R d\theta$$

$$= \frac{1}{2\pi R}\int_0^{2\pi} \frac{1}{2}R^2\cos\theta - \frac{R^2}{6}(1 + \cos 2\theta)\, d\theta$$

$$= \frac{1}{2\pi R}\left(\frac{1}{2}R^2\sin\theta - \frac{R^2}{6}\left(\theta + \frac{1}{2}\sin 2\theta\right)\bigg|_0^{2\pi}\right) = \frac{1}{2\pi R}\cdot -\frac{R^2}{6}(2\pi) = -\frac{R}{6}$$

$$\bar{y} = \frac{1}{V}\iiint_W y\, dV = \frac{1}{2\pi HR}\int_0^{2\pi}\int_0^R\int_0^{H(1-r\cos\theta/R)} r\sin\theta\, dz\, dr\, d\theta$$

$$= \frac{1}{2\pi HR}\int_0^{2\pi}\int_0^R r\sin\theta\cdot z\bigg|_0^{H(1-r\cos\theta/R)} dr\, d\theta$$

$$= \frac{H}{2\pi HR}\int_0^{2\pi}\int_0^R r\sin\theta\left(1 - \frac{r\cos\theta}{R}\right) dr\, d\theta$$

$$= \frac{1}{2\pi R}\int_0^{2\pi}\int_0^R r\sin\theta - \frac{1}{R}r^2\sin\theta\cos\theta\, dr\, d\theta$$

$$= \frac{1}{2\pi R}\int_0^{2\pi} \frac{1}{2}r^2\sin\theta - \frac{1}{3R}r^3\sin\theta\cos\theta\bigg|_0^R d\theta$$

$$= \frac{1}{2\pi R}\int_0^{2\pi} \frac{1}{2}R^2\sin\theta - \frac{R^2}{3}\sin\theta\cos\theta\, d\theta$$

$$= \frac{1}{2\pi R}\left(-\frac{1}{2}R^2\cos\theta - \frac{R^2}{6}\sin^2\theta\bigg|_0^{2\pi}\right) = 0$$

$$\bar{z} = \frac{1}{V}\iiint_W z\, dV = \frac{1}{2\pi HR}\int_0^{2\pi}\int_0^R\int_0^{H(1-r\cos\theta/R)} z\, dz\, dr\, d\theta$$

$$= \frac{1}{2\pi HR}\int_0^{2\pi}\int_0^R \frac{1}{2}z^2\bigg|_0^{H(1-r\cos\theta/R)} dr\, d\theta$$

$$= \frac{H^2}{4\pi HR}\int_0^{2\pi}\int_0^R \left(1 - \frac{r\cos\theta}{R}\right)^2 dr\, d\theta$$

$$= \frac{H}{4\pi R}\int_0^{2\pi}\int_0^R 1 - \frac{2r\cos\theta}{R} + \frac{r^2\cos^2\theta}{R^2}\, dr\, d\theta$$

$$= \frac{H}{4\pi R} \int_0^{2\pi} r - \frac{r^2 \cos\theta}{R} + \frac{r^3 \cos^2\theta}{3R^2} \bigg|_0^R d\theta$$

$$= \frac{H}{4\pi R} \int_0^{2\pi} R - R\cos\theta + \frac{1}{6}R(1 + \cos 2\theta) d\theta$$

$$= \frac{H}{4\pi R} \left(R\theta - R\sin\theta + \frac{1}{6}R\left(\theta + \frac{1}{2}\sin 2\theta\right) \bigg|_0^{2\pi} \right)$$

$$= \frac{H}{4\pi R} \left(2\pi R + \frac{1}{3}\pi R \right) = \frac{H}{4\pi R} \cdot \frac{7\pi R}{3} = \frac{7}{12}H$$

The coordinates of the centroid are $(-R/6, 0, 7H/12)$.

47. Find the centroid of solid (A) in Figure 4 defined by $x^2 + y^2 \le R^2$, $0 \le z \le H$, and $\frac{\pi}{6} \le \theta \le 2\pi$, where θ is the polar angle of (x, y).

SOLUTION Since the mass distribution is uniform, we may assume that $\rho(x, y, z) = 1$, hence the center of mass is

$$x_{CM} = \frac{1}{V}\iiint_{\mathcal{W}} x\, dV, \quad y_{CM} = \frac{1}{V}\iiint_{\mathcal{W}} y\, dV, \quad z_{CM} = \frac{1}{V}\iiint_{\mathcal{W}} z\, dV$$

The inequalities describing $\mathcal{W}$ in cylindrical coordinates are

$$\mathcal{W}: 0 \le \theta \le \frac{\pi}{6}, 0 \le r \le R, 0 \le z \le H$$

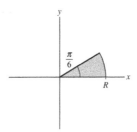

The entire cylinder has a total volume $\pi R^2 H$. The region $\mathcal{W}$ has the fraction $(2\pi - \frac{\pi}{6})/(2\pi)$ of this total volume, so

$$V = \frac{(2\pi - \frac{\pi}{6})}{2\pi}(\pi R^2 H) = \frac{11\pi R^2 H}{12}$$

We use cylindrical coordinates to compute the triple integrals:

$$x_{CM} = \frac{1}{V}\int_{\pi/6}^{2\pi}\int_0^R\int_0^H (r\cos\theta) r\, dz\, dr\, d\theta = \frac{12}{11\pi R^2 H}\left(\int_{\pi/6}^{2\pi} \cos\theta\, d\theta\right)\left(\int_0^R r^2\, dr\right)\left(\int_0^H dz\right)$$

$$= \frac{12}{11\pi R^2 H}\left(-\frac{1}{2}\right)\left(\frac{R^3}{3}\right)(H) = -\frac{2R}{11\pi}$$

$$y_{CM} = \frac{1}{V}\int_{\pi/6}^{2\pi}\int_0^R\int_0^H (r\sin\theta) r\, dz\, dr\, d\theta = \frac{12}{11\pi R^2 H}\left(\int_{\pi/6}^{2\pi} \sin\theta\, d\theta\right)\left(\int_0^R r^2\, dr\right)\left(\int_0^H dz\right)$$

$$= \frac{12}{11\pi R^2 H}\left(\frac{-2 + \sqrt{3}}{2}\right)\left(\frac{R^3}{3}\right)(H) = \frac{2R}{11\pi}(\sqrt{3} - 2)$$

$$z_{CM} = \frac{1}{V}\int_{\pi/6}^{2\pi}\int_0^R\int_0^H zr\, dz\, dr\, d\theta = \frac{12}{11\pi R^2 H}\left(\int_{\pi/6}^{2\pi} d\theta\right)\left(\int_0^R r\, dr\right)\left(\int_0^H z\, dz\right)$$

$$= \frac{12}{11\pi R^2 H}\left(\frac{11\pi}{6}\right)\left(\frac{R^2}{2}\right)\left(\frac{H^2}{2}\right) = \frac{H}{2}$$

Therefore, the center of mass is the following point:

$$\left(-\frac{2R}{11\pi}, \frac{2R}{11\pi}(\sqrt{3} - 2), \frac{H}{2}\right)$$

49. Find the center of mass of the cylinder $x^2 + y^2 = 1$ for $0 \le z \le 1$, assuming a mass density of $\delta(x, y, z) = z$.

SOLUTION

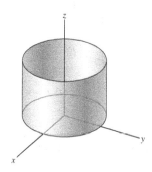

The cylinder $\mathcal{W}$ is defined by the inequalities

$$-1 \le x \le 1, \quad -\sqrt{1-x^2} \le y \le \sqrt{1-x^2}, \quad 0 \le z \le 1$$

In cylindrical coordinates, this is

$$0 \le r \le 1, \; 0 \le \theta \le 2\pi, \; 0 \le z \le 1$$

We first compute the total mass of $\mathcal{W}$ by converting to cylindrical coordinates:

$$\begin{aligned}
M = \iiint_{\mathcal{W}} z\, dV &= \int_0^1 \int_0^{2\pi} \int_0^1 zr\, dz\, d\theta\, dr \\
&= \int_0^1 \int_0^{2\pi} \left. \frac{1}{2} z^2 r \right|_0^1 d\theta\, dr \\
&= \int_0^1 \int_0^{2\pi} \frac{1}{2} r\, d\theta\, dr \\
&= \pi \int_0^1 r\, dr \\
&= \pi \cdot \left. \frac{1}{2} r^2 \right|_0^1 = \frac{\pi}{2}
\end{aligned}$$

Since both the mass density function and the region are symmetric with respect to the z-axis, we have $\bar{x} = \bar{y} = 0$. To calculate $\bar{z}$, we calculate M_{xy}:

$$\begin{aligned}
M_{xy} = \iiint_{\mathcal{W}} z \cdot \delta\, dV &= \int_0^1 \int_0^{2\pi} \int_0^1 z^2 r\, dz\, d\theta\, dr \\
&= \int_0^1 \int_0^{2\pi} \left. \frac{1}{3} z^3 r \right|_0^1 d\theta\, dr \\
&= \frac{1}{3} \int_0^1 \int_0^{2\pi} r\, d\theta\, dr \\
&= \frac{2\pi}{3} \int_0^1 r\, dr \\
&= \frac{2\pi}{3} \cdot \left. \frac{1}{2} r^2 \right|_0^1 = \frac{\pi}{3}
\end{aligned}$$

The center of mass is

$$(\bar{x}, \bar{y}, \bar{z}) = \left(0, 0, \frac{M_{xy}}{M}\right) = \left(0, 0, \frac{2}{3}\right)$$

51. Find the center of mass of the part of the ball $x^2 + y^2 + z^2 = 1$ in the first octant, assuming a mass density of $\delta(x, y, z) = x$.

SOLUTION

(a) The solid is the part of the unit sphere in the first octant. The inequalities defining the projection of the solid onto the xy-plane are

$$\mathcal{D}: 0 \leq y \leq 1, \ 0 \leq x \leq \sqrt{1-y^2}$$

$\mathcal{W}$ is the region bounded by $\mathcal{D}$ and the sphere $z = \sqrt{1-x^2-y^2}$, hence $\mathcal{W}$ is defined by the inequalities

$$\mathcal{W}: 0 \leq y \leq 1, \ 0 \leq x \leq \sqrt{1-y^2}, \ 0 \leq z \leq \sqrt{1-x^2-y^2} \tag{1}$$

We first must compute the mass of the solid. Using the mass as a triple integral, we have

$$M = \int_0^1 \int_0^{\sqrt{1-y^2}} \int_0^{\sqrt{1-x^2-y^2}} x \, dz \, dx \, dy = \int_0^1 \int_0^{\sqrt{1-y^2}} xz \bigg|_{z=0}^{\sqrt{1-x^2-y^2}} dx \, dy$$

$$= \int_0^1 \int_0^{\sqrt{1-y^2}} x\sqrt{1-x^2-y^2} \, dx \, dy$$

We compute the inner integral using the substitution $u = \sqrt{1-x^2-y^2}$, $du = -\frac{x}{u} dx$, or $x \, dx = -u \, du$. We get

$$\int_0^{\sqrt{1-y^2}} x\sqrt{1-x^2-y^2} \, dx = \int_{\sqrt{1-y^2}}^0 u(-u \, du) = \int_0^{\sqrt{1-y^2}} u^2 \, du = \frac{u^3}{3}\bigg|_0^{\sqrt{1-y^2}} = \frac{(1-y^2)^{3/2}}{3} \tag{2}$$

We substitute in (2) and compute the resulting integral substituting $y = \sin t$, $dy = \cos t \, dt$:

$$M = \int_0^1 \frac{(1-y^2)^{3/2}}{3} dy = \frac{1}{3} \int_0^{\pi/2} (1 - \sin^2 t)^{3/2} \cos t \, dt = \frac{1}{3} \int_0^{\pi/2} \cos^4 t \, dt$$

$$= \frac{1}{3} \left(\frac{\cos^3 t \sin t}{4} + \frac{3}{4}\left(\frac{t}{2} + \frac{\sin 2t}{4}\right) \right)\bigg|_0^{\pi/2} = \frac{1}{3} \cdot \frac{\pi}{4} \cdot \frac{\pi}{4} = \frac{\pi}{16}$$

That is, $M = \frac{\pi}{16}$. We now find the coordinates of the center of mass. To compute x_{CM} we use the definition of $\mathcal{D}$ as a vertically simple region to obtain a simpler integral. That is, we write the inequalities for $\mathcal{W}$ as

$$\mathcal{W}: 0 \leq x \leq 1, \ 0 \leq y \leq \sqrt{1-x^2}, \ 0 \leq z \leq \sqrt{1-x^2-y^2} \tag{3}$$

Thus,

$$x_{CM} = \frac{1}{M} \iiint_{\mathcal{W}} x \rho \, dV = \frac{16}{\pi} \int_0^1 \int_0^{\sqrt{1-x^2}} \int_0^{\sqrt{1-x^2-y^2}} x^2 \, dz \, dy \, dx = \frac{16}{\pi} \int_0^1 \int_0^{\sqrt{1-x^2}} x^2 z \bigg|_{z=0}^{\sqrt{1-x^2-y^2}} dy \, dx$$

$$= \frac{16}{\pi} \int_0^1 \int_0^{\sqrt{1-x^2}} x^2 \sqrt{1-x^2-y^2} \, dy \, dx = \frac{16}{\pi} \int_0^1 x^2 \left(\int_0^{\sqrt{1-x^2}} \sqrt{1-x^2-y^2} \, dy \right) dx \tag{4}$$

We compute the inner integral using Integration Formulas:

$$\int_0^{\sqrt{1-x^2}} \sqrt{1-x^2-y^2} \, dy = \frac{y}{2}\sqrt{1-x^2-y^2} + \frac{1-x^2}{2} \sin^{-1} \frac{y}{\sqrt{1-x^2}} \bigg|_{y=0}^{\sqrt{1-x^2}}$$

$$= \frac{1-x^2}{2} \sin^{-1} 1 = \frac{1-x^2}{2} \cdot \frac{\pi}{2} = \frac{\pi}{4}(1-x^2)$$

Substituting in (4) gives

$$x_{CM} = \frac{16}{\pi} \int_0^1 x^2 \cdot \frac{\pi}{4}(1-x^2) \, dx = 4 \int_0^1 (x^2 - x^4) \, dx = 4\left(\frac{x^3}{3} - \frac{x^5}{5} \right)\bigg|_0^1 = 4\left(\frac{1}{3} - \frac{1}{5} \right) = \frac{8}{15}$$

(b) We compute the y-coordinate of the center of mass, using (1):

$$y_{CM} = \frac{1}{M} \iiint_W y\rho\, dV = \frac{16}{\pi} \int_0^1 \int_0^{\sqrt{1-y^2}} \int_0^{\sqrt{1-x^2-y^2}} yx\, dz\, dx\, dy = \frac{16}{\pi} \int_0^1 \int_0^{\sqrt{1-y^2}} yxz \Big|_{z=0}^{\sqrt{1-x^2-y^2}} dx\, dy$$

$$= \frac{16}{\pi} \int_0^1 \int_0^{\sqrt{1-y^2}} yx\sqrt{1-x^2-y^2}\, dx\, dy = \frac{16}{\pi} \int_0^1 y \left(\int_0^{\sqrt{1-y^2}} x\sqrt{1-x^2-y^2}\, dx \right) dy$$

The inner integral was computed in (2), therefore,

$$y_{CM} = \frac{16}{\pi} \int_0^1 y \cdot \frac{(1-y^2)^{3/2}}{3}\, dy = \frac{16}{3\pi} \int_0^1 y(1-y^2)^{3/2}\, dy$$

We compute the integral using the substitution $u = 1 - y^2$, $du = -2y\, dy$. We get

$$y_{CM} = \frac{16}{3\pi} \int_1^0 u^{3/2} \cdot \left(-\frac{du}{2}\right) = \frac{8}{3\pi} \int_0^1 u^{3/2}\, du = \frac{8}{3\pi} \cdot \frac{2}{5} \cdot u^{5/2} \Big|_0^1 = \frac{16}{15\pi}$$

Finally we find the z-coordinate of the center of mass, using (1):

$$z_{CM} = \frac{1}{M} \iiint_W z\rho\, dV = \frac{16}{\pi} \int_0^1 \int_0^{\sqrt{1-y^2}} \int_0^{\sqrt{1-x^2-y^2}} zx\, dz\, dx\, dy = \frac{16}{\pi} \int_0^1 \int_0^{\sqrt{1-y^2}} \frac{z^2 x}{2} \Big|_{z=0}^{\sqrt{1-x^2-y^2}} dx\, dy$$

$$= \frac{16}{\pi} \int_0^1 \int_0^{\sqrt{1-y^2}} \frac{x}{2}(1 - x^2 - y^2)\, dx\, dy = \frac{8}{\pi} \int_0^1 \int_0^{\sqrt{1-y^2}} (x - x^3 - xy^2)\, dx\, dy$$

$$= \frac{8}{\pi} \int_0^1 \frac{x^2}{2} - \frac{x^4}{4} - \frac{x^2 y^2}{2} \Big|_{x=0}^{\sqrt{1-y^2}} dy = \frac{8}{\pi} \int_0^1 \left(\frac{1-y^2}{2} - \frac{(1-y^2)^2}{4} - \frac{(1-y^2)y^2}{2} \right) dy$$

$$= \frac{2}{\pi} \int_0^1 (y^4 - 2y^2 + 1)\, dy = \frac{2}{\pi} \left(\frac{y^5}{5} - \frac{2y^3}{3} + y \right) \Big|_{y=0}^1 = \frac{2}{\pi} \left(\frac{1}{5} - \frac{2}{3} + 1 \right) = \frac{16}{15\pi}$$

The center mass is the following point:

$$P = \left(\frac{8}{15}, \frac{16}{15\pi}, \frac{16}{15\pi} \right)$$

53. Calculate $P(3X + 2Y \geq 6)$ for the probability density in Exercise 52.

SOLUTION Previously we found $p(x, y) = \frac{1}{33}(4x - y + 3)$. Then using $P(3X + 2Y \geq 6)$ we want to find $1 - P(3X + 2Y \leq 6)$. Hence we need to integrate the following:

$$P(3X + 2Y \leq 6) = \int_{x=0}^2 \int_{y=0}^{3-3/2x} \frac{1}{33}(4x - y + 3)\, dy\, dx$$

$$= \frac{1}{33} \int_0^2 4xy - \frac{1}{2}y^2 + 3y \Big|_0^{3-3/2x} dx$$

$$= \frac{1}{33} \int_0^2 4x\left(3 - \frac{3}{2}x\right) - \frac{1}{2}\left(3 - \frac{3}{2}x\right)^2 + 3\left(3 - \frac{3}{2}x\right) dx$$

$$= \frac{1}{33} \int_0^2 -\frac{57}{8}x^2 + 12x + \frac{9}{2}\, dx$$

$$= \frac{1}{33} \left(-\frac{57}{24}x^3 + 6x^2 + \frac{9}{2}x \right) \Big|_0^2$$

$$= \frac{1}{33}(-19 + 24 + 9) = \frac{14}{33}$$

Thus we have:

$$P(3X + 2Y \geq 6) = 1 - P(3X + 2Y \leq 6) = 1 - \frac{14}{33} = \frac{19}{33}$$

55. An insurance company issues two kinds of policies: A and B. Let X be the time until the next claim of type A is filed, and let Y be the time (in days) until the next claim of type B is filed. The random variables have joint probability density

$$p(x, y) = 12e^{-4x-3y}$$

Find the probability that $X \leq Y$.

SOLUTION We must compute

$$P(X \leq Y) = \int_{x=0}^{\infty} \int_{y=x}^{\infty} p(x, y) \, dy \, dx$$

Now evaluating we get:

$$P(X \leq Y) = \int_{x=0}^{\infty} \int_{y=x}^{\infty} 12e^{-4x-3y} \, dy \, dx = 12 \int_{0}^{\infty} \int_{x}^{\infty} e^{-4x} e^{-3y} \, dy \, dx$$

$$= 12 \int_{0}^{\infty} e^{-4x} \left(-\frac{1}{3} e^{-3y} \right) \bigg|_{x}^{\infty} dx = -4 \int_{0}^{\infty} e^{-4x} \left(\lim_{t \to \infty} e^{-3t} - e^{-3x} \right) dx$$

$$= 4 \int_{0}^{\infty} e^{-4x} \cdot e^{-3x} \, dx = 4 \int_{0}^{\infty} e^{-7x} \, dx$$

$$= -\frac{4}{7} \left(e^{-7x} \right) \bigg|_{0}^{\infty} = -\frac{4}{7} \lim_{t \to \infty} \left(e^{-7t} - 1 \right) = \frac{4}{7}$$

57. Find a linear mapping $\Phi(u, v)$ that maps the unit square to the parallelogram in the xy-plane spanned by the vectors $\langle 3, -1 \rangle$ and $\langle 1, 4 \rangle$. Then use the Jacobian to find the area of the image of the rectangle $\mathcal{R} = [0, 4] \times [0, 3]$ under Φ.

SOLUTION We denote the linear map by

$$\Phi(u, v) = (Au + Cv, Bu + Dv) \tag{1}$$

The image of the unit square is the quadrangle whose vertices are the images of the vertices of the square. Therefore, we ask that

$$\Phi(0, 0) = (A \cdot 0 + C \cdot 0, B \cdot 0 + D \cdot 0) = (0, 0) \qquad (0, 0) = (0, 0)$$

$$\Phi(1, 0) = (A \cdot 1 + C \cdot 0, B \cdot 1 + D \cdot 0) = (3, -1) \quad \Rightarrow \quad (A, B) = (3, -1)$$

$$\Phi(0, 1) = (A \cdot 0 + C \cdot 1, B \cdot 0 + D \cdot 1) = (1, 4) \qquad (C, D) = (1, 4)$$

These equations imply that $A = 3$, $B = -1$, $C = 1$, and $D = 4$. Substituting in (1) we obtain the following map:

$$\Phi(u, v) = (3u + v, -u + 4v)$$

The area of the rectangle $\mathcal{R} = [0, 4] \times [0, 3]$ is $4 \cdot 3 = 12$, therefore the area of the image of $\mathcal{R}$ under Φ is

$$|\text{Jac}(\Phi)| \cdot 12.$$

The Jacobian of the linear map Φ is

$$\text{Jac}(\Phi) = \begin{vmatrix} A & C \\ B & D \end{vmatrix} = \begin{vmatrix} 3 & 1 \\ -1 & 4 \end{vmatrix} = 12 - (-1) = 13$$

Therefore the area in question is

$$|\text{Jac}(\Phi)| \cdot 12 = 13 \cdot 12 = 156$$

59. Let $\mathcal{D}$ be the shaded region in Figure 6, and let F be the map
$$u = y + x^2, \qquad v = y - x^3$$

(a) Show that F maps $\mathcal{D}$ to a rectangle $\mathcal{R}$ in the uv-plane.

(b) Apply Eq. (7) in Section 15.6 with $P = (1, 7)$ to estimate Area$(\mathcal{D})$.

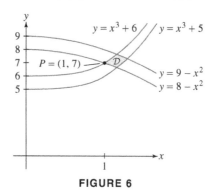

FIGURE 6

SOLUTION

(a) Note that the appropriate map should be $u = y + x^2$ rather than $u = -y + x^2$. We examine the images of the boundary curves of $\mathcal{D}$ under the map $(u, v) = \Phi(x, y) = (x^2 + y, y - x^3)$. The curves $y = x^3 + 6$ and $y = x^3 + 5$ can be rewritten as $y - x^3 = 6$ and $y - x^3 = 5$. Since $v = y - x^3$, these curves are mapped to the horizontal lines $v = 6$ and $v = 5$, respectively. The curves $y = 8 - x^2$ and $y = 9 - x^2$ can be rewritten as $y + x^2 = 8$ and $y + x^2 = 9$. Since $u = y + x^2$, these curves are mapped to the vertical lines $u = 8$ and $u = 9$, respectively. We conclude that $\mathcal{D}$ is mapped to the rectangle $\mathcal{R} = [8, 9] \times [5, 6]$ in the (u, v)-plane.

(b) We use Eq. (5) in section 16.5, where this time Φ is a mapping from the (x, y)-plane to the (u, v)-plane, and $P = (1, 7)$ is a point in $\mathcal{D}$:
$$\text{Area}\Phi(\mathcal{D}) \approx |\text{Jac}(\Phi)(P)|\text{Area}(\mathcal{D})$$

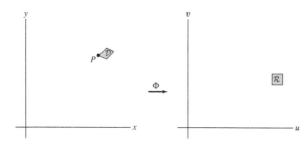

Here, Area$\Phi(\mathcal{D})$ = Area$(\mathcal{R}) = 1^2 = 1$, therefore we get
$$1 \approx |\text{Jac}(\Phi)(P)|\text{Area}(\mathcal{D})$$

or
$$\text{Area}(\mathcal{D}) \approx |\text{Jac}(\Phi)(P)|^{-1} \tag{1}$$

We compute the Jacobian of $\Phi(x, y) = (u, v) = (y + x^2, y - x^3)$ at $P = (1, 7)$:

$$\text{Jac}(\Phi) = \begin{vmatrix} \dfrac{\partial u}{\partial x} & \dfrac{\partial u}{\partial y} \\ \dfrac{\partial v}{\partial x} & \dfrac{\partial v}{\partial y} \end{vmatrix} = \begin{vmatrix} 2x & 1 \\ -3x^2 & 1 \end{vmatrix} = 2x + 3x^2 \quad \Rightarrow \quad \text{Jac}(\Phi)(P) = 2 \cdot 1 + 3 \cdot 7^2 = 149$$

Combining with (1) gives

$$\text{Area}(\mathcal{D}) \approx (149)^{-1} = \frac{1}{149}$$

61. Sketch the region $\mathcal{D}$ bounded by the curves $y = 2/x$, $y = 1/(2x)$, $y = 2x$, $y = x/2$ in the first quadrant. Let F be the map $u = xy$, $v = y/x$ from the xy-plane to the uv-plane.

(a) Find the image of $\mathcal{D}$ under F.

(b) Let $\Phi = F^{-1}$. Show that $|\text{Jac}(\Phi)| = \dfrac{1}{2|v|}$.

(c) Apply the Change of Variables Formula to prove the formula

$$\iint_{\mathcal{D}} f\left(\frac{y}{x}\right) dx\, dy = \frac{3}{4} \int_{1/2}^{2} \frac{f(v)\, dv}{v}$$

(d) Apply (c) to evaluate $\displaystyle\iint_{\mathcal{D}} \frac{y e^{y/x}}{x}\, dx\, dy$.

SOLUTION

(a) The region $\mathcal{D}$ is shown in the figure:

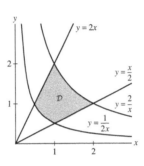

We rewrite the equations of the boundary curves as $xy = 2$, $xy = \frac{1}{2}$, $\frac{y}{x} = 2$, and $\frac{y}{x} = \frac{1}{2}$. These curves are mapped by Φ to the lines $u = 2$, $u = \frac{1}{2}$, $v = 2$, and $v = \frac{1}{2}$. Therefore, the image of $\mathcal{D}$ is the rectangle $\mathcal{R} = \left[\frac{1}{2}, 2\right] \times \left[\frac{1}{2}, 2\right]$ in the (u, v)-plane.

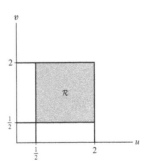

(b) We use the Jacobian of the inverse map:

$$\text{Jac}(\Phi^{-1}) = (\text{Jac}(\Phi))^{-1}$$

We compute the Jacobian of $\Phi(x, y) = (u, v) = \left(xy, \frac{y}{x}\right)$:

$$\text{Jac}(\Phi) = \begin{vmatrix} \dfrac{\partial u}{\partial x} & \dfrac{\partial u}{\partial y} \\ \dfrac{\partial v}{\partial x} & \dfrac{\partial v}{\partial y} \end{vmatrix} = \begin{vmatrix} y & x \\ -\dfrac{y}{x^2} & \dfrac{1}{x} \end{vmatrix} = \frac{y}{x} + \frac{yx}{x^2} = \frac{2y}{x} = 2v$$

(Note that everything is positive, so we don't need absolute values!) Thus,

$$\text{Jac}(\Phi^{-1}) = (\text{Jac}(\Phi))^{-1} = \frac{1}{2v} = \frac{1}{|2v|}$$

(c) The general change of variables formula is

$$\iint_D f(x,y)\,dA = \iint_R f(x(u,v), y(u,v))\,|\text{Jac}(\Phi^{-1})(u,v)|\,du\,dv$$

Here, $f\left(\frac{y}{x}\right) = f(v)$, $\mathcal{R} = \left[\frac{1}{2}, 2\right] \times \left[\frac{1}{2}, 2\right]$ in the (u,v)-plane and $|\text{Jac}(\Phi^{-1})(u,v)| = \left|\frac{1}{2v}\right| = \frac{1}{2v}$ ($v > 0$ in $\mathcal{R}$). Therefore, we have

$$\iint_D f\left(\frac{y}{x}\right) dA = \int_{1/2}^{2}\int_{1/2}^{2} f(v) \cdot \frac{1}{2v}\,du\,dv = \left(\int_{1/2}^{2} 1\,du\right)\left(\int_{1/2}^{2} \frac{f(v)}{2v}\,dv\right) = \frac{3}{4}\int_{1/2}^{2} \frac{f(v)}{v}\,dv$$

(d) We use part (c) with $f\left(\frac{y}{x}\right) = \frac{y}{x} \cdot e^{y/x}$. We have $f(v) = v \cdot e^v$, hence

$$\iint_D \frac{ye^{y/x}}{x}\,dx\,dy = \frac{3}{4}\int_{1/2}^{2} \frac{ve^v}{v}\,dv = \frac{3}{4}\int_{1/2}^{2} e^v\,dv = \frac{3}{4}e^v\bigg|_{1/2}^{2} = \frac{3}{4}(e^2 - e^{1/2}) = \frac{3}{4}\left(e^2 - \sqrt{e}\right)$$

16 LINE AND SURFACE INTEGRALS

16.1 Vector Fields (LT Section 17.1)

Preliminary Questions

1. Which of the following is a unit vector field in the plane?

(a) $\mathbf{F} = \langle y, x \rangle$

(b) $\mathbf{F} = \left\langle \dfrac{y}{\sqrt{x^2+y^2}}, \dfrac{x}{\sqrt{x^2+y^2}} \right\rangle$

(c) $\mathbf{F} = \left\langle \dfrac{y}{x^2+y^2}, \dfrac{x}{x^2+y^2} \right\rangle$

SOLUTION

(a) The length of the vector $\langle y, x \rangle$ is

$$\|\langle y, x \rangle\| = \sqrt{y^2 + x^2}$$

This value is not 1 for all points, hence it is not a unit vector field.

(b) We have

$$\left\|\left\langle \dfrac{y}{\sqrt{x^2+y^2}}, \dfrac{x}{\sqrt{x^2+y^2}} \right\rangle\right\| = \sqrt{\left(\dfrac{y}{\sqrt{x^2+y^2}}\right)^2 + \left(\dfrac{x}{\sqrt{x^2+y^2}}\right)^2}$$

$$= \sqrt{\dfrac{y^2}{x^2+y^2} + \dfrac{x^2}{x^2+y^2}} = \sqrt{\dfrac{y^2+x^2}{x^2+y^2}} = 1$$

Hence the field is a unit vector field, for $(x, y) \neq (0, 0)$.

(c) We compute the length of the vector:

$$\left\|\left\langle \dfrac{y}{x^2+y^2}, \dfrac{x}{x^2+y^2} \right\rangle\right\| = \sqrt{\left(\dfrac{y}{x^2+y^2}\right)^2 + \left(\dfrac{x}{x^2+y^2}\right)^2} = \sqrt{\dfrac{y^2+x^2}{(x^2+y^2)^2}} = \sqrt{\dfrac{1}{x^2+y^2}}$$

Since the length is not identically 1, the field is not a unit vector field.

2. Sketch an example of a nonconstant vector field in the plane in which each vector is parallel to $\langle 1, 1 \rangle$.

SOLUTION The nonconstant vector $\langle x, x \rangle$ is parallel to the vector $\langle 1, 1 \rangle$.

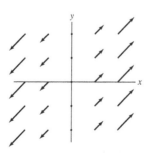

3. Show that the vector field $\mathbf{F} = \langle -z, 0, x \rangle$ is orthogonal to the position vector $\overrightarrow{OP}$ at each point P. Give an example of another vector field with this property.

SOLUTION The position vector at $P = (x, y, z)$ is $\langle x, y, z \rangle$. We must show that the following dot product is zero:

$$\langle x, y, z \rangle \cdot \langle -z, 0, x \rangle = x \cdot (-z) + y \cdot 0 + z \cdot x = 0$$

Therefore, the vector field $\mathbf{F} = \langle -z, 0, x \rangle$ is orthogonal to the position vector. Another vector field with this property is $\mathbf{F} = \langle 0, -z, y \rangle$, since

$$\langle 0, -z, y \rangle \cdot \langle x, y, z \rangle = 0 \cdot x + (-z) \cdot y + y \cdot z = 0$$

4. Show that $f(x, y, z) = xyz$ is a potential function for $\langle yz, xz, xy \rangle$ and give an example of a potential function other than f.

SOLUTION Since any two potential functions of a gradient vector field differ by a constant, a potential function for the given field other than $f(x, y, z) = xyz$ is, for instance, $f_1(x, y, z) = xyz + 1$.

Exercises

1. Compute and sketch the vector assigned to the points $P = (1, 2)$ and $Q = (-1, -1)$ by the vector field $\mathbf{F} = \langle x^2, x \rangle$.

SOLUTION The vector assigned to $P = (1, 2)$ is obtained by substituting $x = 1$ in $\mathbf{F}$, that is,

$$\mathbf{F}(1, 2) = \langle 1^2, 1 \rangle = \langle 1, 1 \rangle$$

Similarly,

$$\mathbf{F}(-1, -1) = \langle (-1)^2, -1 \rangle = \langle 1, -1 \rangle$$

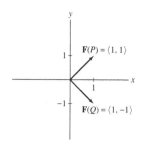

3. Compute and sketch the vector assigned to the points $P = (0, 1, 1)$ and $Q = (2, 1, 0)$ by the vector field $\mathbf{F} = \langle xy, z^2, x \rangle$.

SOLUTION To find the vector assigned to the point $P = (0, 1, 1)$, we substitute $x = 0$, $y = 1$, $z = 1$ in $\mathbf{F} = \langle xy, z^2, x \rangle$. We get

$$\mathbf{F}(P) = \langle 0 \cdot 1, 1^2, 0 \rangle = \langle 0, 1, 0 \rangle$$

Similarly, $\mathbf{F}(Q)$ is obtained by substituting $x = 2$, $y = 1$, $z = 0$ in $\mathbf{F}$. That is,

$$\mathbf{F}(Q) = \langle 2 \cdot 1, 0^2, 2 \rangle = \langle 2, 0, 2 \rangle$$

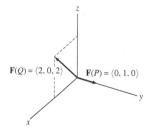

In Exercises 5–12, sketch the following planar vector fields by drawing the vectors attached to points with integer coordinates in the rectangle $-3 \le x \le 3$, $-3 \le y \le 3$. Instead of drawing the vectors with their true lengths, scale them if necessary to avoid overlap.

5. $\mathbf{F} = \langle 1, 0 \rangle$

SOLUTION The constant vector field $\langle 1, 0 \rangle$ is shown in the figure:

7. $\mathbf{F} = x\mathbf{i}$

SOLUTION The vector field $\mathbf{F}(x, y) = x\mathbf{i} = \langle x, 0 \rangle$ is sketched in the following figure:

9. $\mathbf{F} = \langle 0, x \rangle$

SOLUTION We sketch the vector field $\mathbf{F}(x, y) = \langle 0, x \rangle$:

11. $\mathbf{F} = \left\langle \dfrac{x}{x^2 + y^2}, \dfrac{y}{x^2 + y^2} \right\rangle$

SOLUTION

In Exercises 13–16, match each of the following planar vector fields with the corresponding plot in Figure 10.

1372 CHAPTER 16 | LINE AND SURFACE INTEGRALS (LT CHAPTER 17)

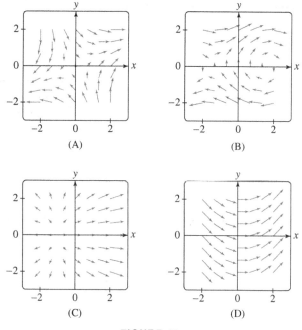

FIGURE 10

13. $\mathbf{F} = \langle 2, x \rangle$

SOLUTION The x-coordinate of the vector field $\langle 2, x \rangle$ is constant, and the y-coordinate is the x value. This matches only with Plot (D).

15. $\mathbf{F} = \langle y, \cos x \rangle$

SOLUTION We compute the images of the point $(0, 2)$, for instance, and identify the corresponding graph accordingly:

$$\mathbf{F}(x, y) = \langle y, \cos x \rangle \quad \Rightarrow \quad \mathbf{F}(0, 2) = \langle 2, 1 \rangle \quad \Rightarrow \quad \text{Plot(B)}$$

In Exercises 17–20, match each three-dimensional vector field with the corresponding plot in Figure 11.

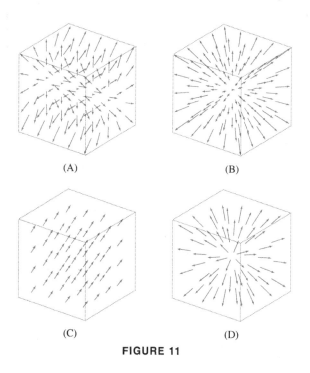

FIGURE 11

17. $\mathbf{F} = \langle 1, 1, 1 \rangle$

SOLUTION The constant vector field $\langle 1, 1, 1 \rangle$ is shown in plot (C).

19. $\mathbf{F} = \langle x, y, z \rangle$

SOLUTION $\langle x, y, z \rangle$ is shown in plot (B). Note that the vectors are pointing away from the origin and are of increasing magnitude.

21. A river 200 meters wide is modeled by the region in the xy-plane given by $-100 \le x \le 100$. The velocity vector field on the surface of the river is given by $\mathbf{F} = \langle -0.05x, 20 - 0.0001x^2 \rangle$ in meters per second. Determine the coordinates of those points that have the maximum speed.

SOLUTION The speed is the magnitude of the velocity vector, so it is

$$\|\mathbf{F}\| = \sqrt{(-0.05x)^2 + (20 - 0.0001x^2)^2} = \sqrt{400 - 0.0375x^2 + 10^{-6}x^4}$$

Maximizing $\|\mathbf{F}\|$ is the same as maximizing its square, so we wish to maximize $s(x) = 400 - 0.0375x^2 + 10^{-6}x^4$. Differentiating gives

$$s'(x) = -0.075x + 4 \cdot 10^{-6}x^3$$

setting equal to zero and solving gives $x = 0$ and $x \approx 136.931$. To see which of these is a maximum, look at $s''(x)$:

$$s''(x) = -0.075 + 12 \cdot 10^{-6}x^2$$

At $x = 0$, we see that $s''(x)$ is negative, while at $x \approx 136.931$,

$$s''(136.931) \approx -0.075 + 12 \cdot 10^{-6} \cdot 136.931^2 \approx 0.15 > 0$$

Therefore, the maximum speed occurs in the center of the river, at $x = 0$, with velocity $\langle 0, 20 \rangle$.

In Exercises 23–30, calculate div($\mathbf{F}$) *and* curl($\mathbf{F}$).

23. $\mathbf{F} = \langle x, y, z \rangle$

SOLUTION The divergence of $\mathbf{F}$ is

$$\text{div}(\mathbf{F}) = \frac{\partial}{\partial x}(x) + \frac{\partial}{\partial y}(y) + \frac{\partial}{\partial z}(z) = 1 + 1 + 1 = 3$$

The curl of $\mathbf{F}$ is

$$\text{curl}(\mathbf{F}) = \left\langle \frac{\partial F_3}{\partial y} - \frac{\partial F_2}{\partial z}, \frac{\partial F_1}{\partial z} - \frac{\partial F_3}{\partial x}, \frac{\partial F_2}{\partial x} - \frac{\partial F_1}{\partial y} \right\rangle$$

$$= \left\langle \frac{\partial z}{\partial y} - \frac{\partial y}{\partial z}, \frac{\partial x}{\partial z} - \frac{\partial z}{\partial x}, \frac{\partial y}{\partial x} - \frac{\partial x}{\partial y} \right\rangle$$

$$= \langle 0, 0, 0 \rangle$$

25. $\mathbf{F} = \langle x - 2zx^2, z - xy, z^2x^2 \rangle$

SOLUTION

$$\text{div}(\mathbf{F}) = \frac{\partial}{\partial x}(x - 2zx^2) + \frac{\partial}{\partial y}(z - xy) + \frac{\partial}{\partial z}(z^2x^2) = (1 - 4zx) + (-x) + (2zx^2) = 1 - 4zx - x + 2zx^2$$

The curl of $\mathbf{F}$ is

$$\text{curl}(\mathbf{F}) = \left\langle \frac{\partial F_3}{\partial y} - \frac{\partial F_2}{\partial z}, \frac{\partial F_1}{\partial z} - \frac{\partial F_3}{\partial x}, \frac{\partial F_2}{\partial x} - \frac{\partial F_1}{\partial y} \right\rangle$$

$$= \left\langle 0 - 1, -2x^2 - 2z^2x, -y - 0 \right\rangle = \left\langle -1, -2x(x + z^2), -y \right\rangle$$

27. $\mathbf{F} = \langle yz, xz, xy \rangle$

SOLUTION The divergence of $\mathbf{F}$ is

$$\text{div}(\mathbf{F}) = \frac{\partial}{\partial x}(yz) + \frac{\partial}{\partial y}(xz) + \frac{\partial}{\partial z}(xy) = 0 + 0 + 0 = 0$$

The curl of **F** is

$$\text{curl}(\mathbf{F}) = \left\langle \frac{\partial F_3}{\partial y} - \frac{\partial F_2}{\partial z}, \frac{\partial F_1}{\partial z} - \frac{\partial F_3}{\partial x}, \frac{\partial F_2}{\partial x} - \frac{\partial F_1}{\partial y} \right\rangle$$

$$= \left\langle \frac{\partial}{\partial y}(xy) - \frac{\partial}{\partial z}(xz), \frac{\partial}{\partial z}(yz) - \frac{\partial}{\partial x}(xy), \frac{\partial}{\partial x}(xz) - \frac{\partial}{\partial y}(yz) \right\rangle$$

$$= \langle 0, 0, 0 \rangle$$

29. $\mathbf{F} = \langle e^y, \sin x, \cos x \rangle$

SOLUTION The divergence of **F** is

$$\text{div}(\mathbf{F}) = \frac{\partial}{\partial x} e^y + \frac{\partial}{\partial y} \sin x + \frac{\partial}{\partial z} \cos x = 0$$

The curl of **F** is

$$\text{curl}(\mathbf{F}) = \begin{vmatrix} \mathbf{i} & \mathbf{j} & \mathbf{k} \\ \frac{\partial}{\partial x} & \frac{\partial}{\partial y} & \frac{\partial}{\partial z} \\ e^y & \sin x & \cos x \end{vmatrix} = 0\mathbf{i} - (-\sin x)\mathbf{j} + (\cos x - e^y)\mathbf{k} = \langle 0, \sin x, \cos x - e^y \rangle$$

In Exercises 31–37, prove the identities assuming that the appropriate partial derivatives exist and are continuous.

31. $\text{div}(\mathbf{F} + \mathbf{G}) = \text{div}(\mathbf{F}) + \text{div}(\mathbf{G})$

SOLUTION Suppose $\mathbf{F} = \langle F_1, F_2, F_3 \rangle$ and $\mathbf{G} = \langle G_1, G_2, G_3 \rangle$. Then

$$\text{div}(\mathbf{F} + \mathbf{G}) = \text{div}(\langle F_1, F_2, F_3 \rangle + \langle G_1, G_2, G_3 \rangle) = \text{div}(\langle F_1 + G_1, F_2 + G_2, F_3 + G_3 \rangle)$$

$$= \frac{\partial}{\partial x}(F_1 + G_1) + \frac{\partial}{\partial y}(F_2 + G_2) + \frac{\partial}{\partial z}(F_3 + G_3)$$

$$= \frac{\partial}{\partial x} F_1 + \frac{\partial}{\partial x} G_1 + \frac{\partial}{\partial y} F_2 + \frac{\partial}{\partial y} G_2 + \frac{\partial}{\partial z} F_3 + \frac{\partial}{\partial z} G_3$$

$$= \left(\frac{\partial}{\partial x} F_1 + \frac{\partial}{\partial y} F_2 + \frac{\partial}{\partial z} F_3 \right) + \left(\frac{\partial}{\partial x} G_1 + \frac{\partial}{\partial y} G_2 + \frac{\partial}{\partial z} G_3 \right)$$

$$= \text{div}(\mathbf{F}) + \text{div}(\mathbf{G})$$

33. $\text{div curl}(\mathbf{F}) = 0$

SOLUTION Let $\mathbf{F} = \langle F_1, F_2, F_3 \rangle$. Then since the necessary partials are continuous, mixed partials are equal, so that

$$\text{div curl}(\mathbf{F}) = \text{div}\left\langle \frac{\partial F_3}{\partial y} - \frac{\partial F_2}{\partial z}, \frac{\partial F_1}{\partial z} - \frac{\partial F_3}{\partial x}, \frac{\partial F_2}{\partial x} - \frac{\partial F_1}{\partial y} \right\rangle$$

$$= \frac{\partial}{\partial x}\left(\frac{\partial F_3}{\partial y} - \frac{\partial F_2}{\partial z} \right) + \frac{\partial}{\partial y}\left(\frac{\partial F_1}{\partial z} - \frac{\partial F_3}{\partial x} \right) + \frac{\partial}{\partial z}\left(\frac{\partial F_2}{\partial x} - \frac{\partial F_1}{\partial y} \right)$$

$$= \frac{\partial F_3}{\partial x \partial y} - \frac{\partial F_2}{\partial x \partial z} + \frac{\partial F_1}{\partial y \partial z} - \frac{\partial F_3}{\partial y \partial x} + \frac{\partial F_2}{\partial z \partial x} - \frac{\partial F_1}{\partial z \partial y}$$

$$= 0$$

35. If f is a scalar function, then $\text{div}(f\mathbf{F}) = f\text{div}(\mathbf{F}) + \mathbf{F} \cdot \nabla f$.

SOLUTION Let $\mathbf{F} = \langle F_1, F_2, F_3 \rangle$. Then

$$\text{div}(f\mathbf{F}) = \text{div}\langle fF_1, fF_2, fF_3 \rangle$$

$$= \frac{\partial}{\partial x}(fF_1) + \frac{\partial}{\partial y}(fF_2) + \frac{\partial}{\partial z}(fF_3)$$

$$= f\frac{\partial}{\partial x}F_1 + F_1\frac{\partial}{\partial x}f + f\frac{\partial}{\partial y}F_2 + F_2\frac{\partial}{\partial y}f + f\frac{\partial}{\partial z}F_3 + F_3\frac{\partial}{\partial z}f$$

$$= f\left(\frac{\partial}{\partial x}F_1 + \frac{\partial}{\partial y}F_2 + \frac{\partial}{\partial z}F_3\right) + \langle F_1, F_2, F_3\rangle \cdot \left(\frac{\partial}{\partial x}f, \frac{\partial}{\partial y}f, \frac{\partial}{\partial z}f\right)$$

$$= f\,\text{div}(\mathbf{F}) + \mathbf{F}\cdot\nabla f$$

37. $\text{div}(\nabla f \times \nabla g) = 0$

SOLUTION From Exercise 34,

$$\text{div}(\nabla f \times \nabla g) = \nabla g \cdot \text{curl}(\nabla f) - \nabla f \cdot \text{curl}(\nabla g)$$

But by Theorem 1, $\text{curl}(\nabla f) = \text{curl}(\nabla g) = 0$, so that $\text{div}(\nabla f \times \nabla g) = 0$.

In Exercises 39–47, find a potential function for the vector field $\mathbf{F}$ by inspection or show that one does not exist.

39. $\mathbf{F} = \langle x, y\rangle$

SOLUTION We must find a function $\varphi(x, y)$ such that $\frac{\partial \varphi}{\partial x} = x$ and $\frac{\partial \varphi}{\partial y} = y$. We choose the following function:

$$\varphi(x, y) = \frac{1}{2}x^2 + \frac{1}{2}y^2$$

41. $\mathbf{F} = \langle y^2 z, 1 + 2xyz, xy^2\rangle$

SOLUTION We want a function whose x derivative is $y^2 z$, so try $xy^2 z$. The z derivative of this function is xy^2, which matches; however, its y derivative is $2xyz$. To fix this, use the function $\varphi(x, y, z) = y + xy^2 z$.

43. $\mathbf{F} = \langle ye^{xy}, xe^{xy}\rangle$

SOLUTION The function $\varphi(x, y) = e^{xy}$ satisfies $\frac{\partial \varphi}{\partial x} = ye^{xy}$ and $\frac{\partial \varphi}{\partial y} = xe^{xy}$, hence φ is a potential function for the given vector field.

45. $\mathbf{F} = \langle yz^2, xz^2, 2xyz\rangle$

SOLUTION We choose a function $\varphi(x, y, z)$ such that

$$\frac{\partial \varphi}{\partial x} = yz^2, \quad \frac{\partial \varphi}{\partial y} = xz^2, \quad \frac{\partial \varphi}{\partial z} = xyz$$

The function $\varphi(x, y, z) = xyz^2$ is a potential function for the given field.

47. $\mathbf{F} = \langle yz\cos(xyz), xz\cos(xyz), xy\cos(xyz)\rangle$

SOLUTION A calculation shows that $\text{curl}(\mathbf{F}) = 0$, so that this vector field is conservative. Integrating the first component with respect to x gives $\varphi(x) = \sin(xyz)$, and it is easily verified that $\text{div}\,\varphi = \mathbf{F}$ so that φ is a potential function for $\mathbf{F}$.

49. Show that $\mathbf{F} = \langle 3, 1, 2\rangle$ is conservative. Then prove more generally that any constant vector field $\mathbf{F} = \langle a, b, c\rangle$ is conservative.

SOLUTION $\mathbf{F} = \nabla\varphi$ for $\varphi(x, y, z) = 3x + y + 2z$. Further for $\mathbf{F} = \langle a, b, c\rangle$, $\mathbf{F} = \nabla\varphi$ for $\varphi(x, y, z) = ax + by + cz$.

51. For $P = (a, b)$, we define the unit radial vector field based at P:

$$\mathbf{e}_P = \frac{\langle x - a, y - b\rangle}{\sqrt{(x-a)^2 + (y-b)^2}}$$

(a) Verify that $\mathbf{e}_P$ is a unit vector field.
(b) Calculate $\mathbf{e}_P(1, 1)$ for $P = (3, 2)$.
(c) Find a potential function for $\mathbf{e}_P$.

SOLUTION
(a)

$$\mathbf{e}_P \cdot \mathbf{e}_P = \frac{(x-a)^2 + (y-b)^2}{\left(\sqrt{(x-a)^2 + (y-b)^2}\right)^2} = 1$$

(b)

$$\mathbf{e}_{(3,2)}(1,1) = \frac{\langle 1-3, 1-2\rangle}{\sqrt{(1-3)^2 + (1-2)^2}} = \frac{\langle -2, -1\rangle}{\sqrt{5}}$$

(c) Let $\varphi(x,y) = \sqrt{(x-a)^2 + (y-b)^2}$. Then,

$$\nabla\varphi = \frac{1}{2}((x-a)^2 + (y-b)^2)^{-1/2}\langle 2(x-a), 2(y-b)\rangle = \mathbf{e}_P$$

53. Which of (A) or (B) in Figure 13 is the contour plot of a potential function for the vector field **F**?

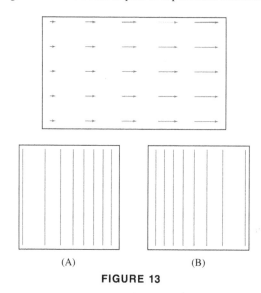

FIGURE 13

SOLUTION Since $\nabla\varphi = \mathbf{F}$, **F** is perpendicular to the level curves of φ, as seen in both the contour plots (A) and (B). The plot of **F** shows that **F** has the form $\mathbf{F} = \langle f(x), 0\rangle$ for an increasing function $f(x)$. Therefore, $\frac{\partial \varphi}{\partial x} = f(x)$ is increasing, implying that the rate of change of φ with respect to x is increasing. Hence, the density of the vertical lines is greater in the direction of growing x. We conclude that (A) is the contour plot of φ.

55. In this exercise, we show that the vector field **F** in Figure 15 is not conservative. Explain the following statements:

(a) If a potential function f for **F** exists, then the level curves of f must be vertical lines.

(b) If a potential function f for **F** exists, then the level curves of f must grow farther apart as y increases.

(c) Explain why (a) and (b) are incompatible, and hence f cannot exist.

FIGURE 15

SOLUTION

(a) If f is a potential function for **F**, then $\nabla f = \mathbf{F}$. Therefore, **F** is orthogonal to the level curves of f. The plot shows that the vectors $\mathbf{F}(x,y)$ are horizontal, hence the level curves of f are vertical lines.

(b) As indicated by the graph of the vector field $\mathbf{F} = \langle F_1, F_2\rangle$, $F_2 = 0$ and F_1 is decreasing as y increases. Since $\mathbf{F} = \frac{\partial f}{\partial x}$, it follows that $\frac{\partial f}{\partial x}$ is decreasing as y increases, or the rate of change of f with respect to x is decreasing as y increases. Therefore, as y increases, the level curves of f are getting farther apart.

(c) By (a), the level curves of f are vertical lines, hence the distance between any two level curves is constant rather than increasing with y as concluded in (b).

Further Insights and Challenges

57. Let $\mathcal{D}$ be a disk in $\mathbf{R}^2$. This exercise shows that if
$$\nabla f(x,y) = \mathbf{0}$$
for all (x,y) in $\mathcal{D}$, then f is constant. Consider points $P = (a,b)$, $Q = (c,d)$, and $R = (c,b)$ as in Figure 16.

(a) Use single-variable calculus to show that f is constant along the segments $\overline{PR}$ and $\overline{RQ}$.

(b) Conclude that $f(P) = f(Q)$ for any two points $P, Q \in \mathcal{D}$.

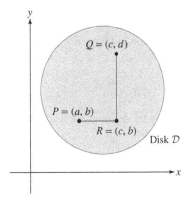

FIGURE 16

SOLUTION Given any two points $P = (a,b)$ and $Q = (c,d)$ in $\mathcal{D}$, we must show that
$$f(P) = f(Q)$$
We consider the point $R = (c,b)$ and the segments $\overline{PR}$ and $\overline{RQ}$. (We assume that (c,b) is in $\mathcal{D}$; if not, just use $R' = (a,d)$.)

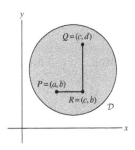

Since $\frac{\partial f}{\partial x}(x,y) = 0$ in $\mathcal{D}$, in particular $\frac{\partial f}{\partial x}(x,b) = 0$ for $a \le x \le c$. Therefore, for $a \le x \le c$ we have
$$f(x,b) = \int_a^x \frac{\partial f}{\partial u}(u,b)\,du + f(a,b) = \int_a^x 0\,du + f(a,b) = k + f(a,b)$$
Substituting $x = a$ determines $k = 0$. Hence,
$$f(x,b) = f(a,b) \quad \text{for} \quad a \le x \le c$$
In particular,
$$f(c,b) = f(a,b) \quad \Rightarrow \quad f(R) = f(P) \tag{1}$$
Similarly, since $\frac{\partial f}{\partial y}(x,y) = 0$ in $\mathcal{D}$, we have $\frac{\partial f}{\partial y}(c,y) = 0$ for $b \le y \le d$. Therefore for $b \le y \le d$ we have
$$f(c,y) = \int_b^y \frac{\partial f}{\partial v}(c,v)\,dv + f(c,b) = \int_b^y 0\,dv + f(c,b) = k + f(c,b)$$
Substituting $y = b$ gives $f(c,b) = k + f(c,b)$ or $k = 0$. Therefore,
$$f(c,y) = f(c,b) \quad \text{for} \quad b \le y \le d$$
In particular,
$$f(c,d) = f(c,b) \quad \Rightarrow \quad f(Q) = f(R) \tag{2}$$
Combining (1) and (2) we obtain the desired equality $f(P) = f(Q)$. Since P and Q are any two points in $\mathcal{D}$, we conclude that f is constant on $\mathcal{D}$.

16.2 Line Integrals (LT Section 17.2)

Preliminary Questions

1. What is the line integral of the constant function $f(x, y, z) = 10$ over a curve C of length 5?

SOLUTION Since the length of C is the line integral $\int_C 1\, ds = 5$, we have

$$\int_C 10\, ds = 10 \int_C 1\, ds = 10 \cdot 5 = 50$$

2. Which of the following have a zero line integral over the vertical segment from $(0, 0)$ to $(0, 1)$?

(a) $f(x, y) = x$ (b) $f(x, y) = y$
(c) $\mathbf{F} = \langle x, 0 \rangle$ (d) $\mathbf{F} = \langle y, 0 \rangle$
(e) $\mathbf{F} = \langle 0, x \rangle$ (f) $\mathbf{F} = \langle 0, y \rangle$

SOLUTION The vertical segment from $(0, 0)$ to $(0, 1)$ has the parametrization

$$\mathbf{r}(t) = (0, t), \quad 0 \le t \le 1$$

Therefore, $\mathbf{r}'(t) = \langle 0, 1 \rangle$ and $\|\mathbf{r}'(t)\| = 1$. The line integrals are thus computed by

$$\int_C f(x, y)\, ds = \int_0^1 f(\mathbf{r}(t)) \|\mathbf{r}'(t)\|\, dt \tag{1}$$

$$\int_C \mathbf{F} \cdot d\mathbf{s} = \int_0^1 \mathbf{F}(\mathbf{r}(t)) \cdot \mathbf{r}'(t)\, dt \tag{2}$$

(a) We have $f(\mathbf{r}(t)) = x = 0$. Therefore by (1) the line integral is zero.
(b) By (1), the line integral is

$$\int_C f(x, y)\, ds = \int_0^1 t \cdot 1\, dt = \frac{1}{2}t^2 \Big|_0^1 = \frac{1}{2} \ne 0$$

(c) This vector line integral is computed using (2). Since $\mathbf{F}(\mathbf{r}(t)) = \langle x, 0 \rangle = \langle 0, 0 \rangle$, the vector line integral is zero.

(d) By (2) we have

$$\int_C \mathbf{F} \cdot d\mathbf{r} = \int_0^1 \langle t, 0 \rangle \cdot \langle 0, 1 \rangle\, dt = \int_0^1 0\, dt = 0$$

(e) The vector integral is computed using (2). Since $\mathbf{F}(\mathbf{r}(t)) = \langle 0, x \rangle = \langle 0, 0 \rangle$, the line integral is zero.

(f) For this vector field we have

$$\int_C \mathbf{F} \cdot d\mathbf{r} = \int_0^1 \mathbf{F}(\mathbf{r}(t)) \cdot \mathbf{r}'(t)\, dt = \int_0^1 \langle 0, t \rangle \cdot \langle 0, 1 \rangle\, dt = \int_0^1 t\, dt = \frac{t^2}{2} \Big|_0^1 = \frac{1}{2} \ne 0$$

So, we conclude that (a), (c), (d), and (e) have an integral of zero.

3. State whether each statement is true or false. If the statement is false, give the correct statement.
(a) The scalar line integral does not depend on how you parametrize the curve.
(b) If you reverse the orientation of the curve, neither the vector line integral nor the scalar line integral changes sign.

SOLUTION
(a) True: It can be shown that any two parametrizations of the curve yield the same value for the scalar line integral, hence the statement is true.
(b) False: For the definition of the scalar line integral, there is no need to specify a direction along the path, hence reversing the orientation of the curve does not change the sign of the integral. However, reversing the orientation of the curve changes the sign of the vector line integral.

4. Suppose that C has length 5. What is the value of $\int_C \mathbf{F} \cdot d\mathbf{r}$ if:
(a) $\mathbf{F}(P)$ is normal to C at all points P on C?
(b) $\mathbf{F}(P)$ is a unit vector pointing in the negative direction along the curve?

SOLUTION

(a) The vector line integral is the integral of the tangential component of the vector field along the curve. Since $\mathbf{F}(P)$ is normal to C at all points P on C, the tangential component is zero, hence the line integral $\int_C \mathbf{F} \cdot d\mathbf{r}$ is zero.

(b) In this case we have

$$\mathbf{F}(P) \cdot \mathbf{T}(P) = (-\mathbf{T}(P)) \cdot \mathbf{T}(P) = -\|\mathbf{T}(P)\|^2 = -1$$

Therefore,

$$\int_C \mathbf{F} \cdot d\mathbf{r} = \int_C (\mathbf{F} \cdot \mathbf{T}) \, ds = \int_C (-1) \, ds = -\text{Length of } C = -5$$

Exercises

1. Let $f(x, y, z) = x + yz$, and let C be the line segment from $P = (0, 0, 0)$ to $(6, 2, 2)$.

(a) Calculate $f(\mathbf{r}(t))$ and $ds = \|\mathbf{r}'(t)\| \, dt$ for the parametrization $\mathbf{r}(t) = \langle 6t, 2t, 2t \rangle$ for $0 \le t \le 1$.

(b) Evaluate $\int_C f(x, y, z) \, ds$.

SOLUTION

(a) We substitute $x = 6t$, $y = 2t$, $z = 2t$ in the function $f(x, y, z) = x + yz$ to find $f(\mathbf{r}(t))$:

$$f(\mathbf{r}(t)) = 6t + (2t)(2t) = 6t + 4t^2$$

We differentiate the vector $\mathbf{r}(t)$ and compute the length of the derivative vector:

$$\mathbf{r}'(t) = \frac{d}{dt}\langle 6t, 2t, 2t \rangle = \langle 6, 2, 2 \rangle \quad \Rightarrow \quad \mathbf{r}'(t) = \sqrt{6^2 + 2^2 + 2^2} = \sqrt{44} = 2\sqrt{11}$$

Hence,

$$ds = \|\mathbf{r}'(t)\| \, dt = 2\sqrt{11} \, dt$$

(b) Computing the scalar line integral, we obtain

$$\int_C f(x, y, z) \, ds = \int_0^1 f(\mathbf{r}(t)) \|\mathbf{r}'(t)\| \, dt = \int_0^1 (6t + 4t^2) \cdot 2\sqrt{11} \, dt$$

$$= 2\sqrt{11} \left(3t^2 + \frac{4}{3}t^3 \right) \Big|_0^1 = 2\sqrt{11} \left(3 + \frac{4}{3} \right) = \frac{26\sqrt{11}}{3}$$

3. Let $\mathbf{F} = \langle y^2, x^2 \rangle$, and let C be the curve $y = x^{-1}$ for $1 \le x \le 2$, oriented from left to right.

(a) Calculate $\mathbf{F}(\mathbf{r}(t))$ and $d\mathbf{r} = \mathbf{r}'(t) \, dt$ for the parametrization of C given by $\mathbf{r}(t) = \langle t, t^{-1} \rangle$.

(b) Calculate the dot product $\mathbf{F}(\mathbf{r}(t)) \cdot \mathbf{r}'(t) \, dt$ and evaluate $\int_C \mathbf{F} \cdot d\mathbf{r}$.

SOLUTION

(a) We calculate $\mathbf{F}(\mathbf{r}(t))$ by substituting $x = t$, $y = t^{-1}$ in $\mathbf{F} = \langle y^2, x^2 \rangle$. We get

$$\mathbf{F}(\mathbf{r}(t)) = \langle (t^{-1})^2, t^2 \rangle = \langle t^{-2}, t^2 \rangle$$

We compute $\mathbf{r}'(t)$:

$$\mathbf{r}'(t) = \frac{d}{dt}\langle t, t^{-1} \rangle = \langle 1, -t^{-2} \rangle \quad \Rightarrow \quad d\mathbf{r} = \langle 1, -t^{-2} \rangle \, dt$$

(b) We compute the dot product:

$$\mathbf{F}(\mathbf{r}(t)) \cdot \mathbf{r}'(t) = \langle t^{-2}, t^2 \rangle \cdot \langle 1, -t^{-2} \rangle = t^{-2} \cdot 1 + t^2 \cdot (-t^{-2}) = t^{-2} - 1$$

Computing the vector line integral, we obtain

$$\int_C \mathbf{F} \cdot d\mathbf{r} = \int_1^2 \mathbf{F}(\mathbf{r}(t)) \cdot \mathbf{r}'(t)\,dt = \int_1^2 (t^{-2} - 1)\,dt = -t^{-1} - t \Big|_1^2 = \left(-\frac{1}{2} - 2\right) - (-1 - 1) = -\frac{1}{2}$$

In Exercises 5–8, compute the integral of the scalar function or vector field over $\mathbf{r}(t) = \langle \cos t, \sin t, t \rangle$ *for* $0 \le t \le \pi$.

5. $f(x, y, z) = x^2 + y^2 + z^2$

SOLUTION

Step 1. Compute $\|\mathbf{r}'(t)\|$. We differentiate $\mathbf{r}(t)$:

$$\mathbf{r}'(t) = \frac{d}{dt} \langle \cos t, \sin t, t \rangle = \langle -\sin t, \cos t, 1 \rangle$$

Hence,

$$\|\mathbf{r}'(t)\| = \sqrt{(-\sin t)^2 + \cos^2 t + 1^2} = \sqrt{\sin^2 t + \cos^2 t + 1} = \sqrt{2}$$

$$ds = \|\mathbf{r}'(t)\|\,dt = \sqrt{2}\,dt$$

Step 2. Write out $f(\mathbf{r}(t))$. We substitute $x = \cos t$, $y = \sin t$, $z = t$ in $f(x, y, z) = x^2 + y^2 + z^2$ to obtain

$$f(\mathbf{r}(t)) = \cos^2 t + \sin^2 t + t^2 = 1 + t^2$$

Step 3. Compute the line integral. Using the Theorem on Scalar Line Integrals we obtain

$$\int_C (x^2 + y^2 + z^2)\,ds = \int_0^\pi f(\mathbf{r}(t)) \|\mathbf{r}'(t)\|\,dt = \int_0^\pi (1 + t^2)\sqrt{2}\,dt = \sqrt{2}\left(t + \frac{t^3}{3}\right)\Big|_0^\pi = \sqrt{2}\left(\pi + \frac{\pi^3}{3}\right)$$

7. $\mathbf{F}(x, y, z) = \langle x, y, z^2 \rangle$

SOLUTION

Step 1. Calculate the integrand. We write out the vectors:

$$\mathbf{r}(t) = \langle \cos t, \sin t, t \rangle$$

$$\mathbf{F}(\mathbf{r}(t)) = \langle x, y, z^2 \rangle = \langle \cos t, \sin t, t^2 \rangle$$

$$\mathbf{r}'(t) = \langle -\sin t, \cos t, 1 \rangle$$

The integrand is the dot product:

$$\mathbf{F}(\mathbf{r}(t)) \cdot \mathbf{r}'(t) = \langle \cos t, \sin t, t^2 \rangle \cdot \langle -\sin t, \cos t, 1 \rangle = -\cos t \sin t + \sin t \cos t + t^2 = t^2$$

Step 2. Evaluate the integral. We use the Theorem on Vector Line Integrals to evaluate the integral:

$$\int_C \mathbf{F}\,d\mathbf{r} = \int_0^\pi \mathbf{F}(\mathbf{r}(t)) \cdot \mathbf{r}'(t)\,dt = \int_0^\pi t^2\,dt = \frac{1}{3}t^3 \Big|_0^\pi = \frac{\pi^3}{3}$$

In Exercises 9–16, compute $\int_C f\,ds$ *for the curve specified.*

9. $f(x, y) = \sqrt{1 + 9xy}$, $y = x^3$ for $0 \le x \le 2$

SOLUTION The curve is parametrized by $\mathbf{r}(t) = \langle t, t^3 \rangle$ for $0 \le t \le 2$.

Step 1. Compute $\|\mathbf{r}'(t)\|$. We have

$$\mathbf{r}'(t) = \frac{d}{dt}\langle t, t^3 \rangle = \langle 1, 3t^2 \rangle \quad \Rightarrow \quad \|\mathbf{r}'(t)\| = \sqrt{1 + 9t^4}$$

Step 2. Write out $f(\mathbf{r}(t))$. We substitute $x = t$, $y = t^3$ in $f(x, y) = \sqrt{1 + 9xy}$ to obtain

$$f(\mathbf{r}(t)) = \sqrt{1 + 9t \cdot t^3} = \sqrt{1 + 9t^4}$$

Step 3. Compute the line integral. We use the Theorem on Scalar Line Integrals to write

$$\int_C f(x, y)\, ds = \int_0^2 f(\mathbf{r}(t))\, \|\mathbf{r}'(t)\|\, dt = \int_0^2 \sqrt{1 + 9t^4}\, \sqrt{1 + 9t^4}\, dt = \int_0^2 \left(1 + 9t^4\right) dt$$

$$= t + \frac{9t^5}{5}\bigg|_0^2 = \frac{298}{5} = 59.6$$

11. $f(x, y, z) = z^2$, $\mathbf{r}(t) = \langle 2t, 3t, 4t \rangle$ for $0 \le t \le 2$

SOLUTION

Step 1. Compute $\|\mathbf{r}'(t)\|$ We have

$$\mathbf{r}'(t) = \frac{d}{dt}\langle 2t, 3t, 4t \rangle = \langle 2, 3, 4 \rangle \quad \Rightarrow \quad \|\mathbf{r}'(t)\| = \sqrt{2^2 + 3^2 + 4^2} = \sqrt{29}$$

Step 2. Write out $f(\mathbf{r}(t))$ We substitute $z = 4t$ in $f(x, y, z) = z^2$ to obtain:

$$f(\mathbf{r}(t)) = 16t^2$$

Step 3. Compute the line integral. By the Theorem on Scalar Line Integrals we have

$$\int_C f(x, y, z)\, ds = \int_0^2 f(\mathbf{r}(t))\, \|\mathbf{r}'(t)\|\, dt = \int_0^2 16t^2 \cdot \sqrt{29}\, dt = \sqrt{29} \cdot \frac{16}{3}t^3 \bigg|_0^2 = \frac{128\sqrt{29}}{3} \approx 229.8$$

13. $f(x, y, z) = xe^{z^2}$, piecewise linear path from $(0, 0, 1)$ to $(0, 2, 0)$ to $(1, 1, 1)$

SOLUTION Let C_1 be the segment joining the points $(0, 0, 1)$ and $(0, 2, 0)$ and C_2 be the segment joining the points $(0, 2, 0)$ and $(1, 1, 1)$. We parametrize C_1 and C_2 by the following parametrization:

$$C_1: \mathbf{r}_1(t) = (0, 2t, 1 - t),\ 0 \le t \le 1$$

$$C_2: \mathbf{r}_2(t) = (t, 2 - t, t),\ 0 \le t \le 1$$

For $C = C_1 + C_1$ we have

$$\int_C f(x, y, z)\, ds = \int_{C_1} f(x, y, z)\, ds + \int_{C_2} f(x, y, z)\, ds \tag{1}$$

We compute the integrals on the right-hand side.

- The integral over C_1: We have

$$\mathbf{r}_1'(t) = \frac{d}{dt}\langle 0, 2t, 1 - t \rangle = \langle 0, 2, -1 \rangle \quad \Rightarrow \quad \|\mathbf{r}_1'(t)\| = \sqrt{0 + 4 + 1} = \sqrt{5}$$

$$f(\mathbf{r}_1(t)) = xe^{z^2} = 0 \cdot e^{(1-t)^2} = 0$$

Hence,

$$\int_{C_1} f(x, y, z)\, ds = \int_0^1 f(\mathbf{r}_1(t))\, \|\mathbf{r}_1'(t)\|\, dt = \int_0^1 0\, dt = 0 \tag{2}$$

- The integral over C_2: We have

$$\mathbf{r}_2'(t) = \frac{d}{dt}\langle t, 2 - t, t \rangle = \langle 1, -1, 1 \rangle \quad \Rightarrow \quad \|\mathbf{r}_2'(t)\| = \sqrt{1 + 1 + 1} = \sqrt{3}$$

$$f(\mathbf{r}_2(t)) = xe^{z^2} = te^{t^2}$$

Hence,

$$\int_{C_2} f(x, y, z)\, ds = \int_0^1 te^{t^2} \sqrt{3}\, dt \tag{3}$$

Using the substitution $u = t^2$ we find that

$$\int_{C_2} f(x,y,z)\,ds = \int_0^1 \frac{\sqrt{3}}{2} e^u\,du = \frac{\sqrt{3}}{2}(e-1) \approx 1.488$$

Hence,

$$\int_C f(x,y,z)\,ds \approx 1.488$$

15. $f(x,y,z) = 2x^2 + 8z$, $\mathbf{r}(t) = \langle e^t, t^2, t \rangle$, $0 \le t \le 1$

SOLUTION

Step 1. Compute $\|\mathbf{r}'(t)\|$.

$$\mathbf{r}'(t) = \frac{d}{dt}\langle e^t, t^2, t \rangle = \langle e^t, 2t, 1 \rangle \quad \Rightarrow \quad \|\mathbf{r}'(t)\| = \sqrt{e^{2t} + 4t^2 + 1}$$

Step 2. Write out $f(\mathbf{r}(t))$. We substitute $x = e^t$, $y = t^2$, $z = t$ in $f(x,y,z) = 2x^2 + 8z$ to obtain:

$$f(\mathbf{r}(t)) = 2e^{2t} + 8t$$

Step 3. Compute the line integral. We have

$$\int_C f(x,y,z)\,ds = \int_0^1 f(\mathbf{r}(t))\|\mathbf{r}'(t)\|\,dt = \int_0^1 (2e^{2t} + 8t)\sqrt{e^{2t} + 4t^2 + 1}\,dt$$

We compute the integral using the substitution $u = e^{2t} + 4t^2 + 1$, $du = 2e^{2t} + 8t\,dt$. We get:

$$\int_C f(x,y,z)\,ds = \int_2^{e^2+5} u^{1/2}\,du = \frac{2}{3}u^{3/2}\Big|_2^{e^2+5} = \frac{2}{3}\left((e^2+5)^{3/2} - 2^{3/2}\right)$$

17. Calculate $\int_C 1\,ds$, where the curve C is parametrized by $\mathbf{r}(t) = \langle 4t, -3t, 12t \rangle$ for $2 \le t \le 5$. What does this integral represent?

SOLUTION Compute $\|\mathbf{r}'(t)\|$.

$$\mathbf{r}'(t) = \frac{d}{dt} < 4t, -3t, 12t > = < 4, -3, 12 > \quad \Rightarrow \quad \|\mathbf{r}'(t)\| = \sqrt{4^2 + (-3)^2 + (12)^2} = 13$$

Compute the line integral. We have

$$\int_C 1\,ds = \int_2^5 \|\mathbf{r}'(t)\|\,dt = \int_2^5 13\,dt = 13(5-2) = 39$$

This represents the distance from the point $(8, -6, 24)$ to the point $(20, -15, 60)$.

In Exercises 19–26, compute $\int_C \mathbf{F} \cdot d\mathbf{r}$ for the oriented curve specified.

19. $\mathbf{F}(x,y) = \langle 1 + x^2, xy^2 \rangle$, line segment from $(0,0)$ to $(1,3)$

SOLUTION Parametrize the line by $\mathbf{r}(t) = \langle t, 3t \rangle$, $0 \le t \le 1$. We compute the integrand:

$$\mathbf{F}(\mathbf{r}(t)) = \langle 1 + t^2, t \cdot (3t)^2 \rangle = \langle 1 + t^2, 9t^3 \rangle$$

$$\mathbf{r}'(t) = \langle 1, 3 \rangle$$

$$\mathbf{F}(\mathbf{r}(t)) \cdot \mathbf{r}'(t) = \langle 1 + t^2, 9t^3 \rangle \cdot \langle 1, 3 \rangle = 1 + t^2 + 27t^3$$

The vector line integral is the following integral:

$$\int_C \mathbf{F} \cdot d\mathbf{r} = \int_0^1 \mathbf{F}(\mathbf{r}(t)) \cdot \mathbf{r}'(t)\,dt = \int_0^1 1 + t^2 + 27t^3\,dt = t + \frac{1}{3}t^3 + \frac{27}{4}t^4 \Big|_0^1 = \frac{97}{12}$$

21. $\mathbf{F}(x,y) = \langle x^2, xy \rangle$, part of circle $x^2 + y^2 = 9$ with $x \le 0$, $y \ge 0$, oriented clockwise

SOLUTION   A plot of the curve is below:

The oriented path is parametrized by

$$\mathbf{r}(t) = (3\cos t, 3\sin t); \quad t \text{ is changing from } \pi \text{ to } \frac{\pi}{2}$$

We compute the integrand:

$$\mathbf{F}(\mathbf{r}(t)) = \langle x^2, xy \rangle = \langle 9\cos^2 t, 9\cos t \sin t \rangle$$

$$\mathbf{r}'(t) = \langle -3\sin t, 3\cos t \rangle$$

$$\mathbf{F}(\mathbf{r}(t)) \cdot \mathbf{r}'(t) = \langle 9\cos^2 t, 9\cos t \sin t \rangle \cdot \langle -3\sin t, 3\cos t \rangle = -27\cos^2 t \sin t + 27\cos^2 t \sin t = 0$$

Hence,

$$\int_C \mathbf{F} \cdot d\mathbf{r} = \int_\pi^{\pi/2} \mathbf{F}(\mathbf{r}(t)) \cdot \mathbf{r}'(t)\, dt = \int_\pi^{\pi/2} 0\, dt = 0$$

23. $\mathbf{F}(x,y) = \langle 3zy^{-1}, 4x, -y \rangle$, $\mathbf{r}(t) = \langle e^t, e^t, t \rangle$ for $-1 \le t \le 1$

SOLUTION
Step 1. Calculate the integrand. We write out the vectors and compute the integrand:

$$\mathbf{r}(t) = \langle e^t, e^t, t \rangle$$

$$\mathbf{F}(\mathbf{r}(t)) = \langle 3zy^{-1}, 4x, -y \rangle = \langle 3te^{-t}, 4e^t, -e^t \rangle$$

$$\mathbf{r}'(t) = \langle e^t, e^t, 1 \rangle$$

The integrand is the dot product:

$$\mathbf{F}(\mathbf{r}(t)) \cdot \mathbf{r}'(t) = \langle 3te^{-t}, 4e^t, -e^t \rangle \cdot \langle e^t, e^t, 1 \rangle = 3te^{-t} \cdot e^t + 4e^t \cdot e^t - e^t \cdot 1 = 3t + 4e^{2t} - e^t$$

Step 2. Evaluate the integral. The vector line integral is:

$$\int_C \mathbf{F} \cdot d\mathbf{r} = \int_{-1}^{1} \mathbf{F}(\mathbf{r}(t)) \cdot \mathbf{r}'(t)\, dt = \int_{-1}^{1} \left(3t + 4e^{2t} - e^t \right) dt = 0 + \int_{-1}^{1} \left(4e^{2t} - e^t \right) dt = 2e^{2t} - e^t \Big|_{-1}^{1}$$

$$= \left(2e^2 - e \right) - \left(2e^{-2} - e^{-1} \right) = 2\left(e^2 - e^{-2} \right) - \left(e - e^{-1} \right) \approx 12.157$$

25. $\mathbf{F}(x,y,z) = \left\langle \dfrac{1}{y^3+1}, \dfrac{1}{z+1}, 1 \right\rangle$, $\mathbf{r}(t) = \langle t^3, 2, t^2 \rangle$ for $0 \le t \le 1$

SOLUTION
Step 1. Calculate the integrand. We have

$$\mathbf{r}(t) = \left(t^3, 2, t^2 \right)$$

$$\mathbf{F}(\mathbf{r}(t)) = \left\langle \dfrac{1}{y^3+1}, \dfrac{1}{z+1}, 1 \right\rangle = \left\langle \dfrac{1}{2^3+1}, \dfrac{1}{t^2+1}, 1 \right\rangle = \left\langle \dfrac{1}{9}, \dfrac{1}{t^2+1}, 1 \right\rangle$$

$$\mathbf{r}'(t) = \langle 3t^2, 0, 2t \rangle$$

Hence,

$$\mathbf{F}(\mathbf{r}(t)) \cdot \mathbf{r}'(t) = \left\langle \dfrac{1}{9}, \dfrac{1}{t^2+1}, 1 \right\rangle \cdot \langle 3t^2, 0, 2t \rangle = \dfrac{1}{3}t^2 + 2t$$

Step 2. Evaluate the integral. Using the Theorem of vector line integrals we obtain:

$$\int_C \mathbf{F} \cdot d\mathbf{r} = \int_0^1 \mathbf{F}(\mathbf{r}(t)) \cdot \mathbf{r}'(t)\, dt = \int_0^1 \left(\frac{1}{3}t^2 + 2t\right) dt = \left(\frac{1}{9}t^3 + t^2\right)\bigg|_0^1 = \frac{10}{9}$$

In Exercises 27–34, evaluate the line integral.

27. $\int_C x\, dx$, over $y = x^3$ for $0 \le x \le 3$

SOLUTION Parametrize C by $x(t) = t$, $y(t) = t^3$ for $0 \le t \le 3$. Then $\frac{dx}{dt} = 1$ and $\frac{dy}{dt} = 3t^2$, so that

$$\int_C x\, dx = \int_0^3 (t \cdot 1\, dt + 0 \cdot 3t^2\, dt) = \int_0^3 t\, dt = \frac{1}{2}t^2\bigg|_0^3 = \frac{9}{2}$$

29. $\int_C y\, dx - x\, dy$, parabola $y = x^2$ for $0 \le x \le 2$

SOLUTION Parametrize C by $x(t) = t$, $y(t) = t^2$ for $0 \le t \le 2$. Then $\frac{dx}{dt} = 1$ and $\frac{dy}{dt} = 2t$, so that

$$\int_C y\, dx - x\, dy = \int_0^2 (t^2\, dt - t(2t\, dt)) = \int_0^2 (-t^2)\, dt = \left(-\frac{1}{3}t^3\right)\bigg|_0^2 = -\frac{8}{3}$$

31. $\int_C (x-y)\, dx + (y-z)\, dy + z\, dz$, line segment from $(0,0,0)$ to $(1,4,4)$

SOLUTION Parametrize the oriented line segment from $(0, 0, 0)$ to $(1, 4, 4)$ by

$$x(t) = t,\ y(t) = 4t,\ z(t) = 4t,\ 0 \le t \le 1$$

Then

$$dx = dt,\ dy = 4\, dt,\ dz = 4\, dt$$

so that

$$\int_C (x-y)\, dx + (y-z)\, dy + z\, dz = \int_0^1 ((t - 4t) \cdot 1 + (4t - 4t) \cdot 4 + 4t \cdot 4)\, dt$$

$$= \int_0^1 13t\, dt = \left(\frac{13}{2}t^2\right)\bigg|_0^1 = \frac{13}{2}$$

33. $\int_C \frac{-y\, dx + x\, dy}{x^2 + y^2}$, segment from $(1, 0)$ to $(0, 1)$

SOLUTION

Step 1. Calculate the integrand.

$$\mathbf{r}(t) = (1 - t, t)\quad (0 \le t \le 1)$$

$$\mathbf{F}(\mathbf{r}(t)) = \frac{1}{x^2 + y^2}\langle -y, x\rangle = \frac{1}{(1-t)^2 + t^2}\langle -t, 1-t\rangle$$

$$\mathbf{r}'(t) = \langle -1, 1\rangle$$

The integrand is the dot product

$$\mathbf{F}(\mathbf{r}(t)) \cdot \mathbf{r}'(t) = \frac{1}{(1-t)^2 + t^2}\langle -t, 1-t\rangle \cdot \langle -1, 1\rangle = \frac{t + 1 - t}{(1-t)^2 + t^2} = \frac{1}{2t^2 - 2t + 1}$$

Step 2.

$$\int_C \frac{-y\, dx + x\, dy}{x^2 + y^2} = \int_0^1 \frac{dt}{2t^2 - 2t + 1} = \frac{1}{2}\int_0^1 \frac{dt}{\left(t - \frac{1}{2}\right)^2 + \frac{1}{4}}$$

We use the trigonometric substitution $t = \frac{1}{2} + \frac{1}{2}\tan\theta \Rightarrow dt = \frac{1}{2}\sec^2\theta\, d\theta$.

$$= \frac{1}{2}\int_{-\frac{\pi}{4}}^{\frac{\pi}{4}} \frac{\frac{1}{2}\sec^2\theta\, d\theta}{\frac{1}{4}(\tan^2\theta + 1)} = \int_{-\frac{\pi}{4}}^{\frac{\pi}{4}} d\theta = \frac{\pi}{2}$$

35. **CAS** Let $f(x, y, z) = x^{-1}yz$, and let C be the curve parametrized by $\mathbf{r}(t) = \langle \ln t, t, t^2 \rangle$ for $2 \leq t \leq 4$. Use a computer algebra system to calculate $\int_C f(x, y, z)\, ds$ to four decimal places.

SOLUTION Note that $\mathbf{r}'(t) = \langle 1/t, 1, 2t \rangle$, so $\|\mathbf{r}'(t)\| = \sqrt{1/t^2 + 1 + 4t^2}$. Our line integral is

$$\int_2^4 f(\ln t, t, t^2)\, \sqrt{1/t^2 + 1 + 4t^2}\, dt$$

which we calculate to be 339.5587.

In Exercises 37 and 38, calculate the line integral of $\mathbf{F}(x, y, z) = \langle e^z, e^{x-y}, e^y \rangle$ over the given path.

37. The blue path from P to Q in Figure 15

FIGURE 15

SOLUTION

Let C_1, C_2, C_3 denote the oriented segments from P to R, from R to S and S to Q, respectively. These paths have the following parametrizations (see figure):

$$\begin{array}{lll} C_1: & \mathbf{r}_1(t) = (0, 0, t) \quad 0 \leq t \leq 1 & \mathbf{r}_1'(t) = \langle 0, 0, 1 \rangle \\ C_2: & \mathbf{r}_2(t) = (0, t, 1) \quad 0 \leq t \leq 1 \quad \Rightarrow & \mathbf{r}_2'(t) = \langle 0, 1, 0 \rangle \\ C_3: & \mathbf{r}_3(t) = (t, 1, 1) \quad 0 \leq t \leq 1 & \mathbf{r}_3'(t) = \langle 1, 0, 0 \rangle \end{array}$$

Since $C = C_1 + C_2 + C_3$ we have

$$\int_C \mathbf{F} \cdot d\mathbf{r} = \int_{C_1} \mathbf{F} \cdot d\mathbf{r} + \int_{C_2} \mathbf{F} \cdot d\mathbf{r} + \int_{C_3} \mathbf{F} \cdot d\mathbf{r} \tag{1}$$

We compute each integral on the right-hand side separately.

$$\int_{C_1} \mathbf{F} \cdot d\mathbf{r} = \int_0^1 \mathbf{F}(\mathbf{r}_1(t)) \cdot \mathbf{r}_1'(t)\, dt = \int_0^1 \langle e^t, e^{0-0}, e^0 \rangle \cdot \langle 0, 0, 1 \rangle\, dt = \int_0^1 1\, dt = 1$$

$$\int_{C_2} \mathbf{F} \cdot d\mathbf{r} = \int_0^1 \mathbf{F}(\mathbf{r}_2(t)) \cdot \mathbf{r}_2'(t)\, dt = \int_0^1 \langle e^1, e^{0-t}, e^t \rangle \cdot \langle 0, 1, 0 \rangle\, dt = \int_0^1 e^{-t}\, dt = -e^{-t}\Big|_0^1 = 1 - e^{-1}$$

$$\int_{C_3} \mathbf{F} \cdot d\mathbf{r} = \int_0^1 \mathbf{F}(\mathbf{r}_3(t)) \cdot \mathbf{r}_3'(t)\, dt = \int_0^1 \langle e^1, e^{t-1}, e^1 \rangle \cdot \langle 1, 0, 0 \rangle\, dt = \int_0^1 e\, dt = e$$

Substituting these integrals in (1) gives

$$\int_C \mathbf{F} \cdot d\mathbf{r} = 1 + (1 - e^{-1}) + e = e + 2 - e^{-1}$$

In Exercises 39 and 40, C is the path from P to Q in Figure 17 that traces C_1, C_2, and C_3 in the orientation indicated, and $\mathbf{F}$ is a vector field such that

$$\int_C \mathbf{F} \cdot d\mathbf{r} = 5, \qquad \int_{C_1} \mathbf{F} \cdot d\mathbf{r} = 8, \qquad \int_{C_3} \mathbf{F} \cdot d\mathbf{r} = 8$$

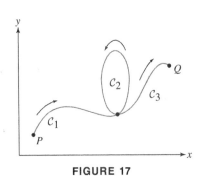

FIGURE 17

39. Determine:

(a) $\displaystyle\int_{-C_3} \mathbf{F} \cdot d\mathbf{r}$ \qquad (b) $\displaystyle\int_{C_2} \mathbf{F} \cdot d\mathbf{r}$ \qquad (c) $\displaystyle\int_{-C_1 - C_3} \mathbf{F} \cdot d\mathbf{r}$

SOLUTION

(a) If the orientation of the path is reversed, the line integral changes sign, thus:

$$\int_{-C_3} \mathbf{F} \cdot d\mathbf{r} = -\int_{C_3} \mathbf{F} \cdot d\mathbf{r} = -8$$

(b) By additivity of line integrals, we have

$$\int_C \mathbf{F} \cdot d\mathbf{r} = \int_{C_1} \mathbf{F} \cdot d\mathbf{r} + \int_{C_2} \mathbf{F} \cdot d\mathbf{r} + \int_{C_3} \mathbf{F} \cdot d\mathbf{r}$$

Substituting the given values we obtain

$$5 = 8 + \int_{C_2} \mathbf{F} \cdot d\mathbf{r} + 8$$

or

$$\int_{C_2} \mathbf{F} \cdot d\mathbf{r} = 5 - 16 = -11$$

(c) Using properties of line integrals gives

$$\int_{-C_1 - C_3} \mathbf{F} \cdot d\mathbf{r} = \int_{-C_1} \mathbf{F} \cdot d\mathbf{r} + \int_{-C_3} \mathbf{F} \cdot d\mathbf{r} = -\int_{C_1} \mathbf{F} \cdot d\mathbf{r} - \int_{C_3} \mathbf{F} \cdot d\mathbf{r} = -8 - 8 = -16$$

41. The values of a function $f(x, y, z)$ and vector field $\mathbf{F}(x, y, z)$ are given at six sample points along the path ABC in Figure 18. Estimate the line integrals of f and $\mathbf{F}$ along ABC.

Point	$f(x,y,z)$	$\mathbf{F}(x,y,z)$
$(1, \frac{1}{6}, 0)$	3	$\langle 1, 0, 2 \rangle$
$(1, \frac{1}{2}, 0)$	3.3	$\langle 1, 1, 3 \rangle$
$(1, \frac{5}{6}, 0)$	3.6	$\langle 2, 1, 5 \rangle$
$(1, 1, \frac{1}{6})$	4.2	$\langle 3, 2, 4 \rangle$
$(1, 1, \frac{1}{2})$	4.5	$\langle 3, 3, 3 \rangle$
$(1, 1, \frac{5}{6})$	4.2	$\langle 5, 3, 3 \rangle$

FIGURE 18

SOLUTION

We write the integrals as sum of integrals and estimate each integral by a Riemann Sum. That is,

$$\int_{ABC} f(x,y,z)\,ds = \int_{AB} f(x,y,z)\,ds + \int_{BC} f(x,y,z)\,d\mathbf{r} \approx \sum_{i=1}^{3} f(P_i)\Delta s_i + \sum_{i=4}^{6} f(P_i)\Delta s_i \quad (1)$$

$$\int_{ABC} \mathbf{F} \cdot d\mathbf{r} = \int_{AB} \mathbf{F} \cdot d\mathbf{r} + \int_{BC} \mathbf{F} \cdot d\mathbf{r} = \int_{AB} (\mathbf{F} \cdot \mathbf{T})\,d\mathbf{r} + \int_{BC} (\mathbf{F} \cdot \mathbf{T})\,d\mathbf{r}$$

On AB, the unit tangent vector is $\mathbf{T} = \langle 0, 1, 0 \rangle$, hence $\mathbf{F} \cdot \mathbf{T} = F_2$. On BC, the unit tangent vector is $\mathbf{T} = \langle 0, 0, 1 \rangle$, hence $\mathbf{F} \cdot \mathbf{T} = F_3$. Therefore,

$$\int_{ABC} \mathbf{F}\,ds = \int_{AB} F_1\,ds + \int_{BC} F_3\,ds \approx \sum_{i=1}^{3} F_1(P_i)\Delta s_i + \sum_{i=4}^{6} F_3(P_i)\Delta s_i \quad (2)$$

We consider the partitions of AB and BC to three subarcs with equal length $\Delta s_i = \frac{1}{3}$, therefore (1) and (2) give

$$\int_{ABC} f(x,y,z)\,ds \approx \frac{1}{3}\left(f(P_1) + f(P_2) + f(P_3) + f(P_4) + f(P_5) + f(P_6)\right)$$

$$\int_{ABC} \mathbf{F}\,d\mathbf{r} \approx \frac{1}{3}\left(F_2(P_1) + F_2(P_2) + F_2(P_3) + F_3(P_4) + F_3(P_5) + F_3(P_6)\right)$$

We now substitute the values of the functions at the sample points to obtain the following approximations:

$$\int_{ABC} f(x,y,z)\,ds \approx \frac{1}{3}(3 + 3.3 + 3.6 + 4.2 + 4.5 + 4.2) = 7.6$$

$$\int_{ABC} \mathbf{F} \cdot d\mathbf{r} \approx \frac{1}{3}(0 + 1 + 1 + 4 + 3 + 3) = 4$$

43. Determine whether the line integrals of the vector fields around the circle (oriented counterclockwise) in Figure 20 are positive, negative, or zero.

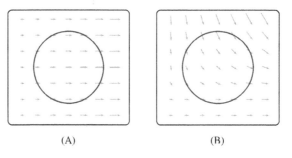

(A)   (B)

(C)

FIGURE 20

SOLUTION The vector line integral of F is the integral of the tangential component of F along the curve. The positive direction of a curve is counterclockwise.

For the vector field in (A), the line integral around the circle is zero because the contribution of the negative tangential components from the upper part of the circle is the same as the contribution of the positive tangential components from the lower part. For the vector in (B) the contribution of the negative tangential component appears to dominate over the positive contribution, hence the line integral is negative. In (C), the vector field is orthogonal to the unit tangent vector at each point, hence the line integral is zero.

45. Calculate the total mass of a circular piece of wire of radius 4 cm centered at the origin whose mass density is $\rho(x, y) = x^2$ g/cm.

SOLUTION The total mass is the following integral:

$$M = \int_C x^2 \, ds$$

We use the following parametrization of the wire:

$$\mathbf{r}(t) = \langle 4\cos t, 4\sin t \rangle, \quad 0 \le t \le 2\pi$$

Hence,

$$\mathbf{r}'(t) = \langle -4\sin t, 4\cos t \rangle \quad \Rightarrow \quad \|\mathbf{r}'(t)\| = \sqrt{(-4\sin t)^2 + (4\cos t)^2} = 4$$

We compute the line integral using the Theorem on Scalar Line Integrals. We get

$$M = \int_0^{2\pi} \rho(\mathbf{r}(t)) \|\mathbf{r}'(t)\| \, dt = \int_0^{2\pi} (4\cos t)^2 \cdot 4 \, dt$$

$$= 64 \int_0^{2\pi} \cos^2 t \, dt = 64 \left(\frac{t}{2} + \frac{\sin 2t}{4} \right) \Big|_0^{2\pi} = 64 \cdot \frac{2\pi}{2} = 64\pi \text{ g}$$

47. Find the total charge on the curve $y = x^{4/3}$ for $1 \le x \le 8$ (in centimeters) assuming a charge density of $\rho(x, y) = x/y$ (in units of 10^{-6} C/cm).

SOLUTION We parametrize the curve by $\mathbf{r}(t) = \left(t, t^{\frac{4}{3}}\right)$ $(1 \le t \le 8)$. Then

$$\mathbf{r}'(t) = \left\langle 1, \frac{4}{3}t^{\frac{1}{3}} \right\rangle \quad \Rightarrow \quad \|\mathbf{r}'(t)\| = \sqrt{1 + \frac{16}{9}t^{\frac{2}{3}}}$$

$$\rho(\mathbf{r}(t)) = \frac{x}{y} = \frac{t}{t^{\frac{4}{3}}}$$

Therefore the total charge will be

$$\int_C \frac{x}{y} \, ds = \int_1^8 \frac{t}{t^{\frac{4}{3}}} \sqrt{1 + \frac{16}{9}t^{\frac{2}{3}}} \, dt = \int_1^8 \sqrt{1 + \frac{16}{9}t^{\frac{2}{3}}} \, t^{-\frac{1}{3}} \, dt$$

Using the substitution $u = 1 + \frac{16}{9}t^{\frac{2}{3}} \Rightarrow du = \frac{32}{27}t^{-\frac{1}{3}} dt$, we calculate the total charge as

$$\int_{\frac{25}{9}}^{\frac{73}{9}} \sqrt{u} \, \frac{27}{32} \, du = \frac{27}{32} \cdot \frac{2}{3} u^{\frac{3}{2}} \bigg|_{\frac{25}{9}}^{\frac{73}{9}} = \frac{1}{48}\left(73^{\frac{3}{2}} - 25^{\frac{3}{2}}\right) \approx 10.39$$

Thus the total charge is 10.39×10^{-6} C.

In Exercises 49–52, use Eq. (6) to compute the electric potential $V(P)$ at the point P for the given charge density (in units of 10^{-6} C).

49. Calculate $V(P)$ at $P = (0, 0, 12)$ if the electric charge is distributed along the quarter circle of radius 4 centered at the origin with charge density $\rho(x, y, z) = xy$.

SOLUTION We parametrize the curve by $\mathbf{r}(t) = (4\cos t, 4\sin t, 0)$, $(0 \le t \le \frac{\pi}{2})$. Then $\mathbf{r}'(t) = (-4\sin t, 4\cos t, 0) \Rightarrow \|\mathbf{r}'(t)\| = 4$. The distance from the point $(0, 0, 12)$ to $\mathbf{r}(t)$ is

$$r_P(t) = \sqrt{(0 - 4\cos t)^2 + (0 - 4\sin t)^2 + (12 - 0)^2} = \sqrt{16 + 144} = 4\sqrt{10}$$

while the charge density along the curve is

$$\rho(\mathbf{r}(t)) = xy = 4\cos t \, 4\sin t = 16\sin t \cos t = 8\sin 2t$$

Therefore

$$V(P) = k \int_C \frac{\rho}{r_P} \, ds = k \int_0^{\frac{\pi}{2}} \frac{8\sin 2t}{4\sqrt{10}} 4 \, dt = \frac{8k}{\sqrt{10}} \cdot \frac{-\cos 2t}{2} \bigg|_0^{\frac{\pi}{2}}$$

$$= \frac{4k}{\sqrt{10}}(-\cos \pi + \cos 0) = \frac{8k}{\sqrt{10}}$$

Thus the electric potential is $\frac{8k}{\sqrt{10}} \times 10^{-6}$ C ≈ 22743.1 volts.

51. Calculate $V(P)$ at $P = (2, 0, 2)$ if the negative charge is distributed along the y-axis for $1 \le y \le 3$ with charge density $\rho(x, y, z) = -y$.

SOLUTION A parametrization for the curve is $\mathbf{r}(t) = (0, t, 0)$ $(1 \le t \le 3)$. Then $\mathbf{r}'(t) = \langle 0, 1, 0 \rangle \Rightarrow \|\mathbf{r}'(t)\| = 1$, and the charge density along the curve is $\rho(\mathbf{r}(t)) = -y = -t$. The distance from the origin to $\mathbf{r}(t)$ is

$$r_P(t) = \sqrt{(2 - 0)^2 + (0 - t)^2 + (2 - 0)^2} = \sqrt{8 + t^2}$$

Therefore,

$$V(P) = k \int_C \frac{\rho}{r_P} \, ds = k \int_1^3 \frac{-t}{\sqrt{8 + t^2}} \cdot 1 \, dt$$

Using the substitution $u = 8 + t^2 \Rightarrow du = 2t \, dt$, we have

$$V(P) = -k \int_9^{17} u^{-\frac{1}{2}} \frac{1}{2} \, du = -\frac{k}{2} \cdot 2u^{\frac{1}{2}} \bigg|_9^{17} = -\frac{k}{2}\left(17^{\frac{1}{2}} - 9^{\frac{1}{2}}\right)$$

Thus the electric potential is $-\frac{k}{2}\left(17^{\frac{1}{2}} - 9^{\frac{1}{2}}\right) \times 10^{-6}$ C ≈ -10097 volts.

53. Calculate the work done by a field $\mathbf{F} = \langle x+y, x-y \rangle$ when an object moves from $(0,0)$ to $(1,1)$ along each of the paths $y = x^2$ and $x = y^2$.

SOLUTION We calculate the work done by $\mathbf{F} = \langle x+y, x-y \rangle$ along the path $y = x^2$ from $(0,0)$ to $(1,1)$. We use the parametrization:

$$\mathbf{r}_1(t) = (t, t^2), \quad 0 \leq t \leq 1$$

We have

$$\mathbf{F}(\mathbf{r}_1(t)) = \langle t + t^2, t - t^2 \rangle$$
$$\mathbf{r}_1'(t) = \langle 1, 2t \rangle$$
$$\mathbf{F}(\mathbf{r}(t)) \cdot \mathbf{r}'(t) = \langle t + t^2, t - t^2 \rangle \cdot \langle 1, 2t \rangle = t + t^2 + 2t^2 - 2t^3 = -2t^3 + 3t^2 + t$$

The work is the following line integral:

$$W = \int_{C_1} \mathbf{F} \cdot d\mathbf{r} = \int_0^1 \mathbf{F}(\mathbf{r}_1(t)) \cdot \mathbf{r}_1'(t)\, dt = \int_0^1 \left(-2t^3 + 3t^2 + t\right) dt = -\frac{1}{2}t^4 + t^3 + \frac{1}{2}t^2 \bigg|_0^1 = 1$$

We now compute the work along the path $x = y^2$. We parametrize the path by:

$$\mathbf{r}_2(t) = (t^2, t), \quad 0 \leq t \leq 1$$

Then

$$\mathbf{F}(\mathbf{r}_2(t)) = \langle t^2 + t, t^2 - t \rangle$$
$$\mathbf{r}_2'(t) = \langle 2t, 1 \rangle$$
$$\mathbf{F}(\mathbf{r}_2(t)) \cdot \mathbf{r}_2'(t) = \langle t^2 + t, t^2 - t \rangle \cdot \langle 2t, 1 \rangle = 2t^3 + 2t^2 + t^2 - t = 2t^3 + 3t^2 - t$$

The work is the line integral

$$W = \int_{C_2} \mathbf{F} \cdot d\mathbf{r} = \int_0^1 \mathbf{F}(\mathbf{r}_2(t)) \cdot \mathbf{r}_2'(t)\, dt = \int_0^1 \left(2t^3 + 3t^2 - t\right) dt = \frac{1}{2}t^4 + t^3 - \frac{1}{2}t^2 \bigg|_0^1 = \frac{1}{2} + 1 - \frac{1}{2} = 1$$

We obtain the same work along the two paths.

In Exercises 54–56, calculate the work done by the field $\mathbf{F}$ when the object moves along the given path from the initial point to the final point.

55. $\mathbf{F}(x, y, z) = \langle xy, yz, xz \rangle$, $\mathbf{r} = \langle t, t^2, t^3 \rangle$ for $0 \leq t \leq 1$

SOLUTION The work done by the force field $\mathbf{F}$ is the line integral:

$$W = \int_C \mathbf{F} \cdot d\mathbf{r}$$

We compute the integrand:

$$\mathbf{F}(\mathbf{r}(t)) = \langle xy, yz, xz \rangle = \langle t^3, t^5, t^4 \rangle$$

$$\mathbf{r}'(t) = \frac{d}{dt}\langle t, t^2, t^3\rangle = \langle 1, 2t, 3t^2\rangle$$

$$\mathbf{F}(\mathbf{r}(t))\cdot\mathbf{r}'(t) = \langle t^3, t^5, t^4\rangle\cdot\langle 1, 2t, 3t^2\rangle = t^3 + 2t^6 + 3t^6 = 5t^6 + t^3$$

We obtain the following integral:

$$W = \int_0^1 \mathbf{F}(\mathbf{r}(t))\cdot\mathbf{r}'(t)\,dt = \int_0^1 (5t^6 + t^3)\,dt = \left(\frac{5}{7}t^7 + \frac{1}{4}t^4\right)\Big|_0^1 = \frac{27}{28}$$

57. Figure 22 shows a force field $\mathbf{F}$.

(a) Over which of the two paths, *ADC* or *ABC*, does $\mathbf{F}$ perform less work?

(b) If you have to work against $\mathbf{F}$ to move an object from *C* to *A*, which of the paths, *CBA* or *CDA*, requires less work?

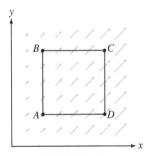

FIGURE 22

SOLUTION

(a) The work performed along either path is the integral of the tangential component of $\mathbf{F}$ along the path. Since the vertical component of the arrows along *AB* are smaller than those along *DC*, and the horizontal components of the arrows along *AD* and *BC* appear about the same, $\mathbf{F}$ will perform less work along *ABC* than it would along *ADC*.

(b) By the same argument as in part (a), since the tangential components along *BA* are smaller (and point against you) than those along *CD* and the tangential components along *CB* and *DA* are about the same, you would perform less work moving the object along *CBA* than along *CDA*.

59. Show that work performed by a constant force field $\mathbf{F}$ over any path C from P to Q is equal to $\mathbf{F}\cdot\overrightarrow{PQ}$.

SOLUTION We denote by $\mathbf{r}(t) = (x(t), y(t), c(t))$, $t_0 \le t \le t_1$ a parametrization of the oriented path from P to Q (then $\mathbf{r}(t_0) = P$ and $\mathbf{r}(t_1) = Q$). Let $\mathbf{F} = \langle a, b, c\rangle$ be a constant vector field. Then,

$$\mathbf{F}(\mathbf{r}(t))\cdot\mathbf{r}'(t) = \langle a, b, c\rangle\cdot\langle x'(t), y'(t), z'(t)\rangle = ax'(t) + by'(t) + cz'(t)$$

The vector line integral is, thus,

$$\int_C \mathbf{F}\cdot d\mathbf{r} = \int_{t_0}^{t_1} \mathbf{F}(\mathbf{r}(t))\cdot\mathbf{r}'(t)\,dt = \int_{t_0}^{t_1} (ax'(t) + by'(t) + cz'(t))\,dt$$

$$= a\int_{t_0}^{t_1} x'(t)\,dt + b\int_{t_0}^{t_1} y'(t)\,dt + c\int_{t_0}^{t_1} z'(t)\,dt = ax(t)\Big|_{t=t_0}^{t_1} + by(t)\Big|_{t=t_0}^{t_1} + cz(t)\Big|_{t=t_0}^{t_1}$$

$$= a(x(t_1) - x(t_0)) + b(y(t_1) - y(t_0)) + c(z(t_1) - z(t_0))$$

$$= \langle a, b, c\rangle\cdot\langle x(t_1) - x(t_0), y(t_1) - y(t_0), z(t_1) - z(t_0)\rangle$$

Since $P = \langle x(t_0), y(t_0), z(t_0)\rangle$ and $Q = \langle x(t_1), y(t_1), z(t_1)\rangle$ we conclude that,

$$\int_C \mathbf{F}\cdot d\mathbf{r} = \langle a, b, c\rangle\cdot\overrightarrow{PQ} = \mathbf{F}\cdot\overrightarrow{PQ}$$

61. Charge is distributed along the spiral with polar equation $r = \theta$ for $0 \le \theta \le 2\pi$. The charge density is $\rho(r, \theta) = r$ (assume distance is in centimeters and charge in units of 10^{-6} C/cm). Use the result of Exercise 60(a) to compute the total charge.

SOLUTION Following Exercise 60(a), $f(\theta) = \theta$, and $f'(\theta) = 1$. Thus $\|\mathbf{r}'(\theta)\| = \sqrt{\theta^2 + 1}$. The total charge will be

$$\int_C \rho \, ds = \int_0^{2\pi} \theta \sqrt{\theta^2 + 1} \, d\theta$$

Substituting $u = \theta^2 + 1 \Rightarrow du = 2\theta \, d\theta$, we have

$$\int_C \rho \, ds = \int_1^{4\pi^2+1} \sqrt{u} \, \frac{1}{2} \, du = \frac{1}{2} \cdot \frac{2}{3} u^{\frac{3}{2}} \Big|_1^{4\pi^2+1} = \frac{1}{2}\left((4\pi^2 + 1)^{\frac{3}{2}} - 1\right) \approx 85.5$$

Thus the total charge is 85.5×10^{-6} C.

In Exercises 62 and 63, let $\mathbf{F}$ be the vortex field (Figure 24):

$$\mathbf{F}(x, y) = \left\langle \frac{-y}{x^2 + y^2}, \frac{x}{x^2 + y^2} \right\rangle$$

FIGURE 24

63. Let C be a curve in polar form $r = f(\theta)$ for $\theta_1 \leq \theta \leq \theta_2$ [Figure 25(B)], parametrized by $\mathbf{r}(\theta) = (f(\theta)\cos\theta, f(\theta)\sin\theta)$ as in Exercise 60.

(a) Show that the vortex field in polar coordinates is written $\mathbf{F}(r, \theta) = r^{-1} \langle -\sin\theta, \cos\theta \rangle$.

(b) Show that $\mathbf{F} \cdot \mathbf{r}'(\theta) \, d\theta = d\theta$.

(c) Show that $\int_C \mathbf{F} \cdot d\mathbf{r} = \theta_2 - \theta_1$.

SOLUTION

(a) Letting $x = r\cos(\theta)$ and $y = r\sin(\theta)$ we have

$$\mathbf{F} = \left\langle \frac{-y}{x^2 + y^2}, \frac{x}{x^2 + y^2} \right\rangle = \left\langle \frac{-r\sin(\theta)}{(r\cos(\theta))^2 + (r\sin(\theta))^2}, \frac{r\cos(\theta)}{(r\cos(\theta))^2 + (r\sin(\theta))^2} \right\rangle$$

$$= \frac{r}{r^2} \left\langle \frac{-\sin(\theta)}{\cos^2(\theta) + \sin^2(\theta)}, \frac{\cos(\theta)}{\cos^2(\theta) + \sin^2(\theta)} \right\rangle = r^{-1} \langle -\sin\theta, \cos\theta \rangle$$

(b) From the solution to Exercise 60(a) we have

$$\mathbf{r}'(\theta) = \langle f'(\theta)\cos(\theta) - f(\theta)\sin(\theta), f'(\theta)\sin(\theta) + f(\theta)\cos(\theta) \rangle$$

Substituting $r = f(\theta)$ into the previous part, we have

$$\mathbf{F} \cdot \mathbf{r}'(\theta) \, d\theta = \frac{1}{f(\theta)} \langle -\sin(\theta), \cos(\theta) \rangle \cdot \langle f'(\theta)\cos(\theta) - f(\theta)\sin(\theta), f'(\theta)\sin(\theta) + f(\theta)\cos(\theta) \rangle \, d\theta$$

$$= \frac{1}{f(\theta)} (-f'(\theta)\sin(\theta)\cos(\theta) + f(\theta)\sin^2(\theta) + f'(\theta)\cos(\theta)\sin(\theta) + f(\theta)\cos^2(\theta)) \, d\theta$$

$$= \frac{1}{f(\theta)} f(\theta)(\sin^2(\theta) + \cos^2(\theta)) \, d\theta = d\theta$$

(c)

$$\int_C \mathbf{F} \cdot d\mathbf{s} = \int_{\theta_1}^{\theta_2} \mathbf{F} \cdot \mathbf{r}'(\theta) \, d\theta = \int_{\theta_1}^{\theta_2} d\theta = \theta_2 - \theta_1$$

In Exercises 64–67, use Eq. (9) to calculate the flux of the vector field across the curve specified.

65. $\mathbf{F}(x,y) = \langle x^2, y^2 \rangle$; segment from $(3,0)$ to $(0,3)$, oriented upward

SOLUTION The curve is parametrized by $\mathbf{r}(t) = (3-t, t)$ $(0 \leq t \leq 3) \Rightarrow \mathbf{r}'(t) = \langle -1, 1 \rangle$. Then
$$\mathbf{F}(\mathbf{r}(t)) = \langle x^2, y^2 \rangle = \langle (3-t)^2, t^2 \rangle$$

Therefore the flux is
$$\int_C F_1 \, dy - F_2 \, dx = \int_0^3 (3-t)^2(1) - (t^2)(-1) \, dt$$
$$= \int_0^3 2t^2 - 6t + 9 \, dt = \frac{2t^3}{3} - 3t^2 + 9t \bigg|_0^3 = 18$$

67. $\mathbf{F}(x,y) = \langle e^y, 2x-1 \rangle$; parabola $y = x^2$ for $0 \leq x \leq 1$, oriented left to right

SOLUTION The curve is parametrized by $\mathbf{r}(t) = (t, t^2)$ $(0 \leq t \leq 1) \Rightarrow \mathbf{r}'(t) = \langle 1, 2t \rangle$. Then
$$\mathbf{F}(\mathbf{r}(t)) = \langle e^y, 2x-1 \rangle = \langle e^{t^2}, 2t-1 \rangle$$

Therefore the flux is
$$\int_C F_1 \, dy - F_2 \, dx = \int_0^1 \left(e^{t^2}\right)(2t) - (2t-1)(1) \, dt = e^{t^2} - t^2 + t \bigg|_0^1 = e - 1$$

Further Insights and Challenges

69. Let $\mathbf{F}(x,y) = \langle x, 0 \rangle$. Prove that if C is any path from (a,b) to (c,d), then
$$\int_C \mathbf{F} \cdot d\mathbf{r} = \frac{1}{2}(c^2 - a^2)$$

SOLUTION

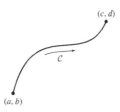

We denote the parametrization of the path by,
$$\mathbf{r}(t) = (x(t), y(t)), \quad t_0 \leq t \leq t_1, \quad \mathbf{r}(t_0) = (a,b), \quad \mathbf{r}(t_1) = (c,d)$$

By the Theorem on vector line integrals we have,
$$\int_C \mathbf{F} \cdot d\mathbf{r} = \int_{t_0}^{t_1} \mathbf{F}(\mathbf{r}(t)) \cdot \mathbf{r}'(t) \, dt = \int_{t_0}^{t_1} \langle x(t), 0 \rangle \cdot \langle x'(t), y'(t) \rangle \, dt = \int_{t_0}^{t_1} x(t) x'(t) \, dt$$

We use the hint to compute the integral, obtaining
$$\int_C \mathbf{F} \cdot d\mathbf{r} = \frac{1}{2} x(t)^2 \bigg|_{t_0}^{t_1} = \frac{1}{2} \left(x(t_1)^2 - x(t_0)^2 \right) = \frac{1}{2} (c^2 - a^2)$$

Proof of the hint: By the Chain Rule for differentiation we have
$$\frac{d}{dt} f^2(t) = 2f(t)f'(t) \quad \Rightarrow \quad f(t)f'(t) = \frac{1}{2} \frac{d}{dt} f^2(t)$$

Applying the Fundamental Theorem of calculus we obtain
$$\int_{t_0}^{t_1} f(t) f'(t) \, dt = \frac{1}{2} \int_{t_0}^{t_1} \frac{d}{dt} \left(f^2(t) \right) dt = \frac{1}{2} \left(f^2(t_1) - f^2(t_0) \right)$$

Alternatively we can evaluate the integral $\int f(t)f'(t)\,dt$ using the substitution $u = f(t)$, $du = f'(t)\,dt$. We get

$$\int f(t)f'(t)\,dt = \int u\,du = \frac{1}{2}u^2 + c = \frac{1}{2}f^2(t) + c$$

71. We wish to define the **average value** $\mathrm{Av}(f)$ of a continuous function f along a curve C of length L. Divide C into N consecutive arcs $C_1, \ldots, C_N$, each of length L/N, and let P_i be a sample point in C_i (Figure 26). The sum

$$\frac{1}{N}\sum_{i=1}^{N} f(P_i)$$

may be considered an approximation to $\mathrm{Av}(f)$, so we define

$$\mathrm{Av}(f) = \lim_{N \to \infty} \frac{1}{N}\sum_{i=1}^{N} f(P_i)$$

Prove that

$$\mathrm{Av}(f) = \frac{1}{L}\int_C f(x,y,z)\,ds \qquad \boxed{10}$$

Hint: Show that $\dfrac{L}{N}\sum_{i=1}^{N} f(P_i)$ is a Riemann sum approximation to the line integral of f along C.

FIGURE 26

SOLUTION The Riemann sum approximation to the line integral is:

$$\sum_{i=1}^{N} f(P_i)\,\Delta S_i$$

If the consecutive arcs $C_1, \ldots, C_2$ have equal lengths $\frac{L}{N}$, the corresponding Riemann sum is,

$$\sum_{i=1}^{N} f(P_i) \cdot \frac{L}{N} = \frac{L}{N}\sum_{i=1}^{N} f(P_i)$$

We let $N \to \infty$,

$$\int_C f(x,y,z)\,ds = \lim_{N \to \infty} \frac{L}{N}\sum_{i=1}^{N} f(P_i) = L \lim_{N \to \infty} \frac{1}{N}\sum_{i=1}^{N} f(P_i) = L\mathrm{Av}(f)$$

That is,

$$\mathrm{Av}(f) = \frac{1}{L}\int_C f(x,y,z)\,ds$$

73. Use Eq. (10) to calculate the average value of $f(x,y) = x$ along the curve $y = x^2$ for $0 \le x \le 1$.

SOLUTION The average value is

$$\mathrm{Av}(f) = \frac{1}{L}\int_C x\,ds \qquad (1)$$

We parametrize the curve by the parametrization,

$$\mathbf{r}(t) = \langle t, t^2 \rangle, \quad 0 \le t \le 1$$

Hence,
$$\mathbf{r}'(t) = \langle 1, 2t \rangle \quad \Rightarrow \quad \|\mathbf{r}'(t)\| = \sqrt{1+4t^2}$$

We first must calculate the length of the path. That is,
$$L = \int_C \|\mathbf{r}'(t)\| \, dt = \int_0^1 \sqrt{1+4t^2} \, dt$$
$$= \frac{1}{2}t\sqrt{1+4t^2} + \frac{1}{4}\ln\left(2t+\sqrt{1+4t^2}\right)\Big|_0^1$$
$$= \frac{\sqrt{5}}{2} + \frac{1}{4}\ln(2+\sqrt{5})$$
$$= \frac{2\sqrt{5}+\ln(2+\sqrt{5})}{4}$$

We compute the line integral in (1):
$$\int_C x \, ds = \int_0^1 t\|\mathbf{r}'(t)\| \, dt = \int_0^1 t\sqrt{1+4t^2} \, dt = \frac{1}{12}(1+4t^2)^{3/2}\Big|_0^1 = \frac{1}{12}(5\sqrt{5}-1)$$

Then
$$\text{Av}(f) = \frac{4}{2\sqrt{5}+\ln(2+\sqrt{5})} \cdot \frac{5\sqrt{5}-1}{12} = \frac{5\sqrt{5}-1}{6\sqrt{5}+3\ln(2+\sqrt{5})}$$

75. The value of a scalar line integral does not depend on the choice of parametrization (because it is defined without reference to a parametrization). Prove this directly. That is, suppose that $\mathbf{r}_1(t)$ and $\mathbf{r}(t)$ are two parametrizations such that $\mathbf{r}_1(t) = \mathbf{r}(\varphi(t))$, where $\varphi(t)$ is an increasing function. Use the Change of Variables Formula to verify that
$$\int_c^d f(\mathbf{r}_1(t))\|\mathbf{r}_1'(t)\| \, dt = \int_a^b f(\mathbf{r}(t))\|\mathbf{r}'(t)\| \, dt$$
where $a = \varphi(c)$ and $b = \varphi(d)$.

SOLUTION We compute the integral $\int_a^b f(\mathbf{r}(t))\|\mathbf{r}'(t)\| \, dt$ using the substitution $t = \varphi(u)$, $a = \varphi(c)$, $b = \varphi(d)$. We get:
$$\int_a^b f(\mathbf{r}(t))\|\mathbf{r}'(t)\| \, dt = \int_{\varphi^{-1}(a)}^{\varphi^{-1}(b)} f(\mathbf{r}(\varphi(u)))\|\mathbf{r}'(\varphi(u))\|\varphi'(u) \, du \quad (1)$$

Since φ is an increasing function, $\varphi'(u) > 0$ for all u, therefore:
$$\|\mathbf{r}'(\varphi(u))\|\varphi'(u) = \|\mathbf{r}'(\varphi(u))\varphi'(u)\| \quad (2)$$

By the Chain Rule for vector valued functions, we have
$$\frac{d}{du}\mathbf{r}(\varphi(u)) = \varphi'(u)\mathbf{r}'(\varphi(u)) \quad (3)$$

Combining (2) and (3) gives:
$$\|\mathbf{r}'(\varphi(u))\|\varphi'(u) = \left\|\frac{d}{du}\mathbf{r}(\varphi(u))\right\| = \left\|\frac{d}{du}\mathbf{r}_1(u)\right\| = \|\mathbf{r}_1'(u)\| \quad (4)$$

We substitute (4) in (1) to obtain:
$$\int_a^b f(\mathbf{r}(t))\|\mathbf{r}'(t)\| \, dt = \int_c^d f(\mathbf{r}_1(u))\|\mathbf{r}_1'(u)\| \, du = \int_c^d f(\mathbf{r}_1(t))\|\mathbf{r}_1'(t)\| \, dt$$

The last step is simply replacing the dummy variable of integration u by t.

16.3 Conservative Vector Fields (LT Section 17.3)

Preliminary Questions

1. The following statement is false. *If* **F** *is a gradient vector field, then the line integral of* **F** *along every curve is zero*. Which single word must be added to make it true?

SOLUTION The missing word is "closed" (curve). The line integral of a gradient vector field along every closed curve is zero.

2. Which of the following statements are true for all vector fields, and which are true only for conservative vector fields?
(a) The line integral along a path from P to Q does not depend on which path is chosen.
(b) The line integral over an oriented curve C does not depend on how C is parametrized.
(c) The line integral around a closed curve is zero.
(d) The line integral changes sign if the orientation is reversed.
(e) The line integral is equal to the difference of a potential function at the two endpoints.
(f) The line integral is equal to the integral of the tangential component along the curve.
(g) The cross-partial derivatives of the components are equal.

SOLUTION
(a) This statement is true only for conservative vector fields.
(b) This statement is true for all vector fields.
(c) This statement holds only for conservative vector fields.
(d) This is a property of all vector fields.
(e) Only conservative vector fields have a potential function, and the line integral is computed by using the potential function as stated.
(f) All vector fields' line integrals share this property.
(g) The cross partial of the components of a conservative field are equal. For other fields, the cross-partial of the components may or may not equal.

3. Let **F** be a vector field on an open, connected domain $\mathcal{D}$ with continuous second partial derivatives. Which of the following statements are always true, and which are true under additional hypotheses on $\mathcal{D}$?
(a) If **F** has a potential function, then **F** is conservative.
(b) If **F** is conservative, then the cross-partial derivatives of **F** are equal.
(c) If the cross-partial derivatives of **F** are equal, then **F** is conservative.

SOLUTION
(a) This statement is always true, since every gradient vector field is conservative.
(b) If **F** is conservative on a connected domain $\mathcal{D}$, then **F** has a potential function $\mathcal{D}$ and consequently the cross partials of **F** are equal in $\mathcal{D}$.
(c) If the cross-partial derivatives of **F** are equal in a simply connected region $\mathcal{D}$, then **F** is a gradient vector field in $\mathcal{D}$.

4. Let C, $\mathcal{D}$, and $\mathcal{E}$ be the oriented curves in Figure 16, and let $\mathbf{F} = \nabla f$ be a gradient vector field such that $\int_C \mathbf{F} \cdot d\mathbf{r} = 4$. What are the values of the following integrals?

(a) $\int_{\mathcal{D}} \mathbf{F} \cdot d\mathbf{r}$ **(b)** $\int_{\mathcal{E}} \mathbf{F} \cdot d\mathbf{r}$

FIGURE 16

SOLUTION Since **F** is a gradient vector field, the integrals over closed paths are zero. Therefore, by the equivalent conditions for path independence we have:
(a) $\int_{\mathcal{D}} \mathbf{F} \cdot d\mathbf{r} = \int_C \mathbf{F} \cdot d\mathbf{r} = 4$
(b) $\int_{\mathcal{E}} \mathbf{F} \cdot d\mathbf{r} = \int_{-C} \mathbf{F} \cdot d\mathbf{r} = -\int_C \mathbf{F} \cdot d\mathbf{r} = -4$

Exercises

1. Let $f(x, y, z) = xy\sin(yz)$ and $\mathbf{F} = \nabla f$. Evaluate $\int_C \mathbf{F} \cdot d\mathbf{r}$, where C is any path from $(0, 0, 0)$ to $(1, 1, \pi)$.

SOLUTION By the Fundamental Theorem for Gradient Vector Fields, we have:

$$\int_C \nabla f \cdot d\mathbf{r} = f(1, 1, \pi) - f(0, 0, 0) = 1 \cdot 1 \sin \pi - 0 = 0$$

In Exercises 3–6, verify that $\mathbf{F} = \nabla f$ and evaluate the line integral of $\mathbf{F}$ over the given path.

3. $\mathbf{F}(x, y) = \langle 3, 6y \rangle$, $\quad f(x, y) = 3x + 3y^2$; $\quad \mathbf{r}(t) = \langle t, 2t^{-1} \rangle$ on the interval $1 \le t \le 4$

SOLUTION The gradient of $f = 3x + 3y^2$ is:

$$\nabla f = \left\langle \frac{\partial f}{\partial x}, \frac{\partial f}{\partial y} \right\rangle = \langle 3, 6y \rangle = \mathbf{F}$$

Using the Fundamental Theorem for Gradient Vector Fields, we have:

$$\int_C \mathbf{F} \cdot d\mathbf{r} = f(\mathbf{r}(4)) - f(\mathbf{r}(1)) = f\left(4, \frac{1}{2}\right) - f(1, 2) = \left(3 \cdot 4 + 3 \cdot \frac{1}{4}\right) - (3 \cdot 1 + 3 \cdot 4) = -\frac{9}{4}$$

5. $\mathbf{F}(x, y, z) = ye^z\mathbf{i} + xe^z\mathbf{j} + xye^z\mathbf{k}$, $\quad f(x, y, z) = xye^z$; $\quad \mathbf{r}(t) = \langle t^2, t^3, t-1 \rangle$ for $1 \le t \le 2$

SOLUTION We verify that $\mathbf{F}$ is the gradient of f:

$$\nabla f = \frac{\partial f}{\partial x}\mathbf{i} + \frac{\partial f}{\partial y}\mathbf{j} + \frac{\partial f}{\partial z}\mathbf{k} = ye^z\mathbf{i} + xe^z\mathbf{j} + xye^z\mathbf{k} = \mathbf{F}$$

We use the Fundamental Theorem for Gradient Vectors with the initial point $\mathbf{r}(1) = (1, 1, 0)$ and terminal point $\mathbf{r}(2) = (4, 8, 1)$, to obtain:

$$\int_C \mathbf{F} \cdot d\mathbf{r} = f(4, 8, 1) - f(1, 1, 0) = 32e - 1$$

In Exercises 7–18, find a potential function for $\mathbf{F}$ or determine that $\mathbf{F}$ is not conservative.

7. $\mathbf{F} = \langle x, y, z \rangle$

SOLUTION We examine whether $\mathbf{F}$ satisfies the cross-partials condition:

$$\left. \begin{array}{l} \dfrac{\partial F_1}{\partial y} = \dfrac{\partial}{\partial y}(x) = 0 \\[6pt] \dfrac{\partial F_2}{\partial x} = \dfrac{\partial}{\partial x}(y) = 0 \end{array} \right\} \Rightarrow \dfrac{\partial F_1}{\partial y} = \dfrac{\partial F_2}{\partial x}$$

$$\left. \begin{array}{l} \dfrac{\partial F_2}{\partial z} = \dfrac{\partial}{\partial z}(y) = 0 \\[6pt] \dfrac{\partial F_3}{\partial y} = \dfrac{\partial}{\partial y}(z) = 0 \end{array} \right\} \Rightarrow \dfrac{\partial F_2}{\partial z} = \dfrac{\partial F_3}{\partial y}$$

$$\left. \begin{array}{l} \dfrac{\partial F_3}{\partial x} = \dfrac{\partial}{\partial x}(z) = 0 \\[6pt] \dfrac{\partial F_1}{\partial z} = \dfrac{\partial}{\partial z}(x) = 0 \end{array} \right\} \Rightarrow \dfrac{\partial F_3}{\partial x} = \dfrac{\partial F_1}{\partial z}$$

$\mathbf{F}$ satisfies the cross-partials condition everywhere. Hence, $\mathbf{F}$ is conservative. If $f(x, y, z)$ is a potential function, then it satisfies

$$f(x, y, z) = \int x\, dx = \frac{1}{2}x^2 + f(y, z)$$

$$f(x, y, z) = \int y\, dy = \frac{1}{2}y^2 + g(x, z)$$

$$f(x, y, z) = \int z\, dz = \frac{1}{2}z^2 + h(x, y)$$

These three ways of writing $f(x, y, z)$ must be equal, so

$$\frac{1}{2}x^2 + f(y, z) = \frac{1}{2}y^2 + g(x, z) = \frac{1}{2}z^2 + h(x, y)$$

If we choose

$$f(y, z) = \frac{1}{2}y^2 + \frac{1}{2}z^2, \quad g(x, z) = \frac{1}{2}x^2 + \frac{1}{2}z^2, \quad h(x, y) = \frac{1}{2}x^2 + \frac{1}{2}y^2$$

the equalities hold, so that a potential function is

$$f(x, y, z) = \frac{1}{2}x^2 + \frac{1}{2}y^2 + \frac{1}{2}z^2 + C$$

for any constant C.

9. $\mathbf{F} = \langle z, x, y \rangle$

SOLUTION We examine whether $\mathbf{F}$ satisfies the cross-partials condition:

$$\frac{\partial F_1}{\partial y} = \frac{\partial}{\partial y}(z) = 0$$
$$\frac{\partial F_2}{\partial x} = \frac{\partial}{\partial x}(x) = 1$$
$$\Rightarrow \quad \frac{\partial F_1}{\partial y} \neq \frac{\partial F_2}{\partial x}$$

$\mathbf{F}$ does not satisfy the cross-partials condition, so $\mathbf{F}$ is not conservative.

11. $\mathbf{F} = y^2 \mathbf{i} + (2xy + e^z)\mathbf{j} + ye^z \mathbf{k}$

SOLUTION We examine whether $\mathbf{F}$ satisfies the cross-partials condition:

$$\frac{\partial F_1}{\partial y} = \frac{\partial}{\partial y}(y^2) = 2y$$
$$\frac{\partial F_2}{\partial x} = \frac{\partial}{\partial x}(2xy + e^z) = 2y$$
$$\Rightarrow \quad \frac{\partial F_1}{\partial y} = \frac{\partial F_2}{\partial x}$$

$$\frac{\partial F_2}{\partial z} = \frac{\partial}{\partial z}(2xy + e^z) = e^z$$
$$\frac{\partial F_3}{\partial y} = \frac{\partial}{\partial y}(ye^z) = e^z$$
$$\Rightarrow \quad \frac{\partial F_2}{\partial z} = \frac{\partial F_3}{\partial y}$$

$$\frac{\partial F_3}{\partial x} = \frac{\partial}{\partial x}(ye^z) = 0$$
$$\frac{\partial F_1}{\partial z} = \frac{\partial}{\partial z}(y^2) = 0$$
$$\Rightarrow \quad \frac{\partial F_3}{\partial x} = \frac{\partial F_1}{\partial z}$$

$\mathbf{F}$ satisfies the cross-partials condition everywhere. Hence, $\mathbf{F}$ is conservative. If $f(x, y, z)$ is a potential function, then it satisfies

$$f(x, y, z) = \int y^2\, dx \quad\quad = xy^2 + f(y, z)$$

$$f(x, y, z) = \int (2xy + e^z)\, dy = xy^2 + ye^z + g(x, z)$$

$$f(x, y, z) = \int ye^z\, dz \quad\quad = ye^z + h(x, y).$$

These three ways of writing $f(x, y, z)$ must be equal, so

$$xy^2 + f(y, z) = xy^2 + ye^z + g(x, z) = ye^z + h(x, y)$$

If we choose $f(y, z) = ye^z$, $g(x, z) = 0$, and $h(x, y) = xy^2$, the equalities hold, so that a potential function is

$$f(x, y, z) = xy^2 + ye^z + C$$

for any constant C.

13. $\mathbf{F} = \langle \cos(xz), \sin(yz), xy \sin z \rangle$

SOLUTION Since $\frac{\partial F_2}{\partial z} = \frac{\partial}{\partial z}(\sin(yz)) = y\cos(yz)$ and $\frac{\partial F_3}{\partial y} = \frac{\partial}{\partial y}(xy \sin z) = x \sin z$, we have $\frac{\partial F_2}{\partial z} \neq \frac{\partial F_3}{\partial y}$. The cross-partials condition is not satisfied; therefore the vector field is not conservative.

15. $\mathbf{F} = \langle z \sec^2 x, z, y + \tan x \rangle$

SOLUTION We examine whether $\mathbf{F}$ satisfies the cross-partials condition:

$$\frac{\partial F_1}{\partial y} = \frac{\partial}{\partial y}(z \sec^2 x) = 0$$
$$\frac{\partial F_2}{\partial x} = \frac{\partial}{\partial x}(z) = 0 \quad \Rightarrow \quad \frac{\partial F_1}{\partial y} = \frac{\partial F_2}{\partial x}$$

$$\frac{\partial F_2}{\partial z} = \frac{\partial}{\partial z}(z) = 1$$
$$\frac{\partial F_3}{\partial y} = \frac{\partial}{\partial y}(y + \tan x) = 1 \quad \Rightarrow \quad \frac{\partial F_2}{\partial z} = \frac{\partial F_3}{\partial y}$$

$$\frac{\partial F_3}{\partial x} = \frac{\partial}{\partial x}(y + \tan x) = \sec^2 x$$
$$\frac{\partial F_1}{\partial z} = \frac{\partial}{\partial z}(z \sec^2 x) = \sec^2 x \quad \Rightarrow \quad \frac{\partial F_3}{\partial x} = \frac{\partial F_1}{\partial z}$$

$\mathbf{F}$ satisfies the cross-partials condition everywhere. Hence, $\mathbf{F}$ is conservative. If $f(x, y, z)$ is a potential function, then it satisfies

$$f(x, y, z) = \int z \sec^2 x \, dx = z \tan x + f(y, z)$$

$$f(x, y, z) = \int z \, dy = yz + g(x, z)$$

$$f(x, y, z) = \int (y + \tan x) \, dz = yz + z \tan x + h(x, y)$$

These three ways of writing $f(x, y, z)$ must be equal, so

$$z \tan x + f(y, z) = yz + g(x, z) = yz + z \tan x + h(x, y)$$

If we choose $f(y, z) = yz$, $g(x, y) = z \tan x$, and $h(x, y) = 0$, the equalities hold, so that a potential function is

$$f(x, y, z) = yz + z \tan x + C$$

for any constant C.

17. $\mathbf{F} = \langle 2xy + 5, x^2 - 4z, -4y \rangle$

SOLUTION We examine whether $\mathbf{F}$ satisfies the cross-partials condition:

$$\frac{\partial F_1}{\partial y} = \frac{\partial}{\partial y}(2xy + 5) = 2x$$
$$\frac{\partial F_2}{\partial x} = \frac{\partial}{\partial x}(x^2 - 4z) = 2x \quad \Rightarrow \quad \frac{\partial F_1}{\partial y} = \frac{\partial F_2}{\partial x}$$

$$\frac{\partial F_2}{\partial z} = \frac{\partial}{\partial z}(x^2 - 4z) = -4$$
$$\frac{\partial F_3}{\partial y} = \frac{\partial}{\partial y}(-4y) = -4 \quad \Rightarrow \quad \frac{\partial F_2}{\partial z} = \frac{\partial F_3}{\partial y}$$

$$\frac{\partial F_3}{\partial x} = \frac{\partial}{\partial x}(-4y) = 0$$
$$\frac{\partial F_1}{\partial z} = \frac{\partial}{\partial z}(2xy + 5) = 0 \quad \Rightarrow \quad \frac{\partial F_3}{\partial x} = \frac{\partial F_1}{\partial z}$$

$\mathbf{F}$ satisfies the cross-partials condition everywhere. Hence, $\mathbf{F}$ is conservative. If $f(x, y, z)$ is a potential function, then it satisfies

$$f(x, y, z) = \int (2xy + 5) \, dx = x^2 y + 5x + f(y, z)$$

$$f(x, y, z) = \int (x^2 - 4z)\, dy = x^2 y - 4yz + g(x, z)$$

$$f(x, y, z) = \int -4y\, dz \quad = -4yz + h(x, y)$$

These three ways of writing $f(x, y, z)$ must be equal, so

$$x^2 y + 5x + f(y, z) = x^2 y - 4yz + g(x, z) = -4yz + h(x, y)$$

If we choose $f(y, z) = -4yz$, $g(x, y) = 5x$, and $h(x, y) = x^2 y + 5x$, the equalities hold, so that a potential function is

$$f(x, y, z) = x^2 y + 5x - 4yz + C$$

for any constant C.

19. Evaluate

$$\int_C 2xyz\, dx + x^2 z\, dy + x^2 y\, dz$$

over the path $\mathbf{r}(t) = (t^2, \sin(\pi t/4), e^{t^2 - 2t})$ for $0 \le t \le 2$.

SOLUTION A potential function is

$$f(x, y, z) = x^2 yz$$

The path begins at $\mathbf{r}(0) = (0, 0, 1)$ and ends at $\mathbf{r}(2) = (4, 1, 1)$ so the line integral is

$$f(4, 1, 1) - f(0, 0, 1) = 16 - 16 = 16$$

In Exercises 21–22, let $\mathbf{F} = \nabla f$, and determine directly $\int_C \mathbf{F} \cdot d\mathbf{r}$ for each of the two paths given, showing that they both give the same answer, which is $f(Q) - f(P)$.

21. $f = x^2 y - z$, $\mathbf{r}_1 = \langle t, t, 0 \rangle$ for $0 \le t \le 1$, and $\mathbf{r}_2 = \langle t, t^2, 0 \rangle$ for $0 \le t \le 1$

SOLUTION We have

$$\mathbf{F} = \nabla f = \left\langle \frac{\partial f}{\partial x}, \frac{\partial f}{\partial y}, \frac{\partial f}{\partial z} \right\rangle = \langle 2xy, x^2, -1 \rangle$$

Also,

$$\mathbf{r}_1'(t) = \langle 1, 1, 0 \rangle, \quad \mathbf{F}(\mathbf{r}_1(t)) = \langle 2t^2, t^2, -1 \rangle$$

$$\mathbf{r}_2'(t) = \langle 1, 2t, 0 \rangle, \quad \mathbf{F}(\mathbf{r}_2(t)) = \langle 2t^3, t^2, -1 \rangle$$

Then

$$\int_0^1 \mathbf{F}(\mathbf{r}_1(t)) \cdot \mathbf{r}_1'(t)\, dt = \int_0^1 3t^2\, dt = t^3 \Big|_0^1 = 1$$

$$\int_0^1 \mathbf{F}(\mathbf{r}_2(t)) \cdot \mathbf{r}_2'(t)\, dt = \int_0^1 4t^3\, dt = t^4 \Big|_0^1 = 1$$

The integrals along the two paths are indeed equal.

23. A vector field $\mathbf{F}$ and contour lines of a potential function for $\mathbf{F}$ are shown in Figure 17. Calculate the common value of $\int_C \mathbf{F} \cdot d\mathbf{r}$ for the curves shown in Figure 17 oriented in the direction from P to Q.

FIGURE 17

SOLUTION

$$\int_C \mathbf{F} \cdot d\mathbf{s} = \int_C \nabla V \cdot d\mathbf{s} = V(Q) - V(P) = 8 - 2 = 6$$

25. Calculate the work expended when a particle is moved from O to Q along segments $\overline{OP}$ and $\overline{PQ}$ in Figure 19 in the presence of the force field $\mathbf{F} = \langle x^2, y^2 \rangle$. How much work is expended moving in a complete circuit around the square?

FIGURE 19

SOLUTION

Since $\frac{\partial F_1}{\partial y} = \frac{\partial}{\partial y}(x^2) = 0$ and $\frac{\partial F_2}{\partial x} = \frac{\partial}{\partial x}(y^2) = 0$, we have $\frac{\partial F_1}{\partial y} = \frac{\partial F_2}{\partial x}$. That is, $\mathbf{F}$ satisfies the cross-partials condition, therefore $\mathbf{F}$ is conservative. We choose the function $\frac{x^3}{3} + \frac{y^3}{3}$, such that $\mathbf{F}$ is the gradient of the function. The potential energy is, thus, $f = -\frac{x^3}{3} - \frac{y^3}{3}$. The work done against $\mathbf{F}$ is computed by the Fundamental Theorem for Gradient vectors:

$$\text{Work against } \mathbf{F} = -\int_C \mathbf{F} \cdot d\mathbf{r} = f(Q) - f(O) = f(1,1) - f(0) = -\frac{2}{3} - 0 = -\frac{2}{3}$$

The negative sign is to be expected, as our force field is actually helping us move along OP and OQ. The line integral of a conservative field along a closed curve is zero; therefore the integral of $\mathbf{F}$ along the complete square is zero, and we get:

$$W = -\int_{OPQR} \mathbf{F} \cdot d\mathbf{r} = 0$$

27. Compute the work W against the earth's gravitational field required to move a satellite of mass $m = 1000$ kg along any path from an orbit of altitude 4000 km to an orbit of altitude 6000 km.

SOLUTION Work against gravity is calculated with the integral

$$W = -\int_C m\mathbf{F} \cdot d\mathbf{s} = 1000 \int_C \nabla V \cdot d\mathbf{s} = 1000(V(r_2) - V(r_1))$$

Since r_1 and r_2 are measured from the center of the earth,
$$r_1 = 4 \times 10^6 + 6.4 \times 10^6 = 10.4 \times 10^6 \text{ meters}$$
$$r_2 = 6 \times 10^6 + 6.4 \times 10^6 = 12.4 \times 10^6 \text{ meters}$$
$$V(r) = -\frac{k}{r} \quad \Rightarrow \quad W = -\frac{1000k}{10^6}\left(\frac{1}{12.4} - \frac{1}{10.4}\right) \approx 6.2 \times 10^9 \text{ J}$$

29. On the surface of the earth, the gravitational field (with z as vertical coordinate measured in meters) is $\mathbf{F} = \langle 0, 0, -g \rangle$.

(a) Find a potential function for $\mathbf{F}$.

(b) Beginning at rest, a ball of mass $m = 2$ kg moves under the influence of gravity (without friction) along a path from $P = (3, 2, 400)$ to $Q = (-21, 40, 50)$. Find the ball's velocity when it reaches Q.

SOLUTION The force of gravity is $\mathbf{F} = \langle 0, 0, -mg \rangle$, therefore $\mathbf{F} = -\nabla f$ for $f(x, y, z) = mgz$. The work performed by the gravitational field is the line integral of $\mathbf{F}$ over the path. Since $\mathbf{F}$ is conservative, the energy is independent of the path connecting the two points. Using the Fundamental Theorem for Gradient Vector Fields we have:
$$W = -\int_C \mathbf{F} \cdot d\mathbf{r} = f(z = 1000) - f(z = 0) = mg \cdot 1000 = 2 \cdot 9.8 \cdot 10^3 = 19{,}600 \text{ joules}$$

31. Let $\mathbf{F} = \left\langle \dfrac{-y}{x^2 + y^2}, \dfrac{x}{x^2 + y^2} \right\rangle$ be the vortex field. Determine $\int_C \mathbf{F} \cdot d\mathbf{r}$ for each of the paths in Figure 20.

FIGURE 20

SOLUTION Since the cross partials of $\mathbf{F}$ are equal, $\mathbf{F}$ has the property,
$$\int_C \mathbf{F} \cdot d\mathbf{r} = 2\pi n$$
where $\mathbf{r}$ is a closed curve not passing through the origin, and n is the number of times $\mathbf{r}$ winds around the origin (n is negative if n winds in the clockwise direction). We use this property to compute the line integrals of $\mathbf{F}$ over the paths in Figure 18:

(A) The path (A) winds around the origin one time in the counterclockwise direction hence the line integral is $2\pi \cdot 1 = 2\pi$.

(B) The point (B) winds around the origin one time in the counterclockwise direction, hence the line integral is $2\pi \cdot 1 = 2\pi$.

(C) The path (C) does not encounter the origin, hence the line integral is $2\pi \cdot 0 = 0$. Notice that there exists a simply connected domain $\mathcal{D}$, not including the origin, so that the path c and the region inside $\mathbf{r}$ are in $\mathcal{D}$. Therefore, Theorem 4 applies in $\mathcal{D}$ and $\mathbf{F}$ is a gradient vector in $\mathcal{D}$. Consequently, the line integral of $\mathbf{F}$ over $\mathbf{r}$ is zero.

(D) This path winds around the origin one time in the clockwise direction, hence $\int_C \mathbf{F} \cdot d\mathbf{r} = 2\pi \cdot (-1) = -2\pi$.
(E) The path winds around the origin twice in the counterclockwise direction, hence the line integral is $2\pi \cdot 2 = 4\pi$.

33. Determine whether or not the vector field $\mathbf{F}(x, y) = \left\langle \dfrac{x^2}{x^2 + y^2}, \dfrac{y^2}{x^2 + y^2} \right\rangle$ has a potential function.

SOLUTION   If $\mathbf{F}$ has a potential function, it must satisfy the cross-partial condition. But

$$\frac{\partial F_2}{\partial x} = \frac{\partial}{\partial x}\left(\frac{y^2}{x^2+y^2}\right) = -\frac{2xy^2}{(x^2+y^2)^2}$$

$$\frac{\partial F_1}{\partial y} = \frac{\partial}{\partial y}\left(\frac{x^2}{x^2+y^2}\right) = -\frac{2x^2 y}{(x^2+y^2)^2}$$

Since the cross partials are unequal, $\mathbf{F}$ does not have a potential function.

Further Insights and Challenges

35. Suppose that $\mathbf{F}$ is defined on $\mathbf{R}^3$ and that $\oint_C \mathbf{F} \cdot d\mathbf{r} = 0$ for all closed paths C in $\mathbf{R}^3$. Prove:

(a) $\mathbf{F}$ is path independent; that is, for any two paths C_1 and C_2 in $\mathcal{D}$ with the same initial and terminal points,

$$\int_{C_1} \mathbf{F} \cdot d\mathbf{r} = \int_{C_2} \mathbf{F} \cdot d\mathbf{r}$$

(b) $\mathbf{F}$ is conservative.

SOLUTION   A closed path C can be represented as a difference

$$C = C_1 - C_2$$

as follows: choose two distinct points P and Q along C and let C_1 and C_2 be the two paths from P to Q as in Figure 21. Thus,

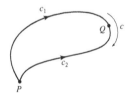

FIGURE 21 The closed loop C is represented as $C_1 - C_2$.

$$\oint_C \mathbf{F} \cdot d\mathbf{r} = \int_{C_1} \mathbf{F} \cdot d\mathbf{r} - \int_{C_2} \mathbf{F} \cdot d\mathbf{r} \qquad (1)$$

If we assume (a), then the right-hand side of Eq. (1) is zero. Thus the left-hand side is zero and (b) holds. On the other hand, Eq. (1) is valid for any two paths C_1 and C_2 from P to Q. If we assume (b), then the left-hand side of Eq. (1) is zero. Thus, the right-hand side is zero and (a) holds as well.

16.4 Parametrized Surfaces and Surface Integrals
(LT Section 17.4)

Preliminary Questions

1. What is the surface integral of the function $f(x, y, z) = 10$ over a surface of total area 5?

SOLUTION Using Surface Integral and Surface Area we have:

$$\iint_S f(x,y,z)\,dS = \iint_D f(\Phi(u,v))\|\mathbf{N}(u,v)\|\,du\,dv = \iint_D 10\|\mathbf{N}(u,v)\|\,du\,dv$$
$$= 10\iint_D \|\mathbf{N}(u,v)\|\,du\,dv = 10\,\text{Area}(S) = 10 \cdot 5 = 50$$

2. What interpretation can we give to the length $\|\mathbf{N}\|$ of the normal vector for a parametrization $\Phi(u, v)$?

SOLUTION The approximation:

$$\text{Area}(S_{ij}) \approx \|\mathbf{N}(u_{ij}, v_{ij})\|\text{Area}(R_{ij})$$

tells that $\|\mathbf{N}\|$ is a distortion factor that indicates how much the area of a small rectangle R_{ij} is altered under the map ϕ.

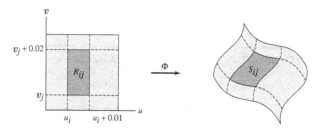

3. A parametrization maps a rectangle of size 0.01×0.02 in the uv-plane onto a small patch S of a surface. Estimate area(S) if $\mathbf{T}_u \times \mathbf{T}_v = \langle 1, 2, 2 \rangle$ at a sample point in the rectangle.

SOLUTION We use the estimation

$$\text{Area}(S) \approx \|\mathbf{N}(u,v)\|\text{Area}(R)$$

where $\mathbf{N}(u, v) = \mathbf{T}_u \times \mathbf{T}_v$ at a sample point in R. We get:

$$\text{Area}(S) \approx \|\langle 1,2,2\rangle\| \cdot 0.01 \cdot 0.02 = \sqrt{1^2 + 2^2 + 2^2} \cdot 0.0002 = 0.0006$$

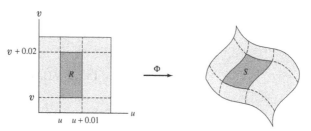

4. A small surface S is divided into three small pieces, each of area 0.2. Estimate $\iint_S f(x,y,z)\,dS$ if $f(x, y, z)$ takes the values 0.9, 1, and 1.1 at sample points in these three pieces.

SOLUTION We use the approximation obtained by the Riemann Sum:

$$\iint_S f(x,y,z)\,dS \approx \sum_{ij} f(P_{ij})\text{Area}(S_{ij}) = 0.9 \cdot 0.2 + 1 \cdot 0.2 + 1.1 \cdot 0.2 = 0.6$$

5. A surface S has a parametrization whose domain is the square $0 \le u, v \le 2$ such that $\|\mathbf{N}(u, v)\| = 5$ for all (u, v). What is area(S)?

SOLUTION Writing the surface area as a surface integral where $\mathcal{D}$ is the square $[0, 2] \times [0, 2]$ in the uv-plane, we have:

$$\text{Area}(S) = \iint_{\mathcal{D}} \|\mathbf{N}(u, v)\| \, du \, dv = \iint_{\mathcal{D}} 5 \, du \, dv = 5 \iint_{\mathcal{D}} 1 \, du \, dv = 5 \text{Area}(\mathcal{D}) = 5 \cdot 2^2 = 20$$

6. What is the outward-pointing unit normal to the sphere of radius 3 centered at the origin at $P = (2, 2, 1)$?

SOLUTION The outward-pointing normal to the sphere of radius $R = 3$ centered at the origin is the following vector:

$$\langle \cos\theta \sin\phi, \sin\theta \sin\phi, \cos\phi \rangle \tag{1}$$

We compute the values in (1) corresponding to $P = (2, 2, 1)$: $x = y = 2, z = 1$ hence $0 \le \theta \le \frac{\pi}{2}$ and $0 < \phi < \frac{\pi}{2}$. We get:

$$\cos\phi = \frac{z}{\rho} = \frac{1}{3} \quad \Rightarrow \quad \sin\phi = \sqrt{1 - \left(\frac{1}{3}\right)^2} = \frac{2\sqrt{2}}{3}$$

$$\cos\theta = \frac{x}{\rho \sin\phi} = \frac{2}{3 \cdot \frac{2\sqrt{2}}{3}} = \frac{1}{\sqrt{2}} \quad \Rightarrow \quad \sin\theta = \sqrt{1 - \frac{1}{2}} = \frac{1}{\sqrt{2}}$$

Substituting in (1) we get the following unit normal:

$$\left\langle \frac{1}{\sqrt{2}} \cdot \frac{2\sqrt{2}}{3}, \frac{1}{\sqrt{2}} \cdot \frac{2\sqrt{2}}{3}, \frac{1}{3} \right\rangle = \left\langle \frac{2}{3}, \frac{2}{3}, \frac{1}{3} \right\rangle$$

Exercises

1. Match each parametrization with the corresponding surface in Figure 16.

(a) $(u, \cos v, \sin v)$

(b) $(u, u + v, v)$

(c) (u, v^3, v)

(d) $(\cos u \sin v, 3 \cos u \sin v, \cos v)$

(e) $(u, u(2 + \cos v), u(2 + \sin v))$

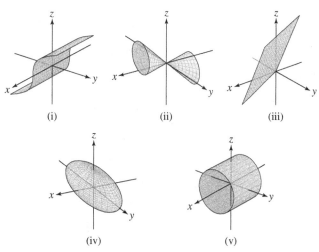

FIGURE 16

SOLUTION (a) = (v), because the y and z coordinates describe a circle with fixed radius.
(b) = (iii), because the coordinates are all linear in u and v.
(c) = (i), because the parametrization gives $y = z^3$.
(d) = (iv), an ellipsoid.
(e) = (ii), because the y and z coordinates describe a circle with varying radius.

3. Show that $\Phi(u, v) = (2u + 1, u - v, 3u + v)$ parametrizes the plane $2x - y - z = 2$. Then
(a) Calculate $\mathbf{T}_u$, $\mathbf{T}_v$, and $\mathbf{N}(u, v)$.
(b) Find the area of $S = \Phi(\mathcal{D})$, where
$\mathcal{D} = \{(u, v) : 0 \leq u \leq 2, 0 \leq v \leq 1\}$.
(c) Express $f(x, y, z) = yz$ in terms of u and v, and evaluate $\iint_S f(x, y, z)\, dS$.

SOLUTION We show that $x = 2u + 1$, $y = u - v$, and $z = 3u + v$ satisfy the equation of the plane,

$$2x - y - z = 2(2u + 1) - (u - v) - (3u + v) = 4u + 2 - u + v - 3u - v = 2$$

Moreover, for any x, y, z satisfying $2x - y - z = z$, there are values of u and v such that $x = 2u + 1$, $y = u - v$, and $z = 3u + v$, since the following equations can be solved for u and v:

$$x = 2u + 1$$
$$y = u - v$$
$$z = 3u + v$$
$$2x - y - z = 2$$

$$\Rightarrow \quad u = \frac{x-1}{2}, \quad v = \frac{x-1}{2} - y$$

We conclude that $\Phi(u, v)$ parametrizes the whole plane $2x - y - z = 2$.

(a) The tangent vectors $\mathbf{T}_u$ and $\mathbf{T}_v$ are:

$$\mathbf{T}_u = \frac{\partial \phi}{\partial u} = \frac{\partial}{\partial u}(2u + 1, u - v, 3u + v) = \langle 2, 1, 3 \rangle$$

$$\mathbf{T}_v = \frac{\partial \phi}{\partial v} = \frac{\partial}{\partial v}(2u + 1, u - v, 3u + v) = \langle 0, -1, 1 \rangle$$

The normal vector is the following cross product:

$$\mathbf{N}(u, v) = \mathbf{T}_u \times \mathbf{T}_v = \begin{vmatrix} \mathbf{i} & \mathbf{j} & \mathbf{k} \\ 2 & 1 & 3 \\ 0 & -1 & 1 \end{vmatrix} = \begin{vmatrix} 1 & 3 \\ -1 & 1 \end{vmatrix} \mathbf{i} - \begin{vmatrix} 2 & 3 \\ 0 & 1 \end{vmatrix} \mathbf{j} + \begin{vmatrix} 2 & 1 \\ 0 & -1 \end{vmatrix} \mathbf{k}$$

$$= 4\mathbf{i} - 2\mathbf{j} - 2\mathbf{k} = \langle 4, -2, -2 \rangle$$

(b) That area of $S = \Phi(\mathcal{D})$ is the following surface integral:

$$\text{Area}(S) = \iint_\mathcal{D} \|\mathbf{N}(u, v)\|\, du\, dv = \iint_\mathcal{D} \|\langle 4, -2, -2 \rangle\|\, du\, dv = \sqrt{24} \iint_\mathcal{D} 1\, du\, dv$$

$$= \sqrt{24}\, \text{Area}(\mathcal{D}) = \sqrt{24} \cdot 2 \cdot 1 = 4\sqrt{6}$$

(c) We express $f(x, y, z) = yz$ in terms of the parameters u and v:

$$f(\phi(u, v)) = (u - v)(3u + v) = 3u^2 - 2uv - v^2$$

Using the Theorem on Surface Integrals we have:

$$\iint_S f(x, y, z)\, dS = \iint_\mathcal{D} f(\phi(u, v))\, \|\mathbf{N}(u, v)\|\, du\, dv = \iint_\mathcal{D} (3u^2 - 2uv - v^2)\, \|\langle 4, -2, -2 \rangle\|\, du\, dv$$

$$= \sqrt{24} \int_0^1 \int_0^2 (3u^2 - 2uv - v^2)\, du\, dv = \sqrt{24} \int_0^1 \left. (u^3 - u^2v - v^2 u) \right|_{u=0}^{2} dv$$

$$= \sqrt{24} \int_0^1 (8 - 4v - 2v^2)\, dv = \sqrt{24} \left. \left(8v - 2v^2 - \frac{2}{3}v^3\right) \right|_0^1 = \frac{32\sqrt{6}}{3}$$

5. Let $\Phi(x, y) = (x, y, xy)$.
(a) Calculate $\mathbf{T}_x$, $\mathbf{T}_y$, and $\mathbf{N}(x, y)$.

(b) Let S be the part of the surface with parameter domain $\mathcal{D} = \{(x,y) : x^2 + y^2 \leq 1, x \geq 0, y \geq 0\}$. Verify the following formula and evaluate using polar coordinates:

$$\iint_S 1\, dS = \iint_\mathcal{D} \sqrt{1 + x^2 + y^2}\, dx\, dy$$

(c) Verify the following formula and evaluate:

$$\iint_S z\, dS = \int_0^{\pi/2} \int_0^1 (\sin\theta \cos\theta) r^3 \sqrt{1 + r^2}\, dr\, d\theta$$

SOLUTION

(a) The tangent vectors are:

$$\mathbf{T}_x = \frac{\partial \phi}{\partial x} = \frac{\partial}{\partial x}(x, y, xy) = \langle 1, 0, y \rangle$$

$$\mathbf{T}_y = \frac{\partial \phi}{\partial y} = \frac{\partial}{\partial y}(x, y, xy) = \langle 0, 1, x \rangle$$

The normal vector is the cross product:

$$\mathbf{N}(x,y) = \mathbf{T}_x \times \mathbf{T}_y = \begin{vmatrix} \mathbf{i} & \mathbf{j} & \mathbf{k} \\ 1 & 0 & y \\ 0 & 1 & x \end{vmatrix} = \begin{vmatrix} 0 & y \\ 1 & x \end{vmatrix} \mathbf{i} - \begin{vmatrix} 1 & y \\ 0 & x \end{vmatrix} \mathbf{j} + \begin{vmatrix} 1 & 0 \\ 0 & 1 \end{vmatrix} \mathbf{k}$$

$$= -y\mathbf{i} - x\mathbf{j} + \mathbf{k} = \langle -y, -x, 1 \rangle$$

(b) Using the Theorem on Evaluating Surface Integrals we have:

$$\iint_S 1\, dS = \iint_\mathcal{D} \|\mathbf{N}(x,y)\|\, dx\, dy = \iint_\mathcal{D} \|\langle -y, -x, 1\rangle\|\, dx\, dy = \iint_\mathcal{D} \sqrt{y^2 + x^2 + 1}\, dx\, dy$$

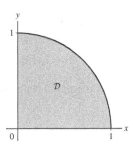

We convert the integral to polar coordinates $x = r\cos\theta$, $y = r\sin\theta$. The new region of integration is:

$$0 \leq r \leq 1, \quad 0 \leq \theta \leq \frac{\pi}{2}$$

We get:

$$\iint_S 1\, dS = \int_0^{\pi/2} \int_0^1 \sqrt{r^2 + 1} \cdot r\, dr\, d\theta = \int_0^{\pi/2} \left(\int_0^1 \sqrt{r^2 + 1} \cdot r\, dr \right) d\theta$$

$$= \int_0^{\pi/2} \left(\int_1^2 \frac{\sqrt{u}}{2}\, du \right) d\theta = \int_0^{\pi/2} \frac{2\sqrt{2} - 1}{3}\, d\theta = \frac{(2\sqrt{2} - 1)\pi}{6}$$

(c) The function z expressed in terms of the parameters x, y is $f(\Phi(x,y)) = xy$. Therefore,

$$\iint_S z\, dS = \iint_\mathcal{D} xy \cdot \|\mathbf{N}(x,y)\|\, dx\, dy = \iint_\mathcal{D} xy \sqrt{1 + x^2 + y^2}\, dx\, dy$$

We compute the double integral by converting it to polar coordinates. We get:

$$\iint_S z\, dS = \int_0^{\pi/2} \int_0^1 (r\cos\theta)(r\sin\theta) \sqrt{1 + r^2} \cdot r\, dr\, d\theta = \int_0^{\pi/2} \int_0^1 (\sin\theta \cos\theta) r^3 \sqrt{1 + r^2}\, dr\, d\theta$$

$$= \left(\int_0^{\pi/2} (\sin\theta \cos\theta)\, d\theta \right) \left(\int_0^1 r^3 \sqrt{1 + r^2}\, dr \right) \qquad (1)$$

We compute each integral in (1). Using the substitution $u = 1 + r^2$, $du = 2r\,dr$ we get:

$$\int_0^1 r^3 \sqrt{1+r^2}\,dr = \int_0^1 r^2 \sqrt{1+r^2} \cdot r\,dr = \int_1^2 \left(u^{3/2} - u^{1/2}\right) \frac{du}{2} = \left.\frac{u^{5/2}}{5} - \frac{u^{3/2}}{3}\right|_1^2 = \frac{2\left(\sqrt{2}+1\right)}{15}$$

Also,

$$\int_0^{\pi/2} \sin\theta \cos\theta\,d\theta = \int_0^{\pi/2} \frac{\sin 2\theta}{2}\,d\theta = \left.-\frac{\cos 2\theta}{4}\right|_0^{\pi/2} = \frac{1}{2}$$

We substitute the integrals in (1) to obtain the following solution:

$$\iint_S z\,dS = \frac{1}{2} \cdot \frac{2\left(\sqrt{2}+1\right)}{15} = \frac{\sqrt{2}+1}{15}$$

In Exercises 7–10, calculate $\mathbf{T}_u$, $\mathbf{T}_v$, and $\mathbf{N}(u,v)$ for the parametrized surface at the given point. Then find the equation of the tangent plane to the surface at that point.

7. $\Phi(u,v) = (2u+v, u-4v, 3u);\qquad u=1,\quad v=4$

SOLUTION The tangent vectors are the following vectors,

$$\mathbf{T}_u = \frac{\partial \Phi}{\partial u} = \frac{\partial}{\partial u}(2u+v, u-4v, 3u) = \langle 2, 1, 3\rangle$$

$$\mathbf{T}_v = \frac{\partial \Phi}{\partial v} = \frac{\partial}{\partial v}(2u+v, u-4v, 3u) = \langle 1, -4, 0\rangle$$

The normal is the cross product:

$$\mathbf{N}(u,v) = \mathbf{T}_u \times \mathbf{T}_v = \begin{vmatrix} \mathbf{i} & \mathbf{j} & \mathbf{k} \\ 2 & 1 & 3 \\ 1 & -4 & 0 \end{vmatrix} = \begin{vmatrix} 1 & 3 \\ -4 & 0 \end{vmatrix}\mathbf{i} - \begin{vmatrix} 2 & 3 \\ 1 & 0 \end{vmatrix}\mathbf{j} + \begin{vmatrix} 2 & 1 \\ 1 & -4 \end{vmatrix}\mathbf{k}$$

$$= 12\mathbf{i} + 3\mathbf{j} - 9\mathbf{k} = 3\langle 4, 1, -3\rangle$$

The equation of the plane passing through the point P: $\Phi(1,4) = (6, -15, 3)$ with the normal vector $\langle 4, 1, -3\rangle$ is:

$$\langle x-6, y+15, z-3\rangle \cdot \langle 4, 1, -3\rangle = 0$$

or

$$4(x-6) + y + 15 - 3(z-3) = 0$$

$$4x + y - 3z = 0$$

9. $\Phi(\theta,\phi) = (\cos\theta\sin\phi, \sin\theta\sin\phi, \cos\phi);\qquad \theta = \frac{\pi}{2},\quad \phi = \frac{\pi}{4}$

SOLUTION We compute the tangent vectors:

$$\mathbf{T}_\theta = \frac{\partial \Phi}{\partial \theta} = \frac{\partial}{\partial \theta}(\cos\theta\sin\phi, \sin\theta\sin\phi, \cos\phi) = \langle -\sin\theta\sin\phi, \cos\theta\sin\phi, 0\rangle$$

$$\mathbf{T}_\phi = \frac{\partial \Phi}{\partial \phi} = \frac{\partial}{\partial \phi}(\cos\theta\sin\phi, \sin\theta\sin\phi, \cos\phi) = \langle \cos\theta\cos\phi, \sin\theta\cos\phi, -\sin\phi\rangle$$

The normal vector is the cross product:

$$\mathbf{N}(\theta,\phi) = \mathbf{T}_\theta \times \mathbf{T}_\phi = \begin{vmatrix} \mathbf{i} & \mathbf{j} & \mathbf{k} \\ -\sin\theta\sin\phi & \cos\theta\sin\phi & 0 \\ \cos\theta\cos\phi & \sin\theta\cos\phi & -\sin\phi \end{vmatrix}$$

$$= \left(-\cos\theta\sin^2\phi\right)\mathbf{i} - \left(\sin\theta\sin^2\phi\right)\mathbf{j} + \left(-\sin^2\theta\sin\phi\cos\phi - \cos^2\theta\cos\phi\sin\phi\right)\mathbf{k}$$

$$= -\left(\cos\theta\sin^2\phi\right)\mathbf{i} - \left(\sin\theta\sin^2\phi\right)\mathbf{j} - (\sin\phi\cos\phi)\mathbf{k}$$

The tangency point and the normal at this point are,

$$P = \Phi\left(\frac{\pi}{2}, \frac{\pi}{4}\right) = \left(\cos\frac{\pi}{2}\sin\frac{\pi}{4}, \sin\frac{\pi}{2}\sin\frac{\pi}{4}, \cos\frac{\pi}{4}\right) = \left(0, \frac{\sqrt{2}}{2}, \frac{\sqrt{2}}{2}\right)$$

$$\mathbf{N}\left(\frac{\pi}{2}, \frac{\pi}{4}\right) = -\frac{1}{2}\mathbf{j} - \frac{1}{2}\mathbf{k} = -\frac{1}{2}(\mathbf{j} + \mathbf{k}) = -\frac{1}{2}\langle 0, 1, 1\rangle$$

The equation of the plane orthogonal to the vector $\langle 0, 1, 1\rangle$ and passing through $P = \left(0, \frac{\sqrt{2}}{2}, \frac{\sqrt{2}}{2}\right)$ is:

$$\left\langle x, y - \frac{\sqrt{2}}{2}, z - \frac{\sqrt{2}}{2}\right\rangle \cdot \langle 0, 1, 1\rangle = 0$$

or

$$y - \frac{\sqrt{2}}{2} + z - \frac{\sqrt{2}}{2} = 0$$

$$y + z = \sqrt{2}$$

11. Use the normal vector computed in Exercise 8 to estimate the area of the small patch of the surface $\Phi(u, v) = (u^2 - v^2, u + v, u - v)$ defined by

$$2 \le u \le 2.1, \quad 3 \le v \le 3.2$$

SOLUTION We denote the rectangle $\mathcal{D} = \{(u, v) : 2 \le u \le 2.1, 3 \le v \le 3.2\}$. Using the sample point corresponding to $u = 2$, $v = 3$ we obtain the following estimation for the area of $S = \Phi(\mathcal{D})$:

$$\text{Area}(S) \approx \|\mathbf{N}(2, 3)\|\text{Area}(\mathcal{D}) = \|\mathbf{N}(2, 3)\| \cdot 0.1 \cdot 0.2 = 0.02\|\mathbf{N}(2, 3)\| \quad (1)$$

In Exercise 8 we found that $\mathbf{N}(2, 3) = 2\langle -1, -1, 5\rangle$. Therefore,

$$\|\mathbf{N}(2, 3)\| = 2\sqrt{1^2 + 1^2 + 5^2} = 2\sqrt{27}$$

Substituting in (1) gives the following estimation:

$$\text{Area}(S) \approx 0.02 \cdot 2 \cdot \sqrt{27} \approx 0.2078$$

In Exercises 13–26, calculate $\iint_S f(x, y, z)\,dS$ for the given surface and function.

13. $\Phi(u, v) = (u\cos v, u\sin v, u), \quad 0 \le u \le 1, \quad 0 \le v \le 1; \; f(x, y, z) = z(x^2 + y^2)$

SOLUTION
Step 1. Compute the tangent and normal vectors. We have:

$$\mathbf{T}_u = \frac{\partial \Phi}{\partial u} = \frac{\partial}{\partial u}(u\cos v, u\sin v, u) = \langle \cos v, \sin v, 1\rangle$$

$$\mathbf{T}_v = \frac{\partial \Phi}{\partial v} = \frac{\partial}{\partial v}(u\cos v, u\sin v, u) = \langle -u\sin v, u\cos v, 0\rangle$$

The normal vector is the cross product:

$$\mathbf{n} = \mathbf{T}_u \times \mathbf{T}_v = \begin{vmatrix} \mathbf{i} & \mathbf{j} & \mathbf{k} \\ \cos v & \sin v & 1 \\ -u\sin v & u\cos v & 0 \end{vmatrix}$$

$$= (-u\cos v)\mathbf{i} - (u\sin v)\mathbf{j} + \left(u\cos^2 v + u\sin^2 v\right)\mathbf{k}$$

$$= (-u\cos v)\mathbf{i} - (u\sin v)\mathbf{j} + u\mathbf{k} = \langle -u\cos v, -u\sin v, u\rangle$$

We compute the length of $\mathbf{N}$:

$$\|\mathbf{N}\| = \sqrt{(-u\cos v)^2 + (-u\sin v)^2 + u^2} = \sqrt{u^2\left(\cos^2 v + \sin^2 v + 1\right)} = \sqrt{u^2 \cdot 2} = \sqrt{2}|u| = \sqrt{2}u$$

Notice that in the region of integration $u \ge 0$, therefore $|u| = u$.

Step 2. Calculate the surface integral. We express the function $f(x, y, z) = z(x^2 + y^2)$ in terms of the parameters u, v:

$$f(\Phi, (u, v)) = u\left(u^2 \cos^2 v + u^2 \sin^2 v\right) = u \cdot u^2 = u^3$$

We obtain the following integral:

$$\iint_S f(x, y, z)\, dS = \int_0^1 \int_0^1 f(\Phi, (u, v)) \|\mathbf{N}\|\, du\, dv = \int_0^1 \int_0^1 u^3 \cdot \sqrt{2}u\, du\, dv$$

$$= \left(\int_0^1 \sqrt{2}\, dv\right)\left(\int_0^1 u^4\, du\right) = \sqrt{2} \cdot \left.\frac{u^5}{5}\right|_0^1 = \frac{\sqrt{2}}{5}$$

15. $y = 4 - z^2$, $0 \le x \le 2$, $0 \le z \le 2$; $f(x, y, z) = 3z$

SOLUTION Let $y = g(x, z) = 4 - z^2$. Then $g_x = 0$ and $g_z = -2z$, so that

$$dS = \sqrt{1 + g_x^2 + g_z^2} = \sqrt{4z^2 + 1}$$

With $f(x, y, z) = 3z$, using the formula for surface integral over a graph we get

$$\iint_D f(x, y, z)\, dS = \int_0^2 \int_0^2 3z\sqrt{4z^2 + 1}\, dx\, dz = \int_0^2 6z\sqrt{4z^2 + 1}\, dz$$

$$= \frac{3}{4}\int_0^2 8z\sqrt{4z^2 + 1}\, dz = \frac{3}{4} \cdot \frac{2}{3}(4z^2 + 1)^{3/2}\bigg|_0^2 = \frac{1}{2}(17\sqrt{17} - 1)$$

17. $x^2 + y^2 + z^2 = 1$, $x, y, z \ge 0$; $f(x, y, z) = x^2$

SOLUTION The octant of the unit sphere centered at the origin, where x, y, $z \ge 0$ has the following parametrization in spherical coordinates:

$$\Phi(\theta, \phi) = (\cos\theta \sin\phi, \sin\theta \sin\phi, \cos\phi), \quad 0 \le \theta \le \frac{\pi}{2},\quad 0 \le \phi \le \frac{\pi}{2}$$

The length of the normal vector is:

$$\|\mathbf{N}\| = \sin\phi$$

The function x^2 expressed in terms of the parameters is $\cos^2\theta \sin^2\phi$. Using the theorem on computing surface integrals we obtain,

$$\iint_S x^2\, dS = \int_0^{\pi/2}\int_0^{\pi/2} \left(\cos^2\theta \sin^2\phi\right)(\sin\phi)\, d\phi\, d\theta = \int_0^{\pi/2}\int_0^{\pi/2} \cos^2\theta \sin^3\phi\, d\phi\, d\theta$$

$$= \left(\int_0^{\pi/2} \cos^2\theta\, d\theta\right)\left(\int_0^{\pi/2} \sin^3\phi\, d\phi\right) = \left(\frac{\theta}{2} + \frac{\sin 2\theta}{4}\right)\bigg|_{\theta=0}^{\pi/2} \cdot \left(-\frac{\sin^2\phi \cos\phi}{3} - \frac{2}{3}\cos\phi\right)\bigg|_{\phi=0}^{\pi/2}$$

$$= \frac{\pi}{4} \cdot \frac{2}{3} = \frac{\pi}{6}$$

19. $x^2 + y^2 = 4$, $0 \le z \le 4$; $f(x, y, z) = e^{-z}$

SOLUTION The cylinder has the following parametrization in cylindrical coordinates:

$$\Phi(\theta, z) = (2\cos\theta, 2\sin\theta, z),\ 0 \le \theta \le 2\pi,\ 0 \le z \le 4$$

Step 1. Compute the tangent and normal vectors. The tangent vectors are the partial derivatives:

$$\mathbf{T}_\theta = \frac{\partial \Phi}{\partial \theta} = \frac{\partial}{\partial \theta}(2\cos\theta, 2\sin\theta, z) = \langle -2\sin\theta, 2\cos\theta, 0\rangle$$

$$\mathbf{T}_z = \frac{\partial}{\partial z}(2\cos\theta, 2\sin\theta, z) = \langle 0, 0, 1\rangle$$

The normal vector is their cross product:

$$\mathbf{N}(\theta, z) = \mathbf{T}_\theta \times \mathbf{T}_z = \begin{vmatrix} \mathbf{i} & \mathbf{j} & \mathbf{k} \\ -2\sin\theta & 2\cos\theta & 0 \\ 0 & 0 & 1 \end{vmatrix} = (2\cos\theta)\mathbf{i} + (2\sin\theta)\mathbf{j} = \langle 2\cos\theta, 2\sin\theta, 0\rangle$$

The length of the normal vector is thus

$$\|\mathbf{N}(\theta, z)\| = \sqrt{(2\cos\theta)^2 + (2\sin\theta)^2 + 0} = \sqrt{4(\cos^2\theta + \sin^2\theta)} = \sqrt{4} = 2$$

Step 2. Calculate the surface integral. The surface integral equals the following double integral:

$$\iint_S f(x, y, z)\, dS = \iint_D f(\Phi(\theta, z))\, \|\mathbf{N}\|\, d\theta\, dz = \int_0^{2\pi} \int_0^4 e^{-z} \cdot 2\, d\theta\, dz$$

$$= \left(\int_0^{2\pi} 2\, d\theta\right)\left(\int_0^4 e^{-z}\, dz\right) = 4\pi \cdot (-e^{-z})\Big|_0^4 = 4\pi\left(1 - e^{-4}\right)$$

21. Part of the plane $x + y + z = 1$, where $x, y, z \geq 0$;
$f(x, y, z) = z$

SOLUTION We let $z = g(x, y) = 1 - x - y$ and use the formula for the surface integral over the graph of $z = g(x, y)$, where $\mathcal{D}$ is the parameter domain in the xy-plane. That is:

$$\iint_S f(x, y, z)\, dS = \iint_D f(x, y, g(x, y)) \sqrt{1 + g_x^2 + g_y^2}\, dx\, dy \tag{1}$$

We have, $g_x = -1$ and $g_y = -1$ therefore:

$$\sqrt{1 + g_x^2 + g_y^2} = \sqrt{1 + (-1)^2 + (-1)^2} = \sqrt{3}$$

We express the function $f(x, y, z) = z$ in terms of the parameters x and y:

$$f(x, y, g(x, y)) = z = 1 - x - y$$

The domain of integration is the triangle $\mathcal{D}$ in the xy-plane shown in the figure.

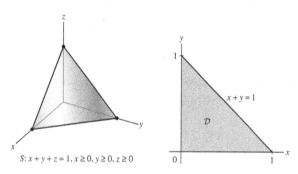

$S: x+y+z=1, x\geq 0, y\geq 0, z\geq 0$

By (1) we get:

$$\iint_S f(x, y, z)\, dS = \int_0^1 \int_0^{1-y} (1 - x - y)\sqrt{3}\, dx\, dy = \sqrt{3} \int_0^1 x - \frac{x^2}{2} - yx\Big|_{x=0}^{1-y} dy$$

$$= \sqrt{3} \int_0^1 \left((1-y)^2 - \frac{(1-y)^2}{2}\right) dy = \frac{\sqrt{3}}{2} \int_0^1 \left(1 - 2y + y^2\right) dy$$

$$= \frac{\sqrt{3}}{2}\left(y - y^2 + \frac{y^3}{3}\right)\Big|_0^1 = \frac{\sqrt{3}}{6}$$

23. $x^2 + y^2 + z^2 = 4$, $1 \leq z \leq 2$; $f(x, y, z) = z^2(x^2 + y^2 + z^2)^{-1}$

SOLUTION We use spherical coordinates to parametrize the cap S.

$$\Phi(\theta, \phi) = (2\cos\theta \sin\phi, 2\sin\theta \sin\phi, 2\cos\phi)$$

$$\mathcal{D}: 0 \leq \theta \leq 2\pi,\ 0 \leq \phi \leq \phi_0$$

The angle ϕ_0 is determined by $\cos\phi_0 = \frac{1}{2}$, that is, $\phi_0 = \frac{\pi}{3}$. The length of the normal vector in spherical coordinates is:

$$\|\mathbf{N}\| = R^2 \sin\phi = 4\sin\phi$$

We express the function $f(x,y,z) = z^2(x^2+y^2+z^2)^{-1}$ in terms of the parameters:

$$f(\Phi(\theta,\phi)) = (2\cos\phi)^2 4^{-1} = \cos^2\phi$$

Using the Theorem on Computing Surface Integral, we get:

$$\iint_S f(x,y,z)\,dS = \iint_D f(\Phi(\theta,\phi))\|\mathbf{N}\|\,d\phi\,d\theta = \int_0^{2\pi}\int_0^{\pi/3}(\cos^2\phi)\cdot 4\sin\phi\,d\phi\,d\theta$$

$$= \left(\int_0^{2\pi} 4\,d\theta\right)\left(\int_0^{\pi/3}\cos^2\phi\sin\phi\,d\phi\right) = 8\pi\left(-\frac{\cos^3\phi}{3}\right)\bigg|_0^{\pi/3}$$

$$= 8\pi\left(\left(\frac{1}{8}-\frac{1}{6}\right)-\left(-\frac{1}{3}\right)\right) = 8\pi\cdot\frac{7}{24} = \frac{7\pi}{3}$$

25. Part of the surface $z = x^3$, where $0 \le x \le 1, 0 \le y \le 1$; $f(x,y,z) = z$

SOLUTION   Write $z = g(x,y) = x^3$; then $g_x = 3x^2$ and $g_y = 0$; further, $f(x,y,g(x,y)) = x^3$. Then using the formula for the surface integral over a graph, we get

$$\int_0^1\int_0^1 f(x,y,g(x,y))\sqrt{1+g_x^2+g_y^2}\,dy\,dx = \int_0^1\int_0^1 x^3\sqrt{1+9x^4}\,dy\,dx$$

$$= \int_0^1 x^3\sqrt{1+9x^4}\,dx$$

$$= \frac{1}{54}(1+9x^4)^{3/2}\bigg|_0^1 = \frac{1}{54}(10\sqrt{10}-1)$$

27. A surface S has a parametrization $\Phi(u,v)$ with rectangular domain $0 \le u \le 2, 0 \le v \le 4$ such that the following partial derivatives are constant:

$$\frac{\partial\Phi}{\partial u} = \langle 2,0,1\rangle, \qquad \frac{\partial\Phi}{\partial v} = \langle 4,0,3\rangle$$

What is the surface area of S?

SOLUTION   Since the partial derivatives are constant, the normal vector is also constant. We find it by computing the cross product:

$$\mathbf{N} = \mathbf{T}_u\times\mathbf{T}_v = \frac{\partial\Phi}{\partial u}\times\frac{\partial\Phi}{\partial v} = \begin{vmatrix} \mathbf{i} & \mathbf{j} & \mathbf{k} \\ 2 & 0 & 1 \\ 4 & 0 & 3 \end{vmatrix} = -2\mathbf{j} = \langle 0,-2,0\rangle \Rightarrow \|\mathbf{N}\| = 2$$

We denote the rectangle $\mathcal{D} = \{(u,v) : 0 \le u \le 2, 0 \le v \le 4\}$, and use the surface area to compute the area of $S = \Phi(\mathcal{D})$. We obtain:

$$\text{Area}(S) = \iint_\mathcal{D} \|\mathbf{N}\|\,du\,dv = \iint_\mathcal{D} 2\,du\,dv = 2\iint_\mathcal{D} 1\,du\,dv = 2\cdot\text{Area}(\mathcal{D}) = 2\cdot 2\cdot 4 = 16$$

29. Calculate $\displaystyle\iint_S (xy+e^z)\,dS$, where S is the triangle in Figure 18 with vertices $(0,0,3)$, $(1,0,2)$, and $(0,4,1)$.

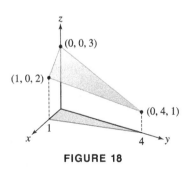

FIGURE 18

SOLUTION   We find the equation of the plane through the points $A = (0,0,3)$, $B = (0,4,1)$, and $C = (1,0,2)$.

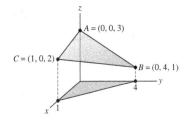

A normal to the plane is the cross product:

$$\vec{AB} \times \vec{AC} = \langle 0, 4, -2 \rangle \times \langle 1, 0, -1 \rangle = \begin{vmatrix} \mathbf{i} & \mathbf{j} & \mathbf{k} \\ 0 & 4 & -2 \\ 1 & 0 & -1 \end{vmatrix} = -4\mathbf{i} - 2\mathbf{j} - 4\mathbf{k} = -2 \langle 2, 1, 2 \rangle$$

The equation of the plane passing through $A = (0, 0, 3)$ and perpendicular to the vector $\langle 2, 1, 2 \rangle$ is:

$$\langle x - 0, y - 0, z - 3 \rangle \cdot \langle 2, 1, 2 \rangle = 0$$
$$2x + y + 2(z - 3) = 0$$
$$2x + y + 2z = 6$$

or

$$z = g(x, y) = -x - \frac{1}{2}y + 3$$

We compute the surface integral of $f(x, y, z) = xy + e^z$ over the triangle ABC using the formula for the surface integral over a graph. The parameter domain $\mathcal{D}$ is the projection of the triangle ABC onto the xy-plane (see figure).

We have:

$$\iint_S f(x, y, z)\, dS = \iint_\mathcal{D} f(x, y, g(x, y)) \sqrt{1 + g_x^2 + g_z^2}\, dx\, dy \tag{1}$$

We compute the functions in the integrand. Since $z = g(x, y) = -x - \frac{y}{2} + 3$, we have:

$$g_x = -1, g_y = -\frac{1}{2} \quad \Rightarrow \quad \sqrt{1 + g_x^2 + g_z^2} = \sqrt{1 + (-1)^2 + \left(-\frac{1}{2}\right)^2} = \frac{3}{2}$$

$$f(x, y, g(x, y, z)) = xy + e^z = xy + e^{-x - \frac{y}{2} + 3}$$

Substituting in (1) gives:

$$\iint_S f(x, y, z)\, dS = \int_0^1 \int_0^{-4x+4} \left(xy + e^{-x - y/2 + 3}\right) \cdot \frac{3}{2}\, dy\, dx = \int_0^1 \frac{3xy^2}{4} - 3e^{-x - y/2 + 3} \Big|_{y=0}^{-4x+4} dx$$

$$= \int_0^1 \left(\frac{3x(-4x + 4)^2}{4} - 3e^{-x - (-4x+4)/2 + 3} + 3e^{-x+3} \right) dx$$

$$= \int_0^1 \left(12x^3 - 24x^2 + 12x - 3e^{x+1} + 3e^{-x+3} \right) dx = 3x^4 - 8x^3 + 6x^2 - 3e^{x+1} - 3e^{-x+3} \Big|_0^1$$

$$= (1 - 3e^2 - 3e^2) - (-3e - 3e^3) = 3e^3 - 6e^2 + 3e + 1 \approx 25.08$$

31. Use cylindrical coordinates to compute the surface area of a sphere of radius R.

SOLUTION In cylindrical coordinates, $x = r\cos\theta$, $y = r\sin\theta$, $z = z$. The sphere is defined by

$$x^2 + y^2 + z^2 = R^2 \quad \Rightarrow \quad z^2 = R^2 - x^2 - y^2 = R^2 - r^2$$

To compute the surface area, we will compute the surface area of the hemisphere S where $z \geq 0$ and double it; then we can parametrize z as $z = \sqrt{R^2 - r^2}$. To compute the surface area of the hemisphere S, we first must find the tangent vectors and the normal vector. That is,

$$\mathbf{T}_r = \frac{\partial \Phi}{\partial r} = \frac{\partial}{\partial r}\left(r\cos\theta, r\sin\theta, \sqrt{R^2 - r^2}\right) = \left\langle \cos\theta, \sin\theta, -\frac{r}{\sqrt{R^2 - r^2}} \right\rangle$$

$$\mathbf{T}_\theta = \frac{\partial \Phi}{\partial \theta} = \frac{\partial}{\partial \theta}\left(r\cos\theta, r\sin\theta, \sqrt{R^2 - r^2}\right) = \langle -r\sin\theta, r\cos\theta, 0 \rangle$$

The normal vector is the cross product:

$$\mathbf{N} = \mathbf{T}_r \times \mathbf{T}_\theta = \begin{vmatrix} \mathbf{i} & \mathbf{j} & \mathbf{k} \\ \cos\theta & \sin\theta & \frac{-r}{\sqrt{R^2 - r^2}} \\ -r\sin\theta & r\cos\theta & 0 \end{vmatrix}$$

$$= \left(\frac{r^2 \cos\theta}{\sqrt{R^2 - r^2}}\right)\mathbf{i} + \left(\frac{r^2 \sin\theta}{\sqrt{R^2 - r^2}}\right)\mathbf{j} + \left(r\cos^2\theta + r\sin^2\theta\right)\mathbf{k}$$

$$= \left(\frac{r^2 \cos\theta}{\sqrt{R^2 - r^2}}\right)\mathbf{i} + \left(\frac{r^2 \sin\theta}{\sqrt{R^2 - r^2}}\right)\mathbf{j} + r\mathbf{k}$$

The length of the normal vector is thus

$$\|\mathbf{N}\| = \sqrt{\frac{r^4 \cos^2\theta}{R^2 - r^2} + \frac{r^4 \sin^2\theta}{R^2 - r^2} + r^2} = \sqrt{\frac{r^4}{R^2 - r^2}(\cos^2\theta + \sin^2\theta) + r^2} = \sqrt{\frac{r^4}{R^2 - r^2} + r^2} = \frac{rR}{\sqrt{R^2 - r^2}}$$

We now compute the surface area as the following surface integral:

$$\text{Area}(S) = \iint_D \|\mathbf{N}\| \, dr \, d\theta = \int_0^{2\pi} \int_0^R \frac{rR}{\sqrt{R^2 - r^2}} \, dr \, d\theta$$

$$= \left(\int_0^{2\pi} R \, d\theta\right)\left(\int_0^R \frac{r}{\sqrt{R^2 - r^2}} \, dr\right) = 2\pi R \int_0^R \frac{r}{\sqrt{R^2 - r^2}} \, dr$$

We compute the integral using the substitution $t = R^2 - r^2$, $dt = -2r \, dr$. We get:

$$\text{Area}(S) = 2\pi R \int_{R^2}^0 \frac{-1}{2t^{1/2}} \, dt = 2\pi R^2$$

Therefore the area of the sphere is $2\text{Area}(S) = 4\pi r^2$.

33. **CAS** Let S be the surface $z = \ln(5 - x^2 - y^2)$ for $0 \leq x \leq 1$, $0 \leq y \leq 1$. Using a computer algebra system:

(a) calculate the surface area of S to four decimal places.

(b) calculate $\iint_S x^2 y^3 \, dS$ to four decimal places.

SOLUTION

(a) Using that $z_x = -2x/(5 - x^2 - y^2)$ and $z_y = -2y/(5 - x^2 - y^2)$, we calculate $\|n\|$ to be

$$\|n\| = \sqrt{1 + (z_x)^2 + (z_y)^2} = \frac{2\sqrt{x^2 + y^2}}{5 - x^2 - y^2}$$

Thus, the surface area is

$$\text{Area}(S) = \int_0^1 \int_0^1 \frac{2\sqrt{x^2 + y^2}}{5 - x^2 - y^2} \, dx \, dy \approx 0.3698$$

(b) We calculate $\iint_S x^2 y^3 \, dS$ as follows:

$$\iint_S x^2 y^3 \, dS = \int_0^1 \int_0^1 x^2 y^3 \frac{2\sqrt{x^2+y^2}}{5-x^2-y^2} \, dx\, dy \approx 0.0508$$

35. Use a surface integral to compute the area of that part of the plane $ax + by + cz = d$ corresponding to $0 \le x, y \le 1$.

SOLUTION We assume $c \ne 0$, since otherwise the plane is parallel to the z-axis, and the surface area is infinite. Then we can rewrite the equation of the plane as:

$$z = g(x,y) = -\frac{a}{c}x - \frac{b}{c}y - \frac{d}{c} \tag{1}$$

Using the integral for surface area and the surface integral over a graph we have:

$$\text{Area}(S) = \iint_S 1 \, dS = \iint_D \sqrt{1 + g_x^2 + g_y^2} \, dx\, dy \tag{2}$$

By (1) we have:

$$g_x = -\frac{a}{c}, \; g_z = -\frac{b}{c} \quad \Rightarrow \quad \sqrt{1 + g_x^2 + g_z^2} = \sqrt{1 + \frac{a^2}{c^2} + \frac{b^2}{c^2}} = \frac{1}{c}\sqrt{1 + a^2 + b^2}$$

We substitute in (2) to obtain:

$$\text{Area}(S) = \iint_D \frac{1}{c} \sqrt{1 + a^2 + b^2} \, dx\, dy = \int_0^1 \int_0^1 \frac{1}{c} \sqrt{1 + a^2 + b^2} \, dx\, dy = \frac{1}{c}\sqrt{1 + a^2 + b^2}$$

37. Find the surface area of the portion S of the cone $z^2 = x^2 + y^2$, where $z \ge 0$, contained within the cylinder $y^2 + z^2 \le 1$.

SOLUTION We rewrite the equation of the cone as $x = \pm\sqrt{z^2 - y^2}$. The projection of the cone onto the yz-plane is obtained by setting $x = 0$ in the equation of the cone, that is,

$$x = 0 = \sqrt{z^2 - y^2} \quad \Rightarrow \quad z = \pm y$$

Since on S, $z \ge 0$, we get $z = |y|$. We conclude that the projection of the upper part of the cone $x^2 + y^2 = z^2$ onto the yz-plane is the region between the lines $z = y$ and $z = -y$ on the upper part of the yz-plane. Therefore, the projection $\mathcal{D}$ of S onto the yz-plane is the region shown in the figure:

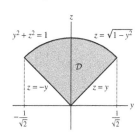

The area of S is the surface integral:

$$\text{Area}(S) = \iint_S dS$$

We compute the integral using a surface integral over a graph. Since $x = g(y,z) = \pm\sqrt{z^2 - y^2}$ we have

$$g_z = \pm \frac{z}{\sqrt{z^2 - y^2}}, \; g_y = \pm \frac{y}{\sqrt{z^2 - y^2}}$$

Hence (notice that $z \ge 0$ on S):

$$\sqrt{1 + g_y^2 + g_z^2} = \sqrt{1 + \frac{z^2}{z^2 - y^2} + \frac{y^2}{z^2 - y^2}} = \sqrt{\frac{2z^2}{z^2 - y^2}} = \frac{z\sqrt{2}}{\sqrt{z^2 - y^2}}$$

We obtain the following integral:

$$\text{Area}(S) = \iint_D \sqrt{1 + g_y^2 + g_z^2}\, dy\, dz = \iint_D \frac{z\sqrt{2}}{\sqrt{z^2 - y^2}}\, dz\, dy$$

Using symmetry gives:

$$\text{Area}(S) = 2 \int_0^{1/\sqrt{2}} \int_y^{\sqrt{1-y^2}} \frac{z\sqrt{2}}{\sqrt{z^2 - y^2}}\, dz\, dy = 2\sqrt{2} \int_0^{1/\sqrt{2}} \left(\int_y^{\sqrt{1-y^2}} \frac{z\, dz}{\sqrt{z^2 - y^2}} \right) dy \quad (1)$$

We compute the inner integral using the substitution $u = \sqrt{z^2 - y^2}$, $du = \frac{z}{u}\, dz$. We get:

$$\int_y^{\sqrt{1-y^2}} \frac{z\, dz}{\sqrt{z^2 - y^2}} = \int_0^{\sqrt{1-y^2}} \frac{u\, du}{u} = \int_0^{\sqrt{1-2y^2}} du = \sqrt{1 - 2y^2}$$

We substitute in (1) and compute the resulting integral using the substitution $t = \sqrt{2}y$. We get:

$$\text{Area}(S) = \int_0^{1/\sqrt{2}} \sqrt{1 - 2y^2}\, dy = 2\sqrt{2} \int_0^1 \sqrt{1 - t^2}\, \frac{dt}{\sqrt{2}} = 2 \int_0^1 \sqrt{1 - t^2}\, dt = 2 \cdot \frac{\pi}{4} = \frac{\pi}{2}$$

39. Calculate $\iint_S x^2 z\, dS$, where S is the cylinder (including the top and bottom) $x^2 + y^2 = 4$, $0 \le z \le 3$.

SOLUTION We calculate the surface integral for each of the three surfaces. We begin with the bottom.

$$S_1 : \phi(x, y) = (x, y, 0)$$

$$\iint_{S_1} x^2 z\, dS_1 = \iint_D x^2(0)\|\mathbf{n}_1\|\, dx\, dy = 0$$

Then the top

$$S_2 : \phi(x, y) = (x, y, 3)$$

$$\mathbf{T}_x = \langle 1, 0, 0 \rangle,\ \mathbf{T}_y \langle 0, 1, 0 \rangle \Rightarrow \mathbf{n}_2 = \langle 0, 0, 1 \rangle$$

$$\iint_{S_2} x^2 z\, dS_2 = \iint_D x^2(3)\|\mathbf{n}_2\|\, dx\, dy = 3 \iint_D x^2\, dx\, dy$$

The domain, $\mathcal{D}$, is the disk of radius 2. Changing to polar coordinates,

$$= 3 \int_0^{2\pi} \int_0^2 (r\cos\theta)^2\, r\, dr\, d\theta = 3 \int_0^{2\pi} \cos^2\theta\, d\theta \cdot \int_0^2 r^3\, dr$$

$$= 3 \left(\frac{1}{2} + \frac{\sin 2\theta}{2} \right) \Big|_0^{2\pi} \cdot \frac{r^4}{4} \Big|_0^2 = 12\pi$$

Finally the side,

$$S_3 : \phi(r, \theta) = (2\cos\theta, 2\sin\theta, z)$$

$$\mathbf{T}_\theta = \langle -2\sin\theta, 2\cos\theta, 0 \rangle,\ \mathbf{T}_z \langle 0, 0, 1 \rangle$$

$$\Rightarrow \mathbf{n}_3 = \langle 2\cos\theta, 2\sin\theta, 0 \rangle \Rightarrow \|\mathbf{n}_3\| = 2$$

$$\iint_{S_3} x^2 z\, dS_2 = \int_0^{2\pi} \int_0^3 (2\cos\theta)^2 z\, 2\, dz\, d\theta$$

$$= 8 \int_0^{2\pi} \cos^2\theta\, d\theta \cdot \int_0^3 z\, dz = 8 \left(\frac{1}{2} + \frac{\sin 2\theta}{2} \right) \Big|_0^{2\pi} \cdot \frac{z^2}{2} \Big|_0^3 = 36\pi$$

The total surface integral is thus

$$\iint_\Phi x^2 z\, dS = 0 + 12\pi + 36\pi = 48\pi$$

41. Prove a famous result of Archimedes: The surface area of the portion of the sphere of radius R between two horizontal planes $z = a$ and $z = b$ is equal to the surface area of the corresponding portion of the circumscribed cylinder (Figure 21).

FIGURE 21

SOLUTION We compute the area of the portion of the sphere between the planes a and b. The portion S_1 of the sphere has the parametrization,

$$\Phi(\theta, \phi) = (r\cos\theta\sin\phi, r\sin\theta\sin\phi, r\cos\phi)$$

where,

$$\mathcal{D}_1 : 0 \leq \theta \leq 2\pi,\ \phi_0 \leq \phi \leq \phi_1$$

If we assume $0 < a < b$, then the angles ϕ_0 and ϕ_1 are determined by,

$$\cos\phi_0 = \frac{b}{r} \quad \Rightarrow \quad \phi_0 = \cos^{-1}\frac{b}{r}$$

$$\cos\phi_1 = \frac{a}{r} \quad \Rightarrow \quad \phi_1 = \cos^{-1}\frac{a}{r}$$

The length of the normal vector is $\|\mathbf{N}\| = r^2\sin\phi$. We obtain the following integral:

$$\text{Area}(S_1) = \iint_{\mathcal{D}_1} \|\mathbf{N}\|\, d\phi\, d\theta = \int_0^{2\pi}\int_{\phi_0}^{\phi_1} r^2 \sin\phi\, d\phi\, d\theta = \left(\int_0^{2\pi} r^2\, d\phi\right)\left(\int_{\phi_1}^{\phi_2} \sin\phi\, d\phi\right)$$

$$= 2\pi r^2\left(-\cos\phi\bigg|_{\phi=\cos^{-1}\frac{b}{r}}^{\cos^{-1}\frac{a}{r}}\right) = 2\pi r^2\left(-\frac{a}{r} + \frac{b}{r}\right) = 2\pi r(b-a)$$

The area of the part S_2 of the cylinder of radius r between the planes $z = a$ and $z = b$ is

$$\text{Area}(S_2) = 2\pi r \cdot (b-a)$$

We see that the two areas are equal:

$$\text{Area}(S_1) = \text{Area}(S_2)$$

Further Insights and Challenges

43. Use Eq. (14) to compute the surface area of $z = 4 - y^2$ for $0 \leq y \leq 2$ rotated about the z-axis.

SOLUTION Since $g(y) = 4 - y^2$, we have $g'(y) = -2y$. By Eq. (14) we obtain the following integral,

$$\text{Area}(S) = 2\pi \int_0^2 |y| \sqrt{1 + (-2y)^2}\, dy = 2\pi \int_0^2 y \cdot \sqrt{1 + 4y^2}\, dy$$

We compute the integral using the substitution $u = 1 + 4y^2$, $du = 8y\, dy$. We get:

$$\text{Area}(S) = 2\pi \int_1^{17} u^{1/2} \cdot \frac{du}{8} = 2\pi \frac{2}{3} \cdot \frac{u^{3/2}}{8}\Big|_1^{17} = \frac{\pi}{6}\left(17\sqrt{17} - 1\right) \approx 36.18$$

45. Area of a Torus Let $\mathcal{T}$ be the torus obtained by rotating the circle in the yz-plane of radius a centered at $(0, b, 0)$ about the z-axis (Figure 23). We assume that $b > a > 0$.

(a) Use Eq. (14) to show that

$$\text{area}(\mathcal{T}) = 4\pi \int_{b-a}^{b+a} \frac{ay}{\sqrt{a^2 - (b - y)^2}}\, dy$$

(b) Show that $\text{area}(\mathcal{T}) = 4\pi^2 ab$.

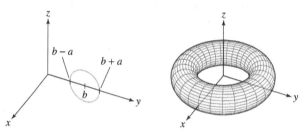

FIGURE 23 The torus obtained by rotating a circle of radius a.

SOLUTION

(a) Using symmetry, the area of the surface obtained by rotating the upper part of the circle is half the area of the torus.

The rotated graph is $z = g(y) = \sqrt{a^2 - (y - b)^2}$, $b - a \leq y \leq b + a$. So, we have,

$$g'(y) = \frac{-2(y - b)}{2\sqrt{a^2 - (y - b)^2}} = -\frac{y - b}{\sqrt{a^2 - (y - b)^2}}$$

$$\sqrt{1 + g'(y)^2} = \sqrt{1 + \frac{(y - b)^2}{a^2 - (y - b)^2}} = \sqrt{\frac{a^2 - (y - b)^2 + (y - b)^2}{a^2 - (y - b)^2}} = \frac{a}{\sqrt{a^2 - (y - b)^2}}$$

We now use symmetry and Eq. (14) to obtain the following area of the torus (we assume that $b - a > 0$, hence $y > 0$):

$$\text{Area}(\mathbf{T}) = 2 \cdot 2\pi \int_{b-a}^{b+a} |y| \sqrt{1 + g'(y)^2}\, dy = 4\pi \int_{b-a}^{b+a} \frac{ay}{\sqrt{a^2 - (y - b)^2}}\, dy \qquad (1)$$

(b) We compute the integral using the substitution $u = \frac{y-b}{a}$, $du = \frac{1}{a} dy$. We get:

$$\int_{b-a}^{b+a} \frac{ay}{\sqrt{a^2-(y-b)^2}} dy = \int_{-1}^{1} \frac{a^2u + ab}{\sqrt{a^2 - a^2u^2}} a\, du = \int_{-1}^{1} \frac{a^2u + ab}{\sqrt{1-u^2}} du = \int_{-1}^{1} \frac{a^2u}{\sqrt{1-u^2}} du + \int_{-1}^{1} \frac{ab}{\sqrt{1-u^2}} du$$

The first integral is zero since the integrand is an odd function. We get:

$$\int_{b-a}^{b+a} \frac{ay}{\sqrt{a^2-(y-b)^2}} dy = 2\int_0^1 \frac{ab}{\sqrt{1-u^2}} du = 2ab \sin^{-1} u \Big|_0^1 = 2ab\left(\frac{\pi}{2} - 0\right) = \pi ab$$

Substituting in (1) gives the following area:

$$\text{Area}(\mathbf{T}) = 4\pi \cdot \pi ab = 4\pi^2 ab$$

47. Compute the surface area of the torus in Exercise 45 using Pappus's Theorem.

SOLUTION The generating curve is the circle of radius a in the (y,z)-plane centered at the point $(0,b,0)$. The length of the generating curve is $L = \pi a$.

The center of mass of the circle is at the center $(\bar{y}, \bar{z}) = (b, 0)$, and it traverses a circle of radius b centered at the origin. Therefore, the center of mass makes a distance of $2\pi b$. Using Pappus's Theorem, the area of the torus is:

$$L \cdot 2\pi a = 2\pi a \cdot 2\pi b = 4\pi^2 ab$$

49. Calculate the gravitational potential V for a hemisphere of radius R with uniform mass distribution.

SOLUTION In Exercise 48(b) we expressed the potential φ for a sphere of radius R. To find the potential for a hemisphere of radius R, we need only to modify the limits of the angle ϕ to $0 \le \phi \le \frac{\pi}{2}$. This gives the following integral:

$$\varphi(0,0,r) = \varphi(r) = -\frac{Gm}{4\pi} \int_0^{\pi/2} \int_0^{2\pi} \frac{\sin\phi\, d\theta\, d\phi}{\sqrt{R^2 + r^2 - 2Rr\cos\phi}} = -\frac{Gm}{4\pi} \cdot 2\pi \int_0^{\pi/2} \frac{\sin\phi\, d\phi}{\sqrt{R^2 + r^2 - 2Rr\cos\phi}}$$

$$= -\frac{Gm}{4\pi} \int_0^{\pi/2} \frac{\sin\phi\, d\phi}{\sqrt{R^2 + r^2 - 2Rr\cos\phi}}$$

We compute the integral using the substitution $u = R^2 + r^2 - 2Rr\cos\phi$, $du = 2Rr\sin\phi\, d\phi$. We obtain:

$$\varphi(r) = -\frac{Gm}{2} \int_{(R-r)^2}^{R^2+r^2} \frac{\frac{du}{2Rr}}{\sqrt{u}} = -\frac{Gm}{4Rr} \int_{(R-r)^2}^{R^2+r^2} u^{-1/2} du = -\frac{Gm}{4Rr} \cdot 2u^{1/2} \Big|_{u=(R-r)^2}^{R^2+r^2}$$

$$= -\frac{Gm}{2Rr}\left((R^2+r^2)^{1/2} - ((R-r)^2)^{1/2}\right) = -\frac{Gm}{2Rr}\left(\sqrt{R^2+r^2} - |R-r|\right)$$

51. Let S be the part of the graph $z = g(x,y)$ lying over a domain $\mathcal{D}$ in the xy-plane. Let $\phi = \phi(x,y)$ be the angle between the normal to S and the vertical. Prove the formula

$$\text{area}(S) = \iint_{\mathcal{D}} \frac{dA}{|\cos\phi|}$$

SOLUTION

Using the Surface Integral over a Graph we have:

$$\text{Area}(S) = \iint_S 1\, dS = \iint_D \sqrt{1 + g_x^2 + g_y^2}\, dA \tag{1}$$

In parametrizing the surface by $\phi(x, y) = (x, y, g(x, y))$, $(x, y) = D$, we have:

$$\mathbf{T}_x = \frac{\partial \Phi}{\partial x} = \langle 1, 0, g_x \rangle$$

$$\mathbf{T}_y = \frac{\partial \Phi}{\partial y} = \langle 0, 1, g_y \rangle$$

Hence,

$$\mathbf{N} = \mathbf{T}_x \times \mathbf{T}_y = \begin{vmatrix} \mathbf{i} & \mathbf{j} & \mathbf{k} \\ 1 & 0 & g_x \\ 0 & 1 & g_y \end{vmatrix} = -g_x \mathbf{i} - g_y \mathbf{j} + \mathbf{k} = \langle -g_x, -g_y, 1 \rangle$$

$$\|\mathbf{N}\| = \sqrt{g_x^2 + g_y^2 + 1}$$

There are two adjacent angles between the normal $\mathbf{N}$ and the vertical, and the cosines of these angles are opposite numbers. Therefore we take the absolute value of $\cos \phi$ to obtain a positive value for $\text{Area}(S)$. Using the Formula for the cosine of the angle between two vectors we get:

$$|\cos \phi| = \frac{|\mathbf{N} \cdot \mathbf{k}|}{\|\mathbf{N}\|\|\mathbf{k}\|} = \frac{|\langle -g_x, -g_y, 1 \rangle \cdot \langle 0, 0, 1 \rangle|}{\sqrt{1 + g_x^2 + g_y^2} \cdot 1} = \frac{1}{\sqrt{1 + g_x^2 + g_y^2}}$$

Substituting in (1) we get:

$$\text{Area}(S) = \iint_D \frac{dA}{|\cos \phi|}$$

16.5 Surface Integrals of Vector Fields (LT Section 17.5)

Preliminary Questions

1. Let $\mathbf{F}$ be a vector field and $\Phi(u, v)$ a parametrization of a surface S, and set $\mathbf{N} = \mathbf{T}_u \times \mathbf{T}_v$. Which of the following is the normal component of $\mathbf{F}$?

(a) $\mathbf{F} \cdot \mathbf{N}$ (b) $\mathbf{F} \cdot \mathbf{n}$

SOLUTION The normal component of $\mathbf{F}$ is $\mathbf{F} \cdot \mathbf{n}$ rather than $\mathbf{F} \cdot \mathbf{N}$.

2. The vector surface integral $\iint_S \mathbf{F} \cdot d\mathbf{S}$ is equal to the scalar surface integral of the function (choose the correct answer):

(a) $\|\mathbf{F}\|$.

(b) $\mathbf{F} \cdot \mathbf{N}$, where $\mathbf{N}$ is a normal vector.

(c) $\mathbf{F} \cdot \mathbf{n}$, where $\mathbf{n}$ is the unit normal vector.

SOLUTION The vector surface integral $\iint_S \mathbf{F} \cdot d\mathbf{S}$ is defined as the scalar surface integral of the normal component of $\mathbf{F}$ on the oriented surface. That is, $\iint_S \mathbf{F} \cdot d\mathbf{S} = \iint_S (\mathbf{F} \cdot \mathbf{n})\, dS$ as stated in (c).

3. $\iint_S \mathbf{F} \cdot d\mathbf{S}$ is zero if (choose the correct answer):

(a) $\mathbf{F}$ is tangent to S at every point.

(b) $\mathbf{F}$ is perpendicular to S at every point.

SOLUTION Since $\iint_S \mathbf{F} \cdot d\mathbf{S}$ is equal to the scalar surface integral of the normal component of $\mathbf{F}$ on S, this integral is zero when the normal component is zero at every point, that is, when $\mathbf{F}$ is tangent to S at every point as stated in (a).

4. If $\mathbf{F}(P) = \mathbf{n}(P)$ at each point on S, then $\iint_S \mathbf{F} \cdot d\mathbf{S}$ is equal to which of the following?

(a) Zero **(b)** Area(S) **(c)** Neither

SOLUTION If $\mathbf{F}(P) = \mathbf{n}(P)$ at each point on S, then,:

$$\iint_S \mathbf{F} \cdot d\mathbf{S} = \iint_S (\mathbf{n} \cdot \mathbf{n})\, dS = \iint_S \|\mathbf{n}\|^2\, dS = \iint_S 1\, dS = \text{Area}(S)$$

Therefore, (b) is the correct answer.

5. Let S be the disk $x^2 + y^2 \leq 1$ in the xy-plane oriented with normal in the positive z-direction. Determine $\iint_S \mathbf{F} \cdot d\mathbf{S}$ for each of the following vector constant fields:

(a) $\mathbf{F} = \langle 1, 0, 0 \rangle$ **(b)** $\mathbf{F} = \langle 0, 0, 1 \rangle$ **(c)** $\mathbf{F} = \langle 1, 1, 1 \rangle$

SOLUTION The unit normal vector to the oriented disk is $\mathbf{n} = \langle 0, 0, 1 \rangle$.

(a) Since $\mathbf{F} \cdot \mathbf{n} = \langle 1, 0, 0 \rangle \cdot \langle 0, 0, 1 \rangle = 0$, $\mathbf{F}$ is perpendicular to the unit normal vector at every point on S, therefore $\iint_S \mathbf{F} \cdot d\mathbf{S} = 0$.

(b) Since $\mathbf{F} = \mathbf{n}$ at every point on S, we have:

$$\iint_S \mathbf{F} \cdot d\mathbf{S} = \iint_S (\mathbf{n} \cdot \mathbf{n})\, dS = \iint_S \|\mathbf{n}\|^2\, dS = \iint_S 1\, dS = \text{Area}(S) = \pi$$

(c) For $\mathbf{F} = \langle 1, 1, 1 \rangle$ we have:

$$\iint_S \mathbf{F} \cdot d\mathbf{S} = \iint_S (\mathbf{F} \cdot \mathbf{n})\, dS = \iint_S \langle 1, 1, 1 \rangle \cdot \langle 0, 0, 1 \rangle\, dS = \iint_S 1\, dS = \text{Area}(S) = \pi$$

6. Estimate $\iint_S \mathbf{F} \cdot d\mathbf{S}$, where S is a tiny oriented surface of area 0.05 and the value of $\mathbf{F}$ at a sample point in S is a vector of length 2 making an angle $\dfrac{\pi}{4}$ with the normal to the surface.

SOLUTION

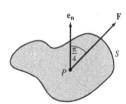

Since S is a tiny surface, we may assume that the dot product $\mathbf{F} \cdot \mathbf{n}$ on S is equal to the dot product at the sample point. This gives the following approximation:

$$\iint_S \mathbf{F} \cdot d\mathbf{S} = \iint_S (\mathbf{F} \cdot \mathbf{n})\, dS \approx \iint_S (\mathbf{F}(P) \cdot \mathbf{n}(P))\, dS = \mathbf{F}(P) \cdot \mathbf{n}(P) \iint_S 1\, dS = \mathbf{F}(P) \cdot \mathbf{n}\,\text{Area}(S)$$

That is,

$$\iint_S \mathbf{F} \cdot d\mathbf{S} \approx \mathbf{F}(P) \cdot \mathbf{n}(P)\text{Area}(S) \tag{1}$$

We are given that Area(S) = 0.05. We compute the dot product:

$$\mathbf{F}(P) \cdot \mathbf{n}(P) = \|\mathbf{F}(P)\|\|\mathbf{n}(P)\| \cos \frac{\pi}{4} = 2 \cdot 1 \cdot \frac{1}{\sqrt{2}} = \sqrt{2}$$

Combining with (1) gives the following estimation:

$$\iint_S \mathbf{F} \cdot d\mathbf{S} \approx 0.05 \sqrt{2} \approx 0.0707$$

7. A small surface S is divided into three pieces of area 0.2. Estimate $\iint_S \mathbf{F} \cdot d\mathbf{S}$ if $\mathbf{F}$ is a unit vector field making angles of 85°, 90°, and 95° with the normal at sample points in these three pieces.

SOLUTION

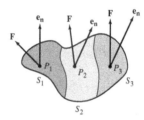

We estimate the vector surface integral by the following sum:

$$\iint_S \mathbf{F} \cdot d\mathbf{S} = \mathbf{F}(P_1) \cdot \mathbf{n}(P_1) \text{Area}(S_1) + \mathbf{F}(P_2) \cdot \mathbf{n}(P_2) \text{Area}(S_2) + \mathbf{F}(P_3) \cdot \mathbf{n}(P_3) \text{Area}(S_3)$$
$$= 0.2 \left(\mathbf{F}(P_1) \cdot \mathbf{n}(P_1) + \mathbf{F}(P_2) \cdot \mathbf{n}(P_2) + \mathbf{F}(P_3) \cdot \mathbf{n}(P_3) \right)$$

We compute the dot product. Since $\mathbf{F}$ and $\mathbf{n}$ are unit vectors, we have:

$$\mathbf{F}(P_1) \cdot \mathbf{n}(P_1) = \cos 85° \approx 0.0872$$
$$\mathbf{F}(P_2) \cdot \mathbf{n}(P_2) = \cos 90° = 0$$
$$\mathbf{F}(P_3) \cdot \mathbf{n}(P_3) = \cos 95° \approx -0.0872$$

Substituting gives the following estimation:

$$\iint_S \mathbf{F} \cdot d\mathbf{S} \approx 0.2(0.0872 + 0 - 0.0872) = 0$$

Exercises

1. Let $\mathbf{F} = \langle z, 0, y \rangle$, and let S be the oriented surface parametrized by $\Phi(u, v) = (u^2 - v, u, v^2)$ for $0 \le u \le 2$, $-1 \le v \le 4$. Calculate:

(a) $\mathbf{N}$ and $\mathbf{F} \cdot \mathbf{N}$ as functions of u and v
(b) The normal component of $\mathbf{F}$ to the surface at $P = (3, 2, 1) = \Phi(2, 1)$
(c) $\iint_S \mathbf{F} \cdot d\mathbf{S}$

SOLUTION

(a) The tangent vectors are

$$\mathbf{T}_u = \frac{\partial G}{\partial u} = \frac{\partial}{\partial u}\left(u^2 - v, u, v^2\right) = \langle 2u, 1, 0 \rangle$$
$$\mathbf{T}_v = \frac{\partial G}{\partial v} = \frac{\partial}{\partial v}\left(u^2 - v, u, v^2\right) = \langle -1, 0, 2v \rangle$$

The normal vector is their cross product:

$$\mathbf{N} = \mathbf{T}_u \times \mathbf{T}_v = \begin{vmatrix} \mathbf{i} & \mathbf{j} & \mathbf{k} \\ 2u & 1 & 0 \\ -1 & 0 & 2v \end{vmatrix} = (2v)\mathbf{i} - (4uv)\mathbf{j} + \mathbf{k} = \langle 2v, -4uv, 1 \rangle$$

We write $\mathbf{F} = \langle z, 0, y \rangle$ in terms of the parameters $x = u^2 - v$, $y = u$, $z = v^2$ and then compute $\mathbf{F} \cdot \mathbf{N}$:

$$\mathbf{F}(G(u, v)) = \langle z, 0, y \rangle = \langle v^2, 0, u \rangle$$

$$\mathbf{F}(G(u, v)) \cdot \mathbf{N}(u, v) = \langle v^2, 0, u \rangle \cdot \langle 2v, -4uv, 1 \rangle = 2v \cdot v^2 - 4uv \cdot 0 + 1 \cdot u^2 = 2v^3 + u$$

(b) At the point $P = (3, 2, 1) = G(2, 1)$ we have:

$$\mathbf{F}(P) = \langle 1, 0, 2 \rangle$$

$$\mathbf{N}(P) = \langle 2 \cdot 1, -4 \cdot 2 \cdot 1, 1 \rangle = \langle 2, -8, 1 \rangle$$

$$\mathbf{n}(P) = \frac{\mathbf{N}(P)}{\|\mathbf{N}(P)\|} = \frac{\langle 2, -8, 1 \rangle}{\sqrt{4 + 64 + 1}} = \frac{1}{\sqrt{69}} \langle 2, -8, 1 \rangle$$

Hence, the normal component of $\mathbf{F}$ to the surface at P is the dot product:

$$\mathbf{F}(P) \cdot \mathbf{n}(P) = \langle 1, 0, 2 \rangle \cdot \frac{1}{\sqrt{69}} \langle 2, -8, 1 \rangle = \frac{1}{\sqrt{69}}(2 - 0 + 2) = \frac{4}{\sqrt{69}}$$

(c) Using the definition of the vector surface integral and the dot product in part (a), we have:

$$\iint_S \mathbf{F} \cdot d\mathbf{S} = \iint_{\mathcal{D}} \mathbf{F}(G(u, v)) \cdot \mathbf{N}(u, v) \, du \, dv$$

$$= \int_0^2 \int_{-1}^4 \left(u + 2v^3 \right) dv \, du$$

$$= \int_0^2 \left(uv + \frac{1}{2} v^4 \right) \Big|_{v=-1}^4 du$$

$$= \int_0^2 \left(5u + \frac{255}{2} \right) du$$

$$= \left(\frac{5}{2} u^2 + \frac{255}{2} u \right) \Big|_0^2 = 265$$

3. Let S be the unit square in the xy-plane shown in Figure 14, oriented with the normal pointing in the positive z-direction. Estimate

$$\iint_S \mathbf{F} \cdot d\mathbf{S}$$

where $\mathbf{F}$ is a vector field whose values at the labeled points are

$$\mathbf{F}(A) = \langle 2, 6, 4 \rangle, \qquad \mathbf{F}(B) = \langle 1, 1, 7 \rangle$$

$$\mathbf{F}(C) = \langle 3, 3, -3 \rangle, \qquad \mathbf{F}(D) = \langle 0, 1, 8 \rangle$$

SOLUTION The unit normal vector to S is $\mathbf{n} = \langle 0, 0, 1 \rangle$. We estimate the vector surface integral $\iint_S \mathbf{F} \cdot d\mathbf{S}$ using the division and sample points given in Figure 12.

Each subsquare has area $\frac{1}{4}$; therefore, we obtain the following estimation:

$$\iint_S \mathbf{F} \cdot d\mathbf{S} \approx (\mathbf{F}(A) \cdot \mathbf{n} + \mathbf{F}(B) \cdot \mathbf{n} + \mathbf{F}(C) \cdot \mathbf{n} + \mathbf{F}(D) \cdot \mathbf{n}) \cdot \frac{1}{4}$$

$$= (\langle 2, 6, 4 \rangle \cdot \langle 0, 0, 1 \rangle + \langle 1, 1, 7 \rangle \cdot \langle 0, 0, 1 \rangle + \langle 3, 3, -3 \rangle \cdot \langle 0, 0, 1 \rangle + \langle 0, 1, 8 \rangle \cdot \langle 0, 0, 1 \rangle) \cdot \frac{1}{4}$$

$$= (4 + 7 - 3 + 8) \cdot \frac{1}{4} = 4$$

In Exercises 5–17, compute $\iint_S \mathbf{F} \cdot d\mathbf{S}$ *for the given oriented surface.*

5. $\mathbf{F} = \langle y, z, x \rangle$, plane $3x - 4y + z = 1$,
$0 \leq x \leq 1, 0 \leq y \leq 1$, upward-pointing normal

SOLUTION We rewrite the equation of the plane as $z = 1 - 3x + 4y$, and parametrize the plane by:

$$\Phi(x, y) = (x, y, 1 - 3x + 4y)$$

Here, the parameter domain is the square $\mathcal{D} = \{(x, y) : 0 \leq x, y \leq 1\}$ in the xy-plane.

Step 1. Compute the tangent and normal vectors.

$$\mathbf{T}_x = \frac{\partial \Phi}{\partial x} = \frac{\partial}{\partial x}(x, y, 1 - 3x + 4y) = \langle 1, 0, -3 \rangle$$

$$\mathbf{T}_y = \frac{\partial \Phi}{\partial y} = \frac{\partial}{\partial y}(x, y, 1 - 3x + 4y) = \langle 0, 1, 4 \rangle$$

$$\mathbf{T}_x \times \mathbf{T}_y = \begin{vmatrix} \mathbf{i} & \mathbf{j} & \mathbf{k} \\ 1 & 0 & -3 \\ 0 & 1 & 4 \end{vmatrix} = 3\mathbf{i} - 4\mathbf{j} + \mathbf{k} = \langle 3, -4, 1 \rangle$$

Since the plane is oriented with upward-pointing normal, the normal vector $\mathbf{N}$ is:

$$\mathbf{N} = \langle 3, -4, 1 \rangle$$

Step 2. Evaluate the dot product $\mathbf{F} \cdot \mathbf{N}$. We write $\mathbf{F}$ in terms of the parameters:

$$\mathbf{F}(\Phi(x, y)) = \langle y, z, x \rangle = \langle y, 1 - 3x + 4y, x \rangle$$

The dot product $\mathbf{F} \cdot \mathbf{N}$ is thus

$$\mathbf{F}(\Phi(x, y)) \cdot \mathbf{N} = \langle y, 1 - 3x + 4y, x \rangle \cdot \langle 3, -4, 1 \rangle = 3y - 4(1 - 3x + 4y) + x = 13x - 13y - 4$$

Step 3. Evaluate the surface integral. The surface integral is equal to the following double integral:

$$\iint_S \mathbf{F} \cdot d\mathbf{S} = \iint_{\mathcal{D}} \mathbf{F}(\Phi(x, y)) \cdot \mathbf{N}(x, y) \, dx \, dy = \int_0^1 \int_0^1 (13x - 13y - 4) \, dx \, dy$$

$$= \int_0^1 \left. \frac{13x^2}{2} - 13yx - 4x \right|_{x=0}^1 dy = \int_0^1 \left(\frac{13}{2} - 13y - 4 \right) dy = \left. \frac{5y}{2} - \frac{13y^2}{2} \right|_0^1 = -4$$

7. $\mathbf{F} = \langle 0, 3, x \rangle$, part of sphere $x^2 + y^2 + z^2 = 9$,
where $x \geq 0, y \geq 0, z \geq 0$, outward-pointing normal

SOLUTION We parametrize the octant S by:

$$\Phi(\theta, \phi) = (3 \cos\theta \sin\phi, 3 \sin\theta \sin\phi, 3 \cos\phi), \, 0 \leq \theta \leq \frac{\pi}{2}, 0 \leq \phi \leq \frac{\pi}{2}.$$

Step 1. Compute the normal vector. As seen in the text, the normal vector that points to the outside of the given region is:

$$\mathbf{N} = \mathbf{T}_\phi \times \mathbf{T}_\theta = \sin\phi \langle \cos\theta \sin\phi, \sin\theta \sin\phi, \cos\phi \rangle$$

For $0 \leq \phi \leq \frac{\pi}{2}$ we have $\sin\phi \cos\phi \geq 0$, therefore $\mathbf{N}$ points to the outside of the region.

Step 2. Evaluate the dot product $\mathbf{F} \cdot \mathbf{N}$. We express the vector field in terms of the parameters:

$$\mathbf{F}(\Phi(\theta, \phi)) = \langle 0, 3, x \rangle = \langle 0, 3, 3\cos\theta\sin\phi \rangle$$

Hence:

$$\mathbf{F}(\Phi(\theta, \phi)) \cdot \mathbf{N}(\theta, \phi) = \langle 0, 3, 3\cos\theta\sin\phi \rangle \cdot \sin\phi \langle \cos\theta\sin\phi, \sin\theta\sin\phi, \cos\phi \rangle$$

$$= \sin\phi \left(3\sin\theta\sin\phi + 3\cos\theta\sin\phi\cos\phi\right)$$

$$= 3\sin\theta\sin^2\phi + 3\cos\theta\sin^2\phi\cos\phi$$

Step 3. Evaluate the surface integral. The surface integral is equal to the following double integral:

$$\iint_S \mathbf{F} \cdot d\mathbf{S} = \iint_\mathcal{D} \mathbf{F}(\Phi(\theta, \phi)) \cdot \mathbf{N}(\theta, \phi) \, d\theta \, d\phi$$

$$= \int_0^{\pi/2} \int_0^{\pi/2} \left(3\sin\theta\sin^2\phi + 3\cos\theta\sin^2\phi\cos\phi\right) d\theta \, d\phi$$

$$= \int_0^{\pi/2} \left(-3\cos\theta\sin^2\phi + 3\sin\theta\sin^2\phi\cos\phi\right)\Big|_{\theta=0}^{\pi/2} d\phi$$

$$= \int_0^{\pi/2} \left(3\sin^2\phi + 3\sin^2\phi\cos\phi\right) d\phi$$

$$= \int_0^{\pi/2} \left(\frac{3}{2} - \frac{3}{2}\cos 2\phi + 3\sin^2\phi\cos\phi\right) d\phi$$

$$= \left(\frac{3}{2}\phi - \frac{3}{4}\sin 2\phi + \sin^3\phi\right)\Big|_0^{\pi/2}$$

$$= \frac{3\pi}{4} + 1$$

9. $\mathbf{F} = \langle z, z, x \rangle$, $z = 9 - x^2 - y^2$, $x \geq 0, y \geq 0, z \geq 0$, upward-pointing normal

SOLUTION

Step 1. Find a parametrization. We use x and y as parameters and parametrize the surface by:

$$\Phi(x, y) = \left(x, y, 9 - x^2 - y^2\right), \quad x \geq 0, y \geq 0, z \geq 0$$

The parameter domain $\mathcal{D}$ is determined by the condition $z = 9 - x^2 - y^2 \geq 0$, or $x^2 + y^2 \leq 9$, together with $x, y \geq 0$. That is:

$$\mathcal{D} = \left\{(x, y) : x^2 + y^2 \leq 9, \ x \geq 0, \ y \geq 0\right\}$$

This is the first-quadrant portion of a circle of radius 3 centered at the origin.

Step 2. Compute the tangent and normal vectors. We have:

$$\mathbf{T}_x = \frac{\partial \Phi}{\partial x} = \frac{\partial}{\partial x}\left(x, y, 9 - x^2 - y^2\right) = \langle 1, 0, -2x \rangle$$

$$\mathbf{T}_y = \frac{\partial \Phi}{\partial y} = \frac{\partial}{\partial y}\left(x, y, 9 - x^2 - y^2\right) = \langle 0, 1, -2y \rangle$$

We compute the cross product of the tangent vectors:

$$\mathbf{T}_x \times \mathbf{T}_y = \begin{vmatrix} \mathbf{i} & \mathbf{j} & \mathbf{k} \\ 1 & 0 & -2x \\ 0 & 1 & -2y \end{vmatrix} = (2x)\mathbf{i} + (2y)\mathbf{j} + \mathbf{k} = \langle 2x, 2y, 1 \rangle$$

Since the z-component is positive, the vector points upward, and we have

$$\mathbf{N} = \langle 2x, 2y, 1 \rangle$$

Step 3. Evaluate the dot product $\mathbf{F} \cdot \mathbf{N}$. We first express the vector field in terms of the parameters x and y, by setting $z = 9 - x^2 - y^2$. We get:

$$\mathbf{F}(\Phi(x,y)) = \langle z, z, x \rangle = \langle 9 - x^2 - y^2, 9 - x^2 - y^2, x \rangle$$

We now compute the dot product:

$$\mathbf{F}(\Phi(x,y)) \cdot \mathbf{N}(x,y) = \langle 9 - x^2 - y^2, 9 - x^2 - y^2, x \rangle \cdot \langle 2x, 2y, 1 \rangle = 2x(9 - x^2 - y^2) + 2y(9 - x^2 - y^2) + x$$

Step 4. Evaluate the surface integral. The surface integral is equal to the following double integral:

$$\iint_S \mathbf{F} \cdot d\mathbf{S} = \iint_D \mathbf{F}(\Phi(x,y)) \cdot \mathbf{N}(x,y) \, dx \, dy = \iint_D \left(2x(9 - x^2 - y^2) + 2y\left(9 - (x^2 + y^2)\right) + x\right) dx \, dy$$

We convert the integral to polar coordinates to obtain:

$$\iint_S \mathbf{F} \cdot d\mathbf{S} = \int_0^3 \int_0^{\pi/2} \left(2r\cos\theta \cdot (9 - r^2) + 2r\sin\theta \cdot (9 - r^2) + r\cos\theta\right) \cdot r \, d\theta \, dr$$

$$= \int_0^3 \int_0^{\pi/2} \left(\left(2r^2(9 - r^2)\right)(\cos\theta + \sin\theta) + r^2 \cos\theta\right) d\theta \, dr$$

$$= \int_0^3 \left((2r^2(9 - r^2))(\sin\theta - \cos\theta) + r^2 \sin\theta\right)\Big|_0^{\pi/2} dr$$

$$= \int_0^3 \left(4r^2(9 - r^2) + r^2\right) dr$$

$$= \int_0^3 (37r^2 - 4r^4) \, dr = \left(\frac{37}{3}r^3 - \frac{4}{5}r^5\right)\Big|_0^3 = \frac{693}{5}$$

11. $\mathbf{F} = y^2 \mathbf{i} + 2\mathbf{j} - x\mathbf{k}$, portion of the plane $x + y + z = 1$ in the octant $x, y, z \geq 0$, upward-pointing normal

SOLUTION

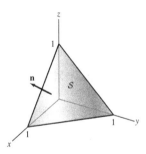

We parametrize the surface by:

$$\Phi(x, y) = (x, y, 1 - x - y)$$

using the parameter domain $\mathcal{D}$ shown in the figure.

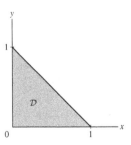

Step 1. Compute the tangent and normal vectors. We have:

$$\mathbf{T}_x = \frac{\partial \Phi}{\partial x} = \frac{\partial}{\partial x}(x, y, 1 - x - y) = \langle 1, 0, -1 \rangle$$

$$\mathbf{T}_y = \frac{\partial \Phi}{\partial y} = \frac{\partial}{\partial y}(x, y, 1 - y) = \langle 0, 1, -1 \rangle$$

$$\mathbf{T}_x \times \mathbf{T}_y = \begin{vmatrix} \mathbf{i} & \mathbf{j} & \mathbf{k} \\ 1 & 0 & -1 \\ 0 & 1 & -1 \end{vmatrix} = \mathbf{i} + \mathbf{j} + \mathbf{k} = \langle 1, 1, 1 \rangle$$

Since the normal points downward, the z-component must be negative, hence:

$$\mathbf{N} = \langle -1, -1, -1 \rangle$$

Step 2. Evaluate the dot product $\mathbf{F} \cdot \mathbf{N}$.

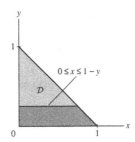

We compute the dot product:

$$\mathbf{F}(\Phi(x, y)) \cdot \mathbf{N} = \langle y^2, 2, -x \rangle \cdot \langle -1, -1, -1 \rangle = -y^2 - 2 + x$$

Step 3. Evaluate the surface integral. The surface integral is equal to the following double integral:

$$\iint_S \mathbf{F} \cdot d\mathbf{S} = \iint_\mathcal{D} \mathbf{F}(\Phi(x,y)) \cdot \mathbf{N}\, dx\, dy = \int_0^1 \int_0^{1-y} \left(-y^2 - 2 + x \right) dx\, dy = \int_0^1 \left. -y^2 x - 2x + \frac{x^2}{2} \right|_{x=0}^{1-y} dy$$

$$= \int_0^1 \left(-y^2(1-y) - 2(1-y) + \frac{(1-y)^2}{2} \right) dy = \int_0^1 \left(y^3 - y^2 + 2(y-1) + \frac{(y-1)^2}{2} \right) dy$$

$$= \left. \frac{y^4}{4} - \frac{y^3}{3} + (y-1)^2 + \frac{(y-1)^3}{6} \right|_0^1 = \left(\frac{1}{4} - \frac{1}{3} \right) - \left(1 - \frac{1}{6} \right) = -\frac{11}{12}$$

13. $\mathbf{F} = \langle xz, yz, z^{-1} \rangle$, disk of radius 3 at height 4 parallel to the xy-plane, upward-pointing normal

SOLUTION

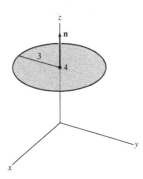

We parametrize the surface S by:

$$\Phi(\theta, r) = (r\cos\theta, r\sin\theta, 4)$$

with the parameter domain:

$$\mathcal{D} = \{(\theta, r) : 0 \le \theta \le 2\pi, 0 \le r \le 3\}$$

Step 1. Compute the tangent and normal vectors. We have:

$$\mathbf{T}_\theta = \frac{\partial \Phi}{\partial \theta} = \frac{\partial}{\partial \theta}(r\cos\theta, r\sin\theta, 4) = \langle -r\sin\theta, r\cos\theta, 0 \rangle$$

$$\mathbf{T}_r = \frac{\partial \Phi}{\partial r} = \frac{\partial}{\partial r}(r\cos\theta, r\sin\theta, 4) = \langle \cos\theta, \sin\theta, 0 \rangle$$

1428 CHAPTER 16 | LINE AND SURFACE INTEGRALS (LT CHAPTER 17)

$$\mathbf{T}_\theta \times \mathbf{T}_r = \begin{vmatrix} \mathbf{i} & \mathbf{j} & \mathbf{k} \\ -r\sin\theta & r\cos\theta & 0 \\ \cos\theta & \sin\theta & 0 \end{vmatrix} = \left(-r\sin^2\theta - r\cos^2\theta\right)\mathbf{k} = -r\mathbf{k} = \langle 0, 0, -r\rangle$$

Since the orientation of S is with an upward-pointing normal, the z-coordinate of $\mathbf{N}$ must be positive. Hence:

$$\mathbf{N} = \langle 0, 0, r\rangle$$

Step 2. Evaluate the dot product $\mathbf{F} \cdot \mathbf{N}$. We first express $\mathbf{F}$ in terms of the parameters:

$$\mathbf{F}(\Phi(\theta, r)) = \langle xz, yz, z^{-1}\rangle = \langle r\cos\theta \cdot 4, r\sin\theta \cdot 4, 4^{-1}\rangle = \langle 4r\cos\theta, 4r\sin\theta, \frac{1}{4}\rangle$$

We now compute the dot product:

$$\mathbf{F}(\Phi(\theta, r)) \cdot \mathbf{N}(\theta, r) = \langle 4r\cos\theta, 4r\sin\theta, \frac{1}{4}\rangle \cdot \langle 0, 0, r\rangle = \frac{r}{4}$$

Step 3. Evaluate the surface integral. The surface integral is equal to the following double integral:

$$\iint_S \mathbf{F} \cdot d\mathbf{S} = \iint_\mathcal{D} \mathbf{F}(\Phi(\theta, r)) \cdot \mathbf{N}(\theta, r)\, dr\, d\theta = \int_0^{2\pi}\!\int_0^3 \frac{r}{4}\, dr\, d\theta = 2\pi \int_0^3 \frac{r}{4}\, dr = 2\pi \cdot \frac{r^2}{8}\Big|_0^3 = \frac{9\pi}{4}$$

15. $\mathbf{F} = \langle 0, 0, e^{y+z}\rangle$, boundary of unit cube $0 \le x \le 1, 0 \le y \le 1, 0 \le z \le 1$, outward-pointing normal

SOLUTION

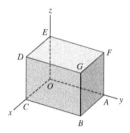

We denote the faces of the cube by:

$$S_1 = \text{Face } OABC \quad S_2 = \text{Face } DGEF \quad S_3 = \text{Face } ABGF$$
$$S_4 = \text{Face } OCDE \quad S_5 = \text{Face } BCDG \quad S_6 = \text{Face } OAFE$$

- On S_1

$$\Phi_1(x, y) = (x, y, 0)$$

and $\mathbf{N}_1 = \langle 0, 0, -1\rangle$. Thus,

$$\mathbf{F}(\Phi_1(x, y)) \cdot \mathbf{N}_1 = \langle 0, 0, e^y\rangle \cdot \langle 0, 0, -1\rangle = -e^y$$

- On S_2

$$\Phi_2(x, y) = (x, y, 1)$$

and $\mathbf{N}_2 = \langle 0, 0, 1\rangle$. Thus,

$$\mathbf{F}(\Phi_2(x, y)) \cdot \mathbf{N}_2 = \langle 0, 0, e^{y+1}\rangle \cdot \langle 0, 0, 1\rangle = e^{y+1}$$

- On any other surface $S_i, 3 \le i \le 6$, we have

$$\mathbf{F}(\Phi_1(x, y)) \cdot \mathbf{N}_i = 0,$$

because the z-component of $\mathbf{N}_i = 0$ and the x, y components of $\mathbf{F}$ equal 0. Thus,

$$\iint_S \mathbf{F} \cdot d\mathbf{S} = \iint_{S_1} \mathbf{F} \cdot d\mathbf{S} + \iint_{S_2} \mathbf{F} \cdot d\mathbf{S} = \int_0^1\!\int_0^1 -e^y\, dx\, dy + \int_0^1\!\int_0^1 e^{y+1}\, dx\, dy$$

$$= \int_0^1\!\int_0^1 \left(e^{y+1} - e^y\right) dx\, dy = \int_0^1 \left(e^{y+1} - e^y\right) dy$$

$$= \int_0^1 e^y(e-1)\,dy = (e-1)e^y\Big|_0^1 = (e-1)^2$$

17. $\mathbf{F} = \langle y, z, 0 \rangle$, $\Phi(u, v) = (u^3 - v, u + v, v^2)$, $0 \le u \le 2$, $0 \le v \le 3$, downward-pointing normal

SOLUTION

Step 1. Compute the tangent and normal vectors. We have,

$$\mathbf{T}_u = \frac{\partial \Phi}{\partial u} = \frac{\partial}{\partial u}\left(u^3 - v, u + v, v^2\right) = \langle 3u^2, 1, 0 \rangle$$

$$\mathbf{T}_v = \frac{\partial \Phi}{\partial v} = \frac{\partial}{\partial v}\left(u^3 - v, u + v, v^2\right) = \langle -1, 1, 2v \rangle$$

$$\mathbf{T}_u \times \mathbf{T}_v = \begin{vmatrix} \mathbf{i} & \mathbf{j} & \mathbf{k} \\ 3u^2 & 1 & 0 \\ -1 & 1 & 2v \end{vmatrix} = (2v)\mathbf{i} - \left(6u^2v\right)\mathbf{j} + \left(3u^2 + 1\right)\mathbf{k} = \langle 2v, -6u^2v, 3u^2 + 1 \rangle$$

Since the normal is pointing downward, the z-coordinate is negative, hence,

$$\mathbf{N} = \langle -2v, 6u^2v, -3u^2 - 1 \rangle$$

Step 2. Evaluate the dot product $\mathbf{F} \cdot \mathbf{N}$. We first express $\mathbf{F}$ in terms of the parameters:

$$\mathbf{F}(\Phi(u, v)) = \langle y, z, 0 \rangle = \langle u + v, v^2, 0 \rangle$$

We compute the dot product:

$$\mathbf{F}(\Phi(u, v)) \cdot \mathbf{N}(u, v) = \langle u + v, v^2, 0 \rangle \cdot \langle -2v, 6u^2v, -3u^2 - 1 \rangle$$

$$= -2v(u + v) + 6u^2v \cdot v^2 + 0 = -2vu - 2v^2 + 6u^2v^3$$

Step 3. Evaluate the surface integral. The surface integral is equal to the following double integral:

$$\iint_S \mathbf{F} \cdot d\mathbf{S} = \iint_D \mathbf{F}(\Phi(u, v)) \cdot \mathbf{N}(u, v)\,du\,dv = \int_0^3 \int_0^2 \left(-2uv - 2v^2 + 6u^2v^3\right)du\,dv$$

$$= \int_0^3 -u^2v - 2v^2u + 2u^3v^3\Big|_{u=0}^2\,dv = \int_0^3 \left(16v^3 - 4v^2 - 4v\right)dv = 4v^4 - \frac{4}{3}v^3 - 2v^2\Big|_0^3 = 270$$

19. Let $\mathbf{e}_r = \langle x/r, y/r, z/r \rangle$ be the unit radial vector, where $r = \sqrt{x^2 + y^2 + z^2}$. Calculate the integral of $\mathbf{F} = e^{-r}\mathbf{e}_r$ over:

(a) the upper hemisphere of $x^2 + y^2 + z^2 = 9$, outward-pointing normal.
(b) the octant $x \ge 0, y \ge 0, z \ge 0$ of the unit sphere centered at the origin.

SOLUTION

(a) We parametrize the upper-hemisphere by

$$\Phi : x = 3\cos\theta\sin\phi, y = 3\sin\theta\sin\phi, z = 3\cos\phi$$

with the parameter domain:

$$\mathcal{D} = \left\{(\theta, \phi) : 0 \le \theta < 2\pi, 0 \le \phi < \frac{\pi}{2}\right\}$$

The outward pointing normal is (see Eq. (4) in Section 17.4):

$$\mathbf{N} = 9\sin\phi\,\mathbf{e}_r$$

We compute the dot product $\mathbf{F} \cdot \mathbf{N}$ on the sphere. On the sphere $r = 3$, hence,

$$\mathbf{F} \cdot \mathbf{N} = e^{-r}\mathbf{e}_r \cdot \mathbf{N} = e^{-3}\mathbf{e}_r \cdot 9\sin\phi\,\mathbf{e}_r = 9e^{-3}\sin\phi\,\mathbf{e}_r \cdot \mathbf{e}_r = 9\,e^{-3}\sin\phi$$

We obtain the following integral:

$$\iint_S \mathbf{F} \cdot d\mathbf{S} = \iint_\mathcal{D} (\mathbf{F} \cdot \mathbf{N})\,d\phi\,d\theta = \int_0^{2\pi}\int_0^{\pi/2} 9e^{-3}\sin\phi\,d\phi\,d\theta$$

$$= 18\pi e^{-3}\int_0^{\pi/2}\sin\phi\,d\phi = 18\pi e^{-3}\left(-\cos\phi\Big|_0^{\pi/2}\right) = 18\pi e^{-3}$$

(b) We parametrize the first octant of the sphere by,

$$\Phi : x = \cos\theta \sin\phi, y = \sin\theta \sin\phi, z = \cos\phi$$

with the parameter domain:

$$\mathcal{D} = \left\{(\theta, \phi) : 0 \le \theta < \frac{\pi}{2}, 0 \le \phi < \frac{\pi}{2}\right\}$$

The outward pointing normal is (as seen above):

$$\mathbf{N} = 1 \sin\phi \mathbf{e}_r$$

We compute the dot product $\mathbf{F} \cdot \mathbf{N}$ on the sphere. On the sphere $r = 1$, hence,

$$\mathbf{F} \cdot \mathbf{N} = e^{-r}\mathbf{e}_r \cdot \mathbf{N} = e^{-1}\mathbf{e}_r \cdot \sin\phi \mathbf{e}_r = e^{-1}\sin\phi \mathbf{e}_r \cdot \mathbf{e}_r = e^{-1}\sin\phi$$

We obtain the following integral:

$$\iint_S \mathbf{F} \cdot d\mathbf{S} = \iint_{\mathcal{D}} (\mathbf{F} \cdot \mathbf{N})\, d\phi\, d\theta = \int_0^{\pi/2} \int_0^{\pi/2} e^{-1}\sin\phi\, d\phi\, d\theta$$

$$= \frac{\pi}{2}e^{-1} \int_0^{\pi/2} \sin\phi\, d\phi = \frac{\pi}{2}e^{-1}\left(-\cos\phi\Big|_0^{\pi/2}\right) = \frac{\pi}{2}e^{-1}$$

21. The electric field due to a point charge located at the origin in $\mathbf{R}^3$ is $\mathbf{E} = k\dfrac{\mathbf{e}_r}{r^2}$, where $r = \sqrt{x^2 + y^2 + z^2}$ and k is a constant. Calculate the flux of $\mathbf{E}$ through the disk D of radius 2 parallel to the xy-plane with center $(0, 0, 3)$.

SOLUTION Let $r = \sqrt{x^2 + y^2 + z^2}$ and $\hat{r} = \sqrt{x^2 + y^2}$. We parametrize the disc by:

$$\Phi(\hat{r}, \theta) = (\hat{r}\cos\theta, \hat{r}\sin\theta, 3)$$

$$\mathbf{T}_{\hat{r}} = \frac{\partial \Phi}{\partial \hat{r}} = \langle \cos\theta, \sin\theta, 0 \rangle$$

$$\mathbf{T}_\theta = \frac{\partial \Phi}{\partial \theta} = \langle -\hat{r}\sin\theta, \hat{r}\cos\theta, 0 \rangle$$

$$\mathbf{N} = \mathbf{T}_{\hat{r}} \times \mathbf{T}_\theta = \begin{vmatrix} \mathbf{i} & \mathbf{j} & \mathbf{k} \\ \cos\theta & \sin\theta & 0 \\ -\hat{r}\sin\theta & \hat{r}\cos\theta & 0 \end{vmatrix} = \langle 0, 0, \hat{r} \rangle$$

Now,

$$\mathbf{E} \cdot \mathbf{N} = k\frac{\mathbf{e}_r}{r^2} \cdot \langle 0, 0, \hat{r} \rangle = \frac{k\hat{r}}{r^3}\langle x, y, z \rangle \cdot \langle 0, 0, 1 \rangle = \frac{zk\hat{r}}{r^3}$$

Since on the disk $z = 3$, we get:

$$\mathbf{E} \cdot \mathbf{N} = 3k\frac{\hat{r}}{r^3} \text{ and } r = \sqrt{\hat{r}^2 + 9}$$

so $\mathbf{E} \cdot \mathbf{N} = 3k\dfrac{\hat{r}}{\left(\sqrt{\hat{r}^2+9}\right)^3}$.

$$\iint_{\mathcal{D}} \mathbf{E} \cdot d\mathbf{S} = \int_0^{2\pi} \int_0^2 \frac{3k\hat{r}}{(\hat{r}^2+9)^{3/2}}\, d\hat{r}\, d\theta = 6\pi k \int_0^2 \frac{\hat{r}}{(\hat{r}^2+9)^{3/2}}\, d\hat{r}$$

Substituting $u = \hat{r}^2 + 9$ and $\frac{1}{2}du = \hat{r}\, d\hat{r}$, we get:

$$\iint_{\mathcal{D}} \mathbf{E} \cdot d\mathbf{S} = 3\pi k \int_9^{13} \frac{du}{u^{3/2}} = -6\pi k u^{-1/2}\Big|_9^{13} = \left(2 - \frac{6}{\sqrt{13}}\right)\pi k$$

23. Let $\mathbf{v} = z\mathbf{k}$ be the velocity field (in meters per second) of a fluid in $\mathbf{R}^3$. Calculate the flow rate (in cubic meters per second) through the upper hemisphere ($z \ge 0$) of the sphere $x^2 + y^2 + z^2 = 1$.

SOLUTION We use the spherical coordinates:

$$x = \cos\theta \sin\phi, \quad y = \sin\theta \sin\phi, \quad z = \cos\phi$$

with the parameter domain

$$0 \leq \theta < 2\pi, \quad 0 \leq \phi \leq \frac{\pi}{2}$$

The normal vector is:

$$\mathbf{N} = \mathbf{T}_\phi \times \mathbf{T}_\theta = \sin\phi \langle \cos\theta \sin\phi, \sin\theta \sin\phi, \cos\phi \rangle$$

We express the function in terms of the parameters:

$$\mathbf{v} = \langle 0, 0, z \rangle = \langle 0, 0, \cos\phi \rangle$$

Hence,

$$\mathbf{v} \cdot \mathbf{n} = \langle 0, 0, \cos\phi \rangle \cdot \sin\phi \langle \cos\theta \sin\phi, \sin\theta \sin\phi, \cos\phi \rangle = \sin\phi \cos^2\phi$$

The flow rate of the fluid through the upper hemisphere S is equal to the flux of the velocity vector through S. That is,

$$\iint_S \mathbf{v} \cdot d\mathbf{S} = \int_0^{\pi/2} \int_0^{2\pi} \sin\phi \cos^2\phi \, d\theta \, d\phi$$

$$= 2\pi \int_0^{\pi/2} \sin\phi \cos^2\phi \, d\phi$$

$$= 2\pi \left(-\frac{1}{3} \cos^3\phi \right) \Big|_0^{\pi/2} = \frac{2\pi}{3} \text{ m}^3/\text{s}$$

In Exercises 25–28, a net is dipped in a river. Determine the flow rate of water across the net if the velocity vector field for the river is given by $\mathbf{v}$ *and the net is described by the given equations.*

25. $\mathbf{v} = \langle x - y, z + y + 4, z^2 \rangle$, net given by $x^2 + z^2 \leq 1, y = 0$, oriented in the positive y-direction

SOLUTION We use modified cylindrical coordinates:

$$x = r\cos\theta, \quad y = y, \quad z = r\sin\theta$$

so that the surface is given by

$$S : x = r\cos\theta, \quad y = 0, \quad z = r\sin\theta$$

$$0 \leq \theta \leq 2\pi, \quad 0 \leq r \leq 1$$

We first compute the normal vector:

$$\mathbf{T}_r = \frac{\partial S}{\partial r} = \frac{\partial}{\partial r}(r\cos\theta, 0, r\sin\theta) = \langle \cos\theta, 0, \sin\theta \rangle$$

$$\mathbf{T}_\theta = \frac{\partial S}{\partial \theta} = \frac{\partial}{\partial \theta}(r\cos\theta, 0, r\sin\theta) = \langle -r\sin\theta, 0, r\cos\theta \rangle$$

$$\mathbf{T}_r \times \mathbf{T}_\theta = \begin{vmatrix} \mathbf{i} & \mathbf{j} & \mathbf{k} \\ \cos\theta & 0 & \sin\theta \\ -r\sin\theta & 0 & r\cos\theta \end{vmatrix} = -\left(r\cos^2\theta + r\sin^2\theta\right)\mathbf{j} = -r\mathbf{j}$$

Since we want the normal to point in the positive y-direction, we take $\mathbf{N} = r\mathbf{j}$. Now express the flow in terms of the parameters:

$$\mathbf{v} = \langle x - y, z + y + 4, z^2 \rangle = \langle r\cos\theta, r\sin\theta + 4, r^2\sin^2\theta \rangle$$

Hence

$$\mathbf{v} \cdot \mathbf{N} = \langle r\cos\theta, r\sin\theta + 4, r^2\sin^2\theta \rangle \cdot \langle 0, r, 0 \rangle = r^2\sin\theta + 4r$$

The flow rate of the water through the net is equal to the flux of the vector field across the surface, which is

$$\iint_S \mathbf{v} \cdot d\mathbf{S} = \int_0^1 \int_0^{2\pi} \mathbf{v} \cdot \mathbf{N} \, d\theta \, dr = \int_0^1 \int_0^{2\pi} (r^2 \sin\theta + 4r) \, d\theta \, dr$$

$$= \int_0^1 (-r^2 \cos\theta + 4r\theta)\Big|_{\theta=0}^{2\pi} dr = \int_0^1 8\pi r \, dr = (4\pi r^2)\Big|_0^1 = 4\pi$$

27. $\mathbf{v} = \langle x - y, z + y + 4, z^2 \rangle$, net given by $y = \sqrt{1 - x^2 - z^2}$, $y \geq 0$, oriented in the positive y-direction

SOLUTION Parametrize the net by a variant of cylindrical coordinates $x = r\cos\theta$, $z = r\sin\theta$, $y = y$, just as in the previous two exercises. The parameter domain $\mathcal{D}$ is the same — the unit circle of radius 1 in the xz-plane centered at the origin, which in our coordinate system is $r = 1, 0 \leq \theta \leq 2\pi$. Then the net is described by

$$G(r, \theta) = (r\cos\theta, \sqrt{1 - r^2}, r\sin\theta), \ 0 \leq r \leq 1, \ 0 \leq \theta \leq 2\pi$$

We first compute the normal vector:

$$\mathbf{T}_r = \frac{\partial G}{\partial r} = \frac{\partial}{\partial r}(r\cos\theta, \sqrt{1 - r^2}, r\sin\theta) = \left\langle \cos\theta, -\frac{r}{\sqrt{1 - r^2}}, \sin\theta \right\rangle$$

$$\mathbf{T}_\theta = \frac{\partial G}{\partial \theta} = \frac{\partial}{\partial \theta}(r\cos\theta, \sqrt{1 - r^2}, r\sin\theta) = \langle -r\sin\theta, 0, r\cos\theta \rangle$$

$$\mathbf{T}_r \times \mathbf{T}_\theta = \begin{vmatrix} \mathbf{i} & \mathbf{j} & \mathbf{k} \\ \cos\theta & -\frac{r}{\sqrt{1-r^2}} & \sin\theta \\ -r\sin\theta & 0 & r\cos\theta \end{vmatrix} = \left\langle -\frac{r^2}{\sqrt{1-r^2}}\cos\theta, -r, -\frac{r^2}{\sqrt{1-r^2}}\sin\theta \right\rangle$$

Since we want the normal to point in the positive y-direction, we let

$$\mathbf{N} = -(\mathbf{T}_r \times \mathbf{T}_\theta) = \left\langle \frac{r^2}{\sqrt{1-r^2}}\cos\theta, r, \frac{r^2}{\sqrt{1-r^2}}\sin\theta \right\rangle$$

Now express the flow in terms of the parameters:

$$\mathbf{v} = \langle x - y, z + y + 4, z^2 \rangle = \langle r\cos\theta - \sqrt{1 - r^2}, r\sin\theta + \sqrt{1 - r^2} + 4, r^2 \sin^2\theta \rangle$$

Hence

$$\mathbf{v} \cdot \mathbf{N} = \langle r\cos\theta - \sqrt{1 - r^2}, r\sin\theta + \sqrt{1 - r^2} + 4, r^2 \sin^2\theta \rangle \cdot \left\langle \frac{r^2}{\sqrt{1-r^2}}\cos\theta, r, \frac{r^2}{\sqrt{1-r^2}}\sin\theta \right\rangle$$

$$= \frac{r^3}{\sqrt{1-r^2}}\cos^2\theta - r^2\cos\theta + r^2\sin\theta + r\sqrt{1 - r^2} + 4r + \frac{r^4}{\sqrt{1-r^2}}\sin^3\theta$$

The flow rate of water through the net is equal to the flux of the vector field across the surface of the net:

$$\iint_S \mathbf{v} \cdot d\mathbf{S} = \int_0^1 \int_0^{2\pi} \mathbf{v} \cdot \mathbf{N} \, d\theta \, dr$$

$$= \int_0^1 \int_0^{2\pi} \left(\frac{r^3}{\sqrt{1-r^2}}\cos^2\theta - r^2\cos\theta + r^2\sin\theta + r\sqrt{1 - r^2} + 4r + \frac{r^4}{\sqrt{1-r^2}}\sin^3\theta \right) d\theta \, dr$$

$$= \int_0^1 \int_0^{2\pi} \left(\frac{r^3}{\sqrt{1-r^2}}\left(\frac{1}{2} + \frac{1}{2}\cos 2\theta\right) - r^2\cos\theta + r^2\sin\theta + r\sqrt{1 - r^2} + 4r \right.$$

$$\left. + \frac{r^4}{\sqrt{1-r^2}}\sin\theta(1 - \cos^2\theta) \right) d\theta \, dr$$

$$= \int_0^1 \left(\frac{r^3}{\sqrt{1-r^2}}\left(\frac{1}{2}\theta + \frac{1}{4}\sin 2\theta\right) - r^2\sin\theta - r^2\cos\theta + r\theta\sqrt{1 - r^2} + 4r\theta \right.$$

$$\left. + \frac{r^4}{\sqrt{1-r^2}}\left(\frac{1}{3}\cos^3\theta - \cos\theta\right) \right)\Bigg|_{\theta=0}^{2\pi} dr$$

$$= \int_0^1 \left(\frac{r^3}{\sqrt{1-r^2}} \cdot \frac{1}{2} \cdot 2\pi + 2\pi r \sqrt{1-r^2} + 8\pi r \right) dr$$

$$= \pi \int_0^1 \left(\frac{r^3}{\sqrt{1-r^2}} + 2r\sqrt{1-r^2} + 8r \right) dr$$

$$= \pi \left(\int_0^1 \frac{r^3}{\sqrt{1-r^2}} \, dr + \int_0^1 2r\sqrt{1-r^2} \, dr + \int_0^1 8r \, dr \right)$$

For the first integral, use the substitution $u = 1 - r^2$; then $-2r \, dr = du$, and we also substitute $r^2 = 1 - u$. For the second, use the substitution $v = \sqrt{1-r^2}$. Then we get

$$\pi \left(\int_0^1 \frac{r^3}{\sqrt{1-r^2}} \, dr + \int_0^1 2r\sqrt{1-r^2} \, dr + \int_0^1 8r \, dr \right)$$

$$= \pi \left(-\frac{1}{2} \int_1^0 \frac{1-u}{\sqrt{u}} \, du - \int_1^0 \sqrt{v} \, dv + 4r^2 \Big|_0^1 \right)$$

$$= \pi \left(-\frac{1}{2} \int_1^0 \left(u^{-1/2} - u^{1/2} \right) du - \frac{2}{3} v^{3/2} \Big|_1^0 + 4 \right)$$

$$= \pi \left(-\frac{1}{2} \left(2u^{1/2} - \frac{2}{3} u^{3/2} \right) \Big|_1^0 + \frac{2}{3} + 4 \right)$$

$$= \pi \left(\frac{1}{2} \left(2 - \frac{2}{3} \right) + \frac{14}{3} \right)$$

$$= \frac{16}{3} \pi$$

In Exercises 29–30, let $\mathcal{T}$ be the triangular region with vertices $(1, 0, 0)$, $(0, 1, 0)$, and $(0, 0, 1)$ oriented with upward-pointing normal vector (Figure 16). Assume distances are in meters.

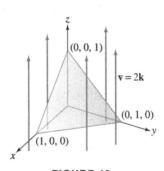

FIGURE 16

29. A fluid flows with constant velocity field $\mathbf{v} = 2\mathbf{k}$ (meters per second). Calculate:

(a) the flow rate through $\mathcal{T}$.

(b) the flow rate through the projection of $\mathcal{T}$ onto the xy-plane [the triangle with vertices $(0, 0, 0)$, $(1, 0, 0)$, and $(0, 1, 0)$].

SOLUTION

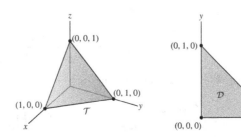

The equation of the plane through the three vertices is $x + y + z = 1$, hence the upward-pointing normal vector is:

$$\mathbf{N} = \langle 1, 1, 1 \rangle$$

and the unit normal is:

$$\mathbf{e_n} = \left\langle \frac{1}{\sqrt{3}}, \frac{1}{\sqrt{3}}, \frac{1}{\sqrt{3}} \right\rangle$$

We compute the dot product $\mathbf{v} \cdot \mathbf{e_n}$:

$$\mathbf{v} \cdot \mathbf{e_n} = \langle 0, 0, 2 \rangle \cdot \left\langle \frac{1}{\sqrt{3}}, \frac{1}{\sqrt{3}}, \frac{1}{\sqrt{3}} \right\rangle = \frac{2}{\sqrt{3}}$$

The flow rate through $\mathcal{T}$ is equal to the flux of $\mathbf{v}$ through $\mathcal{T}$. That is,

$$\iint_S \mathbf{v} \cdot d\mathbf{S} = \iint_S (\mathbf{v} \cdot \mathbf{e_n}) \, dS = \iint_S \frac{2}{\sqrt{3}} \, dS = \frac{2}{\sqrt{3}} \iint_S 1 \, dS = \frac{2}{\sqrt{3}} \cdot \text{Area}(S)$$

The area of the equilateral triangle $\mathcal{T}$ is $\frac{(\sqrt{2})^2 \cdot \sqrt{3}}{4} = \frac{\sqrt{3}}{2}$. Therefore,

$$\iint_S \mathbf{v} \cdot d\mathbf{S} = \frac{2}{\sqrt{3}} \cdot \frac{\sqrt{3}}{2} = 1$$

Let $\mathcal{D}$ denote the projection of $\mathcal{T}$ onto the xy-plane. Then the upward-pointing normal is $\mathbf{N} = \langle 0, 0, 1 \rangle$. We compute the dot product $\mathbf{v} \cdot \mathbf{N}$:

$$\mathbf{v} \cdot \mathbf{N} = \langle 0, 0, 2 \rangle \cdot \langle 0, 0, 1 \rangle = 2$$

The flow rate through $\mathcal{D}$ is equal to the flux of $\mathbf{v}$ through $\mathcal{D}$. That is,

$$\iint_\mathcal{D} \mathbf{v} \cdot d\mathbf{S} = \iint_\mathcal{D} (\mathbf{v} \cdot \mathbf{N}) \, dS = \iint_\mathcal{D} 2 \, dS = 2 \iint_\mathcal{D} 1 \, dS = 2 \cdot \text{Area}(\mathcal{D}) = 2 \cdot \frac{1 \cdot 1}{2} = 1$$

31. Prove that if S is the part of a graph $z = g(x, y)$ lying over a domain $\mathcal{D}$ in the xy-plane, then

$$\iint_S \mathbf{F} \cdot d\mathbf{S} = \iint_\mathcal{D} \left(-F_1 \frac{\partial g}{\partial x} - F_2 \frac{\partial g}{\partial y} + F_3 \right) dx \, dy$$

SOLUTION

Step 1. Find a parametrization. We parametrize the surface by

$$\Phi(x, y) = (x, y, g(x, y)), (x, y) \in \mathcal{D}$$

Step 2. Compute the tangent and normal vectors. We have,

$$\mathbf{T}_x = \frac{\partial \Phi}{\partial x} = \frac{\partial}{\partial x}(x, y, g(x, y)) = \left\langle 1, 0, \frac{\partial g}{\partial x} \right\rangle$$

$$\mathbf{T}_y = \frac{\partial \Phi}{\partial y} = \frac{\partial}{\partial y}(x, y, g(x, y)) = \left\langle 0, 1, \frac{\partial g}{\partial y} \right\rangle$$

$$\mathbf{N} = \mathbf{T}_x \times \mathbf{T}_y = \begin{vmatrix} \mathbf{i} & \mathbf{j} & \mathbf{k} \\ 1 & 0 & \frac{\partial g}{\partial x} \\ 0 & 1 & \frac{\partial g}{\partial y} \end{vmatrix} = -\frac{\partial g}{\partial x}\mathbf{i} - \frac{\partial g}{\partial x}\mathbf{j} + \mathbf{k} = \left\langle -\frac{\partial g}{\partial x}, -\frac{\partial g}{\partial y}, 1 \right\rangle$$

Step 3. Evaluate the dot product $\mathbf{F} \cdot \mathbf{N}$.

$$\mathbf{F} \cdot \mathbf{N} = \langle F_1, F_2, F_3 \rangle \cdot \left\langle -\frac{\partial g}{\partial x}, -\frac{\partial g}{\partial y}, 1 \right\rangle = -F_1 \frac{\partial g}{\partial x} - F_2 \frac{\partial g}{\partial y} + F_3$$

Step 4. Evaluate the surface integral. The surface integral is equal to the following double integral:

$$\iint_S \mathbf{F} \cdot d\mathbf{S} = \iint_\mathcal{D} (\mathbf{F} \cdot \mathbf{N}) \, dx \, dy = \iint_\mathcal{D} \left(-F_1 \frac{\partial g}{\partial x} - F_2 \frac{\partial g}{\partial y} + F_3 \right) dx \, dy$$

In Exercises 32–33, a varying current i(t) flows through a long, straight wire in the xy-plane as in Example 5. The current produces a magnetic field **B** *whose magnitude at a distance r from the wire is* $B = \dfrac{\mu_0 i}{2\pi r}$ *T, where* $\mu_0 = 4\pi \cdot 10^{-7}$ *T-m/A. Furthermore,* **B** *points into the page at points P in the xy-plane.*

33. Assume that $i = 10e^{-0.1t}$ A (t in seconds). Calculate the flux $\Phi(t)$, at time t, of **B** through the isosceles triangle of base 12 cm and height 6 cm whose bottom edge is 3 cm from the wire, as in Figure 17. Assume the triangle is oriented with normal vector pointing out of the page. Use Faraday's Law to determine the voltage drop around the triangular loop (the boundary of the triangle) at time t.

FIGURE 17

SOLUTION The magnetic field is $\mathbf{B} = \dfrac{-\mu_0 i}{2\pi r}\mathbf{k}$ and the unit normal on the triangle points out of the page, hence $\mathbf{N} = \mathbf{n} = \mathbf{k}$.

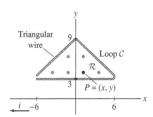

Also, the distance from a point $P = (x, y)$ in $\mathcal{R}$ to the wire is $r = y$. Hence:

$$\mathbf{B} \cdot \mathbf{N} = \frac{-\mu_0 i}{2\pi y}\mathbf{k} \cdot \mathbf{k} = \frac{-\mu_0 i}{2\pi y}$$

The flux $\Phi(t)$ of **B** through $\mathcal{R}$ is the following integral:

$$\Phi(t) = \iint_\mathcal{R} \mathbf{B} \cdot \mathbf{N}\, dx\, dy = \frac{-\mu_0 i}{2\pi} \iint_\mathcal{R} \frac{1}{y}\, dx\, dy$$

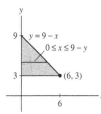

Using symmetry we have:

$$\Phi(t) = \frac{-\mu_0 i}{\pi} \int_3^9 \int_0^{9-y} \frac{1}{y}\, dx\, dy = \frac{-\mu_0 i}{\pi} \int_3^9 \frac{x}{y}\bigg|_{x=0}^{9-y} dy = \frac{-\mu_0 i}{\pi} \int_3^9 \frac{9-y}{y}\, dy$$

$$= \frac{-\mu_0 i}{\pi} \int_3^9 \left(\frac{9}{y} - 1\right) dy = \frac{-\mu_0 i}{\pi}\left(9\ln y - y\bigg|_{y=3}^9\right) = \frac{-\mu_0 i}{\pi}\left((9\ln 9 - 9) - (9\ln 3 - 3)\right)$$

$$= \frac{-\mu_0 i}{\pi}(9\ln 3 - 6) = \frac{-4\pi \cdot 10^{-7}}{\pi}(9\ln 3 - 6)i = -1.56 \cdot 10^{-6} i = -1.56 \cdot 10^{-5} \cdot e^{-0.1t}$$

Using Faraday's Law, the voltage drop around the triangular loop C (oriented counterclockwise):

$$\int_C \mathbf{E} \cdot d\mathbf{S} = -\frac{d\phi}{dt} = -\frac{d}{dt}\left(-1.56 \cdot 10^{-5} \cdot e^{-0.1t}\right) = -1.56 \cdot 10^{-6} \cdot e^{-0.1t} \text{ Volts}$$

In Exercises 34–35, a solid material that has thermal conductivity K in kilowatts per meter-kelvin and temperature given at each point by $w(x, y, z)$ has heat flow given by the vector field $\mathbf{F} = -K\nabla w$ and rate of heat flow across a surface S within the solid given by $-K\iint_S \nabla w \, dS$.

35. An insulated cylinder of solid gold ($K = 310$ kW/m-k) of radius $\sqrt{2}$ m and height 5 m is heated at one end until the temperature at each point in the cylinder is given by $w(x, y, z) = (30 - z^2)(2 - (x^2 + y^2))$. Determine the rate of heat flow across each horizontal disk given by $z = 1$, $z = 2$, and $z = 3$, identifying which has the greatest rate of heat flow across it.

SOLUTION We assume the cylinder is oriented with its axis equal to the z-axis, so that it lies over the circle $x^2 + y^2 = 2$ in the xy-plane and extends from $z = 0$ to $z = 5$. Parametrize it using cylindrical coordinates. Then the disk $z = a$ at height a is parametrized as

$$\Phi(r, \theta) = \langle r\cos\theta, r\sin\theta, a\rangle, \ 0 \le r \le \sqrt{2}, \ 0 \le \theta \le 2\pi$$

We first compute the normal vector:

$$\mathbf{T}_r = \frac{\partial \Phi}{\partial r} = \frac{\partial}{\partial r}(r\cos\theta, r\sin\theta, a) = \langle\cos\theta, \sin\theta, 0\rangle$$

$$\mathbf{T}_\theta = \frac{\partial \Phi}{\partial \theta} = \frac{\partial}{\partial \theta}(r\cos\theta, r\sin\theta, a) = \langle-r\sin\theta, r\cos\theta, 0\rangle$$

$$\mathbf{T}_r \times \mathbf{T}_\theta = \begin{vmatrix} \mathbf{i} & \mathbf{j} & \mathbf{k} \\ \cos\theta & \sin\theta & 0 \\ -r\sin\theta & r\cos\theta & 0 \end{vmatrix} = (r\cos^2\theta + r\sin^2\theta)\mathbf{k} = r\mathbf{k}$$

Since we want the normal to point in the positive y-direction, we let

$$\mathbf{N} = \mathbf{T}_r \times \mathbf{T}_\theta = r\mathbf{k}$$

Next, with w as given, we have

$$\nabla w = \left\langle \frac{\partial w}{\partial x}, \frac{\partial w}{\partial y}, \frac{\partial w}{\partial z} \right\rangle = \langle(30 - z^2)(-2x), (30 - z^2)(-2y), -2z(2 - (x^2 + y^2))\rangle$$

$$= \langle 2x(z^2 - 30), 2y(z^2 - 30), 2z((x^2 + y^2) - 2)\rangle$$

Expressing ∇w in terms of the parameters gives

$$\nabla w = \langle 2r(a^2 - 30)\cos\theta, 2r(a^2 - 30)\sin\theta, 2a(r^2 - 2)\rangle$$

Thus

$$\nabla w \cdot \mathbf{N} = \langle 2r(a^2 - 30)\cos\theta, 2r(a^2 - 30)\sin\theta, 2a(r^2 - 2)\rangle \cdot \langle 0, 0, r\rangle = 2ar(r^2 - 2)$$

Then the rate of heat flow across the horizontal plate at $z = a$ is

$$-K\iint_S \nabla w \, dS = -310 \int_0^{\sqrt{2}} \int_0^{2\pi} 2ar(r^2 - 2) \, d\theta \, dr$$

$$= -620a \cdot 2\pi \int_0^{\sqrt{2}} (r^3 - 2r) \, dr$$

$$= -1240a\pi \left(\frac{1}{4}r^4 - r^2\right)\Big|_0^{\sqrt{2}}$$

$$= 1240a\pi$$

Then the rate of heat flow across the three disks is:

$$z = 1: \ 1240\pi, \quad z = 2: \ 2480\pi, \quad z = 3: \ 3720\pi$$

The rate of heat flow is greatest across $z = 3$.

Further Insights and Challenges

In Exercises 37 and 38, let S be the surface with parametrization

$$\Phi(u, v) = \left(\left(1 + v\cos\frac{u}{2}\right)\cos u, \left(1 + v\cos\frac{u}{2}\right)\sin u, v\sin\frac{u}{2}\right)$$

for $0 \leq u \leq 2\pi$, $-\frac{1}{2} \leq v \leq \frac{1}{2}$

37. *CAS* Use a computer algebra system.

(a) Plot S and confirm visually that S is a Möbius strip.

(b) The intersection of S with the xy-plane is the unit circle $\Phi(u, 0) = (\cos u, \sin u, 0)$. Verify that the normal vector along this circle is

$$\mathbf{N}(u, 0) = \left\langle \cos u \sin\frac{u}{2}, \sin u \sin\frac{u}{2}, -\cos\frac{u}{2} \right\rangle$$

(c) As u varies from 0 to 2π, the point $\Phi(u, 0)$ moves once around the unit circle, beginning and ending at $\Phi(0, 0) = \Phi(2\pi, 0) = (1, 0, 0)$. Verify that $\mathbf{N}(u, 0)$ is a unit vector that varies continuously but that $\mathbf{N}(2\pi, 0) = -\mathbf{N}(0, 0)$. This shows that S is not orientable—that is, it is not possible to choose a nonzero normal vector at each point on S in a continuously varying manner (if it were possible, the unit normal vector would return to itself rather than to its negative when carried around the circle).

SOLUTION

(a) We use a computer algebra system to graph the plot of S, and it is indeed a Möbius strip.

(b) To find the normal vector along the unit circle, we use our computer to first find the cross product $\frac{\partial \mathbf{N}}{\partial u} \times \frac{\partial \mathbf{N}}{\partial v}$ and simplify; we get the very ugly expression

$$\mathbf{N}(u, v) = \frac{\partial \mathbf{N}}{\partial u} \times \frac{\partial \mathbf{N}}{\partial v} = \left\langle \frac{1}{2}\left(-v\cos\left(\frac{u}{2}\right) + 2\cos u + v\cos\left(\frac{3u}{2}\right)\right)\sin\left(\frac{u}{2}\right),\right.$$

$$\frac{1}{4}\left(v + 2\cos\left(\frac{u}{2}\right) + 2v\cos(u) - 2\cos\left(\frac{3u}{2}\right) - v\cos(2u)\right),$$

$$\left. -\cos\left(\frac{u}{2}\right)\left(1 + v\cos\left(\frac{u}{2}\right)\right)\right\rangle$$

(Different computer algebra systems may produce different simplifications.) When we replace v with 0 and simplify, we find that:

$$\mathbf{N}(u, 0) = \left\langle \cos u \sin\frac{u}{2}, \frac{1}{2}\left(\cos\frac{u}{2} - \cos\frac{3u}{2}\right), -\cos\frac{u}{2} \right\rangle$$

This is almost, but not quite, what we want. Let's examine that middle term a bit more.

$$\frac{1}{2}\left(\cos\frac{u}{2} - \cos\frac{3u}{2}\right) = \frac{1}{2}\left(\cos\frac{u}{2} - (\cos u \cos\frac{u}{2} - \sin u \sin\frac{u}{2})\right) = \frac{1}{2}\left(\cos\frac{u}{2}(1 - \cos u) + \sin u \sin\frac{u}{2}\right)$$

$$= \frac{1}{2}\left(\cos\frac{u}{2} \cdot 2\sin^2\frac{u}{2} + \sin u \sin\frac{u}{2}\right) = \frac{1}{2}\sin\frac{u}{2}\left(2\sin\frac{u}{2}\cos\frac{u}{2} + \sin u\right)$$

$$= \frac{1}{2}\sin\frac{u}{2}(\sin u + \sin u) = \sin u \sin\frac{u}{2}$$

which is what we expect. Thus, we see that

$$\mathbf{N}(u, 0) = \left\langle \cos u \sin\frac{u}{2}, \sin u \sin\frac{u}{2}, -\cos\frac{u}{2} \right\rangle$$

(c) To verify that $\mathbf{N}(u, 0)$ is a unit vector, we note that

$$\|\mathbf{N}(u, 0)\| = \sqrt{\left(\cos u \sin\frac{u}{2}\right)^2 + \left(\sin u \sin\frac{u}{2}\right)^2 + \left(\cos\frac{u}{2}\right)^2}$$

$$= \sqrt{\cos^2 u \sin^2\frac{u}{2} + \sin^2 u \sin^2\frac{u}{2} + \cos^2\frac{u}{2}} = \sqrt{\sin^2\frac{u}{2} + \cos^2\frac{u}{2}} = \sqrt{1} = 1$$

It is clear that $\mathbf{N}(u, 0)$ varies continuously with u, as each of its three components are nonconstant continuous functions of u. Finally, we note that $\mathbf{N}(0, 0) = \langle 0, 0, -1 \rangle$ but $\mathbf{N}(2\pi, 0) = \langle 0, 0, 1 \rangle$, so indeed $\mathbf{N}(2\pi, 0) = -\mathbf{N}(0, 0)$.

CHAPTER REVIEW EXERCISES

1. Compute the vector assigned to the point $P = (-3, 5)$ by the vector field:

(a) $\mathbf{F}(x, y) = \langle xy, y - x \rangle$

(b) $\mathbf{F}(x, y) = \langle 4, 8 \rangle$

(c) $\mathbf{F}(x, y) = \langle 3^{x+y}, \log_2(x + y) \rangle$

SOLUTION

(a) Substituting $x = -3$, $y = 5$ in $\mathbf{F} = \langle xy, y - x \rangle$ we obtain:

$$\mathbf{F} = \langle -3 \cdot 5, 5 - (-3) \rangle = \langle -15, 8 \rangle$$

(b) The constant vector field $\mathbf{F} = \langle 4, 8 \rangle$ assigns the vector $\langle 4, 8 \rangle$ to all the vectors. Thus:

$$\mathbf{F}(-3, 5) = \langle 4, 8 \rangle$$

(c) Substituting $x = -3$, $y = 5$ in $\mathbf{F} = \langle 3^{x+y}, \log_2(x + y) \rangle$ we obtain

$$\mathbf{F} = \langle 3^{-3+5}, \log_2(-3 + 5) \rangle = \langle 3^2, \log_2(2) \rangle = \langle 9, 1 \rangle$$

In Exercises 3–6, sketch the vector field.

3. $\mathbf{F}(x, y) = \langle y, 1 \rangle$

SOLUTION Notice that the vector field is constant along horizontal lines.

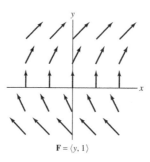

$\mathbf{F} = \langle y, 1 \rangle$

5. ∇f, where $f(x, y) = x^2 - y$

SOLUTION The gradient of $V(x, y) = x^2 - y$ is the following vector:

$$\mathbf{F}(x, y) = \left\langle \frac{\partial V}{\partial x}, \frac{\partial V}{\partial y} \right\rangle = \langle 2x, -1 \rangle$$

This vector is sketched in the following figure:

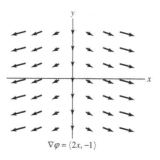

$\nabla \varphi = \langle 2x, -1 \rangle$

In Exercises 7–14, calculate div(**F**) *and* curl(**F**).

7. $\mathbf{F} = \langle x^2, y^2, z^2 \rangle$

SOLUTION The divergence of **F** is

$$\text{div}(\mathbf{F}) = \frac{\partial}{\partial x}(x^2) + \frac{\partial}{\partial y}(y^2) + \frac{\partial}{\partial z}(z^2) = 2x + 2y + 2z$$

The curl of **F** is

$$\text{curl}(F) = \nabla \times \mathbf{F} = \begin{vmatrix} \mathbf{i} & \mathbf{j} & \mathbf{k} \\ \frac{\partial}{\partial x} & \frac{\partial}{\partial y} & \frac{\partial}{\partial z} \\ x^2 & y^2 & z^2 \end{vmatrix}$$

$$= \left(\frac{\partial}{\partial y}(z^2) - \frac{\partial}{\partial z}(y^2)\right)\mathbf{i} - \left(\frac{\partial}{\partial x}(z^2) - \frac{\partial}{\partial z}(x^2)\right)\mathbf{j} + \left(\frac{\partial}{\partial x}(y^2) - \frac{\partial}{\partial y}(x^2)\right)\mathbf{k}$$

$$= \mathbf{0}$$

9. $\mathbf{F} = \langle x^3 y, xz^2, y^2 z \rangle$

SOLUTION The divergence of **F** is

$$\text{div}(\mathbf{F}) = \frac{\partial}{\partial x}(x^3 y) + \frac{\partial}{\partial y}(xz^2) + \frac{\partial}{\partial z}(y^2 z) = 3x^2 y + y^2$$

The curl of **F** is

$$\text{curl}(F) = \nabla \times \mathbf{F} = \begin{vmatrix} \mathbf{i} & \mathbf{j} & \mathbf{k} \\ \frac{\partial}{\partial x} & \frac{\partial}{\partial y} & \frac{\partial}{\partial z} \\ x^3 y & xz^2 & y^2 z \end{vmatrix}$$

$$= \left(\frac{\partial}{\partial y}(y^2 z) - \frac{\partial}{\partial z}(xz^2)\right)\mathbf{i} - \left(\frac{\partial}{\partial x}(y^2 z) - \frac{\partial}{\partial z}(x^3 y)\right)\mathbf{j} + \left(\frac{\partial}{\partial x}(xz^2) - \frac{\partial}{\partial y}(x^3 y)\right)\mathbf{k}$$

$$= (2yz - 2xz)\mathbf{i} + (z^2 - x^3)\mathbf{k}$$

11. $\mathbf{F} = y\mathbf{i} - z\mathbf{k}$

SOLUTION The divergence of **F** is

$$\text{div}(\mathbf{F}) = \frac{\partial}{\partial x}(y) + \frac{\partial}{\partial y}(0) + \frac{\partial}{\partial z}(-z) = 0 + 0 - 1 = -1$$

The curl of **F** is

$$\text{curl}(\mathbf{F}) = \begin{vmatrix} \mathbf{i} & \mathbf{j} & \mathbf{k} \\ \frac{\partial}{\partial x} & \frac{\partial}{\partial y} & \frac{\partial}{\partial z} \\ y & 0 & -z \end{vmatrix}$$

$$= \left(\frac{\partial}{\partial y}(-z) - \frac{\partial}{\partial z}(0)\right)\mathbf{i} - \left(\frac{\partial}{\partial x}(-z) - \frac{\partial}{\partial z}(y)\right)\mathbf{j} + \left(\frac{\partial}{\partial x}(0) - \frac{\partial}{\partial y}(y)\right)\mathbf{k}$$

$$= 0\mathbf{i} + 0\mathbf{j} - 1\mathbf{k} = -\mathbf{k}$$

13. $\mathbf{F} = \nabla(e^{-x^2 - y^2 - z^2})$

SOLUTION Since **F** is the gradient of a function, its curl is zero. To compute its divergence, we have

$$\mathbf{F} = \nabla e^{-x^2 - y^2 - z^2} = \left\langle \frac{\partial}{\partial x} e^{-x^2 - y^2 - z^2}, \frac{\partial}{\partial y} e^{-x^2 - y^2 - z^2}, \frac{\partial}{\partial z} e^{-x^2 - y^2 - z^2} \right\rangle = -\left\langle 2xe^{-x^2 - y^2 - z^2}, 2ye^{-x^2 - y^2 - z^2}, 2ze^{-x^2 - y^2 - z^2} \right\rangle$$

Then

$$\text{div}\mathbf{F} = \frac{\partial}{\partial x}\left(-2xe^{-x^2 - y^2 - z^2}\right) + \frac{\partial}{\partial y}\left(-2ye^{-x^2 - y^2 - z^2}\right) + \frac{\partial}{\partial z}\left(-2ze^{-x^2 - y^2 - z^2}\right)$$

$$= -2e^{-x^2 - y^2 - z^2} + 4x^2 e^{-x^2 - y^2 - z^2} - 2e^{-x^2 - y^2 - z^2} + 4y^2 e^{-x^2 - y^2 - z^2} - 2e^{-x^2 - y^2 - z^2} + 4z^2 e^{-x^2 - y^2 - z^2}$$

$$= e^{-x^2 - y^2 - z^2}(4x^2 + 4y^2 + 4z^2 - 6)$$

15. Show that if F_1, F_2, and F_3 are differentiable functions of one variable, then
$$\text{curl}\,(\langle F_1(x), F_2(y), F_3(z)\rangle) = \mathbf{0}$$
Use this to calculate the curl of
$$\mathbf{F}(x,y,z) = \langle x^2 + y^2, \ln y + z^2, z^3 \sin(z^2)e^{z^3}\rangle$$

SOLUTION To prove the stated property, simply compute the curl of the given vector field:

$$\text{curl}\,(\langle F_1(x), F_2(y), F_3(z)\rangle) = \begin{vmatrix} \mathbf{i} & \mathbf{j} & \mathbf{k} \\ \dfrac{\partial}{\partial x} & \dfrac{\partial}{\partial y} & \dfrac{\partial}{\partial z} \\ F_1(x) & F_2(y) & F_3(z) \end{vmatrix}$$

$$= \left(\dfrac{\partial}{\partial y}(F_3(z)) - \dfrac{\partial}{\partial z}(F_2(y))\right)\mathbf{i} - \left(\dfrac{\partial}{\partial x}(F_3(z)) - \dfrac{\partial}{\partial z}(F_1(x))\right)\mathbf{j}$$
$$+ \left(\dfrac{\partial}{\partial x}(F_2(y)) - \dfrac{\partial}{\partial y}(F_1(x))\right)\mathbf{k}$$
$$= \mathbf{0}$$

Now use the linearity of the curl and the property above to compute the curl of $\mathbf{F}$:

$$\text{curl}\,\mathbf{F} = \text{curl}\,\langle x^2 + y^2, \ln y + z^2, z^3 \sin(z^2)e^{z^3}\rangle$$
$$= \text{curl}\,\langle x^2, \ln y, z^3 \sin(z^2)e^{z^3}\rangle + \text{curl}\,\langle y^2, z^2, 0\rangle$$
$$= \mathbf{0} + \text{curl}\,\langle y^2, z^2, 0\rangle$$
$$= \begin{vmatrix} \mathbf{i} & \mathbf{j} & \mathbf{k} \\ \dfrac{\partial}{\partial x} & \dfrac{\partial}{\partial y} & \dfrac{\partial}{\partial z} \\ y^2 & z^2 & 0 \end{vmatrix}$$
$$= \left\langle \dfrac{\partial}{\partial y}(0) - \dfrac{\partial}{\partial z}(z^2), -\left(\dfrac{\partial}{\partial x}(0) - \dfrac{\partial}{\partial z}(y^2)\right), \dfrac{\partial}{\partial x}z^2 - \dfrac{\partial}{\partial y}y^2 \right\rangle$$
$$= \langle -2z, 0, -2y\rangle$$

17. Verify the identity div(curl$\mathbf{F}$) = 0 for the vector fields $\mathbf{F} = \langle xz, ye^x, yz\rangle$ and $\mathbf{G} = \langle z^2, xy^3, x^2 y\rangle$.

SOLUTION We first show div(curl $\mathbf{F}$)) = 0.

$$\text{curl}\,\mathbf{F} = \begin{vmatrix} \mathbf{i} & \mathbf{j} & \mathbf{k} \\ \dfrac{\partial}{\partial x} & \dfrac{\partial}{\partial y} & \dfrac{\partial}{\partial z} \\ xz & ye^x & yz \end{vmatrix}$$
$$= (z - 0)\mathbf{i} - (0 - x)\mathbf{j} + (ye^x - 0)\mathbf{k} = z\mathbf{i} + x\mathbf{j} + ye^x\mathbf{k}$$

Then
$$\text{div}(\text{curl}\,\mathbf{F}) = (z)_x + (x)_y + (ye^x)_z = 0$$

Likewise, for $\mathbf{G} = \langle z^2, xy^3, x^2 y\rangle$,

$$\text{curl}\,\mathbf{G} = \begin{vmatrix} \mathbf{i} & \mathbf{j} & \mathbf{k} \\ \dfrac{\partial}{\partial x} & \dfrac{\partial}{\partial y} & \dfrac{\partial}{\partial z} \\ z^2 & xy^3 & x^2 y \end{vmatrix}$$
$$= (x^2 - 0)\mathbf{i} - (2xy - 2z)\mathbf{j} + (y^3 - 0)\mathbf{k} = x^2\mathbf{i} + (2z - 2xy)\mathbf{j} + y^3\mathbf{k}$$

Then
$$\text{div}(\text{curl}\,\mathbf{G}) = (x^2)_x + (2z - 2xy)_y + (y^3)_z = 2x - 2x = 0$$

In Exercises 18–26, determine whether the vector field is conservative, and if so, find a potential function.

19. $\mathbf{F}(x, y) = \langle 4x^3y^5, 5x^4y^4 \rangle$

SOLUTION We examine the cross partials of $\mathbf{F}$. We have $F_1 = 4x^3y^5$, $F_2 = 5x^4y^4$, hence:

$$\frac{\partial F_1}{\partial y} = 20x^3y^4$$
$$\frac{\partial F_2}{\partial x} = 20x^3y^4$$
$$\Rightarrow \quad \frac{\partial F_1}{\partial y} = \frac{\partial F_2}{\partial x}$$

Since the cross partials are equal, $\mathbf{F}$ is conservative. We denote the potential field by $V(x, y)$. Then

$$V(x, y) = \int 4x^3y^5 \, dx = x^4y^5 + f(y)$$

$$V(x, y) = \int 5x^4y^4 \, dy = x^4y^5 + g(x)$$

Since V must satisfy both of these equations, we take $f(x) = g(y) = 0$ to find the potential function $V(x, y) = x^4y^5$.

21. $\mathbf{F}(x, y, z) = \langle 2, 4, e^z \rangle$

SOLUTION We examine the cross partials of $\mathbf{F}$. We have, $F_1 = 2$, $F_2 = 4$, $F_3 = e^z$, hence:

$$\frac{\partial F_1}{\partial y} = 0$$
$$\frac{\partial F_2}{\partial x} = 0$$
$$\Rightarrow \quad \frac{\partial F_1}{\partial y} = \frac{\partial F_2}{\partial x}$$

$$\frac{\partial F_2}{\partial z} = 0$$
$$\frac{\partial F_3}{\partial y} = 0$$
$$\Rightarrow \quad \frac{\partial F_2}{\partial z} = \frac{\partial F_3}{\partial y}$$

$$\frac{\partial F_3}{\partial x} = 0$$
$$\frac{\partial F_1}{\partial z} = 0$$
$$\Rightarrow \quad \frac{\partial F_3}{\partial x} = \frac{\partial F_1}{\partial z}$$

Since the cross partials are equal, $\mathbf{F}$ is conservative. We denote the potential field by $V(x, y, z)$. Then

$$V(x, y, z) = \int 2 \, dx = 2x + f(y, z)$$

$$V(x, y, z) = \int 4 \, dy = 4y + g(x, z)$$

$$V(x, y, z) = \int e^z \, dz = e^z + h(x, y)$$

Since V must satisfy all three of these equations, we may take

$$f(y, z) = 4y + e^z, \quad g(x, z) = 2x + e^z, \quad h(x, y) = 2x + 4y$$

so that $V(x, y, z) = 2x + 4y + e^z$ is a potential function.

23. $\mathbf{F}(x, y) = \langle y^4x^3, x^4y^3 \rangle$

SOLUTION Examine the cross partials of $\mathbf{F}$:

$$\frac{\partial F_1}{\partial y} = \frac{\partial}{\partial y}\left(y^4x^3\right) = 4y^3x^3$$

$$\frac{\partial F_2}{\partial x} = \frac{\partial}{\partial x}\left(x^4y^3\right) = 4y^3x^3$$

Since the cross partials are equal, **F** is conservative. We denote the potential field by $V(x, y)$. Then

$$V(x, y) = \int y^4 x^3 \, dx = \frac{1}{4} x^4 y^4 + f(y)$$

$$V(x, y) = \int x^4 y^3 \, dy = \frac{1}{4} x^4 y^4 + g(x)$$

Since V must satisfy both of these equations, we may take $f(y) = g(x) = 0$, so that $V(x, y) = \frac{1}{4} x^4 y^4$ is a potential function.

25. $\mathbf{F}(x, y, z) = \left\langle \dfrac{2xy}{x^2 + z}, \ln(x^2 + z), \dfrac{y}{x^2 + z} \right\rangle$

SOLUTION We examine the cross partials of **F**. Since $F_1 = \frac{2xy}{x^2+z}$, $F_2 = \ln(x^2 + z)$, $F_3 = \frac{y}{x^2+z}$ we have:

$$\frac{\partial F_1}{\partial y} = \frac{2x}{x^2 + z}$$
$$\frac{\partial F_2}{\partial x} = \frac{2x}{x^2 + z}$$
$$\Rightarrow \quad \frac{\partial F_1}{\partial y} = \frac{\partial F_2}{\partial x}$$

$$\frac{\partial F_2}{\partial z} = \frac{1}{x^2 + z}$$
$$\frac{\partial F_3}{\partial y} = \frac{1}{x^2 + z}$$
$$\Rightarrow \quad \frac{\partial F_2}{\partial z} = \frac{\partial F_3}{\partial y}$$

$$\frac{\partial F_3}{\partial x} = -\frac{2xy}{(x^2 + z)^2}$$
$$\frac{\partial F_1}{\partial z} = -\frac{2xy}{(x^2 + z)^2}$$
$$\Rightarrow \quad \frac{\partial F_3}{\partial x} = \frac{\partial F_1}{\partial z}$$

Since the cross partials are equal, **F** is conservative. We denote the potential field by $V(x, y, z)$. Then

$$V(x, y, z) = \int \frac{2xy}{x^2 + z} \, dx = y \int \frac{2x}{x^2 + z} \, dx = y \ln(x^2 + z) + f(y, z)$$

$$V(x, y, z) = \int \ln(x^2 + z) \, dy = y \ln(x^2 + z) + g(x, z)$$

$$V(x, y, z) = \int \frac{y}{x^2 + z} \, dz = y \ln(x^2 + z) + h(x, y)$$

Since V must satisfy all three of these equations, we may take $f(y, z) = g(x, z) = h(x, y) = 0$. Then $V(x, y, z) = y \ln(x^2 + z)$ is a potential function.

27. Find a conservative vector field of the form $\mathbf{F} = \langle g(y), h(x) \rangle$ such that $\mathbf{F}(0, 0) = \langle 1, 1 \rangle$, where $g(y)$ and $h(x)$ are differentiable functions. Determine all such vector fields.

SOLUTION We need to find a scalar function $V(x, y)$ such that $\mathbf{F} = \nabla V$. That is:

$$\frac{\partial V}{\partial x} = g(y) \quad (1)$$

$$\frac{\partial V}{\partial y} = h(x) \quad (2)$$

Integrating (1) with respect to x, treating y as a constant, we get:

$$V(x, y) = xg(y) + f(y) \quad (3)$$

We differentiate $V(x, y)$ with respect to y and equate with (2). This gives:

$$\frac{\partial V}{\partial y} = xg'(y) + f'(y) = h(x)$$

This equation can hold only when $g'(y)$ and $f'(y)$ are constants. That is, $g'(y) = c_1$ and $f'(y) = c_2$, yielding $g(y) = c_1 y + d_1$ and $f(y) = c_2 y + d_2$. Substituting in (3) gives:

$$V(x, y) = x(c_1 y + d_1) + c_2 y + d_2$$

Or
$$V(x, y) = d_1 x + c_2 y + c_1 xy + d_2 \quad \Rightarrow \quad \nabla V = \langle d_1 + c_1 y, c_2 + c_1 x \rangle$$

We conclude that $\mathbf{F} = \langle g(y), h(x) \rangle$ are gradient vector fields if and only if $g(y) = a + by$ and $h(x) = c + bx$ for any constants a, b, c. That is:
$$\mathbf{F} = \langle a + by, c + bx \rangle$$

We also want that $\mathbf{F}(0, 0) = \langle 1, 1 \rangle$, hence:
$$\mathbf{F}(0, 0) = \langle a, c \rangle = \langle 1, 1 \rangle \quad \Rightarrow \quad a = c = 1$$

Therefore, all the gradient field of the form $\mathbf{F} = \langle g(y), h(x) \rangle$ such that $\mathbf{F}(0, 0) = \langle 1, 1 \rangle$ are:
$$\mathbf{F} = \langle 1 + by, 1 + bx \rangle$$

In Exercises 28–31, compute the line integral $\int_C f(x, y)\, ds$ for the given function and path or curve.

29. $f(x, y) = x - y$, the unit semicircle $x^2 + y^2 = 1, y \geq 0$

SOLUTION

The semi unit circle above the x-axis is parametrized by:
$$\mathbf{r}(\theta) = \langle \cos \theta, \sin \theta \rangle, 0 \leq \theta \leq \pi$$

Step 1. Compute $ds = \|\mathbf{r}'(\theta)\|\, d\theta$. We have:
$$\mathbf{r}'(\theta) = \frac{d}{d\theta} \langle \cos \theta, \sin \theta \rangle = \langle -\sin \theta, \cos \theta \rangle$$
$$\|\mathbf{r}'(\theta)\| = \sqrt{(-\sin \theta)^2 + \cos^2 \theta} = 1$$

Hence:
$$ds = \|\mathbf{r}'(\theta)\|\, d\theta = d\theta$$

Step 2. Write out $f(\mathbf{r}(\theta))$ and evaluate the integral.
$$f(\mathbf{r}(\theta)) = x - y = \cos \theta - \sin \theta$$

We use the Theorem on Scalar Line Integrals to obtain:
$$\int_C f(x, y)\, ds = \int_0^\pi f(\mathbf{r}(\theta))\, \|\mathbf{r}'(\theta)\|\, dt = \int_0^\pi (\cos \theta - \sin \theta)\, d\theta$$
$$= \sin \theta + \cos \theta \Big|_0^\pi = (0 - 1) - (0 + 1) = -2$$

31. $f(x, y, z) = x + 2y + z$, the helix $\mathbf{r}(t) = \langle \cos t, \sin t, t \rangle$ for $0 \leq t \leq \pi/2$

SOLUTION We have:
$$\mathbf{r}'(t) = \frac{d}{dt} \langle \cos t, \sin t, t \rangle = \langle -\sin t, \cos t, 1 \rangle$$
$$\|\mathbf{r}'(t)\| = \sqrt{(-\sin t)^2 + \cos^2 t + 1} = \sqrt{1 + 1} = \sqrt{2}$$

We write out $f(\mathbf{r}(t))$:
$$f(\mathbf{r}(t)) = x + 2y + z = \cos t + 2 \sin t + t$$

The scalar line integral is the following integral:

$$\int_C f(x,y,z)\,ds = \int_{-1}^{3} f(\mathbf{r}(t))\,\|\mathbf{r}'(t)\|\,dt = \int_{-1}^{3} (\cos t + 2\sin t + t)\sqrt{2}\,dt$$

$$= \sqrt{2}\left(\sin t - 2\cos t + \frac{t^2}{2}\right)\Big|_{-1}^{3}$$

$$= \sqrt{2}\left(\left(\sin 3 - 2\cos 3 + \frac{9}{2}\right) - \left(\sin(-1) - 2\cos(-1) + \frac{1}{2}\right)\right)$$

$$= \sqrt{2}(\sin 3 - 2\cos 3 + \sin 1 + 2\cos 1 + 4) \approx 11.375$$

33. Calculate $\mathbf{F} = \nabla f$, where $f(x,y,z) = xye^z$, and compute $\int_C \mathbf{F}\cdot d\mathbf{r}$, where:

(a) C is any curve from $(1,1,0)$ to $(3,e,-1)$.

(b) C is the boundary of the square $0 \le x \le 1$, $0 \le y \le 1$ oriented counterclockwise.

SOLUTION The gradient of $V(x,y,z) = xye^z$ is the following vector:

$$\mathbf{F} = \left\langle \frac{\partial V}{\partial x}, \frac{\partial V}{\partial y}, \frac{\partial V}{\partial z} \right\rangle = \langle ye^z, xe^z, xye^z \rangle$$

(a) By the Fundamental Theorem for Gradient Vector Fields, we have:

$$\int_C \mathbf{F}\cdot d\mathbf{r} = \int_C \nabla V \cdot d\mathbf{r} = V(3,e,-1) - V(1,1,0) = 3\cdot e\cdot e^{-1} - 1\cdot 1 e^0 = 3 - 1 = 2$$

(b) Since $\mathbf{F}$ is the gradient of a function, $\mathbf{F}$ is conservative. That is, the line integral of $\mathbf{F}$ over any closed curve is zero. Therefore, the line integral $\int_C \mathbf{F}\cdot d\mathbf{r}$ is zero, where C is the boundary of a square.

35. Let $\mathbf{F}(x,y) = \langle 9y - y^3, e^{\sqrt{y}}(x^2 - 3x)\rangle$, and let C_2 be the oriented curve in Figure 1(B).

(a) Show that $\mathbf{F}$ is not conservative.

(b) Show that $\int_{C_2} \mathbf{F}\cdot d\mathbf{r} = 0$ without explicitly computing the integral. *Hint:* Show that $\mathbf{F}$ is orthogonal to the edges along the square.

SOLUTION

(a) We show that the cross partials of $\mathbf{F}(x,y) = \langle 9y - y^3, e^{\sqrt{y}}(x^2 - 3x)\rangle$ are not equal.

We have $F_1 = 9y - y^3$ and $F_2 = e^{\sqrt{y}}(x^2 - 3x)$, therefore:

$$\frac{\partial F_1}{\partial y} = 9 - 3y^2$$

$$\frac{\partial F_2}{\partial x} = e^{\sqrt{y}}(2x - 3)$$

The cross partials are not equal, hence $\mathbf{F}$ is not conservative.

(b) On $\overline{OA}$, $y = 0$ hence

$$\mathbf{F} = \mathbf{F}(x,0) = \langle 0, x^2 - 3x \rangle$$

Therefore the tangential component of $\mathbf{F}$ along the segment $\overline{OA}$ is zero, resulting in $\int_{\overline{OA}} \mathbf{F}\cdot d\mathbf{r} = 0$. On $\overline{AB}$, $x = 3$ hence

$$\mathbf{F}(3,y) = \langle 9y - y^3, e^{\sqrt{y}}(3^2 - 3\cdot 3)\rangle = \langle 9y - y^3, 0\rangle$$

We see that **F** is orthogonal to the segment AB, resulting in $\int_{\overline{AB}} \mathbf{F} \cdot d\mathbf{r} = 0$. On $\overline{BC}$, $y = 3$ hence

$$\mathbf{F}(x, 3) = \left\langle 9 \cdot 3 - 3^3, e^{\sqrt{3}}\left(x^2 - 3x\right) \right\rangle = \left\langle 0, e^{\sqrt{3}}\left(x^2 - 3x\right) \right\rangle$$

We see that **F** is orthogonal to the segment $\overline{BC}$, therefore $\int_{\overline{BC}} \mathbf{F} \cdot d\mathbf{r} = 0$. Finally, on CO we have $x = 0$, hence $\mathbf{F}(0, y) = \left\langle 9y - y^3, 0 \right\rangle$. It follows that **F** is orthogonal to the segment $\overline{CO}$, therefore the tangential component of **F** along $\overline{CO}$ is zero. We conclude that $\int_{\overline{CO}} \mathbf{F} \cdot d\mathbf{r} = 0$. Combining these integrals we conclude that:

$$\int_{C_2} \mathbf{F} \cdot d\mathbf{r} = \int_{\overline{OA}} \mathbf{F} \cdot d\mathbf{r} + \int_{\overline{AB}} \mathbf{F} \cdot d\mathbf{r} + \int_{\overline{BC}} \mathbf{F} \cdot d\mathbf{r} + \int_{\overline{CO}} \mathbf{F} \cdot d\mathbf{r} = 0 + 0 + 0 + 0 = 0$$

In Exercises 36–39, compute the line integral $\int_c \mathbf{F} \cdot d\mathbf{r}$ for the given vector field and path.

37. $\mathbf{F}(x, y) = \langle 2xy, x^2 + y^2 \rangle$, the part of the unit circle in the first quadrant oriented counterclockwise

SOLUTION

The path is parametrized by:

$$\mathbf{r}(\theta) = \langle \cos\theta, \sin\theta \rangle, \quad 0 \le \theta \le \frac{\pi}{2}$$

Step 1. Calculate the integrand $\mathbf{F}(\mathbf{r}(\theta)) \cdot \mathbf{r}'(\theta)$.

$$\mathbf{r}'(\theta) = \langle -\sin\theta, \cos\theta \rangle$$

$$\mathbf{F}(\mathbf{r}(\theta)) = \langle 2xy, x^2 + y^2 \rangle = \langle 2\cos\theta\sin\theta, \cos^2\theta + \sin^2\theta \rangle = \langle 2\cos\theta\sin\theta, 1 \rangle$$

The integrand is the dot product:

$$\mathbf{F}(\mathbf{r}(\theta)) \cdot \mathbf{r}'(\theta) = \langle 2\cos\theta\sin\theta, 1 \rangle \cdot \langle -\sin\theta, \cos\theta \rangle = -2\cos\theta\sin^2\theta + \cos\theta$$

Step 2. Evaluate the line integral. The vector line integral is:

$$\int_C \mathbf{F} \cdot d\mathbf{r} = \int_0^{\pi/2} \mathbf{F}(\mathbf{r}(\theta)) \cdot \mathbf{r}'(\theta)\, d\theta = \int_0^{\pi/2} \left(-2\cos\theta\sin^2\theta + \cos\theta\right) d\theta$$

$$= -2\int_0^{\pi/2} \sin^2\theta \cos\theta\, d\theta + \int_0^{\pi/2} \cos\theta\, d\theta = \frac{-2\sin^3\theta}{3}\bigg|_0^{\pi/2} + \sin\theta\bigg|_0^{\pi/2} = -\frac{2}{3} + 1 = \frac{1}{3}$$

39. $\mathbf{F} = \nabla f$, where $f(x, y, z) = 4x^2 \ln(1 + y^4 + z^2)$, the path $\mathbf{r}(t) = \left\langle t^3, \ln(1 + t^2), e^t \right\rangle$ for $0 \le t \le 1$

SOLUTION We use the Fundamental Theorem for Conservative Vector Fields to write:

$$\int_C \nabla f \cdot d\mathbf{r} = f(c(1)) - f(c(0)) \tag{1}$$

We compute the values on the right-hand side:

$$c(1) = \left(1^3, \ln(1 + 1^2), e^1\right) = (1, \ln 2, e)$$

$$c(0) = \left(0^3, \ln(1 + 0^2), e^0\right) = (0, 0, 1)$$

Hence:

$$f(c(1)) = 4 \cdot 1^2 \ln\left(1 + \ln^4 2 + e^2\right) = 4\ln\left(1 + e^2 + \ln^4 2\right)$$

$$f(c(0)) = 0$$

Combining with (1) gives:

$$\int_C \nabla f \cdot d\mathbf{r} = 4\ln\left(1 + e^2 + \ln^4 2\right)$$

41. Calculate the work required to move an object from $P = (1, 1, 1)$ to $Q = (3, -4, -2)$ against the force field $\mathbf{F}(x, y, z) = -12r^{-4}\langle x, y, z\rangle$ (distance in meters, force in newtons), where $r = \sqrt{x^2 + y^2 + z^2}$. *Hint:* Find a potential function for $\mathbf{F}$.

SOLUTION The work performed against $\mathbf{F}$ is given by the line integral:

$$W = -\int_{PQ} \mathbf{F} \cdot d\mathbf{r} \tag{1}$$

We notice that $\mathbf{F} = -\frac{12}{(x^2+y^2+z^2)^2}\langle x, y, z\rangle$ is the gradient of the function $V(x, y, z) = \frac{6}{x^2+y^2+z^2}$ since:

$$\nabla V = \left\langle \frac{\partial V}{\partial x}, \frac{\partial V}{\partial y}, \frac{\partial V}{\partial z}\right\rangle = \left\langle \frac{6 \cdot (-2x)}{(x^2+y^2+z^2)^2}, \frac{6 \cdot (-2y)}{(x^2+y^2+z^2)^2}, \frac{6 \cdot (-2z)}{(x^2+y^2+z^2)^2}\right\rangle$$

$$= -\frac{12}{(x^2+y^2+z^2)^2}\langle x, y, z\rangle = \mathbf{F}$$

We now use the Fundamental Theorem for Gradient Vector Field to compute the line integral (1):

$$W = -\int_{PQ} \mathbf{F} \cdot d\mathbf{r} = -\int_{PQ} \nabla V \cdot d\mathbf{r} = -(V(Q) - V(P)) = -(V(3, -4, -2) - V(1, 1, 1))$$

$$= -\left(\frac{6}{9 + 16 + 4} - \frac{6}{1 + 1 + 1}\right) = \frac{52}{29} \approx 1.79$$

43. Calculate the integral of $f(x, y, z) = e^z$ over the portion of the plane $x + 2y + 2z = 3$, where $x, y, z \geq 0$.

SOLUTION We consider the surface a graph of the form $x = g(y, z) = 3 - 2y - 2z$. The requirement that $x \geq 0$ means $3 - 2y - 2z \geq 0 \Rightarrow y + z \leq \frac{3}{2}$. Combined with the requirements that $y, z \geq 0$ we have that our parameter domain, $\mathcal{D}$, is the triangle bounded by the y-axis, the z-axis, and the line $y + z = \frac{3}{2}$.

Calculating the magnitude of the normal vector,

$$\Phi(y, z) = (3 - 2y - 2z, y, z) \quad (x, y) \in \mathcal{D}$$
$$\mathbf{T}_y = \langle -2, 1, 0\rangle$$
$$\mathbf{T}_z = \langle -2, 0, 1\rangle$$
$$\mathbf{n} = \langle 1, 2, 2\rangle \quad \Rightarrow \quad \|\mathbf{n}\| = \sqrt{1^2 + 2^2 + 2^2} = 3$$

Our integral is

$$\iint_S f(x, y, z)\, dS = \iint_\mathcal{D} e^z \|\mathbf{n}\|\, dz\, dy = \int_0^{3/2} \int_0^{-y+3/2} e^z\, 3\, dz\, dy$$

$$= \int_0^{3/2} \left(e^{-y+3/2} - 1\right) 3\, dy = -3e^{-y+3/2} - 3y \Big|_0^{3/2}$$

$$= -3 + 3e^{3/2} - \frac{9}{2} = 3e^{3/2} - \frac{15}{2} \approx 5.945$$

45. *CAS* Plot the surface with parametrization

$$\Phi(u, v) = (u + 4v, 2u - v, 5uv)$$

for $-1 \leq v \leq 1$, $-1 \leq u \leq 1$. Express the surface area as a double integral and use a computer algebra system to compute the area numerically.

SOLUTION The surface is shown in the following plot:

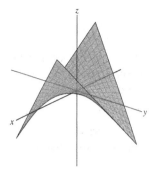

We compute the area of the surface, using the Integral for the Surface Area. We find the tangent and normal vectors:

$$\mathbf{T}_u = \frac{\partial \phi}{\partial u} = \frac{\partial}{\partial u}\langle u + 4v, 2u - v, 5uv\rangle = \langle 1, 2, 5v\rangle$$

$$\mathbf{T}_v = \frac{\partial \phi}{\partial v} = \frac{\partial}{\partial v}\langle u + 4v, 2u - v, 5uv\rangle = \langle 4, -1, 5u\rangle$$

$$\mathbf{N} = \mathbf{T}_u \times \mathbf{T}_v = \begin{vmatrix} \mathbf{i} & \mathbf{j} & \mathbf{k} \\ 1 & 2 & 5v \\ 4 & -1 & 5u \end{vmatrix} = \begin{vmatrix} 2 & 5v \\ -1 & 5u \end{vmatrix}\mathbf{i} - \begin{vmatrix} 1 & 5v \\ 4 & 5u \end{vmatrix}\mathbf{j} + \begin{vmatrix} 1 & 2 \\ 4 & -1 \end{vmatrix}\mathbf{k}$$

$$= (10u + 5v)\mathbf{i} - (5u - 20v)\mathbf{j} + (-1 - 8)\mathbf{k} = \langle 10u + 5v, -5u + 20v, -9\rangle$$

Next, we calculate the length of the normal vector:

$$\|\mathbf{N}\| = \sqrt{(10u + 5v)^2 + (-5u + 20v)^2 + (-9)^2}$$

$$= \sqrt{100u^2 + 100uv + 25v^2 + 25u^2 - 200uv + 400v^2 + 81}$$

$$= \sqrt{125u^2 - 100uv + 425v^2 + 81}$$

We obtain the following integral, which we compute by a CAS:

$$\text{Area}(S) = \iint_{\mathcal{D}} \|\mathbf{N}(u,v)\|\,du\,dv = \int_{-1}^{1}\int_{-1}^{1} \sqrt{125u^2 - 100uv + 425v^2 + 81}\,du\,dv = 62.911$$

47. Evaluate $\iint_S x^2 y\,dS$, where S is the surface $z = \sqrt{3}x + y^2$, $-1 \le x \le 1$, $0 \le y \le 1$

SOLUTION We use the Surface Integral over a Graph with $g(x,y) = \sqrt{3}x + y^2$. The partial derivatives are $g_x = \sqrt{3}$ and $g_y = 2y$. Therefore

$$\sqrt{1 + g_x^2 + g_y^2} = \sqrt{1 + (\sqrt{3})^2 + (2y)^2} = \sqrt{4 + 4y^2} = 2\sqrt{1 + y^2}$$

We obtain the following integral:

$$\iint_S x^2 y \cdot dS = \iint_{\mathcal{D}} x^2 y \sqrt{1 + g_x^2 + g_y^2}\,dx\,dy = \int_0^1 \int_{-1}^1 x^2 y \cdot 2\sqrt{1 + y^2}\,dx\,dy = \left(\int_{-1}^1 x^2\,dx\right)\left(\int_0^1 \sqrt{1 + y^2} \cdot 2y\,dy\right)$$

$$= \left(\frac{x^3}{3}\bigg|_{x=-1}^1\right)\left(\int_0^1 \sqrt{1 + y^2} \cdot 2y\,dy\right) = \frac{2}{3}\int_0^1 \sqrt{1 + y^2} \cdot 2y\,dy$$

We compute the integral using the substitution $u = 1 + y^2$, $du = 2y\,dy$. We get:

$$\iint_S x^2 y \cdot dS = \frac{2}{3}\int_1^2 u^{1/2}\,du = \frac{2}{3}\cdot\frac{2}{3}u^{3/2}\bigg|_1^2 = \frac{4}{9}(2^{3/2} - 1) = \frac{4}{9}(2\sqrt{2} - 1)$$

49. Let S be the upper hemisphere $x^2 + y^2 + z^2 = 1$, $z \ge 0$. For each of the functions (a)–(d), determine whether $\iint_S f\,dS$ is positive, zero, or negative (without evaluating the integral). Explain your reasoning.

(a) $f(x,y,z) = y^3$ \hspace{3em} (b) $f(x,y,z) = z^3$

(c) $f(x,y,z) = xyz$

(d) $f(x,y,z) = z^2 - 2$

SOLUTION

(a) Since $f(x,y,z) = y^3$ is an odd function of y, and the upper hemisphere is symmetric with respect to the (x,z)-plane, the surface integrals over the parts where $y \geq 0$ and $y \leq 0$ cancel each other. Therefore the surface integral is zero.

(b) The function $f(x,y,z) = z^3$ is non-negative on $\mathcal{S}$, hence the surface integral is positive.

(c) Since $f(-x,y,z) = -xyz = -f(x,y,z)$ and since the upper hemisphere is symmetric with respect to (y,z)-plane, the surface integrals over the parts of $\mathcal{S}$ where $x \geq 0$ and $x \leq 0$ cancel each other to obtain a zero surface integral.

(d) On $\mathcal{S}$ we have $z^2 = 1 - x^2 - y^2 \leq 1$, hence $z^2 - 2 < 0$. That is, $f(x,y,z) = z^2 - 2$ is negative on $\mathcal{S}$, therefore the surface integral is negative.

51. The upper half of the sphere $x^2 + y^2 + z^2 = 9$ has parametrization $\Phi(r, \theta) = (r\cos\theta, r\sin\theta, \sqrt{9-r^2})$ in cylindrical coordinates (Figure 3).

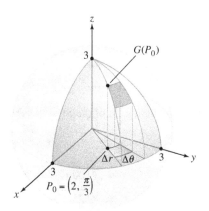

FIGURE 3

(a) Calculate the normal vector $\mathbf{N} = \mathbf{T}_r \times \mathbf{T}_\theta$ at the point $\Phi(2, \frac{\pi}{3})$.

(b) Use Eq. (3) in Section 16.4 to estimate the surface area of $\Phi(\mathcal{R})$, where $\mathcal{R}$ is the small domain defined by

$$2 \leq r \leq 2.1, \qquad \frac{\pi}{3} \leq \theta \leq \frac{\pi}{3} + 0.05$$

SOLUTION

(a) We first find the tangent vectors at the given point:

$$\mathbf{T}_\theta = \frac{\partial \Phi}{\partial \theta} = \frac{\partial}{\partial \theta}\left\langle r\cos\theta, r\sin\theta, \sqrt{9-r^2}\right\rangle = \langle -r\sin\theta, r\cos\theta, 0\rangle$$

$$\Rightarrow \mathbf{T}_\theta\left(2, \frac{\pi}{3}\right) = \left\langle -2\sin\frac{\pi}{3}, 2\cos\frac{\pi}{3}, 0\right\rangle = \langle -\sqrt{3}, 1, 0\rangle$$

$$\mathbf{T}_r = \frac{\partial \Phi}{\partial r} = \frac{\partial}{\partial r}\left\langle r\cos\theta, r\sin\theta, \sqrt{9-r^2}\right\rangle = \left\langle \cos\theta, \sin\theta, -\frac{r}{\sqrt{9-r^2}}\right\rangle$$

$$\Rightarrow \mathbf{T}_r\left(2, \frac{\pi}{3}\right) = \left\langle \cos\frac{\pi}{3}, \sin\frac{\pi}{3}, -\frac{2}{\sqrt{9-2^2}}\right\rangle = \left\langle \frac{1}{2}, \frac{\sqrt{3}}{2}, -\frac{2}{\sqrt{5}}\right\rangle$$

The normal vector is the cross product:

$$\mathbf{N} = \mathbf{T}_r \times \mathbf{T}_\theta = \begin{vmatrix} \mathbf{i} & \mathbf{j} & \mathbf{k} \\ \frac{1}{2} & \frac{\sqrt{3}}{2} & -\frac{2}{\sqrt{5}} \\ -\sqrt{3} & 1 & 0 \end{vmatrix} = \frac{2}{\sqrt{5}}\mathbf{i} + \frac{2\sqrt{3}}{\sqrt{5}}\mathbf{j} + \left(\frac{1}{2} + \frac{3}{2}\right)\mathbf{k} = \left\langle \frac{2}{\sqrt{5}}, \frac{2\sqrt{3}}{\sqrt{5}}, 2\right\rangle$$

(b) We use Eq. (1) with the sample point $(u, v) = \left(2, \frac{\pi}{3}\right)$.

In part (a) we found that the normal to $\Phi(\mathcal{R})$ at this point is $\mathbf{N} = \left\langle \frac{2}{\sqrt{5}}, \frac{2\sqrt{3}}{\sqrt{5}}, 2 \right\rangle$. Therefore, $\|\mathbf{N}\| = \sqrt{\frac{4}{5} + \frac{12}{5} + 4} = \frac{6}{\sqrt{5}}$. We get:

$$\text{Area}(\Phi(\mathcal{R})) \approx \left\| \mathbf{N}\left(2, \frac{\pi}{3}\right) \right\| \text{Area}(\mathcal{R}) = \frac{6}{\sqrt{5}} \cdot 0.1 \cdot 0.05 = \frac{0.03}{\sqrt{5}} \approx 0.0134$$

In Exercises 52–57, compute $\iint_S \mathbf{F} \cdot d\mathbf{S}$ for the given oriented surface or parametrized surface.

53. $\mathbf{F}(x, y, z) = \langle -y, z, -x \rangle$, $\Phi(u, v) = (u + 3v, v - 2u, 2v + 5)$, $0 \le u \le 1, 0 \le v \le 1$, upward-pointing normal

SOLUTION

Step 1. Compute the tangent and normal vectors.

$$\mathbf{T}_u = \frac{\partial \Phi}{\partial u} = \frac{\partial}{\partial u} \langle u + 3v, v - 2u, 2v + 5 \rangle = \langle 1, -2, 0 \rangle$$

$$\mathbf{T}_v = \frac{\partial \Phi}{\partial v} = \frac{\partial}{\partial v} \langle u + 3v, v - 2u, 2v + 5 \rangle = \langle 3, 1, 2 \rangle$$

$$\mathbf{T}_u \times \mathbf{T}_v = \begin{vmatrix} \mathbf{i} & \mathbf{j} & \mathbf{k} \\ 1 & -2 & 0 \\ 3 & 1 & 2 \end{vmatrix} = \begin{vmatrix} -2 & 0 \\ 1 & 2 \end{vmatrix} \mathbf{i} - \begin{vmatrix} 1 & 0 \\ 3 & 2 \end{vmatrix} \mathbf{j} + \begin{vmatrix} 1 & -2 \\ 3 & 1 \end{vmatrix} \mathbf{k} = -4\mathbf{i} - 2\mathbf{j} + 7\mathbf{k} = \langle -4, -2, 7 \rangle$$

Since the normal is pointing upward, the z-component is positive. Therefore the normal vector is $\mathbf{N} = \langle -4, -2, 7 \rangle$.

Step 2. Evaluate the dot product $\mathbf{F} \cdot \mathbf{N}$. We write $\mathbf{F}$ in terms of the parameters. Since $x = u + 3v$, $y = v - 2u$, $z = 2v + 5$, we have:

$$\mathbf{F}(\Phi(u, v)) = \langle -y, z, -x \rangle = \langle 2u - v, 2v + 5, -u - 3v \rangle$$

The dot product is, thus:

$$\mathbf{F}(\Phi(u, v)) \cdot \mathbf{N} = \langle 2u - v, 2v + 5, -u - 3v \rangle \cdot \langle -4, -2, 7 \rangle$$
$$= -4(2u - v) - 2(2v + 5) + 7(-u - 3v) = -15 - 21v - 10$$

Step 3. Evaluate the surface integral. The surface integral is equal to the following double integral:

$$\iint_S \mathbf{F} \cdot d\mathbf{S} = \iint_D \mathbf{F}(\Phi(u, v)) \cdot \mathbf{N} \, du \, dv = \int_0^1 \int_0^1 (-15u - 21v - 10) \, du \, dv$$

$$= \int_0^1 -\frac{15u^2}{2} - 21vu - 10u \bigg|_{u=0}^1 dv = \int_0^1 \left(-\frac{35}{2} - 21v \right) dv$$

$$= -\frac{35v}{2} - \frac{21v^2}{2} \bigg|_0^1 = -\frac{35}{2} - \frac{21}{2} = -28$$

55. $\mathbf{F}(x, y, z) = \langle z, 0, z^2 \rangle$, $\Phi(u, v) = (v \cosh u, v \sinh u, v)$, $0 \le u \le 1, 0 \le v \le 1$, upward-pointing normal

SOLUTION

Step 1. Compute the tangent and normal vectors.

$$\mathbf{T}_u = \frac{\partial \Phi}{\partial u} = \frac{\partial}{\partial u} \langle v \cosh(u), v \sinh(u), v \rangle = \langle v \sinh(u), v \cosh(u), 0 \rangle$$

$$\mathbf{T}_v = \frac{\partial \Phi}{\partial v} = \frac{\partial}{\partial v} \langle v\cosh(u), v\sinh(u), v \rangle = \langle \cosh(u), \sinh(u), 1 \rangle$$

$$\mathbf{T}_u \times \mathbf{T}_v = \begin{vmatrix} \mathbf{i} & \mathbf{j} & \mathbf{k} \\ v\sinh(u) & v\cosh(u) & 0 \\ \cosh(u) & \sinh(u) & 1 \end{vmatrix}$$

$$= \begin{vmatrix} v\cosh(u) & 0 \\ \sinh(u) & 1 \end{vmatrix} \mathbf{i} - \begin{vmatrix} v\sinh(u) & 0 \\ \cosh(u) & 1 \end{vmatrix} \mathbf{j} + \begin{vmatrix} v\sinh(u) & v\cosh(u) \\ \cosh(u) & \sinh(u) \end{vmatrix} \mathbf{k}$$

$$= (v\cosh(u))\mathbf{i} - (v\sinh(u))\mathbf{j} + \left(v\sinh^2(u) - v\cosh^2(u)\right)\mathbf{k}$$

$$= \langle v\cosh(u), -v\sinh(u), -v \rangle$$

Since the normal points upward, the z-component is positive. Therefore the normal vector is (notice that $v \geq 0$):

$$\mathbf{N} = \langle -v\cosh(u), v\sinh(u), v \rangle$$

Step 2. Evaluate the dot product $\mathbf{F} \cdot \mathbf{N}$. We write $\mathbf{F}$ in terms of the parameters:

$$\mathbf{F}(\Phi(u,v)) = \langle z, 0, z^2 \rangle = \langle v, 0, v^2 \rangle$$

Hence:

$$\mathbf{F}(\Phi(u,v)) \cdot \mathbf{N}(u,v) = \langle v, 0, v^2 \rangle \cdot \langle -v\cosh(u), v\sinh(u), v \rangle = -v^2\cosh(u) + v^3$$

Step 3. Evaluate the surface integral. The surface integral is equal to the following double integral:

$$\iint_S \mathbf{F} \cdot d\mathbf{S} = \iint_{\mathcal{D}} \mathbf{F}(\Phi(u,v)) \cdot \mathbf{N} \, du \, dv = \int_0^1 \int_0^1 \left(-v^2 \cosh(u) + v^3\right) du \, dv$$

$$= \int_0^1 \int_0^1 -v^2 \cosh(u) \, du \, dv + \int_0^1 \int_0^1 v^3 \, du \, dv = \left(\int_0^1 -v^2 \, dv\right)\left(\int_0^1 \cosh(u)\, du\right) + \int_0^1 v^3 \, dv$$

$$= \left(-\frac{v^3}{3}\bigg|_{v=0}^1\right)\left(\sinh(u)\bigg|_{u=0}^1\right) + \frac{v^4}{4}\bigg|_{v=0}^1 = -\frac{1}{3}\sinh(1) + \frac{1}{4} \approx -0.1417$$

57. $\mathbf{F}(x,y,z) = \langle 0, 0, z \rangle$, $3x^2 + 2y^2 + z^2 = 1$, $z \geq 0$,
upward-pointing normal

SOLUTION We use modified spherical coordinates to parametrize the ellipsoid:

$$\Phi(\theta, \phi) = \left(\frac{1}{\sqrt{3}} \cos\theta \sin\phi, \frac{1}{\sqrt{2}} \sin\theta \sin\phi, \cos\phi\right)$$

where the parameter domain is:

$$\mathcal{D} = \left\{(\theta, \phi) : 0 \leq \theta \leq 2\pi, 0 \leq \phi \leq \frac{\pi}{2}\right\}$$

Step 1. Compute the tangent and normal vectors.

$$\mathbf{T}_\theta = \frac{\partial \Phi}{\partial \theta} = \left\langle -\frac{1}{\sqrt{3}} \sin\theta \sin\phi, \frac{1}{\sqrt{2}} \cos\theta \sin\phi, 0 \right\rangle$$

$$\mathbf{T}_\phi = \frac{\partial \Phi}{\partial \phi} = \left\langle \frac{1}{\sqrt{3}} \cos\theta \cos\phi, \frac{1}{\sqrt{2}} \sin\theta \cos\phi, -\sin\phi \right\rangle$$

$$\mathbf{T}_\theta \times \mathbf{T}_\phi = \begin{vmatrix} \mathbf{i} & \mathbf{j} & \mathbf{k} \\ -\frac{1}{\sqrt{3}} \sin\theta \sin\phi & \frac{1}{\sqrt{2}} \cos\theta \sin\phi & 0 \\ \frac{1}{\sqrt{3}} \cos\theta \cos\phi & \frac{1}{\sqrt{2}} \sin\theta \cos\phi & -\sin\phi \end{vmatrix}$$

$$= \left(-\frac{1}{\sqrt{2}} \cos\theta \sin^2\phi\right) \mathbf{i} - \left(\frac{1}{\sqrt{3}} \sin\theta \sin^2\phi\right) \mathbf{j} - \left(\frac{1}{\sqrt{6}} \sin^2\theta \sin\phi \cos\phi + \frac{1}{\sqrt{6}} \cos^2\theta \sin\phi \cos\phi\right) \mathbf{k}$$

$$= \left(-\frac{1}{\sqrt{2}} \cos\theta \sin^2\phi\right) \mathbf{i} - \left(\frac{1}{\sqrt{3}} \sin\theta \sin^2\phi\right) \mathbf{j} - \left(\frac{1}{\sqrt{6}} \sin\phi \cos\phi\right) \mathbf{k}$$

Since the normal points upward, the z-component is positive. For $0 \leq \phi \leq \frac{\pi}{2}$, $\sin\phi\cos\phi \geq 0$, therefore the normal vector is:

$$\mathbf{N} = \left\langle \frac{1}{\sqrt{2}} \cos\theta \sin^2\phi, \frac{1}{\sqrt{3}} \sin\theta \sin^2\phi, \frac{1}{\sqrt{6}} \sin\phi \cos\phi \right\rangle$$

Step 2. Compute the dot product $\mathbf{F} \cdot \mathbf{N}$. We have:

$$\mathbf{F}(\Phi(\theta,\phi)) = \langle 0, 0, \cos\phi \rangle \cdot \left\langle \frac{1}{\sqrt{2}} \cos\theta \sin^2\phi, \frac{1}{\sqrt{3}} \sin\theta \sin^2\phi, \frac{1}{\sqrt{6}} \sin\phi \cos\phi \right\rangle = \frac{1}{\sqrt{6}} \sin\theta \cos^2\phi$$

Step 3. Evaluate the surface integral. The surface integral is equal to the following double integral:

$$\iint_S \mathbf{F} \cdot d\mathbf{S} = \iint_\mathcal{D} \mathbf{F}(\Phi(\theta,\phi)) \cdot \mathbf{N}(\theta,\phi)\, d\phi\, d\theta = \int_0^{2\pi} \int_0^{\pi/2} \frac{1}{\sqrt{6}} \sin\phi \cos^2\phi\, d\phi\, d\theta$$

$$= \frac{2\pi}{\sqrt{6}} \int_0^{\pi/2} \cos^2\phi \sin\phi\, d\phi = \frac{2\pi}{\sqrt{6}} \left(-\frac{\cos^3\phi}{3} \Big|_{\phi=0}^{\pi/2} \right) = \frac{2\pi}{\sqrt{6}} \left(\frac{0+1}{3} \right) = \frac{2\pi}{3\sqrt{6}} = \frac{\sqrt{2}}{3\sqrt{3}} \pi$$

59. Find the flow rate of a fluid with velocity field $\mathbf{v} = \langle 2x, y, xy \rangle$ m/s across the part of the cylinder $x^2 + y^2 = 9$ where $x \geq 0, y \geq 0$, and $0 \leq z \leq 4$ (distance in meters).

SOLUTION The flow rate of a fluid with velocity field $\mathbf{v}$ through the part S of the cylinder is the surface integral:

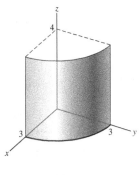

$$\iint_S \mathbf{v} \cdot d\mathbf{S} \tag{1}$$

We parametrize S by,

$$\Phi(\theta, z) = (3\cos\theta, 3\sin\theta, z), \quad 0 \leq \theta \leq \frac{\pi}{2}, \quad 0 \leq z \leq 4$$

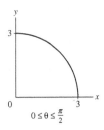

Step 1. Compute the tangent and normal vectors.

$$\mathbf{T}_\theta = \frac{\partial \Phi}{\partial \theta} = \frac{\partial}{\partial \theta} \langle 3\cos\theta, 3\sin\theta, z \rangle = \langle -3\sin\theta, 3\cos\theta, 0 \rangle = (-3\sin\theta)\mathbf{i} + (3\cos\theta)\mathbf{j}$$

$$\mathbf{T}_z = \frac{\partial \Phi}{\partial z} = \frac{\partial}{\partial z} \langle 3\cos\theta, 3\sin\theta, z \rangle = \langle 0, 0, 1 \rangle = \mathbf{k}$$

$$\mathbf{N} = \mathbf{T}_\theta \times \mathbf{T}_z = ((-3\sin\theta)\mathbf{i} + (3\cos\theta)\mathbf{j}) \times \mathbf{k} = (3\sin\theta)\mathbf{j} + (3\cos\theta)\mathbf{i} = \langle 3\cos\theta, 3\sin\theta, 0 \rangle$$

Step 2. Compute the dot product $\mathbf{v} \cdot \mathbf{N}$.

$$\mathbf{v}(\Phi(\theta,z)) \cdot \mathbf{N} = \langle 2 \cdot 3\cos\theta, 3\sin\theta, 9\cos\theta\sin\theta \rangle \cdot \langle 3\cos\theta, 3\sin\theta, 0 \rangle$$

$$= 18\cos^2\theta + 9\sin^2\theta = 9\left(\cos^2\theta + \sin^2\theta\right) + 9\cos^2\theta = 9\cos^2\theta + 9$$

Step 3. Evaluate the flux of **v**. The flux of **v** in (1) is equal to the following double integral (we use the equality $\cos^2\theta = \frac{1}{2} + \frac{1}{2}\cos 2\theta$):

$$\iint_S \mathbf{v}\cdot d\mathbf{S} = \iint_D \mathbf{v}(\Phi(\theta,z))\cdot \mathbf{N}\, d\theta\, dz = \int_0^4 \int_0^{\pi/2} \left(9\cos^2\theta + 9\right) d\theta\, dz$$

$$= 4\int_0^{\pi/2} \left(9\cos^2\theta + 9\right) d\theta = 36\int_0^{\pi/2} \left(\cos^2\theta + 1\right) d\theta$$

$$= 36\int_0^{\pi/2} \left(\frac{3}{2} + \frac{1}{2}\cos 2\theta\right) d\theta = 36\left(\frac{3}{2}\theta + \frac{1}{4}\sin 2\theta\right)\Big|_{\theta=0}^{\pi/2} = 27\pi$$

61. Calculate the flux of the vector field $\mathbf{E}(x,y,z) = \langle 0,0,x\rangle$ through the part of the ellipsoid

$$4x^2 + 9y^2 + z^2 = 36$$

where $z \geq 3$, $x \geq 0$, $y \geq 0$. *Hint:* Use the parametrization

$$\Phi(r,\theta) = \left(3r\cos\theta, 2r\sin\theta, 6\sqrt{1-r^2}\right)$$

SOLUTION The flux of the vector field **E** through the part S of the ellipsoid is the surface integral:

$$\iint_S \mathbf{E}\cdot d\mathbf{S} \tag{1}$$

We use the following parametrization for S:

$$\Phi(\theta,r) = \left(3r\cos\theta, 2r\sin\theta, 6\sqrt{1-r^2}\right)$$

With the parameter domain:

$$\mathcal{D} = \left\{(\theta,r): 0 \leq \theta \leq 2\pi,\ 0 \leq r \leq \frac{\sqrt{3}}{2}\right\}$$

We justify this parametrization. First, we verify that $x = 3r\cos\theta$, $y = 2r\sin\theta$, and $z = 6\sqrt{1-r^2}$ satisfy the equation of the ellipsoid $4x^2 + 9y^2 + z^2 = 36$.

$$4x^2 + 9y^2 + z^2 = 4(3r\cos\theta)^2 + 9(2r\sin\theta)^2 + \left(6\sqrt{1-r^2}\right)^2 = 36r^2\cos^2\theta + 36r^2\sin^2\theta + 36(1-r^2)$$

$$= 36r^2\left(\cos^2\theta + \sin^2\theta\right) + 36 - 36r^2 = 36r^2 + 36 - 36r^2 = 36$$

Next, the part $z \geq 3$ of the ellipsoid corresponds to the values of r such that $6\sqrt{1-r^2} \geq 3$. We solve for $r \geq 0$:

$$6\sqrt{1-r^2} \geq 3 \quad\Rightarrow\quad \sqrt{1-r^2} \geq \frac{1}{2} \quad\Rightarrow\quad 1-r^2 \geq \frac{1}{4} \quad\Rightarrow\quad r^2 \leq \frac{3}{4} \quad\Rightarrow\quad 0 \leq r \leq \frac{\sqrt{3}}{2}$$

Step 1. Compute the tangent and normal vectors.

$$\mathbf{T}_\theta = \frac{\partial \Phi}{\partial \theta} = \frac{\partial}{\partial \theta}\left\langle 3r\cos\theta, 2r\sin\theta, 6\sqrt{1-r^2}\right\rangle = \langle -3r\sin\theta, 2r\cos\theta, 0\rangle$$

$$\mathbf{T}_r = \frac{\partial \Phi}{\partial r} = \frac{\partial}{\partial r}\left\langle 3r\cos\theta, 2r\sin\theta, 6\sqrt{1-r^2}\right\rangle = \left\langle 3\cos\theta, 2\sin\theta, -\frac{6r}{\sqrt{1-r^2}}\right\rangle$$

The outward pointing normal is:

$$\mathbf{N} = -\mathbf{T}_\theta \times \mathbf{T}_r = ((3r\sin\theta)\mathbf{i} - (2r\cos\theta)\mathbf{j}) \times \left((3\cos\theta)\mathbf{i} + (2\sin\theta)\mathbf{j} - \frac{6r}{\sqrt{1-r^2}}\mathbf{k}\right)$$

$$= 6r\sin^2\theta\,\mathbf{k} + \frac{18r^2\sin\theta}{\sqrt{1-r^2}}\mathbf{j} + \left(6r\cos^2\theta\right)\mathbf{k} + \frac{12r^2\cos\theta}{\sqrt{1-r^2}}\mathbf{i} = \frac{12r^2\cos\theta}{\sqrt{1-r^2}}\mathbf{i} + \frac{18r^2\sin\theta}{\sqrt{1-r^2}}\mathbf{j} + 6r\mathbf{k}$$

Step 2. Compute the dot product $\mathbf{E}\cdot\mathbf{N}$.

$$\mathbf{E}\left(\Phi(\theta,r)\right)\cdot\mathbf{N} = \langle 0,0,3r\cos\theta\rangle \cdot \left\langle \frac{12r^2\cos\theta}{\sqrt{1-r^2}}, \frac{18r^2\sin\theta}{\sqrt{1-r^2}}, 6r\right\rangle = 18r^2\cos\theta$$

Step 3. Evaluate the flux of **E**. The flux of **E** in (1) is equal to the following double integral:

$$\iint_S \mathbf{E} \cdot d\mathbf{S} = \iint_D \mathbf{E}(\Phi(\theta, r)) \cdot \mathbf{N}\, d\theta\, dr$$

$$= \int_0^{\sqrt{3}/2} \int_0^{2\pi} 18r^2 \cos\theta\, d\theta\, dr$$

$$= \int_0^{\sqrt{3}/2} 18r^2 \sin\theta \Big|_0^{2\pi} dr$$

$$= 0$$

17 FUNDAMENTAL THEOREMS OF VECTOR ANALYSIS

17.1 Green's Theorem (LT Section 18.1)

Preliminary Questions

1. Which vector field $\mathbf{F}$ is being integrated in the line integral $\oint x^2\, dy - e^y\, dx$?

SOLUTION The line integral can be rewritten as $\oint -e^y\, dx + x^2\, dy$. This is the line integral of $\mathbf{F} = \langle -e^y, x^2 \rangle$ along the curve.

2. Draw a domain in the shape of an ellipse and indicate with an arrow the boundary orientation of the boundary curve. Do the same for the annulus (the region between two concentric circles).

SOLUTION The orientation on C is counterclockwise, meaning that the region enclosed by C lies to the left in traversing C.

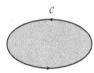

For the annulus, the inner boundary is oriented clockwise and the outer boundary is oriented counterclockwise. The region between the circles lies to the left while traversing each circle.

3. The circulation of a conservative vector field around a closed curve is zero. Is this fact consistent with Green's Theorem? Explain.

SOLUTION Green's Theorem asserts that

$$\int_C \mathbf{F} \cdot d\mathbf{r} = \int_C P\, dx + Q\, dy = \iint_D \left(\frac{\partial Q}{\partial x} - \frac{\partial P}{\partial y} \right) dA \qquad (1)$$

If $\mathbf{F}$ is conservative, the cross partials are equal, that is,

$$\frac{\partial P}{\partial y} = \frac{\partial Q}{\partial x} \quad \Rightarrow \quad \frac{\partial Q}{\partial x} - \frac{\partial P}{\partial y} = 0 \qquad (2)$$

Combining (1) and (2) we obtain $\int_C \mathbf{F} \cdot d\mathbf{r} = 0$. That is, Green's Theorem implies that the integral of a conservative vector field around a simple closed curve is zero.

4. Indicate which of the following vector fields possess this property: For every simple closed curve C, $\oint_C \mathbf{F} \cdot d\mathbf{r}$ is equal to the area enclosed by C.

(a) $\mathbf{F}(x, y) = \langle -y, 0 \rangle$
(b) $\mathbf{F}(x, y) = \langle x, y \rangle$
(c) $\mathbf{F}(x, y) = \langle \sin(x^2), x + e^{y^2} \rangle$

SOLUTION By Green's Theorem,

$$\int_C \mathbf{F} \cdot d\mathbf{r} = \iint_D \left(\frac{\partial Q}{\partial x} - \frac{\partial P}{\partial y} \right) dx\, dy \qquad (1)$$

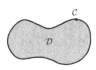

We compute the curl of each one of the given fields.

(a) Here, $P = -y$ and $Q = 0$, hence $\frac{\partial Q}{\partial x} - \frac{\partial P}{\partial y} = 0 - (-1) = 1$. Therefore, by (1),

$$\int_C \mathbf{F} \cdot d\mathbf{r} = \iint_D 1 \, dx \, dy = \text{Area}(D)$$

(b) We have $P = x$ and $Q = y$, therefore $\frac{\partial Q}{\partial x} - \frac{\partial P}{\partial y} = 0 - 0 = 0$. By (1) we get

$$\int_C \mathbf{F} \cdot d\mathbf{r} = \iint_D 0 \, dx \, dy = 0 \neq \text{Area}(D)$$

(c) In this vector field we have $P = \sin(x^2)$ and $Q = x + e^{y^2}$. Therefore,

$$\frac{\partial Q}{\partial x} - \frac{\partial P}{\partial y} = 1 - 0 = 1$$

By (1) we obtain

$$\int_C \mathbf{F} \cdot d\mathbf{r} = \iint_D 1 \, dx \, dy = \text{Area}(D)$$

5. Let A be the area enclosed by a simple closed curve C, and assume that C is oriented counterclockwise. Indicate whether the value of each integral is 0, $-A$, or A.

(a) $\oint_C x \, dx$ (b) $\oint_C y \, dx$

(c) $\oint_C y \, dy$ (d) $\oint_C x \, dy$

SOLUTION Let C be the boundary of the domain D. Then:

(a) In Green's Theorem, we have $F_1 = x$ and $F_2 = 0$, so that

$$\oint_C x \, dx = \iint_D \left(\frac{\partial}{\partial x}(0) - \frac{\partial}{\partial y}(x) \right) dA = \iint_D 0 \, dA = 0$$

(b) In Green's Theorem, we have $F_1 = y$ and $F_2 = 0$, so that

$$\oint_C y \, dx = \iint_D \left(\frac{\partial}{\partial x}(0) - \frac{\partial}{\partial y}(y) \right) dA = \iint_D (-1) \, dA = -A$$

(c) In Green's Theorem, we have $F_1 = 0$ and $F_2 = y$, so that

$$\oint_C y \, dy = \iint_D \left(\frac{\partial}{\partial x}(y) - \frac{\partial}{\partial y}(0) \right) dA = \iint_D 0 \, dA = 0$$

(d) In Green's Theorem, we have $F_1 = 0$ and $F_2 = x$, so that

$$\oint_C y \, dy = \iint_D \left(\frac{\partial}{\partial x}(x) - \frac{\partial}{\partial y}(0) \right) dA = \iint_D 1 \, dA = A$$

Exercises

1. Verify Green's Theorem for the line integral $\oint_C xy \, dx + y \, dy$, where C is the unit circle, oriented counterclockwise.

SOLUTION

Step 1. Evaluate the line integral. We use the parametrization $\gamma(\theta) = \langle \cos\theta, \sin\theta \rangle$, $0 \le \theta \le 2\pi$ of the unit circle. Then

$$dx = -\sin\theta\, d\theta, \quad dy = \cos\theta\, d\theta$$

and

$$xy\, dx + y\, dy = \cos\theta \sin\theta(-\sin\theta\, d\theta) + \sin\theta \cos\theta\, d\theta = \left(-\cos\theta \sin^2\theta + \sin\theta \cos\theta\right) d\theta$$

The line integral is thus

$$\int_C xy\, dx + y\, dy = \int_0^{2\pi} \left(-\cos\theta \sin^2\theta + \sin\theta \cos\theta\right) d\theta$$

$$= \int_0^{2\pi} -\cos\theta \sin^2\theta\, d\theta + \int_0^{2\pi} \sin\theta \cos\theta\, d\theta = -\frac{\sin^3\theta}{3}\bigg|_0^{2\pi} - \frac{\cos 2\theta}{4}\bigg|_0^{2\pi} = 0 \quad (1)$$

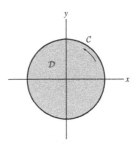

Step 2. Evaluate the double integral. Since $P = xy$ and $Q = y$, we have

$$\frac{\partial Q}{\partial x} - \frac{\partial P}{\partial y} = 0 - x = -x$$

We compute the double integral in Green's Theorem:

$$\iint_{\mathcal{D}} \left(\frac{\partial Q}{\partial x} - \frac{\partial P}{\partial y}\right) dx\, dy = \iint_{\mathcal{D}} -x\, dx\, dy = -\iint_{\mathcal{D}} x\, dx\, dy$$

The integral of x over the disk $\mathcal{D}$ is zero, since by symmetry the positive and negative values of x cancel each other. Therefore,

$$\iint_{\mathcal{D}} \left(\frac{\partial Q}{\partial x} - \frac{\partial P}{\partial y}\right) dx\, dy = 0 \quad (2)$$

Step 3. Compare. The line integral in (1) is equal to the double integral in (2), as stated in Green's Theorem.

In Exercises 3–12, use Green's Theorem to evaluate the line integral. Orient the curve counterclockwise unless otherwise indicated.

3. $\oint_C y^2\, dx + x^2\, dy$, where C is the boundary of the square that is given by $0 \le x \le 1, 0 \le y \le 1$

SOLUTION

We have $P = y^2$ and $Q = x^2$, therefore

$$\frac{\partial Q}{\partial x} - \frac{\partial P}{\partial y} = 2x - 2y$$

Using Green's Theorem we obtain

$$\int_C y^2\,dx + x^2\,dy = \iint_D \frac{\partial Q}{\partial x} - \frac{\partial P}{\partial y}\,dA = \iint_D (2x - 2y)\,dx\,dy$$

$$= \int_0^1 \int_0^1 (2x - 2y)\,dx\,dy = \int_0^1 (x^2 - 2xy)\Big|_0^1 dy = \int_0^1 (1 - 2y)\,dy = (y - y^2)\Big|_0^1 = 0$$

5. $\oint_C 5y\,dx + 2x\,dy$, where C is the triangle with vertices $(-1, 0)$, $(1, 0)$, and $(0, 1)$

SOLUTION

We have $P = 5y$ and $Q = 2x$, therefore

$$\frac{\partial Q}{\partial x} - \frac{\partial P}{\partial y} = 2 - 5 = -3$$

Using Green's Theorem we obtain

$$\int_C 5y\,dx + 2x\,dy = \iint_D \frac{\partial Q}{\partial x} - \frac{\partial P}{\partial y}\,dA = \iint_D (-3)\,dA = -3\iint_D 1\,dA$$

This is just -3 times the area of the triangle. Since the area of the triangle is 1, the value of the integral is -3.

7. $\oint_C x^2 y\,dx$, where C is the unit circle centered at the origin

SOLUTION

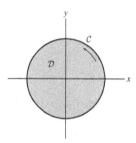

In this function $P = x^2 y$ and $Q = 0$. Therefore,

$$\frac{\partial Q}{\partial x} - \frac{\partial P}{\partial y} = 0 - x^2 = -x^2$$

We obtain the following integral:

$$I = \int_C x^2 y\,dx = \iint_D \left(\frac{\partial Q}{\partial x} - \frac{\partial P}{\partial y}\right) dA = \iint_D -x^2\,dA$$

We convert the integral to polar coordinates. This gives

$$I = \int_0^{2\pi}\int_0^1 -r^2 \cos^2\theta \cdot r\,dr\,d\theta = \int_0^{2\pi}\int_0^1 -r^3 \cos^2\theta\,dr\,d\theta$$

$$= \left(\int_0^{2\pi} \cos^2\theta\,d\theta\right)\left(\int_0^1 -r^3\,dr\right) = \left(\frac{\theta}{2} + \frac{\sin 2\theta}{4}\Big|_{\theta=0}^{2\pi}\right)\left(-\frac{r^4}{4}\Big|_{r=0}^1\right) = \pi \cdot \left(-\frac{1}{4}\right) = -\frac{\pi}{4}$$

9. $\oint_C \mathbf{F} \cdot d\mathbf{r}$, where $\mathbf{F}(x,y) = \langle x^2, x^2 \rangle$ and C consists of the arcs $y = x^2$ and $y = x$ for $0 \le x \le 1$

SOLUTION By Green's Theorem,

$$I = \int_C \mathbf{F} \cdot d\mathbf{r} = \iint_D \left(\frac{\partial Q}{\partial x} - \frac{\partial P}{\partial y} \right) dA$$

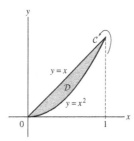

We have $P = Q = x^2$, therefore

$$\frac{\partial Q}{\partial x} - \frac{\partial P}{\partial y} = 2x - 0 = 2x$$

Hence,

$$I = \iint_D 2x\, dA = \int_0^1 \int_{x^2}^x 2x\, dy\, dx = \int_0^1 2xy\Big|_{y=x^2}^x dx = \int_0^1 2x(x - x^2)\, dx = \int_0^1 (2x^2 - 2x^3)\, dx$$

$$= \frac{2}{3}x^3 - \frac{1}{2}x^4 \Big|_0^1 = \frac{2}{3} - \frac{1}{2} = \frac{1}{6}$$

11. The line integral of $\mathbf{F}(x,y) = \langle e^{x+y}, e^{x-y} \rangle$ along the curve (oriented clockwise) consisting of the line segments by joining the points $(0,0)$, $(2,2)$, $(4,2)$, $(2,0)$, and back to $(0,0)$ (Note the orientation.)

SOLUTION We compute the curl of $\mathbf{F} = \langle e^{x+y}, e^{x-y} \rangle$. Here, $P = e^{x+y}$ and $Q = e^{x-y}$, hence

$$\frac{\partial Q}{\partial x} - \frac{\partial P}{\partial y} = e^{x-y} - e^{x+y} = e^x(e^{-y} - e^y)$$

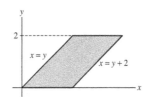

Using Green's Theorem we obtain

$$\int_C \mathbf{F} \cdot d\mathbf{r} = \iint_D e^x(e^{-y} - e^y)\, dx\, dy = \int_0^2 \int_y^{y+2} e^x(e^{-y} - e^y)\, dx\, dy = \int_0^2 e^x(e^{-y} - e^y)\Big|_{x=y}^{y+2} dy$$

$$= \int_0^2 (e^{y+2} - e^y)(e^{-y} - e^y)\, dy = \int_0^2 (e^2 - 1)(1 - e^{2y})\, dy = (e^2 - 1)\left(y - \frac{e^{2y}}{2} \right)\Big|_{y=0}^2$$

$$= (e^2 - 1)\left(2 - \frac{e^4}{2} - \left(-\frac{1}{2} \right) \right) = \frac{(e^2 - 1)(5 - e^4)}{2}$$

13. Let $\mathbf{F}(x,y) = \langle 2xe^y, x + x^2 e^y \rangle$ and let C be the quarter-circle path from A to B in Figure 18. Evaluate $I = \oint_C \mathbf{F} \cdot d\mathbf{r}$ as follows:

(a) Find a function $f(x,y)$ such that $\mathbf{F} = \mathbf{G} + \nabla f$, where $\mathbf{G} = \langle 0, x \rangle$.
(b) Show that the line integrals of $\mathbf{G}$ along the segments $\overline{OA}$ and $\overline{OB}$ are zero.
(c) Evaluate I. *Hint:* Use Green's Theorem to show that

$$I = f(B) - f(A) + 4\pi$$

FIGURE 18

SOLUTION

(a) We need to find a potential function $f(x, y)$ for the difference

$$\mathbf{F} - \mathbf{G} = \langle 2xe^y, x + x^2 e^y \rangle - \langle 0, x \rangle = \langle 2xe^y, x^2 e^y \rangle$$

We let $f(x, y) = x^2 e^y$.

(b) We use the parametrizations $\overline{AO} : \langle t, 0 \rangle, 0 \le t \le 4$ and $\overline{OB} : \langle 0, t \rangle, 0 \le t \le 4$ to evaluate the integrals of $\mathbf{G} = \langle 0, x \rangle$. We get

$$\int_{\overline{OA}} \mathbf{G} \cdot d\mathbf{r} = \int_0^4 \langle 0, t \rangle \cdot \langle 1, 0 \rangle \, dt = \int_0^4 0 \, dt = 0$$

$$\int_{\overline{OB}} \mathbf{G} \cdot d\mathbf{r} = \int_0^4 \langle 0, 0 \rangle \cdot \langle 0, 1 \rangle \, dt = \int_0^4 0 \, dt = 0$$

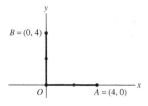

(c) Since $\mathbf{F} - \mathbf{G} = \nabla f$, we have

$$\int_C (\mathbf{F} - \mathbf{G}) \cdot d\mathbf{r} = f(B) - f(A) = \int_C \mathbf{F} \cdot d\mathbf{r} - \int_C \mathbf{G} \cdot d\mathbf{r} = I - \int_C \mathbf{G} \cdot d\mathbf{r}$$

That is,

$$I = f(B) - f(A) + \int_C \mathbf{G} \cdot d\mathbf{r} \qquad (1)$$

To compute the line integral on the right-hand side, we rewrite it as

$$\int_C \mathbf{G} \cdot d\mathbf{r} = \int_{\overline{BO} + \overline{OA} + C} \mathbf{G} \cdot d\mathbf{r} - \int_{\overline{BO}} \mathbf{G} \cdot d\mathbf{r} - \int_{\overline{OA}} \mathbf{G} \cdot d\mathbf{r}$$

Using part (b) we may write

$$\int_C \mathbf{G} \cdot d\mathbf{r} = \int_{\overline{BO} + \overline{OA} + C} \mathbf{G} \cdot d\mathbf{r} \qquad (2)$$

We now use Green's Theorem. Since $\mathbf{G} = \langle 0, x \rangle$, we have $P = 0$ and $Q = x$, hence $\frac{\partial Q}{\partial x} - \frac{\partial P}{\partial y} = 1 - 0 = 1$. Thus,

$$\int_{\overline{BO} + \overline{OA} + C} \mathbf{G} \cdot d\mathbf{r} = \iint_{\mathcal{D}} 1 \, dA = \text{Area}(\mathcal{D}) = \frac{\pi \cdot 4^2}{4} = 4\pi \qquad (3)$$

Combining (1), (2), and (3), we obtain

$$I = f(B) - f(A) + 4\pi$$

Since $f(x, y) = x^2 e^y$, we conclude that

$$I = f(0, 4) - f(4, 0) + 4\pi = 0 - 4^2 e^0 + 4\pi = 4\pi - 16$$

15. Evaluate $I = \int_C (\sin x + y)\,dx + (3x + y)\,dy$ for the nonclosed path $ABCD$ in Figure 20. Use the method of Exercise 14.

FIGURE 20

SOLUTION

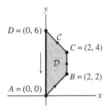

Let $\mathbf{F} = \langle \sin x + y, 3x + y \rangle$, hence $P = \sin x + y$ and $Q = 3x + y$. We denote by C_1 the closed path determined by C and the segment $\overline{DA}$. Then by Green's Theorem,

$$\int_{C_1} P\,dx + Q\,dy = \iint_\mathcal{D} \left(\frac{\partial Q}{\partial x} - \frac{\partial P}{\partial y}\right) dA = \iint_\mathcal{D} (3-1)\,dA = 2\iint_\mathcal{D} dA = 2\,\text{Area}(\mathcal{D}) \quad (1)$$

The area of $\mathcal{D}$ is the area of the trapezoid $ABCD$, that is,

$$\text{Area}(\mathcal{D}) = \frac{\left(\overline{BC} + \overline{AD}\right)h}{2} = \frac{(2+6)\cdot 2}{2} = 8$$

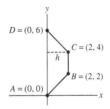

Combining with (1) we get

$$\int_{C_1} P\,dx + Q\,dy = 2 \cdot 8 = 16$$

Using properties of line integrals, we have

$$\int_C P\,dx + Q\,dy + \int_{\overline{DA}} P\,dx + Q\,dy = 16 \quad (2)$$

We compute the line integral over $\overline{DA}$, using the parametrization

$$\overline{DA}: x = 0,\ y = t, t \text{ varies from 6 to 0}$$

We get

$$\int_{\overline{DA}} P\,dx + Q\,dy = \int_6^0 F(0,t) \cdot \frac{d}{dt}\langle 0,t\rangle\,dt = \int_6^0 \langle \sin 0 + t, 3\cdot 0 + t\rangle \cdot \langle 0,1\rangle\,dt$$

$$= \int_6^0 \langle t,t\rangle \cdot \langle 0,1\rangle\,dt = \int_6^0 t\,dt = \left.\frac{t^2}{2}\right|_{t=6}^0 = -18$$

We substitute in (2) and solve for the required integral:

$$\int_C P\,dx + Q\,dy - 18 = 16 \quad \text{or} \quad \int_C P\,dx + Q\,dy = 34$$

17. Use $\dfrac{1}{2}\oint_C x\,dy - y\,dx$ to compute the area of the ellipse $\left(\dfrac{x}{a}\right)^2 + \left(\dfrac{y}{b}\right)^2 = 1$.

SOLUTION As in Example 3, parametrize the boundary of the ellipse by

$$x = a\cos\theta, \quad y = b\sin\theta, \quad 0 \le \theta \le 2\pi$$
$$dx = -a\sin\theta\,d\theta, \quad dy = b\cos\theta\,d\theta$$

$$A = \frac{1}{2}\oint_C x\,dy - y\,dx = \frac{1}{2}\int_0^{2\pi}(ab\cos^2\theta + ab\sin^2\theta)\,d\theta = \frac{1}{2}\int_0^{2\pi} ab\,d\theta = \pi ab$$

In Exercises 18–21, use one of the formulas in Eq. (6) to calculate the area of the given region.

19. The triangle with vertices $(0,0)$, $(1,0)$, and $(1,1)$

SOLUTION We parametrize the segments by

$$\overline{OA} : \langle t, 0\rangle,\ t = 0\text{ to }t = 1$$
$$\overline{AB} : \langle 1, t\rangle,\ t = 0\text{ to }t = 1$$
$$\overline{BO} : \langle t, t\rangle,\ t = 1\text{ to }t = 0$$

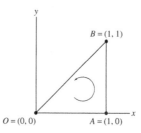

Using Eq. (6), we obtain the following area of the triangle:

$$A = \frac{1}{2}\int_C x\,dy - y\,dx = \frac{1}{2}\int_{\overline{OA}}\langle -y,x\rangle\cdot ds + \frac{1}{2}\int_{\overline{AB}}\langle -y,x\rangle\cdot ds + \frac{1}{2}\int_{\overline{BO}}\langle -y,x\rangle\cdot ds$$

$$= \frac{1}{2}\int_0^1 \langle 0,t\rangle\cdot\langle 1,0\rangle\,dt + \frac{1}{2}\int_0^1 \langle -t,1\rangle\cdot\langle 0,1\rangle\,dt + \frac{1}{2}\int_1^0 \langle -t,t\rangle\cdot\langle 1,1\rangle\,dt$$

$$= \frac{1}{2}\int_0^1 0\,dt + \frac{1}{2}\int_0^1 dt + \frac{1}{2}\int_1^0 0\,dt = \frac{1}{2}\int_0^1 dt = \frac{1}{2}\cdot 1 = \frac{1}{2}$$

21. The region between the graph of $y = x^2$ and the x-axis for $0 \le x \le 2$

SOLUTION The boundary of the region consists of the curve Γ and the segments $\overline{OA}$ and $\overline{AB}$ shown in the figure.

By Eq. (6), the area A of the region is given by

$$A = \frac{1}{2}\int_{\overline{OA}} x\,dy - y\,dx + \frac{1}{2}\int_{\overline{AB}} x\,dy - y\,dx + \frac{1}{2}\int_{\Gamma} x\,dy - y\,dx \tag{1}$$

We compute each integral separately. We use the following parametrizations:

$$\overline{OA}: c_1(t) = (t, 0), \text{ for } 0 \le t \le 2 \quad \Rightarrow \quad c_1'(t) = \langle 1, 0 \rangle$$
$$\overline{AB}: c_2(t) = (2, t), \text{ for } 0 \le t \le 4 \quad \Rightarrow \quad c_2'(t) = \langle 0, 1 \rangle$$
$$\Gamma: c_3(t) = (t, t^2) \text{ for } t \text{ from 2 to 0} \quad \Rightarrow \quad c_3'(t) = \langle 1, 2t \rangle$$

The line integrals in (1) are thus

$$\int_{\overline{OA}} x\,dy - y\,dx = \int_{\overline{OA}} \langle -y, x \rangle \cdot ds = \int_0^2 \langle 0, t \rangle \cdot \langle 1, 0 \rangle\,dt = 0$$

$$\int_{\overline{AB}} x\,dy - y\,dx = \int_{\overline{AB}} \langle -y, x \rangle \cdot ds = \int_0^4 \langle -t, 2 \rangle \cdot \langle 0, 1 \rangle\,dt = \int_0^4 2\,dt = 8$$

$$\int_{\Gamma} x\,dy - y\,dx = \int_{\Gamma} \langle -y, x \rangle \cdot ds = \int_2^0 \langle -t^2, t \rangle \cdot \langle 1, 2t \rangle\,dt = \int_2^0 (-t^2 + 2t^2) \cdot dt$$

$$= \int_0^2 -t^2\,dt = -\frac{t^3}{3}\bigg|_0^2 = -\frac{8}{3}$$

Substituting the integrals in (1) we obtain the following area:

$$A = \frac{1}{2} \cdot 0 + \frac{1}{2} \cdot 8 + \frac{1}{2} \cdot \left(-\frac{8}{3}\right) = \frac{8}{3}$$

23. Let $x^3 + y^3 = 3xy$ be the **folium of Descartes** (Figure 22).

(a) Show that the folium has a parametrization in terms of $t = y/x$ given by

$$x = \frac{3t}{1 + t^3}, \qquad y = \frac{3t^2}{1 + t^3} \qquad (-\infty < t < \infty) \quad (t \ne -1)$$

(b) Show that

$$x\,dy - y\,dx = \frac{9t^2}{(1 + t^3)^2}\,dt$$

Hint: By the Quotient Rule,

$$x^2\,d\left(\frac{y}{x}\right) = x\,dy - y\,dx$$

(c) Find the area of the loop of the folium. *Hint:* The limits of integration are 0 and ∞.

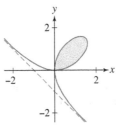

FIGURE 22 Folium of Descartes.

SOLUTION

(a) We show that $x = \frac{3t}{1+t^3}$, $y = \frac{3t^2}{1+t^3}$ satisfy the equation $x^3 + y^3 - 3xy = 0$ of the folium:

$$x^3 + y^3 - 3xy = \left(\frac{3t}{1+t^3}\right)^3 + \left(\frac{3t^2}{1+t^3}\right)^3 - 3 \cdot \frac{3t}{1+t^3} \cdot \frac{3t^2}{1+t^3}$$

$$= \frac{27t^3 + 27t^6}{(1+t^3)^3} - \frac{27t^3(1+t^3)}{(1+t^3)^3} = \frac{27t^3\left(1+t^3 - (1+t^3)\right)}{(1+t^3)^3} = \frac{0}{(1+t^3)^3} = 0$$

This proves that the curve parametrized by $x = \frac{3t}{1+t^3}$, $y = \frac{3t^2}{1+t^3}$ lies on the folium of Descartes. This parametrization parametrizes the whole folium since the two equations can be solved for t in terms of x and y. That is,

$$\begin{aligned} x &= \frac{3t}{1+t^3} \\ y &= \frac{3t^2}{1+t^3} \end{aligned} \quad \Rightarrow \quad t = \frac{y}{x}$$

A glance at the graph of the folium shows that any line $y = tx$, with slope t, intersects the folium exactly once. Thus, there is a one-to-one relationship between the values of t and the points on the graph.

(b) We differentiate the two sides of $t = \frac{y}{x}$ with respect to t. Using the Quotient Rule gives

$$1 = \frac{x\frac{dy}{dt} - y\frac{dx}{dt}}{x^2}$$

or

$$x\frac{dy}{dt} - y\frac{dx}{dt} = x^2 = \left(\frac{3t}{1+t^3}\right)^2$$

This equality can be written in the form

$$x\,dy - y\,dx = \frac{9t^2}{(1+t^3)^2}\,dt$$

(c) We use the formula for the area enclosed by a closed curve and the result of part (b) to find the required area. That is,

$$A = \frac{1}{2}\int_C x\,dy - y\,dx = \frac{1}{2}\int_0^\infty \frac{9t^2}{(1+t^3)^2}\,dt$$

From our earlier discussion on the parametrization of the folium, we see that the loop is traced when the parameter t is increasing along the interval $0 \le t < \infty$. We compute the improper integral using the substitution $u = 1 + t^3$, $du = 3t^2\,dt$. This gives

$$A = \frac{1}{2}\lim_{R\to\infty}\int_0^R \frac{9t^2}{(1+t^3)^2}\,dt = \frac{1}{2}\lim_{R\to\infty}\int_1^{1+R^3}\frac{3\,du}{u^2} = \frac{3}{2}\lim_{R\to\infty}-\frac{1}{u}\bigg|_{u=1}^{1+R^3}$$

$$= \frac{3}{2}\lim_{R\to\infty}\left(1 - \frac{1}{1+R^3}\right) = \frac{3}{2}(1-0) = \frac{3}{2}$$

25. The Centroid via Boundary Measurements The centroid (see Section 15.5) of a domain $\mathcal{D}$ enclosed by a simple closed curve C is the point with coordinates $(\bar{x}, \bar{y}) = (M_y/M, M_x/M)$, where M is the area of $\mathcal{D}$ and the moments are defined by

$$M_x = \iint_\mathcal{D} y\,dA, \qquad M_y = \iint_\mathcal{D} x\,dA$$

Show that $M_x = \oint_C xy\,dy$. Find a similar expression for M_y.

SOLUTION Consider the moment $M_x = \iint_D y\,dA$, we know from Green's Theorem:

$$\iint_D \left(\frac{\partial F_2}{\partial x} - \frac{\partial F_1}{\partial y}\right) dA = \oint_C F_1\,dx + F_2\,dy$$

So then we need

$$\frac{\partial F_2}{dx} - \frac{\partial F_1}{\partial y} = y$$

If we set $F_2 = xy$ and $F_1 = 0$, then $\frac{\partial F_2}{dx} - \frac{\partial F_1}{\partial y} = y$ and

$$M_x = \iint_D y\,dA = \oint_C xy\,dy$$

Similarly, consider the moment $M_y = \iint_D x\,dA$. We will now use Green's Theorem, stated above. Here we need

$$\frac{\partial F_2}{dx} - \frac{\partial F_1}{\partial y} = x$$

If we set $F_1 = -xy$ and $F_2 = 0$ then $\frac{\partial F_2}{\partial x} - \frac{\partial F_1}{\partial y} = x$ and

$$M_y = \iint_D x\,dA = \oint_C -xy\,dx$$

27. Let C_R be the circle of radius R centered at the origin. Use the general form of Green's Theorem to determine $\oint_{C_2} \mathbf{F} \cdot d\mathbf{r}$, where $\mathbf{F}$ is a vector field such that $\oint_{C_1} \mathbf{F} \cdot d\mathbf{r} = 9$ and $\frac{\partial F_2}{\partial x} - \frac{\partial F_1}{\partial y} = x^2 + y^2$ for (x, y) in the annulus $1 \le x^2 + y^2 \le 4$.

SOLUTION We use Green's Theorem for the annulus $\mathcal{D}$ between the circles $C(1)$ and $C(2)$ oriented as shown in the figure.

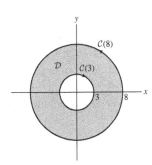

That is,

$$\int_{C(2)} \mathbf{F} \cdot d\mathbf{r} - \int_{C(1)} \mathbf{F} \cdot d\mathbf{r} = \iint_D \mathrm{curl}(\mathbf{F})\,dx\,dy$$

Substituting the given information, we get

$$\int_{C(2)} \mathbf{F} \cdot d\mathbf{r} - 9 = \iint_D (x^2 + y^2)\,dx\,dy$$

or

$$\int_{C(2)} \mathbf{F} \cdot d\mathbf{r} = 9 + \iint_D (x^2 + y^2)\,dx\,dy$$

We compute the double integral by converting it to polar coordinates:

$$\int_{C(2)} \mathbf{F} \cdot d\mathbf{r} = 9 + \int_0^{2\pi} \int_1^2 r^2 \cdot r\, dr\, d\theta = 9 + 2\pi \int_1^2 r^3\, dr = 9 + 2\pi \cdot \frac{r^4}{4}\Big|_1^2 = 9 + 2\pi \left(\frac{2^4 - 1^4}{4}\right) = 9 + \frac{15\pi}{2}$$

29. Referring to Figure 25, suppose that

$$\oint_{C_2} \mathbf{F} \cdot d\mathbf{r} = 3\pi, \qquad \oint_{C_3} \mathbf{F} \cdot d\mathbf{r} = 4\pi$$

Use Green's Theorem to determine the circulation of $\mathbf{F}$ around C_1, assuming that $\frac{\partial F_2}{\partial x} - \frac{\partial F_1}{\partial y} = 9$ on the shaded region.

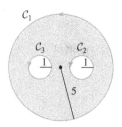

FIGURE 25

SOLUTION We must calculate $\int_{C_1} \mathbf{F} \cdot d\mathbf{r}$. We use Green's Theorem for the region $\mathcal{D}$ between the three circles C_1, C_2, and C_3. Because of orientation, the line integrals $\int_{-C_2} \mathbf{F} \cdot d\mathbf{r} = -\int_{C_2} \mathbf{F} \cdot d\mathbf{r}$ and $\int_{-C_3} \mathbf{F} \cdot d\mathbf{r} = -\int_{C_3} \mathbf{F} \cdot d\mathbf{r}$ must be used in applying Green's Theorem. That is,

$$\int_{C_1} \mathbf{F} \cdot d\mathbf{r} - \int_{C_2} \mathbf{F} \cdot d\mathbf{r} - \int_{C_3} \mathbf{F} \cdot d\mathbf{r} = \iint_{\mathcal{D}} \left(\frac{\partial F_2}{\partial x} - \frac{\partial F_1}{\partial y}\right) dA$$

We substitute the given information to obtain

$$\int_{C_1} \mathbf{F} \cdot d\mathbf{r} - 3\pi - 4\pi = \iint_{\mathcal{D}} 9\, dA = 9 \iint_{\mathcal{D}} 1 \cdot dA = 9\, \text{Area}(\mathcal{D}) \tag{1}$$

The area of $\mathcal{D}$ is computed as the difference of areas of discs. That is,

$$\text{Area}(\mathcal{D}) = \pi \cdot 5^2 - \pi \cdot 1^2 - \pi \cdot 1^2 = 23\pi$$

We substitute in (1) and compute the desired circulation:

$$\int_{C_1} \mathbf{F} \cdot d\mathbf{r} - 7\pi = 9 \cdot 23\pi$$

or

$$\int_{C_1} \mathbf{F} \cdot d\mathbf{r} = 214\pi$$

In Exercises 31–34, we refer to the integrand that occurs in Green's Theorem and that appears as

$$\text{curl}_z(\mathbf{F}) = \frac{\partial F_2}{\partial x} - \frac{\partial F_1}{\partial y}$$

31. For the vector fields (A)–(D) in Figure 27, state whether curl_z at the origin appears to be positive, negative, or zero.

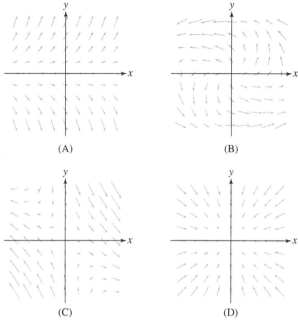

FIGURE 27

SOLUTION The vector field (A) does not have spirals, nor it is a "shear flow." Therefore, the curl appears to be zero. The vector field (B) rotates in the counterclockwise direction, hence we expect the curl to be positive. The vector field (C) perhaps rotates more strongly clockwise than counterclockwise around the origin, indicating a negative curl (however, this is not completely clear, and concluding that it has a zero curl is also reasonable). Finally, in vector field (D) the fluid flows straight toward the origin without spiraling. We expect the curl to be zero.

33. Estimate $\oint_C \mathbf{F} \cdot d\mathbf{r}$, where $\mathbf{F}(x, y) = \langle x + 0.1y^2, y - 0.1x^2 \rangle$ and C encloses a small region of area 0.25 containing the point $P = (1, 1)$.

SOLUTION Use the following estimation:

$$\mathbf{F} \cdot d\mathbf{s} = \oint_C F_1\, dx + F_2\, dy \approx \operatorname{curl}_z(\mathbf{F})(P) \cdot \operatorname{Area}(\mathcal{D})$$

First computing curl$\mathbf{F}$ we have:

$$\operatorname{curl}\mathbf{F} = \begin{vmatrix} \mathbf{i} & \mathbf{j} & \mathbf{k} \\ \partial x & \partial y & \partial z \\ x + 0.1y^2 & y - 0.1x^2 & 0 \end{vmatrix} = \langle 0, 0, -0.2x - 0.2y \rangle$$

Thus $\operatorname{curl}_z(\mathbf{F}) = -0.2x - 0.2y$ and $\operatorname{curl}_z(\mathbf{F})(1, 1) = -0.2 - 0.2 = -0.4$. Also, we are given the area of the region is 0.25. Hence, we see:

$$\oint \mathbf{F} \cdot d\mathbf{s} \approx (-0.4)(0.25) = -0.10$$

35. Let C_R be the circle of radius R centered at the origin. Use Green's Theorem to find the value of R that maximizes $\oint_{C_R} y^3\, dx + x\, dy$.

SOLUTION Using Green's Theorem we can write:

$$\oint_{C_R} y^3\, dx + x\, dy = \iint_{\mathcal{D}} \frac{\partial F_2}{\partial x} - \frac{\partial F_1}{\partial y}\, dA$$

$$= \iint_{\mathcal{D}} 1 - 3y^2\, dA$$

Then we have the following, using polar coordinates:

$$\oint_{C_R} y^3\, dx + x\, dy = \iint_{\mathcal{D}} \frac{\partial F_2}{\partial x} - \frac{\partial F_1}{\partial y}\, dA$$

$$= \iint_D 1 - 3y^2 \, dA$$

$$= \int_0^{2\pi} \int_0^R (1 - 3r^2 \sin^2 \theta)(r) \, dr \, d\theta$$

$$= \int_0^{2\pi} \int_0^R (r - 3r^3 \sin^2 \theta) \, dr \, d\theta$$

$$= \int_0^{2\pi} \left(\frac{1}{2}r^2 - \frac{3}{4}r^4 \sin^2 \theta \right) \Big|_0^R d\theta$$

$$= \int_0^{2\pi} \left(\frac{1}{2}R^2 - \frac{3}{8}R^4(1 - \cos 2\theta) \right) d\theta$$

$$= \int_0^{2\pi} \left(\frac{3R^4}{8}(\cos 2\theta) + \frac{1}{2}R^2 - \frac{3}{8}R^4 \right) d\theta$$

$$= \left(\frac{3R^4}{16} \sin 2\theta + \left(\frac{1}{2}R^2 - \frac{3}{8}R^4 \right)\theta \right) \Big|_0^{2\pi}$$

$$= 0 + 2\pi \left(\frac{1}{2}R^2 - \frac{3}{8}R^4 \right) = \pi \left(R^2 - \frac{3}{4}R^4 \right)$$

Now to maximize this quantity, we need to let $f(R) = \pi(R^2 - 3/4R^4)$ and take the first derivative.

$$f'(R) = \pi(2R - 3R^3) = 0 \quad \Rightarrow R = 0, \pm \sqrt{2/3}$$

This quantity is maximized when $R = \pm \sqrt{\frac{2}{3}}$ (that is, $R = 0$ is a minimum).

37. Use the result of Exercise 36 to compute the areas of the polygons in Figure 28. Check your result for the area of the triangle in (A) using geometry.

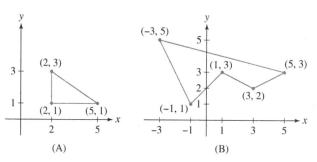

FIGURE 28

SOLUTION

(a) The vertices of the triangle are

$$(x_1, y_1) = (x_4, y_4) = (2, 1), \quad (x_2, y_2) = (5, 1), \quad (x_3, y_3) = (2, 3)$$

Using the formula obtained in Exercise 28, the area of the triangle is the following sum:

$$A = \frac{1}{2}((x_1 y_2 - x_2 y_1) + (x_2 y_3 - x_3 y_2) + (x_3 y_1 - x_1 y_3))$$

$$= \frac{1}{2}((2 \cdot 1 - 5 \cdot 1) + (5 \cdot 3 - 2 \cdot 1) + (2 \cdot 1 - 2 \cdot 3)) = \frac{1}{2}(-3 + 13 - 4) = 3$$

We verify our result using the formula for the area of a triangle:

$$A = \frac{1}{2}bh = \frac{1}{2} \cdot (5-2) \cdot (3-1) = 3$$

(b) The vertices of the polygon are

$$(x_1, y_1) = (x_6, y_6) = (-1, 1)$$
$$(x_2, y_2) = (1, 3)$$
$$(x_3, y_3) = (3, 2)$$
$$(x_4, y_4) = (5, 3)$$
$$(x_5, y_5) = (-3, 5)$$

Using the formula in part (a), the area of the polygon is the following sum:

$$A = \frac{1}{2}((x_1 y_2 - x_2 y_1) + (x_2 y_3 - x_3 y_2) + (x_3 y_4 - x_4 y) + (x_4 y_5 - x_5 y_4) + (x_5 y_1 - x_1 y_5))$$

$$= \frac{1}{2}\Big((-1 \cdot 3 - 1 \cdot 1) + (1 \cdot 2 - 3 \cdot 3) + (3 \cdot 3 - 5 \cdot 2) + (5 \cdot 5 - (-3) \cdot 3) + (-3 \cdot 1 - (-1) \cdot 5)\Big)$$

$$= \frac{1}{2}(-4 - 7 - 1 + 34 + 2) = 12$$

In Exercises 38–43, compute the flux $\oint \mathbf{F} \cdot \mathbf{n}\, ds$ of $\mathbf{F}$ across the curve C for the given vector field and curve using the vector form of Green's Theorem.

39. $\mathbf{F}(x, y) = \langle xy, x - y \rangle$ across the boundary of the square $-1 \le x \le 1$, $-1 \le y \le 1$

SOLUTION Let $\mathcal{D}$ be the given square (including its interior). The divergence of $\mathbf{F}$ is

$$\operatorname{div}(\mathbf{F}) = \frac{\partial}{\partial x}(xy) + \frac{\partial}{\partial y}(x - y) = y - 1$$

so that the flux is

$$\oint \mathbf{F} \cdot \mathbf{n}\, ds = \iint_{\mathcal{D}} \operatorname{div} \mathbf{F}\, ds = \int_{-1}^{1} \int_{-1}^{1} (y - 1)\, dy\, dx = \int_{-1}^{1} \left(\frac{1}{2}y^2 - y\right)\Big|_{-1}^{1} dx = \int_{-1}^{1} (-2)\, dx = -4$$

41. $\mathbf{F}(x, y) = \langle 2x + y^3, 3y - x^4 \rangle$ across the unit circle

SOLUTION Let $\mathcal{D}$ be the unit circle together with its interior. The divergence of $\mathbf{F}$ is

$$\operatorname{div}(\mathbf{F}) = \frac{\partial}{\partial x}(2x + y^3) + \frac{\partial}{\partial y}(3y - x^4) = 2 + 3 = 5$$

so the flux is

$$\oint \mathbf{F} \cdot \mathbf{n}\, ds = \iint_{\mathcal{D}} \operatorname{div} \mathbf{F}\, ds = 5 \iint_{\mathcal{D}} 1\, ds$$

This integral is five times the area of the unit circle, so the answer is 5π.

43. $\mathbf{F}(x, y) = \langle xy^2 + 2x, x^2 y - 2y \rangle$ across the simple closed curve that is the boundary of the half-disk given by $x^2 + y^2 \le 3$, $y \ge 0$

SOLUTION Let $\mathcal{D}$ be the given half-disk. The divergence of **F** is

$$\text{div}(\mathbf{F}) = \frac{d}{dx}(xy^2 + 2x) + \frac{d}{dy}(x^2y - 2y) = y^2 + 2 + x^2 - 2 = x^2 + y^2$$

so that the flux is

$$\oint \mathbf{F} \cdot \mathbf{n}\, ds = \iint_{\mathcal{D}} \text{div}\mathbf{F}\, ds = \iint_{\mathcal{D}} x^2 + y^2\, ds$$

Converting to polar coordinates, the limits of integration are $0 \le r \le 3$ and $0 \le \theta \le \pi$; we get

$$\iint_{\mathcal{D}} x^2 + y^2\, dx = \int_0^3 \int_0^\pi r^2 \cdot r\, d\theta\, dr = \pi \int_0^3 r^3\, dr = \pi \left(\frac{1}{4}r^4\right)\bigg|_0^3 = \frac{81}{4}\pi$$

45. A buffalo stampede (Figure 29) is described by a velocity vector field $\mathbf{F} = \langle xy - y^3, x^2 + y \rangle$ kilometers per hour in the region $\mathcal{D}$ defined by $2 \le x \le 3$, $2 \le y \le 3$ in units of kilometers (Figure 30). Assuming a density is $\rho = 500$ buffalo per square kilometer, use Eq. (13) to determine the net number of buffalo leaving or entering $\mathcal{D}$ per minute (equal to ρ times the flux of **F** across the boundary of $\mathcal{D}$).

FIGURE 29 Buffalo stampede.

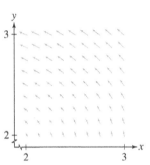

FIGURE 30 The vector field $\mathbf{F} = \langle xy - y^3, x^2 + y \rangle$.

SOLUTION We have

$$\text{div}(\mathbf{F}) = \frac{\partial}{\partial x}(xy - y^3) + \frac{\partial}{\partial y}(x^2 + y) = y + 1$$

Then using Green's Theorem, the net number of buffalo crossing the boundary of $\mathcal{D}$ per minute is ρ times the flux, or

$$\rho \int_{\partial \mathcal{D}} \mathbf{F} \cdot \mathbf{n}\, ds = \rho \iint_{\mathcal{D}} \text{div}(\mathbf{F})\, dA = 500 \int_2^3 \int_2^3 (y+1)\, dy\, dx$$

$$= 500 \int_2^3 \left(\frac{1}{2}y^2 + y\right)\bigg|_2^3 dx = 500 \int_2^3 \frac{7}{2}\, dx = 1750 \text{ buffalo/min}.$$

Further Insights and Challenges

In Exercises 46–49, the **Laplace operator** Δ is defined by

$$\Delta \varphi = \frac{\partial^2 \varphi}{\partial x^2} + \frac{\partial^2 \varphi}{\partial y^2} \qquad \boxed{12}$$

For any vector field $\mathbf{F} = \langle F_1, F_2 \rangle$, define the conjugate vector field $\mathbf{F}^* = \langle -F_2, F_1 \rangle$.

47. Let **n** be the outward-pointing unit normal vector to a simple closed curve C. The **normal derivative** of a function φ, denoted $\dfrac{\partial \varphi}{\partial \mathbf{n}}$, is the directional derivative $D_{\mathbf{n}}(\varphi) = \nabla \varphi \cdot \mathbf{n}$. Prove that

$$\oint_C \frac{\partial \varphi}{\partial \mathbf{n}}\, ds = \iint_{\mathcal{D}} \Delta \varphi\, dA$$

where $\mathcal{D}$ is the domain enclosed by a simple closed curve C. *Hint:* Let $\mathbf{F} = \nabla \varphi$. Show that $\dfrac{\partial \varphi}{\partial \mathbf{n}} = \mathbf{F}^* \cdot \mathbf{T}$, where **T** is the unit tangent vector, and apply Green's Theorem.

SOLUTION Note that $\mathbf{F} \cdot \mathbf{F}^* = \langle F_1, F_2 \rangle \cdot \langle -F_2, F_1 \rangle = 0$, so that $\mathbf{F}$ and $\mathbf{F}^*$ are orthogonal. Looking at the coordinates of $\mathbf{F}$ and $\mathbf{F}^*$, we see that $\mathbf{F}^*$ is just a rotation of $\mathbf{F}$ counterclockwise by $\frac{\pi}{2}$ (to see this more formally, think of $\mathbf{F}$ and $\mathbf{F}^*$ as lying in the xy-plane, and show that their cross product points in the positive z direction, so that $\mathbf{F}$ and $\mathbf{F}^*$ form a right-hand system). Additionally, the unit tangent $\mathbf{e}_n$ is a rotation of $\mathbf{n}$ by $\frac{\pi}{2}$ counterclockwise.

These properties imply that the angle θ between $\mathbf{F}$ and $\mathbf{n}$ is equal to the angle between $\mathbf{F}^*$ and $\mathbf{e}_n$, and $\|\mathbf{F}\| = \|\mathbf{F}^*\|$. Therefore,

$$\mathbf{F} \cdot \mathbf{n} = \|\mathbf{F}\|\|\mathbf{n}\| \cos\theta = \|\mathbf{F}\| \cos\theta$$
$$\mathbf{F}^* \cdot \mathbf{e}_n = \|\mathbf{F}^*\|\|\mathbf{e}_n\| \cos\theta = \|\mathbf{F}\| \cos\theta$$
$$\Rightarrow \mathbf{F} \cdot \mathbf{n} = \mathbf{F}^* \cdot \mathbf{e}_n$$

Now, if $\mathbf{F} = \nabla \varphi$, then

$$\frac{\partial \varphi}{\partial \mathbf{n}} = \nabla \varphi \cdot \mathbf{n} = \mathbf{F} \cdot \mathbf{n} = \mathbf{F}^* \cdot \mathbf{e}_n$$

By the definition of the vector line integral $\int_C \mathbf{F}^* \cdot d\mathbf{r} = \int_C (\mathbf{F}^* \cdot \mathbf{e}_n) \, ds$. Therefore,

$$\int_C \frac{\partial \varphi}{\partial \mathbf{n}} \, ds = \int_C (\mathbf{F}^* \cdot \mathbf{e}_n) \, ds = \int_C \mathbf{F}^* \cdot d\mathbf{r}$$

Using Green's Theorem and the equality $\text{curl}(\mathbf{F}^*) = \Delta\varphi$ obtained in Exercise 46, we get

$$\int_C \frac{\partial \varphi}{\partial \mathbf{n}} \, ds = \int_C \mathbf{F}^* \cdot d\mathbf{r} = \iint_D \text{curl}(\mathbf{F}^*) \, dA = \iint_D \Delta\varphi \, dA$$

49. Prove that $m(r) \leq I_\varphi(r) \leq M(r)$, where $m(r)$ and $M(r)$ are the minimum and maximum values of φ on C_r. Then use the continuity of φ to prove that $\lim_{r \to 0} I_\varphi(r) = \varphi(P)$.

SOLUTION $I_\varphi(r)$ is defined by

$$I_\varphi(r) = \frac{1}{2\pi} \int_0^{2\pi} \varphi(a + r\cos\theta, b + r\sin\theta) \, d\theta$$

The points on C_r have the form $(a + r\cos\theta, b + r\sin\theta)$. Therefore, since $m(r)$ and $M(r)$ are the minimum and maximum values of φ on C_r, we have for all $0 \leq \theta \leq 2\pi$,

$$m(r) \leq \varphi(a + r\cos\theta, b + r\sin\theta) \leq M(r)$$

Using properties of integrals (Eq. (6) in Section 5.2), we conclude that

$$2\pi m(r) \leq \int_0^{2\pi} \varphi(a + r\cos\theta + b + r\sin\theta) \leq 2\pi M(r)$$

Dividing by 2π we obtain

$$m(r) \leq I_\varphi(r) \leq M(r) \tag{1}$$

Now, since φ is continuous and the functions $\sin\theta$ and $\cos\theta$ are bounded for all $0 \leq \theta \leq 2\pi$, the following holds:

$$\lim_{r \to 0} \varphi(a + r\cos\theta, b + r\sin\theta) = \varphi\left(\lim_{r \to 0}(a + r\cos\theta, b + r\sin\theta)\right) = \varphi(a, b)$$

which means that for $\epsilon > 0$ there exists $\delta > 0$ so that

$$|\varphi(a + r\cos\theta, b + r\sin\theta) - \varphi(a, b)| < \epsilon$$

for all $0 \le \theta \le 2\pi$, whenever $0 < r < \delta$. Hence also

$$\lim_{r \to 0} m(r) = \lim_{r \to 0} M(r) = \varphi(a,b) \tag{2}$$

Combining (1), (2), and the Squeeze Theorem, we obtain the following conclusion:

$$\lim_{r \to 0} I_\varphi(r) = \varphi(a,b)$$

In Exercises 50 and 51, let $\mathcal{D}$ be the region bounded by a simple closed curve C. A function $\varphi(x,y)$ on $\mathcal{D}$ (whose second-order partial derivatives exist and are continuous) is called **harmonic** if $\Delta \varphi = 0$, where $\Delta \varphi$ is the Laplace operator defined in Eq. (12).

51. Show that $f(x,y) = x^2 - y^2$ is harmonic. Verify the mean-value property for $f(x,y)$ directly [expand $f(a + r\cos\theta, b + r\sin\theta)$ as a function of θ and compute $I_\varphi(r)$]. Show that $x^2 + y^2$ is not harmonic and does not satisfy the mean-value property.

SOLUTION We show that the function $f(x,y) = x^2 - y^2$ is harmonic by showing that $\Delta f = \frac{\partial^2 f}{\partial x^2} + \frac{\partial^2 f}{\partial y^2} = 0$. We have

$$\frac{\partial f}{\partial x} = 2x, \quad \frac{\partial f}{\partial y} = -2y$$

$$\frac{\partial^2 f}{\partial x^2} = 2, \quad \frac{\partial^2 f}{\partial y^2} = -2$$

Hence,

$$\Delta f = \frac{\partial^2 f}{\partial x^2} + \frac{\partial^2 f}{\partial y^2} = 2 - 2 = 0$$

We now verify the mean-value property for f. That is, we show that for all r,

$$I_f(r) = \frac{1}{2\pi} \int_0^{2\pi} f(a + r\cos\theta, b + r\sin\theta)\, d\theta = f(a,b)$$

We compute the integrand:

$$f(a + r\cos\theta, b + r\sin\theta) = x^2 - y^2 = (a + r\cos\theta)^2 - (b + r\sin\theta)^2$$
$$= a^2 + 2ar\cos\theta + r^2\cos^2\theta - \left(b^2 + 2br\sin\theta + r^2\sin^2\theta\right)$$
$$= a^2 - b^2 + 2r(a\cos\theta - b\sin\theta) + r^2\cos 2\theta$$

We compute the integral:

$$2\pi I_f(r) = \int_0^{2\pi} \left(a^2 - b^2 + 2r(a\cos\theta - b\sin\theta) + r^2\cos 2\theta\right) d\theta$$
$$= (a^2 - b^2)\theta + 2ar\sin\theta + 2br\cos\theta + \frac{r^2}{2}\sin 2\theta \Big|_{\theta=0}^{2\pi} = 2\pi(a^2 - b^2)$$

Hence,

$$I_f(r) = a^2 - b^2$$

However, we have $f(a,b) = a^2 - b^2$. Hence, for all r, $I_f(r) = f(a,b)$, which proves the mean-value property for f. For $g(x,y) = x^2 + y^2$ we have

$$g_{xx} = 2, \quad g_{yy} = 2, \quad \text{and} \quad \Delta g = 2 + 2 = 4 \ne 0$$

We check the mean value property:

$$I_g(r) = \frac{1}{2\pi} \int_0^{2\pi} g(a + r\cos\theta, b + r\sin\theta)\, d\theta = \frac{1}{2\pi} \int_0^{2\pi} (a + r\cos\theta)^2 + (b + r\sin\theta)^2\, d\theta$$
$$= \frac{1}{2\pi} \int_0^{2\pi} a^2 + b^2 + 2r(a\cos\theta + b\sin\theta) + r^2\, d\theta = a^2 + b^2 + r^2 \ne a^2 + b^2 = \varphi(a,b)$$

The mean value property does not hold for g.

17.2 Stokes' Theorem (LT Section 18.2)

Preliminary Questions

1. Indicate with an arrow the boundary orientation of the boundary curves of the surfaces in Figure 14, oriented by the outward-pointing normal vectors.

FIGURE 14

SOLUTION The indicated orientation is defined so that if the normal vector is moving along the boundary curve, the surface lies to the left. Since the surfaces are oriented by the outward-pointing normal vectors, the induced orientation is as shown in the figure:

2. Let $\mathbf{F} = \text{curl}(\mathbf{A})$. Which of the following are related by Stokes' Theorem?
(a) The circulation of $\mathbf{A}$ and flux of $\mathbf{F}$
(b) The circulation of $\mathbf{F}$ and flux of $\mathbf{A}$

SOLUTION Stokes' Theorem states that the circulation of $\mathbf{A}$ is equal to the flux of $\mathbf{F}$. The correct answer is (a).

3. What is the definition of a vector potential?

SOLUTION A vector field $\mathbf{A}$ such that $\mathbf{F} = \text{curl}(\mathbf{A})$ is a vector potential for $\mathbf{F}$.

4. Which of the following statements is correct?
(a) The flux of $\text{curl}(\mathbf{A})$ through every oriented surface is zero.
(b) The flux of $\text{curl}(\mathbf{A})$ through every closed, oriented surface is zero.

SOLUTION Statement (b) is the correct statement. The flux of $\text{curl}(\mathbf{F})$ through an oriented surface is not necessarily zero, unless the surface is closed.

5. Which condition on $\mathbf{F}$ guarantees that the flux through S_1 is equal to the flux through S_2 for any two oriented surfaces S_1 and S_2 with the same oriented boundary?

SOLUTION If $\mathbf{F}$ has a vector potential $\mathbf{A}$, then by a corollary of Stokes' Theorem,

$$\iint_S \mathbf{F} \cdot d\mathbf{S} = \int_C \mathbf{A} \cdot d\mathbf{r}$$

Therefore, if two oriented surfaces S_1 and S_2 have the same oriented boundary curve, C, then

$$\iint_{S_1} \mathbf{F} \cdot d\mathbf{S} = \int_C \mathbf{A} \cdot d\mathbf{r} \quad \text{and} \quad \iint_{S_2} \mathbf{F} \cdot d\mathbf{S} = \int_C \mathbf{A} \cdot d\mathbf{r}$$

Hence,

$$\iint_{S_1} \mathbf{F} \cdot d\mathbf{S} = \iint_{S_2} \mathbf{F} \cdot d\mathbf{S}$$

Exercises

In Exercises 1–4, verify Stokes' Theorem for the given vector field and surface, oriented with an upward-pointing normal.

1. $\mathbf{F} = \langle 2xy, x, y + z \rangle$, the surface $z = 1 - x^2 - y^2$ for $x^2 + y^2 \leq 1$

SOLUTION We must show that

$$\int_C \mathbf{F} \cdot d\mathbf{r} = \iint_S \text{curl}(\mathbf{F}) \cdot d\mathbf{S}$$

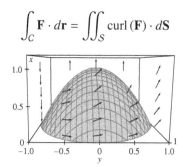

Step 1. Compute the line integral around the boundary curve. The boundary curve C is the unit circle oriented in the counterclockwise direction. We parametrize C by

$$\gamma(t) = (\cos t, \sin t, 0), \quad 0 \leq t \leq 2\pi$$

Then,

$$\mathbf{F}(\gamma(t)) = \langle 2 \cos t \sin t, \cos t, \sin t \rangle$$

$$\gamma'(t) = \langle -\sin t, \cos t, 0 \rangle$$

$$\mathbf{F}(\gamma(t)) \cdot \gamma'(t) = \langle 2 \cos t \sin t, \cos t, \sin t \rangle \cdot \langle -\sin t, \cos t, 0 \rangle = -2 \cos t \sin^2 t + \cos^2 t$$

We obtain the following integral:

$$\int_C \mathbf{F} \, d\mathbf{r} = \int_0^{2\pi} (-2 \cos t \sin^2 t + \cos^2 t) \, dt = -\frac{2 \sin^3 t}{3} + \frac{t}{2} + \frac{\sin 2t}{4} \Big|_0^{2\pi} = \pi \tag{1}$$

Step 2. Compute the flux of the curl through the surface. We parametrize the surface by

$$\Phi(\theta, t) = (t \cos \theta, t \sin \theta, 1 - t^2), \quad 0 \leq t \leq 1, \quad 0 \leq \theta \leq 2\pi$$

We compute the normal vector:

$$\mathbf{T}_\theta = \frac{\partial \Phi}{\partial \theta} = \langle -t \sin \theta, t \cos \theta, 0 \rangle$$

$$\mathbf{T}_t = \frac{\partial \Phi}{\partial t} = \langle \cos \theta, \sin \theta, -2t \rangle$$

$$\mathbf{T}_\theta \times \mathbf{T}_t = \begin{vmatrix} \mathbf{i} & \mathbf{j} & \mathbf{k} \\ -t \sin \theta & t \cos \theta & 0 \\ \cos \theta & \sin \theta & -2t \end{vmatrix} = (-2t^2 \cos \theta)\mathbf{i} - (2t^2 \sin \theta)\mathbf{j} - t(\sin^2 \theta + \cos^2 \theta)\mathbf{k}$$

$$= (-2t^2 \cos \theta)\mathbf{i} - (2t^2 \sin \theta)\mathbf{j} - t\mathbf{k}$$

Since the normal is always supposed to be pointing upward, the z-coordinate of the normal vector must be positive. Therefore, the normal vector is

$$\mathbf{n} = \langle 2t^2 \cos \theta, 2t^2 \sin \theta, t \rangle$$

We compute the curl:

$$\text{curl}(\mathbf{F}) = \begin{vmatrix} \mathbf{i} & \mathbf{j} & \mathbf{k} \\ \dfrac{\partial}{\partial x} & \dfrac{\partial}{\partial y} & \dfrac{\partial}{\partial z} \\ 2xy & x & y + z \end{vmatrix} = \mathbf{i} + (1 - 2x)\mathbf{k} = \langle 1, 0, 1 - 2x \rangle$$

We compute the curl in terms of the parameters:

$$\text{curl}(\mathbf{F}) = \langle 1, 0, 1 - 2t\cos\theta \rangle$$

Hence,

$$\text{curl}(\mathbf{F}) \cdot \mathbf{n} = \langle 1, 0, 1 - 2t\cos\theta \rangle \cdot \langle 2t^2\cos\theta, 2t^2\sin\theta, t \rangle = 2t^2\cos\theta + t - 2t^2\cos\theta = t$$

The surface integral is thus

$$\iint_S \text{curl}(\mathbf{F}) \cdot d\mathbf{S} = \int_0^{2\pi} \int_0^1 t\, dt\, d\theta = 2\pi \int_0^1 t\, dt = 2\pi \cdot \frac{t^2}{2}\bigg|_0^1 = \pi \qquad (2)$$

The values of the integrals in (1) and (2) are equal, as stated in Stokes' Theorem.

3. $\mathbf{F} = \langle e^{y-z}, 0, 0 \rangle$, the square with vertices $(1, 0, 1)$, $(1, 1, 1)$, $(0, 1, 1)$, and $(0, 0, 1)$

SOLUTION

Step 1. Compute the integral around the boundary curve. The boundary consists of four segments C_1, C_2, C_3, and C_4 shown in the figure:

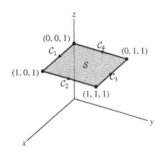

We parametrize the segments by

$$C_1 : \gamma_1(t) = (t, 0, 1), \quad 0 \le t \le 1$$
$$C_2 : \gamma_2(t) = (1, t, 1), \quad 0 \le t \le 1$$
$$C_3 : \gamma_3(t) = (1 - t, 1, 1), \quad 0 \le t \le 1$$
$$C_4 : \gamma_4(t) = (0, 1 - t, 1), \quad 0 \le t \le 1$$

We compute the following values:

$$\mathbf{F}(\gamma_1(t)) = \langle e^{y-z}, 0, 0 \rangle = \langle e^{-1}, 0, 0 \rangle$$
$$\mathbf{F}(\gamma_2(t)) = \langle e^{y-z}, 0, 0 \rangle = \langle e^{t-1}, 0, 0 \rangle$$
$$\mathbf{F}(\gamma_3(t)) = \langle e^{y-z}, 0, 0 \rangle = \langle 1, 0, 0 \rangle$$
$$\mathbf{F}(\gamma_4(t)) = \langle e^{y-z}, 0, 0 \rangle = \langle e^{-t-1}, 0, 0 \rangle$$

Hence,

$$\mathbf{F}(\gamma_1(t)) \cdot \gamma_1'(t) = \langle e^{-1}, 0, 0 \rangle \cdot \langle 1, 0, 0 \rangle = e^{-1}$$
$$\mathbf{F}(\gamma_2(t)) \cdot \gamma_2'(t) = \langle e^{t-1}, 0, 0 \rangle \cdot \langle 0, 1, 0 \rangle = 0$$
$$\mathbf{F}(\gamma_3(t)) \cdot \gamma_3'(t) = \langle 1, 0, 0 \rangle \cdot \langle -1, 0, 0 \rangle = -1$$
$$\mathbf{F}(\gamma_4(t)) \cdot \gamma_4'(t) = \langle e^{-t-1}, 0, 0 \rangle \cdot \langle 0, -1, 0 \rangle = 0$$

We obtain the following integral:

$$\int_C \mathbf{F} \cdot d\mathbf{r} = \sum_{i=1}^4 \int_{C_i} \mathbf{F} \cdot d\mathbf{r} = \int_0^1 e^{-1}\, dt + 0 + \int_0^1 (-1)\, dt + 0 = e^{-1} - 1$$

Step 2. Compute the curl.

$$\text{curl}(\mathbf{F}) = \begin{vmatrix} \mathbf{i} & \mathbf{j} & \mathbf{k} \\ \frac{\partial}{\partial x} & \frac{\partial}{\partial y} & \frac{\partial}{\partial z} \\ e^{y-z} & 0 & 0 \end{vmatrix} = -e^{y-z}\,\mathbf{j} - e^{y-z}\,\mathbf{k} = \langle 0, -e^{y-z}, -e^{y-z} \rangle$$

Step 3. Compute the flux of the curl through the surface. We parametrize the surface by

$$\Phi(x,y) = (x,y,1), \quad 0 \le x, y \le 1$$

The upward pointing normal is $\mathbf{n} = \langle 0, 0, 1 \rangle$. We express curl($\mathbf{F}$) in terms of the parameters x and y:

$$\text{curl}(\mathbf{F})(\Phi(x,y)) = \langle 0, -e^{y-1}, -e^{y-1} \rangle$$

Hence,

$$\text{curl}(\mathbf{F}) \cdot \mathbf{n} = \langle 0, -e^{y-1}, -e^{y-1} \rangle \cdot \langle 0, 0, 1 \rangle = -e^{y-1}$$

The surface integral is thus

$$\iint_S \text{curl}(\mathbf{F}) \cdot d\mathbf{S} = \iint_D -e^{y-1}\, dA = \int_0^1 \int_0^1 -e^{y-1}\, dy\, dx = \int_0^1 -e^{y-1}\, dy = -e^{y-1}\Big|_0^1$$

$$= -1 + e^{-1} = e^{-1} - 1 \tag{1}$$

We see that the integrals in (1) and (2) are equal.

In Exercises 5–10, calculate curl($\mathbf{F}$) *and then apply Stokes' Theorem to compute the flux of* curl($\mathbf{F}$) *through the given surface using a line integral.*

5. $\mathbf{F} = \langle e^{z^2} - y, e^{z^3} + x, \cos(xz) \rangle$, the upper half of the unit sphere $x^2 + y^2 + z^2 = 1$, $z \ge 0$ with outward-pointing normal

SOLUTION curl($\mathbf{F}$) is

$$\text{curl}(\mathbf{F}) = \begin{vmatrix} \mathbf{i} & \mathbf{j} & \mathbf{k} \\ \partial x & \partial y & \partial z \\ e^{z^2} - y & e^{z^3} + x & \cos(xz) \end{vmatrix} = \langle -3z^2 e^{z^3}, 2ze^{z^2} + z\sin(xz), 2 \rangle$$

We compute the flux of the curl through the surface by using Stokes' Theorem and computing the line integral around the boundary curve. The boundary curve is the unit circle oriented in the counterclockwise direction, using the orientation rule described at the start of this section. We use the parametrization $\gamma(t) = \langle \cos t, \sin t, 0 \rangle$, $0 \le t \le 2\pi$. Then

$$\mathbf{F}(\gamma(t)) \cdot \gamma'(t) = \langle e^0 - \sin t, e^0 + \cos t, \cos(0) \rangle \cdot \langle -\sin t, \cos t, 0 \rangle$$

$$= \langle 1 - \sin t, 1 + \cos t, 1 \rangle \cdot \langle -\sin t, \cos t, 0 \rangle$$

$$= -\sin t(1 - \sin t) + \cos t(1 + \cos t) + 0$$

$$= -\sin t + \sin^2 t + \cos t + \cos^2 t$$

$$= 1 - \sin t + \cos t$$

and the line integral is:

$$\int_C \mathbf{F} \cdot d\mathbf{s} = \int_0^{2\pi} (1 - \sin t + \cos t)\, dt = t + \cos t + \sin t \Big|_0^{2\pi} = 2\pi$$

7. $\mathbf{F} = \langle 3z, 5x, -2y \rangle$, that part of the paraboloid $z = x^2 + y^2$ that lies below the plane $z = 4$ with upward-pointing unit normal

SOLUTION curl($\mathbf{F}$) is

$$\text{curl}(\mathbf{F}) = \begin{vmatrix} \mathbf{i} & \mathbf{j} & \mathbf{k} \\ \partial x & \partial y & \partial z \\ 3z & 5x & -2y \end{vmatrix} = \langle -2, 3, 5 \rangle$$

We compute the flux of the curl through the surface by using Stokes' Theorem and computing the line integral around the boundary curve. The boundary curve is the intersection of the paraboloid with the plane $z = 4$, which is the circle $x^2 + y^2 = 4$ of radius 2 in the plane $z = 4$. The orientation of the boundary, given an upward-pointing normal, is counterclockwise, using the orientation rule described at the start of this section. We use the parametrization $\gamma(t) = \langle 2\cos t, 2\sin t, 4 \rangle, 0 \leq t \leq 2\pi$. Then

$$\mathbf{F}(\gamma(t)) \cdot \gamma'(t) = \langle 12, 10\cos t, -4\sin t \rangle \cdot \langle -2\sin t, 2\cos t, 0 \rangle = -24\sin t + 20\cos^2 t$$

and the line integral is:

$$\int_C \mathbf{F} \cdot d\mathbf{s} = \int_0^{2\pi} (-24\sin t + 20\cos^2 t) \, dt = \int_0^{2\pi} (-24\sin t + 10 + 10\cos 2t) \, dt$$

$$= (24\cos t + 10t + 5\sin 2t)\Big|_0^{2\pi} = 20\pi$$

9. $\mathbf{F} = \langle yz, xz, xy \rangle$, that part of the cylinder $x^2 + y^2 = 1$ that lies between the two planes $z = 1$ and $z = 4$ with outward-pointing unit normal vector

SOLUTION curl($\mathbf{F}$) is

$$\text{curl}(\mathbf{F}) = \begin{vmatrix} \mathbf{i} & \mathbf{j} & \mathbf{k} \\ \partial x & \partial y & \partial z \\ yz & xz & xy \end{vmatrix} = \langle 0, 0, 0 \rangle$$

Therefore the flux of the curl is zero. We compute the flux of the curl through the surface using Stokes' Theorem by computing the line integral around the boundary curve. The boundary curve consists of two circles. The two boundaries are each circles $x^2 + y^2 = 1$, lying in the planes $z = 1$ and $z = 4$. As for orientation, using the orientation rule described at the start of this section shows that the orientation of the circle at $z = 4$ is clockwise while the orientation of the circle at $z = 1$ is counterclockwise. Use the following parametrizations for the boundary circles:

$$C_4 \text{ (Upper): } \gamma_4(t) = \langle \cos t, -\sin t, 4 \rangle, \qquad C_1 \text{ (Lower): } \gamma_1(t) = \langle \cos t, \sin t, 1 \rangle, \quad 0 \leq t \leq 2\pi$$

Then

$$\mathbf{F}(\gamma_4(t)) \cdot \gamma_4'(t) = \langle -4\sin t, 4\cos t, -\sin t \cos t \rangle \cdot \langle -\sin t, -\cos t, 0 \rangle = 4\sin^2 t - 4\cos^2 t = -4\cos 2t$$

$$\mathbf{F}(\gamma_1(t)) \cdot \gamma_1'(t) = \langle \sin t, \cos t, \sin t \cos t \rangle \cdot \langle -\sin t, \cos t, 0 \rangle = -\sin^2 t + \cos^2 t = \cos 2t$$

So the line integrals along the boundary are

$$\int_{C_1} \mathbf{F} \cdot d\mathbf{s} + \int_{C_4} \mathbf{F} \cdot d\mathbf{s} = \int_0^{2\pi} \cos 2t \, dt + \int_0^{2\pi} (-4\cos 2t) \, dt = -3\int_0^{2\pi} \cos 2t \, dt = -\frac{3}{2}\sin 2t \Big|_0^{2\pi} = 0$$

The integral is zero, as expected.

In Exercises 11–16, apply Stokes' Theorem to evaluate $\oint_C \mathbf{F} \cdot d\mathbf{r}$ by finding the flux of curl($\mathbf{F}$) across an appropriate surface.

11. $\mathbf{F} = \langle 3y, -2x, 3y \rangle$, C is the circle $x^2 + y^2 = 9, z = 2$, oriented counterclockwise as viewed from above.

SOLUTION Choose for the surface S the interior of the given circle, and parametrize S by

$$\mathbf{r}(r, \theta) = \langle r\cos\theta, r\sin\theta, 2 \rangle, \; 0 \leq r \leq 3, \; 0 \leq \theta \leq 2\pi$$

Now, curl($\mathbf{F}$) is

$$\text{curl}(\mathbf{F}) = \begin{vmatrix} \mathbf{i} & \mathbf{j} & \mathbf{k} \\ \partial x & \partial y & \partial z \\ 3y & -2x & 3y \end{vmatrix} = \langle 3, 0, -5 \rangle$$

so in the given parametrization it is also curl($\mathbf{F}$) = $\langle 3, 0, -5 \rangle$. Compute a normal to the surface:

$$\mathbf{r}_r = \langle \cos\theta, \sin\theta, 0 \rangle$$

$$\mathbf{r}_\theta = \langle -r\sin\theta, r\cos\theta, 0 \rangle$$

$$\mathbf{r}_r \times \mathbf{r}_\theta = \begin{vmatrix} \mathbf{i} & \mathbf{j} & \mathbf{k} \\ \cos\theta & \sin\theta & 0 \\ -r\sin\theta & r\cos\theta & 0 \end{vmatrix} = \langle 0, 0, r \rangle$$

Since the boundary orientation is counterclockwise, we want an upward-pointing normal, so we choose $\mathbf{N} = \langle 0, 0, r \rangle$. Then the integral is

$$\iint_S \text{curl}(\mathbf{F}) \cdot d\mathbf{r} = \int_0^3 \int_0^{2\pi} \langle 3, 0, -5 \rangle \cdot \langle 0, 0, r \rangle \, d\theta \, dr$$

$$= \int_0^3 \int_0^{2\pi} -5r \, d\theta \, dr = \int_0^3 (-10\pi r) \, dr = -5\pi r^2 \Big|_0^3 = -45\pi$$

13. $\mathbf{F} = \langle xz, xy, yz \rangle$, C is the rectangle with vertices $(0,0,0)$, $(0,0,2)$, $(3,0,2)$, and $(3,0,0)$, oriented counterclockwise as viewed from the positive y-axis.

SOLUTION Choose for the surface S the interior of the given rectangle, and parametrize S by

$$\mathbf{r}(x, z) = \langle x, 0, z \rangle, \ 0 \le x \le 3, \ 0 \le z \le 2$$

Now, curl($\mathbf{F}$) is

$$\text{curl}(\mathbf{F}) = \begin{vmatrix} \mathbf{i} & \mathbf{j} & \mathbf{k} \\ \partial x & \partial y & \partial z \\ xz & xy & yz \end{vmatrix} = \langle z, x, y \rangle$$

so in the given parametrization it is curl($\mathbf{F}$) = $\langle z, x, 0 \rangle$. Compute a normal to the surface:

$$\mathbf{r}_x = \langle 1, 0, 0 \rangle$$

$$\mathbf{r}_z = \langle 0, 0, 1 \rangle$$

$$\mathbf{r}_x \times \mathbf{r}_z = \begin{vmatrix} \mathbf{i} & \mathbf{j} & \mathbf{k} \\ 1 & 0 & 0 \\ 0 & 0 & 1 \end{vmatrix} = \langle 0, -1, 0 \rangle$$

Since the boundary orientation is counterclockwise as viewed from the positive y-axis, we want a normal that points that way, so we choose $\mathbf{N} = \langle 0, 1, 0 \rangle$. Then the integral is

$$\iint_S \text{curl}(\mathbf{F}) \cdot d\mathbf{r} = \int_0^3 \int_0^2 \langle z, x, 0 \rangle \cdot \langle 0, 1, 0 \rangle \, dz \, dx = \int_0^3 \int_0^2 x \, dz \, dx = \int_0^3 2x \, dx = x^2 \Big|_0^3 = 9$$

15. $\mathbf{F} = \langle y, z, x \rangle$, C is the triangle with vertices $(0, 0, 0)$, $(3, 0, 0)$, and $(0, 3, 3)$, oriented counterclockwise as viewed from above.

SOLUTION To determine the equation of the plane the triangle lies in, we compute its normal vector:

$$\begin{vmatrix} \mathbf{i} & \mathbf{j} & \mathbf{k} \\ 3 & 0 & 0 \\ 0 & 3 & 3 \end{vmatrix} = \langle 0, -9, 9 \rangle$$

So the plane in which the triangle lies is $-9y + 9z = 0$, or $y = z$. Choose for the surface S the interior of the given triangle, and parametrize S by

$$\mathbf{r}(x, y) = \langle x, y, y \rangle, \ 0 \le x \le 3, \ 0 \le y \le 3 - x$$

Now, curl($\mathbf{F}$) is

$$\text{curl}(\mathbf{F}) = \begin{vmatrix} \mathbf{i} & \mathbf{j} & \mathbf{k} \\ \partial x & \partial y & \partial z \\ y & z & x \end{vmatrix} = \langle -1, -1, -1 \rangle$$

Compute a normal to the surface:

$$\mathbf{r}_x = \langle 1, 0, 0 \rangle$$

$$\mathbf{r}_y = \langle 0, 1, 1 \rangle$$

$$\mathbf{r}_x \times \mathbf{r}_y = \begin{vmatrix} \mathbf{i} & \mathbf{j} & \mathbf{k} \\ 1 & 0 & 0 \\ 0 & 1 & 1 \end{vmatrix} = \langle 0, -1, 1 \rangle$$

Since the boundary orientation is counterclockwise, we want an upward-pointing normal, so we choose $\mathbf{N} = \langle 0, -1, 1 \rangle$. Then the integral is

$$\iint_S \text{curl}(\mathbf{F}) \cdot d\mathbf{r} = \int_0^3 \int_0^{3-x} \langle -1, -1, -1 \rangle \cdot \langle 0, -1, 1 \rangle \, dy \, dx = \int_0^3 \int_0^{3-x} 0 \, dy \, dx = 0$$

17. Let S be the surface of the cylinder (not including the top and bottom) of radius 2 for $1 \le z \le 6$, oriented with outward-pointing normal (Figure 16).

FIGURE 16

(a) Indicate with an arrow the orientation of ∂S (the top and bottom circles).
(b) Verify Stokes' Theorem for S and $\mathbf{F} = \langle yz^2, 0, 0 \rangle$.

SOLUTION
(a) The induced orientation is defined so that as the normal vector travels along the boundary curve, the surface lies to its left. Therefore, the boundary circles on top and bottom have opposite orientations, which are shown in the figure.

(b) We verify Stokes' Theorem for S and $\mathbf{F} = \langle yz^2, 0, 0 \rangle$.
Step 1. Compute the integral around the boundary circles. We use the following parametrizations:

$$C_1 : \gamma_1(t) = (2\cos t, 2\sin t, 6), \quad t \text{ from } 2\pi \text{ to } 0$$
$$C_2 : \gamma_2(t) = (2\cos t, 2\sin t, 1), \quad t \text{ from } 0 \text{ to } 2\pi$$

We compute the following values:

$$\mathbf{F}(\gamma_1(t)) = \langle yz^2, 0, 0 \rangle = \langle 72 \sin t, 0, 0 \rangle,$$
$$\gamma_1'(t) = \langle -2 \sin t, 2 \cos t, 0 \rangle$$
$$\mathbf{F}(\gamma_1(t)) \cdot \gamma_1'(t) = \langle 72 \sin t, 0, 0 \rangle \cdot \langle -2 \sin t, 2 \cos t, 0 \rangle = -144 \sin^2 t$$
$$\mathbf{F}(\gamma_2(t)) = \langle yz^2, 0, 0 \rangle = \langle 2 \sin t, 0, 0 \rangle,$$
$$\gamma_2'(t) = \langle -2 \sin t, 2 \cos t, 0 \rangle$$
$$\mathbf{F}(\gamma_2(t)) \cdot \gamma_2'(t) = \langle 2 \sin t, 0, 0 \rangle \cdot \langle -2 \sin t, 2 \cos t, 0 \rangle = -4 \sin^2 t$$

The line integral is thus

$$\int_C \mathbf{F} \cdot d\mathbf{r} = \int_{C_1} \mathbf{F} \cdot d\mathbf{r} + \int_{C_2} \mathbf{F} \cdot d\mathbf{r} = \int_{2\pi}^0 (-144 \sin^2 t) \, dt + \int_0^{2\pi} (-4 \sin^2 t) \, dt$$
$$= \int_0^{2\pi} 140 \sin^2 t \, dt = 140 \int_0^{2\pi} \frac{1 - \cos 2t}{2} \, dt = 70 \cdot 2\pi - \left. \frac{70 \sin 2t}{2} \right|_0^{2\pi} = 140\pi$$

Step 2. Compute the curl

$$\text{curl}(\mathbf{F}) = \begin{vmatrix} \mathbf{i} & \mathbf{j} & \mathbf{k} \\ \frac{\partial}{\partial x} & \frac{\partial}{\partial y} & \frac{\partial}{\partial z} \\ yz^2 & 0 & 0 \end{vmatrix} = (2yz)\mathbf{j} - z^2\mathbf{k} = \langle 0, 2yz, -z^2 \rangle$$

Step 3. Compute the flux of the curl through the surface. We parametrize S by

$$\Phi(\theta, z) = (2\cos\theta, 2\sin\theta, z), \quad 0 \leq \theta \leq 2\pi, \quad 1 \leq z \leq 6$$

In Example 2 of Chapter 17.4 it is shown that the outward pointing normal is

$$\mathbf{n} = \langle 2\cos\theta, 2\sin\theta, 0 \rangle$$

We compute the dot product:

$$\text{curl}(\mathbf{F})\,(\Phi(\theta, z)) \cdot \mathbf{n} = \langle 0, 4z\sin\theta, -z^2 \rangle \cdot \langle 2\cos\theta, 2\sin\theta, 0 \rangle = 8z\sin^2\theta$$

We obtain the following integral (and use the integral we computed before):

$$\iint_S \text{curl}(\mathbf{F}) \cdot d\mathbf{S} = \int_1^6 \int_0^{2\pi} 8z\sin^2\theta\, d\theta\, dz = \left(\int_1^6 8z\, dz \right)\left(\int_0^{2\pi} \sin^2\theta\, d\theta \right) = 4z^2 \Big|_1^6 \cdot \pi = 140\pi$$

The line integral and the flux have the same value. This verifies Stokes' Theorem.

19. Let I be the flux of $\mathbf{F} = \langle e^y, 2xe^{x^2}, z^2 \rangle$ through the upper hemisphere S of the unit sphere.

(a) Let $\mathbf{G} = \langle e^y, 2xe^{x^2}, 0 \rangle$. Find a vector field $\mathbf{A}$ such that $\text{curl}(\mathbf{A}) = \mathbf{G}$.

(b) Use Stokes' Theorem to show that the flux of $\mathbf{G}$ through S is zero. *Hint:* Calculate the circulation of $\mathbf{A}$ around ∂S.

(c) Calculate I. *Hint:* Use (b) to show that I is equal to the flux of $\langle 0, 0, z^2 \rangle$ through S.

SOLUTION

(a) We search for a vector field $\mathbf{A}$ so that $\mathbf{G} = \text{curl}(\mathbf{A})$. That is,

$$\left\langle \frac{\partial A_3}{\partial y} - \frac{\partial A_2}{\partial z}, \frac{\partial A_1}{\partial z} - \frac{\partial A_3}{\partial x}, \frac{\partial A_2}{\partial x} - \frac{\partial A_1}{\partial y} \right\rangle = \langle e^y, 2xe^{x^2}, 0 \rangle$$

We note that the third coordinate of this curl vector must be zero; this can be satisfied if $A_1 = 0$ and $A_2 = 0$. With this in mind, we let $\mathbf{A} = \langle 0, 0, e^y - e^{x^2} \rangle$. The vector field $\mathbf{A} = \langle 0, 0, e^y - e^{x^2} \rangle$ satisfies this equality. Indeed,

$$\frac{\partial A_3}{\partial y} - \frac{\partial A_2}{\partial z} = e^y, \quad \frac{\partial A_1}{\partial z} - \frac{\partial A_3}{\partial x} = 2xe^{x^2}, \quad \frac{\partial A_2}{\partial x} - \frac{\partial A_1}{\partial y} = 0$$

(b) We found that $\mathbf{G} = \text{curl}(\mathbf{A})$, where $\mathbf{A} = \langle 0, 0, e^y - e^{x^2} \rangle$. We compute the flux of $\mathbf{G}$ through S. By Stokes' Theorem,

$$\iint_S \mathbf{G} \cdot d\mathbf{S} = \iint_S \text{curl}(\mathbf{A}) \cdot d\mathbf{S} = \int_C \mathbf{A} \cdot d\mathbf{r}$$

The boundary C is the circle $x^2 + y^2 = 1$, parametrized by

$$\gamma(t) = (\cos t, \sin t, 0), \quad 0 \leq t \leq 2\pi$$

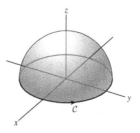

We compute the following values:

$$\mathbf{A}(\gamma(t)) = \left\langle 0, 0, e^y - e^{x^2} \right\rangle = \left\langle 0, 0, e^{\sin t} - e^{\cos^2 t} \right\rangle$$

$$\gamma'(t) = \langle -\sin t, \cos t, 0 \rangle$$

$$\mathbf{A}(\gamma(t)) \cdot \gamma'(t) = \left\langle 0, 0, e^{\sin t} - e^{\cos^2 t} \right\rangle \cdot \langle -\sin t, \cos t, 0 \rangle = 0$$

Therefore,

$$\int_C \mathbf{A} \cdot d\mathbf{s} = \int_0^{2\pi} 0\, dt = 0$$

(c) We rewrite the vector field $\mathbf{F} = \left\langle e^y, 2xe^{x^2}, z^2 \right\rangle$ as

$$\mathbf{F} = \left\langle e^y, 2xe^{x^2}, z^2 \right\rangle = \left\langle e^y, 2xe^{x^2}, 0 \right\rangle + \left\langle 0, 0, z^2 \right\rangle = \text{curl}(\mathbf{A}) + \left\langle 0, 0, z^2 \right\rangle$$

Therefore,

$$\iint_S \mathbf{F} \cdot d\mathbf{S} = \iint_S \text{curl}(\mathbf{A}) \cdot d\mathbf{S} + \iint_S \left\langle 0, 0, z^2 \right\rangle \cdot d\mathbf{S} \quad (1)$$

In part (b) we showed that the first integral on the right-hand side is zero. Therefore,

$$\iint_S \mathbf{F} \cdot d\mathbf{S} = \iint_S \left\langle 0, 0, z^2 \right\rangle \cdot d\mathbf{S} \quad (2)$$

The upper hemisphere is parametrized by

$$\Phi(\theta, \phi) = (\cos\theta \sin\phi, \sin\theta \sin\phi, \cos\phi), \quad 0 \le \theta \le 2\pi, \quad 0 \le \phi \le \frac{\pi}{2}$$

with the outward pointing normal

$$\mathbf{n} = \sin\phi \left\langle \cos\theta \sin\phi, \sin\theta \sin\phi, \cos\phi \right\rangle$$

See Example 4, Section 17.4. We have

$$\left\langle 0, 0, \cos^2\phi \right\rangle \cdot \mathbf{n} = \sin\phi \cos^3\phi$$

Therefore,

$$\iint_S \left\langle 0, 0, z^2 \right\rangle \cdot d\mathbf{S} = \int_0^{2\pi} \int_0^{\pi/2} \sin\phi \cos^3\phi\, d\phi\, d\theta = 2\pi \int_0^{\pi/2} \sin\phi \cos^3\phi\, d\phi$$

$$= 2\pi \frac{-\cos^4\phi}{4} \bigg|_0^{\pi/2} = -\frac{\pi}{2}(0 - 1) = \frac{\pi}{2}$$

Combining with (2) we obtain the solution

$$\iint_S \mathbf{F} \cdot d\mathbf{S} = \frac{\pi}{2}$$

21. Let $\mathbf{A}$ be the vector potential and $\mathbf{B}$ the magnetic field of the infinite solenoid of radius R in Example 4. Use Stokes' Theorem to compute:
(a) The flux of $\mathbf{B}$ through a circle in the xy-plane of radius $r < R$
(b) The circulation of $\mathbf{A}$ around the boundary C of a surface lying outside the solenoid

SOLUTION
(a) In Example 4 it is shown that $\mathbf{B} = \text{curl}(\mathbf{A})$, where

$$\mathbf{A} = \begin{cases} \dfrac{1}{2} R^2 B \left\langle -\dfrac{y}{r^2}, \dfrac{x}{r^2}, 0 \right\rangle & \text{if } r > R \\ \dfrac{1}{2} B \langle -y, x, 0 \rangle & \text{if } r < R \end{cases} \quad (1)$$

Therefore, using Stokes' Theorem, we have (S is the disk of radius r in the xy-plane)

$$\iint_S \mathbf{B} \cdot d\mathbf{S} = \iint_S \text{curl}(\mathbf{A}) \cdot d\mathbf{S} = \int_{\partial S} \mathbf{A} \cdot d\mathbf{s} \quad (2)$$

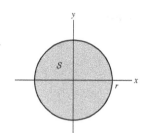

We parametrize the circle $C = \partial S$ by $\mathbf{c}(t) = \langle r\cos t, r\sin t, 0\rangle$, $0 \le t \le 2\pi$. Then
$$\mathbf{c}'(t) = \langle -r\sin t, r\cos t, 0\rangle$$
By (1) for $r < R$,
$$\mathbf{A}(\mathbf{c}(t)) = \frac{1}{2}B\langle -r\sin t, r\cos t, 0\rangle$$
Hence,
$$\mathbf{A}(\mathbf{c}(t))\cdot \mathbf{c}'(t) = \frac{1}{2}B\langle -r\sin t, r\cos t, 0\rangle \cdot \langle -r\sin t, r\cos t, 0\rangle = \frac{1}{2}B\left(r^2\sin^2 t + r^2\cos^2 t\right) = \frac{1}{2}r^2 B$$
Now, by (2) we get
$$\iint_S \mathbf{B}\cdot d\mathbf{S} = \int_{\partial S}\mathbf{A}\cdot d\mathbf{S} = \int_0^{2\pi}\frac{1}{2}r^2\mathbf{B}\,dt = \frac{1}{2}r^2\mathbf{B}\int_0^{2\pi}dt = r^2\mathbf{B}\pi$$

(b) Outside the solenoid $\mathbf{B}$ is the zero field, hence $\mathbf{B} = \mathbf{0}$ on every domain lying outside the solenoid. Therefore, Stokes' Theorem implies that
$$\int_{\partial S}\mathbf{A}\cdot d\mathbf{S} = \iint_S \text{curl}(\mathbf{A})\cdot d\mathbf{S} = \iint_S \mathbf{B}\cdot d\mathbf{S} = \iint_S \mathbf{0}\cdot d\mathbf{S} = 0$$

23. A uniform magnetic field $\mathbf{B}$ has constant strength b in the z-direction [i.e., $\mathbf{B} = \langle 0, 0, b\rangle$].
(a) Verify that $\mathbf{A} = \frac{1}{2}\mathbf{B}\times \mathbf{r}$ is a vector potential for $\mathbf{B}$, where $\mathbf{r} = \langle x, y, 0\rangle$.
(b) Calculate the flux of $\mathbf{B}$ through the rectangle with vertices A, B, C, and D in Figure 19.

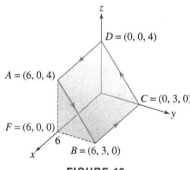

FIGURE 19

SOLUTION

(a) We compute the vector $\mathbf{A} = \frac{1}{2}\mathbf{B}\times \mathbf{r}$. Since $\mathbf{B} = b\mathbf{k}$ and $\mathbf{r} = x\mathbf{i} + y\mathbf{j}$, we have
$$\mathbf{A} = \frac{1}{2}\mathbf{B}\times \mathbf{r} = \frac{1}{2}b\mathbf{k}\times(x\mathbf{i}+y\mathbf{j}) = \frac{1}{2}b(x\mathbf{k}\times\mathbf{i}+y\mathbf{k}\times\mathbf{j}) = \frac{1}{2}b(x\mathbf{j}-y\mathbf{i}) = \left\langle -\frac{by}{2}, \frac{bx}{2}, 0\right\rangle$$

We now show that $\text{curl}(\mathbf{A}) = \mathbf{B}$. We compute the curl of $\mathbf{A}$:
$$\text{curl}(\mathbf{A}) = \begin{vmatrix} \mathbf{i} & \mathbf{j} & \mathbf{k} \\ \dfrac{\partial}{\partial x} & \dfrac{\partial}{\partial y} & \dfrac{\partial}{\partial z} \\ -\dfrac{by}{2} & \dfrac{bx}{2} & 0 \end{vmatrix} = \left\langle 0, 0, \frac{b}{2}+\frac{b}{2}\right\rangle = \langle 0, 0, b\rangle = \mathbf{B}$$

Therefore, $\mathbf{A}$ is a vector potential for $\mathbf{B}$.

(b) Let S be the rectangle $\square ABCD$ and let C be the boundary of S. Since $\mathbf{B} = \text{Curl}(\mathbf{A})$, we see that $\mathbf{B}$ has a vector potential. It follows, as explained in this section, that the flux of $\mathbf{B}$ through rectangle S is equal to the flux of $\mathbf{B}$ through any surface with the same boundary C. Let S' be the wedge-shaped box with four sides and open top. Since the boundary of S' is also C, we have

$$\iint_S \mathbf{B} \cdot d\mathbf{S} = \iint_{S'} \mathbf{B} \cdot d\mathbf{S}$$

The vector field $\mathbf{B}$ points in the $\mathbf{k}$ direction, so it has zero flux through the three vertical sides of S'. On the other hand, the unit normal vector to the bottom face of S' is $\mathbf{k}$, so the normal component of $\mathbf{B}$ along the bottom face is equal to b. We obtain

$$\iint_{S'} \mathbf{B} \cdot d\mathbf{S} = \iint_{\text{Bottom Face of } S'} b\, dA$$

$$= b(\text{Area of Bottom Face of } S') = 18b$$

25. Let $\mathbf{F} = \langle y^2, 2z + x, 2y^2 \rangle$. Use Stokes' Theorem to find a plane with equation $ax + by + cz = 0$ (where a, b, c are not all zero) such that $\oint_C \mathbf{F} \cdot d\mathbf{r} = 0$ for every closed C lying in the plane. *Hint:* Choose a, b, c so that $\text{curl}(\mathbf{F})$ lies in the plane.

SOLUTION Since we are interested in $\oint_C \mathbf{F} \cdot d\mathbf{s}$, we can also consider $\iint \text{curl}\mathbf{F} \cdot d\mathbf{S}$, by Stokes' Theorem. The curl is $\langle 4y - 2, 0, 1 - 2y \rangle$ and the normal to the plane is $\mathbf{n} = \langle a, b, c \rangle$. They are orthogonal if

$$\langle 4y - 2, 0, 1 - 2y \rangle \cdot \langle a, b, c \rangle = a(4y - 2) + c(1 - 2y) = 0$$

which means:

$$4ay - 2a + c - 2cy = 0 \quad \Rightarrow \quad (4a - 2c) = 0, (c - 2a) = 0$$

This yields $c = 2a$ and b is arbitrary.

27. Let $\mathbf{F} = \langle y^2, x^2, z^2 \rangle$. Show that

$$\int_{C_1} \mathbf{F} \cdot d\mathbf{r} = \int_{C_2} \mathbf{F} \cdot d\mathbf{r}$$

for any two closed curves going around a cylinder whose central axis is the z-axis as shown in Figure 21.

FIGURE 21

SOLUTION We denote by S the part of the cylinder for which C_1 and C_2 are boundary curves. Using Stokes' Theorem (notice that C_1 and C_2 have the same orientations), we have

$$\int_{C_1} \mathbf{F} \cdot d\mathbf{r} - \int_{C_2} \mathbf{F} \cdot d\mathbf{r} = \iint_S \text{curl}(\mathbf{F}) \cdot d\mathbf{S} \quad (1)$$

We compute the curl:

$$\text{curl}(\mathbf{F}) = \left\langle \frac{\partial F_3}{\partial y} - \frac{\partial F_2}{\partial z}, \frac{\partial F_1}{\partial z} - \frac{\partial F_3}{\partial x}, \frac{\partial F_2}{\partial x} - \frac{\partial F_1}{\partial y} \right\rangle = \langle 0, 0, 2x - 2y \rangle$$

We parametrize S by

$$\Phi(\theta, z) = \langle R\cos\theta, R\sin\theta, z \rangle$$

where (θ, z) varies in a certain parameter domain $\mathcal{D}$. The outward-pointing normal is

$$\mathbf{n} = \langle R\cos\theta, R\sin\theta, 0\rangle$$

We compute curl($\mathbf{F}$) in terms of the parameters:

$$\text{curl}(\mathbf{F}) = \langle 0, 0, 2x - 2y\rangle = \langle 0, 0, 2R\cos\theta - 2R\sin\theta\rangle$$

We compute the dot product:

$$\text{curl}(\mathbf{F}) \cdot \mathbf{n} = 2R\langle 0, 0, \cos\theta - \sin\theta\rangle \cdot R\langle\cos\theta, \sin\theta, 0\rangle = 2R^2(0 + 0 + 0) = 0$$

Combining with (1) gives

$$\int_{C_1} \mathbf{F}\cdot d\mathbf{r} - \int_{C_2} \mathbf{F}\cdot d\mathbf{r} = \iint_S \text{curl}(\mathbf{F})\cdot d\mathbf{S} = \iint_{\mathcal{D}} 0\, d\theta\, dr = 0$$

or

$$\int_{C_1} \mathbf{F}\cdot d\mathbf{r} = \int_{C_2} \mathbf{F}\cdot d\mathbf{r}$$

29. You know two things about a vector field $\mathbf{F}$:

(i) $\mathbf{F}$ has a vector potential $\mathbf{A}$ (but $\mathbf{A}$ is unknown).

(ii) The circulation of $\mathbf{A}$ around the unit circle (oriented counterclockwise) is 25.

Determine the flux of $\mathbf{F}$ through the surface S in Figure 22, oriented with an upward-pointing normal.

SOLUTION Since $\mathbf{F}$ has a vector potential—that is, $\mathbf{F}$ is the curl of a vector field—the flux of $\mathbf{F}$ through a surface depends only on the boundary curve C. Now, the surface S and the unit disc S_1 in the xy-plane share the same boundary C. Therefore,

$$\iint_S \mathbf{F}\cdot d\mathbf{S} = \iint_{S_1} \mathbf{F}\cdot d\mathbf{S} \quad (1)$$

We compute the flux of $\mathbf{F}$ through S_1, using the parametrization

$$S_1: \Phi(r, \theta) = (r\cos\theta, r\sin\theta, 0), \quad 0 \leq r \leq 1, \quad 0 \leq \theta \leq 2\pi$$

$$\mathbf{n} = \langle 0, 0, 1\rangle$$

By the given information, we have

$$\mathbf{F}(\Phi(r, \theta)) = \mathbf{F}(r\cos\theta, r\sin\theta, 0) = \langle 0, 0, 1\rangle$$

Hence,

$$\mathbf{F}(\Phi(r,\theta))\cdot \mathbf{n} = \langle 0, 0, 1\rangle \cdot \langle 0, 0, 1\rangle = 1$$

We obtain the following integral:

$$\iint_{S_1} \mathbf{F}\cdot d\mathbf{S} = \int_0^{2\pi}\int_0^1 \mathbf{F}(\Phi(r,\theta))\cdot \mathbf{n}\, dr\, d\theta = \int_0^{2\pi}\int_0^1 1\, dr\, d\theta = 2\pi$$

Combining with (1) we obtain

$$\iint_S \mathbf{F}\cdot d\mathbf{S} = 2\pi$$

31. Prove that $\text{curl}(f\mathbf{a}) = \nabla f \times \mathbf{a}$, where f is a differentiable function and $\mathbf{a}$ is a constant vector.

SOLUTION Start with the identity

$$\text{curl}(f\mathbf{a}) = f\,\text{curl}(\mathbf{a}) + \nabla f \times \mathbf{a}$$

Since $\mathbf{a}$ is a constant vector, all the partial derivatives of the components of $\mathbf{a}$ are zero, hence the curl of $\mathbf{a}$ is the zero vector:

$$\text{curl}(\mathbf{a}) = \mathbf{0}$$

Thus we obtain

$$\text{curl}(f\mathbf{a}) = \nabla f \times \mathbf{a}$$

33. Prove the following Product Rule:

$$\text{curl}(f\mathbf{F}) = f\,\text{curl}(\mathbf{F}) + \nabla f \times \mathbf{F}$$

SOLUTION We evaluate the curl of $f\mathbf{F}$. Since $f\mathbf{F} = \langle fF_1, fF_2, fF_3 \rangle$, using the Product Rule for scalar functions we have

$$\text{curl}(f\mathbf{F}) = \left\langle \frac{\partial}{\partial y}(fF_3) - \frac{\partial}{\partial z}(fF_2),\ \frac{\partial}{\partial z}(fF_1) - \frac{\partial}{\partial x}(fF_3),\ \frac{\partial}{\partial x}(fF_2) - \frac{\partial}{\partial y}(fF_1) \right\rangle$$

$$= \left\langle \frac{\partial f}{\partial y}F_3 + f\frac{\partial F_3}{\partial y} - \frac{\partial f}{\partial z}F_2 - f\frac{\partial F_2}{\partial z},\ \frac{\partial f}{\partial z}F_1 + f\frac{\partial F_1}{\partial z} - \frac{\partial f}{\partial x}F_3 - f\frac{\partial F_3}{\partial x},\right.$$

$$\left. \frac{\partial f}{\partial x}F_2 + f\frac{\partial F_2}{\partial x} - \frac{\partial f}{\partial y}F_1 - f\frac{\partial F_1}{\partial y} \right\rangle$$

$$= f\left\langle \frac{\partial F_3}{\partial y} - \frac{\partial F_2}{\partial z},\ \frac{\partial F_1}{\partial z} - \frac{\partial F_3}{\partial x},\ \frac{\partial F_2}{\partial x} - \frac{\partial F_1}{\partial y} \right\rangle$$

$$+ \left\langle \frac{\partial f}{\partial y}F_3 - \frac{\partial f}{\partial z}F_2,\ \frac{\partial f}{\partial z}F_1 - \frac{\partial f}{\partial x}F_3,\ \frac{\partial f}{\partial x}F_2 - \frac{\partial f}{\partial y}F_1 \right\rangle \qquad (1)$$

The vector in the first term is $\text{curl}(\mathbf{F})$. We show that the second term is the cross product $\nabla f \times \mathbf{F}$. We compute the cross product:

$$\nabla f \times \mathbf{F} = \begin{vmatrix} \mathbf{i} & \mathbf{j} & \mathbf{k} \\ \frac{\partial f}{\partial x} & \frac{\partial f}{\partial y} & \frac{\partial f}{\partial z} \\ F_1 & F_2 & F_3 \end{vmatrix} = \left(\frac{\partial f}{\partial y}F_3 - \frac{\partial f}{\partial z}F_2\right)\mathbf{i} - \left(\frac{\partial f}{\partial x}F_3 - \frac{\partial f}{\partial z}F_1\right)\mathbf{j} + \left(\frac{\partial f}{\partial x}F_2 - \frac{\partial f}{\partial y}F_1\right)\mathbf{k}$$

$$= \left\langle \frac{\partial f}{\partial y}F_3 - \frac{\partial f}{\partial z}F_2,\ \frac{\partial f}{\partial z}F_1 - \frac{\partial f}{\partial x}F_3,\ \frac{\partial f}{\partial x}F_2 - \frac{\partial f}{\partial y}F_1 \right\rangle$$

Therefore, (1) gives

$$\text{curl}(f\mathbf{F}) = f\,\text{curl}(\mathbf{F}) + \nabla f \times \mathbf{F}$$

35. Verify that $\mathbf{B} = \text{curl}(\mathbf{A})$ for $r > R$ in the setting of Example 4.

SOLUTION As observed in the example,

$$\text{curl}(\langle f, g, 0 \rangle) = \langle -g_z, f_z, g_x - f_y \rangle$$

and recall $r = x^2 + y^2$. For $r > R$, this yields

$$\text{curl}(\mathbf{A}) = \frac{1}{2}R^2 B \left\langle 0, 0, \frac{\partial}{\partial x}(xr^{-2}) - \frac{\partial}{\partial y}(-yr^{-2}) \right\rangle$$

The z-component on the right is also zero:

$$\frac{\partial}{\partial x}(xr^{-2}) + \frac{\partial}{\partial y}(yr^{-2}) = \frac{\partial}{\partial x}\left(\frac{x}{x^2+y^2}\right) + \frac{\partial}{\partial y}\left(\frac{y}{x^2+y^2}\right)$$

$$= \frac{(x^2+y^2) - x(2x)}{(x^2+y^2)^2} + \frac{(x^2+y^2) - y(2y)}{(x^2+y^2)^2}$$

$$= 0$$

Thus, $\text{curl}(\mathbf{A}) = \mathbf{0}$ when $r > R$ as required.

Further Insights and Challenges

37. In this exercise, we use the notation of the proof of Theorem 1 and prove

$$\oint_C F_3(x, y, z)\mathbf{k} \cdot d\mathbf{r} = \iint_S \operatorname{curl}(F_3(x, y, z)\mathbf{k}) \cdot d\mathbf{S} \quad \boxed{10}$$

In particular, S is the graph of $z = f(x, y)$ over a domain $\mathcal{D}$, and C is the boundary of S with parametrization $(x(t), y(t), f(x(t), y(t)))$.

(a) Use the Chain Rule to show that

$$F_3(x, y, z)\mathbf{k} \cdot d\mathbf{r} = F_3(x(t), y(t), f(x(t), y(t))\big(f_x(x(t), y(t))x'(t) + f_y(x(t), y(t))y'(t)\big)\, dt$$

and verify that

$$\oint_C F_3(x, y, z)\mathbf{k} \cdot d\mathbf{r} = \oint_{C_0} \langle F_3(x, y, z)f_x(x, y), F_3(x, y, z)f_y(x, y) \rangle \cdot d\mathbf{r}$$

where C_0 has parametrization $(x(t), y(t))$.

(b) Apply Green's Theorem to the line integral over C_0 and show that the result is equal to the right-hand side of Eq. (10).

SOLUTION

(a) With $\mathbf{r}(t)$ as given, we have

$$\mathbf{r}'(t) = \left\langle x'(t), y'(t), \frac{d}{dt}f(x(t), y(t)) \right\rangle.$$

Using the Chain Rule gives

$$\left\langle x'(t), y'(t), \frac{d}{dt}f(x(t), y(t)) \right\rangle = \left\langle x'(t), y'(t), f_x(x(t), y(t))x'(t) + f_y(x(t), y(t))y'(t) \right\rangle$$

Assume that C is parametrized as given for $a \le t \le b$. Then

$$\oint_C F_3(x, y, z)\,\mathbf{k} \cdot d\mathbf{r} = \int_a^b \langle 0, 0, F_3(x, y, z) \rangle \cdot \left\langle x'(t), y'(t), f_x(x(t), y(t))x'(t) + f_y(x(t), y(t))y'(t) \right\rangle dt$$

$$= \int_a^b F_3(x, y, z)\big(f_x(x(t), y(t))x'(t) + f_y(x(t), y(t))y'(t)\big)\, dt$$

$$= \int_a^b \langle F_3(x, y, z)f_x(x, y), F_3(x, y, z)f_y(x, y) \rangle \cdot \langle x'(t), y'(t) \rangle\, dt \quad (*)$$

But the curve C projects to the curve C_0 in the xy-plane (see figure below), and therefore the plane curve C_0 has the parametrization (which we also call $\mathbf{r}$)

$$\mathbf{r}(t) = \langle x(t), y(t) \rangle, \quad a \le t \le b$$

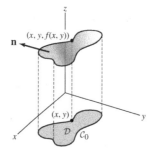

We can now reinterpret the integral $(*)$:

$$\int_a^b \langle F_3(x, y, z)f_x(x, y), F_3(x, y, z)f_y(x, y) \rangle \cdot \langle x'(t), y'(t) \rangle\, dt =$$

$$\oint_{C_0} \langle F_3(x, y, z)f_x(x, y), F_3(x, y, z)f_y(x, y) \rangle \cdot d\mathbf{r}$$

Thus

$$\oint_C F_3(x,y,z)\mathbf{k} \cdot d\mathbf{r} = \oint_{C_0} \langle F_3(x,y,z)f_x(x,y), F_3(x,y,z)f_y(x,y)\rangle \cdot d\mathbf{r}$$

(b) By Green's Theorem applied to the curve C_0 and the two-dimensional vector field $\langle F_3(x,y,z)f_x(x,y), F_3(x,y,z)f_y(x,y)\rangle$ from part **(a)**, we get

$$\oint_C F_3(x,y,z)\mathbf{k} \cdot d\mathbf{r} = \oint_{C_0} \langle F_3(x,y,z)f_x(x,y), F_3(x,y,z)f_y(x,y)\rangle \cdot d\mathbf{r}$$

$$= \iint_D \mathrm{curl}\left(\langle F_3(x,y,z)f_x(x,y), F_3(x,y,z)f_y(x,y)\rangle\right) dA$$

$$= \iint_D \left(\frac{\partial}{\partial x}(F_3(x,y,z)f_y(x,y)) - \frac{\partial}{\partial y}(F_3(x,y,z)f_x(x,y))\right) dA$$

$$= \iint_D \left(\frac{\partial F_3(x,y,z)}{\partial x}f_y(x,y) + F_3(x,y,z)f_{xy}(x,y) \right.$$
$$\left. - \frac{\partial (F_3(x,y,z))}{\partial y}f_x(x,y) - F_3(x,y,z)f_{yx}(x,y)\right) dA$$

$$= \iint_D \left(\frac{\partial F_3(x,y,z)}{\partial x}f_y(x,y) - \frac{\partial (F_3(x,y,z))}{\partial y}f_x(x,y)\right) dA$$

Now compute the surface integral in (1). The normal vector of $z = f(x,y)$ is $\langle -f_x(x,y), -f_y(x,y), 1\rangle$, and the curl of $F_3(x,y,z)\mathbf{k}$ is

$$\mathrm{curl}(F_3(x,y,z)\mathbf{k}) = \begin{vmatrix} \mathbf{i} & \mathbf{j} & \mathbf{k} \\ \frac{\partial}{\partial x} & \frac{\partial}{\partial y} & \frac{\partial}{\partial z} \\ 0 & 0 & F_3(x,y,z) \end{vmatrix} = \left\langle \frac{\partial F_3(x,y,z)}{\partial y}, -\frac{\partial F_3(x,y,z)}{\partial x}, 0\right\rangle$$

Thus the surface integral is

$$\iint_S \mathrm{curl}(F_3(x,y,z)\mathbf{k}) \cdot d\mathbf{S} = \iint_D \left\langle \frac{\partial F_3}{\partial y}(x,y,f(x,y)) - \frac{\partial F_3}{\partial x}(x,y,f(x,y)), 0\right\rangle$$
$$\cdot \langle -f_x(x,y), -f_y(x,y), 1\rangle \, dA$$

$$= \iint_D \left(-\frac{\partial F_3(x,y,f(x,y))}{\partial y}f_x(x,y) + \frac{\partial F_3(x,y,f(x,y))}{\partial x}f_y(x,y)\right) dA$$

The two sides of (1) are equal, and we are done.

17.3 Divergence Theorem (LT Section 18.3)

Preliminary Questions

1. What is the flux of $\mathbf{F} = \langle 1,0,0\rangle$ through a closed surface?

SOLUTION The divergence of $\mathbf{F} = \langle 1,0,0\rangle$ is $\mathrm{div}(\mathbf{F}) = \frac{\partial P}{\partial x} + \frac{\partial Q}{\partial y} + \frac{\partial R}{\partial z} = 0$, therefore the Divergence Theorem implies that the flux of $\mathbf{F}$ through a closed surface $\mathcal{S}$ is

$$\iint_S \mathbf{F} \cdot d\mathbf{S} = \iiint_W \mathrm{div}(\mathbf{F})\, dV = \iiint_W 0\, dV = 0$$

2. Justify the following statement: The flux of $\mathbf{F} = \langle x^3, y^3, z^3\rangle$ through every closed surface is positive.

SOLUTION The divergence of $\mathbf{F} = \langle x^3, y^3, z^3\rangle$ is

$$\mathrm{div}(\mathbf{F}) = 3x^2 + 3y^2 + 3z^2$$

Therefore, by the Divergence Theorem, the flux of **F** through a closed surface S is

$$\iint_S \mathbf{F} \cdot d\mathbf{S} = \iiint_W (3x^2 + 3y^2 + 3z^2) \, dV$$

Since the integrand is positive for all $(x, y, z) \neq (0, 0, 0)$, the triple integral, hence also the flux, is positive.

3. Which of the following expressions are meaningful (where **F** is a vector field and f is a function)? Of those that are meaningful, which are automatically zero?

(a) div(∇f) (b) curl(∇f) (c) ∇curl(f)
(d) div(curl(**F**)) (e) curl(div(**F**)) (f) ∇(div(**F**))

SOLUTION

(a) The divergence is defined on vector fields. The gradient is a vector field, hence div($\nabla \varphi$) is defined. It is not automatically zero since for $\varphi = x^2 + y^2 + z^2$ we have

$$\text{div}(\nabla \varphi) = \text{div}\langle 2x, 2y, 2z \rangle = 2 + 2 + 2 = 6 \neq 0$$

(b) The curl acts on vector valued functions, and $\nabla \varphi$ is such a function. Therefore, curl($\nabla \varphi$) is defined. Since the gradient field $\nabla \varphi$ is conservative, the cross partials of $\nabla \varphi$ are equal, or equivalently, curl($\nabla \varphi$) is the zero vector.

(c) The curl is defined on vector fields rather than on scalar functions. Therefore, curl(φ) is undefined. Obviously, ∇curl(φ) is also undefined.

(d) The curl is defined on the vector field **F** and the divergence is defined on the vector field curl(**F**). Therefore the expression div(curl(**F**)) is meaningful. We show that this vector is automatically zero:

$$\text{div}(\text{curl}(\mathbf{F})) = \text{div}\left\langle \frac{\partial F_3}{\partial y} - \frac{\partial F_2}{\partial z}, \frac{\partial F_1}{\partial z} - \frac{\partial F_3}{\partial x}, \frac{\partial F_2}{\partial x} - \frac{\partial F_1}{\partial y} \right\rangle$$

$$= \frac{\partial}{\partial x}\left(\frac{\partial F_3}{\partial y} - \frac{\partial F_2}{\partial z}\right) + \frac{\partial}{\partial y}\left(\frac{\partial F_1}{\partial z} - \frac{\partial F_3}{\partial x}\right) + \frac{\partial}{\partial z}\left(\frac{\partial F_2}{\partial x} - \frac{\partial F_1}{\partial y}\right)$$

$$= \frac{\partial^2 F_3}{\partial x \partial y} - \frac{\partial^2 F_2}{\partial x \partial z} + \frac{\partial^2 F_1}{\partial y \partial z} - \frac{\partial^2 F_3}{\partial y \partial x} + \frac{\partial^2 F_2}{\partial z \partial x} - \frac{\partial^2 F_1}{\partial z \partial y}$$

$$= \left(\frac{\partial^2 F_3}{\partial x \partial y} - \frac{\partial^2 F_3}{\partial y \partial x}\right) + \left(\frac{\partial^2 F_2}{\partial z \partial x} - \frac{\partial^2 F_2}{\partial x \partial z}\right) + \left(\frac{\partial^2 F_1}{\partial y \partial z} - \frac{\partial^2 F_1}{\partial z \partial y}\right)$$

$$= 0 + 0 + 0 = 0$$

(e) The curl acts on vector valued functions, whereas div(**F**) is a scalar function. Therefore the expression curl(div(**F**)) has no meaning.

(f) div(**F**) is a scalar function, hence ∇(div**F**) is meaningful. It is not necessarily the zero vector as shown in the following example:

$$\mathbf{F} = \langle x^2, y^2, z^2 \rangle$$
$$\text{div}(\mathbf{F}) = 2x + 2y + 2z$$
$$\nabla(\text{div}\mathbf{F}) = \langle 2, 2, 2 \rangle \neq \langle 0, 0, 0 \rangle$$

4. Which of the following statements is correct (where **F** is a continuously differentiable vector field defined everywhere)?

(a) The flux of curl(**F**) through all surfaces is zero.

(b) If $\mathbf{F} = \nabla \varphi$, then the flux of **F** through all surfaces is zero.

(c) The flux of curl(**F**) through all closed surfaces is zero.

SOLUTION

(a) This statement holds only for conservative fields. If **F** is not conservative, there exist closed curves such that $\int_C \mathbf{F} \cdot d\mathbf{r} \neq 0$, hence by Stokes' Theorem $\iint_S \text{curl}(\mathbf{F}) \cdot d\mathbf{S} \neq 0$.

(b) This statement is false. Consider the unit sphere S in the three-dimensional space and the function $\varphi(x, y, z) = x^2 + y^2 + z^2$. Then $\mathbf{F} = \nabla\varphi = \langle 2x, 2y, 2z \rangle$ and $\operatorname{div}(\mathbf{F}) = 2 + 2 + 2 = 6$. Using the Divergence Theorem, we have ($\mathcal{W}$ is the unit ball in R^3)

$$\iint_S \mathbf{F} \cdot d\mathbf{S} = \iiint_\mathcal{W} \operatorname{div}(\mathbf{F}) \, dV = \iiint_\mathcal{W} 6 \, dV = 6 \iiint_\mathcal{W} dV = 6 \operatorname{Vol}(\mathcal{W})$$

(c) This statement is correct, as stated in the corollary of Stokes' Theorem in Section 18.2.

5. How does the Divergence Theorem imply that the flux of the vector field $\mathbf{F} = \langle x^2, y - e^z, y - 2zx \rangle$ through a closed surface is equal to the enclosed volume?

SOLUTION By the Divergence Theorem, the flux is

$$\iint_S \mathbf{F} \cdot d\mathbf{S} = \iiint_\mathcal{W} \operatorname{div}(\mathbf{F}) \, dV = \iiint_\mathcal{W} (2x + 1 - 2x) \, dV = \iiint_\mathcal{W} 1 \, dV = \operatorname{Volume}(\mathcal{W})$$

Therefore the statement is true.

Exercises

In Exercises 1–4, verify the Divergence Theorem for the vector field and region.

1. $\mathbf{F}(x, y, z) = \langle z, x, y \rangle$, the box $[0, 4] \times [0, 2] \times [0, 3]$

SOLUTION Let S be the surface of the box and $\mathcal{R}$ the region enclosed by S.

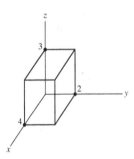

We first compute the surface integral in the Divergence Theorem:

$$\iint_S \mathbf{F} \cdot d\mathbf{S} = \iiint_\mathcal{R} \operatorname{div}(\mathbf{F}) \, dV \tag{1}$$

We denote by S_i, $i = 1, \ldots, 6$, the faces of the box, starting at the face on the xz-plane and moving counterclockwise, then moving to the bottom and the top. We use parametrizations

$$S_1 : \Phi_1(x, z) = (x, 0, z), \quad 0 \le x \le 4, \quad 0 \le z \le 3$$
$$\mathbf{n} = \langle 0, -1, 0 \rangle$$
$$S_2 : \Phi_2(y, z) = (0, y, z), \quad 0 \le y \le 2, \quad 0 \le z \le 3$$
$$\mathbf{n} = \langle -1, 0, 0 \rangle$$
$$S_3 : \Phi_3(x, z) = (x, 2, z), \quad 0 \le x \le 4, \quad 0 \le z \le 3$$
$$\mathbf{n} = \langle 0, 1, 0 \rangle$$
$$S_4 : \Phi_4(y, z) = (4, y, z), \quad 0 \le y \le 2, \quad 0 \le z \le 3$$
$$\mathbf{n} = \langle 1, 0, 0 \rangle$$

$$S_5 : \Phi_5(x, y) = (x, y, 0), \quad 0 \le x \le 4, \quad 0 \le y \le 2$$
$$\mathbf{n} = \langle 0, 0, -1 \rangle$$
$$S_6 : \Phi_6(x, y) = (x, y, 3), \quad 0 \le x \le 4, \quad 0 \le y \le 2$$
$$\mathbf{n} = \langle 0, 0, 1 \rangle$$

Then,

$$\iint_{S_1} \mathbf{F} \cdot d\mathbf{S} = \int_0^3 \int_0^4 \mathbf{F}(\Phi_1(x,z)) \cdot \langle 0, -1, 0 \rangle \, dx \, dz = \int_0^3 \int_0^4 \langle z, x, 0 \rangle \cdot \langle 0, -1, 0 \rangle \, dx \, dz$$
$$= \int_0^3 \int_0^4 -x \, dx \, dz = 3 \left. \frac{-x^2}{2} \right|_0^4 = -24$$

$$\iint_{S_2} \mathbf{F} \cdot d\mathbf{S} = \int_0^3 \int_0^2 \mathbf{F}(\Phi_2(y,z)) \cdot \langle -1, 0, 0 \rangle \, dy \, dz = \int_0^3 \int_0^2 \langle z, 0, y \rangle \cdot \langle -1, 0, 0 \rangle \, dy \, dz$$
$$= \int_0^3 \int_0^2 -z \, dy \, dz = 2 \cdot \left. \frac{-z^2}{2} \right|_0^3 = -9$$

$$\iint_{S_3} \mathbf{F} \cdot d\mathbf{S} = \int_0^3 \int_0^4 \mathbf{F}(\Phi_3(x,z)) \cdot \langle 0, 1, 0 \rangle \, dx \, dz = \int_0^3 \int_0^4 \langle z, x, 2 \rangle \cdot \langle 0, 1, 0 \rangle \, dx \, dz$$
$$= \int_0^3 \int_0^4 x \, dx \, dz = 3 \cdot \left. \frac{x^2}{2} \right|_0^4 = 24$$

$$\iint_{S_4} \mathbf{F} \cdot d\mathbf{S} = \int_0^3 \int_0^2 \mathbf{F}(\Phi_4(y,z)) \cdot \langle 1, 0, 0 \rangle \, dy \, dz = \int_0^3 \int_0^2 \langle z, 4, y \rangle \cdot \langle 1, 0, 0 \rangle \, dy \, dz$$
$$= \int_0^3 \int_0^2 z \, dy \, dz = 2 \cdot \left. \frac{z^2}{2} \right|_0^3 = 9$$

$$\iint_{S_5} \mathbf{F} \cdot d\mathbf{S} = \int_0^2 \int_0^4 \mathbf{F}(\Phi_5(x,y)) \cdot \langle 0, 0, -1 \rangle \, dx \, dy = \int_0^2 \int_0^4 \langle 0, x, y \rangle \cdot \langle 0, 0, -1 \rangle \, dx \, dy$$
$$= \int_0^2 \int_0^4 -y \, dx \, dy = 4 \cdot \left. \frac{-y^2}{2} \right|_0^2 = -8$$

$$\iint_{S_6} \mathbf{F} \cdot d\mathbf{S} = \int_0^2 \int_0^4 \mathbf{F}(\Phi_6(x,y)) \cdot \mathbf{n} \, dx \, dy = \int_0^2 \int_0^4 \langle 3, x, y \rangle \cdot \langle 0, 0, 1 \rangle \, dx \, dy$$
$$= \int_0^2 \int_0^4 y \, dx \, dy = 4 \cdot \left. \frac{y^2}{2} \right|_0^2 = 8$$

We add the integrals to obtain the surface integral

$$\iint_S \mathbf{F} \cdot d\mathbf{S} = \sum_{i=1}^6 \iint_{S_i} \mathbf{F} \cdot d\mathbf{S} = -24 - 9 + 24 + 9 - 8 + 8 = 0 \quad (2)$$

We now evaluate the triple integral in (1). We compute the divergence of $\mathbf{F} = \langle z, x, y \rangle$:

$$\text{div}(\mathbf{F}) = \frac{\partial}{\partial x}(z) + \frac{\partial}{\partial y}(x) + \frac{\partial}{\partial z}(y) = 0$$

Hence,

$$\iiint_{\mathcal{R}} \text{div}(\mathbf{F}) \, dV = \iiint_{\mathcal{R}} 0 \, dV = 0 \quad (3)$$

The equality of the integrals in (2) and (3) verifies the Divergence Theorem.

3. $\mathbf{F}(x, y, z) = \langle 2x, 3z, 3y \rangle$, the region $x^2 + y^2 \leq 1$, $0 \leq z \leq 2$

SOLUTION

Let S be the surface of the cylinder and $\mathcal{R}$ the region enclosed by S. We compute the two sides of the Divergence Theorem:

$$\iint_S \mathbf{F} \cdot d\mathbf{S} = \iiint_{\mathcal{R}} \operatorname{div}(\mathbf{F}) \, dV \tag{1}$$

We first calculate the surface integral.

Step 1. Integral over the side of the cylinder. The side of the cylinder is parametrized by

$$\Phi(\theta, z) = (\cos\theta, \sin\theta, z), \quad 0 \leq \theta \leq 2\pi, \quad 0 \leq z \leq 2$$

$$\mathbf{N} = \langle \cos\theta, \sin\theta, 0 \rangle$$

Then,

$$\mathbf{F}(\Phi(\theta, z)) \cdot \mathbf{N} = \langle 2\cos\theta, 3z, 3\sin\theta \rangle \cdot \langle \cos\theta, \sin\theta, 0 \rangle = 2\cos^2\theta + 3z\sin\theta$$

We obtain the integral

$$\iint_{\text{side}} \mathbf{F} \cdot d\mathbf{S} = \int_0^2 \int_0^{2\pi} \left(2\cos^2\theta + 3z\sin\theta \right) d\theta \, dz = 4 \int_0^{2\pi} \cos^2\theta \, d\theta + \left(\int_0^2 3z \, dz \right) \left(\int_0^{2\pi} \sin\theta \, d\theta \right)$$

$$= 4 \cdot \left(\frac{\theta}{2} + \frac{\sin 2\theta}{4} \Big|_0^{2\pi} \right) + 0 = 4\pi$$

Step 2. Integral over the top of the cylinder. The top of the cylinder is parametrized by

$$\Phi(x, y) = (x, y, 2)$$

with parameter domain $\mathcal{D} = \{(x, y) : x^2 + y^2 \leq 1\}$. The upward pointing normal is

$$\mathbf{N} = \mathbf{T}_x \times \mathbf{T}_y = \langle 1, 0, 0 \rangle \times \langle 0, 1, 0 \rangle = \mathbf{i} \times \mathbf{j} = \mathbf{k} = \langle 0, 0, 1 \rangle$$

Also,

$$\mathbf{F}(\Phi(x, y)) \cdot \mathbf{N} = \langle 2x, 6, 3y \rangle \cdot \langle 0, 0, 1 \rangle = 3y$$

Hence,

$$\iint_{\text{top}} \mathbf{F} \cdot d\mathbf{S} = \iint_{\mathcal{D}} 3y \, dA = 0$$

The last integral is zero due to symmetry.

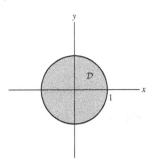

Step 3. Integral over the bottom of the cylinder. We parametrize the bottom by

$$\Phi(x, y) = (x, y, 0), \quad (x, y) \in \mathcal{D}$$

The downward pointing normal is $\mathbf{N} = \langle 0, 0, -1 \rangle$. Then

$$\mathbf{F}(\Phi(x, y)) \cdot \mathbf{N} = \langle 2x, 0, 3y \rangle \cdot \langle 0, 0, -1 \rangle = -3y$$

We obtain the following integral, which is zero due to symmetry:

$$\iint_{\text{bottom}} \mathbf{F} \cdot d\mathbf{S} = \iint_{\mathcal{D}} -3y \, dA = 0$$

Adding the integrals we get

$$\iint_S \mathbf{F} \cdot d\mathbf{S} = \iint_{\text{side}} \mathbf{F} \cdot d\mathbf{S} + \iint_{\text{top}} \mathbf{F} \cdot d\mathbf{S} + \iint_{\text{bottom}} \mathbf{F} \cdot d\mathbf{S} = 4\pi + 0 + 0 = 4\pi \quad (2)$$

Step 4. Compare with integral of divergence.

$$\text{div}(\mathbf{F}) = \text{div} \langle 2x, 3z, 3y \rangle = \frac{\partial}{\partial x}(2x) + \frac{\partial}{\partial y}(3z) + \frac{\partial}{\partial z}(3y) = 2$$

$$\iiint_{\mathcal{R}} \text{div}(\mathbf{F}) \, dV = \iiint_{\mathcal{R}} 2 \, dV = 2 \iiint_{\mathcal{R}} dV = 2 \, \text{Vol}(\mathcal{R}) = 2 \cdot \pi \cdot 2 = 4\pi \quad (3)$$

The equality of (2) and (3) verifies the Divergence Theorem.

In Exercises 5–16, use the Divergence Theorem to evaluate the flux $\iint_S \mathbf{F} \cdot d\mathbf{S}$.

5. $\mathbf{F}(x, y, z) = \langle 0, 0, z^3/3 \rangle$, S is the sphere $x^2 + y^2 + z^2 = 1$.

SOLUTION We compute the divergence of $\mathbf{F} = \langle 0, 0, z^3/3 \rangle$:

$$\text{div}\mathbf{F} = \frac{\partial}{\partial x}(0) + \frac{\partial}{\partial y}(0) + \frac{\partial}{\partial z}(z^3/3) = z^2$$

Hence, by the Divergence Theorem ($\mathcal{W}$ is the unit ball),

$$\iint_S \mathbf{F} \cdot d\mathbf{S} = \iiint_{\mathcal{W}} \text{div}(\mathbf{F}) \, dV = \iiint_{\mathcal{W}} z^2 \, dV$$

Computing this integral we see:

$$\iiint_{\mathcal{W}} z^2 \, dV = \int_0^{2\pi} \int_0^{\pi} \int_0^1 \rho^2 \cos^2 \phi \cdot \rho^2 \sin \phi \, d\rho \, d\phi \, d\theta$$

$$= \int_0^{2\pi} d\theta \cdot \int_0^{\pi} \cos^2 \phi \sin \phi \, d\phi \cdot \int_0^1 \rho^4 \, d\rho$$

$$= (2\pi) \cdot \left(-\frac{\cos^3 \phi}{3} \Big|_0^{\pi} \right) \cdot \left(\frac{\rho^5}{5} \Big|_0^1 \right)$$

$$= 2\pi \left(-\frac{1}{3}(-1 - 1) \right) \left(\frac{1}{5} \right)$$

$$= 2\pi \left(\frac{2}{3} \right) \left(\frac{1}{5} \right) = \frac{4\pi}{15}$$

7. $\mathbf{F}(x, y, z) = \langle xy^2, yz^2, zx^2 \rangle$, S is the boundary of the cylinder given by $x^2 + y^2 \leq 4$, $0 \leq z \leq 3$.

SOLUTION First compute div $\mathbf{F}$:

$$\text{div } \mathbf{F} = \frac{\partial}{\partial x}(xy^2) + \frac{\partial}{\partial y}(yz^2) + \frac{\partial}{\partial z}(zx^2) = x^2 + y^2 + z^2$$

Let $\mathcal{W}$ be the interior of the cylinder together with its boundary. Then by the Divergence Theorem,

$$\iint_S \mathbf{F} \cdot d\mathbf{S} = \iiint_{\mathcal{W}} \operatorname{div} \mathbf{F}\, dV.$$

Parametrize $\mathcal{W}$ using cylindrical coordinates; then div $\mathbf{F}$ becomes $r^2 + z^2$, and we get for the integral

$$\iiint_{\mathcal{W}} \operatorname{div} \mathbf{F}\, dV = \int_0^3 \int_0^2 \int_0^{2\pi} (r^2 + z^2) r\, d\theta\, dr\, dz$$

$$= 2\pi \int_0^3 \int_0^2 (r^3 + rz^2)\, dr\, dz$$

$$= 2\pi \int_0^3 \left(\frac{1}{4} r^4 + \frac{1}{2} r^2 z^2 \right) \Big|_{r=0}^2 dz$$

$$= 2\pi \int_0^3 (4 + 2z^2)\, dz$$

$$= 2\pi \left(4z + \frac{2}{3} z^3 \right) \Big|_0^3 = 60\pi$$

9. $\mathbf{F}(x, y, z) = \langle x + z^2, xz + y^2, zx - y \rangle$, $\mathcal{S}$ is the surface that bounds the solid region with boundary given by the parabolic cylinder $z = 1 - x^2$, and the planes $z = 0$, $y = 0$, and $z + y = 5$.

SOLUTION A diagram of the situation is below:

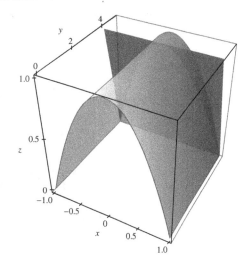

Computing div $\mathbf{F}$ gives

$$\operatorname{div} \mathbf{F} = \frac{\partial}{\partial x}(x + z^2) + \frac{\partial}{\partial y}(xz + y^2) + \frac{\partial}{\partial z}(zx - y) = 1 + 2y + x$$

Let $\mathcal{W}$ be the interior of the given surface together with its boundary. Then by the Divergence Theorem,

$$\iint_S \mathbf{F} \cdot d\mathbf{S} = \iiint_{\mathcal{W}} \operatorname{div} \mathbf{F}\, dV = \int_{-1}^1 \int_0^{1-x^2} \int_0^{5-z} (1 + 2y + x)\, dy\, dz\, dx$$

$$= \int_{-1}^1 \int_0^{1-x^2} (y + y^2 + xy) \Big|_{y=0}^{5-z} dz\, dx$$

$$= \int_{-1}^1 \int_0^{1-x^2} ((5-z) + (5-z)^2 + x(5-z))\, dz\, dx$$

$$= \int_{-1}^1 \int_0^{1-x^2} (z^2 - 11z + 30 - xz + 5x)\, dz\, dx$$

$$= \int_{-1}^1 \left(\frac{1}{3} z^3 - \frac{11}{2} z^2 + 30z - \frac{1}{2} xz^2 + 5xz \right) \Big|_{z=0}^{1-x^2} dx$$

$$= \int_{-1}^{1}\left(\frac{1}{3}(1-x^2)^3 - \frac{11}{2}(1-x^2)^2 + 30(1-x^2) - \frac{1}{2}x(1-x^2)^2 + 5x(1-x^2)\right)dx$$

$$= \int_{-1}^{1}\left(-\frac{1}{3}x^6 - \frac{1}{2}x^5 - \frac{9}{2}x^4 - 4x^3 - 20x^2 + \frac{9}{2}x + \frac{149}{6}\right)dx$$

$$= \left(-\frac{1}{21}x^7 - \frac{1}{12}x^6 - \frac{9}{10}x^5 - x^4 - \frac{20}{3}x^3 + \frac{9}{4}x^2 + \frac{149}{6}x\right)\Big|_{-1}^{1}$$

$$= \frac{3616}{105}$$

11. $\mathbf{F}(x,y,z) = \langle x^3, 0, z^3 \rangle$, $\mathcal{S}$ is the boundary of the region in the first octant of space given by $x^2+y^2+z^2 \le 4$, $x \ge 0, y \ge 0, z \ge 0$.

SOLUTION We compute the divergence of $\mathbf{F} = \langle x^3, 0, z^3 \rangle$:

$$\text{div}(\mathbf{F}) = \frac{\partial}{\partial x}(x^3) + \frac{\partial}{\partial y}(0) + \frac{\partial}{\partial z}(z^3) = 3x^2 + 3z^2 = 3(x^2+z^2)$$

Using the Divergence Theorem we obtain ($\mathcal{W}$ is the region inside the sphere in the first octant)

$$\iint_{\mathcal{S}} \mathbf{F} \cdot d\mathbf{S} = \iiint_{\mathcal{W}} \text{div}(\mathbf{F}) \, dV = \iiint_{\mathcal{W}} 3(x^2+z^2) \, dV$$

We convert the integral to spherical coordinates. We have

$$x^2 + z^2 = \rho^2 \cos^2\theta \sin^2\phi + \rho^2 \cos^2\phi = \rho^2 \cos^2\theta \sin^2\phi + \rho^2(1-\sin^2\phi)$$
$$= -\rho^2 \sin^2\phi(1-\cos^2\theta) + \rho^2 = -\rho^2 \sin^2\phi \sin^2\theta + \rho^2 = \rho^2(1-\sin^2\phi \sin^2\theta)$$

We obtain the following integral:

$$\iint_{\mathcal{S}} \mathbf{F} \cdot d\mathbf{S} = 3\int_0^{\pi/2}\int_0^{\pi/2}\int_0^2 \rho^2(1-\sin^2\phi \sin^2\theta) \cdot \rho^2 \sin\phi \, d\rho \, d\phi \, d\theta$$

$$= 3\int_0^{\pi/2}\int_0^{\pi/2}\int_0^2 \rho^4(\sin\phi - \sin^3\phi \sin^2\theta) \, d\rho \, d\phi \, d\theta$$

$$= 3\int_0^{\pi/2}\int_0^{\pi/2}\left(\frac{1}{5}\rho^5(\sin\phi - \sin^3\phi \sin^2\theta)\right)\Big|_{\rho=0}^{2} d\phi \, d\theta$$

$$= \frac{96}{5}\int_0^{\pi/2}\int_0^{\pi/2}(\sin\phi - \sin^3\phi \sin^2\theta) \, d\phi \, d\theta$$

$$= \frac{96}{5}\int_0^{\pi/2}\int_0^{\pi/2}(\sin\phi - \sin\phi(1-\cos^2\phi)\sin^2\theta) \, d\phi \, d\theta$$

$$= \frac{96}{5}\int_0^{\pi/2}\left(-\cos\phi + \left(\cos\phi - \frac{1}{3}\cos^3\phi\right)\sin^2\theta\right)\Big|_{\phi=0}^{\pi/2} d\theta$$

$$= \frac{96}{5}\int_0^{\pi/2}\left(1 - \frac{2}{3}\sin^2\theta\right)d\theta$$

$$= \frac{32}{5}\int_0^{\pi/2}(3 - 2\sin^2\theta) \, d\theta$$

$$= \frac{32}{5}\int_0^{\pi/2}(2 + \cos 2\theta) \, d\theta$$

$$= \frac{32}{5}\left(2\theta + \frac{1}{2}\sin 2\theta\right)\Big|_0^{\pi/2} = \frac{32}{5}\pi$$

13. $\mathbf{F}(x,y,z) = \langle x, y^2, z+y \rangle$, $\mathcal{S}$ is the boundary of the region contained in the cylinder $x^2+y^2 = 4$ between the planes $z = x$ and $z = 8$.

SOLUTION Let $\mathcal{W}$ be the region enclosed by $\mathcal{S}$.

We compute the divergence of $\mathbf{F} = \langle x, y^2, z+y \rangle$:

$$\text{div}(\mathbf{F}) = \frac{\partial}{\partial x}(x) + \frac{\partial}{\partial y}(y^2) + \frac{\partial}{\partial z}(z+y) = 1 + 2y + 1 = 2 + 2y$$

By the Divergence Theorem we have

$$\iint_S \mathbf{F} \cdot d\mathbf{S} = \iiint_W \text{div}(\mathbf{F})\, dV = \iiint_W (2 + 2y)\, dV$$

We compute the triple integral. Denoting by $\mathcal{D}$ the disk $x^2 + y^2 \leq 4$ in the xy-plane, we have

$$\iint_S \mathbf{F} \cdot d\mathbf{S} = \iint_\mathcal{D} \int_x^8 (2 + 2y)\, dz\, dx\, dy = \iint_\mathcal{D} (2 + 2y)z\Big|_{z=x}^{8} dx\, dy = \iint_\mathcal{D} (2 + 2y)(8 - x)\, dx\, dy$$

We convert the integral to polar coordinates:

$$\iint_S \mathbf{F} \cdot d\mathbf{S} = \int_0^{2\pi} \int_0^2 (2 + 2r\sin\theta)(8 - r\cos\theta)\, r\, dr\, d\theta$$

$$= \int_0^{2\pi} \int_0^2 \left(16r + 2r^2(8\sin\theta - \cos\theta) - r^3 \sin 2\theta\right) dr\, d\theta$$

$$= \int_0^{2\pi} 8r^2 + \frac{2}{3}r^3(8\sin\theta - \cos\theta) - \frac{r^4}{4}\sin 2\theta\bigg|_{r=0}^{2} d\theta$$

$$= \int_0^{2\pi} \left(32 + \frac{16}{3}(8\sin\theta - \cos\theta) - 4\sin 2\theta\right) d\theta$$

$$= 64\pi + \frac{128}{3}\int_0^{2\pi} \sin\theta\, d\theta - \frac{16}{3}\int_0^{2\pi} \cos\theta\, d\theta - \int_0^{2\pi} 4\sin 2\theta\, d\theta = 64\pi$$

15. $\mathbf{F}(x, y, z) = \langle x + y, z, z - x \rangle$, $\mathcal{S}$ is the boundary of the region between the paraboloid $z = 9 - x^2 - y^2$ and the xy-plane.

SOLUTION We compute the divergence of $\mathbf{F} = \langle x + y, z, z - x \rangle$,

$$\text{div}(\mathbf{F}) = \frac{\partial}{\partial x}(x + y) + \frac{\partial}{\partial y}(z) + \frac{\partial}{\partial z}(z - x) = 1 + 0 + 1 = 2$$

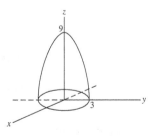

Using the Divergence Theorem we have

$$\iint_S \mathbf{F} \cdot d\mathbf{S} = \iiint_W \text{div}(\mathbf{F})\, dV = \iiint_W 2\, dV$$

We compute the triple integral:

$$\iint_S \mathbf{F} \cdot d\mathbf{S} = \iiint_W 2\, dV = \iint_D \int_0^{9-x^2-y^2} 2\, dz\, dx\, dy = \iint_D 2z \Big|_0^{9-x^2-y^2} dx\, dy$$

$$= \iint_W 2(9 - x^2 - y^2)\, dx\, dy$$

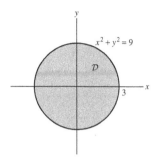

We convert the integral to polar coordinates:

$$x = r\cos\theta, \quad y = r\sin\theta, \quad 0 \le r \le 3, \quad 0 \le \theta \le 2\pi$$

$$\iint_S \mathbf{F} \cdot d\mathbf{S} = \int_0^{2\pi} \int_0^3 2\left(9 - r^2\right) r\, dr\, d\theta = 4\pi \int_0^3 (9r - r^3)\, dr = 4\pi \left(\frac{9r^2}{2} - \frac{r^4}{4} \Big|_0^3 \right) = 81\pi$$

17. Calculate the flux of the vector field $\mathbf{F} = 2xy\mathbf{i} - y^2\mathbf{j} + \mathbf{k}$ through the surface S in Figure 18. *Hint:* Apply the Divergence Theorem to the closed surface consisting of S and the unit disk.

SOLUTION From the diagram in the book, S is the surface in question bounded by the unit circle. Let T be the union of S and the unit disk D. Then T is a closed surface, and we may apply the Divergence Theorem:

$$\iint_S \mathbf{F} \cdot d\mathbf{S} + \iint_D \mathbf{F} \cdot d\mathbf{S} = \iint_T \mathbf{F} \cdot d\mathbf{S} = \iiint_W \operatorname{div}(\mathbf{F}) \cdot d\mathbf{S}$$

where W is the region enclosed by T. Now we observe that $\mathbf{F} = \langle 2xy, -y^2, 1 \rangle$ and we compute the divergence of $\mathbf{F}$:

$$\operatorname{div}(\mathbf{F}) = \frac{\partial}{\partial x}(2xy) + \frac{\partial}{\partial y}(-y^2) + \frac{\partial}{\partial z}(1) = 2y - 2y + 0 = 0$$

Therefore, the triple integral is zero and we obtain:

$$\iint_S \mathbf{F} \cdot d\mathbf{S} = -\iint_D \mathbf{F} \cdot d\mathbf{S} \tag{1}$$

where D is oriented with a downward pointing normal. Let $\Phi(r, \theta) = (r\cos\theta, r\sin\theta, 0)$ be the parametrization of D with polar coordinates. Then

$$\mathbf{F}(\Phi(r, \theta)) = \left\langle 2r^2 \cos\theta \sin\theta, -r^2 \sin^2\theta, 1 \right\rangle$$

Furthermore,

$$\Phi_r(r, \theta) = \langle \cos\theta, \sin\theta, 0 \rangle, \quad \Phi_\theta(r, \theta) = \langle -r\sin\theta, r\cos\theta, 0 \rangle$$

and $\Phi_r \times \Phi_\theta = \langle 0, 0, r \rangle$ is an upward pointing normal. Finally,

$$\mathbf{F} \cdot d\mathbf{S} = \mathbf{F}(\Phi(r, \theta)) \cdot (\Phi_r \times \Phi_\theta)\, dr\, d\theta = r\, dr\, d\theta$$

The integral on the right in (1) uses a downward pointing normal, so we may drop the minus sign and use the upward-pointing normal to obtain:

$$\iint_S \mathbf{F} \cdot d\mathbf{S} = \int_0^{2\pi} \int_0^1 r\, dr\, d\theta = \pi$$

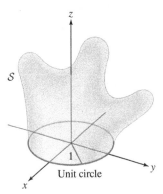

FIGURE 18 Surface S whose boundary is the unit circle.

19. Let S be the half-cylinder $x^2 + y^2 = 1$, $x \geq 0$, $0 \leq z \leq 1$. Assume that $\mathbf{F}$ is a horizontal vector field (the z-component is zero) such that $\mathbf{F}(0, y, z) = zy^2 \mathbf{i}$. Let $\mathcal{W}$ be the solid region enclosed by S, and assume that

$$\iiint_{\mathcal{W}} \text{div}(\mathbf{F}) \, dV = 4$$

Find the flux of $\mathbf{F}$ through the curved side of S.

SOLUTION The flux through the top and bottom are zero. The flux through the flat side (with outward normal $-\mathbf{i}$) is

$$-\int_{z=0}^{1} \int_{y=-1}^{1} zy^2 = -\frac{1}{2}\left(\frac{2}{3}\right) = -\frac{1}{3}$$

The flux through the curved side is $4 + \frac{1}{3} = \frac{13}{3}$.

21. Use Eq. (9) to calculate the volume of the unit ball as a surface integral over the unit sphere.

SOLUTION Let S be the unit sphere and $\mathcal{W}$ is the unit ball. By Eq. (9) we have

$$\text{Volume}(\mathcal{W}) = \frac{1}{3} \iint_S \mathbf{F} \cdot d\mathbf{S}, \quad \mathbf{F} = \langle x, y, z \rangle$$

To compute the surface integral, we parametrize S by

$$\Phi(\theta, \phi) = (\cos\theta \sin\phi, \sin\theta \sin\phi, \cos\phi), \quad 0 \leq \theta \leq 2\pi, \quad 0 \leq \phi \leq \pi$$

$$\mathbf{N} = \sin\phi \langle \cos\theta \sin\phi, \sin\theta \sin\phi, \cos\phi \rangle$$

Then

$$\mathbf{F}(\Phi(\theta, \phi)) \cdot \mathbf{N} = \langle \cos\theta \sin\phi, \sin\theta \sin\phi, \cos\phi \rangle \cdot \langle \cos\theta \sin^2\phi, \sin\theta \sin^2\phi, \cos\phi \sin\phi \rangle$$

$$= \cos^2\theta \sin^3\phi + \sin^2\theta \sin^3\phi + \cos^2\phi \sin\phi = \sin^3\phi(\cos^2\theta + \sin^2\theta) + \cos^2\phi \sin\phi$$

$$= \sin^3\phi + \cos^2\phi \sin\phi = \sin^3\phi + (1 - \sin^2\phi)\sin\phi = \sin\phi$$

We obtain the following integral:

$$\text{Volume}(\mathcal{W}) = \frac{1}{3} \int_0^{2\pi} \int_0^{\pi} \sin\phi \, d\phi \, d\theta = \frac{1}{3} \cdot 2\pi \int_0^{\pi} \sin\phi \, d\phi = \frac{2\pi}{3}\left(-\cos\phi \Big|_0^{\pi}\right) = \frac{2\pi}{3}(1+1) = \frac{4\pi}{3}$$

23. Let $\mathcal{W}$ be the region in Figure 19 bounded by the cylinder $x^2 + y^2 = 4$, the plane $z = x + 1$, and the xy-plane. Use the Divergence Theorem to compute the flux of $\mathbf{F}(x, y, z) = \langle z, x, y + z^2 \rangle$ through the boundary of $\mathcal{W}$.

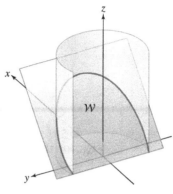

FIGURE 19

SOLUTION We compute the divergence of $\mathbf{F} = \langle z, x, y + z^2 \rangle$:

$$\text{div}(\mathbf{F}) = \frac{\partial}{\partial x}(z) + \frac{\partial}{\partial y}(x) + \frac{\partial}{\partial z}(y + z^2) = 2z$$

By the Divergence Theorem we have

$$\iint_S \mathbf{F} \cdot d\mathbf{S} = \iiint_\mathcal{W} \text{div}(\mathbf{F})\, dV = \iiint_\mathcal{W} 2z\, dV$$

To compute the triple integral, we identify the projection $\mathcal{D}$ of the region on the xy-plane. $\mathcal{D}$ is the region in the xy-plane enclosed by the circle $x^2 + y^2 = 4$ and the line $0 = x + 1$ or $x = -1$. We obtain the following integral:

$$\iint_S \mathbf{F} \cdot d\mathbf{S} = \iiint_\mathcal{W} 2z\, dV = \iint_\mathcal{D} \int_0^{x+1} 2z\, dz\, dx\, dy = \iint_\mathcal{D} z^2 \Big|_{z=0}^{x+1} dx\, dy = \iint_\mathcal{D} (x+1)^2\, dx\, dy$$

The domain $\mathcal{D}$ is shown below:

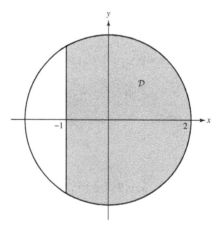

We obtain

$$\iint_S \mathbf{F} \cdot d\mathbf{S} = \iint_\mathcal{D} (x+1)^2\, dA$$

To compute this integral, we will use polar coordinates. Note first that the line $x = -1$ meets the circle at $(-1, \pm\sqrt{3})$; in polar coordinates, these points are $(r, \theta) = \left(2, \frac{2\pi}{3}\right)$ and $\left(2, \frac{4\pi}{3}\right)$. The line $x = -1$ has polar equation $r\cos\theta = -1$, or $r = -\sec\theta$. Further, both the region and the integrand are symmetric about the x-axis, so we can compute the integral by computing it above the x-axis and doubling the result. Finally, the integral over the portion of the shaded area above the x-axis consists of the integral from $\theta = 0$ to $\theta = \frac{2\pi}{3}$, where $0 \le r \le 2$, plus the integral from $\theta = \frac{2\pi}{3}$ to $\theta = \pi$, where $0 \le r \le -\sec\theta$. This gives

$$\iint_D (x+1)^2\, dA = 2\left(\int_0^{2\pi/3}\int_0^2 (r\cos\theta + 1)^2 r\, dr\, d\theta + \int_{2\pi/3}^{\pi}\int_0^{-\sec\theta} (r\cos\theta + 1)^2 r\, dr\, d\theta\right)$$

$$= 2\left(\int_0^{2\pi/3}\int_0^2 (r^3\cos^2\theta + 2r^2\cos\theta + r)\, dr\, d\theta + \int_{2\pi/3}^{\pi}\int_0^{-\sec\theta} (r^3\cos^2\theta + 2r^2\cos\theta + r)\, dr\, d\theta\right)$$

$$= 2\left(\int_0^{2\pi/3} \left(\frac{1}{4}r^4\cos^2\theta + \frac{2}{3}r^3\cos\theta + \frac{1}{2}r^2\right)\Big|_{r=0}^{2}\, d\theta\right.$$
$$\left. + \int_{2\pi/3}^{\pi} \left(\frac{1}{4}r^4\cos^2\theta + \frac{2}{3}r^3\cos\theta + \frac{1}{2}r^2\right)\Big|_{r=0}^{-\sec\theta}\, d\theta\right)$$

$$= 2\left(\int_0^{2\pi/3} \left(4\cos^2\theta + \frac{16}{3}\cos\theta + 2\right) d\theta + \int_{2\pi/3}^{\pi} \left(\frac{1}{4}\sec^2\theta - \frac{2}{3}\sec^2\theta + \frac{1}{2}\sec^2\theta\right) d\theta\right)$$

$$= 2\left(\int_0^{2\pi/3} \left(2\cos 2\theta + \frac{16}{3}\cos\theta + 4\right) d\theta + \frac{1}{12}\int_{2\pi/3}^{\pi} \sec^2\theta\, d\theta\right)$$

$$= 2\left(\left(\sin 2\theta + \frac{16}{3}\sin\theta + 4\theta\right)\Big|_0^{2\pi/3} + \frac{1}{2}\tan\theta\Big|_{2\pi/3}^{\pi}\right)$$

$$= 2\left(-\frac{\sqrt{3}}{2} + \frac{16}{3}\cdot\frac{\sqrt{3}}{2} + \frac{8\pi}{3} + \frac{1}{12}\sqrt{3}\right)$$

$$= \frac{9\sqrt{3}}{2} + \frac{16\pi}{3}$$

25. The velocity field of a fluid **v** (in meters per second) has divergence $\operatorname{div}(\mathbf{v})(P) = 3$ at the point $P = (2, 2, 2)$. Estimate the flow rate out of the sphere of radius 0.5 centered at P.

SOLUTION The flow rate out of the sphere is

$$\iiint_W \operatorname{div}\mathbf{F}\, dV$$

which may be approximated by

$$\operatorname{div}\mathbf{F}(P)\cdot \operatorname{Vol}(W) = 3\cdot\frac{4}{3}\pi\cdot 0.5^3 = 0.5\pi \approx 1.5708$$

27. The electric field due to a unit electric dipole oriented in the **k**-direction is $\mathbf{E} = \nabla(z/r^3)$, where $r = (x^2 + y^2 + z^2)^{1/2}$ (Figure 20). Let $\mathbf{e}_r = r^{-1}\langle x, y, z\rangle$.
(a) Show that $\mathbf{E} = r^{-3}\mathbf{k} - 3zr^{-4}\mathbf{e}_r$.
(b) Calculate the flux of **E** through a sphere centered at the origin.
(c) Calculate div(**E**).
(d) Can we use the Divergence Theorem to compute the flux of **E** through a sphere centered at the origin?

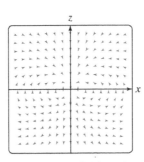

FIGURE 20 The dipole vector field restricted to the xz-plane.

SOLUTION

(a) We first compute the partial derivatives of r:

$$\frac{\partial r}{\partial x} = \frac{1}{2}(x^2 + y^2 + z^2)^{-1/2} \cdot 2x = \frac{x}{r}$$

$$\frac{\partial r}{\partial y} = \frac{1}{2}(x^2 + y^2 + z^2)^{-1/2} \cdot 2y = \frac{y}{r}$$

$$\frac{\partial r}{\partial z} = \frac{1}{2}(x^2 + y^2 + z^2)^{-1/2} \cdot 2z = \frac{z}{r} \tag{1}$$

We compute the partial derivatives of $\frac{z}{r^3}$, using the Chain Rule and the partial derivatives in (1):

$$\frac{\partial}{\partial x}\left(\frac{z}{r^3}\right) = z\frac{\partial}{\partial x}(r^{-3}) = z \cdot (-3)r^{-4}\frac{\partial r}{\partial x} = -3z \cdot r^{-4}\frac{x}{r} = -\frac{3zx}{r^5} = -3zr^{-5}x$$

$$\frac{\partial}{\partial y}\left(\frac{z}{r^3}\right) = z\frac{\partial}{\partial y}(r^{-3}) = z \cdot (-3)r^{-4}\frac{\partial r}{\partial y} = -3z \cdot r^{-4}\frac{y}{r} = -3zr^{-5}y$$

$$\frac{\partial}{\partial z}\left(\frac{z}{r^3}\right) = \frac{\partial}{\partial z}(z \cdot r^{-3}) = 1 \cdot r^{-3} + z \cdot (-3)r^{-4}\frac{\partial r}{\partial z} = r^{-3} - 3z \cdot r^{-4} \cdot \frac{z}{r} = r^{-3} - 3z^2 r^{-5}$$

Therefore,

$$\mathbf{E} = \nabla\left(\frac{z}{r^3}\right) = -3zr^{-5}x\mathbf{i} - 3zr^{-5}y\mathbf{j} + (r^{-3} - 3z^2 r^{-5})\mathbf{k}$$

$$= r^{-3}\mathbf{k} - 3zr^{-4} \cdot r^{-1}(x\mathbf{i} + y\mathbf{j} + z\mathbf{k}) = r^{-3}\mathbf{k} - 3zr^{-4}\mathbf{e}_r$$

(b) To compute the flux $\iint_S \mathbf{E} \cdot d\mathbf{S}$ we use the parametrization $\Phi(\theta, \phi) = (R\cos\theta \sin\phi, R\sin\theta \sin\phi, R\cos\phi)$, $0 \le \theta \le 2\pi$, $0 \le \phi \le \pi$:

$$\mathbf{N} = R^2 \sin\phi \, \mathbf{e}_r$$

We compute $\mathbf{E}(\Phi(\theta, \phi)) \cdot \mathbf{N}$. Since $r = R$ on S, we get

$$\mathbf{E}(\Phi(\theta, \phi)) \cdot \mathbf{N} = \left(R^{-3}\mathbf{k} - 3zR^{-4}\mathbf{e}_r\right) \cdot R^2 \sin\phi \, \mathbf{e}_r = R^{-1}\sin\phi \, \mathbf{k} \cdot \mathbf{e}_r - 3zR^{-2}\sin\phi$$

$$= R^{-1}\sin\phi \, \mathbf{k} \cdot R^{-1}(x\mathbf{i} + y\mathbf{j} + z\mathbf{k}) - 3zR^{-2}\sin\phi$$

$$= R^{-2}z\sin\phi - 3zR^{-2}\sin\phi = -2zR^{-2}\sin\phi$$

$$= -2R\cos\phi \cdot R^{-2}\sin\phi = -R^{-1}\sin 2\phi$$

Hence,

$$\iint_S \mathbf{E} \cdot d\mathbf{S} = \int_0^{2\pi}\int_0^\pi -R^{-1}\sin 2\phi \, d\phi \, d\theta = -\frac{2\pi}{R}\int_0^\pi \sin 2\phi \, d\phi = \frac{\pi}{R}\cos 2\phi \Big|_{\phi=0}^\pi = 0$$

(c) We use part (a) to write the vector $\mathbf{E}$ componentwise:

$$\mathbf{E} = r^{-3}\mathbf{k} - 3zr^{-4}\mathbf{e}_r = r^{-3}\mathbf{k} - 3zr^{-4}r^{-1}\langle x, y, z\rangle = \langle -3zr^{-5}x, -3zr^{-5}y, -3z^2 r^{-5} + r^{-3}\rangle$$

To find $\text{div}(\mathbf{E})$ we compute the following derivatives, using (1) and the laws of differentiation. This gives

$$\frac{\partial}{\partial x}(-3zr^{-5}x) = -3z\frac{\partial}{\partial x}(r^{-5}x) = -3z\left(-5r^{-6}\frac{\partial r}{\partial x}x + r^{-5} \cdot 1\right)$$

$$= -3z\left(-5r^{-6}x\frac{x}{r} + r^{-5}\right) = 3zr^{-7}(5x^2 - r^2)$$

Similarly,

$$\frac{\partial}{\partial y}(-3zr^{-5}y) = 3zr^{-7}(5y^2 - r^2)$$

and

$$\frac{\partial}{\partial z}(-3z^2 r^{-5} + r^{-3}) = -6zr^{-5} - 3z^2(-5)r^{-6}\frac{\partial r}{\partial z} - 3r^{-4}\frac{\partial r}{\partial z}$$

$$= -6zr^{-5} + 15z^2 r^{-6}\frac{z}{r} - 3r^{-4}\frac{z}{r} = 3zr^{-7}(5z^2 - 3r^2)$$

Hence,

$$\text{div}(\mathbf{E}) = 3zr^{-7}(5x^2 - r^2 + 5y^2 - r^2 + 5z^2 - 3r^2) = 15zr^{-7}(x^2 + y^2 + z^2 - r^2)$$

$$= 15zr^{-7}(r^2 - r^2) = 0$$

(d) Since $\mathbf{E}$ is not defined at the origin, which is inside the ball $\mathcal{W}$, we cannot use the Divergence Theorem to compute the flux of $\mathbf{E}$ through the sphere.

29. Let $\mathcal{W}$ be the region between the sphere of radius 4 and the cube of side 1, both centered at the origin. What is the flux through the boundary $\mathcal{S} = \partial \mathcal{W}$ of a vector field $\mathbf{F}$ whose divergence has the constant value $\text{div}(\mathbf{F}) = -4$?

SOLUTION Recall,

$$\text{flux} = \iiint_{\mathcal{W}} \text{div}(\mathbf{F})\, dV$$

Using this fact we see:

$$\text{flux} = \iiint_{\mathcal{W}} (-4)\, dV = -4 \cdot V(\mathcal{W}) = (-4)\left(\frac{256\pi}{3} - 1\right)$$

31. Let f be a scalar function and $\mathbf{F}$ be a vector field. Prove the following Product Rule for Divergence:

$$\text{div}(f\mathbf{F}) = f\text{div}(\mathbf{F}) + \nabla f \cdot \mathbf{F}$$

SOLUTION Let $\mathbf{F} = \langle F_1, F_2, F_3 \rangle$. Then using the Product Rule,

$$\text{div}(f\mathbf{F}) = \frac{\partial}{\partial x}(fF_1) + \frac{\partial}{\partial y}(fF_2) + \frac{\partial}{\partial z}(fF_3)$$

$$= f\frac{\partial}{\partial x}(F_1) + F_1\frac{\partial}{\partial x}(f) + f\frac{\partial}{\partial y}(F_2) + F_2\frac{\partial}{\partial y}(f) + f\frac{\partial}{\partial z}(F_3) + F_3\frac{\partial}{\partial z}(f)$$

$$= f\left(\frac{\partial}{\partial x}(F_1) + \frac{\partial}{\partial y}(F_2) + \frac{\partial}{\partial z}(F_3)\right) + \left\langle \frac{\partial}{\partial x}(f), \frac{\partial}{\partial y}(f), \frac{\partial}{\partial z}(f) \right\rangle \cdot \langle F_1, F_2, F_3 \rangle$$

$$= f\text{div}(\mathbf{F}) + \nabla f \cdot \mathbf{F}$$

*In Exercises 33 and 34, use the product rules in Exercises 31 and 32. A vector field $\mathbf{F}$ is **incompressible** if $\text{div}(\mathbf{F}) = 0$ and is **irrotational** if $\text{curl}(\mathbf{F}) = \mathbf{0}$.*

33. Let $\mathbf{F}$ be an incompressible vector field that is everywhere tangent to level surfaces of f. Prove that $f\mathbf{F}$ is incompressible.

SOLUTION From Exercise 31,

$$\text{div}(f\mathbf{F}) = f\text{div}(\mathbf{F}) + \nabla f \cdot \mathbf{F}$$

Since $\mathbf{F}$ is incompressible, $\text{div}(\mathbf{F}) = 0$. Since $\mathbf{F}$ is tangent to the level surfaces of f, and ∇f is orthogonal to the level surfaces (see Theorem 4 in Section 14.5), it follows that $\nabla f \cdot \mathbf{F} = 0$. Therefore, both terms on the right-hand side are zero, so that $\text{div}(f\mathbf{F}) = 0$ and $f\mathbf{F}$ is incompressible.

In Exercises 35–38, Δ denotes the Laplace operator defined by

$$\Delta \varphi = \frac{\partial^2 \varphi}{\partial x^2} + \frac{\partial^2 \varphi}{\partial y^2} + \frac{\partial^2 \varphi}{\partial z^2}$$

35. Prove the identity

$$\text{div}(\nabla\varphi) = \Delta\varphi$$

SOLUTION Compute the left-hand side:

$$\text{div}(\nabla\varphi) = \text{div}\left\langle \frac{\partial\varphi}{\partial x}, \frac{\partial\varphi}{\partial y}, \frac{\partial\varphi}{\partial z} \right\rangle = \frac{\partial^2\varphi}{\partial x^2} + \frac{\partial^2\varphi}{\partial y^2} + \frac{\partial^2\varphi}{\partial z^2} = \Delta\varphi$$

37. A function φ satisfying $\Delta\varphi = 0$ is called **harmonic**.
 (a) Show that $\Delta\varphi = \text{div}(\nabla\varphi)$ for any function φ.
 (b) Show that φ is harmonic if and only if $\text{div}(\nabla\varphi) = 0$.
 (c) Show that if $\mathbf{F}$ is the gradient of a harmonic function, then $\text{curl}(F) = 0$ and $\text{div}(F) = 0$.
 (d) Show that $\mathbf{F}(x, y, z) = \left\langle xz, -yz, \frac{1}{2}(x^2 - y^2) \right\rangle$ is the gradient of a harmonic function. What is the flux of $\mathbf{F}$ through a closed surface?

SOLUTION
(a) We compute the divergence of $\nabla\varphi$:

$$\text{div}(\nabla\varphi) = \text{div}\left(\left\langle \frac{\partial\varphi}{\partial x}, \frac{\partial\varphi}{\partial y}, \frac{\partial\varphi}{\partial z} \right\rangle\right) = \frac{\partial^2\varphi}{\partial x^2} + \frac{\partial^2\varphi}{\partial y^2} + \frac{\partial^2\varphi}{\partial z^2} = \Delta\varphi$$

(b) In part (a) we showed that $\Delta\varphi = \text{div}(\nabla\varphi)$. Therefore $\Delta\varphi = 0$ if and only if $\text{div}(\nabla\varphi) = 0$. That is, φ is harmonic if and only if $\nabla\varphi$ is divergence free.

(c) We are given that $\mathbf{F} = \nabla\varphi$, where $\Delta\varphi = 0$. In part (b) we showed that

$$\text{div}(\mathbf{F}) = \text{div}(\nabla\varphi) = 0$$

We now show that $\text{curl}(\mathbf{F}) = 0$. We have

$$\text{curl}(\mathbf{F}) = \text{curl}(\nabla\varphi) = \text{curl}\left\langle \varphi_x, \varphi_y, \varphi_z \right\rangle = \begin{vmatrix} \mathbf{i} & \mathbf{j} & \mathbf{k} \\ \frac{\partial}{\partial x} & \frac{\partial}{\partial y} & \frac{\partial}{\partial z} \\ \varphi_x & \varphi_y & \varphi_z \end{vmatrix}$$

$$= \left\langle \varphi_{zy} - \varphi_{yz}, \varphi_{xz} - \varphi_{zx}, \varphi_{yx} - \varphi_{xy} \right\rangle = \langle 0, 0, 0 \rangle = \mathbf{0}$$

The last equality is due to the equality of the mixed partials.

(d) We first show that $\mathbf{F} = \left\langle xz, -yz, \frac{x^2 - y^2}{2} \right\rangle$ is the gradient of a harmonic function. We let $\varphi = \frac{x^2 z}{2} - \frac{y^2 z}{2}$ such that $\mathbf{F} = \nabla\varphi$. Indeed,

$$\nabla\varphi = \left\langle \frac{\partial\varphi}{\partial x}, \frac{\partial\varphi}{\partial y}, \frac{\partial\varphi}{\partial z} \right\rangle = \left\langle xz, -yz, \frac{x^2 - y^2}{2} \right\rangle = \mathbf{F}$$

We show that φ is harmonic, that is, $\Delta\varphi = 0$. We compute the partial derivatives:

$$\frac{\partial\varphi}{\partial x} = xz \quad \Rightarrow \quad \frac{\partial^2\varphi}{\partial x^2} = z$$

$$\frac{\partial\varphi}{\partial y} = -yz \quad \Rightarrow \quad \frac{\partial^2\varphi}{\partial y^2} = -z$$

$$\frac{\partial\varphi}{\partial z} = \frac{x^2 - y^2}{2} \quad \Rightarrow \quad \frac{\partial^2\varphi}{\partial z^2} = 0$$

Therefore,

$$\Delta\varphi = \frac{\partial^2\varphi}{\partial x^2} + \frac{\partial^2\varphi}{\partial y^2} + \frac{\partial^2\varphi}{\partial z^2} = z - z + 0 = 0$$

Since $\mathbf{F}$ is the gradient of a harmonic function, we know by part (c) that $\text{div}(\mathbf{F}) = 0$. Therefore, by the Divergence Theorem, the flux of $\mathbf{F}$ through a closed surface is zero:

$$\iint_S \mathbf{F} \cdot d\mathbf{S} = \iiint_W \text{div}(\mathbf{F}) \, dV = \iiint_W 0 \, dV = 0$$

Further Insights and Challenges

39. Let S be the boundary surface of a region $\mathcal{W}$ in $\mathbf{R}^3$, and let $D_\mathbf{n}\varphi$ denote the directional derivative of φ, where $\mathbf{n}$ is the outward unit normal vector. Let Δ be the Laplace operator defined earlier.

(a) Use the Divergence Theorem to prove that

$$\iint_S D_\mathbf{n}\varphi \, dS = \iiint_\mathcal{W} \Delta\varphi \, dV$$

(b) Show that if φ is a harmonic function (defined in Exercise 37), then

$$\iint_S D_\mathbf{n}\varphi \, dS = 0$$

SOLUTION

(a) By the theorem on evaluating directional derivatives, $D_{\mathbf{e}_n}\varphi = \nabla\varphi \cdot \mathbf{e}_n$, hence,

$$\iint_S D_{\mathbf{e}_n}\varphi \, dS = \iint_S \nabla\varphi \cdot \mathbf{e}_n \, dS \tag{1}$$

By the definition of the vector surface integral, we have

$$\iint_S \nabla\varphi \cdot d\mathbf{S} = \iint_S (\nabla\varphi \cdot \mathbf{e}_n) \, dS$$

Combining with (1) gives

$$\iint_S D_{\mathbf{e}_n}\varphi \, dS = \iint_S \nabla\varphi \cdot d\mathbf{S}$$

We now apply the Divergence Theorem and the identity $\text{div}(\nabla\varphi) = \Delta\varphi$ shown in part (a) of Exercise 27, to write

$$\iint_S D_{\mathbf{e}_n}\varphi \, dS = \iint_S \nabla\varphi \cdot d\mathbf{S} = \iiint_\mathcal{W} \text{div}(\nabla\varphi) \, dV = \iiint_\mathcal{W} \Delta\varphi \, dV$$

(b) If φ is harmonic, then $\Delta\varphi = 0$; therefore, by the equality of part (a) we have

$$\iint_S D_{\mathbf{e}_n}\varphi \, dS = \iiint_\mathcal{W} \Delta\varphi \cdot dV = \iiint_\mathcal{W} 0 \, dV = 0$$

41. Let $\mathbf{F} = \langle P, Q, R \rangle$ be a vector field defined on $\mathbf{R}^3$ such that $\text{div}(\mathbf{F}) = 0$. Use the following steps to show that $\mathbf{F}$ has a vector potential.

(a) Let $\mathbf{A} = \langle f, 0, g \rangle$. Show that

$$\text{curl}(\mathbf{A}) = \left\langle \frac{\partial g}{\partial y}, \frac{\partial f}{\partial z} - \frac{\partial g}{\partial x}, -\frac{\partial f}{\partial y} \right\rangle$$

(b) Fix any value y_0 and show that if we define

$$f(x, y, z) = -\int_{y_0}^{y} R(x, t, z) \, dt + \alpha(x, z)$$

$$g(x, y, z) = \int_{y_0}^{y} P(x, t, z) \, dt + \beta(x, z)$$

where α and β are any functions of x and z, then $\partial g/\partial y = P$ and $-\partial f/\partial y = R$.

(c) It remains for us to show that α and β can be chosen so $Q = \partial f/\partial z - \partial g/\partial x$. Verify that the following choice works (for any choice of z_0):

$$\alpha(x, z) = \int_{z_0}^{z} Q(x, y_0, t) \, dt, \qquad \beta(x, z) = 0$$

Hint: You will need to use the relation $\text{div}(\mathbf{F}) = 0$.

SOLUTION

(a) If $\mathbf{A} = \langle f, 0, g \rangle$, then the curl of $\mathbf{A}$ is the following vector field:

$$\text{curl}(\mathbf{A}) = \begin{vmatrix} \mathbf{i} & \mathbf{j} & \mathbf{k} \\ \dfrac{\partial}{\partial x} & \dfrac{\partial}{\partial y} & \dfrac{\partial}{\partial z} \\ f & 0 & g \end{vmatrix} = \left(\dfrac{\partial g}{\partial y} - 0\right)\mathbf{i} - \left(\dfrac{\partial g}{\partial x} - \dfrac{\partial f}{\partial z}\right)\mathbf{j} + \left(0 - \dfrac{\partial f}{\partial y}\right)\mathbf{k} = \left\langle \dfrac{\partial g}{\partial y}, \dfrac{\partial f}{\partial z} - \dfrac{\partial g}{\partial x}, -\dfrac{\partial f}{\partial y} \right\rangle$$

(b) Using the Fundamental Theorem of Calculus, we have

$$\dfrac{\partial g}{\partial y}(x, y, z) = \dfrac{\partial}{\partial y}\int_{y_0}^{y} P(x, t, z)\, dt + \dfrac{\partial}{\partial y}\beta(x, z) = P(x, y, z) + 0 = P(x, y, z)$$

$$-\dfrac{\partial f}{\partial y}(x, y, z) = \dfrac{\partial}{\partial y}\int_{y_0}^{y} R(x, t, z)\, dt + \dfrac{\partial}{\partial y}\alpha(x, z) = R(x, y, z) + 0 = R(x, y, z)$$

(c) We verify that the functions

$$\alpha(x, z) = \int_{z_0}^{z} Q(x, y_0, t)\, dt, \quad \beta(x, z) = 0$$

satisfy the equality

$$Q = \dfrac{\partial f}{\partial z} - \dfrac{\partial g}{\partial x}$$

We differentiate to obtain

$$\dfrac{\partial f}{\partial z} - \dfrac{\partial g}{\partial x} = -\int_{y_0}^{y} R_z(x, t, z)\, dt + \alpha_z(x, z) - \int_{y_0}^{y} P_x(x, t, z)\, dz - \beta_x(x, z)$$

$$= -\int_{y_0}^{y} (P_x(x, t, z) + R_z(x, t, z))\, dt + \alpha_z(x, z) \qquad (1)$$

By the Fundamental Theorem of Calculus,

$$\alpha_z(x, z) = \dfrac{\partial}{\partial z}\int_{z_0}^{z} Q(x, y_0, t)\, dt = Q(x, y_0, z) \qquad (2)$$

Also, since $\text{div}(\mathbf{F}) = 0$, we have

$$\text{div}(\mathbf{F}) = P_x + Q_y + R_z = 0 \quad \Rightarrow \quad P_x + R_z = -Q_y \qquad (3)$$

Substituting (2) and (3) in (1) gives

$$\dfrac{\partial f}{\partial z} - \dfrac{\partial g}{\partial x} = \int_{y_0}^{y} Q_y(x, t, z)\, dt + Q(x, y_0, z) = Q(x, y, z) - Q(x, y_0, z) + Q(x, y_0, z) = Q(x, y, z)$$

Parts (a)–(c) prove that $\mathbf{F} = \text{curl}(\mathbf{A})$, or $\mathbf{A}$ is a vector potential for $\mathbf{F}$.

43. Show that

$$\mathbf{F}(x, y, z) = \langle 2ye^z - xy, y, yz - z \rangle$$

has a vector potential and find one.

SOLUTION As shown in Exercise 41, if $\mathbf{F}$ is divergence free, then $\mathbf{F}$ has a vector potential. We show that $\text{div}(\mathbf{F}) = 0$:

$$\text{div}(\mathbf{F}) = \dfrac{\partial}{\partial x}(2ye^z - xy) + \dfrac{\partial}{\partial y}(y) + \dfrac{\partial}{\partial z}(yz - z) = -y + 1 + y - 1 = 0$$

We find a vector potential $\mathbf{A}$, using the result in Exercise 41:

$$\mathbf{A} = \langle f, 0, g \rangle \qquad (1)$$

Using $z_0 = 0$, we have

$$f(x, y, z) = -\int_{y_0}^{y} R(x, t, z)\, dt + \int_{0}^{z} Q(x, y_0, t)\, dt$$

$$g(x, y, z) = \int_{y_0}^{y} P(x, t, z)\, dt$$

Hence, $P(x, y, z) = 2ye^z - xy$, $Q(x, y, z) = y$, and $R(x, y, z) = yz - z$. We choose $y_0 = 0$ and compute the functions f and g:

$$f(x, y, z) = -\int_0^y (tz - z)\, dt + \int_0^z 0\, dt = -\left(\frac{t^2 z}{2} - zt\right)\bigg|_{t=0}^y = zy - \frac{y^2 z}{2} = z\left(y - \frac{y^2}{2}\right)$$

$$g(x, y, z) = \int_0^y (2te^z - xt)\, dt = t^2 e^z - \frac{xt^2}{2}\bigg|_{t=0}^y = y^2 e^z - \frac{xy^2}{2} = y^2\left(e^z - \frac{x}{2}\right)$$

Substituting in (1) we obtain

$$\mathbf{A} = \left\langle z\left(y - \frac{y^2}{2}\right), 0, y^2\left(e^z - \frac{x}{2}\right)\right\rangle$$

CHAPTER REVIEW EXERCISES

1. Let $\mathbf{F}(x, y) = \langle x + y^2, x^2 - y\rangle$, and let C be the unit circle, oriented counterclockwise. Evaluate $\oint_C \mathbf{F} \cdot d\mathbf{r}$ directly as a line integral and using Green's Theorem.

SOLUTION We parametrize the unit circle by $\mathbf{c}(t) = (\cos t, \sin t)$, $0 \le t \le 2\pi$. Then, $\mathbf{c}'(t) = \langle -\sin t, \cos t\rangle$ and $\mathbf{F}(\mathbf{c}(t)) = (\cos t + \sin^2 t, \cos^2 t - \sin t)$. We compute the dot product:

$$\mathbf{F}(\mathbf{c}(t)) \cdot \mathbf{c}'(t) = \left\langle\cos t + \sin^2 t, \cos^2 t - \sin t\right\rangle \cdot \langle -\sin t, \cos t\rangle$$

$$= (-\sin t)(\cos t + \sin^2 t) + \cos t(\cos^2 t - \sin t)$$

$$= \cos^3 t - \sin^3 t - 2\sin t \cos t$$

The line integral is thus

$$\int_C \mathbf{F}(\mathbf{c}(t)) \cdot \mathbf{c}'(t)\, dt = \int_0^{2\pi} \left(\cos^3 t - \sin^3 t - 2\sin t \cos t\right) dt$$

$$= \int_0^{2\pi} \cos^3 t\, dt - \int_0^{2\pi} \sin^3 t\, dt - \int_0^{2\pi} \sin 2t\, dt$$

$$= \frac{\cos^2 t \sin t}{3} + \frac{2\sin t}{3}\bigg|_0^{2\pi} + \left(\frac{\sin^2 t \cos t}{3} + \frac{2\cos t}{3}\right)\bigg|_0^{2\pi} + \frac{\cos 2t}{2}\bigg|_0^{2\pi} = 0$$

We now compute the integral using Green's Theorem. We compute the curl of $\mathbf{F}$. Since $P = x + y^2$ and $Q = x^2 - y$, we have

$$\frac{\partial Q}{\partial x} - \frac{\partial P}{\partial y} = 2x - 2y$$

Thus,

$$\int_C \mathbf{F} \cdot d\mathbf{r} = \iint_\mathcal{D} (2x - 2y)\, dx\, dy$$

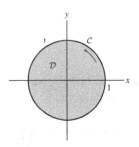

We compute the double integral by converting to polar coordinates. We get

$$\int_C \mathbf{F} \cdot d\mathbf{r} = \int_0^{2\pi} \int_0^1 (2r\cos\theta - 2r\sin\theta) r \, dr \, d\theta = \int_0^{2\pi} \int_0^1 2r^2(\cos\theta - \sin\theta) \, dr \, d\theta$$

$$= \left(\int_0^1 2r^2 \, dr\right)\left(\int_0^{2\pi} (\cos\theta - \sin\theta) \, d\theta\right) = \left(\frac{2}{3}r^3 \Big|_0^1\right)\left(\sin\theta + \cos\theta \Big|_0^{2\pi}\right) = \frac{2}{3}(1 - 1) = 0$$

In Exercises 3–6, use Green's Theorem to evaluate the line integral around the given closed curve.

3. $\oint_C xy^3 \, dx + x^3 y \, dy$, where C is the boundary of the rectangle $-1 \le x \le 2$, $-2 \le y \le 3$, oriented counterclockwise

SOLUTION

Since $P = xy^3$, $Q = x^3 y$ the curl of $\mathbf{F}$ is

$$\frac{\partial Q}{\partial x} - \frac{\partial P}{\partial y} = 3x^2 y - 3xy^2$$

By Green's Theorem we obtain

$$\int_C xy^3 \, dx + x^3 y \, dy = \iint_{\mathcal{D}} (3x^2 y - 3xy^2) \, dx \, dy = \int_{-2}^3 \int_{-1}^2 (3x^2 y - 3xy^2) \, dx \, dy$$

$$= \int_{-2}^3 x^3 y - \frac{3x^2 y^2}{2}\Big|_{x=-1}^2 \, dy = \int_{-2}^3 \left((8y - 6y^2) - \left(-y - \frac{3y^2}{2}\right)\right) dy$$

$$= \int_{-2}^3 \left(-\frac{9y^2}{2} + 9y\right) dy = -\frac{3y^3}{2} + \frac{9y^2}{2}\Big|_{-2}^3 = \left(-\frac{81}{2} + \frac{81}{2}\right) - (12 + 18) = -30$$

5. $\oint_C y^2 \, dx - x^2 \, dy$, where C consists of the arcs $y = x^2$ and $y = \sqrt{x}$, $0 \le x \le 1$, oriented clockwise

SOLUTION We compute the curl of $\mathbf{F}$.

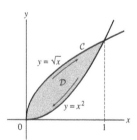

We have $P = y^2$ and $Q = -x^2$, hence

$$\frac{\partial Q}{\partial x} - \frac{\partial P}{\partial y} = -2x - 2y$$

We now compute the line integral using Green's Theorem. Since the curve is oriented clockwise, we consider the negative of the double integrals. We get

$$\int_C y^2 \, dx - x^2 \, dy = -\iint_{\mathcal{D}} (-2x - 2y) \, dA = -\int_0^1 \int_{x^2}^{\sqrt{x}} (-2x - 2y) \, dy \, dx$$

$$= \int_0^1 2xy + y^2 \Big|_{y=x^2}^{\sqrt{x}} dx = \int_0^1 \left(\left(2x\sqrt{x} + x\right) - \left(2x \cdot x^2 + x^4\right) \right) dx$$

$$= \int_0^1 (-x^4 - 2x^3 + 2x^{3/2} + x) \, dx = -\frac{x^5}{5} - \frac{x^4}{2} + \frac{4x^{5/2}}{5} + \frac{x^2}{2} \Big|_0^1$$

$$= -\frac{1}{5} - \frac{1}{2} + \frac{4}{5} + \frac{1}{2} = \frac{3}{5}$$

7. Let $\mathbf{r}(t) = \langle t^2(1-t), t(t-1)^2 \rangle$.

(a) [GU] Plot the path $\mathbf{r}(t)$ for $0 \le t \le 1$.

(b) Calculate the area A of the region enclosed by $\mathbf{r}(t)$ for $0 \le t \le 1$ using the formula $A = \frac{1}{2} \oint_C (x\, dy - y\, dx)$.

SOLUTION

(a) The path $\mathbf{r}(t)$ for $0 \le t \le 1$ is shown in the figure:

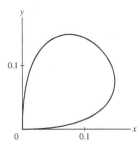

Note that the path is traced out clockwise as t goes from 0 to 1.

(b) We use the formula for the area enclosed by a closed curve,

$$A = \frac{1}{2} \int_C (x\, dy - y\, dx)$$

We compute the line integral. Since $x = t^2(1-t)$ and $y = t(t-1)^2$, we have

$$dx = \left(2t(1-t) - t^2\right) dt = \left(2t - 3t^2\right) dt$$

$$dy = (t-1)^2 + t \cdot 2(t-1) = (t-1)(3t-1)\, dt$$

Therefore,

$$x\, dy - y\, dx = t^2(1-t) \cdot (t-1)(3t-1)\, dt - t(t-1)^2 \cdot (2t - 3t^2)\, dt = t^2(t-1)^2\, dt$$

We obtain the following integral (note that the path must be counterclockwise):

$$A = \frac{1}{2} \int_1^0 -t^2(t-1)^2\, dt = \frac{1}{2} \int_0^1 (t^4 - 2t^3 + t^2)\, dt = \frac{1}{2} \left(\frac{t^5}{5} - \frac{t^4}{2} + \frac{t^3}{3} \Big|_0^1 \right) = \frac{1}{60}$$

9. Calculate the area of the region bounded by the two curves $y = x^2$ and $y = \sqrt{x}$ for $x \ge 0$ using the formula $A = \oint_C x\, dy$.

SOLUTION The boundary of the region consists of two pieces, one along $y = x^2$ and the other along $y = \sqrt{x}$. These two pieces meet at the points $(0, 0)$ and $(1, 1)$. Parametrize them so the boundary is oriented counterclockwise as follows:

$$C_1: (t, t^2),\ 0 \le t \le 1, \qquad C_2: ((1-t)^2, 1-t),\ 0 \le t \le 1$$

Then we have

$$C_1: \quad x = t \quad dy = 2t\, dt, \qquad C_2: \quad x = (1-t)^2 \quad dy = -1\, dt$$

Thus the area of the region is

$$\int_{C_1} x\, dy + \int_{C_2} x\, dy = \int_0^1 t \cdot 2t\, dt + \int_0^1 (1-t)^2(-1)\, dt = \frac{2}{3} t^3 \Big|_0^1 + \frac{1}{3}(1-t)^3 \Big|_0^1 = \frac{1}{3}$$

This agrees with the answer one gets by integrating directly:

$$\int_0^1 (\sqrt{x} - x^2)\,dx = \left(\frac{2}{3}x^{3/2} - \frac{1}{3}x^3\right)\bigg|_0^1 = \frac{1}{3}$$

11. Calculate the area of the region bounded by the two curves $y = x^2$ and $y = \sqrt{x}$ for $x \geq 0$ using the formula $A = \oint_C -y\,dx$.

SOLUTION The boundary of the region consists of two pieces, one along $y = x^2$ and the other along $y = \sqrt{x}$. These two pieces meet at the points $(0, 0)$ and $(1, 1)$. Parametrize them so the boundary is oriented counterclockwise as follows:

$$C_1 : (t, t^2),\ 0 \leq t \leq 1, \qquad C_2 : ((1-t)^2, 1-t),\ 0 \leq t \leq 1$$

Then we have

$$C_1 : \quad y = t^2 \quad dx = dt, \qquad C_2 : \quad y = 1-t \quad dx = -2(1-t)\,dt$$

Thus the area of the region is

$$\int_{C_1} (-y)\,dx + \int_{C_2} (-y)\,dx = -\int_0^1 t^2\,dt + \int_0^1 (1-t)\cdot 2(1-t)\,dt$$

$$= -\frac{1}{3}t^3\bigg|_0^1 + 2\int_0^1 (1-t)^2\,dt = -\frac{1}{3} - \frac{2}{3}(1-t)^3\bigg|_0^1 = \frac{1}{3}$$

This agrees with the answer in Exercise 9 as well as the one you get by integrating directly:

$$\int_0^1 (\sqrt{x} - x^2)\,dx = \left(\frac{2}{3}x^{3/2} - \frac{1}{3}x^3\right)\bigg|_0^1 = \frac{1}{3}$$

13. Let $\mathbf{F}(x, y) = \langle x^2 y, xy^2 \rangle$ be the velocity vector field for a fluid in the plane. Find all points where the angular velocity of a small paddle wheel inserted into the fluid would be 0.

SOLUTION The angular velocity of such a paddle wheel is approximately $\frac{1}{2}\mathrm{curl}(\mathbf{F})(P)$. Now,

$$\mathrm{curl}(\mathbf{F}) = \begin{vmatrix} \mathbf{i} & \mathbf{j} & \mathbf{k} \\ \partial x & \partial y & \partial z \\ x^2 y & xy^2 & 0 \end{vmatrix} = \langle 0, 0, y^2 - x^2 \rangle$$

so that $\mathrm{curl}(\mathbf{F})$ vanishes when $y = \pm x$; that is, at any point along the lines $y = x$ or $y = -x$.

15. Compute the flux $\oint_{\partial D} \mathbf{F} \cdot \mathbf{n}\,ds$ of $\mathbf{F}(x, y) = \langle x^3 + 2x, y^3 + y \rangle$ across the circle $\mathcal{D}$ given by $x^2 + y^2 = 4$ using the vector form of Green's Theorem.

SOLUTION We have

$$\mathrm{div}\,\mathbf{F} = \frac{\partial}{\partial x}(x^3 + 2x) + \frac{\partial}{\partial y}(y^3 + y) = 3x^2 + 3y^2 + 3$$

Therefore the flux of $\mathbf{F}$ out of the unit square is given by

$$\oint_{\partial D} \mathbf{F} \cdot \mathbf{n}\,ds = \iint_D \mathrm{div}\,\mathbf{F}\,dA = \iint_D 3(x^2 + y^2 + 1)\,dA$$

To compute the integral, convert to polar coordinates, giving

$$\iint_D 3(x^2 + y^2 + 1)\,dA = 3\int_0^2 \int_0^{2\pi} (r^2 + 1)r\,d\theta\,dr = 6\pi \int_0^2 (r^3 + r)\,dr = 6\pi \left(\frac{1}{4}r^4 + \frac{1}{2}r^2\right)\bigg|_0^2 = 36\pi$$

17. Prove that if $\mathbf{F}$ is a gradient vector field, then the flux of $\mathrm{curl}(\mathbf{F})$ through a smooth surface $\mathcal{S}$ (whether closed or not) is equal to zero.

SOLUTION If **F** is a gradient vector field, then **F** is conservative; therefore the line integral of **F** over any closed curve is zero. Combining with Stokes' Theorem yields

$$\iint_S \text{curl}(\mathbf{F}) \cdot d\mathbf{S} = \int_{\partial S} \mathbf{F} \cdot d\mathbf{r} = 0$$

19. Let $\mathbf{F}(x, y, z) = \langle z^2, x + z, y^2 \rangle$, and let S be the upper half of the ellipsoid

$$\frac{x^2}{4} + y^2 + z^2 = 1$$

oriented by outward-pointing normals. Use Stokes' Theorem to compute $\iint_S \text{curl}(\mathbf{F}) \cdot d\mathbf{S}$.

SOLUTION We compute the curl of $\mathbf{F} = \langle z^2, x + z, y^2 \rangle$:

$$\text{curl}(\mathbf{F}) = \begin{vmatrix} \mathbf{i} & \mathbf{j} & \mathbf{k} \\ \frac{\partial}{\partial x} & \frac{\partial}{\partial y} & \frac{\partial}{\partial z} \\ z^2 & x+z & y^2 \end{vmatrix} = (2y - 1)\mathbf{i} - (0 - 2z)\mathbf{j} + (1 - 0)\mathbf{k} = \langle 2y - 1, 2z, 1 \rangle$$

Let C denote the boundary of S, that is, the ellipse $\frac{x^2}{4} + y^2 = 1$ in the xy-plane, oriented counterclockwise. Then by Stoke's Theorem we have

$$\iint_S \text{curl}(\mathbf{F}) \cdot d\mathbf{S} = \int_C \mathbf{F} \cdot d\mathbf{r} \tag{1}$$

We parametrize C by

$$C : r(t) = (2\cos t, \sin t, 0), \quad 0 \le t \le 2\pi$$

Then

$$\mathbf{F}(r(t)) \cdot r'(t) = \langle 0, 2\cos t, \sin^2 t \rangle \cdot \langle -2\sin t, \cos t, 0 \rangle = 2\cos^2 t$$

Combining with (1) gives

$$\iint_S \text{curl}(\mathbf{F}) \cdot d\mathbf{S} = \int_0^{2\pi} 2\cos^2 t \, dt = t + \frac{\sin 2t}{2} \Big|_0^{2\pi} = 2\pi$$

21. Let S be the side of the cylinder $x^2 + y^2 = 4$, $0 \le z \le 2$ (not including the top and bottom of the cylinder). Use Stokes' Theorem to compute the flux of $\mathbf{F}(x, y, z) = \langle 0, y, -z \rangle$ through S (with outward-pointing normal) by finding a vector potential **A** such that $\text{curl}(\mathbf{A}) = \mathbf{F}$.

SOLUTION We can write $\mathbf{F} = \text{curl}(\mathbf{A})$ where $\mathbf{A} = \langle yz, 0, 0 \rangle$. The flux of **F** through S is equal to the line integral of A around the oriented boundary which consists of two circles of radius 2 with center on the z-axis (one at height $z = 0$ and one at height $z = 2$).

However, the line integrals of A about both circles are zero. This is clear for the circle at $z = 0$ because then $A = 0$, but it is also true at $z = 2$ because the vector field $A = \langle 2y, 0, 0 \rangle$ integrates to zero around the circle.

In Exercises 23–26, use the Divergence Theorem to calculate $\iint_S \mathbf{F} \cdot d\mathbf{S}$ *for the given vector field and surface.*

23. $\mathbf{F}(x, y, z) = \langle xy, yz, x^2z + z^2 \rangle$, S is the boundary of the box $[0, 1] \times [2, 4] \times [1, 5]$.

SOLUTION

We compute the divergence of $\mathbf{F} = \langle xy, yz, x^2z + z^2 \rangle$:

$$\text{div}(\mathbf{F}) = \frac{\partial}{\partial x}xy + \frac{\partial}{\partial y}yz + \frac{\partial}{\partial z}(x^2z + z^2) = y + z + x^2 + 2z = x^2 + y + 3z$$

The Divergence Theorem gives

$$\iint_S \langle xy, yz, x^2z + z^2 \rangle \cdot d\mathbf{S} = \int_1^5 \int_2^4 \int_0^1 (x^2 + y + 3z)\,dx\,dy\,dz = \int_1^5 \int_2^4 \frac{x^3}{3} + (y + 3z)x \Big|_{x=0}^1 dy\,dz$$

$$= \int_1^5 \int_1^4 \left(\frac{1}{3} + y + 3z\right) dy\,dz = \int_1^5 \frac{1}{3}y + \frac{1}{2}y^2 + 3zy \Big|_{y=1}^4 dz$$

$$= \int_1^5 \left(\left(\frac{4}{3} + \frac{16}{2} + 12z\right) - \left(\frac{1}{3} + \frac{1}{2} + 3z\right)\right) dz = \int_1^5 \left(\frac{17}{2} + 9z\right) dz$$

$$= \frac{17z}{2} + \frac{9z^2}{2} \Big|_1^5 = \frac{(17 \cdot 5 + 9 \cdot 25) - (17 + 9)}{2} = 142$$

25. $\mathbf{F}(x, y, z) = \langle xyz + xy, \frac{1}{2}y^2(1 - z) + e^x, e^{x^2+y^2} \rangle$, S is the boundary of the solid bounded by the cylinder $x^2 + y^2 = 16$ and the planes $z = 0$ and $z = y - 4$.

SOLUTION We compute the divergence of $\mathbf{F}$:

$$\text{div}(\mathbf{F}) = \frac{\partial}{\partial x}(xyz + xy) + \frac{\partial}{\partial y}\left(\frac{y^2}{2}(1 - z) + e^x\right) + \frac{\partial}{\partial z}(e^{x^2+y^2}) = yz + y + y(1 - z) = 2y$$

Let S denote the surface of the solid $\mathcal{W}$. The Divergence Theorem gives

$$\iint_S \mathbf{F} \cdot d\mathbf{S} = \iiint_{\mathcal{W}} \text{div}(\mathbf{F})\,dV = \iiint_{\mathcal{W}} 2y\,dV = \iint_D \int_{y-4}^0 2y\,dz\,dx\,dy$$

$$= \iint_D 2yz \Big|_{z=y-4}^0 dx\,dy = \iint_D 2y(0 - (y - 4))\,dx\,dy = \iint_D (8y - 2y^2)\,dx\,dy$$

We convert the integral to polar coordinates:

$$\iint_S \mathbf{F} \cdot d\mathbf{S} = \int_0^{2\pi} \int_0^4 (8r\cos\theta - 2r^2\cos^2\theta)r\,dr\,d\theta$$

$$= 8\left(\int_0^4 r^2\,dr\right)\left(\int_0^{2\pi} \cos\theta\,d\theta\right) - \left(\int_0^4 r^3\,dr\right)\left(\int_0^{2\pi} 2\cos^2\theta\,d\theta\right)$$

$$= 0 - \left(\frac{r^4}{4}\Big|_0^4\right)\left(\theta + \frac{\sin 2\theta}{2}\Big|_0^{2\pi}\right) = -\frac{4^4}{4} \cdot 2\pi = -128\pi$$

27. Find the volume of a region $\mathcal{W}$ if

$$\iint_{\partial \mathcal{W}} \left\langle x + xy + z, x + 3y - \frac{1}{2}y^2, 4z \right\rangle \cdot d\mathbf{S} = 16$$

SOLUTION Let $\mathbf{F} = \langle x + xy + z, x + 3y - \frac{1}{2}y^2, 4z \rangle$. We compute the divergence of $\mathbf{F}$:

$$\text{div}(\mathbf{F}) = \frac{\partial}{\partial x}(x + xy + z) + \frac{\partial}{\partial y}\left(x + 3y - \frac{1}{2}y^2\right) + \frac{\partial}{\partial z}(4z) = 1 + y + 3 - y + 4 = 8$$

Using the Divergence Theorem and the given information, we obtain

$$16 = \iint_S \mathbf{F} \cdot d\mathbf{S} = \iiint_W \text{div}(\mathbf{F}) \, dV = \iiint_W 8 \, dV = 8 \iiint_W 1 \, dV = 8 \text{ Volume}(W)$$

That is,

$$16 = 8 \text{ Volume}(W)$$

or

$$\text{Volume}(W) = 2$$

In Exercises 29–32, let $\mathbf{F}$ be a vector field whose curl and divergence at the origin are

$$\text{curl}(\mathbf{F})(0, 0, 0) = \langle 2, -1, 4 \rangle, \quad \text{div}(\mathbf{F})(0, 0, 0) = -2$$

29. Estimate $\oint_C \mathbf{F} \cdot d\mathbf{r}$, where C is the circle of radius 0.03 in the xy-plane centered at the origin.

SOLUTION We use the estimation

$$\int_C \mathbf{F} \cdot d\mathbf{r} \approx (\text{curl}(\mathbf{F})(\mathbf{0}) \cdot \mathbf{e}_n) \text{ Area}(\mathcal{R})$$

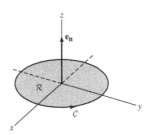

The unit normal vector to the disk $\mathcal{R}$ is $\mathbf{e}_n = \mathbf{k} = \langle 0, 0, 1 \rangle$. The area of the disk is

$$\text{Area}(\mathcal{R}) = \pi \cdot 0.03^2 = 0.0009\pi$$

Using the given curl at the origin, we have

$$\int_C \mathbf{F} \cdot d\mathbf{r} \approx \langle 2, -1, 4 \rangle \cdot \langle 0, 0, 1 \rangle \cdot 0.0009\pi = 4 \cdot 0.0009\pi \approx 0.0113$$

31. Suppose that $\mathbf{v}$ is the velocity field of a fluid and imagine placing a small paddle wheel at the origin. Find the equation of the plane in which the paddle wheel should be placed to make it rotate as quickly as possible.

SOLUTION The paddle wheel has the maximum spin when the circulation of the velocity field $\mathbf{F}$ around the wheel is maximum. The maximum circulation occurs when $\mathbf{e}_n$, and the curl of $\mathbf{F}$ at the origin (i.e., the vector $\langle 2, -1, 4 \rangle$) point in the same direction. Therefore, the plane in which the paddle wheel should be placed is the plane through the origin with the normal $\langle 2, -1, 4 \rangle$. This plane has the equation, $2x - y + 4z = 0$.

33. The velocity vector field of a fluid (in meters per second) is

$$\mathbf{F}(x, y, z) = \langle x^2 + y^2, 0, z^2 \rangle$$

Let $\mathcal{W}$ be the region between the hemisphere

$$\mathcal{S} = \{(x, y, z) : x^2 + y^2 + z^2 = 1, \quad x, y, z \geq 0\}$$

and the disk $\mathcal{D} = \{(x, y, 0) : x^2 + y^2 \leq 1\}$ in the xy-plane. Recall that the flow rate of a fluid across a surface is equal to the flux of $\mathbf{F}$ through the surface.

(a) Show that the flow rate across $\mathcal{D}$ is zero.

(b) Use the Divergence Theorem to show that the flow rate across $\mathcal{S}$, oriented with an outward-pointing normal, is equal to $\iiint_{\mathcal{W}} \text{div}(\mathbf{F})\,dV$. Then compute this triple integral.

SOLUTION

(a) To show that no fluid flows across $\mathcal{D}$, we show that the normal component of $\mathbf{F}$ at each point on $\mathcal{D}$ is zero. At each point $P = (x, y, 0)$ on the xy-plane,

$$\mathbf{F}(P) = \langle x^2 + y^2, 0, 0^2 \rangle = \langle x^2 + y^2, 0, 0 \rangle$$

Moreover, the unit normal vector to the xy-plane is $\mathbf{e}_n = (0, 0, 1)$. Therefore,

$$\mathbf{F}(P) \cdot \mathbf{e}_n = \langle x^2 + y^2, 0, 0 \rangle \cdot \langle 0, 0, 1 \rangle = 0$$

Since $\mathcal{D}$ is contained in the xy-plane, we conclude that the normal component of $\mathbf{F}$ at each point on $\mathcal{D}$ is zero. Therefore, no fluid flows across $\mathcal{D}$.

(b) By the Divergence Theorem and the linearity of the flux we have

$$\iint_{\mathcal{S}} \mathbf{F} \cdot d\mathbf{S} + \iint_{\mathcal{D}} \mathbf{F} \cdot d\mathbf{S} = \iiint_{\mathcal{W}} \text{div}(\mathbf{F})\,dV$$

Since the flux through the disk $\mathcal{D}$ is zero, we have

$$\iint_{\mathcal{S}} \mathbf{F} \cdot d\mathbf{S} = \iiint_{\mathcal{W}} \text{div}(\mathbf{F})\,dV \quad (1)$$

To compute the triple integral, we first compute $\text{div}(\mathbf{F})$:

$$\text{div}(\mathbf{F}) = \frac{\partial}{\partial x}(x^2 + y^2) + \frac{\partial}{\partial y}(0) + \frac{\partial}{\partial z}(z^2) = 2x + 2z = 2(x + z)$$

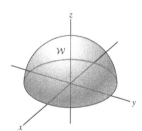

Using spherical coordinate we get

$$\iiint_{\mathcal{W}} \text{div}(\mathbf{F})\,dV = 2\int_0^{\pi/2}\int_0^{2\pi}\int_0^1 (\rho \sin\phi \cos\theta + \rho\cos\phi)\rho^2 \sin\phi\,d\rho\,d\theta\,d\phi$$

$$= 2\int_0^1 \rho^3 d\rho\left(\left(\int_0^{\pi/2}\sin^2\phi\,d\phi\right)\left(\int_0^{2\pi}\cos\theta\,d\theta\right) + 2\pi\int_0^{\pi/2}\cos\phi\sin\phi\,d\phi\right)$$

$$= \frac{1}{2}\left(0 + \pi\int_0^{\pi/2}\sin 2\phi\,d\phi\right) = \frac{\pi}{2}\left(-\frac{\cos 2\phi}{2}\right)\Big|_0^{\pi/2} = -\frac{\pi}{4}(-1-1) = \frac{\pi}{2}$$

Combining with (1) we obtain the flux:

$$\iint_{\mathcal{S}} \mathbf{F}\cdot d\mathbf{S} = \frac{\pi}{2}$$

35. Let $f(x, y) = x + \dfrac{x}{x^2 + y^2}$. The vector field $\mathbf{F} = \nabla f$ (Figure 5) provides a model in the plane of the velocity field of an incompressible, irrotational fluid flowing past a cylindrical obstacle (in this case, the obstacle is the unit circle $x^2 + y^2 = 1$).

(a) Verify that **F** is irrotational [by definition, **F** is irrotational if curl(**F**) = **0**].

FIGURE 5 The vector field ∇f for $f(x,y) = x + \dfrac{x}{x^2 + y^2}$.

(b) Verify that **F** is tangent to the unit circle at each point along the unit circle except $(1,0)$ and $(-1,0)$ (where **F** = **0**).
(c) What is the circulation of **F** around the unit circle?
(d) Calculate the line integral of **F** along the upper and lower halves of the unit circle separately.

SOLUTION
(a) In Exercise 8, we proved the identity curl($\nabla \varphi$) = **0**. Since **F** is a gradient vector field, it is irrotational; that is, curl(**F**) = **0** for $(x,y) \neq (0,0)$, where **F** is defined.
(b) We compute **F** explicitly:

$$\mathbf{F} = \nabla \varphi = \left\langle \frac{\partial \varphi}{\partial x}, \frac{\partial \varphi}{\partial y} \right\rangle = \left\langle 1 + \frac{y^2 - x^2}{(x^2 + y^2)^2}, -\frac{2xy}{(x^2 + y^2)^2} \right\rangle$$

Now, using $x = \cos t$ and $y = \sin t$ as a parametrization of the circle, we see that

$$\mathbf{F} = \left\langle 1 + \sin^2 t - \cos^2 t, -2\cos t \sin t \right\rangle = \left\langle 2\sin^2 t, -2\cos t \sin t \right\rangle$$

and so

$$\mathbf{F} = 2\sin t \langle \sin t, -\cos t \rangle = 2\sin t \langle y, -x \rangle$$

which is clearly perpendicular to the radial vector $\langle x, y \rangle$ for the circle.

(c) We use our expression of **F** from part (b):

$$\mathbf{F} = \nabla \varphi = \left\langle 1 + \frac{y^2 - x^2}{(x^2 + y^2)^2}, -\frac{2xy}{(x^2 + y^2)^2} \right\rangle$$

Now, **F** is not defined at the origin and therefore we cannot use Green's Theorem to compute the line integral along the unit circle. We thus compute the integral directly, using the parametrization

$$\mathbf{c}(t) = (\cos t, \sin t), \quad 0 \le t \le 2\pi$$

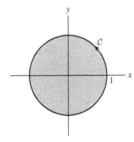

Then,

$$\mathbf{F}(\mathbf{c}(t)) \cdot \mathbf{c}'(t) = \left\langle 1 + \frac{\sin^2 t - \cos^2 t}{(\cos^2 t + \sin^2 t)^2}, -\frac{2\cos t \sin t}{(\cos^2 t + \sin^2 t)^2} \right\rangle \cdot \langle -\sin t, \cos t \rangle$$

$$= \left\langle 1 + \sin^2 t - \cos^2 t, -2\cos t \sin t \right\rangle \cdot \langle -\sin t, \cos t \rangle = \left\langle 2\sin^2 t, -2\cos t \sin t \right\rangle \cdot \langle -\sin t, \cos t \rangle$$

$$= -2\sin^3 t - 2\cos^2 t \sin t = -2\sin t(\sin^2 t + \cos^2 t) = -2\sin t$$

Hence,

$$\int_C \mathbf{F} \cdot d\mathbf{r} = \int_0^{2\pi} -2\sin t \, dt = 0$$

(d) We denote by C_1 and C_2 the upper and lower halves of the unit circle. Using part (c) we have

$$\int_{C_1} \mathbf{F} \cdot d\mathbf{r} + \int_{C_2} \mathbf{F} \cdot d\mathbf{r} = 0 \quad \Rightarrow \quad \int_{C_2} \mathbf{F} \cdot d\mathbf{r} = -\int_{C_1} \mathbf{F} \cdot d\mathbf{r} \qquad (1)$$

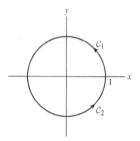

To compute the circulation along C_1, we compute the integral as in part (c), only that the limits of integration are now $t = 0$ and $t = \pi$. Using the computations in part (c) we obtain

$$\int_{C_1} \mathbf{F} \cdot d\mathbf{r} = \int_0^{\pi} -2\sin^2 t \, dt = -4$$

Therefore, by (1),

$$\int_{C_2} \mathbf{F} \cdot d\mathbf{r} = 4$$

37. In Section 17.1, we showed that if C is a simple closed curve, oriented counterclockwise, then the area enclosed by C is given by

$$\text{area enclosed by } C = \frac{1}{2} \oint_C x \, dy - y \, dx \qquad \boxed{1}$$

Suppose that C is a path from P to Q that is not closed but has the property that every line through the origin intersects C in at most one point, as in Figure 7. Let $\mathcal{R}$ be the region enclosed by C and the two radial segments joining P and Q to the origin. Show that the line integral in Eq. (1) is equal to the area of $\mathcal{R}$. *Hint:* Show that the line integral of $\mathbf{F} = \langle -y, x \rangle$ along the two radial segments is zero and apply Green's Theorem.

FIGURE 7

SOLUTION

Let $\mathbf{F} = \langle -y, x \rangle$. Then $P = -y$ and $Q = x$, and $\frac{\partial Q}{\partial x} - \frac{\partial P}{\partial y} = 2$. By Green's Theorem, we have

$$\int_C -y\,dx + x\,dy + \int_{\overline{QO}} -y\,dx + x\,dy + \int_{\overline{OP}} -y\,dx + x\,dy = \iint_{\mathcal{R}} 2\,dA = 2\iint_{\mathcal{R}} dA$$

Denoting by A the area of the region $\mathcal{R}$, we obtain

$$A = \frac{1}{2}\int_C -y\,dx + x\,dy + \frac{1}{2}\int_{\overline{QO}} -y\,dx + x\,dy + \frac{1}{2}\int_{\overline{OP}} -y\,dx + x\,dy \tag{1}$$

We parametrize the two segments by

$$\overline{QO}: \mathbf{c}(t) = (t, t\tan\beta) \quad\Rightarrow\quad \mathbf{c}'(t) = \langle 1, \tan\beta \rangle$$
$$\overline{OP}: \mathbf{d}(t) = (t, t\tan\alpha) \quad\Rightarrow\quad \mathbf{d}'(t) = \langle 1, \tan\alpha \rangle$$

Then,

$$\mathbf{F}(\mathbf{c}(t)) \cdot \mathbf{c}'(t) = \langle -t\tan\beta, t \rangle \cdot \langle 1, \tan\beta \rangle = -t\tan\beta + t\tan\beta = 0$$
$$\mathbf{F}(\mathbf{d}(t)) \cdot \mathbf{d}'(t) = \langle -t\tan\alpha, t \rangle \cdot \langle 1, \tan\alpha \rangle = -t\tan\alpha + t\tan\alpha = 0$$

Therefore,

$$\int_{\overline{QO}} \mathbf{F}\cdot d\mathbf{r} = \int_{\overline{OP}} \mathbf{F}\cdot d\mathbf{r} = 0$$

Combining with (1) gives

$$A = \frac{1}{2}\int_C -y\,dx + x\,dy$$

39. Prove the following generalization of Eq. (1). Let C be a simple closed curve in the plane S with equation $ax + by + cz + d = 0$ (Figure 8). Then the area of the region R enclosed by C is equal to

$$\frac{1}{2\|\mathbf{N}\|} \oint_C (bz - cy)\,dx + (cx - az)\,dy + (ay - bx)\,dz$$

where $\mathbf{N} = \langle a, b, c \rangle$ is the normal to S, and C is oriented as the boundary of $\mathcal{R}$ (relative to the normal vector $\mathbf{N}$). *Hint:* Apply Stokes' Theorem to $\mathbf{F} = \langle bz - cy, cx - az, ay - bx \rangle$.

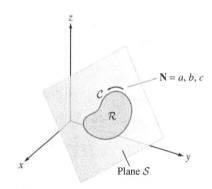

FIGURE 8

SOLUTION By Stokes' Theorem,

$$\iint_S \text{curl}(\mathbf{F})\cdot d\mathbf{S} = \iint_S (\text{curl}(\mathbf{F})\cdot \mathbf{e}_n)\,dS = \int_C \mathbf{F}\cdot d\mathbf{r} \tag{1}$$

We compute the curl of $\mathbf{F}$:

$$\text{curl}(\mathbf{F}) = \begin{vmatrix} \mathbf{i} & \mathbf{j} & \mathbf{k} \\ \frac{\partial}{\partial x} & \frac{\partial}{\partial y} & \frac{\partial}{\partial z} \\ bz - cy & cx - az & ay - bx \end{vmatrix} = 2a\mathbf{i} + 2b\mathbf{j} + 2c\mathbf{k} = 2\langle a, b, c \rangle$$

The unit normal to the plane $ax + by + cz + d = 0$ is

$$\mathbf{e}_n = \frac{\langle a, b, c \rangle}{\sqrt{a^2 + b^2 + c^2}}$$

Therefore,

$$\text{curl}(\mathbf{F}) \cdot \mathbf{e}_n = 2\langle a, b, c \rangle \cdot \frac{1}{\sqrt{a^2 + b^2 + c^2}} \langle a, b, c \rangle$$

$$= \frac{2}{\sqrt{a^2 + b^2 + c^2}}(a^2 + b^2 + c^2) = 2\sqrt{a^2 + b^2 + c^2}$$

Hence,

$$\iint_S \text{curl}(\mathbf{F}) \cdot d\mathbf{S} = \iint_S \text{curl}(\mathbf{F}) \cdot \mathbf{e}_n \, dS = \iint_S 2\sqrt{a^2 + b^2 + c^2} \, dS = 2\sqrt{a^2 + b^2 + c^2} \iint_S 1 \, dS \quad (2)$$

The sign of $\iint_S 1 \, dS$ is determined by the orientation of S. Since the area is a positive value, we have

$$\left| \iint_S 1 \, ds \right| = \text{Area}(S)$$

Therefore, (2) gives

$$\left| \iint_S \text{curl}(\mathbf{F}) \cdot d\mathbf{S} \right| = 2\sqrt{a^2 + b^2 + c^2} \, \text{Area}(S)$$

Combining with (1) we obtain

$$2\sqrt{a^2 + b^2 + c^2} \, \text{Area}(S) = \left| \int_C \mathbf{F} \cdot d\mathbf{r} \right|$$

or

$$\text{Area}(S) = \frac{1}{2\sqrt{a^2 + b^2 + c^2}} = \frac{1}{2\|\mathbf{N}\|} \cdot \left| \int_C (bz - cy)\,dx + (cx - az)\,dy + (ay - bx)\,dz \right|$$

41. Show that $\Phi(\theta, \phi) = (a\cos\theta\sin\phi, b\sin\theta\sin\phi, c\cos\phi)$ is a parametrization of the ellipsoid

$$\left(\frac{x}{a}\right)^2 + \left(\frac{y}{b}\right)^2 + \left(\frac{z}{c}\right)^2 = 1$$

Then calculate the volume of the ellipsoid as the surface integral of $\mathbf{F} = \frac{1}{3}\langle x, y, z \rangle$ (this surface integral is equal to the volume by the Divergence Theorem).

SOLUTION For the given parametrization,

$$x = a\cos\theta\sin\phi, \quad y = b\sin\theta\sin\phi, \quad z = c\cos\phi \quad (1)$$

We show that it satisfies the equation of the ellipsoid

$$\left(\frac{x}{a}\right)^2 + \left(\frac{y}{b}\right)^2 + \left(\frac{z}{c}\right)^2 = \left(\frac{a\cos\theta\sin\phi}{a}\right)^2 + \left(\frac{b\sin\theta\sin\phi}{b}\right)^2 + \left(\frac{c\cos\phi}{c}\right)^2$$

$$= \cos^2\theta\sin^2\phi + \sin^2\theta\sin^2\phi + \cos^2\phi$$

$$= \sin^2\phi(\cos^2\theta + \sin^2\theta) + \cos^2\phi$$

$$= \sin^2\phi + \cos^2\phi = 1$$

Conversely, for each (x, y, z) on the ellipsoid, there exists θ and ϕ so that (1) holds. Therefore $\Phi(\theta, \phi)$ parametrizes the whole ellipsoid. Let $\mathcal{W}$ be the interior of the ellipsoid S. Then by Eq. (9):

$$\text{Volume}(\mathcal{W}) = \frac{1}{3}\iint_S \mathbf{F} \cdot d\mathbf{S}, \quad \mathbf{F} = \langle x, y, z \rangle$$

We compute the surface integral, using the given parametrization. We first compute the normal vector:

$$\frac{\partial \Phi}{\partial \theta} = \langle -a\sin\theta\sin\phi, b\cos\theta\sin\phi, 0 \rangle$$

$$\frac{\partial \Phi}{\partial \phi} = \langle a \cos\theta \cos\phi, b \sin\theta \cos\phi, -c \sin\phi \rangle$$

$$\frac{\partial \Phi}{\partial \theta} \times \frac{\partial \Phi}{\partial \phi} = -ab\sin^2\theta \sin\phi\cos\phi \mathbf{k} - ac\sin\theta\sin^2\phi \mathbf{j} - ab\cos^2\theta\sin\phi\cos\phi \mathbf{k} - bc\cos\theta\sin^2\phi \mathbf{i}$$

$$= \langle -bc\cos\theta\sin^2\phi, -ac\sin\theta\sin^2\phi, -ab\sin\phi\cos\phi \rangle$$

Hence, the outward pointing normal is

$$\mathbf{N} = \langle bc\cos\theta\sin^2\phi, ac\sin\theta\sin^2\phi, ab\sin\phi\cos\phi \rangle$$

$$\mathbf{F}(\Phi(\theta,\phi)) \cdot \mathbf{N} = \langle a\cos\theta\sin\phi, b\sin\theta\sin\phi, c\cos\phi \rangle \cdot \langle bc\cos\theta\sin^2\phi, ac\sin\theta\sin^2\phi, ab\sin\phi\cos\phi \rangle$$

$$= abc\cos^2\theta\sin^3\phi + abc\sin^2\theta\sin^3\phi + abc\sin\phi\cos^2\phi$$

$$= abc\sin^3\phi(\cos^2\theta + \sin^2\theta) + abc\sin\phi\cos^2\phi$$

$$= abc\sin^3\phi + abc\sin\phi\cos^2\phi = abc\sin^3\phi + abc\sin\phi(1 - \sin^2\phi)$$

$$= abc\sin\phi$$

We obtain the following integral:

$$\text{Volume}(\mathcal{W}) = \frac{1}{3}\iint_S \mathbf{F} \cdot d\mathbf{S} = \frac{1}{3}\int_0^{2\pi}\int_0^{\pi} abc\sin\phi\, d\phi\, d\theta$$

$$= \frac{2\pi abc}{3}\int_0^{\pi}\sin\phi\, d\varphi = \frac{2\pi abc}{3}\left(-\cos\phi\Big|_0^{\pi}\right) = \frac{4\pi abc}{3}$$